W9-CCR-951

To the Student

As you begin, you may feel anxious about the number of theorems, definitions, procedures, and equations. You may wonder if you can learn it all in time. Don't worry–your concerns are normal. This textbook was written with you in mind. If you attend class, work hard, and read and study this text, you will build the knowledge and skills you need to be successful. Here's how you can use the text to your benefit.

Read Carefully

When you get busy, it's easy to skip reading and go right to the problems. Don't ... the text has a large number of examples and clear explanations to help you break down the mathematics into easy-to-understand steps. Reading will provide you with a clearer understanding, beyond simple memorization. Read before class (not after) so you can ask questions about anything you didn't understand. You'll be amazed at how much more you'll get out of class if you do this.

Use the Features

I use many different methods in the classroom to communicate. Those methods, when incorporated into the text, are called "features." The features serve many purposes, from providing timely review of material you learned before (just when you need it) to providing organized review sessions to help you prepare for quizzes and tests. Take advantage of the features and you will master the material.

To make this easier, we've provided a brief guide to getting the most from this text. Refer to "Prepare for Class," "Practice," and "Review" at the front of the text. Spend fifteen minutes reviewing the guide and familiarizing yourself with the features by flipping to the page numbers provided. Then, as you read, use them. This is the best way to make the most of your text.

Please do not hesitate to contact me through Pearson Education, with any questions, comments, or suggestions for improving this text. I look forward to hearing from you, and good luck with all of your studies.

Best Wishes!

Michael Sullivan

NC

NEOSHO COUNTY
COMMUNITY COLLEGE

NCCC Bookstore
800 W. 14th St.
Chanute, KS 66720
620.432.0400

NCCC Bookstore

TORE:00001   REG:101   TRAN#:3848
ASHIER:NOEMI P

ustomer ID: 5395143
INZIE
LEAVER

ULLIVA/COLLEGE AL
00000000030185   N
1 @ 0.00)   0.00
 Rental - Due Date: 12/15/2023
OTAL   **0.00**

Returns or Refunds within 14 days
Receipt Required To Return Purchases
NO CASH REFUNDS - Acct Credit ONLY
Exclusions may apply for Returns
Thank You!

225.67   08/31/2023 02:14PM

CUSTOMER COPY

Neosho County
COMMUNITY COLLEGE

NCCC Bookstore
800 W. 14th St.
Chanute, KS 66720
620.432.0400

NCCC Bookstore

STORE:00001   REG:101   TRAN#:3848
CASHIER:NOEMI P

Customer ID: 6395143
INZIE
LEAVER

SULLIVAN/COLLEGE AL
000000003018S   N.
1 @ 0:00)   0.00
Rental - Due Date: 12/15/2023
TOTAL   0.00

Returns or Refunds within 14 days
Receipt Required To Return Purchases
NO CASH REFUNDS - Acct Credit ONLY
Exclusions may apply For Returns
Thank You!

226.67   08/31/2023 02:14PM

CUSTOMER COPY

Prepare for Class "Read the Book"

Feature	Description	Benefit	Page
Every Chapter Opener begins with . . .			
Chapter-Opening Topic & Project	Each chapter begins with a discussion of a topic of current interest and ends with a related project.	The Project lets you apply what you learned to solve a problem related to the topic.	414
Internet-Based Projects	The projects allow for the integration of spreadsheet technology that you will need to be a productive member of the workforce.	The projects give you an opportunity to collaborate and use mathematics to deal with issues of current interest.	516
Every Section begins with . . .			
LEARNING OBJECTIVES 2	Each section begins with a list of objectives. Objectives also appear in the text where the objective is covered.	These focus your study by emphasizing what's most important and where to find it.	435
Sections contain . . .			
PREPARING FOR THIS SECTION	Most sections begin with a list of key concepts to review with page numbers.	Ever forget what you've learned? This feature highlights previously learned material to be used in this section. Review it, and you'll always be prepared to move forward.	435
Now Work the 'Are You Prepared?' Problems	Problems that assess whether you have the prerequisite knowledge for the upcoming section.	Not sure you need the Preparing for This Section review? Work the 'Are You Prepared?' problems. If you get one wrong, you'll know exactly what you need to review and where to review it!	435, 446
Now Work PROBLEMS	These follow most examples and direct you to a related exercise.	We learn best by doing. You'll solidify your understanding of examples if you try a similar problem right away, to be sure you understand what you've just read.	442, 447
WARNING	Warnings are provided in the text.	These point out common mistakes and help you to avoid them.	469
Exploration and Seeing the Concept	These graphing utility activities foreshadow a concept or solidify a concept just presented.	You will obtain a deeper and more intuitive understanding of theorems and definitions.	430, 455
In Words	These provide alternative descriptions of select definitions and theorems.	Does math ever look foreign to you? This feature translates math into plain English.	452
Calculus	These appear next to information essential for the study of calculus.	Pay attention–if you spend extra time now, you'll do better later!	210, 419, 442
SHOWCASE EXAMPLES	These examples provide "how-to" instruction by offering a guided, step-by-step approach to solving a problem.	With each step presented on the left and the mathematics displayed on the right, you can immediately see how each step is used.	381
Model It! Examples and Problems	These examples and problems require you to build a mathematical model from either a verbal description or data. The homework Model It! problems are marked by purple headings.	It is rare for a problem to come in the form "*Solve the following equation.*" Rather, the equation must be developed based on an explanation of the problem. These problems require you to develop models to find a solution to the problem.	459, 488
NEW! Need to Review?	These margin notes provide a just-in-time reminder of a concept needed now, but covered in an earlier section of the book. Each note is back-referenced to the chapter, section and page where the concept was originally discussed.	Sometimes as you read, you encounter a word or concept you know you've seen before, but don't remember exactly what it means. This feature will point you to where you first learned the word or concept. A quick review now will help you see the connection to what you are learning for the first time and make remembering easier the next time.	428

Practice "Work the Problems"

Feature	Description	Benefit	Page
'Are You Prepared?' Problems	These assess your retention of the prerequisite material you'll need. Answers are given at the end of the section exercises. This feature is related to the Preparing for This Section feature.	Do you always remember what you've learned? Working these problems is the best way to find out. If you get one wrong, you'll know exactly what you need to review and where to review it!	452, 460
Concepts and Vocabulary	These short-answer questions, mainly Fill-in-the-Blank, Multiple-Choice and True/False items, assess your understanding of key definitions and concepts in the current section.	It is difficult to learn math without knowing the language of mathematics. These problems test your understanding of the formulas and vocabulary.	446
Skill Building	Correlated with section examples, these problems provide straightforward practice.	It's important to dig in and develop your skills. These problems provide you with ample opportunity to do so.	446–448
Applications and Extensions	These problems allow you to apply your skills to real-world problems. They also allow you to extend concepts learned in the section.	You will see that the material learned within the section has many uses in everyday life.	449–451
NEW! Challenge Problems	These problems have been added in most sections and appear at the end of the Application and Extensions exercises. They are intended to be thought-provoking, requiring some ingenuity to solve.	Are you a student who likes being challenged? Then the Challenge Problems are for you! Your professor might also choose to assign a challenge problem as a group project. The ability to work with a team is a highly regarded skill in the working world.	451
Explaining Concepts: Discussion and Writing	"Discussion and Writing" problems are colored red. They support class discussion, verbalization of mathematical ideas, and writing and research projects.	To verbalize an idea, or to describe it clearly in writing, shows real understanding. These problems nurture that understanding. Many are challenging, but you'll get out what you put in.	451
Retain Your Knowledge	These problems allow you to practice content learned earlier in the course.	Remembering how to solve all the different kinds of problems that you encounter throughout the course is difficult. This practice helps you remember.	451
Now Work PROBLEMS	Many examples refer you to a related homework problem. These related problems are marked by a pencil and orange numbers.	If you get stuck while working problems, look for the closest Now Work problem, and refer to the related example to see if it helps.	444, 447, 448
Review Exercises	Every chapter concludes with a comprehensive list of exercises to pratice. Use the list of objectives to determine the objective and examples that correspond to the problems.	Work these problems to ensure that you understand all the skills and concepts of the chapter. Think of it as a comprehensive review of the chapter.	511–514

Review "Study for Quizzes and Tests"

Feature	Description	Benefit	Page
The Chapter Review at the end of each chapter contains . . .			
Things to Know	A detailed list of important theorems, formulas, and definitions from the chapter.	Review these and you'll know the most important material in the chapter!	509–510
You Should Be Able to . . .	Contains a complete list of objectives by section, examples that illustrate the objective, and practice exercises that test your understanding of the objective.	Do the recommended exercises and you'll have mastered the key material. If you get something wrong, go back and work through the objective listed and try again.	510–511
Review Exercises	These provide comprehensive review and practice of key skills, matched to the Learning Objectives for each section.	Practice makes perfect. These problems combine exercises from all sections, giving you a comprehensive review in one place.	511–514
Chapter Test	About 15–20 problems that can be taken as a Chapter Test. Be sure to take the Chapter Test under test conditions—no notes!	Be prepared. Take the sample practice test under test conditions. This will get you ready for your instructor's test. If you get a problem wrong, you can watch the Chapter Test Prep Video.	514
Cumulative Review	These problem sets appear at the end of each chapter, beginning with Chapter 2. They combine problems from previous chapters, providing an ongoing cumulative review. When you use them in conjunction with the Retain Your Knowledge problems, you will be ready for the final exam.	These problem sets are really important. Completing them will ensure that you are not forgetting anything as you go. This will go a long way toward keeping you primed for the final exam.	515
Chapter Projects	The Chapter Projects apply to what you've learned in the chapter. Additional projects are available on the Instructor's Resource Center (IRC).	The Chapter Projects give you an opportunity to use what you've learned in the chapter to the opening topic. If your instructor allows, these make excellent opportunities to work in a group, which is often the best way to learn math.	516
Internet-Based Projects	In selected chapters, a Web-based project is given.	These projects give you an opportunity to collaborate and use mathematics to deal with issues of current interest by using the Internet to research and collect data.	516

Dedicated to the memory of Mary

College Algebra

Eleventh Edition

Michael Sullivan
Chicago State University

 Pearson

Director, Portfolio Management: *Anne Kelly*
Senior Portfolio Management Analyst: *Dawn Murrin*
Portfolio Management Administrator: *Joseph Colella*
Manager, Courseware QA: *Mary Durnwald*
VP, Production & Digital Studio: *Ruth Berry*
Manager Producer: *Vicki Dreyfus*
Associate Producer: *Stacey Miller*
Manager, Content Development: *Kristina Evans*
Senior Content Developer: *Megan M. Burns*
Managing Producer: *Scott Disanno*
Content Producer: *Peggy McMahon*
Product Marketing Director: *Erin Kelly*

Product Marketer for Precalculus: *Stacey Sveum*
Product Marketing Assistant: *Shannon McCormack*
Field Marketing Manager: *Peggy Lucas*
Senior Publishing Services Analyst, Author Support: *Joe Vetere*
Manager, Rights/ Permissions: *Gina Cheselka*
Manager Design: *Blair Brown*
Manufacturing Buyer: *Carol Melville, LSC Communications*
Cover Illustration: *Tamara Newnam*
Cover Design: *Jerilyn Bockorick, Pearson CSC*
Full Service Vendor: *Pearson, CSC*
Full Service Project Management: *Pearson CSC, Rose Kernan*
Composition: *Pearson CSC*

Copyright © 2020, 2016, 2012 by Pearson Education, Inc. 221 River Street, Hoboken, NJ 07030.

All Rights Reserved. Printed in the United States of America. This publication is protected by copyright, and permission should be obtained from the publisher prior to any prohibited reproduction, storage in a retrieval system, or transmission in any form or by any means, electronic, mechanical, photocopying, recording, or otherwise. For information regarding permissions, request forms and the appropriate contacts within the Pearson Education Global Rights & Permissions department, please visit www.pearsoned.com/permissions/.

Attributions of third party content appear on page C-1, which constitutes an extension of this copyright page.

MICROSOFT® AND WINDOWS® ARE REGISTERED TRADEMARKS OF THE MICROSOFT CORPORATION IN THE U.S.A. AND OTHER COUNTRIES. SCREEN SHOTS AND ICONS REPRINTED WITH PERMISSION FROM THE MICROSOFT CORPORATION. THIS BOOK IS NOT SPONSORED OR ENDORSED BY OR AFFILIATED WITH THE MICROSOFT CORPORATION.

MICROSOFT AND/OR ITS RESPECTIVE SUPPLIERS MAKE NO REPRESENTATIONS ABOUT THE SUITABILITY OF THE INFORMATION CONTAINED IN THE DOCUMENTS AND RELATED GRAPHICS PUBLISHED AS PART OF THE SERVICES FOR ANY PURPOSE. ALL SUCH DOCUMENTS AND RELATED GRAPHICS ARE PROVIDED "AS IS" WITHOUT WARRANTY OF ANY KIND. MICROSOFT AND/OR ITS RESPECTIVE SUPPLIERS HEREBY DISCLAIM ALL WARRANTIES AND CONDITIONS WITH REGARD TO THIS INFORMATION, INCLUDING ALL WARRANTIES AND CONDITIONS OF MERCHANTABILITY, WHETHER EXPRESS, IMPLIED OR STATUTORY, FITNESS FOR A PARTICULAR PURPOSE, TITLE AND NON-INFRINGEMENT. IN NO EVENT SHALL MICROSOFT AND/OR ITS RESPECTIVE SUPPLIERS BE LIABLE FOR ANY SPECIAL, INDIRECT OR CONSEQUENTIAL DAMAGES OR ANY DAMAGES WHATSOEVER RESULTING FROM LOSS OF USE, DATA OR PROFITS, WHETHER IN AN ACTION OF CONTRACT, NEGLIGENCE OR OTHER TORTIOUS ACTION, ARISING OUT OF OR IN CONNECTION WITH THE USE OR PERFORMANCE OF INFORMATION AVAILABLE FROM THE SERVICES. THE DOCUMENTS AND RELATED GRAPHICS CONTAINED HEREIN COULD INCLUDE TECHNICAL INACCURACIES OR TYPOGRAPHICAL ERRORS. CHANGES ARE PERIODICALLY ADDED TO THE INFORMATION HEREIN. MICROSOFT AND/OR ITS RESPECTIVE SUPPLIERS MAY MAKE IMPROVEMENTS AND/OR CHANGES IN THE PRODUCT(S) AND/OR THE PROGRAM(S) DESCRIBED HEREIN AT ANY TIME. PARTIAL SCREEN SHOTS MAY BE VIEWED IN FULL WITHIN THE SOFTWARE VERSION SPECIFIED.

PEARSON, ALWAYS LEARNING, and MYLAB™MATH are exclusive trademarks owned by Pearson Education, Inc. or its affiliates in the U.S. and/or other countries.

Unless otherwise indicated herein, any third-party trademarks that may appear in this work are the property of their respective owners and any references to third-party trademarks, logos or other trade dress are for demonstrative or descriptive purposes only. Such references are not intended to imply any sponsorship, endorsement, authorization, or promotion of Pearson's products by the owners of such marks, or any relationship between the owner and Pearson Education, Inc. or its affiliates, authors, licensees or distributors.

The student edition of this book has been cataloged by the Library of Congress as follows:
Library of Congress Cataloging-in-Publication Data

Names: Sullivan, Michael, 1942- author.
Title: College algebra / Michael Sullivan (Chicago State University).
Description: Eleventh edition. | Hoboken, NJ : Pearson, [2020] | Includes index.
Identifiers: LCCN 2018060244 | ISBN 9780135163047
Subjects: LCSH: Algebra--Textbooks. | Algebra--Study and teaching (Higher)
Classification: LCC QA154.3 .S763 2020 | DDC 512.9--dc23 LC record available at https://lccn.loc.gov/2018060244

6 2020

About the Cover:
The image on this book's cover was inspired by a talk given by Michael Sullivan III: *Is Mathematical Talent Overrated?*

The answer is yes. In mathematics, innate talent plays a much smaller role than grit and motivation as you work toward your goal. If you put in the time and hard work, you can succeed in your math course—just as an athlete must work to medal in their sport.

ISBN 10: **0135163048**
ISBN-13: **9780135163047**

Contents

7 Analytic Geometry — 517

Three Distinct Series

Students have different goals, learning styles, and levels of preparation. Instructors have different teaching philosophies, styles, and techniques. Rather than write one series to fit all, the Sullivans have written three distinct series. All share the same goal—to develop a high level of mathematical understanding and an appreciation for the way mathematics can describe the world around us. The manner of reaching that goal, however, differs from series to series.

Flagship Series, Eleventh Edition

The Flagship Series is the most traditional in approach yet modern in its treatment of precalculus mathematics. In each text, needed review material is included, and is referenced when it is used. Graphing utility coverage is optional and can be included or excluded at the discretion of the instructor: *College Algebra, Algebra & Trigonometry, Trigonometry: A Unit Circle Approach, Precalculus.*

Enhanced with Graphing Utilities Series, Seventh Edition

This series provides a thorough integration of graphing utilities into topics, allowing students to explore mathematical concepts and encounter ideas usually studied in later courses. Many examples show solutions using algebra side-by-side with graphing techniques. Using technology, the approach to solving certain problems differs from the Flagship Series, while the emphasis on understanding concepts and building strong skills is maintained: *College Algebra, Algebra & Trigonometry, Precalculus.*

Concepts through Functions Series, Fourth Edition

This series differs from the others, utilizing a functions approach that serves as the organizing principle tying concepts together. Functions are introduced early in various formats. The approach supports the Rule of Four, which states that functions can be represented symbolically, numerically, graphically, and verbally. Each chapter introduces a new type of function and then develops all concepts pertaining to that particular function. The solutions of equations and inequalities, instead of being developed as stand-alone topics, are developed in the context of the underlying functions. Graphing utility coverage is optional and can be included or excluded at the discretion of the instructor: *College Algebra; Precalculus, with a Unit Circle Approach to Trigonometry; Precalculus, with a Right Triangle Approach to Trigonometry.*

The Flagship Series

College Algebra, Eleventh Edition

This text provides a contemporary approach to college algebra, with three chapters of review material preceding the chapters on functions. Graphing calculator usage is provided, but is optional. After completing this book, a student will be adequately prepared for trigonometry, finite mathematics, and business calculus.

Algebra & Trigonometry, Eleventh Edition

This text contains all the material in *College Algebra*, but also develops the trigonometric functions using a right triangle approach and shows how it relates to the unit circle approach. Graphing techniques are emphasized, including a thorough discussion of polar coordinates, parametric equations, and conics using polar coordinates. Vectors in the plane, sequences, induction, and the binomial theorem are also presented. Graphing calculator usage is provided, but is optional. After completing this book, a student will be adequately prepared for finite mathematics, business calculus, and engineering calculus.

Precalculus, Eleventh Edition

This text contains one review chapter before covering the traditional precalculus topics of polynomial, rational, exponential, and logarithmic functions and their graphs. The trigonometric functions are introduced using a unit circle approach and showing how it relates to the right triangle approach. Graphing techniques are emphasized, including a thorough discussion of polar coordinates, parametric equations, and conics using polar coordinates. Vectors in the plane and in space, including the dot and cross products, sequences, induction, and the binomial theorem are also presented. Graphing calculator usage is provided, but is optional. The final chapter provides an introduction to calculus, with a discussion of the limit, the derivative, and the integral of a function. After completing this book, a student will be adequately prepared for finite mathematics, business calculus, and engineering calculus.

Trigonometry: a Unit Circle Approach, Eleventh Edition

This text, designed for stand-alone courses in trigonometry, develops the trigonometric functions using a unit circle approach and shows how it relates to the right triangle approach. Vectors in the plane and in space, including the dot and cross products, are presented. Graphing techniques are emphasized, including a thorough discussion of polar coordinates, parametric equations, and conics using polar coordinates. Graphing calculator usage is provided, but is optional. After completing this book, a student will be adequately prepared for finite mathematics, business calculus, and engineering calculus.

Preface to the Instructor

As a professor of mathematics at an urban public university for 35 years, I understand the varied needs of college algebra students. Students range from being underprepared with little mathematical background and a fear of mathematics, to being highly prepared and motivated. For some, this is their final course in mathematics. For others, it is preparation for future mathematics courses. I have written this text with both groups in mind.

A tremendous benefit of authoring a successful series is the broad-based feedback I receive from instructors. and students who have used previous editions. I am sincerely grateful for their support. Virtually every change to this edition is the result of their thoughtful comments and suggestions. I hope that I have been able to take their ideas and, building upon a successful foundation of the tenth edition, make this series an even better learning and teaching tool for students and instructors.

Features in the Eleventh Edition

A descriptive list of the many special features of *College Algebra* can be found on the endpapers in the front of this text. This list places the features in their proper context, as building blocks of an overall learning system that has been carefully crafted over the years to help students get the most out of the time they put into studying. Please take the time to review it and to discuss it with your students at the beginning of your course. My experience has been that when students use these features, they are more successful in the course.

- **Updated! Retain Your Knowledge Problems** These problems, which were new to the previous edition, are based on the article *"To Retain New Learning, Do the Math,"* published in the *Edurati Review.* In this article, Kevin Washburn suggests that "the more students are required to recall new content or skills, the better their memory will be." The Retain Your Knowledge problems were so well received that they have been expanded in this edition. Moreover, while the focus remains to help students maintain their skills, in most sections, problems were chosen that preview skills required to succeed in subsequent sections or in calculus. These are easily identified by the calculus icon ($\mathcal{J}$). All answers to Retain Your Knowledge problems are given in the back of the text and all are assignable in MyLab Math.

- **Guided Lecture Notes** Ideal for online, emporium/redesign courses, inverted classrooms, or traditional lecture classrooms. These lecture notes help students take thorough, organized, and understandable notes as they watch the Author in Action videos. They ask students to complete definitions, procedures, and examples based on the content of the videos and text. In addition, experience suggests that students learn by doing and understanding the why/how of the concept or property. Therefore, many

sections will have an exploration activity to motivate student learning. These explorations introduce the topic and/or connect it to either a real-world application or a previous section. For example, when the vertical-line test is discussed in Section 3.2, after the theorem statement, the notes ask the students to explain why the vertical-line test works by using the definition of a function. This challenge helps students process the information at a higher level of understanding.

- **Illustrations** Many of the figures have captions to help connect the illustrations to the explanations in the body of the text.

- **Graphing Utility Screen Captures** In several instances we have added Desmos screen captures along with the TI-84 Plus C screen captures. These updated screen captures provide alternate ways of visualizing concepts and making connections between equations, data and graphs in full color.

- **Chapter Projects,** which apply the concepts of each chapter to a real-world situation, have been enhanced to give students an up-to-the-minute experience. Many of these projects are new requiring the student to research information online in order to solve problems.

- **Exercise Sets** The exercises in the text have been reviewed and analyzed some have been removed, and new ones have been added. All time-sensitive problems have been updated to the most recent information available. The problem sets remain classified according to purpose.

  The *'Are You Prepared?'* problems have been improved to better serve their purpose as a just-in-time review of concepts that the student will need to apply in the upcoming section.

  The *Concepts and Vocabulary* problems have been expanded to cover each objective of the section. These multiple-choice, fill-in-the-blank, and True/False exercises have been written to also serve as reading quizzes.

  Skill Building problems develop the student's computational skills with a large selection of exercises that are directly related to the objectives of the section. *Mixed Practice* problems offer a comprehensive assessment of skills that relate to more than one objective. Often these require skills learned earlier in the course.

  Applications and Extensions problems have been updated. Further, many new application-type exercises have been added, especially ones involving information and data drawn from sources the student will recognize, to improve relevance and timeliness.

  At the end of Applications and Extensions, we have a collection of one or more *Challenge Problems.* These problems, as the title suggests, are intended to be thought-provoking, requiring some ingenuity to solve. They can be used for group work or to challenge students. At the end of the Annotated Instructor's

Edition and in the online Instructor's Solutions Manual, we have provided solutions to all these problems.

The ***Explaining Concepts: Discussion and Writing*** exercises provide opportunity for classroom discussion and group projects.

Updated! ***Retain Your Knowledge*** has been improved and expanded. The problems are based on material learned earlier in the course. They serve to keep information that has already been learned "fresh" in the mind of the student. Answers to all these problems appear in the Student Edition.

Need to Review? These margin notes provide a just-in-time reminder of a concept needed now, but covered in an earlier section of the book. Each note includes a reference to the chapter, section and page where the concept was originally discussed.

Content Changes to the 11ᵗʰ edition

- ***Challenge Problems*** have been added in most sections at the end of the Application and Extensions exercises. Challenge Problems are intended to be thought-provoking problems that require some ingenuity to solve. They can be used to challenge students or for group work. Solutions to Challenge Problems are at available in the Annotated Instructor's Edition and the online Instructors Solutions Manual.

- **Need to Review?** These margin notes provide a just-in-time review for a concept needed now, but covered in an earlier section of the book. Each note is back-referenced to the chapter, section and page where the concept was originally discussed.

- Additional **Retain Your Knowledge** exercises, whose purpose is to keep learned material fresh in a student's mind, have been added to each section. Many of these new problems preview skills required for calculus or for concepts needed in subsequent sections.

- **Desmos** screen captures have been added throughout the text. This is done to recognize that graphing technology expands beyond graphing calculators.

- Examples and exercises throughout the text have been augmented to reflect a broader selection of STEM applications.

- Concepts and Vocabulary exercises have been expanded to cover each objective of a section.

- Skill building exercises have been expanded to assess a wider range of difficulty.

- Applied problems and those based on real data have been updated where appropriate.

Chapter R

- Section R.8 Objective 3 now includes rationalizing the numerator
  - NEW Example 6 Rationalizing Numerators
  - Problems 69-76 provide practice.
- Section R.8 Exercises now include more practice in simplifying radicals

Chapter 1

- NEW Section 1.2 Objective 2 Solve a Quadratic Equation Using the Square Root Method.

Chapter 2

- NEW Section 2.2 Example 9 Testing an Equation for Symmetry
- Section 2.3 has been reorganized to treat the slope-intercept form of the equation of a line before finding an equation of a line using two points.

Chapter 3

- NEW Section 3.1 Objective 1 Describe a Relation
- NEW Section 3.2 Example 4 Expending Energy
- NEW Section 3.4 Example 4 Analyzing a Piecewise-defined Function
- NEW Example 1 Describing a Relation demonstrates using the Rule of Four to express a relation numerically, as a mapping, and graphically given a verbal description.

Chapter 4

- Section 4.3 introduces the concept of concavity for a quadratic function
- NEW Section 4.3 Example 3 Graphing a Quadratic Function Using Its Vertex, Axis, and Intercepts
- Section 4.3 Example 8 Analyzing the Motion of a Projectile (formerly in Section 4.4)
- NEW Section 4.4 Example 4 Fitting a Quadratic Function to Data

Chapter 5

- Section 5.1 has been revised and split into two sections:
  - 5.1 Polynomial Functions
  - 5.2 Graphing Polynomial Functions; Models
- NEW Section 5.2 Example 2 Graphing a Polynomial Function (a 4th degree polynomial function)

Chapter 6

- Section 6.2 now finds and verifies inverse functions analytically and graphically

Chapter 8

- NEW Section 8.5 Example 1 Identifying Proper and Improper Rational Expressions

Using the Eleventh Edition Effectively with Your Syllabus

To meet the varied needs of diverse syllabi, this text contains more content than is likely to be covered in an *College Algebra* course. As the chart illustrates, this text has been organized with flexibility of use in mind. Within a given chapter, certain sections are optional (see the details that follow the figure below) and can be omitted without loss of continuity.

Chapter R Review

This chapter consists of review material. It may be used as the first part of the course or later as a just-in-time review when the content is required. Specific references to this chapter occur throughout the text to assist in the review process.

Chapter 1 Equations and Inequalities

Primarily a review of Intermediate Algebra topics, this material is a prerequisite for later topics. The coverage of complex numbers and quadratic equations with a negative discriminant is optional and may be postponed or skipped entirely without loss of continuity.

Chapter 2 Graphs

This chapter lays the foundation for functions. Section 2.5 is optional.

Chapter 3 Functions and Their Graphs

Perhaps the most important chapter. Section 3.6 is optional.

Chapter 4 Linear and Quadratic Functions

Topic selection depends on your syllabus. Sections 4.2 and 4.4 may be omitted without loss of continuity.

Chapter 5 Polynomial and Rational Functions

Topic selection depends on your syllabus.

Chapter 6 Exponential and Logarithmic Functions

Sections 6.1–6.6 follow in sequence. Sections 6.7, 6.8, and 6.9 are optional.

Chapter 7 Analytic Geometry

Sections 7.1–7.4 follow in sequence.

Chapter 8 Systems of Equations and Inequalities

Sections 8.2–8.7 may be covered in any order, but each requires Section 8.1. Section 8.8 requires Section 8.7.

Chapter 9 Sequences; Induction; The Binomial Theorem

There are three independent parts: Sections 9.1–9.3; Section 9.4; and Section 9.5.

Chapter 10 Counting and Probability

The sections follow in sequence.

Acknowledgments

Textbooks are written by authors, but evolve from an idea to final form through the efforts of many people. It was Don Dellen who first suggested this text and series to me. Don is remembered for his extensive contributions to publishing and mathematics.

Thanks are due to the following people for their assistance and encouragement to the preparation of this edition:

- From Pearson Education: Anne Kelly for her substantial contributions, ideas, and enthusiasm; Dawn Murrin, for her unmatched talent at getting the details right; Joseph Colella for always getting the reviews and pages to me on time; Peggy McMahon for directing the always difficult production process; Rose Kernan for handling liaison between the compositor and author; Peggy Lucas and Stacey Sveum for their genuine interest in marketing this text. Marcia Horton for her continued support and genuine interest; Paul Corey for his leadership and commitment to excellence; and the Pearson Sales team, for their continued confidence and personal support of Sullivan texts.

- Accuracy checkers: C. Brad Davis who read the entire manuscript and accuracy checked answers. His attention to detail is amazing; Timothy Britt, for creating the Solutions Manuals; and Kathleen Miranda and Pamela Trim for accuracy checking answers.

Finally, I offer my grateful thanks to the dedicated users and reviewers of my texts, whose collective insights form the backbone of each textbook revision.

James Africh, College of DuPage
Steve Agronsky, Cal Poly State University
Gererdo Aladro, Florida International University
Grant Alexander, Joliet Junior College
Dave Anderson, South Suburban College
Wes Anderson, Northwest Vista College
Richard Andrews, Florida A&M University
Joby Milo Anthony, University of Central Florida
James E. Arnold, University of Wisconsin-Milwaukee

Adel Arshaghi, Center for Educational Merit
Carolyn Autray, University of West Georgia
Agnes Azzolino, Middlesex County College
Wilson P. Banks, Illinois State University
Sudeshna Basu, Howard University
Timothy Bayer, Virginia Western CC
Dale R. Bedgood, East Texas State University
Beth Beno, South Suburban College
Carolyn Bernath, Tallahassee Community College
Rebecca Berthiaume, Edison State College
William H. Beyer, University of Akron

Annette Blackwelder, Florida State University
Richelle Blair, Lakeland Community College
Kevin Bodden, Lewis and Clark College
Jeffrey Boerner, University of Wisconsin-Stout
Connie Booker, Owensboro Community and Technical College
Barry Booten, Florida Atlantic University
Laurie Boudreaux, Nicholls State University
Larry Bouldin, Roane State Community College
Bob Bradshaw, Ohlone College
Trudy Bratten, Grossmont College

Tim Bremer, Broome Community College
Tim Britt, Jackson State Community College
Holly Broesamle, Oakland CC-Auburn Hills
Michael Brook, University of Delaware
Timothy Brown, Central Washington
University
Joanne Brunner, Joliet Junior College
Warren Burch, Brevard Community College
Mary Butler, Lincoln Public Schools
Melanie Butler, West Virginia University
Jim Butterbach, Joliet Junior College
Roberto Cabezas, Miami Dade College
William J. Cable, University of
Wisconsin-Stevens Point
Lois Calamia, Brookdale Community College
Jim Campbell, Lincoln Public Schools
Roger Carlsen, Moraine Valley Community
College
Elena Catoiu, Joliet Junior College
Mathews Chakkanakuzhi, Palomar College
Tim Chappell, Penn Valley Community
College
John Collado, South Suburban College
Amy Collins, Northwest Vista College
Alicia Collins, Mesa Community College
Nelson Collins, Joliet Junior College
Rebecca Connell, Troy University
Jim Cooper, Joliet Junior College
Denise Corbett, East Carolina University
Carlos C. Corona, San Antonio College
Theodore C. Coskey, South Seattle
Community College
Rebecca Connell, Troy University
Donna Costello, Plano Senior High School
Rebecca Courter, Pasadena City College
Garrett Cox, The University of Texas at San
Antonio
Paul Crittenden, University of Nebraska at
Lincoln
John Davenport, East Texas State University
Faye Dang, Joliet Junior College
Antonio David, Del Mar College
Stephanie Deacon, Liberty University
Duane E. Deal, Ball State University
Jerry DeGroot, Purdue North Central
Timothy Deis, University of Wisconsin-
Platteville
Joanna DelMonaco, Middlesex Community
College
Vivian Dennis, Eastfield College
Deborah Dillon, R. L. Turner High School
Guesna Dohrman, Tallahassee Community
College
Cheryl Doolittle, Iowa State University
Karen R. Dougan, University of Florida
Jerrett Dumouchel, Florida Community
College at Jacksonville
Louise Dyson, Clark College
Paul D. East, Lexington Community College
Don Edmondson, University of Texas-Austin
Erica Egizio, Joliet Junior College
Jason Eltrevoog, Joliet Junior College
Christopher Ennis, University of Minnesota
Kathy Eppler, Salt Lake Community College
Ralph Esparza, Jr., Richland College
Garret J. Etgen, University of Houston
Scott Fallstrom, Shoreline Community College
Pete Falzone, Pensacola Junior College
Arash Farahmand, Skyline College
Said Fariabli, San Antonio College
W.A. Ferguson, University of Illinois-Urbana/
Champaign
Iris B. Fetta, Clemson University
Mason Flake, student at Edison Community
College
Timothy W. Flood, Pittsburg State University
Robert Frank, Westmoreland County
Community College

Merle Friel, Humboldt State University
Richard A. Fritz, Moraine Valley
Community College
Dewey Furness, Ricks College
Mary Jule Gabiou, North Idaho College
Randy Gallaher, Lewis and Clark College
Tina Garn, University of Arizona
Dawit Getachew, Chicago State University
Wayne Gibson, Rancho Santiago College
Loran W. Gierhart, University of Texas at San
Antonio and Palo Alto College
Robert Gill, University of Minnesota Duluth
Nina Girard, University of Pittsburgh at
Johnstown
Sudhir Kumar Goel, Valdosta State University
Adrienne Goldstein, Miami Dade College,
Kendall Campus
Joan Goliday, Sante Fe Community College
Lourdes Gonzalez, Miami Dade College,
Kendall Campus
Frederic Gooding, Goucher College
Donald Goral, Northern Virginia Community
College
Sue Graupner, Lincoln Public Schools
Mary Beth Grayson, Liberty University
Jennifer L. Grimsley, University of Charleston
Ken Gurganus, University of North Carolina
Igor Halfin, University of Texas-San Antonio
James E. Hall, University of Wisconsin-
Madison
Judy Hall, West Virginia University
Edward R. Hancock, DeVry Institute of
Technology
Julia Hassett, DeVry Institute, Dupage
Christopher Hay-Jahans, University of South
Dakota
Michah Heibel, Lincoln Public Schools
LaRae Helliwell, San Jose City College
Celeste Hernandez, Richland College
Gloria P. Hernandez, Louisiana State
University at Eunice
Brother Herron, Brother Rice High School
Robert Hoburg, Western Connecticut State
University
Lynda Hollingsworth, Northwest Missouri
State University
Deltrye Holt, Augusta State University
Charla Holzbog, Denison High School
Lee Hruby, Naperville North High School
Miles Hubbard, St. Cloud State University
Kim Hughes, California State College-San
Bernardino
Stanislav, Jabuka, University of Nevada, Reno
Ron Jamison, Brigham Young University
Richard A. Jensen, Manatee Community
College
Glenn Johnson, Middlesex Community
College
Sandra G. Johnson, St. Cloud State University
Tuesday Johnson, New Mexico State University
Susitha Karunaratne, Purdue University North
Central
Moana H. Karsteter, Tallahassee Community
College
Donna Katula, Joliet Junior College
Arthur Kaufman, College of Staten Island
Thomas Kearns, North Kentucky University
Jack Keating, Massasoit Community College
Shelia Kellenbarger, Lincoln Public Schools
Rachael Kenney, North Carolina State
University
Penelope Kirby, Florida State University
John B. Klassen, North Idaho College
Debra Kopcso, Louisiana State University
Lynne Kowski, Raritan Valley Community
College
Yelena Kravchuk, University of Alabama at
Birmingham

Ray S. Kuan, Skyline College
Keith Kuchar, Manatee Community College
Tor Kwembe, Chicago State University
Linda J. Kyle, Tarrant Country Jr. College
H.E. Lacey, Texas A & M University
Darren Lacoste, Valencia College-West
Campus
Harriet Lamm, Coastal Bend College
James Lapp, Fort Lewis College
Matt Larson, Lincoln Public Schools
Christopher Lattin, Oakton Community College
Julia Ledet, Lousiana State University
Wayne Lee, St. Phillips CC
Adele LeGere, Oakton Community College
Kevin Leith, University of Houston
JoAnn Lewin, Edison College
Jeff Lewis, Johnson County Community
College
Janice C. Lyon, Tallahassee Community
College
Jean McArthur, Joliet Junior College
Virginia McCarthy, Iowa State University
Karla McCavit, Albion College
Michael McClendon, University of Central
Oklahoma
Tom McCollow, DeVry Institute of
Technology
Marilyn McCollum, North Carolina State
University
Jill McGowan, Howard University
Will McGowant, Howard University
Angela McNulty, Joliet Junior College
Lisa Meads, College of the Albemarle
Laurence Maher, North Texas State University
Jay A. Malmstrom, Oklahoma City
Community College
Rebecca Mann, Apollo High School
Lynn Marecek, Santa Ana College
Sherry Martina, Naperville North High School
Ruby Martinez, San Antonio College
Alec Matheson, Lamar University
Nancy Matthews, University of Oklahoma
James Maxwell, Oklahoma State
University-Stillwater
Marsha May, Midwestern State University
James McLaughlin, West Chester University
Judy Meckley, Joliet Junior College
David Meel, Bowling Green State University
Carolyn Meitler, Concordia University
Samia Metwali, Erie Community College
Rich Meyers, Joliet Junior College
Eldon Miller, University of Mississippi
James Miller, West Virginia University
Michael Miller, Iowa State University
Kathleen Miranda, SUNY at Old Westbury
Chris Mirbaha, The Community College of
Baltimore County
Val Mohanakumar, Hillsborough Community
College
Thomas Monaghan, Naperville North High
School
Miguel Montanez, Miami Dade College,
Wolfson Campus
Maria Montoya, Our Lady of the Lake
University
Susan Moosai, Florida Atlantic University
Craig Morse, Naperville North High School
Samad Mortabit, Metropolitan State University
Pat Mower, Washburn University
Tammy Muhs, University of Central Florida
A. Muhundan, Manatee Community College
Jane Murphy, Middlesex Community College
Richard Nadel, Florida International University
Gabriel Nagy, Kansas State University
Bill Naegele, South Suburban College
Karla Neal, Lousiana State University
Lawrence E. Newman, Holyoke Community
College

Dwight Newsome, Pasco-Hernando Community College
Denise Nunley, Maricopa Community Colleges
James Nymann, University of Texas-El Paso
Mark Omodt, Anoka-Ramsey Community College
Seth F. Oppenheimer, Mississippi State University
Leticia Oropesa, University of Miami
Linda Padilla, Joliet Junior College
Sanja Pantic, University of Illinois at Chicago
E. James Peake, Iowa State University
Kelly Pearson, Murray State University
Dashamir Petrela, Florida Atlantic University
Philip Pina, Florida Atlantic University
Charlotte Pisors, Baylor University
Michael Prophet, University of Northern Iowa
Laura Pyzdrowski, West Virginia University
Carrie Quesnell, Weber State University
Neal C. Raber, University of Akron
Thomas Radin, San Joaquin Delta College
Aibeng Serene Radulovic, Florida Atlantic University
Ken A. Rager, Metropolitan State College
Traci Reed, St. Johns River State College
Kenneth D. Reeves, San Antonio College
Elsi Reinhardt, Truckee Meadows Community College
Jose Remesar, Miami Dade College, Wolfson Campus
Jane Ringwald, Iowa State University
Douglas F. Robertson, University of Minnesota, MPLS
Stephen Rodi, Austin Community College
William Rogge, Lincoln Northeast High School
Howard L. Rolf, Baylor University
Mike Rosenthal, Florida International University
Phoebe Rouse, Lousiana State University
Edward Rozema, University of Tennessee at Chattanooga
Dennis C. Runde, Manatee Community College
Paul Runnion, Missouri University of Science and Technology
Amit Saini, University of Nevada-Reno
Laura Salazar, Northwest Vista College

Alan Saleski, Loyola University of Chicago
Susan Sandmeyer, Jamestown Community College
Brenda Santistevan, Salt Lake Community College
Linda Schmidt, Greenville Technical College
Ingrid Scott, Montgomery College
A.K. Shamma, University of West Florida
Zachery Sharon, University of Texas at San Antonio
Joshua Shelor, Virginia Western CC
Martin Sherry, Lower Columbia College
Carmen Shershin, Florida International University
Tatrana Shubin, San Jose State University
Anita Sikes, Delgado Community College
Timothy Sipka, Alma College
Charlotte Smedberg, University of Tampa
Lori Smellegar, Manatee Community College
Gayle Smith, Loyola Blakefield
Cindy Soderstrom, Salt Lake Community College
Leslie Soltis, Mercyhurst College
John Spellman, Southwest Texas State University
Karen Spike, University of North Carolina
Rajalakshmi Sriram, Okaloosa-Walton Community College
Katrina Staley, North Carolina Agricultural and Technical State University
Becky Stamper, Western Kentucky University
Judy Staver, Florida Community College-South
Robin Steinberg, Pima Community College
Neil Stephens, Hinsdale South High School
Sonya Stephens, Florida A&M Univeristy
Patrick Stevens, Joliet Junior College
John Sumner, University of Tampa
Matthew TenHuisen, University of North Carolina, Wilmington
Christopher Terry, Augusta State University
Diane Tesar, South Suburban College
Tommy Thompson, Brookhaven College
Martha K. Tietze, Shawnee Mission Northwest High School

Richard J. Tondra, Iowa State University
Florentina Tone, University of West Florida
Suzanne Topp, Salt Lake Community College
Marilyn Toscano, University of Wisconsin, Superior
Marvel Townsend, University of Florida
Jim Trudnowski, Carroll College
David Tseng, Miami Dade College, Kendall Campus
Robert Tuskey, Joliet Junior College
Mihaela Vajiac, Chapman University-Orange
Julia Varbalow, Thomas Nelson Community College-Leesville
Richard G. Vinson, University of South Alabama
Jorge Viola-Prioli, Florida Atlantic University
Mary Voxman, University of Idaho
Jennifer Walsh, Daytona Beach Community College
Donna Wandke, Naperville North High School
Timothy L. Warkentin, Cloud County Community College
Melissa J. Watts, Virginia State University
Hayat Weiss, Middlesex Community College
Kathryn Wetzel, Amarillo College
Darlene Whitkenack, Northern Illinois University
Suzanne Williams, Central Piedmont Community College
Larissa Williamson, University of Florida
Christine Wilson, West Virginia University
Brad Wind, Florida International University
Anna Wiodarczyk, Florida International University
Mary Wolyniak, Broome Community College
Canton Woods, Auburn University
Tamara S. Worner, Wayne State College
Terri Wright, New Hampshire Community Technical College, Manchester
Rob Wylie, Carl Albert State College
Aletheia Zambesi, University of West Florida
George Zazi, Chicago State University
Loris Zucca, Lone Star College-Kingwood
Steve Zuro, Joliet Junior College

Michael Sullivan

Get the Most Out of
MyLab Math

Pearson
MyLab

Math courses are continuously evolving to help today's students succeed. It's more challenging than ever to support students with a wide range of backgrounds, learner styles, and math anxieties. The flexibility to build a course that fits instructors' individual course formats—with a variety of content options and multimedia resources all in one place—has made MyLab Math the market-leading solution for teaching and learning mathematics since its inception.

Preparedness

One of the biggest challenges in College Algebra, Trigonometry, and Precalculus is making sure students are adequately prepared with prerequisite knowledge. For a student, having the essential algebra skills upfront in this course can dramatically increase success.

- **MyLab Math with Integrated Review** can be used in corequisite courses, or simply to help students who enter without a full understanding of prerequisite skills and concepts. **Integrated Review** provides videos on review topics with a corresponding worksheet, along with premade, assignable skills-check quizzes and personalized review homework assignments. **Integrated Review** is now available within all Sullivan 11th Edition MyLab Math courses.

Assignments

10/18/19 11:59pm	◆	Chapter 4 Skills Check
10/18/19 11:59pm	⚑ ●	Chapter 4 Skills Review Homework
04/01/20 11:59pm	◆	Chapter 5 Skills Check
04/01/20 11:59pm	⚑ ●	Chapter 5 Skills Review Homework
09/14/20 11:59pm	◆	Chapter 6 Skills Check
09/14/20 11:59pm	⚑ ●	Chapter 6 Skills Review Homework

pearson.com/mylab/math

Resources for Success

Pearson
MyLab

MyLab Math Online Course for College Algebra, 11th Edition by Michael Sullivan (access code required)

MyLab™ Math is tightly integrated with each author's style, offering a range of author-created multimedia resources, so your students have a consistent experience.

Video Program and Resources

Author in Action Videos are actual classroom lectures by Michael Sullivan III with fully worked-out examples.

- **Video assessment** questions are available to assign in MyLab Math for key videos.
- Updated! The corresponding **Guided Lecture Notes** assist students in taking thorough, organized, and understandable notes while watching Author in Action videos.

EXAMPLE

Finding the Exact Value of a Logarithmic Expression

$(a)\ \log_3 81 = 4$ $(b)\ \log_2 \dfrac{1}{8}$

$y = \log_a x$ means $a^y = x$

$(b)\ y = \log_2 \dfrac{1}{8}$

$2^y = 2$

$2^y = \dfrac{1}{8}$

03:27 / 04:07   info   Speed   CC

Guided Visualizations

New! **Guided Visualizations**, created in GeoGebra by Michael Sullivan III, bring mathematical concepts to life, helping students visualize the concept through directed exploration and purposeful manipulation. Assignable in MyLab Math with assessment questions to check students' conceptual understanding.

☑ Show Unit Circle   ☐ Show Key Points on Graph of the Cosine Function
                           ☑ Show Graph of the Cosine Function

Angle in Radians $= 0\ rad$
Angle in Degrees $= 0°$
Note : Rounded to Nearest Degree

(0, 1)

$y = \cos x$ (1, 0) θ

The cosine function takes as input an angle θ and outputs the x-coordinate of the unit circle.
$\cos 0\ rad = 1$

Retain Your Knowledge Exercises

Updated! Retain Your Knowledge Exercises, assignable in MyLab Math, improve students' recall of concepts learned earlier in the course. New for the 11th Edition, additional exercises will be included that will have an emphasis on content that students will build upon in the immediate upcoming section.

Retain Your Knowledge

Problems 154–162 are based on material learned earlier in the course. The purpose of these problems is to keep the material fresh in your mind so that you are better prepared for the final exam.

154. Simplify $\left(\dfrac{x^2 y^{-3}}{x^4 y^3}\right)^{-2}$. Assume $x \neq 0$ and $y \neq 0$. Express the answer so that all exponents are positive. $x^4 y^{16}$

155. The lengths of the legs of a right triangle are $a = 8$ and $b = 15$. Find the hypotenuse. 17

156. Solve the equation: $(x - 3)^2 + 25 = 49$
$\{3 - 2\sqrt 6, 3 + 2\sqrt 6\}$

157. Solve $|2x - 5| + 7 < 10$. Express the answer using set notation or interval notation. Graph the solution set.

158. Determine the domain of the variable x in the expression:
$\sqrt{8 - \dfrac{2}{3}x}$ $(-\infty, 12]$

159. Determine what number should be added to complete the square:
$x^2 + \dfrac{3}{4}x \quad \dfrac{9}{64}$

160. Multiply and simplify the result:
$\dfrac{x^2 - 16}{x^2 + 6x + 8} \cdot \dfrac{x + 2}{16 - 4x} \quad -\dfrac{1}{4}$

161. Rationalize the denominator:
$\dfrac{\sqrt{x + 1} + \sqrt x}{\sqrt{x + 1} - \sqrt x} \quad 2x + 1 + 2\sqrt{x(x + 1)}$

162. Solve: $x - 5\sqrt x + 6 = 0 \quad \{4, 9\}$

pearson.com/mylab/math

Resources for Success

Pearson
MyLab

Instructor Resources

Online resources can be downloaded from www.pearson.com, or hardcopy resources can be ordered from your sales representative.

Annotated Instructor's Edition
College Algebra, 11th Edition
ISBN – 013516320X / 9780135163207
Shorter answers are on the page beside the exercises. Longer answers are in the back of the text.

Instructor's Solutions Manual
ISBN – 0135163722 / 9780135163726
Includes fully worked solutions to all exercises in the text.

Learning Catalytics Question Library
Questions written by Michael Sullivan III are available within MyLab Math to deliver through Learning Catalytics to engage students in your course.

Powerpoint® Lecture Slides
Fully editable slides correlate to the textbook.

Mini Lecture Notes
Includes additional examples and helpful teaching tips, by section.

Testgen®
TestGen (www.pearsoned.com/testgen) enables instructors to build, edit, print, and administer tests using a computerized bank of questions developed to cover all the objectives of the text.

Online Chapter Projects
Additional projects that give students an opportunity to apply what they learned in the chapter.

Student Resources

Additional resources to enhance student success.

Lecture Video
Author in Action videos are actual classroom lectures with fully worked out examples presented by Michael Sullivan, III. All video is assignable within MyLab Math.

Chapter Test Prep Videos
Students can watch instructors work through step-by-step solutions to all chapter test exercises from the text. These are available in MyLab Math and on YouTube.

Student's Solutions Manual
ISBN - 013516317X / 9780135163177
Provides detailed worked-out solutions to odd-numbered exercises.

Guided Lecture Notes
ISBN – 0135163188 / 9780135163184
These lecture notes assist students in taking thorough, organized, and understandable notes while watching Author in Action videos. Students actively participate in learning the how/why of important concepts through explorations and activities. The Guided Lecture Notes are available as PDF's and customizable Word files in MyLab Math. They can also be packaged with the text and the MyLab Math access code.

Algebra Review
ISBN: 0131480065 / 9780131480063
Four printed chapters of Intermediate Algebra review available. Perfect for a corequisite course or for individual review.

Applications Index

Calculus

absolute maximum/minimum in, 233, 551
area under a curve, 267, 612
average rate of change in, 236, 353, 473, 528, 570, 662
composite functions in, 419
concavity test, 311, 644
critical numbers, 662
difference quotient in, 210, 217, 267, 324, 346, 451, 473, 490, 538, 629
discontinuous functions, 379
e^x in, 443, 661
expression as single quotient in, 77
expressions with rational exponents in, 77
factoring in, 53, 57, 80, 465, 570, 644
functions approximated by polynomial functions in, 353
increasing/decreasing functions in, 231, 311, 346, 637
Intermediate Value Theorem, 396, 637
inverse functions, 551
local maxima/minima in, 233, 291, 501
partial fraction decomposition, 668, 685, 702, 711
perpendicular lines, 595, 620
quadratic equations in, 99–100
rationalizing numerators, 595
reducing expression to lowest terms in, 71
secant line in, 236, 291, 490
second derivative, 702
Simpson's rule, 320
trigonometric expressions and functions, 685

Acoustics

amplifying sound, 512
loudness of sound, 463, 514
whispering galleries, 536

Aeronautics

sonic boom, 550

Agriculture

farm management, 643
farm workers in U.S., 500
field enclosure, 628
milk production, 507
minimizing cost, 643
watering a field, 102

Air travel

distance between two planes, 269
parking at O'Hare International Airport, 251

Archaeology

age of ancient tools, 493–494
age of fossil, 499
age of tree, 499
date of prehistoric man's death, 513

Architecture

brick staircase, 668, 693
Burj Khalifa building, 31
floor design, 666, 693
football stadium seating, 667
mosaic design, 668, 693
Norman window, 37, 318
parabolic arch, 318
racetrack design, 537
special window, 318, 326
stadium construction, 668
vertically circular building, 190
window design, 318
window dimensions, 102

Art

framing a painting, 146

Astronomy

distance from Earth to its moon, 29
distances of planets from Sun, 661
light-year, 29
planetary orbits, 537

Aviation

orbital launches, 567

Biology

alcohol and driving, 459, 464
bacterial growth, 492–493, 506
  E-coli, 240, 282
blood types, 701
bone length, 326–327
cricket chirp rate and temperature, 319
healing of wounds, 449, 463
maternal age versus Down syndrome, 297
yeast biomass as function of time, 505

Business

advertising, 183, 298, 327
automobile production, 421, 583
blending coffee, 141
candy bar size, 103
checkout lines, 720
clothing store, 723
commissions, 326
cookie orders, 648
copying machines, 146
cost
  of can, 375, 378
  of commodity, 421
  of manufacturing, 29, 141, 386, 636
  marginal, 311, 326
  minimizing, 326, 643, 648
  of printing, 350–351
  of production, 240, 421, 610, 648
  of transporting goods, 252
cost equation, 182, 195
cost function, 290
customer wait times, 377
demand
  for candy, 196
demand equation, 326, 412
depreciation, 414, 464
discount pricing, 91, 92, 422
drive-thru rate
  at Burger King, 445
  at Citibank, 449, 463
  at McDonald's, 449–450
equipment depreciation, 678
expense computation, 142
farm workers in U.S., 500
inventory management, 272
Jiffy Lube's car arrival rate, 449, 463
managing a meat market, 643
milk production, 507
mixing candy, 141
mixing nuts, 141
orange juice production, 583
precision ball bearings, 29
presale orders, 568
price markup, 91
product design, 644
production scheduling, 643
product promotion, 183
profit, 610
  maximizing, 641–642, 643–644
profit function, 218
rate of return on, 488
restaurant management, 568
revenue, 141, 311, 324, 327, 506
  advertising, 508
  airline, 644
  of clothing store, 600
  daily, 311
  from digital music, 266
  from football seating, 679
  maximizing, 311, 317–318
  monthly, 311
  theater, 569
revenue equation, 195
RV rental, 327
salary, 422, 668
  gross, 217
  increases in, 678, 693
sales
  commission on, 128
  of movie theater ticket, 556, 561, 567
  net, 157
salvage value, 513
straight-line depreciation, 285–286, 289
supply and demand, 286–287, 289
tax, 386
theater attendance, 92
toy truck manufacturing, 636
transporting goods, 637
truck rentals, 182
unemployment, 723

Review

R

A Look Ahead →

Chapter R, as the title states, contains review material. Your instructor may choose to cover all or part of it as a regular chapter at the beginning of your course or later as a just-in-time review when the content is required. Regardless, when information in this chapter is needed, a specific reference to this chapter will be made so you can review.

Outline

R.1 Real Numbers

PREPARING FOR THIS TEXT *Before getting started, read "To the Student" at the front of this text.*

OBJECTIVES
1 Work with Sets (p. 2)
2 Classify Numbers (p. 4)
3 Evaluate Numerical Expressions (p. 8)
4 Work with Properties of Real Numbers (p. 9)

1 Work with Sets

A **set** is a well-defined collection of distinct objects. The objects of a set are called its **elements**. By **well-defined**, we mean that there is a rule that enables us to determine whether a given object is an element of the set. If a set has no elements, it is called the **empty set**, or **null set**, and is denoted by the symbol $\varnothing$.

For example, the set of **digits** consists of the collection of numbers 0, 1, 2, 3, 4, 5, 6, 7, 8, and 9. If we use the symbol D to denote the set of digits, then we can write

$$D = \{0, 1, 2, 3, 4, 5, 6, 7, 8, 9\}$$

In this notation, the braces { } are used to enclose the objects, or **elements**, in the set. This method of denoting a set is called the **roster method**. A second way to denote a set is to use **set-builder notation**, where the set D of digits is written as

$$D = \{ \quad x \quad | \quad x \text{ is a digit}\}$$

Read as "D is the set of all x such that x is a digit."

| EXAMPLE 1 | **Using Set-builder Notation and the Roster Method** |

(a) $E = \{x | x \text{ is an even digit}\} = \{0, 2, 4, 6, 8\}$
(b) $O = \{x | x \text{ is an odd digit}\} = \{1, 3, 5, 7, 9\}$

Because the elements of a set are distinct, we never repeat elements. For example, we would never write $\{1, 2, 3, 2\}$; the correct listing is $\{1, 2, 3\}$. Because a set is a collection, the order in which the elements are listed is immaterial. $\{1, 2, 3\}, \{1, 3, 2\}, \{2, 1, 3\}$, and so on, all represent the same set.

If every element of a set A is also an element of a set B, then A is a **subset** of B, which is denoted $A \subseteq B$. If two sets A and B have the same elements, then A **equals** B, which is denoted $A = B$.

For example, $\{1, 2, 3\} \subseteq \{1, 2, 3, 4, 5\}$ and $\{1, 2, 3\} = \{2, 3, 1\}$.

DEFINITION Intersection and Union of Two Sets

If A and B are sets, the **intersection** of A with B, denoted $A \cap B$, is the set consisting of elements that belong to both A and B. The **union** of A with B, denoted $A \cup B$, is the set consisting of elements that belong to either A or B, or both.

| EXAMPLE 2 | **Finding the Intersection and Union of Sets** |

Let $A = \{1, 3, 5, 8\}, B = \{3, 5, 7\}$, and $C = \{2, 4, 6, 8\}$. Find:

(a) $A \cap B$ (b) $A \cup B$ (c) $B \cap (A \cup C)$

Solution
(a) $A \cap B = \{1, 3, 5, 8\} \cap \{3, 5, 7\} = \{3, 5\}$

(b) $A \cup B = \{1, 3, 5, 8\} \cup \{3, 5, 7\} = \{1, 3, 5, 7, 8\}$

(c) $B \cap (A \cup C) = \{3, 5, 7\} \cap [\{1, 3, 5, 8\} \cup \{2, 4, 6, 8\}]$
$= \{3, 5, 7\} \cap \{1, 2, 3, 4, 5, 6, 8\} = \{3, 5\}$

Now Work PROBLEM 15

Usually, in working with sets, we designate a **universal set** U, the set consisting of all the elements that we wish to consider. Once a universal set has been designated, we can consider elements of the universal set not found in a given set.

DEFINITION Complement of a Set

If A is a set, the **complement** of A, denoted $\overline{A}$, is the set consisting of all the elements in the universal set that are not in A.*

EXAMPLE 3 Finding the Complement of a Set

If the universal set is $U = \{1, 2, 3, 4, 5, 6, 7, 8, 9\}$ and if $A = \{1, 3, 5, 7, 9\}$, then $\overline{A} = \{2, 4, 6, 8\}$.

It follows from the definition of complement that $A \cup \overline{A} = U$ and $A \cap \overline{A} = \varnothing$. Do you see why?

Now Work PROBLEM 19

It is often helpful to draw pictures of sets. Such pictures, called **Venn diagrams**, represent sets as circles enclosed in a rectangle, which represents the universal set. Such diagrams often help us to visualize various relationships among sets. See Figure 1.

If we know that $A \subseteq B$, we might use the Venn diagram in Figure 2(a). If we know that A and B have no elements in common—that is, if $A \cap B = \varnothing$—we might use the Venn diagram in Figure 2(b). The sets A and B in Figure 2(b) are said to be **disjoint**.

Universal set

Figure 1 Venn diagram

Figure 2

(a) $A \subseteq B$
subset

(b) $A \cap B = \varnothing$
disjoint sets

Figures 3(a), 3(b), and 3(c) use Venn diagrams to illustrate intersection, union, and complement, respectively.

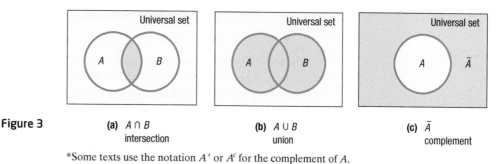

Figure 3

(a) $A \cap B$
intersection

(b) $A \cup B$
union

(c) $\overline{A}$
complement

*Some texts use the notation A' or A^c for the complement of A.

2 Classify Numbers

It is helpful to classify the various kinds of numbers that we deal with as sets. The **counting numbers**, or **natural numbers**, are the numbers in the set $\{1, 2, 3, 4, \ldots\}$. (The three dots, called an **ellipsis**, indicate that the pattern continues indefinitely.) As their name implies, these numbers are often used to count things. For example, there are 26 letters in our alphabet; there are 100 cents in a dollar. The **whole numbers** are the numbers in the set $\{0, 1, 2, 3, \ldots\}$ —that is, the counting numbers together with 0. The set of counting numbers is a subset of the set of whole numbers.

DEFINITION Integers

The **integers** are the set of numbers $\{\ldots, -3, -2, -1, 0, 1, 2, 3, \ldots\}$.

These numbers are useful in many situations. For example, if your checking account has $10 in it and you write a check for $15, you can represent the current balance as $-\$5$.

Each time we expand a number system, such as from the whole numbers to the integers, we do so in order to be able to handle new, and usually more complicated, problems. The integers enable us to solve problems requiring both positive and negative counting numbers, such as profit/loss, height above/below sea level, temperature above/below 0°F, and so on.

But integers alone are not sufficient for *all* problems. For example, they do not answer the question "What part of a dollar is 38 cents?" To answer such a question, we enlarge our number system to include *rational numbers*. For example, $\dfrac{38}{100}$ answers the question "What part of a dollar is 38 cents?"

DEFINITION Rational Number

A **rational number** is a number that can be expressed as a quotient $\dfrac{a}{b}$ of two integers. The integer a is called the **numerator**, and the integer b, which cannot be 0, is called the **denominator**. The rational numbers are the numbers in the set $\left\{ x \,\middle|\, x = \dfrac{a}{b}, \text{ where } a, b \text{ are integers and } b \neq 0 \right\}$.

Examples of rational numbers are $\dfrac{3}{4}, \dfrac{5}{2}, \dfrac{0}{4}, -\dfrac{2}{3}$, and $\dfrac{100}{3}$. Since $\dfrac{a}{1} = a$ for any integer a, it follows that the set of integers is a subset of the set of rational numbers.

Rational numbers may be represented as **decimals**. For example, the rational numbers $\dfrac{3}{4}, \dfrac{5}{2}, -\dfrac{2}{3}$, and $\dfrac{7}{66}$ may be represented as decimals by merely carrying out the indicated division:

$$\frac{3}{4} = 0.75 \qquad \frac{5}{2} = 2.5 \qquad -\frac{2}{3} = -0.666\ldots = -0.\overline{6} \qquad \frac{7}{66} = 0.1060606\ldots = 0.1\overline{06}$$

Notice that the decimal representations of $\dfrac{3}{4}$ and $\dfrac{5}{2}$ terminate, or end. The decimal representations of $-\dfrac{2}{3}$ and $\dfrac{7}{66}$ do not terminate, but they do exhibit a pattern of repetition. For $-\dfrac{2}{3}$, the 6 repeats indefinitely, as indicated by the bar over the 6; for $\dfrac{7}{66}$, the block 06 repeats indefinitely, as indicated by the bar over the 06. It can be shown that every rational number may be represented by a decimal that either terminates or is nonterminating with a repeating block of digits, and vice versa.

On the other hand, some decimals do not fit into either of these categories. Such decimals represent **irrational numbers**. Every irrational number may be represented by a decimal that neither repeats nor terminates. In other words, irrational numbers cannot be written in the form $\dfrac{a}{b}$, where a, b are integers and $b \neq 0$.

Irrational numbers occur naturally. For example, consider the isosceles right triangle whose legs are each of length 1. See Figure 4. The length of the hypotenuse is $\sqrt{2}$, an irrational number.

Also, the number that equals the ratio of the circumference C to the diameter d of any circle, denoted by the symbol π (the Greek letter pi), is an irrational number. See Figure 5.

Figure 4

Figure 5 $\pi = \dfrac{C}{d}$

DEFINITION Real Numbers

The set of **real numbers** is the union of the set of rational numbers with the set of irrational numbers.

Figure 6 shows the relationship of various types of numbers.*

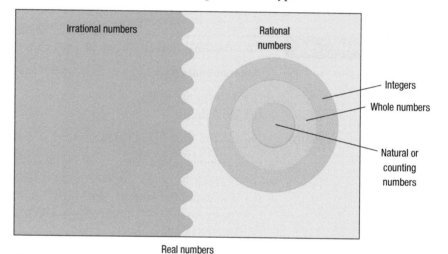

Figure 6

| EXAMPLE 4 | Classifying the Numbers in a Set |

List the numbers in the set

$$\left\{ -3, \frac{4}{3}, 0.12, \sqrt{2}, \pi, 10, 2.151515\ldots \text{ (where the block 15 repeats)} \right\}$$

that are

(a) Natural numbers   (b) Integers   (c) Rational numbers

(d) Irrational numbers   (e) Real numbers

Solution

(a) 10 is the only natural number.

(b) -3 and 10 are integers.

(c) $-3, 10, \dfrac{4}{3}, 0.12,$ and $2.151515\ldots$ are rational numbers.

(d) $\sqrt{2}$ and π are irrational numbers.

(e) All the numbers listed are real numbers.

➤ **Now Work** PROBLEM 25

*The set of real numbers is a subset of the set of complex numbers. We discuss complex numbers in Chapter 1, Section 1.3.

Approximations

Every decimal may be represented by a real number (either rational or irrational), and every real number may be represented by a decimal.

In practice, the decimal representation of an irrational number is given as an approximation. For example, using the symbol ≈ (read as "approximately equal to"), we can write

$$\sqrt{2} \approx 1.4142 \qquad \pi \approx 3.1416$$

In approximating decimals, we either *round* or *truncate* to a given number of decimal places.* The number of places establishes the location of the *final digit* in the decimal approximation.

> **Truncation:** Drop all of the digits that follow the specified final digit in the decimal.
>
> **Rounding:** Identify the specified final digit in the decimal. If the next digit is 5 or more, add 1 to the final digit; if the next digit is 4 or less, leave the final digit as it is. Then truncate following the final digit.

EXAMPLE 5 **Approximating a Decimal to Two Places**

Approximate 20.98752 to two decimal places by

(a) Truncating
(b) Rounding

Solution For 20.98752, the final digit is 8, since it is two decimal places from the decimal point.

(a) To truncate, we remove all digits following the final digit 8. The truncation of 20.98752 to two decimal places is 20.98.
(b) The digit following the final digit 8 is the digit 7. Since 7 is 5 or more, we add 1 to the final digit 8 and truncate. The rounded form of 20.98752 to two decimal places is 20.99.

EXAMPLE 6 **Approximating a Decimal to Two and Four Places**

Number	Rounded to Two Decimal Places	Rounded to Four Decimal Places	Truncated to Two Decimal Places	Truncated to Four Decimal Places
(a) 3.14159	3.14	3.1416	3.14	3.1415
(b) 0.056128	0.06	0.0561	0.05	0.0561
(c) 893.46125	893.46	893.4613	893.46	893.4612

Now Work PROBLEM 29

Calculators and Graphing Utilities

Calculators are incapable of displaying decimals that contain a large number of digits. For example, some calculators are capable of displaying only eight digits. When a number requires more than eight digits, the calculator either truncates or rounds.

*Sometimes we say "correct to a given number of decimal places" instead of "truncate."

To see how your calculator handles decimals, divide 2 by 3. How many digits do you see? Is the last digit a 6 or a 7? If it is a 6, your calculator truncates; if it is a 7, your calculator rounds.

There are different kinds of calculators. An **arithmetic** calculator can only add, subtract, multiply, and divide numbers; therefore, this type is not adequate for this course. **Scientific** calculators have all the capabilities of arithmetic calculators and also contain **function keys** labeled ln, log, sin, cos, tan, x^y, inv, and so on. As you proceed through this text, you will discover how to use many of the function keys. **Graphing** calculators have all the capabilities of scientific calculators and contain a screen on which graphs can be displayed. We use the term **graphing utilities** to refer generically to all graphing calculators and computer software graphing packages.

For those who have access to a graphing utility, we have included comments, examples, and exercises marked with a ▦, indicating that a graphing utility is required. We have also included an appendix that explains some of the capabilities of graphing utilities. The ▦ comments, examples, and exercises may be omitted without loss of continuity, if so desired.

Operations

In algebra, we use letters such as x, y, a, b, and c to represent numbers. The symbols used in algebra for the operations of addition, subtraction, multiplication, and division are $+$, $-$, $\cdot$, and $/$. The words used to describe the results of these operations are **sum**, **difference**, **product**, and **quotient**. Table 1 summarizes these ideas.

Table 1

Operation	Symbol	Words
Addition	$a + b$	Sum: a plus b
Subtraction	$a - b$	Difference: a minus b
Multiplication	$a \cdot b$, $(a) \cdot b$, $a \cdot (b)$, $(a) \cdot (b)$, ab, $(a)b$, $a(b)$, $(a)(b)$	Product: a times b
Division	a/b or $\dfrac{a}{b}$	Quotient: a divided by b

In algebra, we generally avoid using the multiplication sign $\times$ and the division sign $\div$ so familiar in arithmetic. Notice also that when two expressions are placed next to each other without an operation symbol, as in ab, or in parentheses, as in $(a)(b)$, it is understood that the expressions, called **factors**, are to be multiplied.

We also prefer not to use mixed numbers in algebra. When mixed numbers are used, addition is understood; for example, $2\dfrac{3}{4}$ means $2 + \dfrac{3}{4}$. In algebra, use of a mixed number may be confusing because the absence of an operation symbol between two terms is generally taken to mean multiplication. The expression $2\dfrac{3}{4}$ is therefore written instead as 2.75 or as $\dfrac{11}{4}$.

The symbol $=$, called an **equal sign** and read as "equals" or "is," is used to express the idea that the number or expression on the left of the equal sign is equivalent to the number or expression on the right.

EXAMPLE 7

Writing Statements Using Symbols

(a) The sum of 2 and 7 equals 9. In symbols, this statement is written as $2 + 7 = 9$.

(b) The product of 3 and 5 is 15. In symbols, this statement is written as $3 \cdot 5 = 15$.

▬▬▶ **Now Work** PROBLEM 41

3 Evaluate Numerical Expressions

Consider the expression $2 + 3 \cdot 6$. It is not clear whether we should add 2 and 3 to get 5, and then multiply by 6 to get 30; or first multiply 3 and 6 to get 18, and then add 2 to get 20. To avoid this ambiguity, we have the following agreement.

In Words

Multiply first, then add.

We agree that whenever the two operations of addition and multiplication separate three numbers, the multiplication operation is always performed first, followed by the addition operation.

For $2 + 3 \cdot 6$, then, we have

$$2 + 3 \cdot 6 = 2 + 18 = 20$$

EXAMPLE 8

Finding the Value of an Expression

Evaluate each expression.

(a) $3 + 4 \cdot 5$ (b) $8 \cdot 2 + 1$ (c) $2 + 2 \cdot 2$

Solution

(a) $3 + 4 \cdot 5 = 3 + 20 = 23$ (b) $8 \cdot 2 + 1 = 16 + 1 = 17$

          ↑   ↑

   Multiply first.   Multiply first.

(c) $2 + 2 \cdot 2 = 2 + 4 = 6$

➤ **Now Work** PROBLEM 53

When we want to indicate adding 3 and 4 and then multiplying the result by 5, we use parentheses and write $(3 + 4) \cdot 5$. Whenever parentheses appear in an expression, it means "perform the operations within the parentheses first!"

EXAMPLE 9

Finding the Value of an Expression

(a) $(5 + 3) \cdot 4 = 8 \cdot 4 = 32$

(b) $(4 + 5) \cdot (8 - 2) = 9 \cdot 6 = 54$

When we divide two expressions, as in

$$\frac{2 + 3}{4 + 8}$$

it is understood that the division bar acts like parentheses; that is,

$$\frac{2 + 3}{4 + 8} = \frac{(2 + 3)}{(4 + 8)}$$

Rules for the Order of Operations

1. Begin with the innermost parentheses and work outward. Remember that in dividing two expressions, we treat the numerator and denominator as if they were enclosed in parentheses.
2. Perform multiplications and divisions, working from left to right.
3. Perform additions and subtractions, working from left to right.

| EXAMPLE 10 | **Finding the Value of an Expression** |

Evaluate each expression.

(a) $8 \cdot 2 + 3$ (b) $5 \cdot (3 + 4) + 2$

(c) $\dfrac{2 + 5}{2 + 4 \cdot 7}$ (d) $2 + [\, 4 + 2 \cdot (10 + 6) \,]$

Solution

(a) $8 \cdot 2 + 3 = 16 + 3 = 19$

                    ↑

            Multiply first.

(b) $5 \cdot (3 + 4) + 2 = 5 \cdot 7 + 2 = 35 + 2 = 37$

                 ↑   ↑

       Parentheses first   **Multiply before adding.**

(c) $\dfrac{2 + 5}{2 + 4 \cdot 7} = \dfrac{2 + 5}{2 + 28} = \dfrac{7}{30}$

(d) $2 + [\, 4 + 2 \cdot (10 + 6) \,] = 2 + [\, 4 + 2 \cdot (16) \,]$

                        $= 2 + [\, 4 + 32 \,] = 2 + [\, 36 \,] = 38$

Be careful if you use a calculator. For Example 10(c), you need to use parentheses. See Figure 7.* If you don't, the calculator will compute the expression

$$2 + \frac{5}{2} + 4 \cdot 7 = 2 + 2.5 + 28 = 32.5$$

```
NORMAL FLOAT AUTO REAL RADIAN MP
(2+5)/(2+4*7)
               .2333333333
Ans▶Frac
                        7
                       ──
                       30
```

Figure 7

giving a wrong answer.

✏— **Now Work** PROBLEMS 59 AND 67

4 Work with Properties of Real Numbers

An equal sign is used to mean that one expression is equivalent to another. Four important properties of equality are listed next. In this list, a, b, and c represent real numbers.

- The **reflexive property** states that a number equals itself; that is, $a = a$.
- The **symmetric property** states that if $a = b$, then $b = a$.
- The **transitive property** states that if $a = b$ and $b = c$, then $a = c$.
- The **principle of substitution** states that if $a = b$, then we may substitute b for a in any expression containing a.

Now, let's consider some other properties of real numbers.

| EXAMPLE 11 | **Commutative Properties** |

(a) $3 + 5 = 8$ (b) $2 \cdot 3 = 6$

     $5 + 3 = 8$ $3 \cdot 2 = 6$

     $3 + 5 = 5 + 3$ $2 \cdot 3 = 3 \cdot 2$

This example illustrates the *commutative property* of real numbers, which states that the order in which addition or multiplication takes place does not affect the final result.

---

*Notice that we converted the decimal to its fraction form. Another option, when using a TI-84 Plus C, is to use the fraction template under the MATH button to enter the expression as it appears in Example 10(c). Consult your manual to see how to enter such expressions on your calculator.

Commutative Properties

$$a + b = b + a \tag{1a}$$

$$a \cdot b = b \cdot a \tag{1b}$$

Here, and in the properties listed next and on pages 11–13, a, b, and c represent real numbers.

EXAMPLE 12 **Associative Properties**

(a) $2 + (3 + 4) = 2 + 7 = 9$ (b) $2 \cdot (3 \cdot 4) = 2 \cdot 12 = 24$

$\quad (2 + 3) + 4 = 5 + 4 = 9$ $(2 \cdot 3) \cdot 4 = 6 \cdot 4 = 24$

$\quad 2 + (3 + 4) = (2 + 3) + 4$ $2 \cdot (3 \cdot 4) = (2 \cdot 3) \cdot 4$

The way we add or multiply three real numbers does not affect the final result. Expressions such as $2 + 3 + 4$ and $3 \cdot 4 \cdot 5$ present no ambiguity, even though addition and multiplication are performed on one pair of numbers at a time. This property is called the *associative property*.

Associative Properties

$$a + (b + c) = (a + b) + c = a + b + c \tag{2a}$$

$$a \cdot (b \cdot c) = (a \cdot b) \cdot c = a \cdot b \cdot c \tag{2b}$$

Distributive Property

$$a \cdot (b + c) = a \cdot b + a \cdot c \tag{3a}$$

$$(a + b) \cdot c = a \cdot c + b \cdot c \tag{3b}$$

The distributive property may be used in two different ways.

EXAMPLE 13 **Distributive Property**

(a) $2 \cdot (x + 3) = 2 \cdot x + 2 \cdot 3 = 2x + 6$ Use to remove parentheses.

(b) $3x + 5x = (3 + 5)x = 8x$ Use to combine two expressions.

(c) $(x + 2)(x + 3) = x(x + 3) + 2(x + 3) = (x^2 + 3x) + (2x + 6)$

$\qquad\qquad\qquad\quad = x^2 + (3x + 2x) + 6 = x^2 + 5x + 6$

✏️— **Now Work** PROBLEM 89

The real numbers 0 and 1 have unique properties called the *identity properties*.

EXAMPLE 14 **Identity Properties**

(a) $4 + 0 = 0 + 4 = 4$ (b) $3 \cdot 1 = 1 \cdot 3 = 3$

Identity Properties

$$0 + a = a + 0 = a \tag{4a}$$

$$a \cdot 1 = 1 \cdot a = a \tag{4b}$$

We call 0 the **additive identity** and 1 the **multiplicative identity**.

For each real number a, there is a real number $-a$, called the **additive inverse** of a, having the following property:

Additive Inverse Property

$$a + (-a) = -a + a = 0 \tag{5a}$$

EXAMPLE 15 **Finding an Additive Inverse**

(a) The additive inverse of 6 is -6, because $6 + (-6) = 0$.
(b) The additive inverse of -8 is 8, because $-8 + 8 = 0$.

The additive inverse of a, that is, $-a$, is often called the *negative* of a. The use of the term 'negative' can be dangerous, because it suggests that the additive inverse is a negative number, which may not be the case. For example, the additive inverse of -3 equals 3, a positive number.

For each *nonzero* real number a, there is a real number $\dfrac{1}{a}$, called the **multiplicative inverse** of a, having the following property:

Multiplicative Inverse Property

$$a \cdot \frac{1}{a} = \frac{1}{a} \cdot a = 1 \qquad \text{if } a \neq 0 \tag{5b}$$

The multiplicative inverse $\dfrac{1}{a}$ of a nonzero real number a is also referred to as the **reciprocal** of a. The number 0 has no reciprocal.

EXAMPLE 16 **Finding a Reciprocal**

(a) The reciprocal of 6 is $\dfrac{1}{6}$, because $6 \cdot \dfrac{1}{6} = 1$.

(b) The reciprocal of -3 is $\dfrac{1}{-3}$, because $-3 \cdot \dfrac{1}{-3} = 1$.

(c) The reciprocal of $\dfrac{2}{3}$ is $\dfrac{3}{2}$, because $\dfrac{2}{3} \cdot \dfrac{3}{2} = 1$.

With these properties for adding and multiplying real numbers, we can define the operations of subtraction and division as follows:

DEFINITION Difference

The **difference** $a - b$, also read "a less b" or "a minus b," is defined as

$$a - b = a + (-b) \tag{6}$$

In Words
To subtract b from a, add the additive inverse of b to a.

> **DEFINITION Quotient**
>
> If b is a nonzero real number, the **quotient** $\dfrac{a}{b}$, also read as "a divided by b" or "the ratio of a to b," is defined as
>
> $$\frac{a}{b} = a \cdot \frac{1}{b} \qquad \text{if } b \neq 0 \tag{7}$$

EXAMPLE 17 **Working with Differences and Quotients**

(a) $8 - 5 = 8 + (-5) = 3$

(b) $4 - 9 = 4 + (-9) = -5$

(c) $\dfrac{5}{8} = 5 \cdot \dfrac{1}{8}$

For any number a, the product of a times 0 is always 0; that is,

In Words
The result of multiplying by zero is zero.

> **Multiplication by Zero**
>
> $$a \cdot 0 = 0 \tag{8}$$

For a nonzero number a,

> **Division Properties**
>
> $$\frac{0}{a} = 0 \qquad \frac{a}{a} = 1 \quad \text{if } a \neq 0 \tag{9}$$

NOTE Division by 0 is not defined. One reason is to avoid the following difficulty: $\dfrac{2}{0} = x$ means to find x such that $0 \cdot x = 2$. But $0 \cdot x$ equals 0 for all x, so there is no number x for which $\dfrac{2}{0} = x$. ∎

> **Rules of Signs**
>
> $$a(-b) = -(ab) \qquad (-a)b = -(ab) \qquad\qquad (-a)(-b) = ab$$
> $$-(-a) = a \qquad\qquad \frac{a}{-b} = \frac{-a}{b} = -\frac{a}{b} \text{ if } b \neq 0 \qquad \frac{-a}{-b} = \frac{a}{b} \text{ if } b \neq 0 \tag{10}$$

EXAMPLE 18 **Applying the Rules of Signs**

(a) $2(-3) = -(2 \cdot 3) = -6$ (b) $(-3)(-5) = 3 \cdot 5 = 15$

(c) $\dfrac{3}{-2} = \dfrac{-3}{2} = -\dfrac{3}{2}$ (d) $\dfrac{-4}{-9} = \dfrac{4}{9}$

(e) $\dfrac{x}{-2} = \dfrac{1}{-2} \cdot x = -\dfrac{1}{2}x$

Cancellation Properties

$$ac = bc \quad \text{implies} \quad a = b \quad \text{if } c \neq 0$$

$$\frac{ac}{bc} = \frac{a}{b} \qquad\qquad \text{if } b \neq 0, c \neq 0 \tag{11}$$

EXAMPLE 19 Using the Cancellation Properties

(a) If $2x = 6$, then

$$\begin{aligned} 2x &= 6 \\ 2x &= 2 \cdot 3 &&\text{Factor 6.} \\ x &= 3 &&\text{Cancel the 2's.} \end{aligned}$$

NOTE We follow the common practice of using slash marks to indicate cancellations. ∎

(b) $\dfrac{18}{12} = \dfrac{3 \cdot \cancel{6}}{2 \cdot \cancel{6}} = \dfrac{3}{2}$

Cancel the 6's.

In Words
If a product equals 0, then one or both of the factors is 0.

Zero-Product Property

$$\text{If } ab = 0, \text{ then } a = 0, \text{ or } b = 0, \text{ or both.} \tag{12}$$

EXAMPLE 20 Using the Zero-Product Property

If $2x = 0$, then either $2 = 0$ or $x = 0$. Since $2 \neq 0$, it follows that $x = 0$.

Arithmetic of Quotients

$$\frac{a}{b} + \frac{c}{d} = \frac{ad}{bd} + \frac{bc}{bd} = \frac{ad + bc}{bd} \qquad \text{if } b \neq 0, d \neq 0 \tag{13}$$

$$\frac{a}{b} \cdot \frac{c}{d} = \frac{ac}{bd} \qquad\qquad\qquad \text{if } b \neq 0, d \neq 0 \tag{14}$$

$$\frac{\dfrac{a}{b}}{\dfrac{c}{d}} = \frac{a}{b} \cdot \frac{d}{c} = \frac{ad}{bc} \qquad\qquad \text{if } b \neq 0, c \neq 0, d \neq 0 \tag{15}$$

EXAMPLE 21 Adding, Subtracting, Multiplying, and Dividing Quotients

(a) $\dfrac{2}{3} + \dfrac{5}{2} = \dfrac{2 \cdot 2}{3 \cdot 2} + \dfrac{3 \cdot 5}{3 \cdot 2} = \dfrac{2 \cdot 2 + 3 \cdot 5}{3 \cdot 2} = \dfrac{4 + 15}{6} = \dfrac{19}{6}$

↑ By equation (13)

(b) $\dfrac{3}{5} - \dfrac{2}{3} = \dfrac{3}{5} + \left(-\dfrac{2}{3}\right) = \dfrac{3}{5} + \dfrac{-2}{3}$

↑ By equation (6) ↑ By equation (10)

$$= \frac{3 \cdot 3 + 5 \cdot (-2)}{5 \cdot 3} = \frac{9 + (-10)}{15} = \frac{-1}{15} = -\frac{1}{15}$$

↑ By equation (13)

NOTE Slanting the slash marks in different directions for different factors, as shown here, is a good practice to follow, since it will help in checking for errors. ∎

(c) $\dfrac{8}{3} \cdot \dfrac{15}{4} = \dfrac{8 \cdot 15}{3 \cdot 4} = \dfrac{2 \cdot \cancel{4} \cdot \cancel{3} \cdot 5}{\cancel{3} \cdot \cancel{4} \cdot 1} = \dfrac{2 \cdot 5}{1} = 10$

       ↑ **By equation (14)**   ↑ **By equation (11)**

(d) $\dfrac{\frac{3}{5}}{\frac{7}{9}} = \dfrac{3}{5} \cdot \dfrac{9}{7} = \dfrac{3 \cdot 9}{5 \cdot 7} = \dfrac{27}{35}$

  ↑   ↑ **By equation (14)**

By equation (15)

NOTE In writing quotients, we follow the usual convention and write the quotient in lowest terms. That is, we write it so that any common factors of the numerator and the denominator have been removed using the Cancellation Properties, equation (11). As examples,

$$\frac{90}{24} = \frac{15 \cdot \cancel{6}}{4 \cdot \cancel{6}} = \frac{15}{4}$$

$$\frac{24x^2}{18x} = \frac{4 \cdot \cancel{6} \cdot x \cdot \cancel{x}}{3 \cdot \cancel{6} \cdot \cancel{x}} = \frac{4x}{3} \qquad x \neq 0$$

∎

Now Work PROBLEMS 69, 73, AND 83

Sometimes it is easier to add two fractions using *least common multiples* (LCM). The LCM of two numbers is the smallest number that each has as a common multiple.

EXAMPLE 22 **Finding the Least Common Multiple of Two Numbers**

Find the least common multiple of 15 and 12.

Solution To find the LCM of 15 and 12, we look at multiples of 15 and 12.

    15, 30, 45, **60**, 75, 90, 105, **120**, . . .

    12, 24, 36, 48, **60**, 72, 84, 96, 108, **120**, . . .

The *common* multiples are in blue. The *least* common multiple is 60.

EXAMPLE 23 **Using the Least Common Multiple to Add Two Fractions**

Find: $\dfrac{8}{15} + \dfrac{5}{12}$

Solution We use the LCM of the denominators of the fractions and rewrite each fraction using the LCM as a common denominator. The LCM of the denominators (12 and 15) is 60. Rewrite each fraction using 60 as the denominator.

$$\frac{8}{15} + \frac{5}{12} = \frac{8}{15} \cdot \frac{4}{4} + \frac{5}{12} \cdot \frac{5}{5}$$

$$= \frac{32}{60} + \frac{25}{60}$$

$$= \frac{32 + 25}{60}$$

$$= \frac{57}{60}$$

$$= \frac{19}{20} \qquad \text{Reduce the fraction.}$$

Now Work PROBLEM 77

Historical Feature

The real number system has a history that stretches back at least to the ancient Babylonians (1800 BC). It is remarkable how much the ancient Babylonian approach resembles our own. As we stated in the text, the fundamental difficulty with irrational numbers is that they cannot be written as quotients of integers or, equivalently, as repeating or terminating decimals. The Babylonians wrote their numbers in a system based on 60 in the same way that we write ours based on 10. They would carry as many places for π as the accuracy of the problem demanded, just as we now use

$$\pi \approx 3\frac{1}{7} \quad \text{or} \quad \pi \approx 3.1416 \quad \text{or} \quad \pi \approx 3.14159$$

$$\text{or} \quad \pi \approx 3.14159265358979$$

depending on how accurate we need to be.

Things were very different for the Greeks, whose number system allowed only rational numbers. When it was discovered that $\sqrt{2}$ was not a rational number, this was regarded as a fundamental flaw in the number concept. So serious was the matter that the Pythagorean Brotherhood (an early mathematical society) is said to have drowned one of its members for revealing this terrible secret.

Greek mathematicians then turned away from the number concept, expressing facts about whole numbers in terms of line segments.

In astronomy, however, Babylonian methods, including the Babylonian number system, continued to be used. Simon Stevin (1548-1620), probably using the Babylonian system as a model, invented the decimal system, complete with rules of calculation, in 1585. [Others, for example, al-Kashi of Samarkand (d. 1429), had made some progress in the same direction.] The decimal system so effectively conceals the difficulties that the need for more logical precision began to be felt only in the early 1800s. Around 1880, Georg Cantor (1845-1918) and Richard Dedekind (1831-1916) gave precise definitions of real numbers. Cantor's definition, although more abstract and precise, has its roots in the decimal (and hence Babylonian) numerical system.

Sets and set theory were a spin-off of the research that went into clarifying the foundations of the real number system. Set theory has developed into a large discipline of its own, and many mathematicians regard it as the foundation upon which modern mathematics is built. Cantor's discoveries that infinite sets can also be counted and that there are different sizes of infinite sets are among the most astounding results of modern mathematics.

R.1 Assess Your Understanding

Concepts and Vocabulary

1. The numbers in the set $\left\{ x \mid x = \dfrac{a}{b}, \text{where } a, b \text{ are integers} \right.$ and $b \neq 0 \left.\right\}$ are called _____ numbers.

2. The value of the expression $4 + 5 \cdot 6 - 3$ is _____.

3. The fact that $2x + 3x = (2 + 3)x$ is a consequence of the _____ Property.

4. *Multiple Choice* Which of the following represents "the product of 5 and $x + 3$ equals 6"?
   (a) $5 + (x + 3)$ (b) $5 \cdot x + 3 = 6$
   (c) $5(x + 3) = 6$ (d) None of these

5. *Multiple Choice* The intersection of sets A and B is denoted by which of the following?
   (a) $A \cap B$ (b) $A \cup B$ (c) $A \subseteq B$ (d) $A \varnothing B$

6. *Multiple Choice* Choose the correct name for the set of numbers $\{0, 1, 2, 3, \dots\}$.
   (a) Counting numbers   (b) Whole numbers
   (c) Integers   (d) Irrational numbers

7. *True or False* Rational numbers have decimals that either terminate or are nonterminating with a repeating block of digits.

8. *True or False* The Zero-Product Property states that the product of any number and zero equals zero.

9. *True or False* The least common multiple of 12 and 18 is 6.

10. *True or False* No real number is both rational and irrational.

Skill Building

In Problems 11–22, use U = universal set = $\{0, 1, 2, 3, 4, 5, 6, 7, 8, 9\}$, $A = \{1, 3, 4, 5, 9\}$, $B = \{2, 4, 6, 7, 8\}$, and $C = \{1, 3, 4, 6\}$ to find each set.

11. $A \cup B$ 12. $A \cup C$ 13. $A \cap B$ 14. $A \cap C$

15. $(A \cup B) \cap C$ 16. $(A \cap B) \cup C$ 17. $\overline{A}$ 18. $\overline{C}$

19. $\overline{A \cap B}$ 20. $\overline{B \cup C}$ 21. $\overline{A} \cup \overline{B}$ 22. $\overline{B} \cap \overline{C}$

In Problems 23–28, list the numbers in each set that are (a) Natural numbers, (b) Integers, (c) Rational numbers, (d) Irrational numbers, (e) Real numbers.

23. $A = \left\{ -6, \dfrac{1}{2}, -1.333 \dots \text{(the 3's repeat)}, \pi, 2, 5 \right\}$

24. $B = \left\{ -\dfrac{5}{3}, 2.060606 \dots \text{(the block 06 repeats)}, 1.25, 0, 1, \sqrt{5} \right\}$

25. $C = \left\{ 0, 1, \dfrac{1}{2}, \dfrac{1}{3}, \dfrac{1}{4} \right\}$

26. $D = \{ -1, -1.1, -1.2, -1.3 \}$

27. $E = \left\{ \sqrt{2}, \pi, \sqrt{2} + 1, \pi + \dfrac{1}{2} \right\}$

28. $F = \left\{ -\sqrt{2}, \pi + \sqrt{2}, \dfrac{1}{2} + 10.3 \right\}$

In Problems 29–40, approximate each number (a) rounded and (b) truncated to three decimal places.

29. 18.9526   **30.** 25.86134   **31.** 28.65319   **32.** 99.05249   **33.** 0.06291   **34.** 0.05388

35. 9.9985   **36.** 1.0006   **37.** $\dfrac{3}{7}$ **38.** $\dfrac{5}{9}$ **39.** $\dfrac{521}{15}$ **40.** $\dfrac{81}{5}$

In Problems 41–50, write each statement using symbols.

41. The sum of 3 and 2 equals 5.

42. The product of 5 and 2 equals 10.

43. The sum of x and 2 is the product of 3 and 4.

44. The sum of 3 and y is the sum of 2 and 2.

45. The product of 3 and y is the sum of 1 and 2.

46. The product of 2 and x is the product of 4 and 6.

47. The difference x less 2 equals 6.

48. The difference 2 less y equals 6.

49. The quotient x divided by 2 is 6.

50. The quotient 2 divided by x is 6.

In Problems 51–88, evaluate each expression.

51. $9 - 4 + 2$

52. $6 - 4 + 3$

53. $-6 + 4 \cdot 3$

54. $8 - 4 \cdot 2$

55. $18 - 5 \cdot 2$

56. $100 - 10 \cdot 2$

57. $4 + \dfrac{1}{3}$

58. $2 - \dfrac{1}{2}$

59. $6 - [3 \cdot 5 + 2 \cdot (3 - 2)]$

60. $2 \cdot [8 - 3(4 + 2)] - 3$

61. $4 \cdot (9 + 5) - 6 \cdot 7 + 3$

62. $1 - (4 \cdot 3 - 2 + 2)$

63. $10 - [6 - 2 \cdot 2 + (8 - 3)] \cdot 2$

64. $2 - 5 \cdot 4 - [6 \cdot (3 - 4)]$

65. $(5 - 3)\dfrac{1}{2}$

66. $(5 + 4)\dfrac{1}{3}$

67. $\dfrac{4 + 8}{5 - 3}$

68. $\dfrac{2 - 4}{5 - 3}$

69. $\dfrac{3}{5} \cdot \dfrac{10}{21}$

70. $\dfrac{5}{9} \cdot \dfrac{3}{10}$

71. $\dfrac{6}{25} \cdot \dfrac{10}{27}$

72. $\dfrac{21}{25} \cdot \dfrac{100}{3}$

73. $\dfrac{3}{4} + \dfrac{2}{5}$

74. $\dfrac{4}{3} + \dfrac{1}{2}$

75. $\dfrac{7}{8} + \dfrac{4}{7}$

76. $\dfrac{8}{9} + \dfrac{15}{2}$

77. $\dfrac{5}{18} + \dfrac{1}{12}$

78. $\dfrac{2}{15} + \dfrac{8}{9}$

79. $\dfrac{5}{24} - \dfrac{8}{15}$

80. $\dfrac{3}{14} - \dfrac{2}{21}$

81. $\dfrac{3}{20} - \dfrac{2}{15}$

82. $\dfrac{6}{35} - \dfrac{3}{14}$

83. $\dfrac{\dfrac{5}{18}}{\dfrac{11}{27}}$

84. $\dfrac{\dfrac{5}{21}}{\dfrac{2}{35}}$

85. $\dfrac{1}{3} \cdot \dfrac{4}{7} + \dfrac{17}{21}$

86. $\dfrac{2}{3} + \dfrac{4}{5} \cdot \dfrac{1}{6}$

87. $2 \cdot \dfrac{3}{4} + \dfrac{3}{8}$

88. $3 \cdot \dfrac{5}{6} - \dfrac{1}{2}$

In Problems 89–100, use the Distributive Property to remove the parentheses.

89. $6(x + 4)$

90. $4(2x - 1)$

91. $x(x - 4)$

92. $4x(x + 3)$

93. $2\left(\dfrac{3}{4}x - \dfrac{1}{2}\right)$

94. $3\left(\dfrac{2}{3}x + \dfrac{1}{6}\right)$

95. $(x + 2)(x + 4)$

96. $(x + 5)(x + 1)$

97. $(x + 9)(2x - 7)$

98. $(3x - 1)(x + 5)$

99. $(x - 8)(x - 2)$

100. $(x - 4)(x - 2)$

101. Challenge Problem Find k if $3x(x - 5k) = 3x^2 - 60x$.

102. Challenge Problem Find k if $(x - k)(x + 3k) = x^2 + 4x - 12$.

Explaining Concepts: Discussion and Writing

103. Explain to a friend how the Distributive Property is used to justify the fact that $2x + 3x = 5x$.

104. Explain to a friend why $2 + 3 \cdot 4 = 14$, whereas $(2 + 3) \cdot 4 = 20$.

105. Explain why $2(3 \cdot 4)$ is not equal to $(2 \cdot 3) \cdot (2 \cdot 4)$.

106. Explain why $\dfrac{4 + 3}{2 + 5}$ is not equal to $\dfrac{4}{2} + \dfrac{3}{5}$.

107. Is subtraction commutative? Support your conclusion with an example.

108. Is subtraction associative? Support your conclusion with an example.

109. Is division commutative? Support your conclusion with an example.

110. Is division associative? Support your conclusion with an example.

111. If $2 = x$, why does $x = 2$?

112. If $x = 5$, why does $x^2 + x = 30$?

113. Are there any real numbers that are both rational and irrational? Are there any real numbers that are neither? Explain your reasoning.

114. Explain why the sum of a rational number and an irrational number must be irrational.

115. A rational number is defined as the quotient of two integers. When written as a decimal, the decimal will either repeat or terminate. By looking at the denominator of the rational number, there is a way to tell in advance whether its decimal representation will repeat or terminate. Make a list of rational numbers and their decimals. See if you can discover the pattern. Confirm your conclusion by consulting books on number theory at the library. Write a brief essay on your findings.

116. The current time is 12 noon CST. What time (CST) will it be 12,997 hours from now?

117. Both $\frac{a}{0}$ ($a \neq 0$) and $\frac{0}{0}$ are undefined, but for different reasons. Write a paragraph or two explaining the different reasons.

R.2 Algebra Essentials

OBJECTIVES 1 Graph Inequalities (p. 18)
2 Find Distance on the Real Number Line (p. 19)
3 Evaluate Algebraic Expressions (p. 20)
4 Determine the Domain of a Variable (p. 21)
5 Use the Laws of Exponents (p. 21)
6 Evaluate Square Roots (p. 23)
7 Use a Calculator to Evaluate Exponents (p. 24)
8 Use Scientific Notation (p. 24)

The Real Number Line

Real numbers can be represented by points on a line called the **real number line**. There is a one-to-one correspondence between real numbers and points on a line. That is, every real number corresponds to a point on the line, and each point on the line has a unique real number associated with it.

Pick a point on a line somewhere in the center, and label it O. This point, called the **origin**, corresponds to the real number 0. See Figure 8. The point 1 unit to the right of O corresponds to the number 1. The distance between 0 and 1 determines the **scale** of the number line. For example, the point associated with the number 2 is twice as far from O as 1. Notice that an arrowhead on the right end of the line indicates the direction in which the numbers increase. Points to the left of the origin correspond to the real numbers $-1, -2$, and so on. Figure 8 also shows the points associated with the rational numbers $-\frac{1}{2}$ and $\frac{1}{2}$ and with the irrational numbers $\sqrt{2}$ and π.

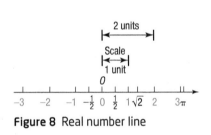

Figure 8 Real number line

DEFINITION Coordinate; Real Number Line

The real number associated with a point P is called the **coordinate** of P, and the line whose points have been assigned coordinates is called the **real number line**.

- Now Work PROBLEM 13

The real number line consists of three classes of real numbers, as shown in Figure 9.

Figure 9

- The **negative real numbers** are the coordinates of points to the left of the origin O.
- The real number **zero** is the coordinate of the origin O.
- The **positive real numbers** are the coordinates of points to the right of the origin O.

Multiplication Properties of Positive and Negative Numbers

- The product of two positive numbers is a positive number.
- The product of two negative numbers is a positive number.
- The product of a positive number and a negative number is a negative number.

1 Graph Inequalities

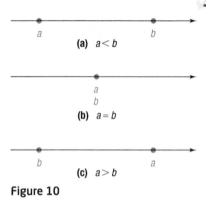

(a) $a < b$

(b) $a = b$

(c) $a > b$

Figure 10

An important property of the real number line follows from the fact that, given two numbers a and b, either a is to the left of b, or a is at the same location as b, or a is to the right of b. See Figure 10.

If a is to the left of b, then "a is less than b," which is written $a < b$. If a is to the right of b, then "a is greater than b," which is written $a > b$. If a is at the same location as b, then $a = b$. If a is either less than or equal to b, then $a \leq b$. Similarly, $a \geq b$ means that a is either greater than or equal to b. Collectively, the symbols $<, >, \leq,$ and $\geq$ are called **inequality symbols**.

Note that $a < b$ and $b > a$ mean the same thing. It does not matter whether we write $2 < 3$ or $3 > 2$.

Furthermore, if $a < b$ or if $b > a$, then the difference $b - a$ is positive. Do you see why?

EXAMPLE 1 **Using Inequality Symbols**

(a) $3 < 7$ (b) $-8 > -16$ (c) $-6 < 0$

(d) $-8 < -4$ (e) $4 > -1$ (f) $8 > 0$

In Example 1(a), we conclude that $3 < 7$ either because 3 is to the left of 7 on the real number line or because the difference, $7 - 3 = 4$, is a positive real number.

Similarly, we conclude in Example 1(b) that $-8 > -16$ either because -8 lies to the right of -16 on the real number line or because the difference, $-8 - (-16) = -8 + 16 = 8$, is a positive real number.

Look again at Example 1. Note that the inequality symbol always points in the direction of the smaller number.

An **inequality** is a statement in which two expressions are related by an inequality symbol. The expressions are referred to as the **sides** of the inequality. Inequalities of the form $a < b$ or $b > a$ are called **strict inequalities**, whereas inequalities of the form $a \leq b$ or $b \geq a$ are called **nonstrict inequalities**.

Based on the discussion so far, we conclude that

$$
\begin{array}{ll}
a > 0 & \text{is equivalent to}\quad a \text{ is positive} \\
a < 0 & \text{is equivalent to}\quad a \text{ is negative}
\end{array}
$$

We sometimes read $a > 0$ by saying that "a is positive." If $a \geq 0$, then either $a > 0$ or $a = 0$, and we may read this as "a is nonnegative."

Now Work PROBLEMS **17** AND **27**

EXAMPLE 2

Graphing Inequalities

(a) On the real number line, graph all numbers x for which $x > 4$.
(b) On the real number line, graph all numbers x for which $x \leq 5$.

Solution

Figure 11 $x > 4$

Figure 12 $x \leq 5$

(a) See Figure 11. Notice that we use a left parenthesis to indicate that the number 4 is *not* part of the graph.

(b) See Figure 12. Notice that we use a right bracket to indicate that the number 5 *is* part of the graph.

↗ **Now Work** PROBLEM 33

2 Find Distance on the Real Number Line

Figure 13

The *absolute value* of a number a is the distance from 0 to a on the number line. For example, -4 is 4 units from 0, and 3 is 3 units from 0. See Figure 13. That is, the absolute value of -4 is 4, and the absolute value of 3 is 3.

A more formal definition of absolute value is given next.

> **DEFINITION** **Absolute Value**
>
> The **absolute value** of a real number a, denoted by the symbol $|a|$, is defined by the rules
>
> $$|a| = a \quad \text{if } a \geq 0 \qquad \text{and} \qquad |a| = -a \quad \text{if } a < 0$$

For example, because $-4 < 0$, the second rule must be used to get

$$|-4| = -(-4) = 4$$

EXAMPLE 3

Computing Absolute Value

(a) $|8| = 8$ (b) $|0| = 0$ (c) $|-15| = -(-15) = 15$

Look again at Figure 13. The distance from -4 to 3 is 7 units. This distance is the difference $3 - (-4)$, obtained by subtracting the smaller coordinate from the larger. However, since $|3 - (-4)| = |7| = 7$ and $|-4 - 3| = |-7| = 7$, we can use absolute value to calculate the distance between two points without being concerned about which is smaller.

> **DEFINITION** **Distance Between Two Points**
>
> If P and Q are two points on a real number line with coordinates a and b, respectively, the **distance between P and Q**, denoted by $d(P, Q)$, is
>
> $$d(P, Q) = |b - a|$$

Since $|b - a| = |a - b|$, it follows that $d(P, Q) = d(Q, P)$.

EXAMPLE 4

Finding Distance on a Number Line

Let P, Q, and R be points on a real number line with coordinates -5, 7, and -3, respectively. Find the distance

(a) between P and Q (b) between Q and R

Solution See Figure 14.

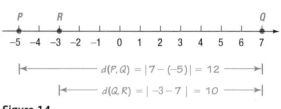

Figure 14

(a) $d(P, Q) = |7 - (-5)| = |12| = 12$

(b) $d(Q, R) = |-3 - 7| = |-10| = 10$

➤ **Now Work** PROBLEM 39

3 Evaluate Algebraic Expressions

Remember, in algebra we use letters such as x, y, a, b, and c to represent numbers. If a letter used is to represent *any* number from a given set of numbers, it is called a **variable**. A **constant** is either a fixed number, such as 5 or $\sqrt{3}$, or a letter that represents a fixed (possibly unspecified) number.

Constants and variables are combined using the operations of addition, subtraction, multiplication, and division to form *algebraic expressions*. Examples of algebraic expressions include

$$x + 3 \qquad \frac{3}{1 - t} \qquad 7x - 2y$$

To evaluate an algebraic expression, substitute a numerical value for each variable.

EXAMPLE 5

Evaluating an Algebraic Expression

Evaluate each expression if $x = 3$ and $y = -1$.

(a) $x + 3y$ (b) $5xy$ (c) $\dfrac{3y}{2 - 2x}$ (d) $|-4x + y|$

Solution (a) Substitute 3 for x and -1 for y in the expression $x + 3y$.

$$x + 3y = \underset{\underset{x=3,\, y=-1}{\uparrow}}{3} + 3(-1) = 3 + (-3) = 0$$

(b) If $x = 3$ and $y = -1$, then

$$5xy = 5 \cdot 3 \cdot (-1) = -15$$

(c) If $x = 3$ and $y = -1$, then

$$\frac{3y}{2 - 2x} = \frac{3(-1)}{2 - 2 \cdot 3} = \frac{-3}{2 - 6} = \frac{-3}{-4} = \frac{3}{4}$$

(d) If $x = 3$ and $y = -1$, then

$$|-4x + y| = |-4 \cdot 3 + (-1)| = |-12 + (-1)| = |-13| = 13$$

➤ **Now Work** PROBLEMS 41 AND 49

4 Determine the Domain of a Variable

In working with expressions or formulas involving variables, the variables may be allowed to take on values from only a certain set of numbers. For example, in the formula for the area A of a circle of radius r, $A = \pi r^2$, the variable r is necessarily restricted to the positive real numbers. In the expression $\dfrac{1}{x}$, the variable x cannot take on the value 0, since division by 0 is not defined.

> **DEFINITION Domain of a Variable**
>
> The set of values that a variable may assume is called the **domain of the variable**.

| EXAMPLE 6 | **Finding the Domain of a Variable** |

The domain of the variable x in the expression

$$\frac{5}{x - 2}$$

is $\{x | x \neq 2\}$ since, if $x = 2$, the denominator becomes 0, which is not defined.

| EXAMPLE 7 | **Circumference of a Circle** |

In the formula for the circumference C of a circle of radius r,

$$C = 2\pi r$$

the domain of the variable r, representing the radius of the circle, is the set of positive real numbers, $\{r | r > 0\}$. The domain of the variable C, representing the circumference of the circle, is also the set of positive real numbers, $\{C | C > 0\}$.

In describing the domain of a variable, we may use either set notation or words, whichever is more convenient.

─ **Now Work** PROBLEM 59

5 Use the Laws of Exponents

Integer exponents provide a shorthand notation for representing repeated multiplications of a real number. For example,

$$3^4 = 3 \cdot 3 \cdot 3 \cdot 3 = 81$$

Additionally, many formulas have exponents. For example,

- The formula for the horsepower rating H of an engine is

$$H = \frac{D^2 N}{2.5}$$

  where D is the diameter of a cylinder and N is the number of cylinders.

- A formula for the resistance R of blood flowing in a blood vessel is

$$R = C\frac{L}{r^4}$$

  where L is the length and r is the radius of the blood vessel, and C is a positive constant.

DEFINITION a^n

If a is a real number and n is a positive integer, then the symbol a^n represents the product of n factors of a. That is,

$$a^n = \underbrace{a \cdot a \cdot \ldots \cdot a}_{n \text{ factors}} \qquad \text{(1)}$$

WARNING Be careful with minus signs and exponents.

$$-2^4 = -1 \cdot 2^4 = -16$$

whereas

$$(-2)^4 = (-2)(-2)(-2)(-2) = 16 \ \blacksquare$$

In the definition it is understood that $a^1 = a$. Furthermore, $a^2 = a \cdot a$, $a^3 = a \cdot a \cdot a$, and so on. In the expression a^n, a is called the **base** and n is called the **exponent**, or **power**. We read a^n as "a raised to the power n" or as "a to the nth power." We usually read a^2 as "a squared" and a^3 as "a cubed."

In working with exponents, the operation of *raising to a power* is performed before any other operation. As examples,

$$4 \cdot 3^2 = 4 \cdot 9 = 36 \qquad 2^2 + 3^2 = 4 + 9 = 13$$

$$-2^4 = -16 \qquad 5 \cdot 3^2 + 2 \cdot 4 = 5 \cdot 9 + 2 \cdot 4 = 45 + 8 = 53$$

Parentheses are used to indicate operations to be performed first. For example,

$$(-2)^4 = (-2)(-2)(-2)(-2) = 16 \qquad (2 + 3)^2 = 5^2 = 25$$

DEFINITION a^0

If $a \neq 0$, then

$$a^0 = 1$$

DEFINITION a^{-n}

If $a \neq 0$ and if n is a positive integer, then

$$a^{-n} = \frac{1}{a^n}$$

Whenever you encounter a negative exponent, think "reciprocal."

EXAMPLE 8 **Evaluating Expressions Containing Negative Exponents**

(a) $2^{-3} = \dfrac{1}{2^3} = \dfrac{1}{8}$ (b) $x^{-4} = \dfrac{1}{x^4}$ (c) $\left(\dfrac{1}{5}\right)^{-2} = \dfrac{1}{\left(\dfrac{1}{5}\right)^2} = \dfrac{1}{\dfrac{1}{25}} = 25$

- **Now Work** PROBLEMS **77** AND **97**

The following properties, called the **Laws of Exponents**, can be proved using the preceding definitions. In the list, a and b are real numbers, and m and n are integers.

THEOREM **Laws of Exponents**

$$a^m a^n = a^{m+n} \qquad (a^m)^n = a^{mn} \qquad (ab)^n = a^n b^n$$

$$\frac{a^m}{a^n} = a^{m-n} = \frac{1}{a^{n-m}} \ \text{if } a \neq 0 \qquad \left(\frac{a}{b}\right)^n = \frac{a^n}{b^n} \ \text{if } b \neq 0$$

EXAMPLE 9

Using the Laws of Exponents

(a) $x^{-3} \cdot x^5 = x^{-3+5} = x^2 \quad x \neq 0$

(b) $(x^{-3})^2 = x^{-3 \cdot 2} = x^{-6} = \dfrac{1}{x^6} \quad x \neq 0$

(c) $(2x)^3 = 2^3 \cdot x^3 = 8x^3$

(d) $\left(\dfrac{2}{3}\right)^4 = \dfrac{2^4}{3^4} = \dfrac{16}{81}$

NOTE Always write the final answer using positive exponents. ∎

(e) $\dfrac{x^{-2}}{x^{-5}} = x^{-2-(-5)} = x^3 \quad x \neq 0$

Now Work PROBLEM 79

EXAMPLE 10

Using the Laws of Exponents

Write each expression so that all exponents are positive.

(a) $\dfrac{x^5 y^{-2}}{x^3 y} \quad x \neq 0, \quad y \neq 0$

(b) $\left(\dfrac{x^{-3}}{3y^{-1}}\right)^{-2} \quad x \neq 0, \quad y \neq 0$

Solution

(a) $\dfrac{x^5 y^{-2}}{x^3 y} = \dfrac{x^5}{x^3} \cdot \dfrac{y^{-2}}{y} = x^{5-3} \cdot y^{-2-1} = x^2 y^{-3} = x^2 \cdot \dfrac{1}{y^3} = \dfrac{x^2}{y^3}$

(b) $\left(\dfrac{x^{-3}}{3y^{-1}}\right)^{-2} = \dfrac{(x^{-3})^{-2}}{(3y^{-1})^{-2}} = \dfrac{x^6}{3^{-2}(y^{-1})^{-2}} = \dfrac{x^6}{\dfrac{1}{9}y^2} = \dfrac{9x^6}{y^2}$

Now Work PROBLEM 89

6 Evaluate Square Roots

In Words
$\sqrt{36}$ means "what is the nonnegative number whose square is 36?"

A real number is squared when it is raised to the power 2. The inverse of squaring is finding a **square root**. For example, since $6^2 = 36$ and $(-6)^2 = 36$, the numbers 6 and -6 are square roots of 36.

The symbol $\sqrt{}$, called a **radical sign**, is used to denote the **principal**, or nonnegative, square root. For example, $\sqrt{36} = 6$.

DEFINITION Principal Square Root

If a is a nonnegative real number, the nonnegative number b for which $b^2 = a$ is the **principal square root** of a, and is denoted by $b = \sqrt{a}$.

The following comments are noteworthy:

- Negative numbers do not have square roots (in the real number system), because the square of any real number is *nonnegative*. For example, $\sqrt{-4}$ is not a real number, because there is no real number whose square is -4.
- The principal square root of 0 is 0, since $0^2 = 0$. That is, $\sqrt{0} = 0$.
- The principal square root of a positive number is positive.
- If $c \geq 0$, then $(\sqrt{c})^2 = c$. For example, $(\sqrt{2})^2 = 2$ and $(\sqrt{3})^2 = 3$.

EXAMPLE 11

Evaluating Square Roots

(a) $\sqrt{64} = 8$

(b) $\sqrt{\dfrac{1}{16}} = \dfrac{1}{4}$

(c) $(\sqrt{1.4})^2 = 1.4$

Examples 11(a) and (b) are examples of square roots of perfect squares, since $64 = 8^2$ and $\dfrac{1}{16} = \left(\dfrac{1}{4}\right)^2$.

Consider the expression $\sqrt{a^2}$. Since $a^2 \geq 0$, the principal square root of a^2 is defined whether $a > 0$ or $a < 0$. However, since the principal square root is nonnegative, we need an absolute value to ensure the nonnegative result. That is,

$$\sqrt{a^2} = |a| \quad a \text{ any real number} \tag{2}$$

EXAMPLE 12 **Using Equation (2)**

(a) $\sqrt{(2.3)^2} = |2.3| = 2.3$

(b) $\sqrt{(-2.3)^2} = |-2.3| = 2.3$

(c) $\sqrt{x^2} = |x|$

 Now Work PROBLEM 85

7 Use a Calculator to Evaluate Exponents

Your calculator has either a caret key, $\boxed{\wedge}$, or an $\boxed{x^y}$ key, that is used for computations involving exponents.

EXAMPLE 13 **Exponents on a Graphing Calculator**

Evaluate: $(2.3)^5$

Solution Figure 15 shows the result using a TI-84 Plus C graphing calculator.

 Now Work PROBLEM 115

NORMAL FLOAT AUTO REAL RADIAN MP

2.3⁵

64.36343

Figure 15

8 Use Scientific Notation

Measurements of physical quantities can range from very small to very large. For example, the mass of a proton is approximately 0.00000000000000000000000000167 kilogram and the mass of Earth is about 5,980,000,000,000,000,000,000,000 kilograms. These numbers obviously are tedious to write down and difficult to read, so we use exponents to rewrite them.

DEFINITION Scientific Notation

A number written as the product of a number x, where $1 \leq x < 10$, times a power of 10, is written in **scientific notation**.

In scientific notation,

$$\text{Mass of a proton} = 1.67 \times 10^{-27} \text{ kilogram}$$
$$\text{Mass of Earth} = 5.98 \times 10^{24} \text{ kilograms}$$

Converting a Decimal to Scientific Notation

To change a positive number into scientific notation:

- Count the number N of places that the decimal point must be moved to arrive at a number x, where $1 \leq x < 10$.
- If the original number is greater than or equal to 1, the scientific notation is $x \times 10^N$. If the original number is between 0 and 1, the scientific notation is $x \times 10^{-N}$.

EXAMPLE 14

Using Scientific Notation

Write each number in scientific notation.

(a) 9582   (b) 1.245   (c) 0.285   (d) 0.000561

Solution   (a) The decimal point in 9582 follows the 2. Count left from the decimal point

$$9\quad 5\quad 8\quad 2\quad .$$

$$\underset{3\quad 2\quad 1}{}$$

stopping after three moves, because 9.582 is a number between 1 and 10. Since 9582 is greater than 1, we write

$$9582 = 9.582 \times 10^3$$

(b) The decimal point in 1.245 is between the 1 and 2. Since the number is already between 1 and 10, the scientific notation for it is $1.245 \times 10^0 = 1.245$.

(c) The decimal point in 0.285 is between the 0 and the 2. We count

$$0\quad .\quad 2\quad 8\quad 5$$

$$\underset{1}{}$$

stopping after one move, because 2.85 is a number between 1 and 10. Since 0.285 is between 0 and 1, we write

$$0.285 = 2.85 \times 10^{-1}$$

(d) The decimal point in 0.000561 is moved as follows:

$$0\quad .\quad 0\quad 0\quad 0\quad 5\quad 6\quad 1$$

$$\underset{1\quad 2\quad 3\quad 4}{}$$

As a result,

$$0.000561 = 5.61 \times 10^{-4}$$

—Now Work PROBLEM **121**

EXAMPLE 15

Changing from Scientific Notation to Decimals

Write each number as a decimal.

(a) 2.1×10^4 (b) 3.26×10^{-5} (c) 1×10^{-2}

Solution   (a) $2.1 \times 10^4 = 2\quad .\quad 1\quad 0\quad 0\quad 0\quad \times 10^4 = 21{,}000$

$$\underset{1\quad 2\quad 3\quad 4}{}$$

(b) $3.26 \times 10^{-5} = 0\quad 0\quad 0\quad 0\quad 0\quad 3\quad .\quad 2\quad 6 \times 10^{-5} = 0.0000326$

$$\underset{5\quad 4\quad 3\quad 2\quad 1}{}$$

(c) $1 \times 10^{-2} = 0\quad 0\quad 1\quad .\quad 0 \times 10^{-2} = 0.01$

$$\underset{2\quad 1}{}$$

—Now Work PROBLEM **129**

EXAMPLE 16	Using Scientific Notation

(a) The diameter of the smallest living cell is only about 0.00001 centimeter (cm).* Express this number in scientific notation.

(b) The surface area of Earth is about 1.97×10^8 square miles.† Express the surface area as a whole number.

Solution

(a) 0.00001 cm $= 1 \times 10^{-5}$ cm because the decimal point is moved five places and the number is less than 1.

(b) 1.97×10^8 square miles $= 197{,}000{,}000$ square miles.

➤ **Now Work** PROBLEM 155

COMMENT On a calculator, a number such as 3.615×10^{12} is usually displayed as $\boxed{3.615E12.}$ ■

* *Powers of Ten*, Philip and Phylis Morrison.

† *2011 Information Please Almanac.*

Historical Feature

The word *algebra* is derived from the Arabic word *al-jabr*. This word is a part of the title of a ninth-century work, "Hisâb al-jabr w'al-muqâbalah," written by Mohammed ibn Músâ al-Khwârizmî. The word *al-jabr* means "a restoration," a reference to the fact that if a number is added to one side of an equation, then it must also be added to the other side in order to "restore" the equality. The title of the work, freely translated, is "The Science of Reduction and Cancellation." Of course, today, algebra has come to mean a great deal more.

R.2 Assess Your Understanding

Concepts and Vocabulary

1. A(n) _____ is a letter used in algebra to represent any number from a given set of numbers.

2. On the real number line, the real number zero is the coordinate of the _____.

3. An inequality of the form $a > b$ is called a(n) _____ inequality.

4. In the expression 2^4, the number 2 is called the _____ and 4 is called the _____.

5. In scientific notation, $1234.5678 =$ _____.

6. *Multiple Choice* If a is a nonnegative real number, then which inequality statement best describes a?
   (a) $a < 0$ (b) $a > 0$ (c) $a \le 0$ (d) $a \ge 0$

7. *Multiple Choice* The set of values that a variable may assume is called the _____ of the variable.
   (a) domain (b) range (c) coordinate (d) none of these

8. *Multiple Choice* Let a and b be non-zero real numbers and m and n be integers. Which of the following is not a law of exponents?
   (a) $\left(\dfrac{a}{b}\right)^n = \dfrac{a^n}{b^n}$ (b) $(a^m)^n = a^{m+n}$
   (c) $\dfrac{a^m}{a^n} = a^{m-n}$ (d) $(ab)^n = a^n b^n$

9. *True or False* The distance between two distinct points on the real number line is always greater than zero.

10. *True or False* The absolute value of a real number is always greater than zero.

11. *True or False* When a number is expressed in scientific notation, it is expressed as the product of a number x, $0 \le x < 1$, and a power of 10.

12. *True or False* The inverse of squaring is finding a square root.

Skill Building

13. On the real number line, label the points with coordinates $0, 1, -1, \dfrac{5}{2}, -2.5, \dfrac{3}{4}$, and 0.25.

14. Repeat Problem 13 for the coordinates $0, -2, 2, -1.5, \dfrac{3}{2}, \dfrac{1}{3}$, and $\dfrac{2}{3}$.

In Problems 15–24, replace the question mark by $<$, $>$, or $=$, whichever is correct.

15. $\dfrac{1}{2}$? 0

16. 5 ? 6

17. -1 ? -2

18. -3 ? $-\dfrac{5}{2}$

19. π ? 3.14

20. $\sqrt{2}$? 1.41

21. $\dfrac{1}{2}$? 0.5

22. $\dfrac{1}{3}$? 0.33

23. $\dfrac{2}{3}$? 0.67

24. $\dfrac{1}{4}$? 0.25

In Problems 25–30, write each statement as an inequality.

25. x is positive

26. z is negative

27. x is less than 2

28. y is greater than -5

29. x is less than or equal to 1

30. x is greater than or equal to 2

In Problems 31–34, graph the numbers x on the real number line.

31. $x \geq -2$

32. $x < 4$

33. $x > -1$

34. $x \leq 7$

In Problems 35–40, use the given real number line to compute each distance.

$$\overset{\text{A}}{} \quad \overset{\text{B}}{} \overset{\text{C}}{} \overset{\text{D}}{} \quad \overset{\text{E}}{}$$

$$-4 \;\; -3 \;\; -2 \;\; -1 \;\; 0 \;\; 1 \;\; 2 \;\; 3 \;\; 4 \;\; 5 \;\; 6$$

35. $d(C, D)$

36. $d(C, A)$

37. $d(D, E)$

38. $d(C, E)$

39. $d(A, E)$

40. $d(D, B)$

In Problems 41–48, evaluate each expression if x = −2 and y = 3.

41. $x + 2y$

42. $3x + y$

43. $5xy + 2$

44. $-2x + xy$

45. $\dfrac{2x}{x - y}$

46. $\dfrac{x + y}{x - y}$

47. $\dfrac{3x + 2y}{2 + y}$

48. $\dfrac{2x - 3}{y}$

In Problems 49–58, find the value of each expression if x = 3 and y = −2.

49. $|x + y|$

50. $|x - y|$

51. $|x| + |y|$

52. $|x| - |y|$

53. $\dfrac{|x|}{x}$

54. $\dfrac{|y|}{y}$

55. $|4x - 5y|$

56. $|3x + 2y|$

57. $||4x| - |5y||$

58. $3|x| + 2|y|$

In Problems 59–66, determine which of the values (a) through (d), if any, must be excluded from the domain of the variable in each expression.

(a) $x = 0$ (b) $x = 1$ (c) $x = 0$ (d) $x = -1$

59. $\dfrac{x^2 - 1}{x}$

60. $\dfrac{x^2 + 1}{x}$

61. $\dfrac{x}{x^2 - 9}$

62. $\dfrac{x}{x^2 + 9}$

63. $\dfrac{x^2}{x^2 + 1}$

64. $\dfrac{x^3}{x^2 - 1}$

65. $\dfrac{x^2 + 5x - 10}{x^3 - x}$

66. $\dfrac{-9x^2 - x + 1}{x^3 + x}$

In Problems 67–70, determine the domain of the variable x in each expression.

67. $\dfrac{4}{x - 5}$

68. $\dfrac{-6}{x + 4}$

69. $\dfrac{x}{x + 4}$

70. $\dfrac{x - 2}{x - 6}$

In Problems 71–74, use the formula $C = \dfrac{5}{9}(F - 32)$ for converting degrees Fahrenheit into degrees Celsius to find the Celsius measure of each Fahrenheit temperature.

71. $F = 32°$

72. $F = 212°$

73. $F = 77°$

74. $F = -4°$

In Problems 75–86, simplify each expression.

75. $(-9)^2$

76. -4^2

77. 4^{-2}

78. -4^{-2}

79. $3^{-6} \cdot 3^4$

80. $4^{-2} \cdot 4^3$

81. $(4^{-3})^{-1}$

82. $(2^{-1})^{-3}$

83. $\sqrt{100}$

84. $\sqrt{36}$

85. $\sqrt{(-4)^2}$

86. $\sqrt{(-3)^2}$

In Problems 87–96, simplify each expression. Express the answer so that all exponents are positive. Whenever an exponent is 0 or negative, we assume that the base is not 0.

87. $(9x^4)^2$

88. $(-4x^2)^{-1}$

89. $(x^2 y^{-1})^2$

90. $(x^{-1} y)^3$

91. $\dfrac{x^2 y^5}{x^3 y^4}$

92. $\dfrac{x^{-2} y}{xy^2}$

93. $\dfrac{(-4)^2 y^5 (xz)^3}{(-3)^3 xy^7 z^2}$

94. $\dfrac{4x^{-2}(yz)^{-1}}{2^3 x^4 y}$

95. $\left(\dfrac{2x^{-3}}{3y^{-1}}\right)^{-2}$

96. $\left(\dfrac{5x^{-2}}{6y^{-2}}\right)^{-3}$

In Problems 97–108, find the value of each expression if $x = 2$ and $y = -1$.

97. $2xy^{-1}$

98. $-3x^{-1}y$

99. $x^2 + y^2$

100. x^2y^2

101. $(xy)^2$

102. $(x + y)^2$

103. $\sqrt{x^2}$

104. $(\sqrt{x})^2$

105. $\sqrt{x^2 + y^2}$

106. $\sqrt{x^2} + \sqrt{y^2}$

107. x^y

108. y^x

109. Find the value of the expression $2x^3 - 3x^2 + 5x - 4$ if $x = 2$. What is the value if $x = 1$?

110. Find the value of the expression $4x^3 + 3x^2 - x + 2$ if $x = 1$. What is the value if $x = 2$?

111. What is the value of $\dfrac{(666)^4}{(222)^4}$?

112. What is the value of $(0.1)^3 \cdot (20)^3$?

In Problems 113–120, use a calculator to evaluate each expression. Round your answer to three decimal places.

113. $(8.2)^6$

114. $(3.7)^5$

115. $(6.1)^{-3}$

116. $(2.2)^{-5}$

117. $(-2.8)^6$

118. $-(2.8)^6$

119. $(-8.11)^{-4}$

120. $-(8.11)^{-4}$

In Problems 121–128, write each number in scientific notation.

121. 454.2

122. 32.14

123. 0.013

124. 0.00421

125. 32,155

126. 21,210

127. 0.000423

128. 0.0514

In Problems 129–136, write each number as a decimal.

129. 6.15×10^4

130. 9.7×10^3

131. 1.214×10^{-3}

132. 9.88×10^{-4}

133. 1.1×10^8

134. 4.112×10^2

135. 8.1×10^{-2}

136. 6.453×10^{-1}

Applications and Extensions

In Problems 137–146, express each statement as an equation involving the indicated variables.

137. Area of a Rectangle The area A of a rectangle is the product of its length l and its width w.

138. Perimeter of a Rectangle The perimeter P of a rectangle is twice the sum of its length l and its width w.

139. Circumference of a Circle The circumference C of a circle is the product of π and its diameter d.

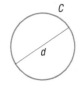

140. Area of a Triangle The area A of a triangle is one-half the product of its base b and its height h.

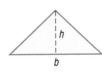

141. Area of an Equilateral Triangle The area A of an equilateral triangle is $\dfrac{\sqrt{3}}{4}$ times the square of the length x of one side.

142. Perimeter of an Equilateral Triangle The perimeter P of an equilateral triangle is 3 times the length x of one side.

143. Volume of a Sphere The volume V of a sphere is $\dfrac{4}{3}$ times π times the cube of the radius r.

144. Surface Area of a Sphere The surface area S of a sphere is 4 times π times the square of the radius r.

145. Volume of a Cube The volume V of a cube is the cube of the length x of a side.

146. Surface Area of a Cube The surface area S of a cube is 6 times the square of the length x of a side.

147. Manufacturing Cost The weekly production cost C of manufacturing x watches is given by the formula $C = 4000 + 2x$, where the variable C is in dollars.
(a) What is the cost of producing 1000 watches?
(b) What is the cost of producing 2000 watches?

148. Balancing a Checking Account At the beginning of the month, Mike had a balance of $210 in his checking account. During the next month, he deposited $80, made an ATM withdrawal for $120, made another deposit of $25, and made two electronic bill payments: one for $60 and the other for $32. He was also assessed a monthly service charge of $5. What was his balance at the end of the month?

In Problems 149 and 150, write an inequality using an absolute value to describe each statement.

149. x is at least 6 units from 4.

150. x is more than 5 units from 2.

151. U.S. Voltage In the United States, normal household voltage is 110 volts. It is acceptable for the actual voltage x to differ from normal by at most 5 volts. A formula that describes this is

$$|x - 110| \leq 5$$

(a) Show that a voltage of 108 volts is acceptable.
(b) Show that a voltage of 104 volts is not acceptable.

152. Foreign Voltage In other countries, normal household voltage is 220 volts. It is acceptable for the actual voltage x to differ from normal by at most 8 volts. A formula that describes this is

$$|x - 220| \leq 8$$

(a) Show that a voltage of 214 volts is acceptable.
(b) Show that a voltage of 209 volts is not acceptable.

153. Making Precision Ball Bearings The FireBall Company manufactures ball bearings for precision equipment. One of its products is a ball bearing with a stated radius of 3 centimeters (cm). Only ball bearings with a radius within 0.01 cm of this stated radius are acceptable. If x is the radius of a ball bearing, a formula describing this situation is

$$|x - 3| \leq 0.01$$

(a) Is a ball bearing of radius $x = 2.999$ acceptable?
(b) Is a ball bearing of radius $x = 2.89$ acceptable?

154. Body Temperature Normal human body temperature is 98.6°F. A temperature x that differs from normal by at least 1.5°F is considered unhealthy. A formula that describes this is

$$|x - 98.6| \geq 1.5$$

(a) Show that a temperature of 97°F is unhealthy.
(b) Show that a temperature of 100°F is not unhealthy.

155. Distance from Earth to Its Moon The distance from Earth to the Moon is about 4×10^8 meters.* Express this distance as a whole number.

156. Height of Mt. Everest The height of Mt. Everest is 8848 meters.* Express this height in scientific notation.

157. Wavelength of Visible Light The wavelength of visible light is about 5×10^{-7} meter.* Express this wavelength as a decimal.

158. Diameter of an Atom The diameter of an atom is about 1×10^{-10} meter.* Express this diameter as a decimal.

159. Diameter of Wire The diameter of 18-gauge wire is about 0.0403 inch. Express this diameter using scientific notation.

160. Tiny Motor The world's tiniest rotary motor is only 0.00004 millimeter tall. Express this measurement using scientific notation.

Source: Herkewitz, William. "Scientists Build the World's Tiniest Rotary Motor," Popular Mechanics. Feb. 19, 2016.

161. Astronomy One light-year is defined by astronomers to be the distance that a beam of light will travel in 1 year (365 days). If the speed of light is 186,000 miles per second, how many miles are in a light-year? Express your answer in scientific notation.

162. Astronomy How long does it take a beam of light to reach Earth from the Sun when the Sun is 93,000,000 miles from Earth? Express your answer in seconds, using scientific notation.

163. Does $\dfrac{1}{3}$ equal 0.333? If not, which is larger? By how much?

164. Does $\dfrac{2}{3}$ equal 0.666? If not, which is larger? By how much?

165. *Challenge Problem* Find the product and write the answer in scientific notation.

$$(5.24 \times 10^6)(6.5 \times 10^{13})$$

166. *Challenge Problem* Find the quotient and write the answer in scientific notation.

$$\frac{1.62 \times 10^{-4}}{4.5 \times 10^{-10}}$$

Explaining Concepts: Discussion and Writing

167. Is there a positive real number "closest" to 0?

168. Number Game I'm thinking of a number! It lies between 1 and 10; its square is rational and lies between 1 and 10. The number is larger than π. Correct to two decimal places (that is, truncated to two decimal places), name the number. Now think of your own number, describe it, and challenge a fellow student to name it.

169. Write a brief paragraph that illustrates the similarities and differences between "less than" ($<$) and "less than or equal to" ($\leq$).

170. Give a reason why the statement $5 < 8$ is true.

Powers of Ten, Philip and Phylis Morrison.

R.3 Geometry Essentials

OBJECTIVES 1 Use the Pythagorean Theorem and Its Converse (p. 30)
2 Know Geometry Formulas (p. 31)
3 Understand Congruent Triangles and Similar Triangles (p. 32)

1 Use the Pythagorean Theorem and Its Converse

Figure 16 A right triangle

The *Pythagorean Theorem* is a statement about *right triangles*. A **right triangle** is one that contains a **right angle**—that is, an angle of 90°. The side of the triangle opposite the 90° angle is called the **hypotenuse**; the remaining two sides are called **legs**. In Figure 16 we have used c to represent the length of the hypotenuse and a and b to represent the lengths of the legs. Notice the use of the symbol ⌐ to show the 90° angle. We now state the Pythagorean Theorem.

> **PYTHAGOREAN THEOREM**
>
> In a right triangle, the square of the length of the hypotenuse is equal to the sum of the squares of the lengths of the legs. That is, in the right triangle shown in Figure 16,
>
> $$c^2 = a^2 + b^2 \qquad (1)$$

A proof of the Pythagorean Theorem is given at the end of this section.

EXAMPLE 1 **Finding the Hypotenuse of a Right Triangle**

In a right triangle, one leg has length 4 and the other has length 3. What is the length of the hypotenuse?

Solution Since the triangle is a right triangle, we use the Pythagorean Theorem with $a = 4$ and $b = 3$ to find the length c of the hypotenuse. From equation (1),

$$c^2 = a^2 + b^2$$
$$c^2 = 4^2 + 3^2 = 16 + 9 = 25$$
$$c = \sqrt{25} = 5$$

Now Work PROBLEM 15

The converse of the Pythagorean Theorem is also true.

> **CONVERSE OF THE PYTHAGOREAN THEOREM**
>
> In a triangle, if the square of the length of one side equals the sum of the squares of the lengths of the other two sides, the triangle is a right triangle. The 90° angle is opposite the longest side.

A proof is given at the end of this section.

EXAMPLE 2 **Verifying That a Triangle Is a Right Triangle**

Show that a triangle whose sides are of lengths 5, 12, and 13 is a right triangle. Identify the hypotenuse.

Solution Square the lengths of the sides.

$$5^2 = 25 \qquad 12^2 = 144 \qquad 13^2 = 169$$

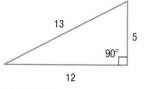

Figure 17

Notice that the sum of the first two squares (25 and 144) equals the third square (169). That is, because $5^2 + 12^2 = 13^2$, the triangle is a right triangle. The longest side, 13, is the hypotenuse. See Figure 17.

━━ **Now Work** PROBLEM 23

| **EXAMPLE 3** | **Applying the Pythagorean Theorem** |

The tallest building in the world is Burj Khalifa in Dubai, United Arab Emirates, at 2717 feet and 163 floors. The observation deck is 1483 feet above ground level. How far can a person standing on the observation deck see (with the aid of a telescope)? Use 3960 miles for the radius of Earth, and assume Earth is a sphere.
Source: *Council on Tall Buildings and Urban Habitat.*

Solution   From the center of Earth, draw two radii: one through Burj Khalifa and the other to the farthest point a person can see from the observation deck. See Figure 18. Apply the Pythagorean Theorem to the right triangle.

Since 1 mile = 5280 feet, 1483 feet = $\dfrac{1483}{5280}$ mile. Then

$$d^2 + (3960)^2 = \left(3960 + \frac{1483}{5280}\right)^2$$

$$d^2 = \left(3960 + \frac{1483}{5280}\right)^2 - (3960)^2 \approx 2224.58$$

$$d \approx 47.17$$

A person can see more than 47 miles from the observation deck.

1483 ft d

3960 mi

Figure 18

━━ **Now Work** PROBLEM 55

2 Know Geometry Formulas

Certain formulas from geometry are useful in solving algebra problems.
For a rectangle of length l and width w,

$$\text{Area} = lw \qquad \text{Perimeter} = 2l + 2w$$

For a triangle with base b and altitude h,

$$\text{Area} = \frac{1}{2}bh$$

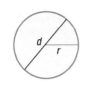

For a circle of radius r (diameter $d = 2r$),

$$\text{Area} = \pi r^2 \qquad \text{Circumference} = 2\pi r = \pi d$$

For a closed rectangular box of length l, width w, and height h,

$$\text{Volume} = lwh \qquad \text{Surface area} = 2lh + 2wh + 2lw$$

For a sphere of radius r,

$$\text{Volume} = \frac{4}{3}\pi r^3 \qquad \text{Surface area} = 4\pi r^2$$

For a closed right circular cylinder of height h and radius r,

$$\text{Volume} = \pi r^2 h \qquad \text{Surface area} = 2\pi r^2 + 2\pi rh$$

- Now Work PROBLEM 31

| EXAMPLE 4 | **Using Geometry Formulas** |

A Christmas tree ornament is in the shape of a semicircle on top of a triangle. How many square centimeters (cm^2) of copper is required to make the ornament if the height of the triangle is 6 cm and the base is 4 cm?

Solution

See Figure 19. The amount of copper required equals the shaded area. This area is the sum of the areas of the triangle and the semicircle. The triangle has height $h = 6$ and base $b = 4$. The semicircle has diameter $d = 4$, so its radius is $r = 2$.

$$\text{Total area} = \text{Area of triangle} + \text{Area of semicircle} \quad b = 4; h = 6; r = 2$$
$$= \frac{1}{2}bh + \frac{1}{2}\pi r^2 = \frac{1}{2}(4)(6) + \frac{1}{2}\pi \cdot 2^2$$
$$= 12 + 2\pi \approx 18.28 \ \text{cm}^2$$

Figure 19

About 18.28 cm^2 of copper is required.

- Now Work PROBLEM 49

3 Understand Congruent Triangles and Similar Triangles

In Words
Two triangles are congruent if they have the same size and shape.

Throughout the text we will make reference to triangles. We begin with a discussion of *congruent* triangles. According to dictionary.com, the word **congruent** means "coinciding exactly when superimposed." For example, two angles are congruent if they have the same measure, and two line segments are congruent if they have the same length.

> **DEFINITION Congruent Triangles**
>
> Two triangles are **congruent** if each pair of corresponding angles have the same measure and each pair of corresponding sides are the same length.

In Figure 20, corresponding angles are equal and the corresponding sides are equal in length: $a = d$, $b = e$, and $c = f$. As a result, these triangles are congruent.

Figure 20 Congruent triangles

It is not necessary to verify that all three angles and all three sides are the same measure to determine whether two triangles are congruent.

Determining Congruent Triangles

- **Angle–Side–Angle Case** Two triangles are congruent if two of the angles are equal and the lengths of the corresponding sides between the two angles are equal.

  For example, in Figure 21(a), the two triangles are congruent because two angles and the included side are equal.
- **Side–Side–Side Case** Two triangles are congruent if the lengths of the corresponding sides of the triangles are equal.

  For example, in Figure 21(b), the two triangles are congruent because the three corresponding sides are all equal.
- **Side–Angle–Side Case** Two triangles are congruent if the lengths of two corresponding sides are equal and the angles between the two sides are the same.

  For example, in Figure 21(c), the two triangles are congruent because two sides and the included angle are equal.

Figure 21

We contrast congruent triangles with *similar* triangles.

DEFINITION Similar Triangles

Two triangles are **similar** if the corresponding angles are equal and the lengths of the corresponding sides are proportional.

In Words

Two triangles are similar if they have the same shape, but (possibly) different sizes.

For example, the triangles in Figure 22 (on the next page) are similar because the corresponding angles are equal. In addition, the lengths of the corresponding sides are proportional because each side in the triangle on the right is twice as long as each corresponding side in the triangle on the left. That is, the ratio of the corresponding sides is a constant: $\dfrac{d}{a} = \dfrac{e}{b} = \dfrac{f}{c} = 2$.

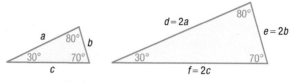

Figure 22 Similar triangles

It is not necessary to verify that all three angles are equal and all three sides are proportional to determine whether two triangles are similar.

Determining Similar Triangles

- **Angle-Angle Case** Two triangles are similar if two of the corresponding angles are equal.
  For example, in Figure 23(a), the two triangles are similar because two angles are equal.
- **Side–Side–Side Case** Two triangles are similar if the lengths of all three sides of each triangle are proportional.
  For example, in Figure 23(b), the two triangles are similar because

$$\frac{10}{30} = \frac{5}{15} = \frac{6}{18} = \frac{1}{3}$$

- **Side–Angle–Side Case** Two triangles are similar if two corresponding sides are proportional and the angles between the two sides are equal.

  For example, in Figure 23(c), the two triangles are similar because $\frac{4}{6} = \frac{12}{18} = \frac{2}{3}$ and the angles between the sides are equal.

Figure 23

EXAMPLE 5   **Using Similar Triangles**

Given that the triangles in Figure 24 are similar, find the missing length x and the angles $A, B,$ and C.

Figure 24

Solution

Because the triangles are similar, corresponding angles are equal. So $A = 90°$, $B = 60°$, and $C = 30°$. Also, the corresponding sides are proportional. That is, $\frac{3}{5} = \frac{6}{x}$. We solve this equation for x.

$$\frac{3}{5} = \frac{6}{x}$$

$$5x \cdot \frac{3}{5} = 5x \cdot \frac{6}{x} \qquad \text{Multiply both sides by } 5x.$$

$$3x = 30 \qquad \text{Simplify.}$$

$$x = 10 \qquad \text{Divide both sides by 3.}$$

The missing length is 10 units.

Now Work PROBLEM 43

Proof of the Pythagorean Theorem Begin with a square, each side of length $a + b$. In this square, form four right triangles, each having legs equal in length to a and b. See Figure 25. All these triangles are congruent (two sides and their included angle are equal). As a result, the hypotenuse of each is the same, say c, and the pink shading in Figure 25 indicates a square with an area equal to c^2.

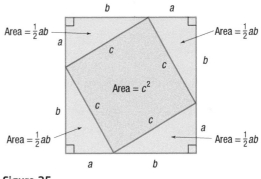

Figure 25

The area of the original square with sides $a + b$ equals the sum of the areas of the four triangles (each of area $\frac{1}{2}ab$) plus the area of the square with side c. That is,

$$(a + b)^2 = \frac{1}{2}ab + \frac{1}{2}ab + \frac{1}{2}ab + \frac{1}{2}ab + c^2$$

$$a^2 + 2ab + b^2 = 2ab + c^2$$

$$a^2 + b^2 = c^2$$

The proof is complete. ∎

Proof of the Converse of the Pythagorean Theorem Begin with two triangles: one a right triangle with legs a and b and the other a triangle with sides a, b, and c for which $c^2 = a^2 + b^2$. See Figure 26. By the Pythagorean Theorem, the length x of the third side of the first triangle is

$$x^2 = a^2 + b^2$$

But $c^2 = a^2 + b^2$. Then,

$$x^2 = c^2$$
$$x = c$$

The two triangles have the same sides and are therefore congruent. This means corresponding angles are equal, so the angle opposite side c of the second triangle equals 90°.

The proof is complete. ∎

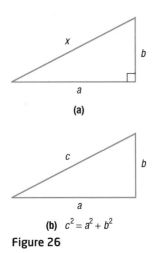

(a)

(b) $c^2 = a^2 + b^2$

Figure 26

R.3 Assess Your Understanding

Concepts and Vocabulary

1. A(n) _____ triangle is one that contains an angle of 90 degrees. The longest side is called the _____.

2. For a triangle with base b and altitude h, a formula for the area A is _____.

3. The formula for the circumference C of a circle of radius r is _____.

4. Two triangles are _____ if corresponding angles are equal and the lengths of the corresponding sides are proportional.

5. *Multiple Choice* Which of the following is not a case for determining congruent triangles?
   (a) Angle–Side–Angle   (b) Side–Angle–Side
   (c) Angle–Angle–Angle   (d) Side-Side-Side

6. *Multiple Choice* Choose the formula for the volume of a sphere of radius r.
   (a) $\frac{4}{3}\pi r^2$ (b) $\frac{4}{3}\pi r^3$ (c) $4\pi r^3$ (d) $4\pi r^2$

7. *True or False* In a right triangle, the square of the length of the longest side equals the sum of the squares of the lengths of the other two sides.

8. *True or False* The triangle with sides of lengths 6, 8, and 10 is a right triangle.

9. *True or False* The surface area of a sphere of radius r is $\frac{4}{3}\pi r^2$.

10. *True or False* The triangles shown are congruent.

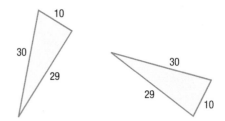

11. *True or False* The triangles shown are similar.

12. *True or False* The triangles shown are similar.

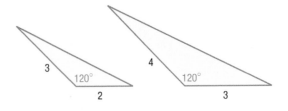

Skill Building

In Problems 13–18, the lengths of the legs of a right triangle are given. Find the hypotenuse.

13. $a = 5$, $b = 12$
14. $a = 6$, $b = 8$
15. $a = 10$, $b = 24$
16. $a = 4$, $b = 3$
17. $a = 7$, $b = 24$
18. $a = 14$, $b = 48$

In Problems 19–26, the lengths of the sides of a triangle are given. Determine which are right triangles. For those that are, identify the hypotenuse.

19. 3, 4, 5
20. 6, 8, 10
21. 4, 5, 6
22. 2, 2, 3
23. 7, 24, 25
24. 10, 24, 26
25. 6, 4, 3
26. 5, 4, 7

27. Find the area A of a rectangle with length 6 inches and width 7 inches.

28. Find the area A of a rectangle with length 9 centimeters and width 4 centimeters.

29. Find the area A of a triangle with height 14 inches and base 4 inches.

30. Find the area A of a triangle with height 9 centimeters and base 4 centimeters.

31. Find the area A and circumference C of a circle of radius 5 meters.

32. Find the area A and circumference C of a circle of radius 2 feet.

33. Find the volume V and surface area S of a closed rectangular box with length 6 feet, width 8 feet, and height 5 feet.

34. Find the volume V and surface area S of a closed rectangular box with length 9 inches, width 4 inches, and height 8 inches.

35. Find the volume V and surface area S of a sphere of radius 5 centimeters.

36. Find the volume V and surface area S of a sphere of radius 3 feet.

37. Find the volume V and surface area S of a closed right circular cylinder with radius 9 inches and height 8 inches.

38. Find the volume V and surface area S of a closed right circular cylinder with radius 8 inches and height 9 inches.

In Problems 39–42, find the area of the shaded region.

39. 2   **40.** 2   **41.**   **42.**

In Problems 43–46, the triangles in each pair are similar. Find the missing length x and the missing angles A, B, and C.

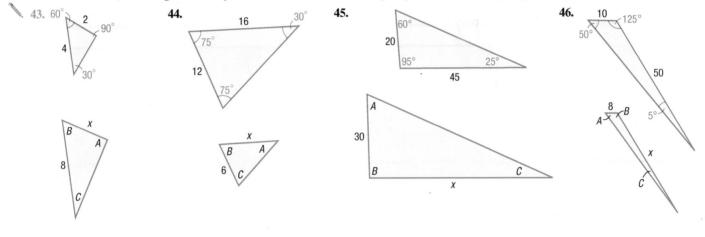

43.   **44.**   **45.**   **46.**

Applications and Extensions

47. How many feet has a wheel with a diameter of 16 inches traveled after four revolutions?

48. How many revolutions will a circular disk with a diameter of 4 feet have completed after it has rolled 20 feet?

49. In the figure shown, *ABCD* is a square, with each side of length 6 feet. The width of the border (shaded portion) between the outer square *EFGH* and *ABCD* is 2 feet. Find the area of the border.

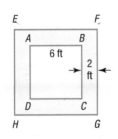

50. Refer to the figure. Square *ABCD* has an area of 100 square feet; square *BEFG* has an area of 16 square feet. What is the area of the triangle *CGF*?

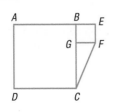

51. Architecture A **Norman window** consists of a rectangle surmounted by a semicircle. Find the area of the Norman window shown in the illustration. How much wood frame is needed to enclose the window?

52. Construction A circular swimming pool that is 20 feet in diameter is enclosed by a wooden deck that is 3 feet wide. What is the area of the deck? How much fence is required to enclose the deck?

53. How Tall Is the Great Pyramid? The ancient Greek philosopher Thales of Miletus is reported on one occasion to have visited Egypt and calculated the height of the Great Pyramid of Cheops by means of shadow reckoning. Thales knew that each side of the base of the pyramid was 252 paces and that his own height was 2 paces. He measured the length of the pyramid's shadow to be 114 paces and determined the length of his shadow to be 3 paces. See the illustration. Using similar triangles, determine the height of the Great Pyramid in terms of the number of paces.

Source: Diggins, Julie E, *String, Straightedge, and Shadow: The Story of Geometry,* 2003, Whole Spirit Press, http://wholespiritpress.com.

54. The Bermuda Triangle Karen is doing research on the Bermuda Triangle which she defines roughly by Hamilton, Bermuda; San Juan, Puerto Rico; and Fort Lauderdale, Florida. On her atlas Karen measures the straight-line distances from Hamilton to Fort Lauderdale, Fort Lauderdale to San Juan, and San Juan to Hamilton to be approximately 57 millimeters (mm), 58 mm, and 53.5 mm, respectively. If the actual distance from Fort Lauderdale to San Juan is 1046 miles, approximate the actual distances from San Juan to Hamilton and from Hamilton to Fort Lauderdale.

In Problems 55–57, use the facts that the radius of Earth is 3960 miles and 1 mile = 5280 feet.

55. How Far Can You See? The conning tower of the U.S.S. *Silversides,* a World War II submarine now permanently stationed in Muskegon, Michigan, is approximately 20 feet above sea level. How far can you see from the conning tower?

56. How Far Can You See? A person who is 6 feet tall is standing on the beach in Fort Lauderdale, Florida, and looks out onto the Atlantic Ocean. Suddenly, a ship appears on the horizon. How far is the ship from shore?

57. How Far Can You See? The deck of a destroyer is 100 feet above sea level. How far can you see from the deck? How far can you see from the bridge, which is 150 feet above sea level?

58. Suppose that m and n are positive integers with $m > n$. If $a = m^2 - n^2$, $b = 2mn$, and $c = m^2 + n^2$, show that a, b, and c are the lengths of the sides of a right triangle. (This formula can be used to find the sides of a right triangle that are integers, such as 3, 4, 5; 5, 12, 13; and so on. Such triplets of integers are called **Pythagorean triples**.)

59. *Challenge Problem* If the water depth of the pool in Problem 52 is uniformly 4.5 feet, how many gallons of water are in the pool? Round to the nearest gallon.
Hint: $1 \text{ ft}^3 \approx 7.48052$ gal

60. *Challenge Problem* A cylindrical oil tank holds 10,000 barrels of oil. If the diameter of the tank is 50 feet, what is its height? Round to the nearest tenth of a foot.
Hint: 1 oil barrel $\approx 5.61458 \text{ ft}^3$

Explaining Concepts: Discussion and Writing

61. If the radius of a circle is doubled, does the area of the circle also double? Explain.

62. If the radius of a sphere is doubled, does the volume of the sphere also double? Explain.

63. You have 1000 feet of flexible pool siding and intend to construct a swimming pool. Experiment with rectangular-shaped pools with perimeters of 1000 feet. How do their areas vary? What is the shape of the rectangle with the largest area? Now compute the area enclosed by a circular pool with a perimeter (circumference) of 1000 feet. What would be your choice of shape for the pool? If rectangular, what is your preference for dimensions? Justify your choice. If your only consideration is to have a pool that encloses the most area, what shape should you use?

64. The Gibb's Hill Lighthouse, Southampton, Bermuda, in operation since 1846, stands 117 feet high on a hill 245 feet high, so its beam of light is 362 feet above sea level. A brochure states that the light itself can be seen on the horizon about 26 miles distant. Verify the accuracy of this information. The brochure further states that ships 40 miles away can see the light and that planes flying at 10,000 feet can see it 120 miles away. Verify the accuracy of these statements. What assumption did the brochure make about the height of the ship?

R.4 Polynomials

OBJECTIVES
1 Recognize Monomials (p. 39)
2 Recognize Polynomials (p. 39)
3 Add and Subtract Polynomials (p. 41)
4 Multiply Polynomials (p. 41)
5 Know Formulas for Special Products (p. 42)
6 Divide Polynomials Using Long Division (p. 44)
7 Work with Polynomials in Two Variables (p. 46)

We have described algebra as a generalization of arithmetic in which letters are used to represent real numbers. From now on, we shall use the letters at the end of the alphabet, such as x, y, and z, to represent variables and use the letters at the beginning of the alphabet, such as a, b, and c, to represent constants. In the expressions $3x + 5$ and $ax + b$, it is understood that x is a variable and that a and b are constants, even though the constants a and b are unspecified. As you will find out, the context usually makes the intended meaning clear.

1 Recognize Monomials

DEFINITION Monomial

NOTE The nonnegative integers are the whole numbers 0, 1, 2, 3, ... ■

A **monomial** in one variable is the product of a constant and a variable raised to a nonnegative integer power. A monomial is of the form

$$ax^k$$

where a is a constant, x is a variable, and $k \geq 0$ is an integer. The constant a is called the **coefficient** of the monomial. If $a \neq 0$, then k is called the **degree** of the monomial.

EXAMPLE 1   **Examples of Monomials**

Monomial	Coefficient	Degree	
$6x^2$	6	2	
$-\sqrt{2}x^3$	$-\sqrt{2}$	3	
3	3	0	Since $3 = 3 \cdot 1 = 3x^0$, $x \neq 0$
$-5x$	-5	1	Since $-5x = -5x^1$
x^4	1	4	Since $x^4 = 1 \cdot x^4$

EXAMPLE 2   **Examples of Expressions that are Not Monomials**

(a) $3x^{1/2}$ is not a monomial, since the exponent of the variable x is $\frac{1}{2}$, and $\frac{1}{2}$ is not a nonnegative integer.

(b) $4x^{-3}$ is not a monomial, since the exponent of the variable x is -3, and -3 is not a nonnegative integer.

━━━━ **Now Work** PROBLEM 9

2 Recognize Polynomials

Two monomials with the same variable raised to the same power are called **like terms**. For example, $2x^4$ and $-5x^4$ are like terms. In contrast, the monomials $2x^3$ and $2x^5$ are not like terms.

We can add or subtract like terms using the Distributive Property. For example,

$$2x^2 + 5x^2 = (2 + 5)x^2 = 7x^2 \quad \text{and} \quad 8x^3 - 5x^3 = (8 - 5)x^3 = 3x^3$$

The sum or difference of two monomials having different degrees is called a **binomial**. The sum or difference of three monomials with three different degrees is called a **trinomial**. For example,

- $x^2 - 2$ is a binomial.
- $x^3 - 3x + 5$ is a trinomial.
- $2x^2 + 5x^2 + 2 = 7x^2 + 2$ is a binomial.

DEFINITION Polynomial

A **polynomial** in one variable is an algebraic expression of the form

$$a_n x^n + a_{n-1} x^{n-1} + \cdots + a_1 x + a_0 \qquad (1)$$

where $a_n, a_{n-1}, \ldots, a_1, a_0$ are constants,[*] called the **coefficients** of the polynomial, $n \geq 0$ is an integer, and x is a variable. If $a_n \neq 0$, it is called the **leading coefficient**, $a_n x^n$ is called the **leading term**, and n is the **degree** of the polynomial.

In Words

A polynomial is a sum of monomials.

The monomials that make up a polynomial are called its **terms**. If all of the coefficients are 0, the polynomial is called the **zero polynomial**, which has no degree.

Polynomials are usually written in **standard form**, beginning with the nonzero term of highest degree and continuing with terms in descending order according to degree. If a power of x is missing, it is because its coefficient is zero.

EXAMPLE 3	**Examples of Polynomials**		

Polynomial	**Coefficients**	**Degree**
$-8x^3 + 4x^2 - 6x + 2$	$-8, 4, -6, 2$	3
$3x^2 - 5 = 3x^2 + 0 \cdot x - 5$	$3, 0, -5$	2
$8 - 2x + x^2 = 1 \cdot x^2 - 2x + 8$	$1, -2, 8$	2
$5x + \sqrt{2} = 5x^1 + \sqrt{2}$	$5, \sqrt{2}$	1
$3 = 3 \cdot 1 = 3 \cdot x^0$	3	0
0	0	No degree

Although we have been using x to represent the variable, letters such as y and z are also commonly used.

- $3x^4 - x^2 + 2$ is a polynomial (in x) of degree 4.
- $9y^3 - 2y^2 + y - 3$ is a polynomial (in y) of degree 3.
- $z^5 + \pi$ is a polynomial (in z) of degree 5.

Algebraic expressions such as

$$\frac{1}{x} \quad \text{and} \quad \frac{x^2 + 1}{x + 5}$$

are not polynomials. The first is not a polynomial because $\dfrac{1}{x} = x^{-1}$ has an exponent that is not a nonnegative integer. The second expression is not a polynomial because the quotient cannot be simplified to a sum of monomials.

Now Work PROBLEM 19

---

[*]The notation a_n is read as "a sub n." The number n is called a **subscript** and should not be confused with an exponent. We use subscripts to distinguish one constant from another when a large or undetermined number of constants are required.

3 Add and Subtract Polynomials

Polynomials are added and subtracted by combining like terms.

EXAMPLE 4   **Adding Polynomials**

Find the sum of the polynomials:

$$8x^3 - 2x^2 + 6x - 2 \quad \text{and} \quad 3x^4 - 2x^3 + x^2 + x$$

Solution   We shall find the sum in two ways.

Horizontal Addition: The idea here is to group the like terms and then combine them.

$$(8x^3 - 2x^2 + 6x - 2) + (3x^4 - 2x^3 + x^2 + x)$$
$$= 3x^4 + (8x^3 - 2x^3) + (-2x^2 + x^2) + (6x + x) - 2$$
$$= 3x^4 + 6x^3 - x^2 + 7x - 2$$

Vertical Addition: The idea here is to vertically line up the like terms in each polynomial and then add the coefficients.

$$
\begin{array}{cccccc}
x^4 & x^3 & x^2 & x^1 & x^0 \\
 & 8x^3 & - 2x^2 & + 6x & - 2 \\
+ 3x^4 & - 2x^3 & + x^2 & + x & \\
\hline
3x^4 & + 6x^3 & - x^2 & + 7x & - 2
\end{array}
$$

We can subtract two polynomials horizontally or vertically as well.

EXAMPLE 5   **Subtracting Polynomials**

Find the difference: $(3x^4 - 4x^3 + 6x^2 - 1) - (2x^4 - 8x^2 - 6x + 5)$

Solution   *Horizontal Subtraction:*

$$(3x^4 - 4x^3 + 6x^2 - 1) - (2x^4 - 8x^2 - 6x + 5)$$
$$= 3x^4 - 4x^3 + 6x^2 - 1 + \underbrace{(-2x^4 + 8x^2 + 6x - 5)}$$

Be sure to change the sign of each term in the second polynomial.

$$= (3x^4 - 2x^4) + (-4x^3) + (6x^2 + 8x^2) + 6x + (-1 - 5)$$
↑
Group like terms.
$$= x^4 - 4x^3 + 14x^2 + 6x - 6$$

COMMENT Vertical subtraction will be used when we divide polynomials. ∎

Vertical Subtraction: We line up like terms, change the sign of each coefficient of the second polynomial, and add.

$$
\begin{array}{ccccc}
x^4 & x^3 & x^2 & x^1 & x^0 \\
3x^4 & - 4x^3 & + 6x^2 & & - 1 \\
- [2x^4 & & - 8x^2 & - 6x & + 5]
\end{array}
\quad = \quad + \quad
\begin{array}{ccccc}
x^4 & x^3 & x^2 & x^1 & x^0 \\
3x^4 & - 4x^3 & + 6x^2 & & - 1 \\
-2x^4 & & + 8x^2 & + 6x & - 5 \\
\hline
x^4 & - 4x^3 & + 14x^2 & + 6x & - 6
\end{array}
$$

Which of these methods to use for adding and subtracting polynomials is up to you. To save space, we shall most often use the horizontal format.

Now Work PROBLEM 31

4 Multiply Polynomials

Two monomials may be multiplied using the Laws of Exponents and the Commutative and Associative Properties. For example,

$$(2x^3) \cdot (5x^4) = (2 \cdot 5) \cdot (x^3 \cdot x^4) = 10x^{3+4} = 10x^7$$

Products of polynomials are found by repeated use of the Distributive Property and the Laws of Exponents. Again, you have a choice of horizontal or vertical format.

EXAMPLE 6 | **Multiplying Polynomials**

Find the product: $(2x + 5)(x^2 - x + 2)$

Solution | **Horizontal Multiplication:**

$$(2x + 5)(x^2 - x + 2) = 2x(x^2 - x + 2) + 5(x^2 - x + 2)$$
↑
Distributive Property
$$= (2x \cdot x^2 - 2x \cdot x + 2x \cdot 2) + (5 \cdot x^2 - 5 \cdot x + 5 \cdot 2)$$
↑
Distributive Property
$$= (2x^3 - 2x^2 + 4x) + (5x^2 - 5x + 10)$$
↑
Law of Exponents
$$= 2x^3 + 3x^2 - x + 10$$
↑
Combine like terms.

Vertical Multiplication: The idea here is very much like multiplying a two-digit number by a three-digit number.

$$
\begin{array}{r}
x^2 - x + 2 \\
2x + 5 \\
\hline
2x^3 - 2x^2 + 4x \\
(+) \quad 5x^2 - 5x + 10 \\
\hline
2x^3 + 3x^2 - x + 10
\end{array}
$$

This line is $2x(x^2 - x + 2)$.
This line is $5(x^2 - x + 2)$.
Sum of the above two lines

◢

Now Work PROBLEM 47

---

5 Know Formulas for Special Products

Certain products, which we call **special products**, occur frequently in algebra. We can calculate them easily using the **FOIL** (*First, Outer, Inner, Last*) method of multiplying two binomials.

$$
\begin{array}{c}
\overset{\text{First}\quad\text{Outer}\quad\text{Inner}\quad\text{Last}}{(ax + b)(cx + d) = \overbrace{ax \cdot cx} + \overbrace{ax \cdot d} + \overbrace{b \cdot cx} + \overbrace{b \cdot d}} \\
= acx^2 + adx + bcx + bd \\
= acx^2 + (ad + bc)x + bd
\end{array}
$$

EXAMPLE 7 | **Using FOIL**

(a) $(x - 3)(x + 3) = x^2 + 3x - 3x - 9 = x^2 - 9$
       F   O   I   L
(b) $(x + 2)^2 = (x + 2)(x + 2) = x^2 + 2x + 2x + 4 = x^2 + 4x + 4$
(c) $(x - 3)^2 = (x - 3)(x - 3) = x^2 - 3x - 3x + 9 = x^2 - 6x + 9$
(d) $(x + 3)(x + 1) = x^2 + x + 3x + 3 = x^2 + 4x + 3$
(e) $(2x + 1)(3x + 4) = 6x^2 + 8x + 3x + 4 = 6x^2 + 11x + 4$

◢

Notice the factors in part (a). The first binomial is a difference and the second one is a sum. Now notice that the outer product O and the inner product I are additive inverses; their sum is zero. So the product is a difference of two squares.

Now Work PROBLEMS 49 AND 57

Some products have been given special names because of their form. The following special products are based on Examples 7(a), (b), and (c).

Difference of Two Squares

$$(x - a)(x + a) = x^2 - a^2 \qquad \text{(2)}$$

Squares of Binomials, or Perfect Squares

$$(x + a)^2 = x^2 + 2ax + a^2 \qquad \text{(3a)}$$
$$(x - a)^2 = x^2 - 2ax + a^2 \qquad \text{(3b)}$$

EXAMPLE 8

Using Special Product Formulas

(a) $(x - 5)(x + 5) = x^2 - 5^2 = x^2 - 25$ Difference of two squares

(b) $(x + 7)^2 = x^2 + 2 \cdot 7 \cdot x + 7^2 = x^2 + 14x + 49$ Square of a binomial

(c) $(2x + 1)^2 = (2x)^2 + 2 \cdot 1 \cdot 2x + 1^2 = 4x^2 + 4x + 1$ Notice that we used $2x$ in place of x in formula (3a).

(d) $(3x - 4)^2 = (3x)^2 - 2 \cdot 4 \cdot 3x + 4^2 = 9x^2 - 24x + 16$ Replace x by $3x$ in formula (3b).

Now Work PROBLEMS **67, 69,** AND **71**

Let's look at some more examples that lead to general formulas.

EXAMPLE 9

Cubing a Binomial

(a) $(x + 2)^3 = (x + 2)(x + 2)^2 = (x + 2)(x^2 + 4x + 4)$ Formula (3a)
$$= (x^3 + 4x^2 + 4x) + (2x^2 + 8x + 8)$$
$$= x^3 + 6x^2 + 12x + 8$$

(b) $(x - 1)^3 = (x - 1)(x - 1)^2 = (x - 1)(x^2 - 2x + 1)$ Formula (3b)
$$= (x^3 - 2x^2 + x) - (x^2 - 2x + 1)$$
$$= x^3 - 3x^2 + 3x - 1$$

Cubes of Binomials, or Perfect Cubes

$$(x + a)^3 = x^3 + 3ax^2 + 3a^2x + a^3 \qquad \text{(4a)}$$
$$(x - a)^3 = x^3 - 3ax^2 + 3a^2x - a^3 \qquad \text{(4b)}$$

Now Work PROBLEM **87**

EXAMPLE 10

Forming the Difference of Two Cubes

$$(x - 1)(x^2 + x + 1) = x(x^2 + x + 1) - 1(x^2 + x + 1)$$
$$= x^3 + x^2 + x - x^2 - x - 1$$
$$= x^3 - 1$$

EXAMPLE 11

Forming the Sum of Two Cubes

$$(x + 2)(x^2 - 2x + 4) = x(x^2 - 2x + 4) + 2(x^2 - 2x + 4)$$
$$= x^3 - 2x^2 + 4x + 2x^2 - 4x + 8$$
$$= x^3 + 8$$

Examples 10 and 11 lead to two more special products.

Difference of Two Cubes

$$(x - a)(x^2 + ax + a^2) = x^3 - a^3 \qquad \textbf{(5)}$$

Sum of Two Cubes

$$(x + a)(x^2 - ax + a^2) = x^3 + a^3 \qquad \textbf{(6)}$$

6 Divide Polynomials Using Long Division

The procedure for dividing two polynomials is similar to the procedure for dividing two integers.

EXAMPLE 12

Dividing Two Integers

Divide 842 by 15.

Solution

$$
\begin{array}{r}
56 \quad \leftarrow \textbf{Quotient} \\
\text{Divisor} \rightarrow 15\overline{)842} \quad \leftarrow \textbf{Dividend} \\
75 \quad \leftarrow 5 \cdot 15 \\
\overline{92} \quad \leftarrow \textbf{Subtract and bring down the 2.} \\
90 \quad \leftarrow 6 \cdot 15 \\
\overline{2} \quad \leftarrow \textbf{Subtract; the remainder is 2.}
\end{array}
$$

So, $\dfrac{842}{15} = 56 + \dfrac{2}{15}$.

In the division problem detailed in Example 12, the number 15 is called the **divisor**, the number 842 is called the **dividend**, the number 56 is called the **quotient**, and the number 2 is called the **remainder**.

To check the answer obtained in a division problem, multiply the quotient by the divisor and add the remainder. The answer should be the dividend.

Quotient · Divisor + Remainder = Dividend

For example, we can check the results obtained in Example 12 as follows:

$$56 \cdot 15 + 2 = 840 + 2 = 842$$

NOTE Remember, a polynomial is in standard form when its terms are written in descending powers of x. ∎

To divide two polynomials, we first write each polynomial in standard form. The process then follows a pattern similar to that of Example 12. The next example illustrates the procedure.

EXAMPLE 13   Dividing Two Polynomials

Find the quotient and the remainder when

$$3x^3 + 4x^2 + x + 7 \quad \text{is divided by} \quad x^2 + 1$$

Solution   Each polynomial is in standard form. The dividend is $3x^3 + 4x^2 + x + 7$, and the divisor is $x^2 + 1$.

STEP 1: Divide the leading term of the dividend, $3x^3$, by the leading term of the divisor, x^2. Enter the result, $3x$, over the term $3x^3$, as follows:

$$
\begin{array}{r}
3x \\
x^2 + 1 \overline{)3x^3 + 4x^2 + x + 7}
\end{array}
$$

STEP 2: Multiply $3x$ by $x^2 + 1$, and enter the result below the dividend.

$$
\begin{array}{r}
3x \\
x^2 + 1 \overline{)3x^3 + 4x^2 + x + 7} \\
3x^3 + 3x
\end{array}
$$
$\leftarrow 3x \cdot (x^2 + 1) = 3x^3 + 3x$

Align the 3x term under the x to make the next step easier.

STEP 3: Subtract and bring down the remaining terms.

$$
\begin{array}{r}
3x \\
x^2 + 1 \overline{)3x^3 + 4x^2 + x + 7} \\
\underline{3x^3 + 3x } \\
4x^2 - 2x + 7
\end{array}
$$
$\leftarrow$ Subtract (change the signs and add).
$\leftarrow$ Bring down the $4x^2$ and the 7.

STEP 4: Repeat Steps 1–3 using $4x^2 - 2x + 7$ as the dividend.

$$
\begin{array}{r}
3x + 4 \\
x^2 + 1 \overline{)3x^3 + 4x^2 + x + 7} \\
\underline{3x^3 + 3x } \\
4x^2 - 2x + 7 \\
\underline{4x^2 + 4} \\
-2x + 3
\end{array}
$$
$\leftarrow$ Divide $4x^2$ by x^2 to get 4.
$\leftarrow$ Multiply $(x^2 + 1)$ by 4; subtract.

COMMENT When the degree of the divisor is greater than the degree of the dividend, the process ends. ∎

Since x^2 does not divide $-2x$ evenly (that is, the result is not a monomial), the process ends. The quotient is $3x + 4$, and the remainder is $-2x + 3$.

✓ **Check:** Quotient · Divisor + Remainder

$$
\begin{aligned}
&= (3x + 4)(x^2 + 1) + (-2x + 3) \\
&= 3x^3 + 3x + 4x^2 + 4 - 2x + 3 \\
&= 3x^3 + 4x^2 + x + 7 = \text{Dividend}
\end{aligned}
$$

Then

$$\frac{3x^3 + 4x^2 + x + 7}{x^2 + 1} = 3x + 4 + \frac{-2x + 3}{x^2 + 1}$$

The next example combines the steps involved in long division.

EXAMPLE 14   Dividing Two Polynomials

Find the quotient and the remainder when

$$x^4 - 3x^3 + 2x - 5 \quad \text{is divided by} \quad x^2 - x + 1$$

Solution   In setting up this division problem, it is necessary to leave a space for the missing x^2 term in the dividend.

$$
\begin{array}{r}
x^2 - 2x\ - 3 \quad\leftarrow \text{Quotient}\\
x^2 - x + 1\overline{)x^4 - 3x^3 \qquad\ + 2x - 5}\quad\leftarrow \text{Dividend}\\
\underline{x^4 -\ x^3 +\ x^2}\\
-2x^3 -\ x^2 + 2x - 5\\
\underline{-2x^3 + 2x^2 - 2x}\\
-3x^2 + 4x - 5\\
\underline{-3x^2 + 3x - 3}\\
x - 2 \quad\leftarrow \text{Remainder}
\end{array}
$$

Divisor $\rightarrow$ Subtract $\rightarrow$ Subtract $\rightarrow$ Subtract $\rightarrow$

✓ **Check:** Quotient · Divisor + Remainder

$$= (x^2 - 2x - 3)(x^2 - x + 1) + x - 2$$

$$= x^4 - x^3 + x^2 - 2x^3 + 2x^2 - 2x - 3x^2 + 3x - 3 + x - 2$$

$$= x^4 - 3x^3 + 2x - 5 = \text{Dividend}$$

As a result,

$$\frac{x^4 - 3x^3 + 2x - 5}{x^2 - x + 1} = x^2 - 2x - 3 + \frac{x - 2}{x^2 - x + 1}$$

The process of dividing two polynomials leads to the following result:

THEOREM

Let Q be a polynomial of positive degree, and let P be a polynomial whose degree is greater than or equal to the degree of Q. The remainder after dividing P by Q is either the zero polynomial or a polynomial whose degree is less than the degree of the divisor Q.

— Now Work PROBLEM 95

7 Work with Polynomials in Two Variables

A **monomial in two variables** x and y has the form $ax^n y^m$, where a is a constant, x and y are variables, and n and m are nonnegative integers. The **degree** of a monomial is the sum of the powers of the variables.

For example,

$$2xy^3, \quad x^2y^2, \quad \text{and} \quad x^3y$$

are all monomials that have degree 4.

A **polynomial in two variables** x and y is the sum of one or more monomials in two variables. The **degree of a polynomial** in two variables is the highest degree of all the monomials with nonzero coefficients.

EXAMPLE 15   Examples of Polynomials in Two Variables

$$3x^2 + 2x^3y + 5 \qquad \pi x^3 - y^2 \qquad x^4 + 4x^3y - xy^3 + y^4$$

Two variables,   Two variables,   Two variables,
degree is 4.   degree is 3.   degree is 4.

We multiply polynomials in two variables the same way as we multiply polynomials in one variable.

| EXAMPLE 16 | **Using a Special Product Formula** |

To multiply $(2x - y)^2$, use the Squares of Binomials formula (3b) with $2x$ instead of x and with y instead of a.

$$(2x - y)^2 = (2x)^2 - 2 \cdot y \cdot 2x + y^2$$
$$= 4x^2 - 4xy + y^2$$

⌐

———— **Now Work** PROBLEM 81

R.4 Assess Your Understanding

Concepts and Vocabulary

1. The polynomial $3x^4 - 2x^3 + 13x^2 - 5$ is of degree ____. The leading coefficient is ____.

2. $(x^2 - 4)(x^2 + 4) =$ _____.

3. $(x - 2)(x^2 + 2x + 4) =$ _____

4. *Multiple Choice* The monomials that make up a polynomial are called which of the following?
(a) terms **(b)** variables **(c)** factors **(d)** coefficients

5. *Multiple Choice* Choose the degree of the monomial $3x^4 y^2$.
(a) 3 **(b)** 8 **(c)** 6 **(d)** 2

6. *True or False* $4x^{-2}$ is a monomial of degree -2.

7. *True or False* The degree of the product of two nonzero polynomials equals the sum of their degrees.

8. *True or False* To check division, the divisor should equal Quotient · Dividend + Remainder.

Skill Building

In Problems 9–18, is the expression a monomial? If it is, name the variable(s) and the coefficient, and give the degree of the monomial. If it is not a monomial, state why not.

9. $2x^3$

10. $-4x^2$

11. $\dfrac{8}{x}$

12. $-2x^{-3}$

13. $-2xy^2$

14. $5x^2 y^3$

15. $\dfrac{8x}{y}$

16. $-\dfrac{2x^2}{y^3}$

17. $x^2 + y^2$

18. $3x^2 + 4$

In Problems 19–28, is the expression a polynomial? If it is, give its degree. If it is not, state why not.

19. $3x^2 - 5$

20. $1 - 4x$

21. 5

22. $-\pi$

23. $3x^2 - \dfrac{5}{x}$

24. $\dfrac{3}{x} + 2$

25. $2y^3 - \sqrt{2}$

26. $10z^2 + z$

27. $\dfrac{x^2 + 5}{x^3 - 1}$

28. $\dfrac{3x^3 + 2x - 1}{x^2 + x + 1}$

In Problems 29–48, add, subtract, or multiply, as indicated. Express your answer as a single polynomial in standard form.

29. $(x^2 + 6x + 8) + (3x^2 - 4x + 7)$

30. $(x^3 + 3x^2 + 2) + (x^2 - 4x + 4)$

31. $(x^3 - 2x^2 + 5x + 10) - (2x^2 - 4x + 3)$

32. $(x^2 - 3x - 4) - (x^3 - 3x^2 + x + 5)$

33. $(6x^5 + x^3 + x) + (5x^4 - x^3 + 3x^2)$

34. $(10x^5 - 8x^2) + (3x^3 - 2x^2 + 6)$

35. $(x^2 - 6x + 4) + 3(2x^2 + x - 5)$

36. $-2(x^2 + x + 1) + (-5x^2 - x + 2)$

37. $6(x^3 + x^2 - 3) - 4(2x^3 - 3x^2)$

38. $8(4x^3 - 3x^2 - 1) - 6(4x^3 + 8x - 2)$

39. $(x^2 - x + 2) + (2x^2 - 3x + 5) - (x^2 + 1)$

40. $(x^2 + 1) - (4x^2 + 5) + (x^2 + x - 2)$

41. $7(y^2 - 5y + 3) - 4(3 - y^2)$

42. $8(1 - y^3) + 4(1 + y + y^2 + y^3)$

43. $x^2(x^2 + 2x - 5)$

44. $4x^2(x^3 - x + 2)$

45. $-2x^2(4x^3 + 5)$

46. $5x^3(3x - 4)$

47. $(x + 1)(x^2 + 2x - 4)$

48. $(2x - 3)(x^2 + x + 1)$

In Problems 49–66, multiply the polynomials using the FOIL method. Express your answer as a single polynomial in standard form.

49. $(x + 2)(x + 4)$

50. $(x + 3)(x + 5)$

51. $(2x + 7)(x + 5)$

52. $(3x + 1)(2x + 1)$

53. $(x - 4)(x + 2)$

54. $(x + 4)(x - 2)$

55. $(x - 6)(x - 3)$

56. $(x - 5)(x - 1)$

57. $(2x + 3)(x - 2)$

58. $(2x - 4)(3x + 1)$

59. $(-3x + 4)(x - 2)$

60. $(-3x - 1)(x + 1)$

61. $(-x - 5)(-2x - 7)$

62. $(-2x - 3)(3 - x)$

63. $(x - 2y)(x + y)$

64. $(2x + 3y)(x - y)$

65. $(-2x - 3y)(3x + 2y)$

66. $(x - 3y)(-2x + y)$

In Problems 67–90, multiply the polynomials using the special product formulas. Express your answer as a single polynomial in standard form.

67. $(x - 7)(x + 7)$

68. $(x - 1)(x + 1)$

69. $(2x + 3)(2x - 3)$

70. $(3x + 2)(3x - 2)$

71. $(x + 4)^2$

72. $(x + 5)^2$

73. $(x - 4)^2$

74. $(x - 5)^2$

75. $(3x + 4)(3x - 4)$

76. $(5x - 3)(5x + 3)$

77. $(2x - 3)^2$

78. $(3x - 4)^2$

79. $(x + y)(x - y)$

80. $(x + 3y)(x - 3y)$

81. $(3x + y)(3x - y)$

82. $(3x + 4y)(3x - 4y)$

83. $(x + y)^2$

84. $(x - y)^2$

85. $(x - 2y)^2$

86. $(2x + 3y)^2$

87. $(x - 2)^3$

88. $(x + 1)^3$

89. $(2x + 1)^3$

90. $(3x - 2)^3$

In Problems 91–106, find the quotient and the remainder. Check your work by verifying that

$$Quotient \cdot Divisor + Remainder = Dividend$$

91. $4x^3 - 3x^2 + x + 1$ divided by $x + 2$

92. $3x^3 - x^2 + x - 2$ divided by $x + 2$

93. $4x^3 - 3x^2 + x + 1$ divided by x^2

94. $3x^3 - x^2 + x - 2$ divided by x^2

95. $5x^4 - 3x^2 + x + 1$ divided by $x^2 + 2$

96. $5x^4 - x^2 + x - 2$ divided by $x^2 + 2$

97. $4x^5 - 3x^2 + x + 1$ divided by $2x^3 - 1$

98. $3x^5 - x^2 + x - 2$ divided by $3x^3 - 1$

99. $2x^4 - 3x^3 + x + 1$ divided by $2x^2 + x + 1$

100. $3x^4 - x^3 + x - 2$ divided by $3x^2 + x + 1$

101. $-4x^3 + x^2 - 4$ divided by $x - 1$

102. $-3x^4 - 2x - 1$ divided by $x - 1$

103. $1 - x^2 + x^4$ divided by $x^2 + x + 1$

104. $1 - x^2 + x^4$ divided by $x^2 - x + 1$

105. $x^3 - a^3$ divided by $x - a$

106. $x^5 - a^5$ divided by $x - a$

107. **Challenge Problem** Find the numbers k so that
$$(3x - 2k)(4x + 3k) = 12x^2 + kx - 96$$

108. **Challenge Problem** Without actually multiplying, determine the number of terms the simplified product will have
$$(x + y)(z + w)(x - y)(z - w)$$

Explaining Concepts: Discussion and Writing

109. Explain why the degree of the product of two nonzero polynomials equals the sum of their degrees.

110. Explain why the degree of the sum of two polynomials of different degrees equals the larger of their degrees.

111. Give a careful statement about the degree of the sum of two polynomials of the same degree.

112. Do you prefer adding two polynomials using the horizontal method or the vertical method? Write a brief position paper defending your choice.

113. Do you prefer to memorize the rule for the square of a binomial $(x + a)^2$ or to use FOIL to obtain the product? Write a brief position paper defending your choice.

R.5 Factoring Polynomials

OBJECTIVES
1 Factor the Difference of Two Squares and the Sum and Difference of Two Cubes (p. 50)
2 Factor Perfect Squares (p. 50)
3 Factor a Second-Degree Polynomial: $x^2 + Bx + C$ (p. 51)
4 Factor by Grouping (p. 53)
5 Factor a Second-Degree Polynomial: $Ax^2 + Bx + C, A \neq 1$ (p. 53)
6 Complete the Square (p. 55)

Consider the following product:

$$(2x + 3)(x - 4) = 2x^2 - 5x - 12$$

The two polynomials on the left side are called **factors** of the polynomial on the right side. Expressing a given polynomial as a product of other polynomials—that is, finding the factors of a polynomial—is called **factoring**.

We restrict our discussion here to factoring polynomials in one variable into products of polynomials in one variable, where all coefficients are integers. We call this **factoring over the integers**.

Any polynomial can be written as the product of 1 times itself or as -1 times its additive inverse. If a polynomial cannot be written as the product of two other polynomials (excluding 1 and -1), then the polynomial is **prime**. When a polynomial has been written as a product consisting only of prime factors, it is **factored completely**. Examples of prime polynomials (over the integers) are

$$2, \quad 3, \quad 5, \quad x, \quad x + 1, \quad x - 1, \quad 3x + 4, \quad x^2 + 4$$

COMMENT Over the real numbers, $3x + 4$ factors into $3\left(x + \frac{4}{3}\right)$. It is the fraction $\frac{4}{3}$ that causes $3x + 4$ to be prime over the integers. ■

The first factor to look for in a factoring problem is a common monomial factor present in each term of the polynomial. If one is present, use the Distributive Property to factor it out. Continue factoring out monomial factors until none are left.

EXAMPLE 1 Identifying Common Monomial Factors

Polynomial	Common Monomial Factor	Remaining Factor	Factored Form
$2x + 4$	2	$x + 2$	$2x + 4 = 2(x + 2)$
$3x - 6$	3	$x - 2$	$3x - 6 = 3(x - 2)$
$2x^2 - 4x + 8$	2	$x^2 - 2x + 4$	$2x^2 - 4x + 8 = 2(x^2 - 2x + 4)$
$8x - 12$	4	$2x - 3$	$8x - 12 = 4(2x - 3)$
$x^2 + x$	x	$x + 1$	$x^2 + x = x(x + 1)$
$x^3 - 3x^2$	x^2	$x - 3$	$x^3 - 3x^2 = x^2(x - 3)$
$6x^2 + 9x$	$3x$	$2x + 3$	$6x^2 + 9x = 3x(2x + 3)$

Notice that once all common monomial factors have been removed from a polynomial, the remaining factor is either a prime polynomial of degree 1 or a polynomial of degree 2 or higher. (Do you see why?)

═══════ Now Work PROBLEM 9

1 Factor the Difference of Two Squares and the Sum and Difference of Two Cubes

When you factor a polynomial, first check for common monomial factors. Then see if it is one of the special products discussed in the previous section.

Difference of Two Squares	$x^2 - a^2 = (x - a)(x + a)$
Perfect Squares	$x^2 + 2ax + a^2 = (x + a)^2$
	$x^2 - 2ax + a^2 = (x - a)^2$
Sum of Two Cubes	$x^3 + a^3 = (x + a)(x^2 - ax + a^2)$
Difference of Two Cubes	$x^3 - a^3 = (x - a)(x^2 + ax + a^2)$

EXAMPLE 2　**Factoring the Difference of Two Squares**

Factor completely: $x^2 - 4$

Solution　Note that $x^2 - 4$ is the difference of two squares, x^2 and 2^2.

$$x^2 - 4 = (x - 2)(x + 2)$$

EXAMPLE 3　**Factoring the Difference of Two Cubes**

Factor completely: $x^3 - 1$

Solution　Because $x^3 - 1$ is the difference of two cubes, x^3 and 1^3,

$$x^3 - 1 = (x - 1)(x^2 + x + 1)$$

EXAMPLE 4　**Factoring the Sum of Two Cubes**

Factor completely: $x^3 + 8$

Solution　Because $x^3 + 8$ is the sum of two cubes, x^3 and 2^3,

$$x^3 + 8 = (x + 2)(x^2 - 2x + 4)$$

EXAMPLE 5　**Factoring the Difference of Two Squares**

Factor completely: $x^4 - 16$

Solution　Because $x^4 - 16$ is the difference of two squares, $x^4 = (x^2)^2$ and $16 = 4^2$,

$$x^4 - 16 = (x^2 - 4)(x^2 + 4)$$

But $x^2 - 4$ is also the difference of two squares. Then,

$$x^4 - 16 = (x^2 - 4)(x^2 + 4) = (x - 2)(x + 2)(x^2 + 4)$$

Now Work PROBLEMS **19** AND **37**

2 Factor Perfect Squares

When the first term and third term of a trinomial are both positive and are perfect squares, such as x^2, $9x^2$, 1, and 4, check to see whether the trinomial is a perfect square.

EXAMPLE 6　**Factoring a Perfect Square**

Factor completely: $x^2 + 6x + 9$

Solution　The first term, x^2, and the third term, $9 = 3^2$, are perfect squares. Because the middle term, $6x$, is twice the product of x and 3, we have a perfect square.

$$x^2 + 6x + 9 = (x + 3)^2$$

EXAMPLE 7

Factoring a Perfect Square

Factor completely: $9x^2 - 6x + 1$

Solution

The first term, $9x^2 = (3x)^2$, and the third term, $1 = 1^2$, are perfect squares. Because the middle term, $-6x$, is -2 times the product of $3x$ and 1, we have a perfect square.

$$9x^2 - 6x + 1 = (3x - 1)^2$$

EXAMPLE 8

Factoring a Perfect Square

Factor completely: $25x^2 + 30x + 9$

Solution

The first term, $25x^2 = (5x)^2$, and the third term, $9 = 3^2$, are perfect squares. Because the middle term, $30x$, is twice the product of $5x$ and 3, we have a perfect square.

$$25x^2 + 30x + 9 = (5x + 3)^2$$

Now Work PROBLEMS 29 AND 103

If a trinomial is not a perfect square, it may be possible to factor it using the technique discussed next.

3 Factor a Second-Degree Polynomial: $x^2 + Bx + C$

The idea behind factoring a second-degree polynomial like $x^2 + Bx + C$ is to see whether it can be made equal to the product of two (possibly equal) first-degree polynomials.

For example, consider

$$(x + 3)(x + 4) = x^2 + 7x + 12$$

The factors of $x^2 + 7x + 12$ are $x + 3$ and $x + 4$. Notice the following:

$$x^2 + 7x + 12 = (x + 3)(x + 4)$$

— 12 is the product of 3 and 4.

— 7 is the sum of 3 and 4.

In general, if $x^2 + Bx + C = (x + a)(x + b) = x^2 + (a + b)x + ab$, then $ab = C$ and $a + b = B$.

To factor a second-degree polynomial $x^2 + Bx + C$, find integers whose product is C and whose sum is B. That is, if there are numbers a, b, where $ab = C$ and $a + b = B$, then

$$x^2 + Bx + C = (x + a)(x + b)$$

EXAMPLE 9

Factoring a Trinomial

Factor completely: $x^2 + 7x + 10$

Solution

First determine all pairs of integers whose product is 10, and then compute their sums.

Integers whose product is 10	1, 10	−1, −10	2, 5	−2, −5
Sum	11	−11	7	−7

The integers 2 and 5 have a product of 10 and add up to 7, the coefficient of the middle term. As a result,

$$x^2 + 7x + 10 = (x + 2)(x + 5)$$

EXAMPLE 10

Factoring a Trinomial

Factor completely: $x^2 - 6x + 8$

Solution First determine all pairs of integers whose product is 8, and then compute each sum.

Integers whose product is 8	1, 8	−1, −8	2, 4	−2, −4
Sum	9	−9	6	−6

Since −6 is the coefficient of the middle term,

$$x^2 - 6x + 8 = (x - 2)(x - 4)$$

EXAMPLE 11

Factoring a Trinomial

Factor completely: $x^2 - x - 12$

Solution First determine all pairs of integers whose product is −12, and then compute each sum.

Integers whose product is −12	1, −12	−1, 12	2, −6	−2, 6	3, −4	−3, 4
Sum	−11	11	−4	4	−1	1

Since −1 is the coefficient of the middle term,

$$x^2 - x - 12 = (x + 3)(x - 4)$$

EXAMPLE 12

Factoring a Trinomial

Factor completely: $x^2 + 4x - 12$

Solution The integers −2 and 6 have a product of −12 and have the sum 4. So,

$$x^2 + 4x - 12 = (x - 2)(x + 6)$$

To avoid errors in factoring, always check your answer by multiplying it out to see whether the result equals the original expression.
When none of the possibilities works, the polynomial is prime.

EXAMPLE 13

Identifying a Prime Polynomial

Show that $x^2 + 9$ is prime.

Solution First list the pairs of integers whose product is 9, and then compute their sums.

Integers whose product is 9	1, 9	−1, −9	3, 3	−3, −3
Sum	10	−10	6	−6

Since the coefficient of the middle term in $x^2 + 9 = x^2 + 0x + 9$ is 0 and none of the sums equals 0, we conclude that $x^2 + 9$ is prime.

Example 13 demonstrates a more general result:

THEOREM

Any polynomial of the form $x^2 + a^2$, a a real number, is prime.

Now Work PROBLEMS 43 AND 87

4 Factor by Grouping

Sometimes a common factor does not occur in every term of the polynomial but does occur in each of several groups of terms that together make up the polynomial. When this happens, the common factor can be factored out of each group by means of the Distributive Property. This technique is called **factoring by grouping**.

EXAMPLE 14

Factoring by Grouping

Factor completely by grouping: $(x^2 + 2)x + (x^2 + 2) \cdot 3$

Solution

Notice the common factor $x^2 + 2$. Using the Distributive Property yields

$$(x^2 + 2)x + (x^2 + 2) \cdot 3 = (x^2 + 2)(x + 3)$$

Since $x^2 + 2$ and $x + 3$ are prime, the factorization is complete.

The next example shows a factoring problem that occurs in calculus.

EXAMPLE 15

Factoring by Grouping

Factor completely by grouping: $3(x - 1)^2(x + 2)^4 + 4(x - 1)^3(x + 2)^3$

Solution

Here, $(x - 1)^2(x + 2)^3$ is a common factor of both $3(x - 1)^2(x + 2)^4$ and $4(x - 1)^3(x + 2)^3$. As a result,

$$3(x - 1)^2(x + 2)^4 + 4(x - 1)^3(x + 2)^3 = (x - 1)^2(x + 2)^3[3(x + 2) + 4(x - 1)]$$

$$= (x - 1)^2(x + 2)^3[3x + 6 + 4x - 4]$$

$$= (x - 1)^2(x + 2)^3(7x + 2)$$

EXAMPLE 16

Factoring by Grouping

Factor completely by grouping: $x^3 - 4x^2 + 2x - 8$

Solution

To see whether factoring by grouping will work, group the first two terms and the last two terms. Then look for a common factor in each group. In this example, factor x^2 from $x^3 - 4x^2$ and 2 from $2x - 8$. The remaining factor in each case is the same, $x - 4$. This means that factoring by grouping will work, as follows:

$$x^3 - 4x^2 + 2x - 8 = (x^3 - 4x^2) + (2x - 8)$$

$$= x^2(x - 4) + 2(x - 4)$$

$$= (x - 4)(x^2 + 2)$$

Since $x^2 + 2$ and $x - 4$ are prime, the factorization is complete.

-**Now Work** PROBLEMS 55 AND 131

5 Factor a Second-Degree Polynomial: $Ax^2 + Bx + C, A \neq 1$

To factor a second-degree polynomial $Ax^2 + Bx + C$, when $A \neq 1$ and A, B, and C have no common factors, follow these steps:

Steps for Factoring $Ax^2 + Bx + C$,
When $A \neq 1$ and A, B, and C Have No Common Factors

STEP 1: Find the value of AC.
STEP 2: Find a pair of integers whose product is AC and that add up to B.
   That is, find a and b so that $ab = AC$ and $a + b = B$.
STEP 3: Write $Ax^2 + Bx + C = Ax^2 + ax + bx + C$.
STEP 4: Factor this last expression by grouping.

| EXAMPLE 17 | **Factoring a Trinomial** |

Factor completely: $2x^2 + 5x + 3$

Solution   Comparing $2x^2 + 5x + 3$ to $Ax^2 + Bx + C$, we find that $A = 2$, $B = 5$, and $C = 3$.

STEP 1: The value of AC is $2 \cdot 3 = 6$.
STEP 2: Determine the pairs of integers whose product is $AC = 6$ and compute their sums.

Integers whose product is 6	1, 6	−1, −6	2, 3	−2, −3
Sum	7	−7	5	−5

The integers whose product is 6 that add up to $B = 5$ are 2 and 3.

STEP 3: Write $2x^2 + 5x + 3 = 2x^2 + 2x + 3x + 3$

STEP 4: Factor by grouping.

$$2x^2 + 2x + 3x + 3 = (2x^2 + 2x) + (3x + 3)$$
$$= 2x(x + 1) + 3(x + 1)$$
$$= (x + 1)(2x + 3)$$

As a result,

$$2x^2 + 5x + 3 = (x + 1)(2x + 3)$$

| EXAMPLE 18 | **Factoring a Trinomial** |

Factor completely: $2x^2 - x - 6$

Solution   Comparing $2x^2 - x - 6$ to $Ax^2 + Bx + C$, we find that $A = 2$, $B = -1$, and $C = -6$.

STEP 1: The value of AC is $2 \cdot (-6) = -12$.
STEP 2: Determine the pairs of integers whose product is $AC = -12$ and compute their sums.

Integers whose product is − 12	1, −12	−1, 12	2, −6	−2, 6	3, −4	−3, 4
Sum	−11	11	−4	4	−1	1

The integers whose product is −12 that add up to $B = -1$ are −4 and 3.

STEP 3: Write $2x^2 - x - 6 = 2x^2 - 4x + 3x - 6$

STEP 4: Factor by grouping.

$$2x^2 - 4x + 3x - 6 = (2x^2 - 4x) + (3x - 6)$$
$$= 2x(x - 2) + 3(x - 2)$$
$$= (x - 2)(2x + 3)$$

As a result,

$$2x^2 - x - 6 = (x - 2)(2x + 3)$$

Now Work PROBLEM 61

SUMMARY

Type of Polynomial	Method	Example
Any polynomial	Look for common monomial factors. (Always do this first!)	$6x^2 + 9x = 3x(2x + 3)$
Binomials of degree 2 or higher	Check for a special product: Difference of two squares, $x^2 - a^2$ Difference of two cubes, $x^3 - a^3$ Sum of two cubes, $x^3 + a^3$	$x^2 - 16 = (x - 4)(x + 4)$ $x^3 - 64 = (x - 4)(x^2 + 4x + 16)$ $x^3 + 27 = (x + 3)(x^2 - 3x + 9)$
Trinomials of degree 2	Check for a perfect square, $(x \pm a)^2$ Factoring $x^2 + Bx + C$ (p. 51) Factoring $Ax^2 + Bx + C$ (p. 53)	$x^2 + 8x + 16 = (x + 4)^2$ $x^2 - 10x + 25 = (x - 5)^2$ $x^2 - x - 2 = (x - 2)(x + 1)$ $6x^2 + x - 1 = (2x + 1)(3x - 1)$
Four or more terms	Grouping	$2x^3 - 3x^2 + 4x - 6 = (2x - 3)(x^2 + 2)$

6 Complete the Square

The idea behind completing the square in one variable is to "adjust" an expression of the form $x^2 + bx$ to make it a perfect square. Perfect squares are trinomials of the form

$$x^2 + 2ax + a^2 = (x + a)^2 \text{ or } x^2 - 2ax + a^2 = (x - a)^2$$

For example, $x^2 + 6x + 9$ is a perfect square because $x^2 + 6x + 9 = (x + 3)^2$. And $p^2 - 12p + 36$ is a perfect square because $p^2 - 12p + 36 = (p - 6)^2$.

So how do we "adjust" $x^2 + bx$ to make it a perfect square? We do it by adding a number. For example, to make $x^2 + 6x$ a perfect square, add 9. But how do we know to add 9? If we divide the coefficient of the first-degree term, 6, by 2, and then square the result, we obtain 9. This approach works in general.

> **Completing the Square of $x^2 + bx$**
>
> - Identify the coefficient of the first-degree term, namely b.
> - Multiply b by $\dfrac{1}{2}$ and then square the result. That is, compute $\left(\dfrac{1}{2}b\right)^2$.
> - Add $\left(\dfrac{1}{2}b\right)^2$ to $x^2 + bx$ to get $x^2 + bx + \left(\dfrac{1}{2}b\right)^2 = \left(x + \dfrac{b}{2}\right)^2$

WARNING To use $\left(\frac{1}{2}b\right)^2$ to complete the square, the coefficient of the x^2 term must be 1. ∎

EXAMPLE 19   **Completing the Square**

Determine the number that must be added to each expression to complete the square. Then factor the expression.

Start	Add	Result	Factored Form
$y^2 + 8y$	$\left(\dfrac{1}{2} \cdot 8\right)^2 = 16$	$y^2 + 8y + 16$	$(y + 4)^2$
$x^2 + 12x$	$\left(\dfrac{1}{2} \cdot 12\right)^2 = 36$	$x^2 + 12x + 36$	$(x + 6)^2$
$a^2 - 20a$	$\left(\dfrac{1}{2} \cdot (-20)\right)^2 = 100$	$a^2 - 20a + 100$	$(a - 10)^2$
$p^2 - 5p$	$\left(\dfrac{1}{2} \cdot (-5)\right)^2 = \dfrac{25}{4}$	$p^2 - 5p + \dfrac{25}{4}$	$\left(p - \dfrac{5}{2}\right)^2$

Notice that the factored form of a perfect square is either

$$x^2 + bx + \left(\frac{b}{2}\right)^2 = \left(x + \frac{b}{2}\right)^2 \quad \text{or} \quad x^2 - bx + \left(\frac{b}{2}\right)^2 = \left(x - \frac{b}{2}\right)^2$$

Now Work PROBLEM 73

Figure 27

Are you wondering why we refer to making an expression a perfect square as "completing the square"? Look at the square in Figure 27. Its area is $(y + 4)^2$. The yellow area is y^2 and each orange area is $4y$ (for a total area of $8y$). The sum of these areas is $y^2 + 8y$. To complete the square, we need to add the area of the green region, which is $4 \cdot 4 = 16$. As a result, $y^2 + 8y + 16 = (y + 4)^2$.

R.5 Assess Your Understanding

Concepts and Vocabulary

1. If factored completely, $3x^3 - 12x =$ _____.

2. If a polynomial cannot be written as the product of two other polynomials (excluding 1 and -1), then the polynomial is said to be _____.

3. *Multiple Choice* For $x^2 + Bx + C = (x + a)(x + b)$, which of the following must be true?
   (a) $ab = B$ and $a + b = C$
   (b) $a + b = C$ and $a - b = B$
   (c) $ab = C$ and $a + b = B$
   (d) $ab = B$ and $a - b = C$

4. *Multiple Choice* Choose the best description of $x^2 - 64$.
   (a) Prime
   (b) Difference of two squares
   (c) Difference of two cubes
   (d) Perfect Square

5. *Multiple Choice* Choose the complete factorization of
   $$4x^2 - 8x - 60$$
   (a) $2(x + 3)(x - 5)$
   (b) $4(x^2 - 2x - 15)$
   (c) $(2x + 6)(2x - 10)$
   (d) $4(x + 3)(x - 5)$

6. *Multiple Choice* To complete the square of $x^2 + bx$, add which of the following?
   (a) $(2b)^2$ (b) $2b^2$ (c) $\left(\frac{1}{2}b\right)^2$ (d) $\frac{1}{2}b^2$

7. *True or False* The polynomial $x^2 + 4$ is prime.

8. *True or False* $3x^3 - 2x^2 - 6x + 4 = (3x - 2)(x^2 + 2)$.

Skill Building

In Problems 9–18, factor each polynomial by factoring out the common monomial factor.

9. $3x + 6$
10. $7x - 14$
11. $ax^2 + a$
12. $ax - a$
13. $x^3 + x^2 + x$
14. $x^3 - x^2 + x$
15. $2x^2 - 2x$
16. $3x^2 - 3x$
17. $3x^2y - 6xy^2 + 12xy$
18. $60x^2y - 48xy^2 + 72x^3y$

In Problems 19–26, factor the difference of two squares.

19. $x^2 - 1$
20. $x^2 - 4$
21. $4x^2 - 1$
22. $9x^2 - 1$
23. $x^2 - 16$
24. $x^2 - 25$
25. $25x^2 - 4$
26. $36x^2 - 9$

In Problems 27–36, factor the perfect squares.

27. $x^2 + 2x + 1$
28. $x^2 - 4x + 4$
29. $x^2 + 4x + 4$
30. $x^2 - 2x + 1$
31. $x^2 - 10x + 25$
32. $x^2 + 10x + 25$
33. $4x^2 + 4x + 1$
34. $9x^2 + 6x + 1$
35. $16x^2 + 8x + 1$
36. $25x^2 + 10x + 1$

In Problems 37–42, factor the sum or difference of two cubes.

37. $x^3 - 27$
38. $x^3 + 125$
39. $x^3 + 27$
40. $27 - 8x^3$
41. $8x^3 + 27$
42. $64 - 27x^3$

In Problems 43–54, factor each polynomial.

43. $x^2 + 5x + 6$
44. $x^2 + 6x + 8$
45. $x^2 + 7x + 6$
46. $x^2 + 9x + 8$
47. $x^2 + 7x + 10$
48. $x^2 + 11x + 10$
49. $x^2 - 10x + 16$
50. $x^2 - 17x + 16$
51. $x^2 - 7x - 8$
52. $x^2 - 2x - 8$
53. $x^2 + 7x - 8$
54. $x^2 + 2x - 8$

In Problems 55–60, factor by grouping.

55. $2x^2 + 4x + 3x + 6$
56. $3x^2 - 3x + 2x - 2$
57. $5x^2 - 15x + x - 3$
58. $3x^2 + 6x - x - 2$
59. $6x^2 + 21x + 8x + 28$
60. $9x^2 - 6x + 3x - 2$

In Problems 61–72, factor each polynomial.

61. $3x^2 + 4x + 1$ **62.** $2x^2 + 3x + 1$ **63.** $2z^2 + 9z + 7$ **64.** $6z^2 + 5z + 1$

65. $5x^2 + 6x - 8$ **66.** $3x^2 + 10x + 8$ **67.** $5x^2 - 6x - 8$ **68.** $3x^2 - 10x + 8$

69. $5x^2 + 22x + 8$ **70.** $3x^2 - 14x + 8$ **71.** $5x^2 + 18x - 8$ **72.** $3x^2 - 10x - 8$

In Problems 73–78, determine what number should be added to complete the square of each expression. Then factor each expression.

73. $x^2 + 10x$ **74.** $p^2 + 14p$ **75.** $y^2 - 6y$

76. $x^2 - 4x$ **77.** $x^2 - \dfrac{1}{2}x$ **78.** $x^2 + \dfrac{1}{3}x$

Mixed Practice *In Problems 79–126, factor each polynomial completely. If the polynomial cannot be factored, say it is prime.*

79. $x^2 - 36$ **80.** $x^2 - 9$ **81.** $2 - 8x^2$ **82.** $3 - 27x^2$

83. $x^2 + 11x + 10$ **84.** $x^2 + 5x + 4$ **85.** $x^2 - 10x + 21$ **86.** $x^2 - 6x + 8$

87. $4x^2 - 8x + 32$ **88.** $3x^2 - 12x + 15$ **89.** $x^2 + 4x + 16$ **90.** $x^2 + 12x + 36$

91. $15 + 2x - x^2$ **92.** $14 + 6x - x^2$ **93.** $3x^2 - 12x - 36$ **94.** $x^3 + 8x^2 - 20x$

95. $y^4 + 11y^3 + 30y^2$ **96.** $3y^3 - 18y^2 - 48y$ **97.** $4x^2 + 12x + 9$ **98.** $9x^2 - 12x + 4$

99. $6x^2 + 8x + 2$ **100.** $8x^2 + 6x - 2$ **101.** $x^4 - 81$ **102.** $x^4 - 1$

103. $x^6 - 2x^3 + 1$ **104.** $x^6 + 2x^3 + 1$ **105.** $x^7 - x^5$ **106.** $x^8 - x^5$

107. $16x^2 + 24x + 9$ **108.** $9x^2 - 24x + 16$ **109.** $5 + 16x - 16x^2$ **110.** $5 + 11x - 16x^2$

111. $4y^2 - 16y + 15$ **112.** $9y^2 + 9y - 4$ **113.** $1 - 8x^2 - 9x^4$ **114.** $4 - 14x^2 - 8x^4$

115. $x(x + 3) - 6(x + 3)$ **116.** $5(3x - 7) + x(3x - 7)$ **117.** $(x + 2)^2 - 5(x + 2)$

118. $(x - 1)^2 - 2(x - 1)$ **119.** $(3x - 2)^3 - 27$ **120.** $(5x + 1)^3 - 1$

121. $3(x^2 + 10x + 25) - 4(x + 5)$ **122.** $7(x^2 - 6x + 9) + 5(x - 3)$ **123.** $x^3 + 2x^2 - x - 2$

124. $x^3 - 3x^2 - x + 3$ **125.** $x^4 - x^3 + x - 1$ **126.** $x^4 + x^3 + x + 1$

Applications and Extensions

In Problems 127–136, expressions that occur in calculus are given. Factor each expression completely.

127. $2(3x + 4)^2 + (2x + 3) \cdot 2(3x + 4) \cdot 3$

128. $5(2x + 1)^2 + (5x - 6) \cdot 2(2x + 1) \cdot 2$

129. $2x(2x + 5) + x^2 \cdot 2$

130. $3x^2(8x - 3) + x^3 \cdot 8$

131. $2(x + 3)(x - 2)^3 + (x + 3)^2 \cdot 3(x - 2)^2$

132. $4(x + 5)^3(x - 1)^2 + (x + 5)^4 \cdot 2(x - 1)$

133. $(4x - 3)^2 + x \cdot 2(4x - 3) \cdot 4$

134. $3x^2(3x + 4)^2 + x^3 \cdot 2(3x + 4) \cdot 3$

135. $2(3x - 5) \cdot 3(2x + 1)^3 + (3x - 5)^2 \cdot 3(2x + 1)^2 \cdot 2$

136. $3(4x + 5)^2 \cdot 4(5x + 1)^2 + (4x + 5)^3 \cdot 2(5x + 1) \cdot 5$

137. *Challenge Problem* Show that $x^2 + 4$ is prime.

138. *Challenge Problem* Show that $x^2 + x + 1$ is prime.

Explaining Concepts: Discussion and Writing

139. Make up a polynomial that factors into a perfect square.

140. Explain to a fellow student what you look for first when presented with a factoring problem. What do you do next?

R.6 Synthetic Division

OBJECTIVE 1 Divide Polynomials Using Synthetic Division (p. 57)

1 Divide Polynomials Using Synthetic Division

To find the quotient as well as the remainder when a polynomial of degree 1 or higher is divided by $x - c$, a shortened version of long division, called **synthetic division**, makes the task simpler.

To see how synthetic division works, first consider long division for dividing the polynomial $2x^3 - x^2 + 3$ by $x - 3$.

$$
\begin{array}{r}
2x^2 + 5x\ + 15 \quad \leftarrow \text{Quotient} \\
x - 3\overline{\smash{)}2x^3 -\ x^2 \qquad\quad +\ 3} \\
\underline{2x^3 - 6x^2} \\
5x^2 \\
\underline{5x^2 - 15x} \\
15x +\ 3 \\
\underline{15x - 45} \\
48 \quad \leftarrow \text{Remainder}
\end{array}
$$

✓ **Check:** Divisor · Quotient + Remainder

$$
\begin{aligned}
&= (x - 3)(2x^2 + 5x + 15) + 48 \\
&= 2x^3 + 5x^2 + 15x - 6x^2 - 15x - 45 + 48 \\
&= 2x^3 - x^2 + 3
\end{aligned}
$$

The process of synthetic division arises from rewriting the long division in a more compact form, using simpler notation. For example, in the long division above, the terms in blue are not really necessary because they are identical to the terms directly above them. With these terms removed, we have

$$
\begin{array}{r}
2x^2 + 5x\ + 15 \\
x - 3\overline{\smash{)}2x^3 -\ x^2 \qquad\quad +\ 3} \\
-\ 6x^2 \\
\underline{} \\
5x^2 \\
-\ 15x \\
\underline{} \\
15x \\
-\ 45 \\
\underline{} \\
48
\end{array}
$$

Most of the x's that appear in this process can also be removed, provided that we are careful about positioning each coefficient. In this regard, we will need to use 0 as the coefficient of x in the dividend, because that power of x is missing. Now we have

$$
\begin{array}{r}
2x^2 + 5x + 15 \\
x - 3\overline{\smash{)}2\quad -\ 1 \qquad 0 \qquad 3} \\
-\ 6 \\
\underline{} \\
5 \\
-\ 15 \\
\underline{} \\
15 \\
-\ 45 \\
\underline{} \\
48
\end{array}
$$

We can make this display more compact by moving the lines up until the numbers in blue align horizontally.

$$
\begin{array}{rl}
2x^2 + 5x + 15 & \text{Row 1} \\
x - 3\overline{\smash{)}2 \quad -\ 1 \qquad 0 \qquad 3} & \text{Row 2} \\
-\ 6 \ -\ 15 - 45 & \text{Row 3} \\
\underline{} & \\
\bigcirc \quad 5 \qquad 15 \quad 48 & \text{Row 4}
\end{array}
$$

Because the leading coefficient of the divisor is always 1, the leading coefficient of the dividend will also be the leading coefficient of the quotient. So we place the leading coefficient of the quotient, 2, in the circled position. Now, the first three numbers in row 4 are precisely the coefficients of the quotient, and the last number in row 4 is the remainder. Since row 1 is not really needed, we can compress the process to three rows, where the bottom row contains both the coefficients of the quotient and the remainder.

$$x - 3\overline{)2 \quad -1 \quad 0 \quad 3} \quad \text{Row 1}$$
$$\underline{-6 - 15 - 45} \quad \text{Row 2 (subtract)}$$
$$2 \quad 5 \quad 15 \quad 48 \quad \text{Row 3}$$

Recall that the entries in row 3 are obtained by subtracting the entries in row 2 from those in row 1. Rather than subtracting the entries in row 2, we can change the sign of each entry and add. With this modification, our display will look like this:

$$x - 3\overline{)2 \quad -1 \quad 0 \quad 3} \quad \text{Row 1}$$
$$\underline{6 \quad 15 \quad 45} \quad \text{Row 2 (add)}$$
$$2 \quad 5 \quad 15 \quad 48 \quad \text{Row 3}$$

Notice that the entries in row 2 are three times the prior entries in row 3. Our last modification to the display replaces the $x - 3$ by 3. The entries in row 3 give the quotient and the remainder, as shown next.

$$3\overline{)2 \quad -1 \quad 0 \quad 3} \quad \text{Row 1}$$
$$\underline{6 \quad 15 \quad 45} \quad \text{Row 2 (add)}$$
$$2 \quad 5 \quad 15 \quad 48 \quad \text{Row 3}$$
$$\underbrace{2x^2 + 5x + 15}_{\text{Quotient}} \quad \underbrace{48}_{\text{Remainder}}$$

Let's go through an example step by step.

EXAMPLE 1   **Using Synthetic Division to Find the Quotient and Remainder**

Use synthetic division to find the quotient and remainder when

$$x^3 - 4x^2 - 5 \quad \text{is divided by} \quad x - 3$$

Solution   **STEP 1:** Write the dividend in descending powers of x. Then copy the coefficients, remembering to insert a 0 for any missing powers of x.

$$1 \quad -4 \quad 0 \quad -5 \quad \text{Row 1}$$

STEP 2: Insert the usual division symbol. In synthetic division, the divisor is of the form $x - c$, and c is the number placed to the left of the division symbol. Here, since the divisor is $x - 3$, insert 3 to the left of the division symbol.

$$3\overline{)1 \quad -4 \quad 0 \quad -5} \quad \text{Row 1}$$

STEP 3: Bring the 1 down two rows, and enter it in row 3.

$$3\overline{)1 \quad -4 \quad 0 \quad -5} \quad \text{Row 1}$$
$$\downarrow \qquad\qquad\qquad \text{Row 2}$$
$$1 \qquad\qquad\qquad \text{Row 3}$$

STEP 4: Multiply the latest entry in row 3 by 3, and place the result in row 2, one column over to the right.

$$3\overline{)1 \quad -4 \quad 0 \quad -5} \quad \text{Row 1}$$
$$\underline{3} \qquad\qquad\qquad \text{Row 2}$$
$$1 \qquad\qquad\qquad \text{Row 3}$$

STEP 5: Add the entry in row 2 to the entry above it in row 1, and enter the sum in row 3.

$$3\overline{)1 \quad -4 \quad 0 \quad -5} \quad \text{Row 1}$$
$$\underline{3} \qquad\qquad\qquad \text{Row 2}$$
$$1 \quad -1 \qquad\qquad \text{Row 3}$$

STEP 6: Repeat Steps 4 and 5 until no more entries are available in row 1.

$$3\overline{)1 \quad -4 \quad 0 \quad -5} \quad \text{Row 1}$$
$$\underline{ \quad 3 \quad -3 \quad -9} \quad \text{Row 2}$$
$$1 \quad -1 \quad -3 \quad -14 \quad \text{Row 3}$$

STEP 7: The final entry in row 3, the -14, is the remainder; the other entries in row 3, the 1, -1, and -3, are the coefficients (in descending order) of a polynomial whose degree is 1 less than that of the dividend. This is the quotient. That is,

$$\text{Quotient} = x^2 - x - 3 \qquad \text{Remainder} = -14$$

✓ **Check:** Divisor · Quotient + Remainder

$$= (x - 3)(x^2 - x - 3) + (-14)$$
$$= (x^3 - x^2 - 3x - 3x^2 + 3x + 9) + (-14)$$
$$= x^3 - 4x^2 - 5 = \text{Dividend}$$

Let's do an example in which all seven steps are combined.

EXAMPLE 2 **Using Synthetic Division to Verify a Factor**

Use synthetic division to show that $x + 3$ is a factor of

$$2x^5 + 5x^4 - 2x^3 + 2x^2 - 2x + 3$$

Solution The divisor is $x + 3 = x - (-3)$, so place -3 to the left of the division symbol. Then the row 3 entries will be multiplied by -3, entered in row 2, and added to row 1.

$$-3\overline{)2 \quad 5 \quad -2 \quad 2 \quad -2 \quad 3} \quad \text{Row 1}$$
$$\underline{ \quad -6 \quad 3 \quad -3 \quad 3 \quad -3} \quad \text{Row 2}$$
$$2 \quad -1 \quad 1 \quad -1 \quad 1 \quad 0 \quad \text{Row 3}$$

Because the remainder is 0, we have

Divisor · Quotient + Remainder

$$= (x + 3)(2x^4 - x^3 + x^2 - x + 1) = 2x^5 + 5x^4 - 2x^3 + 2x^2 - 2x + 3$$

As we see, $x + 3$ is a factor of $2x^5 + 5x^4 - 2x^3 + 2x^2 - 2x + 3$.

As Example 2 illustrates, the remainder after division gives information about whether the divisor is, or is not, a factor. We shall have more to say about this in Chapter 5.

✏ **Now Work** PROBLEMS **9** AND **19**

R.6 Assess Your Understanding

Concepts and Vocabulary

1. To check division, the expression that is being divided, the dividend, should equal the product of the _____ and the _____ plus the _____.

2. To divide $2x^3 - 5x + 1$ by $x + 3$ using synthetic division, the first step is to write ___$\overline{)}$.

3. *Multiple Choice* Choose the division problem that cannot be done using synthetic division.
   (a) $2x^3 - 4x^2 + 6x - 8$ is divided by $x - 8$
   (b) $x^4 - 3$ is divided by $x + 1$
   (c) $x^5 + 3x^2 - 9x + 2$ is divided by $x + 10$
   (d) $x^4 - 5x^3 + 3x^2 - 9x + 13$ is divided by $x^2 + 5$

4. *Multiple Choice* Choose the correct conclusion based on the following synthetic division:
   $$-5\overline{)2 \quad 3 \quad -38 \quad -15}$$
   $$\underline{ \quad -10 \quad 35 \quad 15}$$
   $$2 \quad -7 \quad -3 \quad 0$$
   (a) $x + 5$ is a factor of $2x^3 + 3x^2 - 38x - 15$
   (b) $x - 5$ is a factor of $2x^3 + 3x^2 - 38x - 15$
   (c) $x + 5$ is not a factor of $2x^3 + 3x^2 - 38x - 15$
   (d) $x - 5$ is not a factor of $2x^3 + 3x^2 - 38x - 15$

5. *True or False* In using synthetic division, the divisor is always a polynomial of degree 1, whose leading coefficient is 1.

6. *True or False* $-2\overline{)5\quad 3\quad 2\quad 1}$

$$\underline{\quad -10\quad 14\quad -32\quad}\ \text{means}\quad \frac{5x^3 + 3x^2 + 2x + 1}{x + 2} = 5x^2 - 7x + 16 + \frac{-31}{x + 2}.$$

$$\quad 5\quad -7\quad 16\quad -31$$

Skill Building

In Problems 7–18, use synthetic division to find the quotient and remainder when:

7. $x^3 - 7x^2 + 5x + 10$ is divided by $x - 2$

8. $x^3 + 2x^2 - 3x + 1$ is divided by $x + 1$

9. $3x^3 + 2x^2 - x + 3$ is divided by $x - 3$

10. $-4x^3 + 2x^2 - x + 1$ is divided by $x + 2$

11. $x^5 - 4x^3 + x$ is divided by $x + 3$

12. $x^4 + x^2 + 2$ is divided by $x - 2$

13. $4x^6 - 3x^4 + x^2 + 5$ is divided by $x - 1$

14. $x^5 + 5x^3 - 10$ is divided by $x + 1$

15. $0.1x^3 + 0.2x$ is divided by $x + 1.1$

16. $0.1x^2 - 0.2$ is divided by $x + 2.1$

17. $x^5 - 32$ is divided by $x - 2$

18. $x^5 + 1$ is divided by $x + 1$

In Problems 19–28, use synthetic division to determine whether $x - c$ is a factor of the given polynomial.

19. $4x^3 - 3x^2 - 8x + 4$; $x - 2$

20. $-4x^3 + 5x^2 + 8$; $x + 3$

21. $2x^4 - 6x^3 - 7x + 21$; $x - 3$

22. $4x^4 - 15x^2 - 4$; $x - 2$

23. $5x^6 + 43x^3 + 24$; $x + 2$

24. $2x^6 - 18x^4 + x^2 - 9$; $x + 3$

25. $x^5 - 16x^3 - x^2 + 19$; $x + 4$

26. $x^6 - 16x^4 + x^2 - 16$; $x + 4$

27. $3x^4 - x^3 + 6x - 2$; $x - \dfrac{1}{3}$

28. $3x^4 + x^3 - 3x + 1$; $x + \dfrac{1}{3}$

Applications and Extensions

29. Find the sum of $a, b, c,$ and d if

$$\frac{x^3 - 2x^2 + 3x + 5}{x + 2} = ax^2 + bx + c + \frac{d}{x + 2}$$

30. *Challenge Problem* Use synthetic division to divide
$x^4 + (3 - h)\,x^3 - 2hx^2 - 2h^2x + h^3$ by $x - h$.

31. *Challenge Problem* Use synthetic division to determine
whether $x + y$ is a factor of $x^4 + 3x^3y - 3x^2y^2 - xy^3 + 4y^4$.

Explaining Concepts: Discussion and Writing

32. When dividing a polynomial by $x - c$, do you prefer to use long division or synthetic division? Does the value of c make a difference to you in choosing? Give reasons.

R.7 Rational Expressions

OBJECTIVES **1** Reduce a Rational Expression to Lowest Terms (p. 61)

 2 Multiply and Divide Rational Expressions (p. 62)

 3 Add and Subtract Rational Expressions (p. 63)

 4 Use the Least Common Multiple Method (p. 65)

 5 Simplify Complex Rational Expressions (p. 67)

1 Reduce a Rational Expression to Lowest Terms

If we form the quotient of two polynomials, the result is called a **rational expression**. Some examples of rational expressions are

(a) $\dfrac{x^3 + 1}{x}$ (b) $\dfrac{3x^2 + x - 2}{x^2 + 5}$ (c) $\dfrac{x}{x^2 - 1}$, (d) $\dfrac{xy^2}{(x - y)^2}$

Expressions (a), (b), and (c) are rational expressions in one variable, x, whereas (d) is a rational expression in two variables, x and y.

Rational expressions are described in the same manner as rational numbers. In expression (a), the polynomial $x^3 + 1$ is the **numerator**, and x is the **denominator**. When the numerator and denominator of a rational expression contain no common factors (except 1 and -1), we say that the rational expression is **reduced to lowest terms**, or **simplified**.

The polynomial in the denominator of a rational expression cannot be equal to 0 because division by 0 is not defined. For example, for the expression $\dfrac{x^3 + 1}{x}$, x cannot take on the value 0. The domain of the variable x is $\{x \mid x \neq 0\}$.

A rational expression is reduced to lowest terms by factoring the numerator and the denominator completely and canceling any common factors using the Cancellation Property:

Cancellation Property

$$\frac{a\cancel{c}}{b\cancel{c}} = \frac{a}{b} \qquad \text{if } b \neq 0, c \neq 0 \tag{1}$$

EXAMPLE 1 **Reducing a Rational Expression to Lowest Terms**

Reduce to lowest terms: $\dfrac{x^2 + 4x + 4}{x^2 + 3x + 2}$

Solution Begin by factoring the numerator and the denominator.

$$x^2 + 4x + 4 = (x + 2)(x + 2)$$
$$x^2 + 3x + 2 = (x + 2)(x + 1)$$

Since a common factor, $x + 2$, appears, the original expression is not in lowest terms. To reduce it to lowest terms, use the Cancellation Property:

$$\frac{x^2 + 4x + 4}{x^2 + 3x + 2} = \frac{\cancel{(x + 2)}(x + 2)}{\cancel{(x + 2)}(x + 1)} = \frac{x + 2}{x + 1} \qquad x \neq -2, -1$$

WARNING Use the Cancellation Property only with rational expressions written in factored form. Be sure to cancel only common factors, not common terms! ∎

EXAMPLE 2 **Reducing Rational Expressions to Lowest Terms**

Reduce each rational expression to lowest terms.

(a) $\dfrac{x^3 - 8}{x^3 - 2x^2}$ 　　　　(b) $\dfrac{8 - 2x}{x^2 - x - 12}$

Solution (a) $\dfrac{x^3 - 8}{x^3 - 2x^2} = \dfrac{\cancel{(x - 2)}(x^2 + 2x + 4)}{x^2\cancel{(x - 2)}} = \dfrac{x^2 + 2x + 4}{x^2} \qquad x \neq 0, 2$

(b) $\dfrac{8 - 2x}{x^2 - x - 12} = \dfrac{2(4 - x)}{(x - 4)(x + 3)} = \dfrac{2(-1)\cancel{(x - 4)}}{\cancel{(x - 4)}(x + 3)} = \dfrac{-2}{x + 3} \qquad x \neq -3, 4$

- Now Work PROBLEM 7

2 Multiply and Divide Rational Expressions

The rules for multiplying and dividing rational expressions are the same as the rules for multiplying and dividing rational numbers.

If $\dfrac{a}{b}$ and $\dfrac{c}{d}$ are two rational expressions, then

$$\frac{a}{b} \cdot \frac{c}{d} = \frac{ac}{bd} \qquad \text{if } b \neq 0, d \neq 0 \qquad \text{(2)}$$

$$\frac{\dfrac{a}{b}}{\dfrac{c}{d}} = \frac{a}{b} \cdot \frac{d}{c} = \frac{ad}{bc} \qquad \text{if } b \neq 0, c \neq 0, d \neq 0 \qquad \text{(3)}$$

In using equations (2) and (3) with rational expressions, be sure first to factor each polynomial completely so that common factors can be canceled. Leave your answer in factored form.

EXAMPLE 3

Multiplying and Dividing Rational Expressions

Perform the indicated operation and simplify the result. Leave your answer in factored form.

(a) $\dfrac{x^2 - 2x + 1}{x^3 + x} \cdot \dfrac{4x^2 + 4}{x^2 + x - 2}$

(b) $\dfrac{\dfrac{x + 3}{x^2 - 4}}{\dfrac{x^2 - x - 12}{x^3 - 8}}$

Solution

(a) $\dfrac{x^2 - 2x + 1}{x^3 + x} \cdot \dfrac{4x^2 + 4}{x^2 + x - 2} = \dfrac{(x - 1)^2}{x(x^2 + 1)} \cdot \dfrac{4(x^2 + 1)}{(x + 2)(x - 1)}$

$$= \frac{(x - 1)^2 (4) \cancel{(x^2 + 1)}}{x \cancel{(x^2 + 1)} (x + 2) \cancel{(x - 1)}}$$

$$= \frac{4(x - 1)}{x(x + 2)} \qquad x \neq -2, 0, 1$$

(b) $\dfrac{\dfrac{x + 3}{x^2 - 4}}{\dfrac{x^2 - x - 12}{x^3 - 8}} = \dfrac{x + 3}{x^2 - 4} \cdot \dfrac{x^3 - 8}{x^2 - x - 12}$

$$= \frac{x + 3}{(x - 2)(x + 2)} \cdot \frac{(x - 2)(x^2 + 2x + 4)}{(x - 4)(x + 3)}$$

$$= \frac{\cancel{(x + 3)}\,\cancel{(x - 2)}\,(x^2 + 2x + 4)}{\cancel{(x - 2)}\,(x + 2)(x - 4)\,\cancel{(x + 3)}}$$

$$= \frac{x^2 + 2x + 4}{(x + 2)(x - 4)} \qquad x \neq -3, -2, 2, 4$$

━━━ **Now Work** PROBLEMS **19** AND **27**

In Words

To add (or subtract) two rational expressions with the same denominator, keep the common denominator and add (or subtract) the numerators.

3 Add and Subtract Rational Expressions

The rules for adding and subtracting rational expressions are the same as the rules for adding and subtracting rational numbers. If the denominators of two rational expressions to be added (or subtracted) are equal, then add (or subtract) the numerators and keep the common denominator.

If $\dfrac{a}{b}$ and $\dfrac{c}{b}$ are two rational expressions, then

$$\frac{a}{b} + \frac{c}{b} = \frac{a+c}{b} \qquad \frac{a}{b} - \frac{c}{b} = \frac{a-c}{b} \qquad \text{if } b \neq 0 \qquad \textbf{(4)}$$

EXAMPLE 4 **Adding and Subtracting Rational Expressions with Equal Denominators**

Perform the indicated operation and simplify the result. Leave your answer in factored form.

(a) $\dfrac{2x^2 - 4}{2x + 5} + \dfrac{x + 3}{2x + 5} \qquad x \neq -\dfrac{5}{2}$ (b) $\dfrac{x}{x - 3} - \dfrac{3x + 2}{x - 3} \qquad x \neq 3$

Solution (a) $\dfrac{2x^2 - 4}{2x + 5} + \dfrac{x + 3}{2x + 5} = \dfrac{(2x^2 - 4) + (x + 3)}{2x + 5}$

$$= \frac{2x^2 + x - 1}{2x + 5} = \frac{(2x - 1)(x + 1)}{2x + 5}$$

(b) $\dfrac{x}{x - 3} - \dfrac{3x + 2}{x - 3} = \dfrac{x - (3x + 2)}{x - 3} = \dfrac{x - 3x - 2}{x - 3}$

$$= \frac{-2x - 2}{x - 3} = \frac{-2(x + 1)}{x - 3}$$

EXAMPLE 5 **Adding Rational Expressions Whose Denominators Are Additive Inverses of Each Other**

Perform the indicated operation and simplify the result. Leave your answer in factored form.

$$\frac{2x}{x - 3} + \frac{5}{3 - x} \qquad x \neq 3$$

Solution Notice that the denominators of the two rational expressions are different. However, the denominator of the second expression is the additive inverse of the denominator of the first. That is,

$$3 - x = -x + 3 = -1 \cdot (x - 3) = -(x - 3)$$

Then

$$\frac{2x}{x - 3} + \frac{5}{3 - x} = \frac{2x}{x - 3} + \frac{5}{-(x - 3)} = \frac{2x}{x - 3} + \frac{-5}{x - 3}$$

$$3 - x = -(x - 3) \qquad\qquad \frac{a}{-b} = \frac{-a}{b}$$

$$= \frac{2x + (-5)}{x - 3} = \frac{2x - 5}{x - 3}$$

Now Work PROBLEMS **39** AND **45**

If the denominators of two rational expressions to be added or subtracted are not equal, we can use the general formulas for adding and subtracting rational expressions.

If $\dfrac{a}{b}$ and $\dfrac{c}{d}$ are two rational expressions, then

$$\frac{a}{b} + \frac{c}{d} = \frac{a \cdot d}{b \cdot d} + \frac{b \cdot c}{b \cdot d} = \frac{ad + bc}{bd} \qquad \text{if } b \neq 0, d \neq 0 \qquad \textbf{(5a)}$$

$$\frac{a}{b} - \frac{c}{d} = \frac{a \cdot d}{b \cdot d} - \frac{b \cdot c}{b \cdot d} = \frac{ad - bc}{bd} \qquad \text{if } b \neq 0, d \neq 0 \qquad \textbf{(5b)}$$

EXAMPLE 6

Adding and Subtracting Rational Expressions with Unequal Denominators

Perform the indicated operation and simplify the result. Leave your answer in factored form.

(a) $\dfrac{x - 3}{x + 4} + \dfrac{x}{x - 2} \qquad x \neq -4, 2$ (b) $\dfrac{x^2}{x^2 - 4} - \dfrac{1}{x} \qquad x \neq -2, 0, 2$

Solution

(a) $\dfrac{x - 3}{x + 4} + \dfrac{x}{x - 2} \underset{\underset{\text{(5a)}}{\uparrow}}{=} \dfrac{x - 3}{x + 4} \cdot \dfrac{x - 2}{x - 2} + \dfrac{x + 4}{x + 4} \cdot \dfrac{x}{x - 2}$

$$= \frac{(x - 3)(x - 2) + (x + 4)(x)}{(x + 4)(x - 2)}$$

$$= \frac{x^2 - 5x + 6 + x^2 + 4x}{(x + 4)(x - 2)} = \frac{2x^2 - x + 6}{(x + 4)(x - 2)}$$

(b) $\dfrac{x^2}{x^2 - 4} - \dfrac{1}{x} \underset{\underset{\text{(5b)}}{\uparrow}}{=} \dfrac{x^2}{x^2 - 4} \cdot \dfrac{x}{x} - \dfrac{x^2 - 4}{x^2 - 4} \cdot \dfrac{1}{x} = \dfrac{x^2(x) - (x^2 - 4)(1)}{(x^2 - 4) \cdot x}$

$$= \frac{x^3 - x^2 + 4}{x(x - 2)(x + 2)}$$

Now Work PROBLEM **49**

4 Use the Least Common Multiple Method

If the denominators of two rational expressions to be added (or subtracted) have common factors, we usually do not use the general rules given by equations (5a) and (5b). Just as with fractions, we use the **least common multiple (LCM) method**. The LCM method uses the polynomial of least degree that has each denominator polynomial as a factor.

The LCM Method for Adding or Subtracting Rational Expressions

The Least Common Multiple (LCM) Method requires four steps:

STEP 1: Factor completely the polynomial in the denominator of each rational expression.

STEP 2: The LCM of the denominators is the product of each unique factor, with each of these factors raised to a power equal to the greatest number of times that the factor occurs in any denominator.

STEP 3: Write each rational expression using the LCM as the common denominator.

STEP 4: Add or subtract the rational expressions using equation (4).

We begin with an example that requires only Steps 1 and 2.

EXAMPLE 7

Finding the Least Common Multiple

Find the least common multiple of the following pair of polynomials:

$$x(x-1)^2(x+1) \quad \text{and} \quad 4(x-1)(x+1)^3$$

Solution   **STEP 1:** The polynomials are already factored completely as

$$x(x-1)^2(x+1) \quad \text{and} \quad 4(x-1)(x+1)^3$$

STEP 2: Start by writing the factors of the left-hand polynomial. (Or you could start with the one on the right.)

$$x(x-1)^2(x+1)$$

Now look at the right-hand polynomial. Its first factor, 4, does not appear in our list, so we insert it.

$$4x(x-1)^2(x+1)$$

The next factor, $x-1$, is already in our list, so no change is necessary. The final factor is $(x+1)^3$. Since our list has $x+1$ to the first power only, we replace $x+1$ in the list by $(x+1)^3$. The LCM is

$$4x(x-1)^2(x+1)^3$$

Notice that the LCM is, in fact, the polynomial of least degree that contains $x(x-1)^2(x+1)$ and $4(x-1)(x+1)^3$ as factors. ⌐

━━━ **Now Work** PROBLEM 55

EXAMPLE 8

Using the Least Common Multiple to Add Rational Expressions

Perform the indicated operation and simplify the result. Leave your answer in factored form.

$$\frac{x}{x^2+3x+2} + \frac{2x-3}{x^2-1} \quad x \neq -2, -1, 1$$

Solution   **STEP 1:** Factor completely the polynomials in the denominators.

$$x^2 + 3x + 2 = (x+2)(x+1)$$
$$x^2 - 1 = (x-1)(x+1)$$

STEP 2: The LCM is $(x+2)(x+1)(x-1)$. Do you see why?

STEP 3: Write each rational expression using the LCM as the denominator.

$$\frac{x}{x^2+3x+2} = \frac{x}{(x+2)(x+1)} = \frac{x}{(x+2)(x+1)} \cdot \frac{x-1}{x-1} = \frac{x(x-1)}{(x+2)(x+1)(x-1)}$$

↑ Multiply numerator and denominator by $x-1$ to get the LCM in the denominator.

$$\frac{2x-3}{x^2-1} = \frac{2x-3}{(x-1)(x+1)} = \frac{2x-3}{(x-1)(x+1)} \cdot \frac{x+2}{x+2} = \frac{(2x-3)(x+2)}{(x-1)(x+1)(x+2)}$$

↑ Multiply numerator and denominator by $x+2$ to get the LCM in the denominator.

STEP 4: Now add by using equation (4).

$$\frac{x}{x^2 + 3x + 2} + \frac{2x - 3}{x^2 - 1} = \frac{x(x - 1)}{(x + 2)(x + 1)(x - 1)} + \frac{(2x - 3)(x + 2)}{(x + 2)(x + 1)(x - 1)}$$

$$= \frac{(x^2 - x) + (2x^2 + x - 6)}{(x + 2)(x + 1)(x - 1)}$$

$$= \frac{3x^2 - 6}{(x + 2)(x + 1)(x - 1)} = \frac{3(x^2 - 2)}{(x + 2)(x + 1)(x - 1)}$$

EXAMPLE 9

Using the Least Common Multiple to Subtract Rational Expressions

Perform the indicated operation and simplify the result. Leave your answer in factored form.

$$\frac{3}{x^2 + x} - \frac{x + 4}{x^2 + 2x + 1} \qquad x \neq -1, 0$$

Solution   **STEP 1:** Factor completely the polynomials in the denominators.

$$x^2 + x = x(x + 1)$$

$$x^2 + 2x + 1 = (x + 1)^2$$

STEP 2: The LCM is $x(x + 1)^2$.

STEP 3: Write each rational expression using the LCM as the denominator.

$$\frac{3}{x^2 + x} = \frac{3}{x(x + 1)} = \frac{3}{x(x + 1)} \cdot \frac{x + 1}{x + 1} = \frac{3(x + 1)}{x(x + 1)^2}$$

$$\frac{x + 4}{x^2 + 2x + 1} = \frac{x + 4}{(x + 1)^2} = \frac{x + 4}{(x + 1)^2} \cdot \frac{x}{x} = \frac{x(x + 4)}{x(x + 1)^2}$$

STEP 4: Subtract, using equation (4).

$$\frac{3}{x^2 + x} - \frac{x + 4}{x^2 + 2x + 1} = \frac{3(x + 1)}{x(x + 1)^2} - \frac{x(x + 4)}{x(x + 1)^2}$$

$$= \frac{3(x + 1) - x(x + 4)}{x(x + 1)^2}$$

$$= \frac{3x + 3 - x^2 - 4x}{x(x + 1)^2}$$

$$= \frac{-x^2 - x + 3}{x(x + 1)^2}$$

— **Now Work** PROBLEM **65**

5 Simplify Complex Rational Expressions

When sums and/or differences of rational expressions appear as the numerator and/or denominator of a quotient, the quotient is called a **complex rational expression**.* For example,

$$\frac{1 + \dfrac{1}{x}}{1 - \dfrac{1}{x}} \qquad \text{and} \qquad \frac{\dfrac{x^2}{x^2 - 4} - 3}{\dfrac{x - 3}{x + 2} - 1}$$

* Some texts use the term **complex fraction**.

are complex rational expressions. To **simplify** a complex rational expression means to write it as a rational expression reduced to lowest terms. This can be accomplished in either of two ways.

Simplifying a Complex Rational Expression

OPTION 1: Treat the numerator and denominator of the complex rational expression separately, performing whatever operations are indicated and simplifying the results. Follow this by simplifying the resulting rational expression.

OPTION 2: Find the LCM of the denominators of all rational expressions that appear in the complex rational expression. Multiply the numerator and denominator of the complex rational expression by the LCM and simplify the result.

We use both options in the next example. By carefully studying each option, you can discover situations in which one may be easier to use than the other.

EXAMPLE 10 **Simplifying a Complex Rational Expression**

Simplify: $\dfrac{\dfrac{1}{2} + \dfrac{3}{x}}{\dfrac{x+3}{4}}$ $x \neq -3, 0$

Solution **Option 1:** First, we perform the indicated operation in the numerator, and then we divide.

$$\dfrac{\dfrac{1}{2} + \dfrac{3}{x}}{\dfrac{x+3}{4}} = \dfrac{\dfrac{1 \cdot x + 2 \cdot 3}{2 \cdot x}}{\dfrac{x+3}{4}} = \dfrac{\dfrac{x+6}{2x}}{\dfrac{x+3}{4}} = \dfrac{x+6}{2x} \cdot \dfrac{4}{x+3}$$

↑ **Rule for adding quotients**   ↑ **Rule for dividing quotients**

$$= \dfrac{(x+6) \cdot 4}{2 \cdot x \cdot (x+3)} = \dfrac{2 \cdot 2 \cdot (x+6)}{2 \cdot x \cdot (x+3)} = \dfrac{2(x+6)}{x(x+3)}$$

↑ **Rule for multiplying quotients**

Option 2: The rational expressions that appear in the complex rational expression are

$$\dfrac{1}{2}, \quad \dfrac{3}{x}, \quad \dfrac{x+3}{4}$$

The LCM of their denominators is $4x$. We multiply the numerator and denominator of the complex rational expression by $4x$ and then simplify.

$$\dfrac{\dfrac{1}{2} + \dfrac{3}{x}}{\dfrac{x+3}{4}} = \dfrac{4x \cdot \left(\dfrac{1}{2} + \dfrac{3}{x}\right)}{4x \cdot \left(\dfrac{x+3}{4}\right)} = \dfrac{4x \cdot \dfrac{1}{2} + 4x \cdot \dfrac{3}{x}}{\dfrac{4x \cdot (x+3)}{4}}$$

↑ **Multiply the numerator and denominator by 4x.**   ↑ **Use the Distributive Property in the numerator.**

$$= \dfrac{2 \cdot 2x \cdot \dfrac{1}{2} + 4x \cdot \dfrac{3}{x}}{\dfrac{4x \cdot (x+3)}{4}} = \dfrac{2x + 12}{x(x+3)} = \dfrac{2(x+6)}{x(x+3)}$$

↑ **Simplify.**   ↑ **Factor.**

EXAMPLE 11   Simplifying a Complex Rational Expression

Simplify: $\dfrac{\dfrac{x^2}{x-4}+2}{\dfrac{2x-2}{x}-1}$ $x \neq 0, 2, 4$

Solution   We use Option 1.

$$\dfrac{\dfrac{x^2}{x-4}+2}{\dfrac{2x-2}{x}-1} = \dfrac{\dfrac{x^2}{x-4}+\dfrac{2(x-4)}{x-4}}{\dfrac{2x-2}{x}-\dfrac{x}{x}} = \dfrac{\dfrac{x^2+2x-8}{x-4}}{\dfrac{2x-2-x}{x}}$$

$$= \dfrac{\dfrac{(x+4)(x-2)}{x-4}}{\dfrac{x-2}{x}} = \dfrac{(x+4)\cancel{(x-2)}}{x-4} \cdot \dfrac{x}{\cancel{x-2}}$$

$$= \dfrac{x(x+4)}{x-4}$$

━━━━ **Now Work** PROBLEM 75

Application

EXAMPLE 12   Solving an Application in Electricity

An electrical circuit contains two resistors connected in parallel, as shown in Figure 28. If these two resistors provide resistance of R_1 and R_2 ohms, respectively, their combined resistance R is given by the formula

$$R = \dfrac{1}{\dfrac{1}{R_1}+\dfrac{1}{R_2}}$$

Figure 28 Resistors in parallel

(a) Express R as a rational expression; that is, simplify the right-hand side of this formula.
(b) Evaluate the rational expression if $R_1 = 6$ ohms and $R_2 = 10$ ohms.

Solution   (a) We use Option 2 to simplify R. If we consider 1 as the fraction $\dfrac{1}{1}$, the rational expressions in the complex rational expression are

$$\dfrac{1}{1}, \quad \dfrac{1}{R_1}, \quad \dfrac{1}{R_2}$$

The LCM of the denominators is $R_1 R_2$. We multiply the numerator and denominator of the complex rational expression by $R_1 R_2$ and simplify.

$$\dfrac{1}{\dfrac{1}{R_1}+\dfrac{1}{R_2}} = \dfrac{1 \cdot R_1 R_2}{\left(\dfrac{1}{R_1}+\dfrac{1}{R_2}\right) \cdot R_1 R_2} = \dfrac{R_1 R_2}{\dfrac{1}{R_1} \cdot R_1 R_2 + \dfrac{1}{R_2} \cdot R_1 R_2} = \dfrac{R_1 R_2}{R_2 + R_1}$$

So,

$$R = \dfrac{R_1 R_2}{R_2 + R_1}$$

(b) If $R_1 = 6$ and $R_2 = 10$, then

$$R = \dfrac{6 \cdot 10}{10 + 6} = \dfrac{60}{16} = \dfrac{15}{4} \text{ ohms}$$

R.7 Assess Your Understanding

Concepts and Vocabulary

1. When the numerator and denominator of a rational expression contain no common factors (except 1 and −1), the rational expression is in _____ _____.

2. LCM is an abbreviation for _____ _____ _____.

3. *Multiple Choice* Choose the statement that is not true. Assume $b \neq 0, c \neq 0$, and $d \neq 0$ as necessary.

   (a) $\dfrac{ac}{bc} = \dfrac{a}{b}$

   (b) $\dfrac{a}{b} + \dfrac{c}{b} = \dfrac{a+c}{b}$

   (c) $\dfrac{a}{b} - \dfrac{c}{d} = \dfrac{ad - bc}{bd}$

   (d) $\dfrac{\dfrac{a}{b}}{\dfrac{c}{d}} = \dfrac{ac}{bd}$

4. *Multiple Choice* Choose the rational expression that simplifies to −1.

   (a) $\dfrac{a-b}{b-a}$

   (b) $\dfrac{a-b}{a-b}$

   (c) $\dfrac{a+b}{a-b}$

   (d) $\dfrac{b-a}{b+a}$

5. *True or False* The complex rational expression $\dfrac{\dfrac{1}{3} + \dfrac{1}{x}}{\dfrac{1}{5} - \dfrac{1}{x}}$ simplifies to $\dfrac{5(x+3)}{3(x-5)}$.

6. *True or False* The LCM of $2x^3 + 6x^2$ and $6x^4 + 4x^3$ is $4x^3(x+1)$.

Skill Building

In Problems 7–18, reduce each rational expression to lowest terms.

7. $\dfrac{3x+9}{x^2-9}$

8. $\dfrac{4x^2+8x}{12x+24}$

9. $\dfrac{x^2-2x}{3x-6}$

10. $\dfrac{15x^2+24x}{3x^2}$

11. $\dfrac{24x^2}{12x^2-6x}$

12. $\dfrac{x^2+4x+4}{x^2-4}$

13. $\dfrac{y^2-49}{3y^2-18y-21}$

14. $\dfrac{3y^2-y-2}{3y^2+5y+2}$

15. $\dfrac{x^2+4x-12}{x^2-4x+4}$

16. $\dfrac{x-x^2}{x^2+x-2}$

17. $\dfrac{x^2+x-20}{4-x}$

18. $\dfrac{2x^2+5x-3}{1-2x}$

In Problems 19–36, perform the indicated operation and simplify the result. Leave your answer in factored form.

19. $\dfrac{3x+6}{5x^2} \cdot \dfrac{x}{x^2-4}$

20. $\dfrac{3}{2x} \cdot \dfrac{x^2}{6x+10}$

21. $\dfrac{4x^2}{x^2-16} \cdot \dfrac{x^3-64}{2x}$

22. $\dfrac{12}{x^2+x} \cdot \dfrac{x^3+1}{4x-2}$

23. $\dfrac{4x-8}{-3x} \cdot \dfrac{12}{12-6x}$

24. $\dfrac{6x-27}{5x} \cdot \dfrac{2}{4x-18}$

25. $\dfrac{x^2-4x-12}{x^2+2x-48} \cdot \dfrac{x^2+4x-32}{x^2+10x+16}$

26. $\dfrac{x^2+x-6}{x^2+4x-5} \cdot \dfrac{x^2-25}{x^2+2x-15}$

27. $\dfrac{\dfrac{6x}{x^2-4}}{\dfrac{3x-9}{2x+4}}$

28. $\dfrac{\dfrac{12x}{5x+20}}{\dfrac{4x^2}{x^2-16}}$

29. $\dfrac{\dfrac{8x}{x^2-1}}{\dfrac{10x}{x+1}}$

30. $\dfrac{\dfrac{x-2}{4x}}{\dfrac{x^2-4x+4}{12x}}$

31. $\dfrac{\dfrac{4-x}{4+x}}{\dfrac{4x}{x^2-16}}$

32. $\dfrac{\dfrac{3+x}{3-x}}{\dfrac{x^2-9}{9x^3}}$

33. $\dfrac{\dfrac{x^2+7x+12}{x^2-7x+12}}{\dfrac{x^2+x-12}{x^2-x-12}}$

34. $\dfrac{\dfrac{x^2+7x+6}{x^2+x-6}}{\dfrac{x^2+5x-6}{x^2+5x+6}}$

35. $\dfrac{\dfrac{5x^2-7x-6}{2x^2+3x-5}}{\dfrac{15x^2+14x+3}{2x^2+13x+20}}$

36. $\dfrac{\dfrac{9x^2+3x-2}{12x^2+5x-2}}{\dfrac{9x^2-6x+1}{8x^2-10x-3}}$

In Problems 37–54, perform the indicated operation and simplify the result. Leave your answer in factored form.

37. $\dfrac{x}{2} + \dfrac{5}{2}$

38. $\dfrac{3}{x} - \dfrac{6}{x}$

39. $\dfrac{x^2}{2x-3} - \dfrac{4}{2x-3}$

40. $\dfrac{3x^2}{2x-1} - \dfrac{9}{2x-1}$

41. $\dfrac{x+5}{x-4} + \dfrac{3x-2}{x-4}$

42. $\dfrac{2x-5}{3x+2} + \dfrac{x+4}{3x+2}$

43. $\dfrac{3x + 5}{2x - 1} - \dfrac{2x - 4}{2x - 1}$

44. $\dfrac{5x - 4}{3x + 4} - \dfrac{x + 1}{3x + 4}$

45. $\dfrac{4}{x - 2} + \dfrac{x}{2 - x}$

46. $\dfrac{6}{x - 1} - \dfrac{x}{1 - x}$

47. $\dfrac{7}{x - 3} - \dfrac{3}{x + 1}$

48. $\dfrac{2}{x + 5} - \dfrac{5}{x - 5}$

49. $\dfrac{x}{x + 1} + \dfrac{2x - 3}{x - 1}$

50. $\dfrac{3x}{x - 4} + \dfrac{2x}{x + 3}$

51. $\dfrac{x - 3}{x + 2} - \dfrac{x + 4}{x - 2}$

52. $\dfrac{2x - 3}{x - 1} - \dfrac{2x + 1}{x + 1}$

53. $\dfrac{x}{x^2 - 4} + \dfrac{1}{x}$

54. $\dfrac{x - 1}{x^3} + \dfrac{x}{x^2 + 1}$

In Problems 55–62, find the LCM of the given polynomials.

55. $x^2 - 4, \quad x^2 - x - 2$

56. $x^2 - x - 12, \quad x^2 - 8x + 16$

57. $x^3 - x, \quad x^2 - x$

58. $3x^2 - 27, \quad 2x^2 - x - 15$

59. $4x^3 - 4x^2 + x, \quad 2x^3 - x^2, \quad x^3$

60. $x - 3, \quad x^2 + 3x, \quad x^3 - 9x$

61. $x^3 - x, \quad x^3 - 2x^2 + x, \quad x^3 - 1$

62. $x^2 + 4x + 4, \quad x^3 + 2x^2, \quad (x + 2)^3$

In Problems 63–74, perform the indicated operations and simplify the result. Leave your answer in factored form.

63. $\dfrac{x}{x^2 - 7x + 6} - \dfrac{x}{x^2 - 2x - 24}$

64. $\dfrac{x}{x - 3} - \dfrac{x + 1}{x^2 + 5x - 24}$

65. $\dfrac{4x}{x^2 - 4} - \dfrac{2}{x^2 + x - 6}$

66. $\dfrac{3x}{x - 1} - \dfrac{x - 4}{x^2 - 2x + 1}$

67. $\dfrac{3}{(x - 1)^2(x + 1)} + \dfrac{2}{(x - 1)(x + 1)^2}$

68. $\dfrac{2}{(x + 2)^2(x - 1)} - \dfrac{6}{(x + 2)(x - 1)^2}$

69. $\dfrac{x + 4}{x^2 - x - 2} - \dfrac{2x + 3}{x^2 + 2x - 8}$

70. $\dfrac{2x - 3}{x^2 + 8x + 7} - \dfrac{x - 2}{(x + 1)^2}$

71. $\dfrac{1}{x} - \dfrac{2}{x^2 + x} + \dfrac{3}{x^3 - x^2}$

72. $\dfrac{x}{(x - 1)^2} + \dfrac{2}{x} - \dfrac{x + 1}{x^3 - x^2}$

73. $\dfrac{1}{h}\left(\dfrac{1}{x + h} - \dfrac{1}{x}\right)$

74. $\dfrac{1}{h}\left[\dfrac{1}{(x + h)^2} - \dfrac{1}{x^2}\right]$

In Problems 75–86, perform the indicated operations and simplify the result. Leave your answer in factored form.

75. $\dfrac{1 + \dfrac{1}{x}}{1 - \dfrac{1}{x}}$

76. $\dfrac{4 + \dfrac{1}{x^2}}{3 - \dfrac{1}{x^2}}$

77. $\dfrac{2 - \dfrac{x + 1}{x}}{3 + \dfrac{x - 1}{x + 1}}$

78. $\dfrac{1 - \dfrac{x}{x + 1}}{2 - \dfrac{x - 1}{x}}$

79. $\dfrac{\dfrac{x + 4}{x - 2} - \dfrac{x - 3}{x + 1}}{x + 1}$

80. $\dfrac{\dfrac{x - 2}{x + 1} - \dfrac{x}{x - 2}}{x + 3}$

81. $\dfrac{\dfrac{x - 2}{x + 2} + \dfrac{x - 1}{x + 1}}{\dfrac{x}{x + 1} - \dfrac{2x - 3}{x}}$

82. $\dfrac{\dfrac{2x + 5}{x} - \dfrac{x}{x - 3}}{\dfrac{x^2}{x - 3} - \dfrac{(x + 1)^2}{x + 3}}$

83. $1 - \dfrac{1}{1 - \dfrac{1}{x}}$

84. $1 - \dfrac{1}{1 - \dfrac{1}{1 - x}}$

85. $\dfrac{2(x - 1)^{-1} + 3}{3(x - 1)^{-1} + 2}$

86. $\dfrac{4(x + 2)^{-1} - 3}{3(x + 2)^{-1} - 1}$

In Problems 87–94, expressions that occur in calculus are given. Reduce each expression to lowest terms.

87. $\dfrac{(2x + 3) \cdot 3 - (3x - 5) \cdot 2}{(3x - 5)^2}$

88. $\dfrac{(4x + 1) \cdot 5 - (5x - 2) \cdot 4}{(5x - 2)^2}$

89. $\dfrac{x \cdot 2x - (x^2 + 1) \cdot 1}{(x^2 + 1)^2}$

90. $\dfrac{x \cdot 2x - (x^2 - 4) \cdot 1}{(x^2 - 4)^2}$

91. $\dfrac{(3x + 1) \cdot 2x - x^2 \cdot 3}{(3x + 1)^2}$

92. $\dfrac{(2x - 5) \cdot 3x^2 - x^3 \cdot 2}{(2x - 5)^2}$

93. $\dfrac{(x^2 + 1) \cdot 3 - (3x + 4) \cdot 2x}{(x^2 + 1)^2}$

94. $\dfrac{(x^2 + 9) \cdot 2 - (2x - 5) \cdot 2x}{(x^2 + 9)^2}$

Applications and Extensions

95. The Lensmaker's Equation The focal length f of a lens with index of refraction n is

$$\frac{1}{f} = (n - 1)\left[\frac{1}{R_1} + \frac{1}{R_2}\right]$$

where R_1 and R_2 are the radii of curvature of the front and back surfaces of the lens. Express f as a rational expression. Evaluate the rational expression for $n = 1.5$, $R_1 = 0.1$ meter, and $R_2 = 0.2$ meter.

97. Challenge Problem Find k so that

$$\frac{x^2 + 3kx - 3x - 9k}{x^2 + 2x - 15} = \frac{x + 12}{x + 5}$$

96. Electrical Circuits An electrical circuit contains three resistors connected in parallel. If these three resistors provide resistance of R_1, R_2, and R_3 ohms, respectively, their combined resistance R is given by the formula

$$\frac{1}{R} = \frac{1}{R_1} + \frac{1}{R_2} + \frac{1}{R_3}$$

Express R as a rational expression. Evaluate R for $R_1 = 5$ ohms, $R_2 = 4$ ohms, and $R_3 = 10$ ohms.

98. Challenge Problem Find A so that

$$\frac{A}{x + 2} + \frac{7}{x - 3} = \frac{12x - 1}{x^2 - x - 6}$$

Explaining Concepts: Discussion and Writing

99. The following expressions are called **continued fractions**:

$$1 + \frac{1}{x} \qquad 1 + \cfrac{1}{1 + \cfrac{1}{x}} \qquad 1 + \cfrac{1}{1 + \cfrac{1}{1 + \cfrac{1}{x}}} \cdots$$

Each simplifies to an expression of the form $\dfrac{ax + b}{bx + c}$

Trace the successive values of $a, b,$ and c as you "continue" the fraction. Can you discover the pattern of these values? Go to the library and research Fibonacci numbers. Write a report on your findings.

100. Explain to a fellow student when you would use the LCM method to add two rational expressions. Give two examples of adding two rational expressions: one in which you use the LCM and the other in which you do not.

101. Which of the two options given in the text for simplifying complex rational expressions do you prefer? Write a brief paragraph stating the reasons for your choice.

R.8 *n*th Roots; Rational Exponents

PREPARING FOR THIS SECTION *Before getting started, review the following:*

- Exponents, Square Roots (Section R.2, pp. 21–24)

  Now Work the 'Are You Prepared?' problems on page 78.

  OBJECTIVES 1 Work with *n*th Roots (p. 72)
  2 Simplify Radicals (p. 73)
  3 Rationalize Denominators and Numerators (p. 74)
  4 Simplify Expressions with Rational Exponents (p. 76)

1 Work with *n*th Roots

DEFINITION Principal *n*th Root

The **principal *n*th root of a real number** a, $n \geq 2$ an integer, symbolized by $\sqrt[n]{a}$, is defined as follows:

$$\sqrt[n]{a} = b \quad \text{means} \quad a = b^n$$

where $a \geq 0$ and $b \geq 0$ if n is even and a, b are any real numbers if n is odd.

Notice that if a is negative and n is even, then $\sqrt[n]{a}$ is not defined. When it is defined, the principal nth root of a number is unique.

The symbol $\sqrt[n]{a}$ for the principal nth root of a is called a **radical**; the integer n is called the **index**, and a is called the **radicand**. If the index of a radical is 2, we call $\sqrt[2]{a}$ the **square root** of a and omit the index 2 by simply writing $\sqrt{a}$. If the index is 3, we call $\sqrt[3]{a}$ the **cube root** of a.

> **In Words**
> The symbol $\sqrt[n]{a}$ means "what is the number that, when raised to the power n, equals a."

EXAMPLE 1 **Simplifying Principal nth Roots**

(a) $\sqrt[3]{8} = \sqrt[3]{2^3} = 2$

(b) $\sqrt[3]{-64} = \sqrt[3]{(-4)^3} = -4$

(c) $\sqrt[4]{\dfrac{1}{16}} = \sqrt[4]{\left(\dfrac{1}{2}\right)^4} = \dfrac{1}{2}$

(d) $\sqrt[6]{(-2)^6} = |-2| = 2$

These are examples of **perfect roots**, since each simplifies to a rational number. Notice the absolute value in Example 1(d). If n is even, then the principal nth root must be nonnegative.

In general, if $n \geq 2$ is an integer and a is a real number, we have

$$\sqrt[n]{a^n} = a \qquad \text{if } n \geq 3 \text{ is odd} \tag{1a}$$

$$\sqrt[n]{a^n} = |a| \qquad \text{if } n \geq 2 \text{ is even} \tag{1b}$$

━━━━ **Now Work** PROBLEM **11**

Radicals provide a way of representing many irrational real numbers. For example, it can be shown that there is no rational number whose square is 2. Using radicals, we can say that $\sqrt{2}$ is the positive number whose square is 2.

EXAMPLE 2 **Using a Calculator to Approximate Roots**

Use a calculator to approximate $\sqrt[5]{16}$.

Solution Figure 29 shows the result using a TI-84 Plus C graphing calculator.

━━━━ **Now Work** PROBLEM **127**

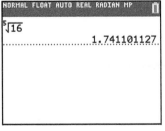

Figure 29

2 Simplify Radicals

Let $n \geq 2$ and $m \geq 2$ denote integers, and let a and b represent real numbers. Assuming that all radicals are defined, we have the following properties:

Properties of Radicals

$$\sqrt[n]{ab} = \sqrt[n]{a}\,\sqrt[n]{b} \tag{2a}$$

$$\sqrt[n]{\dfrac{a}{b}} = \dfrac{\sqrt[n]{a}}{\sqrt[n]{b}} \qquad b \neq 0 \tag{2b}$$

$$\sqrt[n]{a^m} = \left(\sqrt[n]{a}\right)^m \tag{2c}$$

When used in reference to radicals, the direction to "simplify" means to remove from the radicals any perfect roots that occur as factors.

EXAMPLE 3 **Simplifying Radicals**

(a) $\sqrt{32} = \sqrt{16 \cdot 2} = \sqrt{16} \cdot \sqrt{2} = 4\sqrt{2}$

    Factor out 16,   (2a)
    a perfect square.

(b) $\sqrt[3]{16} = \sqrt[3]{8 \cdot 2} = \sqrt[3]{8} \cdot \sqrt[3]{2} = \sqrt[3]{2^3} \cdot \sqrt[3]{2} = 2\sqrt[3]{2}$

    Factor out 8,   (2a)
    a perfect cube.

(c) $\sqrt[3]{-16x^4} = \sqrt[3]{-8 \cdot 2 \cdot x^3 \cdot x} = \sqrt[3]{(-8x^3)(2x)}$

    Factor perfect   Group perfect
    cubes inside radical. cubes.

$\qquad\qquad = \sqrt[3]{(-2x)^3 \cdot 2x} = \sqrt[3]{(-2x)^3} \cdot \sqrt[3]{2x} = -2x\sqrt[3]{2x}$

(d) $\sqrt[4]{\dfrac{16x^5}{81}} = \sqrt[4]{\dfrac{2^4 x^4 x}{3^4}} = \sqrt[4]{\left(\dfrac{2x}{3}\right)^4 \cdot x} = \sqrt[4]{\left(\dfrac{2x}{3}\right)^4} \cdot \sqrt[4]{x} = \left|\dfrac{2x}{3}\right| \sqrt[4]{x}$

Now Work PROBLEMS 15 AND 27

Two or more radicals can be combined, provided that they have the same index and the same radicand. Such radicals are called **like radicals**.

EXAMPLE 4 **Combining Like Radicals**

(a) $-8\sqrt{12} + \sqrt{3} = -8\sqrt{4 \cdot 3} + \sqrt{3}$

$\qquad\qquad\qquad = -8 \cdot \sqrt{4}\,\sqrt{3} + \sqrt{3}$

$\qquad\qquad\qquad = -16\sqrt{3} + \sqrt{3} = -15\sqrt{3}$

(b) $\sqrt[3]{8x^4} + \sqrt[3]{-x} + 4\sqrt[3]{27x} = \sqrt[3]{2^3 x^3 x} + \sqrt[3]{-1 \cdot x} + 4\sqrt[3]{3^3 x}$

$\qquad\qquad\qquad\qquad = \sqrt[3]{(2x)^3} \cdot \sqrt[3]{x} + \sqrt[3]{-1} \cdot \sqrt[3]{x} + 4\sqrt[3]{3^3} \cdot \sqrt[3]{x}$

$\qquad\qquad\qquad\qquad = 2x\sqrt[3]{x} - 1 \cdot \sqrt[3]{x} + 12\sqrt[3]{x}$

$\qquad\qquad\qquad\qquad = (2x + 11)\sqrt[3]{x}$

Now Work PROBLEM 45

3 Rationalize Denominators and Numerators

When radicals occur in quotients, it is customary to rewrite the quotient so that the new denominator contains no radicals. This process is referred to as **rationalizing the denominator**.

The idea is to multiply by an appropriate expression so that the new denominator contains no radicals. For example:

If a Denominator Contains the Factor	Multiply by	To Obtain a Denominator Free of Radicals
$\sqrt{3}$	$\sqrt{3}$	$(\sqrt{3})^2 = 3$
$\sqrt{3} + 1$	$\sqrt{3} - 1$	$(\sqrt{3})^2 - 1^2 = 3 - 1 = 2$
$\sqrt{2} - 3$	$\sqrt{2} + 3$	$(\sqrt{2})^2 - 3^2 = 2 - 9 = -7$
$\sqrt{5} - \sqrt{3}$	$\sqrt{5} + \sqrt{3}$	$(\sqrt{5})^2 - (\sqrt{3})^2 = 5 - 3 = 2$
$\sqrt[3]{4}$	$\sqrt[3]{2}$	$\sqrt[3]{4} \cdot \sqrt[3]{2} = \sqrt[3]{8} = 2$

In rationalizing the denominator of a quotient, be sure to multiply both the numerator and the denominator by the expression.

EXAMPLE 5

Rationalizing Denominators

Rationalize the denominator of each expression:

(a) $\dfrac{1}{\sqrt{3}}$ (b) $\dfrac{5}{4\sqrt{2}}$ (c) $\dfrac{\sqrt{2}}{\sqrt{3} - 3\sqrt{2}}$

Solution

(a) The denominator contains the factor $\sqrt{3}$, so we multiply the numerator and denominator by $\sqrt{3}$ to obtain

$$\frac{1}{\sqrt{3}} = \frac{1}{\sqrt{3}} \cdot \frac{\sqrt{3}}{\sqrt{3}} = \frac{\sqrt{3}}{(\sqrt{3})^2} = \frac{\sqrt{3}}{3}$$

(b) The denominator contains the factor $\sqrt{2}$, so we multiply the numerator and denominator by $\sqrt{2}$ to obtain

$$\frac{5}{4\sqrt{2}} = \frac{5}{4\sqrt{2}} \cdot \frac{\sqrt{2}}{\sqrt{2}} = \frac{5\sqrt{2}}{4(\sqrt{2})^2} = \frac{5\sqrt{2}}{4 \cdot 2} = \frac{5\sqrt{2}}{8}$$

(c) The denominator contains the factor $\sqrt{3} - 3\sqrt{2}$, so we multiply the numerator and denominator by $\sqrt{3} + 3\sqrt{2}$ to obtain

$$\frac{\sqrt{2}}{\sqrt{3} - 3\sqrt{2}} = \frac{\sqrt{2}}{\sqrt{3} - 3\sqrt{2}} \cdot \frac{\sqrt{3} + 3\sqrt{2}}{\sqrt{3} + 3\sqrt{2}} = \frac{\sqrt{2}(\sqrt{3} + 3\sqrt{2})}{(\sqrt{3})^2 - (3\sqrt{2})^2}$$

$$= \frac{\sqrt{2}\sqrt{3} + 3(\sqrt{2})^2}{3 - 18} = \frac{\sqrt{6} + 6}{-15} = -\frac{6 + \sqrt{6}}{15}$$

Now Work PROBLEM 59

In calculus, it is sometimes necessary to **rationalize the numerator** of a quotient. To do this, multiply by an appropriate expression so that the new numerator contains no radicals.

EXAMPLE 6

Rationalizing Numerators

Rationalize the numerator of the following expression.

$$\frac{\sqrt{x + h} - \sqrt{x}}{h} \qquad h \neq 0$$

Solution

The numerator contains the factor $\sqrt{x + h} - \sqrt{x}$, so we multiply the numerator and denominator by $\sqrt{x + h} + \sqrt{x}$ to obtain

$$\frac{\sqrt{x + h} - \sqrt{x}}{h} = \frac{\sqrt{x + h} - \sqrt{x}}{h} \cdot \frac{\sqrt{x + h} + \sqrt{x}}{\sqrt{x + h} + \sqrt{x}} = \frac{(\sqrt{x + h})^2 - (\sqrt{x})^2}{h(\sqrt{x + h} + \sqrt{x})}$$

$$= \frac{x + h - x}{h(\sqrt{x + h} + \sqrt{x})} = \frac{h}{h(\sqrt{x + h} + \sqrt{x})} = \frac{1}{\sqrt{x + h} + \sqrt{x}}$$

Now Work PROBLEM 69

4 Simplify Expressions with Rational Exponents

Radicals are used to define rational exponents.

DEFINITION $a^{1/n}$

If a is a real number and $n \geq 2$ is an integer, then

$$a^{1/n} = \sqrt[n]{a} \tag{3}$$

provided that $\sqrt[n]{a}$ exists.

Note that if n is even and $a < 0$, then $\sqrt[n]{a}$ and $a^{1/n}$ do not exist.

EXAMPLE 7 | **Writing Expressions Containing Fractional Exponents as Radicals**

(a) $4^{1/2} = \sqrt{4} = 2$

(b) $8^{1/2} = \sqrt{8} = 2\sqrt{2}$

(c) $(-27)^{1/3} = \sqrt[3]{-27} = -3$

(d) $16^{1/3} = \sqrt[3]{16} = 2\sqrt[3]{2}$

DEFINITION $a^{m/n}$

If a is a real number and m and n are integers containing no common factors, with $n \geq 2$, then

$$a^{m/n} = \sqrt[n]{a^m} = \left(\sqrt[n]{a}\right)^m \tag{4}$$

provided that $\sqrt[n]{a}$ exists.

We have two comments about equation (4):

- The exponent $\dfrac{m}{n}$ must be in lowest terms, and $n \geq 2$ must be positive.
- In simplifying the rational expression $a^{m/n}$, either $\sqrt[n]{a^m}$ or $\left(\sqrt[n]{a}\right)^m$ may be used, the choice depending on which is easier to simplify. Generally, taking the root first, as in $\left(\sqrt[n]{a}\right)^m$, is easier.

EXAMPLE 8 | **Simplifying Expressions With Rational Exponents**

(a) $4^{3/2} = \left(\sqrt{4}\right)^3 = 2^3 = 8$

(b) $(-8)^{4/3} = \left(\sqrt[3]{-8}\right)^4 = (-2)^4 = 16$

(c) $(32)^{-2/5} = \left(\sqrt[5]{32}\right)^{-2} = 2^{-2} = \dfrac{1}{4}$

(d) $25^{6/4} = 25^{3/2} = \left(\sqrt{25}\right)^3 = 5^3 = 125$

Now Work PROBLEM 77

It can be shown that the Laws of Exponents hold for rational exponents. The next example illustrates using the Laws of Exponents to simplify.

EXAMPLE 9 | **Simplifying Expressions With Rational Exponents**

Simplify each expression. Express your answer so that only positive exponents occur. Assume that the variables are positive.

(a) $\left(x^{2/3}y\right)\left(x^{-2}y\right)^{1/2}$

(b) $\left(\dfrac{2x^{1/3}}{y^{2/3}}\right)^{-3}$

(c) $\left(\dfrac{9x^2y^{1/3}}{x^{1/3}y}\right)^{1/2}$

Solution

(a) $\left(x^{2/3}y\right)\left(x^{-2}y\right)^{1/2} = x^{2/3} \cdot y \cdot (x^{-2})^{1/2} \cdot y^{1/2}$ $(ab)^n = a^n b^n$

$= x^{2/3} \cdot y \cdot x^{-1} \cdot y^{1/2}$ $(a^m)^n = a^{mn}$

$= x^{2/3} \cdot x^{-1} \cdot y \cdot y^{1/2}$

$= x^{-1/3} \cdot y^{3/2}$ $a^m \cdot a^n = a^{m+n}$

$= \dfrac{y^{3/2}}{x^{1/3}}$ $a^{-n} = \dfrac{1}{a^n}$

(b) $\left(\dfrac{2x^{1/3}}{y^{2/3}}\right)^{-3} = \left(\dfrac{y^{2/3}}{2x^{1/3}}\right)^{3} = \dfrac{(y^{2/3})^{3}}{(2x^{1/3})^{3}} = \dfrac{y^{2}}{2^{3}(x^{1/3})^{3}} = \dfrac{y^{2}}{8x}$

(c) $\left(\dfrac{9x^{2}y^{1/3}}{x^{1/3}y}\right)^{1/2} = \left(\dfrac{9x^{2-(1/3)}}{y^{1-(1/3)}}\right)^{1/2} = \left(\dfrac{9x^{5/3}}{y^{2/3}}\right)^{1/2} = \dfrac{9^{1/2}(x^{5/3})^{1/2}}{(y^{2/3})^{1/2}} = \dfrac{3x^{5/6}}{y^{1/3}}$

Now Work PROBLEM 97

The next two examples illustrate some algebra that you will need to know for certain calculus problems.

EXAMPLE 10 **Writing an Expression as a Single Quotient**

Write the following expression as a single quotient in which only positive exponents appear.

$$(x^{2} + 1)^{1/2} + x \cdot \dfrac{1}{2}(x^{2} + 1)^{-1/2} \cdot 2x$$

Solution

$(x^{2} + 1)^{1/2} + x \cdot \dfrac{1}{2}(x^{2} + 1)^{-1/2} \cdot 2x = (x^{2} + 1)^{1/2} + \dfrac{x^{2}}{(x^{2} + 1)^{1/2}}$

$\qquad = \dfrac{(x^{2} + 1)^{1/2}(x^{2} + 1)^{1/2} + x^{2}}{(x^{2} + 1)^{1/2}}$

$\qquad = \dfrac{(x^{2} + 1) + x^{2}}{(x^{2} + 1)^{1/2}}$

$\qquad = \dfrac{2x^{2} + 1}{(x^{2} + 1)^{1/2}}$

Now Work PROBLEM 103

EXAMPLE 11 **Factoring an Expression Containing Rational Exponents**

Factor and simplify: $\dfrac{4}{3}x^{1/3}(2x + 1) + 2x^{4/3}$

Solution Begin by writing $2x^{4/3}$ as a fraction with 3 as the denominator.

$\dfrac{4}{3}x^{1/3}(2x + 1) + 2x^{4/3} = \dfrac{4x^{1/3}(2x + 1)}{3} + \dfrac{6x^{4/3}}{3} = \dfrac{4x^{1/3}(2x + 1) + 6x^{4/3}}{3}$

$\qquad\qquad\qquad\qquad\qquad\qquad\qquad\qquad\qquad$ Add the two fractions.

$\qquad = \dfrac{2x^{1/3}[2(2x + 1) + 3x]}{3} = \dfrac{2x^{1/3}(7x + 2)}{3}$

$\qquad\quad$ 2 and $x^{1/3}$ are common factors. $\qquad$ Simplify.

Now Work PROBLEM 115

Historical Note

The radical sign, $\sqrt{}$, was first used in print by Christoff Rudolff in 1525. It is thought to be the manuscript form of the letter r (for the Latin word *radix* = *root*), although this has not been quite conclusively confirmed. It took a long time for $\sqrt{}$ to become the standard symbol for a square root and much longer to standardize $\sqrt[3]{}$, $\sqrt[4]{}$, $\sqrt[5]{}$, and so on. The indexes of the root were placed in every conceivable position, with

$$\sqrt[3]{8}, \quad \sqrt[③]{8}, \quad \text{and} \quad \underset{3}{\sqrt{}}8$$

all being variants for $\sqrt[3]{8}$. The notation $\sqrt{}\sqrt{16}$ was popular for $\sqrt[4]{16}$. By the 1700s, the index had settled where we now put it.

The bar on top of the present radical symbol, as follows,

$$\sqrt{a^2 + 2ab + b^2}$$

is the last survivor of the **vinculum**, a bar placed atop an expression to indicate what we would now indicate with parentheses. For example,

$$\overline{ab + c} = a(b + c)$$

R.8 Assess Your Understanding

'Are You Prepared?' *Answers are given at the end of these exercises. If you get a wrong answer, read the pages in red.*

1. $(-3)^2 = \underline{\quad}; -3^2 = \underline{\quad}$ (pp. 21–24)

2. $\sqrt{16} = \underline{\quad}; \sqrt{(-4)^2} = \underline{\quad}$ (pp. 21–24)

Concepts and Vocabulary

3. In the symbol $\sqrt[n]{a}$, the integer n is called the _____.

4. We call $\sqrt[3]{a}$ the _____ _____ of a.

5. *Multiple Choice* Let $n \geq 2$ and $m \geq 2$ be integers, and let a and b be real numbers. Which of the following is not a property of radicals? Assume all radicals are defined.

   (a) $\sqrt[n]{\dfrac{a}{b}} = \dfrac{\sqrt[n]{a}}{\sqrt[n]{b}}$

   (b) $\sqrt[n]{a + b} = \sqrt[n]{a} + \sqrt[n]{b}$

   (c) $\sqrt[n]{ab} = \sqrt[n]{a}\sqrt[n]{b}$

   (d) $\sqrt[n]{a^m} = \left(\sqrt[n]{a}\right)^m$

6. *Multiple Choice* If a is a real number and $n \geq 2$ is an integer, then which of the following expressions is equivalent to $\sqrt[n]{a}$, provided that it exists?

   (a) a^{-n} (b) a^n (c) $\dfrac{1}{a^n}$ (d) $a^{1/n}$

7. *Multiple Choice* Which of the following phrases best defines like radicals?
   (a) Radical expressions that have the same index
   (b) Radical expressions that have the same radicand
   (c) Radical expressions that have the same index and the same radicand
   (d) Radical expressions that have the same variable

8. *Multiple Choice* To rationalize the denominator of the expression $\dfrac{\sqrt{2}}{1 - \sqrt{3}}$, multiply both the numerator and the denominator by which of the following?

   (a) $\sqrt{3}$ (b) $\sqrt{2}$ (c) $1 + \sqrt{3}$ (d) $1 - \sqrt{3}$

9. *True or False* $\sqrt[5]{-32} = -2$

10. *True or False* $\sqrt[4]{(-3)^4} = -3$

Skill Building

In Problems 11–54, simplify each expression. Assume that all variables are positive when they appear.

11. $\sqrt[3]{27}$

12. $\sqrt[4]{16}$

13. $\sqrt[3]{-8}$

14. $\sqrt[3]{-1}$

15. $\sqrt{8}$

16. $\sqrt{75}$

17. $\sqrt{700}$

18. $\sqrt{45x^3}$

19. $\sqrt[3]{32}$

20. $\sqrt[3]{54}$

21. $\sqrt[3]{-8x^4}$

22. $\sqrt[3]{192x^5}$

23. $\sqrt[4]{243}$

24. $\sqrt[4]{48x^5}$

25. $\sqrt[4]{x^{12}y^8}$

26. $\sqrt[5]{x^{10}y^5}$

27. $\sqrt[4]{\dfrac{x^9 y^7}{xy^3}}$

28. $\sqrt[3]{\dfrac{3xy^2}{81x^4y^2}}$

29. $\sqrt{64x}$

30. $\sqrt{9x^5}$

31. $\sqrt[4]{162x^9 y^{12}}$

32. $\sqrt[3]{-40x^{14} y^{10}}$

33. $\sqrt{15x^2}\sqrt{5x}$

34. $\sqrt{5x}\sqrt{20x^3}$

35. $\left(\sqrt{5}\sqrt[3]{9}\right)^2$

36. $\left(\sqrt[3]{3}\sqrt{10}\right)^4$

37. $(3\sqrt{6})(2\sqrt{2})$

38. $(5\sqrt{8})(-3\sqrt{3})$

39. $3\sqrt{2} + 4\sqrt{2}$

40. $6\sqrt{5} - 4\sqrt{5}$

41. $-\sqrt{48} + 5\sqrt{12}$

42. $2\sqrt{12} - 3\sqrt{27}$

43. $(\sqrt{3} + 3)(\sqrt{3} - 1)$

44. $(\sqrt{5} - 2)(\sqrt{5} + 3)$

45. $5\sqrt[3]{2} - 2\sqrt[3]{54}$

46. $9\sqrt[3]{24} - \sqrt[3]{81}$

47. $(\sqrt{x} - 1)^2$

48. $(\sqrt{x} + \sqrt{5})^2$

49. $\sqrt[3]{16x^4} - \sqrt[3]{2x}$

50. $\sqrt[4]{32x} + \sqrt[4]{2x^5}$

51. $\sqrt{8x^3} - 3\sqrt{50x}$

52. $3x\sqrt{9y} + 4\sqrt{25y}$

53. $\sqrt[3]{16x^4y} - 3x\sqrt[3]{2xy} + 5\sqrt[3]{-2xy^4}$

54. $8xy - \sqrt{25x^2y^2} + \sqrt[3]{8x^3y^3}$

In Problems 55–68, rationalize the denominator of each expression. Assume that all variables are positive when they appear.

55. $\dfrac{1}{\sqrt{2}}$

56. $\dfrac{2}{\sqrt{3}}$

57. $\dfrac{-\sqrt{3}}{\sqrt{5}}$

58. $\dfrac{-\sqrt{3}}{\sqrt{8}}$

59. $\dfrac{\sqrt{3}}{5 - \sqrt{2}}$

60. $\dfrac{\sqrt{2}}{\sqrt{7} + 2}$

61. $\dfrac{2 - \sqrt{5}}{2 + 3\sqrt{5}}$

62. $\dfrac{\sqrt{3} - 1}{2\sqrt{3} + 3}$

63. $\dfrac{5}{\sqrt{2} - 1}$

64. $\dfrac{-3}{\sqrt{5} + 4}$

65. $\dfrac{5}{\sqrt[3]{2}}$

66. $\dfrac{-2}{\sqrt[3]{9}}$

67. $\dfrac{\sqrt{x + h} - \sqrt{x}}{\sqrt{x + h} + \sqrt{x}}$

68. $\dfrac{\sqrt{x + h} + \sqrt{x - h}}{\sqrt{x + h} - \sqrt{x - h}}$

In Problems 69–76, rationalize the numerator of each expression. Assume that all variables are positive when they appear.

69. $\dfrac{\sqrt{11} + 1}{2}$

70. $\dfrac{5 - \sqrt{43}}{3}$

71. $\dfrac{\sqrt{6} - \sqrt{15}}{\sqrt{15}}$

72. $\dfrac{\sqrt{5} + \sqrt{3}}{\sqrt{5}}$

73. $\dfrac{\sqrt{x} - \sqrt{c}}{x - c} \quad x \neq c$

74. $\dfrac{\sqrt{x} - 2}{x - 4} \quad x \neq 4$

75. $\dfrac{\sqrt{x - 7} - 1}{x - 8} \quad x \neq 8$

76. $\dfrac{4 - \sqrt{x - 9}}{x - 25} \quad x \neq 25$

In Problems 77–92, simplify each expression.

77. $8^{2/3}$

78. $4^{3/2}$

79. $(-64)^{1/3}$

80. $16^{3/4}$

81. $100^{3/2}$

82. $25^{3/2}$

83. $4^{-3/2}$

84. $16^{-3/2}$

85. $\left(\dfrac{9}{8}\right)^{3/2}$

86. $\left(\dfrac{27}{8}\right)^{2/3}$

87. $\left(\dfrac{8}{9}\right)^{-3/2}$

88. $\left(\dfrac{8}{27}\right)^{-2/3}$

89. $(-1000)^{-1/3}$

90. $-25^{-1/2}$

91. $\left(-\dfrac{64}{125}\right)^{-2/3}$

92. $-81^{-3/4}$

In Problems 93–100, simplify each expression. Express your answer so that only positive exponents occur. Assume that the variables are positive.

93. $x^{3/4}x^{1/3}x^{-1/2}$

94. $x^{2/3}x^{1/2}x^{-1/4}$

95. $(x^3 y^6)^{1/3}$

96. $(x^4 y^8)^{3/4}$

97. $\dfrac{(x^2 y)^{1/3}(xy^2)^{2/3}}{x^{2/3}y^{2/3}}$

98. $\dfrac{(xy)^{1/4}(x^2 y^2)^{1/2}}{(x^2 y)^{3/4}}$

99. $\dfrac{(16x^2 y^{-1/3})^{3/4}}{(xy^2)^{1/4}}$

100. $\dfrac{(4x^{-1}y^{1/3})^{3/2}}{(xy)^{3/2}}$

Applications and Extensions

In Problems 101–114, expressions that occur in calculus are given. Write each expression as a single quotient in which only positive exponents and/or radicals appear.

101. $\dfrac{x}{(1 + x)^{1/2}} + 2(1 + x)^{1/2} \quad x > -1$

102. $\dfrac{1 + x}{2x^{1/2}} + x^{1/2} \quad x > 0$

103. $2x(x^2 + 1)^{1/2} + x^2 \cdot \dfrac{1}{2}(x^2 + 1)^{-1/2} \cdot 2x$

104. $(x + 1)^{1/3} + x \cdot \dfrac{1}{3}(x + 1)^{-2/3} \quad x \neq -1$

105. $\sqrt{4x + 3} \cdot \dfrac{1}{2\sqrt{x - 5}} + \sqrt{x - 5} \cdot \dfrac{1}{5\sqrt{4x + 3}} \quad x > 5$

106. $\dfrac{\sqrt[3]{8x + 1}}{3\sqrt[3]{(x - 2)^2}} + \dfrac{\sqrt[3]{x - 2}}{24\sqrt[3]{(8x + 1)^2}} \quad x \neq 2, x \neq -\dfrac{1}{8}$

107. $\dfrac{\sqrt{1 + x} - x \cdot \dfrac{1}{2\sqrt{1 + x}}}{1 + x} \quad x > -1$

108. $\dfrac{\sqrt{x^2 + 1} - x \cdot \dfrac{2x}{2\sqrt{x^2 + 1}}}{x^2 + 1}$

109. $\dfrac{(x + 4)^{1/2} - 2x(x + 4)^{-1/2}}{x + 4} \quad x > -4$

110. $\dfrac{(9 - x^2)^{1/2} + x^2(9 - x^2)^{-1/2}}{9 - x^2} \quad -3 < x < 3$

111. $\dfrac{\dfrac{x^2}{(x^2 - 1)^{1/2}} - (x^2 - 1)^{1/2}}{x^2} \quad x < -1 \ \text{ or } \ x > 1$

112. $\dfrac{(x^2 + 4)^{1/2} - x^2(x^2 + 4)^{-1/2}}{x^2 + 4}$

113. $\dfrac{\dfrac{1 + x^2}{2\sqrt{x}} - 2x\sqrt{x}}{(1 + x^2)^2}$ $x > 0$

114. $\dfrac{2x(1 - x^2)^{1/3} + \dfrac{2}{3}x^3(1 - x^2)^{-2/3}}{(1 - x^2)^{2/3}}$ $x \neq -1, x \neq 1$

In Problems 115–124, expressions that occur in calculus are given. Factor each expression. Express your answer so that only positive exponents occur.

115. $(x + 1)^{3/2} + x \cdot \dfrac{3}{2}(x + 1)^{1/2}$ $x \geq -1$

116. $(x^2 + 4)^{4/3} + x \cdot \dfrac{4}{3}(x^2 + 4)^{1/3} \cdot 2x$

117. $6x^{1/2}(x^2 + x) - 8x^{3/2} - 8x^{1/2}$ $x \geq 0$

118. $6x^{1/2}(2x + 3) + x^{3/2} \cdot 8$ $x \geq 0$

119. $3(x^2 + 4)^{4/3} + x \cdot 4(x^2 + 4)^{1/3} \cdot 2x$

120. $2x(3x + 4)^{4/3} + x^2 \cdot 4(3x + 4)^{1/3}$

121. $4(3x + 5)^{1/3}(2x + 3)^{3/2} + 3(3x + 5)^{4/3}(2x + 3)^{1/2}$ $x \geq -\dfrac{3}{2}$

122. $6(6x + 1)^{1/3}(4x - 3)^{3/2} + 6(6x + 1)^{4/3}(4x - 3)^{1/2}$ $x \geq \dfrac{3}{4}$

123. $3x^{-1/2} + \dfrac{3}{2}x^{1/2}$ $x > 0$

124. $8x^{1/3} - 4x^{-2/3}$ $x \neq 0$

In Problems 125–132, use a calculator to approximate each radical. Round your answer to two decimal places.

125. $\sqrt{2}$

126. $\sqrt{7}$

127. $\sqrt[3]{4}$

128. $\sqrt[3]{-5}$

129. $\dfrac{2 + \sqrt{3}}{3 - \sqrt{5}}$

130. $\dfrac{\sqrt{5} - 2}{\sqrt{2} + 4}$

131. $\dfrac{3\sqrt[3]{5} - \sqrt{2}}{\sqrt{3}}$

132. $\dfrac{2\sqrt{3} - \sqrt[3]{4}}{\sqrt{2}}$

133. Calculating the Amount of Gasoline in a Tank A Shell station stores its gasoline in underground tanks that are right circular cylinders lying on their sides. See the illustration. The volume V of gasoline in the tank (in gallons) is given by the formula

$$V = 40h^2 \sqrt{\dfrac{96}{h} - 0.608}$$

where h is the height of the gasoline (in inches) as measured on a depth stick.
(a) If $h = 12$ inches, how many gallons of gasoline are in the tank?
(b) If $h = 1$ inch, how many gallons of gasoline are in the tank?

134. Inclined Planes The final velocity v of an object in feet per second (ft/s) after it slides down a frictionless inclined plane of height h feet is

$$v = \sqrt{64h + v_0^2}$$

where v_0 is the initial velocity (in ft/s) of the object.
(a) What is the final velocity v of an object that slides down a frictionless inclined plane of height 4 feet? Assume that the initial velocity is 0.
(b) What is the final velocity v of an object that slides down a frictionless inclined plane of height 16 feet? Assume that the initial velocity is 0.
(c) What is the final velocity v of an object that slides down a frictionless inclined plane of height 2 feet with an initial velocity of 4 ft/s?

Problems 135 and 136 require the following information.

Period of a Pendulum The period T, in seconds, of a pendulum of length l, in feet, may be approximated using the formula

$$T = 2\pi \sqrt{\dfrac{l}{32}}$$

In Problems 135 and 136, express your answer both as a square root and as a decimal approximation.

135. Find the period T of a pendulum whose length is 64 feet.

136. Find the period T of a pendulum whose length is 16 feet.

137. *Challenge Problem* Use synthetic division to find the quotient and remainder when $x^3 - 4x^2 - 3x + 13$ is divided by $x - \sqrt{3}$.

138. *Challenge Problem* Use synthetic division to determine whether $x - (1 + \sqrt{2})$ is a factor of $x^3 - 9x^2 + 13x + 7$.

Explaining Concepts: Discussion and Writing

139. Give an example to show that $\sqrt{a^2}$ is not equal to a. Use it to explain why $\sqrt{a^2} = |a|$.

'Are You Prepared?' Answers

1. $9; -9$ **2.** $4; 4$

Equations and Inequalities

Financing a Purchase

Whenever we make a major purchase, such as an automobile or a house, we often need to finance the purchase by borrowing money from a lending institution, such as a bank. Have you ever wondered how the bank determines the monthly payment? How much total interest will be paid over the course of the loan? What roles do the rate of interest and the length of the loan play?

 —See the Internet-based Chapter Project I—

→ A Look Ahead

Chapter 1, Equations and Inequalities, reviews many topics covered in Intermediate Algebra. If your instructor decides to exclude complex numbers from the course, don't be alarmed. This text has been designed so that the topic of complex numbers can be included or excluded without any confusion later on.

Outline

1.1 Linear Equations

PREPARING FOR THIS SECTION *Before getting started, review the following:*

- Properties of Real Numbers (Section R.1, pp. 9–13)
- Domain of a Variable (Section R.2, p. 21)

Now Work the 'Are You Prepared?' problems on page 90.

OBJECTIVES 1 Solve a Linear Equation (p. 84)
2 Solve Equations That Lead to Linear Equations (p. 86)
3 Solve Problems That Can Be Modeled by Linear Equations (p. 87)

An **equation in one variable** is a statement in which two expressions, at least one containing the variable, are set equal. The expressions are called the **sides** of the equation. Since an equation is a statement, it may be true or false, depending on the value of the variable. Unless otherwise restricted, the admissible values of the variable are those in the domain of the variable. The admissible values of the variable, if any, that result in a true statement are called **solutions**, or **roots**, of the equation. To **solve an equation** means to find all the solutions of the equation.

For example, the following are all equations in one variable, x:

$$x + 5 = 9 \qquad x^2 + 5x = 2x - 2 \qquad \frac{x^2 - 4}{x + 1} = 0 \qquad \sqrt{x^2 + 9} = 5$$

The first of these statements, $x + 5 = 9$, is true when $x = 4$ and false for any other choice of x. That is, 4 is the solution of the equation $x + 5 = 9$. We also say that 4 **satisfies** the equation $x + 5 = 9$, because, when 4 is substituted for x, a true statement results.

Sometimes an equation will have more than one solution. For example, the equation

$$\frac{x^2 - 4}{x + 1} = 0$$

has -2 and 2 as solutions.

Usually, we write the solutions of an equation as a set, called the **solution set** of the equation. For example, the solution set of the equation $x^2 - 9 = 0$ is $\{-3, 3\}$.

Some equations have no real solution. For example, $x^2 + 9 = 5$ has no real solution, because there is no real number whose square, when added to 9, equals 5.

An equation that is satisfied for every value of the variable for which both sides are defined is called an **identity**. For example, the equation

$$3x + 5 = x + 3 + 2x + 2$$

is an identity, because this statement is true for any real number x.

Solving Equations Algebraically

One method for solving an equation is to replace the original equation by a succession of **equivalent equations**, equations having the same solution set, until an equation with an obvious solution set is obtained.

For example, consider the following succession of equivalent equations:

$$2x + 3 = 13$$
$$2x = 10$$
$$x = 5$$

We conclude that the solution set of the original equation is $\{5\}$.

How do we obtain equivalent equations? In general, there are five ways.

Procedures That Result in Equivalent Equations

- Interchange the two sides of the equation:

  Replace $\quad 3 = x \quad$ by $\quad x = 3$

- Simplify the sides of the equation by combining like terms, eliminating parentheses, and so on:

  Replace $\quad (x + 2) + 6 = 2x + (x + 1)$

  by $\quad\quad\quad x + 8 = 3x + 1$

- Add or subtract the same expression on both sides of the equation:

  Replace $\quad\quad\quad\quad 3x - 5 = 4$

  by $\quad\quad (3x - 5) + 5 = 4 + 5$

- Multiply or divide both sides of the equation by the same nonzero expression:

  Replace $\quad\quad \dfrac{3x}{x - 1} = \dfrac{6}{x - 1} \quad x \neq 1$

  by $\quad \dfrac{3x}{x - 1} \cdot (x - 1) = \dfrac{6}{x - 1} \cdot (x - 1)$

- If one side of the equation is 0 and the other side can be factored, then write it as the product of factors:

  Replace $\quad\quad x^2 - 3x = 0$

  by $\quad\quad x(x - 3) = 0$

WARNING Squaring both sides of an equation does not necessarily lead to an equivalent equation. For example, $x = 3$ has one solution, but $x^2 = 9$ has two solutions, -3 and 3. ∎

Whenever it is possible to solve an equation in your head, do so. For example,

- The solution of $2x = 8$ is 4.
- The solution of $3x - 15 = 0$ is 5.

──**Now Work** PROBLEM 11

EXAMPLE 1

Solving an Equation

Solve the equation: $3x - 5 = 4$

Solution

Replace the original equation by a succession of equivalent equations.

$$3x - 5 = 4$$
$$(3x - 5) + 5 = 4 + 5 \quad\quad \text{Add 5 to both sides.}$$
$$3x = 9 \quad\quad \text{Simplify.}$$
$$\frac{3x}{3} = \frac{9}{3} \quad\quad \text{Divide both sides by 3.}$$
$$x = 3 \quad\quad \text{Simplify.}$$

The last equation, $x = 3$, has the solution 3. All these equations are equivalent, so 3 is the only solution of the original equation, $3x - 5 = 4$.

(continued)

✓ **Check:** Check the solution by substituting 3 for x in the original equation.

$$3x - 5 = 4$$
$$3 \cdot 3 - 5 \overset{?}{=} 4$$
$$9 - 5 \overset{?}{=} 4$$
$$4 = 4$$

The solution checks. The solution set is $\{3\}$.

Now Work PROBLEM 25

Steps for Solving Equations

STEP 1: List any restrictions on the domain of the variable.

STEP 2: Simplify the equation by replacing the original equation by a succession of equivalent equations using the procedures listed earlier.

STEP 3: If the result of Step 2 is a product of factors equal to 0 use the Zero-Product Property to set each factor equal to 0 and solve the resulting equations.

STEP 4: Check your solution(s).

Need to Review?

The Zero-Product Property is discussed in Section R.1, p. 13.

1 Solve a Linear Equation

Linear equations are equations such as

$$3x + 12 = 0 \qquad -2x + 5 = 0 \qquad \frac{1}{2}x - \sqrt{3} = 0$$

DEFINITION Linear Equation in One Variable

A **linear equation in one variable** is an equation equivalent in form to

$$ax + b = 0 \qquad a \neq 0$$

where a and b are real numbers.

Sometimes a linear equation is called a **first-degree equation**, because the left side is a polynomial in x of degree 1.

To solve a linear equation, we *isolate* the variable:

$$ax + b = 0 \qquad a \neq 0$$
$$ax = -b \qquad \text{Subtract } b \text{ from both sides.}$$
$$x = \frac{-b}{a} \qquad \text{Divide both sides by } a, a \neq 0$$

The linear equation $ax + b = 0$, $a \neq 0$, has the single solution $x = -\dfrac{b}{a}$.

EXAMPLE 2   Solving a Linear Equation

Solve the equation: $\dfrac{1}{2}(x + 5) - 4 = \dfrac{1}{3}(2x - 1)$

Solution   To clear the equation of fractions, multiply both sides by 6, the least common multiple (LCM) of the denominators of the fractions $\frac{1}{2}$ and $\frac{1}{3}$.

$$\frac{1}{2}(x+5) - 4 = \frac{1}{3}(2x-1)$$

$$6\left[\frac{1}{2}(x+5) - 4\right] = 6\left[\frac{1}{3}(2x-1)\right] \qquad \text{Multiply both sides by 6, the LCM of 2 and 3.}$$

$$3(x+5) - 24 = 2(2x-1) \qquad \begin{array}{l}\text{Use the Distributive Property on the left}\\\text{and the Associative Property on the right.}\end{array}$$

$$3x + 15 - 24 = 4x - 2 \qquad \text{Use the Distributive Property.}$$

$$3x - 9 = 4x - 2 \qquad \text{Combine like terms.}$$

$$3x - 9 + 9 = 4x - 2 + 9 \qquad \text{Add 9 to both sides.}$$

$$3x = 4x + 7 \qquad \text{Simplify.}$$

$$3x - 4x = 4x + 7 - 4x \qquad \text{Subtract 4x from both sides.}$$

$$-x = 7 \qquad \text{Simplify.}$$

$$x = -7 \qquad \text{Multiply both sides by } -1.$$

✔ **Check:** Substitute -7 for x in the expressions on the left and right sides of the original equation, and simplify. If the two expressions are equal, the solution checks.

$$\frac{1}{2}(x+5) - 4 = \frac{1}{2}(-7+5) - 4 = \frac{1}{2}(-2) - 4 = -1 - 4 = -5$$

$$\frac{1}{3}(2x-1) = \frac{1}{3}[2(-7) - 1] = \frac{1}{3}(-14 - 1) = \frac{1}{3}(-15) = -5$$

Since the two expressions are equal, the solution checks. The solution set is $\{-7\}$.

━━━━ **Now Work** PROBLEM 35

EXAMPLE 3   **Solving a Linear Equation Using a Calculator**

Solve the equation: $2.78x + \dfrac{2}{17.931} = 54.06$

Round the answer to two decimal places.

Solution   To avoid rounding errors, solve for x before using a calculator.

$$2.78x + \frac{2}{17.931} = 54.06$$

$$2.78x = 54.06 - \frac{2}{17.931} \qquad \text{Subtract } \frac{2}{17.931} \text{ from both sides.}$$

$$x = \frac{54.06 - \dfrac{2}{17.931}}{2.78} \qquad \text{Divide both sides by 2.78.}$$

Now use your calculator. The solution, rounded to two decimal places, is 19.41.

✔ **Check:** Store the calculator solution 19.40592134 in memory, and proceed to evaluate $2.78x + \dfrac{2}{17.931}$.

$$2.78 \cdot 19.40592134 + \frac{2}{17.931} = 54.06$$

━━━━ **Now Work** PROBLEM 69

2 Solve Equations That Lead to Linear Equations

| EXAMPLE 4 | Solving an Equation That Leads to a Linear Equation |

Solve the equation: $(2y + 1)(y - 1) = (y + 5)(2y - 5)$

Solution

$$(2y + 1)(y - 1) = (y + 5)(2y - 5)$$

$$2y^2 - y - 1 = 2y^2 + 5y - 25 \qquad \text{Multiply and combine like terms.}$$

$$-y - 1 = 5y - 25 \qquad \text{Subtract } 2y^2 \text{ from both sides.}$$

$$-y = 5y - 24 \qquad \text{Add 1 to both sides.}$$

$$-6y = -24 \qquad \text{Subtract 5y from both sides.}$$

$$y = 4 \qquad \text{Divide both sides by } -6.$$

✓ Check: $(2y + 1)(y - 1) = (2 \cdot 4 + 1)(4 - 1) = (8 + 1) \cdot 3 = 9 \cdot 3 = 27$

$(y + 5)(2y - 5) = (4 + 5)(2 \cdot 4 - 5) = 9 \cdot (8 - 5) = 9 \cdot 3 = 27$

The two expressions are equal, so the solution checks. The solution set is {4}.

| EXAMPLE 5 | Solving an Equation That Leads to a Linear Equation |

Solve the equation: $\dfrac{3}{x - 2} = \dfrac{1}{x - 1} + \dfrac{7}{(x - 1)(x - 2)}$

Need to Review?

Least common multiples are discussed in Section R.7, pp. 65–67.

Solution First, notice that the domain of the variable is $\{x | x \neq 1, x \neq 2\}$. Clear the equation of fractions by multiplying both sides by the least common multiple of the denominators of the three fractions, $(x - 1)(x - 2)$.

$$\frac{3}{x - 2} = \frac{1}{x - 1} + \frac{7}{(x - 1)(x - 2)}$$

$$(x - 1)(x - 2) \frac{3}{x - 2} = (x - 1)(x - 2)\left[\frac{1}{x - 1} + \frac{7}{(x - 1)(x - 2)} \right] \qquad \begin{array}{l} \text{Multiply both sides by} \\ (x - 1)(x - 2). \end{array}$$

$$3x - 3 = (x - 1)(x - 2) \frac{1}{x - 1} + (x - 1)(x - 2) \frac{7}{(x - 1)(x - 2)} \qquad \begin{array}{l} \text{Use the Distributive} \\ \text{Property on both sides.} \end{array}$$

$$3x - 3 = (x - 2) + 7 \qquad \text{Simplify.}$$

$$3x - 3 = x + 5 \qquad \text{Combine like terms.}$$

$$2x = 8 \qquad \begin{array}{l} \text{Add 3 to both sides;} \\ \text{subtract x from both sides.} \end{array}$$

$$x = 4 \qquad \text{Divide both sides by 2.}$$

✓ Check: $\dfrac{3}{x - 2} = \dfrac{3}{4 - 2} = \dfrac{3}{2}$

$$\frac{1}{x - 1} + \frac{7}{(x - 1)(x - 2)} = \frac{1}{4 - 1} + \frac{7}{(4 - 1)(4 - 2)} = \frac{1}{3} + \frac{7}{3 \cdot 2} = \frac{2}{6} + \frac{7}{6} = \frac{9}{6} = \frac{3}{2}$$

The solution checks. The solution set is {4}.

───── **Now Work** PROBLEM 63

| EXAMPLE 6 | An Equation with No Solution |

Solve the equation: $\dfrac{3x}{x - 1} + 2 = \dfrac{3}{x - 1}$

Solution

First, note that the domain of the variable is $\{x | x \neq 1\}$. Since the two quotients in the equation have the same denominator, $x - 1$, simplify by multiplying both sides by $x - 1$. The resulting equation is equivalent to the original equation, since we are multiplying by $x - 1$, which is not 0. (Remember, $x \neq 1$.)

$$\frac{3x}{x - 1} + 2 = \frac{3}{x - 1}$$

$$\left(\frac{3x}{x - 1} + 2\right) \cdot (x - 1) = \frac{3}{x - 1} \cdot (x - 1) \quad \text{Multiply both sides by } x - 1,$$

$$\frac{3x}{x - 1} \cdot (x - 1) + 2 \cdot (x - 1) = 3 \quad \begin{array}{l}\text{Use the Distributive Property}\\ \text{on the left side,}\end{array}$$

$$3x + 2x - 2 = 3 \quad \text{Simplify.}$$

$$5x - 2 = 3 \quad \text{Combine like terms.}$$

$$5x = 5 \quad \text{Add 2 to both sides.}$$

$$x = 1 \quad \text{Divide both sides by 5.}$$

NOTE Example 6 illustrates an *extraneous solution*. This is discussed more in Section 1.4. ■

The solution appears to be 1. But recall that 1 is not in the domain of the variable, so this value must be discarded. The equation has no solution. The solution set is $\varnothing$.

━ **Now Work** PROBLEM 53

EXAMPLE 7

Converting Temperature to Fahrenheit from Celsius

In the United States we measure temperature in both degrees Fahrenheit (°F) and degrees Celsius (°C), which are related by the formula $C = \frac{5}{9}(F - 32)$. What are the Fahrenheit temperatures corresponding to Celsius temperatures of 0°, 10°, 20°, and 30°C?

Solution

We could solve four equations for F by replacing C each time by 0, 10, 20, and 30. Instead, it is much easier and faster to first solve the equation $C = \frac{5}{9}(F - 32)$ for F and then to substitute in the values of C.

$$C = \frac{5}{9}(F - 32)$$

$$9C = 5(F - 32) \quad \text{Multiply both sides by 9.}$$

$$9C = 5F - 160 \quad \text{Use the Distributive Property.}$$

$$5F - 160 = 9C \quad \text{Interchange sides.}$$

$$5F = 9C + 160 \quad \text{Add 160 to both sides.}$$

$$F = \frac{9}{5}C + 32 \quad \text{Divide both sides by 5.}$$

Now do the required arithmetic.

$$0°C: \quad F = \frac{9}{5} \cdot 0 + 32 = 32°F$$

$$10°C: \quad F = \frac{9}{5} \cdot 10 + 32 = 50°F$$

$$20°C: \quad F = \frac{9}{5} \cdot 20 + 32 = 68°F$$

NOTE The icon is a Model It! icon. It indicates that the discussion or problem involves modeling. ■

$$30°C: \quad F = \frac{9}{5} \cdot 30 + 32 = 86°F$$

3 Solve Problems That Can Be Modeled by Linear Equations

Although each situation has its unique features, we can provide an outline of the steps to follow when solving applied problems.

Steps for Solving Applied Problems

STEP 1: Read the problem carefully, perhaps two or three times. Pay particular attention to the question being asked in order to identify what you are looking for. Identify any relevant formulas you may need ($d = rt$, $A = \pi r^2$, etc.). If you can, determine realistic possibilities for the answer.

NOTE It is a good practice to choose a variable that reminds you of the unknown. For example, use t for time. ∎

STEP 2: Assign a letter (variable) to represent what you are looking for, and, if necessary, express any remaining unknown quantities in terms of this variable.

STEP 3: Make a list of all the known facts, and translate them into mathematical expressions. These may take the form of an equation (or, later, an inequality) involving the variable. The equation (or inequality) is called the **model**. If possible, draw an appropriately labeled diagram to assist you. Sometimes a table or chart helps.

STEP 4: Solve the equation for the variable, and then answer the question, using a complete sentence.

STEP 5: Check the answer with the facts in the problem. If it agrees, congratulations! If it does not agree, review your work and try again.

EXAMPLE 8   Investments

A total of $18,000 is invested, some in stocks and some in bonds. If the amount invested in bonds is half that invested in stocks, how much is invested in each category?

Step-by-Step Solution

Step 1: Determine what you are looking for.

We are being asked to find the amount of two investments. These amounts must total $18,000. (Do you see why?)

Step 2: Assign a variable to represent what you are looking for. If necessary, express any remaining unknown quantities in terms of this variable.

If s equals the amount invested in stocks, then the rest of the money, $18{,}000 - s$, is the amount invested in bonds.

Step 3: Translate the English into mathematical statements. It may be helpful to draw a figure that represents the situation. Sometimes a table can be used to organize the information. Use the information to build your model.

Set up a table:

Amount in Stocks	Amount in Bonds	Reason
s	$18{,}000 - s$	Total invested is $18,000.

We also know that:

Amount invested in bonds	is	one-half that in stocks
$18{,}000 - s$	$=$	$\dfrac{1}{2}s$

Step 4: Solve the equation and answer the original question.

$$18{,}000 - s = \frac{1}{2}s \qquad \text{The Model}$$

$$18{,}000 = s + \frac{1}{2}s \qquad \text{Add } s \text{ to both sides.}$$

$$18{,}000 = \frac{3}{2}s \qquad \text{Simplify.}$$

$$\frac{2}{3} \cdot 18{,}000 = \frac{2}{3} \cdot \frac{3}{2}s \qquad \text{Multiply both sides by } \frac{2}{3}.$$

$$12{,}000 = s \qquad \text{Simplify.}$$

So $12,000 is invested in stocks, and $18{,}000 - \$12{,}000 = \$6000$ is invested in bonds.

Step 5: Check your answer with the facts presented in the problem.

The total invested is $12,000 + $6000 = $18,000, and the amount in bonds, $6000, is half that in stocks, $12,000.

━━━━ **Now Work** PROBLEM 85

| EXAMPLE 9 | Determining an Hourly Wage |

Shannon grossed $928 one week by working 52 hours. Her employer pays time-and-a-half for all hours worked in excess of 40 hours. With this information, determine Shannon's regular hourly wage.

Solution **STEP 1:** We are looking for an hourly wage. Our answer will be expressed in dollars per hour.

STEP 2: Let x represent the regular hourly wage, measured in dollars per hour. Then $1.5x$ is the overtime hourly wage.

STEP 3: Set up a table:

	Hours Worked	Hourly Wage	Salary
Regular	40	x	$40x$
Overtime	12	$1.5x$	$12(1.5x) = 18x$

The sum of regular salary plus overtime salary will equal $928. From the table, $40x + 18x = 928$.

STEP 4:
$$40x + 18x = 928 \qquad \text{The Model}$$
$$58x = 928$$
$$x = 16.00$$

Shannon's regular hourly wage is $16.00 per hour.

STEP 5: Forty hours yields a salary of $40 \cdot 16.00 = \$640$, and 12 hours of overtime yields a salary of $12 \cdot 1.5 \cdot 16.00 = \$288$, for a total of $928.

Now Work PROBLEM 87

SUMMARY

Steps for Solving a Linear Equation

To solve a linear equation, follow these steps:

STEP 1: List any restrictions on the variable.

STEP 2: If necessary, clear the equation of fractions by multiplying both sides by the least common multiple (LCM) of the denominators of all the fractions.

STEP 3: Remove all parentheses and simplify.

STEP 4: Collect all terms containing the variable on one side and all remaining terms on the other side.

STEP 5: Simplify and solve.

STEP 6: Check your solution(s).

Historical Feature

Solving equations is among the oldest of mathematical activities, and efforts to systematize this activity determined much of the shape of modern mathematics.

Consider the following problem and its solution using only words: Solve the problem of how many apples Jim has, given that

"Bob's five apples and Jim's apples together make twelve apples" by thinking,

"Jim's apples are all twelve apples less Bob's five apples" and then concluding,

"Jim has seven apples."

The mental steps translated into algebra are
$$5 + x = 12$$
$$x = 12 - 5$$
$$= 7$$

The solution of this problem using only words is the earliest form of algebra. Such problems were solved exactly this way in Babylonia in 1800 BC. We know almost nothing of mathematical work before this date, although most authorities believe the sophistication of the earliest known texts indicates that there must have been a long period of previous development. The method of writing out equations in words persisted for thousands of years, and although it now seems extremely cumbersome, it was used very effectively by many generations of mathematicians. The Arabs developed a good deal of the theory of cubic equations while writing out all the equations in words. About AD 1500, the tendency to abbreviate words in the written equations began to lead in the direction of modern notation; for example, the Latin word *et* (meaning "and") developed into the plus sign, +. Although the occasional use of letters to represent variables dates back to AD 1200, the practice did not become common until about AD 1600. Development thereafter was rapid, and by 1635 algebraic notation did not differ essentially from what we use now.

1.1 Assess Your Understanding

'Are You Prepared?' *Answers are given at the end of these exercises. If you get a wrong answer, read the pages listed in* red.

1. The fact that $2(x + 3) = 2x + 6$ is attributable to the _____ Property. (pp. 9–13)

2. The fact that $3x = 0$ implies that $x = 0$ is a result of the _____ Property. (pp. 9–13)

3. The domain of the variable in the expression $\dfrac{x}{x - 4}$ is _____. (p. 21)

Concepts and Vocabulary

4. *True or False* Multiplying both sides of an equation by any number results in an equivalent equation.

5. An equation that is satisfied for every value of the variable for which both sides are defined is called a(n) _____.

6. An equation of the form $ax + b = 0$ is called a(n) _____ equation or a(n) _____ equation.

7. *True or False* The solution of the equation $3x - 8 = 0$ is $\dfrac{3}{8}$.

8. *True or False* Some equations have no solution.

9. *Multiple Choice* An admissible value for the variable that makes the equation a true statement is called a(n) _____ of the equation.
   (a) identity   **(b)** solution   **(c)** degree   **(d)** model

10. *Multiple Choice* A chemist mixes 10 liters of a 20% solution with x liters of a 35% solution. Which of the following expressions represents the total number of liters in the mixture?
   (a) x **(b)** $20 - x$ **(c)** $\dfrac{35}{x}$ **(d)** $10 + x$

Skill Building

In Problems 11–18, mentally solve each equation.

11. $7x = 21$

12. $6x = -24$

13. $3x + 15 = 0$

14. $6x + 18 = 0$

15. $2x - 3 = 0$

16. $3x + 4 = 0$

17. $\dfrac{1}{4}x = \dfrac{7}{20}$

18. $\dfrac{2}{3}x = \dfrac{9}{2}$

In Problems 19–68, solve each equation, if possible.

19. $3x + 4 = x$

20. $2x + 9 = 5x$

21. $2t - 6 = 3 - t$

22. $5y + 6 = -18 - y$

23. $6 - x = 2x + 9$

24. $3 - 2x = 2 - x$

25. $3 + 2n = 4n + 7$

26. $6 - 2m = 3m + 1$

27. $3(5 + 3x) = 8(x - 1)$

28. $3(2 - x) = 2x - 1$

29. $8x - (3x + 2) = 3x - 10$

30. $7 - (2x - 1) = 10$

31. $\dfrac{3}{2}x + 2 = \dfrac{1}{2} - \dfrac{1}{2}x$

32. $\dfrac{1}{3}x = 2 - \dfrac{2}{3}x$

33. $\dfrac{1}{2}x - 5 = \dfrac{3}{4}x$

34. $1 - \dfrac{1}{2}x = 6$

35. $\dfrac{2}{3}p = \dfrac{1}{2}p + \dfrac{1}{3}$

36. $\dfrac{1}{2} - \dfrac{1}{3}p = \dfrac{4}{3}$

37. $0.2m = 0.9 + 0.5m$

38. $0.9t = 1 + t$

39. $\dfrac{x + 1}{3} + \dfrac{x + 2}{7} = 2$

40. $\dfrac{2x + 1}{3} + 16 = 3x$

41. $\dfrac{5}{8}(p + 3) - 2 = \dfrac{1}{4}(2p - 3) + \dfrac{11}{16}$

42. $\dfrac{1}{3}(w + 1) - 3 = \dfrac{2}{5}(w - 4) - \dfrac{2}{15}$

43. $\dfrac{2}{y} + \dfrac{4}{y} = 3$

44. $\dfrac{4}{y} - 5 = \dfrac{5}{2y}$

45. $\dfrac{1}{2} + \dfrac{2}{x} = \dfrac{3}{4}$

46. $\dfrac{3}{x} - \dfrac{1}{3} = \dfrac{1}{6}$

47. $(x + 7)(x - 1) = (x + 1)^2$

48. $(x + 2)(x - 3) = (x + 3)^2$

49. $x(2x - 3) = (2x + 1)(x - 4)$

50. $x(1 + 2x) = (2x - 1)(x - 2)$

51. $p(p^2 + 3) = 12 + p^3$

52. $w(4 - w^2) = 8 - w^3$

53. $\dfrac{x}{x - 2} + 3 = \dfrac{2}{x - 2}$

54. $\dfrac{2x}{x + 3} = \dfrac{-6}{x + 3} - 2$

55. $\dfrac{2x}{x^2 - 4} = \dfrac{4}{x^2 - 4} - \dfrac{3}{x + 2}$

56. $\dfrac{x}{x^2 - 9} + \dfrac{4}{x + 3} = \dfrac{3}{x^2 - 9}$

57. $\dfrac{x}{x + 2} = \dfrac{3}{2}$

58. $\dfrac{3x}{x - 1} = 2$

59. $\dfrac{7}{3x + 10} = \dfrac{2}{x - 3}$

60. $\dfrac{-4}{x + 4} = \dfrac{-3}{x + 6}$

61. $\dfrac{6t + 7}{4t - 1} = \dfrac{3t + 8}{2t - 4}$

62. $\dfrac{8w + 5}{10w - 7} = \dfrac{4w - 3}{5w + 7}$

63. $\dfrac{4}{x - 2} = \dfrac{-3}{x + 5} + \dfrac{7}{(x + 5)(x - 2)}$

64. $\dfrac{-4}{2x+3} + \dfrac{1}{x-1} = \dfrac{1}{(2x+3)(x-1)}$

65. $\dfrac{2}{y+3} + \dfrac{3}{y-4} = \dfrac{5}{y+6}$

66. $\dfrac{5}{5z-11} + \dfrac{4}{2z-3} = \dfrac{-3}{5-z}$

67. $\dfrac{x}{x^2-9} - \dfrac{x-4}{x^2+3x} = \dfrac{10}{x^2-3x}$

68. $\dfrac{x+1}{x^2+2x} - \dfrac{x+4}{x^2+x} = \dfrac{-3}{x^2+3x+2}$

In Problems 69–72, use a calculator to solve each equation. Round the solution to two decimal places.

69. $3.2x + \dfrac{21.3}{65.871} = 19.23$

70. $6.2x - \dfrac{19.1}{83.72} = 0.195$

71. $14.72 - 21.58x = \dfrac{18}{2.11}x + 2.4$

72. $18.63x - \dfrac{21.2}{2.6} = \dfrac{14}{2.32}x - 20$

Applications and Extensions

In Problems 73–76, solve each equation. The letters a, b, and c are constants.

73. $ax - b = c, \quad a \neq 0$

74. $1 - ax = b, \quad a \neq 0$

75. $\dfrac{x}{a} + \dfrac{x}{b} = c, a \neq 0, b \neq 0, a \neq -b$

76. $\dfrac{a}{x} + \dfrac{b}{x} = c, \quad c \neq 0$

77. Find the number a for which $x = 4$ is a solution of the equation
$$x + 2a = 16 + ax - 6a$$

78. Find the number b for which $x = 2$ is a solution of the equation
$$x + 2b = x - 4 + 2bx$$

Problems 79–84 list some formulas that occur in applications. Solve each formula for the indicated variable.

79. Electricity $\dfrac{1}{R} = \dfrac{1}{R_1} + \dfrac{1}{R_2}$ for R

80. Finance $A = P(1 + rt)$ for r

81. Mechanics $F = \dfrac{mv^2}{R}$ for R

82. Chemistry $PV = nRT$ for T

83. Mathematics $S = \dfrac{a}{1-r}$ for r

84. Mechanics $v = -gt + v_0$ for t

85. Finance A total of $20,000 is to be invested, some in bonds and some in certificates of deposit (CDs). If the amount invested in bonds is to exceed that in CDs by $3000, how much will be invested in each type of investment?

86. Finance A total of $10,000 is to be divided between Sean and George, with George to receive $3000 less than Sean. How much will each receive?

87. Computing Hourly Wages Kim is paid time-and-a-half for hours worked in excess of 40 hours and had gross weekly wages of $910 for 48 hours worked. What is her regular hourly rate?

88. Computing Hourly Wages Leigh is paid time-and-a-half for hours worked in excess of 40 hours and double-time for hours worked on Sunday. If Leigh had gross weekly wages of $1083 for working 50 hours, 4 of which were on Sunday, what is her regular hourly rate?

89. Computing Grades Going into the final exam, which will count as two tests, Brooke has test scores of 80, 83, 71, 61, and 95. What score does Brooke need on the final in order to have an average score of 80?

90. Computing Grades Going into the final exam, which will count as two-thirds of the final grade, Mike has test scores of 86, 80, 84, and 90. What minimum score does Mike need on the final in order to earn a B, which requires an average score of 80? What does he need to earn an A, which requires an average of 90?

91. Business: Discount Pricing A store sells refurbished iPhones that cost 12% less than the original price. If the new price of a refurbished iPhone is $572, what was the original price? How much is saved by purchasing the refurbished phone?

92. Business: Discount Pricing A car dealer, at a year-end clearance, reduces the list price of last year's models by 15%. If a certain four-door model has a discounted price of $18,000, what was its list price? How much can be saved by purchasing last year's model?

93. Business: Concession Markup A movie theater marks up the candy it sells by 275%. If a box of candy sells for $4.50 at the theater, how much did the theater pay for the box?

94. Personal Finance: Cost of a Car The suggested list price of a new car is $24,000. The dealer's cost is 85% of list. How much will you pay if the dealer is willing to accept $300 over cost for the car?

95. Business: Theater Attendance The manager of the Coral Theater wants to know whether the majority of its patrons are adults or children. One day in July, 5200 tickets were sold and the receipts totaled $29,961. The adult admission is $7.50, and the children's admission is $4.50. How many adult patrons were there?

96. Business: Discount Pricing A pair of leather boots, discounted by 30% for a clearance sale, has a price tag of $399. What was the original price?

97. Geometry The perimeter of a rectangle is 60 feet. Find its length and width if the length is 8 feet longer than the width.

98. Geometry The perimeter of a rectangle is 42 meters. Find its length and width if the length is twice the width.

99. Counting Calories Herschel uses an app on his smartphone to keep track of his daily calories from meals. One day his calories from breakfast were 125 more than his calories from lunch, and his calories from dinner were 300 less than twice his calories from lunch. If his total caloric intake from meals was 2025, determine his calories for each meal.

100. Counting Calories Tyshira tracks her net calories (calories taken in minus calories burned) as part of her fitness program. For one particular day, her net intake was 1480 calories. Her lunch calories were half her breakfast calories, and her dinner calories were 200 more than her breakfast calories. She ate 120 less calories in snacks than for breakfast, and she burned 700 calories by exercising on her elliptical. How many calories did she take in from snacks?

101. _Challenge Problem_ Sharing the Cost of a Pizza Judy and Tom agree to share the cost of an $18 pizza based on how much each ate. If Tom ate $\frac{2}{3}$ the amount that Judy ate, how much should each pay?

[**Hint**: Some pizza may be left.]

Tom's portion

Judy's portion

102. _Challenge Problem_ Find the largest perimeter of an isosceles triangle whose sides are of lengths $4x + 10$, $2x + 40$, and $3x + 18$.

103. _Challenge Problem_ Solve:

$$\frac{3}{4}x - \frac{1}{5}\left(\frac{1}{2} - 3x\right) + 1 = \frac{1}{4}\left(\frac{1}{20}x + 6\right) - \frac{4}{5}$$

104. _Challenge Problem_ A regular hexagon is inscribed in a circle. Find the radius of the circle if the perimeter of the hexagon is 10 inches more than the radius.

Explaining Concepts: Discussion and Writing

*105. **What Is Wrong?** One step in the following list contains an error. Identify it and explain what is wrong.

$x = 2$	**(1)**
$3x - 2x = 2$	**(2)**
$3x = 2x + 2$	**(3)**
$x^2 + 3x = x^2 + 2x + 2$	**(4)**
$x^2 + 3x - 10 = x^2 + 2x - 8$	**(5)**
$(x - 2)(x + 5) = (x - 2)(x + 4)$	**(6)**
$x + 5 = x + 4$	**(7)**
$1 = 0$	**(8)**

106. The equation

$$\frac{5}{x + 3} + 3 = \frac{8 + x}{x + 3}$$

has no solution, yet when we go through the process of solving it, we obtain $x = -3$. Write a brief paragraph to explain what causes this to happen.

107. Make up an equation that has no solution and give it to a fellow student to solve. Ask the fellow student to write a critique of your equation.

'Are You Prepared?' Answers

1. Distributive   **2.** Zero-Product   **3.** $\{x | x \neq 4\}$

1.2 Quadratic Equations

PREPARING FOR THIS SECTION *Before getting started, review the following:*

- Factoring Polynomials (Section R.5, pp. 49–55)
- Zero-Product Property (Section R.1, p. 13)
- Square Roots (Section R.2, pp. 23–24)
- Complete the Square (Section R.5, pp. 55–56)

Now Work the 'Are You Prepared?' problems on page 101.

OBJECTIVES
1. Solve a Quadratic Equation by Factoring (p. 93)
2. Solve a Quadratic Equation Using the Square Root Method (p. 94)
3. Solve a Quadratic Equation by Completing the Square (p. 95)
4. Solve a Quadratic Equation Using the Quadratic Formula (p. 96)
5. Solve Problems That Can Be Modeled by Quadratic Equations (p. 99)

Quadratic equations are equations such as

$$2x^2 + x + 8 = 0$$
$$3x^2 - 5x + 6 = 0$$
$$x^2 - 9 = 0$$

> **DEFINITION** Quadratic Equation
>
> A **quadratic equation** is an equation equivalent to one of the form
>
> $$ax^2 + bx + c = 0 \qquad a \neq 0 \qquad \qquad \textbf{(1)}$$
>
> where a, b, and c are real numbers.

A quadratic equation written in the form $ax^2 + bx + c = 0$ is said to be in **standard form**.

Sometimes, a quadratic equation is called a **second-degree equation** because, when it is in standard form, the left side is a polynomial of degree 2. We shall discuss four ways of solving quadratic equations: by factoring, by using the square root method, by completing the square, and by using the quadratic formula.

1 Solve a Quadratic Equation by Factoring

When a quadratic equation is written in standard form, it may be possible to factor the expression on the left side into the product of two first-degree polynomials. Then, using the Zero-Product Property and setting each factor equal to 0, the resulting linear equations are solved to obtain the solutions of the quadratic equation.

EXAMPLE 1   **Solving a Quadratic Equation by Factoring**

Solve the equations:

(a) $x^2 + 6x = 0$ (b) $2x^2 = x + 3$

Solution   (a) The equation is in standard form. The left side may be factored as

$$x^2 + 6x = 0$$
$$x(x + 6) = 0 \quad \text{Factor.}$$

Using the Zero-Product Property, set each factor equal to 0 and solve the first-degree equations.

$$x = 0 \quad \text{or} \quad x + 6 = 0 \qquad \text{Zero-Product Property}$$
$$x = 0 \quad \text{or} \qquad x = -6 \qquad \text{Solve.}$$

The solution set is $\{0, -6\}$.

(b) Place the equation $2x^2 = x + 3$ in standard form by subtracting x and 3 from both sides.

$$2x^2 = x + 3$$
$$2x^2 - x - 3 = 0 \qquad \text{Subtract } x \text{ and 3 from both sides.}$$

The left side may now be factored as

$$(2x - 3)(x + 1) = 0 \qquad \text{Factor.}$$

so that

$$2x - 3 = 0 \quad \text{or} \quad x + 1 = 0 \qquad \text{Zero-Product Property}$$
$$x = \frac{3}{2} \qquad \qquad x = -1 \quad \text{Solve.}$$

The solution set is $\left\{ -1, \dfrac{3}{2} \right\}$.

When the left side factors into two linear equations with the same solution, the quadratic equation is said to have a **repeated solution**. This solution is also called a **root of multiplicity 2**, or a **double root**.

EXAMPLE 2   **Solving a Quadratic Equation by Factoring**

Solve the equation: $9x^2 - 6x + 1 = 0$

Solution   This equation is already in standard form, and the left side can be factored.

$$9x^2 - 6x + 1 = 0$$
$$(3x - 1)(3x - 1) = 0$$

so

$$x = \frac{1}{3} \quad \text{or} \quad x = \frac{1}{3}$$

This equation has only the repeated solution $\frac{1}{3}$. The solution set is $\left\{\frac{1}{3}\right\}$.

Now Work PROBLEMS 13 AND 23

2 Solve a Quadratic Equation Using the Square Root Method

Suppose that we wish to solve the quadratic equation

$$x^2 = p \tag{2}$$

where $p > 0$ is a real number. Proceeding as in the earlier examples,

$$x^2 - p = 0 \qquad \text{Put in standard form.}$$
$$(x - \sqrt{p})(x + \sqrt{p}) = 0 \qquad \text{Factor (over the real numbers).}$$
$$x = \sqrt{p} \quad \text{or} \quad x = -\sqrt{p} \qquad \text{Solve.}$$

We have the following result:

> If $x^2 = p$ and $p > 0$, then $x = \sqrt{p}$ or $x = -\sqrt{p}$. $\tag{3}$

When statement (3) is used, it is called the **Square Root Method**. In statement (3), note that if $p > 0$ the equation $x^2 = p$ has two solutions, $x = \sqrt{p}$ and $x = -\sqrt{p}$. We usually abbreviate these solutions as $x = \pm\sqrt{p}$, which is read as "x equals plus or minus the square root of p."

For example, the two solutions of the equation

$$x^2 = 4$$

are

$$x = \pm\sqrt{4} \qquad \text{Use the Square Root Method.}$$

and, since $\sqrt{4} = 2$, we have

$$x = \pm 2$$

The solution set is $\{-2, 2\}$.

EXAMPLE 3   **Solving a Quadratic Equation Using the Square Root Method**

Solve each equation.

(a) $x^2 = 5$ (b) $(x - 2)^2 = 16$

Solution   (a) Use the Square Root Method:

$$x^2 = 5$$
$$x = \pm\sqrt{5} \qquad \text{Use the Square Root Method.}$$
$$x = \sqrt{5} \quad \text{or} \quad x = -\sqrt{5}$$

The solution set is $\{-\sqrt{5}, \sqrt{5}\}$.

(b) Use the Square Root Method:

$$(x - 2)^2 = 16$$

$$x - 2 = \pm\sqrt{16} \qquad \text{Use the Square Root Method.}$$

$$x - 2 = \pm 4 \qquad \qquad \sqrt{16} = 4$$

$$x - 2 = 4 \quad \text{or} \quad x - 2 = -4$$

$$x = 6 \quad \text{or} \qquad x = -2$$

The solution set is $\{-2, 6\}$.

Now Work PROBLEM 33

3 Solve a Quadratic Equation by Completing the Square

EXAMPLE 4

Solving a Quadratic Equation by Completing the Square

Solve by completing the square: $x^2 + 5x + 4 = 0$

Solution

Always begin by writing the equation with the constant on the right side.

$$x^2 + 5x + 4 = 0$$

$$x^2 + 5x = -4$$

NOTE If the coefficient of the square term is not 1, divide both sides by the coefficient of the square term before attempting to complete the square. ∎

Since the coefficient of x^2 is 1, complete the square on the left side by adding $\left(\dfrac{1}{2} \cdot 5\right)^2 = \dfrac{25}{4}$. Of course, in an equation, whatever is added to the left side must also be added to the right side. So add $\dfrac{25}{4}$ to *both* sides.

$$x^2 + 5x + \frac{25}{4} = -4 + \frac{25}{4} \qquad \text{Add } \frac{25}{4} \text{ to both sides.}$$

$$\left(x + \frac{5}{2}\right)^2 = \frac{9}{4} \qquad \qquad \text{Factor on the left; simplify on the right.}$$

$$x + \frac{5}{2} = \pm\sqrt{\frac{9}{4}} \qquad \qquad \text{Use the Square Root Method.}$$

$$x + \frac{5}{2} = \pm\frac{3}{2} \qquad \qquad \sqrt{\frac{9}{4}} = \frac{\sqrt{9}}{\sqrt{4}} = \frac{3}{2}$$

$$x = -\frac{5}{2} \pm \frac{3}{2}$$

$$x = -\frac{5}{2} + \frac{3}{2} = -1 \quad \text{or} \quad x = -\frac{5}{2} - \frac{3}{2} = -4$$

The solution set is $\{-4, -1\}$.

THE SOLUTION OF THE EQUATION IN EXAMPLE 4 CAN ALSO BE OBTAINED BY FACTORING. REWORK EXAMPLE 4 USING FACTORING.

The next example illustrates an equation that cannot be solved by factoring.

EXAMPLE 5

Solving a Quadratic Equation by Completing the Square

Solve by completing the square: $2x^2 - 8x - 5 = 0$

Solution

First, rewrite the equation so that the constant is on the right side.

$$2x^2 - 8x - 5 = 0$$

$$2x^2 - 8x = 5 \quad \text{Add 5 to both sides.}$$

(continued)

Next, divide both sides by 2 so that the coefficient of x^2 is 1. (This enables us to complete the square at the next step.)

$$x^2 - 4x = \frac{5}{2}$$

Finally, complete the square by adding $\left[\frac{1}{2}(-4)\right]^2 = 4$ to both sides.

$$x^2 - 4x + 4 = \frac{5}{2} + 4 \qquad \text{Add 4 to both sides.}$$

$$(x - 2)^2 = \frac{13}{2} \qquad \text{Factor on the left; simplify on the right.}$$

$$x - 2 = \pm\sqrt{\frac{13}{2}} \qquad \text{Use the Square Root Method.}$$

$$x - 2 = \pm\frac{\sqrt{26}}{2} \qquad \sqrt{\frac{13}{2}} = \frac{\sqrt{13}}{\sqrt{2}} = \frac{\sqrt{13}}{\sqrt{2}} \cdot \frac{\sqrt{2}}{\sqrt{2}} = \frac{\sqrt{26}}{2}$$

$$x = 2 \pm \frac{\sqrt{26}}{2}$$

NOTE If we wanted an approximation, say rounded to two decimal places, of these solutions, we would use a calculator to get $\{-0.55, 4.55\}$. ∎

The solution set is $\left\{2 - \frac{\sqrt{26}}{2}, 2 + \frac{\sqrt{26}}{2}\right\}$.

➤ **Now Work** PROBLEM 37

4 Solve a Quadratic Equation Using the Quadratic Formula

We can use the method of completing the square to obtain a general formula for solving any quadratic equation

$$ax^2 + bx + c = 0 \qquad a \neq 0$$

NOTE We can assume that $a > 0$, since if $a < 0$ we can multiply by -1 to obtain an equivalent equation with a positive leading coefficient. ∎

As in Examples 4 and 5, rearrange the terms as

$$ax^2 + bx = -c \quad a > 0$$

Since $a > 0$, divide both sides by a to get

$$x^2 + \frac{b}{a}x = -\frac{c}{a}$$

Now the coefficient of x^2 is 1. To complete the square on the left side, add the square of $\frac{1}{2}$ the coefficient of x; that is, add

$$\left(\frac{1}{2} \cdot \frac{b}{a}\right)^2 = \frac{b^2}{4a^2}$$

to both sides. Then

$$x^2 + \frac{b}{a}x + \frac{b^2}{4a^2} = \frac{b^2}{4a^2} - \frac{c}{a}$$

$$\left(x + \frac{b}{2a}\right)^2 = \frac{b^2 - 4ac}{4a^2} \qquad \frac{b^2}{4a^2} - \frac{c}{a} = \frac{b^2}{4a^2} - \frac{4ac}{4a^2} = \frac{b^2 - 4ac}{4a^2} \qquad \textbf{(4)}$$

Provided that $b^2 - 4ac \geq 0$, we can now use the Square Root Method to get

$$x + \frac{b}{2a} = \pm\sqrt{\frac{b^2 - 4ac}{4a^2}}$$

$$x + \frac{b}{2a} = \frac{\pm\sqrt{b^2 - 4ac}}{2a} \qquad \begin{array}{l}\text{The square root of a quotient equals} \\ \text{the quotient of the square roots.} \\ \text{Also, } \sqrt{4a^2} = 2a \text{ since } a > 0.\end{array}$$

$$x = -\frac{b}{2a} \pm \frac{\sqrt{b^2 - 4ac}}{2a} \qquad \text{Add } -\frac{b}{2a} \text{ to both sides.}$$

$$= \frac{-b \pm \sqrt{b^2 - 4ac}}{2a} \qquad \text{Combine the quotients on the right.}$$

What if $b^2 - 4ac$ is negative? Then equation (4) states that the left expression (a real number squared) equals the right expression (a negative number). Since this is impossible for real numbers, we conclude that if $b^2 - 4ac < 0$, the quadratic equation has no *real* solution. (We discuss quadratic equations for which the quantity $b^2 - 4ac < 0$ in detail in the next section.)

THEOREM **Quadratic Formula**

Consider the quadratic equation

$$ax^2 + bx + c = 0 \qquad a \neq 0$$

- If $b^2 - 4ac < 0$, this equation has no real solution.
- If $b^2 - 4ac \geq 0$, the real solution(s) of this equation is (are) given by the **quadratic formula**:

$$x = \frac{-b \pm \sqrt{b^2 - 4ac}}{2a} \qquad\qquad (5)$$

The quantity $b^2 - 4ac$ is called the **discriminant** of the quadratic equation, because its value tells us whether the equation has real solutions. In fact, it also tells us how many solutions to expect.

Discriminant of a Quadratic Equation

For a quadratic equation $ax^2 + bx + c = 0$, $a \neq 0$:

- If $b^2 - 4ac > 0$, there are two unequal real solutions.
- If $b^2 - 4ac = 0$, there is a repeated real solution, a double root.
- If $b^2 - 4ac < 0$, there is no real solution.

When asked to find the real solutions of a quadratic equation, always evaluate the discriminant first to see if there are any real solutions.

EXAMPLE 6 **Solving a Quadratic Equation Using the Quadratic Formula**

Use the quadratic formula to find the real solutions, if any, of the equation

$$3x^2 - 5x + 1 = 0$$

Solution The equation is in standard form, so compare it to $ax^2 + bx + c = 0$ to find a, b, and c.

$$3x^2 - 5x + 1 = 0$$

$$ax^2 + bx + c = 0 \qquad a = 3, b = -5, c = 1$$

(continued)

With $a = 3$, $b = -5$, and $c = 1$, evaluate the discriminant $b^2 - 4ac$.

$$b^2 - 4ac = (-5)^2 - 4 \cdot 3 \cdot 1 = 25 - 12 = 13$$

Since $b^2 - 4ac > 0$, there are two real solutions, which can be found using the quadratic formula.

$$x = \frac{-b \pm \sqrt{b^2 - 4ac}}{2a} = \frac{-(-5) \pm \sqrt{13}}{2 \cdot 3} = \frac{5 \pm \sqrt{13}}{6}$$

The solution set is $\left\{ \dfrac{5 - \sqrt{13}}{6}, \dfrac{5 + \sqrt{13}}{6} \right\}$.

EXAMPLE 7

Solving a Quadratic Equation Using the Quadratic Formula

Use the quadratic formula to find the real solutions, if any, of the equation

$$\frac{25}{2}x^2 - 30x + 18 = 0$$

Solution The equation is given in standard form. However, to simplify the arithmetic, clear the fraction.

$$\frac{25}{2}x^2 - 30x + 18 = 0$$

$25x^2 - 60x + 36 = 0$ **Multiply both sides by 2 to clear the fraction.**

$ax^2 + bx + c = 0$ **Compare to standard form.**

With $a = 25$, $b = -60$, and $c = 36$, evaluate the discriminant.

$$b^2 - 4ac = (-60)^2 - 4 \cdot 25 \cdot 36 = 3600 - 3600 = 0$$

The equation has a repeated real solution, which is found by using the quadratic formula.

$$x = \frac{-b \pm \sqrt{b^2 - 4ac}}{2a} = \frac{60 \pm \sqrt{0}}{50} = \frac{60}{50} = \frac{6}{5}$$

The solution set is $\left\{ \dfrac{6}{5} \right\}$.

EXAMPLE 8

Solving a Quadratic Equation Using the Quadratic Formula

Use the quadratic formula to find the real solutions, if any, of the equation

$$3x^2 + 2 = 4x$$

Solution The equation, as given, is not in standard form.

$$3x^2 + 2 = 4x$$

$3x^2 - 4x + 2 = 0$ **Put in standard form.**

$ax^2 + bx + c = 0$ **Compare to standard form.**

With $a = 3$, $b = -4$, and $c = 2$, the discriminant is

$$b^2 - 4ac = (-4)^2 - 4 \cdot 3 \cdot 2 = 16 - 24 = -8$$

Since $b^2 - 4ac < 0$, the equation has no real solution.

Now Work PROBLEMS 47 AND 57

EXAMPLE 9 **Solving an Equation Using the Quadratic Formula**

Find the real solutions, if any, of the equation: $9 + \dfrac{3}{x} - \dfrac{2}{x^2} = 0$, $x \neq 0$

Solution In its present form, the equation

$$9 + \frac{3}{x} - \frac{2}{x^2} = 0$$

is not a quadratic equation. However, it can be transformed into one by multiplying both sides by x^2. The result is

$$9x^2 + 3x - 2 = 0, \quad x \neq 0$$

Although we multiplied both sides by x^2, we know that $x^2 \neq 0$ (do you see why?), so this quadratic equation is equivalent to the original equation.

Using $a = 9$, $b = 3$, and $c = -2$, the discriminant is

$$b^2 - 4ac = 3^2 - 4 \cdot 9 \cdot (-2) = 9 + 72 = 81$$

Since $b^2 - 4ac > 0$, the new equation has two real solutions.

$$x = \frac{-b \pm \sqrt{b^2 - 4ac}}{2a} = \frac{-3 \pm \sqrt{81}}{2 \cdot 9} = \frac{-3 \pm 9}{18}$$

$$x = \frac{-3 + 9}{18} = \frac{6}{18} = \frac{1}{3} \quad \text{or} \quad x = \frac{-3 - 9}{18} = \frac{-12}{18} = -\frac{2}{3}$$

The solution set is $\left\{ -\dfrac{2}{3}, \dfrac{1}{3} \right\}$.

SUMMARY

Steps for Solving a Quadratic Equation

To solve a quadratic equation, first put it in standard form:

$$ax^2 + bx + c = 0$$

Then:

STEP 1: Identify a, b, and c.

STEP 2: Evaluate the discriminant, $b^2 - 4ac$.

STEP 3: • If the discriminant is negative, the equation has no real solution.

   • If the discriminant is zero, the equation has one real solution, a double root.

   • If the discriminant is positive, the equation has two distinct real solutions.

If you can easily spot factors, use the factoring method to solve the equation. Otherwise, use the quadratic formula or the method of completing the square.

5 Solve Problems That Can Be Modeled by Quadratic Equations

Many applied problems require the solution of a quadratic equation. Let's look at one that you will probably see again in a slightly different form if you study calculus.

EXAMPLE 10 **Constructing a Box**

From each corner of a square piece of sheet metal, remove a square with sides of length 9 centimeters. Turn up the edges to form an open box. If the box is to hold 144 cubic centimeters (cm^3), what should be the dimensions of the piece of sheet metal?

(continued)

Solution Use Figure 1 as a guide. We have labeled by x the length of a side of the square piece of sheet metal. The box will be of height 9 centimeters, and its square base will measure $x - 18$ on each side. The volume V (Length $\times$ Width $\times$ Height) of the box is therefore

$$V = (x - 18)(x - 18) \cdot 9 = 9(x - 18)^2$$

Figure 1

Since the volume of the box is to be 144 cm^3, we have

$$\begin{aligned} 9(x - 18)^2 &= 144 && V = 144 \\ (x - 18)^2 &= 16 && \text{Divide both sides by 9.} \\ x - 18 &= \pm 4 && \text{Use the Square Root Method.} \\ x &= 18 \pm 4 \\ x &= 22 \quad \text{or} \quad x = 14 \end{aligned}$$

✔ **Check:** If we take a piece of sheet metal 22 cm by 22 cm, cut out a 9-centimeter square from each corner, and fold up the edges, we get a box whose dimensions are 9 cm by 4 cm by 4 cm, with volume $9 \cdot 4 \cdot 4 = 144$ cm^3, as required.

Discard the solution $x = 14$ (do you see why?) and conclude that the sheet metal should be 22 centimeters by 22 centimeters.

━━━ **Now Work** PROBLEM 99

Historical Feature

Problems using quadratic equations are found in the oldest known mathematical literature. Babylonians and Egyptians were solving such problems before 1800 BC. Euclid solved quadratic equations geometrically in his *Data* (300 BC), and the Hindus and Arabs gave rules for solving any quadratic equation with real roots. Because negative numbers were not freely used before AD 1500, there were several different types of quadratic equations, each with its own rule. Thomas Harriot (1560-1621) introduced the method of factoring to obtain solutions, and François Viète (1540-1603) introduced a method that is essentially completing the square.

Until modern times it was usual to neglect the negative roots (if there were any), and equations involving square roots of negative quantities were regarded as unsolvable until the 1500s.

Historical Problems

1. *One of al-Khwārízmí solutions* Solve $x^2 + 12x = 85$ by drawing the square shown. The area of the four white rectangles and the yellow square is $x^2 + 12x$. We then set this expression equal to 85 to get the equation $x^2 + 12x = 85$. If we add the four blue squares, we will have a larger square of known area. Complete the solution.

2. *Viète's method* Solve $x^2 + 12x - 85 = 0$ by letting $x = u + z$. Then

$$(u + z)^2 + 12(u + z) - 85 = 0$$
$$u^2 + (2z + 12)u + (z^2 + 12z - 85) = 0$$

Now select z so that $2z + 12 = 0$ and finish the solution.

3. *Another method to get the quadratic formula* Look at equation (4) on page 96. Rewrite the right side as $\left(\dfrac{\sqrt{b^2 - 4ac}}{2a}\right)^2$ and then subtract it from both sides. The right side is now 0 and the left side is a difference of two squares. If you factor this difference of two squares, you will easily be able to get the quadratic formula, and moreover, the quadratic expression is factored, which is sometimes useful.

1.2 Assess Your Understanding

'Are You Prepared?' *Answers are given at the end of these exercises. If you get a wrong answer, read the pages listed in* red.

1. Factor: $x^2 - 5x - 6$ (pp. 49–55)

2. Factor: $2x^2 - x - 3$ (pp. 49–55)

3. The solution set of the equation $(x - 3)(3x + 5) = 0$ is _____. (p. 13)

4. *True or False* $\sqrt{x^2} = |x|$. (pp. 23–24)

5. Complete the square of $x^2 + 5x$. Factor the new expression. (pp. 55–56)

Concepts and Vocabulary

6. The quantity $b^2 - 4ac$ is called the _____ of a quadratic equation. If it is _____, the equation has no real solution.

7. *True or False* Quadratic equations always have two real solutions.

8. *True or False* If $x^2 = p$ and $p > 0$, then $x = \sqrt{p}$.

9. *Multiple Choice* A quadratic equation is sometimes called a _____ equation.

   (a) first-degree   **(b)** second-degree

   (c) third-degree   **(d)** fourth-degree

10. *Multiple Choice* Which of the following quadratic equations is in standard form?

   (a) $x^2 - 7x = 5$ **(b)** $9 = x^2$

   (c) $(x + 5)(x - 4) = 0$ **(d)** $0 = 5x^2 - 6x - 1$

Skill Building

In Problems 11–30, solve each equation by factoring.

11. $x^2 - 9x = 0$

12. $x^2 + 4x = 0$

13. $x^2 - 25 = 0$

14. $x^2 - 9 = 0$

15. $z^2 + z - 6 = 0$

16. $v^2 + 7v + 6 = 0$

17. $2x^2 - 5x - 3 = 0$

18. $3x^2 + 5x + 2 = 0$

19. $-5w^2 + 180 = 0$

20. $2y^2 - 50 = 0$

21. $x(x + 3) - 10 = 0$

22. $x(x + 4) = 12$

23. $4x^2 + 9 = 12x$

24. $25x^2 + 16 = 40x$

25. $6(p^2 - 1) = 5p$

26. $2(2u^2 - 4u) + 3 = 0$

27. $6x - 5 = \dfrac{6}{x}$

28. $x + \dfrac{12}{x} = 7$

29. $\dfrac{4(x - 2)}{x - 3} + \dfrac{3}{x} = \dfrac{-3}{x(x - 3)}$

30. $\dfrac{5}{x + 4} = 4 + \dfrac{3}{x - 2}$

In Problems 31–36, solve each equation by the Square Root Method.

31. $x^2 = 25$

32. $x^2 = 36$

33. $(x - 1)^2 = 4$

34. $(x + 2)^2 = 1$

35. $\left(\dfrac{1}{3}h + 4\right)^2 = 16$

36. $(3z - 2)^2 = 4$

In Problems 37–42, solve each equation by completing the square.

37. $x^2 + 4x = 21$

38. $x^2 - 6x = 13$

39. $x^2 - \dfrac{1}{2}x - \dfrac{3}{16} = 0$

40. $x^2 + \dfrac{2}{3}x - \dfrac{1}{3} = 0$

41. $3x^2 + x - \dfrac{1}{2} = 0$

42. $2x^2 - 3x - 1 = 0$

In Problems 43–66, find the real solutions, if any, of each equation. Use the quadratic formula.

43. $x^2 + 2x - 13 = 0$

44. $x^2 + 4x + 2 = 0$

45. $x^2 - 4x - 1 = 0$

46. $x^2 + 6x + 1 = 0$

47. $2x^2 - 5x + 3 = 0$

48. $2x^2 + 5x + 3 = 0$

49. $4y^2 - y + 2 = 0$

50. $4t^2 + t + 1 = 0$

51. $9x^2 + 8x = 5$

52. $2x^2 = 1 - 2x$

53. $4x^2 = 9x$

54. $5x = 4x^2$

55. $9t^2 - 6t + 1 = 0$

56. $4u^2 - 6u + 9 = 0$

57. $\dfrac{3}{4}x^2 - \dfrac{1}{4}x = \dfrac{1}{2}$

58. $\frac{2}{3}x^2 - x - 3 = 0$

59. $\frac{5}{3}x^2 - x = \frac{1}{3}$

60. $\frac{3}{5}x^2 - x = \frac{1}{5}$

61. $2x(x + 2) = 3$

62. $3x(x + 2) = 1$

63. $4 + \frac{1}{x} - \frac{1}{x^2} = 0$

64. $2 + \frac{8}{x} + \frac{3}{x^2} = 0$

65. $\frac{3x}{x - 2} + \frac{1}{x} = 4$

66. $\frac{2x}{x - 3} + \frac{1}{x} = 4$

In Problems 67–72, find the real solutions, if any, of each equation. Use the quadratic formula and a calculator. Express any solutions rounded to two decimal places.

67. $x^2 - 4.1x + 2.2 = 0$

68. $x^2 + 3.9x + 1.8 = 0$

69. $x^2 + \sqrt{3}x - 3 = 0$

70. $x^2 + \sqrt{2}x - 2 = 0$

71. $\pi x^2 - x - \pi = 0$

72. $\pi x^2 + \pi x - 2 = 0$

In Problems 73–78, use the discriminant to determine whether each quadratic equation has two unequal real solutions, a repeated real solution (a double root), or no real solution, without solving the equation.

73. $2x^2 - 6x + 7 = 0$

74. $x^2 + 4x + 7 = 0$

75. $9x^2 - 30x + 25 = 0$

76. $25x^2 - 20x + 4 = 0$

77. $3x^2 + 5x - 8 = 0$

78. $2x^2 - 3x - 7 = 0$

Mixed Practice In Problems 79–92, find the real solutions, if any, of each equation. Use any method.

79. $x^2 - 5 = 0$

80. $x^2 - 6 = 0$

81. $16x^2 - 8x + 1 = 0$

82. $9x^2 - 12x + 4 = 0$

83. $10x^2 - 19x - 15 = 0$

84. $6x^2 + 7x - 20 = 0$

85. $2 + z = 6z^2$

86. $2 = y + 6y^2$

87. $x^2 + \sqrt{2}x = \frac{1}{2}$

88. $\frac{1}{2}x^2 = \sqrt{2}x + 1$

89. $x^2 + x = 4$

90. $x^2 + x = 1$

91. $\frac{x}{x - 2} + \frac{2}{x + 1} = \frac{7x + 1}{x^2 - x - 2}$

92. $\frac{3x}{x + 2} + \frac{1}{x - 1} = \frac{4 - 7x}{x^2 + x - 2}$

Applications and Extensions

93. Pythagorean Theorem How many right triangles have a hypotenuse that measures $2x + 3$ meters and legs that measure $2x - 5$ meters and $x + 7$ meters? What are their dimensions?

94. Pythagorean Theorem How many right triangles have a hypotenuse that measures $4x + 5$ inches and legs that measure $3x + 13$ inches and x inches? What are the dimensions of the triangle(s)?

95. Dimensions of a Window The area of the opening of a rectangular window is to be 143 square feet. If the length is to be 2 feet more than the width, what are the dimensions?

96. Dimensions of a Window The area of a rectangular window is to be 306 square centimeters. If the length exceeds the width by 1 centimeter, what are the dimensions?

97. Geometry Find the dimensions of a rectangle whose perimeter is 26 meters and whose area is 40 square meters.

98. Watering a Square Field An adjustable water sprinkler that sprays water in a circular pattern is placed at the center of a square field whose area is 1250 square feet (see the figure). What is the shortest radius setting that can be used if the field is to be completely enclosed within the circle?

99. Constructing a Box An open box is to be constructed from a square piece of sheet metal by removing a square with sides of length 1 foot from each corner and turning up the edges. If the box is to hold 4 cubic feet, what should be the dimensions of the sheet metal?

100. Constructing a Box Rework Problem 99 if the piece of sheet metal is a rectangle whose length is twice its width.

101. Physics A ball is thrown vertically upward from the top of a building 96 feet tall with an initial velocity of 80 feet per second. The distance s (in feet) of the ball from the ground after t seconds is $s = 96 + 80t - 16t^2$.
(a) After how many seconds does the ball strike the ground?
(b) After how many seconds will the ball pass the top of the building on its way down?

102. Physics An object is propelled vertically upward with an initial velocity of 20 meters per second. The distance s (in meters) of the object from the ground after t seconds is $s = -4.9t^2 + 20t$.
(a) When will the object be 15 meters above the ground?
(b) When will it strike the ground?
(c) Will the object reach a height of 100 meters?

103. Reducing the Size of a Candy Bar A jumbo chocolate bar with a rectangular shape measures 12 centimeters in length, 7 centimeters in width, and 3 centimeters in thickness. Due to escalating costs of cocoa, management decides to reduce the volume of the bar by 10%. To accomplish this reduction, management decides that the new bar should have the same 3-centimeter thickness, but the length and width of each should be reduced an equal number of centimeters. What should be the dimensions of the new candy bar?

104. Reducing the Size of a Candy Bar Rework Problem 103 if the reduction is to be 20%.

105. Constructing a Border around a Pool A circular pool measures 10 feet across. One cubic yard of concrete is to be used to create a circular border of uniform width around the pool. If the border is to have a depth of 3 inches, how wide will the border be? (1 cubic yard = 27 cubic feet) See the illustration.

106. Constructing a Border around a Pool Rework Problem 105 if the depth of the border is 4 inches.

107. Constructing a Border around a Garden A landscaper, who just completed a rectangular flower garden measuring 6 feet by 10 feet, orders 1 cubic yard of premixed cement, all of which is to be used to create a border of uniform width around the garden. If the border is to have a depth of 3 inches, how wide will the border be? (**Hint:** 1 cubic yard = 27 cubic feet)

108. Dimensions of a Patio A contractor orders 8 cubic yards of premixed cement, all of which is to be used to pour a patio that will be 4 inches thick. If the length of the patio is specified to be twice the width, what will be the patio dimensions? (1 cubic yard = 27 cubic feet)

109. Comparing Tablets The screen size of a tablet is determined by the length of the diagonal of the rectangular screen. The 12.9-inch iPad Pro™ comes in a 16:9 format, which means that the ratio of the length to the width of the rectangular screen is 16:9. What is the area of the iPad's screen? What is the area of a 12.3-inch Microsoft Surface Pro™ if its screen is in a 3:2 format? Which screen is larger? (**Hint:** If x is the length of a 4:3 format screen, then $\frac{3}{4}x$ is the width.)

110. Comparing Tablets Refer to Problem 109. Find the screen area of a 7.9-inch iPad mini 4™ in a 4:3 format, and compare it with an 8-inch Amazon Fire HD 8™ if its screen is in a 16:10 format. Which screen is larger?

111. Field Design A football field, 50 feet wide, is sloped from the center toward the sides for drainage. The height h, in feet, of the field, x feet from the side, is given by $h = -0.00025x^2 + 0.04x$. Find the distance from the side when the height is 1.1 feet. Round to the nearest tenth of a foot.

112. College Value The difference, d, in median earnings, in $1000s, between high school graduates and college graduates can be approximated by $d = -0.012x^2 + 0.828x + 15.750$, where x is the number of years after 1980. Based on this model, estimate to the nearest year when the difference in median earnings was $25,000.
Source: Current Population Survey

113. Student Working A study found that a student's GPA, g, is related to the number of hours worked each week, h, by the equation $g = -0.0006h^2 + 0.015h + 3.04$. Estimate the number of hours worked each week for a student with a GPA of 2.97. Round to the nearest whole hour.

114. Fraternity Purchase A fraternity wants to buy a new LED Smart TV that costs $1470. If 7 members of the fraternity are not able to contribute, the share for the remaining members increases by $5. How many members are in the fraternity?

115. Discretionary Income Individuals are able to save money when their discretionary income (income after all bills are paid) exceeds their consumption. Suppose the model, $y = -25a^2 + 2400a - 30,700$, represents the average discretionary income for an individual who is a years old. If the consumption model for an individual is given by $y = 160a + 7840$, at what age is the individual able to start saving money? Round to the nearest year.

116. Living at Home A study found that Millennials are more likely to live at home than GenXers at the same age. The model $P = 0.003x^2 - 0.034x + 8.086$ gives the approximate percentage of 25-to-35-year-olds living in their parents' home, x years after 1960. Use the model to estimate when the percentage of 25-to-35-year-olds living in their parents' home was 10%. Round to the nearest year.
Data: Pew Research Center

117. The sum of the consecutive integers 1, 2, 3, ..., n is given by the formula $\frac{1}{2}n(n + 1)$. How many consecutive integers, starting with 1, must be added to get a sum of 703?

118. Geometry If a polygon of n sides has $\frac{1}{2}n(n - 3)$ diagonals, how many sides will a polygon with 65 diagonals have? Is there a polygon with 80 diagonals?

119. Show that the sum of the roots of a quadratic equation is $-\frac{b}{a}$.

120. Show that the product of the roots of a quadratic equation is $\frac{c}{a}$.

121. Find k so that the equation $kx^2 + x + k = 0$ has a repeated real solution.

122. Find k so that the equation $x^2 - kx + 4 = 0$ has a repeated real solution.

123. *Challenge Problem* Show that the real solutions of the equation $ax^2 + bx + c = 0, a \neq 0$, are the negatives of the real solutions of the equation $ax^2 - bx + c = 0$. Assume that $b^2 - 4ac \geq 0$.

124. *Challenge Problem* Show that the real solutions of the equation $ax^2 + bx + c = 0, a \neq 0$, are the reciprocals of the real solutions of the equation $cx^2 + bx + a = 0$, $c \neq 0$. Assume that $b^2 - 4ac \geq 0$.

125. *Challenge Problem* A rectangular piece of paper with area 1 m² is to be folded in half along its longest side. If the ratio of side lengths is to be the same for the folded paper as for the original paper, what are the dimensions of the original paper?

Explaining Concepts: Discussion and Writing

126. Which of the following pairs of equations are equivalent? Explain.
(a) $x^2 = 9$; $x = 3$ (b) $x = \sqrt{9}$; $x = 3$
(c) $(x - 1)(x - 2) = (x - 1)^2$; $x - 2 = x - 1$

127. Describe three ways that you might solve a quadratic equation. State your preferred method; explain why you chose it.

128. Explain the benefits of evaluating the discriminant of a quadratic equation before attempting to solve it.

129. Create three quadratic equations: one having two distinct real solutions, one having no real solution, and one having exactly one real solution.

130. The word *quadratic* seems to imply four (*quad*), yet a quadratic equation is an equation that involves a polynomial of degree 2. Investigate the origin of the term *quadratic* as it is used in the expression *quadratic equation*. Write a brief essay on your findings.

'Are You Prepared?' Answers

1. $(x - 6)(x + 1)$ **2.** $(2x - 3)(x + 1)$ **3.** $\left\{-\frac{5}{3}, 3\right\}$ **4.** True   **5.** $x^2 + 5x + \frac{25}{4} = \left(x + \frac{5}{2}\right)^2$

1.3 Complex Numbers; Quadratic Equations in the Complex Number System*

PREPARING FOR THIS SECTION *Before getting started, review the following:*

- Classification of Numbers (Section R.1, pp. 4–5)
- Laws of Exponents (Section R.2, pp. 21–23)
- Rationalizing Denominators and Numerators (Section R.8, pp. 74–75)

Now Work the 'Are You Prepared?' problems on page 111.

OBJECTIVES 1 Add, Subtract, Multiply, and Divide Complex Numbers (p. 105)
2 Solve Quadratic Equations in the Complex Number System (p. 109)

Complex Numbers

One property of a real number is that its square is nonnegative. For example, there is no real number x for which

$$x^2 = -1$$

To remedy this situation, we introduce a new number called the *imaginary unit*.

> **DEFINITION** The Imaginary Unit
>
> The **imaginary unit**, which we denote by i, is the number whose square is -1. That is,
>
> $$i^2 = -1$$

*This section may be omitted without any loss of continuity.

This should not surprise you. If our universe were to consist only of integers, there would be no number x for which $2x = 1$. This was remedied by introducing numbers such as $\frac{1}{2}$ and $\frac{2}{3}$, the *rational numbers*. If our universe were to consist only of rational numbers, there would be no x whose square equals 2. That is, there would be no number x for which $x^2 = 2$. To remedy this, we introduced numbers such as $\sqrt{2}$ and $\sqrt[3]{5}$, the *irrational numbers*. Recall that the *real numbers* consist of the rational numbers and the irrational numbers. Now, if our universe were to consist only of real numbers, then there would be no number x whose square is -1. To remedy this, we introduce the number i, whose square is -1.

In the progression outlined, each time we encountered a situation that was unsuitable, a new number system was introduced to remedy the situation. The number system that results from introducing the number i is called the **complex number system**.

DEFINITION **The Complex Number System**

Complex numbers are numbers of the form $a + bi$, where a and b are real numbers. The real number a is called the **real part** of the number $a + bi$; the real number b is called the **imaginary part** of $a + bi$; and i is the imaginary unit, so $i^2 = -1$.

For example, the complex number $-5 + 6i$ has the real part -5 and the imaginary part 6.

When a complex number is written in the form $a + bi$, where a and b are real numbers, it is in **standard form**. However, if the imaginary part of a complex number is negative, such as in the complex number $3 + (-2)i$, we agree to write it instead in the form $3 - 2i$.

Also, the complex number $a + 0i$ is usually written merely as a. This serves to remind us that the real numbers are a subset of the complex numbers. Similarly, the complex number $0 + bi$ is usually written as bi. Sometimes the complex number bi is called a **pure imaginary number**.

1 Add, Subtract, Multiply, and Divide Complex Numbers

Equality, addition, subtraction, and multiplication of complex numbers are defined so as to preserve the familiar rules of algebra for real numbers. Two complex numbers are equal if and only if their real parts are equal and their imaginary parts are equal.

Equality of Complex Numbers

$$a + bi = c + di \quad \text{if and only if} \quad a = c \text{ and } b = d \qquad (1)$$

Two complex numbers are added by forming the complex number whose real part is the sum of the real parts and whose imaginary part is the sum of the imaginary parts.

Sum of Complex Numbers

$$(a + bi) + (c + di) = (a + c) + (b + d)i \qquad (2)$$

To subtract two complex numbers, use this rule:

Difference of Complex Numbers

$$(a + bi) - (c + di) = (a - c) + (b - d)i \qquad (3)$$

EXAMPLE 1

Adding and Subtracting Complex Numbers

(a) $(3 + 5i) + (-2 + 3i) = [3 + (-2)] + (5 + 3)i = 1 + 8i$

(b) $(6 + 4i) - (3 + 6i) = (6 - 3) + (4 - 6)i = 3 + (-2)i = 3 - 2i$

---Now Work PROBLEM 15

Products of complex numbers are calculated as illustrated in Example 2.

EXAMPLE 2

Multiplying Complex Numbers

$$(5 + 3i)(2 + 7i) = 5(2 + 7i) + 3i(2 + 7i) \qquad \text{Distributive Property}$$
$$= 10 + 35i + 6i + 21i^2 \qquad \text{Distributive Property}$$
$$= 10 + 41i + 21(-1) \qquad i^2 = -1$$
$$= -11 + 41i \qquad \text{Simplify.}$$

Based on the procedure of Example 2, the **product** of two complex numbers is defined as follows:

Product of Complex Numbers

$$(a + bi)(c + di) = (ac - bd) + (ad + bc)i \qquad \textbf{(4)}$$

Do not bother to memorize formula (4). Instead, whenever it is necessary to multiply two complex numbers, follow the usual rules for multiplying two binomials, as we did in Example 2. Just remember that $i^2 = -1$. For example,

$$(2i)(2i) = 4i^2 = 4(-1) = -4$$
$$(2 + i)(1 - i) = 2 - 2i + i - i^2 = 3 - i$$

---Now Work PROBLEM 21

Algebraic properties for addition and multiplication, such as the commutative, associative, and distributive properties, hold for complex numbers. The property that every nonzero complex number has a multiplicative inverse, or reciprocal, requires a closer look.

Conjugates

In Words
The conjugate of a complex number is found by changing the sign of the imaginary part.

DEFINITION Complex Conjugate

If $z = a + bi$ is a complex number, then its **conjugate**, denoted by $\bar{z}$, is defined as

$$\bar{z} = \overline{a + bi} = a - bi$$

For example, $\overline{2 + 3i} = 2 - 3i$ and $\overline{-6 - 2i} = -6 + 2i$.

EXAMPLE 3

Multiplying a Complex Number by Its Conjugate

Find the product of the complex number $z = 3 + 4i$ and its conjugate $\bar{z}$.

Solution   Since $\bar{z} = 3 - 4i$, we have

$$z\bar{z} = (3 + 4i)(3 - 4i) = 9 - 12i + 12i - 16i^2 = 9 + 16 = 25$$

The result obtained in Example 3 has an important generalization.

THEOREM

The product of a complex number and its conjugate is a nonnegative real number. That is, if $z = a + bi$, then

$$z\bar{z} = a^2 + b^2 \tag{5}$$

Proof If $z = a + bi$, then

$$z\bar{z} = (a + bi)(a - bi) = a^2 - abi + abi - (bi)^2 = a^2 - b^2 i^2$$
$$= a^2 - b^2(-1) = a^2 + b^2 \qquad \blacksquare$$

━━ **Now Work** PROBLEM 89

To express the reciprocal of a nonzero complex number z in standard form, multiply the numerator and denominator of $\dfrac{1}{z}$ by $\bar{z}$. That is, if $z = a + bi$ is a nonzero complex number, then

$$\frac{1}{a + bi} = \frac{1}{z} = \frac{1}{z} \cdot \frac{\bar{z}}{\bar{z}} = \frac{\bar{z}}{z\bar{z}} = \underset{\underset{\text{Use (5).}}{\uparrow}}{\frac{a - bi}{a^2 + b^2}} = \frac{a}{a^2 + b^2} - \frac{b}{a^2 + b^2}i$$

EXAMPLE 4 | **Writing the Reciprocal of a Complex Number in Standard Form**

Write $\dfrac{1}{3 + 4i}$ in standard form; that is, find the reciprocal of $3 + 4i$.

Solution The idea is to multiply the numerator and denominator by the conjugate of $3 + 4i$, that is, by the complex number $3 - 4i$. The result is

$$\frac{1}{3 + 4i} = \frac{1}{3 + 4i} \cdot \frac{3 - 4i}{3 - 4i} = \frac{3 - 4i}{9 + 16} = \frac{3}{25} - \frac{4}{25}i$$

To express the quotient of two complex numbers in standard form, multiply the numerator and denominator of the quotient by the conjugate of the denominator.

EXAMPLE 5 | **Writing the Quotient of Two Complex Numbers in Standard Form**

Write each quotient in standard form.

(a) $\dfrac{1 + 4i}{5 - 12i}$
(b) $\dfrac{2 - 3i}{4 - 3i}$

Solution (a) $\dfrac{1 + 4i}{5 - 12i} = \dfrac{1 + 4i}{5 - 12i} \cdot \dfrac{5 + 12i}{5 + 12i} = \dfrac{5 + 12i + 20i + 48i^2}{25 + 144}$

$$= \frac{-43 + 32i}{169} = -\frac{43}{169} + \frac{32}{169}i$$

(b) $\dfrac{2 - 3i}{4 - 3i} = \dfrac{2 - 3i}{4 - 3i} \cdot \dfrac{4 + 3i}{4 + 3i} = \dfrac{8 + 6i - 12i - 9i^2}{16 + 9} = \dfrac{17 - 6i}{25} = \dfrac{17}{25} - \dfrac{6}{25}i$

━━ **Now Work** PROBLEM 29

EXAMPLE 6 | **Writing Other Expressions in Standard Form**

If $z = 2 - 3i$ and $w = 5 + 2i$, write each expression in standard form.

(a) $\dfrac{z}{w}$
(b) $\overline{z + w}$
(c) $z + \bar{z}$

(continued)

Solution (a) $\dfrac{z}{w} = \dfrac{z \cdot \overline{w}}{w \cdot \overline{w}} = \dfrac{(2 - 3i)(5 - 2i)}{(5 + 2i)(5 - 2i)} = \dfrac{10 - 4i - 15i + 6i^2}{25 + 4}$

$\qquad\qquad = \dfrac{4 - 19i}{29} = \dfrac{4}{29} - \dfrac{19}{29}i$

(b) $\overline{z + w} = \overline{(2 - 3i) + (5 + 2i)} = \overline{7 - i} = 7 + i$

(c) $z + \overline{z} = (2 - 3i) + (2 + 3i) = 4$

The conjugate of a complex number has certain general properties that will be useful later.

For a real number $a = a + 0i$, the conjugate is $\overline{a} = \overline{a + 0i} = a - 0i = a$.

> **THEOREM**
>
> The conjugate of a real number is the real number itself.

Other properties that are direct consequences of the definition of the conjugate are given next. In each statement, z and w represent complex numbers.

> **THEOREMS**
>
> - The conjugate of the conjugate of a complex number is the complex number itself.
>
> $$\overline{\overline{z}} = z \qquad\qquad (6)$$
>
> - The conjugate of the sum of two complex numbers equals the sum of their conjugates.
>
> $$\overline{z + w} = \overline{z} + \overline{w} \qquad\qquad (7)$$
>
> - The conjugate of the product of two complex numbers equals the product of their conjugates.
>
> $$\overline{z \cdot w} = \overline{z} \cdot \overline{w} \qquad\qquad (8)$$

The proofs of equations (6), (7), and (8) are left as exercises.

Powers of i

The **powers of i** follow a pattern that is useful to know.

$i^1 = i$ $\qquad\qquad\qquad\qquad$ $i^5 = i^4 \cdot i = 1 \cdot i = i$

$i^2 = -1$ $\qquad\qquad\qquad\qquad$ $i^6 = i^4 \cdot i^2 = -1$

$i^3 = i^2 \cdot i = -1 \cdot i = -i$ $\qquad\qquad$ $i^7 = i^4 \cdot i^3 = -i$

$i^4 = i^2 \cdot i^2 = (-1)(-1) = 1$ $\qquad\quad$ $i^8 = i^4 \cdot i^4 = 1$

And so on. The powers of i repeat with every fourth power.

EXAMPLE 7 **Evaluating Powers of i**

(a) $i^{27} = i^{24} \cdot i^3 = (i^4)^6 \cdot i^3 = 1^6 \cdot i^3 = -i$

(b) $i^{101} = i^{100} \cdot i^1 = (i^4)^{25} \cdot i = 1^{25} \cdot i = i$

| EXAMPLE 8 | **Writing the Power of a Complex Number in Standard Form** |

Write $(2 + i)^3$ in standard form.

Solution Use the special product formula for $(x + a)^3$.

$$(x + a)^3 = x^3 + 3ax^2 + 3a^2x + a^3$$

NOTE Another way to find $(2 + i)^3$ is to multiply out $(2 + i)^2(2 + i)$. ■

Using this special product formula,

$$(2 + i)^3 = 2^3 + 3 \cdot i \cdot 2^2 + 3 \cdot i^2 \cdot 2 + i^3$$
$$= 8 + 12i + 6(-1) + (-i)$$
$$= 2 + 11i$$

 Now Work PROBLEMS 35 AND 43

2 Solve Quadratic Equations in the Complex Number System

Quadratic equations with a negative discriminant have no real number solution. However, if we extend our number system to allow complex numbers, quadratic equations always have solutions. Since the solution to a quadratic equation involves the square root of the discriminant, we begin with a discussion of square roots of negative numbers.

> **DEFINITION Principal Square Root of $-N$**
>
> If N is a positive real number, we define the **principal square root of $-N$**, denoted by $\sqrt{-N}$, as
>
> $$\sqrt{-N} = \sqrt{N}\,i$$
>
> where i is the imaginary unit and $i^2 = -1$.

WARNING In writing $\sqrt{-N} = \sqrt{N}\,i$, be sure to place i outside the $\sqrt{}$ symbol. ■

| EXAMPLE 9 | **Evaluating the Square Root of a Negative Number** |

(a) $\sqrt{-1} = \sqrt{1}\,i = i$

(b) $\sqrt{-16} = \sqrt{16}\,i = 4i$

(c) $\sqrt{-8} = \sqrt{8}\,i = 2\sqrt{2}\,i$

 Now Work PROBLEM 51

| EXAMPLE 10 | **Solving Equations** |

Solve each equation in the complex number system.

(a) $x^2 = 4$ (b) $x^2 = -9$

Solution (a) $x^2 = 4$

$x = \pm\sqrt{4} = \pm 2$

The equation has two solutions, -2 and 2. The solution set is $\{-2, 2\}$.

(b) $x^2 = -9$

$x = \pm\sqrt{-9} = \pm\sqrt{9}\,i = \pm 3i$

The equation has two solutions, $-3i$ and $3i$. The solution set is $\{-3i, 3i\}$.

 Now Work PROBLEM 59

WARNING When working with square roots of negative numbers, do not set the square root of a product equal to the product of the square roots (which can be done with positive real numbers). To see why, look at this calculation: We know that $\sqrt{100} = 10$. However, it is also true that $100 = (-25)(-4)$, so

$$10 = \sqrt{100} = \sqrt{(-25)(-4)} = \sqrt{-25}\sqrt{-4} = \sqrt{25}\,i \cdot \sqrt{4}\,i = 5i \cdot 2i = 10i^2 = -10$$

↑
Here is the error. ■

Because we have defined the square root of a negative number, we now restate the quadratic formula without restriction.

THEOREM Quadratic Formula

In the complex number system, the solutions of the quadratic equation $ax^2 + bx + c = 0$, where a, b, and c are real numbers and $a \neq 0$, are given by the formula

$$x = \frac{-b \pm \sqrt{b^2 - 4ac}}{2a} \tag{9}$$

EXAMPLE 11

Solving a Quadratic Equation in the Complex Number System

Solve the equation $x^2 - 4x + 8 = 0$ in the complex number system.

Solution Here $a = 1, b = -4, c = 8$, and $b^2 - 4ac = (-4)^2 - 4 \cdot 1 \cdot 8 = -16$. Using equation (9), we find that

$$x = \frac{-(-4) \pm \sqrt{-16}}{2 \cdot 1} = \frac{4 \pm \sqrt{16}\,i}{2} = \frac{4 \pm 4i}{2} = \frac{2(2 \pm 2i)}{2} = 2 \pm 2i$$

The equation has the solution set $\{2 - 2i, 2 + 2i\}$.

✓ **Check:** $2 + 2i$: $(2 + 2i)^2 - 4(2 + 2i) + 8 = 4 + 8i + 4i^2 - 8 - 8i + 8$
$$= 4 + 4i^2$$
$$= 4 - 4 = 0$$

  $2 - 2i$: $(2 - 2i)^2 - 4(2 - 2i) + 8 = 4 - 8i + 4i^2 - 8 + 8i + 8$
$$= 4 - 4 = 0$$

> **Now Work** PROBLEM 65

The discriminant $b^2 - 4ac$ of a quadratic equation still serves as a way to determine the character of the solutions.

Character of the Solutions of a Quadratic Equation

In the complex number system, consider a quadratic equation $ax^2 + bx + c = 0$, where $a \neq 0$ and a, b, and c are real numbers.

- If $b^2 - 4ac > 0$, the equation has two unequal real solutions.
- If $b^2 - 4ac = 0$, the equation has a repeated real solution, a double root.
- If $b^2 - 4ac < 0$, the equation has two complex solutions that are not real. The solutions are conjugates of each other.

The third conclusion in the display is a consequence of the fact that if $b^2 - 4ac = -N < 0$, then by the quadratic formula, the solutions are

$$x = \frac{-b + \sqrt{b^2 - 4ac}}{2a} = \frac{-b + \sqrt{-N}}{2a} = \frac{-b + \sqrt{N}\,i}{2a} = -\frac{b}{2a} + \frac{\sqrt{N}}{2a}\,i$$

and

$$x = \frac{-b - \sqrt{b^2 - 4ac}}{2a} = \frac{-b - \sqrt{-N}}{2a} = \frac{-b - \sqrt{N}\,i}{2a} = -\frac{b}{2a} - \frac{\sqrt{N}}{2a}\,i$$

which are conjugates of each other.

EXAMPLE 12

Determining the Character of the Solutions of a Quadratic Equation

Without solving, determine the character of the solutions of each equation.

(a) $3x^2 + 4x + 5 = 0$ (b) $2x^2 + 4x + 1 = 0$ (c) $9x^2 - 6x + 1 = 0$

Solution

(a) Here $a = 3$, $b = 4$, and $c = 5$, so $b^2 - 4ac = 16 - 4 \cdot 3 \cdot 5 = -44$. The solutions are two complex numbers that are not real and are conjugates of each other.

(b) Here $a = 2$, $b = 4$, and $c = 1$, so $b^2 - 4ac = 16 - 8 = 8$. The solutions are two unequal real numbers.

(c) Here $a = 9$, $b = -6$, and $c = 1$, so $b^2 - 4ac = 36 - 4 \cdot 9 \cdot 1 = 0$. The solution is a repeated real number—that is, a double root.

Now Work PROBLEM 79

1.3 Assess Your Understanding

'Are You Prepared?' *Answers are given at the end of these exercises. If you get a wrong answer, read the pages listed in* red.

1. Name the integers and the rational numbers in the set $\left\{ -3, 0, \sqrt{2}, \dfrac{6}{5}, \pi \right\}$. (pp. 4–5)

2. *True or False* If a is a real number, and m and n are integers, then $a^m \cdot a^n = a^{m+n}$. (pp. 21–23)

3. Rationalize the denominator of $\dfrac{3}{2 + \sqrt{3}}$. (pp. 74–75)

Concepts and Vocabulary

4. In the complex number $5 + 2i$, the number 5 is called the _____ part; the number 2 is called the _____ part; the number i is called the _____ _____.

5. *True or False* The conjugate of $2 + 5i$ is $-2 - 5i$.

6. *True or False* All real numbers are complex numbers.

7. *True or False* If $2 - 3i$ is a solution of a quadratic equation with real coefficients, then $-2 + 3i$ is also a solution.

8. *Multiple Choice* Which of the following is the principal square root of -4?
   (a) $-2i$ (b) $2i$ (c) -2 (d) 2

9. *Multiple Choice* Which operation involving complex numbers requires the use of a conjugate?
   (a) division (b) multiplication
   (c) subtraction (d) addition

10. *Multiple Choice* Powers of i repeat every _____ power.
   (a) second (b) third (c) fourth (d) fifth

Skill Building

In Problems 11–58, perform the indicated operation, and write each expression in the standard form $a + bi$.

11. $(2 - 3i) + (6 + 8i)$

12. $(4 + 5i) + (-8 + 2i)$

13. $(-3 + 2i) - (4 - 4i)$

14. $(3 - 4i) - (-3 - 4i)$

15. $(2 - 5i) - (8 + 6i)$

16. $(-8 + 4i) - (2 - 2i)$

17. $3(2 - 6i)$

18. $-4(2 + 8i)$

19. $-3i(7 + 6i)$

20. $3i(-3 + 4i)$

21. $(3 - 4i)(2 + i)$

22. $(5 + 3i)(2 - i)$

23. $(-5 - i)(-5 + i)$

24. $(-3 + i)(3 + i)$

25. $\dfrac{10}{3 - 4i}$

26. $\dfrac{13}{5 - 12i}$

27. $\dfrac{2 + i}{i}$

28. $\dfrac{2 - i}{-2i}$

29. $\dfrac{6 - i}{1 + i}$

30. $\dfrac{2 + 3i}{1 - i}$

31. $\left(\dfrac{1}{2} + \dfrac{\sqrt{3}}{2}i \right)^2$

32. $\left(\dfrac{\sqrt{3}}{2} - \dfrac{1}{2}i \right)^2$

33. $(1 + i)^2$

34. $(1 - i)^2$

35. i^{23} **36.** i^{14} **37.** i^{-20} **38.** i^{-23} **39.** $i^6 - 5$

40. $4 + i^3$ **41.** $6i^3 - 4i^5$ **42.** $4i^3 - 2i^2 + 1$ **43.** $(1 + i)^3$ **44.** $(3i)^4 + 1$

45. $i^7(1 + i^2)$ **46.** $2i^4(1 + i^2)$ **47.** $i^8 + i^6 - i^4 - i^2$ **48.** $i^7 + i^5 + i^3 + i$ **49.** $\sqrt{-4}$

50. $\sqrt{-9}$ **51.** $\sqrt{-25}$ **52.** $\sqrt{-64}$ **53.** $\sqrt{-12}$ **54.** $\sqrt{-18}$

55. $\sqrt{-200}$ **56.** $\sqrt{-45}$ **57.** $\sqrt{(3 + 4i)(4i - 3)}$ **58.** $\sqrt{(4 + 3i)(3i - 4)}$

In Problems 59–78, solve each equation in the complex number system.

59. $x^2 + 4 = 0$ **60.** $x^2 - 4 = 0$ **61.** $x^2 - 16 = 0$ **62.** $x^2 + 25 = 0$

63. $x^2 - 6x + 13 = 0$ **64.** $x^2 + 4x + 8 = 0$ **65.** $x^2 - 6x + 10 = 0$ **66.** $x^2 - 2x + 5 = 0$

67. $25x^2 - 10x + 2 = 0$ **68.** $10x^2 + 6x + 1 = 0$ **69.** $5x^2 + 1 = 2x$ **70.** $13x^2 + 1 = 6x$

71. $x^2 + x + 1 = 0$ **72.** $x^2 - x + 1 = 0$ **73.** $x^3 - 64 = 0$ **74.** $x^3 + 27 = 0$

75. $x^4 = 16$ **76.** $x^4 = 1$ **77.** $x^4 + 13x^2 + 36 = 0$ **78.** $x^4 + 3x^2 - 4 = 0$

In Problems 79–84, without solving, determine the character of the solutions of each equation in the complex number system.

79. $3x^2 - 3x + 4 = 0$ **80.** $2x^2 - 4x + 1 = 0$ **81.** $2x^2 + 3x = 4$

82. $x^2 + 6 = 2x$ **83.** $9x^2 - 12x + 4 = 0$ **84.** $4x^2 + 12x + 9 = 0$

85. $2 + 3i$ is a solution of a quadratic equation with real coefficients. Find the other solution.

86. $4 - i$ is a solution of a quadratic equation with real coefficients. Find the other solution.

In Problems 87–90, $z = 3 - 4i$ and $w = 8 + 3i$. Write each expression in the standard form $a + bi$.

87. $z + \bar{z}$ **88.** $w - \bar{w}$ **89.** $z\bar{z}$ **90.** $\overline{z - w}$

Applications and Extensions

91. Electrical Circuits The impedance Z, in ohms, of a circuit element is defined as the ratio of the phasor voltage V, in volts, across the element to the phasor current I, in amperes, through the element. That is, $Z = \dfrac{V}{I}$. If the voltage across a circuit element is $18 + i$ volts and the current through the element is $3 - 4i$ amperes, determine the impedance.

92. Parallel Circuits In an ac circuit with two parallel pathways, the total impedance Z, in ohms, satisfies the formula $\dfrac{1}{Z} = \dfrac{1}{Z_1} + \dfrac{1}{Z_2}$, where Z_1 is the impedance of the first pathway and Z_2 is the impedance of the second pathway. Determine the total impedance if the impedances of the two pathways are $Z_1 = 2 + i$ ohms and $Z_2 = 4 - 3i$ ohms.

93. Use $z = a + bi$ to show that $z + \bar{z} = 2a$ and $z - \bar{z} = 2bi$.

94. Use $z = a + bi$ to show that $\bar{\bar{z}} = z$.

95. Use $z = a + bi$ and $w = c + di$ to show that $\overline{z + w} = \bar{z} + \bar{w}$.

96. Use $z = a + bi$ and $w = c + di$ to show that $\overline{z \cdot w} = \bar{z} \cdot \bar{w}$.

97. Challenge Problem Find a complex number, $a + bi$, if its square equals the opposite of the square of its conjugate.

98. Challenge Problem Find the exact solutions of $x^3 + 2 = 0$.

99. Challenge Problem Find $x - y$ if x and y are real numbers, and $(x + 5)(y - 5) = (x + y)^2$.

Hint: let $u = x + 5$ and $v = y - 5$.

Explaining Concepts: Discussion and Writing

100. Explain to a friend how you would add two complex numbers and how you would multiply two complex numbers. Explain any differences between the two explanations.

101. Write a brief paragraph that compares the method used to rationalize the denominator of a radical expression and the method used to write the quotient of two complex numbers in standard form.

102. Use an Internet search engine to investigate the origins of complex numbers. Write a paragraph describing what you find, and present it to the class.

103. Explain how the method of multiplying two complex numbers is related to multiplying two binomials.

104. What Went Wrong? A student multiplied $\sqrt{-9}$ and $\sqrt{-9}$ as follows:

$$\sqrt{-9} \cdot \sqrt{-9} = \sqrt{(-9)(-9)} = \sqrt{81} = 9$$

The instructor marked the problem incorrect. Why?

'Are You Prepared?' Answers

1. Integers: $\{-3, 0\}$; rational numbers: $\left\{-3, 0, \dfrac{6}{5}\right\}$ **2.** True   **3.** $3(2 - \sqrt{3})$

1.4 Radical Equations; Equations Quadratic in Form; Factorable Equations

PREPARING FOR THIS SECTION *Before getting started, review the following:*

- Square Roots (Section R.2, pp. 23–24)
- Factoring Polynomials (Section R.5, pp. 49–55)
- *n*th Roots; Rational Exponents (Section R.8, pp. 72–77)

Now Work the 'Are You Prepared?' problems on page 117.

OBJECTIVES 1 Solve Radical Equations (p. 113)
2 Solve Equations Quadratic in Form (p. 114)
3 Solve Equations by Factoring (p. 116)

1 Solve Radical Equations

When the variable in an equation occurs in a square root, cube root, and so on—that is, when it occurs in a radical—the equation is called a **radical equation**. Sometimes a suitable operation will change a radical equation to one that is linear or quadratic. A commonly used procedure is to isolate the most complicated radical on one side of the equation and then eliminate it by raising both sides to a power equal to the index of the radical. Care must be taken, however, because apparent solutions that are not, in fact, solutions of the original equation may result. These are called **extraneous solutions**. Therefore, we need to check all answers when working with radical equations, and we check them in the *original* equation.

EXAMPLE 1 **Solving a Radical Equation**

Find the real solutions of the equation: $\sqrt[3]{2x - 4} - 2 = 0$

Solution The equation contains a radical whose index is 3. Isolate it on the left side.

$$\sqrt[3]{2x - 4} - 2 = 0$$
$$\sqrt[3]{2x - 4} = 2 \quad \text{Add 2 to both sides.}$$

Now raise both sides to the third power (the index of the radical is 3) and solve.

$$(\sqrt[3]{2x - 4})^3 = 2^3 \qquad \text{Raise both sides to the power 3.}$$
$$2x - 4 = 8 \qquad \text{Simplify.}$$
$$2x = 12 \qquad \text{Add 4 to both sides.}$$
$$x = 6 \qquad \text{Divide both sides by 2.}$$

✓ **Check:** $\sqrt[3]{2 \cdot 6 - 4} - 2 = \sqrt[3]{12 - 4} - 2 = \sqrt[3]{8} - 2 = 2 - 2 = 0$

The solution set is {6}.

Now Work PROBLEM 9

EXAMPLE 2 **Solving a Radical Equation**

Find the real solutions of the equation: $\sqrt{x - 1} = x - 7$

Solution Square both sides since the index of a square root is 2.

$$\sqrt{x - 1} = x - 7$$
$$(\sqrt{x - 1})^2 = (x - 7)^2 \qquad \text{Square both sides.}$$
$$x - 1 = x^2 - 14x + 49 \qquad \text{Multiply out.}$$
$$x^2 - 15x + 50 = 0 \qquad \text{Put in standard form.}$$

(continued)

$$(x - 10)(x - 5) = 0 \quad \text{Factor.}$$

$$x = 10 \quad \text{or} \quad x = 5 \quad \text{Use the Zero-Product Property and solve.}$$

✔ **Check:**

$$x = 10: \quad \sqrt{x - 1} = \sqrt{10 - 1} = \sqrt{9} = 3 \text{ and } x - 7 = 10 - 7 = 3$$

$$x = 5: \quad \sqrt{x - 1} = \sqrt{5 - 1} = \sqrt{4} = 2 \text{ and } x - 7 = 5 - 7 = -2$$

The solution 5 does not check, so it is extraneous; the only solution of the equation is 10. The solution set is {10}.

➤ **Now Work** PROBLEM 19

Sometimes it is necessary to raise both sides to a power more than once to solve a radical equation.

EXAMPLE 3

Solving a Radical Equation

Find the real solutions of the equation: $\sqrt{2x + 3} - \sqrt{x + 2} = 2$

Solution

First, isolate the more complicated radical expression (in this case, $\sqrt{2x + 3}$) on the left side.

$$\sqrt{2x + 3} = \sqrt{x + 2} + 2$$

Now square both sides (the index of the isolated radical is 2).

$$(\sqrt{2x + 3})^2 = (\sqrt{x + 2} + 2)^2 \qquad \text{Square both sides.}$$

$$2x + 3 = (\sqrt{x + 2})^2 + 4\sqrt{x + 2} + 4 \qquad \text{Multiply out.}$$

$$2x + 3 = x + 2 + 4\sqrt{x + 2} + 4 \qquad \text{Simplify.}$$

$$2x + 3 = x + 6 + 4\sqrt{x + 2} \qquad \text{Combine like terms.}$$

Because the equation still contains a radical, isolate the remaining radical on the right side and again square both sides.

$$x - 3 = 4\sqrt{x + 2} \qquad \text{Isolate the radical on the right side.}$$

$$(x - 3)^2 = 16(x + 2) \qquad \text{Square both sides.}$$

$$x^2 - 6x + 9 = 16x + 32 \qquad \text{Multiply out.}$$

$$x^2 - 22x - 23 = 0 \qquad \text{Put in standard form.}$$

$$(x - 23)(x + 1) = 0 \qquad \text{Factor.}$$

$$x = 23 \quad \text{or} \quad x = -1 \qquad \text{Use the Zero-Product Property and solve.}$$

The original equation appears to have the solution set {−1, 23}. However, we have not yet checked.

✔ **Check:**

$$x = 23: \quad \sqrt{2x + 3} - \sqrt{x + 2} = \sqrt{2 \cdot 23 + 3} - \sqrt{23 + 2} = \sqrt{49} - \sqrt{25} = 7 - 5 = 2$$

$$x = -1: \quad \sqrt{2x + 3} - \sqrt{x + 2} = \sqrt{2(-1) + 3} - \sqrt{-1 + 2} = \sqrt{1} - \sqrt{1} = 1 - 1 = 0$$

The equation has only one solution, 23; the apparent solution −1 is extraneous. The solution set is {23}.

➤ **Now Work** PROBLEM 31

2 Solve Equations Quadratic in Form

The equation $x^4 + x^2 - 12 = 0$ is not quadratic in x, but it is quadratic in x^2. That is, if we let $u = x^2$, we get $u^2 + u - 12 = 0$, a quadratic equation. This equation can be solved for u, and in turn, by using $u = x^2$, we can find the solutions x of the original equation.

In general, if an appropriate substitution u transforms an equation into one of the form

$$au^2 + bu + c = 0 \qquad a \neq 0$$

then the original equation is called an **equation of the quadratic type** or an **equation quadratic in form**.

The difficulty of solving such an equation lies in the determination that the equation is, in fact, quadratic in form. After you are told an equation is quadratic in form, it is easy enough to see it, but some practice is needed to enable you to recognize such equations on your own.

EXAMPLE 4

Solving an Equation Quadratic in Form

Find the real solutions of the equation: $(x + 2)^2 + 11(x + 2) - 12 = 0$

Solution   For this equation, let $u = x + 2$. Then $u^2 = (x + 2)^2$, and the original equation,

$$(x + 2)^2 + 11(x + 2) - 12 = 0$$

becomes

$$u^2 + 11u - 12 = 0 \quad \text{Let } u = x + 2. \text{ Then } u^2 = (x + 2)^2.$$
$$(u + 12)(u - 1) = 0 \quad \text{Factor.}$$
$$u = -12 \quad \text{or} \quad u = 1 \quad \text{Solve.}$$

WARNING Do not stop after finding values for u. Remember to finish solving for the original variable. ∎

But we want to solve for x. Because $u = x + 2$, we have

$$x + 2 = -12 \quad \text{or} \quad x + 2 = 1$$
$$x = -14 \qquad\qquad x = -1$$

✓ **Check:** $x = -14$: $(-14 + 2)^2 + 11(-14 + 2) - 12$
$$= (-12)^2 + 11(-12) - 12 = 144 - 132 - 12 = 0$$
$x = -1$: $(-1 + 2)^2 + 11(-1 + 2) - 12 = 1 + 11 - 12 = 0$

The original equation has the solution set $\{-14, -1\}$.

EXAMPLE 5

Solving an Equation Quadratic in Form

Find the real solutions of the equation: $(x^2 - 1)^2 + (x^2 - 1) - 12 = 0$

Solution   Let $u = x^2 - 1$ so that $u^2 = (x^2 - 1)^2$. Then the original equation,

$$(x^2 - 1)^2 + (x^2 - 1) - 12 = 0$$

becomes

$$u^2 + u - 12 = 0 \quad \text{Let } u = x^2 - 1. \text{ Then } u^2 = (x^2 - 1)^2.$$
$$(u + 4)(u - 3) = 0 \quad \text{Factor.}$$
$$u = -4 \quad \text{or} \quad u = 3 \quad \text{Solve.}$$

But remember that we want to solve for x. Because $u = x^2 - 1$, we have

$$x^2 - 1 = -4 \quad \text{or} \quad x^2 - 1 = 3$$
$$x^2 = -3 \qquad\qquad x^2 = 4$$

The first of these has no real solution; the second has the solution set $\{-2, 2\}$.

✓ **Check:** $x = -2$: $((-2)^2 - 1)^2 + ((-2)^2 - 1) - 12 = 9 + 3 - 12 = 0$
$x = 2$: $(2^2 - 1)^2 + (2^2 - 1) - 12 = 9 + 3 - 12 = 0$

The original equation has the solution set $\{-2, 2\}$.

EXAMPLE 6 **Solving an Equation Quadratic in Form**

Find the real solutions of the equation: $x + 2\sqrt{x} - 3 = 0$

Solution For the equation $x + 2\sqrt{x} - 3 = 0$, let $u = \sqrt{x}$. Then $u^2 = x$, and the original equation,

$$x + 2\sqrt{x} - 3 = 0$$

becomes

$$u^2 + 2u - 3 = 0 \quad \text{Let } u = \sqrt{x}. \text{ Then } u^2 = x.$$
$$(u + 3)(u - 1) = 0 \quad \text{Factor.}$$
$$u = -3 \quad \text{or} \quad u = 1 \quad \text{Solve.}$$

Since $u = \sqrt{x}$, we have $\sqrt{x} = -3$ or $\sqrt{x} = 1$. The first of these, $\sqrt{x} = -3$, has no real solution, since the principal square root of a real number is never negative. The second, $\sqrt{x} = 1$, has the solution $x = 1$.

✓ **Check:** $1 + 2\sqrt{1} - 3 = 1 + 2 - 3 = 0$

The original equation has the solution set {1}.

━━━━━ ANOTHER METHOD FOR SOLVING EXAMPLE 6 WOULD BE TO TREAT IT AS A RADICAL EQUATION. SOLVE IT THIS WAY FOR PRACTICE.

If an equation contains an expression and that same expression squared, make a substitution for the expression. You may get a quadratic equation.

━━━━━ **Now Work** PROBLEM 53

3 Solve Equations by Factoring

We have already solved certain quadratic equations using factoring. Let's look at examples of other kinds of equations that can be solved by factoring.

EXAMPLE 7 **Solving an Equation by Factoring**

Solve the equation: $x^4 = 4x^2$

Solution Begin by collecting all terms on one side. This results in 0 on one side and an expression to be factored on the other.

$$x^4 = 4x^2$$
$$x^4 - 4x^2 = 0$$
$$x^2(x^2 - 4) = 0 \quad \text{Factor.}$$
$$x^2 = 0 \quad \text{or} \quad x^2 - 4 = 0 \quad \text{Use the Zero-Product Property.}$$
$$x = 0 \quad \text{or} \quad x^2 = 4$$
$$x = 0 \quad \text{or} \quad x = -2 \quad \text{or} \quad x = 2 \quad \text{Use the Square Root Method.}$$

✓ **Check:** $x = -2$: $(-2)^4 = 16$ and $4(-2)^2 = 16$ $\quad -2$ is a solution.
$\quad\quad\quad\quad x = 0$: $\quad\quad 0^4 = 0$ and $4 \cdot 0^2 = 0$ $\quad\quad$ 0 is a solution.
$\quad\quad\quad\quad x = 2$: $\quad\quad 2^4 = 16$ and $4 \cdot 2^2 = 16$ $\quad$ 2 is a solution.

The solution set is $\{-2, 0, 2\}$.

EXAMPLE 8	Solving an Equation by Factoring

Solve the equation: $x^3 - x^2 - 4x + 4 = 0$

Solution

Do you recall the method of factoring by grouping? Group the terms of $x^3 - x^2 - 4x + 4 = 0$ as follows:

$$(x^3 - x^2) - (4x - 4) = 0$$

Factor x^2 from the first grouping and 4 from the second.

$$x^2(x - 1) - 4(x - 1) = 0$$

> **Need to Review?**
> *Factoring by Grouping* is discussed in Section R.5, p. 53.

This reveals the common factor $(x - 1)$, so we have

$$(x^2 - 4)(x - 1) = 0$$
$$(x - 2)(x + 2)(x - 1) = 0 \qquad \text{Factor again.}$$
$$x - 2 = 0 \quad \text{or} \quad x + 2 = 0 \quad \text{or} \quad x - 1 = 0 \quad \text{Use the Zero-Product Property.}$$
$$x = 2 \qquad\qquad x = -2 \qquad\qquad x = 1 \quad \text{Solve.}$$

✓ **Check:**

$x = -2$: $\quad (-2)^3 - (-2)^2 - 4(-2) + 4 = -8 - 4 + 8 + 4 = 0 \quad -2$ **is a solution.**

$x = 1$: $\qquad 1^3 - 1^2 - 4 \cdot 1 + 4 = 1 - 1 - 4 + 4 = 0 \qquad\qquad 1$ **is a solution.**

$x = 2$: $\qquad 2^3 - 2^2 - 4 \cdot 2 + 4 = 8 - 4 - 8 + 4 = 0 \qquad\qquad 2$ **is a solution.**

The solution set is $\{-2, 1, 2\}$.

──── **Now Work** PROBLEM 81

1.4 Assess Your Understanding

'Are You Prepared?' *Answers are given at the end of these exercises. If you get a wrong answer, read the pages listed in* red.

1. ***True or False*** The principal square root of any nonnegative real number is always nonnegative. (pp. 23–24)

2. $\left(\sqrt[3]{x}\right)^3 = $ _____. (p. 73)

3. Factor $6x^3 - 2x^2$ (pp. 49–55)

Concepts and Vocabulary

4. ***True or False*** Factoring can only be used to solve quadratic equations or equations that are quadratic in form.

5. If u is an expression that involves x, then the equation $au^2 + bu + c = 0, a \neq 0$, is called an equation _____ _____ _____.

6. ***True or False*** Radical equations sometimes have extraneous solutions.

7. ***Multiple Choice*** An apparent solution that does not satisfy the original equation is called a(n) _____ solution.

   (a) extraneous   **(b)** radical
   (c) imaginary   **(d)** conditional

8. ***Multiple Choice*** Which equation is likely to require squaring both sides more than once?

   (a) $\sqrt{x + 2} = \sqrt{3x - 5}$ **(b)** $x^4 - 3x^2 = 10$
   (c) $\sqrt{x + 1} + \sqrt{x - 4} = 8$ **(d)** $\sqrt{3x + 1} = 5$

Skill Building

In Problems 9–42, find the real solutions, if any, of each equation.

9. $\sqrt{2t - 1} = 1$

10. $\sqrt{3t + 4} = 2$

11. $\sqrt{3t + 4} = -6$

12. $\sqrt{5t + 3} = -2$

13. $\sqrt[3]{1 - 2x} - 3 = 0$

14. $\sqrt[3]{1 - 2x} - 1 = 0$

15. $\sqrt[5]{x^2 + 2x} = -1$

16. $\sqrt[4]{x^2 + 16} = \sqrt{5}$

17. $x = 8\sqrt{x}$

18. $x = 3\sqrt{x}$

19. $\sqrt{15 - 2x} = x$

20. $\sqrt{12 - x} = x$

21. $x = 2\sqrt{x - 1}$

22. $x = 2\sqrt{-x - 1}$

23. $\sqrt{x^2 - x - 4} = x + 2$

24. $\sqrt{3 - x + x^2} = x - 2$

25. $3 + \sqrt{3x + 1} = x$

26. $2 + \sqrt{12 - 2x} = x$

27. $\sqrt{3(x + 10)} - 4 = x$

28. $\sqrt{1 - x} - 3 = x + 2$

29. $\sqrt{3x - 5} - \sqrt{x + 7} = 2$

30. $\sqrt{3x + 7} + \sqrt{x + 2} = 1$

31. $\sqrt{3x + 1} - \sqrt{x - 1} = 2$

32. $\sqrt{2x + 3} - \sqrt{x + 1} = 1$

33. $\sqrt{3 - 2\sqrt{x}} = \sqrt{x}$

34. $\sqrt{10 + 3\sqrt{x}} = \sqrt{x}$

35. $(3x + 1)^{1/2} = 4$

36. $(3x - 5)^{1/2} = 2$

37. $(5x - 2)^{1/3} = 2$

38. $(2x + 1)^{1/3} = -1$

39. $(x^2 + 9)^{1/2} = 5$

40. $(x^2 - 16)^{1/2} = 9$

41. $x^{3/2} - 3x^{1/2} = 0$

42. $x^{3/4} - 9x^{1/4} = 0$

In Problems 43–74, find the real solutions of each equation.

43. $x^4 - 5x^2 + 4 = 0$

44. $x^4 - 10x^2 + 25 = 0$

45. $6x^4 - 5x^2 - 1 = 0$

46. $2x^4 - 5x^2 - 12 = 0$

47. $x^6 + 7x^3 - 8 = 0$

48. $x^6 - 7x^3 - 8 = 0$

49. $(x + 2)^2 + 7(x + 2) + 12 = 0$

50. $(2x + 5)^2 - (2x + 5) - 6 = 0$

51. $(4x - 9)^2 - 10(4x - 9) + 25 = 0$

52. $(2 - x)^2 + (2 - x) - 20 = 0$

53. $2(s + 1)^2 - 5(s + 1) = 3$

54. $3(1 - y)^2 + 5(1 - y) + 2 = 0$

55. $x - 4x\sqrt{x} = 0$

56. $x + 8\sqrt{x} = 0$

57. $x + \sqrt{x} = 20$

58. $x + \sqrt{x} = 6$

59. $t^{1/2} - 2t^{1/4} + 1 = 0$

60. $z^{1/2} - 4z^{1/4} + 4 = 0$

61. $x^{1/2} - 3x^{1/4} + 2 = 0$

62. $4x^{1/2} - 9x^{1/4} + 4 = 0$

63. $\sqrt[3]{5x^2 - 6} = x$

64. $\sqrt[4]{4 - 5x^2} = x$

65. $x^2 + 3x + \sqrt{x^2 + 3x} = 6$

66. $x^2 - 3x - \sqrt{x^2 - 3x} = 2$

67. $\dfrac{1}{(x + 1)^2} = \dfrac{1}{x + 1} + 2$

68. $\dfrac{1}{(x - 1)^2} + \dfrac{1}{x - 1} = 12$

69. $3x^{-2} - 7x^{-1} - 6 = 0$

70. $2x^{-2} - 3x^{-1} - 4 = 0$

71. $2x^{2/3} - 5x^{1/3} - 3 = 0$

72. $3x^{4/3} + 5x^{2/3} - 2 = 0$

73. $\left(\dfrac{v}{v + 2}\right)^2 + \dfrac{3v}{v + 2} = 10$

74. $\left(\dfrac{y}{y - 1}\right)^2 = 6\left(\dfrac{y}{y - 1}\right) + 16$

In Problems 75–90, find the real solutions of each equation by factoring.

75. $x^3 - 9x = 0$

76. $x^4 - x^2 = 0$

77. $4x^3 = 3x^2$

78. $x^5 = 4x^3$

79. $x^3 + x^2 - 20x = 0$

80. $x^3 + 6x^2 - 7x = 0$

81. $x^3 + x^2 - x - 1 = 0$

82. $x^3 + 4x^2 - x - 4 = 0$

83. $x^3 - 3x^2 - 16x + 48 = 0$

84. $x^3 - 3x^2 - x + 3 = 0$

85. $2x^3 + 4 = x^2 + 8x$

86. $3x^3 + 4x^2 = 27x + 36$

87. $3x^3 + 12x = 7x^2 + 28$

88. $3x^3 + 12x = 5x^2 + 20$

89. $x(x^2 - 3x)^{1/3} + 2(x^2 - 3x)^{4/3} = 0$

90. $3x(x^2 + 2x)^{1/2} - 2(x^2 + 2x)^{3/2} = 0$

In Problems 91–96, find the real solutions of each equation. Use a calculator to express any solutions rounded to two decimal places.

91. $x - 4x^{1/2} + 2 = 0$

92. $x^{2/3} + 4x^{1/3} + 2 = 0$

93. $x^4 + \sqrt{3}x^2 - 3 = 0$

94. $x^4 + \sqrt{2}x^2 - 2 = 0$

95. $\pi(1 + t)^2 = \pi + 1 + t$

96. $\pi(1 + r)^2 = 2 + \pi(1 + r)$

Mixed Practice *In Problems 97–110, find the real solutions, if any, of each equation.*

97. $3x^2 + 7x - 20 = 0$

98. $2x^2 - 13x + 21 = 0$

99. $5a^3 - 45a = -2a^2 + 18$

100. $3z^3 - 12z = -5z^2 + 20$

101. $4(w - 3) = w + 3$

102. $6(k + 3) - 2k = 12$

103. $\left(\dfrac{v}{v + 1}\right)^2 + \dfrac{2v}{v + 1} = 8$

104. $\left(\dfrac{y}{y - 1}\right)^2 = \dfrac{6y}{y - 1} + 7$

105. $\sqrt{2x + 5} - x = 1$

106. $\sqrt{3x + 1} - 2x = -6$

107. $3m^2 + 6m = -1$

108. $4y^2 - 8y = 3$

109. $\sqrt[4]{5x^2 - 6} = x$

110. $\sqrt[4]{4 - 3x^2} = x$

111. If $k = \dfrac{x + 3}{x - 3}$ and $k^2 - k = 12$, find x.

112. If $k = \dfrac{x + 3}{x - 4}$ and $k^2 - 3k = 28$, find x.

Applications

113. Physics: Using Sound to Measure Distance The distance to the surface of the water in a well can sometimes be found by dropping an object into the well and measuring the time elapsed until a sound is heard. If t_1 is the time (measured in seconds) that it takes for the object to strike the water, then t_1 will obey the equation $s = 16t_1^2$, where s is the distance (measured in feet). It follows that $t_1 = \dfrac{\sqrt{s}}{4}$. Suppose that t_2 is the time that it takes for the sound of the impact to reach your ears. Because sound waves are known to travel at a speed of approximately 1100 feet per second, the time t_2 to travel the distance s will be $t_2 = \dfrac{s}{1100}$. See the illustration.

Now $t_1 + t_2$ is the total time that elapses from the moment that the object is dropped to the moment that a sound is heard. We have the equation

$$\text{Total time elapsed} = \frac{\sqrt{s}}{4} + \frac{s}{1100}$$

Find the distance to the water's surface if the total time elapsed from dropping a rock to hearing it hit water is 4 seconds.

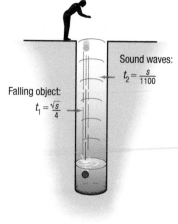

Sound waves:
$t_2 = \frac{s}{1100}$

Falling object:
$t_1 = \frac{\sqrt{s}}{4}$

114. **Crushing Load** A civil engineer relates the thickness T, in inches, and height H, in feet, of a square wooden pillar to its crushing load L, in tons, using the model $T = \sqrt[4]{\dfrac{LH^2}{25}}$. If a square wooden pillar is 4 inches thick and 10 feet high, what is its crushing load?

115. **Foucault's Pendulum** The period of a pendulum is the time it takes the pendulum to make one full swing back and forth. The period T, in seconds, is given by the formula $T = 2\pi\sqrt{\dfrac{l}{32}}$ where l is the length, in feet, of the pendulum. In 1851, Jean Bernard Leon Foucault win Paris. The period of Foucault's pendulum was approximately 16.5 seconds. What was its length?

116. **Challenge Problem** Solve:
$$\sqrt{3x + 5} - \sqrt{x - 2} = \sqrt{x + 3}$$

117. **Challenge Problem** Solve:
$$\sqrt[4]{\sqrt[3]{\sqrt{x - 7}} - 10} + 18 = 2$$

118. **Challenge Problem** Find all the real solutions of $12x^{7/5} + 3x^{2/5} = 13x^{9/10}$.

119. **Challenge Problem** Find all the complex solutions of $z^6 + 28z^3 + 27 = 0$.

Explaining Concepts: Discussion and Writing

120. Make up a radical equation that has no solution.

121. Make up a radical equation that has an extraneous solution.

122. Discuss the step in the solving process for radical equations that leads to the possibility of extraneous solutions. Why is there no such possibility for linear and quadratic equations?

123. **What Went Wrong?** On an exam, Jane solved the equation $\sqrt{2x + 3} - x = 0$ and wrote that the solution set was $\{-1, 3\}$. Jane received 3 out of 5 points for the problem. Jane asks you why she received 3 out of 5 points. Provide an explanation.

'Are You Prepared?' Answers

1. True   2. x 3. $2x^2(3x - 1)$

1.5 Solving Inequalities

PREPARING FOR THIS SECTION *Before getting started, review the following:*

- Graphing Inequalities (Section R.2, pp. 17–19)

Now Work the 'Are You Prepared?' problems on page 127.

OBJECTIVES   **1** Use Interval Notation (p. 120)
2 Use Properties of Inequalities (p. 121)
3 Solve Inequalities (p. 123)
4 Solve Combined Inequalities (p. 124)

Suppose that a and b are two real numbers and $a < b$. The notation $a < x < b$ means that x is a number *between* a and b. The expression $a < x < b$ is equivalent to the two inequalities $a < x$ and $x < b$. Similarly, the expression $a \le x \le b$ is equivalent to the two inequalities $a \le x$ and $x \le b$. The remaining two possibilities, $a \le x < b$ and $a < x \le b$, are defined similarly.

Although it is acceptable to write $3 \ge x \ge 2$, it is preferable to reverse the inequality symbols and write instead $2 \le x \le 3$ so that the values go from smaller to larger, reading from left to right.

A statement such as $2 \le x \le 1$ is false because there is no number x for which $2 \le x$ and $x \le 1$. Finally, never mix inequality symbols, as in $2 \le x \ge 3$.

1 Use Interval Notation

Let a and b represent two real numbers with $a < b$.

> **DEFINITION** Intervals
>
> - An **open interval**, denoted by **(a, b)**, consists of all real numbers x for which $a < x < b$.
> - A **closed interval**, denoted by **$[a, b]$**, consists of all real numbers x for which $a \le x \le b$.
> - The **half-open**, or **half-closed**, **intervals** are **$(a, b]$**, consisting of all real numbers x for which $a < x \le b$, and **$[a, b)$**, consisting of all real numbers x for which $a \le x < b$.

In Words
The notation $[a, b]$ means all real numbers between a and b, inclusive. The notation (a, b) means all real numbers between a and b, not including a and b.

In each of these definitions, a is called the **left endpoint** and b the **right endpoint** of the interval.

The symbol ∞ (read as "infinity") is not a real number, but notation used to indicate unboundedness in the positive direction. The symbol $-\infty$ (read as "negative infinity") also is not a real number, but notation used to indicate unboundedness in the negative direction. The symbols ∞ and $-\infty$ are used to define five other kinds of intervals:

> $[a, \infty)$ consists of all real numbers x for which $x \ge a$.
> (a, ∞) consists of all real numbers x for which $x > a$.
> $(-\infty, a]$ consists of all real numbers x for which $x \le a$.
> $(-\infty, a)$ consists of all real numbers x for which $x < a$.
> $(-\infty, \infty)$ consists of all real numbers.

In Words
An interval is a nonempty set of real numbers.

Note that ∞ and $-\infty$ are never included as endpoints, since neither is a real number.

Table 1 summarizes interval notation, corresponding inequality notation, and their graphs.

Table 1

Interval	Inequality	Graph
The open interval (a, b)	$a < x < b$	
The closed interval $[a, b]$	$a \le x \le b$	
The half-open interval $[a, b)$	$a \le x < b$	
The half-open interval $(a, b]$	$a < x \le b$	
The interval $[a, \infty)$	$x \ge a$	
The interval (a, ∞)	$x > a$	
The interval $(-\infty, a]$	$x \le a$	
The interval $(-\infty, a)$	$x < a$	
The interval $(-\infty, \infty)$	All real numbers	

EXAMPLE 1   Writing Inequalities Using Interval Notation

Write each inequality using interval notation.

(a) $1 \le x \le 3$ (b) $-4 < x < 0$ (c) $x > 5$ (d) $x \le 1$

Solution
(a) $1 \leq x \leq 3$ describes all real numbers x between 1 and 3, inclusive. In interval notation, we write $[1, 3]$.
(b) In interval notation, $-4 < x < 0$ is written $(-4, 0)$.
(c) In interval notation, $x > 5$ is written $(5, \infty)$.
(d) In interval notation, $x \leq 1$ is written $(-\infty, 1]$.

EXAMPLE 2 **Writing Intervals Using Inequality Notation**

Write each interval as an inequality involving x.

(a) $[1, 4)$ (b) $(2, \infty)$ (c) $[2, 3]$ (d) $(-\infty, -3]$

Solution
(a) $[1, 4)$ consists of all real numbers x for which $1 \leq x < 4$.
(b) $(2, \infty)$ consists of all real numbers x for which $x > 2$.
(c) $[2, 3]$ consists of all real numbers x for which $2 \leq x \leq 3$.
(d) $(-\infty, -3]$ consists of all real numbers x for which $x \leq -3$.

Now Work PROBLEMS 13, 25, AND 33

2 Use Properties of Inequalities

The product of two positive real numbers is positive, the product of two negative real numbers is positive, and the product of 0 and 0 is 0. For any real number a, the value of a^2 is 0 or positive; that is, a^2 is nonnegative. This is called the **nonnegative property**.

In Words
The square of a real number is never negative.

> **Nonnegative Property**
> For any real number a,
>
> $$a^2 \geq 0 \qquad\qquad (1)$$

When the same number is added to both sides of an inequality, an equivalent inequality is obtained. For example, since $3 < 5$, then $3 + 4 < 5 + 4$ or $7 < 9$. This is called the **addition property** of inequalities.

In Words
The addition property states that the sense, or direction, of an inequality remains unchanged if the same number is added to both sides.

> **Addition Property of Inequalities**
> For real numbers a, b, and c,
>
> - If $a < b$, then $a + c < b + c$. $\qquad\qquad$ **(2a)**
> - If $a > b$, then $a + c > b + c$. $\qquad\qquad$ **(2b)**

Figure 2 illustrates the addition property (2a). In Figure 2(a), we see that a lies to the left of b. If c is positive, then $a + c$ and $b + c$ lie c units to the right of a and c units to the right of b, respectively. Consequently, $a + c$ must lie to the left of $b + c$; that is, $a + c < b + c$. Figure 2(b) illustrates the situation if c is negative.

(a) If $a < b$ and $c > 0$, then $a + c < b + c$.

(b) If $a < b$ and $c < 0$, then $a + c < b + c$.

Figure 2 Addition property of inequalities

DRAW AN ILLUSTRATION SIMILAR TO FIGURE 2 THAT ILLUSTRATES THE ADDITION PROPERTY (2b).

EXAMPLE 3 **Addition Property of Inequalities**

(a) If $x < -5$, then $x + 5 < -5 + 5$ or $x + 5 < 0$.

(b) If $x > 2$, then $x + (-2) > 2 + (-2)$ or $x - 2 > 0$.

━━━ **Now Work** PROBLEM 41

Now let's see what happens when both sides of an inequality are multiplied by a nonzero number.

EXAMPLE 4 **Multiplying an Inequality by a Positive Number**

Express as an inequality the result of multiplying both sides of the inequality $3 < 7$ by 2.

Solution Begin with

$$3 < 7$$

Multiplying both sides by 2 yields the numbers 6 and 14, so we have

$$6 < 14$$

EXAMPLE 5 **Multiplying an Inequality by a Negative Number**

Express as an inequality the result of multiplying both sides of the inequality $9 > 2$ by -4.

Solution Begin with

$$9 > 2$$

Multiplying both sides by -4 yields the numbers -36 and -8, so we have

$$-36 < -8$$

In Words
Multiplying by a negative number reverses the inequality.

Note that the effect of multiplying both sides of $9 > 2$ by the negative number -4 is that the direction of the inequality symbol is reversed.

Examples 4 and 5 illustrate the following general **multiplication properties** for inequalities:

In Words
The multiplication properties state that the sense, or direction, of an inequality *remains the same* if both sides are multiplied by a *positive* real number, whereas the direction is *reversed* if both sides are multiplied by a *negative* real number.

Multiplication Properties for Inequalities
For real numbers a, b, and c,

- If $a < b$ and if $c > 0$, then $ac < bc$.
- If $a < b$ and if $c < 0$, then $ac > bc$. **(3a)**

- If $a > b$ and if $c > 0$, then $ac > bc$.
- If $a > b$ and if $c < 0$, then $ac < bc$. **(3b)**

EXAMPLE 6 **Multiplication Property of Inequalities**

(a) If $2x < 6$, then $\dfrac{1}{2} \cdot 2x < \dfrac{1}{2} \cdot 6$ or $x < 3$.

(b) If $\dfrac{x}{-3} > 12$, then $-3\left(\dfrac{x}{-3}\right) < -3 \cdot 12$ or $x < -36$.

(c) If $-4x < -8$, then $\dfrac{-4x}{-4} > \dfrac{-8}{-4}$ or $x > 2$.

(d) If $-x > 8$, then $(-1)(-x) < (-1) \cdot 8$ or $x < -8$.

━━━ **Now Work** PROBLEM 47

Reciprocal Properties for Inequalities

- If $a > 0$, then $\dfrac{1}{a} > 0$. **(4a)**

- If $a < 0$, then $\dfrac{1}{a} < 0$. **(4b)**

- If $b > a > 0$, then $\dfrac{1}{a} > \dfrac{1}{b} > 0$. **(4c)**

- If $a < b < 0$, then $\dfrac{1}{b} < \dfrac{1}{a} < 0$. **(4d)**

---

3 Solve Inequalities

An **inequality in one variable** is a statement involving two expressions, at least one containing the variable, separated by one of the inequality symbols: $<$, $\leq$, $>$, or $\geq$. To **solve an inequality** means to find all values of the variable for which the statement is true. These values are called **solutions** of the inequality.

For example, the following are all inequalities involving one variable x:

$$x + 5 < 8 \qquad 2x - 3 \geq 4 \qquad x^2 - 1 \leq 3 \qquad \frac{x + 1}{x - 2} > 0$$

As with equations, one method for solving an inequality is to replace it by a series of equivalent inequalities until an inequality with an obvious solution, such as $x < 3$, is obtained. Equivalent inequalities are obtained by applying some of the same properties that are used to find equivalent equations. The addition property and the multiplication properties for inequalities form the basis for the following procedures.

Procedures That Leave the Inequality Symbol Unchanged

- Simplify both sides of the inequality by combining like terms and eliminating parentheses:

  Replace $\quad x + 2 + 6 > 2x + 5(x + 1)$

  by $\qquad x + 8 > 7x + 5$

- Add or subtract the same expression on both sides of the inequality:

  Replace $\qquad 3x - 5 < 4$

  by $\quad (3x - 5) + 5 < 4 + 5$

- Multiply or divide both sides of the inequality by the same *positive* expression:

  Replace $\quad 4x > 16 \quad$ by $\quad \dfrac{4x}{4} > \dfrac{16}{4}$

Procedures That Reverse the Sense or Direction of the Inequality Symbol

- Interchange the two sides of the inequality:

  Replace $\quad 3 < x \quad$ by $\quad x > 3$

- Multiply or divide both sides of the inequality by the same *negative* expression:

  Replace $\quad -2x > 6 \quad$ by $\quad \dfrac{-2x}{-2} < \dfrac{6}{-2}$

As the examples that follow illustrate, we solve inequalities using many of the same steps that are used to solve equations. In writing the solution of an inequality, either set notation or interval notation may be used, whichever is more convenient.

EXAMPLE 7 **Solving an Inequality**

Solve the inequality $3 - 2x < 5$, and graph the solution set.

Solution

$$3 - 2x < 5$$

$$3 - 2x - 3 < 5 - 3 \quad \text{Subtract 3 from both sides.}$$

$$-2x < 2 \quad \text{Simplify.}$$

$$\frac{-2x}{-2} > \frac{2}{-2} \quad \begin{array}{l}\text{Divide both sides by } -2. \text{ (The sense of}\\ \text{the inequality symbol is reversed.)}\end{array}$$

$$x > -1 \quad \text{Simplify.}$$

Figure 3 $x > -1$

The solution set is $\{x \mid x > -1\}$ or, using interval notation, all numbers in the interval $(-1, \infty)$. See Figure 3 for the graph.

EXAMPLE 8 **Solving an Inequality**

Solve the inequality $4x + 7 \geq 2x - 3$, and graph the solution set.

Solution

$$4x + 7 \geq 2x - 3$$

$$4x + 7 - 7 \geq 2x - 3 - 7 \quad \text{Subtract 7 from both sides.}$$

$$4x \geq 2x - 10 \quad \text{Simplify.}$$

$$4x - 2x \geq 2x - 10 - 2x \quad \text{Subtract 2x from both sides.}$$

$$2x \geq -10 \quad \text{Simplify.}$$

$$\frac{2x}{2} \geq \frac{-10}{2} \quad \begin{array}{l}\text{Divide both sides by 2. (The direction}\\ \text{of the inequality symbol is unchanged.)}\end{array}$$

$$x \geq -5 \quad \text{Simplify.}$$

Figure 4 $x \geq -5$

The solution set is $\{x \mid x \geq -5\}$ or, using interval notation, all numbers in the interval $[-5, \infty)$. See Figure 4 for the graph.

Now Work PROBLEM 59

---

4 Solve Combined Inequalities

EXAMPLE 9 **Solving a Combined Inequality**

Solve the inequality $-5 < 3x - 2 < 1$, and graph the solution set.

Solution Recall that the inequality

$$-5 < 3x - 2 < 1$$

is equivalent to the two inequalities

$$-5 < 3x - 2 \quad \text{and} \quad 3x - 2 < 1$$

Solve each of these inequalities separately.

$$-5 < 3x - 2 \qquad\qquad\qquad\qquad 3x - 2 < 1$$

$$-5 + 2 < 3x - 2 + 2 \quad \text{Add 2 to both sides.} \qquad 3x - 2 + 2 < 1 + 2$$

$$-3 < 3x \qquad\qquad \text{Simplify.} \qquad\qquad\qquad 3x < 3$$

$$\frac{-3}{3} < \frac{3x}{3} \qquad\qquad \text{Divide both sides by 3.} \qquad\qquad \frac{3x}{3} < \frac{3}{3}$$

$$-1 < x \qquad\qquad \text{Simplify.} \qquad\qquad\qquad\qquad x < 1$$

The solution set of the original pair of inequalities consists of all x for which

$$-1 < x \quad \text{and} \quad x < 1$$

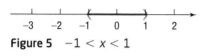

Figure 5 $-1 < x < 1$

This may be written more compactly as $\{x \mid -1 < x < 1\}$. In interval notation, the solution is $(-1, 1)$. See Figure 5 for the graph.

Observe in Example 9 above that solving each of the two inequalities required exactly the same steps. A shortcut to solving the original inequality algebraically is to deal with the two inequalities at the same time, as follows:

$$-5 < \quad 3x - 2 \quad < 1$$

$$-5 + 2 < \ 3x - 2 + 2 \ < 1 + 2 \quad \text{Add 2 to each part.}$$

$$-3 < \qquad 3x \qquad < 3 \qquad \text{Simplify.}$$

$$\frac{-3}{3} < \qquad \frac{3x}{3} \qquad < \frac{3}{3} \qquad \text{Divide each part by 3.}$$

$$-1 < \qquad x \qquad < 1 \qquad \text{Simplify.}$$

EXAMPLE 10	**Solving a Combined Inequality**

Solve the inequality $-1 \le \dfrac{3 - 5x}{2} \le 9$, and graph the solution set.

Solution

$$-1 \le \quad \frac{3 - 5x}{2} \quad \le 9$$

$$2(-1) \le \ 2 \cdot \frac{3 - 5x}{2} \ \le 2 \cdot 9 \qquad \text{Multiply each part by 2 to remove the denominator.}$$

$$-2 \le \quad 3 - 5x \quad \le 18 \qquad \text{Simplify.}$$

$$-2 - 3 \le 3 - 5x - 3 \le 18 - 3 \qquad \text{Subtract 3 from each part to isolate the term containing } x.$$

$$-5 \le \quad -5x \quad \le 15 \qquad \text{Simplify.}$$

$$\frac{-5}{-5} \ge \quad \frac{-5x}{-5} \quad \ge \frac{15}{-5} \qquad \text{Divide each part by } -5 \text{ (reverse the direction of each inequality symbol).}$$

$$1 \ge \quad x \quad \ge -3 \qquad \text{Simplify.}$$

$$-3 \le \quad x \quad \le 1 \qquad \text{Reverse the order so that the numbers get larger as you read from left to right.}$$

Figure 6 $-3 \le x \le 1$

The solution set is $\{x \mid -3 \le x \le 1\}$, that is, all real numbers in the interval $[-3, 1]$. Figure 6 illustrates the graph.

- **Now Work** PROBLEM 79

EXAMPLE 11   **Using a Reciprocal Property to Solve an Inequality**

Solve the inequality $(4x - 1)^{-1} > 0$, and graph the solution set.

Solution   Recall that $(4x - 1)^{-1} = \dfrac{1}{4x - 1}$. Reciprocal Property (4a) states that if $a > 0$, then its reciprocal is greater than zero.

$$(4x - 1)^{-1} > 0$$

$$\frac{1}{4x - 1} > 0$$

$$4x - 1 > 0 \quad \text{Reciprocal Property (4a)}$$

$$4x > 1 \quad \text{Add 1 to both sides.}$$

$$x > \frac{1}{4} \quad \text{Divide both sides by 4.}$$

Figure 7 $x > \dfrac{1}{4}$

The solution set is $\left\{ x \mid x > \dfrac{1}{4} \right\}$, that is, all real numbers in the interval $\left(\dfrac{1}{4}, \infty \right)$. Figure 7 illustrates the graph.

━━━━━━ **Now Work** PROBLEM 89

EXAMPLE 12   **Creating Equivalent Inequalities**

If $-1 < x < 4$, find a and b so that $a < 2x + 1 < b$.

Solution   The idea here is to change the middle part of the combined inequality from x to $2x + 1$, using properties of inequalities.

$$-1 < x < 4$$

$$-2 < 2x < 8 \quad \text{Multiply each part by 2.}$$

$$-1 < 2x + 1 < 9 \quad \text{Add 1 to each part.}$$

Now we see that $a = -1$ and $b = 9$.

━━━━━━ **Now Work** PROBLEM 99

Application

EXAMPLE 13   **Physics: Ohm's Law**

In electricity, Ohm's law states that $E = IR$, where E is the voltage (in volts), I is the current (in amperes), and R is the resistance (in ohms). An air-conditioning unit is rated at a resistance of 10 ohms. If the voltage varies from 110 to 120 volts, inclusive, what corresponding range of current will the air conditioner draw?

Solution   The voltage lies between 110 and 120, inclusive, so

$$110 \leq E \leq 120$$

$$110 \leq IR \leq 120 \quad \text{Ohm's law, } E = IR$$

$$110 \leq I \cdot 10 \leq 120 \quad R = 10$$

$$\frac{110}{10} \leq \frac{I \cdot 10}{10} \leq \frac{120}{10} \quad \text{Divide each part by 10.}$$

$$11 \leq I \leq 12 \quad \text{Simplify.}$$

The air conditioner will draw between 11 and 12 amperes of current, inclusive.

1.5 Assess Your Understanding

'Are You Prepared?' *Answers are given at the end of these exercises. If you get a wrong answer, read the pages listed in* red.

   1. Graph the inequality: $x \geq -2$ (pp. 17–19)   **2.** *True or False* $-5 > -3$ (pp. 17–19)

Concepts and Vocabulary

   3. A(n) _____ _____, denoted $[a, b]$, consists of all real numbers x for which $a \leq x \leq b$.

   4. The _____ _____ state that the sense, or direction, of an inequality remains the same if both sides are multiplied by a positive number, while the direction is reversed if both sides are multiplied by a negative number.

In Problems 5–8, assume that $a < b$ and $c < 0$.

   5. *True or False* $a + c < b + c$

   6. *True or False* $a - c < b - c$

   7. *True or False* $ac > bc$

   8. *True or False* $\dfrac{a}{c} < \dfrac{b}{c}$

   9. *True or False* The square of any real number is always nonnegative.

   10. *True or False* A half-closed interval must have an endpoint of either $-\infty$ or ∞.

   11. *Multiple Choice* Which of the following will *not* change the direction, or sense, of an inequality?
   (a) Dividing both sides by a negative number
   (b) Interchanging sides
   (c) Taking the reciprocal of both sides
   (d) Subtracting a positive number from both sides

   12. *Multiple Choice* Which pair of inequalities is equivalent to $0 < x \leq 3$?
   (a) $x > 0$ and $x \geq 3$
   (b) $x < 0$ and $x \geq 3$
   (c) $x > 0$ and $x \leq 3$
   (d) $x < 0$ and $x \leq 3$

Skill Building

In Problems 13–18, express the graph shown in blue using interval notation. Also express each as an inequality involving x.

   13.

   14.

   15.

   16.

   17.

   18.

In Problems 19–24, an inequality is given. Write the inequality obtained by:

   (a) Adding 3 to both sides of the given inequality.
   (c) Multiplying both sides of the given inequality by 3.

   (b) Subtracting 5 from both sides of the given inequality.
   (d) Multiplying both sides of the given inequality by -2.

   19. $3 < 5$ **20.** $2 > 1$ **21.** $4 > -3$ **22.** $-3 > -5$ **23.** $2x + 1 < 2$ **24.** $1 - 2x > 5$

In Problems 25–32, write each inequality using interval notation, and graph each inequality on the real number line.

   25. $0 \leq x \leq 4$ **26.** $-1 < x < 5$ **27.** $4 \leq x < 6$ **28.** $-2 < x < 0$

   29. $x \geq -3$ **30.** $x \leq 5$ **31.** $x < -4$ **32.** $x > 1$

In Problems 33–40, write each interval as an inequality involving x, and graph each inequality on the real number line.

   33. $[2, 5]$ **34.** $(1, 2)$ **35.** $(-4, 3]$ **36.** $[0, 1)$

   37. $[4, \infty)$ **38.** $(-\infty, 2]$ **39.** $(-\infty, -3)$ **40.** $(-8, \infty)$

In Problems 41–58, fill in the blank to form a correct inequality statement.

   41. If $x < 5$, then $x - 5$ ____ 0.

   43. If $x > -4$, then $x + 4$ ____ 0.

   45. If $x \geq -4$, then $3x$ ____ -12.

   47. If $x > 6$, then $-2x$ ____ -12.

   49. If $x \geq 5$, then $-4x$ ____ -20.

   51. If $8x > 40$, then x ____ 5.

   53. If $-\dfrac{1}{2}x \leq 3$, then x ____ -6.

   55. If $0 < 5 < x$, then $0 < \dfrac{1}{\boxed{}} < \dfrac{1}{\boxed{}}$.

   57. If $-5 < x < 0$, then $\dfrac{1}{\boxed{}} < \dfrac{1}{\boxed{}} < 0$.

   42. If $x < -4$, then $x + 4$ ____ 0.

   44. If $x > 6$, then $x - 6$ ____ 0.

   46. If $x \leq 3$, then $2x$ ____ 6.

   48. If $x > -2$, then $-4x$ ____ 8.

   50. If $x \leq -4$, then $-3x$ ____ 12.

   52. If $3x \leq 12$, then x ____ 4.

   54. If $-\dfrac{1}{4}x > 1$, then x ____ -4.

   56. If $x \leq -4 < 0$, then $\dfrac{1}{\boxed{}} \leq \dfrac{1}{\boxed{}} < 0$.

   58. If $0 < x \leq 10$, then $0 < \dfrac{1}{\boxed{}} \leq \dfrac{1}{\boxed{}}$.

In Problems 59–96, solve each inequality. Express your answer using set notation or interval notation. Graph the solution set.

59. $x + 1 < 5$

60. $x - 6 < 1$

61. $3 - 5x \leq -7$

62. $2 - 3x \leq 5$

63. $3x - 7 > 2$

64. $2x + 5 > 1$

65. $3x - 1 \geq 3 + x$

66. $2x - 2 \geq 3 + x$

67. $-2(x + 3) < 8$

68. $-3(1 - x) < 12$

69. $4 - 3(1 - x) \leq 3$

70. $8 - 4(2 - x) \leq -2x$

71. $\frac{1}{2}(x - 4) > x + 8$

72. $3x + 4 > \frac{1}{3}(x - 2)$

73. $\frac{x}{2} \geq 1 - \frac{x}{4}$

74. $\frac{x}{3} \geq 2 + \frac{x}{6}$

75. $0 < 3x - 7 \leq 5$

76. $4 \leq 2x + 2 \leq 10$

77. $-5 \leq 4 - 3x \leq 2$

78. $-3 \leq 3 - 2x \leq 9$

79. $-3 < \frac{2x - 1}{4} < 0$

80. $0 < \frac{3x + 2}{2} < 4$

81. $1 < 1 - \frac{1}{2}x < 4$

82. $0 < 1 - \frac{1}{3}x < 1$

83. $(x + 2)(x - 3) > (x - 1)(x + 1)$

84. $(x - 1)(x + 1) > (x - 3)(x + 4)$

85. $x(4x + 3) \leq (2x + 1)^2$

86. $x(9x - 5) \leq (3x - 1)^2$

87. $\frac{1}{2} \leq \frac{x + 1}{3} < \frac{3}{4}$

88. $\frac{1}{3} < \frac{x + 1}{2} \leq \frac{2}{3}$

89. $(4x + 2)^{-1} < 0$

90. $(2x - 1)^{-1} > 0$

91. $(1 - 4x)^{-1} \geq 7$

92. $2(3x + 5)^{-1} \leq -3$

93. $0 < \frac{2}{x} < \frac{3}{5}$

94. $0 < \frac{4}{x} < \frac{2}{3}$

95. $0 < (2x - 4)^{-1} < \frac{1}{2}$

96. $0 < (3x + 6)^{-1} < \frac{1}{3}$

Applications and Extensions

In Problems 97–106, find a and b.

97. If $-1 < x < 1$, then $a < x + 4 < b$.

98. If $-3 < x < 2$, then $a < x - 6 < b$.

99. If $2 < x < 3$, then $a < -4x < b$.

100. If $-4 < x < 0$, then $a < \frac{1}{2}x < b$.

101. If $0 < x < 4$, then $a < 2x + 3 < b$.

102. If $-3 < x < 3$, then $a < 1 - 2x < b$.

103. If $-3 < x < 0$, then $a < \frac{1}{x + 4} < b$.

104. If $2 < x < 4$, then $a < \frac{1}{x - 6} < b$.

105. If $6 < 3x < 12$, then $a < x^2 < b$.

106. If $0 < 2x < 6$, then $a < x^2 < b$.

107. What is the domain of the variable in the expression $\sqrt{3x + 6}$?

108. What is the domain of the variable in the expression $\sqrt{8 + 2x}$?

109. A young adult may be defined as someone older than 21 but less than 30 years of age. Express this statement using inequalities.

110. Middle-aged may be defined as being 40 or more and less than 60. Express this statement using inequalities.

111. **Life Expectancy** The Social Security Administration determined that an average 30-year-old male in 2018 could expect to live at least 52.2 more years and that an average 30-year-old female in 2018 could expect to live at least 55.8 more years.

    (a) To what age could an average 30-year-old male expect to live? Express your answer as an inequality.

    (b) To what age could an average 30-year-old female expect to live? Express your answer as an inequality.

    (c) Who can expect to live longer, a male or a female? By how many years?

Source: Social Security Administration, 2018

112. **General Chemistry** For a certain ideal gas, the volume V (in cubic centimeters) equals 20 times the temperature T (in degrees Celsius). If the temperature varies from 80° to 120°C, inclusive, what is the corresponding range of the volume of the gas?

113. **Real Estate** A real estate agent agrees to sell an apartment complex according to the following commission schedule: $45,000 plus 25% of the selling price in excess of $900,000. Assuming that the complex will sell at some price between $900,000 and $1,100,000, inclusive, over what range does the agent's commission vary? How does the commission vary as a percent of selling price?

114. **Sales Commission** A used car salesperson is paid a commission of $25 plus 40% of the selling price in excess of owner's cost. The owner claims that used cars typically sell for at least owner's cost plus $200 and at most owner's cost plus $3000. For each sale made, over what range can the salesperson expect the commission to vary?

115. Federal Tax Withholding The percentage method of withholding for federal income tax (2018) states that a single person whose weekly wages, after subtracting withholding allowances, are over $815, but not over $1658, shall have $85.62 plus 22% of the excess over $815 withheld. Over what range does the amount withheld vary if the weekly wages vary from $900 to $1100, inclusive?

Source: Employer's Tax Guide. Internal Revenue Service, 2018

116. Exercising Sue wants to lose weight. For healthy weight loss, the American College of Sports Medicine (ACSM) recommends 200 to 300 minutes of exercise per week. For the first six days of the week, Sue exercised 40, 45, 0, 50, 25, and 35 minutes. How long should Sue exercise on the seventh day in order to stay within the ACSM guidelines?

117. Electricity Rates During summer months in 2018, Omaha Public Power District charged residential customers a monthly service charge of $25, plus a usage charge of 10.06¢ per kilowatt-hour (kWh). If one customer's monthly summer bills ranged from a low of $140.69 to a high of $231.23, over what range did usage vary (in kWh)?

Source: Omaha Public Power District, 2018

118. Sewer Bills The village of Oak Lawn charges homeowners $23.55 per quarter-year for sewer usage, plus $0.40 per 1000 gallons of water metered. In 2018, one homeowner's quarterly bill ranged from a high of $36.75 to a low of $30.35. Over what range did metered water usage vary?

Source: Village of Oak Lawn, Illinois, January 2018

119. Fat Content Suppose that you order a small McCafe™ chocolate shake (15 g of fat) and an Artisan Grilled Chicken Sandwich™ (7 g of fat) at McDonald's. How many oatmeal raisin cookies can you eat (5 g of fat) and still keep the total fat content of your meal to no more than 47 g?

Source: McDonald's Corporation

120. Sodium Content Suppose that you order a Garden Side Salad (95 mg of sodium) and a Strawberry Banana Smoothie (50 mg of sodium) at Burger King. How many hamburgers can you eat (380 mg of sodium) and still keep the total sodium content of your meal to no more than 1285 mg?

Source: Burger King

121. Computing Grades In your Economics 101 class, you have scores of 68, 82, 87, and 89 on the first four of five tests. To get a grade of B, the average of the first five test scores must be greater than or equal to 80 and less than 90.
(a) Solve an inequality to find the least score you can get on the last test and still earn a B.

(b) What score do you need if the fifth test counts double?

What do I need to get a B?

122. IQ Tests A standard intelligence test has an average score of 100. According to statistical theory, of the people who take the test, the 2.5% with the highest scores will have scores of more than 1.96σ above the average, where σ (sigma, a number called the **standard deviation**) depends on the nature of the test. If $\sigma = 12$ for this test and there is (in principle) no upper limit to the score possible on the test, write the interval of possible test scores of the people in the top 2.5%.

123. Arithmetic Mean If $a < b$, show that $a < \dfrac{a + b}{2} < b$. The number $\dfrac{a + b}{2}$ is called the **arithmetic mean** of a and b.

124. Refer to Problem 123. Show that the arithmetic mean of a and b is equidistant from a and b.

125. Geometric Mean If $0 < a < b$, show that $a < \sqrt{ab} < b$. The number $\sqrt{ab}$ is called the **geometric mean** of a and b.

126. Refer to Problems 123 and 125. Show that the geometric mean of a and b is less than the arithmetic mean of a and b.

127. *Challenge Problem* Harmonic Mean For $0 < a < b$, let h be defined by

$$\frac{1}{h} = \frac{1}{2}\left(\frac{1}{a} + \frac{1}{b}\right)$$

Show that $a < h < b$. The number h is called the **harmonic mean** of a and b.

128. *Challenge Problem* Refer to Problems 123, 125, and 127. Show that the harmonic mean of a and b equals the geometric mean squared, divided by the arithmetic mean.

129. *Challenge Problem* Solve: $x - 4 < 2x - 3 \le \dfrac{x + 5}{3}$

130. *Challenge Problem* If $\dfrac{2}{3} \le \dfrac{5 - x}{3} \le 3$, find the largest possible value of $2x^2 - 3$.

Explaining Concepts: Discussion and Writing

131. How would you explain to a fellow student the underlying reason for the multiplication properties for inequalities (page 122)? That is, the sense or direction of an inequality remains the same if both sides are multiplied by a positive real number, whereas the direction is reversed if both sides are multiplied by a negative real number.

132. Make up an inequality that has no solution. Make up an inequality that has exactly one solution.

133. The inequality $x^2 + 1 < -5$ has no real solution. Explain why.

134. Do you prefer to use inequality notation or interval notation to express the solution to an inequality? Give your reasons. Are there particular circumstances when you prefer one to the other? Cite examples.

'Are You Prepared?' Answers

1. [number line with marks at −4, −2, 0]   **2.** False

1.6 Equations and Inequalities Involving Absolute Value

PREPARING FOR THIS SECTION *Before getting started, review the following:*

- Absolute Value (Section R.2, p. 19)

Now Work the 'Are You Prepared?' problems on page 132.

OBJECTIVES 1 Solve Equations Involving Absolute Value (p. 130)
2 Solve Inequalities Involving Absolute Value (p. 130)

1 Solve Equations Involving Absolute Value

Recall that on the real number line, the absolute value of a equals the distance from the origin to the point whose coordinate is a. For example, there are two points whose distance from the origin is 5 units, -5 and 5. So the equation $|x| = 5$ will have the solution set $\{-5, 5\}$. This leads to the following result:

THEOREM

If a is a positive real number and if u is any algebraic expression, then

$$|u| = a \quad \text{is equivalent to} \quad u = a \quad \text{or} \quad u = -a \qquad \textbf{(1)}$$

Statement (1) requires that a be a positive number. If $a = 0$, then $|u| = 0$ which is equivalent to $u = 0$. If a is less than zero, the equation has no real solution. To see why, consider the equation $|x| = -2$. Because there is no real number whose distance from 0 on the real number line is negative, this equation has no real solution.

EXAMPLE 1 **Solving an Equation Involving Absolute Value**

Solve the equations:

(a) $|x + 4| = 13$ (b) $|2x - 3| + 2 = 7$

Solution   (a) Apply statement (1), where $u = x + 4$. There are two possibilities:

$$x + 4 = 13 \quad \text{or} \quad x + 4 = -13$$
$$x = 9 \quad \text{or} \quad x = -17$$

The solution set is $\{-17, 9\}$.

(b) The equation $|2x - 3| + 2 = 7$ is not in the form of statement (1). Put it in the form of (1) by subtracting 2 from both sides of the equation.

$$|2x - 3| + 2 = 7$$
$$|2x - 3| = 5 \qquad \text{Subtract 2 from both sides.}$$
$$2x - 3 = 5 \quad \text{or} \quad 2x - 3 = -5 \qquad \text{Use (1).}$$
$$2x = 8 \quad \text{or} \quad 2x = -2 \qquad \text{Add 3 to both sides.}$$
$$x = 4 \quad \text{or} \quad x = -1 \qquad \text{Divide both sides by 2.}$$

The solution set is $\{-1, 4\}$.

Now Work PROBLEM 11

2 Solve Inequalities Involving Absolute Value

EXAMPLE 2 **Solving an Inequality Involving Absolute Value**

Solve the inequality: $|x| < 4$

Solution

Less than 4 units
from origin O

O

-5 -4 -3 -2 -1 0 1 2 3 4

Figure 8 $|x| < 4$

We are looking for all points whose coordinate x is a distance less than 4 units from the origin. See Figure 8 for an illustration. Because any number x between -4 and 4 satisfies the condition $|x| < 4$, the solution set consists of all numbers x for which $-4 < x < 4$, that is, all real numbers in the interval $(-4, 4)$.

THEOREM

If a is a positive number and if u is an algebraic expression, then

- $|u| < a$ is equivalent to $-a < u < a$ **(2)**
- $|u| \le a$ is equivalent to $-a \le u \le a$ **(3)**

In other words, $|u| < a$ is equivalent to $-a < u$ and $u < a$.

$-a$ 0 a

Figure 9 $|u| \le a, a > 0$

See Figure 9 for an illustration of statement (3).

EXAMPLE 3

Solving an Inequality Involving Absolute Value

Solve the inequality $|2x + 4| \le 3$, and graph the solution set.

Solution

$$|2x + 4| \le 3 \qquad \text{This follows the form of statement (3)} \\ \text{where } u = 2x + 4.$$

$$-3 \le \quad 2x + 4 \quad \le 3 \qquad \text{Use statement (3).}$$

$$-3 - 4 \le 2x + 4 - 4 \le 3 - 4 \qquad \text{Subtract 4 from each part.}$$

$$-7 \le \quad 2x \quad \le -1 \qquad \text{Simplify.}$$

$$\frac{-7}{2} \le \quad \frac{2x}{2} \quad \le \frac{-1}{2} \qquad \text{Divide each part by 2.}$$

$$-\frac{7}{2} \le \quad x \quad \le -\frac{1}{2} \qquad \text{Simplify.}$$

-5 $-\frac{7}{2}$ -2 $-\frac{1}{2}$ 0   2   4

Figure 10 $|2x + 4| \le 3$

The solution set is $\left\{ x \middle| -\dfrac{7}{2} \le x \le -\dfrac{1}{2} \right\}$, that is, all real numbers in the interval $\left[-\dfrac{7}{2}, -\dfrac{1}{2} \right]$. See Figure 10 for the graph of the solution set.

EXAMPLE 4

Solving an Inequality Involving Absolute Value

Solve the inequality $|1 - 4x| < 5$, and graph the solution set.

Solution

$$|1 - 4x| < 5 \qquad \text{This follows the form of statement (2)} \\ \text{where } u = 1 - 4x.$$

$$-5 < \quad 1 - 4x \quad < 5 \qquad \text{Use statement (2).}$$

$$-5 - 1 < 1 - 4x - 1 < 5 - 1 \qquad \text{Subtract 1 from each part.}$$

$$-6 < \quad -4x \quad < 4 \qquad \text{Simplify.}$$

$$\frac{-6}{-4} > \quad \frac{-4x}{-4} \quad > \frac{4}{-4} \qquad \text{Divide each part by } -4, \text{ which reverses} \\ \text{the direction of the inequality symbols.}$$

$$\frac{3}{2} > \quad x \quad > -1 \qquad \text{Simplify.}$$

$$-1 < \quad x \quad < \frac{3}{2} \qquad \text{Rearrange the ordering.}$$

-5 -4 -3 -2 -1   0   1 $\frac{3}{2}$ 2   3   4

Figure 11 $|1 - 4x| < 5$

The solution set is $\left\{ x \middle| -1 < x < \dfrac{3}{2} \right\}$, that is, all real numbers in the open interval $\left(-1, \dfrac{3}{2} \right)$. See Figure 11 for the graph of the solution set.

Now Work PROBLEM 41

EXAMPLE 5 | Solving an Inequality Involving Absolute Value

Solve the inequality $|x| > 3$, and graph the solution set.

Solution

$$-5 \quad -4 \quad -3 \quad -2 \quad -1 \quad 0 \quad 1 \quad 2 \quad 3 \quad 4$$

Figure 12 $|x| > 3$

We are looking for all points whose coordinate x is a distance greater than 3 units from the origin. Figure 12 illustrates the situation. Any number x less than -3 or greater than 3 satisfies the condition $|x| > 3$. The solution set consists of all numbers x for which $x < -3$ or $x > 3$, that is, all real numbers in $(-\infty, -3) \cup (3, \infty)$.

Need to Review?

The symbol $\cup$ stands for the union of two sets. Refer to page 2 if necessary.

THEOREM

If a is a positive number and u is an algebraic expression, then

- $|u| > a$ is equivalent to $u < -a$ or $u > a$ **(4)**
- $|u| \geq a$ is equivalent to $u \leq -a$ or $u \geq a$ **(5)**

$$-a \qquad 0 \qquad a$$

Figure 13 $|u| \geq a, a > 0$

See Figure 13 for an illustration of statement (5).

EXAMPLE 6 | Solving an Inequality Involving Absolute Value

Solve the inequality $|2x - 5| > 3$, and graph the solution set.

Solution

$\|2x - 5\| > 3$			This follows the form of statement (4) where $u = 2x - 5$.
$2x - 5 < -3$	or	$2x - 5 > 3$	Use statement (4).
$2x - 5 + 5 < -3 + 5$	or	$2x - 5 + 5 > 3 + 5$	Add 5 to each part.
$2x < 2$	or	$2x > 8$	Simplify.
$\dfrac{2x}{2} < \dfrac{2}{2}$	or	$\dfrac{2x}{2} > \dfrac{8}{2}$	Divide each part by 2.
$x < 1$	or	$x > 4$	Simplify.

$$-2 \quad -1 \quad 0 \quad 1 \quad 2 \quad 3 \quad 4 \quad 5 \quad 6 \quad 7$$

Figure 14 $|2x - 5| > 3$

The solution set is $\{x | x < 1 \text{ or } x > 4\}$, that is, all real numbers in $(-\infty, 1) \cup (4, \infty)$. See Figure 14 for the graph of the solution set.

WARNING A common error to be avoided is to attempt to write the solution $x < 1$ or $x > 4$ as the combined inequality $1 > x > 4$, which is incorrect, since there are no numbers x for which $1 > x$ and $x > 4$. Another common error is to "mix" the symbols and write $1 < x > 4$, which makes no sense. ∎

Now Work PROBLEM 45

1.6 Assess Your Understanding

'Are You Prepared?' *Answers are given at the end of these exercises. If you get a wrong answer, read the pages listed in red.*

1. $|-2| =$ _____ (p. 19)

2. *True or False* $|x| \geq 0$ for any real number x. (p. 19)

Concepts and Vocabulary

3. The solution set of the equation $|x| = 5$ is { _____ }.

4. The solution set of the inequality $|x| < 5$ is $\{x |$ _____ $\}$.

5. *True or False* The equation $|x| = -2$ has no solution.

6. *True or False* The inequality $|x| \geq -2$ has the set of real numbers as its solution set.

7. *Multiple Choice* Which of the following pairs of inequalities is equivalent to $|x| > 4$?
   - (a) $x > -4$ and $x < 4$
   - (b) $x < -4$ and $x < 4$
   - (c) $x > -4$ or $x > 4$
   - (d) $x < -4$ or $x > 4$

8. *Multiple Choice* Which of the following has no solution?
   - (a) $|x| < -5$
   - (b) $|x| \leq 0$
   - (c) $|x| > 0$
   - (d) $|x| \geq 0$

Skill Building

In Problems 9–36, find the real solutions, if any, of each equation.

9. $|3x| = 15$

10. $|3x| = 12$

11. $|2x + 3| = 5$

12. $|3x - 1| = 2$

13. $|1 - 4t| + 8 = 13$

14. $|1 - 2z| + 6 = 9$

15. $|-2x| = |8|$

16. $|-x| = |1|$

17. $|-2|x = 4$

18. $|3|x = 9$

19. $\dfrac{8}{7}|x| = 3$

20. $\dfrac{3}{4}|x| = 9$

21. $\left|\dfrac{x}{3} + \dfrac{2}{5}\right| = 2$

22. $\left|\dfrac{x}{2} - \dfrac{1}{3}\right| = 1$

23. $|u - 2| = -\dfrac{1}{2}$

24. $|2 - v| = -1$

25. $5 - |4x| = 4$

26. $5 - \left|\dfrac{1}{2}x\right| = 3$

27. $|x^2 - 9| = 0$

28. $|x^2 - 16| = 0$

29. $|x^2 - 2x| = 3$

30. $|x^2 + x| = 12$

31. $|x^2 + x - 1| = 1$

32. $|x^2 + 3x - 2| = 2$

33. $\left|\dfrac{5x - 3}{3x - 5}\right| = 2$

34. $\left|\dfrac{2x + 1}{3x + 4}\right| = 1$

35. $|x^2 + 3x| = |x^2 - 2x|$

36. $|x^2 - 2x| = |x^2 + 6x|$

In Problems 37–70, solve each inequality. Express your answer using set notation or interval notation. Graph the solution set.

37. $|2x| < 8$

38. $|3x| < 15$

39. $|7x| > 42$

40. $|2x| > 6$

41. $|x - 2| + 2 < 3$

42. $|x + 4| + 3 < 5$

43. $|3t - 2| \le 4$

44. $|2u + 5| \le 7$

45. $|2x - 3| \ge 2$

46. $|3x + 4| \ge 2$

47. $|1 - 4x| - 7 < -2$

48. $|1 - 2x| - 4 < -1$

49. $|5 - 2x| > 7$

50. $|2 - 3x| > 1$

51. $|-4x| + |-5| \le 1$

52. $|-x| - |4| \le 2$

53. $|-2x| > |-3|$

54. $|-x - 2| \ge 1$

55. $-3|2x - 5| \ge -21$

56. $-|1 - 2x| \ge -3$

57. $|9x| < -5$

58. $|3x| \ge 0$

59. $|5x| \ge -1$

60. $|6x| < -2$

61. $\left|\dfrac{2x + 3}{3} - \dfrac{1}{2}\right| < 1$

62. $3 - |x + 1| < \dfrac{1}{2}$

63. $|8 - 4x| \le -13$

64. $|7x + 4| < -9$

65. $\left|\dfrac{7 - 2x}{3}\right| \le 0$

66. $-\left|\dfrac{4x - 15}{6}\right| \ge 0$

67. $\left|\dfrac{3x - 7}{5}\right| > 4$

68. $\left|\dfrac{5 - 2x}{9}\right| \ge 8$

69. $5 + |x - 1| > \dfrac{1}{2}$

70. $\left|\dfrac{2x - 3}{2} + \dfrac{1}{3}\right| > 1$

Applications and Extensions

71. Body Temperature "Normal" human body temperature is 98.6°F. If a temperature x that differs from normal by at least 1.5° is considered unhealthy, write the condition for an unhealthy temperature x as an inequality involving an absolute value, and solve for x.

72. Tolerance Specifications for a rod in an internal combustion engine call for a length of 5.375 inches. Lengths within 0.0025 inch of this length are acceptable. Express this situation as an inequality involving an absolute value. Use x as the actual rod length and solve for x.

73. Fake News A Pew Research Center poll conducted in December 2016 found that 64% of adult Americans believe that fake news is causing confusion about basic facts of current issues and events. Suppose that the researchers are 99% confident that the result from the poll is off by fewer than 3.9 percentage points from the actual percentage x of adult Americans who feel this way. Express this situation

as an inequality involving absolute value, then solve the inequality for x to determine the interval in which the actual percentage is likely to fall. [**Note:** In statistics, this interval is called a 99% **confidence interval.**]

74. Speed of Sound According to data from the Hill Aerospace Museum (Hill Air Force Base, Utah), the speed of sound varies depending on altitude, barometric pressure, and temperature. For example, at 20,000 feet, 13.75 inches of mercury, and $-12.3°F$, the speed of sound is about 707 miles per hour, but the speed can vary from this result by as much as 55 miles per hour as conditions change.

(a) Using x for the speed of sound, express this situation as an inequality involving an absolute value.

(b) Solve for x to find an interval for the speed of sound.

75. Express the fact that x differs from 3 by less than $\dfrac{1}{2}$ as an inequality involving an absolute value. Solve for x.

76. Express the fact that x differs from -4 by less than 1 as an inequality involving an absolute value. Solve for x.

77. Express the fact that x differs from -3 by more than 2 as an inequality involving an absolute value. Solve for x.

78. Express the fact that x differs from 2 by more than 3 as an inequality involving an absolute value. Solve for x.

In Problems 79–84, find a and b.

79. If $|x - 1| < 3$, then $a < x + 4 < b$.

80. If $|x + 2| < 5$, then $a < x - 2 < b$.

81. If $|x + 4| \leq 2$, then $a \leq 2x - 3 \leq b$.

82. If $|x - 3| \leq 1$, then $a \leq 3x + 1 \leq b$.

83. If $|x - 2| \leq 7$, then $a \leq \dfrac{1}{x - 10} \leq b$.

84. If $|x + 1| \leq 3$, then $a \leq \dfrac{1}{x + 5} \leq b$.

85. Show that: if $a > 0, b > 0$, and $\sqrt{a} < \sqrt{b}$, then $a < b$.
[**Hint:** $b - a = (\sqrt{b} - \sqrt{a})(\sqrt{b} + \sqrt{a})$.]

86. Show that $a \leq |a|$.

87. Prove the *triangle inequality* $|a + b| \leq |a| + |b|$.
[**Hint:** Expand $|a + b|^2 = (a + b)^2$, and use the result of Problem 86.]

88. Prove that $|a - b| \geq |a| - |b|$.
[**Hint:** Apply the triangle inequality from Problem 87 to $|a| = |(a - b) + b|$.]

89. If $a > 0$, show that the solution set of the inequality

$$x^2 < a$$

consists of all numbers x for which

$$-\sqrt{a} < x < \sqrt{a}$$

90. If $a > 0$, show that the solution set of the inequality

$$x^2 > a$$

consists of all numbers x for which

$$x < -\sqrt{a} \quad \text{or} \quad x > \sqrt{a}$$

In Problems 91–98, use the results found in Problems 89 and 90 to solve each inequality.

91. $x^2 < 1$

92. $x^2 < 4$

93. $x^2 \geq 9$

94. $x^2 \geq 1$

95. $x^2 \leq 16$

96. $x^2 \leq 9$

97. $x^2 > 4$

98. $x^2 > 16$

99. *Challenge Problem* Solve $|3x - |2x + 1|| = 4$.

100. *Challenge Problem* Solve $|x + |3x - 2|| = 2$.

101. *Challenge Problem* If $|2x - 5| = x + 13$ and $|4 - 3y| = 2$, what is the largest possible value of $\dfrac{y}{x}$?

102. *Challenge Problem* Solve $\left| 1 + \dfrac{|x|}{1 + |x|} \right| \leq \dfrac{3}{2}$.

Explaining Concepts: Discussion and Writing

103. The equation $|x| = -2$ has no solution. Explain why.

104. The inequality $|x| > -0.5$ has all real numbers as solutions. Explain why.

105. The inequality $|x| > 0$ has as its solution set $\{x | x \neq 0\}$. Explain why.

'Are You Prepared?' Answers

1. 2   **2.** True

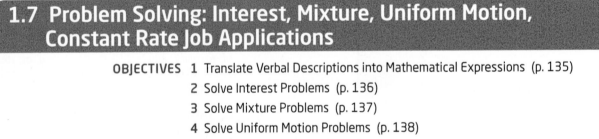

1.7 Problem Solving: Interest, Mixture, Uniform Motion, Constant Rate Job Applications

> **OBJECTIVES** 1 Translate Verbal Descriptions into Mathematical Expressions (p. 135)
>
> 2 Solve Interest Problems (p. 136)
>
> 3 Solve Mixture Problems (p. 137)
>
> 4 Solve Uniform Motion Problems (p. 138)
>
> 5 Solve Constant Rate Job Problems (p. 140)

Applied (word) problems do not come in the form "Solve the equation. . . ." Instead, they supply information using words, a verbal description of the real problem. So, to solve applied problems, we must be able to translate the verbal description into the language of mathematics. This can be done by using variables to represent unknown quantities and then finding relationships (such as equations) that involve these variables. The process of doing all this is called **mathematical modeling**. An equation or inequality that describes a relationship among the variables is called a **model**.

Any solution to the mathematical problem must be checked against the mathematical problem, the verbal description, and the real problem. See Figure 15 for an illustration of the **modeling process**.

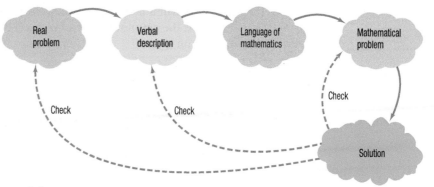

Figure 15 The modeling process

1 Translate Verbal Descriptions into Mathematical Expressions

---

EXAMPLE 1

Translating Verbal Descriptions into Mathematical Expressions

(a) For uniform motion, the average speed of an object equals the distance traveled divided by the time required.

Translation: If r is the speed, d the distance, and t the time, then $r = \dfrac{d}{t}$.

(b) Let x denote a number.

The number 5 times as large as x is $5x$.

The number 3 less than x is $x - 3$.

The number that exceeds x by 4 is $x + 4$.

The number that, when added to x, gives 5 is $5 - x$.

Now Work PROBLEM 9

Always check the units used to measure the variables of an applied problem. In Example 1(a), if r is measured in miles per hour, then the distance d must be expressed in miles, and the time t must be expressed in hours. It is a good practice to check units to be sure that they are consistent and make sense.

The steps for solving applied problems, given earlier, are repeated next.

Steps for Solving Applied Problems

STEP 1: Read the problem carefully, perhaps two or three times. Pay particular attention to the question being asked in order to identify what you are looking for. Identify any relevant formulas you may need ($d = rt$, $A = \pi r^2$, etc.). If you can, determine realistic possibilities for the answer.

STEP 2: Assign a letter (variable) to represent what you are looking for, and if necessary, express any remaining unknown quantities in terms of this variable.

STEP 3: Make a list of all the known facts, and translate them into mathematical expressions. These may take the form of an equation or an inequality involving the variable. If possible, draw an appropriately labeled diagram to assist you. Sometimes, creating a table or chart helps.

STEP 4: Solve for the variable, and then answer the question.

STEP 5: Check the answer with the facts in the problem. If it agrees, congratulations! If it does not agree, review your work and try again.

2 Solve Interest Problems

Interest is money paid for the use of money. The total amount borrowed (whether by an individual from a bank in the form of a loan, or by a bank from an individual in the form of a savings account) is called the **principal**. The **rate of interest**, expressed as a percent, is the amount charged for the use of the principal for a given period of time, usually on a yearly (that is, per annum) basis.

> **Simple Interest Formula**
> If a principal of P dollars is borrowed for a period of t years at a per annum interest rate r, expressed as a decimal, the interest I charged is
>
> $$I = Prt \qquad\qquad\qquad (1)$$

Interest charged according to formula (1) is called **simple interest**. When using formula (1), be sure to express r as a decimal. For example, if the rate of interest is 4%, then $r = 0.04$.

EXAMPLE 2	**Finance: Computing Interest on a Loan**

Suppose that Juanita borrows $500 for 6 months at the simple interest rate of 9% per annum. What is the interest that Juanita will be charged on the loan? How much does Juanita owe after 6 months?

Solution The rate of interest is given per annum, so the actual time that the money is borrowed must be expressed in years. The interest charged would be the principal, $500, times the rate of interest (9% = 0.09), times the time in years, $\frac{1}{2}$:

$$\text{Interest charged} = I = Prt = 500 \cdot 0.09 \cdot \frac{1}{2} = \$22.50$$

After 6 months, Juanita will owe what she borrowed plus the interest:

$$\$500 + \$22.50 = \$522.50$$

EXAMPLE 3	**Financial Planning**

Candy has $70,000 to invest and wants an annual return of $2800, which requires an overall rate of return of 4%. She can invest in a safe, government-insured certificate of deposit, but it pays only 2%. To obtain 4%, she agrees to invest some of her money in noninsured corporate bonds paying 7%. How much should be placed in each investment to achieve her goal?

Solution **STEP 1:** The question is asking for two dollar amounts: the principal to invest in the corporate bonds and the principal to invest in the certificate of deposit.

STEP 2: Let b represent the amount (in dollars) to be invested in the bonds. Then $70,000 - b$ is the amount that will be invested in the certificate. (Do you see why?)

STEP 3: We set up a table:

NOTE: We could have also let c represent the amount invested in the certificate and $70,000 - c$ the amount invested in bonds. ∎

	Principal ($)	Rate	Time (yr)	Interest ($)
Bonds	b	7% = 0.07	1	$0.07b$
Certificate	$70,000 - b$	2% = 0.02	1	$0.02(70,000 - b)$
Total	70,000	4% = 0.04	1	$0.04(70,000) = 2800$

Since the combined interest from the investments is equal to the total interest, we have

Bond interest + Certificate interest = Total interest

$$0.07b + 0.02(70,000 - b) = 2800$$

(Note that the units are consistent: the unit is dollars on both sides.)

Step 4: $0.07b + 1400 - 0.02b = 2800$

$0.05b = 1400$ Simplify.

$b = 28,000$ Divide both sides by 0.05.

Candy should place $28,000 in the bonds and $70,000 - 28,000 = $42,000 in the certificate.

Step 5: The interest on the bonds after 1 year is 0.07($28,000) = $1960; the interest on the certificate after 1 year is 0.02($42,000) = $840. The total annual interest is $2800, the required amount.

— **Now Work** PROBLEM 19

3 Solve Mixture Problems

Oil refineries sometimes produce gasoline that is a blend of two or more types of fuel; bakeries occasionally blend two or more types of flour for their bread. These problems are referred to as **mixture problems** because they involve combining two or more quantities to form a mixture.

EXAMPLE 4	**Blending Coffees**

The manager of a Starbucks store decides to experiment with a new blend of coffee. She will mix some B grade Colombian coffee that sells for $5 per pound with some A grade Arabica coffee that sells for $10 per pound to get 100 pounds of the new blend. The selling price of the new blend is to be $7 per pound, and there is to be no difference in revenue between selling the new blend and selling the other types separately. How many pounds of the B grade Colombian coffee and how many pounds of the A grade Arabica coffees are required?

Solution Let c represent the number of pounds of the B grade Colombian coffee. Then $100 - c$ equals the number of pounds of the A grade Arabica coffee. See Figure 16.

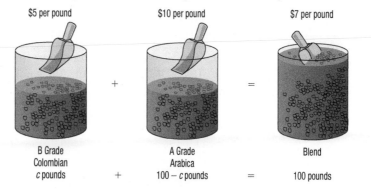

Figure 16

Since there is to be no difference in revenue between selling the A and B grades separately and selling the blend, we have

Revenue from B grade + Revenue from A grade = Revenue from blend

$$\left\{\begin{array}{c}\text{Price per pound}\\\text{of B grade}\end{array}\right\}\left\{\begin{array}{c}\text{Pounds of}\\\text{B grade}\end{array}\right\} + \left\{\begin{array}{c}\text{Price per pound}\\\text{of A grade}\end{array}\right\}\left\{\begin{array}{c}\text{Pounds of}\\\text{A grade}\end{array}\right\} = \left\{\begin{array}{c}\text{Price per pound}\\\text{of blend}\end{array}\right\}\left\{\begin{array}{c}\text{Pounds of}\\\text{blend}\end{array}\right\}$$

$$\$5 \quad \cdot \quad c \quad + \quad \$10 \quad \cdot \quad (100 - c) \quad = \quad \$7 \quad \cdot \quad 100$$

(*continued*)

Now solve the equation:

$$5c + 10(100 - c) = 700$$
$$5c + 1000 - 10c = 700$$
$$-5c = -300$$
$$c = 60$$

The manager should blend 60 pounds of B grade Colombian coffee with $100 - 60 = 40$ pounds of A grade Arabica coffee to get the desired blend.

✓ **Check:** The 60 pounds of B grade coffee would sell for ($5)(60) = $300, and the 40 pounds of A grade coffee would sell for ($10)(40) = $400. The total revenue, $700, equals the revenue obtained from selling the blend, as desired.

━━━━ **Now Work** PROBLEM 23

4 Solve Uniform Motion Problems

Objects that move at a constant speed are said to be in **uniform motion**. When the average speed of an object is known, it can be interpreted as that object's constant speed. For example, a bicyclist traveling at an average speed of 25 miles per hour can be modeled as being in uniform motion with a constant speed of 25 miles per hour.

> **Uniform Motion Formula**
> If an object moves at an average speed (rate) r, the distance d covered in time t is given by the formula
>
> $$d = rt \qquad (2)$$
>
> That is, Distance = Rate · Time.

EXAMPLE 5

Physics: Uniform Motion

Tanya, who is a long-distance runner, runs at an average speed of 8 miles per hour (mi/h). Two hours after Tanya leaves your house, you leave in your Honda and follow the same route. If your average speed is 40 mi/h, how long will it be before you catch up to Tanya? How far will each of you be from your home?

Solution

Refer to Figure 17. We use t to represent the time (in hours) that it takes you to catch up to Tanya. When this occurs, the total time elapsed for Tanya is $t + 2$ hours because she left 2 hours earlier.

Figure 17

Set up the following table:

	Rate (mi/h)	Time (h)	Distance (mi)
Tanya	8	$t + 2$	$8(t + 2)$
Honda	40	t	$40t$

The distance traveled is the same for both, which leads to the equation

$$8(t + 2) = 40t$$

$$8t + 16 = 40t$$

$$32t = 16$$

$$t = \frac{1}{2} \text{ hour}$$

It will take you $\frac{1}{2}$ hour to catch up to Tanya. Each of you will have gone 20 miles.

✓ **Check:** In 2.5 hours, Tanya travels a distance of $2.5 \cdot 8 = 20$ miles. In $\frac{1}{2}$ hour, you travel a distance of $\frac{1}{2} \cdot 40 = 20$ miles.

EXAMPLE 6

Physics: Uniform Motion

A motorboat heads upstream a distance of 24 miles on a river whose current is running at 3 miles per hour (mi/h). The trip up and back takes 6 hours. Assuming that the motorboat maintains a constant speed relative to the water, what is its speed?

Solution

See Figure 18. Use r to represent the constant speed of the motorboat relative to the water. Then the true speed going upstream is $r - 3$ mi/h, and the true speed going downstream is $r + 3$ mi/h. Since Distance = Rate · Time, then Time = $\dfrac{\text{Distance}}{\text{Rate}}$. Set up a table.

Figure 18

	Rate (mi/h)	Distance (mi)	Time (h) = $\dfrac{\text{Distance}}{\text{Rate}}$
Upstream	$r - 3$	24	$\dfrac{24}{r - 3}$
Downstream	$r + 3$	24	$\dfrac{24}{r + 3}$

The total time up and back is 6 hours, which gives the equation

$$\frac{24}{r - 3} + \frac{24}{r + 3} = 6$$

$$\frac{24(r + 3) + 24(r - 3)}{(r - 3)(r + 3)} = 6 \qquad \text{Add the quotients on the left.}$$

$$\frac{48r}{r^2 - 9} = 6 \qquad \text{Simplify.}$$

$$48r = 6(r^2 - 9) \qquad \text{Multiply both sides by } r^2 - 9.$$

$$6r^2 - 48r - 54 = 0 \qquad \text{Write in standard form.}$$

$$r^2 - 8r - 9 = 0 \qquad \text{Divide by 6.}$$

$$(r - 9)(r + 1) = 0 \qquad \text{Factor.}$$

$$r = 9 \quad \text{or} \quad r = -1 \qquad \text{Use the Zero-Product Property and solve.}$$

Discard the solution $r = -1$ mi/h and conclude that the speed of the motorboat relative to the water is 9 mi/h.

━━━━━ **Now Work** PROBLEM 29

5 Solve Constant Rate Job Problems

Here we look at jobs that are performed at a **constant rate**. The assumption is that if a job can be done in t units of time, then $\dfrac{1}{t}$ of the job is done in 1 unit of time. In other words, if a job takes 4 hours, then $\dfrac{1}{4}$ of the job is done in 1 hour.

| EXAMPLE 7 | Working Together to Do a Job |

At 10 AM Danny is asked by his father to weed the garden. From past experience, Danny knows that this will take him 4 hours, working alone. His older brother Mike, when it is his turn to do this job, requires 6 hours. Since Mike wants to go golfing with Danny and has a reservation for 1 PM, he agrees to help Danny. Assuming no gain or loss of efficiency, when will they finish if they work together? Can they make the golf date?

Solution Set up Table 2. In 1 hour, Danny does $\dfrac{1}{4}$ of the job, and in 1 hour, Mike does $\dfrac{1}{6}$ of the job. Let t be the time (in hours) that it takes them to do the job together. In 1 hour, then, $\dfrac{1}{t}$ of the job is completed. Reason as follows:

$$\left(\begin{array}{c}\text{Part done by Danny}\\ \text{in 1 hour}\end{array}\right) + \left(\begin{array}{c}\text{Part done by Mike}\\ \text{in 1 hour}\end{array}\right) = \left(\begin{array}{c}\text{Part done together}\\ \text{in 1 hour}\end{array}\right)$$

Table 2

	Hours to Do Job	Part of Job Done in 1 Hour
Danny	4	$\dfrac{1}{4}$
Mike	6	$\dfrac{1}{6}$
Together	t	$\dfrac{1}{t}$

From Table 2,

$$\frac{1}{4} + \frac{1}{6} = \frac{1}{t} \qquad \text{The Model}$$

$$\frac{3}{12} + \frac{2}{12} = \frac{1}{t} \qquad \text{LCD} = 12 \text{ on the left}$$

$$\frac{5}{12} = \frac{1}{t} \qquad \text{Simplify.}$$

$$5t = 12 \qquad \text{Multiply both sides by } 12t.$$

$$t = \frac{12}{5} \qquad \text{Divide both sides by 5.}$$

Working together, Mike and Danny can do the job in $\dfrac{12}{5}$ hours, or 2 hours, 24 minutes. They should make the golf date, since they will finish at 12:24 PM.

Now Work PROBLEM 35

1.7 Assess Your Understanding

Concepts and Vocabulary

1. The process of using variables to represent unknown quantities and then finding relationships that involve these variables is referred to as _____ _____.

2. The money paid for the use of money is _____.

3. Objects that move at a constant speed are said to be in _____ _____.

4. *True or False* The amount charged for the use of principal for a given period of time is called the rate of interest.

5. *True or False* If an object moves at an average speed r, the distance d covered in time t is given by the formula $d = rt$.

6. *Multiple Choice* Suppose that you want to mix two coffees in order to obtain 100 pounds of a blend. If x represents the number of pounds of coffee A, which algebraic expression represents the number of pounds of coffee B?
   (a) $100 - x$ (b) $x - 100$ (c) $100x$ (d) $100 + x$

7. *Multiple Choice* Which of the following is the simple interest formula?
   (a) $I = \dfrac{rt}{P}$ (b) $I = Prt$ (c) $I = \dfrac{P}{rt}$ (d) $I = P + rt$

8. *Multiple Choice* If it takes 5 hours to complete a job, what fraction of the job is done in 1 hour?
   (a) $\dfrac{4}{5}$ (b) $\dfrac{5}{4}$ (c) $\dfrac{1}{5}$ (d) $\dfrac{1}{4}$

Applications and Extensions

In Problems 9–18, translate each sentence into a mathematical equation. Be sure to identify the meaning of all symbols.

9. **Geometry** The area of a circle is the product of the number π and the square of the radius.

10. **Geometry** The circumference of a circle is the product of the number π and twice the radius.

11. **Geometry** The area of a square is the square of the length of a side.

12. **Geometry** The perimeter of a square is four times the length of a side.

13. **Physics** Force equals the product of mass and acceleration.

14. **Physics** Pressure is force per unit area.

15. **Physics** Work equals force times distance.

16. **Physics** Kinetic energy is one-half the product of the mass and the square of the velocity.

17. **Business** The total variable cost of manufacturing x dishwashers is $150 per dishwasher times the number of dishwashers manufactured.

18. **Business** The total revenue derived from selling x dishwashers is $250 per dishwasher times the number of dishwashers sold.

19. **Financial Planning** Betsy, a recent retiree, requires $6000 per year in extra income. She has $50,000 to invest and can invest in B-rated bonds paying 15% per year or in a certificate of deposit (CD) paying 7% per year. How much money should Betsy invest in each to realize exactly $6000 in interest per year?

20. **Financial Planning** After 2 years, Betsy (see Problem 19) finds that she will now require $7000 per year. Assuming that the remaining information is the same, how should the money be reinvested?

21. **Banking** A bank loaned out $12,000, part of it at the rate of 8% per year and the rest at the rate of 18% per year. If the interest received totaled $1000, how much was loaned at 8%?

22. **Banking** Wendy, a loan officer at a bank, has $1,000,000 to lend and is required to obtain an average return of 18% per year. If she can lend at the rate of 19% or at the rate of 16%, how much can she lend at the 16% rate and still meet her requirement?

23. **Blending Teas** The manager of a store that specializes in selling tea decides to experiment with a new blend. She will mix some Earl Grey tea that sells for $6 per pound with some Orange Pekoe tea that sells for $4 per pound to get 100 pounds of the new blend. The selling price of the new blend is to be $5.50 per pound, and there is to be no difference in revenue between selling the new blend and selling the other types. How many pounds of the Earl Grey tea and of the Orange Pekoe tea are required?

24. **Business: Blending Coffee** A coffee manufacturer wants to market a new blend of coffee that sells for $4.10 per pound by mixing two coffees that sell for $2.75 and $5 per pound, respectively. What amounts of each coffee should be blended to obtain the desired mixture?

    [**Hint**: Assume that the total weight of the desired blend is 100 pounds.]

25. **Business: Mixing Nuts** A nut store normally sells cashews for $9.00 per pound and almonds for $4.50 per pound. But at the end of the month the almonds had not sold well, so, in order to sell 60 pounds of almonds, the manager decided to mix the 60 pounds of almonds with some cashews and sell the mixture for $7.75 per pound. How many pounds of cashews should be mixed with the almonds to ensure no change in the revenue?

26. **Business: Mixing Candy** A candy store sells boxes of candy containing caramels and cremes. Each box sells for $12.50 and holds 30 pieces of candy (all pieces are the same size). If the caramels cost $0.25 to produce and the cremes cost $0.45 to produce, how many of each should be in a box to yield a profit of $3?

27. **Physics: Uniform Motion** A motorboat can maintain a constant speed of 16 miles per hour relative to the water. The boat travels upstream to a certain point in 20 minutes; the return trip takes 15 minutes. What is the speed of the current? See the figure.

28. **Physics: Uniform Motion** A motorboat heads upstream on a river that has a current of 3 miles per hour. The trip upstream takes 5 hours, and the return trip takes 2.5 hours. What is the speed of the motorboat? (Assume that the boat maintains a constant speed relative to the water.)

29. **Physics: Uniform Motion** A motorboat maintained a constant speed of 15 miles per hour relative to the water in going 10 miles upstream and then returning. The total time for the trip was 1.5 hours. Use this information to find the speed of the current.

30. **Physics: Uniform Motion** Two cars enter the Florida Turnpike at Commercial Boulevard at 8:00 AM, each heading for Wildwood. One car's average speed is 10 miles per hour more than the other's. The faster car arrives at Wildwood at 11:00 AM, $\frac{1}{2}$ hour before the other car. What was the average speed of each car? How far did each travel?

31. **Moving Walkways** The speed of a moving walkway is typically about 2.5 feet per second. Walking on such a moving walkway, it takes Karen a total of 48 seconds to travel 50 feet with the movement of the walkway and then back again against the movement of the walkway. What is Karen's normal walking speed?

    Source: Answers.com

32. **High-Speed Walkways** Toronto's Pearson International Airport has a high-speed version of a moving walkway. If Liam walks while riding this moving walkway, he can travel 280 meters in 60 seconds less time than if he stands still on the moving walkway. If Liam walks at a normal rate of 1.5 meters per second, what is the speed of the walkway?

    Source: Answers.com

33. **Tennis** A regulation doubles tennis court has an area of 2808 square feet. If it is 6 feet longer than twice its width, determine the dimensions of the court.

    Source: United States Tennis Association

34. **Laser Printers** It takes a Xerox VersaLink C500 laser printer 9 minutes longer to complete a 1440-page print job by itself than it takes a Brother HL-L8350CDW to complete the same job by itself. Together the two printers can complete the job in 20 minutes. How long does it take each printer to complete the print job alone? What is the speed of each printer?

    Source: Top Ten Reviews

35. **Working Together on a Job** Trent can deliver his newspapers in 30 minutes. It takes Lois 20 minutes to do the same route. How long would it take them to deliver the newspapers if they worked together?

36. **Working Together on a Job** Patrice, by himself, can paint four rooms in 10 hours. If he hires April to help, they can do the same job together in 6 hours. If he lets April work alone, how long will it take her to paint four rooms?

37. **Enclosing a Garden** A gardener has 46 feet of fencing to be used to enclose a rectangular garden that has a border 2 feet wide surrounding it. See the figure.
    (a) If the length of the garden is to be twice its width, what will be the dimensions of the garden?
    (b) What is the area of the garden?
    (c) If the length and width of the garden are to be the same, what will be the dimensions of the garden?
    (d) What will be the area of the square garden?

38. **Construction** A pond is enclosed by a wooden deck that is 3 feet wide. The fence surrounding the deck is 100 feet long.
    (a) If the pond is square, what are its dimensions?
    (b) If the pond is rectangular and the length of the pond is to be three times its width, what are its dimensions?
    (c) If the pond is circular, what is its diameter?
    (d) Which pond has the larger area?

39. **Football** A tight end can run the 100-yard dash in 12 seconds. A defensive back can do it in 10 seconds. The tight end catches a pass at his own 20-yard line with the defensive back at the 15-yard line. (See the figure-top, right.) If no other players are nearby, at what yard line will the defensive back catch up to the tight end?

    [**Hint**: At time $t = 0$, the defensive back is 5 yards behind the tight end.]

40. **Computing Business Expense** Therese, an outside salesperson, uses her car for both business and pleasure. Last year, she traveled 30,000 miles, using 900 gallons of gasoline. Her car gets 40 miles per gallon on the highway and 25 in the city. She can deduct all highway travel, but no city travel, on her taxes. How many miles should Therese deduct as a business expense?

41. **Mixing Water and Antifreeze** How much water should be added to 1 gallon of pure antifreeze to obtain a solution that is 60% antifreeze?

42. **Mixing Water and Antifreeze** The cooling system of a certain foreign-made car has a capacity of 15 liters. If the system is filled with a mixture that is 40% antifreeze, how much of this mixture should be drained and replaced by pure antifreeze so that the system is filled with a solution that is 60% antifreeze?

43. **Chemistry: Salt Solutions** How much water must be evaporated from 32 ounces of a 4% salt solution to make a 6% salt solution?

44. **Chemistry: Salt Solutions** How much water must be evaporated from 240 gallons of a 3% salt solution to produce a 5% salt solution?

45. **Purity of Gold** The purity of gold is measured in karats, with pure gold being 24 karats. Other purities of gold are expressed as proportional parts of pure gold. Thus, 18-karat gold is $\frac{18}{24}$, or 75% pure gold; 12-karat gold is $\frac{12}{24}$, or 50% pure gold; and so on. How much 12-karat gold should be mixed with pure gold to obtain 60 grams of 16-karat gold?

46. **Chemistry: Sugar Molecules** A sugar molecule has twice as many atoms of hydrogen as it does oxygen and one more atom of carbon than of oxygen. If a sugar molecule has a total of 45 atoms, how many are oxygen? How many are hydrogen?

47. **Running a Race** Mike can run the mile in 6 minutes, and Dan can run the mile in 9 minutes. If Mike gives Dan a head start of 1 minute, how far from the start will Mike pass Dan? How long does it take? See the figure.

48. **Range of an Airplane** An air rescue plane averages 300 miles per hour in still air. It carries enough fuel for 5 hours of flying time. If, upon takeoff, it encounters a head wind of 30 mi/h, how far can it fly and return safely? (Assume that the wind remains constant.)

49. **Emptying Oil Tankers** An oil tanker can be emptied by the main pump in 4 hours. An auxiliary pump can empty the tanker in 9 hours. If the main pump is started at 9 AM, when should the auxiliary pump be started so that the tanker is emptied by noon?

50. **Cement Mix** A 20-pound bag of Economy brand cement mix contains 25% cement and 75% sand. How much pure cement must be added to produce a cement mix that is 40% cement?

51. **Filling a Tub** A bathroom tub will fill in 15 minutes with both faucets open and the stopper in place. With both faucets closed and the stopper removed, the tub will empty in 20 minutes. How long will it take for the tub to fill if both faucets are open and the stopper is removed?

52. **Using Two Pumps** A 5-horsepower (hp) pump can empty a pool in 5 hours. A smaller, 2-hp pump empties the same pool in 8 hours. The pumps are used together to begin emptying the pool. After two hours, the 2-hp pump breaks down. How long will it take the larger pump to finish emptying the pool?

53. **A Biathlon** Suppose that you have entered an 87-mile biathlon that consists of a run and a bicycle race. During your run, your average speed is 6 miles per hour, and during your bicycle race, your average speed is 25 miles per hour. You finish the race in 5 hours. What is the distance of the run? What is the distance of the bicycle race?

54. **Cyclists** Two cyclists leave a city at the same time, one going east and the other going west. The westbound cyclist bikes 5 mph faster than the eastbound cyclist. After 6 hours they are 246 miles apart. How fast is each cyclist riding?

55. **Comparing Olympic Heroes** In the 2016 Olympics, Usain Bolt of Jamaica won the gold medal in the 100-meter race with a time of 9.81 seconds. In the 1896 Olympics, Thomas Burke of the United States won the gold medal in the 100-meter race in 12.0 seconds. If they ran in the same race, repeating their respective times, by how many meters would Bolt beat Burke?

56. **Constructing a Coffee Can**
A 30.5-ounce can of Hills Bros.® coffee requires 58.9π square inches of aluminum. If its height is 6.4 inches, what is its radius? [**Hint**: The surface area S of a right cylinder is $S = 2\pi r^2 + 2\pi rh$, where r is the radius and h is the height.]

57. ***Challenge Problem*** **Landscaping** Elaine can complete a landscaping project in 2 hours with the help of either her husband Brian or both her two daughters. If Brian and one of his daughters work together, it would take them 4 hours to complete the project. Assuming the rate of work is constant for each person, and the two daughters work at the same rate, how long would it take Elaine, Brian, and one of their daughters to complete the project?

58. ***Challenge Problem*** **Saltwater** Detty has a vat in her lab that contains x liters of a 20% saltwater solution. Over time, 25% of the pure water evaporates leaving all the salt. She adds 20 liters of salt and 10 liters of water resulting in a concentration of $33\frac{1}{3}\%$. How many liters of solution were initially in the vat?

59. ***Challenge Problem*** **Physics: Uniform Motion** A man is walking at an average speed of 4 miles per hour alongside a railroad track. A freight train, going in the same direction at an average speed of 30 miles per hour, requires 5 seconds to pass the man. How long is the freight train? Give your answer in feet.

Explaining Concepts: Discussion and Writing

60. **Critical Thinking** Make up a word problem that requires solving a linear equation as part of its solution. Exchange problems with a friend. Write a critique of your friend's problem.

61. **Critical Thinking** You are the manager of a clothing store and have just purchased 100 dress shirts for $20.00 each. After 1 month of selling the shirts at the regular price, you plan to have a sale giving 40% off the original selling price. However, you still want to make a profit of $4 on each shirt at the sale price. What should you price the shirts at initially to ensure this? If, instead of 40% off at the sale, you give 50% off, by how much is your profit reduced?

62. **Computing Average Speed** In going from Chicago to Atlanta, a car averages 45 miles per hour, and in going from Atlanta to Miami, it averages 55 miles per hour. If Atlanta is halfway between Chicago and Miami, what is the average

speed from Chicago to Miami? Discuss an intuitive solution. Write a paragraph defending your intuitive solution. Then solve the problem algebraically. Is your intuitive solution the same as the algebraic one? If not, find the flaw.

63. **Speed of a Plane** On a recent flight from Phoenix to Kansas City, a distance of 919 nautical miles, the plane arrived 20 minutes early. On leaving the aircraft, I asked the captain, "What was our tail wind?" He replied, "I don't know, but our ground speed was 550 knots." Has enough information been provided for you to find the tail wind? If possible, find the tail wind. (1 knot = 1 nautical mile per hour)

64. **Critical Thinking** Without solving, explain what is wrong with the following mixture problem: How many liters of 25% ethanol should be added to 20 liters of 48% ethanol to obtain a solution of 58% ethanol? Now go through an algebraic solution. What happens?

Chapter Review

Things to Know

Quadratic formula (pp. 97 and 110)

If $ax^2 + bx + c = 0, a \neq 0$, then $x = \dfrac{-b \pm \sqrt{b^2 - 4ac}}{2a}$.

If $b^2 - 4ac < 0$, there are no real solutions.

Discriminant (pp. 97 and 110)

If $b^2 - 4ac > 0$, there are two unequal real solutions.

If $b^2 - 4ac = 0$, there is one repeated real solution, a double root.

If $b^2 - 4ac < 0$, there are no real solutions, but there are two distinct complex solutions that are not real; the complex solutions are conjugates of each other.

Interval notation (p. 120)

$[a, b]$	$\{x \mid a \leq x \leq b\}$	(a, b)	$\{x \mid a < x < b\}$	$(-\infty, a]$	$\{x \mid x \leq a\}$
$[a, b)$	$\{x \mid a \leq x < b\}$	$[a, \infty)$	$\{x \mid x \geq a\}$	$(-\infty, a)$	$\{x \mid x < a\}$
$(a, b]$	$\{x \mid a < x \leq b\}$	(a, ∞)	$\{x \mid x > a\}$	$(-\infty, \infty)$	All real numbers

Properties of inequalities

Nonnegative property (p. 121)	For any real number a, $a^2 \geq 0$.	
Addition property (p. 121)	If $a < b$, then $a + c < b + c$.	If $a > b$, then $a + c > b + c$.
Multiplication properties (p. 122)	If $a < b$ and if $c > 0$, then $ac < bc$.	If $a > b$ and if $c > 0$, then $ac > bc$.
	If $a < b$ and if $c < 0$, then $ac > bc$.	If $a > b$ and if $c < 0$, then $ac < bc$.
Reciprocal properties (p. 123)	If $a > 0$, then $\dfrac{1}{a} > 0$.	If $a < 0$, then $\dfrac{1}{a} < 0$.
	If $b > a > 0$, then $\dfrac{1}{a} > \dfrac{1}{b} > 0$.	If $a < b < 0$, then $\dfrac{1}{b} < \dfrac{1}{a} < 0$.

Absolute value

If $|u| = a, a > 0$, then $u = -a$ or $u = a$. (p. 130)

If $|u| \leq a, a > 0$, then $-a \leq u \leq a$. (p. 131)

If $|u| \geq a, a > 0$, then $u \leq -a$ or $u \geq a$. (p. 132)

Objectives

Section	You should be able to...	Examples	Review Exercises
1.1	1 Solve a linear equation (p. 84)	1–3	1, 2, 7, 8
	2 Solve equations that lead to linear equations (p. 86)	4–6	3
	3 Solve problems that can be modeled by linear equations (p. 87)	7–9	45, 57
1.2	1 Solve a quadratic equation by factoring (p. 93)	1, 2	6, 9, 22
	2 Solve a quadratic equation using the square root method (p. 94)	3	4
	3 Solve a quadratic equation by completing the square (p. 95)	4, 5	5, 6, 9, 10, 13, 21, 22
	4 Solve a quadratic equation using the quadratic formula (p. 96)	6–9	5, 6, 9, 10, 13, 21, 22
	5 Solve problems that can be modeled by quadratic equations (p. 99)	10	50, 54, 56
1.3	1 Add, subtract, multiply, and divide complex numbers (p. 105)	1–8	35–39
	2 Solve quadratic equations in the complex number system (p. 109)	9–12	40–43
1.4	1 Solve radical equations (p. 113)	1–3	11, 12, 15–19, 23
	2 Solve equations quadratic in form (p. 114)	4–6	14, 20
	3 Solve equations by factoring (p. 116)	7, 8	26, 27
1.5	1 Use interval notation (p. 120)	1, 2	28–34
	2 Use properties of inequalities (p. 121)	3–6	28–34, 47

Section	You should be able to...	Examples	Review Exercises
	3 Solve inequalities (p. 123)	7, 8, 11, 13	28, 47
	4 Solve combined inequalities (p. 124)	9, 10	29, 30
1.6	1 Solve equations involving absolute value (p. 130)	1	24, 25
	2 Solve inequalities involving absolute value (p. 130)	2–6	31–34
1.7	1 Translate verbal descriptions into mathematical expressions (p. 135)	1	44
	2 Solve interest problems (p. 136)	2, 3	45, 60
	3 Solve mixture problems (p. 137)	4	53, 55
	4 Solve uniform motion problems (p. 138)	5, 6	46, 48, 49, 59
	5 Solve constant rate job problems (p. 140)	7	51, 52, 58

Review Exercises

In Problems 1–27, find the real solutions, if any, of each equation. (Where they appear, a, b, m, and n are positive constants.)

1. $2 - \dfrac{x}{3} = 8$

2. $-2(5 - 3x) + 8 = 4 + 5x$

3. $\dfrac{x}{x - 1} = \dfrac{6}{5} \quad x \neq 1$

4. $(2x + 7)^2 = 20$

5. $x(1 - x) = 6$

6. $x(1 + x) = 6$

7. $\dfrac{1}{2}\left(x - \dfrac{1}{3}\right) = \dfrac{3}{4} - \dfrac{x}{6}$

8. $\dfrac{1 - 3x}{4} = \dfrac{x + 6}{3} + \dfrac{1}{2}$

9. $(x - 1)(2x + 3) = 3$

10. $2x + 3 = 4x^2$

11. $\sqrt[3]{x^2 - 1} = 2$

12. $\sqrt{1 + x^3} = 3$

13. $x(x + 1) + 2 = 0$

14. $x^4 - 5x^2 + 4 = 0$

15. $\sqrt{2x - 3} + x = 3$

16. $\sqrt[4]{2x + 3} = 2$

17. $\sqrt{4x^2 + x - 6} = \sqrt{x - 1}$

18. $\sqrt{2x - 1} - \sqrt{x - 5} = 3$

19. $2x^{1/2} - 3 = 0$

20. $x^{-6} - 7x^{-3} - 8 = 0$

21. $x^2 + m^2 = 2mx + (nx)^2 \quad n \neq 1$

22. $10a^2 x^2 - 2abx - 36b^2 = 0$

23. $\sqrt{x^2 + 3x + 7} - \sqrt{x^2 - 3x + 9} + 2 = 0$

24. $|2x + 3| = 7$

25. $|2 - 3x| + 2 = 9$

26. $2x^3 = 3x^2$

27. $2x^3 + 5x^2 - 8x - 20 = 0$

In Problems 28–34, solve each inequality. Express your answer using set notation or interval notation. Graph the solution set.

28. $\dfrac{2x - 3}{5} + 2 \leq \dfrac{x}{2}$

29. $-9 \leq \dfrac{2x + 3}{-4} \leq 7$

30. $2 < \dfrac{3 - 3x}{12} < 6$

31. $|3x + 4| < \dfrac{1}{2}$

32. $|2x - 5| \geq 9$

33. $2 + |2 - 3x| \leq 4$

34. $1 - |2 - 3x| < -4$

In Problems 35–39, use the complex number system and write each expression in the standard form a + bi.

35. $(6 + 3i) - (2 - 4i)$

36. $4(3 - i) + 3(-5 + 2i)$

37. $\dfrac{3}{3 + i}$

38. i^{50}

39. $(2 + 3i)^3$

In Problems 40–43, solve each equation in the complex number system.

40. $x^2 + x + 1 = 0$

41. $2x^2 + x - 2 = 0$

42. $x^2 + 3 = x$

43. $x(1 - x) = 6$

44. Translate the following statement into a mathematical expression: The total cost C of manufacturing x bicycles in one day is $50,000 plus $95 times the number of bicycles manufactured.

45. Financial Planning Steve, a recent retiree, requires $5000 per year in extra income. He has $70,000 to invest and can invest in A-rated bonds paying 8% per year or in a certificate of deposit (CD) paying 5% per year. How much money should be invested in each to realize exactly $5000 in interest per year?

46. Lightning and Thunder A flash of lightning is seen, and the resulting thunderclap is heard 3 seconds later. If the speed of sound averages 1100 feet per second, how far away is the storm?

47. Physics: Intensity of Light The intensity I (in candlepower) of a certain light source obeys the equation $I = \dfrac{900}{x^2}$, where x is the distance (in meters) from the light. Over what range of distances can an object be placed from this light source so that the range of intensity of light is from 1600 to 3600 candlepower, inclusive?

48. Extent of Search and Rescue A search plane has a cruising speed of 250 miles per hour and carries enough fuel for at most 5 hours of flying. If there is a wind that averages 30 miles per hour and the direction of the search is with the wind one way and against it the other, how far can the search plane travel before it has to turn back?

49. Rescue at Sea A life raft, set adrift from a sinking ship 150 miles offshore, travels directly toward a Coast Guard station at the rate of 5 miles per hour. At the time that the raft is set adrift, a rescue helicopter is dispatched from the Coast Guard station. If the helicopter's average speed is 90 miles per hour, how long will it take the helicopter to reach the life raft?

90 mi/h

5 mi/h

|← 150 mi →|

50. Physics An object is thrown down from the top of a building 1280 feet tall with an initial velocity of 32 feet per second. The distance s (in feet) of the object from the ground after t seconds is $s = 1280 - 32t - 16t^2$.
(a) When will the object strike the ground?
(b) What is the height of the object after 4 seconds?

51. Working Together to Get a Job Done Clarissa and Shawna, working together, can paint the exterior of a house in 6 days. Clarissa by herself can complete this job in 5 days less than Shawna. How long will it take Clarissa to complete the job by herself?

52. Emptying a Tank Two pumps of different sizes, working together, can empty a fuel tank in 5 hours. The larger pump can empty this tank in 4 hours less than the smaller one. If the larger pump is out of order, how long will it take the smaller one to do the job alone?

53. Chemistry: Salt Solutions How much water should be added to 64 ounces of a 10% salt solution to make a 2% salt solution?

54. Geometry The diagonal of a rectangle measures 10 inches. If the length is 2 inches more than the width, find the dimensions of the rectangle.

55. Chemistry: Mixing Acids A laboratory has 60 cubic centimeters (cm³) of a solution that is 40% HCl acid. How many cubic centimeters of a 15% solution of HCl acid should be mixed with the 60 cm³ of 40% acid to obtain a solution of 25% HCl? How much of the 25% solution is there?

56. Framing a Painting An artist has 50 inches of oak trim to frame a painting. The frame is to have a border 3 inches wide surrounding the painting.
(a) If the painting is square, what are its dimensions? What are the dimensions of the frame?
(b) If the painting is rectangular with a length twice its width, what are the dimensions of the painting? What are the dimensions of the frame?

57. Finance An inheritance of $900,000 is to be divided among Scott, Alice, and Tricia in the following manner: Alice is to receive $\dfrac{3}{4}$ of what Scott gets, while Tricia gets $\dfrac{1}{2}$ of what Scott gets. How much does each receive?

58. Utilizing Copying Machines A new copying machine can do a certain job in 1 hour less than an older copier. Together they can do this job in 72 minutes. How long would it take the older copier by itself to do the job?

59. Evening Up a Race In a 100-meter race, Todd crosses the finish line 5 meters ahead of Scott. To even things up, Todd suggests to Scott that they race again, this time with Todd lining up 5 meters behind the start.
(a) Assuming that Todd and Scott run at the same pace as before, does the second race end in a tie?
(b) If not, who wins?
(c) By how many meters does he win?
(d) How far back should Todd start so that the race ends in a tie?

After running the race a second time, Scott, to even things up, suggests to Todd that he (Scott) line up 5 meters in front of the start.
(e) Assuming again that they run at the same pace as in the first race, does the third race result in a tie?
(f) If not, who wins?
(g) By how many meters?
(h) How far ahead should Scott start so that the race ends in a tie?

60. Student Loan Lenah accumulated $24,000 in simple interest school loans by the time she graduated. She deferred payments for three years while she looked for a permanent full-time job. If she owed $27,060 at the end of the three years when she began repayment, what was her interest rate?

Chapter Test

CHAPTER
Test Prep
VIDEOS

The Chapter Test Prep Videos include step-by-step solutions to all chapter test exercises. These videos are available in MyLab™ Math, or on this text's YouTube Channel. Refer to the Preface for a link to the YouTube channel.

In Problems 1–7, find the real solutions of each equation.

1. $\dfrac{2x}{3} - \dfrac{x}{2} = \dfrac{5}{12}$

2. $x(x - 1) = 6$

3. $x^4 - 3x^2 - 4 = 0$

4. $\sqrt{2x - 5} + 2 = 4$

5. $|2x - 3| + 7 = 10$

6. $3x^3 + 2x^2 - 12x - 8 = 0$

7. $3x^2 - x + 1 = 0$

In Problems 8–10, solve each inequality. Express your answer using interval notation. Graph the solution set.

8. $-3 \le \dfrac{3x - 4}{2} \le 6$

9. $|3x + 4| < 8$

10. $2 + |2x - 5| \ge 9$

11. Write $\dfrac{-2}{3 - i}$ in the standard form $a + bi$.

12. Solve the equation $4x^2 - 4x + 5 = 0$ in the complex number system.

13. Blending Coffee A coffee house has 20 pounds of a coffee that sells for $4 per pound. How many pounds of a coffee that sells for $8 per pound should be mixed with the 20 pounds of $4-per-pound coffee to obtain a blend that will sell for $5 per pound? How much of the $5-per-pound coffee is there to sell?

Chapter Projects

(((•))) **Internet-based Project**

I. **Financing a Purchase** At some point in your life, you are likely to need to borrow money to finance a purchase. For example, most of us will finance the purchase of a car or a home. What is the mathematics behind financing a purchase? When you borrow money from a bank, the bank uses a rather complex equation (or formula) to determine how much you need to pay each month to repay the loan. There are a number of variables that determine the monthly payment. These variables include the amount borrowed, the interest rate, and the length of the loan. The interest rate is based on current economic conditions, the length of the loan, the type of item being purchased, and your credit history.

The formula (top, right) gives the monthly payment P required to pay off a loan amount L at an annual interest rate r, expressed as a decimal, but usually given as a percent.

The time t, measured in months, is the length of the loan. For example, a 30-year loan requires $12 \times 30 = 360$ monthly payments.

$$P = L\left[\dfrac{\dfrac{r}{12}}{1 - \left(1 + \dfrac{r}{12}\right)^{-t}}\right]$$

P = monthly payment
L = loan amount
r = annual rate of interest expressed as a decimal
t = length of loan, in months

1. Interest rates change daily. Many websites post current interest rates on loans. Go to *www.bankrate.com* (or some other website that posts lenders' interest rates) and find the current best interest rate on a 60-month new-car purchase loan. Use this rate to determine the monthly payment on a $30,000 automobile loan.

2. Determine the total amount paid for the loan by multiplying the loan payment by the term of the loan.

3. Determine the total amount of interest paid by subtracting the loan amount from the total amount paid from question 2.

4. More often than not, we decide how much of a payment we can afford and use that information to determine the loan amount. Suppose you can afford a monthly payment of $500. Use the interest rate from question 1 to determine the maximum amount you can borrow. If you have $5000 to put down on the car, what is the maximum value of a car you can purchase?

5. Repeat questions 1 through 4 using a 72-month new-car purchase loan, a 60-month used-car purchase loan, and a 72-month used-car purchase loan.

6. We can use the power of a spreadsheet, such as Excel, to create a loan amortization schedule. A loan amortization schedule is a list of the monthly payments, a breakdown of interest and principal, along with a current loan balance. Create a loan amortization schedule for each of the four loan scenarios discussed on the previous page, using the following as a guide. You may want to use an Internet search engine to research specific keystrokes for creating an amortization schedule in a spreadsheet. We supply a sample spreadsheet with formulas included as a guide. Use the spreadsheet to verify your results from questions 1 through 5.

	A	B	C	D	E	F	G	H
1	**Loan Information**		**Payment Number**	**Payment Amount**	**Interest**	**Principal**	**Balance**	**Total Interest Paid**
2	Loan Amount	$30,000	1	=PMT($B$3/12,$B$5,-$B$2,0)	=B2*$B$3/12	=D2-E2	=B2-F2	=B2*$B$3/12
3	Annual Interest Rate	0.045	2	=PMT($B$3/12,$B$5,-$B$2,0)	=G2*$B$3/12	=D3-E3	=G2-F3	=H2+E3
4	Length of Loan (years)	5	3	=PMT($B$3/12,$B$5,-$B$2,0)	=G3*$B$3/12	=D4-E4	=G3-F4	=H3+E4
5	Number of Payments	=B4*12	⋮	⋮	⋮	⋮	⋮	⋮

7. Go to an online automobile website such as *www.cars.com, www.edmunds.com,* or *www.autobytel.com.* Research the types of vehicles you can afford for a monthly payment of $500. Decide on a vehicle you would purchase based on your analysis in questions 1–6. Be sure to justify your decision, and include the impact the term of the loan has on your decision. You might consider other factors in your decision, such as expected maintenance costs and insurance costs.

Citation: Excel ©2018 Microsoft Corporation. Used with permission from Microsoft.

The following project is also available on the Instructor's Resource Center (IRC):

II. **Project at Motorola** **How Many Cellular Phones Can I Make?** An industrial engineer uses a model involving equations to be sure production levels meet customer demand.

Graphs

How to Value a House

Two things to consider in valuing a home: (1) How does it compare to similar nearby homes that have sold recently? (2) What value do you place on the advertised features and amenities?

The Zestimate® home value is a good starting point in figuring out the value of a home. It shows you how the home compares relative to others in the area, but you then need to add in all the other qualities that only someone who has seen the house knows.

Looking at "comps"

Knowing whether an asking price is fair will be important when you're ready to make an offer on a house. It will be even more important when your mortgage lender hires an appraiser to determine whether the house is worth the loan you're after.

Check on Zillow to see recent sales of similar, or comparable, homes in the area. Print them out and keep these "comps." You'll be referring to them quite a bit.

Note that "recent sales" usually means within the past six months. A sales price from a year ago probably bears little or no relation to what is going on in your area right now. In fact, some lenders will not accept comps older than three months.

Market activity also determines how easy or difficult it is to find accurate comps. In a "hot" or busy market, you're likely to have lots of comps to choose from. In a less active market finding reasonable comps becomes harder. And if the home you're looking at has special design features, finding a comparable property is harder still. It's also necessary to know what's going on in a given sub-segment. Maybe large, high-end homes are selling like hotcakes, but owners of smaller houses are staying put, or vice versa.

Source: *http://luthersanchez.com/2016/03/09/how-to-value-a-house/*

 —*See the Internet-based Chapter Project*—

← A Look Back

Chapter R and Chapter 1 review skills from intermediate algebra.

A Look Ahead →

Here we connect algebra and geometry using the rectangular coordinate system. In the 1600s, algebra had developed to the point that René Descartes (1596-1650) and Pierre de Fermat (1601-1665) were able to use rectangular coordinates to translate geometry problems into algebra problems, and vice versa. This enabled both geometers and algebraists to gain new insights into their subjects, which had been thought to be separate but now were seen as connected.

Outline

2.1 The Distance and Midpoint Formulas

PREPARING FOR THIS SECTION *Before getting started, review the following:*

- Algebra Essentials (Section R.2, pp. 17–26)
- Geometry Essentials (Section R.3, pp. 30–35)

Now Work the 'Are You Prepared?' problems on page 154.

OBJECTIVES 1 Use the Distance Formula (p. 151)
2 Use the Midpoint Formula (p. 153)

Rectangular Coordinates

We locate a point on the real number line by assigning it a single real number, called the *coordinate of the point*. For work in a two-dimensional plane, we locate points by using two numbers.

Begin with two real number lines located in the same plane: one horizontal and the other vertical. The horizontal line is called the **x-axis**, the vertical line the **y-axis**, and the point of intersection the **origin O**. See Figure 1. Assign coordinates to every point on these number lines using a convenient scale. In mathematics, we usually use the same scale on each axis, but in applications, different scales appropriate to the application may be used.

The origin O has a value of 0 on both the x-axis and the y-axis. Points on the x-axis to the right of O are associated with positive real numbers, and those to the left of O are associated with negative real numbers. Points on the y-axis above O are associated with positive real numbers, and those below O are associated with negative real numbers. In Figure 1, the x-axis and y-axis are labeled as x and y, respectively, and an arrow at the end of each axis is used to denote the positive direction.

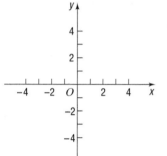

Figure 1 *xy*-Plane

The coordinate system described here is called a **rectangular** or **Cartesian**[*] **coordinate system**. The x-axis and y-axis lie in a *plane* called the **xy-plane**, and the x-axis and y-axis are referred to as the **coordinate axes**.

Any point P in the xy-plane can be located by using an **ordered pair** (x, y) of real numbers. Let x denote the signed distance of P from the y-axis (*signed* means that if P is to the right of the y-axis, then $x > 0$, and if P is to the left of the y-axis, then $x < 0$); and let y denote the signed distance of P from the x-axis. The ordered pair (x, y), also called the **coordinates** of P, gives us enough information to locate the point P in the plane.

For example, to locate the point whose coordinates are $(-3, 1)$, go 3 units along the x-axis to the left of O and then go straight up 1 unit. We **plot** this point by placing a dot at this location. See Figure 2, in which the points with coordinates $(-3, 1)$, $(-2, -3)$, $(3, -2)$, and $(3, 2)$ are plotted.

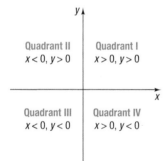

Figure 2

The origin has coordinates $(0, 0)$. Any point on the x-axis has coordinates of the form $(x, 0)$, and any point on the y-axis has coordinates of the form $(0, y)$.

If (x, y) are the coordinates of a point P, then x is called the **x-coordinate**, or **abscissa**, of P, and y is the **y-coordinate**, or **ordinate**, of P. We identify the point P by its coordinates (x, y) by writing $P = (x, y)$. Usually, we will simply say "the point (x, y)" rather than "the point whose coordinates are (x, y)."

The coordinate axes partition the xy-plane into four sections called **quadrants**, as shown in Figure 3. In quadrant I, both the x-coordinate and the y-coordinate of all points are positive; in quadrant II, x is negative and y is positive; in quadrant III, both x and y are negative; and in quadrant IV, x is positive and y is negative. Points on the coordinate axes belong to no quadrant.

Quadrant II	Quadrant I
$x < 0, y > 0$	$x > 0, y > 0$
Quadrant III	Quadrant IV
$x < 0, y < 0$	$x > 0, y < 0$

Figure 3

Now Work PROBLEM 15

---

*Named after René Descartes (1596–1650), a French mathematician, philosopher, and theologian.

COMMENT On a graphing calculator, you can set the scale on each axis. Once this has been done, you obtain the **viewing rectangle**. See Figure 4 for a typical viewing rectangle. You should now read Section 1, *The Viewing Rectangle*, in the Appendix.

Figure 4 TI-84 Plus C Standard Viewing Rectangle ∎

1 Use the Distance Formula

If the same units of measurement (such as inches, centimeters, and so on) are used for both the *x*-axis and *y*-axis, then all distances in the *xy*-plane can be measured using this unit of measurement.

EXAMPLE 1

Finding the Distance between Two Points

Find the distance *d* between the points $(1, 3)$ and $(5, 6)$.

Solution

First plot the points $(1, 3)$ and $(5, 6)$ and connect them with a line segment. See Figure 5(a). To find the length *d*, begin by drawing a horizontal line segment from $(1, 3)$ to $(5, 3)$ and a vertical line segment from $(5, 3)$ to $(5, 6)$, forming a right triangle, as shown in Figure 5(b). One leg of the triangle is of length 4 (since $|5 - 1| = 4$), and the other is of length 3 (since $|6 - 3| = 3$). By the Pythagorean Theorem, the square of the distance *d* that we seek is

$$d^2 = 4^2 + 3^2 = 16 + 9 = 25$$
$$d = \sqrt{25} = 5$$

> **Need to Review?**
>
> The Pythagorean Theorem and its converse are discussed in Section R.3, pp. 30–31.

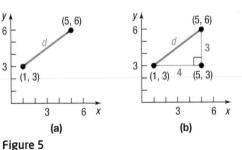

(a)   **(b)**

Figure 5

The **distance formula** provides a straightforward method for computing the distance between two points.

THEOREM Distance Formula

The distance between two points $P_1 = (x_1, y_1)$ and $P_2 = (x_2, y_2)$, denoted by $d(P_1, P_2)$, is

$$d(P_1, P_2) = \sqrt{(x_2 - x_1)^2 + (y_2 - y_1)^2} \tag{1}$$

> **In Words**
>
> To compute the distance between two points, find the difference of the x-coordinates, square it, and add this to the square of the difference of the y-coordinates. The square root of this sum is the distance.

Proof of the Distance Formula Let (x_1, y_1) denote the coordinates of point P_1 and let (x_2, y_2) denote the coordinates of point P_2.

- Assume that the line joining P_1 and P_2 is neither horizontal nor vertical. Refer to Figure 6(a) on page 152. The coordinates of P_3 are (x_2, y_1). The horizontal distance from P_1 to P_3 equals the absolute value of the difference of the x-coordinates, $|x_2 - x_1|$. The vertical distance from P_3 to P_2 equals the absolute value of the difference of the y-coordinates, $|y_2 - y_1|$. See Figure 6(b). The distance $d(P_1, P_2)$ is the length of the hypotenuse of the right triangle, so, by the Pythagorean Theorem, it follows that

$$[d(P_1, P_2)]^2 = |x_2 - x_1|^2 + |y_2 - y_1|^2$$
$$= (x_2 - x_1)^2 + (y_2 - y_1)^2$$
$$d(P_1, P_2) = \sqrt{(x_2 - x_1)^2 + (y_2 - y_1)^2}$$

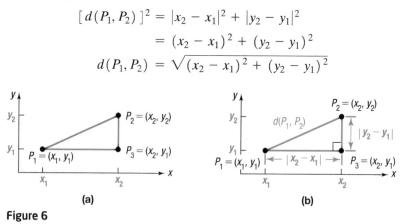

(a)　　　　　　　　　　　　　　(b)

Figure 6

- If the line joining P_1 and P_2 is horizontal, then the y-coordinate of P_1 equals the y-coordinate of P_2; that is, $y_1 = y_2$. Refer to Figure 7(a). In this case, the distance formula (1) still works, because for $y_1 = y_2$, it reduces to

$$d(P_1, P_2) = \sqrt{(x_2 - x_1)^2 + 0^2} = \sqrt{(x_2 - x_1)^2} = |x_2 - x_1|$$

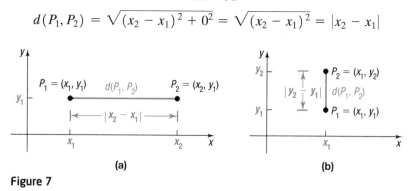

(a)　　　　　　　　　　　　　　(b)

Figure 7

- A similar argument holds if the line joining P_1 and P_2 is vertical. See Figure 7(b).

EXAMPLE 2 | **Using the Distance Formula**

Find the distance d between the points $(-4, 5)$ and $(3, 2)$.

Solution　Using the distance formula, equation (1), reveals that the distance d is

$$d = \sqrt{[3 - (-4)]^2 + (2 - 5)^2} = \sqrt{7^2 + (-3)^2}$$
$$= \sqrt{49 + 9} = \sqrt{58} \approx 7.62$$

Now Work PROBLEMS 19 AND 23

The distance between two points $P_1 = (x_1, y_1)$ and $P_2 = (x_2, y_2)$ is never a negative number. Also, the distance between two points is 0 only when the points are identical—that is, when $x_1 = x_2$ and $y_1 = y_2$. And, because $(x_2 - x_1)^2 = (x_1 - x_2)^2$ and $(y_2 - y_1)^2 = (y_1 - y_2)^2$, it makes no difference whether the distance is computed from P_1 to P_2 or from P_2 to P_1; that is, $d(P_1, P_2) = d(P_2, P_1)$.

The introduction to this chapter mentioned that rectangular coordinates enable us to translate geometry problems into algebra problems, and vice versa. The next example shows how algebra (the distance formula) can be used to solve geometry problems.

EXAMPLE 3

Using Algebra to Solve a Geometry Problem

Consider the three points $A = (-2, 1)$, $B = (2, 3)$, and $C = (3, 1)$.

(a) Plot each point and form the triangle ABC.
(b) Find the length of each side of the triangle.
(c) Show that the triangle is a right triangle.
(d) Find the area of the triangle.

Solution

(a) Figure 8 shows the points A, B, C and the triangle ABC.

(b) To find the length of each side of the triangle, use the distance formula, equation (1).

$$d(A, B) = \sqrt{[2 - (-2)]^2 + (3 - 1)^2} = \sqrt{16 + 4} = \sqrt{20} = 2\sqrt{5}$$
$$d(B, C) = \sqrt{(3 - 2)^2 + (1 - 3)^2} = \sqrt{1 + 4} = \sqrt{5}$$
$$d(A, C) = \sqrt{[3 - (-2)]^2 + (1 - 1)^2} = \sqrt{25 + 0} = 5$$

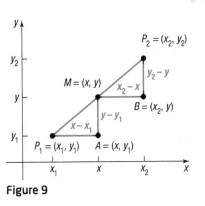
Figure 8

(c) If the sum of the squares of the lengths of two of the sides equals the square of the length of the third side, then the triangle is a right triangle. Looking at Figure 8, it seems reasonable to conjecture that the angle at vertex B might be a right angle. We shall check to see whether

$$[d(A, B)]^2 + [d(B, C)]^2 = [d(A, C)]^2$$

Using the results in part (b) yields

$$[d(A, B)]^2 + [d(B, C)]^2 = (2\sqrt{5})^2 + (\sqrt{5})^2$$
$$= 20 + 5 = 25 = [d(A, C)]^2$$

It follows from the converse of the Pythagorean Theorem that triangle ABC is a right triangle.

(d) Because the right angle is at vertex B, the sides AB and BC form the base and height of the triangle. Its area is

$$\text{Area} = \frac{1}{2} \cdot \text{Base} \cdot \text{Height} = \frac{1}{2} \cdot 2\sqrt{5} \cdot \sqrt{5} = 5 \text{ square units}$$

Now Work PROBLEM 33

2 Use the Midpoint Formula

We now derive a formula for the coordinates of the **midpoint of a line segment**. Let $P_1 = (x_1, y_1)$ and $P_2 = (x_2, y_2)$ be the endpoints of a line segment, and let $M = (x, y)$ be the point on the line segment that is the same distance from P_1 as it is from P_2. See Figure 9. The triangles P_1AM and MBP_2 are congruent. [Do you see why? $d(P_1, M) = d(M, P_2)$ is given; also, $\angle AP_1M = \angle BMP_2$[*] and $\angle P_1MA = \angle MP_2B$. So, we have angle–side–angle.] Because triangles P_1AM and MBP_2 are congruent, corresponding sides are equal in length. That is,

$$x - x_1 = x_2 - x \quad \text{and} \quad y - y_1 = y_2 - y$$
$$2x = x_1 + x_2 \qquad\qquad 2y = y_1 + y_2$$
$$x = \frac{x_1 + x_2}{2} \qquad\qquad y = \frac{y_1 + y_2}{2}$$

Figure 9

*A postulate from geometry states that the transversal $\overline{P_1P_2}$ forms congruent corresponding angles with the parallel line segments $\overline{P_1A}$ and $\overline{MB}$.

┌─ THEOREM Midpoint Formula

In Words

To find the midpoint of a line segment, average the x-coordinates of the endpoints, and average the y-coordinates of the endpoints.

The midpoint $M = (x, y)$ of the line segment from $P_1 = (x_1, y_1)$ to $P_2 = (x_2, y_2)$ is

$$M = (x, y) = \left(\frac{x_1 + x_2}{2}, \frac{y_1 + y_2}{2} \right)$$ (2)

EXAMPLE 4   **Finding the Midpoint of a Line Segment**

Find the midpoint of the line segment from $P_1 = (-5, 5)$ to $P_2 = (3, 1)$. Plot the points P_1 and P_2 and their midpoint.

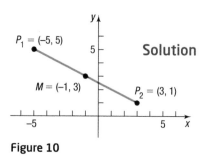

Figure 10

Solution   Use the midpoint formula (2) with $x_1 = -5, y_1 = 5$, $x_2 = 3$, and $y_2 = 1$. The coordinates (x, y) of the midpoint M are

$$x = \frac{x_1 + x_2}{2} = \frac{-5 + 3}{2} = -1 \quad \text{and} \quad y = \frac{y_1 + y_2}{2} = \frac{5 + 1}{2} = 3$$

That is, $M = (-1, 3)$. See Figure 10.

─────── **Now Work** PROBLEM 39

2.1 Assess Your Understanding

'Are You Prepared?' *Answers are given at the end of these exercises. If you get a wrong answer, read the pages listed in* red.

1. On the real number line, the origin is assigned the number _____. (p. 17)

2. If -3 and 5 are the coordinates of two points on the real number line, the distance between these points is _____. (pp. 19–20)

3. If 3 and 4 are the legs of a right triangle, the hypotenuse is _____. (p. 30)

4. Use the converse of the Pythagorean Theorem to show that a triangle whose sides are of lengths 11, 60, and 61 is a right triangle. (pp. 30–31)

5. The area A of a triangle whose base is b and whose altitude is h is $A =$ _____. (p. 31)

6. *True or False* Two triangles are congruent if two angles and the included side of one equals two angles and the included side of the other. (pp. 32–33)

Concepts and Vocabulary

7. If (x, y) are the coordinates of a point P in the xy-plane, then x is called the _____ of P, and y is the _____ of P.

8. The coordinate axes partition the xy-plane into four sections called _____.

9. If three distinct points P, Q, and R all lie on a line, and if $d(P, Q) = d(Q, R)$, then Q is called the _____ of the line segment from P to R.

10. *True or False* The distance between two points is sometimes a negative number.

11. *True or False* The point $(-1, 4)$ lies in quadrant IV of the Cartesian plane.

12. *True or False* The midpoint of a line segment is found by averaging the x-coordinates and averaging the y-coordinates of the endpoints.

13. *Multiple Choice* Which of the following statements is true for a point (x, y) that lies in quadrant III?
    (a) Both x and y are positive.
    (b) Both x and y are negative.
    (c) x is positive, and y is negative.
    (d) x is negative, and y is positive.

14. *Multiple Choice* Choose the expression that equals the distance between two points (x_1, y_1) and (x_2, y_2).
    (a) $\sqrt{(x_2 - x_1)^2 + (y_2 - y_1)^2}$
    (b) $\sqrt{(x_2 + x_1)^2 - (y_2 + y_1)^2}$
    (c) $\sqrt{(x_2 - x_1)^2 - (y_2 - y_1)^2}$
    (d) $\sqrt{(x_2 + x_1)^2 + (y_2 + y_1)^2}$

Skill Building

In Problems 15 and 16, plot each point in the xy-plane. State which quadrant or on what coordinate axis each point lies.

15. **(a)** $A = (-3, 2)$ **(d)** $D = (6, 5)$
    (b) $B = (6, 0)$ **(e)** $E = (0, -3)$
    (c) $C = (-2, -2)$ **(f)** $F = (6, -3)$

16. **(a)** $A = (1, 4)$ **(d)** $D = (4, 1)$
    (b) $B = (-3, -4)$ **(e)** $E = (0, 1)$
    (c) $C = (-3, 4)$ **(f)** $F = (-3, 0)$

17. Plot the points $(2, 0), (2, -3), (2, 4), (2, 1)$, and $(2, -1)$. Describe the set of all points of the form $(2, y)$, where y is a real number.

18. Plot the points $(0, 3), (1, 3), (-2, 3), (5, 3)$, and $(-4, 3)$. Describe the set of all points of the form $(x, 3)$, where x is a real number.

In Problems 19–32, find the distance d between the points P_1 and P_2.

19.

20.

21.

22.

23. $P_1 = (3, -4)$; $P_2 = (5, 4)$ 24. $P_1 = (-1, 0)$; $P_2 = (2, 4)$

25. $P_1 = (-7, 3)$; $P_2 = (4, 0)$ 26. $P_1 = (2, -3)$; $P_2 = (4, 2)$

27. $P_1 = (5, -2)$; $P_2 = (6, 1)$ 28. $P_1 = (-4, -3)$; $P_2 = (6, 2)$

29. $P_1 = (-0.2, 0.3)$; $P_2 = (2.3, 1.1)$ 30. $P_1 = (1.2, 2.3)$; $P_2 = (-0.3, 1.1)$

31. $P_1 = (a, b)$; $P_2 = (0, 0)$ 32. $P_1 = (a, a)$; $P_2 = (0, 0)$

In Problems 33–38, plot each point and form the triangle ABC. Show that the triangle is a right triangle. Find its area.

33. $A = (-2, 5)$; $B = (1, 3)$; $C = (-1, 0)$ 34. $A = (-2, 5)$; $B = (12, 3)$; $C = (10, -11)$

35. $A = (-5, 3)$; $B = (6, 0)$; $C = (5, 5)$ 36. $A = (-6, 3)$; $B = (3, -5)$; $C = (-1, 5)$

37. $A = (4, -3)$; $B = (0, -3)$; $C = (4, 2)$ 38. $A = (4, -3)$; $B = (4, 1)$; $C = (2, 1)$

In Problems 39–46, find the midpoint of the line segment joining the points P_1 and P_2.

39. $P_1 = (3, -4)$; $P_2 = (5, 4)$ 40. $P_1 = (-2, 0)$; $P_2 = (2, 4)$

41. $P_1 = (-1, 4)$; $P_2 = (8, 0)$ 42. $P_1 = (2, -3)$; $P_2 = (4, 2)$

43. $P_1 = (7, -5)$; $P_2 = (9, 1)$ 44. $P_1 = (-4, -3)$; $P_2 = (2, 2)$

45. $P_1 = (a, b)$; $P_2 = (0, 0)$ 46. $P_1 = (a, a)$; $P_2 = (0, 0)$

Applications and Extensions

47. If the point $(2, 5)$ is shifted 3 units to the right and 2 units down, what are its new coordinates?

48. If the point $(-1, 6)$ is shifted 2 units to the left and 4 units up, what are its new coordinates?

49. Find all points having an x-coordinate of 3 whose distance from the point $(-2, -1)$ is 13.

    (a) By using the Pythagorean Theorem.
    (b) By using the distance formula.

50. Find all points having a y-coordinate of -6 whose distance from the point $(1, 2)$ is 17.

    (a) By using the Pythagorean Theorem.
    (b) By using the distance formula.

51. Find all points on the x-axis that are 6 units from the point $(4, -3)$.

52. Find all points on the y-axis that are 6 units from the point $(4, -3)$.

53. Suppose that $A = (2, 5)$ are the coordinates of a point in the xy-plane.

    (a) Find the coordinates of the point if A is shifted 3 units to the left and 4 units down.
    (b) Find the coordinates of the point if A is shifted 2 units to the left and 8 units up.

54. Plot the points $A = (-1, 8)$ and $M = (2, 3)$ in the xy-plane. If M is the midpoint of a line segment AB, find the coordinates of B.

55. The midpoint of the line segment from P_1 to P_2 is $(-1, 4)$. If $P_1 = (-3, 6)$, what is P_2?

56. The midpoint of the line segment from P_1 to P_2 is $(5, -4)$. If $P_2 = (7, -2)$, what is P_1?

57. Geometry The **medians** of a triangle are the line segments from each vertex to the midpoint of the opposite side (see the figure). Find the lengths of the medians of the triangle with vertices at $A = (0, 0)$, $B = (6, 0)$, and $C = (4, 4)$.

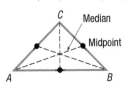

58. Geometry An **equilateral triangle** has three sides of equal length. If two vertices of an equilateral triangle are $(0, 4)$ and $(0, 0)$ find the third vertex. How many of these triangles are possible?

*In Problems 59–62, find the length of each side of the triangle determined by the three points P_1, P_2, and P_3. State whether the triangle is an isosceles triangle, a right triangle, neither of these, or both. (An **isosceles triangle** is one in which at least two of the sides are of equal length.)*

59. $P_1 = (2, 1);\ \ P_2 = (-4, 1);\ \ P_3 = (-4, -3)$

60. $P_1 = (-1, 4);\ \ P_2 = (6, 2);\ \ P_3 = (4, -5)$

61. $P_1 = (-2, -1);\ \ P_2 = (0, 7);\ \ P_3 = (3, 2)$

62. $P_1 = (7, 2);\ \ P_2 = (-4, 0);\ \ P_3 = (4, 6)$

63. Baseball A major league baseball "diamond" is actually a square 90 feet on a side (see the figure). What is the distance directly from home plate to second base (the diagonal of the square)?

64. Little League Baseball The layout of a Little League playing field is a square 60 feet on a side. How far is it directly from home plate to second base (the diagonal of the square)?

Source: 2018 Little League Baseball Official Regulations, Playing Rules, and Operating Policies

65. Baseball Refer to Problem 63. Overlay a rectangular coordinate system on a major league baseball diamond so that the origin is at home plate, the positive x-axis lies in the direction from home plate to first base, and the positive y-axis lies in the direction from home plate to third base.
  (a) What are the coordinates of first base, second base, and third base? Use feet as the unit of measurement.
  (b) If the right fielder is located at $(310, 15)$ how far is it from the right fielder to second base?
  (c) If the center fielder is located at $(300, 300)$, how far is it from the center fielder to third base?

66. Little League Baseball Refer to Problem 64. Overlay a rectangular coordinate system on a Little League baseball diamond so that the origin is at home plate, the positive x-axis lies in the direction from home plate to first base, and the positive y-axis lies in the direction from home plate to third base.
  (a) What are the coordinates of first base, second base, and third base? Use feet as the unit of measurement.
  (b) If the right fielder is located at $(180, 20)$, how far is it from the right fielder to second base?
  (c) If the center fielder is located at $(220, 220)$, how far is it from the center fielder to third base?

67. Distance between Moving Objects A Ford Focus and a Freightliner Cascadia truck leave an intersection at the same time. The Focus heads east at an average speed of 60 miles per hour, while the Cascadia heads south at an average speed of 45 miles per hour. Find an expression for their distance apart d (in miles) at the end of t hours.

68. Distance of a Moving Object from a Fixed Point A hot-air balloon, headed due east at an average speed of 15 miles per hour and at a constant altitude of 100 feet, passes over an intersection (see the figure). Find an expression for the distance d (measured in feet) from the balloon to the intersection t seconds later.

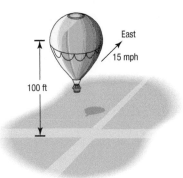

69. Drafting Error When a draftsman draws three lines that are to intersect at one point, the lines may not intersect as intended and subsequently will form an **error triangle**. If this error triangle is long and thin, one estimate for the location of the desired point is the midpoint of the shortest side. The figure shows one such error triangle.

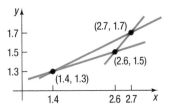

  (a) Find an estimate for the desired intersection point.
  (b) Find the distance from $(1.4, 1.3)$ to the midpoint found in part (a).

70. Net Sales The figure illustrates the net sales growth of Costco Wholesale Corporation from 2013 through 2017. Use the midpoint formula to estimate the net sales of Costco Wholesale Corporation in 2015. How does your result compare to the reported value of $113.67 billion?

Source: Costco Wholesale Corporation 2017 Annual Report

Costco Wholesale Corporation

71. Poverty Threshold Poverty thresholds are determined by the U.S. Census Bureau. A poverty threshold represents the minimum annual household income for a family not to be considered poor. In 2009, the poverty threshold for a family of four with two children under the age of 18 years was $21,756. In 2017, the poverty threshold for a family of four with two children under the age of 18 years was $24,858.

Assuming that poverty thresholds increase in a straight-line fashion, use the midpoint formula to estimate the poverty threshold for a family of four with two children under the age of 18 in 2013. How does your result compare to the actual poverty threshold in 2013 of $23,624?

Source: U.S. Census Bureau

72. Challenge Problem Geometry Verify that the points $(0,0), (a,0)$, and $\left(\dfrac{a}{2}, \dfrac{\sqrt{3}a}{2}\right)$ are the vertices of an equilateral triangle. Then show that the midpoints of the three sides are the vertices of a second equilateral triangle.

73. Challenge Problem Geometry Find the midpoint of each diagonal of a square with side of length s. Draw the conclusion that the diagonals of a square intersect at their midpoints.

[**Hint:** Use $(0, 0), (0, s), (s, 0)$, and (s, s) as the vertices of the square.]

74. Challenge Problem Geometry A point P is equidistant from $(-5, 1)$ and $(4, -4)$. Find the coordinates of P if its y-coordinate is twice its x-coordinate.

75. Challenge Problem Geometry For any parallelogram, prove that the sum of the squares of the lengths of the sides equals the sum of the squares of the lengths of the diagonals.

[**Hint:** Use $(0, 0), (a, 0), (a + b, c)$, and (b, c) as the vertices of the parallelogram.]

Explaining Concepts: Discussion and Writing

76. Write a paragraph that describes a Cartesian plane. Then write a second paragraph that describes how to plot points in the Cartesian plane. Your paragraphs should include the terms "coordinate axes," "ordered pair," "coordinates," "plot," "x-coordinate," and "y-coordinate."

Retain Your Knowledge

Problems 77–86 are based on material learned earlier in the course. The purpose of these problems is to keep the material fresh in your mind so that you are better prepared for the final exam.

77. Determine the domain of the variable x in the expression:
$$\frac{3x + 1}{2x - 5}$$

78. Find the value of $\dfrac{x^2 - 3xy + 2}{5x - 2y}$ if $x = 4$ and $y = 7$.

79. Express $(5x - 2)(3x + 7)$ as a polynomial in standard form.

80. Reduce the rational expression to lowest terms:
$$\frac{3x^2 + 7x + 2}{3x^2 - 11x - 4}$$

81. Factor the expression completely:
$$6(x + 1)^3(2x - 5)^7 - 5(x + 1)^4(2x - 5)^6$$

In Problems 82–84, find the real solution(s), if any, of each equation.

82. $3x^2 - 7x - 20 = 0$

83. $\dfrac{x}{x + 3} + \dfrac{1}{x - 3} = 1$

84. $|7x - 4| = 31$

85. Solve the inequality $5(x - 3) + 2x \geq 6(2x - 3) - 7$. Express the solution using interval notation. Graph the solution set.

86.[†] Multiply $(7 + 3i)(1 - 2i)$. Write the answer in the form $a + bi$.

[†]This problem is based on content from Section 1.3, which is optional.

'Are You Prepared?' Answers

1. 0   **2.** 8   **3.** 5   **4.** $11^2 + 60^2 = 121 + 3600 = 3721 = 61^2$ **5.** $\dfrac{1}{2}bh$ **6.** True

2.2 Graphs of Equations in Two Variables; Intercepts; Symmetry

PREPARING FOR THIS SECTION *Before getting started, review the following:*

- Solving Linear Equations (Section 1.1, pp. 82–85)
- Solve a Quadratic Equation by Factoring (Section 1.2, pp. 93–94)

Now Work the 'Are You Prepared?' problems on page 166.

OBJECTIVES **1** Graph Equations by Plotting Points (p. 158)
    2 Find Intercepts from a Graph (p. 160)
    3 Find Intercepts from an Equation (p. 161)
    4 Test an Equation for Symmetry with Respect to the x-Axis, the y-Axis, and the Origin (p. 161)
    5 Know How to Graph Key Equations (p. 164)

1 Graph Equations by Plotting Points

An **equation in two variables**, say x and y, is a statement in which two expressions involving x and y are equal. The expressions are called the **sides** of the equation. Since an equation is a statement, it may be true or false, depending on the value of the variables. Any values of x and y that result in a true statement are said to **satisfy** the equation.

For example, the following are all equations in two variables x and y:

$$x^2 + y^2 = 5 \qquad 2x - y = 6 \qquad y = 2x + 5 \qquad x^2 = y$$

The first of these, $x^2 + y^2 = 5$, is satisfied for $x = 1, y = 2$, since $1^2 + 2^2 = 5$. Other choices of x and y, such as $x = -1, y = -2$, also satisfy this equation. It is not satisfied for $x = 2$ and $y = 3$, since $2^2 + 3^2 = 4 + 9 = 13 \neq 5$.

The **graph of an equation in two variables** x and y consists of the set of points in the xy-plane whose coordinates (x, y) satisfy the equation.

Graphs play an important role in helping us to visualize the relationships that exist between two variables or quantities. Table 1 shows the average price of gasoline in the United States for the years 1991–2017 (adjusted for inflation). If we plot these data using year as the x-coordinate and price as the y-coordinate, and then connect the points (year, price), we obtain Figure 11.

Table 1   Average Price of Gasoline

Year	Price	Year	Price	Year	Price
1991	1.98	2000	2.11	2009	2.68
1992	1.90	2001	1.97	2010	3.12
1993	1.81	2002	1.83	2011	3.84
1994	1.78	2003	2.07	2012	3.87
1995	1.78	2004	2.40	2013	3.68
1996	1.87	2005	2.85	2014	3.48
1997	1.83	2006	3.13	2015	2.51
1998	1.55	2007	3.31	2016	2.19
1999	1.67	2008	3.70	2017	2.38

Source: U.S. Energy Information Administration *(https://www.eia.gov/ dnav/pet/pet_pri_gnd_dcus_nus_a.htm)*

Figure 11

EXAMPLE 1

Determining Whether a Point Is on the Graph of an Equation

Determine whether the points are on the graph of the equation $2x - y = 6$.

(a) $(2, 3)$ (b) $(2, -2)$

Solution (a) For the point $(2, 3)$, check to see whether $x = 2, y = 3$ satisfies the equation $2x - y = 6$.

$$2x - y = 2 \cdot 2 - 3 = 4 - 3 = 1 \neq 6$$

The equation is not satisfied, so the point $(2, 3)$ is not on the graph of $2x - y = 6$.

(b) For the point $(2, -2)$,

$$2x - y = 2 \cdot 2 - (-2) = 4 + 2 = 6$$

The equation is satisfied, so the point $(2, -2)$ is on the graph of $2x - y = 6$.

Now Work PROBLEM 13

EXAMPLE 2

Graphing an Equation by Plotting Points

Graph the equation: $y = 2x + 5$

Solution The graph consists of all points (x, y) that satisfy the equation. To locate some of these points (and get an idea of the pattern of the graph), assign some numbers to x, and find corresponding values for y.

If	Then	Point on Graph
$x = 0$	$y = 2 \cdot 0 + 5 = 5$	$(0, 5)$
$x = 1$	$y = 2 \cdot 1 + 5 = 7$	$(1, 7)$
$x = -5$	$y = 2 \cdot (-5) + 5 = -5$	$(-5, -5)$
$x = 10$	$y = 2 \cdot 10 + 5 = 25$	$(10, 25)$

By plotting these points and then connecting them, we obtain the graph (a *line*) of the equation $y = 2x + 5$, as shown in Figure 12.

Figure 12 $y = 2x + 5$

EXAMPLE 3

Graphing an Equation by Plotting Points

Graph the equation: $y = x^2$

Solution Table 2 provides several points on the graph of $y = x^2$. Plotting these points and connecting them with a smooth curve gives the graph (a *parabola*) shown in Figure 13.

Table 2

x	$y = x^2$	(x, y)
-4	16	$(-4, 16)$
-3	9	$(-3, 9)$
-2	4	$(-2, 4)$
-1	1	$(-1, 1)$
0	0	$(0, 0)$
1	1	$(1, 1)$
2	4	$(2, 4)$
3	9	$(3, 9)$
4	16	$(4, 16)$

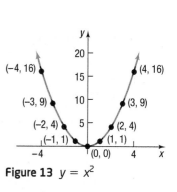

Figure 13 $y = x^2$

The graphs of the equations shown in Figures 12 and 13 do not show all points. For example, in Figure 12, the point (20, 45) is a part of the graph of $y = 2x + 5$, but it is not shown. Since the graph of $y = 2x + 5$ can be extended out indefinitely, we use arrows to indicate that the pattern shown continues. It is important, when showing a graph, to present enough of the graph so that any viewer of the illustration will "see" the rest of it as an obvious continuation of what is actually there. This is referred to as a **complete graph**.

COMMENT Another way to obtain the graph of an equation is to use a graphing utility. Read Section 2, *Using a Graphing Utility to Graph Equations*, in the Appendix. ∎

One way to obtain the complete graph of an equation is to plot enough points on the graph for a pattern to become evident. Then these points are connected with a smooth curve following the suggested pattern. But how many points are sufficient? Sometimes knowledge about the equation tells us. For example, we will learn in the next section that if an equation is of the form $y = mx + b$, then its graph is a line. In this case, only two points are needed to obtain the complete graph.

One purpose of this text is to investigate the properties of equations in order to decide whether a graph is complete. Sometimes we shall graph equations by plotting points. Shortly, we shall investigate various techniques that will enable us to graph an equation without plotting so many points.

Two techniques that sometimes reduce the number of points required to graph an equation involve finding *intercepts* and checking for *symmetry*.

2 Find Intercepts from a Graph

The points, if any, at which a graph crosses or touches the coordinate axes are called the **intercepts** of the graph. See Figure 14. The x-coordinate of a point at which the graph crosses or touches the x-axis is an ***x*-intercept**, and the y-coordinate of a point at which the graph crosses or touches the y-axis is a ***y*-intercept**.

In Words

Intercepts are points (ordered pairs). An x-intercept or a y-intercept is a number. For example, the point (3, 0) is an intercept; the number 3 is an x-intercept.

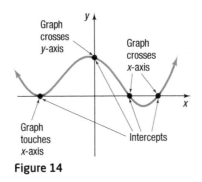

Figure 14

| EXAMPLE 4 | **Finding Intercepts from a Graph** |

Find the intercepts of the graph in Figure 15. What are its x-intercepts? What are its y-intercepts?

Solution   The intercepts of the graph are the points

$$(-3, 0), \quad (0, 3), \quad \left(\frac{3}{2}, 0\right), \quad \left(0, -\frac{4}{3}\right), \quad (0, -3.5), \quad (4.5, 0)$$

The x-intercepts are $-3, \frac{3}{2}$, and 4.5; the y-intercepts are $-3.5, -\frac{4}{3}$, and 3.

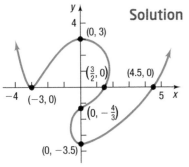

Figure 15

In Example 4, notice that intercepts are listed as ordered pairs, and the x-intercepts and the y-intercepts are listed as numbers. We use this distinction throughout the text.

Now Work PROBLEM 41(a)

3 Find Intercepts from an Equation

The intercepts of a graph can be found from its equation by using the fact that points on the *x*-axis have *y*-coordinates equal to 0, and points on the *y*-axis have *x*-coordinates equal to 0.

COMMENT For many equations, finding intercepts may not be so easy. In such cases, a graphing utility can be used. Read the first part of Section 3, *Using a Graphing Utility to Locate Intercepts and Check for Symmetry*, in the Appendix, to find out how to locate intercepts using a graphing utility. ∎

Procedure for Finding Intercepts

- To find the *x*-intercept(s), if any, of the graph of an equation, let $y = 0$ in the equation and solve for *x*, where *x* is a real number.
- To find the *y*-intercept(s), if any, of the graph of an equation, let $x = 0$ in the equation and solve for *y*, where *y* is a real number.

EXAMPLE 5

Finding Intercepts from an Equation

Find the *x*-intercept(s) and the *y*-intercept(s) of the graph of $y = x^2 - 4$. Then graph $y = x^2 - 4$ by plotting points.

Solution

To find the *x*-intercept(s), let $y = 0$ and obtain the equation

$$x^2 - 4 = 0 \quad y = x^2 - 4 \text{ with } y = 0$$
$$(x + 2)(x - 2) = 0 \quad \text{Factor.}$$
$$x + 2 = 0 \quad \text{or} \quad x - 2 = 0 \quad \text{Use the Zero-Product Property.}$$
$$x = -2 \quad \text{or} \quad x = 2 \quad \text{Solve.}$$

The equation has two solutions, -2 and 2. The *x*-intercepts are -2 and 2.

To find the *y*-intercept(s), let $x = 0$ in the equation.

$$y = x^2 - 4$$
$$= 0^2 - 4 = -4$$

The *y*-intercept is -4.

Since $x^2 \geq 0$ for all *x*, we deduce from the equation $y = x^2 - 4$ that $y \geq -4$ for all *x*. This information, the intercepts, and the points from Table 3 enable us to graph $y = x^2 - 4$. See Figure 16.

Table 3

x	$y = x^2 - 4$	(x, y)
-3	5	$(-3, 5)$
-1	-3	$(-1, -3)$
1	-3	$(1, -3)$
3	5	$(3, 5)$

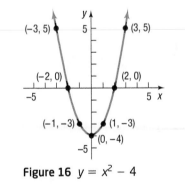

Figure 16 $y = x^2 - 4$

Now Work PROBLEM 23

4 Test an Equation for Symmetry with Respect to the *x*-Axis, the *y*-Axis, and the Origin

Another helpful tool for graphing equations by hand involves *symmetry*, particularly symmetry with respect to the *x*-axis, the *y*-axis, and the origin.

Symmetry often occurs in nature. Consider the picture of the butterfly. Do you see the symmetry?

Figure 17 Symmetry with respect to the *x*-axis

> **DEFINITION** **Symmetry with Respect to the *x*-Axis**
>
> A graph is **symmetric with respect to the *x*-axis** if, for every point (x, y) on the graph, the point $(x, -y)$ is also on the graph.

Figure 17 illustrates the definition. Note that when a graph is symmetric with respect to the *x*-axis, the part of the graph above the *x*-axis is a reflection (or mirror image) of the part below it, and vice versa.

EXAMPLE 6 **Points Symmetric with Respect to the *x*-Axis**

If a graph is symmetric with respect to the *x*-axis, and the point $(3, 2)$ is on the graph, then the point $(3, -2)$ is also on the graph.

Figure 18 Symmetry with respect to the *y*-axis

> **DEFINITION** **Symmetry with Respect to the *y*-Axis**
>
> A graph is **symmetric with respect to the *y*-axis** if, for every point (x, y) on the graph, the point $(-x, y)$ is also on the graph.

Figure 18 illustrates the definition. When a graph is symmetric with respect to the *y*-axis, the part of the graph to the right of the *y*-axis is a reflection of the part to the left of it, and vice versa.

EXAMPLE 7 **Points Symmetric with Respect to the *y*-Axis**

If a graph is symmetric with respect to the *y*-axis and the point $(5, 8)$ is on the graph, then the point $(-5, 8)$ is also on the graph.

> **DEFINITION** **Symmetry with Respect to the Origin**
>
> A graph is **symmetric with respect to the origin** if, for every point (x, y) on the graph, the point $(-x, -y)$ is also on the graph.

Figure 19 Symmetry with respect to the origin

Figure 19 illustrates the definition. Symmetry with respect to the origin may be viewed in three ways:

- As a reflection about the *y*-axis, followed by a reflection about the *x*-axis
- As a projection along a line through the origin so that the distances from the origin are equal
- As half of a complete revolution about the origin

EXAMPLE 8 **Points Symmetric with Respect to the Origin**

If a graph is symmetric with respect to the origin, and the point $(4, -2)$ is on the graph, then the point $(-4, 2)$ is also on the graph.

Now Work PROBLEMS 31 AND 41(b)

When the graph of an equation is symmetric with respect to a coordinate axis or the origin, the number of points that you need to plot in order to see the pattern is reduced. For example, if the graph of an equation is symmetric with respect to the *y*-axis, then once points to the right of the *y*-axis are plotted, an equal number of points on the graph can be obtained by reflecting them about the *y*-axis. Because of this, before we graph an equation, we should first determine whether it has any symmetry. The following tests are used for this purpose.

> **Tests for Symmetry**
>
> To test the graph of an equation for symmetry with respect to the
>
> - *x-Axis* Replace y by $-y$ in the equation and simplify. If an equivalent equation results, the graph of the equation is symmetric with respect to the x-axis.
> - *y-Axis* Replace x by $-x$ in the equation and simplify. If an equivalent equation results, the graph of the equation is symmetric with respect to the y-axis.
> - *Origin* Replace x by $-x$ and y by $-y$ in the equation and simplify. If an equivalent equation results, the graph of the equation is symmetric with respect to the origin.

EXAMPLE 9 **Testing an Equation for Symmetry**

Test $4x^2 + 9y^2 = 36$ for symmetry.

Solution *x-Axis:* To test for symmetry with respect to the x-axis, replace y by $-y$. Since $4x^2 + 9(-y)^2 = 36$ is equivalent to $4x^2 + 9y^2 = 36$, the graph of the equation is symmetric with respect to the x-axis.

y-Axis: To test for symmetry with respect to the y-axis, replace x by $-x$. Since $4(-x)^2 + 9y^2 = 36$ is equivalent to $4x^2 + 9y^2 = 36$, the graph of the equation is symmetric with respect to the y-axis.

Origin: To test for symmetry with respect to the origin, replace x by $-x$ and y by $-y$. Since $4(-x)^2 + 9(-y)^2 = 36$ is equivalent to $4x^2 + 9y^2 = 36$, the graph of the equation is symmetric with respect to the origin.

EXAMPLE 10 **Testing an Equation for Symmetry**

Test $y = \dfrac{4x^2}{x^2 + 1}$ for symmetry.

Solution *x-Axis:* To test for symmetry with respect to the x-axis, replace y by $-y$. Since $-y = \dfrac{4x^2}{x^2 + 1}$ is not equivalent to $y = \dfrac{4x^2}{x^2 + 1}$, the graph of the equation is not symmetric with respect to the x-axis.

y-Axis: To test for symmetry with respect to the y-axis, replace x by $-x$. Since $y = \dfrac{4(-x)^2}{(-x)^2 + 1} = \dfrac{4x^2}{x^2 + 1}$ is equivalent to $y = \dfrac{4x^2}{x^2 + 1}$, the graph of the equation is symmetric with respect to the y-axis.

Origin: To test for symmetry with respect to the origin, replace x by $-x$ and y by $-y$.

$$-y = \frac{4(-x)^2}{(-x)^2 + 1} \qquad \text{Replace } x \text{ by } -x \text{ and } y \text{ by } -y.$$

$$-y = \frac{4x^2}{x^2 + 1} \qquad \text{Simplify.}$$

$$y = -\frac{4x^2}{x^2 + 1} \qquad \text{Multiply both sides by } -1.$$

Since the result is not equivalent to the original equation, the graph of the equation $y = \dfrac{4x^2}{x^2 + 1}$ is not symmetric with respect to the origin.

Seeing the Concept

Figure 20 shows the graph of $4x^2 + 9y^2 = 36$ using Desmos. Do you see the symmetry with respect to the x-axis, the y-axis, and the origin?

Figure 20 $4x^2 + 9y^2 = 36$

Seeing the Concept

Figure 21 shows the graph of $y = \dfrac{4x^2}{x^2 + 1}$ using a TI-84 Plus C graphing calculator. Do you see the symmetry with respect to the y-axis?

Figure 21 $y = \dfrac{4x^2}{x^2 + 1}$

Now Work PROBLEM 61

5 Know How to Graph Key Equations

The next three examples use intercepts, symmetry, and point plotting to obtain the graphs of key equations. It is important to know the graphs of these key equations because we use them later. The first of these is $y = x^3$.

EXAMPLE 11

Graphing the Equation $y = x^3$ by Finding Intercepts, Checking for Symmetry, and Plotting Points

Graph the equation $y = x^3$ by plotting points. Find any intercepts and check for symmetry first.

Solution

First, find the intercepts. When $x = 0$, then $y = 0$; and when $y = 0$, then $x = 0$. The origin $(0, 0)$ is the only intercept. Now test for symmetry.

x-Axis: Replace y by $-y$. Since $-y = x^3$ is not equivalent to $y = x^3$, the graph is not symmetric with respect to the x-axis.

y-Axis: Replace x by $-x$. Since $y = (-x)^3 = -x^3$ is not equivalent to $y = x^3$, the graph is not symmetric with respect to the y-axis.

Origin: Replace x by $-x$ and y by $-y$. Since $-y = (-x)^3 = -x^3$ is equivalent to $y = x^3$ (multiply both sides by -1), the graph is symmetric with respect to the origin.

To graph $y = x^3$, use the equation to obtain several points on the graph. Because of the symmetry, we need to locate only points on the graph for which $x \geq 0$. See Table 4. Since $(1, 1)$ is on the graph, and the graph is symmetric with respect to the origin, the point $(-1, -1)$ is also on the graph. Plot the points from Table 4 and use the symmetry. Figure 22 shows the graph.

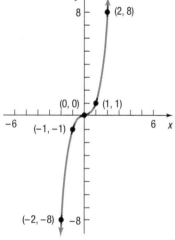

Figure 22 $y = x^3$

Table 4

x	y = x³	(x, y)
0	0	(0, 0)
1	1	(1, 1)
2	8	(2, 8)
3	27	(3, 27)

EXAMPLE 12

Graphing the Equation $x = y^2$

(a) Graph the equation $x = y^2$. Find any intercepts and check for symmetry first.

(b) Graph $x = y^2$, $y \geq 0$.

Solution
(a) The lone intercept is $(0, 0)$. The graph is symmetric with respect to the x-axis. (Do you see why? Replace y by $-y$.) Figure 23(a) shows the graph.

(b) If we restrict y so that $y \geq 0$, the equation $x = y^2$, $y \geq 0$, may be written equivalently as $y = \sqrt{x}$. The portion of the graph of $x = y^2$ in quadrant I is therefore the graph of $y = \sqrt{x}$. See Figure 23(b).

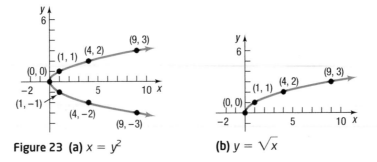

Figure 23 (a) $x = y^2$ **(b)** $y = \sqrt{x}$

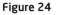

Figure 24

⬛ **COMMENT** To see the graph of the equation $x = y^2$ on a graphing calculator, you will need to graph two equations: $Y_1 = \sqrt{x}$ and $Y_2 = -\sqrt{x}$. We discuss why in Chapter 3. See Figure 24. ∎

EXAMPLE 13

Graphing the Equation $y = \dfrac{1}{x}$

Graph the equation $y = \dfrac{1}{x}$. First, find any intercepts and check for symmetry.

Solution
Check for intercepts first. If we let $x = 0$, we obtain 0 in the denominator, which makes y undefined. We conclude that there is no y-intercept. If we let $y = 0$, we get the equation $\dfrac{1}{x} = 0$, which has no solution. We conclude that there is no x-intercept.

The graph of $y = \dfrac{1}{x}$ does not cross or touch the coordinate axes.

Next check for symmetry:

x-Axis: Replacing y by $-y$ yields $-y = \dfrac{1}{x}$, which is not equivalent to $y = \dfrac{1}{x}$.

y-Axis: Replacing x by $-x$ yields $y = \dfrac{1}{-x} = -\dfrac{1}{x}$, which is not equivalent to $y = \dfrac{1}{x}$.

Origin: Replacing x by $-x$ and y by $-y$ yields $-y = -\dfrac{1}{x}$, which is equivalent to $y = \dfrac{1}{x}$. The graph is symmetric with respect to the origin.

Now set up Table 5, listing several points on the graph. Because of the symmetry with respect to the origin, we use only positive values of x. From Table 5 we infer that if x is a large and positive number, then $y = \dfrac{1}{x}$ is a positive number close to 0. We also infer that if x is a positive number close to 0, then $y = \dfrac{1}{x}$ is a large and positive number. Armed with this information, we can graph the equation.

Figure 25 illustrates some of these points and the graph of $y = \dfrac{1}{x}$. Observe how the absence of intercepts and the existence of symmetry with respect to the origin were utilized. ⌐

⬛ **COMMENT** Refer to Example 2 in the Appendix, Section 3, for the graph of $y = \dfrac{1}{x}$ using a graphing calculator. ∎

Table 5

x	$y = \dfrac{1}{x}$	(x, y)
$\dfrac{1}{10}$	10	$\left(\dfrac{1}{10}, 10\right)$
$\dfrac{1}{3}$	3	$\left(\dfrac{1}{3}, 3\right)$
$\dfrac{1}{2}$	2	$\left(\dfrac{1}{2}, 2\right)$
1	1	$(1, 1)$
2	$\dfrac{1}{2}$	$\left(2, \dfrac{1}{2}\right)$
3	$\dfrac{1}{3}$	$\left(3, \dfrac{1}{3}\right)$
10	$\dfrac{1}{10}$	$\left(10, \dfrac{1}{10}\right)$

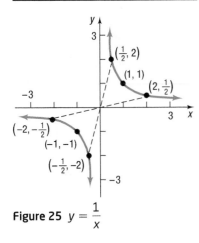

Figure 25 $y = \dfrac{1}{x}$

2.2 Assess Your Understanding

'Are You Prepared?' *Answers are given at the end of these exercises. If you get a wrong answer, read the pages listed in* red.

1. Solve the equation $2(x + 3) - 1 = -7$. (pp. 82–85)

2. Solve the equation $x^2 - 9 = 0$. (pp. 93–94)

Concepts and Vocabulary

3. The points, if any, at which a graph crosses or touches the coordinate axes are called _____.

4. The *x*-intercepts of the graph of an equation are those *x*-values for which _____.

5. If for every point (x, y) on the graph of an equation the point $(-x, y)$ is also on the graph, then the graph is symmetric with respect to the _____.

6. If the graph of an equation is symmetric with respect to the *y*-axis and -4 is an *x*-intercept of this graph, then ___ is also an *x*-intercept.

7. If the graph of an equation is symmetric with respect to the origin and $(3, -4)$ is a point on the graph, then _____ is also a point on the graph.

8. *True or False* To find the *y*-intercepts of the graph of an equation, let $x = 0$ and solve for *y*.

9. *True or False* The *y*-coordinate of a point at which the graph crosses or touches the *x*-axis is an *x*-intercept.

10. *True or False* If a graph is symmetric with respect to the *x*-axis, then it cannot be symmetric with respect to the *y*-axis.

11. *Multiple Choice* Given that the intercepts of a graph are $(-4, 0)$ and $(0, 5)$, choose the statement that is true.

(a) The *y*-intercept is -4, and the *x*-intercept is 5.
(b) The *y*-intercepts are -4 and 5.
(c) The *x*-intercepts are -4 and 5.
(d) The *x*-intercept is -4, and the *y*-intercept is 5.

12. *Multiple Choice* To test whether the graph of an equation is symmetric with respect to the origin, replace _____ in the equation and simplify. If an equivalent equation results, then the graph is symmetric with respect to the origin.

(a) *x* by $-x$
(b) *y* by $-y$
(c) *x* by $-x$ and *y* by $-y$
(d) *x* by $-y$ and *y* by $-x$

Skill Building

In Problems 13–18, determine which of the given points are on the graph of the equation.

13. Equation: $y = x^4 - \sqrt{x}$
Points: $(0, 0); (1, 1); (2, 4)$

14. Equation: $y = x^3 - 2\sqrt{x}$
Points: $(0, 0); (1, 1); (1, -1)$

15. Equation: $y^2 = x^2 + 9$
Points: $(0, 3); (3, 0); (-3, 0)$

16. Equation: $y^3 = x + 1$
Points: $(1, 2); (0, 1); (-1, 0)$

17. Equation: $x^2 + y^2 = 4$
Points: $(0, 2); (-2, 2); (\sqrt{2}, \sqrt{2})$

18. Equation: $x^2 + 4y^2 = 4$
Points: $(0, 1); (2, 0); \left(2, \frac{1}{2}\right)$

In Problems 19–30, find the intercepts and graph each equation by plotting points. Be sure to label the intercepts.

19. $y = x + 2$
20. $y = x - 6$
21. $y = 2x + 8$
22. $y = 3x - 9$

23. $y = x^2 - 1$
24. $y = x^2 - 9$
25. $y = -x^2 + 4$
26. $y = -x^2 + 1$

27. $2x + 3y = 6$
28. $5x + 2y = 10$
29. $9x^2 + 4y = 36$
30. $4x^2 + y = 4$

In Problems 31–40, plot each point. Then plot the point that is symmetric to it with respect to (a) the x-axis; (b) the y-axis; (c) the origin.

31. $(3, 4)$
32. $(5, 3)$
33. $(-2, 1)$
34. $(4, -2)$
35. $(5, -2)$

36. $(-1, -1)$
37. $(-3, -4)$
38. $(4, 0)$
39. $(0, -3)$
40. $(-3, 0)$

In Problems 41–52, the graph of an equation is given. (a) Find the intercepts. (b) Indicate whether the graph is symmetric with respect to the x-axis, the y-axis, the origin or none of these.

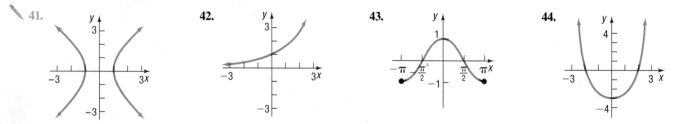

41.　　　**42.**　　　**43.**　　　**44.**

45. **46.** **47.** **48.**

49. **50.** **51.** **52.**

In Problems 53–56, draw a complete graph so that it has the type of symmetry indicated.

53. *y*-axis   **54.** *x*-axis   **55.** Origin   **56.** *y*-axis

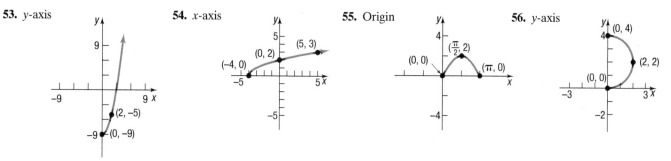

In Problems 57–72, list the intercepts and test for symmetry.

57. $y^2 = x + 16$

58. $y^2 = x + 9$

59. $y = \sqrt[3]{x}$

60. $y = \sqrt[5]{x}$

61. $x^2 + y - 9 = 0$

62. $x^2 - y - 4 = 0$

63. $25x^2 + 4y^2 = 100$

64. $4x^2 + y^2 = 4$

65. $y = x^3 - 64$

66. $y = x^4 - 1$

67. $y = x^2 - 2x - 8$

68. $y = x^2 + 4$

69. $y = \dfrac{4x}{x^2 + 16}$

70. $y = \dfrac{x^2 - 4}{2x}$

71. $y = \dfrac{-x^3}{x^2 - 9}$

72. $y = \dfrac{x^4 + 1}{2x^5}$

In Problems 73–76, graph each equation.

73. $y = x^3$

74. $x = y^2$

75. $y = \sqrt{x}$

76. $y = \dfrac{1}{x}$

77. If $(a, 4)$ is a point on the graph of $y = x^2 + 3x$, what is a?   **78.** If $(a, -5)$ is a point on the graph of $y = x^2 + 6x$, what is a?

Applications and Extensions

79. Given that the point $(1, 2)$ is on the graph of an equation that is symmetric with respect to the origin, what other point is on the graph?

80. If the graph of an equation is symmetric with respect to the *y*-axis and 6 is an *x*-intercept of this graph, name another *x*-intercept.

81. If the graph of an equation is symmetric with respect to the origin and -4 is an *x*-intercept of this graph, name another *x*-intercept.

82. If the graph of an equation is symmetric with respect to the *x*-axis and 2 is a *y*-intercept, name another *y*-intercept.

83. Microphones In studios and on stages, cardioid microphones are often preferred for the richness they add to voices and for their ability to reduce the level of sound from the sides

and rear of the microphone. Suppose one such cardioid pattern is given by the equation $(x^2 + y^2 - x)^2 = x^2 + y^2$.

(a) Find the intercepts of the graph of the equation.
(b) Test for symmetry with respect to the *x*-axis, the *y*-axis, and the origin.

Source: www.notaviva.com

84. Solar Energy The solar electric generating systems at Kramer Junction, California, use parabolic troughs to heat a heat-transfer fluid to a high temperature. This fluid is used to generate steam that drives a power conversion system to produce electricity. For troughs 7.5 feet wide, an equation for the cross section is $16y^2 = 120x - 225$.

(a) Find the intercepts of the graph of the equation.

(b) Test for symmetry with respect to the x-axis, the y-axis, and the origin.

Source: U.S. Department of Energy

85. Challenge Problem Lemniscate For a nonzero constant a, find the intercepts of the graph of $(x^2 + y^2)^2 = a^2(x^2 - y^2)$. Then test for symmetry with respect to the x-axis, the y-axis, and the origin.

86. Challenge Problem Limaçon For nonzero constants a and b, find the intercepts of the graph of

$$(x^2 + y^2 - ax)^2 = b^2(x^2 + y^2)$$

Then test for symmetry with respect to the x-axis, the y-axis, and the origin.

Explaining Concepts: Discussion and Writing

87. **(a)** Graph $y = \sqrt{x^2}$, $y = x$, $y = |x|$, and $y = (\sqrt{x})^2$, noting which graphs are the same.

(b) Explain why the graphs of $y = \sqrt{x^2}$ and $y = |x|$ are the same.

(c) Explain why the graphs of $y = x$ and $y = (\sqrt{x})^2$ are not the same.

(d) Explain why the graphs of $y = \sqrt{x^2}$ and $y = x$ are not the same.

88. Explain what is meant by a complete graph.

89. Draw a graph of an equation that contains two x-intercepts; at one the graph crosses the x-axis, and at the other the graph touches the x-axis.

90. Make up an equation with the intercepts (2, 0), (4, 0), and (0, 1). Compare your equation with a friend's equation. Comment on any similarities.

91. Draw a graph that contains the points

$$(-2, -1), (0, 1), (1, 3), \text{ and } (3, 5)$$

Compare your graph with those of other students. Are most of the graphs almost straight lines? How many are "curved"? Discuss the various ways in which these points might be connected.

92. An equation is being tested for symmetry with respect to the x-axis, the y-axis, and the origin. Explain why, if two of these symmetries are present, the remaining one must also be present.

93. Draw a graph that contains the points $(-2, 5)$, $(-1, 3)$, and $(0, 2)$ and is symmetric with respect to the y-axis. Compare your graph with those of other students; comment on any similarities. Can a graph contain these points and be symmetric with respect to the x-axis? the origin? Why or why not?

Retain Your Knowledge

Problems 94–103 are based on material learned earlier in the course. The purpose of these problems is to keep the material fresh in your mind so that you are better prepared for the final exam.

94. Find the value of $\dfrac{x + y}{x - y}$ if $x = 6$ and $y = -2$.

95. Factor $3x^2 - 30x + 75$ completely.

96. Simplify: $(2x^2 y^3)^3 (3x^3 y)^2$

97. Use synthetic division to find the quotient and remainder when $2x^4 - 7x^3 + 7x + 6$ is divided by $x - 3$.

98. Solve: $\sqrt{2x + 5} - 1 = x$

99. Solve the inequality $1 < 3(4 - x) - 5 \le 19$. Express the solution using interval notation. Graph the solution set.

100. Solve $x^2 - 8x + 4 = 0$ by completing the square.

101. Solve $3x^2 + x = 1$ by using the quadratic formula.

102. Water Pipe When mineral deposits formed a coating 2 millimeters thick on the inside of a circular pipe, the area through which water can flow was reduced by 30%. Find the original diameter inside the pipe. Round to the nearest hundredth.

103.[†] Simplify: $\sqrt{-196}$

[†]This problem is based on content from Section 1.3, which is optional.

'Are You Prepared?' Answers

1. $\{-6\}$ **2.** $\{-3, 3\}$

2.3 Lines

In this section we study a certain type of equation that contains two variables, called a *linear equation,* and its graph, a *line.*

1 Calculate and Interpret the Slope of a Line

Figure 26

Consider the staircase illustrated in Figure 26. Each step contains exactly the same horizontal **run** and the same vertical **rise**. The ratio of the rise to the run, called the *slope,* is a numerical measure of the steepness of the staircase. For example, if the run is increased and the rise remains the same, the staircase becomes less steep. If the run is kept the same but the rise is increased, the staircase becomes more steep. This important characteristic of a line is best defined using rectangular coordinates.

DEFINITION Slope

Let $P = (x_1, y_1)$ and $Q = (x_2, y_2)$ be two distinct points. If $x_1 \neq x_2$, the **slope m** of the nonvertical line L containing P and Q is defined by the formula

$$m = \frac{y_2 - y_1}{x_2 - x_1} \qquad x_1 \neq x_2 \qquad \textbf{(1)}$$

If $x_1 = x_2$, then L is a **vertical line** and the slope m of L is **undefined** (since this results in division by 0).

Figure 27(a) illustrates the slope of a nonvertical line; Figure 27(b) illustrates a vertical line.

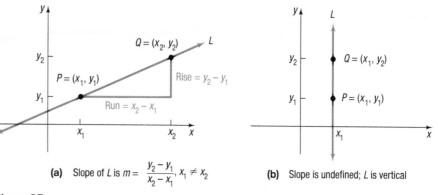

(a) Slope of L is $m = \frac{y_2 - y_1}{x_2 - x_1}, x_1 \neq x_2$ (b) Slope is undefined; L is vertical

Figure 27

In Words
The symbol Δ is the Greek uppercase letter delta. In mathematics, Δ is read "change in," so $\dfrac{\Delta y}{\Delta x}$ is read "change in y divided by change in x."

As Figure 27(a) illustrates, the slope m of a nonvertical line may be viewed as

$$m = \frac{y_2 - y_1}{x_2 - x_1} = \frac{\text{Rise}}{\text{Run}} \quad \text{or as} \quad m = \frac{y_2 - y_1}{x_2 - x_1} = \frac{\text{Change in } y}{\text{Change in } x} = \frac{\Delta y}{\Delta x}$$

That is, the slope m of a nonvertical line measures the amount y changes when x changes from x_1 to x_2. The expression $\dfrac{\Delta y}{\Delta x}$ is called the **average rate of change** of y with respect to x.

Two comments about computing the slope of a nonvertical line may prove helpful:

- Any two distinct points on the line can be used to compute the slope of the line. (See Figure 28 for justification.) Since any two distinct points can be used to compute the slope of a line, the average rate of change of a line is always the same number.

Figure 28
Triangles ABC and PQR are similar (equal angles), so ratios of corresponding sides are proportional. Then the slope using P and Q is

$$\frac{y_2 - y_1}{x_2 - x_1} = \frac{d(B, C)}{d(A, C)},$$

which is the slope using A and B.

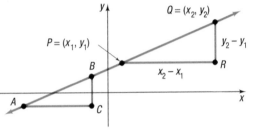

- The slope of a line may be computed from $P = (x_1, y_1)$ to $Q = (x_2, y_2)$ or from Q to P because

$$\frac{y_2 - y_1}{x_2 - x_1} = \frac{y_1 - y_2}{x_1 - x_2}$$

EXAMPLE 1 **Finding and Interpreting the Slope of a Line Given Two Points**

The slope m of the line containing the points $(1, 2)$ and $(5, -3)$ may be computed as

$$m = \frac{-3 - 2}{5 - 1} = \frac{-5}{4} = -\frac{5}{4} \quad \text{or as} \quad m = \frac{2 - (-3)}{1 - 5} = \frac{5}{-4} = -\frac{5}{4}$$

For every 4-unit change in x, y will change by -5 units. That is, if x increases by 4 units, then y will decrease by 5 units. The average rate of change of y with respect to x is $-\dfrac{5}{4}$.

Now Work PROBLEMS 13 AND 19

EXAMPLE 2 **Finding the Slopes of Various Lines Containing the Same Point (2, 3)**

Compute the slopes of the lines L_1, L_2, L_3, and L_4 containing the following pairs of points. Graph all four lines on the same set of coordinate axes.

$$L_1: \quad P = (2, 3) \qquad Q_1 = (-1, -2)$$
$$L_2: \quad P = (2, 3) \qquad Q_2 = (3, -1)$$
$$L_3: \quad P = (2, 3) \qquad Q_3 = (5, 3)$$
$$L_4: \quad P = (2, 3) \qquad Q_4 = (2, 5)$$

Solution Let m_1, m_2, m_3, and m_4 denote the slopes of the lines L_1, L_2, L_3, and L_4, respectively.

Then

Figure 29

$$m_1 = \frac{-2 - 3}{-1 - 2} = \frac{-5}{-3} = \frac{5}{3} \quad \text{A rise of 5 divided by a run of 3}$$

$$m_2 = \frac{-1 - 3}{3 - 2} = \frac{-4}{1} = -4$$

$$m_3 = \frac{3 - 3}{5 - 2} = \frac{0}{3} = 0$$

m_4 is undefined because $x_1 = x_2 = 2$

The graphs of these lines are in Figure 29.

Figure 29 illustrates the following facts:

- When the slope of a line is positive, the line slants upward from left to right (L_1).
- When the slope of a line is negative, the line slants downward from left to right (L_2).
- When the slope is 0, the line is horizontal (L_3).
- When the slope is undefined, the line is vertical (L_4).

Seeing the Concept

On the same screen, graph the following equations:

Figure 30

$Y_1 = 0$	Slope of line is 0.
$Y_2 = \dfrac{1}{4}x$	Slope of line is $\dfrac{1}{4}$.
$Y_3 = \dfrac{1}{2}x$	Slope of line is $\dfrac{1}{2}$.
$Y_4 = x$	Slope of line is 1.
$Y_5 = 2x$	Slope of line is 2.
$Y_6 = 6x$	Slope of line is 6.

See Figure 30.

Seeing the Concept

On the same screen, graph the following equations:

Figure 31

$Y_1 = 0$	Slope of line is 0.
$Y_2 = -\dfrac{1}{4}x$	Slope of line is $-\dfrac{1}{4}$.
$Y_3 = -\dfrac{1}{2}x$	Slope of line is $-\dfrac{1}{2}$.
$Y_4 = -x$	Slope of line is -1.
$Y_5 = -2x$	Slope of line is -2.
$Y_6 = -6x$	Slope of line is -6.

See Figure 31.

Figures 30 and 31 illustrate that the closer the line is to the vertical position, the greater the magnitude of the slope.

2 Graph Lines Given a Point and the Slope

EXAMPLE 3

Graphing a Line Given a Point and a Slope

Draw a graph of the line that contains the point $(3, 2)$ and has a slope of:

(a) $\dfrac{3}{4}$ (b) $-\dfrac{4}{5}$

Solution

(a) Slope $= \dfrac{\text{Rise}}{\text{Run}}$. The slope $\dfrac{3}{4}$ means that for every horizontal change (run) of 4 units to the right, there is a vertical change (rise) of 3 units. Start at the point $(3, 2)$ and move 4 units to the right and 3 units up, arriving at the point $(7, 5)$. Drawing the line through the points $(7, 5)$ and $(3, 2)$ gives the graph. See Figure 32.

Figure 32

(b) A slope of

$$-\frac{4}{5} = \frac{-4}{5} = \frac{\text{Rise}}{\text{Run}}$$

means that for every horizontal change of 5 units to the right, there is a corresponding vertical change of -4 units (a downward movement). Start at the point $(3, 2)$ and move 5 units to the right and then 4 units down, arriving at the point $(8, -2)$. Drawing the line through these points gives the graph. See Figure 33.

Alternatively, consider that

$$-\frac{4}{5} = \frac{4}{-5} = \frac{\text{Rise}}{\text{Run}}$$

so for every horizontal change of -5 units (a movement to the left), there is a corresponding vertical change of 4 units (upward). This approach leads to the point $(-2, 6)$, which is also on the graph of the line in Figure 33.

Figure 33

Now Work PROBLEM 25

3 Find the Equation of a Vertical Line

EXAMPLE 4

Graphing a Line

Graph the equation: $x = 3$

Solution

To graph $x = 3$, we find all points (x, y) in the plane for which $x = 3$. No matter what y-coordinate is used, the corresponding x-coordinate always equals 3. Consequently, the graph of the equation $x = 3$ is a vertical line with x-intercept 3 and an undefined slope. See Figure 34 .

Figure 34 $x = 3$

Example 4 suggests the following result:

> **THEOREM Equation of a Vertical Line**
>
> A vertical line is given by an equation of the form
>
> $$x = a$$
>
> where a is the x-intercept.

COMMENT To graph an equation using most graphing utilities, we need to express the equation in the form y = {*expression in x*}. But x = 3 cannot be put in this form. To overcome this, most graphing utilities have special commands for drawing vertical lines. DRAW, LINE, PLOT, and VERT are among the more common ones. Consult your manual to determine the correct methodology for your graphing utility. ∎

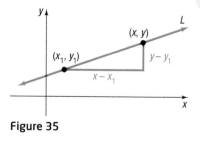

Figure 35

4 Use the Point-Slope Form of a Line; Identify Horizontal Lines

Let L be a nonvertical line with slope m that contains the point (x_1, y_1). See Figure 35. For any other point (x, y) on L, we have

$$m = \frac{y - y_1}{x - x_1} \quad \text{or} \quad y - y_1 = m(x - x_1)$$

> **THEOREM Point-Slope Form of an Equation of a Line**
>
> An equation of a nonvertical line with slope m that contains the point (x_1, y_1) is
>
> $$y - y_1 = m(x - x_1) \qquad \text{(2)}$$

EXAMPLE 5 **Finding the Point-Slope Form of an Equation of a Line**

Find the point-slope form of an equation of the line with slope 4, containing the point $(1, 2)$.

Solution An equation of the line with slope 4 that contains the point $(1, 2)$ can be found by using the point-slope form with $m = 4$, $x_1 = 1$, and $y_1 = 2$.

$$y - y_1 = m(x - x_1)$$
$$y - 2 = 4(x - 1) \qquad m = 4, x_1 = 1, y_1 = 2$$

See Figure 36 for the graph.

Figure 36 $y - 2 = 4(x - 1)$

━━━ **Now Work** PROBLEM 33

EXAMPLE 6 **Finding the Equation of a Horizontal Line**

Find an equation of the horizontal line containing the point $(3, 2)$.

Solution Because all the y-values are equal on a horizontal line, the slope of a horizontal line is 0. To get an equation, use the point-slope form with $m = 0$, $x_1 = 3$, and $y_1 = 2$.

$$y - y_1 = m(x - x_1)$$
$$y - 2 = 0 \cdot (x - 3) \qquad m = 0, x_1 = 3, y_1 = 2$$
$$y - 2 = 0$$
$$y = 2$$

Figure 37 $y = 2$

See Figure 37 for the graph.

Example 6 suggests the following result:

> **THEOREM** **Equation of a Horizontal Line**
>
> A horizontal line is given by an equation of the form
>
> $$y = b$$
>
> where b is the y-intercept.

5 Use the Slope-Intercept Form of a Line

Another useful equation of a line is obtained when the slope m and y-intercept b are known. Then the point $(0, b)$ is on the line. Using the point-slope form, equation (2), we obtain

$$y - b = m(x - 0) \quad \text{or} \quad y = mx + b$$

> **THEOREM** **Slope-Intercept Form of an Equation of a Line**
>
> An equation of a line with slope m and y-intercept b is
>
> $$y = mx + b \qquad\qquad (3)$$

For example, if a line has slope 5 and y-intercept 2, we can write the equation in slope-intercept form as

$$y = 5x + 2$$

slope   y-intercept

Now Work PROBLEMS 53 AND 59 (express answer in slope-intercept form)

Seeing the Concept

To see the role that the slope m plays, graph the following lines on the same screen.

$$Y_1 = 2 \qquad\qquad m = 0$$
$$Y_2 = x + 2 \qquad\; m = 1$$
$$Y_3 = -x + 2 \qquad m = -1$$
$$Y_4 = 3x + 2 \qquad m = 3$$
$$Y_5 = -3x + 2 \qquad m = -3$$

Figure 38 displays the graphs using Desmos. What do you conclude about lines of the form $y = mx + 2$?

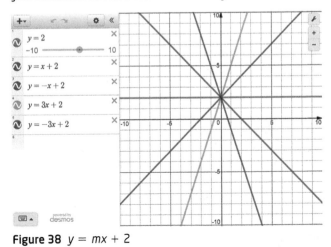

Figure 38 $y = mx + 2$

To see the role of the y-intercept b, graph the following lines on the same screen.

$$Y_1 = 2x \qquad\quad b = 0$$
$$Y_2 = 2x + 1 \qquad b = 1$$
$$Y_3 = 2x - 1 \qquad b = -1$$
$$Y_4 = 2x + 4 \qquad b = 4$$
$$Y_5 = 2x - 4 \qquad b = -4$$

Figure 39 displays the graphs using Desmos. What do you conclude about lines of the form $y = 2x + b$?

Figure 39 $y = 2x + b$

When the equation of a line is written in slope-intercept form, it is easy to find the slope m and y-intercept b of the line. For example, suppose that the equation of a line is

$$y = -2x + 7$$

Compare this equation to $y = mx + b$.

$$y = -2x + 7$$
$$y = mx + b$$

The slope of this line is -2 and its y-intercept is 7.

━━━ **Now Work** PROBLEM 79

| **EXAMPLE 7** | **Finding the Slope and *y*-Intercept of a Line** |

Find the slope m and y-intercept b of the equation $2x + 4y = 8$. Graph the equation.

Solution To find the slope and y-intercept, write the equation in slope-intercept form by solving for y.

$$2x + 4y = 8$$
$$4y = -2x + 8$$
$$y = -\frac{1}{2}x + 2 \quad y = mx + b$$

Figure 40 $y = -\frac{1}{2}x + 2$

The coefficient of x, $-\frac{1}{2}$, is the slope, and the constant, 2, is the y-intercept. To graph the line with y-intercept 2 and with slope $-\frac{1}{2}$, start at the point $(0, 2)$ and move to the right 2 units and then down 1 unit to the point $(2, 1)$. Draw the line through these points. See Figure 40.

━━━ **Now Work** PROBLEM 85

6 Find the Equation of a Line Given Two Points

| **EXAMPLE 8** | **Finding an Equation of a Line Given Two Points** |

Find an equation of the line containing the points $(2, 3)$ and $(-4, 5)$. Graph the line.

Solution First compute the slope of the line.

$$m = \frac{5 - 3}{-4 - 2} = \frac{2}{-6} = -\frac{1}{3} \quad m = \frac{y_2 - y_1}{x_2 - x_1}$$

Use the point $(2, 3)$ and the slope $m = -\frac{1}{3}$ to get the point-slope form of the equation of the line.

$$y - 3 = -\frac{1}{3}(x - 2) \quad y - y_1 = m(x - x_1)$$

Continue to get the slope-intercept form.

$$y - 3 = -\frac{1}{3}x + \frac{2}{3}$$

$$y = -\frac{1}{3}x + \frac{11}{3} \quad y = mx + b$$

Figure 41 $y = -\frac{1}{3}x + \frac{11}{3}$

See Figure 41 for the graph.

In the solution to Example 8, we could have used the other point, $(-4, 5)$, instead of the point $(2, 3)$. The equation that results, when written in slope-intercept form, is the equation that we obtained in the example. (Try it for yourself.)

━━━ **Now Work** PROBLEM 45

7 Graph Lines Written in General Form Using Intercepts

> **DEFINITION General Form**
>
> The equation of a line is in **general form*** when it is written as
> $$Ax + By = C \qquad \qquad (4)$$
> where A, B, and C are real numbers and A and B are not both 0.

If $B = 0$ in equation (4), then $A \neq 0$ and the graph of the equation is a vertical line: $x = \dfrac{C}{A}$.

If $B \neq 0$ in equation (4), then we can solve the equation for y and write the equation in slope-intercept form as we did in Example 7.

One way to graph a line given in general form, equation (4), is to find its intercepts. Remember, the intercepts of the graph of an equation are the points where the graph crosses or touches a coordinate axis.

EXAMPLE 9

Graphing an Equation in General Form Using Its Intercepts

Graph the equation $2x + 4y = 8$ by finding its intercepts.

Solution

To obtain the x-intercept, let $y = 0$ in the equation and solve for x.

$$2x + 4y = 8$$
$$2x + 4 \cdot 0 = 8 \quad \text{Let } y = 0.$$
$$2x = 8$$
$$x = 4 \quad \text{Divide both sides by 2.}$$

The x-intercept is 4, and the point $(4, 0)$ is on the graph of the equation.

To obtain the y-intercept, let $x = 0$ in the equation and solve for y.

$$2x + 4y = 8$$
$$2 \cdot 0 + 4y = 8 \quad \text{Let } x = 0.$$
$$4y = 8$$
$$y = 2 \quad \text{Divide both sides by 4.}$$

The y-intercept is 2, and the point $(0, 2)$ is on the graph of the equation.

Plot the points $(4, 0)$ and $(0, 2)$ and draw the line through the points. See Figure 42.

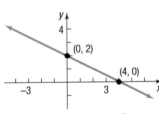

Figure 42 $2x + 4y = 8$

Now Work PROBLEM 99

The equation of every line can be written in general form. For example, a vertical line whose equation is

$$x = a$$

can be written in the general form

$$1 \cdot x + 0 \cdot y = a \quad A = 1, B = 0, C = a$$

A horizontal line whose equation is

$$y = b$$

can be written in the general form

$$0 \cdot x + 1 \cdot y = b \quad A = 0, B = 1, C = b$$

---

*Some texts use the term **standard form**.

Lines that are neither vertical nor horizontal have general equations of the form

$$Ax + By = C \quad A \neq 0 \text{ and } B \neq 0$$

Because the equation of every line can be written in general form, any equation equivalent to equation (4) is called a **linear equation**.

8 Find Equations of Parallel Lines

When two lines (in the plane) do not intersect (that is, they have no points in common), they are **parallel**. Look at Figure 43. We have drawn two parallel lines and have constructed two right triangles by drawing sides parallel to the coordinate axes. The right triangles are similar. (Do you see why? Two angles are equal.) Because the triangles are similar, the ratios of corresponding sides are equal.

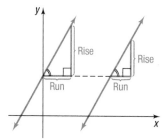

Figure 43 Parallel lines

> **THEOREM Criteria for Parallel Lines**
>
> Two nonvertical lines are parallel if and only if their slopes are equal and they have different y-intercepts.

The use of the phrase "if and only if" in the preceding theorem means that actually two statements are being made, one the converse of the other.

- If two nonvertical lines are parallel, then their slopes are equal and they have different y-intercepts.
- If two nonvertical lines have equal slopes and they have different y-intercepts, then they are parallel.

EXAMPLE 10

Showing That Two Lines Are Parallel

Show that the lines given by the following equations are parallel.

$$L_1: \quad 2x + 3y = 6 \qquad L_2: \quad 4x + 6y = 0$$

Solution

To determine whether these lines have equal slopes and different y-intercepts, write each equation in slope-intercept form.

$$L_1: \quad 2x + 3y = 6 \qquad\qquad\qquad L_2: \quad 4x + 6y = 0$$
$$3y = -2x + 6 \qquad\qquad\qquad 6y = -4x$$
$$y = -\frac{2}{3}x + 2 \qquad\qquad\qquad y = -\frac{2}{3}x$$
$$\text{Slope} = -\frac{2}{3}; \quad y\text{-intercept} = 2 \qquad\qquad \text{Slope} = -\frac{2}{3}; \quad y\text{-intercept} = 0$$

Figure 44 Parallel lines

Because these lines have the same slope, $-\frac{2}{3}$, but different y-intercepts, the lines are parallel. See Figure 44.

EXAMPLE 11

Finding a Line That Is Parallel to a Given Line

Find an equation for the line that contains the point $(2, -3)$ and is parallel to the line $2x + y = 6$.

Solution

Since the two lines are to be parallel, the slope of the line containing the point $(2, -3)$ equals the slope of the line $2x + y = 6$. Begin by writing the equation of the line $2x + y = 6$ in slope-intercept form.

$$2x + y = 6$$
$$y = -2x + 6$$

(continued)

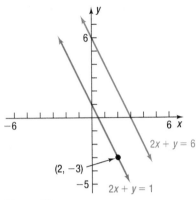

Figure 45

The slope is -2. Since the line containing the point $(2, -3)$ also has slope -2, use the point-slope form to obtain its equation.

$$y - y_1 = m(x - x_1) \qquad \text{Point-slope form}$$
$$y - (-3) = -2(x - 2) \qquad m = -2, x_1 = 2, y_1 = -3$$
$$y + 3 = -2x + 4 \qquad \text{Simplify.}$$
$$y = -2x + 1 \qquad \text{Slope-intercept form}$$
$$2x + y = 1 \qquad \text{General form}$$

This line is parallel to the line $2x + y = 6$ and contains the point $(2, -3)$. See Figure 45.

──── **Now Work** PROBLEM 67

9 Find Equations of Perpendicular Lines

When two lines intersect at a right angle (90°), they are **perpendicular**. See Figure 46.

The following result gives a condition, in terms of their slopes, for two lines to be perpendicular.

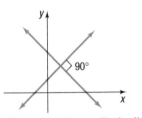

Figure 46 Perpendicular lines

THEOREM **Criterion for Perpendicular Lines**

Two nonvertical lines are perpendicular if and only if the product of their slopes is -1.

Here we prove the "only if" part of the statement:

If two nonvertical lines are perpendicular, then the product of their slopes is -1.

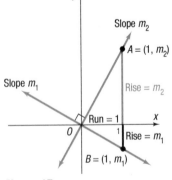

Figure 47

Proof Let m_1 and m_2 denote the slopes of the two lines. There is no loss in generality (that is, neither the angle nor the slopes are affected) if we situate the lines so that they meet at the origin. See Figure 47. The point $A = (1, m_2)$ is on the line having slope m_2, and the point $B = (1, m_1)$ is on the line having slope m_1. (Do you see why this must be true?)

Suppose that the lines are perpendicular. Then triangle OAB is a right triangle. As a result of the Pythagorean Theorem, it follows that

$$[d(O, A)]^2 + [d(O, B)]^2 = [d(A, B)]^2 \qquad \text{(5)}$$

Using the distance formula, the squares of these distances are

$$[d(O, A)]^2 = (1 - 0)^2 + (m_2 - 0)^2 = 1 + m_2^2$$
$$[d(O, B)]^2 = (1 - 0)^2 + (m_1 - 0)^2 = 1 + m_1^2$$
$$[d(A, B)]^2 = (1 - 1)^2 + (m_2 - m_1)^2 = m_2^2 - 2m_1 m_2 + m_1^2$$

Using these facts in equation (5), we get

$$(1 + m_2^2) + (1 + m_1^2) = m_2^2 - 2m_1 m_2 + m_1^2$$

which, upon simplification, can be written as

$$m_1 m_2 = -1$$

If the lines are perpendicular, the product of their slopes is -1. ∎

In Problem 138, you are asked to prove the "if" part of the theorem:

> If two nonvertical lines have slopes whose product is -1, then the lines are perpendicular.

You may find it easier to remember the condition for two nonvertical lines to be perpendicular by observing that the equality $m_1 m_2 = -1$ means that m_1 and m_2 are negative reciprocals of each other; that is,

$$m_1 = -\frac{1}{m_2} \quad \text{and} \quad m_2 = -\frac{1}{m_1}$$

EXAMPLE 12   **Finding the Slope of a Line Perpendicular to Another Line**

If a line has slope $\dfrac{3}{2}$, any line having slope $-\dfrac{2}{3}$ is perpendicular to it.

EXAMPLE 13   **Finding an Equation of a Line Perpendicular to a Given Line**

Find an equation of the line that contains the point $(1, -2)$ and is perpendicular to the line $x + 3y = 6$. Graph the two lines.

Solution   First write the equation of the line $x + 3y = 6$ in slope-intercept form to find its slope.

$$x + 3y = 6$$

$$3y = -x + 6 \qquad \text{Proceed to solve for } y.$$

$$y = -\frac{1}{3}x + 2 \qquad \text{Place in the form } y = mx + b.$$

The slope of the line $x + 3y = 6$ is $-\dfrac{1}{3}$. Any line perpendicular to this line has slope 3.

Because the point $(1, -2)$ is on the line with slope 3, use the point-slope form of the equation of a line.

$$y - y_1 = m(x - x_1) \qquad \text{Point-slope form}$$

$$y - (-2) = 3(x - 1) \qquad m = 3, x_1 = 1, y_1 = -2$$

$$y + 2 = 3(x - 1)$$

To obtain other forms of the equation, proceed as follows:

$$y + 2 = 3x - 3 \qquad \text{Simplify.}$$

$$y = 3x - 5 \qquad \text{Slope-intercept form}$$

$$3x - y = 5 \qquad \text{General form}$$

Figure 48

Figure 48 shows the graphs.

 ✏ **Now Work** PROBLEM 73

WARNING Be sure to use a square screen when you use a graphing calculator to graph perpendicular lines. Otherwise, the angle between the two lines will appear distorted. A discussion of square screens is given in Section 5 of the Appendix. ■

2.3 Assess Your Understanding

Concepts and Vocabulary

1. The slope of a vertical line is _____; the slope of a horizontal line is _____.

2. For the line $2x + 3y = 6$, the x-intercept is _____ and the y-intercept is _____.

3. *True or False* The equation $3x + 4y = 6$ is written in general form.

4. *True or False* The slope of the line $2y = 3x + 5$ is 3.

5. *True or False* The point $(1, 2)$ is on the line $2x + y = 4$.

6. Two nonvertical lines have slopes m_1 and m_2, respectively. The lines are parallel if _____ and the _____ are unequal; the lines are perpendicular if _____.

7. The lines $y = 2x + 3$ and $y = ax + 5$ are parallel if $a =$ _____.

8. The lines $y = 2x - 1$ and $y = ax + 2$ are perpendicular if $a =$ _____.

9. *Multiple Choice* If a line slants downward from left to right, then which of the following describes its slope?
(a) positive   **(b)** zero
(c) negative   **(d)** undefined

10. *Multiple Choice* Choose the formula for finding the slope m of a nonvertical line that contains the two distinct points (x_1, y_1) and (x_2, y_2).

(a) $m = \dfrac{y_2 - x_2}{y_1 - x_1}$ $x_1 \neq y_1$ **(b)** $m = \dfrac{y_2 - x_1}{x_2 - y_1}$ $y_1 \neq x_2$

(c) $m = \dfrac{x_2 - x_1}{y_2 - y_1}$ $y_1 \neq y_2$ **(d)** $m = \dfrac{y_2 - y_1}{x_2 - x_1}$ $x_1 \neq x_2$

11. *Multiple Choice* Choose the correct statement about the graph of the line $y = -3$.
(a) The graph is vertical with x-intercept -3.
(b) The graph is horizontal with y-intercept -3.
(c) The graph is vertical with y-intercept -3.
(d) The graph is horizontal with x-intercept -3.

12. *Multiple Choice* Choose the point-slope equation of a nonvertical line with slope m that passes through the point (x_1, y_1).

(a) $y + y_1 = mx + x_1$ **(b)** $y - y_1 = mx - x_1$

(c) $y + y_1 = m(x + x_1)$ **(d)** $y - y_1 = m(x - x_1)$

Skill Building

In Problems 13–16, (a) find the slope of the line and (b) interpret the slope.

13. (2, 1) (0, 0)

14. (−2, 1) (0, 0)

15. (−2, 2) (1, 1)

16. (2, 2) (−1, 1)

In Problems 17–24, plot each pair of points and determine the slope of the line containing the points. Graph the line.

17. $(2, 3); (4, 0)$ **18.** $(4, 2); (3, 4)$ **19.** $(-2, 3); (2, 1)$ **20.** $(-1, 1); (2, 3)$

21. $(-3, -1); (2, -1)$ **22.** $(4, 2); (-5, 2)$ **23.** $(-1, 2); (-1, -2)$ **24.** $(2, 0); (2, 2)$

In Problems 25–32, graph the line that contains the point P and has slope m.

25. $P = (1, 2); m = 3$ **26.** $P = (2, 1); m = 4$ **27.** $P = (2, 4); m = -\dfrac{3}{4}$ **28.** $P = (1, 3); m = -\dfrac{2}{5}$

29. $P = (-1, 3); m = 0$ **30.** $P = (2, -4); m = 0$ **31.** $P = (0, 3)$; slope undefined   **32.** $P = (-2, 0)$; slope undefined

In Problems 33–38, a point on a line and its slope are given. Find the point-slope form of the equation of the line.

33. $P = (1, 2); m = 3$ **34.** $P = (2, 1); m = 4$ **35.** $P = (2, 4); m = -\dfrac{3}{4}$

36. $P = (1, 3); m = -\dfrac{2}{5}$ **37.** $P = (-1, 3); m = 0$ **38.** $P = (2, -4); m = 0$

In Problems 39–44, the slope and a point on a line are given. Use this information to locate three additional points on the line. Answers may vary. [**Hint**: *It is not necessary to find the equation of the line. See Example 3.*]

39. Slope 4; point $(1, 2)$ **40.** Slope 2; point $(-2, 3)$ **41.** Slope $-\dfrac{3}{2}$; point $(2, -4)$

42. Slope $\dfrac{4}{3}$; point $(-3, 2)$ **43.** Slope -2; point $(-2, -3)$ **44.** Slope -1; point $(4, 1)$

In Problems 45–52, find an equation of the line L.

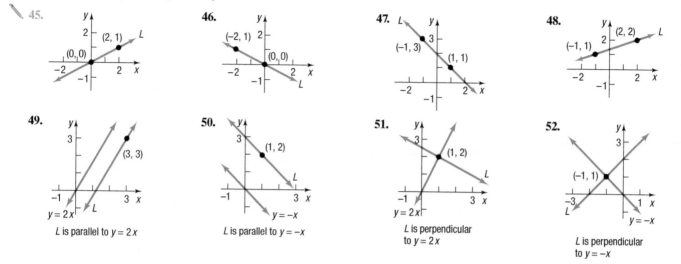

45. (2, 1) *L*, (0, 0), *y*, *x*

46. (−2, 1), (0, 0), *L*, *y*, *x*

47. *L*, (−1, 3), (1, 1), *y*, *x*

48. (2, 2) *L*, (−1, 1), *y*, *x*

49. (3, 3), *y* = 2*x*, *L*
L is parallel to *y* = 2*x*

50. (1, 2), *L*, *y* = −*x*
L is parallel to *y* = −*x*

51. (1, 2), *L*, *y* = 2*x*
L is perpendicular to *y* = 2*x*

52. (−1, 1), *L*, *y* = −*x*
L is perpendicular to *y* = −*x*

In Problems 53–78, find an equation for the line with the given properties. Express your answer using either the general form or the slope-intercept form of the equation of a line, whichever you prefer.

53. Slope = 3; containing the point $(-2, 3)$

54. Slope = 2; containing the point $(4, -3)$

55. Slope = $\dfrac{1}{2}$; containing the point $(3, 1)$

56. Slope = $-\dfrac{2}{3}$; containing the point $(1, -1)$

57. Containing the points $(1, 3)$ and $(-1, 2)$

58. Containing the points $(-3, 4)$ and $(2, 5)$

59. Slope = -3; y-intercept = 3

60. Slope = -2; y-intercept = -2

61. x-intercept = -4; y-intercept = 4

62. x-intercept = 2; y-intercept = -1

63. Slope undefined; containing the point $(2, 4)$

64. Slope undefined; containing the point $(3, 8)$

65. Horizontal; containing the point $(-3, 2)$

66. Vertical; containing the point $(4, -5)$

67. Parallel to the line $y = 2x$; containing the point $(-1, 2)$

68. Parallel to the line $y = -3x$; containing the point $(-1, 2)$

69. Parallel to the line $x - 2y = -5$; containing the point $(0, 0)$

70. Parallel to the line $2x - y = -2$; containing the point $(0, 0)$

71. Parallel to the line $x = 5$; containing the point $(4, 2)$

72. Parallel to the line $y = 5$; containing the point $(4, 2)$

73. Perpendicular to the line $y = \dfrac{1}{2}x + 4$; containing the point $(1, -2)$

74. Perpendicular to the line $y = 2x - 3$; containing the point $(1, -2)$

75. Perpendicular to the line $x - 2y = -5$; containing the point $(0, 4)$

76. Perpendicular to the line $2x + y = 2$; containing the point $(-3, 0)$

77. Perpendicular to the line $x = 8$; containing the point $(3, 4)$

78. Perpendicular to the line $y = 8$; containing the point $(3, 4)$

In Problems 79–98, find the slope and y-intercept of each line. Graph the line.

79. $y = 2x + 3$

80. $y = -3x + 4$

81. $\dfrac{1}{2}y = x - 1$

82. $\dfrac{1}{3}x + y = 2$

83. $y = \dfrac{1}{2}x + 2$

84. $y = 2x + \dfrac{1}{2}$

85. $x + 2y = 4$

86. $-x + 3y = 6$

87. $2x - 3y = 6$

88. $3x + 2y = 6$

89. $x + y = 1$

90. $x - y = 2$

91. $x = -4$

92. $y = -1$

93. $y = 5$

94. $x = 2$

95. $y - x = 0$

96. $x + y = 0$

97. $2y - 3x = 0$

98. $3x + 2y = 0$

In Problems 99–108, (a) find the intercepts of the graph of each equation and (b) graph the equation.

99. $2x + 3y = 6$

100. $3x - 2y = 6$

101. $-4x + 5y = 40$

102. $6x - 4y = 24$

103. $7x + 2y = 21$

104. $5x + 3y = 18$

105. $\dfrac{1}{2}x + \dfrac{1}{3}y = 1$

106. $x - \dfrac{2}{3}y = 4$

107. $0.2x - 0.5y = 1$

108. $-0.3x + 0.4y = 1.2$

109. Find an equation of the x-axis.

110. Find an equation of the y-axis.

In Problems 111–114, the equations of two lines are given. Determine whether the lines are parallel, perpendicular, or neither.

111. $y = 2x - 3$
$y = 2x + 4$

112. $y = \frac{1}{2}x - 3$
$y = -2x + 4$

113. $y = 4x + 5$
$y = -4x + 2$

114. $y = -2x + 3$
$y = -\frac{1}{2}x + 2$

In Problems 115–118, write an equation of each line. Express your answer using either the general form or the slope-intercept form of the equation of a line, whichever you prefer.

115.

116.

117.

118.

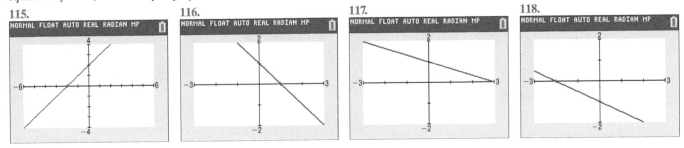

Applications and Extensions

119. Geometry Use slopes to show that the triangle whose vertices are $(-2, 5)$, $(1, 3)$, and $(-1, 0)$ is a right triangle.

120. Geometry Use slopes to show that the quadrilateral whose vertices are $(1, -1)$, $(4, 1)$, $(2, 2)$, and $(5, 4)$ is a parallelogram.

121. Geometry Use slopes to show that the quadrilateral whose vertices are $(-1, 0)$, $(2, 3)$, $(1, -2)$, and $(4, 1)$ is a rectangle.

122. Geometry Use slopes and the distance formula to show that the quadrilateral whose vertices are $(0, 0)$, $(1, 3)$, $(4, 2)$, and $(3, -1)$ is a square.

123. Truck Rentals A truck rental company rents a moving truck for one day by charging $39 plus $0.60 per mile. Write a linear equation that relates the cost C, in dollars, of renting the truck to the number x of miles driven. What is the cost of renting the truck if the truck is driven 110 miles? 230 miles?

124. Cost Equation The **fixed costs** of operating a business are the costs incurred regardless of the level of production. Fixed costs include rent, fixed salaries, and costs of leasing machinery. The **variable costs** of operating a business are the costs that change with the level of output. Variable costs include raw materials, hourly wages, and electricity. Suppose that a manufacturer of jeans has fixed daily costs of $1200 and variable costs of $20 for each pair of jeans manufactured. Write a linear equation that relates the daily cost C, in dollars, of manufacturing the jeans to the number x of jeans manufactured. What is the cost of manufacturing 400 pairs of jeans? 740 pairs?

125. Cost of Driving a Car The annual fixed costs of owning a small sedan are $4252, assuming the car is completely paid for. The cost to drive the car is approximately $0.14 per mile. Write a linear equation that relates the cost C and the number x of miles driven annually.
Source: AAA, 2017

126. Wages of a Car Salesperson Dan receives $525 per week for selling new and used cars. He also receives 5% of the profit on any sales. Write a linear equation that represents Dan's weekly salary S when he has sales that generate a profit of x dollars.

127. Electricity Rates in Florida Florida Power & Light Company supplies electricity to residential customers for a monthly customer charge of $8.01 plus 8.89 cents per kilowatt hour (kWh) for up to 1000 kilowatt hours.

(a) Write a linear equation that relates the monthly charge C, in dollars, to the number x of kilowatt hours used in a month, $0 \le x \le 1000$.

(b) Graph this equation.

(c) What is the monthly charge for using 200 kilowatt hours?

(d) What is the monthly charge for using 500 kilowatt hours?

(e) Interpret the slope of the line.

Source: Florida Power & Light Company, March 2018

128. Natural Gas Rates in Illinois Ameren Illinois supplies natural gas to residential customers for a monthly customer charge of $21.82 plus 64.9 cents per therm of heat energy.

(a) Write a linear equation that relates the monthly charge C, in dollars, to the number x of therms used in a month.

(b) What is the monthly charge for using 90 therms?

(c) What is the monthly charge for using 150 therms?

(d) Graph the equation.

(e) Interpret the slope of the line.

Source: Ameren Illinois, March 2018

129. Measuring Temperature The relationship between Celsius (°C) and Fahrenheit (°F) degrees of measuring temperature is linear. Find a linear equation relating °C and °F if 0°C corresponds to 32°F and 100°C corresponds to 212°F. Use the equation to find the Celsius measure of 70°F.

130. Measuring Temperature The Kelvin (K) scale for measuring temperature is obtained by adding 273 to the Celsius temperature.

(a) Write a linear equation relating K and °C.
(b) Write a linear equation relating K and °F (see Problem 129).

131. Access Ramp A wooden access ramp is being built to reach a platform that sits 30 inches above the floor. The ramp drops 2 inches for every 25-inch run.

(a) Write a linear equation that relates the height y of the ramp above the floor to the horizontal distance x from the platform.
(b) Find and interpret the x-intercept of the graph of your equation.
(c) Design requirements stipulate that the maximum run be 30 feet and that the maximum slope be a drop of 1 inch for each 12 inches of run. Will this ramp meet the requirements? Explain.
(d) What slopes could be used to obtain the 30-inch rise and still meet design requirements?

Source: www.adaptiveaccess.com/wood_ramps.php

132. U.S. Advertising Share A report showed that Internet ads accounted for 19% of all U.S. advertisement spending when print ads (magazines and newspapers) accounted for 26% of the spending. The report further showed that Internet ads accounted for 35% of all advertisement spending when print ads accounted for 16% of the spending.

(a) Write a linear equation that relates that percent y of print ad spending to the percent x of Internet ad spending.

(b) Find the intercepts of the graph of your equation.
(c) Do the intercepts have any meaningful interpretation?
(d) Predict the percent of print ad spending if Internet ads account for 39% of all advertisement spending in the United States.

Source: Marketing Fact Pack 2018. Ad Age Datacenter, December 18, 2017

133. Product Promotion A cereal company finds that the number of people who will buy one of its products in the first month that the product is introduced is linearly related to the amount of money it spends on advertising. If it spends $40,000 on advertising, then 100,000 boxes of cereal will be sold, and if it spends $60,000, then 200,000 boxes will be sold.

(a) Write a linear equation that relates the amount A spent on advertising to the number x of boxes the company aims to sell.
(b) How much expenditure on advertising is needed to sell 300,000 boxes of cereal?
(c) Interpret the slope.

134. The equation $2x - y = C$ defines a **family of lines**, one line for each value of C. On one set of coordinate axes, graph the members of the family when $C = -4$, $C = 0$, and $C = 2$. Can you draw a conclusion from the graph about each member of the family?

135. *Challenge Problem* Find three numbers a for which the lines $x + 2y = 5$, $2x - 3y + 4 = 0$, and $ax + y = 0$ do not form a triangle.

136. *Challenge Problem* Show that the line containing the points (a, b) and (b, a), $a \neq b$, is perpendicular to the line $y = x$. Also show that the midpoint of (a, b) and (b, a) lies on the line $y = x$.

137. *Challenge Problem* Form a triangle using the points $(0, 0)$, $(a, 0)$, and (b, c), where $a > 0$, $b > 0$, and $c > 0$. Find the point of intersection of the three lines joining the midpoint of a side of the triangle to the opposite vertex.

138. *Challenge Problem* Prove that if two nonvertical lines have slopes whose product is -1, then the lines are perpendicular. [**Hint**: Refer to Figure 47 and use the converse of the Pythagorean Theorem.]

Explaining Concepts: Discussion and Writing

139. Which of the following equations might have the graph shown? (More than one answer is possible.)

(a) $2x + 3y = 6$ (b) $-2x + 3y = 6$
(c) $3x - 4y = -12$ (d) $x - y = 1$
(e) $x - y = -1$ (f) $y = 3x - 5$
(g) $y = 2x + 3$ (h) $y = -3x + 3$

140. Which of the following equations might have the graph shown? (More than one answer is possible.)

(a) $2x + 3y = 6$ (b) $2x - 3y = 6$
(c) $3x + 4y = 12$ (d) $x - y = 1$
(e) $x - y = -1$ (f) $y = -2x - 1$
(g) $y = -\dfrac{1}{2}x + 10$ (h) $y = x + 4$

141. The figure shows the graph of two parallel lines. Which of the following pairs of equations might have such a graph?

    (a) $x - 2y = 3$
        $x + 2y = 7$

    (b) $x + y = 2$
        $x + y = -1$

    (c) $x - y = -2$
        $x - y = 1$

    (d) $x - y = -2$
        $2x - 2y = -4$

    (e) $x + 2y = 2$
        $x + 2y = -1$

142. The figure shows the graph of two perpendicular lines. Which of the following pairs of equations might have such a graph?

    (a) $y - 2x = 2$
        $y + 2x = -1$

    (b) $y - 2x = 0$
        $2y + x = 0$

    (c) $2y - x = 2$
        $2y + x = -2$

    (d) $y - 2x = 2$
        $x + 2y = -1$

    (e) $2x + y = -2$
        $2y + x = -2$

143. **m is for Slope** The accepted symbol used to denote the slope of a line is the letter m. Investigate the origin of this practice. Begin by consulting a French dictionary and looking up the French word *monter*. Write a brief essay on your findings.

144. **Grade of a Road** The term *grade* is used to describe the inclination of a road. How is this term related to the notion of slope of a line? Is a 4% grade very steep? Investigate the grades of some mountainous roads and determine their slopes. Write a brief essay on your findings.

145. **Carpentry** Carpenters use the term *pitch* to describe the steepness of staircases and roofs. How is pitch related to slope? Investigate typical pitches used for stairs and for roofs. Write a brief essay on your findings.

146. Can the equation of every line be written in slope-intercept form? Why?

147. Does every line have exactly one x-intercept and one y-intercept? Are there any lines that have no intercepts?

148. What can you say about two lines that have equal slopes and equal y-intercepts?

149. What can you say about two lines with the same x-intercept and the same y-intercept? Assume that the x-intercept is not 0.

150. If two distinct lines have the same slope but different x-intercepts, can they have the same y-intercept?

151. If two distinct lines have the same y-intercept but different slopes, can they have the same x-intercept?

152. Which form of the equation of a line do you prefer to use? Justify your position with an example that shows that your choice is better than another. Have reasons.

153. **What Went Wrong?** A student is asked to find the slope of the line joining $(-3, 2)$ and $(1, -4)$. He states that the slope is $\frac{3}{2}$. Is he correct? If not, what went wrong?

Retain Your Knowledge

Problems 154–162 are based on material learned earlier in the course. The purpose of these problems is to keep the material fresh in your mind so that you are better prepared for the final exam.

154. Simplify $\left(\dfrac{x^2 y^{-3}}{x^4 y^5}\right)^{-2}$. Assume $x \neq 0$ and $y \neq 0$. Express the answer so that all exponents are positive.

155. The lengths of the legs of a right triangle are $a = 8$ and $b = 15$. Find the hypotenuse.

156. Solve the equation: $(x - 3)^2 + 25 = 49$

157. Solve $|2x - 5| + 7 < 10$. Express the answer using set notation or interval notation. Graph the solution set.

158. Determine the domain of the variable x in the expression:

$$\sqrt{8 - \frac{2}{3}x}$$

159. Determine what number should be added to complete the square:

$$x^2 + \frac{3}{4}x$$

160. Multiply and simplify the result.

$$\frac{x^2 - 16}{x^2 + 6x + 8} \cdot \frac{x + 2}{16 - 4x}$$

161. Rationalize the denominator:

$$\frac{\sqrt{x + 1} + \sqrt{x}}{\sqrt{x + 1} - \sqrt{x}}$$

162. Solve: $x - 5\sqrt{x} + 6 = 0$

2.4 Circles

PREPARING FOR THIS SECTION *Before getting started, review the following:*

- Completing the Square (Section R.5, p. 55–56)
- Square Root Method (Section 1.2, pp. 94–95)

Now Work the 'Are You Prepared?' problems on page 188.

OBJECTIVES 1 Write the Standard Form of the Equation of a Circle (p. 185)
2 Graph a Circle (p. 186)
3 Work with the General Form of the Equation of a Circle (p. 187)

1 Write the Standard Form of the Equation of a Circle

One advantage of a coordinate system is that it enables us to translate a geometric statement into an algebraic statement, and vice versa. Consider, for example, the following geometric statement that defines a circle.

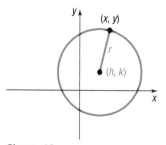

Figure 49
$(x - h)^2 + (y - k)^2 = r^2$

> **DEFINITION Circle**
>
> A **circle** is a set of points in the *xy*-plane that are a fixed distance *r* from a fixed point (h, k). The fixed distance *r* is called the **radius**, and the fixed point (h, k) is called the **center** of the circle.

Figure 49 shows the graph of a circle. To find the equation, let (x, y) represent the coordinates of any point on a circle with radius *r* and center (h, k). Then the distance between the points (x, y) and (h, k) must always equal *r*. That is, by the distance formula,

$$\sqrt{(x - h)^2 + (y - k)^2} = r$$

or, equivalently,

$$(x - h)^2 + (y - k)^2 = r^2$$

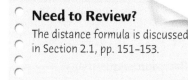

Need to Review?
The distance formula is discussed in Section 2.1, pp. 151–153.

> **DEFINITION Standard Form of an Equation of a Circle**
>
> The **standard form of an equation of a circle** with radius *r* and center (h, k) is
>
> $$(x - h)^2 + (y - k)^2 = r^2 \qquad \textbf{(1)}$$

> **THEOREM**
>
> The standard form of an equation of a circle of radius *r* with center at the origin $(0, 0)$ is
>
> $$x^2 + y^2 = r^2$$

> **DEFINITION Unit Circle**
>
> If the radius $r = 1$, the circle whose center is at the origin is called the **unit circle** and has the equation
>
> $$x^2 + y^2 = 1$$

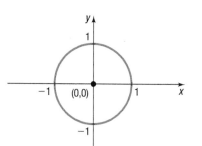

Figure 50
Unit circle $x^2 + y^2 = 1$

See Figure 50. Notice that the graph of the unit circle is symmetric with respect to the *x*-axis, the *y*-axis, and the origin.

EXAMPLE 1

Writing the Standard Form of the Equation of a Circle

Write the standard form of the equation of the circle with radius 5 and center $(-3, 6)$.

Solution   Substitute the values $r = 5, h = -3$, and $k = 6$ into equation (1).

$$(x - h)^2 + (y - k)^2 = r^2 \quad \text{Equation (1)}$$
$$[x - (-3)]^2 + (y - 6)^2 = 5^2$$
$$(x + 3)^2 + (y - 6)^2 = 25$$

Now Work PROBLEM 9

2 Graph a Circle

EXAMPLE 2

Graphing a Circle

Graph the equation: $(x + 3)^2 + (y - 2)^2 = 16$

Solution   The given equation is in the standard form of an equation of a circle. To graph the circle, compare the equation to equation (1). The comparison gives information about the circle.

$$(x + 3)^2 + (y - 2)^2 = 16$$
$$(x - (-3))^2 + (y - 2)^2 = 4^2$$
$$(x - h)^2 + (y - k)^2 = r^2$$

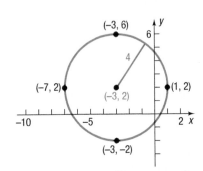

Figure 51 $(x + 3)^2 + (y - 2)^2 = 16$

We see that $h = -3, k = 2$, and $r = 4$. The circle has center $(-3, 2)$ and radius 4. To graph this circle, first plot the center $(-3, 2)$. Since the radius is 4, locate four points on the circle by plotting points 4 units to the left, to the right, up, and down from the center. These four points are then used as guides to obtain the graph. See Figure 51.

Now Work PROBLEMS 27(a) AND (b)

EXAMPLE 3

Finding the Intercepts of a Circle

Find the intercepts, if any, of the graph of the circle $(x + 3)^2 + (y - 2)^2 = 16$.

Solution   This is the equation graphed in Example 2. To find the x-intercepts, if any, let $y = 0$. Then

$$(x + 3)^2 + (y - 2)^2 = 16$$
$$(x + 3)^2 + (0 - 2)^2 = 16 \qquad y = 0$$
$$(x + 3)^2 + 4 = 16 \qquad \text{Simplify.}$$
$$(x + 3)^2 = 12 \qquad \text{Simplify.}$$
$$x + 3 = \pm\sqrt{12} \qquad \text{Use the Square Root Method.}$$
$$x = -3 \pm 2\sqrt{3} \quad \text{Solve for } x.$$

> **In Words**
> The symbol $\pm$ is read "plus or minus." It means to add and subtract the quantity following the $\pm$ symbol. For example, 5 ± 2 means "$5 - 2 = 3$ or $5 + 2 = 7$."

The x-intercepts are $-3 - 2\sqrt{3} \approx -6.46$ and $-3 + 2\sqrt{3} \approx 0.46$.
To find the y-intercepts, if any, let $x = 0$. Then

$$(x + 3)^2 + (y - 2)^2 = 16$$
$$(0 + 3)^2 + (y - 2)^2 = 16$$
$$9 + (y - 2)^2 = 16$$
$$(y - 2)^2 = 7$$
$$y - 2 = \pm\sqrt{7}$$
$$y = 2 \pm \sqrt{7}$$

The y-intercepts are $2 - \sqrt{7} \approx -0.65$ and $2 + \sqrt{7} \approx 4.65$.

Look back at Figure 51 to verify the approximate locations of the intercepts.

Now Work PROBLEM 27(c)

3 Work with the General Form of the Equation of a Circle

If we eliminate the parentheses from the standard form of the equation of the circle given in Example 2, we get

$$(x + 3)^2 + (y - 2)^2 = 16$$
$$x^2 + 6x + 9 + y^2 - 4y + 4 = 16$$

which simplifies to

$$x^2 + y^2 + 6x - 4y - 3 = 0$$

It can be shown that any equation of the form

$$x^2 + y^2 + ax + by + c = 0$$

has a graph that is a circle, is a point, or has no graph at all. For example, the graph of the equation $x^2 + y^2 = 0$ is the single point $(0, 0)$. The equation $x^2 + y^2 + 5 = 0$, or $x^2 + y^2 = -5$, has no graph, because sums of squares of real numbers are never negative.

> **DEFINITION General Form of the Equation of a Circle**
>
> When its graph is a circle, the equation
>
> $$x^2 + y^2 + ax + by + c = 0$$
>
> is the **general form of the equation of a circle**.

Need to Review?

Completing the square is discussed in Section R.5, pp. 55–56.

Now Work PROBLEM 15

If an equation of a circle is in general form, we use the method of completing the square to put the equation in standard form so that we can identify its center and radius.

EXAMPLE 4 | **Graphing a Circle Whose Equation Is in General Form**

Graph the equation: $x^2 + y^2 + 4x - 6y + 12 = 0$

Solution Group the terms involving x, group the terms involving y, and put the constant on the right side of the equation. The result is

$$(x^2 + 4x) + (y^2 - 6y) = -12$$

Next, complete the square of each expression in parentheses. Remember that any number added on the left side of the equation must also be added on the right.

$$(x^2 + 4x + 4) + (y^2 - 6y + 9) = -12 + 4 + 9$$

$$\left(\frac{4}{2}\right)^2 = 4 \qquad \left(\frac{-6}{2}\right)^2 = 9$$

$$(x + 2)^2 + (y - 3)^2 = 1 \quad \text{Factor.}$$

This equation is the standard form of the equation of a circle with radius 1 and center $(-2, 3)$. To graph the equation, use the center $(-2, 3)$ and the radius 1. See Figure 52.

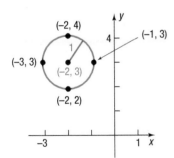

Figure 52 $(x + 2)^2 + (y - 3)^2 = 1$

Now Work PROBLEM 31

| EXAMPLE 5 | Using a Graphing Utility to Graph a Circle |

Graph the equation: $x^2 + y^2 = 4$

Solution

$Y_1 = \sqrt{4 - x^2}$

$Y_2 = -\sqrt{4 - x^2}$

Figure 53 $x^2 + y^2 = 4$

This is the equation of a circle with center at the origin and radius 2. To graph this equation using a graphing utility, begin by solving for y.

$$x^2 + y^2 = 4$$

$$y^2 = 4 - x^2 \qquad \text{Subtract } x^2 \text{ from both sides.}$$

$$y = \pm\sqrt{4 - x^2} \qquad \text{Use the Square Root Method to solve for } y.$$

There are two equations to graph: first graph $Y_1 = \sqrt{4 - x^2}$ and then graph $Y_2 = -\sqrt{4 - x^2}$ on the same square screen. (Your circle will appear oval if you do not use a square screen.*) See Figure 53.

Overview

The discussion in Sections 2.3 and 2.4 about lines and circles deals with two problems that can be generalized as follows:

- Given an equation, classify it and graph it.
- Given a graph, or information about a graph, find its equation.

This text deals with both problems. We shall study various equations, classify them, and graph them.

*The square screen ratio for the TI-84 Plus C graphing calculator is 8:5.

2.4 Assess Your Understanding

'Are You Prepared?' *Answers are given at the end of these exercises. If you get a wrong answer, read the pages listed in* red.

1. To complete the square of $x^2 + 10x$, you would _____ (*add/subtract*) the number _____. (p. 55–56)

2. Use the Square Root Method to solve the equation $(x - 2)^2 = 9$. (pp. 94–95)

Concepts and Vocabulary

3. *True or False* Every equation of the form
$$x^2 + y^2 + ax + by + c = 0$$
has a circle as its graph.

4. For a circle, the _____ is the distance from the center to any point on the circle.

5. *True or False* The radius of the circle $x^2 + y^2 = 9$ is 3.

6. *True or False* The center of the circle
$$(x + 3)^2 + (y - 2)^2 = 13$$
is $(3, -2)$.

7. *Multiple Choice* Choose the equation of a circle with radius 6 and center $(3, -5)$.
 (a) $(x - 3)^2 + (y + 5)^2 = 6$
 (b) $(x + 3)^2 + (y - 5)^2 = 36$
 (c) $(x + 3)^2 + (y - 5)^2 = 6$
 (d) $(x - 3)^2 + (y + 5)^2 = 36$

8. *Multiple Choice* The equation of a circle can be changed from general form to standard from by doing which of the following?
 (a) completing the squares (b) solving for x
 (c) solving for y (d) squaring both sides

Skill Building

In Problems 9–12, find the center and radius of each circle. Write the standard form of the equation.

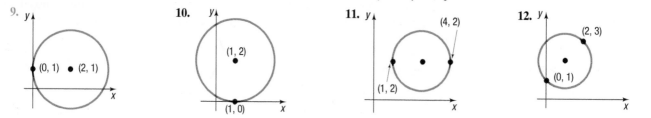

9.

10.

11.

12.

In Problems 13–24, write the standard form of the equation and the general form of the equation of each circle of radius r and center (h, k). Graph each circle.

13. $r = 2$; $(h, k) = (0, 0)$ **14.** $r = 3$; $(h, k) = (0, 0)$ **15.** $r = 2$; $(h, k) = (0, 2)$ **16.** $r = 3$; $(h, k) = (1, 0)$

17. $r = 5$; $(h, k) = (4, -3)$ **18.** $r = 4$; $(h, k) = (2, -3)$ **19.** $r = 4$; $(h, k) = (-2, 1)$ **20.** $r = 7$; $(h, k) = (-5, -2)$

21. $r = \dfrac{1}{2}$; $(h, k) = \left(\dfrac{1}{2}, 0\right)$ **22.** $r = \dfrac{1}{2}$; $(h, k) = \left(0, -\dfrac{1}{2}\right)$ **23.** $r = \sqrt{13}$; $(h, k) = (5, -1)$ **24.** $r = 2\sqrt{5}$; $(h, k) = (-3, 2)$

In Problems 25–38, (a) find the center (h, k) and radius r of each circle; (b) graph each circle; (c) find the intercepts, if any.

25. $x^2 + y^2 = 4$ **26.** $x^2 + (y - 1)^2 = 1$ **27.** $2(x - 3)^2 + 2y^2 = 8$

28. $3(x + 1)^2 + 3(y - 1)^2 = 6$ **29.** $x^2 + y^2 - 2x - 4y - 4 = 0$ **30.** $x^2 + y^2 + 4x + 2y - 20 = 0$

31. $x^2 + y^2 + 4x - 4y - 1 = 0$ **32.** $x^2 + y^2 - 6x + 2y + 9 = 0$ **33.** $x^2 + y^2 - x + 2y + 1 = 0$

34. $x^2 + y^2 + x + y - \dfrac{1}{2} = 0$ **35.** $2x^2 + 2y^2 - 12x + 8y - 24 = 0$ **36.** $2x^2 + 2y^2 + 8x + 7 = 0$

37. $2x^2 + 8x + 2y^2 = 0$ **38.** $3x^2 + 3y^2 - 12y = 0$

In Problems 39–44, find the standard form of the equation of each circle.

39. Center at the origin and containing the point $(-2, 3)$ **40.** Center $(1, 0)$ and containing the point $(-3, 2)$

41. With endpoints of a diameter at $(1, 4)$ and $(-3, 2)$ **42.** With endpoints of a diameter at $(4, 3)$ and $(0, 1)$

43. Center $(2, -4)$ and circumference 16π **44.** Center $(-5, 6)$ and area 49π

In Problems 45–48, match each graph with the correct equation.

(a) $(x - 3)^2 + (y + 3)^2 = 9$ **(b)** $(x + 1)^2 + (y - 2)^2 = 4$ **(c)** $(x - 1)^2 + (y + 2)^2 = 4$ **(d)** $(x + 3)^2 + (y - 3)^2 = 9$

45.   **46.**   **47.**   **48.**

Applications and Extensions

49. Find an equation of the line containing the centers of the two circles

$$(x - 4)^2 + (y + 2)^2 = 25$$

and

$$(x + 1)^2 + (y - 5)^2 = 16$$

50. Find an equation of the line containing the centers of the two circles

$$x^2 + y^2 - 4x + 6y + 4 = 0$$

and

$$x^2 + y^2 + 6x + 4y + 9 = 0$$

51. If a circle of radius 2 rolls along the x-axis in quadrants I and II, what is an equation for the path of the center of the circle?

52. A circle of radius 7 rolls along the y-axis in quadrants I and IV. Find an equation for the path of the center of the circle.

53. Find the area of the square in the figure.

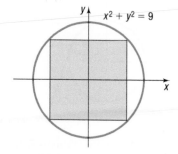
$x^2 + y^2 = 9$

54. Find the area of the blue shaded region in the figure, assuming the quadrilateral inside the circle is a square.

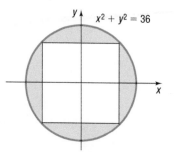
$x^2 + y^2 = 36$

55. Ferris Wheel The original Ferris wheel was built in 1893 by Pittsburgh, Pennsylvania bridge builder George W. Ferris. The Ferris wheel was originally built for the 1893 World's Fair in Chicago, but it was also later reconstructed for the 1904 World's Fair in St. Louis. It had a maximum height of 264 feet and a wheel diameter of 250 feet. Find an equation for the wheel if the center of the wheel is on the y-axis.

Source: guinnessworldrecords.com

56. Ferris Wheel The High Roller observation wheel in Las Vegas has a maximum height of 550 feet and a diameter of 520 feet, with one full rotation taking approximately 30 minutes. Find an equation for the wheel if the center of the wheel is on the y-axis.

Source: Las Vegas Review Journal

57. Vertically Circular Building Located in Al Raha, Abu Dhabi, the headquarters of property developing company Aldar is a vertically circular building with a diameter of 121 meters. The tip of the building is 110 meters aboveground. Find an equation for the building's outline if the center of the building is on the y-axis.

58. Vertically Circular Building The Sunrise Kempinski Hotel in Beijing, China, is a vertically circular building whose outline is described by the equation $x^2 + y^2 - 78y - 1843 = 0$ if the center of the building is on the y-axis. If x and y are in meters, what is the height of the building?

Problems 59–66 require the following discussion.

The **tangent line** to a circle may be defined as the line that intersects the circle in a single point, called the **point of tangency**. See the figure.

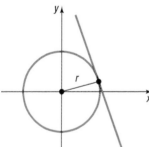

In Problems 59–62, find the standard form of the equation of each circle. (Refer to the preceding discussion).

59. Center $(2, 3)$ and tangent to the x-axis

60. Center $(-3, 1)$ and tangent to the y-axis

61. Center $(-1, 3)$ and tangent to the line $y = 2$

62. Center $(4, -2)$ and tangent to the line $x = 1$

63. Challenge Problem If the equation of a circle is $x^2 + y^2 = r^2$ and the equation of a tangent line is $y = mx + b$, show that:

(a) $r^2(1 + m^2) = b^2$

[**Hint**: The quadratic equation $x^2 + (mx + b)^2 = r^2$ has exactly one solution.]

(b) The point of tangency is $\left(\dfrac{-r^2 m}{b}, \dfrac{r^2}{b} \right)$.

(c) The tangent line is perpendicular to the line containing the center of the circle and the point of tangency.

64. Challenge Problem Refer to Problem 63. The line $x - 2y + 4 = 0$ is tangent to a circle at $(0, 2)$. The line $y = 2x - 7$ is tangent to the same circle at $(3, -1)$. Find the center of the circle.

65. Challenge Problem The Greek Method The Greek method for finding the equation of the tangent line to a circle uses the fact that at any point on a circle, the line containing the center and the tangent line are perpendicular. Use this method to find an equation of the tangent line to the circle $x^2 + y^2 = 9$ at the point $(1, 2\sqrt{2})$.

66. Challenge Problem Use the Greek method described in Problem 65 to find an equation of the tangent line to the circle

$$x^2 + y^2 - 4x + 6y + 4 = 0$$

at the point $(3, 2\sqrt{2} - 3)$.

67. Challenge Problem If (x_1, y_1) and (x_2, y_2) are the endpoints of a diameter of a circle, show that an equation of the circle is

$$(x - x_1)(x - x_2) + (y - y_1)(y - y_2) = 0$$

68. Challenge Problem If $x^2 + y^2 + dx + ey + f = 0$ is the equation of a circle, show that

$$x_0 x + y_0 y + d\left(\frac{x + x_0}{2} \right) + e\left(\frac{y + y_0}{2} \right) + f = 0$$

is an equation of the tangent line to the circle at the point (x_0, y_0).

Explaining Concepts: Discussion and Writing

69. Which of the following equations might have the graph shown? (More than one answer is possible.)

(a) $(x - 2)^2 + (y + 3)^2 = 13$

(b) $(x - 2)^2 + (y - 2)^2 = 8$

(c) $(x - 2)^2 + (y - 3)^2 = 13$

(d) $(x + 2)^2 + (y - 2)^2 = 8$

(e) $x^2 + y^2 - 4x - 9y = 0$

(f) $x^2 + y^2 + 4x - 2y = 0$

(g) $x^2 + y^2 - 9x - 4y = 0$

(h) $x^2 + y^2 - 4x - 4y = 4$

70. Which of the following equations might have the graph shown? (More than one answer is possible.)

(a) $(x - 2)^2 + y^2 = 3$

(b) $(x + 2)^2 + y^2 = 3$

(c) $x^2 + (y - 2)^2 = 3$

(d) $(x + 2)^2 + y^2 = 4$

(e) $x^2 + y^2 + 10x + 16 = 0$

(f) $x^2 + y^2 + 10x - 2y = 1$

(g) $x^2 + y^2 + 9x + 10 = 0$

(h) $x^2 + y^2 - 9x - 10 = 0$

71. Explain how the center and radius of a circle can be used to graph the circle.

72. **What Went Wrong?** A student stated that the center and radius of the graph whose equation is $(x + 3)^2 + (y - 2)^2 = 16$ are $(3, -2)$ and 4, respectively. Why is this incorrect?

─Retain Your Knowledge ───

Problems 73–82 are based on material learned earlier in the course. The purpose of these problems is to keep the material fresh in your mind so that you are better prepared for the final exam.

73. Find the area and circumference of a circle of radius 13 cm.

74. Multiply $(3x - 2)(x^2 - 2x + 3)$. Express the answer as a polynomial in standard form.

75. Solve the equation: $\sqrt{2x^2 + 3x - 1} = x + 1$

76. Aaron can load a delivery van in 22 minutes. Elizabeth can load the same van in 28 minutes. How long would it take them to load the van if they worked together?

77. Write 9.57×10^{-5} as a decimal.

78. Find the quotient and the remainder:
    $8x^5 - 10x^4 + 2x^3 + 11x^2 - 16x + 7$ divided by $2x^2 - 3$

79. Factor completely: $12x^5 - 15x^4 + 84x^3 - 105x^2$

80. Simplify: $\sqrt[4]{405x^{11}y^{20}}$

81. A square with an area of 64 square meters is inscribed inside a circle. What are the circumference and area of the circle?

82. How much pure antifreeze must be added to 2 liters of a 40% solution of antifreeze to obtain a 75% solution?

'Are You Prepared?' Answers

1. add; 25

2. $\{-1, 5\}$

2.5 Variation

OBJECTIVES 1 Construct a Model Using Direct Variation (p. 191)
2 Construct a Model Using Inverse Variation (p. 192)
3 Construct a Model Using Joint Variation (p. 193)

When a mathematical model is developed for a real-world problem, it often involves relationships between quantities that are expressed in terms of proportionality:

- Force is proportional to acceleration.
- When an ideal gas is held at a constant temperature, pressure and volume are inversely proportional.
- The force of attraction between two heavenly bodies is inversely proportional to the square of the distance between them.
- Revenue is directly proportional to sales.

Each of the preceding statements illustrates the idea of **variation**, or how one quantity varies in relation to another quantity. Quantities may vary *directly, inversely,* or *jointly.*

1 Construct a Model Using Direct Variation

DEFINITION Varies Directly, Directly Proportional, and Constant of Proportionality

Let x and y denote two quantities. Then y **varies directly** with x, or y is **directly proportional** to x, if there is a nonzero number k for which

$$y = kx$$

The number k is called the **constant of proportionality**.

Figure 54 $y = kx$; $k > 0$, $x \geq 0$

If y varies directly with x, $x \geq 0$, and k, $k > 0$, is the constant of proportionality, then the graph in Figure 54 illustrates the relationship between y and x. Note that the constant of proportionality is, in fact, the slope of the line.

If two quantities vary directly, then knowing the value of each quantity in one instance enables us to write a formula that is true in all cases.

EXAMPLE 1

Mortgage Payments

The monthly payment p on a mortgage varies directly with the amount borrowed B. If the monthly payment on a 30-year mortgage is \$6.65 for every \$1000 borrowed, find a formula that relates the monthly payment p to the amount borrowed B for a mortgage with these terms. Then find the monthly payment p when the amount borrowed B is \$120,000.

Solution

Because p varies directly with B, we know that

$$p = kB$$

for some constant k. Because $p = 6.65$ when $B = 1000$, it follows that

$$6.65 = k \cdot 1000$$

$$k = 0.00665 \quad \text{Solve for } k.$$

Since $p = kB$,

$$p = 0.00665B \quad \text{The Model}$$

In particular, when $B = \$120,000$,

$$p = 0.00665(\$120,000) = \$798$$

Figure 55

Figure 55 illustrates the relationship between the monthly payment p and the amount borrowed B.

- **Now Work** PROBLEMS 5 AND 23

---

2 Construct a Model Using Inverse Variation

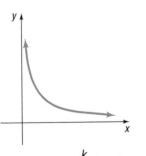

Figure 56 $y = \dfrac{k}{x}$; $k > 0$, $x > 0$

DEFINITION Varies Inversely and Inversely Proportional

Let x and y denote two quantities. Then y **varies inversely** with x, or y is **inversely proportional** to x, if there is a nonzero constant k for which

$$y = \frac{k}{x}$$

The graph in Figure 56 illustrates the relationship between y and x if y varies inversely with x and $k > 0$, $x > 0$.

EXAMPLE 2

Maximum Weight That Can Be Supported by a Piece of Pine

See Figure 57. The maximum weight W that can be safely supported by a 2-inch by 4-inch piece of pine varies inversely with its length l. Experiments indicate that the maximum weight that a 10-foot-long 2-by-4 piece of pine can support is 500 pounds. Write a general formula relating the maximum weight W (in pounds) to length l (in feet). Find the maximum weight W that can be safely supported by a length of 25 feet.

Solution   Because W varies inversely with l, we know that

$$W = \frac{k}{l}$$

for some constant k. Because $W = 500$ when $l = 10$, we have

$$500 = \frac{k}{10}$$
$$k = 5000$$

Figure 57

Since $W = \frac{k}{l}$,

$$W = \frac{5000}{l} \quad \text{The Model}$$

In particular, the maximum weight W that can be safely supported by a piece of pine 25 feet in length is

$$W = \frac{5000}{25} = 200 \text{ pounds}$$

Figure 58 illustrates the relationship between the weight W and the length l.   ⌐

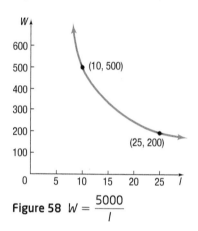

Figure 58 $W = \dfrac{5000}{l}$

───── **Now Work** PROBLEM 33

3 Construct a Model Using Joint Variation

When a variable quantity Q is proportional to two or more other variables, we say that Q **varies jointly** with these quantities. In joint variations, combinations of direct and/or inverse variation may occur.

EXAMPLE 3

Modeling a Joint Variation: Loss of Heat through a Wall

The loss of heat through a wall varies directly with the area of the wall and the difference between the inside and outside temperatures and varies inversely with the thickness of the wall. Write an equation that relates these quantities.

Solution   Begin by assigning symbols to represent the quantities:

L = Heat loss T = Temperature difference
A = Area of wall D = Thickness of wall

Then

$$L = k\frac{AT}{D} \quad \text{The Model}$$

where k is the constant of proportionality.   ⌐

In direct or inverse variation, the quantities that vary may be raised to powers. For example, in the early seventeenth century, Johannes Kepler (1571–1630) discovered that the square of the period of revolution T of a planet around the Sun varies directly with the cube of its mean distance a from the Sun. That is, $T^2 = ka^3$, where k is the constant of proportionality.

EXAMPLE 4

Force of the Wind on a Window

The force F of the wind on a flat surface positioned at a right angle to the direction of the wind varies directly with the area A of the surface and the square of the speed v of the wind. A wind of 30 miles per hour blowing on a window measuring 4 feet by 5 feet has a force of 150 pounds. See Figure 59. What force does a wind of 50 miles per hour exert on a window measuring 3 feet by 4 feet?

Solution

Since F varies directly with A and v^2, we have

$$F = kAv^2$$

where k is the constant of proportionality. We are told that $F = 150$ when $A = 4 \cdot 5 = 20$ and $v = 30$. Then

$$150 = k \cdot 20 \cdot 900 \qquad F = kAv^2, F = 150, A = 20, v = 30$$

$$k = \frac{1}{120}$$

Figure 59

Since $F = kAv^2$,

$$F = \frac{1}{120} Av^2 \qquad \text{The Model}$$

For a wind of 50 miles per hour blowing on a window whose area is $A = 3 \cdot 4 = 12$ square feet, the force F is

$$F = \frac{1}{120} \cdot 12 \cdot 2500 = 250 \text{ pounds}$$

Now Work PROBLEM 41

2.5 Assess Your Understanding

Concepts and Vocabulary

1. If x and y are two quantities, then y is directly proportional to x if there is a nonzero number k for which _____.

2. *True or False* If y varies directly with x, then $y = \dfrac{k}{x}$, where k is a constant.

3. *Multiple Choice* Which equation represents a joint variation model?
(a) $y = 5x$
(b) $y = x^2 + z^2$
(c) $y = \dfrac{5}{x}$
(d) $y = \dfrac{5xz}{w}$

4. *Multiple Choice* Choose the best description for the model $y = \dfrac{kx}{z}$, if k is a nonzero constant.
(a) y varies directly with x and z.
(b) y is inversely proportional to x and z.
(c) y varies directly with x and inversely with z.
(d) y is directly proportion to z and inversely proportional to x.

Skill Building

In Problems 5–16, write a general formula to describe each variation.

5. y varies directly with x; $y = 2$ when $x = 10$

6. v varies directly with t; $v = 16$ when $t = 2$

7. V varies directly with x^3; $V = 36\pi$ when $x = 3$

8. A varies directly with x^2; $A = 4\pi$ when $x = 2$

9. y varies inversely with $\sqrt{x}$; $y = 4$ when $x = 9$

10. F varies inversely with d^2; $F = 10$ when $d = 5$

11. z varies directly with the sum of the squares of x and y; $z = 26$ when $x = 5$ and $y = 12$

12. T varies directly with the cube root of x and the square of d; $T = 18$ when $x = 8$ and $d = 3$

13. M varies directly with the square of d and inversely with the square root of x; $M = 24$ when $x = 9$ and $d = 4$

14. z varies directly with the sum of the cube of x and the square of y; $z = 1$ when $x = 2$ and $y = 3$

15. The square of T varies directly with the cube of a and inversely with the square of d; $T = 2$ when $a = 2$ and $d = 4$

16. The cube of z varies directly with the sum of the squares of x and y; $z = 2$ when $x = 9$ and $y = 4$

Applications and Extensions

In Problems 17–22, write an equation that relates the quantities.

17. **Geometry** The volume V of a sphere varies directly with the cube of its radius r. The constant of proportionality is $\dfrac{4\pi}{3}$.

18. **Geometry** The square of the length of the hypotenuse c of a right triangle varies jointly with the sum of the squares of the lengths of its legs a and b. The constant of proportionality is 1.

19. **Geometry** The perimeter p of a rectangle varies jointly with the sum of the lengths of its sides l and w. The constant of proportionality is 2.

20. **Geometry** The area A of a triangle varies jointly with the product of the lengths of the base b and the height h. The constant of proportionality is $\dfrac{1}{2}$.

21. **Physics: Newton's Law** The force F (in newtons) of attraction between two bodies varies jointly with their masses m and M (in kilograms) and inversely with the square of the distance d (in meters) between them. The constant of proportionality is $G = 6.67 \times 10^{-11}$.

22. **Physics: Simple Pendulum** The **period** of a pendulum is the time required for one oscillation; the pendulum is usually referred to as **simple** when the angle made to the vertical is less than 5°. The period T of a simple pendulum (in seconds) varies directly with the square root of its length l (in feet). The constant of proportionality is $\dfrac{2\pi}{\sqrt{32}}$.

23. **Mortgage Payments** The monthly payment p on a mortgage varies directly with the amount borrowed B. If the monthly payment on a 30-year mortgage is $6.49 for every $1000 borrowed, find a linear equation that relates the monthly payment p to the amount borrowed B for a mortgage with these terms. Then find the monthly payment p when the amount borrowed B is $145,000.

24. **Mortgage Payments** The monthly payment p on a mortgage varies directly with the amount borrowed B. If the monthly payment on a 15-year mortgage is $8.99 for every $1000 borrowed, find a linear equation that relates the monthly payment p to the amount borrowed B for a mortgage with these terms. Then find the monthly payment p when the amount borrowed B is $175,000.

25. **Physics: Falling Objects** The distance s that an object falls is directly proportional to the square of the time t of the fall. If an object falls 16 feet in 1 second, how far will it fall in 3 seconds? How long will it take an object to fall 64 feet?

26. **Physics: Falling Objects** The velocity v of a falling object is directly proportional to the time t of the fall. If, after 2 seconds, the velocity of the object is 64 feet per second, what will its velocity be after 3 seconds?

27. **Physics: Stretching a Spring** The elongation E of a spring balance varies directly with the applied weight W (see the figure). If $E = 3$ when $W = 20$, find E when $W = 15$.

28. **Physics: Vibrating String** The rate of vibration of a string under constant tension varies inversely with the length of the string. If a string is 48 inches long and vibrates 256 times per second, what is the length of a string that vibrates 576 times per second?

29. **Revenue Equation** At the corner Shell station, the revenue R varies directly with the number g of gallons of gasoline sold. If the revenue is $34.08 when the number of gallons sold is 12, find a linear equation that relates revenue R to the number g of gallons of gasoline sold. Then find the revenue R when the number of gallons of gasoline sold is 10.5.

30. **Cost Equation** The cost C of roasted almonds varies directly with the number A of pounds of almonds purchased. If the cost is $23.75 when the number of pounds of roasted almonds purchased is 5, find a linear equation that relates the cost C to the number A of pounds of almonds purchased. Then find the cost C when the number of pounds of almonds purchased is 3.5.

31. Demand Suppose that the demand D for candy at the movie theater is inversely related to the price p.

(a) When the price of candy is \$2.75 per bag, the theater sells 156 bags of candy. Express the demand for candy in terms of its price.

(b) Determine the number of bags of candy that will be sold if the price is raised to \$3 a bag.

32. Driving to School The time t that it takes to get to school varies inversely with your average speed s.

(a) Suppose that it takes you 40 minutes to get to school when your average speed is 30 miles per hour. Express the driving time to school in terms of average speed.

(b) Suppose that your average speed to school is 40 miles per hour. How long will it take you to get to school?

33. Pressure The volume V of a gas held at a constant temperature in a closed container varies inversely with its pressure P. If the volume of a gas is 600 cubic centimeters (cm^3) when the pressure is 150 millimeters of mercury (mm Hg), find the volume when the pressure is 200 mm Hg.

34. Resistance The current i in a circuit is inversely proportional to its resistance Z measured in ohms. Suppose that when the current in a circuit is 30 amperes, the resistance is 8 ohms. Find the current in the same circuit when the resistance is 10 ohms.

35. Weight The weight of an object above the surface of Earth varies inversely with the square of the distance from the center of Earth. If Maria weighs 125 pounds when she is on the surface of Earth (3960 miles from the center), determine Maria's weight when she is at the top of Denali (3.8 miles from the surface of Earth).

36. Weight of a Body The weight of a body above the surface of Earth varies inversely with the square of the distance from the center of Earth. If a certain body weighs 55 pounds when it is 3960 miles from the center of Earth, how much will it weigh when it is 3965 miles from the center?

37. Geometry The volume V of a right circular cylinder varies directly with the square of its radius r and its height h. The constant of proportionality is π. See the figure. Write an equation for V.

38. Geometry The volume V of a right circular cone varies directly with the square of its radius r and its height h. The constant of proportionality is $\dfrac{\pi}{3}$. See the figure. Write an equation for V.

39. Intensity of Light The intensity I of light (measured in foot-candles) varies inversely with the square of the distance from the bulb. Suppose that the intensity of a 100-watt light bulb at a distance of 2 meters is 0.075 foot-candle. Determine the intensity of the bulb at a distance of 5 meters.

40. Force of the Wind on a Window The force exerted by the wind on a plane surface varies directly with the area of the surface and the square of the velocity of the wind. If the force on an area of 20 square feet is 11 pounds when the wind velocity is 22 miles per hour, find the force on a surface area of 47.125 square feet when the wind velocity is 36.5 miles per hour.

41. Horsepower The horsepower (hp) that a shaft can safely transmit varies directly with its speed (in revolutions per minute, rpm) and the cube of its diameter. If a shaft of a certain material 2 inches in diameter can transmit 36 hp at 75 rpm, what diameter must the shaft have in order to transmit 45 hp at 125 rpm?

42. Chemistry: Gas Laws The volume V of an ideal gas varies directly with the temperature T and inversely with the pressure P. Write an equation relating V, T, and P using k as the constant of proportionality. If a cylinder contains oxygen at a temperature of 300 K and a pressure of 15 atmospheres in a volume of 100 liters, what is the constant of proportionality k? If a piston is lowered into the cylinder, decreasing the volume occupied by the gas to 80 liters and raising the temperature to 310 K, what is the gas pressure?

43. Electrical Resistance of a Wire The electrical resistance of a wire varies directly with the length of the wire and inversely with the square of the diameter of the wire. If a wire 432 feet long and 4 millimeters in diameter has a resistance of 1.24 ohms, find the length of a wire of the same material whose resistance is 1.44 ohms and whose diameter is 3 millimeters.

44. Physics: Kinetic Energy The kinetic energy K of a moving object varies directly with its mass m and the square of its velocity v. If an object weighing 25 kilograms and moving with a velocity of 10 meters per second has a kinetic energy of 1250 joules, find its kinetic energy when the velocity is 15 meters per second.

45. Measuring the Stress of Materials The stress in the material of a pipe subject to internal pressure varies directly with the internal pressure and the internal diameter of the pipe and inversely with the thickness of the pipe. The stress is 100 pounds per square inch when the diameter is 5 inches, the thickness is 0.75 inch, and the internal pressure is 25 pounds per square inch. Find the stress when the internal pressure is 40 pounds per square inch if the diameter is 8 inches and the thickness is 0.50 inch.

46. Safe Load for a Beam The maximum safe load for a horizontal rectangular beam varies directly with the width of the beam and the square of the thickness of the beam and inversely with its length. If an 8-foot beam will support up to 750 pounds when the beam is 4 inches wide and 2 inches thick, what is the maximum safe load in a similar beam 10 feet long, 6 inches wide, and 2 inches thick?

Explaining Concepts: Discussion and Writing

47. In the early seventeenth century, Johannes Kepler discovered that the square of the period T of the revolution of a planet around the Sun varies directly with the cube of its mean distance a from the Sun. Research this law and Kepler's other two laws. Write a brief paper about these laws and Kepler's place in history.

48. Using a situation that has not been discussed in the text, write a real-world problem that you think involves two variables that vary directly. Exchange your problem with another student's to solve and critique.

49. Using a situation that has not been discussed in the text, write a real-world problem that you think involves two variables that vary inversely. Exchange your problem with another student's to solve and critique.

50. Using a situation that has not been discussed in the text, write a real-world problem that you think involves three variables that vary jointly. Exchange your problem with another student's to solve and critique.

Retain Your Knowledge

Problems 51–60 are based on material learned earlier in the course. The purpose of these problems is to keep the material fresh in your mind so that you are better prepared for the final exam.

51. Factor $3x^3 + 25x^2 - 12x - 100$ completely.

52. Add $\dfrac{5}{x+3} + \dfrac{x-2}{x^2 + 7x + 12}$ and simplify the result.

53. Simplify: $\left(\dfrac{4}{25}\right)^{3/2}$

54. Rationalize the denominator of $\dfrac{3}{\sqrt{7} - 2}$.

55. Evaluate $\dfrac{c^2 - b^2}{ac - 2b(c - a)}$ for $a = 3, b = 2$, and $c = 5$.

In Problems 56–59, solve each equation.

56. $6x - 2(x - 4) = 24$

57. $\dfrac{6}{x} - \dfrac{1}{10} = \dfrac{1}{5}$

58. $10x^2 = 117 - 19x$

59. $7 - 3|4x - 7| = 4$

60. Solve the inequality. Express the answer in interval notation, and graph the solution set.
$$5(x + 2) - 9x > x - 3(2x + 1) + 7$$

Chapter Review

Things to Know

Formulas

Distance formula (p. 151)	$d = \sqrt{(x_2 - x_1)^2 + (y_2 - y_1)^2}$
Midpoint formula (p. 154)	$(x, y) = \left(\dfrac{x_1 + x_2}{2}, \dfrac{y_1 + y_2}{2}\right)$
Slope of a line (p. 169)	$m = \dfrac{y_2 - y_1}{x_2 - x_1}$ if $x_1 \neq x_2$; undefined if $x_1 = x_2$
Parallel lines (p. 177)	Equal slopes $(m_1 = m_2)$ and different y-intercepts $(b_1 \neq b_2)$
Perpendicular lines (p. 178)	Product of slopes is -1 $(m_1 \cdot m_2 = -1)$
Direct variation (p. 191)	$y = kx, \ k \neq 0$
Inverse variation (p. 192)	$y = \dfrac{k}{x}, \ k \neq 0$

Equations of Lines and Circles

Vertical line (p. 173)	$x = a$; a is the x-intercept
Point-slope form of the equation of a line (p. 173)	$y - y_1 = m(x - x_1)$; m is the slope of the line, (x_1, y_1) is a point on the line
Horizontal line (p. 174)	$y = b$; b is the y-intercept
Slope-intercept form of the equation of a line (p. 174)	$y = mx + b$; m is the slope of the line, b is the y-intercept
General form of the equation of a line (p. 176)	$Ax + By = C$; A, B not both 0

Standard form of the equation of a circle (p. 185) $(x - h)^2 + (y - k)^2 = r^2$; r is the radius of the circle,
(h, k) is the center of the circle

Equation of the unit circle (p. 185) $x^2 + y^2 = 1$

General form of the equation of a circle (p. 187) $x^2 + y^2 + ax + by + c = 0$

Objectives

Section	You should be able to ...	Examples	Review Exercises
2.1	1 Use the distance formula (p. 151)	1–3	1(a)–3(a), 29, 30(a), 31
	2 Use the midpoint formula (p. 153)	4	1(b)–3(b), 31
2.2	1 Graph equations by plotting points (p. 158)	1–3	4
	2 Find intercepts from a graph (p. 160)	4	5
	3 Find intercepts from an equation (p. 161)	5	6–10
	4 Test an equation for symmetry with respect to the x-axis, the y-axis, and the origin (p. 161)	6–10	6–10
	5 Know how to graph key equations (p. 164)	11–13	26, 27
2.3	1 Calculate and interpret the slope of a line (p. 169)	1, 2	1(c)–3(c), 1(d)–3(d), 32
	2 Graph lines given a point and the slope (p. 172)	3	28
	3 Find the equation of a vertical line (p. 172)	4	17
	4 Use the point-slope form of a line; identify horizontal lines (p. 173)	5, 6	16
	5 Use the slope-intercept form of a line (p. 174)	7	16, 18–23
	6 Find an equation of a line given two points (p. 175)	8	18, 19
	7 Graph lines written in general form using intercepts (p. 176)	9	24, 25
	8 Find equations of parallel lines (p. 177)	10, 11	20
	9 Find equations of perpendicular lines (p. 178)	12, 13	21, 30(b)
2.4	1 Write the standard form of the equation of a circle (p. 185)	1	11, 12, 31
	2 Graph a circle (p. 186)	2, 3, 5	13–15
	3 Work with the general form of the equation of a circle (p. 187)	4	14, 15
2.5	1 Construct a model using direct variation (p. 191)	1	33
	2 Construct a model using inverse variation (p. 192)	2	34
	3 Construct a model using joint variation (p. 193)	3, 4	35

Review Exercises

In Problems 1–3, find the following for each pair of points:

   (a) The distance between the points

   (b) The midpoint of the line segment connecting the points

   (c) The slope of the line containing the points

   (d) Interpret the slope found in part (c)

1. $(0, 0)$; $(4, 2)$

2. $(1, -1)$; $(-2, 3)$

3. $(4, -4)$; $(4, 8)$

4. Graph $y = x^2 + 4$ by plotting points.

5. List the intercepts of the graph below.

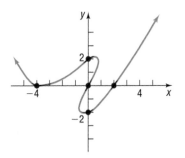

In Problems 6–10, list the intercepts and test for symmetry with respect to the x-axis, the y-axis, and the origin.

6. $2x = 3y^2$

7. $x^2 + 4y^2 = 16$

8. $y = x^4 + 2x^2 + 1$

9. $y = x^3 - x$

10. $x^2 + x + y^2 + 2y = 0$

In Problems 11 and 12, find the standard form of the equation of the circle whose center and radius are given.

11. $(h, k) = (-2, 3); r = 4$

12. $(h, k) = (-1, -2); r = 1$

In Problems 13–15, find the center and radius of each circle. Graph each circle. Find the intercepts, if any, of each circle.

13. $x^2 + (y - 1)^2 = 4$

14. $x^2 + y^2 - 2x + 4y - 4 = 0$

15. $3x^2 + 3y^2 - 6x + 12y = 0$

In Problems 16–21, find an equation of the line having the given characteristics. Express your answer using either the general form or the slope-intercept form of the equation of a line, whichever you prefer.

16. Slope $= -2$; containing the point $(3, -1)$

17. Vertical; containing the point $(-3, 4)$

18. y-intercept $= -2$; containing the point $(5, -3)$

19. Containing the points $(3, -4)$ and $(2, 1)$

20. Parallel to the line $2x - 3y = -4$;
containing the point $(-5, 3)$

21. Perpendicular to the line $x + y = 2$;
containing the point $(4, -3)$

In Problems 22 and 23, find the slope and y-intercept of each line. Graph the line, labeling any intercepts.

22. $4x - 5y = -20$

23. $\dfrac{1}{2}x - \dfrac{1}{3}y = -\dfrac{1}{6}$

In Problems 24 and 25, find the intercepts and graph each line.

24. $2x - 3y = 12$

25. $\dfrac{1}{2}x + \dfrac{1}{3}y = 2$

26. Graph $y = x^3$.

27. Graph $y = \sqrt{x}$.

28. Graph the line with slope $\dfrac{2}{3}$ containing the point $(1, 2)$.

29. Show that the points $A = (3, 4)$, $B = (1, 1)$, and $C = (-2, 3)$ are the vertices of an isosceles triangle.

30. Show that the points $A = (-2, 0)$, $B = (-4, 4)$, and $C = (8, 5)$ are the vertices of a right triangle in two ways:

(a) By using the converse of the Pythagorean Theorem

(b) By using the slopes of the lines joining the vertices

31. The endpoints of the diameter of a circle are $(-3, 2)$ and $(5, -6)$. Find the center and radius of the circle. Write the standard equation of this circle.

32. Show that the points $A = (2, 5)$, $B = (6, 1)$, and $C = (8, -1)$ lie on a line by using slopes.

33. Mortgage Payments The monthly payment p on a mortgage varies directly with the amount borrowed B. If the monthly payment on a 30-year mortgage is $854.00 when $130,000 is borrowed, find an equation that relates the monthly payment p to the amount borrowed B for a mortgage with these terms. Then find the monthly payment p when the amount borrowed B is $165,000.

34. Weight of a Body The weight of a body varies inversely with the square of its distance from the center of Earth. Assuming that the radius of Earth is 3960 miles, how much would a man weigh at an altitude of 1 mile above Earth's surface if he weighs 200 pounds on Earth's surface?

35. Heat Loss The amount of heat transferred per hour through a glass window varies jointly with the surface area of the window and the difference in temperature between the areas separated by the glass. A window with a surface area of 7.5 square feet loses 135 Btu per hour when the temperature difference is 40°F. How much heat is lost per hour for a similar window with a surface area of 12 square feet when the temperature difference is 35°F?

Chapter Test

 CHAPTER **Test Prep** VIDEOS

The Chapter Test Prep Videos include step-by-step solutions to all chapter test exercises. These videos are available in MyLab™ Math, or on this text's YouTube Channel. Refer to the Preface for a link to the YouTube channel.

In Problems 1–3, use $P_1 = (-1, 3)$ and $P_2 = (5, -1)$.

1. Find the distance from P_1 to P_2.

2. Find the midpoint of the line segment joining P_1 and P_2.

3. **(a)** Find the slope of the line containing P_1 and P_2.

   (b) Interpret this slope.

4. Graph $y = x^2 - 9$ by plotting points.

5. Sketch the graph of $y^2 = x$.

6. List the intercepts and test for symmetry: $x^2 + y = 9$

7. Write the slope-intercept form of the line with slope -2 containing the point $(3, -4)$. Graph the line.

8. Write the general form of the circle with center $(4, -3)$ and radius 5.

9. Find the center and radius of the circle
$$x^2 + y^2 + 4x - 2y - 4 = 0$$
Graph this circle.

10. For the line $2x + 3y = 6$, find a line parallel to the given line containing the point $(1, -1)$. Also find a line perpendicular to the given line containing the point $(0, 3)$.

11. **Resistance Due to a Conductor** The resistance (in ohms) of a circular conductor varies directly with the length of the conductor and inversely with the square of the radius of the conductor. If 50 feet of wire with a radius of 6×10^{-3} inch has a resistance of 10 ohms, what would be the resistance of 100 feet of the same wire if the radius were increased to 7×10^{-3} inch?

Cumulative Review

In Problems 1–8, find the real solution(s), if any, of each equation.

1. $3x - 5 = 0$

2. $x^2 - x - 12 = 0$

3. $2x^2 - 5x - 3 = 0$

4. $x^2 - 2x - 2 = 0$

5. $x^2 + 2x + 5 = 0$

6. $\sqrt{2x + 1} = 3$

7. $|x - 2| = 1$

8. $\sqrt{x^2 + 4x} = 2$

In Problems 9 and 10, solve each equation in the complex number system.[†]

9. $x^2 = -9$

10. $x^2 - 2x + 5 = 0$

[†]These problems are based on content from Section 1.3, which is optional.

In Problems 11–14, solve each inequality. Graph the solution set.

11. $2x - 3 \le 7$

12. $-1 < x + 4 < 5$

13. $|x - 2| \le 1$

14. $|2 + x| > 3$

15. Find the distance between the points $P = (-1, 3)$ and $Q = (4, -2)$. Find the midpoint of the line segment from P to Q.

16. Which of the following points are on the graph of $y = x^3 - 3x + 1$?

    (a) $(-2, -1)$ **(b)** $(2, 3)$ **(c)** $(3, 1)$

17. Sketch the graph of $y = x^3$.

18. Find an equation of the line containing the points $(-1, 4)$ and $(2, -2)$. Express your answer in slope-intercept form.

19. Find an equation of the line perpendicular to the line $y = 2x + 1$ and containing the point $(3, 5)$. Express your answer in slope-intercept form and graph the line.

20. Graph the equation $x^2 + y^2 - 4x + 8y - 5 = 0$.

Chapter Project

 Internet-based Project

I. Determining the Selling Price of a Home Determining how much to pay for a home is one of the more difficult decisions that must be made when purchasing a home. There are many factors that play a role in a home's value. Location, size, number of bedrooms, number of bathrooms, lot size, and building materials are just a few. Fortunately, the website Zillow.com has developed its own formula for predicting the selling price of a home. This information is a great tool for predicting the actual sale price. For example, the data below show the "zestimate"—the selling price of a home as predicted by the folks at Zillow—and the actual selling price of the home, for homes in Oak Park, Illinois.

Zestimate ($ thousands)	Sale Price ($ thousands)
291.5	268
320	305
371.5	375
303.5	283
351.5	350
314	275
332.5	356
295	300
313	285
368	385

The graph below, called a scatter plot, shows the points

$$(291.5, 268), (320, 305), \ldots, (368, 385)$$

in a Cartesian plane. From the graph, it appears that the data follow a linear relation.

1. Imagine drawing a line through the data that appears to fit the data well. Do you believe the slope of the line would be positive, negative, or close to zero? Why?

2. Pick two points from the scatter plot. Treat the zestimate as the value of x, and treat the sale price as the corresponding value of y. Find the equation of the line through the two points you selected.

3. Interpret the slope of the line.

4. Use your equation to predict the selling price of a home whose zestimate is $335,000.

5. Do you believe it would be a good idea to use the equation you found in part 2 if the zestimate were $950,000? Why or why not?

6. Choose a location in which you would like to live. Go to zillow.com and randomly select at least ten homes that have recently sold.

   (a) Draw a scatter plot of your data.

   (b) Select two points from the scatter plot and find the equation of the line through the points.

   (c) Interpret the slope.

   (d) Find a home from the Zillow website that interests you under the "Make Me Move" option for which a zestimate is available. Use your equation to predict the sale price based on the zestimate.

3

Functions and Their Graphs

Choosing a Data Plan

When selecting a data plan for a device, most consumers choose a service provider first and then select an appropriate data plan from that provider. The choice as to the type of plan selected depends on your use of the device. For example, is online gaming important? Do you want to stream audio or video? The mathematics learned in this chapter can help you decide what plan is best suited to your particular needs.

—See the Internet-based Chapter Project—

Outline

⟵ A Look Back

So far, our discussion has focused on solving equations and developing techniques for graphing equations containing two variables.

A Look Ahead ⟶

In this chapter, we look at a special type of equation involving two variables called a *function*. This chapter deals with what a function is, how to graph functions, properties of functions, and how functions are used in applications. The word *function* apparently was introduced by René Descartes in 1637. For him, a function was simply any positive integral power of a variable *x*. Gottfried Wilhelm Leibniz (1646–1716), who always emphasized the geometric side of mathematics, used the word *function* to denote any quantity associated with a curve, such as the coordinates of a point on the curve. Leonhard Euler (1707–1783) used the word to mean any equation or formula involving variables and constants. His idea of a function is similar to the one most often seen in courses that precede calculus. Later, the use of functions in investigating heat flow equations led to a very broad definition that originated with Lejeune Dirichlet (1805–1859), which describes a function as a correspondence between two sets. That is the definition used in this text.

3.1 Functions

PREPARING FOR THIS SECTION *Before getting started, review the following:*

- Intervals (Section 1.5, pp. 120–121)
- Solving Inequalities (Section 1.5, pp. 123–126)
- Evaluating Algebraic Expressions, Domain of a Variable (Section R.2, pp. 20–21)
- Rationalizing Denominators and Numerators (Section R.8, pp. 74–75)

Now Work the 'Are You Prepared?' problems on page 215.

OBJECTIVES 1 Describe a Relation (p. 203)
2 Determine Whether a Relation Represents a Function (p. 205)
3 Use Function Notation; Find the Value of a Function (p. 207)
4 Find the Difference Quotient of a Function (p. 210)
5 Find the Domain of a Function Defined by an Equation (p. 211)
6 Form the Sum, Difference, Product, and Quotient of Two Functions (p. 213)

1 Describe a Relation

Often there are situations where one variable is somehow linked to another variable. For example, the price of a gallon of gas is linked to the price of a barrel of oil. A person can be associated to her telephone number(s). The volume V of a sphere depends on its radius R. The force F exerted by an object corresponds to its acceleration a. These are examples of a *relation*, a correspondence between two sets called the *domain* and the *range*.

Figure 1

> ### DEFINITION Relation
>
> A **relation** is a correspondence between two sets: a set X, called the **domain**, and a set Y, called the **range**. In a relation, each element from the domain corresponds to at least one element from the range.

If x is an element of the domain and y is an element of the range, and if a relation exists from x to y, then we say that y **corresponds** to x or that y **depends on** x, and we write $x \rightarrow y$. It is often helpful to think of x as the **input** and y as the **output** of the relation. See Figure 1.

Suppose an astronaut standing on the Moon throws a rock 20 meters up and starts a stopwatch as the rock begins to fall back down. The astronaut measures the height of the rock at 1, 2, 2.5, 3, 4, and 5 seconds and obtains heights of 19.2, 16.8, 15, 12.8, 7.2, and 0 meters, respectively. This is an example of a relation expressed **verbally**. The domain of the relation is the set $\{0, 1, 2, 2.5, 3, 4, 5\}$ and the range of the relation is the set $\{20, 19.2, 16.8, 15, 12.8, 7.2, 0\}$.

The astronaut could also express this relation *numerically*, *graphically*, or *algebraically*.

The relation can be expressed **numerically** using a table of numbers, as in Table 1, or by using a **set of ordered pairs**. Using ordered pairs, the relation is

$$\{(0, 20), (1, 19.2), (2, 16.8), (2.5, 15), (3, 12.8), (4, 7.2), (5, 0)\}$$

where the first element of each pair denotes the time and the second element denotes the height.

Suppose x represents the number of seconds on the stopwatch and y represents the height of the rock in meters. Then the relation can be expressed **graphically** by plotting the points (x, y). See Figure 2 on the next page.

The relation can be represented as a **mapping** by drawing an arrow from an element in the domain to the corresponding element in the range. See Figure 3 on the next page.

Table 1

Time (in seconds)	Height (in meters)
0	20
1	19.2
2	16.8
2.5	15
3	12.8
4	7.2
5	0

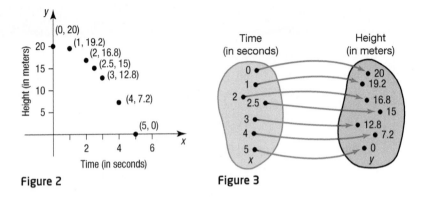

Figure 2

Figure 3

Finally, from physics, the relation can be expressed **algebraically** using the equation

$$y = 20 - 0.8x^2$$

EXAMPLE 1   Describing a Relation

A verbal description of a relation is given below.

The price of First Class U.S. postage stamps has changed over the years. To mail a letter in 2015 cost $0.49. In 2016 it cost $0.49 for part of the year and $0.47 for the rest of the year. In 2017 it cost $0.47 for part of the year and $0.49 for the rest of the year. In 2018 it cost $0.49 for part of the year and $0.50 for the rest of the year.

Using year as input and price as output,

    (a) What is the domain and the range of the relation?

    (b) Express the relation as a set of ordered pairs.

    (c) Express the relation as a mapping.

    (d) Express the relation as a graph.

Solution   The relation establishes a correspondence between the input, year, and the output, price of a First Class U.S. postage stamp.

(a) The domain of the relation is $\{2015, 2016, 2017, 2018\}$. The range of the relation is $\{\$0.47, \$0.49, \$0.50\}$.

(b) The relation expressed as a set of ordered pairs is $\{(2015, \$0.49), (2016, \$0.47), (2016, \$0.49), (2017, \$0.47), (2017, \$0.49), (2018, \$0.49), (2018, \$0.50)\}$

(c) See Figure 4 for the relation expressed as a mapping, using t for year and p for price.

(d) Figure 5 shows a graph of the relation.

Figure 4 First Class stamp price

Figure 5 First Class stamp price (2015–2018)

Now Work PROBLEM 17

2 Determine Whether a Relation Represents a Function

Look back at the relation involving the height of a rock on the Moon described at the beginning of the section. Notice that each input, time, corresponds to exactly one output, height. Given a time, you could tell the exact height of the rock. But that is not the case with the price of stamps. Given the year 2018, you cannot determine the price of a stamp with certainty. It could be $0.49, or it could be $0.50.

Consider the mapping of the relation in Figure 6. It shows a correspondence between a substance and its specific heat. Notice that for each substance you can tell its specific heat with certainty.

The relation associating the time to the height of the rock is a *function*, and the relation associating a given substance to its specific heat is a function. But the relation associating the year to the price of a First Class postage stamp is *not* a function. To be a function, each input must correspond to *exactly one* output.

Figure 6 Specific heat of some common substances

DEFINITION Function

Let X and Y be two nonempty sets.* A **function** from X into Y is a relation that associates with each element of X exactly one element of Y.

The set X is called the **domain** of the function. For each element x in X, the corresponding element y in Y is called the **value** of the function at x, or the **image** of x. The set of all images of the elements in the domain is called the **range** of the function. See Figure 7.

Since there may be some elements in Y that are not the image of some x in X, it follows that the range of a function may be a proper subset of Y, as shown in Figure 7.

The idea behind a function is its certainty. If an input is given, we can use the function to determine the output. This is not possible if a relation is not a function. The requirement of "one output" provides a predictable behavior that is important when using mathematics to model or analyze the real world. It allows doctors to know exactly how much medicine to give a patient, an engineer to determine the material to use in construction, a store manager to choose how many units to keep in stock, etc.

Figure 7

EXAMPLE 2 Determining Whether a Relation Given by a Mapping Is a Function

For each relation, state the domain and range. Then determine whether the relation is a function.

(a) See Figure 8. This relation shows a correspondence between an item on McDonald's $1-$2-$3 menu and its price.
Source: McDonald's Corporation, 2018

(b) See Figure 9. This relation shows a correspondence between activity level and daily calories needed for an individual with a basal metabolic rate (BMR) of 1975 kilocalories per day (kcal/day).

(c) See Figure 10. This relation shows a correspondence between gestation period (in days) and life expectancy (in years) for five animal species.

Figure 8

Figure 9

Figure 10

*The sets X and Y will usually be sets of real numbers, in which case a (real) function results. The two sets can also be sets of complex numbers, and then we have defined a complex function. In the broad definition (proposed by Lejeune Dirichlet), X and Y can be any two sets.

Solution (a) The domain of the relation is {Cheeseburger, McChicken, Bacon McDouble, Triple Cheeseburger, Happy Meal}, and the range of the relation is { $1, $2, $3 }. The relation in Figure 8 is a function because each element in the domain corresponds to exactly one element in the range.

(b) The domain of the relation is { Sedentary, Light, Moderate, Intense }, and the range of the relation is {2370, 2716, 3061, 3407}. The relation in Figure 9 is a function because each element in the domain corresponds to exactly one element in the range.

(c) The domain of the relation is { 122, 201, 240, 284 }, and the range of the relation is {5, 8, 12, 15, 20}. The relation in Figure 10 is not a function because there is an element in the domain, 240, that corresponds to two elements in the range, 12 and 20. If a gestation period of 240 days is selected from the domain, a single life expectancy cannot be associated with it.

━━━ **Now Work** PROBLEM 19

We may also think of a function as a set of ordered pairs (x, y) in which no ordered pairs have the same first element and different second elements. The set of all first elements x is the domain of the function, and the set of all second elements y is its range. Each element x in the domain corresponds to exactly one element y in the range.

EXAMPLE 3 **Determining Whether a Relation Given by a Set of Ordered Pairs Is a Function**

For each relation, state the domain and range. Then determine whether the relation is a function.

(a) { (1, 4), (2, 5), (3, 6), (4, 7) }

(b) { (1, 4), (2, 4), (3, 5), (6, 10) }

(c) { (−3, 9), (−2, 4), (0, 0), (1, 1), (−3, 8) }

Solution (a) The domain of this relation is {1, 2, 3, 4}, and its range is {4, 5, 6, 7}. This relation is a function because there are no ordered pairs with the same first element and different second elements.

(b) The domain of this relation is {1, 2, 3, 6}, and its range is {4, 5, 10}. This relation is a function because there are no ordered pairs with the same first element and different second elements.

(c) The domain of this relation is {−3, −2, 0, 1}, and its range is {0, 1, 4, 8, 9}. This relation is not a function because there are two ordered pairs, (−3, 9) and (−3, 8), that have the same first element and different second elements.

In Example 3(b), notice that 1 and 2 in the domain both have the same image in the range. This does not violate the definition of a function; two different first elements can have the same second element. The definition is violated when two ordered pairs have the same first element and different second elements, as in Example 3(c).

━━━ **Now Work** PROBLEM 23

Up to now we have shown how to identify when a relation is a function for relations defined by mappings (Example 2) and ordered pairs (Example 3). But relations can also be expressed as equations.

It is usually easiest to determine whether an equation, where y depends on x, is a function, when the equation is solved for y. If any value of x in the domain corresponds to more than one y, the equation does not define a function; otherwise, it does define a function.

EXAMPLE 4   Determining Whether an Equation Is a Function

Determine whether the equation $y = 2x - 5$ defines y as a function of x.

Solution The equation tells us to take an input x, multiply it by 2, and then subtract 5. For any input x, these operations yield only one output y, so the equation is a function. For example, if $x = 1$, then $y = 2 \cdot 1 - 5 = -3$. If $x = 3$, then $y = 2 \cdot 3 - 5 = 1$.

The graph of the equation $y = 2x - 5$ is a line with slope 2 and y-intercept -5. The function is called a *linear function*. See Figure 11.

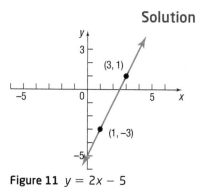

Figure 11 $y = 2x - 5$

EXAMPLE 5   Determining Whether an Equation Is a Function

Determine whether the equation $x^2 + y^2 = 1$ defines y as a function of x.

Solution To determine whether the equation $x^2 + y^2 = 1$, which defines the unit circle, is a function, solve the equation for y.

$$x^2 + y^2 = 1$$
$$y^2 = 1 - x^2$$
$$y = \pm\sqrt{1 - x^2}$$

For values of x for which $-1 < x < 1$, two values of y result. For example, if $x = 0$, then $y = \pm 1$, so two different outputs result from the same input. This means that the equation $x^2 + y^2 = 1$ does not define a function. See Figure 12.

Figure 12 $x^2 + y^2 = 1$

  Now Work PROBLEM 37

3 Use Function Notation; Find the Value of a Function

It is common practice to denote functions by letters such as f, g, F, G, and others. If f is a function, then for each number x in the domain, the corresponding number y in the range is designated by the symbol $f(x)$, read as "f of x," and we write $y = f(x)$. When a function is expressed in this way, we are using **function notation**.

We refer to $f(x)$ as the **value** of the function f at the number x. For example, the function in Example 4 may be written using function notation as $y = f(x) = 2x - 5$. Then $f(1) = -3$ and $f(3) = 1$.

Sometimes it is helpful to think of a function f as a machine that receives as input a number from the domain, manipulates it, and outputs a value in the range. See Figure 13.

The restrictions on this input/output machine are as follows:

- It accepts only numbers from the domain of the function.
- For each input, there is exactly one output (which may be repeated for different inputs).

Figure 13 Input/output machine

In Words
If $y = f(x)$, then x is the input and y is the output corresponding to x.

WARNING The notation $y = f(x)$ denotes a function f. It does NOT mean "f times x." ∎

Figure 14 illustrates some other functions. Notice that in every function, for each x in the domain, there is one corresponding value in the range.

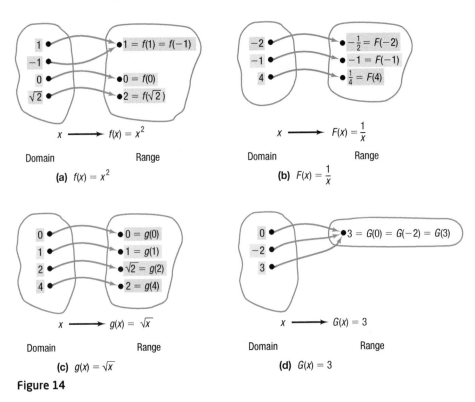

Figure 14

For a function $y = f(x)$, the variable x is called the **independent variable** because it can be assigned any number from the domain. The variable y is called the **dependent variable**, because its value depends on x.

Any symbols can be used to represent the independent and dependent variables. For example, if f is the *cube function*, then f can be given by $f(x) = x^3$ or $f(t) = t^3$ or $f(z) = z^3$. All three functions are the same. Each says to cube the independent variable to get the output. In practice, the symbols used for the independent and dependent variables are based on common usage, such as using t for time and a for acceleration.

The independent variable is also called the **argument** of the function. Thinking of the independent variable as an argument can sometimes make it easier to find the value of a function. For example, if f is the function defined by $f(x) = x^3$, then f tells us to cube the argument. Then $f(2)$ means to cube 2, $f(a)$ means to cube the number a, and $f(x + h)$ means to cube the quantity $x + h$.

EXAMPLE 6

Finding Values of a Function

For the function f defined by $f(x) = 2x^2 - 3x$, evaluate

(a) $f(3)$ (b) $f(x) + f(3)$ (c) $3f(x)$ (d) $f(-x)$

(e) $-f(x)$ (f) $f(3x)$ (g) $f(x + 3)$ (h) $f(x + h)$

Solution

(a) Substitute 3 for x in the equation for f, $f(x) = 2x^2 - 3x$, to get

$$f(3) = 2 \cdot 3^2 - 3 \cdot 3 = 18 - 9 = 9$$

The image of 3 is 9.

(b) $f(x) + f(3) = (2x^2 - 3x) + 9 = 2x^2 - 3x + 9$

(c) Multiply the equation for f by 3.

$$3f(x) = 3(2x^2 - 3x) = 6x^2 - 9x$$

(d) Substitute $-x$ for x in the equation for f and simplify.
$$f(-x) = 2(-x)^2 - 3(-x) = 2x^2 + 3x \quad \text{Notice the use of parentheses here.}$$

(e) $-f(x) = -(2x^2 - 3x) = -2x^2 + 3x$

(f) Substitute $3x$ for x in the equation for f and simplify.
$$f(3x) = 2(3x)^2 - 3 \cdot 3x = 2 \cdot 9x^2 - 9x = 18x^2 - 9x$$

(g) Substitute $x + 3$ for x in the equation for f and simplify.
$$\begin{aligned} f(x + 3) &= 2(x + 3)^2 - 3(x + 3) \\ &= 2(x^2 + 6x + 9) - 3x - 9 \\ &= 2x^2 + 12x + 18 - 3x - 9 \\ &= 2x^2 + 9x + 9 \end{aligned}$$

(h) Substitute $x + h$ for x in the equation for f and simplify.
$$\begin{aligned} f(x + h) &= 2(x + h)^2 - 3(x + h) \\ &= 2(x^2 + 2xh + h^2) - 3x - 3h \\ &= 2x^2 + 4xh + 2h^2 - 3x - 3h \end{aligned}$$

Notice in this example that $f(x + 3) \neq f(x) + f(3)$, $f(-x) \neq -f(x)$, and $3f(x) \neq f(3x)$.

- **Now Work** PROBLEM 43

Most calculators have special keys that allow you to find the value of certain commonly used functions. Examples are keys for the square function $f(x) = x^2$, the square root function $f(x) = \sqrt{x}$, and the reciprocal function $f(x) = \dfrac{1}{x} = x^{-1}$.

EXAMPLE 7   **Finding Values of a Function on a Calculator**

Verify the following results on your calculator.

(a) $f(x) = x^2$ $f(1.234) = 1.234^2 = 1.522756$

(b) $F(x) = \dfrac{1}{x}$ $F(1.234) = \dfrac{1}{1.234} \approx 0.8103727715$

(c) $g(x) = \sqrt{x}$ $g(1.234) = \sqrt{1.234} \approx 1.110855526$

COMMENT Graphing calculators can be used to evaluate a function. Figure 15 shows the function $Y_1 = f(x) = 2x^2 - 3x$ evaluated at 3 on a TI-84 Plus C graphing calculator. Compare this result with the solution to Example 6(a).

Figure 15 Evaluating $f(x) = 2x^2 - 3x$ for $x = 3$

Implicit Form of a Function

⊞ **COMMENT** A graphing calculator requires the explicit form of a function. ∎

In general, when a function f is defined by an equation in x and y, we say that the function f is given **implicitly**. If it is possible to solve the equation for y in terms of x, then we write $y = f(x)$ and say that the function is given **explicitly**. For example,

Implicit Form	**Explicit Form**
$3x + y = 5$	$y = f(x) = -3x + 5$
$x^2 - y = 6$	$y = f(x) = x^2 - 6$
$xy = 4$	$y = f(x) = \dfrac{4}{x}$

SUMMARY

Important Facts about Functions

- For each x in the domain of a function f, there is exactly one image $f(x)$ in the range; however, more than one x in the domain can have the same image in the range.
- f is the symbol that we use to denote the function. It is symbolic of the equation (rule) that we use to get from x in the domain to $f(x)$ in the range.
- If $y = f(x)$, then x is the independent variable, or the argument of f, and y, or $f(x)$, is the dependent variable, or the value of f at x.

⚖ 4 Find the Difference Quotient of a Function

An important concept in calculus involves using a certain quotient. For a function f, the inputs x and $x + h$, $h \neq 0$, result in the values $f(x)$ and $f(x + h)$. The quotient of their differences

$$\frac{f(x + h) - f(x)}{(x + h) - x} = \frac{f(x + h) - f(x)}{h}$$

with $h \neq 0$, is called the *difference quotient of f at x*.

DEFINITION Difference Quotient

The **difference quotient** of a function f at x is given by

$$\frac{f(x + h) - f(x)}{h} \qquad h \neq 0 \qquad \textbf{(1)}$$

The difference quotient is used in calculus to define the *derivative*, which leads to applications such as the velocity of an object and optimization of resources.

When finding a difference quotient, it is necessary to simplify expression (1) so that the h in the denominator can be cancelled.

EXAMPLE 8

Finding the Difference Quotient of a Function

Find the difference quotient of each function.

(a) $f(x) = 2x^2 - 3x$

(b) $f(x) = \dfrac{4}{x}$

(c) $f(x) = \sqrt{x}$

Solution   **(a)** $\dfrac{f(x+h)-f(x)}{h} = \dfrac{[2(x+h)^2 - 3(x+h)] - [2x^2 - 3x]}{h}$

$f(x+h) = 2(x+h)^2 - 3(x+h)$

$$= \dfrac{2(x^2 + 2xh + h^2) - 3x - 3h - 2x^2 + 3x}{h} \quad \text{Simplify.}$$

$$= \dfrac{2x^2 + 4xh + 2h^2 - 3h - 2x^2}{h} \quad \begin{array}{l}\text{Distribute and combine}\\ \text{like terms.}\end{array}$$

$$= \dfrac{4xh + 2h^2 - 3h}{h} \quad \text{Combine like terms.}$$

$$= \dfrac{h(4x + 2h - 3)}{h} \quad \text{Factor out } h.$$

$$= 4x + 2h - 3 \quad \text{Cancel the } h\text{'s.}$$

(b) $\dfrac{f(x+h)-f(x)}{h} = \dfrac{\dfrac{4}{x+h} - \dfrac{4}{x}}{h} \qquad f(x+h) = \dfrac{4}{x+h}$

$$= \dfrac{\dfrac{4x - 4(x+h)}{x(x+h)}}{h} \quad \text{Subtract.}$$

$$= \dfrac{4x - 4x - 4h}{x(x+h)h} \quad \text{Divide and distribute.}$$

$$= \dfrac{-4h}{x(x+h)h} \quad \text{Simplify.}$$

$$= -\dfrac{4}{x(x+h)} \quad \text{Cancel the } h\text{'s.}$$

Need to Review?
Rationalizing Numerators is discussed in Section R.8, pp. 74–75.

(c) When a function involves a square root, we rationalize the numerator of expression (1). After simplifying, the *h*'s cancel.

$$\dfrac{f(x+h)-f(x)}{h} = \dfrac{\sqrt{x+h} - \sqrt{x}}{h} \qquad f(x+h) = \sqrt{x+h}$$

$$= \dfrac{\sqrt{x+h} - \sqrt{x}}{h} \cdot \dfrac{\sqrt{x+h} + \sqrt{x}}{\sqrt{x+h} + \sqrt{x}} \quad \begin{array}{l}\text{Rationalize the}\\ \text{numerator.}\end{array}$$

$$= \dfrac{(\sqrt{x+h})^2 - (\sqrt{x})^2}{h(\sqrt{x+h} + \sqrt{x})} \quad (A - B)(A + B) = A^2 - B^2$$

$$= \dfrac{h}{h(\sqrt{x+h} + \sqrt{x})} \quad \begin{array}{l}(\sqrt{x+h})^2 - (\sqrt{x})^2\\ = x + h - x = h\end{array}$$

$$= \dfrac{1}{\sqrt{x+h} + \sqrt{x}} \quad \text{Cancel the } h\text{'s.}$$

Now Work PROBLEM 83

5 Find the Domain of a Function Defined by an Equation

Often the domain of a function *f* is not specified; instead, only the equation defining the function is given. In such cases, the **domain of *f*** is the largest set of real numbers for which the value $f(x)$ is a real number. The domain of a function *f* is the same as the domain of the variable *x* in the expression $f(x)$.

| EXAMPLE 9 | **Finding the Domain of a Function** |

Find the domain of each function.

(a) $f(x) = x^2 + 5x$

(b) $g(x) = \dfrac{3x}{x^2 - 4}$

(c) $h(t) = \sqrt{4 - 3t}$

(d) $F(x) = \dfrac{\sqrt{3x + 12}}{x - 5}$

Solution

(a) The function $f(x) = x^2 + 5x$ says to sum the square of a number and five times the number. Since these operations can be performed on any real number, the domain of f is the set of all real numbers.

In Words

The domain of g found in Example 9(b) is $\{x \mid x \neq -2, x \neq 2\}$.
This notation is read, "The domain of the function g is the set of all real numbers x such that x does not equal -2 and x does not equal 2."

(b) The function $g(x) = \dfrac{3x}{x^2 - 4}$ says to divide $3x$ by $x^2 - 4$. Since division by 0 is not defined, the denominator $x^2 - 4$ cannot be 0, so x cannot equal -2 or 2. The domain of the function g is $\{x \mid x \neq -2, x \neq 2\}$.

(c) The function $h(t) = \sqrt{4 - 3t}$ says to take the square root of $4 - 3t$. Since only nonnegative numbers have real square roots, the expression under the square root (the radicand) must be nonnegative (greater than or equal to zero). That is,

$$4 - 3t \geq 0$$
$$-3t \geq -4$$
$$t \leq \frac{4}{3}$$

The domain of h is $\left\{ t \,\middle|\, t \leq \dfrac{4}{3} \right\}$, or the interval $\left(-\infty, \dfrac{4}{3} \right]$.

(d) The function $F(x) = \dfrac{\sqrt{3x + 12}}{x - 5}$ says to take the square root of $3x + 12$ and divide the result by $x - 5$. This requires that $3x + 12 \geq 0$, so $x \geq -4$, and also that $x - 5 \neq 0$, so $x \neq 5$. Combining these two restrictions, the domain of F is

$$\{x \mid x \geq -4, x \neq 5\}$$

The following steps may prove helpful for finding the domain of a function that is defined by an equation and whose domain is a subset of the real numbers.

Finding the Domain of a Function Defined by an Equation

- Start with the domain as the set of all real numbers.
- If the equation has a denominator, exclude any numbers for which the denominator is zero.
- If the equation has a radical with an even index, exclude any numbers for which the expression inside the radical (the radicand) is negative.

Now Work PROBLEM 55

We express the domain of a function using interval notation, set notation, or words, whichever is most convenient. If x is in the domain of a function f, we say that f **is defined at x**, or $f(x)$ **exists**. If x is not in the domain of f, we say that f **is not defined at x**, or $f(x)$ **does not exist**. For example, if $f(x) = \dfrac{x}{x^2 - 1}$, then $f(0)$ exists, but $f(1)$ and $f(-1)$ do not exist. (Do you see why?)

When a function is defined by an equation, it can be difficult to find its range unless we also have a graph of the function. So we are usually content to find only the domain.

When we use functions in applications, the domain may be restricted by physical or geometric considerations. For example, the domain of the function f defined by $f(x) = x^2$ is the set of all real numbers. However, if f represents the area of a square whose sides are of length x, the domain of f is restricted to the positive real numbers, since the length of a side can never be 0 or negative.

EXAMPLE 10

Finding the Domain of a Function Used in an Application

Express the area of a circle as a function of its radius. Find the domain.

Solution

See Figure 16. The formula for the area A of a circle of radius r is $A = \pi r^2$. Using r to represent the independent variable and A to represent the dependent variable, the function expressing this relationship is

$$A = A(r) = \pi r^2$$

In this application, the domain is $\{r \mid r > 0\}$. (Do you see why?)

Figure 16 Circle of radius r

Now Work PROBLEM 105

6 Form the Sum, Difference, Product, and Quotient of Two Functions

Functions, like numbers, can be added, subtracted, multiplied, and divided. For example, if $f(x) = x^2 + 9$ and $g(x) = 3x + 5$, then

$$f(x) + g(x) = (x^2 + 9) + (3x + 5) = x^2 + 3x + 14$$

The new function $y = x^2 + 3x + 14$ is called the *sum function $f + g$*. Similarly,

$$f(x) \cdot g(x) = (x^2 + 9)(3x + 5) = 3x^3 + 5x^2 + 27x + 45$$

The new function $y = 3x^3 + 5x^2 + 27x + 45$ is called the *product function $f \cdot g$*. The general definitions are given next.

> **DEFINITION** Sum Function
>
> Given functions f and g, the **sum function** is defined by
>
> $$(f + g)(x) = f(x) + g(x)$$

RECALL

The symbol $\cap$ stands for intersection. It means the set of elements that are common to both sets.

The domain of $f + g$ consists of all real numbers x that are in the domains of both f and g. That is, domain of $f + g$ = domain of $f \cap$ domain of g.

> **DEFINITION** Difference Function
>
> Given functions f and g, the **difference function** is defined by
>
> $$(f - g)(x) = f(x) - g(x)$$

The domain of $f - g$ consists of all real numbers x that are in the domains of both f and g. That is, domain of $f - g$ = domain of $f \cap$ domain of g.

> **DEFINITION** Product Function
>
> Given functions f and g, the **product function** is defined by
>
> $$(f \cdot g)(x) = f(x) \cdot g(x)$$

The domain of $f \cdot g$ consists of all real numbers x that are in the domains of both f and g. That is, domain of $f \cdot g$ = domain of $f \cap$ domain of g.

> **DEFINITION Quotient Function**
>
> Given functions f and g, the **quotient function** is defined by
>
> $$\left(\frac{f}{g}\right)(x) = \frac{f(x)}{g(x)} \qquad g(x) \neq 0$$

The domain of $\dfrac{f}{g}$ consists of all real numbers x for which $g(x) \neq 0$ that are also in the domains of both f and g. That is,

$$\text{domain of } \frac{f}{g} = \{x \mid g(x) \neq 0\} \cap \text{domain of } f \cap \text{domain of } g$$

EXAMPLE 11 | **Operations on Functions**

Let f and g be two functions defined as

$$f(x) = \frac{1}{x+2} \quad \text{and} \quad g(x) = \frac{x}{x-1}$$

Find the following functions, and determine the domain.

(a) $(f+g)(x)$ (b) $(f-g)(x)$ (c) $(f \cdot g)(x)$ (d) $\left(\dfrac{f}{g}\right)(x)$

Solution The domain of f is $\{x \mid x \neq -2\}$ and the domain of g is $\{x \mid x \neq 1\}$.

(a) $(f+g)(x) = f(x) + g(x) = \dfrac{1}{x+2} + \dfrac{x}{x-1}$

$\qquad = \dfrac{x-1}{(x+2)(x-1)} + \dfrac{x(x+2)}{(x+2)(x-1)}$

$\qquad = \dfrac{x-1+x^2+2x}{(x+2)(x-1)} = \dfrac{x^2+3x-1}{(x+2)(x-1)}$

The domain of $f+g$ consists of all real numbers x that are in the domains of both f and g. The domain of $f+g$ is $\{x \mid x \neq -2, x \neq 1\}$.

(b) $(f-g)(x) = f(x) - g(x) = \dfrac{1}{x+2} - \dfrac{x}{x-1}$

$\qquad = \dfrac{x-1}{(x+2)(x-1)} - \dfrac{x(x+2)}{(x+2)(x-1)}$

$\qquad = \dfrac{x-1-x^2-2x}{(x+2)(x-1)} = \dfrac{-x^2-x-1}{(x+2)(x-1)} = -\dfrac{x^2+x+1}{(x+2)(x-1)}$

The domain of $f-g$ consists of all real numbers x that are in the domains of both f and g. The domain of $f-g$ is $\{x \mid x \neq -2, x \neq 1\}$.

(c) $(f \cdot g)(x) = f(x) \cdot g(x) = \dfrac{1}{x+2} \cdot \dfrac{x}{x-1} = \dfrac{x}{(x+2)(x-1)}$

The domain of $f \cdot g$ consists of all real numbers x that are in the domains of both f and g. The domain of $f \cdot g$ is $\{x \mid x \neq -2, x \neq 1\}$.

(d) $\left(\dfrac{f}{g}\right)(x) = \dfrac{f(x)}{g(x)} = \dfrac{\dfrac{1}{x+2}}{\dfrac{x}{x-1}} = \dfrac{1}{x+2} \cdot \dfrac{x-1}{x} = \dfrac{x-1}{x(x+2)}$

The domain of $\dfrac{f}{g}$ consists of all real numbers x for which $g(x) \neq 0$ that are also in the domains of both f and g. Since $g(x) = 0$ when $x = 0$, exclude 0 as well as -2 and 1 from the domain. The domain of $\dfrac{f}{g}$ is $\{x \mid x \neq -2, x \neq 0, x \neq 1\}$.

Now Work PROBLEM 71

In calculus, it is sometimes helpful to view a complicated function as the sum, difference, product, or quotient of simpler functions. For example,

$$F(x) = x^2 + \sqrt{x} \text{ is the sum of } f(x) = x^2 \text{ and } g(x) = \sqrt{x}.$$

$$H(x) = \frac{x^2 - 1}{x^2 + 1} \text{ is the quotient of } f(x) = x^2 - 1 \text{ and } g(x) = x^2 + 1.$$

SUMMARY

Function	• A relation between two sets of real numbers so that each number x in the first set, the domain, corresponds to exactly one number y in the second set.
	• A set of ordered pairs (x, y) or $(x, f(x))$ in which no first element is paired with two different second elements.
	• The range is the set of y-values of the function that are the images of the x-values in the domain.
	• A function f may be defined implicitly by an equation involving x and y or explicitly by writing $y = f(x)$.
Unspecified domain	If a function f is defined by an equation and no domain is specified, then the domain is the largest set of real numbers for which $f(x)$ is a real number.
Function notation	• $y = f(x)$
	• f is a symbol for the function.
	• x is the independent variable, or argument.
	• y is the dependent variable.
	• $f(x)$ is the value of the function at x.

3.1 Assess Your Understanding

'Are You Prepared?' *Answers are given at the end of these exercises. If you get a wrong answer, read the pages listed in* red.

1. The inequality $-1 < x < 3$ can be written in interval notation as _____. (pp. 120–121)

2. If $x = -2$, the value of the expression $3x^2 - 5x + \dfrac{1}{x}$ is _____. (p. 20)

3. The domain of the variable in the expression $\dfrac{x - 3}{x + 4}$ is _____. (p. 21)

4. Solve the inequality: $3 - 2x > 5$. Graph the solution set. (pp. 123–126)

5. To rationalize the denominator of $\dfrac{3}{\sqrt{5} - 2}$, multiply the numerator and denominator by _____. (pp. 74–75)

6. A quotient is considered rationalized if its denominator has no _____. (pp. 74–75)

Concepts and Vocabulary

7. For a function $y = f(x)$, the variable x is the _____ variable, and the variable y is the _____ variable.

8. *Multiple Choice* The set of all images of the elements in the domain of a function is called the _____.

   (a) range (b) domain (c) solution set (d) function

9. *Multiple Choice* The independent variable is sometimes referred to as the _____ of the function.

   (a) range (b) value (c) argument (d) definition

10. *True or False* The domain of $\dfrac{f}{g}$ consists of the numbers x that are in the domains of both f and g.

11. *True or False* Every relation is a function.

12. Four ways of expressing a relation are _____, _____, _____, and _____.

13. *True or False* If no domain is specified for a function f, then the domain of f is the set of real numbers.

14. *True or False* If x is in the domain of a function f, we say that f is not defined at x, or $f(x)$ does not exist.

15. The expression $\dfrac{f(x + h) - f(x)}{h}$ is called the _____ _____ of f.

16. When written as $y = f(x)$, a function is said to be defined _____.

Skill Building

In Problems 17 and 18, a relation expressed verbally is given.

(a) What is the domain and the range of the relation?   (b) Express the relation using a mapping.   (c) Express the relation as a set of ordered pairs.

17. The density of a gas under constant pressure depends on temperature. Holding pressure constant at 14.5 pounds per square inch, a chemist measures the density of an oxygen sample at temperatures of 0, 22, 40, 70, and 100°C and obtains densities of 1.411, 1.305, 1.229, 1.121, and 1.031 kg/m³, respectively.

18. A researcher wants to investigate how weight depends on height among adult males in Europe. She visits five regions in Europe and determines the average heights in those regions to be 1.80, 1.78, 1.77, 1.77, and 1.80 meters. The corresponding average weights are 87.1, 86.9, 83.0, 84.1, and 86.4 kg, respectively.

In Problems 19–30, find the domain and range of each relation. Then determine whether the relation represents a function.

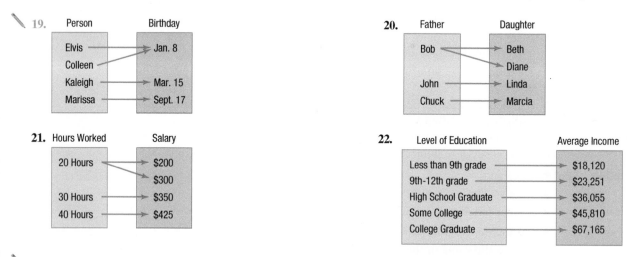

19.
Person	Birthday
Elvis	Jan. 8
Colleen	
Kaleigh	Mar. 15
Marissa	Sept. 17

20.
Father	Daughter
Bob	Beth
	Diane
John	Linda
Chuck	Marcia

21.
Hours Worked	Salary
20 Hours	$200
	$300
30 Hours	$350
40 Hours	$425

22.
Level of Education	Average Income
Less than 9th grade	$18,120
9th-12th grade	$23,251
High School Graduate	$36,055
Some College	$45,810
College Graduate	$67,165

23. $\{(2, 6), (-3, 6), (4, 9), (2, 10)\}$ **24.** $\{(-2, 5), (-1, 3), (3, 7), (4, 12)\}$ **25.** $\{(1, 3), (2, 3), (3, 3), (4, 3)\}$

26. $\{(0, -2), (1, 3), (2, 3), (3, 7)\}$ **27.** $\{(3, 3), (3, 5), (0, 1), (-4, 6)\}$ **28.** $\{(-4, 4), (-3, 3), (-2, 2), (-1, 1), (-4, 0)\}$

29. $\{(-1, 8), (0, 3), (2, -1), (4, 3)\}$ **30.** $\{(-2, 16), (-1, 4), (0, 3), (1, 4)\}$

In Problems 31–42, determine whether the equation defines y as a function of x.

31. $y = 2x^2 - 3x + 4$

32. $y = x^3$

33. $y = \dfrac{1}{x}$

34. $y = |x|$

35. $x^2 = 8 - y^2$

36. $y = \pm\sqrt{1 - 2x}$

37. $x = y^2$

38. $x + y^2 = 1$

39. $y = \sqrt[3]{x}$

40. $y = \dfrac{3x - 1}{x + 2}$

41. $|y| = 2x + 3$

42. $x^2 - 4y^2 = 1$

In Problems 43–50, find the following for each function:

(a) $f(0)$ (b) $f(1)$ (c) $f(-1)$ (d) $f(-x)$ (e) $-f(x)$ (f) $f(x + 1)$ (g) $f(2x)$ (h) $f(x + h)$

43. $f(x) = 3x^2 + 2x - 4$

44. $f(x) = -2x^2 + x - 1$

45. $f(x) = \dfrac{x}{x^2 + 1}$

46. $f(x) = \dfrac{x^2 - 1}{x + 4}$

47. $f(x) = |x| + 4$

48. $f(x) = \sqrt{x^2 + x}$

49. $f(x) = \dfrac{2x + 1}{3x - 5}$

50. $f(x) = 1 - \dfrac{1}{(x + 2)^2}$

In Problems 51–70, find the domain of each function.

51. $f(x) = -5x + 4$

52. $f(x) = x^2 + 2$

53. $f(x) = \dfrac{x + 1}{2x^2 + 8}$

54. $f(x) = \dfrac{x^2}{x^2 + 1}$

55. $g(x) = \dfrac{x}{x^2 - 16}$

56. $h(x) = \dfrac{2x}{x^2 - 4}$

57. $F(x) = \dfrac{x - 2}{x^3 + x}$

58. $G(x) = \dfrac{x + 4}{x^3 - 4x}$

59. $h(x) = \sqrt{3x - 12}$ **60.** $G(x) = \sqrt{1 - x}$ **61.** $p(x) = \dfrac{x}{|2x + 3| - 1}$ **62.** $f(x) = \dfrac{x - 1}{|3x - 1| - 4}$

63. $f(x) = \dfrac{x}{\sqrt{x - 4}}$ **64.** $f(x) = \dfrac{-x}{\sqrt{-x - 2}}$ **65.** $P(t) = \dfrac{\sqrt{t - 4}}{3t - 21}$ **66.** $h(z) = \dfrac{\sqrt{z + 3}}{z - 2}$

67. $f(x) = \sqrt[3]{5x - 4}$ **68.** $g(t) = -t^2 + \sqrt[3]{t^2 + 7t}$ **69.** $M(t) = \sqrt[5]{\dfrac{t + 1}{t^2 - 5t - 14}}$ **70.** $N(p) = \sqrt[5]{\dfrac{p}{2p^2 - 98}}$

In Problems 71–80, for the given functions f and g, find the following. For parts (a)–(d), also find the domain.

(a) $(f + g)(x)$ (b) $(f - g)(x)$ (c) $(f \cdot g)(x)$ (d) $\left(\dfrac{f}{g}\right)(x)$

(e) $(f + g)(3)$ (f) $(f - g)(4)$ (g) $(f \cdot g)(2)$ (h) $\left(\dfrac{f}{g}\right)(1)$

71. $f(x) = 3x + 4;\ \ g(x) = 2x - 3$
73. $f(x) = x - 1;\ \ g(x) = 2x^2$
75. $f(x) = \sqrt{x};\ \ g(x) = 3x - 5$
77. $f(x) = 1 + \dfrac{1}{x};\ \ g(x) = \dfrac{1}{x}$
79. $f(x) = \dfrac{2x + 3}{3x - 2};\ \ g(x) = \dfrac{4x}{3x - 2}$

72. $f(x) = 2x + 1;\ \ g(x) = 3x - 2$
74. $f(x) = 2x^2 + 3;\ \ g(x) = 4x^3 + 1$
76. $f(x) = |x|;\ \ g(x) = x$
78. $f(x) = \sqrt{x - 1};\ \ g(x) = \sqrt{4 - x}$
80. $f(x) = \sqrt{x + 1};\ \ g(x) = \dfrac{2}{x}$

81. Given $f(x) = 3x + 1$ and $(f + g)(x) = 6 - \dfrac{1}{2}x$, find the function g.

82. Given $f(x) = \dfrac{1}{x}$ and $\left(\dfrac{f}{g}\right)(x) = \dfrac{x + 1}{x^2 - x}$, find the function g.

In Problems 83–98, find the difference quotient of f; that is, find $\dfrac{f(x + h) - f(x)}{h}, h \neq 0$, for each function. Be sure to simplify.

83. $f(x) = 4x + 3$ **84.** $f(x) = -3x + 1$ **85.** $f(x) = x^2 - 4$ **86.** $f(x) = 3x^2 + 2$

87. $f(x) = x^2 - x + 4$ **88.** $f(x) = 3x^2 - 2x + 6$ **89.** $f(x) = \dfrac{5}{4x - 3}$ **90.** $f(x) = \dfrac{1}{x + 3}$

91. $f(x) = \dfrac{2x}{x + 3}$ **92.** $f(x) = \dfrac{5x}{x - 4}$ **93.** $f(x) = \sqrt{x - 2}$ **94.** $f(x) = \sqrt{x + 1}$

95. $f(x) = \dfrac{1}{x^2}$ **96.** $f(x) = \dfrac{1}{x^2 + 1}$ **97.** $f(x) = \sqrt{4 - x^2}$ **98.** $f(x) = \dfrac{1}{\sqrt{x + 2}}$

Applications and Extensions

99. If $f(x) = x^2 - 2x + 3$, find the value(s) of x so that $f(x) = 11$.

100. If $f(x) = \dfrac{5}{6}x - \dfrac{3}{4}$, find the value(s) of x so that $f(x) = -\dfrac{7}{16}$.

101. If $f(x) = 2x^3 + Ax^2 + 4x - 5$ and $f(2) = 5$, what is the value of A?

102. If $f(x) = 3x^2 - Bx + 4$ and $f(-1) = 12$, what is the value of B?

103. If $f(x) = \dfrac{3x + 8}{2x - A}$ and $f(0) = 2$, what is the value of A?

104. If $f(x) = \dfrac{2x - B}{3x + 4}$ and $f(2) = \dfrac{1}{2}$, what is the value of B?

105. Geometry Express the area A of a rectangle as a function of the length x if the length of the rectangle is twice its width.

106. Geometry Express the area A of an isosceles right triangle as a function of the length x of one of the two equal sides.

107. Constructing Functions Express the gross wages G of a person who earns $16 per hour as a function of the number x of hours worked.

108. Constructing Functions Ann, a commissioned salesperson, earns $100 base pay plus $10 per item sold. Express her gross salary G as a function of the number x of items sold.

109. Effect of Gravity on Earth If a rock falls from a height of 20 meters on Earth, the height H (in meters) after x seconds is approximately

$$H(x) = 20 - 4.9x^2$$

(a) What is the height of the rock when $x = 1$ second? When $x = 1.1$ seconds? When $x = 1.2$ seconds?

(b) When is the height of the rock 15 meters? When is it 10 meters? When is it 5 meters?

(c) When does the rock strike the ground?

110. Effect of Gravity on Jupiter If a rock falls from a height of 20 meters on the planet Jupiter, its height H (in meters) after x seconds is approximately

$$H(x) = 20 - 13x^2$$

(a) What is the height of the rock when $x = 1$ second? When $x = 1.1$ seconds? When $x = 1.2$ seconds?

(b) When is the height of the rock 15 meters? When is it 10 meters? When is it 5 meters?

(c) When does the rock strike the ground?

111. Cost of Transatlantic Travel A Boeing 747 crosses the Atlantic Ocean (3000 miles) with an airspeed of 500 miles per hour. The cost C (in dollars) per passenger is given by

$$C(x) = 100 + \frac{x}{10} + \frac{36,000}{x}$$

where x is the ground speed (airspeed $\pm$ wind).

(a) What is the cost per passenger for quiescent (no wind) conditions?

(b) What is the cost per passenger with a head wind of 50 miles per hour?

(c) What is the cost per passenger with a tail wind of 100 miles per hour?

(d) What is the cost per passenger with a head wind of 100 miles per hour?

112. Cross-sectional Area The cross-sectional area of a beam cut from a log with radius 1 foot is given by the function $A(x) = 4x\sqrt{1 - x^2}$, where x represents the length, in feet, of half the base of the beam. See the figure. Determine the cross-sectional area of the beam if the length of half the base of the beam is as follows:

(a) One-third of a foot

(b) One-half of a foot

(c) Two-thirds of a foot

$A(x) = 4x\sqrt{1 - x^2}$

113. Economics The **participation rate** is the number of people in the labor force divided by the civilian population (excludes military). Let $L(x)$ represent the size of the labor force in year x, and $P(x)$ represent the civilian population in year x. Determine a function that represents the participation rate R as a function of x.

114. Crimes Suppose that $V(x)$ represents the number of violent crimes committed in year x and $P(x)$ represents the number of property crimes committed in year x. Determine a function T that represents the combined total of violent crimes and property crimes in year x.

115. Health Care Suppose that $P(x)$ represents the percentage of income spent on health care in year x and $I(x)$ represents income in year x. Find a function H that represents total health care expenditures in year x.

116. Income Tax Suppose that $I(x)$ represents the income of an individual in year x before taxes and $T(x)$ represents the individual's tax bill in year x. Find a function N that represents the individual's net income (income after taxes) in year x.

117. Profit Function Suppose that the revenue R, in dollars, from selling x smartphones, in hundreds, is

$$R(x) = -1.2x^2 + 220x$$

The cost C, in dollars, of selling x smartphones, in hundreds, is $C(x) = 0.05x^3 - 2x^2 + 65x + 500$.

(a) Find the profit function, $P(x) = R(x) - C(x)$.

(b) Find the profit if $x = 15$ hundred smartphones are sold.

(c) Interpret $P(15)$.

118. Population as a Function of Age The function

$$P = P(a) = 0.027a^2 - 6.530a + 363.804$$

represents the population P (in millions) of Americans who are at least a years old in 2015.

(a) Identify the dependent and independent variables.

(b) Evaluate $P(20)$. Explain the meaning of $P(20)$.

(c) Evaluate $P(0)$. Explain the meaning of $P(0)$.

Source: U.S. Census Bureau

119. Stopping Distance When the driver of a vehicle observes an impediment, the total stopping distance involves both the reaction distance R (the distance the vehicle travels while the driver moves his or her foot to the brake pedal) and the braking distance B (the distance the vehicle travels once the brakes are applied). For a car traveling at a speed of v miles per hour, the reaction distance R, in feet, can be estimated by $R(v) = 2.2v$. Suppose that the braking distance B, in feet, for a car is given by $B(v) = 0.05v^2 + 0.4v - 15$.

(a) Find the stopping distance function

$$D(v) = R(v) + B(v)$$

(b) Find the stopping distance if the car is traveling at a speed of 60 mph.

(c) Interpret $D(60)$.

120. Some functions f have the property that

$$f(a + b) = f(a) + f(b)$$

for all real numbers a and b. Which of the following functions have this property?

(a) $h(x) = 2x$ (b) $g(x) = x^2$

(c) $F(x) = 5x - 2$ (d) $G(x) = \frac{1}{x}$

121. *Challenge Problem* Find the difference quotient of the function $f(x) = \sqrt[3]{x}$.

(*Hint:* Factor using $a^3 - b^3 = (a - b)(a^2 + ab + b^2)$ with $a = \sqrt[3]{x + h}$ and $b = \sqrt[3]{x}$.)

122. *Challenge Problem* If $f\left(\dfrac{x + 4}{5x - 4}\right) = 3x^2 - 2$, find $f(1)$.

123. *Challenge Problem* Find the domain of $f(x) = \sqrt{\dfrac{x^2 + 1}{7 - |3x - 1|}}$.

Explaining Concepts: Discussion and Writing

124. Are the functions $f(x) = x - 1$ and $g(x) = \dfrac{x^2 - 1}{x + 1}$ the same? Explain.

125. Find a function H that multiplies a number x by 3 and then subtracts the cube of x and divides the result by your age.

126. Investigate when, historically, the use of the function notation $y = f(x)$ first appeared.

Retain Your Knowledge

Problems 127–135 are based on material learned earlier in the course. The purpose of these problems is to keep the material fresh in your mind so that you are better prepared for the final exam.

127. List the intercepts and test for symmetry the graph of
$$(x + 12)^2 + y^2 = 16$$

128. Determine which of the given points are on the graph of the equation $y = 3x^2 - 8\sqrt{x}$.
Points: $(-1, -5)$, $(4, 32)$, $(9, 171)$

129. How many pounds of lean hamburger that is 7% fat must be mixed with 12 pounds of ground chuck that is 20% fat to have a hamburger mixture that is 15% fat?

130. Solve $x^3 - 9x = 2x^2 - 18$.

131. Given $a + bx = ac + d$, solve for a.

132. Rotational Inertia The rotational inertia of an object varies directly with the square of the perpendicular distance from the object to the axis of rotation. If the rotational inertia is $0.4 \text{ kg} \cdot \text{m}^2$ when the perpendicular distance is 0.6 m, what is the rotational inertia of the object if the perpendicular distance is 1.5 m?

133. Find the slope of a line perpendicular to the line
$$3x - 10y = 12$$

134. Simplify $\dfrac{(4x^2 - 7) \cdot 3 - (3x + 5) \cdot 8x}{(4x^2 - 7)^2}$.

135. Determine the degree of the polynomial
$$9x^2(3x - 5)(5x + 1)^4$$

'Are You Prepared?' Answers

1. $(-1, 3)$ **2.** 21.5   **3.** $\{x \mid x \neq -4\}$ **4.** $\{x \mid x < -1\}$ **5.** $\sqrt{5} + 2$ **6.** radicals

3.2 The Graph of a Function

PREPARING FOR THIS SECTION *Before getting started, review the following:*

- Graphs of Equations (Section 2.2, pp. 158–160)
- Intercepts (Section 2.2, pp. 160–161)

Now Work the 'Are You Prepared?' problems on page 223.

OBJECTIVES   **1** Identify the Graph of a Function (p. 220)
          2 Obtain Information from or about the Graph of a Function (p. 220)

In Section 2.2 we saw how a graph can more clearly demonstrate the relationship between two variables. Consider the average gasoline price data and corresponding graph provided again for convenience in Table 2 and Figure 17.

Table 2

Year	Price	Year	Price	Year	Price
1991	1.98	2000	2.11	2009	2.68
1992	1.90	2001	1.97	2010	3.12
1993	1.81	2002	1.83	2011	3.84
1994	1.78	2003	2.07	2012	3.87
1995	1.78	2004	2.40	2013	3.68
1996	1.87	2005	2.85	2014	3.48
1997	1.83	2006	3.13	2015	2.51
1998	1.55	2007	3.31	2016	2.19
1999	1.67	2008	3.70	2017	2.38

Source: U.S. Energy Information Administration

Figure 17 Average retail price of gasoline (2017 dollars)

We can see from the graph that the price of gasoline (adjusted for inflation) stayed roughly the same from 1993 to 1997 and was increasing from 2002 to 2008. The graph also shows that the lowest price occurred in 1998.

Look again at Figure 17. The graph shows that for each date on the horizontal axis, there is only one price on the vertical axis. The graph represents a function, although a rule for getting from date to price is not given.

When a function is defined by an equation in x and y, the **graph of the function** is the graph of the equation; that is, it is the set of points (x, y) in the xy-plane that satisfy the equation.

1 Identify the Graph of a Function

Not every collection of points in the xy-plane represents the graph of a function. Remember, for a function, each number x in the domain has exactly one image y in the range. This means that the graph of a function cannot contain two points with the same x-coordinate and different y-coordinates. Therefore, the graph of a function must satisfy the following *vertical-line test*.

In Words

If any vertical line intersects a graph at more than one point, the graph is not the graph of a function.

THEOREM   Vertical-Line Test

A set of points in the xy-plane is the graph of a function if and only if every vertical line intersects the graph in at most one point.

EXAMPLE 1   Identifying the Graph of a Function

Which of the graphs in Figure 18 are graphs of functions?

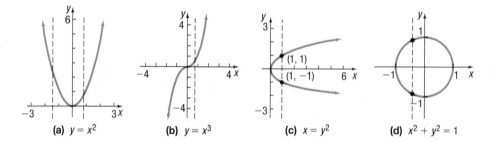

Figure 18   (a) $y = x^2$ (b) $y = x^3$ (c) $x = y^2$ (d) $x^2 + y^2 = 1$

Solution   The graphs in Figures 18(a) and 18(b) are graphs of functions, because every vertical line intersects each graph in at most one point. The graphs in Figures 18(c) and 18(d) are not graphs of functions, because there is a vertical line that intersects each graph in more than one point. Notice in Figure 18(c) that the input 1 corresponds to two outputs, -1 and 1. This is why the graph does not represent a function.

Now Work   PROBLEMS 15 AND 17

2 Obtain Information from or about the Graph of a Function

If (x, y) is a point on the graph of a function f, then y is the value of f at x; that is, $y = f(x)$. Also if $y = f(x)$, then (x, y) is a point on the graph of f. For example, if $(-2, 7)$ is on the graph of f, then $f(-2) = 7$, and if $f(5) = 8$, then the point $(5, 8)$ is on the graph of $y = f(x)$.

EXAMPLE 2   Obtaining Information from the Graph of a Function

Let f be the function whose graph is given in Figure 19. (The graph of f might represent the distance y that the bob of a pendulum is from its *at-rest* position at time x. Negative values of y mean that the pendulum is to the left of the at-rest position, and positive values of y mean that the pendulum is to the right of the at-rest position.)

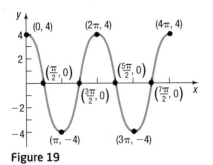

Figure 19

(a) What are $f(0), f\left(\dfrac{3\pi}{2}\right)$, and $f(3\pi)$?

(b) What is the domain of f?

(c) What is the range of f?

(d) List the intercepts. (Recall that these are the points, if any, where the graph crosses or touches the coordinate axes.)

(e) How many times does the line $y = 2$ intersect the graph?

(f) For what values of x does $f(x) = -4$?

(g) For what values of x is $f(x) > 0$?

Solution

(a) Since $(0, 4)$ is on the graph of f, the y-coordinate 4 is the value of f at the x-coordinate 0; that is, $f(0) = 4$. In a similar way, when $x = \dfrac{3\pi}{2}$, then $y = 0$, so $f\left(\dfrac{3\pi}{2}\right) = 0$. When $x = 3\pi$, then $y = -4$, so $f(3\pi) = -4$.

(b) To determine the domain of f, notice that the points on the graph of f have x-coordinates between 0 and 4π, inclusive; and for each number x between 0 and 4π, there is a point $(x, f(x))$ on the graph. The domain of f is $\{x \mid 0 \le x \le 4\pi\}$ or the interval $[0, 4\pi]$.

(c) The points on the graph all have y-coordinates between -4 and 4, inclusive; and for each such number y, there is at least one number x in the domain. The range of f is $\{y \mid -4 \le y \le 4\}$ or the interval $[-4, 4]$.

(d) The intercepts are the points

$$(0, 4), \left(\dfrac{\pi}{2}, 0\right), \left(\dfrac{3\pi}{2}, 0\right), \left(\dfrac{5\pi}{2}, 0\right), \quad \text{and} \quad \left(\dfrac{7\pi}{2}, 0\right)$$

(e) Draw the horizontal line $y = 2$ on the graph in Figure 19. Notice that the line intersects the graph four times.

(f) Since $(\pi, -4)$ and $(3\pi, -4)$ are the only points on the graph for which $y = f(x) = -4$, we have $f(x) = -4$ when $x = \pi$ and $x = 3\pi$.

(g) To determine where $f(x) > 0$, look at Figure 19 and determine the x-values from 0 to 4π for which the y-coordinate is positive. This occurs on $\left[0, \dfrac{\pi}{2}\right) \cup \left(\dfrac{3\pi}{2}, \dfrac{5\pi}{2}\right) \cup \left(\dfrac{7\pi}{2}, 4\pi\right]$. Using inequality notation, $f(x) > 0$ for $0 \le x < \dfrac{\pi}{2}$ or $\dfrac{3\pi}{2} < x < \dfrac{5\pi}{2}$ or $\dfrac{7\pi}{2} < x \le 4\pi$.

RECALL

The symbol $\cup$ stands for union. It means the set of elements that are in either of two sets.

When the graph of a function is given, its domain may be viewed as the shadow created by the graph on the x-axis by vertical beams of light. Its range can be viewed as the shadow created by the graph on the y-axis by horizontal beams of light. Try this technique with the graph given in Figure 19.

━━━ **Now Work** PROBLEM 11

EXAMPLE 3

Obtaining Information about the Graph of a Function

Consider the function: $f(x) = \dfrac{x + 1}{x + 2}$

(a) Find the domain of f.

(b) Is the point $\left(1, \dfrac{1}{2}\right)$ on the graph of f?

(c) If $x = 2$, what is $f(x)$? What point is on the graph of f?

(d) If $f(x) = 2$, what is x? What point is on the graph of f?

(e) What are the x-intercepts of the graph of f (if any)? What corresponding point(s) are on the graph of f?

Solution

(a) The domain of f is $\{x \mid x \neq -2\}$.

(b) When $x = 1$, then

$$f(1) = \frac{1+1}{1+2} = \frac{2}{3} \quad f(x) = \frac{x+1}{x+2}$$

The point $\left(1, \dfrac{2}{3}\right)$ is on the graph of f; the point $\left(1, \dfrac{1}{2}\right)$ is not.

(c) If $x = 2$, then

$$f(2) = \frac{2+1}{2+2} = \frac{3}{4}$$

The point $\left(2, \dfrac{3}{4}\right)$ is on the graph of f.

(d) If $f(x) = 2$, then

$$\frac{x+1}{x+2} = 2 \qquad\qquad f(x) = 2$$

$$x + 1 = 2(x+2) \quad \text{Multiply both sides by } x+2.$$

$$x + 1 = 2x + 4 \qquad \text{Distribute.}$$

$$x = -3 \qquad\qquad \text{Solve for } x.$$

NOTE In Example 3, -2 is not in the domain of f, so $x + 2$ is not zero and we can multiply both sides of the equation by $x + 2$. ∎

If $f(x) = 2$, then $x = -3$. The point $(-3, 2)$ is on the graph of f.

(e) The x-intercepts of the graph of f are the real solutions of the equation $f(x) = 0$ that are in the domain of f.

$$\frac{x+1}{x+2} = 0$$

$$x + 1 = 0 \quad \text{Multiply both sides by } x+2.$$

$$x = -1 \quad \text{Subtract 1 from both sides.}$$

The only real solution of the equation $f(x) = \dfrac{x+1}{x+2} = 0$ is $x = -1$, so -1 is the only x-intercept. Since $f(-1) = 0$, the point $(-1, 0)$ is on the graph of f. ⌝

Now Work PROBLEM 27

EXAMPLE 4 — Energy Expended

For an individual walking, the energy expended E in terms of speed v can be approximated by

$$E(v) = \frac{29}{v} + 0.0053v$$

where E has units of cal/min/kg and v has units of m/min.

(a) Find the energy expended for a speed of $v = 40$ m/min.
(b) Find the energy expended for a speed of $v = 70$ m/min.
(c) Find the energy expended for a speed of $v = 100$ m/min.
(d) Graph the function $E = E(v), 0 < v \leq 200$
(e) Create a TABLE with TblStart $= 1$ and ΔTbl $= 1$. Which value of v minimizes the energy expended?

Source: Ralston, H.J. Int. Z. Angew. Physiol. Einschl. Arbeitsphysiol. (1958) 17: 277.

Solution

(a) The energy expended for a walking speed of $v = 40$ meters per minute is

$$E(40) = \frac{29}{40} + 0.0053 \cdot 40 \approx 0.94 \, \text{cal/min/kg}$$

(b) The energy expended for a walking speed of $v = 70$ meters per minute is

$$E(70) = \frac{29}{70} + 0.0053 \cdot 70 \approx 0.79 \, \text{cal/min/kg}$$

(c) The energy expended for a walking speed of $v = 100$ meters per minute is

$$E(100) = \frac{29}{100} + 0.0053 \cdot 100 = 0.82 \, \text{cal/min/kg}$$

(d) See Figure 20 for the graph of $E = E(v)$ on a TI-84 Plus C.

(e) With the function $E = E(v)$ in Y_1, we create Table 3. Scroll down to find a value of v for which Y_1 is smallest. Table 4 shows that a walking speed of $v = 74$ meters per minute minimizes the expended energy at about 0.784 cal/min/kg.

Table 3 **Table 4**

Figure 20 $E(v) = \dfrac{29}{v} + 0.0053v$

⎯⎯⎯ **Now Work** PROBLEM 35

SUMMARY

- **Graph of a Function** The set of points (x, y) in the xy-plane that satisfy the equation $y = f(x)$.
- **Vertical-Line Test** A set of points in the xy-plane is the graph of a function if and only if every vertical line intersects the graph in at most one point.

3.2 Assess Your Understanding

'Are You Prepared?' *Answers are given at the end of these exercises. If you get a wrong answer, read the pages listed in* red.

1. The intercepts of the equation $x^2 + 4y^2 = 16$ are _____.
   (pp. 160–161)

2. **True or False** The point $(-2, -6)$ is on the graph of the equation $x = 2y - 2$. (pp. 158–160)

Concepts and Vocabulary

3. A set of points in the xy-plane is the graph of a function if and only if every _____ line intersects the graph in at most one point.

4. If the point $(5, -3)$ is a point on the graph of f, then $f(\_\_) = \_\_$.

5. Find a so that the point $(-1, 2)$ is on the graph of $f(x) = ax^2 + 4$.

6. **True or False** Every graph represents a function.

7. **True or False** The graph of a function $y = f(x)$ always crosses the y-axis.

8. **True or False** The y-intercept of the graph of the function $y = f(x)$, whose domain is all real numbers, is $f(0)$.

9. **Multiple Choice** If a function is defined by an equation in x and y, then the set of points (x, y) in the xy-plane that satisfy the equation is called the _____.
   (a) domain of the function (b) range of the function
   (c) graph of the function   (d) relation of the function

10. **Multiple Choice** The graph of a function $y = f(x)$ can have more than one of which type of intercept?
    (a) x-intercept (b) y-intercept (c) both (d) neither

Skill Building

11. Use the given graph of the function f to answer parts (a)–(n).

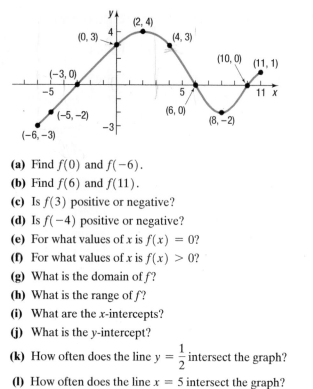

(a) Find $f(0)$ and $f(-6)$.

(b) Find $f(6)$ and $f(11)$.

(c) Is $f(3)$ positive or negative?

(d) Is $f(-4)$ positive or negative?

(e) For what values of x is $f(x) = 0$?

(f) For what values of x is $f(x) > 0$?

(g) What is the domain of f?

(h) What is the range of f?

(i) What are the x-intercepts?

(j) What is the y-intercept?

(k) How often does the line $y = \dfrac{1}{2}$ intersect the graph?

(l) How often does the line $x = 5$ intersect the graph?

(m) For what values of x does $f(x) = 3$?

(n) For what values of x does $f(x) = -2$?

12. Use the given graph of the function f to answer parts (a)–(n).

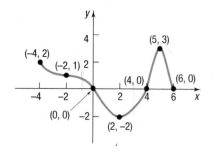

(a) Find $f(0)$ and $f(6)$.

(b) Find $f(2)$ and $f(-2)$.

(c) Is $f(3)$ positive or negative?

(d) Is $f(-1)$ positive or negative?

(e) For what values of x is $f(x) = 0$?

(f) For what values of x is $f(x) < 0$?

(g) What is the domain of f?

(h) What is the range of f?

(i) What are the x-intercepts?

(j) What is the y-intercept?

(k) How often does the line $y = -1$ intersect the graph?

(l) How often does the line $x = 1$ intersect the graph?

(m) For what value of x does $f(x) = 3$?

(n) For what value of x does $f(x) = -2$?

In Problems 13–24, determine whether or not the graph is that of a function by using the vertical-line test. In either case, use the graph to find:
(a) The domain and range　　*(b) The intercepts, if any*　　*(c) Any symmetry with respect to the x-axis, the y-axis, or the origin*

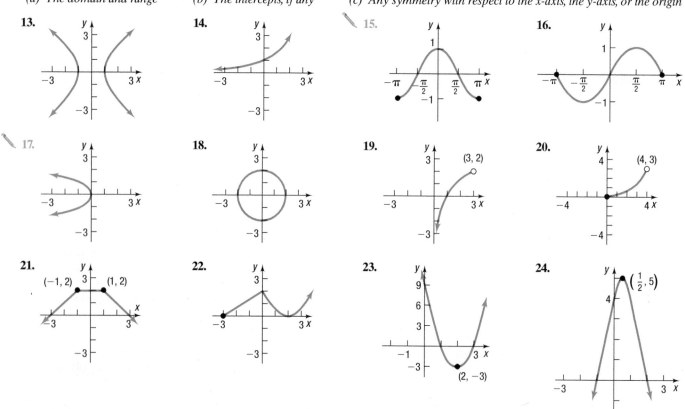

In Problems 25–30, answer the questions about each function.

25. $f(x) = 3x^2 + x - 2$

    (a) Is the point $(1, 2)$ on the graph of f?

    (b) If $x = -2$, what is $f(x)$? What point is on the graph of f?

    (c) If $f(x) = -2$, what is x? What point(s) are on the graph of f?

    (d) What is the domain of f?

    (e) List the x-intercepts, if any, of the graph of f.

    (f) List the y-intercept, if there is one, of the graph of f.

26. $f(x) = -3x^2 + 5x$

    (a) Is the point $(-1, 2)$ on the graph of f?

    (b) If $x = -2$, what is $f(x)$? What point is on the graph of f?

    (c) If $f(x) = -2$, what is x? What point(s) are on the graph of f?

    (d) What is the domain of f?

    (e) List the x-intercepts, if any, of the graph of f.

    (f) List the y-intercept, if there is one, of the graph of f.

27. $f(x) = \dfrac{x + 2}{x - 6}$

    (a) Is the point $(3, 14)$ on the graph of f?

    (b) If $x = 4$, what is $f(x)$? What point is on the graph of f?

    (c) If $f(x) = 2$, what is x? What point(s) are on the graph of f?

    (d) What is the domain of f?

    (e) List the x-intercepts, if any, of the graph of f.

    (f) List the y-intercept, if there is one, of the graph of f.

28. $f(x) = \dfrac{x^2 + 2}{x + 4}$

    (a) Is the point $\left(1, \dfrac{3}{5}\right)$ on the graph of f?

    (b) If $x = 0$, what is $f(x)$? What point is on the graph of f?

    (c) If $f(x) = \dfrac{1}{2}$, what is x? What point(s) are on the graph of f?

    (d) What is the domain of f?

    (e) List the x-intercepts, if any, of the graph of f.

    (f) List the y-intercept, if there is one, of the graph of f.

29. $f(x) = \dfrac{12x^4}{x^2 + 1}$

    (a) Is the point $(-1, 6)$ on the graph of f?

    (b) If $x = 3$, what is $f(x)$? What point is on the graph of f?

    (c) If $f(x) = 1$, what is x? What point(s) are on the graph of f?

    (d) What is the domain of f?

    (e) List the x-intercepts, if any, of the graph of f.

    (f) List the y-intercept, if there is one, of the graph of f.

30. $f(x) = \dfrac{2x}{x - 2}$

    (a) Is the point $\left(\dfrac{1}{2}, -\dfrac{2}{3}\right)$ on the graph of f?

    (b) If $x = 4$, what is $f(x)$? What point is on the graph of f?

    (c) If $f(x) = 1$, what is x? What point(s) are on the graph of f?

    (d) What is the domain of f?

    (e) List the x-intercepts, if any, of the graph of f.

    (f) List the y-intercept, if there is one, of the graph of f.

Applications and Extensions

31. The graphs of two functions, f and g, are illustrated. Use the graphs to answer parts (a)–(f).

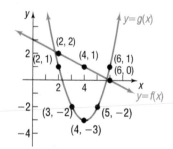

    (a) $(f + g)(2)$ **(b)** $(f + g)(4)$

    (c) $(f - g)(6)$ **(d)** $(g - f)(6)$

    (e) $(f \cdot g)(2)$ **(f)** $\left(\dfrac{f}{g}\right)(4)$

32. Granny Shots The last player in the NBA to use an underhand foul shot (a "granny" shot) was Hall of Fame forward Rick Barry, who retired in 1980. Barry believes that current NBA players could increase their free-throw percentage if they were to use an underhand shot. Since underhand shots are released from a lower position, the angle of the shot must be increased. If a player shoots an underhand foul shot, releasing the ball at a 70-degree angle from a position 3.5 feet above the floor, then the path of the ball can be modeled by the function $h(x) = -\dfrac{136x^2}{v^2} + 2.7x + 3.5$, where h is the height of the ball above the floor, x is the forward distance

of the ball in front of the foul line, and v is the initial velocity with which the ball is shot in feet per second.

    (a) The center of the hoop is 10 feet above the floor and 15 feet in front of the foul line. Determine the initial velocity with which the ball must be shot for the ball to go through the hoop.

    (b) Write the function for the path of the ball using the velocity found in part (a).

    (c) Determine the height of the ball after it has traveled 9 feet in front of the foul line.

    (d) Find additional points and graph the path of the basketball.

Source: The Physics of Foul Shots, Discover, Vol. 21, No. 10, October 2000

33. Free-throw Shots According to physicist Peter Brancazio, the key to a successful foul shot in basketball lies in the arc of the shot. Brancazio determined the optimal angle of the arc from the free-throw line to be 45 degrees. The arc also depends on the velocity with which the ball is shot. If a player shoots a foul shot, releasing the ball at a 45-degree angle from a position 6 feet above the floor, then the path of the ball can be modeled by the function

$$h(x) = -\dfrac{44x^2}{v^2} + x + 6$$

where h is the height of the ball above the floor, x is the forward distance of the ball in front of the foul line, and v is the initial velocity with which the ball is shot in feet per second. Suppose a player shoots a ball with an initial velocity of 28 feet per second.

(continued)

(a) Find the height of the ball after it has traveled 8 feet in front of the foul line.

(b) Find the height of the ball after it has traveled 12 feet in front of the foul line.

(c) Find additional points and graph the path of the basketball.

(d) The center of the hoop is located 10 feet above the floor and 15 feet in front of the foul line. Will the ball go through the hoop? Why or why not? If not, with what initial velocity must the ball be shot in order for the ball to go through the hoop?

Source: The Physics of Foul Shots, Discover, *Vol. 21, No. 10, October 2000*

34. Cross-sectional Area The cross-sectional area of a beam cut from a log with radius 1 foot is given by the function $A(x) = 4x\sqrt{1 - x^2}$, where x represents the length, in feet, of half the base of the beam. See the figure.

(a) Find the domain of A.

(b) Use a graphing utility to graph the function $A = A(x)$.

(c) Create a TABLE with TblStart $= 0$ and ΔTbl $= 0.1$ for $0 \le x \le 1$. Which value of x maximizes the cross-sectional area? What should be the length of the base of the beam to maximize the cross-sectional area?

$A(x) = 4x\sqrt{1 - x^2}$

35. Motion of a Golf Ball A golf ball is hit with an initial velocity of 130 feet per second at an inclination of 45° to the horizontal. In physics, it is established that the height h of the golf ball is given by the function

$$h(x) = \frac{-32x^2}{130^2} + x$$

where x is the horizontal distance that the golf ball has traveled.

(a) Determine the height of the golf ball after it has traveled 100 feet.

(b) What is the height after it has traveled 300 feet?

(c) What is $h(500)$? Interpret this value.

(d) How far was the golf ball hit?

(e) Use a graphing utility to graph the function $h = h(x)$.

(f) Use a graphing utility to determine the distance that the ball has traveled when the height of the ball is 90 feet.

(g) Create a TABLE with TblStart $= 0$ and ΔTbl $= 25$. To the nearest 25 feet, how far does the ball travel before it reaches a maximum height? What is the maximum height?

(h) Adjust the value of ΔTbl until you determine the distance, to within 1 foot, that the ball travels before it reaches its maximum height.

36. Effect of Elevation on Weight If an object weighs m pounds at sea level, then its weight W (in pounds) at a height of h miles above sea level is given approximately by

$$W(h) = m\left(\frac{4000}{4000 + h}\right)^2$$

(a) If Amy weighs 120 pounds at sea level, how much will she weigh on Pikes Peak, which is 14,110 feet above sea level?

(b) Use a graphing utility to graph the function $W = W(h)$. Use $m = 120$ pounds.

(c) Create a TABLE with TblStart $= 0$ and ΔTbl $= 0.5$ to see how the weight W varies as h changes from 0 to 5 miles.

(d) At what height will Amy weigh 119.95 pounds?

(e) Does your answer to part (d) seem reasonable? Explain.

37. Cost of Transatlantic Travel A Boeing 747 crosses the Atlantic Ocean (3000 miles) with an airspeed of 500 miles per hour. The cost C (in dollars) per passenger is given by

$$C(x) = 100 + \frac{x}{10} + \frac{36,000}{x}$$

where x is the groundspeed (airspeed $\pm$ wind).

(a) What is the cost when the groundspeed is 480 miles per hour? 600 miles per hour?

(b) Find the domain of C.

(c) Use a graphing utility to graph the function $C = C(x)$.

(d) Create a TABLE with TblStart $= 0$ and ΔTbl $= 50$.

(e) To the nearest 50 miles per hour, what groundspeed minimizes the cost per passenger?

38. Reading and Interpreting Graphs Let C be the function whose graph is given below. This graph represents the cost C of manufacturing q computers in a day.

(a) Find $C(0)$. Interpret this value.

(b) Find $C(10)$. Interpret this value.

(c) Find $C(50)$. Interpret this value.

(d) What is the domain of C? What does this domain imply in terms of daily production?

(e) Describe the shape of the graph.

(f) The point (30, 32 000) is called an *inflection point*. Describe the behavior of the graph around the inflection point.

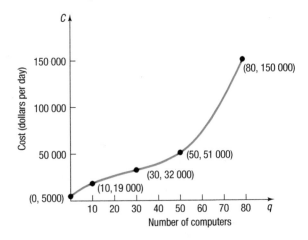

39. Reading and Interpreting Graphs Let C be the function whose graph is given below. This graph represents the cost C of using g gigabytes of data in a month for a data-only plan.

 (a) Find $C(0)$. Interpret this value.

 (b) Find $C(5)$. Interpret this value.

 (c) Find $C(15)$. Interpret this value.

 (d) What is the domain of C? What does this domain imply in terms of the number of gigabytes?

 (e) Describe the shape of the graph.

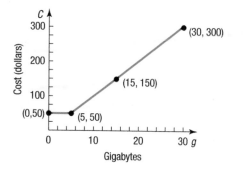

40. Challenge Problem Suppose $f(x) = x^2 - 4x + c$ and $g(x) = \dfrac{f(x)}{3} - 4$. Find $f(3)$ if $g(-2) = 5$.

41. Challenge Problem Suppose $f(x) = \sqrt{x} + 2$ and $g(x) = x^2 + n$. If $f(g(5)) = 4$, what is the value of $g(n)$?

Explaining Concepts: Discussion and Writing

42. Describe how you would find the domain and range of a function if you were given its graph. How would your strategy change if you were given the equation defining the function instead of its graph?

43. How many x-intercepts can the graph of a function have? How many y-intercepts can the graph of a function have? Explain why.

44. Is a graph that consists of a single point the graph of a function? Can you write the equation of such a function?

45. Match each of the following functions with the graph that best describes the situation.
 (a) The cost of building a house as a function of its square footage
 (b) The height of an egg dropped from a 300-foot building as a function of time
 (c) The height of a human as a function of time
 (d) The demand for Big Macs as a function of price
 (e) The height of a child on a swing as a function of time

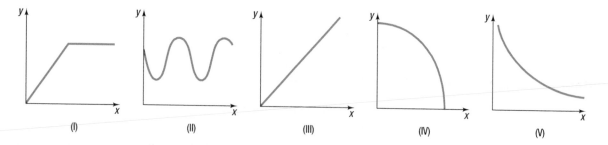

46. Match each of the following functions with the graph that best describes the situation.
 (a) The temperature of a bowl of soup as a function of time
 (b) The number of hours of daylight per day over a 2-year period
 (c) The population of Florida as a function of time
 (d) The distance traveled by a car going at a constant velocity as a function of time
 (e) The height of a golf ball hit with a 7-iron as a function of time

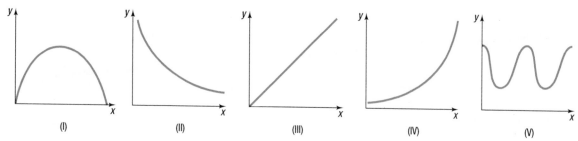

47. Consider the following scenario: Barbara decides to take a walk. She leaves home, walks 2 blocks in 5 minutes at a constant speed, and realizes that she forgot to lock the door. So Barbara runs home in 1 minute. While at her doorstep, it takes her 1 minute to find her keys and lock the door. Barbara walks 5 blocks in 15 minutes and then decides to jog home. It takes her 7 minutes to get home. Draw a graph of Barbara's distance from home (in blocks) as a function of time.

48. Consider the following scenario: Jayne enjoys riding her bicycle through the woods. At the forest preserve, she gets on her bicycle and rides up a 2000-foot incline in 10 minutes. She then travels down the incline in 3 minutes. The next 5000 feet is level terrain, and she covers the distance in 20 minutes. She rests for 15 minutes. Jayne then travels 10,000 feet in 30 minutes. Draw a graph of Jayne's distance traveled (in feet) as a function of time.

49. The graph below represents the distance d (in miles) that Kevin was from home as a function of time t (in hours). Answer the questions by referring to the graph. In parts (a)–(g), how many hours elapsed and how far was Kevin from home during the times listed?

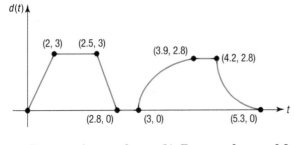

(a) From $t = 0$ to $t = 2$ (b) From $t = 2$ to $t = 2.5$
(c) From $t = 2.5$ to $t = 2.8$ (d) From $t = 2.8$ to $t = 3$
(e) From $t = 3$ to $t = 3.9$ (f) From $t = 3.9$ to $t = 4.2$
(g) From $t = 4.2$ to $t = 5.3$
(h) What is the farthest distance that Kevin was from home?
(i) How many times did Kevin return home?

50. The graph below represents the speed v (in miles per hour) of Michael's car as a function of time t (in minutes).

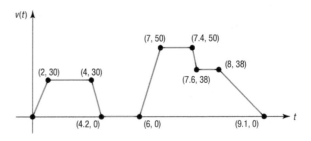

(a) Over what interval of time was Michael traveling fastest?
(b) Over what interval(s) of time was Michael's speed zero?
(c) What was Michael's speed between 0 and 2 minutes?
(d) What was Michael's speed between 4.2 and 6 minutes?
(e) What was Michael's speed between 7 and 7.4 minutes?
(f) When was Michael's speed constant?

51. Graph a function whose domain is

$$\{x \mid -3 \le x \le 8, \quad x \ne 5\}$$

and whose range is

$$\{y \mid -1 \le y \le 2, \quad y \ne 0\}$$

What point(s) in the rectangle $-3 \le x \le 8, -1 \le y \le 2$ cannot be on the graph? Compare your graph with those of other students. What differences do you see?

52. Is there a function whose graph is symmetric with respect to the x-axis? Explain.

53. Explain why the vertical-line test works.

Retain Your Knowledge

Problems 54–63 are based on material learned earlier in the course. The purpose of these problems is to keep the material fresh in your mind so that you are better prepared for the final exam.

54. If $f(x) = -x^2 + x - 3$, find $f(x - 2)$.

55. Find the distance between the points $(3, -6)$ and $(1, 0)$.

56. Write an equation of the line with slope $\frac{2}{3}$ that contains the point $(-6, 4)$.

57. Find the domain of $g(x) = \sqrt[3]{x + 4} - 5$.

58. Find the number that must be added to $m^2 - 12m$ to complete the square.

59. Rationalize the numerator: $\dfrac{\sqrt{x} - \sqrt{6}}{x - 6}$.

60. Two cars leave an intersection at the same time, one traveling north at 25 mph and the other traveling west at 35 mph. How long will it take for the cars to be 40 miles apart?

61. Find the numbers x that satisfy both of the inequalities

$$3x + 4 \le 7 \text{ and } 5 - 2x < 13$$

62. Simplify $(5x^2 - 7x + 2) - (8x - 10)$.

63. Write the inequality $-3 \le x \le 10$ in interval notation.

'Are You Prepared?' Answers

1. $(-4, 0), (4, 0), (0, -2), (0, 2)$ 2. False

3.3 Properties of Functions

PREPARING FOR THIS SECTION *Before getting started, review the following:*

- Intervals (Section 1.5, pp. 120–121)
- Intercepts (Section 2.2, pp. 160–161)
- Slope of a Line (Section 2.3, pp. 169–171)

- Point-Slope Form of a Line (Section 2.3, p. 173)
- Symmetry (Section 2.2, pp. 161–164)

Now Work the 'Are You Prepared?' problems on page 237.

OBJECTIVES
1. Identify Even and Odd Functions from a Graph (p. 229)
2. Identify Even and Odd Functions from an Equation (p. 230)
3. Use a Graph to Determine Where a Function Is Increasing, Decreasing, or Constant (p. 231)
4. Use a Graph to Locate Local Maxima and Local Minima (p. 232)
5. Use a Graph to Locate the Absolute Maximum and the Absolute Minimum (p. 233)
6. Use a Graphing Utility to Approximate Local Maxima and Local Minima and to Determine Where a Function Is Increasing or Decreasing (p. 235)
7. Find the Average Rate of Change of a Function (p. 235)

To obtain the graph of a function $y = f(x)$, it is often helpful to know properties of the function and the impact of these properties on the graph of the function.

1 Identify Even and Odd Functions from a Graph

The words *even* and *odd*, when discussing a function f, describe the symmetry of the graph of the function.

A function f is even if and only if, whenever the point (x, y) is on the graph of f, the point $(-x, y)$ is also on the graph. Using function notation, we define an even function as follows:

DEFINITION Even Function

A function f is **even** if, for every number x in its domain, the number $-x$ is also in the domain and

$$f(-x) = f(x)$$

A function f is odd if and only if, whenever the point (x, y) is on the graph of f, the point $(-x, -y)$ is also on the graph. Using function notation, we define an odd function as follows:

DEFINITION Odd Function

A function f is **odd** if, for every number x in its domain, the number $-x$ is also in the domain and

$$f(-x) = -f(x)$$

Refer to page 163, where the tests for symmetry are listed. The results below follow.

> **THEOREM** Graphs of Even and Odd Functions
>
> - A function is even if and only if its graph is symmetric with respect to the y-axis.
> - A function is odd if and only if its graph is symmetric with respect to the origin.

EXAMPLE 1 Identifying Even and Odd Functions from a Graph

Determine whether each graph given in Figure 21 is the graph of an even function, an odd function, or a function that is neither even nor odd.

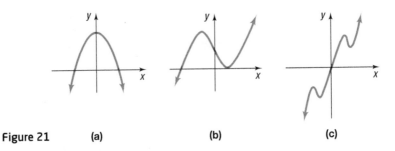

Figure 21   (a)   (b)   (c)

Solution (a) The graph in Figure 21(a) is that of an even function, because the graph is symmetric with respect to the y-axis.

(b) The function whose graph is given in Figure 21(b) is neither even nor odd, because the graph is neither symmetric with respect to the y-axis nor symmetric with respect to the origin.

(c) The function whose graph is given in Figure 21(c) is odd, because its graph is symmetric with respect to the origin.

──**Now Work** PROBLEMS 25(a), (b), AND (d)

2 Identify Even and Odd Functions from an Equation

EXAMPLE 2 Identifying Even and Odd Functions Algebraically

Determine whether each of the following functions is even, odd, or neither. Then determine whether the graph is symmetric with respect to the y-axis, with respect to the origin, or neither.

(a) $f(x) = x^2 - 5$ (b) $g(x) = x^3 - 1$

(c) $h(x) = 5x^3 - x$ (d) $F(x) = |x|$

Solution (a) To determine whether f is even, odd, or neither, replace x by $-x$ in $f(x) = x^2 - 5$.

$$f(-x) = (-x)^2 - 5 = x^2 - 5 = f(x)$$

Since $f(-x) = f(x)$, the function is even, and the graph of f is symmetric with respect to the y-axis.

(b) Replace x by $-x$ in $g(x) = x^3 - 1$.

$$g(-x) = (-x)^3 - 1 = -x^3 - 1$$

Since $g(-x) \neq g(x)$ and $g(-x) \neq -g(x) = -(x^3 - 1) = -x^3 + 1$, the function is neither even nor odd. The graph of g is not symmetric with respect to the y-axis, nor is it symmetric with respect to the origin.

(c) Replace x by $-x$ in $h(x) = 5x^3 - x$.

$$h(-x) = 5(-x)^3 - (-x) = -5x^3 + x = -(5x^3 - x) = -h(x)$$

Since $h(-x) = -h(x)$, h is an odd function, and the graph of h is symmetric with respect to the origin.

(d) Replace x by $-x$ in $F(x) = |x|$.

$$F(-x) = |-x| = |-1| \cdot |x| = |x| = F(x)$$

Since $F(-x) = F(x)$, F is an even function, and the graph of F is symmetric with respect to the y-axis.

Now Work PROBLEM 37

3 Use a Graph to Determine Where a Function Is Increasing, Decreasing, or Constant

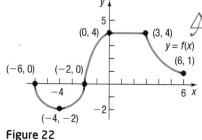

Figure 22

Consider the graph given in Figure 22. If you look from left to right along the graph of the function, you will notice that parts of the graph are going up, parts are going down, and parts are horizontal. In such cases, the function is described as *increasing*, *decreasing*, or *constant*, respectively.

> **DEFINITIONS** Increasing Function, Decreasing Function, Constant Function
>
> - A function f is **increasing** on an interval I if, for any choice of x_1 and x_2 in I, with $x_1 < x_2$, then $f(x_1) < f(x_2)$.
> - A function f is **decreasing** on an interval I if, for any choice of x_1 and x_2 in I, with $x_1 < x_2$, then $f(x_1) > f(x_2)$.
> - A function f is **constant** on an interval I if, for all choices of x in I, the values $f(x)$ are equal.

In Words

- If a function is increasing, then as the values of x get bigger, the values of the function also get bigger.
- If a function is decreasing, then as the values of x get bigger, the values of the function get smaller.
- If a function is constant, then as the values of x get bigger, the values of the function remain unchanged.

Need to Review?

Interval Notation is discussed in Section 1.5, pp. 120–121.

Figure 23 illustrates the definitions. The graph of an increasing function goes up from left to right, the graph of a decreasing function goes down from left to right, and the graph of a constant function remains at a fixed height. The interval I on which a function is increasing, decreasing, or constant may be open, closed, or half-open/half-closed depending on whether the endpoints of the interval satisfy the required inequality or not.

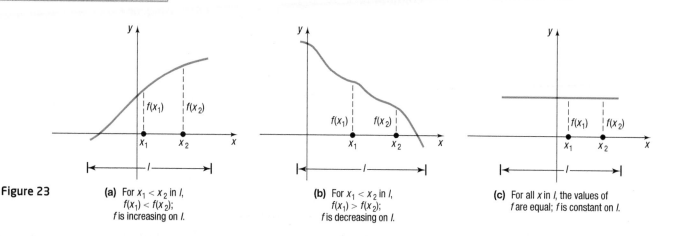

Figure 23

(a) For $x_1 < x_2$ in I, $f(x_1) < f(x_2)$; f is increasing on I.

(b) For $x_1 < x_2$ in I, $f(x_1) > f(x_2)$; f is decreasing on I.

(c) For all x in I, the values of f are equal; f is constant on I.

| EXAMPLE 3 | Determining Where a Function Is Increasing, Decreasing, or Constant from Its Graph |

Determine the values of x for which the function in Figure 24 is increasing. Where is it decreasing? Where is it constant?

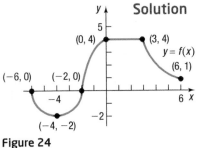

Figure 24

Solution   When determining where a function is increasing, where it is decreasing, and where it is constant, we use intervals involving the independent variable x.

The function whose graph is given in Figure 24 is decreasing on the interval $[-6, -4]$ because for any choice of x_1 and x_2 in the interval for which $x_1 < x_2$, we have $f(x_1) > f(x_2)$. For example, -6 is included in the interval where f is decreasing because if x is any number for which $-6 < x \leq -4$, then $f(-6) > f(x)$.

Similarly, f is increasing on the interval $[-4, 0]$ because for any choice of x_1 and x_2 in the interval for which $x_1 < x_2$, we have $f(x_1) < f(x_2)$.

Finally, the function f is constant on the interval $[0, 3]$ and is decreasing on the interval $[3, 6]$.

───── **Now Work** PROBLEMS 13, 15, 17, AND 25(C)

4 Use a Graph to Locate Local Maxima and Local Minima

Suppose f is a function defined on an open interval I containing c. If the value of f at c is greater than or equal to the other values of f on I, then f has a *local maximum* at c.[†] See Figure 25(a).

If the value of f at c is less than or equal to the other values of f on I, then f has a *local minimum* at c. See Figure 25(b).

NOTE "Maxima" is the plural of "maximum"; "minima" is the plural of "minimum." ■

f has a local maximum at c.

$f(c)$ — — — —

$(c, f(c))$

c

x

$\overleftarrow{\ \ I\ \ }\overrightarrow{}$

(a)

f has a local minimum at c.

$f(c)$ — — — —

$(c, f(c))$

c

x

$\overleftarrow{\ \ I\ \ }\overrightarrow{}$

(b)

Figure 25 Local maximum and local minimum

> **DEFINITIONS** Local Maximum/Minimum
>
> Let f be a function defined on some interval I and let c be a number in I.
> - A function f has a **local maximum** at c if there is an open interval in I containing c so that $f(c) \geq f(x)$ for all x in this open interval. The number $f(c)$ is called a **local maximum value of** f.
> - A function f has a **local minimum** at c if there is an open interval in I containing c so that $f(c) \leq f(x)$ for all x in this open interval. The number $f(c)$ is called a **local minimum value of** f.

If f has a local maximum at c, then the value of f at c is greater than or equal to the values of f near c. If f has a local minimum at c, then the value of f at c is less than or equal to the values of f near c. The word *local* is used to suggest that it is only near c, not necessarily over the entire domain, that the value $f(c)$ has these properties.

[†]Some texts use the term *relative* instead of *local*.

EXAMPLE 4

Finding Local Maxima and Local Minima from the Graph of a Function and Determining Where the Function Is Increasing, Decreasing, or Constant

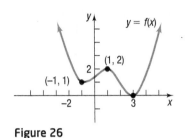

Figure 26

Figure 26 shows the graph of a function f.

(a) At what numbers x, if any, does f have a local maximum? List the local maximum value(s).

(b) At what numbers x, if any, does f have a local minimum? List the local minimum value(s).

(c) Find the intervals on which f is increasing. Find the intervals on which f is decreasing.

Solution

The domain of f is the set of real numbers.

(a) f has a local maximum at 1, since for all x close to 1, we have $f(x) \le f(1)$. The local maximum value is $f(1) = 2$.

(b) f has local minima at -1 and at 3. The local minimum values are $f(-1) = 1$ and $f(3) = 0$.

(c) The function whose graph is given in Figure 26 is increasing on the intervals $[-1, 1]$ and $[3, \infty)$, or for $-1 \le x \le 1$ and $x \ge 3$. The function is decreasing on the intervals $(-\infty, -1]$ and $[1, 3]$, or for $x \le -1$ and $1 \le x \le 3$. ⌐

WARNING The y-value is the local maximum value or local minimum value, and it occurs at some number x. For example, in Figure 26, we say f has a local maximum at 1 and the local maximum value is 2. ∎

━━ **Now Work** PROBLEMS 19 AND 21

5 Use a Graph to Locate the Absolute Maximum and the Absolute Minimum

Look at the graph of the function f given in Figure 27. The domain of f is the closed interval $[a, b]$. Also, the largest value of f is $f(u)$ and the smallest value of f is $f(v)$. These are called, respectively, the *absolute maximum* and the *absolute minimum* of f on $[a, b]$.

domain: $[a, b]$
for all x in $[a, b]$, $f(x) \le f(u)$
for all x in $[a, b]$, $f(x) \ge f(v)$
absolute maximum: $f(u)$
absolute minimum: $f(v)$

Figure 27

DEFINITIONS Absolute Maximum and Absolute Minimum

Let f be a function defined on some interval I.

- If there is a number u in I for which $f(u) \ge f(x)$ for all x in I, then f has an **absolute maximum at u**, and the number $f(u)$ is the **absolute maximum of f on I**.

- If there is a number v in I for which $f(v) \le f(x)$ for all x in I, then f has an **absolute minimum at v**, and the number $f(v)$ is the **absolute minimum of f on I**. ⌐

The absolute maximum and absolute minimum of a function f are sometimes called the **absolute extrema** or the **extreme values** of f on I.

The absolute maximum or absolute minimum of a function f may not exist. Let's look at some examples.

> **EXAMPLE 5**   **Finding the Absolute Maximum and the Absolute Minimum from the Graph of a Function**

For the graph of each function $y = f(x)$ in Figure 28, find the absolute maximum and the absolute minimum, if they exist. Also, find any local maxima or local minima.

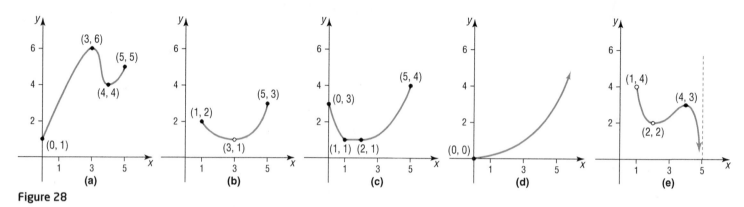

Figure 28

Solution   (a) The function f whose graph is given in Figure 28(a) has the closed interval $[0, 5]$ as its domain. The largest value of f is $f(3) = 6$, the absolute maximum. The smallest value of f is $f(0) = 1$, the absolute minimum. The function has a local maximum value of 6 at $x = 3$ and a local minimum value of 4 at $x = 4$.

(b) The function f whose graph is given in Figure 28(b) has domain $\{x | 1 \le x \le 5, x \ne 3\}$. Note that we exclude 3 from the domain because of the "hole" at $(3, 1)$. The largest value of f on its domain is $f(5) = 3$, the absolute maximum. There is no absolute minimum. Do you see why? As you trace the graph, getting closer to the point $(3, 1)$, there is no single smallest value. [As soon as you claim a smallest value, we can trace closer to $(3, 1)$ and get a smaller value!] The function has no local maxima or minima.

(c) The function f whose graph is given in Figure 28(c) has the interval $[0, 5]$ as its domain. The absolute maximum of f is $f(5) = 4$. The absolute minimum is 1. Notice that the absolute minimum occurs at any number in the interval $[1, 2]$. The function has a local minimum value of 1 at every x in the interval $[1, 2]$, but it has no local maximum.

(d) The function f given in Figure 28(d) has the interval $[0, \infty)$ as its domain. The function has no absolute maximum; the absolute minimum is $f(0) = 0$. The function has no local maxima or local minima.

(e) The function f in Figure 28(e) has domain $\{x | 1 < x < 5, x \ne 2\}$. The function has no absolute maximum and no absolute minimum. Do you see why? The function has a local maximum value of 3 at $x = 4$, but no local minimum value.

In calculus, there is a theorem with conditions that guarantee a function will have an absolute maximum and an absolute minimum.

> **THEOREM   Extreme Value Theorem**
>
> If a function f is continuous* on a closed interval $[a, b]$, then f has an absolute maximum and an absolute minimum on $[a, b]$.

The absolute maximum (minimum) can be found by selecting the largest (smallest) value of f from the following list:

- The values of f at any local maxima or local minima of f in $[a, b]$.
- The value of f at each endpoint of $[a, b]$—that is, $f(a)$ and $f(b)$.

*Although a precise definition requires calculus, we'll agree for now that a function is continuous if its graph has no gaps or holes and can be traced without lifting the pencil from the paper.

For example, the graph of the function f given in Figure 28(a) is continuous on the closed interval $[0, 5]$. The Extreme Value Theorem guarantees that f has extreme values on $[0, 5]$. To find them, we list

- The value of f at the local extrema: $f(3) = 6, f(4) = 4$
- The value of f at the endpoints: $f(0) = 1, f(5) = 5$

The largest of these, 6, is the absolute maximum; the smallest of these, 1, is the absolute minimum.

━━━━▶ Now Work PROBLEM 49

6 Use a Graphing Utility to Approximate Local Maxima and Local Minima and to Determine Where a Function Is Increasing or Decreasing

To locate the exact value at which a function f has a local maximum or a local minimum usually requires calculus. However, a graphing utility may be used to approximate these values using the MAXIMUM and MINIMUM features.

EXAMPLE 6

Using a Graphing Utility to Approximate Local Maxima and Minima and to Determine Where a Function Is Increasing or Decreasing

(a) Use a graphing utility to graph $f(x) = 6x^3 - 12x + 5$ for $-2 \leq x \leq 2$. Approximate where f has a local maximum and where f has a local minimum.

(b) Determine where f is increasing and where it is decreasing.

Solution

(a) Graphing utilities have a feature that finds the maximum or minimum point of a graph within a given interval. Graph the function f for $-2 \leq x \leq 2$. Using MAXIMUM on a TI-84 Plus C, we find that the local maximum value is 11.53 and that it occurs at $x = -0.82$, rounded to two decimal places. See Figure 29(a). Using MINIMUM, we find that the local minimum value is -1.53 and that it occurs at $x = 0.82$, rounded to two decimal places. See Figure 29(b). Figure 29(c) shows the two extrema in Desmos.

Figure 29   **(a)** Local maximum   **(b)** Local minimum   **(c)** Local extrema

(b) Looking at Figure 29, we see that f is increasing on the intervals $[-2, -0.82]$ and $[0.82, 2]$, or for $-2 \leq x \leq -0.82$ and $0.82 \leq x \leq 2$. And f is decreasing on the interval $[-0.82, 0.82]$, or for $-0.82 \leq x \leq 0.82$. ⏝

━━━━▶ Now Work PROBLEM 57

7 Find the Average Rate of Change of a Function

In Section 2.3, we said that the slope of a line can be interpreted as the average rate of change. To find the average rate of change of a function between any two points on its graph, calculate the slope of the line containing the two points.

> **DEFINITION** **Average Rate of Change**
>
> If a and b, $a \neq b$, are in the domain of a function $y = f(x)$, the **average rate of change of f** from a to b is defined as
>
> $$\text{Average rate of change} = \frac{\Delta y}{\Delta x} = \frac{f(b) - f(a)}{b - a} \qquad a \neq b \qquad \textbf{(1)}$$

The symbol Δy in equation (1) is the "change in y," and Δx is the "change in x." The average rate of change of f is the change in y divided by the change in x.

EXAMPLE 7	Finding the Average Rate of Change

Find the average rate of change of $f(x) = 3x^2$:

(a) From 1 to 3   (b) From 1 to 5   (c) From 1 to 7

Solution

(a) The average rate of change of $f(x) = 3x^2$ from 1 to 3 is

$$\frac{\Delta y}{\Delta x} = \frac{f(3) - f(1)}{3 - 1} = \frac{27 - 3}{3 - 1} = \frac{24}{2} = 12$$

(b) The average rate of change of $f(x) = 3x^2$ from 1 to 5 is

$$\frac{\Delta y}{\Delta x} = \frac{f(5) - f(1)}{5 - 1} = \frac{75 - 3}{5 - 1} = \frac{72}{4} = 18$$

(c) The average rate of change of $f(x) = 3x^2$ from 1 to 7 is

$$\frac{\Delta y}{\Delta x} = \frac{f(7) - f(1)}{7 - 1} = \frac{147 - 3}{7 - 1} = \frac{144}{6} = 24$$

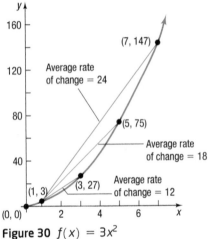

Figure 30 $f(x) = 3x^2$

See Figure 30 for a graph of $f(x) = 3x^2$. The function f is increasing on the interval $[0, \infty)$. The fact that the average rate of change is positive for any $x_1, x_2, x_1 \neq x_2$, in the interval $[1, 7]$ indicates that f is increasing on $1 \leq x \leq 7$.

 Now Work PROBLEM 65

The Secant Line

The average rate of change of a function has an important geometric interpretation. Look at the graph of $y = f(x)$ in Figure 31. Two points are labeled on the graph: $(a, f(a))$ and $(b, f(b))$. The line containing these two points is called a **secant line**; its slope is

$$m_{\text{sec}} = \frac{f(b) - f(a)}{b - a}$$

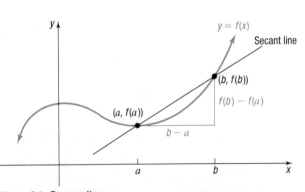

Figure 31 Secant line

⌐ THEOREM **Slope of a Secant Line**

The average rate of change of a function from a to b equals the slope of the secant line containing the two points $(a, f(a))$ and $(b, f(b))$ on its graph. ⌐

EXAMPLE 8 **Finding an Equation of a Secant Line**

Suppose that $g(x) = 3x^2 - 2x + 3$.

(a) Find the average rate of change of g from -2 to 1.

(b) Find an equation of the secant line containing $(-2, g(-2))$ and $(1, g(1))$.

(c) Using a graphing utility, draw the graph of g and the secant line obtained in part (b) on the same screen.

Solution (a) The average rate of change of $g(x) = 3x^2 - 2x + 3$ from -2 to 1 is

$$\text{Average rate of change} = \frac{g(1) - g(-2)}{1 - (-2)}$$

$$= \frac{4 - 19}{3} \qquad \begin{aligned} g(1) &= 3(1)^2 - 2 \cdot 1 + 3 = 4 \\ g(-2) &= 3(-2)^2 - 2(-2) + 3 = 19 \end{aligned}$$

$$= -\frac{15}{3} = -5$$

NORMAL FLOAT AUTO REAL RADIAN MP

(b) The slope of the secant line containing $(-2, g(-2)) = (-2, 19)$ and $(1, g(1)) = (1, 4)$ is $m_{\text{sec}} = -5$. Use the point-slope form to find an equation of the secant line.

$$y - y_1 = m_{\text{sec}}(x - x_1) \qquad \text{Point-slope form of a secant line}$$

$$y - 19 = -5(x - (-2)) \qquad x_1 = -2, y_1 = g(-2) = 19, m_{\text{sec}} = -5$$

$$y - 19 = -5x - 10 \qquad \text{Distribute.}$$

$$y = -5x + 9 \qquad \text{Slope-intercept form of the secant line}$$

Figure 32 Graph of g and the secant line

(c) Figure 32 shows the graph of g along with the secant line $y = -5x + 9$ on a TI-84 Plus C. ⌐

✏ **Now Work** PROBLEM 71

3.3 Assess Your Understanding

'Are You Prepared?' *Answers are given at the end of these exercises. If you get a wrong answer, read the pages listed in* red.

1. The interval $(2, 5)$ can be written as the inequality _____. (pp. 120–121)

2. The slope of the line containing the points $(-2, 3)$ and $(3, 8)$ is _____. (pp. 169–171)

3. Test the equation $y = 5x^2 - 1$ for symmetry with respect to the x-axis, the y-axis, and the origin. (pp. 161–164)

4. Write the point-slope form of the line with slope 5 containing the point $(3, -2)$. (p. 173)

5. The intercepts of the graph of $y = x^2 - 9$ are _____. (pp. 160–161)

Concepts and Vocabulary

6. A function f is _____ on an interval I if, for any choice of x_1 and x_2 in I, with $x_1 < x_2$, then $f(x_1) < f(x_2)$.

7. A(n) _____ function f is one for which $f(-x) = f(x)$ for every x and $-x$ in the domain of f; a(n) _____ function f is one for which $f(-x) = -f(x)$ for every x and $-x$ in the domain of f.

8. *True or False* A function f is decreasing on an interval I if, for any choice of x_1 and x_2 in I, with $x_1 < x_2$, then $f(x_1) > f(x_2)$.

9. *True or False* A function f has a local minimum at c if there is an open interval I containing c so that $f(c) \le f(x)$ for all x in this open interval.

10. *True or False* Even functions have graphs that are symmetric with respect to the origin.

11. *Multiple Choice* An odd function is symmetric with respect to _____.
   (a) the x-axis   (b) the y-axis
   (c) the origin   (d) the line $y = x$

12. *Multiple Choice* A function that is continuous on the interval _____ is guaranteed to have both an absolute maximum and an absolute minimum.
   (a) (a, b) (b) $(a, b]$ (c) $[a, b)$ (d) $[a, b]$

Skill Building

In Problems 13–24, use the graph on the right of the function f.

13. Is *f* increasing on the interval $[-8, -2]$?

14. Is *f* decreasing on the interval $[-8, -4]$?

15. Is *f* increasing on the interval $[-2, 6]$?

16. Is *f* decreasing on the interval $[2, 5]$?

17. List the interval(s) on which *f* is increasing.

18. List the interval(s) on which *f* is decreasing.

19. Is there a local maximum at 2? If yes, what is it?

20. Is there a local maximum at 5? If yes, what is it?

21. List the number(s) at which *f* has a local maximum. What are the local maximum values?

22. List the number(s) at which *f* has a local minimum. What are the local minimum values?

23. Find the absolute minimum of *f* on $[-10, 7]$.

24. Find the absolute maximum of *f* on $[-10, 7]$.

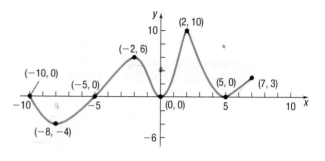

In Problems 25–32, the graph of a function is given. Use the graph to find:
- (a) The intercepts, if any
- (b) The domain and range
- (c) The intervals on which the function is increasing, decreasing, or constant
- (d) Whether the function is even, odd, or neither

In Problems 33–36, the graph of a function f is given. Use the graph to find:
- (a) The numbers, if any, at which f has a local maximum. What are the local maximum values?
- (b) The numbers, if any, at which f has a local minimum. What are the local minimum values?

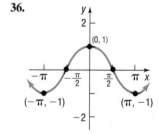

In Problems 37–48, determine algebraically whether each function is even, odd, or neither.

37. $f(x) = 4x^3$

38. $f(x) = 2x^4 - x^2$

39. $g(x) = 10 - x^2$

40. $h(x) = 3x^3 + 5$

41. $F(x) = \sqrt[3]{4x}$

42. $G(x) = \sqrt{x}$

43. $f(x) = x + |x|$

44. $f(x) = \sqrt[3]{2x^2 + 1}$

45. $g(x) = \dfrac{1}{x^2 + 8}$

46. $h(x) = \dfrac{x}{x^2 - 1}$

47. $h(x) = \dfrac{-x^3}{3x^2 - 9}$

48. $F(x) = \dfrac{2x}{|x|}$

In Problems 49–56, for each graph of a function $y = f(x)$, *find the absolute maximum and the absolute minimum, if they exist. Identify any local maximum values or local minimum values.*

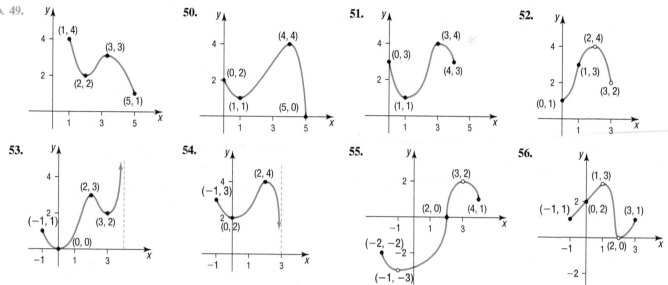

In Problems 57–64, use a graphing utility to graph each function over the indicated interval and approximate any local maximum values and local minimum values. Determine where the function is increasing and where it is decreasing. Round answers to two decimal places.

57. $f(x) = x^3 - 3x + 2 \quad [-2, 2]$

58. $f(x) = x^3 - 3x^2 + 5 \quad [-1, 3]$

59. $f(x) = x^5 - x^3 \quad [-2, 2]$

60. $f(x) = x^4 - x^2 \quad [-2, 2]$

61. $f(x) = -0.2x^3 - 0.6x^2 + 4x - 6 \quad [-6, 4]$

62. $f(x) = -0.4x^3 + 0.6x^2 + 3x - 2 \quad [-4, 5]$

63. $f(x) = 0.25x^4 + 0.3x^3 - 0.9x^2 + 3 \quad [-3, 2]$

64. $f(x) = -0.4x^4 - 0.5x^3 + 0.8x^2 - 2 \quad [-3, 2]$

65. Find the average rate of change of $f(x) = -2x^2 + 4$:
   (a) From 0 to 2   **(b)** From 1 to 3   **(c)** From 1 to 4

66. Find the average rate of change of $f(x) = -x^3 + 1$:
   (a) From 0 to 2   **(b)** From 1 to 3   **(c)** From −1 to 1

67. Find the average rate of change of $g(x) = x^3 - 4x + 7$:
   (a) From −3 to −2   **(b)** From −1 to 1   **(c)** From 1 to 3

68. Find the average rate of change of $h(x) = x^2 - 2x + 3$:
   (a) From −1 to 1   **(b)** From 0 to 2   **(c)** From 2 to 5

69. $f(x) = 5x - 2$
   (a) Find the average rate of change from 1 to 3.
   (b) Find an equation of the secant line containing $(1, f(1))$ and $(3, f(3))$.

70. $f(x) = -4x + 1$
   (a) Find the average rate of change from 2 to 5.
   (b) Find an equation of the secant line containing $(2, f(2))$ and $(5, f(5))$.

71. $g(x) = x^2 - 2$
   (a) Find the average rate of change from −2 to 1.
   (b) Find an equation of the secant line containing $(-2, g(-2))$ and $(1, g(1))$.

72. $g(x) = x^2 + 1$
   (a) Find the average rate of change from −1 to 2.
   (b) Find an equation of the secant line containing $(-1, g(-1))$ and $(2, g(2))$.

73. $h(x) = x^2 - 2x$
   (a) Find the average rate of change from 2 to 4.
   (b) Find an equation of the secant line containing $(2, h(2))$ and $(4, h(4))$.

74. $h(x) = -2x^2 + x$
   (a) Find the average rate of change from 0 to 3.
   (b) Find an equation of the secant line containing $(0, h(0))$ and $(3, h(3))$.

75. *Mixed Practice* $g(x) = x^3 - 27x$
   (a) Determine whether g is even, odd, or neither.
   (b) There is a local minimum value of −54 at 3. Determine the local maximum value.

76. *Mixed Practice* $f(x) = -x^3 + 12x$
   (a) Determine whether f is even, odd, or neither.
   (b) There is a local maximum value of 16 at 2. Determine the local minimum value.

Applications and Extensions

77. $F(x) = -x^4 + 8x^2 + 9$
   (a) Determine whether F is even, odd, or neither.
   (b) There is a local maximum value of 25 at $x = 2$. Find a second local maximum value.
   (c) Suppose the area of the region enclosed by the graph of F and the x-axis between $x = 0$ and $x = 3$ is 50.4 square units. Using the result from (a), determine the area of the region enclosed by the graph of F and the x-axis between $x = -3$ and $x = 0$.

78. $G(x) = -x^4 + 32x^2 + 144$

 (a) Determine whether G is even, odd, or neither.

 (b) There is a local maximum value of 400 at $x = 4$. Find a second local maximum value.

 (c) Suppose the area of the region enclosed by the graph of G and the x-axis between $x = 0$ and $x = 6$ is 1612.8 square units. Using the result from (a), determine the area of the region enclosed by the graph of G and the x-axis between $x = -6$ and $x = 0$.

79. Minimum Average Cost The average cost per hour in dollars, $\overline{C}$, of producing x riding lawn mowers can be modeled by the function

$$\overline{C}(x) = 0.3x^2 + 21x - 251 + \frac{2500}{x}$$

 (a) Use a graphing utility to graph $\overline{C} = \overline{C}(x)$.

 (b) Determine the number of riding lawn mowers to produce in order to minimize average cost.

 (c) What is the minimum average cost?

80. Medicine Concentration The concentration C of a medication in the bloodstream t hours after being administered is modeled by the function

$$C(t) = -0.002t^4 + 0.039t^3 - 0.285t^2 + 0.766t + 0.085$$

 (a) After how many hours will the concentration be highest?

 (b) A woman nursing a child must wait until the concentration is below 0.5 before she can feed him. After taking the medication, how long must she wait before feeding her child?

81. E. coli Growth A strain of E. coli Beu 397-recA441 is placed into a nutrient broth at 30° Celsius and allowed to grow. The data shown in the table are collected. The population is measured in grams and the time in hours. Since population P depends on time t, and each input corresponds to exactly one output, we can say that population is a function of time, so $P(t)$ represents the population at time t.

Time (hours), t	Population (grams), P
0	0.09
2.5	0.18
3.5	0.26
4.5	0.35
6	0.50

 (a) Find the average rate of change of the population from 0 to 2.5 hours.

 (b) Find the average rate of change of the population from 4.5 to 6 hours.

 (c) What is happening to the average rate of change as time passes?

82. National Debt The size of the total debt owed by the United States federal government continues to grow. In fact, according to the Department of the Treasury, the debt per person living in the United States is approximately \$63,720 (or over \$172,000 per U.S. household). The following data represent the U.S. debt for the years 2007–2017. Since the debt D depends on the year y, and each input corresponds to exactly one output, the debt is a function of the year. So $D(y)$ represents the debt for each year y.

Year	Debt (billions of dollars)
2007	9008
2008	10,025
2009	11,910
2010	13,562
2011	14,790
2012	16,066
2013	16,738
2014	17,824
2015	18,151
2016	19,573
2017	20,245

Source: www.treasurydirect.gov

 (a) Plot the points (2007, 9008), (2008, 10 025), and so on.

 (b) Draw a line segment from the point (2007, 9008) to (2012, 16 066). What does the slope of this line segment represent?

 (c) Find the average rate of change of the debt from 2008 to 2010.

 (d) Find the average rate of change of the debt from 2011 to 2013.

 (e) Find the average rate of change of the debt from 2014 to 2016.

 (f) What appears to be happening to the average rate of change as time passes?

83. For the function $f(x) = x^2$, compute the average rate of change:

 (a) From 0 to 1   **(b)** From 0 to 0.5   **(c)** From 0 to 0.1
 (d) From 0 to 0.01   **(e)** From 0 to 0.001

 (f) Use a graphing utility to graph each of the secant lines along with f.

 (g) What do you think is happening to the secant lines?

 (h) What is happening to the slopes of the secant lines? Is there some number that they are getting closer to? What is that number?

84. For the function $f(x) = x^2$, compute the average rate of change:

 (a) From 1 to 2   **(b)** From 1 to 1.5   **(c)** From 1 to 1.1
 (d) From 1 to 1.01   **(e)** From 1 to 1.001

 (f) Use a graphing utility to graph each of the secant lines along with f.

 (g) What do you think is happening to the secant lines?

 (h) What is happening to the slopes of the secant lines? Is there some number that they are getting closer to? What is that number?

⚠ *Problems 85–92 require the following discussion of a secant line. The slope of the secant line containing the two points $(x, f(x))$ and $(x + h, f(x + h))$ on the graph of a function $y = f(x)$ may be given as*

$$m_{\text{sec}} = \frac{f(x + h) - f(x)}{(x + h) - x} = \frac{f(x + h) - f(x)}{h} \qquad h \neq 0$$

(a) *Express the slope of the secant line of each function in terms of x and h. Be sure to simplify your answer.*
(b) *Find m_{sec} for $h = 0.5, 0.1$, and 0.01 at $x = 1$. What value does m_{sec} approach as h approaches 0?*
(c) *Find an equation for the secant line at $x = 1$ with $h = 0.01$.*
(d) *Use a graphing utility to graph f and the secant line found in part (c) in the same viewing window.*

85. $f(x) = 2x + 5$ **86.** $f(x) = -3x + 2$ **87.** $f(x) = x^2 + 2x$ **88.** $f(x) = 2x^2 + x$

89. $f(x) = 2x^2 - 3x + 1$ **90.** $f(x) = -x^2 + 3x - 2$ **91.** $f(x) = \dfrac{1}{x}$ **92.** $f(x) = \dfrac{1}{x^2}$

⚠ **93. *Challenge Problem* Mean Value Theorem** Suppose $f(x) = x^3 + 2x^2 - x + 6$. From calculus, the Mean Value Theorem guarantees that there is at least one number in the open interval $(-1, 2)$ at which the value of the derivative of f, given by $f'(x) = 3x^2 + 4x - 1$, is equal to the average rate of change of f on the interval. Find all such numbers x in the interval.

94. *Challenge Problem* If f is an odd function, determine whether $g(x) = -2f\left(-\dfrac{x}{3}\right)$ is even, odd, or neither.

Explaining Concepts: Discussion and Writing

95. Draw the graph of a function that has the following properties: domain: all real numbers; range: all real numbers; intercepts: $(0, -3)$ and $(3, 0)$; a local maximum value of -2 at -1; a local minimum value of -6 at 2. Compare your graph with those of others. Comment on any differences.

96. Redo Problem 95 with the following additional information: increasing on $(-\infty, -1], [2, \infty)$; decreasing on $[-1, 2]$. Again compare your graph with others and comment on any differences.

97. How many x-intercepts can a function defined on an interval have if it is increasing on that interval? Explain.

98. Suppose that a friend of yours does not understand the idea of increasing and decreasing functions. Provide an explanation, complete with graphs, that clarifies the idea.

99. Can a function be both even and odd? Explain.

100. Using a graphing utility, graph $y = 5$ on the interval $[-3, 3]$. Use MAXIMUM to find the local maximum values on $[-3, 3]$. Comment on the result provided by the graphing utility.

101. A function f has a positive average rate of change on the interval $[2, 5]$. Is f increasing on $[2, 5]$? Explain.

102. Show that a constant function $f(x) = b$ has an average rate of change of 0. Compute the average rate of change of $y = \sqrt{4 - x^2}$ on the interval $[-2, 2]$. Explain how this can happen.

Retain Your Knowledge

Problems 103–112 are based on material learned earlier in the course. The purpose of these problems is to keep the material fresh in your mind so that you are better prepared for the final exam.

103. Simplify: $\sqrt{540}$

104. Multiply: $(4a - b)^2$

105. The daily rental charge for a moving truck is $40 plus a mileage charge of $0.80 per mile. Express the cost C to rent a moving truck for one day as a function of the number x of miles driven.

106. The shelf life of a perishable commodity varies inversely with the storage temperature. If the shelf life at $10°C$ is 33 days, what is the shelf life at $40°C$?

107. Write the standard form of the equation of a circle with center $(3, -2)$ and radius $r = \dfrac{\sqrt{6}}{2}$.

⚠ **108.** Factor completely:

$$(x^2 - 2)^3 \cdot 2(3x^5 + 1) \cdot 15x^4 + (3x^5 + 1)^2 \cdot 3(x^2 - 2)^2 \cdot 2x$$

109. Find the midpoint of the line segment connecting the points $(-2, 1)$ and $\left(\dfrac{3}{5}, -4\right)$.

110. Solve: $|3x + 7| - 3 = 5$

111. Find the real solutions of $x^6 + 7x^3 = 8$.

⚠ **112.** Solve for D if

$$3y^2 \cdot D + 3x^2 - 3xy^2 - 3x^2y \cdot D = 0$$

'Are You Prepared?' Answers

1. $2 < x < 5$ **2.** 1 **3.** symmetric with respect to the y-axis   **4.** $y + 2 = 5(x - 3)$ **5.** $(-3, 0), (3, 0), (0, -9)$

3.4 Library of Functions; Piecewise-defined Functions

PREPARING FOR THIS SECTION *Before getting started, review the following:*

- Intercepts (Section 2.2, pp. 160–161)
- Graphs of Key Equations (Section 2.2: Example 3, p. 159; Examples 11–13, pp. 164–165)

Now Work the 'Are You Prepared?' problems on page 250.

OBJECTIVES 1 Graph the Functions Listed in the Library of Functions (p. 242)
2 Analyze a Piecewise-defined Function (p. 247)

1 Graph the Functions Listed in the Library of Functions

Figure 33 shows the graph of $y = \sqrt{x}$. Based on the Vertical-Line Test, $y = f(x) = \sqrt{x}$ is a function, called the **square root function**. From the graph, we have the following properties:

Figure 33 $y = f(x) = \sqrt{x}$

Properties of $f(x) = \sqrt{x}$

- The domain and the range of $f(x) = \sqrt{x}$ are the set of nonnegative real numbers.
- The x-intercept of the graph of $f(x) = \sqrt{x}$ is 0. The y-intercept of the graph of $f(x) = \sqrt{x}$ is also 0.
- $f(x) = \sqrt{x}$ is neither even nor odd.
- $f(x) = \sqrt{x}$ is increasing on the interval $[0, \infty)$.
- $f(x) = \sqrt{x}$ has an absolute minimum of 0 at $x = 0$.

EXAMPLE 1

Graphing the Cube Root Function

(a) Determine whether $f(x) = \sqrt[3]{x}$ is even, odd, or neither. State whether the graph of f is symmetric with respect to the y-axis, symmetric with respect to the origin, or neither.

(b) Find the intercepts, if any, of the graph of $f(x) = \sqrt[3]{x}$.

(c) Graph $f(x) = \sqrt[3]{x}$.

Solution

(a) Because

$$f(-x) = \sqrt[3]{-x} = -\sqrt[3]{x} = -f(x)$$

the cube root function is odd. The graph of f is symmetric with respect to the origin.

(b) The y-intercept is $f(0) = \sqrt[3]{0} = 0$. The x-intercept is found by solving the equation $f(x) = 0$.

$$f(x) = 0$$
$$\sqrt[3]{x} = 0 \qquad f(x) = \sqrt[3]{x}$$
$$x = 0 \qquad \text{Cube both sides of the equation.}$$

The x-intercept is also 0.

(c) Use the function to form Table 5 and obtain some points on the graph. Because of the symmetry with respect to the origin, we find only points (x, y) for which $x \geq 0$. Figure 34 shows the graph of $f(x) = \sqrt[3]{x}$.

Table 5

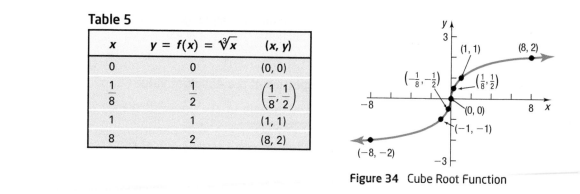

x	y = f(x) = ∛x	(x, y)
0	0	(0, 0)
$\frac{1}{8}$	$\frac{1}{2}$	$\left(\frac{1}{8}, \frac{1}{2}\right)$
1	1	(1, 1)
8	2	(8, 2)

Figure 34 Cube Root Function

From the results of Example 1 and Figure 34, we have the following properties of the cube root function.

Properties of the Cube Root Function $f(x) = \sqrt[3]{x}$

- The domain and the range of $f(x) = \sqrt[3]{x}$ are the set of all real numbers.
- The x-intercept of the graph of $f(x) = \sqrt[3]{x}$ is 0. The y-intercept of the graph of $f(x) = \sqrt[3]{x}$ is also 0.
- $f(x) = \sqrt[3]{x}$ is odd. The graph is symmetric with respect to the origin.
- $f(x) = \sqrt[3]{x}$ is increasing on the interval $(-\infty, \infty)$.
- $f(x) = \sqrt[3]{x}$ does not have any local minima or any local maxima.

EXAMPLE 2   **Graphing the Absolute Value Function**

(a) Determine whether $f(x) = |x|$ is even, odd, or neither. State whether the graph of f is symmetric with respect to the y-axis, symmetric with respect to the origin, or neither.

(b) Determine the intercepts, if any, of the graph of $f(x) = |x|$.

(c) Graph $f(x) = |x|$.

Solution   (a) Because

$$f(-x) = |-x|$$
$$= |x| = f(x)$$

the absolute value function is even. The graph of f is symmetric with respect to the y-axis.

(b) The y-intercept is $f(0) = |0| = 0$. The x-intercept is found by solving the equation $f(x) = |x| = 0$. The x-intercept is 0.

(c) Use the function to form Table 6 and obtain some points on the graph. Because of the symmetry with respect to the y-axis, we only need to find points (x, y) for which $x \geq 0$. Figure 35 shows the graph of $f(x) = |x|$.

Table 6

x	y = f(x) = |x|	(x, y)
0	0	(0, 0)
1	1	(1, 1)
2	2	(2, 2)
3	3	(3, 3)

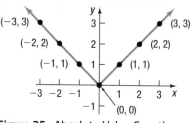

Figure 35 Absolute Value Function

From Example 2 and Figure 35, we have the following properties of the absolute value function.

Properties of the Absolute Value Function $f(x) = |x|$

- The domain of $f(x) = |x|$ is the set of all real numbers. The range of f is $\{y \mid y \geq 0\}$.
- The x-intercept of the graph of $f(x) = |x|$ is 0. The y-intercept of the graph of $f(x) = |x|$ is also 0.
- $f(x) = |x|$ is even. The graph is symmetric with respect to the y-axis.
- $f(x) = |x|$ is decreasing on the interval $(-\infty, 0]$. It is increasing on the interval $[0, \infty)$.
- $f(x) = |x|$ has an absolute minimum of 0 at $x = 0$.

Below is a list of the key functions that we have discussed. In going through this list, pay special attention to the properties of each function, particularly to the domain of each function and to the shape of each graph. Knowing these graphs, along with key points on each graph, lays the foundation for further graphing techniques.

Constant Function

$$f(x) = b \qquad b \text{ is a real number}$$

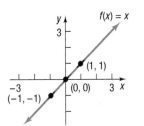

Figure 36 Constant Function

See Figure 36.

The domain of a **constant function** is the set of all real numbers; its range is the set consisting of a single number b. Its graph is a horizontal line whose y-intercept is b. The constant function is an even function.

Identity Function

$$f(x) = x$$

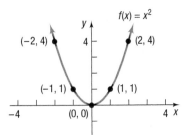

Figure 37 Identity Function

See Figure 37.

The domain and the range of the **identity function** are the set of all real numbers. Its graph is a line with slope 1 and y-intercept 0. The line consists of all points for which the x-coordinate equals the y-coordinate. The identity function is an odd function and is increasing over its domain. Note that the graph bisects quadrants I and III.

Square Function

$$f(x) = x^2$$

Figure 38 Square Function

See Figure 38.

The domain of the **square function** is the set of all real numbers; its range is the set of nonnegative real numbers. The graph of the square function is a parabola whose intercept is at $(0, 0)$. The square function is an even function that is decreasing on the interval $(-\infty, 0]$ and increasing on the interval $[0, \infty)$.

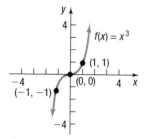

Figure 39 Cube Function

Cube Function

$$f(x) = x^3$$

See Figure 39.

The domain and the range of the **cube function** are the set of all real numbers. The intercept of the graph is at $(0, 0)$. The cube function is odd and is increasing on the interval $(-\infty, \infty)$.

Figure 40 Square Root Function

Square Root Function

$$f(x) = \sqrt{x}$$

See Figure 40.

The domain and the range of the **square root function** are the set of nonnegative real numbers. The intercept of the graph is at $(0, 0)$. The square root function is neither even nor odd and is increasing on the interval $[0, \infty)$.

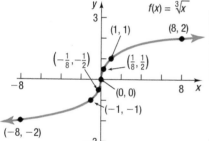

Figure 41 Cube Root Function

Cube Root Function

$$f(x) = \sqrt[3]{x}$$

See Figure 41.

The domain and the range of the **cube root function** are the set of all real numbers. The intercept of the graph is at $(0, 0)$. The cube root function is an odd function that is increasing on the interval $(-\infty, \infty)$.

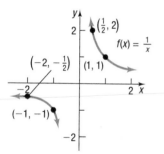

Figure 42 Reciprocal Function

Reciprocal Function

$$f(x) = \frac{1}{x}$$

Refer to Example 13, page 165, for a discussion of the equation $y = \dfrac{1}{x}$. See Figure 42.

The domain and the range of the **reciprocal function** are the set of all nonzero real numbers. The graph has no intercepts. The reciprocal function is decreasing on the intervals $(-\infty, 0)$ and $(0, \infty)$ and is an odd function.

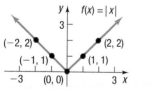

Figure 43 Absolute Value Function

Absolute Value Function

$$f(x) = |x|$$

See Figure 43.

The domain of the **absolute value function** is the set of all real numbers; its range is the set of nonnegative real numbers. The intercept of the graph is at $(0, 0)$. If $x \geq 0$, then $f(x) = x$, and the graph of f is part of the line $y = x$; if $x < 0$, then $f(x) = -x$, and the graph of f is part of the line $y = -x$. The absolute value function is an even function; it is decreasing on the interval $(-\infty, 0]$ and increasing on the interval $[0, \infty)$.

The notation int(x) stands for the greatest integer less than or equal to x. For example,

$$\text{int}(1) = 1 \quad \text{int}(2.5) = 2 \quad \text{int}\!\left(\frac{1}{2}\right) = 0 \quad \text{int}\!\left(-\frac{3}{4}\right) = -1 \quad \text{int}(\pi) = 3$$

Table 7

x	$y = f(x)$ $= \text{int}(x)$	(x, y)
-1	-1	$(-1, -1)$
$-\dfrac{1}{2}$	-1	$\left(-\dfrac{1}{2}, -1\right)$
$-\dfrac{1}{4}$	-1	$\left(-\dfrac{1}{4}, -1\right)$
0	0	$(0, 0)$
$\dfrac{1}{4}$	0	$\left(\dfrac{1}{4}, 0\right)$
$\dfrac{1}{2}$	0	$\left(\dfrac{1}{2}, 0\right)$
$\dfrac{3}{4}$	0	$\left(\dfrac{3}{4}, 0\right)$

> **DEFINITION Greatest Integer Function**
>
> $$f(x) = \text{int}(x) = \text{greatest integer less than or equal to } x^{*}$$

We obtain the graph of $f(x) = \text{int}(x)$ by plotting several points. See Table 7. For values of x, $-1 \le x < 0$, the value of $f(x) = \text{int}(x)$ is -1; for values of x, $0 \le x < 1$, the value of f is 0. See Figure 44 for the graph.

The domain of the **greatest integer function** is the set of all real numbers; its range is the set of integers. The y-intercept of the graph is 0. The x-intercepts lie in the interval $[0, 1)$. The greatest integer function is neither even nor odd. It is constant on every interval of the form $[k, k + 1)$, for k an integer. In Figure 44, a solid dot indicates, for example, that at $x = 1$ the value of f is $f(1) = 1$; an open circle is used to show that the value of f is not 0 at $x = 1$.

Although a precise definition requires calculus, in a rough sense, a function is *continuous* if its graph has no gaps or holes and can be traced without lifting the pencil from the paper on which the graph is drawn. A function is *discontinuous* if its graph has gaps or holes and cannot be traced without lifting the pencil from the paper.

The greatest integer function is also called a **step function**. At $x = 0$, $x = \pm 1, x = \pm 2$, and so on, this function is discontinuous because, at integer values, the graph suddenly "steps" from one value to another without taking on any of the intermediate values. For example, to the immediate left of $x = 3$, the y-coordinates of the points on the graph are 2, and at $x = 3$ and to the immediate right of $x = 3$, the y-coordinates of the points on the graph are 3.

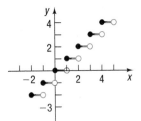

Figure 44 Greatest Integer Function

COMMENT When graphing a function using a graphing utility, typically you can choose either **connected mode**, in which points plotted on the screen are connected, making the graph appear continuous, or **dot mode**, in which only the points plotted appear. When graphing the greatest integer function with a graphing utility, it may be necessary to be in **dot mode**. This is to prevent the utility from "connecting the dots" when $f(x)$ changes from one integer value to the next. However, some utilities will display the gaps even when in "connected" mode. See Figure 45. ∎

(a) TI-83 Plus, connected mode   (b) TI-83 Plus, dot mode   (c) TI-84 Plus C   (d) Desmos

Figure 45 $f(x) = \text{int}(x)$

The functions discussed so far are basic. Whenever you encounter one of them, you should see a mental picture of its graph. For example, if you encounter the function $f(x) = x^2$, you should see in your mind's eye a picture of a parabola.

 Now Work PROBLEMS 11 THROUGH 18

---

*Some texts use the notation $f(x) = [x]$ or call the greatest integer function the "floor function" and use the notation $f(x) = \lfloor x \rfloor$.

2 Analyze a Piecewise-defined Function

Figure 46 Absolute Value Function

Sometimes a function is defined using different equations on different parts of its domain. For example, the absolute value function $f(x) = |x|$ is actually defined by two equations: $f(x) = x$ if $x \geq 0$ and $f(x) = -x$ if $x < 0$. See Figure 46. For convenience, these equations are generally combined into one expression as

$$f(x) = |x| = \begin{cases} x & \text{if } x \geq 0 \\ -x & \text{if } x < 0 \end{cases}$$

When a function is defined by different equations on different parts of its domain, it is called a **piecewise-defined** function.

EXAMPLE 3   **Analyzing a Piecewise-defined Function**

A piecewise-defined function f is defined as

$$f(x) = \begin{cases} -2x & \text{if } x < 0 \\ \sqrt{x} & \text{if } x \geq 0 \end{cases}$$

(a) Find $f(-4)$, $f(0)$, and $f(4)$.

(b) Find the domain of f.

(c) Locate any intercepts.

(d) Graph f.

(e) Use the graph to find the range of f.

Solution   (a) When evaluating a piecewise-defined function, we first must identify which equation should be used in the evaluation. We do this by identifying in which part of the domain the value for the independent variable lies.

● To find $f(-4)$, observe that $-4 < 0$ so $x = -4$ lies in the domain of the first equation. This means that when $x = -4$, the equation for f is $f(x) = -2x$. Then

$$f(-4) = -2(-4) = 8$$

● To find $f(0)$, observe that when $x = 0$, the equation for f is $f(x) = \sqrt{x}$. Then

$$f(0) = \sqrt{0} = 0$$

● To find $f(4)$, observe that when $x = 4$, the equation for f is $f(x) = \sqrt{x}$. Then

$$f(4) = \sqrt{4} = 2$$

(b) The domain of the first equation of the function f is the set of all negative real numbers, $\{x \mid x < 0\}$, or $(-\infty, 0)$ in interval notation. The domain of the second equation of the function f is the set of all nonnegative real numbers, $\{x \mid x \geq 0\}$, or $[0, \infty)$ in interval notation. The domain of the function is the union of the domains of the individual equations. Since

$$(-\infty, 0) \cup [0, \infty) = (-\infty, \infty)$$

the domain of f is the set of all real numbers.

(c) The y-intercept of the graph of the function is $f(0)$. Because the equation for f when $x = 0$ is $f(x) = \sqrt{x}$, the y-intercept is $f(0) = \sqrt{0} = 0$.

(continued)

The *x*-intercepts of the graph of a function are the real solutions to the equation $f(x) = 0$. To find the *x*-intercepts of *f*, solve $f(x) = 0$ for each equation of the function, and then determine which values of *x*, if any, satisfy the domain of the equation.

$$f(x) = 0 \quad x < 0 \qquad\qquad f(x) = 0 \quad x \geq 0$$
$$-2x = 0 \qquad\qquad\qquad \sqrt{x} = 0$$
$$x = 0 \qquad\qquad\qquad x = 0$$

The result on the left $(x = 0)$ is discarded because it does not satisfy the condition $x < 0$. The result on the right $(x = 0)$ is obtained under the condition $x \geq 0$, so 0 is an *x*-intercept. The only intercept is $(0, 0)$.

(d) To graph *f*, graph each equation. First graph the line $y = -2x$ and keep only the part for which $x < 0$. Then graph the square root function $y = \sqrt{x}$ for $x \geq 0$. See Figure 47.

(e) From the graph, we conclude that the range of *f* is the set of nonnegative real numbers, or the interval $[0, \infty)$.

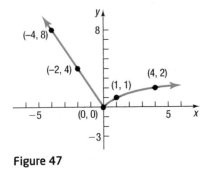

Figure 47

EXAMPLE 4

Analyzing a Piecewise-defined Function

A piecewise-defined function *f* is defined as

$$f(x) = \begin{cases} -2x + 1 & \text{if } -3 \leq x < 1 \\ 2 & \text{if } x = 1 \\ x^2 & \text{if } x > 1 \end{cases}$$

(a) Find $f(-2), f(1)$, and $f(2)$.

(b) Find the domain of *f*.

(c) Locate any intercepts.

(d) Graph *f*.

(e) Use the graph to find the range of *f*.

Solution

(a) To find $f(-2)$, observe that $-3 \leq -2 < 1$, so when $x = -2$, the equation for *f* is given by $f(x) = -2x + 1$. Then

$$f(-2) = -2(-2) + 1 = 5 \quad f(x) = -2x + 1 \quad \text{if } -3 \leq x < 1$$

To find $f(1)$, observe that when $x = 1$, the equation for *f* is $f(x) = 2$. So,

$$f(1) = 2 \quad f(x) = 2 \text{ if } x = 1$$

When $x = 2$, the equation for *f* is $f(x) = x^2$. Then

$$f(2) = 2^2 = 4 \quad f(x) = x^2 \text{ if } x > 1$$

(b) The domain of *f* is the union of the domains of each equation in the piecewise-defined function. So the domain of *f* is $[-3, 1) \cup \{1\} \cup (1, \infty) = [-3, \infty)$. The domain of *f* is $[-3, \infty)$ in interval notation, or $\{x | x \geq -3\}$ in set notation.

(c) The *y*-intercept of the graph of the function is $f(0)$. Because the equation for *f* when $x = 0$ is $f(x) = -2x + 1$, the *y*-intercept is $f(0) = -2 \cdot 0 + 1 = 1$. The *x*-intercepts of the graph of a function *f* are the real solutions of the equation $f(x) = 0$. To find the *x*-intercepts of *f*, solve $f(x) = 0$ for each equation of the function, and then determine what values of *x*, if any, are in the domain of the equation.

$$f(x) = 0 \quad -3 \leq x < 1 \qquad f(x) = 0 \quad x = 1 \qquad f(x) = 0 \quad x > 1$$
$$-2x + 1 = 0 \qquad\qquad\qquad 2 = 0 \qquad\qquad\qquad x^2 = 0$$
$$-2x = -1 \qquad\qquad\qquad \text{No solution} \qquad\qquad x = 0$$
$$x = \frac{1}{2}$$

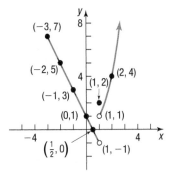

Figure 48

The first potential x-intercept, $x = \dfrac{1}{2}$, satisfies the condition $-3 \le x < 1$, so $x = \dfrac{1}{2}$ is an x-intercept. The second potential x-intercept, $x = 0$, does not satisfy the condition $x > 1$, so we discard it. The only x-intercept is $\dfrac{1}{2}$. The intercepts are $(0, 1)$ and $\left(\dfrac{1}{2}, 0\right)$.

(d) To graph f, first graph the line $y = -2x + 1$ and keep only the part for which $-3 \le x < 1$. When $x = 1, f(x) = 2$, so plot the point $(1, 2)$. Finally, graph the parabola $y = x^2$ and keep only the part for which $x > 1$. See Figure 48. Notice we use open circles at the points $(1, -1)$ and $(1, 1)$ to indicate these points are not part of the graph.

(e) From the graph, we conclude that the range of f is $\{y \mid y > -1\}$, or the interval $(-1, \infty)$.

 Now Work PROBLEM 31

 EXAMPLE 5

Cost of Electricity

In the spring of 2018, Duke Energy Progress supplied electricity to residences in South Carolina for a monthly customer charge of \$8.29 plus 8.6035¢ per kilowatt-hour (kWh) for the first 1000 kWh supplied in the month and 9.1446¢ per kWh for all usage over 1000 kWh in the month.

(a) What is the charge for using 500 kWh in a month?

(b) What is the charge for using 1500 kWh in a month?

(c) If C is the monthly charge for x kWh, develop a model relating the monthly charge and kilowatt-hours used. That is, express C as a function of x.
Source: *Duke Energy Progress, 2018*

Solution

(a) For 500 kWh, the charge is \$8.29 plus $(8.6035¢ = \$0.086035)$ per kWh. That is,

$$\text{Charge} = \$8.29 + \$0.086035 \cdot 500 = \$51.31$$

(b) For 1500 kWh, the charge is \$8.29 plus 8.6035¢ per kWh for the first 1000 kWh plus 9.1446¢ per kWh for the 500 kWh in excess of 1000. That is,

$$\text{Charge} = \$8.29 + \$0.086035 \cdot 1000 + \$0.091446 \cdot 500 = \$140.05$$

(c) Let x represent the number of kilowatt-hours used. If $0 \le x \le 1000$, then the monthly charge C (in dollars) can be found by multiplying x times \$0.086035 and adding the monthly customer charge of \$8.29. So if $0 \le x \le 1000$, then

$$C(x) = 0.086035x + 8.29$$

For $x > 1000$, the charge is $0.086035 \cdot 1000 + 8.29 + 0.091446(x - 1000)$, since $(x - 1000)$ equals the usage in excess of 1000 kWh, which costs \$0.091446 per kWh. That is, if $x > 1000$, then

$$C(x) = 0.086035 \cdot 1000 + 8.29 + 0.091446(x - 1000)$$
$$= 86.035 + 8.29 + 0.091446x - 91.446$$
$$= 0.091446x + 2.879$$

The rule for computing C follows two equations:

$$C(x) = \begin{cases} 0.086035x + 8.29 & \text{if } 0 \le x \le 1000 \\ 0.091446x + 2.879 & \text{if } x > 1000 \end{cases} \quad \text{The model}$$

Figure 49

See Figure 49 for the graph. Note that the two graphs are both lines, but they have different slopes (rates) and intersect at the point $(1000, 94.33)$.

3.4 Assess Your Understanding

'Are You Prepared?' *Answers are given at the end of these exercises. If you get a wrong answer, read the pages listed in* red.

1. Graph $y = \sqrt{x}$. (pp. 164–165)

2. Graph $y = \dfrac{1}{x}$. (p. 165)

3. Find the intercepts of the equation $y = x^3 - 8$. (p. 161)

Concepts and Vocabulary

4. The function $f(x) = x^2$ is decreasing on the interval _____.

5. When functions are defined by more than one equation, they are called _____ functions.

6. *True or False* The cube function is odd and is increasing on the interval $(-\infty, \infty)$.

7. *True or False* The cube root function is odd and is decreasing on the interval $(-\infty, \infty)$.

8. *True or False* The domain and the range of the reciprocal function are the set of all real numbers.

9. *Multiple Choice* Which of the following functions has a graph that is symmetric about the *y*-axis?

(a) $y = \sqrt{x}$ (b) $y = |x|$ (c) $y = x^3$ (d) $y = \dfrac{1}{x}$

10. *Multiple Choice* Consider the following function.

$$f(x) = \begin{cases} 3x - 2 & \text{if} & x < 2 \\ x^2 + 5 & \text{if} & 2 \le x < 10 \\ 3 & \text{if} & x \ge 10 \end{cases}$$

Which expression(s) should be used to find the *y*-intercept?

(a) $3x - 2$ (b) $x^2 + 5$ (c) 3 (d) all three

Skill Building

In Problems 11–18, match each graph to its function.

A. *Constant function*   B. *Identity function*   C. *Square function*   D. *Cube function*
E. *Square root function*   F. *Reciprocal function*   G. *Absolute value function* H. *Cube root function*

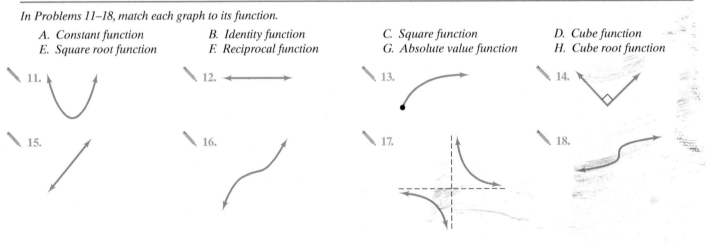

11.   **12.**   **13.**   **14.**

15.   **16.**   **17.**   **18.**

In Problems 19–26, graph each function. Be sure to label three points on the graph.

19. $f(x) = x$ **20.** $f(x) = x^2$ **21.** $f(x) = x^3$ **22.** $f(x) = \sqrt{x}$

23. $f(x) = \dfrac{1}{x}$ **24.** $f(x) = |x|$ **25.** $f(x) = \sqrt[3]{x}$ **26.** $f(x) = 3$

27. If $f(x) = \begin{cases} -x^2 & \text{if } x < 0 \\ 4 & \text{if } x = 0 \\ 3x - 2 & \text{if } x > 0 \end{cases}$ find:

(a) $f(-3)$ (b) $f(0)$ (c) $f(3)$

28. If $f(x) = \begin{cases} -3x & \text{if } x < -1 \\ 0 & \text{if } x = -1 \\ 2x^2 + 1 & \text{if } x > -1 \end{cases}$ find:

(a) $f(-2)$ (b) $f(-1)$ (c) $f(0)$

29. If $f(x) = \begin{cases} 2x + 4 & \text{if } -3 \le x \le 1 \\ x^3 - 1 & \text{if } 1 < x \le 5 \end{cases}$ find:

(a) $f(-2)$ (b) $f(0)$ (c) $f(1)$ (d) $f(3)$

30. If $f(x) = \begin{cases} x^3 & \text{if } -2 \le x < 1 \\ 3x + 2 & \text{if } 1 \le x \le 4 \end{cases}$ find:

(a) $f(-1)$ (b) $f(0)$ (c) $f(1)$ (d) $f(3)$

In Problems 31–42:

(a) *Find the domain of each function.*   (b) *Locate any intercepts.*   (c) *Graph each function.*

(d) *Based on the graph, find the range.*

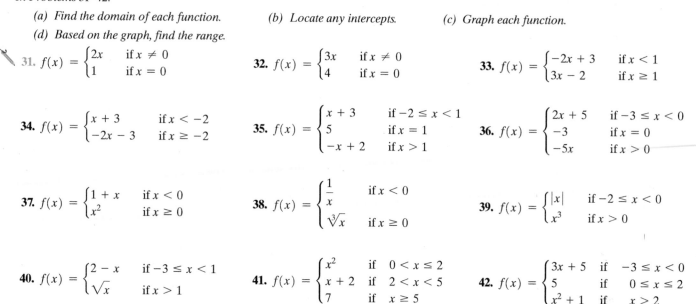

31. $f(x) = \begin{cases} 2x & \text{if } x \neq 0 \\ 1 & \text{if } x = 0 \end{cases}$

32. $f(x) = \begin{cases} 3x & \text{if } x \neq 0 \\ 4 & \text{if } x = 0 \end{cases}$

33. $f(x) = \begin{cases} -2x + 3 & \text{if } x < 1 \\ 3x - 2 & \text{if } x \geq 1 \end{cases}$

34. $f(x) = \begin{cases} x + 3 & \text{if } x < -2 \\ -2x - 3 & \text{if } x \geq -2 \end{cases}$

35. $f(x) = \begin{cases} x + 3 & \text{if } -2 \leq x < 1 \\ 5 & \text{if } x = 1 \\ -x + 2 & \text{if } x > 1 \end{cases}$

36. $f(x) = \begin{cases} 2x + 5 & \text{if } -3 \leq x < 0 \\ -3 & \text{if } x = 0 \\ -5x & \text{if } x > 0 \end{cases}$

37. $f(x) = \begin{cases} 1 + x & \text{if } x < 0 \\ x^2 & \text{if } x \geq 0 \end{cases}$

38. $f(x) = \begin{cases} \dfrac{1}{x} & \text{if } x < 0 \\ \sqrt[3]{x} & \text{if } x \geq 0 \end{cases}$

39. $f(x) = \begin{cases} |x| & \text{if } -2 \leq x < 0 \\ x^3 & \text{if } x > 0 \end{cases}$

40. $f(x) = \begin{cases} 2 - x & \text{if } -3 \leq x < 1 \\ \sqrt{x} & \text{if } x > 1 \end{cases}$

41. $f(x) = \begin{cases} x^2 & \text{if } 0 < x \leq 2 \\ x + 2 & \text{if } 2 < x < 5 \\ 7 & \text{if } x \geq 5 \end{cases}$

42. $f(x) = \begin{cases} 3x + 5 & \text{if } -3 \leq x < 0 \\ 5 & \text{if } 0 \leq x \leq 2 \\ x^2 + 1 & \text{if } x > 2 \end{cases}$

In Problems 43–46, the graph of a piecewise-defined function is given. Write a definition for each function.

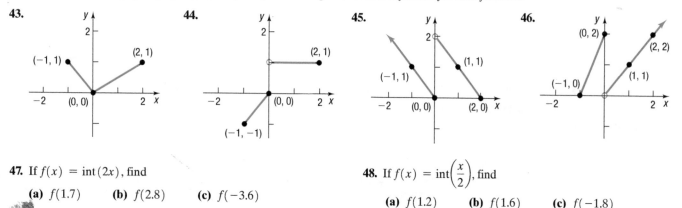

43.

44.

45.

46.

47. If $f(x) = \text{int}(2x)$, find

    (a) $f(1.7)$ (b) $f(2.8)$ (c) $f(-3.6)$

48. If $f(x) = \text{int}\left(\dfrac{x}{2}\right)$, find

    (a) $f(1.2)$ (b) $f(1.6)$ (c) $f(-1.8)$

Applications and Extensions

49. (a) Graph $f(x) = \begin{cases} (x - 1)^2 & \text{if } 0 \leq x < 2 \\ -2x + 10 & \text{if } 2 \leq x \leq 6 \end{cases}$

    (b) Find the domain of f.

    (c) Find the absolute maximum and the absolute minimum, if they exist.

50. (a) Graph $f(x) = \begin{cases} -x + 1 & \text{if } -2 \leq x < 0 \\ 2 & \text{if } x = 0 \\ x + 1 & \text{if } 0 < x \leq 2 \end{cases}$

    (b) Find the domain of f.

    (c) Find the absolute maximum and the absolute minimum, if they exist.

51. **Tablet Service** A monthly tablet plan costs $34.99. It includes 3 gigabytes of data and charges $15 per gigabyte for additional gigabytes. The following function is used to compute the monthly cost for a subscriber.

$$C(x) = \begin{cases} 34.99 & \text{if } 0 \leq x \leq 3 \\ 15x - 10.01 & \text{if } x > 3 \end{cases}$$

Compute the monthly cost for each of the following gigabytes of use.

(a) 2   (b) 5   (c) 13

52. **Parking at O'Hare International Airport** The short-term (no more than 24 hours) parking fee F (in dollars) for parking x hours on a weekday at O'Hare International Airport's main parking garage can be modeled by the function

$$F(x) = \begin{cases} 2 - 3\,\text{int}(1 - x) & \text{if } 0 < x \leq 4 \\ 2 - 9\,\text{int}(3 - x) & \text{if } 4 < x \leq 9 \\ 74 & \text{if } 9 < x \leq 24 \end{cases}$$

Determine the fee for parking in the short-term parking garage for

(a) 2 hours   (b) 7 hours

(c) 15 hours   (d) 8 hours and 24 minutes

Source: O'Hare International Airport

53. Cost of Natural Gas In March 2018, Spire, Inc. had the following rate schedule for natural gas usage in single-family residences.

Monthly service charge	$23.44
Delivery charge	
First 30 therms	$0.91686/therm
Over 30 therms	$0
Natural gas cost	
First 30 therms	$0.26486/therm
Over 30 therms	$0.50897/therm

(a) What is the charge for using 20 therms in a month?

(b) What is the charge for using 150 therms in a month?

(c) Develop a function that models the monthly charge C for x therms of gas.

(d) Graph the function found in part (c).

Source: Spire, Inc.

54. Cost of Natural Gas In April 2018, Nicor Gas had the following rate schedule for natural gas usage in small businesses.

Monthly customer charge	$90.00
Distribution charge	
1st 150 therms	$0.1201/therm
Next 4850 therms	$0.0549/therm
Over 5000 therms	$0.0482/therm
Gas supply charge	$0.35/therm

(a) What is the charge for using 1000 therms in a month?

(b) What is the charge for using 6000 therms in a month?

(c) Develop a function that models the monthly charge C for x therms of gas.

(d) Graph the function found in part (c).

Source: Nicor Gas

55. Federal Income Tax Two 2018 Tax Rate Schedules are given in the accompanying table. If x equals taxable income and y equals the tax due, construct a function $y = f(x)$ for Schedule X.

2018 Tax Rate Schedules										
Schedule X—Single						Schedule Y-1—Married Filing Jointly or Qualified Widow(er)				
If Taxable Income is Over	But Not Over	The Tax is This Amount		Plus This %	Of the Excess Over	If Taxable Income is Over	But Not Over	The Tax is This Amount		Plus This %	Of the Excess Over
$0	$9,525	$0	+	10%	$0	$0	$19,050	$0	+	10%	$0
9,525	38,700	952.50	+	12%	9,525	19,050	77,400	1,905	+	12%	19,050
38,700	82,500	4,453.50	+	22%	38,700	77,400	165,000	8,907.00	+	22%	77,400
82,500	157,500	14,089.50	+	24%	82,500	165,000	315,000	28,179.00	+	24%	165,000
157,500	200,000	32,089.50	+	32%	157,500	315,000	400,000	64,179.00	+	32%	315,000
200,000	500,000	45,689.50	+	35%	200,000	400,000	600,000	91,379.00	+	35%	400,000
500,000	–	150,689.50	+	37%	500,000	600,000	–	161,379.00	+	37%	600,000

56. Federal Income Tax Refer to the 2018 tax rate schedules. If x equals taxable income and y equals the tax due, construct a function $y = f(x)$ for Schedule Y-1.

57. Cost of Transporting Goods A trucking company transports goods between Chicago and New York, a distance of 960 miles. The company's policy is to charge, for each pound, $0.50 per mile for the first 100 miles, $0.40 per mile for the next 300 miles, $0.25 per mile for the next 400 miles, and no charge for the remaining 160 miles.

(a) Graph the relationship between the per-pound cost of transportation in dollars and mileage over the entire 960-mile route.

(b) Find the cost as a function of mileage for hauls between 100 and 400 miles from Chicago.

(c) Find the cost as a function of mileage for hauls between 400 and 800 miles from Chicago.

58. Car Rental Costs An economy car rented in Florida from Enterprise® on a weekly basis costs $185 per week. Extra days cost $37 per day until the day rate exceeds the weekly rate, in which case the weekly rate applies. Also, any part of a day used counts as a full day. Find the cost C of renting an economy car as a function of the number of days used x, where $7 \leq x \leq 14$. Graph this function.

59. Mortgage Fees Fannie Mae charges a loan-level price adjustment (LLPA) on all mortgages, which represents a fee homebuyers seeking a loan must pay. The rate paid depends on the credit score of the borrower, the amount borrowed, and the loan-to-value (LTV) ratio. The LTV ratio is the ratio of amount borrowed to appraised value of the home. For example, a homebuyer who wishes to borrow $250,000 with a credit score of 730 and an LTV ratio of 80% will pay 0.75% (0.0075) of $250,000, or $1875. The table shows the LLPA for various credit scores and an LTV ratio of 80%.

Credit Score	Loan-Level Price Adjustment Rate
≤ 659	3.00%
660–679	2.75%
680–699	1.75%
700–719	1.25%
720–739	0.75%
≥ 740	0.50%

Source: Fannie Mae

(a) Construct a function $C = C(s)$, where C is the loan-level price adjustment (LLPA) and s is the credit score of an individual who wishes to borrow $300,000 with an 80% LTV ratio.

(b) What is the LLPA on a $300,000 loan with an 80% LTV ratio for a borrower whose credit score is 725?

(c) What is the LLPA on a $300,000 loan with an 80% LTV ratio for a borrower whose credit score is 670?

60. Minimum Payments for Credit Cards Holders of credit cards issued by banks, department stores, oil companies, and so on, receive bills each month that state minimum amounts that must be paid by a certain due date. The minimum due depends on the total amount owed. One such credit card company uses the following rules: For a bill of less than $10, the entire amount is due. For a bill of at least $10 but less than $500, the minimum due is $10. A minimum of $30 is due on a bill of at least $500 but less than $1000, a minimum of $50 is due on a bill of at least $1000 but less than $1500, and a minimum of $70 is due on bills of $1500 or more. Find the function f that describes the minimum payment due on a bill of x dollars. Graph f.

61. Wind Chill The wind chill factor represents the air temperature at a standard wind speed that would produce the same heat loss as the given temperature and wind speed. One formula for computing the equivalent temperature is

$$W = \begin{cases} t & 0 \le v < 1.79 \\ 33 - \dfrac{(10.45 + 10\sqrt{v} - v)(33 - t)}{22.04} & 1.79 \le v \le 20 \\ 33 - 1.5958(33 - t) & v > 20 \end{cases}$$

where v represents the wind speed (in meters per second) and t represents the air temperature (°C). Compute the wind chill for the following:

(a) An air temperature of 10°C and a wind speed of 1 meter per second (m/sec)

(b) An air temperature of 10°C and a wind speed of 5 m/sec

(c) An air temperature of 10°C and a wind speed of 15 m/sec

(d) An air temperature of 10°C and a wind speed of 25 m/sec

(e) Explain the physical meaning of the equation corresponding to $0 \le v < 1.79$.

(f) Explain the physical meaning of the equation corresponding to $v > 20$.

62. Wind Chill Redo Problem 61(a)–(d) for an air temperature of -10°C.

63. First-class Mail In 2018 the U.S. Postal Service charged $1.00 postage for certain first-class mail retail flats (such as an 8.5″ by 11″ envelope) weighing up to 1 ounce, plus $0.21 for each additional ounce up to 13 ounces. First-class rates do not apply to flats weighing more than 13 ounces. Develop a model that relates C, the first-class postage charged, for a flat weighing x ounces. Graph the function.

Source: United States Postal Service

64. Challenge Problem Pool Depth Develop a model for the depth of the swimming pool shown below as a function of the distance from the wall on the left.

65. Challenge Problem Find the sum function $(f + g)(x)$ if

$$f(x) = \begin{cases} 2x + 3 & \text{if } x < 2 \\ x^2 + 5x & \text{if } x \ge 2 \end{cases}$$

and

$$g(x) = \begin{cases} -4x + 1 & \text{if } x \le 0 \\ x - 7 & \text{if } x > 0 \end{cases}$$

Explaining Concepts: Discussion and Writing

In Problems 66–73, use a graphing utility.

66. Exploration Graph $y = x^2$. Then on the same screen graph $y = x^2 + 2$, followed by $y = x^2 + 4$, followed by $y = x^2 - 2$. What pattern do you observe? Can you predict the graph of $y = x^2 - 4$? Of $y = x^2 + 5$?

67. Exploration Graph $y = x^2$. Then on the same screen graph $y = (x - 2)^2$, followed by $y = (x - 4)^2$, followed by $y = (x + 2)^2$. What pattern do you observe? Can you predict the graph of $y = (x + 4)^2$? Of $y = (x - 5)^2$?

68. Exploration Graph $y = |x|$. Then on the same screen graph $y = 2|x|$, followed by $y = 4|x|$, followed by $y = \dfrac{1}{2}|x|$. What pattern do you observe? Can you predict the graph of $y = \dfrac{1}{4}|x|$? Of $y = 5|x|$?

69. Exploration Graph $y = x^2$. Then on the same screen graph $y = -x^2$. Now try $y = |x|$ and $y = -|x|$. What do you conclude?

70. Exploration Graph $y = \sqrt{x}$. Then on the same screen graph $y = \sqrt{-x}$. Now try $y = 2x + 1$ and $y = 2(-x) + 1$. What do you conclude?

71. Exploration Graph $y = x^3$. Then on the same screen graph $y = (x - 1)^3 + 2$. Could you have predicted the result?

72. Exploration Graph $y = x^2$, $y = x^4$, and $y = x^6$ on the same screen. What do you notice is the same about each graph? What do you notice is different?

73. Exploration Graph $y = x^3$, $y = x^5$, and $y = x^7$ on the same screen. What do you notice is the same about each graph? What do you notice is different?

74. Consider the equation

$$y = \begin{cases} 1 & \text{if } x \text{ is rational} \\ 0 & \text{if } x \text{ is irrational} \end{cases}$$

Is this a function? What is its domain? What is its range? What is its y-intercept, if any? What are its x-intercepts, if any? Is it even, odd, or neither? How would you describe its graph?

75. Define some functions that pass through $(0, 0)$ and $(1, 1)$ and are increasing for $x \ge 0$. Begin your list with $y = \sqrt{x}$, $y = x$, and $y = x^2$. Can you propose a general result about such functions?

Retain Your Knowledge

Problems 76–85 are based on material learned earlier in the course. The purpose of these problems is to keep the material fresh in your mind so that you are better prepared for the final exam.

76. Simplify: $(x^{-3}y^5)^{-2}$

77. Find the center and radius of the circle $x^2 + y^2 = 6y + 16$.

78. Solve: $4x - 5(2x - 1) = 4 - 7(x + 1)$

79. Ethan has \$60,000 to invest. He puts part of the money in a CD that earns 3% simple interest per year and the rest in a mutual fund that earns 8% simple interest per year. How much did he invest in each if his earned interest the first year was \$3700?

80. Find the quotient and remainder when $x^3 + 3x^2 - 6$ is divided by $x + 2$.

81. What is the conjugate of $\dfrac{3}{2} - 2i$?

[This problem is based on content from Section 1.3, which is optional.]

82. Identify the leading term: $-5x^4 + 8x^2 - 2x^7$

83. Simplify: $\sqrt{(5t^2)^2 + (25t^{7/2})^2}$

84. Find the domain of $h(x) = \sqrt[4]{x + 7} + 7x$.

85. Factor: $3x^3y - 2x^2y^2 + 18x - 12y$

'Are You Prepared?' Answers

1.

2.

3. $(0, -8), (2, 0)$

3.5 Graphing Techniques: Transformations

OBJECTIVES
1. Graph Functions Using Vertical and Horizontal Shifts (p. 254)
2. Graph Functions Using Compressions and Stretches (p. 258)
3. Graph Functions Using Reflections about the *x*-Axis and the *y*-Axis (p. 260)

At this stage, you should be able to quickly graph any of the functions

$$y = x \qquad y = x^2 \qquad y = x^3 \qquad y = \sqrt{x} \qquad y = \sqrt[3]{x} \qquad y = \frac{1}{x} \qquad y = |x|$$

with key points plotted. If necessary, review the previous section, Figures 37 through 43.

Sometimes we are asked to graph a function that is "almost" like one that we already know how to graph. In this section, we develop techniques for graphing such functions. Collectively, these techniques are referred to as **transformations**.

1 Graph Functions Using Vertical and Horizontal Shifts

EXAMPLE 1   **Vertical Shift Up**

Use the graph of $f(x) = x^2$ to obtain the graph of $g(x) = x^2 + 3$. Find the domain and range of *g*.

Solution   The function $g(x) = x^2 + 3$ is basically a square function. In fact, notice that $g(x) = f(x) + 3$. Create a table of values to obtain some points on the graphs of *f* and *g*. For example, when $x = 0$, then $y = f(0) = 0$ and $y = g(0) = 3$. When $x = 1$, then $y = f(1) = 1$ and $y = g(1) = 4$. Table 8 lists these and a few other points on each graph. Notice that each *y*-coordinate of a point on the graph of *g* is 3 units larger than the *y*-coordinate of the corresponding point on the graph of *f*. We conclude that the graph of *g* is identical to that of *f*, except that it is shifted vertically up 3 units. See Figure 50.

Table 8

x	$y = f(x)$ $= x^2$	$y = g(x)$ $= x^2 + 3$
−2	4	7
−1	1	4
0	0	3
1	1	4
2	4	7

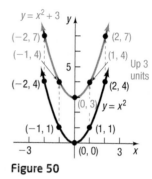

Figure 50

The domain of g is all real numbers, or $(-\infty, \infty)$. The range of g is $[3, \infty)$.

EXAMPLE 2 **Vertical Shift Down**

Use the graph of $f(x) = x^2$ to obtain the graph of $g(x) = x^2 - 4$. Find the domain and range of g.

Solution The function $g(x) = x^2 - 4$ is basically a square function. Table 9 lists some points on the graphs of f and g. Notice that each y-coordinate of g is 4 units less than the corresponding y-coordinate of f. Also notice that $g(x) = f(x) - 4$.

To obtain the graph of g from the graph of f, subtract 4 from each y-coordinate on the graph of f. The graph of g is identical to that of f except that it is shifted down 4 units. See Figure 51.

Table 9

x	$y = f(x)$ $= x^2$	$y = g(x)$ $= x^2 - 4$
−2	4	0
−1	1	−3
0	0	−4
1	1	−3
2	4	0

Figure 51

The domain of g is all real numbers, or $(-\infty, \infty)$. The range of g is $[-4, \infty)$.

A vertical shift affects only the range of a function, not the domain. For example, the range of $f(x) = x^2$ is $[0, \infty)$. In Example 1 the range of $g(x) = f(x) + 3$ is $[3, \infty)$, whereas in Example 2 the range of $g(x) = f(x) - 4$ is $[-4, \infty)$. The domain for all three functions is all real numbers.

Figure 52

Exploration On the same screen, graph each of the following functions:

$$Y_1 = x^2 \qquad Y_2 = x^2 + 2 \qquad Y_3 = x^2 - 2$$

Figure 52 illustrates the graphs using Desmos. You should have observed a general pattern. With $Y_1 = x^2$ on the screen, the graph of $Y_2 = x^2 + 2$ is identical to that of $Y_1 = x^2$, except that it is shifted vertically up 2 units. The graph of $Y_3 = x^2 - 2$ is identical to that of $Y_1 = x^2$, except that it is shifted vertically down 2 units.

In Words
For $y = f(x) + k, k > 0$, add k to each y-coordinate on the graph of $y = f(x)$ to shift the graph up k units.
For $y = f(x) - k, k > 0$, subtract k from each y-coordinate to shift the graph down k units.

We are led to the following conclusions:

Vertical Shifts

- If a positive real number k is added to the output of a function $y = f(x)$, the graph of the new function $y = f(x) + k$ is the graph of f **shifted vertically up** k units.
- If a positive real number k is subtracted from the output of a function $y = f(x)$, the graph of the new function $y = f(x) - k$ is the graph of f **shifted vertically down** k units.

━━━━━━**Now Work** PROBLEM 37

EXAMPLE 3 | **Horizontal Shift to the Right**

Use the graph of $f(x) = \sqrt{x}$ to obtain the graph of $g(x) = \sqrt{x - 2}$. Find the domain and range of g.

Solution The function $g(x) = \sqrt{x - 2}$ is basically a square root function. Table 10 lists some points on the graphs of f and g. Note that when $f(x) = 0$, then $x = 0$, but when $g(x) = 0$, then $x = 2$. Also, when $f(x) = 2$, then $x = 4$, but when $g(x) = 2$, then $x = 6$. The x-coordinates on the graph of g are 2 units larger than the corresponding x-coordinates on the graph of f for any given y-coordinate. Also notice that $g(x) = f(x - 2)$. We conclude that the graph of g is identical to that of f, except that it is shifted horizontally 2 units to the right. See Figure 53.

Table 10

	$y = f(x)$		$y = g(x)$
x	$= \sqrt{x}$	x	$= \sqrt{x - 2}$
0	0	2	0
1	1	3	1
4	2	6	2
9	3	11	3

Figure 53

The domain of g is $[2, \infty)$ and the range of g is $[0, \infty)$.

EXAMPLE 4 | **Horizontal Shift to the Left**

Use the graph of $f(x) = \sqrt{x}$ to obtain the graph of $g(x) = \sqrt{x + 4}$. Find the domain and range of g.

Solution The function $g(x) = \sqrt{x + 4}$ is basically a square root function. Table 11 lists some points on the graphs of f and g. Note that when $f(x) = 0$, then $x = 0$, but when $g(x) = 0$, then $x = -4$. Also, when $f(x) = 2$, then $x = 4$, but when $g(x) = 2$, then $x = 0$. The x-coordinates on the graph of g are 4 units smaller than the corresponding x-coordinates on the graph of f for any given y-coordinate. Also notice that $g(x) = f(x + 4)$. We conclude that the graph of g is identical to that of f, except that it is shifted horizontally 4 units to the left. See Figure 54.

Table 11

	$y = f(x)$		$y = g(x)$
x	$= \sqrt{x}$	x	$= \sqrt{x + 4}$
0	0	-4	0
1	1	-3	1
4	2	0	2
9	3	5	3

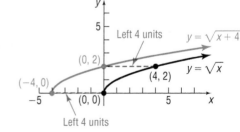

Figure 54

The domain of g is $[-4, \infty)$ and the range of g is $[0, \infty)$.

━━━━━━**Now Work** PROBLEM 41

A horizontal shift affects only the domain of a function, not the range. For example, the domain of $f(x) = \sqrt{x}$ is $[0, \infty)$. In Example 3 the domain of $g(x) = f(x - 2)$ is $[2, \infty)$, whereas in Example 4 the domain of $g(x) = f(x + 4)$ is $[-4, \infty)$. The range for all three functions is $[0, \infty)$.

Figure 55

Exploration On the same screen, graph each of the following functions:

$$Y_1 = x^2 \qquad Y_2 = (x - 3)^2 \qquad Y_3 = (x + 2)^2$$

Figure 55 illustrates the graphs using Desmos.

You should have observed the following pattern. With the graph of $Y_1 = x^2$ on the screen, the graph of $Y_2 = (x - 3)^2$ is identical to that of $Y_1 = x^2$, except that it is shifted horizontally to the right 3 units. The graph of $Y_3 = (x + 2)^2$ is identical to that of $Y_1 = x^2$, except that it is shifted horizontally to the left 2 units.

We are led to the following conclusions:

In Words

For $y = f(x - h)$, $h > 0$, add h to each x-coordinate on the graph of $y = f(x)$ to shift the graph right h units. For $y = f(x + h)$, $h > 0$, subtract h from each x-coordinate on the graph of $y = f(x)$ to shift the graph left h units.

Horizontal Shifts

- If the argument x of a function f is replaced by $x - h$, $h > 0$, the graph of the new function $y = f(x - h)$ is the graph of f **shifted horizontally right h units**.

- If the argument x of a function f is replaced by $x + h$, $h > 0$, the graph of the new function $y = f(x + h)$ is the graph of f **shifted horizontally left h units**.

Observe the distinction between vertical and horizontal shifts. The graph of $f(x) = x^3 + 2$ is obtained by shifting the graph vertically because we evaluate the cube function first and then add 2. The graph of $g(x) = (x + 2)^3$ is obtained by shifting the graph of $y = x^3$ horizontally because we add 2 to x before we evaluate the cube function.

Another way to think of the distinction is to note where the shift occurs. If the shift occurs *outside* the basic function, as is the case with $f(x) = x^3 + 2$, then there is a vertical shift. If the shift occurs *inside* the basic function, as is the case with $f(x) = (x + 2)^3$, then there is a horizontal shift.

The graph of a function can be moved anywhere in the coordinate plane by combining vertical and horizontal shifts.

EXAMPLE 5 **Combining Vertical and Horizontal Shifts**

Graph the function $f(x) = |x + 3| - 5$. Find the domain and range of f.

Solution   We graph f in steps. First, note that f is basically an absolute value function, so begin with the graph of $y = |x|$ as shown in Figure 56(a). Next, to get the graph of $y = |x + 3|$, shift the graph of $y = |x|$ horizontally 3 units to the left. See Figure 56(b). Finally, to get the graph of $y = |x + 3| - 5$, shift the graph of $y = |x + 3|$ vertically down 5 units. See Figure 56(c).

Figure 56

The domain of f is all real numbers, or $(-\infty, \infty)$. The range of f is $[-5, \infty)$.

Note the points plotted on each graph in Figure 56. Using key points can be helpful in keeping track of the transformations that have been made.

In Example 5, if the vertical shift had been done first, followed by the horizontal shift, the final graph would have been the same. Try it for yourself.

Now Work PROBLEMS 43 AND 67

2 Graph Functions Using Compressions and Stretches

EXAMPLE 6

Vertical Stretch

Use the graph of $f(x) = \sqrt{x}$ to obtain the graph of $g(x) = 2\sqrt{x}$.

Solution

To see the relationship between the graphs of f and g, we list points on each graph, as shown in Table 12. For each x, the y-coordinate of a point on the graph of g is 2 times as large as the corresponding y-coordinate on the graph of f. That is, $g(x) = 2f(x)$. The graph of $f(x) = \sqrt{x}$ is vertically stretched by a factor of 2 to obtain the graph of $g(x) = 2\sqrt{x}$. For example, $(1, 1)$ is on the graph of f, but $(1, 2)$ is on the graph of g. See Figure 57.

Table 12

x	$y = f(x)$ $= \sqrt{x}$	$y = g(x)$ $= 2\sqrt{x}$
0	0	0
1	1	2
4	2	4
9	3	6

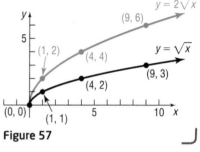

Figure 57

EXAMPLE 7

Vertical Compression

Use the graph of $f(x) = |x|$ to obtain the graph of $g(x) = \dfrac{1}{2}|x|$.

Solution

For each x, the y-coordinate of a point on the graph of g is $\dfrac{1}{2}$ as large as the corresponding y-coordinate on the graph of f. That is, $g(x) = \dfrac{1}{2}f(x)$. The graph of $f(x) = |x|$ is vertically compressed by a factor of $\dfrac{1}{2}$ to obtain the graph of $g(x) = \dfrac{1}{2}|x|$. For example, $(2, 2)$ is on the graph of f, but $(2, 1)$ is on the graph of g. See Table 13 and Figure 58.

Table 13

x	$y = f(x)$ $= |x|$	$y = g(x)$ $= \dfrac{1}{2}|x|$
-2	2	1
-1	1	$\dfrac{1}{2}$
0	0	0
1	1	$\dfrac{1}{2}$
2	2	1

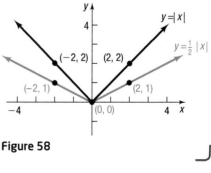

Figure 58

In Words
For $y = af(x)$, $a > 0$, the factor a is "outside" the function, so it affects the y-coordinates. Multiply each y-coordinate on the graph of $y = f(x)$ by a.

Vertical Compression or Stretch

If a function $y = f(x)$ is multiplied by a positive number a, then the graph of the new function $y = af(x)$ is obtained by multiplying each y-coordinate of the graph of $y = f(x)$ by a.

- If $0 < a < 1$, a **vertical compression by a factor of a** results.
- If $a > 1$, a **vertical stretch by a factor of a** results.

━━━━ **Now Work** PROBLEM 45

What happens if the argument x of a function $y = f(x)$ is multiplied by a positive number a, creating a new function $y = f(ax)$? To find the answer, look at the following Exploration.

Exploration On the same screen, graph each of the following functions:

$$Y_1 = f(x) = \sqrt{x} \qquad Y_2 = f(2x) = \sqrt{2x} \qquad Y_3 = f\left(\frac{1}{2}x\right) = \sqrt{\frac{1}{2}x} = \sqrt{\frac{x}{2}}$$

Create a table of values to explore the relation between the x- and y-coordinates of each function.

Result You should have obtained the graphs in Figure 59. Look at Table 14(a). Note that (1, 1), (4, 2), and (9, 3) are points on the graph of $Y_1 = \sqrt{x}$. Also, (0.5, 1), (2, 2), and (4.5, 3) are points on the graph of $Y_2 = \sqrt{2x}$. For a given y-coordinate, the x-coordinate on the graph of Y_2 is $\frac{1}{2}$ of the x-coordinate on Y_1.

Figure 59

Table 14

(a)

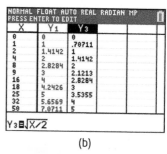

(b)

We conclude that the graph of $Y_2 = \sqrt{2x}$ is obtained by multiplying the x-coordinate of each point on the graph of $Y_1 = \sqrt{x}$ by $\frac{1}{2}$. The graph of $Y_2 = \sqrt{2x}$ is the graph of $Y_1 = \sqrt{x}$ *compressed* horizontally.

Look at Table 14(b). Notice that (1, 1), (4, 2), and (9, 3) are points on the graph of $Y_1 = \sqrt{x}$. Also notice that (2, 1), (8, 2), and (18, 3) are points on the graph of $Y_3 = \sqrt{\frac{x}{2}}$. For a given y-coordinate, the x-coordinate on the graph of Y_3 is 2 times the x-coordinate on Y_1. We conclude that the graph of $Y_3 = \sqrt{\frac{x}{2}}$ is obtained by multiplying the x-coordinate of each point on the graph of $Y_1 = \sqrt{x}$ by 2. The graph of $Y_3 = \sqrt{\frac{x}{2}}$ is the graph of $Y_1 = \sqrt{x}$ *stretched* horizontally.

Based on the Exploration, we have the following result:

In Words
For $y = f(ax)$, $a > 0$, the factor a is "inside" the function, so it affects the x-coordinates. Multiply each x-coordinate on the graph of $y = f(x)$ by $\frac{1}{a}$.

Horizontal Compression or Stretch

If the argument of a function $y = f(x)$ is multiplied by a positive number a, then the graph of the new function $y = f(ax)$ is obtained by multiplying each x-coordinate of the graph of $y = f(x)$ by $\frac{1}{a}$.

- If $a > 1$, a **horizontal compression by a factor of $\frac{1}{a}$** results.
- If $0 < a < 1$, a **horizontal stretch by a factor of $\frac{1}{a}$** results.

EXAMPLE 8 **Graphing Using Stretches and Compressions**

The graph of $y = f(x)$ is given in Figure 60. Use this graph to find the graphs of

(a) $y = 2f(x)$ (b) $y = f(3x)$

Solution (a) Because the 2 is "outside" the function f, the graph of $y = 2f(x)$ is obtained by multiplying each y-coordinate of $y = f(x)$ by 2. See Figure 61.

(b) Because the 3 is "inside" the function f, the graph of $y = f(3x)$ is obtained from the graph of $y = f(x)$ by multiplying each x-coordinate of $y = f(x)$ by $\frac{1}{3}$. See Figure 62.

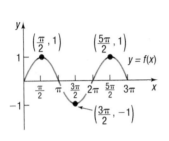

Figure 60 $y = f(x)$

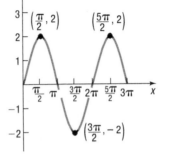

Figure 61 $y = 2f(x)$ Vertical stretch by a factor of 2

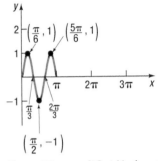

Figure 62 $y = f(3x)$ Horizontal compression by a factor of $\frac{1}{3}$

━━ **Now Work** PROBLEMS 61(e) AND (g)

3 Graph Functions Using Reflections about the *x*-Axis and the *y*-Axis

EXAMPLE 9 **Reflection about the *x*-Axis**

Graph the function $f(x) = -x^2$. Find the domain and range of f.

Solution Note that f is basically a square function, so begin with the graph of $y = x^2$, as shown in black in Figure 63. For each point (x, y) on the graph of $y = x^2$, the point $(x, -y)$ is on the graph of $y = -x^2$, as indicated in Table 15. Draw the graph of $y = -x^2$ by reflecting the graph of $y = x^2$ about the x-axis. See Figure 63.

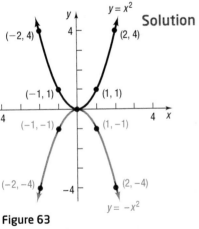

Figure 63

Table 15

x	$y = x^2$	$y = -x^2$
-2	4	-4
-1	1	-1
0	0	0
1	1	-1
2	4	-4

The domain of f is all real numbers, or $(-\infty, \infty)$. The range of f is $(-\infty, 0]$.

Reflection about the *x*-Axis

When a function f is multiplied by -1, the graph of the new function $y = -f(x)$ is the **reflection about the *x*-axis** of the graph of the function f.

━━ **Now Work** PROBLEM 47

| EXAMPLE 10 | **Reflection about the y-Axis** |

Graph the function $f(x) = \sqrt{-x}$. Find the domain and range of f.

Solution To graph $f(x) = \sqrt{-x}$, begin with the graph of $y = \sqrt{x}$, as shown in black in Figure 64. For each point (x, y) on the graph of $y = \sqrt{x}$, the point $(-x, y)$ is on the graph of $y = \sqrt{-x}$, as listed in Table 16. Obtain the graph of $y = \sqrt{-x}$ by reflecting the graph of $y = \sqrt{x}$ about the y-axis. See Figure 64.

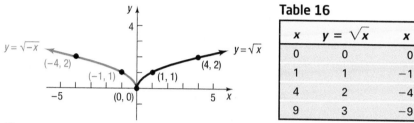

Table 16

x	$y = \sqrt{x}$	x	$y = \sqrt{-x}$
0	0	0	0
1	1	−1	1
4	2	−4	2
9	3	−9	3

Figure 64

The domain of f is $(-\infty, 0]$. The range of f is the set of all nonnegative real numbers, or $[0, \infty)$.

In Words

For $y = -f(x)$, multiply each y-coordinate on the graph of $y = f(x)$ by −1.
For $y = f(-x)$, multiply each x-coordinate by −1.

Reflection about the y-Axis

When the graph of the function f is known, the graph of the new function $y = f(-x)$ is the **reflection about the y-axis** of the graph of the function f.

SUMMARY OF GRAPHING TECHNIQUES

Each graphing technique has a different effect on the graph of a function. Compressions and stretches change the proportions of a graph, and reflections change the orientation, but not its proportions. Vertical and horizontal shifts change the location of the graph, without changing its proportions or orientation.

To Graph:	Draw the Graph of f and:	Functional Change to $f(x)$
Vertical shifts		
$y = f(x) + k, \quad k > 0$	Shift the graph of f up k units.	Add k to $f(x)$.
$y = f(x) - k, \quad k > 0$	Shift the graph of f down k units.	Subtract k from $f(x)$.
Horizontal shifts		
$y = f(x + h), \quad h > 0$	Shift the graph of f to the left h units.	Replace x by $x + h$.
$y = f(x - h), \quad h > 0$	Shift the graph of f to the right h units.	Replace x by $x - h$.
Compressing or stretching		
$y = af(x), \quad a > 0$	Multiply each y-coordinate of $y = f(x)$ by a. Stretch the graph of f vertically if $a > 1$. Compress the graph of f vertically if $0 < a < 1$.	Multiply $f(x)$ by a.
$y = f(ax), \quad a > 0$	Multiply each x-coordinate of $y = f(x)$ by $\dfrac{1}{a}$. Stretch the graph of f horizontally if $0 < a < 1$. Compress the graph of f horizontally if $a > 1$.	Replace x by ax.
Reflection about the x-axis		
$y = -f(x)$	Reflect the graph of f about the x-axis.	Multiply $f(x)$ by −1.
Reflection about the y-axis		
$y = f(-x)$	Reflect the graph of f about the y-axis.	Replace x by $-x$.

EXAMPLE 11 **Determining the Function Obtained from a Series of Transformations**

Find the function that is finally graphed after the following sequence of transformations are applied to the graph of $y = |x|$.

1. Shift left 2 units
2. Shift up 3 units
3. Reflect about the y-axis

Solution

1. Shift left 2 units: Replace x by $x + 2$. $y = |x + 2|$
2. Shift up 3 units: Add 3. $y = |x + 2| + 3$
3. Reflect about the y-axis: Replace x by $-x$. $y = |-x + 2| + 3$

———**Now Work** PROBLEM 29

EXAMPLE 12 **Using Graphing Techniques**

Graph the function $f(x) = \dfrac{3}{x - 2} + 1$. Find the domain and range of f.

Solution

It is helpful to write f as $f(x) = 3 \cdot \dfrac{1}{x - 2} + 1$ Now use the following steps to obtain the graph of f.

HINT: Although the order in which transformations are performed can be altered, consider using the following order for consistency:
1. Horizontal shift
2. Reflections
3. Compressions and stretches
4. Vertical shift

STEP 1: $y = \dfrac{1}{x}$ 　　　　　　　　　　**Reciprocal function**

STEP 2: $y = \dfrac{1}{x - 2}$ 　　　　　　**Replace x by $x - 2$; horizontal shift to the right 2 units.**

STEP 3: $y = 3 \cdot \dfrac{1}{x - 2} = \dfrac{3}{x - 2}$ 　　**Multiply by 3; vertical stretch by a factor of 3.**

STEP 4: $y = \dfrac{3}{x - 2} + 1$ 　　　　　**Add 1; vertical shift up 1 unit.**

See Figure 65.

Figure 65

(a) $y = \dfrac{1}{x}$ Replace x by $x - 2$; Horizontal shift right 2 units

(b) $y = \dfrac{1}{x-2}$ Multiply by 3; Vertical stretch

(c) $y = \dfrac{3}{x-2}$ Add 1; Vertical shift up 1 unit

(d) $y = \dfrac{3}{x-2} + 1$

The domain of $y = \dfrac{1}{x}$ is $\{x | x \neq 0\}$ and its range is $\{y | y \neq 0\}$. Because we shifted right 2 units and up 1 unit to obtain f, the domain of f is $\{x | x \neq 2\}$ and its range is $\{y | y \neq 1\}$.

Other orderings of the steps shown in Example 12 would also result in the graph of f. For example, try this one:

STEP 1: $y = \dfrac{1}{x}$ Reciprocal function

STEP 2: $y = 3 \cdot \dfrac{1}{x} = \dfrac{3}{x}$ Multiply by 3; vertical stretch by a factor of 3.

STEP 3: $y = \dfrac{3}{x - 2}$ Replace x by $x - 2$; horizontal shift to the right 2 units.

STEP 4: $y = \dfrac{3}{x - 2} + 1$ Add 1; vertical shift up 1 unit.

EXAMPLE 13   **Using Graphing Techniques**

Graph the function $f(x) = \sqrt{1 - x} + 2$. Find the domain and range of f.

Solution   Begin by rewriting $f(x)$ as $f(x) = \sqrt{1 - x} + 2 = \sqrt{-x + 1} + 2$. Now use the following steps.

STEP 1: $y = \sqrt{x}$ Square root function

STEP 2: $y = \sqrt{x + 1}$ Replace x by $x + 1$; horizontal shift to the left 1 unit.

STEP 3: $y = \sqrt{-x + 1} = \sqrt{1 - x}$ Replace x by $-x$; reflect about the y-axis.

STEP 4: $y = \sqrt{1 - x} + 2$ Add 2; vertical shift up 2 units.

See Figure 66.

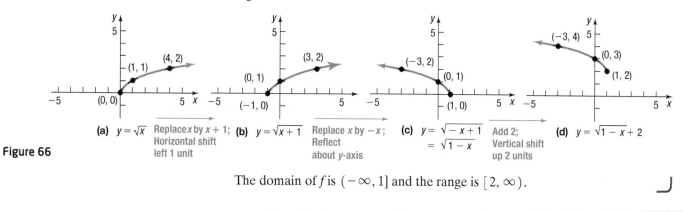

Figure 66

(a) $y = \sqrt{x}$ Replace x by $x + 1$; Horizontal shift left 1 unit   (b) $y = \sqrt{x + 1}$ Replace x by $-x$; Reflect about y-axis   (c) $y = \sqrt{-x + 1} = \sqrt{1 - x}$ Add 2; Vertical shift up 2 units   (d) $y = \sqrt{1 - x} + 2$

The domain of f is $(-\infty, 1]$ and the range is $[2, \infty)$.

— **Now Work** PROBLEM 53

3.5 Assess Your Understanding

Concepts and Vocabulary

1. Suppose the graph of a function f is known. Then the graph of $y = f(x - 2)$ is obtained by a _____ shift of the graph of f to the _____ a distance of 2 units.

2. Suppose the graph of a function f is known. Then the graph of $y = f(-x)$ is a reflection about the _____-axis of the graph of the function $y = f(x)$.

3. *True or False* The graph of $y = \dfrac{1}{3}g(x)$ is the graph of $y = g(x)$ vertically stretched by a factor of 3.

4. *True or False* The graph of $y = -f(x)$ is the reflection about the x-axis of the graph of $y = f(x)$.

5. *Multiple Choice* Which function has a graph that is the graph of $y = \sqrt{x}$ shifted down 3 units?

   (a) $y = \sqrt{x + 3}$ **(b)** $y = \sqrt{x - 3}$

   (c) $y = \sqrt{x} + 3$ **(d)** $y = \sqrt{x} - 3$

6. *Multiple Choice* Which function has a graph that is the graph of $y = f(x)$ horizontally stretched by a factor of 4?

   (a) $y = f(4x)$ **(b)** $y = f\left(\dfrac{1}{4}x\right)$

   (c) $y = 4f(x)$ **(d)** $y = \dfrac{1}{4}f(x)$

Skill Building

In Problems 7–18, match each graph to one of the following functions:

A. $y = x^2 + 2$

B. $y = -x^2 + 2$

C. $y = |x| + 2$

D. $y = -|x| + 2$

E. $y = (x - 2)^2$

F. $y = -(x + 2)^2$

G. $y = |x - 2|$

H. $y = -|x + 2|$

I. $y = 2x^2$

J. $y = -2x^2$

K. $y = 2|x|$

L. $y = -2|x|$

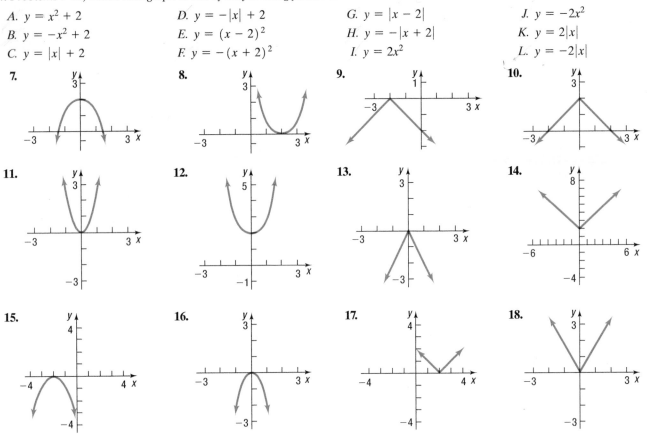

In Problems 19–28, write the function whose graph is the graph of $y = x^3$, but is:

19. Shifted to the right 4 units

20. Shifted to the left 4 units

21. Shifted up 4 units

22. Shifted down 4 units

23. Reflected about the y-axis

24. Reflected about the x-axis

25. Vertically stretched by a factor of 5

26. Horizontally stretched by a factor of 4

27. Horizontally compressed by a factor of $\dfrac{1}{2}$

28. Vertically compressed by a factor of $\dfrac{1}{4}$

In Problems 29–32, find the function that is finally graphed after each of the following transformations is applied to the graph of $y = \sqrt{x}$ in the order stated.

29. (1) Shift up 2 units
    (2) Reflect about the x-axis
    (3) Reflect about the y-axis

30. (1) Reflect about the x-axis
    (2) Shift right 3 units
    (3) Shift down 2 units

31. (1) Vertical stretch by a factor of 3
    (2) Shift up 4 units
    (3) Shift left 5 units

32. (1) Shift up 2 units
    (2) Reflect about the y-axis
    (3) Shift left 3 units

33. If $(3, 6)$ is a point on the graph of $y = f(x)$, which of the following points must be on the graph of $y = -f(x)$?
    (a) $(6, 3)$
    (b) $(6, -3)$
    (c) $(3, -6)$
    (d) $(-3, 6)$

34. If $(3, 6)$ is a point on the graph of $y = f(x)$, which of the following points must be on the graph of $y = f(-x)$?
    (a) $(6, 3)$
    (b) $(6, -3)$
    (c) $(3, -6)$
    (d) $(-3, 6)$

35. If $(1, 3)$ is a point on the graph of $y = f(x)$, which of the following points must be on the graph of $y = 2f(x)$?
    (a) $\left(1, \dfrac{3}{2}\right)$
    (b) $(2, 3)$
    (c) $(1, 6)$
    (d) $\left(\dfrac{1}{2}, 3\right)$

36. If $(4, 2)$ is a point on the graph of $y = f(x)$, which of the following points must be on the graph of $y = f(2x)$?
    (a) $(4, 1)$
    (b) $(8, 2)$
    (c) $(2, 2)$
    (d) $(4, 4)$

In Problems 37–60, graph each function using the techniques of shifting, compressing, stretching, and/or reflecting. Start with the graph of the basic function (for example, $y = x^2$) and show all the steps. Be sure to show at least three key points. Find the domain and the range of each function.

37. $f(x) = x^2 - 1$

38. $f(x) = x^2 + 4$

39. $g(x) = \sqrt{3x}$

40. $g(x) = \sqrt[3]{\frac{1}{2}x}$

41. $h(x) = \sqrt{x + 2}$

42. $h(x) = \sqrt{x + 1}$

43. $f(x) = (x - 1)^3 + 2$

44. $f(x) = (x + 2)^3 - 3$

45. $g(x) = 4\sqrt{x}$

46. $g(x) = \frac{1}{2}\sqrt{x}$

47. $f(x) = -\sqrt[3]{x}$

48. $f(x) = -\sqrt{x}$

49. $f(x) = 2(x + 1)^2 - 3$

50. $f(x) = 3(x - 2)^2 + 1$

51. $g(x) = 2\sqrt{x - 2} + 1$

52. $g(x) = 3|x + 1| - 3$

53. $h(x) = \sqrt{-x} - 2$

54. $h(x) = \frac{4}{x} + 2$

55. $f(x) = -(x + 1)^3 - 1$

56. $f(x) = -4\sqrt{x - 1}$

57. $g(x) = 2|1 - x|$

58. $g(x) = 4\sqrt{2 - x}$

59. $h(x) = \frac{1}{2x}$

60. $h(x) = \sqrt[3]{x - 1} + 3$

In Problems 61–64, the graph of a function f is illustrated. Use the graph of f as the first step toward graphing each of the following functions:

(a) $F(x) = f(x) + 3$

(b) $G(x) = f(x + 2)$

(c) $P(x) = -f(x)$

(d) $H(x) = f(x + 1) - 2$

(e) $Q(x) = \frac{1}{2}f(x)$

(f) $g(x) = f(-x)$

(g) $h(x) = f(2x)$

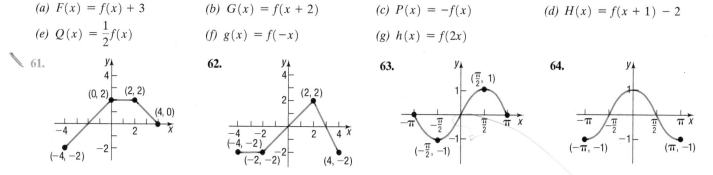

61.

62.

63.

64.

Mixed Practice In Problems 65–72, complete the square of each quadratic expression. Then graph each function using graphing techniques. (If necessary, refer to Chapter R, Section R.5 to review completing the square.)

65. $f(x) = x^2 + 2x$

66. $f(x) = x^2 - 6x$

67. $f(x) = x^2 - 8x + 1$

68. $f(x) = x^2 + 4x + 2$

69. $f(x) = 2x^2 - 12x + 19$

70. $f(x) = 3x^2 + 6x + 1$

71. $f(x) = -3x^2 - 12x - 17$

72. $f(x) = -2x^2 - 12x - 13$

Applications and Extensions

73. Suppose that the x-intercepts of the graph of $y = f(x)$ are -5 and 3.
   (a) What are the x-intercepts of the graph of $y = f(x + 2)$?
   (b) What are the x-intercepts of the graph of $y = f(x - 2)$?
   (c) What are the x-intercepts of the graph of $y = 4f(x)$?
   (d) What are the x-intercepts of the graph of $y = f(-x)$?

74. Suppose that the x-intercepts of the graph of $y = f(x)$ are -8 and 1.
   (a) What are the x-intercepts of the graph of $y = f(x + 4)$?
   (b) What are the x-intercepts of the graph of $y = f(x - 3)$?
   (c) What are the x-intercepts of the graph of $y = 2f(x)$?
   (d) What are the x-intercepts of the graph of $y = f(-x)$?

75. Suppose that the function $y = f(x)$ is increasing on the interval $[-1, 5]$.
   (a) Over what interval is the graph of $y = f(x + 2)$ increasing?
   (b) Over what interval is the graph of $y = f(x - 5)$ increasing?
   (c) Is the graph of $y = -f(x)$ increasing, decreasing, or neither on the interval $[-1, 5]$?
   (d) Is the graph of $y = f(-x)$ increasing, decreasing, or neither on the interval $[-5, 1]$?

76. Suppose that the function $y = f(x)$ is decreasing on the interval $[-2, 7]$.
   (a) Over what interval is the graph of $y = f(x + 2)$ decreasing?
   (b) Over what interval is the graph of $y = f(x - 5)$ decreasing?
   (c) Is the graph of $y = -f(x)$ increasing, decreasing, or neither on the interval $[-2, 7]$?
   (d) Is the graph of $y = f(-x)$ increasing, decreasing, or neither on the interval $[-7, 2]$?

77. The graph of a function f is illustrated in the figure.
   (a) Graph $y = |f(x)|$.
   (b) Graph $y = f(|x|)$.

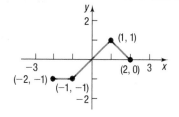

78. The graph of a function f is illustrated in the figure.

　(a) Graph $y = |f(x)|$.

　(b) Graph $y = f(|x|)$.

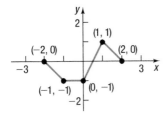

79. Suppose $(1, 3)$ is a point on the graph of $y = f(x)$.

　(a) What point is on the graph of $y = f(x + 3) - 5$?

　(b) What point is on the graph of $y = -2f(x - 2) + 1$?

　(c) What point is on the graph of $y = f(2x + 3)$?

80. Suppose $(-3, 5)$ is a point on the graph of $y = g(x)$.

　(a) What point is on the graph of $y = g(x + 1) - 3$?

　(b) What point is on the graph of $y = -3g(x - 4) + 3$?

　(c) What point is on the graph of $y = g(3x + 9)$?

81. Graph the following functions using transformations.

　(a) $f(x) = \text{int}(-x)$　　　　(b) $g(x) = -\text{int}(x)$

82. Graph the following functions using transformations

　(a) $f(x) = \text{int}(x - 1)$　　　　(b) $g(x) = \text{int}(1 - x)$

83. (a) Graph $f(x) = |x - 3| - 3$ using transformations.

　(b) Find the area of the region that is bounded by f and the x-axis and lies below the x-axis.

84. (a) Graph $f(x) = -2|x - 4| + 4$ using transformations.

　(b) Find the area of the region that is bounded by f and the x-axis and lies above the x-axis.

85. Thermostat Control Energy conservation experts estimate that homeowners can save 5% to 10% on winter heating bills by programming their thermostats 5 to 10 degrees lower while sleeping. In the graph below, the temperature T (in degrees Fahrenheit) of a home is given as a function of time t (in hours after midnight) over a 24-hour period.

　(a) At what temperature is the thermostat set during daytime hours? At what temperature is the thermostat set overnight?

　(b) The homeowner reprograms the thermostat to $y = T(t) - 2$. Explain how this affects the temperature in the house. Graph this new function.

　(c) The homeowner reprograms the thermostat to $y = T(t + 1)$. Explain how this affects the temperature in the house. Graph this new function.

　Source: Roger Albright, 547 Ways to Be Fuel Smart, 2000

86. Digital Music Revenues The total worldwide digital music revenues R, in billions of dollars, for the years 2012 through 2017 can be modeled by the function

$$R(x) = 0.15x^2 - 0.03x + 5.46$$

where x is the number of years after 2012.

　(a) Find $R(0)$, $R(3)$, and $R(5)$ and explain what each value represents.

　(b) Find $r(x) = R(x - 2)$.

　(c) Find $r(2)$, $r(5)$ and $r(7)$ and explain what each value represents.

　(d) In the model $r = r(x)$, what does x represent?

　(e) Would there be an advantage in using the model r when estimating the projected revenues for a given year instead of the model R?

　Source: IFPI Digital Music Report 2017

87. Temperature Measurements The relationship between the Celsius (°C) and Fahrenheit (°F) scales for measuring temperature is given by the equation

$$F = \frac{9}{5}C + 32$$

The relationship between the Celsius (°C) and Kelvin (K) scales is $K = C + 273$. Graph the equation $F = \frac{9}{5}C + 32$ using degrees Fahrenheit on the y-axis and degrees Celsius on the x-axis. Use the techniques introduced in this section to obtain the graph showing the relationship between Kelvin and Fahrenheit temperatures.

88. Period of a Pendulum The period T (in seconds) of a simple pendulum is a function of its length l (in feet) defined by the equation

$$T = 2\pi\sqrt{\frac{l}{g}}$$

where $g \approx 32.2$ feet per second per second is the acceleration due to gravity.

　(a) Use a graphing utility to graph the function $T = T(l)$.

　(b) Now graph the functions $T = T(l + 1)$, $T = T(l + 2)$, and $T = T(l + 3)$.

　(c) Discuss how adding to the length l changes the period T.

　(d) Now graph the functions $T = T(2l)$, $T = T(3l)$, and $T = T(4l)$.

　(e) Discuss how multiplying the length l by factors of 2, 3, and 4 changes the period T.

89. The equation $y = (x - c)^2$ defines a *family of parabolas*, one parabola for each value of c. On one set of coordinate axes, graph the members of the family for $c = 0$, $c = 3$, and $c = -2$.

90. Repeat Problem 89 for the family of parabolas $y = x^2 + c$.

91. *Challenge Problem* If a function f is increasing on the intervals $[-3, 3]$ and $[11, 19]$ and decreasing on the interval $[3, 11]$, determine the interval(s) on which $g(x) = -3f(2x - 5)$ is increasing.

92. Challenge Problem In statistics, the standard normal density function is given by $f(x) = \dfrac{1}{\sqrt{2\pi}} \cdot \exp\left[-\dfrac{x^2}{2}\right]$. This function can be transformed to describe any general normal distribution with mean, μ, and standard deviation, σ. A general normal density function is given by $f(x) = \dfrac{1}{\sqrt{2\pi} \cdot \sigma} \cdot \exp\left[-\dfrac{(x - \mu)^2}{2\sigma^2}\right]$. Describe the transformations needed to get from the graph of the standard normal function to the graph of a general normal function.

Explaining Concepts: Discussion and Writing

93. Suppose that the graph of a function f is known. Explain how the graph of $y = 4f(x)$ differs from the graph of $y = f(4x)$.

94. Suppose that the graph of a function f is known. Explain how the graph of $y = f(x) - 2$ differs from the graph of $y = f(x - 2)$.

95. The area under the curve $y = \sqrt{x}$ bounded from below by the x-axis and on the right by $x = 4$ is $\dfrac{16}{3}$ square units. Using the ideas presented in this section, what do you think is the area under the curve of $y = \sqrt{-x}$ bounded from below by the x-axis and on the left by $x = -4$? Justify your answer.

96. Explain how the range of the function $f(x) = x^2$ compares to the range of $g(x) = f(x) + k$.

97. Explain how the domain of $g(x) = \sqrt{x}$ compares to the domain of $g(x - k)$, where $k \geq 0$.

Retain Your Knowledge

Problems 98–106 are based on material learned earlier in the course. The purpose of these problems is to keep the material fresh in your mind so that you are better prepared for the final exam.

98. Find the slope and y-intercept of the line
$$3x - 5y = 30$$

99. Angie runs 7 mph for the first half of a marathon, 13.1 miles, but twists her knee and must walk 2 mph for the second half. What was her average speed? Round to 2 decimal places.

100. The amount of water used when taking a shower varies directly with the number of minutes the shower is run. If a 4-minute shower uses 7 gallons of water, how much water is used in a 9-minute shower?

101. Find the intercepts and test for symmetry: $y^2 = x + 4$

102. Find the domain of $h(x) = \dfrac{x + 2}{x^2 - 5x - 14}$.

103. **Projectile Motion** A ball is thrown upward from the top of a building. Its height h, in feet, after t seconds is given by the equation $h = -16t^2 + 96t + 200$. How long will it take for the ball to be 88 ft above the ground?

104. Simplify $\sqrt[3]{16x^5y^6z}$.

105. Find the difference quotient of $f(x) = 3x^2 + 2x - 1$.

106. Factor $z^3 + 216$.

3.6 Mathematical Models: Building Functions

OBJECTIVE 1 Build and Analyze Functions (p. 267)

 1 Build and Analyze Functions

Real-world problems often result in mathematical models that involve functions. These functions need to be constructed or built based on the information given. In building functions, we must be able to translate the verbal description into the language of mathematics. This is done by assigning symbols to represent the independent and dependent variables and then by finding the function or rule that relates these variables.

EXAMPLE 1   **Finding the Distance from the Origin to a Point on a Graph**

Let $P = (x, y)$ be a point on the graph of $y = x^2 - 1$.

(a) Graph f. Express the distance d from P to the origin O as a function of x.

(b) What is d if $x = 0$?   (c) What is d if $x = 1$?   (d) What is d if $x = \dfrac{\sqrt{2}}{2}$?

(e) Use a graphing utility to graph the function $d = d(x), x \geq 0$. Rounding to two decimal places, find the value(s) of x at which d has a local minimum. [This gives the point(s) on the graph of $y = x^2 - 1$ closest to the origin.]

Solution

(a) Figure 67 illustrates the graph of $y = x^2 - 1$. The distance d from P to O is

$$d = \sqrt{(x - 0)^2 + (y - 0)^2} = \sqrt{x^2 + y^2}$$

Since P is a point on the graph of $y = x^2 - 1$, substitute $x^2 - 1$ for y. Then

$$d(x) = \sqrt{x^2 + (x^2 - 1)^2} = \sqrt{x^4 - x^2 + 1}$$

The distance d is expressed as a function of x.

(b) If $x = 0$, the distance d is

$$d(0) = \sqrt{0^4 - 0^2 + 1} = \sqrt{1} = 1$$

(c) If $x = 1$, the distance d is

$$d(1) = \sqrt{1^4 - 1^2 + 1} = 1$$

(d) If $x = \dfrac{\sqrt{2}}{2}$, the distance d is

$$d\left(\frac{\sqrt{2}}{2}\right) = \sqrt{\left(\frac{\sqrt{2}}{2}\right)^4 - \left(\frac{\sqrt{2}}{2}\right)^2 + 1} = \sqrt{\frac{1}{4} - \frac{1}{2} + 1} = \frac{\sqrt{3}}{2}$$

(0, 0) d $P = (x, y)$

Figure 67 $y = x^2 - 1$

Figure 68
$d(x) = \sqrt{x^4 - x^2 + 1}$

(e) Figure 68 shows the graph of $Y_1 = \sqrt{x^4 - x^2 + 1}$. Using the MINIMUM feature on a TI-84 Plus C, we find that when $x \approx 0.71$ the value of d is smallest. The local minimum value is $d \approx 0.87$ rounded to two decimal places. Since $d(x)$ is even, it follows by symmetry that when $x \approx -0.71$, the value of d is the same local minimum value. Since $(\pm 0.71)^2 - 1 \approx -0.50$, the points $(-0.71, -0.50)$ and $(0.71, -0.50)$ on the graph of $y = x^2 - 1$ are closest to the origin.

Now Work PROBLEM 1

EXAMPLE 2

Area of a Rectangle

A rectangle has one corner in quadrant I on the graph of $y = 25 - x^2$, another at the origin, a third on the positive y-axis, and the fourth on the positive x-axis. See Figure 69.

(a) Express the area A of the rectangle as a function of x.

(b) What is the domain of A?

(c) Graph $A = A(x)$.

(d) For what value of x is the area A largest?

(x, y)
$y = 25 - x^2$
$(0, 0)$

Figure 69

Solution

(a) The area A of the rectangle is $A = xy$, where $y = 25 - x^2$. Substituting this expression for y, we obtain $A(x) = x(25 - x^2) = 25x - x^3$.

(b) Since (x, y) is in quadrant I, we have $x > 0$. Also, $y = 25 - x^2 > 0$, which implies that $x^2 < 25$, so $-5 < x < 5$. Combining these restrictions, the domain of A is $\{x \mid 0 < x < 5\}$, or $(0, 5)$ using interval notation.

(c) See Figure 70 for the graph of $A = A(x)$ on a TI-84 Plus C.

(d) Using **MAXIMUM**, we find that the maximum area is 48.11 square units at $x = 2.89$ units, each rounded to two decimal places. See Figure 71.

Figure 70 $A(x) = 25x - x^3$ **Figure 71**

➤ **Now Work** PROBLEM 7

EXAMPLE 3   **Close Call?**

Suppose two planes flying at the same altitude are headed toward each other. One plane is flying due south at a groundspeed of 400 miles per hour and is 600 miles from the potential intersection point of the planes. The other plane is flying due west with a groundspeed of 250 miles per hour and is 400 miles from the potential intersection point of the planes. See Figure 72.

(a) Build a model that expresses the distance d between the planes as a function of time t.

(b) Use a graphing utility to graph $d = d(t)$. How close do the planes come to each other? At what time are the planes closest?

Solution

(a) Refer to Figure 72. The distance d between the two planes is the hypotenuse of a right triangle. At any time t, the length of the north/south leg of the triangle is $600 - 400t$. At any time t, the length of the east/west leg of the triangle is $400 - 250t$. Use the Pythagorean Theorem to find that the square of the distance between the two planes is

$$d^2 = (600 - 400t)^2 + (400 - 250t)^2$$

Therefore, the distance between the two planes as a function of time is given by the model

$$d(t) = \sqrt{(600 - 400t)^2 + (400 - 250t)^2}$$

Figure 72

(b) Figure 73(a) shows the graph of $d = d(t)$ on a TI-84 Plus C. Using **MINIMUM**, the minimum distance between the planes is 21.20 miles, and the time at which the planes are closest is after 1.53 hours, each rounded to two decimal places. See Figure 73(b).

(a)   (b)

Figure 73

➤ **Now Work** PROBLEM 19

3.6 Assess Your Understanding

Applications and Extensions

1. Let $P = (x, y)$ be a point on the graph of $y = x^2 - 8$.
   (a) Express the distance d from P to the origin as a function of x.
   (b) What is d if $x = 0$?
   (c) What is d if $x = 1$?
   (d) Use a graphing utility to graph $d = d(x)$.
   (e) For what values of x is d smallest?

2. Let $P = (x, y)$ be a point on the graph of $y = x^2 - 8$.
   (a) Express the distance d from P to the point $(0, -1)$ as a function of x.
   (b) What is d if $x = 0$?
   (c) What is d if $x = -1$?
   (d) Use a graphing utility to graph $d = d(x)$.
   (e) For what values of x is d smallest?

3. Let $P = (x, y)$ be a point on the graph of $y = \sqrt{x}$.
   (a) Express the distance d from P to the point $(1, 0)$ as a function of x.
   (b) Use a graphing utility to graph $d = d(x)$.
   (c) For what values of x is d smallest?
   (d) What is the smallest distance?

4. Let $P = (x, y)$ be a point on the graph of $y = \dfrac{1}{x}$.
   (a) Express the distance d from P to the origin as a function of x.
   (b) Use a graphing utility to graph $d = d(x)$.
   (c) For what values of x is d smallest?
   (d) What is the smallest distance?

5. A right triangle has one vertex on the graph of $y = x^3, x > 0$, at (x, y), another at the origin, and the third on the positive y-axis at $(0, y)$, as shown in the figure. Express the area A of the triangle as a function of x.

6. A right triangle has one vertex on the graph of $y = 9 - x^2, x > 0$, at (x, y), another at the origin, and the third on the positive x-axis at $(x, 0)$. Express the area A of the triangle as a function of x.

7. A rectangle has one corner in quadrant I on the graph of $y = 16 - x^2$, another at the origin, a third on the positive y-axis, and the fourth on the positive x-axis. See the figure.
   (a) Express the area A of the rectangle as a function of x.

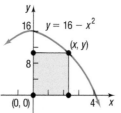

   (b) What is the domain of A?
   (c) Graph $A = A(x)$. For what value of x is A largest?
   (d) What is the largest area?

8. A rectangle is inscribed in a semicircle of radius 2. See the figure. Let $P = (x, y)$ be the point in quadrant I that is a vertex of the rectangle and is on the circle.

   (a) Express the area A of the rectangle as a function of x.
   (b) Express the perimeter p of the rectangle as a function of x.
   (c) Graph $A = A(x)$. For what value of x is A largest?
   (d) Graph $p = p(x)$. For what value of x is p largest?
   (e) What is the largest area? What is the largest perimeter?

9. A rectangle is inscribed in a circle of radius 2. See the figure. Let $P = (x, y)$ be the point in quadrant I that is a vertex of the rectangle and is on the circle.

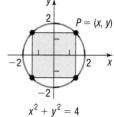

   (a) Express the area A of the rectangle as a function of x.
   (b) Express the perimeter p of the rectangle as a function of x.
   (c) Graph $A = A(x)$. For what value of x is A largest?
   (d) Graph $p = p(x)$. For what value of x is p largest?

10. A circle of radius r is inscribed in a square. See the figure.

    (a) Express the area A of the square as a function of the radius r of the circle.
    (b) Express the perimeter p of the square as a function of r.

11. **Geometry** A wire 10 meters long is to be cut into two pieces. One piece will be shaped as a square, and the other piece will be shaped as a circle. See the figure.

    (a) Express the total area A enclosed by the pieces of wire as a function of the length x of a side of the square.
    (b) What is the domain of A?
    (c) Graph $A = A(x)$. For what value of x is A smallest?

12. **Geometry** A wire 10 meters long is to be cut into two pieces. One piece will be shaped as an equilateral triangle, and the other piece will be shaped as a circle.
    (a) Express the total area A enclosed by the pieces of wire as a function of the length x of a side of the equilateral triangle.
    (b) What is the domain of A?
    (c) Graph $A = A(x)$. For what value of x is A smallest?

13. Geometry A wire of length x is bent into the shape of a circle.

    (a) Express the circumference C of the circle as a function of x.

    (b) Express the area A of the circle as a function of x.

14. Geometry A wire of length x is bent into the shape of a square.

    (a) Express the perimeter p of the square as a function of x.

    (b) Express the area A of the square as a function of x.

15. Geometry A semicircle of radius r is inscribed in a rectangle so that the diameter of the semicircle is the length of the rectangle. See the figure.

    (a) Express the area A of the rectangle as a function of the radius r of the semicircle.

    (b) Express the perimeter p of the rectangle as a function of r.

16. Geometry An equilateral triangle is inscribed in a circle of radius r. See the figure. Express the circumference C of the circle as a function of the length x of a side of the triangle.

$$\left[\textbf{Hint: } \text{First show that } r^2 = \frac{x^2}{3}.\right]$$

17. Geometry An equilateral triangle is inscribed in a circle of radius r. See the figure in Problem 16. Express the area A within the circle, but outside the triangle, as a function of the length x of a side of the triangle.

18. Uniform Motion Two cars leave an intersection at the same time. One is headed south at a constant speed of 30 miles per hour, and the other is headed west at a constant speed of 40 miles per hour (see the figure). Build a model that expresses the distance d between the cars as a function of the time t.

    [**Hint:** At $t = 0$, the cars leave the intersection.]

19. Uniform Motion Two cars are approaching an intersection. One is 2 miles south of the intersection and is moving at a constant speed of 30 miles per hour. At the same time, the other car is 3 miles east of the intersection and is moving at a constant speed of 40 miles per hour.

    (a) Build a model that expresses the distance d between the cars as a function of time t.

        [**Hint:** At $t = 0$, the cars are 2 miles south and 3 miles east of the intersection, respectively.]

    (b) Use a graphing utility to graph $d = d(t)$. For what value of t is d smallest?

20. Inscribing a Cylinder in a Sphere Inscribe a right circular cylinder of height h and radius r in a sphere of fixed radius R. See the figure. Express the volume V of the cylinder as a function of h.

    [**Hint:** $V = \pi r^2 h$. Note also the right triangle.]

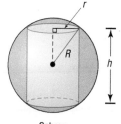

Sphere

21. Inscribing a Cylinder in a Cone Inscribe a right circular cylinder of height h and radius r in a cone of fixed radius R and fixed height H. See the figure. Express the volume V of the cylinder as a function of r.

    [**Hint:** $V = \pi r^2 h$. Note also the similar triangles.]

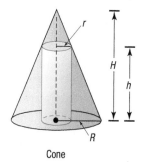

Cone

22. Installing Cable TV MetroMedia Cable is asked to provide service to a customer whose house is located 2 miles from the road along which the cable is buried. The nearest connection box for the cable is located 5 miles down the road. See the figure.

    (a) If the installation cost is $500 per mile along the road and $700 per mile off the road, build a model that expresses the total cost C of installation as a function of the distance x (in miles) from the connection box to the point where the cable installation turns off the road. Find the domain of $C = C(x)$.

    (b) Compute the cost if $x = 1$ mile.

    (c) Compute the cost if $x = 3$ miles.

    (d) Graph the function $C = C(x)$. Use TRACE to see how the cost C varies as x changes from 0 to 5.

    (e) What value of x results in the least cost?

23. Time Required to Go from an Island to a Town An island is 2 miles from the nearest point P on a straight shoreline. A town is 12 miles down the shore from P. See the figure.

(a) If a person can row a boat at an average speed of 3 miles per hour and the same person can walk 5 miles per hour, build a model that expresses the time T that it takes to go from the island to town as a function of the distance x from P to where the person lands the boat.

(b) What is the domain of T?

(c) How long will it take to travel from the island to town if the person lands the boat 4 miles from P?

(d) How long will it take if the person lands the boat 8 miles from P?

24. Constructing an Open Box An open box with a square base is required to have a volume of 10 cubic feet.

(a) Express the amount A of material used to make such a box as a function of the length x of a side of the square base.

(b) How much material is required for such a box with a base 1 foot by 1 foot?

(c) How much material is required for such a box with a base 2 feet by 2 feet?

(d) Use a graphing utility to graph $A = A(x)$. For what value of x is A smallest?

(e) What is the least amount of material needed?

25. Constructing an Open Box An open box with a square base is to be made from a square piece of cardboard 24 inches on a side by cutting out a square from each corner and turning up the sides. See the figure.

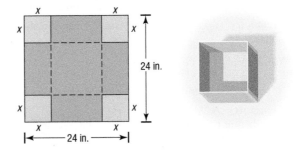

(a) Express the volume V of the box as a function of the length x of the side of the square cut from each corner.

(b) What is the volume if a 3-inch square is cut out?

(c) What is the volume if a 10-inch square is cut out?

(d) Graph $V = V(x)$. For what value of x is V largest?

(e) What is the largest volume?

26. Challenge Problem Filling a Conical Tank Water is poured into a container in the shape of a right circular cone with radius 4 feet and height 16 feet. See the figure. Express the volume V of the water in the cone as a function of the height h of the water.

27. Challenge Problem Inventory Management A retailer buys 600 USB Flash Drives per year from a distributor. The retailer wants to determine how many drives to order, x, per shipment so that her inventory is exhausted just as the next shipment arrives. The processing fee is \$15 per shipment, the yearly storage cost is \$1.60x, and each drive costs the retailer \$4.85.

(a) Express the total yearly cost C as a function of the number x of drives in each shipment.

(b) Use a graphing utility to determine the minimum yearly cost and the number of drives per order that yields the minimum cost.

Retain Your Knowledge

Problems 28–37 are based on material learned earlier in the course. The purpose of these problems is to keep the material fresh in your mind so that you are better prepared for the final exam.

28. Solve: $|2x - 3| - 5 = -2$

29. A 16-foot long Ford Fusion wants to pass a 50-foot truck traveling at 55 mi/h. How fast must the car travel to completely pass the truck in 5 seconds?

30. Find the slope of the line containing the points $(3, -2)$ and $(1, 6)$.

31. Find the missing length x for the given pair of similar triangles.

32. Given $y = \dfrac{x}{x+1}$ and $u = x + 1$, express y in terms of u.

33. Write $\dfrac{x+5}{3x^{2/3}} + x^{1/3}$ as a single quotient with only positive exponents.

34. Solve $-\sqrt{3x-2} \geq 4$.

35. If the point $(3, -2)$ is on the graph of an equation that is symmetric about the origin, what other point must be on the graph?

36. Solve $v = \dfrac{2.6t}{d^2}\sqrt{\dfrac{E}{P}}$ for P.

37. Find the discriminant of the quadratic equation $3x^2 - 7x = 4x - 2$.

Chapter Review

Library of Functions

Constant function (p. 244)

$$f(x) = b$$

The graph is a horizontal line with y-intercept b.

Identity function (p. 244)

$$f(x) = x$$

The graph is a line with slope 1 and y-intercept 0.

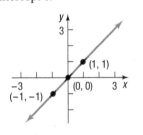

Square function (p. 244)

$$f(x) = x^2$$

The graph is a parabola with intercept at $(0, 0)$

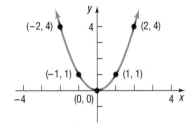

Cube function (p. 245)

$$f(x) = x^3$$

Square root function (p. 245)

$$f(x) = \sqrt{x}$$

Cube root function (p. 245)

$$f(x) = \sqrt[3]{x}$$

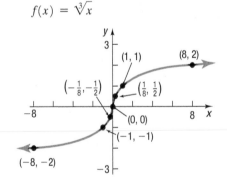

Reciprocal function (p. 245)

$$f(x) = \dfrac{1}{x}$$

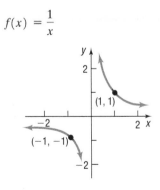

Absolute value function (p. 245)

$$f(x) = |x|$$

Greatest integer function (p. 246)

$$f(x) = \text{int}(x)$$

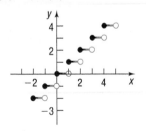

Things to Know

Function (pp. 205–207)	• A relation between two nonempty sets so that each element x in the first set, the domain, has corresponding to it exactly one element y in the second set. The range is the set of images of the elements in the domain.
	• A function can also be described as a set of ordered pairs (x, y) in which no first element is paired with two different second elements.
Function notation (pp. 207–210)	• $y = f(x)$
	• f is a symbol for the function.
	• x is the argument, or independent variable.
	• y is the dependent variable.
	• $f(x)$ is the value of the function at x.
	• A function f may be defined implicitly by an equation involving x and y or explicitly by writing $y = f(x)$.
Difference quotient of f (p. 210)	$\dfrac{f(x + h) - f(x)}{h} \quad h \neq 0$
Domain (pp. 211–213)	If unspecified, the domain of a function f defined by an equation is the largest set of real numbers for which $f(x)$ is a real number.
Vertical-line test (p. 220)	A set of points in the xy-plane is the graph of a function if and only if every vertical line intersects the graph in at most one point.
Even function f (p. 229)	$f(-x) = f(x)$ for every x in the domain ($-x$ must also be in the domain).
Odd function f (p. 229)	$f(-x) = -f(x)$ for every x in the domain ($-x$ must also be in the domain).
Increasing function (p. 231)	A function f is increasing on an interval I if, for any choice of x_1 and x_2 in I, with $x_1 < x_2$, then $f(x_1) < f(x_2)$.
Decreasing function (p. 231)	A function f is decreasing on an interval I if, for any choice of x_1 and x_2 in I, with $x_1 < x_2$, then $f(x_1) > f(x_2)$.
Constant function (p. 231)	A function f is constant on an interval I if, for all choices of x in I, the values of $f(x)$ are equal.
Local maximum (p. 232)	A function f, defined on some interval I, has a local maximum at c if there is an open interval in I containing c so that $f(c) \geq f(x)$, for all x in this open interval. The local maximum value is $f(c)$.
Local minimum (p. 232)	A function f, defined on some interval I, has a local minimum at c if there is an open interval in I containing c so that $f(c) \leq f(x)$, for all x in this open interval. The local minimum value is $f(c)$.
Absolute maximum and absolute minimum (p. 233)	Let f denote a function defined on some interval I.
	• If there is a number u in I for which $f(u) \geq f(x)$ for all x in I, then f has an absolute maximum at u, and the number $f(u)$ is the absolute maximum of f on I.
	• If there is a number v in I for which $f(v) \leq f(x)$, for all x in I, then f has an absolute minimum at v, and the number $f(v)$ is the absolute minimum of f on I.
Average rate of change of a function (pp. 235–236)	The average rate of change of f from a to b is $$\frac{\Delta y}{\Delta x} = \frac{f(b) - f(a)}{b - a} \quad a \neq b$$

Objectives

Section	You should be able to . . .	Examples	Review Exercises
3.1	1 Describe a relation (p. 203)	1	1
	2 Determine whether a relation represents a function (p. 205)	2–5	2–5
	3 Use function notation; find the value of a function (p. 207)	6, 7	6–8, 43
	4 Find the difference quotient of a function (p. 210)	8	18
	5 Find the domain of a function defined by an equation (p. 211)	9, 10	9–14
	6 Form the sum, difference, product, and quotient of two functions (p. 213)	11	15–17

Section	You should be able to . . .	Examples	Review Exercises
3.2	1 Identify the graph of a function (p. 220)	1	31, 32
	2 Obtain information from or about the graph of a function (p. 220)	2–4	19(a)–(e), 20(a), 20(e), 20(g)
3.3	1 Identify even and odd functions from a graph (p. 229)	1	20(f)
	2 Identify even and odd functions from an equation (p. 230)	2	21–24
	3 Use a graph to determine where a function is increasing, decreasing, or constant (p. 231)	3	20(b)
	4 Use a graph to locate local maxima and local minima (p. 232)	4	20(c)
	5 Use a graph to locate the absolute maximum and the absolute minimum (p. 233)	5	20(d)
	6 Use a graphing utility to approximate local maxima and local minima and to determine where a function is increasing or decreasing (p. 235)	6	25, 26, 44(d), 45(b)
	7 Find the average rate of change of a function (p. 235)	7, 8	27–30
3.4	1 Graph the functions listed in the library of functions (p. 242)	1, 2	33, 34
	2 Analyze a piecewise-defined function (p. 247)	3–5	41, 42
3.5	1 Graph functions using vertical and horizontal shifts (p. 254)	1–5, 11–13	19(f), 35, 37–40
	2 Graph functions using compressions and stretches (p. 258)	6–8, 12	19(g), 36, 40
	3 Graph functions using reflections about the x-axis and the y-axis (p. 260)	9, 10, 13	19(h), 36, 38, 40
3.6	1 Build and analyze functions (p. 267)	1–3	44, 45

Review Exercises

1. While shopping online for AA batteries, Masoud found that he could order a pack of 8 batteries for $6.30, a pack of 16 for $13.99, a pack of 20 for $12.32, or a pack of 24 for $13.99. Define a relation using number of batteries as input and price as output.

   (a) What is the domain and range of the relation?

   (b) Express the relation as a set of ordered pairs.

   (c) Express the relation as a mapping.

   (d) Express the relation as a graph.

In Problems 2–5, find the domain and range of each relation. Then determine whether the relation represents a function.

2. $\{(-1, 0), (2, 3), (4, 0)\}$ 3. $\{(4, -1), (2, 1), (4, 2)\}$ 4. $(x - 1)^2 + y^2 = 4$ 5. $y = |-4x - 5| - 3$

In Problems 6–8, find the following for each function:

   (a) $f(2)$ (b) $f(-2)$ (c) $f(-x)$ (d) $-f(x)$ (e) $f(x - 2)$ (f) $f(2x)$

6. $f(x) = \dfrac{3x}{x^2 - 1}$ 7. $f(x) = \sqrt{x^2 - 4}$ 8. $f(x) = \dfrac{x^2 - 4}{x^2}$

In Problems 9–14, find the domain of each function.

9. $f(x) = \dfrac{x}{x^2 - 9}$ 10. $f(x) = \sqrt{2 - x}$ 11. $g(x) = \dfrac{|x|}{x}$

12. $f(x) = \dfrac{x}{x^2 + 2x - 3}$ 13. $f(x) = \dfrac{\sqrt{x + 1}}{x^2 - 4}$ 14. $g(x) = \dfrac{x}{\sqrt{x + 8}}$

In Problems 15–17, find $f + g$, $f - g$, $f \cdot g$, and $\dfrac{f}{g}$ for each pair of functions. State the domain of each of these functions.

15. $f(x) = 2 - x$; $g(x) = 3x + 1$ 16. $f(x) = 3x^2 + x + 1$; $g(x) = 3x$ 17. $f(x) = \dfrac{x + 1}{x - 1}$; $g(x) = \dfrac{1}{x}$

18. Find the difference quotient of $f(x) = -2x^2 + x + 1$;

that is, find $\dfrac{f(x + h) - f(x)}{h}$, $h \neq 0$.

19. Consider the graph of the function f below.
   (a) Find the domain and the range of f.
   (b) List the intercepts.
   (c) Find $f(-2)$.
   (d) For what value of x does $f(x) = -3$?
   (e) Solve $f(x) > 0$.
   (f) Graph $y = f(x - 3)$.
   (g) Graph $y = f\left(\dfrac{1}{2}x\right)$.
   (h) Graph $y = -f(x)$.

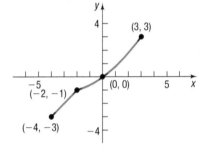

20. Use the graph of the function f shown below to find:
   (a) The domain and the range of f.
   (b) The intervals on which f is increasing, decreasing, or constant.
   (c) The local minimum values and local maximum values.
   (d) The absolute maximum and absolute minimum.
   (e) Whether the graph is symmetric with respect to the x-axis, the y-axis, the origin, or none of these.
   (f) Whether the function is even, odd, or neither.
   (g) The intercepts, if any.

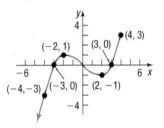

In Problems 21–24, determine (algebraically) whether the given function is even, odd, or neither.

21. $f(x) = x^3 - 4x$

22. $g(x) = \dfrac{4 + x^2}{1 + x^4}$

23. $G(x) = 1 - x + x^3$

24. $f(x) = \dfrac{x}{1 + x^2}$

In Problems 25 and 26, use a graphing utility to graph each function over the indicated interval. Approximate any local maximum values and local minimum values. Determine where the function is increasing and where it is decreasing.

25. $f(x) = 2x^3 - 5x + 1$ $(-3, 3)$

26. $f(x) = 2x^4 - 5x^3 + 2x + 1$ $(-2, 3)$

27. Find the average rate of change of $f(x) = 8x^2 - x$:
   (a) From 1 to 2
   (b) From 0 to 1
   (c) From 2 to 4

In Problems 28 and 29, find the average rate of change from 2 to 3 for each function f. Be sure to simplify.

28. $f(x) = 2 - 5x$

29. $f(x) = 3x - 4x^2$

30. If $f(x) = 3x - 4x^2$, find an equation of the secant line of f from 2 to 3.

In Problems 31 and 32, is the graph shown the graph of a function?

31.   **32.**

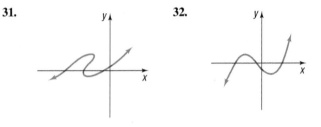

In Problems 33 and 34, graph each function. Be sure to label at least three points.

33. $f(x) = |x|$

34. $f(x) = \sqrt{x}$

In Problems 35–40, graph each function using the techniques of shifting, compressing or stretching, and reflections. Identify any intercepts of the graph. State the domain and, based on the graph, find the range.

35. $F(x) = |x| - 4$

36. $g(x) = -2|x|$

37. $h(x) = \sqrt{x - 1}$

38. $f(x) = \sqrt{1 - x}$

39. $h(x) = (x - 1)^2 + 2$

40. $g(x) = -2(x + 2)^3 - 8$

In Problems 41 and 42:
   (a) Find the domain of each function.
   (b) Locate any intercepts.
   (c) Graph each function.
   (d) Based on the graph, find the range.

41. $f(x) = \begin{cases} 3x & \text{if } -2 < x \leq 1 \\ x + 1 & \text{if } x > 1 \end{cases}$

42. $f(x) = \begin{cases} x & \text{if } -4 \leq x < 0 \\ 1 & \text{if } x = 0 \\ 3x & \text{if } x > 0 \end{cases}$

43. A function f is defined by

$$f(x) = \frac{Ax + 5}{6x - 2}$$

   If $f(1) = 4$, find A.

44. **Constructing a Closed Box** A closed box with a square base is required to have a volume of 10 cubic feet.
   (a) Build a model that expresses the amount A of material used to make such a box as a function of the length x of a side of the square base.
   (b) How much material is required for a base 1 foot by 1 foot?
   (c) How much material is required for a base 2 feet by 2 feet?
   (d) Graph $A = A(x)$. For what value of x is A smallest?

45. **Area of a Rectangle** A rectangle has one vertex in quadrant I on the graph of $y = 10 - x^2$, another at the origin, one on the positive x-axis, and one on the positive y-axis.
   (a) Express the area A of the rectangle as a function of x.
   (b) Find the largest area A that can be enclosed by the rectangle.

Chapter Test

CHAPTER
Test Prep
VIDEOS The Chapter Test Prep Videos include step-by-step solutions to all chapter test exercises. These videos are available in MyLab™ Math, or on this text's YouTube channel. Refer to the Preface for a link to the YouTube channel.

1. Find the domain and range of each relation. Then determine whether each relation represents a function.
   (a) $\{(2,5), (4,6), (6,7), (8,8)\}$
   (b) $\{(1,3), (4,-2), (-3,5), (1,7)\}$
   (c)

   (d)

In Problems 2–4, find the domain of each function and evaluate each function at $x = -1$.

2. $f(x) = \sqrt{4 - 5x}$

3. $g(x) = \dfrac{x + 2}{|x + 2|}$

4. $h(x) = \dfrac{x - 4}{x^2 + 5x - 36}$

5. Consider the graph of the function f below.

   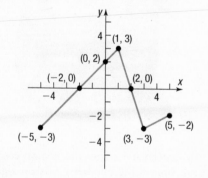

   (a) Find the domain and the range of f.
   (b) List the intercepts.
   (c) Find $f(1)$.
   (d) For what value(s) of x does $f(x) = -3$?
   (e) Solve $f(x) < 0$.

6. Graph the function $f(x) = -x^4 + 2x^3 + 4x^2 - 2$ on the interval $(-5, 5)$ using a graphing utility. Then approximate any local maximum values and local minimum values rounded to two decimal places. Determine where the function is increasing and where it is decreasing.

7. Consider the function $g(x) = \begin{cases} 2x + 1 & \text{if } x < -1 \\ x - 4 & \text{if } x \ge -1 \end{cases}$

   (a) Graph the function.
   (b) List the intercepts.
   (c) Find $g(-5)$.
   (d) Find $g(2)$.

8. For the function $f(x) = 3x^2 - 3x + 4$,
   (a) Find the average rate of change of f from 3 to 4.
   (b) Find an equation of the secant line from 3 to 4.

9. For the functions $f(x) = 2x^2 + 1$ and $g(x) = 3x - 2$, find the following and simplify.
   (a) $(f - g)(x)$
   (b) $(f \cdot g)(x)$
   (c) $f(x + h) - f(x)$

10. Graph each function using the techniques of shifting, compressing or stretching, and reflecting. Start with the graph of the basic function and show all the steps.
    (a) $h(x) = -2(x + 1)^3 + 3$
    (b) $g(x) = |x + 4| + 2$

11. Find the difference quotient of $f(x) = x^2 - 3x$.

12. A community skating rink is in the shape of a rectangle with semicircles attached at the ends. The length of the rectangle is 20 feet less than twice the width. The thickness of the ice is 2 inches.

    (a) Build a model that expresses the ice volume V as a function of the width, x.
    (b) How much ice is in the rink if the width is 90 feet?

13. Determine if the function $f(x) = -x^2 - 7$ is even, odd, or neither.

Cumulative Review

In Problems 1–6, find the real solutions of each equation.

1. $3x - 8 = 10$
2. $3x^2 - x = 0$
3. $x^2 - 8x - 9 = 0$
4. $6x^2 - 5x + 1 = 0$
5. $|2x + 3| = 4$
6. $\sqrt{2x + 3} = 2$

In Problems 7–9, solve each inequality. Graph the solution set.

7. $2 - 3x > 6$
8. $|2x - 5| < 3$
9. $|4x + 1| \geq 7$

10. (a) Find the distance from $P_1 = (-2, -3)$ to $P_2 = (3, -5)$.
    (b) What is the midpoint of the line segment from P_1 to P_2?
    (c) What is the slope of the line containing the points P_1 and P_2?

In Problems 11–14, graph each equation.

11. $3x - 2y = 12$
12. $x = y^2$
13. $x^2 + (y - 3)^2 = 16$
14. $y = \sqrt{x}$

15. For the equation $3x^2 - 4y = 12$, find the intercepts and check for symmetry.

16. Find the slope-intercept form of the equation of the line containing the points $(-2, 4)$ and $(6, 8)$.

In Problems 17–19, graph each function.

17. $f(x) = (x + 2)^2 - 3$
18. $f(x) = \dfrac{1}{x}$
19. $f(x) = \begin{cases} 2 - x & \text{if } x \leq 2 \\ |x| & \text{if } x > 2 \end{cases}$

Chapter Projects

((•)) Internet-based Project

I. **Choosing a Data Plan** Collect information from your family, friends, or consumer agencies such as Consumer Reports. Then decide on a service provider, choosing the company that you feel offers the best service. Once you have selected a service provider, research the various types of individual plans offered by the company by visiting the provider's website. Many cellular providers offer family plans that include unlimited talk, text, and data. However, once a data cap has been reached, service may be slowed, which prevents media from being streamed. So, many customers still purchase data-only plans for devices such as tablets or laptops. The monthly cost is primarily determined by the amount of data used and the number of data-only devices.

1. Suppose you expect to use 10 gigabytes of data for a single tablet. What would be the monthly cost of each plan you are considering?

2. Suppose you expect to use 30 gigabytes of data and want a personal hotspot, but you still have only a single tablet. What would be the monthly cost of each plan you are considering?

3. Suppose you expect to use 20 gigabytes of data with three tablets sharing the data. What would be the monthly cost of each plan you are considering?

4. Suppose you expect to use 20 gigabytes of data with a single tablet and a personal hotspot. What would be the monthly cost of each plan you are considering?

5. Build a model that describes the monthly cost C, in dollars, as a function of the number g of data gigabytes used, assuming a single tablet and a personal hotspot for each plan you are considering.

6. Graph each function from Problem 5.

7. Based on your particular usage, which plan is best for you?

8. Now, develop an Excel spreadsheet to analyze the various plans you are considering. Suppose you want a plan that offers 50 gigabytes of shared data and costs $60 per month. Additional gigabytes of data cost $15 per gigabyte, extra tablets can be added to the plan for $10 each per month, and each hotspot or laptop costs $20 per month. Because these data plans have a cost structure based on piecewise-defined functions, we need an "if/then" statement within Excel to analyze the cost of the plan. Use the accompanying Excel spreadsheet as a guide in developing your spreadsheet. Enter into your spreadsheet a variety of possible amounts of data and various numbers of additional tablets, laptops, and hotspots.

	A	B	C	D	
1					
2	Monthly fee	$60			
3	Allotted data per month (GB)	50			
4	Data used (GB)	12			
5	Cost per additional GB of data	$15			
6					
7	Monthly cost of hotspot or laptop	$20			
8	Number of hotspots or laptops	1			
9	Monthly cost of additional tablet	$10			
10	Number of additional tablets	2			
11					
12	Cost of data	=IF(B4<B3,B2,B2+B5*(B4-B3))			
13	Cost of additional devices/hotspots	=B8*B7+B10*B9			
14					
15	Total Cost	=B12+B13			
16					

9. Write a paragraph supporting the choice in plans that best meets your needs.

10. How are "if/then" loops similar to a piecewise-defined function?

Source: *Excel © 2018 Microsoft Corporation. Used with Permission from Microsoft.*

The following projects are available on the Instructor's Resource Center (IRC).

II. Project at Motorola: ***Wireless Internet Service*** Use functions and their graphs to analyze the total cost of various wireless Internet service plans.

III. Cost of Cable When government regulations and customer preference influence the path of a new cable line, the Pythagorean Theorem can be used to assess the cost of installation.

IV. Oil Spill Functions are used to analyze the size and spread of an oil spill from a leaking tanker.

4

Linear and Quadratic Functions

The Beta of a Stock

Investing in the stock market can be rewarding and fun, but how does one go about selecting which stocks to purchase? Financial investment firms hire thousands of analysts who track individual stocks (equities) and assess the value of the underlying company. One measure the analysts consider is the *beta* of the stock. **Beta** measures the risk of an individual company's equity relative to that of a market basket of stocks, such as the Standard & Poor's 500. But how is beta computed?

—See the Internet-based Chapter Project I—

Outline

← A Look Back

Up to now, our discussion has focused on equations and functions and their graphs. We graphed equations by plotting points, using intercepts, and testing for symmetry. We learned to identify whether a relation represents a function, and we discussed properties of functions, such as domain/range, increasing/decreasing, even/odd, and average rate of change. Using this information, we developed a library of functions, and analyzed piecewise-defined functions. Finally, we used transformations to graph new functions from the graphs of the familiar functions listed in the library of functions.

A Look Ahead →

Now we begin looking at classes of functions. This chapter focuses on linear and quadratic functions, their properties, and their applications.

4.1 Properties of Linear Functions and Linear Models

PREPARING FOR THIS SECTION *Before getting started, review the following:*

- Lines (Section 2.3, pp. 169–179)
- Graphs of Equations in Two Variables; Intercepts; Symmetry (Section 2.2, pp. 158–165)
- Linear Equations (Section 1.1, pp. 82–85)

- Functions (Section 3.1, pp. 205–215)
- The Graph of a Function (Section 3.2, pp. 219–223)
- Properties of Functions (Section 3.3, pp. 229–237)

Now Work the 'Are You Prepared?' problems on page 287.

OBJECTIVES 1 Graph Linear Functions (p. 281)
2 Use Average Rate of Change to Identify Linear Functions (p. 281)
3 Determine Whether a Linear Function Is Increasing, Decreasing, or Constant (p. 284)
4 Build Linear Models from Verbal Descriptions (p. 285)

1 Graph Linear Functions

In Section 2.3 we discussed lines. In particular, for nonvertical lines we developed the slope-intercept form of the equation of a line, $y = mx + b$. When the slope-intercept form of a line is written using function notation, the result is a *linear function*.

DEFINITION Linear Function

A **linear function** is a function of the form

$$f(x) = mx + b$$

The graph of a linear function is a line with slope m and y-intercept b. Its domain is the set of all real numbers.

Functions that are not linear are said to be **nonlinear**.

EXAMPLE 1 **Graphing a Linear Function**

Graph the linear function $f(x) = -3x + 7$. What are the domain and the range of f?

Solution This is a linear function with slope $m = -3$ and y-intercept $b = 7$. To graph this function, plot the point $(0, 7)$, the y-intercept, and use the slope to find an additional point by moving right 1 unit and down 3 units. See Figure 1. The domain and the range of f are each the set of all real numbers.

Alternatively, an additional point can be found by evaluating the function at some $x \neq 0$. For $x = 1$, $f(1) = -3 \cdot 1 + 7 = 4$, so the point $(1, 4)$ lies on the graph.

Figure 1 $f(x) = -3x + 7$

Now Work PROBLEMS 13(a) AND (b)

2 Use Average Rate of Change to Identify Linear Functions

Look at Table 1 on the next page, which shows several values of the independent variable x and corresponding values of the dependent variable y for the function $f(x) = -3x + 7$. Notice that as the value of the independent variable x increases by 1, the value of the dependent variable y decreases by 3. That is, the average rate of change of y with respect to x is a constant, -3.

Table 1

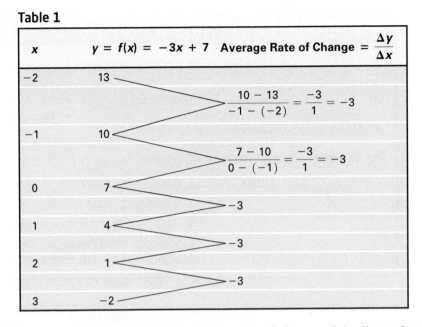

It is not a coincidence that the average rate of change of the linear function $f(x) = -3x + 7$ is the slope of the linear function. That is, $\dfrac{\Delta y}{\Delta x} = m = -3$. The following theorem states this fact.

> **THEOREM Average Rate of Change of a Linear Function**
>
> Linear functions have a constant average rate of change. That is, the average rate of change of a linear function $f(x) = mx + b$ is
>
> $$\frac{\Delta y}{\Delta x} = m$$

Proof The average rate of change of $f(x) = mx + b$ from x_1 to x_2, $x_1 \neq x_2$, is

$$\frac{\Delta y}{\Delta x} = \frac{f(x_2) - f(x_1)}{x_2 - x_1} = \frac{(mx_2 + b) - (mx_1 + b)}{x_2 - x_1}$$

$$= \frac{mx_2 - mx_1}{x_2 - x_1} = \frac{m(x_2 - x_1)}{x_2 - x_1} = m \qquad \blacksquare$$

Based on the theorem just proved, the average rate of change of the function $g(x) = -\dfrac{2}{5}x + 5$ is $-\dfrac{2}{5}$.

Now Work PROBLEM 13(c)

As it turns out, only linear functions have a constant average rate of change. Because of this, the average rate of change can be used to determine whether a function is linear. This is especially useful if the function is defined by a data set.

EXAMPLE 2

Using the Average Rate of Change to Identify Linear Functions

(a) A strain of *E. coli* known as Beu 397-recA441 is placed into a Petri dish at 30° Celsius and allowed to grow. The data shown in Table 2 are collected. The population is measured in grams and the time in hours. Plot the ordered pairs (x, y) in the Cartesian plane, and use the average rate of change to determine whether the function is linear.

(b) The data in Table 3 represent the maximum number of heartbeats that a healthy individual of different ages should have during a 15-second interval of time while exercising. Plot the ordered pairs (x, y) in the Cartesian plane, and use the average rate of change to determine whether the function is linear.

Table 2

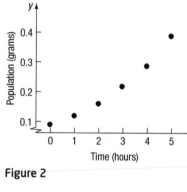

Time (hours), x	Population (grams), y	(x, y)
0	0.09	(0, 0.09)
1	0.12	(1, 0.12)
2	0.16	(2, 0.16)
3	0.22	(3, 0.22)
4	0.29	(4, 0.29)
5	0.39	(5, 0.39)

Table 3

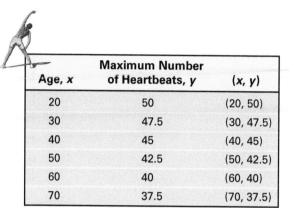

Age, x	Maximum Number of Heartbeats, y	(x, y)
20	50	(20, 50)
30	47.5	(30, 47.5)
40	45	(40, 45)
50	42.5	(50, 42.5)
60	40	(60, 40)
70	37.5	(70, 37.5)

Source: American Heart Association

Solution

Compute the average rate of change of each function. If the average rate of change is constant, the function is linear. If the average rate of change is not constant, the function is nonlinear.

(a) Figure 2 shows the points listed in Table 2 plotted in the Cartesian plane. Note that it is impossible to draw a straight line that contains all the points. Table 4 displays the average rate of change of the population.

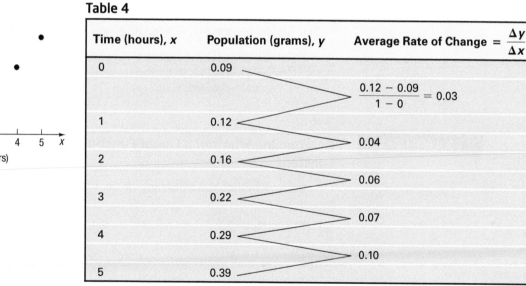

Figure 2

Table 4

Time (hours), x	Population (grams), y	Average Rate of Change $= \dfrac{\Delta y}{\Delta x}$
0	0.09	
		$\dfrac{0.12 - 0.09}{1 - 0} = 0.03$
1	0.12	
		0.04
2	0.16	
		0.06
3	0.22	
		0.07
4	0.29	
		0.10
5	0.39	

Because the average rate of change is not constant, the function is not linear. In fact, because the average rate of change is increasing as the value of the independent variable increases, the function is increasing at an increasing rate. So not only is the population increasing over time, but it is also growing more rapidly as time passes.

(b) Figure 3 on the next page shows the points listed in Table 3 plotted in the Cartesian plane. Table 5 on the next page displays the average rate of change of the maximum number of heartbeats. The average rate of change of the heartbeat data is constant, -0.25 beat per year, so the function is linear, and the points in Figure 3 lie on a line.

Figure 3

Table 5

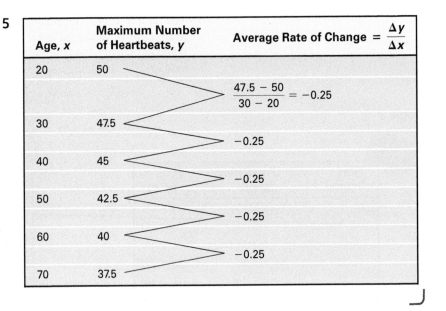

Age, x	Maximum Number of Heartbeats, y	Average Rate of Change $= \dfrac{\Delta y}{\Delta x}$
20	50	
		$\dfrac{47.5 - 50}{30 - 20} = -0.25$
30	47.5	
		-0.25
40	45	
		-0.25
50	42.5	
		-0.25
60	40	
		-0.25
70	37.5	

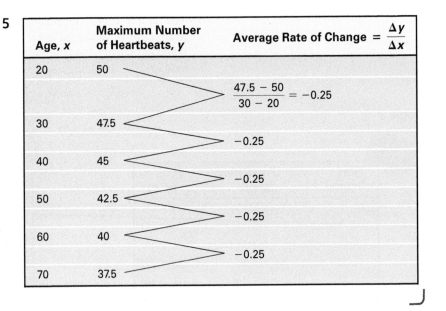

Now Work PROBLEM 21

3 Determine Whether a Linear Function Is Increasing, Decreasing, or Constant

When the slope m of a linear function is positive ($m > 0$), the line slants upward from left to right. When the slope m of a linear function is negative ($m < 0$), the line slants downward from left to right. When the slope m of a linear function is zero ($m = 0$), the line is horizontal.

> **THEOREM Increasing, Decreasing, and Constant Linear Functions**
>
> A linear function $f(x) = mx + b$ is increasing over its domain if its slope, m, is positive. It is decreasing over its domain if its slope, m, is negative. It is constant over its domain if its slope, m, is zero.

EXAMPLE 3

Determining Whether a Linear Function Is Increasing, Decreasing, or Constant

Determine whether the following linear functions are increasing, decreasing, or constant.

(a) $f(x) = 5x - 2$ (b) $g(x) = -2x + 8$

(c) $s(t) = \dfrac{3}{4}t - 4$ (d) $h(z) = 7$

Solution

(a) The linear function $f(x) = 5x - 2$ has slope 5, which is positive. The function f is increasing on the interval $(-\infty, \infty)$.

(b) The linear function $g(x) = -2x + 8$ has slope -2, which is negative. The function g is decreasing on the interval $(-\infty, \infty)$.

(c) The linear function $s(t) = \dfrac{3}{4}t - 4$ has slope $\dfrac{3}{4}$, which is positive. The function s is increasing on the interval $(-\infty, \infty)$.

(d) The linear function h can be written as $h(z) = 0z + 7$. Because the slope is 0, the function h is constant on the interval $(-\infty, \infty)$.

Now Work PROBLEM 13(d)

4 Build Linear Models from Verbal Descriptions

When the average rate of change of a function is constant, a linear function can model the relation between the two variables. For example, if a recycling company pays $0.52 per pound for aluminum cans, then the relation between the price p paid and the pounds recycled x can be modeled as the linear function $p(x) = 0.52x$, with slope $m = \dfrac{0.52 \text{ dollar}}{1 \text{ pound}}$.

In Words
A constant rate of change involving two variables indicates a linear function.

Modeling with a Linear Function

If the average rate of change of a function is a constant m, a linear function f can be used to model the relation between the two variables as follows:

$$f(x) = mx + b$$

where b is the value of f at 0; that is, $b = f(0)$.

EXAMPLE 4　**Straight-line Depreciation**

Book value is the value of an asset that a company uses to create its balance sheet. Some companies depreciate assets using straight-line depreciation so that the value of the asset declines by a fixed amount each year. The amount of the decline depends on the useful life that the company assigns to the asset. Suppose a company just purchased a fleet of new cars for its sales force at a cost of $31,500 per car. The company chooses to depreciate each vehicle using the straight-line method over 7 years. This means that each car will depreciate by $\dfrac{\$31{,}500}{7} = \$4500$ per year.

(a) Write a linear function that expresses the book value V of each car as a function of its age, x, in years.
(b) Graph the linear function.
(c) What is the book value of each car after 3 years?
(d) Interpret the slope.
(e) When will the book value of each car be $9000?

Solution　(a) If we let $V(x)$ represent the value of each car after x years, then $V(0)$ represents the original value of each car, so $V(0) = \$31{,}500$. The y-intercept of the linear function is $31,500. Because each car depreciates by $4500 per year, the slope of the linear function is -4500. The linear function that represents the book value V of each car after x years is

$$V(x) = -4500x + 31{,}500$$

(b) Figure 4 shows the graph of V.
(c) The book value of each car after 3 years is

$$V(3) = -4500 \cdot 3 + 31{,}500$$
$$= \$18{,}000$$

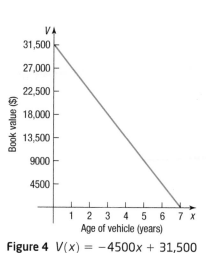

Figure 4 $V(x) = -4500x + 31{,}500$

(d) Since the slope of $V(x) = -4500x + 31{,}500$ is -4500, the average rate of change of the book value is $-\$4500/\text{year}$. So for each additional year that passes, the book value of the car decreases by $4500.

(continued)

(e) To find when the book value will be $9000, solve the equation

$$V(x) = 9000$$

$$-4500x + 31{,}500 = 9000$$

$$-4500x = -22{,}500 \qquad \text{Subtract 31,500 from both sides.}$$

$$x = \frac{-22{,}500}{-4500} = 5 \qquad \text{Divide by } -4500.$$

The car will have a book value of $9000 when it is 5 years old.

Now Work PROBLEM 47

EXAMPLE 5

Supply and Demand

The **quantity supplied** of a good is the amount of a product that a company is willing to make available for sale at a given price. The **quantity demanded** of a good is the amount of a product that consumers are willing to purchase at a given price p. Suppose that the quantity supplied, S, and the quantity demanded, D, of smartphones each month are given by the following functions:

$$S(p) = 15p - 900$$

$$D(p) = -3.75p + 2850$$

where p is the price (in dollars) of the smartphone.

(a) The **equilibrium price** of a product is defined as the price at which quantity supplied equals quantity demanded. That is, the equilibrium price is the price at which $S(p) = D(p)$. Find the equilibrium price of smartphones. What is the **equilibrium quantity**, the amount demanded (or supplied) at the equilibrium price?

(b) Determine the prices for which quantity supplied is greater than quantity demanded. That is, solve the inequality $S(p) > D(p)$.

(c) Graph $S = S(p)$ and $D = D(p)$, and label the **equilibrium point**, the point of intersection of S and D.

Solution

(a) To find the equilibrium price, solve the equation $S(p) = D(p)$.

$$15p - 900 = -3.75p + 2850 \qquad \begin{array}{l} S(p) = 15p - 900; \\ D(p) = -3.75p + 2850 \end{array}$$

$$15p = -3.75p + 3750 \qquad \text{Add 900 to both sides.}$$

$$18.75p = 3750 \qquad \text{Add 3.75p to both sides.}$$

$$p = 200 \qquad \text{Divide both sides by 18.75.}$$

The equilibrium price is $200 per smartphone. To find the equilibrium quantity, evaluate either $S(p)$ or $D(p)$ at $p = 200$.

$$S(200) = 15 \cdot 200 - 900 = 2100$$

The equilibrium quantity is 2100 smartphones. At a price of $200 per phone, the company will produce and sell 2100 phones each month and have no shortages or excess inventory.

(b) The inequality $S(p) > D(p)$ is

$$15p - 900 > -3.75p + 2850 \qquad S(p) > D(p)$$

$$15p > -3.75p + 3750 \qquad \text{Add 900 to both sides.}$$

$$18.75p > 3750 \qquad \text{Add 3.75p to both sides.}$$

$$p > 200 \qquad \text{Divide both sides by 18.75.}$$

If the company charges more than $200 per phone, quantity supplied will exceed quantity demanded. In this case the company will have excess phones in inventory.

(c) Figure 5 shows the graphs of $S = S(p)$ and $D = D(p)$ with the equilibrium point labeled.

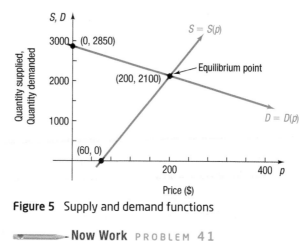

Figure 5 Supply and demand functions

━━━━━ **Now Work** PROBLEM 41

4.1 Assess Your Understanding

'Are You Prepared?' *Answers are given at the end of these exercises. If you get a wrong answer, read the pages listed in* red.

1. Graph $y = 3x - 1$. (pp. 174–175)
2. Find the slope of the line joining the points $(2, 5)$ and $(-1, 3)$. (pp. 169–171)
3. Find the average rate of change of $f(x) = -4x + 3$, from 2 to 4. (pp. 235–236)
4. Solve: $60x - 900 = -15x + 2850$. (pp. 82–85)
5. If $f(x) = 7.5x + 15$, find $f(-2)$. (pp. 207–209)
6. *True or False* The function $f(x) = \dfrac{2}{3}x + 15$ is increasing on the interval $(-\infty, \infty)$. (pp. 231–232)

Concepts and Vocabulary

7. For the graph of the linear function $f(x) = mx + b$, m is the _____ and b is the _____.
8. If the slope m of the graph of a linear function is _____, the function is increasing over its domain.
9. *True or False* The slope of a nonvertical line is the average rate of change of the linear function.
10. *True or False* The average rate of change of $f(x) = 2x + 8$ is 8.

11. *Multiple Choice* What is the only type of function that has a constant average rate of change?
   (a) linear function   **(b)** quadratic function
   (c) step function   **(d)** absolute value function
12. *Multiple Choice* A car has 12,500 miles on its odometer. Say the car is driven an average of 40 miles per day. Choose the model that expresses the number of miles N that will be on its odometer after x days.
   (a) $N(x) = -40x + 12,500$ **(b)** $N(x) = 40x - 12,500$
   (c) $N(x) = 12,500x + 40$ **(d)** $N(x) = 40x + 12,500$

Skill Building

In Problems 13–20, a linear function is given.
   (a) Find the slope and y-intercept of each function.
   (b) Use the slope and y-intercept to graph each function.
   (c) What is the average rate of change of each function?
   (d) Determine whether each function is increasing, decreasing, or constant.

13. $f(x) = 2x + 3$ 14. $g(x) = 5x - 4$ 15. $h(x) = -3x + 4$ 16. $p(x) = -x + 6$

17. $f(x) = \dfrac{1}{4}x - 3$ 18. $h(x) = -\dfrac{2}{3}x + 4$ 19. $F(x) = 4$ 20. $G(x) = -2$

In Problems 21–28, determine whether each function is linear or nonlinear. If it is linear, determine the slope.

21.

x	y = f(x)
-2	4
-1	1
0	-2
1	-5
2	-8

22.

x	y = f(x)
-2	1/4
-1	1/2
0	1
1	2
2	4

23.

x	y = f(x)
-2	-8
-1	-3
0	0
1	1
2	0

24.

x	y = f(x)
-2	-4
-1	0
0	4
1	8
2	12

25.

x	y = f(x)
−2	−26
−1	−4
0	2
1	−2
2	−10

26.

x	y = f(x)
−2	−4
−1	−3.5
0	−3
1	−2.5
2	−2

27.

x	y = f(x)
−2	8
−1	8
0	8
1	8
2	8

28.

x	y = f(x)
−2	0
−1	1
0	4
1	9
2	16

Applications and Extensions

29. Suppose that $f(x) = 4x - 1$ and $g(x) = -2x + 5$.
 (a) Solve $f(x) = 0$.
 (b) Solve $f(x) > 0$.
 (c) Solve $f(x) = g(x)$.
 (d) Solve $f(x) \le g(x)$.
 (e) Graph $y = f(x)$ and $y = g(x)$ and label the point that represents the solution to the equation $f(x) = g(x)$.

30. Suppose that $f(x) = 3x + 5$ and $g(x) = -2x + 15$.
 (a) Solve $f(x) = 0$.
 (b) Solve $f(x) < 0$.
 (c) Solve $f(x) = g(x)$.
 (d) Solve $f(x) \ge g(x)$.
 (e) Graph $y = f(x)$ and $y = g(x)$ and label the point that represents the solution to the equation $f(x) = g(x)$.

31. In parts (a)–(f), use the figure below.

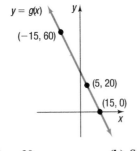

 (a) Solve $f(x) = 50$.
 (b) Solve $f(x) = 80$.
 (c) Solve $f(x) = 0$.
 (d) Solve $f(x) > 50$.
 (e) Solve $f(x) \le 80$.
 (f) Solve $0 < f(x) < 80$.

32. In parts (a)–(f), use the figure below.

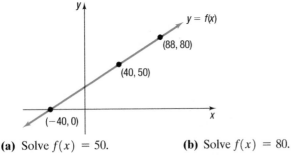

 (a) Solve $g(x) = 20$.
 (b) Solve $g(x) = 60$.
 (c) Solve $g(x) = 0$.
 (d) Solve $g(x) > 20$.
 (e) Solve $g(x) \le 60$.
 (f) Solve $0 < g(x) < 60$.

33. In parts (a) and (b), use the figure below.

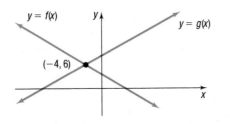

 (a) Solve the equation: $f(x) = g(x)$.
 (b) Solve the inequality: $f(x) > g(x)$.

34. In parts (a) and (b), use the figure below.

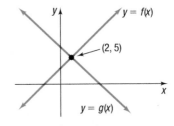

 (a) Solve the equation: $f(x) = g(x)$.
 (b) Solve the inequality: $f(x) \le g(x)$.

35. In parts (a) and (b), use the figure below.

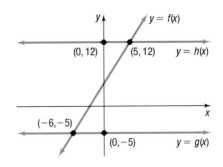

 (a) Solve the equation: $f(x) = g(x)$.
 (b) Solve the inequality: $g(x) \le f(x) < h(x)$.

36. In parts (a) and (b), use the figure below.

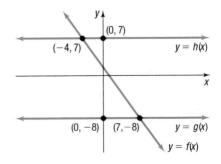

 (a) Solve the equation: $f(x) = g(x)$.
 (b) Solve the inequality: $g(x) < f(x) \le h(x)$.

37. Getting Towed The cost C, in dollars, to tow a car is modeled by the function $C(x) = 2.5x + 85$, where x is the number of miles towed.
 (a) What is the cost of towing a car 40 miles?
 (b) If the cost of towing a car is $245, how many miles was it towed?
 (c) Suppose that you have only $150. What is the maximum number of miles that you can be towed?
 (d) What is the domain of C?

38. Phone Charges The monthly cost C, in dollars, for calls from the United States to Germany on a certain wireless plan is modeled by the function $C(x) = 0.26x + 5$, where x is the number of minutes used.

(a) What is the cost if you talk on the phone for 50 minutes?

(b) Suppose that your monthly bill is $21.64. How many minutes did you use the phone?

(c) Suppose that you budget $50 per month for calls to Germany. What is the maximum number of minutes that you can talk?

(d) What is the domain of C if there are 30 days in the month?

39. Forensic Science The relationship between the height H of an adult male and the length x of his humerus, in centimeters, can be modeled by the linear function $H(x) = 2.89x + 78.10$.

(a) If incomplete skeletal remains of an adult male include a humerus measuring 37.1 centimeters, approximate the height of this male to the nearest tenth.

(b) If an adult male is 175.3 centimeters tall, approximate the length of his humerus to the nearest tenth.

40. Forensic Science The relationship between the height H of an adult female and the length x of her femur, in centimeters, can be modeled by the linear function $H(x) = 2.47x + 54.10$.

(a) If incomplete skeletal remains of an adult female include a femur measuring 46.8 centimeters, approximate the height of this female to the nearest tenth.

(b) If an adult female is 152.4 centimeters tall, approximate the length of her femur to the nearest tenth.

41. Supply and Demand Suppose that the quantity supplied S and the quantity demanded D of T-shirts at a concert are given by the following functions:

$$S(p) = -600 + 50p$$
$$D(p) = 1200 - 25p$$

where p is the price of a T-shirt.

(a) Find the equilibrium price for T-shirts at this concert. What is the equilibrium quantity?

(b) Determine the prices for which quantity demanded is greater than quantity supplied.

(c) What do you think will eventually happen to the price of T-shirts if quantity demanded is greater than quantity supplied?

42. Supply and Demand Suppose that the quantity supplied S and the quantity demanded D of hot dogs at a baseball game are given by the following functions:

$$S(p) = -2000 + 3000p$$
$$D(p) = 10{,}000 - 1000p$$

where p is the price of a hot dog.

(a) Find the equilibrium price for hot dogs at the baseball game. What is the equilibrium quantity?

(b) Determine the prices for which quantity demanded is less than quantity supplied.

(c) What do you think will eventually happen to the price of hot dogs if quantity demanded is less than quantity supplied?

43. Taxes The function $T(x) = 0.12(x - 9525) + 952.50$ represents the tax bill T of a single person whose adjusted gross income is x dollars for income over $9525 but not over $38,700, in 2018.

Source: Internal Revenue Service

(a) What is the domain of this linear function?

(b) What is a single filer's tax bill if adjusted gross income is $20,000?

(c) Which variable is independent and which is dependent?

(d) Graph the linear function over the domain specified in part (a).

(e) What is a single filer's adjusted gross income if the tax bill is $3109.50?

44. Competitive Balance Tax Under the 2017–2021 labor agreement between Major League Baseball and the players, any team whose payroll exceeded $195 million in 2017 had to pay a competitive balance tax of 50%. The linear function $T(p) = 0.50(p - 195)$ describes the competitive balance tax T for a team whose payroll was p (in millions of dollars).

Source: Major League Baseball

(a) What is the domain of this linear function?

(b) What was the competitive balance tax for the New York Yankees, whose 2017 payroll was $209.3 million?

(c) Graph the linear function.

(d) What was the 2017 payroll for the Los Angeles Dodgers who paid a competitive balance tax of $24.5 million?

*The point at which a company's profits equal zero is called the company's **break-even point**. For Problems 45 and 46, let R represent a company's revenue, let C represent the company's costs, and let x represent the number of units produced and sold each day.*

(a) Find the firm's break-even point; that is, find x so that $R = C$.

(b) Solve the inequality $R(x) > C(x)$ to find the units that represent a profit for the company.

45. $R(x) = 8x$
$C(x) = 4.5x + 17{,}500$

46. $R(x) = 12x$
$C(x) = 10x + 15{,}000$

47. Straight-line Depreciation Suppose that a company has just purchased a new computer for $3000. The company chooses to depreciate the computer using the straight-line method over 3 years.

(a) Write a linear model that expresses the book value V of the computer as a function of its age x.

(b) What is the domain of the function found in part (a)?

(c) Graph the linear function.

(d) What is the book value of the computer after 2 years?

(e) When will the computer have a book value of $2000?

48. Straight-line Depreciation Suppose that a company has just purchased a new machine for its manufacturing facility for $120,000. The company chooses to depreciate the machine using the straight-line method over 10 years.

(a) Write a linear model that expresses the book value V of the machine as a function of its age x.

(b) What is the domain of the function found in part (a)?

(c) Graph the linear function.

(d) What is the book value of the machine after 4 years?

(e) When will the machine have a book value of $72,000?

49. Cost Function The simplest cost function C is a linear cost function, $C(x) = mx + b$, where the y-intercept b represents the fixed costs of operating a business and the slope m represents the cost of each item produced. Suppose that a small bicycle manufacturer has daily fixed costs of $1800, and each bicycle costs $90 to manufacture.
 (a) Write a linear model that expresses the cost C of manufacturing x bicycles in a day.
 (b) Graph the model.
 (c) What is the cost of manufacturing 14 bicycles in a day?
 (d) How many bicycles could be manufactured for $3780?

50. Cost Function Refer to Problem 49. Suppose that the landlord of the building increases the bicycle manufacturer's rent by $100 per month.
 (a) Assuming that the manufacturer is open for business 20 days per month, what are the new daily fixed costs?
 (b) Write a linear model that expresses the cost C of manufacturing x bicycles in a day with the higher rent.
 (c) Graph the model.
 (d) What is the cost of manufacturing 14 bicycles in a day?
 (e) How many bicycles can be manufactured for $3780?

51. Hooke's Law According to Hooke's Law, a linear relationship exists between the distance that a spring stretches and the force stretching it. Suppose a weight of 0.5 kilograms causes a spring to stretch 2.75 centimeters and a weight of 1.2 kilograms causes the same spring to stretch 6.6 centimeters.
 (a) Find a linear model that relates the distance d of the stretch and the weight w.
 (b) What stretch is caused by a weight of 2.4 kilograms?
 (c) What weight causes a stretch of 19.8 centimeters?

52. Hooke's Law The distance d between the bottom of a suspended spring and a countertop is a linear function of the weight w attached to the bottom of the spring. The bottom of the spring is 9 inches from the countertop when

the attached weight is 1.5 pounds and 5 inches from the countertop when the attached weight is 2.5 pounds.
 (a) Find a linear model that relates the distance d from the countertop and the weight w.
 (b) Find the distance between the bottom of the spring and the countertop if no weight is attached.
 (c) What is the smallest weight that will make the bottom of the spring reach the countertop? (Ignore the thickness of the weight.)

53. *Challenge Problem* Temperature Conversion The linear function $F(C) = \dfrac{9}{5}C + 32$ converts degrees Celsius to degrees Fahrenheit, and the linear function $R(F) = F + 459.67$ converts degrees Fahrenheit to degrees Rankine. Find a linear function that converts degrees Rankine to degrees Celsius.

54. *Mixed Practice* Building a Linear Model from Data The following data represent the various combinations of soda and hot dogs that Yolanda can buy at a baseball game with $60.

Soda, s	Hot Dogs, h
20	0
15	3
10	6
5	9

 (a) Plot the ordered pairs (s, h) in a Cartesian plane.
 (b) Show that the number h of hot dogs purchased is a linear function of the number s of sodas purchased.
 (c) Determine the linear function that describes the relation between s and h.
 (d) What is the domain of the linear function?
 (e) Graph the linear function in the Cartesian plane drawn in part (a).
 (f) Interpret the slope.
 (g) Interpret the intercepts.

Explaining Concepts: Discussion and Writing

55. Which functions might have the graph shown? (More than one answer is possible.)
 (a) $f(x) = 2x - 7$
 (b) $g(x) = -3x + 4$
 (c) $H(x) = 5$
 (d) $F(x) = 3x + 4$
 (e) $G(x) = \dfrac{1}{2}x + 2$

56. Which functions might have the graph shown? (More than one answer is possible.)
 (a) $f(x) = 3x + 1$
 (b) $g(x) = -2x + 3$
 (c) $H(x) = 3$
 (d) $F(x) = -4x - 1$
 (e) $G(x) = -\dfrac{2}{3}x + 3$

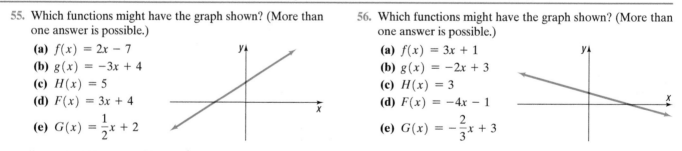

57. Under what circumstances is a linear function $f(x) = mx + b$ odd? Can a linear function ever be even?

58. Explain how the graph of $f(x) = mx + b$ can be used to solve $mx + b > 0$.

Retain Your Knowledge

Problems 59–68 are based on material learned earlier in the course. The purpose of these problems is to keep the material fresh in your mind so that you are better prepared for the final exam.

59. Graph $x^2 - 4x + y^2 + 10y - 7 = 0$.

60. If $f(x) = \dfrac{2x + B}{x - 3}$ and $f(5) = 8$, what is the value of B?

61. Find the average rate of change of $f(x) = 3x^2 - 5x$ from 1 to 3.

62. Graph $g(x) = \begin{cases} x^2 & \text{if } x \le 0 \\ \sqrt{x} + 1 & \text{if } x > 0 \end{cases}$

In Problems 63–64, complete the square for each quadratic function.

63. $f(x) = x^2 - 10x + 7$

64. $g(x) = 3x^2 + 15x + 13$

65. Find the x-intercept(s) and y-intercepts(s) of the graph of $4x^2 + 9y = 72$.

66. Find the domain of $f(x) = \dfrac{\sqrt{x + 2}}{x - 4}$.

67. Suppose $f(x) = x^2 - 3x + 7$. Find an equation of the secant line containing the points $(-1, f(-1))$ and $(2, f(2))$.

68. Use a graphing utility to graph $f(x) = x^3 - 8x^2 + 13x - 2$ over the interval $[-2, 8]$. Then, approximate any local maximum values and local minimum values, and determine where f is increasing and where f is decreasing. Round answers to two decimal places.

'Are You Prepared?' Answers

1.

2. $\dfrac{2}{3}$ **3.** -4 **4.** $\{50\}$ **5.** 0   **6.** True

---

4.2 Building Linear Models from Data

PREPARING FOR THIS SECTION *Before getting started, review the following:*

- Rectangular Coordinates (Section 2.1, pp. 150–151)
- Functions (Section 3.1, pp. 203–213)
- Lines (Section 2.3, pp. 169–179)

Now Work the 'Are You Prepared?' problems on page 295.

OBJECTIVES 1 Draw and Interpret Scatter Plots (p. 291)
2 Distinguish between Linear and Nonlinear Relations (p. 292)
3 Use a Graphing Utility to Find the Line of Best Fit (p. 294)

1 Draw and Interpret Scatter Plots

In Section 4.1, we built linear models from verbal descriptions. Linear models can also be constructed by fitting a linear function to data that can be represented as ordered pairs. The first step is to plot the ordered pairs using rectangular coordinates. The resulting graph is a **scatter plot**.

EXAMPLE 1   **Drawing and Interpreting a Scatter Plot**

In baseball, the on-base percentage for a team represents the percentage of time that the players safely reach base. The data given in Table 6 on the next page represent the number of runs scored y and the on-base percentage x for teams in the National League during the 2017 regular baseball season.

(a) Draw a scatter plot of the data, treating on-base percentage as the independent variable.

(b) Use a graphing utility to draw a scatter plot.

(c) Describe what happens to runs scored as the on-base percentage increases.

Table 6

Team	On-Base Percentage, x	Runs Scored, y	(x, y)
Arizona	32.9	812	(32.9, 812)
Atlanta	32.6	732	(32.6, 732)
Chicago Cubs	33.8	822	(33.8, 822)
Cincinnati	32.9	753	(32.9, 753)
Colorado	33.8	824	(33.8, 824)
LA Dodgers	33.4	770	(33.4, 770)
Miami	33.1	778	(33.1, 778)
Milwaukee	32.2	732	(32.2, 732)
NY Mets	32.0	735	(32.0, 735)
Philadelphia	31.5	690	(31.5, 690)
Pittsburgh	31.8	668	(31.8, 668)
San Diego	29.9	604	(29.9, 604)
San Francisco	30.9	639	(30.9, 639)
St. Louis	33.4	761	(33.4, 761)
Washington	33.2	819	(33.2, 819)

Source: espn.com

Solution

(a) To draw a scatter plot, plot the ordered pairs listed in Table 6, with the on-base percentage as the *x*-coordinate and the runs scored as the *y*-coordinate. See Figure 6(a). Notice that the points in the scatter plot are not connected.

(b) Figure 6(b) shows a scatter plot using a TI-84 Plus C graphing calculator.

(c) The scatter plots show that as the on-base percentage increases, the number of runs scored also increases.

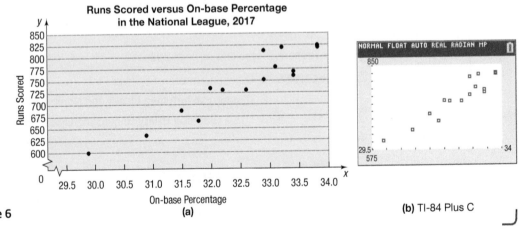

Figure 6

(a) Runs Scored versus On-base Percentage in the National League, 2017

(b) TI-84 Plus C

 Now Work PROBLEM 11(a)

2 Distinguish between Linear and Nonlinear Relations

Notice that the points in Figure 6 do not follow a perfect linear relation. However, the data exhibit a linear pattern. There are numerous possible explanations why the data are not perfectly linear, but one easy explanation is the fact that other variables besides on-base percentage (such as number of home runs hit) play a role in determining runs scored.

Scatter plots are used to help us to see the type of relation that exists between two variables. In this text, we discuss a variety of different relations that may exist between two variables. For now, we concentrate on distinguishing between linear and nonlinear relations. See Figure 7.

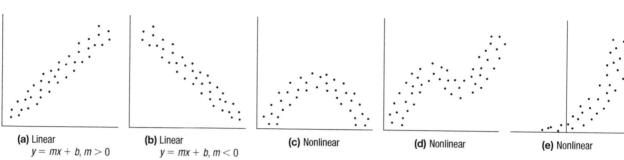

(a) Linear
$y = mx + b, m > 0$

(b) Linear
$y = mx + b, m < 0$

(c) Nonlinear

(d) Nonlinear

(e) Nonlinear

Figure 7

EXAMPLE 2

Distinguishing between Linear and Nonlinear Relations

Determine whether the relation between the two variables in each scatter plot in Figure 8 is linear or nonlinear.

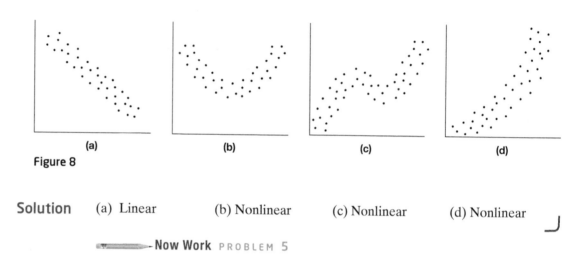

(a)

(b)

(c)

(d)

Figure 8

Solution   (a) Linear   (b) Nonlinear   (c) Nonlinear   (d) Nonlinear

Now Work PROBLEM 5

Suppose that the scatter plot of a set of data indicate a linear relationship, as in Figure 7(a) or (b). We might want to model the data by finding an equation of a line that relates the two variables. One way to obtain a model for such data is to draw a line through two points on the scatter plot and determine an equation of the line.

EXAMPLE 3

Finding a Model for Linearly Related Data

Use the data in Table 6 from Example 1.

(a) Select two points and find an equation of the line containing the points.
(b) Graph the line on the scatter plot obtained in Example 1(a).

Solution   (a) Select two points, say $(31.5, 690)$ and $(33.1, 778)$. The slope of the line joining the points $(31.5, 690)$ and $(33.1, 778)$ is

$$m = \frac{778 - 690}{33.1 - 31.5} = \frac{88}{1.6} = 55$$

An equation of the line with slope 55 and containing the point $(31.5, 690)$ is found using the point-slope form with $m = 55$, $x_1 = 31.5$, and $y_1 = 690$.

$$y - y_1 = m(x - x_1) \qquad \text{Point-slope form of a line}$$

$$y - 690 = 55(x - 31.5) \qquad x_1 = 31.5, y_1 = 690, m = 55$$

$$y - 690 = 55x - 1732.5$$

$$y = 55x - 1042.5 \qquad \text{The model}$$

(continued)

(b) Figure 9 shows the scatter plot with the graph of the line found in part (a).

Figure 9 $y = 55x - 1042.5$

Select two other points and complete the solution. Graph the line on the scatter plot obtained in Figure 6.

Now Work PROBLEMS 11(b) AND (c)

3 Use a Graphing Utility to Find the Line of Best Fit

The model obtained in Example 3 depends on the selection of points, which varies from person to person. So, the model that we found might be different from the model you found. Although the model in Example 3 appears to fit the data well, there may be a model that "fits them better." Do you think your model fits the data better? Is there a *line of best fit*? As it turns out, there is a method for finding a model that best fits linearly related data (called the **line of best fit**).*

EXAMPLE 4

Finding the Line of Best Fit for Linearly Related Data

Use the data in Table 6 from Example 1.

(a) Use a graphing utility to find the line of best fit that models the relation between on-base percentage and runs scored.

(b) Graph the line of best fit on the scatter plot obtained in Example 1(b).

(c) Interpret the slope.

(d) Use the line of best fit to predict the number of runs a team will score if their on-base percentage is 32.2.

Solution

(a) Graphing utilities contain built-in programs that find the line of best fit for a collection of points in a scatter plot. Executing the LINear REGression program on a TI-84 Plus C provides the results shown in Figure 10. This output shows the equation $y = ax + b$, where a is the slope of the line and b is the y-intercept. The line of best fit that relates on-base percentage to runs scored may be expressed as the line

$$y = 57.35x - 1120.75 \quad \text{The model}$$

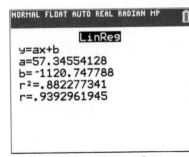

Figure 10 Linear Regression

---

* We do not discuss the underlying mathematics of lines of best fit in this text.

Figure 11 $y = 57.35x - 1120.75$

(b) Figure 11 shows the graph of the line of best fit, along with the scatter plot.

(c) The slope of the line of best fit is 57.35, which means that, for every 1 percent increase in the on-base percentage, runs scored increase by 57.35, on average.

(d) Letting $x = 32.2$ in the equation of the line of best fit, we obtain

$$y = 57.35 \cdot 32.2 - 1120.75 \approx 726 \text{ runs}$$

Now Work PROBLEMS 11(d) AND (f)

Does the line of best fit appear to be a good fit? In other words, does it appear to accurately describe the relation between on-base percentage and runs scored?

And just how "good" is this line of best fit? Look again at Figure 10. The last line of output is $r = 0.939$. This number, called the **correlation coefficient**, $r, -1 \leq r \leq 1$, is a measure of the strength of the linear relation that exists between two variables. The closer $|r|$ is to 1, the more nearly perfect the linear relationship is. If r is close to 0, there is little or no linear relationship between the variables. A negative value of $r, r < 0$, indicates that as x increases, y decreases; a positive value of $r, r > 0$, indicates that as x increases, y does also. The data given in Table 6, which have a correlation coefficient of 0.939, indicate a linear relationship with positive slope.

4.2 Assess Your Understanding

'Are You Prepared?' *Answers are given at the end of these exercises. If you get a wrong answer, read the pages listed in* red.

1. Plot the points $(1, 5), (2, 6), (3, 9), (1, 12)$ in the Cartesian plane. Is the relation $\{(1, 5), (2, 6), (3, 9), (1, 12)\}$ a function? Why? (pp. 150 and 203–207)

2. Find an equation of the line containing the points $(1, 4)$ and $(3, 8)$. (p. 175)

Concepts and Vocabulary

3. A _____ is used to help us to see what type of relation, if any, may exist between two variables.

4. *True or False* The correlation coefficient is a measure of the strength of a linear relation between two variables and must lie between -1 and 1, inclusive.

Skill Building

In Problems 5–10, examine each scatter plot and determine whether the relation is linear or nonlinear.

5.

6.

7.

8. 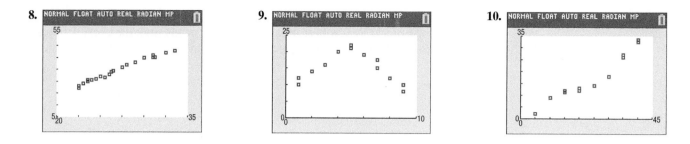 **9.** **10.**

In Problems 11–16:
(a) *Draw a scatter plot.*
(b) *Select two points from the scatter plot, and find an equation of the line containing the points selected.*
(c) *Graph the line found in part (b) on the scatter plot.*
(d) *Use a graphing utility to find the line of best fit.*
(e) *What is the correlation coefficient r?*
(f) *Use a graphing utility to draw the scatter plot and graph the line of best fit on it.*

11.

x	3	4	5	6	7	8	9
y	4	6	7	10	12	14	16

12.

x	3	5	7	9	11	13
y	0	2	3	6	9	11

13.

x	−2	−1	0	1	2
y	−4	0	1	4	5

14.

x	−2	−1	0	1	2
y	7	6	3	2	0

15.

x	−20	−17	−15	−14	−10
y	100	120	118	130	140

16.

x	−30	−27	−25	−20	−14
y	10	12	13	13	18

Applications and Extensions

17. Candy The following data represent the weight (in grams) of various candy bars and the corresponding number of calories.

Candy Bar	Weight, x	Calories, y
Hershey's Milk Chocolate®	44.28	230
Nestle's Crunch®	44.84	230
Butterfinger®	61.30	270
Baby Ruth®	66.45	280
Almond Joy®	47.33	220
Twix® (with caramel)	58.00	280
Snickers®	61.12	280
Heath®	39.52	210

Source: Megan Pocius, student at Joliet Junior College

(a) Draw a scatter plot of the data, treating weight as the independent variable.
(b) What type of relation appears to exist between the weight of a candy bar and the number of calories?
(c) Select two points and find a linear model that contains the points.
(d) Graph the line on the scatter plot drawn in part (a).
(e) Use the linear model to predict the number of calories in a candy bar that weighs 62.3 grams.
(f) Interpret the slope of the line found in part (c).

18. Tornadoes The following data represent the width (in yards) and length (in miles) of various tornadoes.

Width (yards), w	Length (miles), L
200	2.5
350	4.8
180	2.0
300	2.5
500	5.8
400	4.5
500	8.0
800	8.0
100	3.4
50	0.5
700	9.0
600	5.7

Source: NOAA

(a) Draw a scatter plot of the data, treating width as the independent variable.
(b) What type of relation appears to exist between the width and the length of tornadoes?
(c) Select two points and find a linear model that contains the points.
(d) Graph the line on the scatter plot drawn in part (a).
(e) Use the linear model to predict the length of a tornado that has a width of 450 yards.
(f) Interpret the slope of the line found in part (c).

19. Video Games and Grade-Point Average Professor Grant Alexander wanted to find a linear model that relates the number h of hours a student plays video games each week to the cumulative grade-point average G of the student. He randomly selected 10 full-time students at his college and asked each student to disclose the number of hours spent playing video games and the student's cumulative grade-point average.

Hours of Video Games per Week, h	Grade-point Average, G
0	3.49
0	3.05
2	3.24
3	2.82
3	3.19
5	2.78
8	2.31
8	2.54
10	2.03
12	2.51

(a) Explain why the number of hours spent playing video games is the independent variable and cumulative grade-point average is the dependent variable.
(b) Use a graphing utility to draw a scatter plot.
(c) Use a graphing utility to find the line of best fit that models the relation between number of hours of video game playing each week and grade-point average. Express the model using function notation.
(d) Interpret the slope of the line of best fit.
(e) Predict the grade-point average of a student who plays video games for 8 hours each week.
(f) How many hours of video game playing do you think a student plays whose grade-point average is 2.40?

20. Hurricanes The data at the top of the next column represent the atmospheric pressure p (in millibars) and the wind speed w (in knots) measured during various tropical systems in the Atlantic Ocean.

(a) Use a graphing utility to draw a scatter plot of the data, treating atmospheric pressure as the independent variable.
(b) Use a graphing utility to find the line of best fit that models the relation between atmospheric pressure and wind speed. Express the model using function notation.
(c) Interpret the slope.
(d) Predict the wind speed of a tropical storm if the atmospheric pressure measures 990 millibars.
(e) What is the atmospheric pressure of a hurricane if the wind speed is 85 knots?

Atmospheric Pressure (millibars), p	Wind Speed (knots), w
993	50
994	60
997	45
1003	45
1004	40
1000	55
994	55
942	105
1006	40
942	120
986	50
983	70
940	120
966	100
982	55

Source: National Hurricane Center

21. Maternal Age versus Down Syndrome A biologist would like to know how the age of the mother affects the incidence of Down syndrome. The following data represent the age of the mother and the incidence of Down syndrome per 1000 pregnancies. Draw a scatter plot treating age of the mother as the independent variable. Would it make sense to find the line of best fit for these data? Why or why not?

Age of Mother, x	Incidence of Down Syndrome, y
33	2.4
34	3.1
35	4
36	5
37	6.7
38	8.3
39	10
40	13.3
41	16.7
42	22.2
43	28.6
44	33.3
45	50

Source: Hook, E.B., Journal of the American Medical Association, 249, 2034–2038, 1983.

22. U.S. Advertising The following data represent the percentages of U.S. advertising spending for Internet ads, n, and magazine ads, m, over time.

Internet, n	Magazine, m
8.2%	14.5%
10.7	13.9
14.0	12.5
15.5	12.3
17.2	12.0
19.4	11.1
22.0	10.7
24.8	10.1
28.3	9.5
31.7	8.7
35.1	7.8

Source: Marketing Fact Pack 2018. The Ad Age Datacenter, 12/18/2017

(a) Draw a scatter plot of the data, treating percentage of spending on Internet ads as the independent variable. Does the relation appear to be linear?

(b) Use a graphing utility to find the line of best fit that models the relation between the percentages of spending on Internet ads and magazine ads. Express the model using function notation.

(c) What is the correlation coefficient? Does the correlation coefficient support your conclusion from part (a)?

(d) Interpret the slope of the line of best fit.

(e) What is the domain of the function?

(f) Predict the percentage of spending on magazine ads when Internet ads account for 26.0% of ad spending.

Explaining Concepts: Discussion and Writing

23. Find the line of best fit for the ordered pairs $(1, 5)$ and $(3, 8)$. What is the correlation coefficient for these data? Why is this result reasonable?

24. What does a correlation coefficient of 0 imply?

25. Explain why it does not make sense to interpret the y-intercept in Problem 17.

26. Refer to Problem 19. Solve $G(h) = 0$ and provide an interpretation of this result. Find $G(0)$ and provide an interpretation of this result.

Retain Your Knowledge

Problems 27–36 are based on material learned earlier in the course. The purpose of these problems is to keep the material fresh in your mind so that you are better prepared for the final exam.

27. Find an equation for the line containing the points $(-1, 5)$ and $(3, -3)$. Write the equation using either the general form or the slope-intercept form, whichever you prefer.

28. Find the domain of $f(x) = \dfrac{x-1}{x^2-25}$.

29. For $f(x) = 5x - 8$ and $g(x) = x^2 - 3x + 4$, find $(g - f)(x)$.

30. Find a function whose graph is the graph of $y = x^2$, but shifted to the left 3 units and shifted down 4 units.

31. Solve: $x^2 - 4x = 3$

32. Solve. Write the answer in interval notation.
$$5(2x + 7) - 6x \geq 10 - 3(x + 9)$$

33. Determine algebraically whether the function $f(x) = \dfrac{x^2}{5 - 2x^2}$ is even, odd, or neither.

34. Find the x-intercept(s) and y-intercept(s) of the graph of $3x - 8y = 6$.

35. Rationalize the numerator: $\dfrac{6 - \sqrt{x+1}}{x - 35}$ $x \geq -1, x \neq 35$

36. Write $\dfrac{4x}{(2+x)^{1/2}} + 3(2+x)^{1/2}$, $x > -2$, as a single quotient in which only positive exponents appear.

'Are You Prepared?' Answers

1. Not a function because the input, 1, corresponds to two different outputs.

2. $y = 2x + 2$

4.3 Quadratic Functions and Their Properties

PREPARING FOR THIS SECTION *Before getting started, review the following:*

- Intercepts (Section 2.2, pp. 160–161)
- Graphing Techniques: Transformations (Section 3.5, pp. 254–263)
- Completing the Square (Section R.5, pp. 55–56; Section 1.2, Examples 4 and 5, pp. 95–96)
- Quadratic Equations (Section 1.2, pp. 92–100)

Now Work the 'Are You Prepared?' problems on page 308.

OBJECTIVES
1 Graph a Quadratic Function Using Transformations (p. 300)
2 Identify the Vertex and Axis of Symmetry of a Parabola (p. 302)
3 Graph a Quadratic Function Using Its Vertex, Axis, and Intercepts (p. 303)
4 Find a Quadratic Function Given Its Vertex and One Other Point (p. 306)
5 Find the Maximum or Minimum Value of a Quadratic Function (p. 306)

Quadratic Functions

Here are some examples of quadratic functions.

$$f(x) = x^2 \quad F(x) = 3x^2 - 5x + 1 \quad g(x) = -6x^2 + 1 \quad H(x) = \frac{1}{2}x^2 + \frac{2}{3}x$$

> **DEFINITION** **Quadratic Function**
>
> A **quadratic function** is a function of the form
>
> $$f(x) = ax^2 + bx + c$$
>
> where a, b, and c are real numbers and $a \neq 0$. The domain of a quadratic function is the set of all real numbers.

In Words
A quadratic function is a function defined by a second-degree polynomial in one variable.

Many applications require a knowledge of quadratic functions. For example, suppose that Texas Instruments collects the data shown in Table 7, which relate the number of calculators sold to the price p (in dollars) per calculator. Since the price of a product determines the quantity that will be purchased, we treat price as the independent variable. The relationship between the number x of calculators sold and the price p per calculator is given by the linear equation

$$x = 21{,}000 - 150p$$

Table 7

Price p per Calculator, (in dollars)	Number of Calculators, x
60	12,000
65	11,250
70	10,500
75	9,750
80	9,000
85	8,250
90	7,500

Then the revenue R derived from selling x calculators at the price p per calculator is equal to the unit selling price p of the calculator times the number x of units actually sold. That is,

$$R = xp$$

$$R(p) = (21{,}000 - 150p)p \quad x = 21{,}000 - 150p$$

$$= -150p^2 + 21{,}000p$$

So the revenue R is a quadratic function of the price p. Figure 12 illustrates the graph of this revenue function, whose domain is $0 \le p \le 140$, since both x and p must be nonnegative.

Figure 12 $R(p) = -150p^2 + 21{,}000p$

Figure 13 Path of a cannonball

A quadratic function also models the motion of a projectile. Based on Newton's Second Law of Motion (force equals mass times acceleration, $F = ma$), it can be shown that, ignoring air resistance, the path of a projectile propelled upward at an inclination to the horizontal is the graph of a quadratic function. See Figure 13 for an illustration.

1 Graph a Quadratic Function Using Transformations

Figure 14 shows the graph of three functions of the form $f(x) = ax^2, a > 0$, for $a = 1$, $a = \dfrac{1}{2}$, and $a = 3$. Figure 15 shows the graphs of $f(x) = ax^2$ for $a < 0$. Notice that these graphs are reflections about the x-axis of the graphs in Figure 14.

The graphs in Figures 14 and 15 are typical of the graphs of all quadratic functions, which are called **parabolas**.

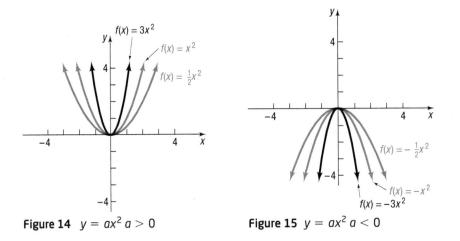

Figure 14 $y = ax^2 \ a > 0$

Figure 15 $y = ax^2 \ a < 0$

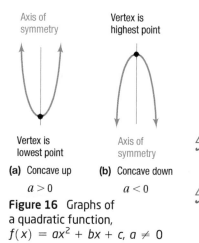

Axis of symmetry

Vertex is highest point

Vertex is lowest point

Axis of symmetry

(a) Concave up

$a > 0$

(b) Concave down

$a < 0$

Figure 16 Graphs of a quadratic function, $f(x) = ax^2 + bx + c, a \neq 0$

Two conclusions can be drawn about the graph of $f(x) = ax^2$.

- As $|a|$ increases, the graph is vertically stretched (becomes "taller"), and as $|a|$ gets closer to zero, the graph is vertically compressed (becomes "shorter").
- If $a > 0$, the graph opens "up," and if $a < 0$, the graph opens "down."

Refer to Figure 16, where two parabolas are pictured.

- The parabola on the left opens up. We describe this by saying the graph is **concave up**. Notice that the graph has a lowest point, where there is an absolute minimum.
- The parabola on the right opens down. We describe this by saying the graph is **concave down**. Notice the graph has a highest point, where there is an absolute maximum.

The lowest or highest point of a parabola is called the **vertex**. The vertical line passing through the vertex in each parabola is called the **axis of symmetry** (usually abbreviated to **axis**) of the parabola. The axis of a parabola can be used to find additional points on the parabola.

The parabolas shown in Figure 16 are the graphs of a quadratic function $f(x) = ax^2 + bx + c, a \neq 0$. Notice that the coordinate axes are not included in the figure. Depending on the numbers $a, b,$ and c, the axes could be placed anywhere. The important fact is that the shape of the graph of a quadratic function will look like one of the parabolas in Figure 16.

In the following example, techniques from Section 3.5 are used to graph a quadratic function $f(x) = ax^2 + bx + c, a \neq 0$. The method of completing the square is used to write the function f in the form $f(x) = a(x - h)^2 + k$.

EXAMPLE 1

Graphing a Quadratic Function Using Transformations

Graph the function $f(x) = 2x^2 + 8x + 5$. Find the vertex and axis of symmetry.

Solution

Recall that to complete the square, the coefficient of x^2 must equal 1. So we factor out 2 on the right-hand side.

> **Need To Review?**
> Completing the square is discussed in Section 1.2, pp. 95–96.

$$f(x) = 2x^2 + 8x + 5$$

$$= 2(x^2 + 4x) + 5 \qquad \text{Factor out the 2 from } 2x^2 + 8x.$$

$$= 2(x^2 + 4x + 4) + 5 - 8 \qquad \text{Complete the square of } x^2 + 4x \text{ by}$$

$$= 2(x + 2)^2 - 3 \qquad \begin{array}{l}\text{adding 4. Notice that the factor of 2}\\ \text{requires that 8 be added and subtracted.}\end{array}$$

The graph of f can be obtained from the graph of $y = x^2$ using transformations as shown in Figure 17. Now compare this graph to the graph in Figure 16(a). The graph of $f(x) = 2x^2 + 8x + 5$ is a parabola that is concave up and has its vertex (lowest point) at $(-2, -3)$. Its axis of symmetry is the line $x = -2$.

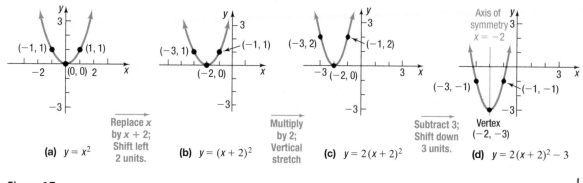

(a) $y = x^2$

Replace x by $x + 2$; Shift left 2 units.

(b) $y = (x + 2)^2$

Multiply by 2; Vertical stretch

(c) $y = 2(x + 2)^2$

Subtract 3; Shift down 3 units.

Vertex $(-2, -3)$

(d) $y = 2(x + 2)^2 - 3$

Figure 17

━ **Now Work** PROBLEM 35

The method used in Example 1 can be used to graph any quadratic function $f(x) = ax^2 + bx + c, a \neq 0$, as follows:

$f(x) = ax^2 + bx + c$

$\qquad = a\left(x^2 + \dfrac{b}{a}x\right) + c$ Factor out a from $ax^2 + bx$.

$\qquad = a\left(x^2 + \dfrac{b}{a}x + \dfrac{b^2}{4a^2}\right) + c - a \cdot \dfrac{b^2}{4a^2}$ Complete the square by adding $\dfrac{b^2}{4a^2}$ in the

$\qquad = a\left(x + \dfrac{b}{2a}\right)^2 + c - \dfrac{b^2}{4a}$ parentheses, and subtracting $a \cdot \dfrac{b^2}{4a^2}$.

Factor; simplify.

$\qquad = a\left(x + \dfrac{b}{2a}\right)^2 + \dfrac{4ac - b^2}{4a}$ $c - \dfrac{b^2}{4a} = c \cdot \dfrac{4a}{4a} - \dfrac{b^2}{4a} = \dfrac{4ac - b^2}{4a}$

These results lead to the following conclusion:

> **DEFINITION** **Vertex Form of a Quadratic Function**
>
> Suppose f is the quadratic function $f(x) = ax^2 + bx + c$. If f is written as
>
> $$f(x) = a(x - h)^2 + k \qquad\qquad (1)$$
>
> where $h = -\dfrac{b}{2a}$ and $k = \dfrac{4ac - b^2}{4a}$, then the quadratic function f is in **vertex form**.

The graph of $f(x) = a(x - h)^2 + k$ is the parabola $y = ax^2$ shifted horizontally h units (replace x by $x - h$) and vertically k units (add k). The vertex of the parabola is (h, k), and the function is concave up if $a > 0$ and is concave down if $a < 0$. The axis of symmetry is the vertical line $x = h$.

For example, compare equation (1) with the solution given in Example 1.

$$f(x) = 2(x + 2)^2 - 3$$
$$= 2(x - (-2))^2 + (-3)$$
$$\qquad\quad \uparrow \qquad\quad \uparrow \qquad\qquad \uparrow$$
$$= a(x - \quad h)^2 \ + \ k$$

Because $a = 2$, the graph is concave up. Also, because $h = -2$ and $k = -3$, its vertex is $(-2, -3)$, the lowest point on the graph.

2 Identify the Vertex and Axis of Symmetry of a Parabola

We do not need to complete the square to identify the vertex of a parabola. It usually easier to obtain the vertex by remembering that its x-coordinate is $h = -\dfrac{b}{2a}$.

The y-coordinate k is then found by evaluating f at $-\dfrac{b}{2a}$. That is, $k = f\left(-\dfrac{b}{2a}\right)$.

> **Properties of the Graph of a Quadratic Function (a Parabola)**
>
> $$f(x) = ax^2 + bx + c \qquad a \neq 0$$
>
> - Vertex $= \left(-\dfrac{b}{2a}, f\left(-\dfrac{b}{2a}\right)\right)$
> - Axis of symmetry: the vertical line $x = -\dfrac{b}{2a}$
> - A parabola is concave up if $a > 0$; the vertex is the minimum point.
> - A parabola is concave down if $a < 0$; the vertex is the maximum point.

EXAMPLE 2

Locating the Vertex of a Parabola without Graphing

Without graphing, locate the vertex and axis of symmetry of the parabola defined by $f(x) = -3x^2 + 6x + 1$. Is it concave up or concave down?

Solution

For this quadratic function, $a = -3$, $b = 6$, and $c = 1$.
The x-coordinate of the vertex is

$$h = -\frac{b}{2a} = -\frac{6}{2(-3)} = 1$$

The y-coordinate of the vertex is

$$k = f\left(-\frac{b}{2a}\right) = f(1) = -3 \cdot 1^2 + 6 \cdot 1 + 1 = 4$$

The vertex is located at the point $(1, 4)$. The axis of symmetry is the line $x = 1$. Because $a = -3 < 0$, the parabola is concave down.

➤ **Now Work** PROBLEM 43(a)

3 Graph a Quadratic Function Using Its Vertex, Axis, and Intercepts

The location of the vertex and intercepts, along with knowledge of whether the graph is concave up or concave down, is usually enough information to graph $f(x) = ax^2 + bx + c, a \neq 0$.

- The y-intercept is the value of f at $x = 0$; that is, the y-intercept is $f(0) = c$.
- The x-intercepts, if there are any, are found by solving the quadratic equation

$$ax^2 + bx + c = 0$$

A quadratic equation has two, one, or no real solutions, depending on whether the discriminant $b^2 - 4ac$ is positive, 0, or negative.

> **The x-Intercepts of a Quadratic Function**
>
> - If the discriminant $b^2 - 4ac > 0$, the graph of $f(x) = ax^2 + bx + c$ has two distinct x-intercepts so it crosses the x-axis in two places.
> - If the discriminant $b^2 - 4ac = 0$, the graph of $f(x) = ax^2 + bx + c$ has one x-intercept so it touches the x-axis at its vertex.
> - If the discriminant $b^2 - 4ac < 0$, the graph of $f(x) = ax^2 + bx + c$ has no x-intercepts so it does not cross or touch the x-axis.

Figure 18 illustrates these possibilities for parabolas that are concave up.

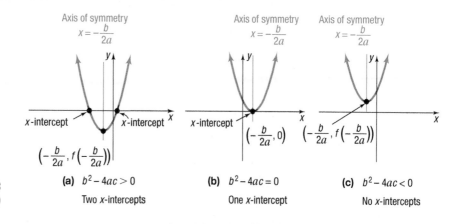

Figure 18
$f(x) = ax^2 + bx + c, a > 0$

(a) $b^2 - 4ac > 0$
Two x-intercepts

(b) $b^2 - 4ac = 0$
One x-intercept

(c) $b^2 - 4ac < 0$
No x-intercepts

| EXAMPLE 3 | Graphing a Quadratic Function Using Its Vertex, Axis, and Intercepts |

(a) Graph $f(x) = -x^2 + 6x - 5$ by determining whether the graph is concave up or down and by finding its vertex, axis of symmetry, y-intercept, and x-intercepts, if any.

(b) Find the domain and the range of f.

(c) Determine where f is increasing and where it is decreasing.

(d) Determine where $f(x) > 0$ and where $f(x) < 0$.

Solution (a) For $f(x) = -x^2 + 6x - 5$, $a = -1$, $b = 6$, and $c = -5$. Because $a = -1 < 0$, the parabola is concave down. The x-coordinate of the vertex is

$$h = -\frac{b}{2a} = -\frac{6}{2(-1)} = 3$$

The y-coordinate of the vertex is

$$k = f(3) = -3^2 + 6 \cdot 3 - 5 = -9 + 18 - 5 = 4$$

The vertex is $(3, 4)$. The axis of symmetry is $x = 3$. The y-intercept is $f(0) = -5$. The x-intercepts are found by solving $f(x) = 0$. This results in the equation $-x^2 + 6x - 5 = 0$, which we solve by factoring:

$$-x^2 + 6x - 5 = 0$$
$$-(x - 1)(x - 5) = 0 \qquad \text{Factor.}$$
$$x - 1 = 0 \quad \text{or} \quad x - 5 = 0 \qquad \text{Use the Zero-Product Property.}$$
$$x = 1 \quad \text{or} \quad x = 5$$

The x-intercepts are 1 and 5.

See Figure 19. Notice that we used the y-intercept and the axis of symmetry, $x = 3$, to obtain the additional point $(6, -5)$ on the graph.

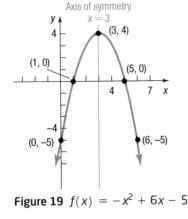

Figure 19 $f(x) = -x^2 + 6x - 5$

(b) The domain of f is the set of all real numbers. Based on the graph, the range of f is the interval $(-\infty, 4]$.

(c) The function f is increasing on the interval $(-\infty, 3]$ and decreasing on the interval $[3, -\infty)$.

(d) Note that $f(x) > 0$ where the graph of f is above the x-axis, and $f(x) < 0$ where the graph of f is below the x-axis. So, $f(x) > 0$ on the interval $(1, 5)$ or for $1 < x < 5$, and $f(x) < 0$ on $(-\infty, 1) \cup (5, \infty)$ or for $x < 1$, $x > 5$. ⌋

✏ **Graph the function in Example 3 by completing the square and using transformations. Which method do you prefer?**

✏ **Now Work** PROBLEMS 27 AND 43 (b)–(f)

If the graph of a quadratic function has only one x-intercept or no x-intercepts, it is usually necessary to plot an additional point to obtain the graph.

| EXAMPLE 4 | Graphing a Quadratic Function Using Its Vertex, Axis, and Intercepts |

(a) Graph $f(x) = x^2 - 6x + 9$ by determining whether the parabola is concave up or concave down and by finding its vertex, axis of symmetry, y-intercept, and x-intercepts, if any.

(b) Find the domain and the range of f.

(c) Determine where f is increasing and where it is decreasing.

(d) Determine where $f(x) > 0$ and where $f(x) < 0$.

Solution (a) For $f(x) = x^2 - 6x + 9$, $a = 1$, $b = -6$, and $c = 9$. Because $a = 1 > 0$, the parabola is concave up. The x-coordinate of the vertex is

$$h = -\frac{b}{2a} = -\frac{-6}{2 \cdot 1} = 3$$

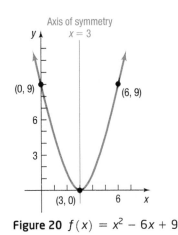

Figure 20 $f(x) = x^2 - 6x + 9$

The y-coordinate of the vertex is

$$k = f(3) = 3^2 - 6 \cdot 3 + 9 = 0$$

The vertex is $(3, 0)$. The axis of symmetry is the line $x = 3$. The y-intercept is $f(0) = 9$. Since the vertex $(3, 0)$ lies on the x-axis, the graph touches the x-axis at the x-intercept. By using the axis of symmetry and the y-intercept at $(0, 9)$, we can locate the additional point $(6, 9)$ on the graph. See Figure 20.

(b) The domain of f is the set of all real numbers. Based on the graph, the range of f is the interval $[0, \infty)$.

(c) The function f is decreasing on the interval $(-\infty, 3]$ and increasing on the interval $[3, \infty)$.

(d) The graph of f is above the x-axis everywhere except at the vertex $(3, 0)$. So, $f(x) > 0$ on $(-\infty, 3) \cup (3, \infty)$ or for $x < 3, x > 3$, and $f(x)$ is never negative.

──── **Now Work** PROBLEM 49

EXAMPLE 5   **Graphing a Quadratic Function Using Its Vertex, Axis, and Intercepts**

(a) Graph $f(x) = 2x^2 + x + 1$ by determining whether the graph is concave up or concave down and by finding its vertex, axis of symmetry, y-intercept, and x-intercepts, if any.

(b) Find the domain and the range of f.

(c) Determine where f is increasing and where it is decreasing.

(d) Determine where $f(x) > 0$ and where $f(x) < 0$.

Solution (a) For $f(x) = 2x^2 + x + 1$, we have $a = 2, b = 1$, and $c = 1$. Because $a = 2 > 0$, the parabola is concave up. The x-coordinate of the vertex is

$$h = -\frac{b}{2a} = -\frac{1}{4}$$

NOTE In Example 5, since the vertex is above the x-axis and the parabola is concave up, we can conclude that the graph of the quadratic function has no x-intercepts. ∎

The y-coordinate of the vertex is

$$k = f\left(-\frac{1}{4}\right) = 2 \cdot \frac{1}{16} + \left(-\frac{1}{4}\right) + 1 = \frac{7}{8}$$

The vertex is $\left(-\frac{1}{4}, \frac{7}{8}\right)$. The axis of symmetry is the line $x = -\frac{1}{4}$. The y-intercept is $f(0) = 1$. The x-intercept(s), if any, satisfy the equation $2x^2 + x + 1 = 0$. The discriminant $b^2 - 4ac = 1^2 - 4 \cdot 2 \cdot 1 = -7 < 0$. This equation has no real solutions, which means the graph has no x-intercepts. Use the point $(0, 1)$ and the axis of symmetry $x = -\frac{1}{4}$ to locate the additional point $\left(-\frac{1}{2}, 1\right)$ on the graph. See Figure 21.

(b) The domain of f is the set of all real numbers. Based on the graph, the range of f is the interval $\left[\frac{7}{8}, \infty\right)$.

(c) The function f is decreasing on the interval $\left(-\infty, -\frac{1}{4}\right]$ and is increasing on the interval $\left[-\frac{1}{4}, \infty\right)$.

(d) The graph of f is always above the x-axis. So, $f(x) > 0$ on the interval $(-\infty, \infty)$ or for all real numbers x.

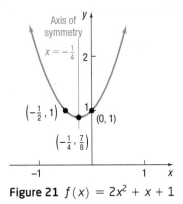

Figure 21 $f(x) = 2x^2 + x + 1$

──── **Now Work** PROBLEM 53

4 Find a Quadratic Function Given Its Vertex and One Other Point

If the vertex (h, k) and one additional point on the graph of a quadratic function $f(x) = ax^2 + bx + c, a \neq 0$, are known, then the vertex form of f,

$$f(x) = a(x - h)^2 + k \tag{2}$$

can be used to obtain the quadratic function.

EXAMPLE 6 **Finding the Quadratic Function Given Its Vertex and One Other Point**

Determine the quadratic function whose vertex is $(1, -5)$ and whose y-intercept is -3.

Solution The vertex is $(1, -5)$, so $h = 1$ and $k = -5$. Substitute these values into equation (2).

$$f(x) = a(x - h)^2 + k \qquad \text{Equation (2)}$$
$$f(x) = a(x - 1)^2 - 5 \qquad h = 1, k = -5$$

To determine the value of a, use the fact that $f(0) = -3$ (the y-intercept).

$$f(x) = a(x - 1)^2 - 5$$
$$-3 = a(0 - 1)^2 - 5 \qquad x = 0, y = f(0) = -3$$
$$-3 = a - 5$$
$$a = 2$$

The quadratic function we seek is

$$f(x) = a(x - h)^2 + k = 2(x - 1)^2 - 5 = 2x^2 - 4x - 3$$

See Figure 22.

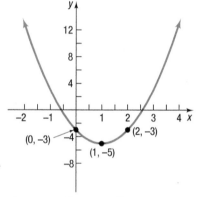

Figure 22 $f(x) = 2x^2 - 4x - 3$

Now Work PROBLEM 59

5 Find the Maximum or Minimum Value of a Quadratic Function

The graph of a quadratic function

$$f(x) = ax^2 + bx + c \qquad a \neq 0$$

is a parabola with vertex $\left(-\dfrac{b}{2a}, f\left(-\dfrac{b}{2a}\right)\right)$. The vertex is the highest point on the graph if $a < 0$ and the lowest point on the graph if $a > 0$. If the vertex is the highest point $(a < 0)$, then $f\left(-\dfrac{b}{2a}\right)$ is the **maximum value** of f. If the vertex is the lowest point $(a > 0)$, then $f\left(-\dfrac{b}{2a}\right)$ is the **minimum value** of f.

EXAMPLE 7 **Finding the Maximum or Minimum Value of a Quadratic Function**

Determine whether the quadratic function

$$f(x) = x^2 - 4x - 5$$

has a maximum or a minimum value. Then find the maximum or minimum value.

Solution For $f(x) = x^2 - 4x - 5$, $a = 1$, $b = -4$, and $c = -5$. Because $a > 0$, the graph of f is concave up, which means the vertex is a minimum point. The minimum occurs at

$$x = -\frac{b}{2a} = -\frac{-4}{2 \cdot 1} = \frac{4}{2} = 2$$
$$\underset{\uparrow}{}$$
$$a = 1, b = -4$$

The minimum value of f is

$$f\left(-\frac{b}{2a}\right) = f(2) = 2^2 - 4 \cdot 2 - 5 = 4 - 8 - 5 = -9$$

-**Now Work** PROBLEM 67

EXAMPLE 8

Analyzing the Motion of a Projectile

A projectile is fired from a cliff 500 feet above the water at an inclination of 45° to the horizontal, with a muzzle velocity of 400 feet per second. From physics, the height h of the projectile above the water can be modeled by

$$h(x) = \frac{-32x^2}{400^2} + x + 500$$

where x is the horizontal distance of the projectile from the base of the cliff. See Figure 23.

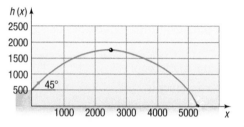

Figure 23

(a) Find the maximum height of the projectile.
(b) How far from the base of the cliff will the projectile strike the water?

Solution

(a) The height of the projectile is given by the quadratic function

$$h(x) = \frac{-32x^2}{400^2} + x + 500 = \frac{-1}{5000}x^2 + x + 500$$

We are looking for the maximum value of h. Because $a < 0$, the vertex is the maximum point and occurs at

$$x = -\frac{b}{2a} = -\frac{1}{2 \cdot \dfrac{-1}{5000}} = \frac{5000}{2} = 2500$$

The maximum height of the projectile is

$$h(2500) = \frac{-1}{5000} \cdot 2500^2 + 2500 + 500 = -1250 + 2500 + 500 = 1750 \text{ ft}$$

(b) The projectile strikes the water when the height h is zero. To find the distance x traveled, solve the equation

$$h(x) = \frac{-1}{5000}x^2 + x + 500 = 0$$

The discriminant of this quadratic equation is

$$b^2 - 4ac = 1^2 - 4 \cdot \frac{-1}{5000} \cdot 500 = 1.4$$

Then

$$x = \frac{-b \pm \sqrt{b^2 - 4ac}}{2a} = \frac{-1 \pm \sqrt{1.4}}{2 \cdot \dfrac{-1}{5000}}$$

$$x \approx -458 \quad \text{or} \quad x \approx 5458$$

Discard the negative solution. The projectile strikes the water about 5458 feet from the base of the cliff.

-**Now Work** PROBLEM 85

Seeing the Concept

Use a graphing utility to graph

$$h(x) = \frac{-1}{5000}x^2 + x + 500$$

$$0 \le x \le 5500$$

Use the appropriate commands to find the maximum height of the projectile and the distance from the base of the cliff to where it strikes the water. Compare your results with those obtained in Example 8.

SUMMARY

Steps for Graphing a Quadratic Function $f(x) = ax^2 + bx + c, a \neq 0$

Option 1

STEP 1: Complete the square in x to write the quadratic function in the vertex form $f(x) = a(x - h)^2 + k$.

STEP 2: Graph the function using transformations.

Option 2

STEP 1: Determine whether the parabola is concave up $(a > 0)$ or down $(a < 0)$.

STEP 2: Find the vertex $\left(-\dfrac{b}{2a}, f\left(-\dfrac{b}{2a} \right) \right)$.

STEP 3: Find the axis of symmetry, $x = -\dfrac{b}{2a}$.

STEP 4: Find the y-intercept, $f(0)$, and the x-intercepts, if any.

    (a) If $b^2 - 4ac > 0$, the graph of the quadratic function has two x-intercepts, which are found by solving the equation $ax^2 + bx + c = 0$.

    (b) If $b^2 - 4ac = 0$, the vertex is the x-intercept.

    (c) If $b^2 - 4ac < 0$, there are no x-intercepts.

STEP 5: Find an additional point using the y-intercept and the axis of symmetry.

STEP 6: Plot the points and draw the graph.

4.3 Assess Your Understanding

'Are You Prepared?' *Answers are given at the end of these exercises. If you get a wrong answer, read the pages listed in* red.

1. Find the intercepts of the equation $y = x^2 - 9$. (p. 161)
2. Find the real solutions of the equation $2x^2 + 7x - 4 = 0$. (pp. 92–100)
3. To complete the square of $x^2 - 5x$, add the number _____ . (pp. 55–56)
4. To graph $y = (x - 4)^2$, shift the graph of $y = x^2$ to the _____ a distance of _____ units. (pp. 254–258)
5. Find the discriminant of $2x^2 - 5x - 8 = 0$. Then identify the number of real solutions of the equation. (p. 97)
6. Complete the square of $3x^2 + 7x$. Factor the new expression. (pp. 55–56)

Concepts and Vocabulary

7. The graph of a quadratic function is called a(n) _____ .
8. The vertical line passing through the vertex of a parabola is called the _____ .
9. The x-coordinate of the vertex of $f(x) = ax^2 + bx + c, a \neq 0$, is _____ .
10. **True or False** The graph of $f(x) = 2x^2 + 3x - 4$ is concave up.
11. **True or False** The y-coordinate of the vertex of $f(x) = -x^2 + 4x + 5$ is $f(2)$.
12. **True or False** If the discriminant $b^2 - 4ac = 0$, the graph of $f(x) = ax^2 + bx + c, a \neq 0$, touches the x-axis at its vertex.

13. **Multiple Choice** If $b^2 - 4ac > 0$, which conclusion can be made about the graph of $f(x) = ax^2 + bx + c, a \neq 0$?

    (a) The graph has two distinct x-intercepts.

    (b) The graph has no x-intercepts.

    (c) The graph has three distinct x-intercepts.

    (d) The graph has one x-intercept.

14. **Multiple Choice** If the graph of $f(x) = ax^2 + bx + c, a \neq 0$, has a maximum value at its vertex, which condition must be true?

    (a) $-\dfrac{b}{2a} > 0$ (b) $-\dfrac{b}{2a} < 0$

    (c) $a > 0$ (d) $a < 0$

Skill Building

In Problems 15–22, match each graph to one the following functions.

15. $f(x) = x^2 - 1$

16. $f(x) = -x^2 - 1$

17. $f(x) = x^2 - 2x + 1$

18. $f(x) = x^2 + 2x + 1$

19. $f(x) = x^2 - 2x + 2$

20. $f(x) = x^2 + 2x$

21. $f(x) = x^2 - 2x$

22. $f(x) = x^2 + 2x + 2$

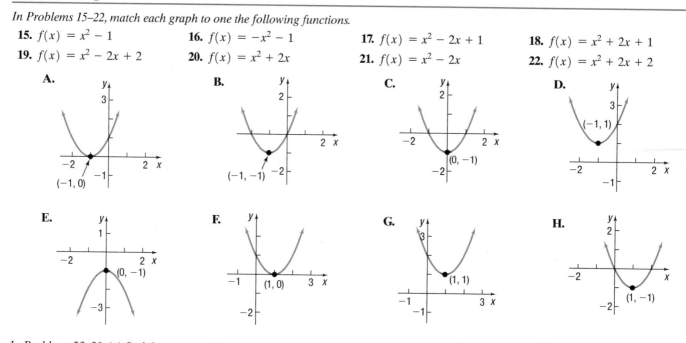

In Problems 23–30, (a) find the vertex and axis of symmetry of each quadratic function. (b) Determine whether the graph is concave up or concave down. (c) Graph the quadratic function.

23. $f(x) = (x - 3)^2 - 2$

24. $f(x) = -(x + 4)^2 - 1$

25. $f(x) = -2(x - 3)^2 + 5$

26. $f(x) = 3(x + 1)^2 - 4$

27. $f(x) = 2(x - 6)^2 + 3$

28. $f(x) = \frac{1}{2}(x + 1)^2 - 3$

29. $f(x) = -\frac{1}{3}\left(x - \frac{1}{2}\right)^2 - \frac{7}{6}$

30. $f(x) = -(x + 5)^2$

In Problems 31–42, graph the function f by starting with the graph of $y = x^2$ and using transformations (shifting, compressing, stretching, and/or reflecting).

[**Hint:** If necessary, write f in the form $f(x) = a(x - h)^2 + k$.]

31. $f(x) = \frac{1}{4}x^2$

32. $f(x) = 2x^2 + 4$

33. $f(x) = (x + 2)^2 - 2$

34. $f(x) = (x - 3)^2 - 10$

35. $f(x) = x^2 + 4x + 2$

36. $f(x) = x^2 - 6x - 1$

37. $f(x) = 2x^2 - 4x + 1$

38. $f(x) = 3x^2 + 6x$

39. $f(x) = -x^2 - 2x$

40. $f(x) = -2x^2 + 6x + 2$

41. $f(x) = \frac{1}{2}x^2 + x - 1$

42. $f(x) = \frac{2}{3}x^2 + \frac{4}{3}x - 1$

In Problems 43–58, (a) find the vertex and the axis of symmetry of each quadratic function, and determine whether the graph is concave up or concave down. (b) Find the y-intercept and the x-intercepts, if any. (c) Use parts (a) and (b) to graph the function. (d) Find the domain and the range of the quadratic function. (e) Determine where the quadratic function is increasing and where it is decreasing. (f) Determine where $f(x) > 0$ and where $f(x) < 0$.

43. $f(x) = x^2 + 2x$

44. $f(x) = x^2 - 4x$

45. $f(x) = -x^2 - 6x$

46. $f(x) = -x^2 + 4x$

47. $f(x) = x^2 + 2x - 8$

48. $f(x) = x^2 - 2x - 3$

49. $f(x) = x^2 + 2x + 1$

50. $f(x) = x^2 + 6x + 9$

51. $f(x) = 2x^2 - x + 2$

52. $f(x) = 4x^2 - 2x + 1$

53. $f(x) = -2x^2 + 2x - 3$

54. $f(x) = -3x^2 + 3x - 2$

55. $f(x) = 3x^2 + 6x + 2$

56. $f(x) = 2x^2 + 5x + 3$

57. $f(x) = -4x^2 - 6x + 2$

58. $f(x) = 3x^2 - 8x + 2$

In Problems 59–64, determine the quadratic function whose graph is given.

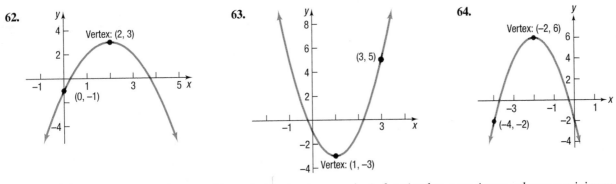

In Problems 65–72, determine, without graphing, whether the given quadratic function has a maximum value or a minimum value, and then find the value.

65. $f(x) = 3x^2 + 24x$

66. $f(x) = -2x^2 + 12x$

67. $f(x) = 2x^2 + 12x - 3$

68. $f(x) = 4x^2 - 8x + 3$

69. $f(x) = -x^2 + 6x - 1$

70. $f(x) = -2x^2 + 8x + 3$

71. $f(x) = -5x^2 + 20x + 3$

72. $f(x) = 4x^2 - 4x$

Applications and Extensions

73. The graph of the function $f(x) = ax^2 + bx + c$ has vertex at $(0, 2)$ and passes through the point $(1, 8)$. Find a, b, and c.

74. The graph of the function $f(x) = ax^2 + bx + c$ has vertex at $(1, 4)$ and passes through the point $(-1, -8)$. Find a, b, and c.

In Problems 75–80, for the given functions f and g:

(a) Graph f and g on the same Cartesian plane.
(b) Solve $f(x) = g(x)$.
(c) Use the result of part (b) to label the points of intersection of the graphs of f and g.
(d) Shade the region for which $f(x) > g(x)$; that is, the region below f and above g.

75. $f(x) = 2x - 1$; $g(x) = x^2 - 4$

76. $f(x) = -2x - 1$; $g(x) = x^2 - 9$

77. $f(x) = -x^2 + 4$; $g(x) = -2x + 1$

78. $f(x) = -x^2 + 9$; $g(x) = 2x + 1$

79. $f(x) = -x^2 + 5x$; $g(x) = x^2 + 3x - 4$

80. $f(x) = -x^2 + 7x - 6$; $g(x) = x^2 + x - 6$

For Problems 81 and 82, use the fact that a quadratic function of the form $f(x) = ax^2 + bx + c$ with $b^2 - 4ac > 0$ may also be written in the form $f(x) = a(x - r_1)(x - r_2)$, where r_1 and r_2 are the x-intercepts of the graph of the quadratic function.

81. (a) Find quadratic functions whose x-intercepts are -3 and 1 with $a = 1$; $a = 2$; $a = -2$; $a = 5$.
   (b) How does the value of a affect the intercepts?
   (c) How does the value of a affect the axis of symmetry?
   (d) How does the value of a affect the vertex?
   (e) Compare the x-coordinate of the vertex with the midpoint of the x-intercepts. What might you conclude?

82. (a) Find quadratic functions whose x-intercepts are -5 and 3 with $a = 1$; $a = 2$; $a = -2$; $a = 5$.
   (b) How does the value of a affect the intercepts?
   (c) How does the value of a affect the axis of symmetry?
   (d) How does the value of a affect the vertex?
   (e) Compare the x-coordinate of the vertex with the midpoint of the x-intercepts. What might you conclude?

83. Suppose that $f(x) = x^2 + 4x - 21$.
   (a) What is the vertex of f?
   (b) What are the x-intercepts of the graph of f?
   (c) Solve $f(x) = -21$ for x. What points are on the graph of f?
   (d) Use the information obtained in parts (a)–(c) to graph $f(x) = x^2 + 4x - 21$.

84. Suppose that $f(x) = x^2 + 2x - 8$.
   (a) What is the vertex of f?
   (b) What are the x-intercepts of the graph of f?
   (c) Solve $f(x) = -8$ for x. What points are on the graph of f?
   (d) Use the information obtained in parts (a)–(c) to graph $f(x) = x^2 + 2x - 8$.

85. Analyzing the Motion of a Projectile A projectile is fired from a cliff 200 feet above the water at an inclination of 45° to the horizontal, with a muzzle velocity of 50 feet per second. The height h of the projectile above the water is modeled by

$$h(x) = \frac{-32x^2}{50^2} + x + 200$$

where x is the horizontal distance of the projectile from the face of the cliff.
   (a) At what horizontal distance from the face of the cliff is the height of the projectile a maximum?
   (b) Find the maximum height of the projectile.
   (c) At what horizontal distance from the face of the cliff will the projectile strike the water?
   (d) Graph the function h, $0 \le x \le 200$.
   (e) Use a graphing utility to verify the solutions found in parts (b) and (c).
   (f) When the height of the projectile is 100 feet above the water, how far is it from the cliff?

86. Analyzing the Motion of a Projectile A projectile is fired at an inclination of 45° to the horizontal, with a muzzle velocity of 100 feet per second. The height h of the projectile is modeled by

$$h(x) = \frac{-32x^2}{100^2} + x$$

where x is the horizontal distance of the projectile from the firing point.

(a) At what horizontal distance from the firing point is the height of the projectile a maximum?

(b) Find the maximum height of the projectile.

(c) At what horizontal distance from the firing point will the projectile strike the ground?

(d) Graph the function h, $0 \le x \le 350$.

(e) Use a graphing utility to verify the results obtained in parts (b) and (c).

(f) When the height of the projectile is 50 feet above the ground, how far has it traveled horizontally?

87. Maximizing Revenue Suppose that the manufacturer of a gas clothes dryer has found that when the unit price is p dollars, the revenue R (in dollars) is

$$R(p) = -4p^2 + 4000p$$

What unit price p maximizes revenue? What is the maximum revenue?

88. Maximizing Revenue A lawn mower manufacturer has found that the revenue, in dollars, from sales of zero-turn mowers is a function of the unit price p, in dollars, that it charges. If the revenue R is

$$R(p) = -\frac{1}{2}p^2 + 2900p$$

what unit price p should be charged to maximize revenue? What is the maximum revenue?

89. Minimizing Marginal Cost The **marginal cost** of a product can be thought of as the cost of producing one additional unit of output. For example, if the marginal cost of producing the 50th product is \$6.20, it costs \$6.20 to increase production from 49 to 50 units of output. Suppose the marginal cost C (in dollars) to produce x thousand digital music players is given by the function

$$C(x) = x^2 - 140x + 7400$$

(a) How many players should be produced to minimize the marginal cost?

(b) What is the minimum marginal cost?

90. Minimizing Marginal Cost (See Problem 89.) The marginal cost C (in dollars) of manufacturing x smartphones (in thousands) is given by

$$C(x) = 5x^2 - 200x + 4000$$

(a) How many smartphones should be manufactured to minimize the marginal cost?

(b) What is the minimum marginal cost?

91. Business The monthly revenue R achieved by selling x wristwatches is $R(x) = 75x - 0.2x^2$. The monthly cost C of selling x wristwatches is

$$C(x) = 32x + 1750$$

(a) How many wristwatches must the firm sell to maximize revenue? What is the maximum revenue?

(b) Profit is given as $P(x) = R(x) - C(x)$. What is the profit function?

(c) How many wristwatches must the firm sell to maximize profit? What is the maximum profit?

(d) Provide a reasonable explanation as to why the answers found in parts (a) and (c) differ. Explain why a quadratic function is a reasonable model for revenue.

92. Business The daily revenue R achieved by selling x boxes of candy is $R(x) = 9.5x - 0.04x^2$. The daily cost C of selling x boxes of candy is $C(x) = 1.25x + 250$.

(a) How many boxes of candy must the firm sell to maximize revenue? What is the maximum revenue?

(b) Profit is given as $P(x) = R(x) - C(x)$. What is the profit function?

(c) How many boxes of candy must the firm sell to maximize profit? What is the maximum profit?

(d) Provide a reasonable explanation as to why the answers found in parts (a) and (c) differ. Explain why a quadratic function is a reasonable model for revenue.

93. Stopping Distance An accepted relationship between stopping distance d (in feet), and the speed v of a car (in mph), is $d = 1.1v + 0.06v^2$ on dry, level concrete.

(a) How many feet will it take a car traveling 45 mph to stop on dry, level concrete?

(b) If an accident occurs 200 feet ahead of you, what is the maximum speed you can be traveling to avoid being involved?

94. Birth Rate of Unmarried Women In the United States, the birth rate B of unmarried women (births per 1000 unmarried women) for women whose age is a is modeled by the function $B(a) = -0.33a^2 + 19.17a - 213.37$.

(a) What is the age of unmarried women with the highest birth rate?

(b) What is the highest birth rate of unmarried women?

(c) Evaluate and interpret $B(40)$.

Source: National Vital Statistics Reports, 1/31/2018

95. Chemical Reactions A self-catalytic chemical reaction results in the formation of a compound that causes the formation ratio to increase. If the reaction rate V is modeled by

$$V(x) = kx(a - x), \qquad 0 \le x \le a$$

where k is a positive constant, a is the initial amount of the compound, and x is the variable amount of the compound, for what value of x is the reaction rate a maximum?

96. *Mixed Practice* Find the distance from the vertex of the parabola $f(x) = 2(x - 3)^2 + 5$ to the center of the circle $(x + 3)^2 + (y - 1)^2 = 4$.

97. *Mixed Practice* Find the distance from the vertex of the parabola $g(x) = -3x^2 + 6x + 1$ to the center of the circle $x^2 + y^2 + 10x + 8y + 32 = 0$.

98. *Challenge Problem* Let $f(x) = ax^2 + bx + c$, where a, b, and c are odd integers. If x is an integer, show that $f(x)$ must be an odd integer.

[**Hint:** x is either an even integer or an odd integer.]

99. *Challenge Problem* Find the point on the line $y = x$ that is closest to the point $(3, 1)$.

100. *Challenge Problem* Find the point on the line $y = x + 1$ that is closest to the point $(4, 1)$.

101. *Challenge Problem* Increasing/Decreasing Function Test Suppose $f(x) = x^3 - 7x^2 - 5x + 35$. From calculus, the derivative of f is given by $f'(x) = 3x^2 - 14x - 5$. The function f is increasing where $f'(x) > 0$ and decreasing where $f'(x) < 0$. Determine where f is increasing and where f is decreasing.

102. *Challenge Problem* Test for Concavity Suppose $f(x) = 3x^4 - 8x^3 + 6x + 1$. From calculus, the second derivative of f is given by $f''(x) = 36x^2 - 48x$. The function f is concave up where $f''(x) > 0$ and concave down where $f''(x) < 0$. Determine where f is concave up and where f is concave down.

Explaining Concepts: Discussion and Writing

103. Make up a quadratic function that opens down and has only one x-intercept. Compare yours with others in the class. What are the similarities? What are the differences?

104. On one set of coordinate axes, graph the family of parabolas $f(x) = x^2 + 2x + c$ for $c = -3$, $c = 0$, and $c = 1$. Describe the characteristics of a member of this family.

105. On one set of coordinate axes, graph the family of parabolas $f(x) = x^2 + bx + 1$ for $b = -4$, $b = 0$, and $b = 4$. Describe the general characteristics of this family.

106. State the circumstances that cause the graph of a quadratic function $f(x) = ax^2 + bx + c$ to have no x-intercepts.

107. Why is the graph of a quadratic function concave up if $a > 0$ and concave down if $a < 0$?

108. Can a quadratic function have a range of $(-\infty, \infty)$? Justify your answer.

109. What are the possibilities for the number of times the graphs of two different quadratic functions intersect?

Retain Your Knowledge

Problems 110–119 are based on material learned earlier in the course. The purpose of these problems is to keep the material fresh in your mind so that you are better prepared for the final exam.

110. Determine whether $x^2 + 4y^2 = 16$ is symmetric respect to the x-axis, the y-axis, and/or the origin.

111. Solve the inequality $27 - x \geq 5x + 3$. Write the solution in both set notation and interval notation.

112. Find the center and radius of the circle
$$x^2 + y^2 - 10x + 4y + 20 = 0$$

113. Find the function whose graph is the graph of $y = \sqrt{x}$, but reflected about the y-axis.

114. Find an equation of the line that contains the point $(14, -3)$ and is parallel to the line $5x + 7y = 35$. Write the equation in slope-intercept form.

115. State the domain and range of the relation given below. Is the relation a function?
$$\{(5, -3), (4, -4), (3, -5), (2, -6), (1, -7)\}$$

116. If $f(x) = 3x^2 - 25x + 28$, find $f(7)$.

117. If $g(x) = \dfrac{2}{3}x - 8$, find $g\left(\dfrac{3}{2}x + 12\right)$.

118. Write $\dfrac{4x^2}{(3x + 5)^{2/3}} + 8x(3x + 5)^{1/3}$ as a single quotient with positive exponents.

119. If $f(x) = x^2 + 5x$, find and simplify $\dfrac{f(x) - f(c)}{x - c}$, $x \neq c$.

'Are You Prepared?' Answers

1. $(0, -9), (-3, 0), (3, 0)$ **2.** $\left\{-4, \dfrac{1}{2}\right\}$ **3.** $\dfrac{25}{4}$ **4.** right; 4   **5.** 89; two real solutions   **6.** $3\left(x^2 + \dfrac{7}{3}x + \dfrac{49}{36}\right) = 3\left(x + \dfrac{7}{6}\right)^2$

4.4 Building Quadratic Models from Verbal Descriptions and from Data

PREPARING FOR THIS SECTION *Before getting started, review the following:*

- Problem Solving (Section 1.7, pp. 134–140)
- Building Linear Models from Data (Section 4.2, pp. 291–295)

Now Work the 'Are You Prepared?' problems on page 317.

OBJECTIVES **1** Build Quadratic Models from Verbal Descriptions (p. 313)
         2 Build Quadratic Models from Data (p. 316)

In this section, we first discuss models that lead to a quadratic function from verbal descriptions. Then we fit a quadratic function to data, which is another form of modeling.

When a quadratic function models a problem, the properties of the graph of the function can provide important information about the model. In particular, we can determine the maximum or minimum value of the function. The fact that the graph of a quadratic function has a maximum or minimum value enables us to answer questions involving **optimization**—that is, finding the maximum or minimum values in models.

1 Build Quadratic Models from Verbal Descriptions

In economics, revenue R, in dollars, is defined as the amount of money received from the sale of an item and is equal to the unit selling price p, in dollars, of the item times the number x of units actually sold. That is,

$$R = xp$$

The **Law of Demand** states that p and x are inversely related: As one increases, the other decreases. The equation that relates p and x is called a **demand equation**. When a demand equation is linear, the revenue model is a quadratic function.

EXAMPLE 1   **Maximizing Revenue**

The marketing department at Texas Instruments has found that when certain calculators are sold at a price of p dollars per unit, the number x of calculators sold is given by the demand equation

$$x = 21{,}000 - 150p$$

(a) Find a model that expresses the revenue R as a function of the price p.
(b) What is the domain of R? Assume revenue is nonnegative.
(c) What unit price should be used to maximize revenue?
(d) If this price is charged, what is the maximum revenue?
(e) How many units are sold at this price?
(f) Graph R.
(g) What price should Texas Instruments charge to collect at least \$675,000 in revenue?

Solution   (a) The revenue is $R = xp$, where $x = 21{,}000 - 150p$.

$$R = xp = (21{,}000 - 150p)p = -150p^2 + 21{,}000p \quad \text{The model}$$

(b) Because x represents the number of calculators sold, we have $x \geq 0$, so $21{,}000 - 150p \geq 0$. Solving this linear inequality gives $p \leq 140$. Also from $R = (21{,}000 - 150p)p$, since R is assumed to be nonnegative and $21{,}000 - 150p \geq 0$, it follows that $p \geq 0$. Combining these inequalities gives the domain of R, which is $\{ p \mid 0 \leq p \leq 140 \}$.

(c) The function R is a quadratic function with $a = -150$, $b = 21{,}000$, and $c = 0$. Because $a < 0$, the vertex is the highest point on the parabola. The revenue R is a maximum when the price p is

$$p = -\frac{b}{2a} \underset{\substack{\uparrow \\ a\,=\,-150,\, b\,=\,21{,}000}}{=} -\frac{21{,}000}{2(-150)} = \$70.00$$

(d) The maximum revenue R is

$$R(70) = -150 \cdot 70^2 + 21{,}000 \cdot 70 = \$735{,}000$$

(e) The number of calculators sold is given by the demand equation $x = 21{,}000 - 150p$. At a price of $p = \$70$,

$$x = 21{,}000 - 150 \cdot 70 = 10{,}500$$

calculators are sold.   *(continued)*

(f) To graph R, plot the intercepts $(0, 0)$ and $(140, 0)$ and the vertex $(70, 735\,000)$. See Figure 24 for the graph.

Figure 24

(g) Graph $R = 675{,}000$ and $R(p) = -150p^2 + 21{,}000p$ on the same Cartesian plane. See Figure 25. We find where the graphs intersect by solving

$$675{,}000 = -150p^2 + 21{,}000p$$

$$150p^2 - 21{,}000p + 675{,}000 = 0 \qquad \text{Add } 150p^2 - 21{,}000p \text{ to both sides.}$$

$$p^2 - 140p + 4500 = 0 \qquad \text{Divide both sides by 150.}$$

$$(p - 50)(p - 90) = 0 \qquad \text{Factor.}$$

$$p = 50 \quad \text{or} \quad p = 90 \qquad \text{Use the Zero-Product Property.}$$

The graphs intersect at $(50, 675\,000)$ and $(90, 675\,000)$. Based on the graph in Figure 25, Texas Instruments should charge between \$50 and \$90 to earn at least \$675,000 in revenue.

Figure 25

 Now Work PROBLEM 3

EXAMPLE 2

Maximizing the Area Enclosed by a Fence

A farmer has 2000 yards of fence to enclose a rectangular field. What are the dimensions of the rectangle that encloses the most area?

Solution

Figure 26

Figure 26 illustrates the situation. The available fence represents the perimeter of the rectangle. If x is the length and w is the width, then

$$2x + 2w = 2000 \tag{1}$$

The area A of the rectangle is

$$A = xw$$

To express A in terms of a single variable, solve equation (1) for w and substitute the result in $A = xw$. Then A involves only the variable x. [You could also solve equation (1) for x and express A in terms of w alone. Try it!]

$$2x + 2w = 2000$$

$$2w = 2000 - 2x$$

$$w = \frac{2000 - 2x}{2} = 1000 - x$$

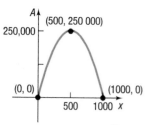

Figure 27 $A(x) = -x^2 + 1000x$

Then the area A is

$$A = xw = x(1000 - x) = -x^2 + 1000x$$

Now, A is a quadratic function of x.

$$A = A(x) = -x^2 + 1000x \qquad a = -1, b = 1000, c = 0$$

Figure 27 shows the graph of $A(x) = -x^2 + 1000x$. Because $a < 0$, the vertex is a maximum point on the graph of A. The maximum value occurs at

$$x = -\frac{b}{2a} = -\frac{1000}{2(-1)} = 500$$

The maximum value of A is

$$A\left(-\frac{b}{2a}\right) = A(500) = -500^2 + 1000 \cdot 500 = -250{,}000 + 500{,}000 = 250{,}000$$

The largest rectangle that can be enclosed by 2000 yards of fence has an area of 250,000 square yards. Its dimensions are 500 yards by 500 yards. ⌐

━━━━━▶ **Now Work** PROBLEM 7

EXAMPLE 3

The Golden Gate Bridge

The Golden Gate Bridge, a suspension bridge, spans the entrance to San Francisco Bay. Its 746-foot-tall towers are 4200 feet apart. The bridge is suspended from two huge cables more than 3 feet in diameter; the 90-foot-wide roadway is 220 feet above the water. The cables are parabolic in shape* and touch the road surface at the center of the bridge. Find the height of the cable above the road at a distance of 1000 feet from the center.

Solution

Begin by choosing the coordinate axes so that the x-axis coincides with the road surface and the origin coincides with the center of the bridge. See Figure 28. As a result, the 746-foot towers will be vertical (height $746 - 220 = 526$ feet above the road) and located 2100 feet from the center. Also, the cable has the shape of a parabola that is concave up. The parabola extends from the towers and has its vertex at $(0, 0)$. This choice of the axes results in the equation of the parabola having the form $y = ax^2$, $a > 0$. Note that the points $(-2100, 526)$ and $(2100, 526)$ are on the graph.

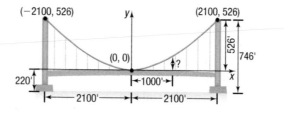

Figure 28

Use these facts to find the value of a in $y = ax^2$.

$$y = ax^2$$
$$526 = a \cdot 2100^2 \qquad x = 2100, y = 526$$
$$a = \frac{526}{2100^2}$$

The equation of the parabola is

$$y = \frac{526}{2100^2}x^2$$

(continued)

---

*A cable suspended from two towers is in the shape of a **catenary**, but when a horizontal roadway is suspended from the cable, the cable takes the shape of a parabola.

When $x = 1000$, the height of the cable is

$$y = \frac{526}{2100^2} \cdot 1000^2 \approx 119.3 \text{ feet}$$

The cable is 119.3 feet above the road at a distance of 1000 feet from the center of the bridge.

── **Now Work** PROBLEM 11

 2 Build Quadratic Models from Data

In Section 4.2, we found the line of best fit for data that appeared to be linearly related. But data may also follow a nonlinear relation. Figures 29(a) and (b) show scatter plots of data that follow a quadratic relation.

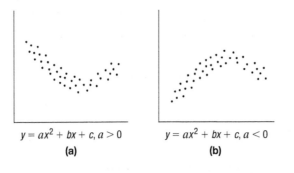

$y = ax^2 + bx + c, a > 0$ $\qquad$ $y = ax^2 + bx + c, a < 0$

Figure 29 $\qquad$ (a) $\qquad\qquad\qquad$ (b)

EXAMPLE 4

Table 8

Age, x	Average Annual Expenditures (in thousands of dollars), E
21.4	34.4
29.6	52.8
39.5	66.4
49.5	71.2
59.4	61.3
68.9	50.9
74.3	45.8

Source: Consumer Expenditure Survey, August 2017

Fitting a Quadratic Function to Data

The data in Table 8 represent the average annual expenditures for consumers of various ages in 2016.

(a) Draw a scatter plot of the data, treating age as the independent variable. Comment on the type of relation that exists between age and average annual expenditures.

(b) Use a graphing utility to find the quadratic function of best fit that models the relation between age and average annual expenditures.

(c) Use the model found in part (b) to approximate the age at which the expenditures is greatest.

(d) Use the model found in part (b) to approximate the highest average annual expenditure.

(e) Use a graphing utility to draw the quadratic function of best fit on the scatter plot.

Solution

(a) Figure 30 shows the scatter plot on a TI-84 Plus C graphing calculator. It appears the data follow a quadratic relation, with $a < 0$.

(b) Use the QUADratic REGression program to obtain the results shown in Figure 31 on the next page. The output shows the equation $y = ax^2 + bx + c$. The quadratic function of best fit that models the relation between age and average annual expenditures is

$$E(x) = -0.0425x^2 + 4.2056x - 34.9639 \quad \text{The model}$$

where x represents age and E represents average annual expenditure in thousands of dollars.

Figure 30 TI-84 Plus C

Figure 31 Quadratic Regression

Figure 32
$E(x) = -0.0425x^2 + 4.2056x - 34.9639$

(c) Based on the quadratic function of best fit, the age of consumers with the greatest expenditures is

$$x = -\frac{b}{2a} = -\frac{4.2056}{2(-0.0425)} \approx 49.5 \text{ years}$$

(d) Evaluate the function $E(x)$ at $x = 49.5$.

$$E(49.5) = -0.0425 \cdot 49.5^2 + 4.2056 \cdot 49.5 - 34.9639 \approx 69.1$$

According to the model, 49.5-year-olds have the greatest expenditures. On average, a 49.5-year-old spends about \$69,100 annually.

(e) Figure 32 shows the graph of the quadratic function found in part (b) drawn on the scatter plot.

Look again at Figure 31. Notice that the output given by the graphing calculator does not include r, the correlation coefficient. Recall that the correlation coefficient is a measure of the strength of a linear relation that exists between two variables. The graphing calculator does not provide an indication of how well the function fits the data in terms of r, since a quadratic function is not linear.

Now Work PROBLEM 17

4.4 Assess Your Understanding

'Are You Prepared?' *Answers are given at the end of these exercises. If you get a wrong answer, read the pages listed in* red.

1. Translate the following sentence into a mathematical equation: The area A of a circle equals the product of the square of its radius r and the constant π. (pp. 134–135)

2. Use a graphing utility to find the line of best fit for the following data: (pp. 294–295)

x	3	5	5	6	7	8
y	10	13	12	15	16	19

Applications and Extensions

3. **Maximizing Revenue** The price p (in dollars) and the quantity x sold of a certain product satisfy the demand equation

$$x = -6p + 600$$

(a) Find a model that expresses the revenue R as a function of p. (Remember, $R = xp$.)
(b) What is the domain of R? Assume R is nonnegative.
(c) What price p maximizes the revenue?
(d) What is the maximum revenue?
(e) How many units are sold at this price?
(f) Graph R.
(g) What price should the company charge to earn at least \$12,600 in revenue?

4. **Maximizing Revenue** The price p (in dollars) and the quantity x sold of a certain product satisfy the demand equation

$$x = -3p + 360$$

(a) Find a model that expresses the revenue R as a function of p.
(b) What is the domain of R? Assume R is nonnegative.

(c) What price p maximizes the revenue?
(d) What is the maximum revenue?
(e) How many units are sold at this price?
(f) Graph R.
(g) What price should the company charge to earn at least \$9600 in revenue?

5. **Maximizing Revenue** The price p (in dollars) and the quantity x sold of a certain product satisfy the demand equation

$$x = -5p + 100$$

(a) Find a model that expresses the revenue R as a function of p.
(b) What is the domain of R? Assume R is nonnegative.
(c) What price p maximizes the revenue?
(d) What is the maximum revenue?
(e) How many units are sold at this price?
(f) Graph R.
(g) What price should the company charge to earn at least \$480 in revenue?

6. **Maximizing Revenue** The price p (in dollars) and the quantity x sold of a certain product satisfy the demand equation

$$x = -20p + 500$$

   (a) Find a model that expresses the revenue R as a function of p.
   (b) What is the domain of R? Assume R is nonnegative.
   (c) What price p maximizes the revenue?
   (d) What is the maximum revenue?
   (e) How many units are sold at this price?
   (f) Graph R.
   (g) What price should the company charge to earn at least \$3000 in revenue?

7. **Enclosing a Rectangular Field** David has 400 yards of fencing and wishes to enclose a rectangular area.
   (a) Express the area A of the rectangle as a function of the width w of the rectangle.
   (b) For what value of w is the area largest?
   (c) What is the maximum area?

8. **Enclosing a Rectangular Field** Beth has 3000 feet of fencing available to enclose a rectangular field.
   (a) Express the area A of the rectangle as a function of x, where x is the length of the rectangle.
   (b) For what value of x is the area largest?
   (c) What is the maximum area?

9. **Enclosing the Most Area with a Fence** A farmer with 4000 meters of fencing wants to enclose a rectangular plot that borders on a river. If the farmer does not fence the side along the river, what is the largest area that can be enclosed? (See the figure.)

$4000 - 2x$

10. **Enclosing the Most Area with a Fence** A farmer with 2000 meters of fencing wants to enclose a rectangular plot that borders on a straight highway. If the farmer does not fence the side along the highway, what is the largest area that can be enclosed?

11. **Suspension Bridge** A suspension bridge with weight uniformly distributed along its length has twin towers that extend 75 meters above the road surface and are 400 meters apart. The cables are parabolic in shape and are suspended from the tops of the towers. The cables touch the road surface at the center of the bridge. Find the height of the cables at a point 100 meters from the center. (Assume that the road is level.)

12. **Architecture** A parabolic arch has a span of 120 feet and a maximum height of 25 feet. Choose suitable rectangular coordinate axes and find the equation of the parabola. Then calculate the height of the arch at points 10 feet, 20 feet, and 40 feet from the center.

13. **Constructing Rain Gutters** A rain gutter is to be made of aluminum sheets that are 12 inches wide by turning up the edges 90°. See the illustration.
   (a) What depth will provide maximum cross-sectional area and allow the most water to flow?
   (b) What depths will allow at least 16 square inches of water to flow?

14. **Norman Windows** A **Norman window** has the shape of a rectangle surmounted by a semicircle of diameter equal to the width of the rectangle. See the figure. If the perimeter of the window is 20 feet, what dimensions will admit the most light (maximize the area)?

15. **Constructing a Stadium** A track-and-field playing area is in the shape of a rectangle with semicircles at each end. See the figure. The inside perimeter of the track is to be 1500 meters. What should the dimensions of the rectangle be so that the area of the rectangle is a maximum?

16. **Architecture** A special window has the shape of a rectangle surmounted by an equilateral triangle. See the figure. If the perimeter of the window is 16 feet, what dimensions will admit the most light?

   [**Hint:** Area of an equilateral triangle $= \dfrac{\sqrt{3}}{4}x^2$, where x is the length of a side of the triangle.]

17. Life Cycle Hypothesis An individual's income varies with his or her age. The following table shows the median income I of males of different age groups within the United States for 2016. For each age group, let the class midpoint represent the independent variable, x. For the class "65 years and older," we will assume that the class midpoint is 69.5.

Age	Class Midpoint, x	Median Income, I
15–24 years	19.5	$12,396
25–34 years	29.5	$38,152
35–44 years	39.5	$51,443
45–54 years	49.5	$51,476
55–64 years	59.5	$47,001
65 years and older	69.5	$31,618

Source: U.S. Census Bureau

(a) Use a graphing utility to draw a scatter plot of the data. Comment on the type of relation that may exist between the two variables.
(b) Use a graphing utility to find the quadratic function of best fit that models the relation between age and median income.
(c) Use the function found in part (b) to determine the age at which an individual can expect to earn the most income.
(d) Use the function found in part (b) to predict the peak income earned.
(e) With a graphing utility, graph the quadratic function of best fit on the scatter plot

18. Height of a Ball A shot-putter throws a ball at an inclination of 45° to the horizontal. The following data represent the height of the ball h, in feet, at the instant that it has traveled x feet horizontally.

Distance, x	Height, h
20	25
40	40
60	55
80	65
100	71
120	77
140	77
160	75
180	71
200	64

(a) Use a graphing utility to draw a scatter plot of the data. Comment on the type of relation that may exist between the two variables.
(b) Use a graphing utility to find the quadratic function of best fit that models the relation between distance and height.
(c) Use the function found in part (b) to determine how far the ball will travel before it reaches its maximum height.
(d) Use the function found in part (b) to find the maximum height of the ball.
(e) With a graphing utility, graph the quadratic function of best fit on the scatter plot.

19. Mixed Practice Which Model? The following data represent the square footage and rents (dollars per month) for apartments in the La Jolla area of San Diego, California.

Square Footage, x	Rent per Month, R
520	$1630
625	$1820
710	$1860
765	$1975
855	$1985
925	$2200
1040	$2360

Source: apartments.com, 2018

(a) Using a graphing utility, draw a scatter plot of the data treating square footage as the independent variable. What type of relation appears to exist between square footage and rent?
(b) Based on your response to part (a), find either a linear or a quadratic model that describes the relation between square footage and rent.
(c) Use your model to predict the rent for an apartment in San Diego that is 875 square feet.

20. Mixed Practice Which Model? A cricket makes a chirping noise by sliding its wings together rapidly. Perhaps you have noticed that the number of chirps seems to increase with the temperature. The following data list the temperature (in degrees Fahrenheit) and the number of chirps per second for the striped ground cricket.

Temperature (°F), x	Chirps per Second, C
88.6	20.0
93.3	19.8
80.6	17.1
69.7	14.7
69.4	15.4
79.6	15.0
80.6	16.0
76.3	14.4
75.2	15.5

Source: Pierce, George W. The Songs of Insects. *Cambridge, MA Harvard University Press, 1949, pp. 12 – 21*

(a) Using a graphing utility, draw a scatter plot of the data, treating temperature as the independent variable. What type of relation appears to exist between temperature and chirps per second?
(b) Based on your response to part (a), find either a linear or a quadratic model that best describes the relation between temperature and chirps per second.
(c) Use your model to predict the chirps per second if the temperature is 80°F.

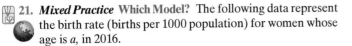

21. *Mixed Practice* **Which Model?** The following data represent the birth rate (births per 1000 population) for women whose age is a, in 2016.

Age, a	Birth Rate, B
16	8.8
19	37.5
22	73.8
27	102.1
32	102.7
37	52.7
42	11.4

Source: National Vital Statistics Reports, 2018

(a) Using a graphing utility, draw a scatter plot of the data, treating age as the independent variable. What type of relation appears to exist between age and birth rate?

(b) Based on your response to part (a), find either a linear or a quadratic model that describes the relation between age and birth rate.

(c) Use your model to predict the birth rate for 35-year-old women.

22. *Challenge Problem* **Simpson's Rule** The figure shows the graph of $y = ax^2 + bx + c$. Suppose that the points $(-h, y_0)$, $(0, y_1)$, and (h, y_2) are on the graph. It can be shown that the area enclosed by the parabola, the x-axis, and the lines $x = -h$ and $x = h$ is

$$\text{Area} = \frac{h}{3}(2ah^2 + 6c)$$

Show that this area may also be given by

$$\text{Area} = \frac{h}{3}(y_0 + 4y_1 + y_2)$$

23. *Challenge Problem* Use the result obtained in Problem 22 to find the area enclosed by $f(x) = -5x^2 + 8$, the x-axis, and the lines $x = -1$ and $x = 1$.

24. *Challenge Problem* Use the result obtained in Problem 22 to find the area enclosed by $f(x) = 2x^2 + 8$, the x-axis, and the lines $x = -2$ and $x = 2$.

25. *Challenge Problem* Use the result obtained in Problem 22 to find the area enclosed by $f(x) = x^2 + 3x + 5$, the x-axis, and the lines $x = -4$ and $x = 4$.

26. *Challenge Problem* Use the result obtained in Problem 22 to find the area enclosed by $f(x) = -x^2 + x + 4$, the x-axis, and the lines $x = -1$ and $x = 1$.

Explaining Concepts: Discussion and Writing

27. Refer to Example 1 in this section. Notice that if the price charged for the calculators is $0 or $140, then the revenue is $0. It is easy to explain why revenue would be $0 if the price charged were $0, but how can revenue be $0 if the price charged is $140?

Retain Your Knowledge

Problems 28–37 are based on material learned earlier in the course. The purpose of these problems is to keep the material fresh in your mind so that you are better prepared for the final exam.

28. Find the distance between the points $P_1 = (4, -7)$ and $P_2 = (-1, 5)$.

29. Find the equation of the circle with center $(-6, 0)$ and radius $r = \sqrt{7}$.

30. Solve: $5x^2 + 8x - 3 = 0$

31. Find the x-intercept and y-intercept of the graph of $5x + 7y = 140$.

32. Solve: $2|3x - 7| - 9 = 21$

33. Find the quotient and remainder:

$x^3 - 7x^2 + 19x - 15$ is divided by $x - 3$

34. Find the domain of $f(x) = \dfrac{5x - 1}{x^3 - 16x}$.

35. Find the function that is finally graphed after all three of the following transformations are applied to the graph of $f(x) = \sqrt{9 - x^2}$:
   (1) Shift left 3 units
   (2) Vertical stretch by a factor of 2
   (3) Shift down 4 units

36. Find the difference quotient of f: $f(x) = \dfrac{3}{x - 1}$

37. Factor completely:

$$4(x + 1)^5(x - 7)^3 + 5(x + 1)^4(x - 7)^4$$

'Are You Prepared?' Answers

1. $A = \pi r^2$ **2.** $y = 1.7826x + 4.0652$

4.5 Inequalities Involving Quadratic Functions

PREPARING FOR THIS SECTION *Before getting started, review the following:*

- Solve Inequalities (Section 1.5, pp. 123–126)
- Use Interval Notation (Section 1.5, pp. 120–121)

Now Work the 'Are You Prepared?' problems on page 323.

OBJECTIVE 1 Solve Inequalities Involving a Quadratic Function (p. 321)

1 Solve Inequalities Involving a Quadratic Function

In this section we solve inequalities that involve quadratic functions.

- To solve the inequality

$$ax^2 + bx + c > 0 \quad a \neq 0$$

graph the quadratic function $f(x) = ax^2 + bx + c$, and, from the graph, determine where the function is above the x-axis—that is, where $f(x) > 0$.

- To solve the inequality

$$ax^2 + bx + c < 0 \quad a \neq 0$$

graph the quadratic function $f(x) = ax^2 + bx + c$, and, from the graph, determine where the function is below the x-axis—that is, where $f(x) < 0$.

- If the inequality is not strict, include the x-intercepts, if any, in the solution.

EXAMPLE 1

Solving an Inequality

Solve the inequality $x^2 - 4x - 12 \leq 0$ and graph the solution set.

Solution

Graph the function $f(x) = x^2 - 4x - 12$.

- y-intercept: $f(0) = -12$ Evaluate f at 0.
- x-intercepts (if any): $x^2 - 4x - 12 = 0$ Solve $f(x) = 0$.

$$(x - 6)(x + 2) = 0 \quad \text{Factor.}$$

$$x - 6 = 0 \quad \text{or} \quad x + 2 = 0 \quad \text{Use the Zero-Product}$$

$$x = 6 \quad \text{or} \quad x = -2 \quad \text{Property.}$$

The y-intercept is -12; the x-intercepts are -2 and 6.

The vertex is at $x = -\dfrac{b}{2a} = -\dfrac{-4}{2} = 2$. Because $f(2) = -16$, the vertex is $(2, -16)$.

Figure 33 $f(x) = x^2 - 4x - 12$

See Figure 33 for the graph.

The graph is below the x-axis for $-2 < x < 6$. Because the original inequality is not strict, the solution includes the x-intercepts. The solution set is $\{x \mid -2 \leq x \leq 6\}$ or, using interval notation, $[-2, 6]$. See Figure 34 for the graph of the solution set.

Figure 34

═══ **Now Work** PROBLEM 9

EXAMPLE 2	Solving an Inequality

Solve the inequality $2x^2 < x + 10$ and graph the solution set.

Solution   ***Option 1***   Rearrange the inequality so that 0 is on the right side.

$$2x^2 < x + 10$$
$$2x^2 - x - 10 < 0 \qquad \text{Subtract } x + 10 \text{ from both sides.}$$

This inequality is equivalent to the original inequality.

Next graph the function $f(x) = 2x^2 - x - 10$ to find where $f(x) < 0$.

- y-intercept: $f(0) = -10$ Evaluate f at 0.
- x-intercepts (if any): $2x^2 - x - 10 = 0$ Solve $f(x) = 0$.

$$(2x - 5)(x + 2) = 0 \qquad \text{Factor.}$$
$$2x - 5 = 0 \quad \text{or} \quad x + 2 = 0 \qquad \text{Use the Zero-Product Property.}$$
$$x = \frac{5}{2} \quad \text{or} \quad x = -2$$

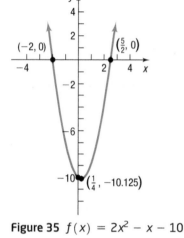

Figure 35 $f(x) = 2x^2 - x - 10$

The y-intercept is -10; the x-intercepts are -2 and $\dfrac{5}{2}$.

The vertex is at $x = -\dfrac{b}{2a} = -\dfrac{-1}{4} = \dfrac{1}{4}$. Because $f\left(\dfrac{1}{4}\right) = -10.125$, the vertex is $\left(\dfrac{1}{4}, -10.125\right)$. See Figure 35 for the graph.

The graph is below the x-axis ($f(x) < 0$) between $x = -2$ and $x = \dfrac{5}{2}$. Because the inequality is strict, the solution set is $\left\{x \middle| -2 < x < \dfrac{5}{2}\right\}$ or, using interval notation, $\left(-2, \dfrac{5}{2}\right)$.

Option 2   If $f(x) = 2x^2$ and $g(x) = x + 10$, then the inequality to be solved is $f(x) < g(x)$. Graph the functions $f(x) = 2x^2$ and $g(x) = x + 10$. See Figure 36. The graphs intersect where $f(x) = g(x)$. Then

$$2x^2 = x + 10 \qquad f(x) = g(x)$$
$$2x^2 - x - 10 = 0$$
$$(2x - 5)(x + 2) = 0 \qquad \text{Factor.}$$
$$2x - 5 = 0 \quad \text{or} \quad x + 2 = 0 \qquad \text{Use the Zero-Product Property.}$$
$$x = \frac{5}{2} \quad \text{or} \qquad x = -2$$

Figure 36

Figure 37

The graphs intersect at the points $(-2, 8)$ and $\left(\dfrac{5}{2}, \dfrac{25}{2}\right)$. Then $f(x) < g(x)$ on the interval $\left(-2, \dfrac{5}{2}\right)$, where the graph of f is below the graph of g.

See Figure 37 for the graph of the solution set.

⌐

━━━━━ **Now Work** PROBLEMS 5 AND 13

EXAMPLE 3	Solving an Inequality

Solve the inequality $x^2 + x + 1 > 0$ and graph the solution set.

Solution

The graph of the function $f(x) = x^2 + x + 1$ has y-intercept 1. There are no x-intercepts (Do you know why? Check the discriminant.). The vertex is $\left(-\dfrac{1}{2}, \dfrac{3}{4}\right)$.

Since $a > 0$, the parabola is concave up and lies above the x-axis for all real numbers x. So $x^2 + x + 1 > 0$ for all real numbers. See Figure 38.

See Figure 39 for the graph of the solution set $(-\infty, \infty)$.

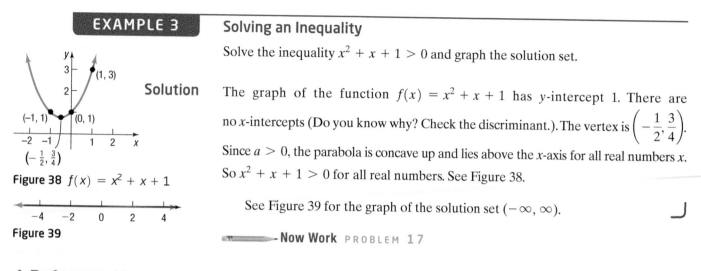

Figure 38 $f(x) = x^2 + x + 1$

Figure 39

━━━━━ **Now Work** PROBLEM 17

4.5 Assess Your Understanding

'Are You Prepared?' *Answers are given at the end of these exercises. If you get a wrong answer, read the pages listed in* red.

1. Solve the inequality $-3x - 2 < 7$. (pp. 123–126)

2. Write $(-2, 7]$ using inequality notation. (pp. 120–121)

Skill Building

In Problems 3–6, use the figure to solve each inequality.

3.

(a) $f(x) > 0$
(b) $f(x) \le 0$

4.

(a) $g(x) < 0$
(b) $g(x) \ge 0$

5.

(a) $g(x) \ge f(x)$
(b) $f(x) > g(x)$

6.

(a) $f(x) < g(x)$
(b) $f(x) \ge g(x)$

In Problems 7–22, solve each inequality.

7. $x^2 - 3x - 10 < 0$

8. $x^2 + 3x - 10 > 0$

9. $x^2 - 4x > 0$

10. $x^2 + 8x > 0$

11. $x^2 - 9 < 0$

12. $x^2 - 1 < 0$

13. $x^2 + x > 12$

14. $x^2 + 7x < -12$

15. $2x^2 < 5x + 3$

16. $6x^2 < 6 + 5x$

17. $x^2 - x + 1 \le 0$

18. $x^2 + 2x + 4 > 0$

19. $4x^2 + 9 < 6x$

20. $25x^2 + 16 < 40x$

21. $6(x^2 - 1) > 5x$

22. $2(2x^2 - 3x) > -9$

Mixed Practice *In Problems 23–30, use the given functions f and g.*
   (a) *Solve* $f(x) = 0$.
   (e) *Solve* $g(x) \le 0$.
   (b) *Solve* $g(x) = 0$.
   (f) *Solve* $f(x) > g(x)$.
   (c) *Solve* $f(x) = g(x)$.
   (g) *Solve* $f(x) \ge 1$.
   (d) *Solve* $f(x) > 0$.

23. $f(x) = x^2 - 1$
$g(x) = 3x + 3$

24. $f(x) = -x^2 + 3$
$g(x) = -3x + 3$

25. $f(x) = -x^2 + 1$
$g(x) = 4x + 1$

26. $f(x) = -x^2 + 4$
$g(x) = -x - 2$

27. $f(x) = x^2 - 4$
$g(x) = -x^2 + 4$

28. $f(x) = x^2 - 2x + 1$
$g(x) = -x^2 + 1$

29. $f(x) = x^2 - x - 2$
$g(x) = x^2 + x - 2$

30. $f(x) = -x^2 - x + 1$
$g(x) = -x^2 + x + 6$

Applications and Extensions

31. What is the domain of the function $f(x) = \sqrt{x^2 - 16}$?

32. What is the domain of the function $f(x) = \sqrt{x - 3x^2}$?

33. Physics A ball is thrown vertically upward with an initial velocity of 80 feet per second. The distance s (in feet) of the ball from the ground after t seconds is $s(t) = 80t - 16t^2$.

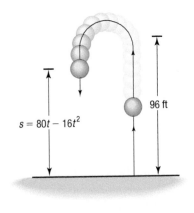

$s = 80t - 16t^2$

96 ft

(a) At what time t will the ball strike the ground?

(b) For what time t is the ball more than 96 feet above the ground?

34. Physics A ball is thrown vertically upward with an initial velocity of 96 feet per second. The distance s (in feet) of the ball from the ground after t seconds is $s(t) = 96t - 16t^2$.

(a) At what time t will the ball strike the ground?

(b) For what times t is the ball more than 128 feet above the ground?

35. Revenue Suppose that the manufacturer of a gas clothes dryer has found that when the unit price is p dollars, the revenue R (in dollars) is

$$R(p) = -4p^2 + 4000p$$

(a) At what prices p is revenue zero?

(b) For what range of prices will revenue exceed $800,000?

36. Revenue The John Deere company has found that the revenue from sales of heavy-duty tractors is a function of the unit price p, in dollars, that it charges. The revenue R, in dollars, is given by

$$R(p) = -\frac{1}{2}p^2 + 1900p$$

(a) At what prices p is revenue zero?

(b) For what range of prices will revenue exceed $1,200,000?

37. Artillery A projectile fired from the point $(0, 0)$ at an angle to the positive x-axis has a trajectory given by

$$y = cx - (1 + c^2)\left(\frac{g}{2}\right)\left(\frac{x}{v}\right)^2$$

where

x = horizontal distance in meters

y = height in meters

v = initial muzzle velocity in meters per second (m/s)

g = acceleration due to gravity = 9.81 meters per second squared (m/s²)

$c > 0$ is a constant determined by the angle of elevation.

A howitzer fires an artillery round with a muzzle velocity of 897 m/s.

(a) If the round must clear a hill 200 meters high at a distance of 2000 meters in front of the howitzer, what c values are permitted in the trajectory equation?

(b) If the goal in part (a) is to hit a target on the ground 75 kilometers away, is it possible to do so? If so, for what values of c? If not, what is the maximum distance the round will travel?

Source: www.answers.com

38. Challenge Problem Runaway Car Using Hooke's Law, we can show that the work W done in compressing a spring a distance of x feet from its at-rest position is $W = \frac{1}{2}kx^2$, where k is a stiffness constant depending on the spring. It can also be shown that the work done by a body in motion before it comes to rest is given by $\widetilde{W} = \frac{w}{2g}v^2$, where

w = weight of the object (in lb), g = acceleration due to gravity (32.2 ft/s²), and v = object's velocity (in ft/s). A parking garage has a spring shock absorber at the end of a ramp to stop runaway cars. The spring has a stiffness constant $k = 9450$ lb/ft and must be able to stop a 4000-lb car traveling at 25 mph. What is the least compression required of the spring? Express your answer using feet to the nearest tenth.

Source: www.sciforums.com

Explaining Concepts: Discussion and Writing

39. Show that the inequality $(x - 4)^2 \leq 0$ has exactly one solution.

40. Show that the inequality $(x - 2)^2 > 0$ has one real number that is not a solution.

41. Explain why the inequality $x^2 + x + 1 > 0$ has all real numbers as the solution set.

42. Explain why the inequality $x^2 - x + 1 < 0$ has the empty set as the solution set.

43. Explain the circumstances under which the x-intercepts of the graph of a quadratic function are included in the solution set of a quadratic inequality.

Retain Your Knowledge

Problems 44–53 are based on material learned earlier in the course. The purpose of these problems is to keep the material fresh in your mind so that you are better prepared for the final exam.

44. Find the domain of $f(x) = \sqrt{10 - 2x}$.

45. Determine algebraically whether $f(x) = \dfrac{-x}{x^2 + 9}$ is even, odd, or neither.

46. Suppose $f(x) = \dfrac{2}{3}x - 6$.

(a) Find the intercepts of the graph of f.

(b) Graph f.

47. Write a general formula to describe the variation: d varies directly with t; $d = 203$ when $t = 3.5$.

48. Find the zeros of $f(x) = x^2 + 6x - 8$.

49. Find the intercepts of the graph of $y = \dfrac{4x^2 - 25}{x^2 - 1}$.

In Problems 50 and 51, if $f(x) = x^2 + 2x - 7$ and $g(x) = 3x - 4$, find:

50. $(g - f)(x)$

51. $(f \cdot g)(x)$

52. Find the difference quotient of f: $f(x) = 3x^2 - 5x$

53. Simplify: $\dfrac{5x^4(2x + 7)^4 - 8x^5(2x + 7)^3}{(2x + 7)^8}$

'Are You Prepared?' Answers

 1. $\{x | x > -3\}$ or $(-3, \infty)$ 2. $-2 < x \leq 7$

Chapter Review

Things to Know

Linear function (p. 281)

$f(x) = mx + b$ Average rate of change of $f = m$

The graph of f is a line with slope m and y-intercept b.

Quadratic function (pp. 299–303)

$f(x) = ax^2 + bx + c, a \neq 0$ The graph of f is a parabola that is concave up if $a > 0$ and is concave down if $a < 0$.

Vertex: $\left(-\dfrac{b}{2a}, f\left(-\dfrac{b}{2a} \right) \right)$

Axis of symmetry: $x = -\dfrac{b}{2a}$

y-intercept: $f(0) = c$

x-intercept(s): If any, found by finding the real solutions of the equation $ax^2 + bx + c = 0$

Objectives

Section	You should be able to...	Examples	Review Exercises
4.1	1 Graph linear functions (p. 281)	1	1(a)–3(a), 1(c)–3(c)
	2 Use average rate of change to identify linear functions (p. 281)	2	1(b)–3(b), 4, 5
	3 Determine whether a linear function is increasing, decreasing, or constant (p. 284)	3	1(d)–3(d)
	4 Build linear models from verbal descriptions (p. 285)	4, 5	21
4.2	1 Draw and interpret scatter plots (p. 291)	1	29(a), 30(a)
	2 Distinguish between linear and nonlinear relations (p. 292)	2, 3	29(b), 30(a)
	3 Use a graphing utility to find the line of best fit (p. 294)	4	29(c)
4.3	1 Graph a quadratic function using transformations (p. 300)	1	6–8
	2 Identify the vertex and axis of symmetry of a parabola (p. 302)	2	9–13
	3 Graph a quadratic function using its vertex, axis, and intercepts (p. 303)	3–5	9–13
	4 Find a quadratic function given its vertex and one other point (p. 306)	6	19, 20
	5 Find the maximum or minimum value of a quadratic function (p. 306)	7–8	14–16, 22–27
4.4	1 Build quadratic models from verbal descriptions (p. 313)	1–3	22–28
	2 Build quadratic models from data (p. 316)	4	30
4.5	1 Solve inequalities involving a quadratic function (p. 321)	1–3	17, 18

Review Exercises

In Problems 1–3:
 (a) Find the slope and y-intercept of each linear function.
 (b) What is the average rate of change of each function?
 (c) Graph each function. Label the intercepts.
 (d) Determine whether the function is increasing, decreasing, or constant.

 1. $f(x) = 2x - 5$ **2.** $h(x) = \dfrac{4}{5}x - 6$ **3.** $G(x) = 4$

In Problems 4 and 5, determine whether the function is linear or nonlinear. If the function is linear, state its slope.

4.

x	y = f(x)
−1	−2
0	3
1	8
2	13
3	18

5.

x	y = g(x)
−1	−3
0	4
1	7
2	6
3	1

In Problems 6–8, graph each quadratic function using transformations (shifting, compressing, stretching, and/or reflecting).

6. $f(x) = (x + 1)^2 - 4$ **7.** $f(x) = -(x - 4)^2$ **8.** $f(x) = -3(x + 2)^2 + 1$

In Problems 9–13, (a) graph each quadratic function by determining whether its graph is concave up or is concave down and by finding its vertex, axis of symmetry, y-intercept, and x-intercepts, if any. (b) Determine the domain and the range of the function. (c) Determine where the function is increasing and where it is decreasing.

9. $f(x) = (x - 2)^2 + 2$ **10.** $f(x) = \frac{1}{4}x^2 - 16$ **11.** $f(x) = -4x^2 + 4x$

12. $f(x) = \frac{9}{2}x^2 + 3x + 1$ **13.** $f(x) = 3x^2 + 4x - 1$

In Problems 14–16, determine whether the given quadratic function has a maximum value or a minimum value, and then find the value.

14. $f(x) = 3x^2 - 6x + 4$ **15.** $f(x) = -x^2 + 8x - 4$

16. $f(x) = -3x^2 + 12x + 4$

In Problems 17 and 18, solve each quadratic inequality.

17. $x^2 + 6x - 16 < 0$ **18.** $3x^2 \geq 14x + 5$

In Problems 19 and 20, find the quadratic function for which:

19. Vertex is $(2, -4)$; y-intercept is -16

20. Vertex is $(-1, 2)$; contains the point $(1, 6)$

21. Sales Commissions Bill was just offered a sales position for a computer company. His salary would be $25,000 per year plus 1% of his total annual sales.

(a) Find a linear function that relates Bill's annual salary, S, to his total annual sales, x.
(b) If Bill's total annual sales were $1,000,000, what would be Bill's salary?
(c) What would Bill have to sell to earn $100,000?
(d) Determine the sales required of Bill for his salary to exceed $150,000.

22. Demand Equation The price p (in dollars) and the quantity x sold of a certain product satisfy the demand equation

$$x = 1500 - 10p$$

(a) Find a model that expresses the revenue R as a function of the price p.
(b) What is the domain of R? Assume R is nonnegative.
(c) What unit price should be used to maximize revenue?
(d) If this price is charged, what is the maximum revenue?
(e) How many units are sold at this price?
(f) What price should be charged to collect at least $56,000 in revenue?

23. Landscaping A landscape engineer has 200 feet of border to enclose a rectangular pond. What dimensions will result in the largest pond?

24. Enclosing the Most Area with a Fence A farmer with 10,000 meters of fencing wants to enclose a rectangular field and then divide it into two plots with a fence parallel to one of the sides. See the figure. What is the largest area that can be enclosed?

25. Architecture A special window in the shape of a rectangle with semicircles at each end is to be constructed so that the outside perimeter is 100 feet. See the illustration. Find the dimensions of the rectangle that maximizes the area of the rectangle.

26. Minimizing Marginal Cost Callaway Golf Company has determined that the marginal cost C of manufacturing x Big Bertha golf clubs may be expressed by the quadratic function

$$C(x) = 4.9x^2 - 617.4x + 19,600$$

(a) How many clubs should be manufactured to minimize the marginal cost?
(b) At this level of production, what is the marginal cost?

27. Maximizing Area A rectangle has one vertex on the line $y = 10 - x$, $x > 0$, another at the origin, one on the positive x-axis, and one on the positive y-axis. Express the area A of the rectangle as a function of x. Find the largest area A that can be enclosed by the rectangle.

28. Parabolic Arch Bridge A horizontal bridge is in the shape of a parabolic arch. Given the information shown in the figure, what is the height h of the arch 2 feet from shore?

29. Bone Length Research performed at NASA, led by Dr. Emily R. Morey-Holton, measured the lengths of the right humerus and right tibia in 11 rats that were sent to space on Spacelab Life Sciences 2. The data shown on the next page were collected.

(a) Draw a scatter plot of the data, treating length of the right humerus as the independent variable.

(b) Based on the scatter plot, do you think that there is a linear relation between the length of the right humerus and the length of the right tibia?

(c) Use a graphing utility to find the line of best fit relating length of the right humerus and length of the right tibia.

(d) Predict the length of the right tibia on a rat whose right humerus is 26.5 millimeters (mm).

Right Humerus (mm), x	Right Tibia (mm), y
24.80	36.05
24.59	35.57
24.59	35.57
24.29	34.58
23.81	34.20
24.87	34.73
25.90	37.38
26.11	37.96
26.63	37.46
26.31	37.75
26.84	38.50

Source: NASA Life Sciences Data Archive

30. Advertising A small manufacturing firm collected the following data on advertising expenditures A (in thousands of dollars) and total revenue R (in thousands of dollars).

Advertising Expenditures ($1000s)	Total Revenue ($1000s)
20	6101
22	6222
25	6350
25	6378
27	6453
28	6423
29	6360
31	6231

(a) Draw a scatter plot of the data. Comment on the type of relation that may exist between the two variables.

(b) The quadratic function of best fit to these data is

$$R(A) = -7.76A^2 + 411.88A + 942.72$$

Use this function to determine the optimal level of advertising.

(c) Use the function to predict the total revenue when the optimal level of advertising is spent.

(d) Use a graphing utility to verify that the function given in part (b) is the quadratic function of best fit.

(e) Use a graphing utility to draw a scatter plot of the data, and then graph the quadratic function of best fit on the scatter plot.

Chapter Test

Test Prep VIDEOS

CHAPTER

The Chapter Test Prep Videos include step-by-step solutions to all chapter test exercises. These videos are available in MyLab™ Math, or on this text's YouTube Channel. Refer to the Preface for a link to the YouTube channel.

1. For the linear function $f(x) = -4x + 3$:
   (a) Find the slope and y-intercept.
   (b) What is the average rate of change of f?
   (c) Determine whether f is increasing, decreasing, or constant.
   (d) Graph f.

In Problems 2 and 3, find the intercepts, if any, of each quadratic function.

2. $f(x) = 3x^2 - 2x - 8$ **3.** $G(x) = -2x^2 + 4x + 1$

4. Suppose $f(x) = x^2 + 3x$ and $g(x) = 5x + 3$.
   (a) Solve $f(x) = g(x)$.
   (b) Graph each function and label the points of intersection.
   (c) Solve the inequality $f(x) < g(x)$ and graph the solution set.

5. Graph $f(x) = (x - 3)^2 - 2$ using transformations.

6. Consider the quadratic function $f(x) = 3x^2 - 12x + 4$.
   (a) Is the graph concave up or concave down?
   (b) Find the vertex.
   (c) Find the axis of symmetry.
   (d) Find the intercepts.
   (e) Use the information from parts (a)–(d) to graph f.

7. Determine whether $f(x) = -2x^2 + 12x + 3$ has a maximum or a minimum. Then find the maximum or minimum value.

8. Solve $x^2 - 10x + 24 \geq 0$.

9. RV Rental The weekly rental cost of a 20-foot recreational vehicle is $129.50 plus $0.15 per mile.
   (a) Find a linear function that expresses the cost C as a function of miles driven m.
   (b) What is the rental cost if 860 miles are driven?
   (c) How many miles were driven if the rental cost is $213.80?

10. Maximizing Revenue The price p (in dollars) and the quantity x sold of a certain product satisfy the demand equation

$$x = -10p + 10,000$$

   (a) Find a model that expresses the revenue R as a function of p.
   (b) What is the domain of R? Assume R is nonnegative.
   (c) What price p maximizes the revenue?
   (d) What is the maximum revenue?
   (e) How many units are sold at this price?
   (f) What price should the company charge to earn at least $1,600,000 in revenue?

Cumulative Review

1. Find the distance between the points $P = (-1, 3)$ and $Q = (4, -2)$. Find the midpoint of the line segment from P to Q.

2. Which points are on the graph of $y = x^3 - 3x + 1$?
   (a) $(-2, -1)$ (b) $(2, 3)$ (c) $(3, 1)$

3. Solve the inequality $5x + 3 \geq 0$ and graph the solution set.

4. Find the equation of the line containing the points $(-1, 4)$ and $(2, -2)$. Express your answer in slope-intercept form and graph the line.

5. Find the equation of the line perpendicular to the line $y = 2x + 1$ and containing the point $(3, 5)$. Express your answer in slope-intercept form and graph both lines.

6. Graph the equation $x^2 + y^2 - 4x + 8y - 5 = 0$.

7. Does the following relation represent a function?
   $\{ (-3, 8), (1, 3), (2, 5), (3, 8) \}$.

8. For the function f defined by $f(x) = x^2 - 4x + 1$, find:
   (a) $f(2)$
   (b) $f(x) + f(2)$
   (c) $f(-x)$
   (d) $-f(x)$
   (e) $f(x + 2)$
   (f) $\dfrac{f(x + h) - f(x)}{h}$, $h \neq 0$

9. Find the domain of $h(z) = \dfrac{3z - 1}{6z - 7}$.

10. Is the following graph the graph of a function?

11. Consider the function $f(x) = \dfrac{x}{x + 4}$.
   (a) Is the point $\left(1, \dfrac{1}{4} \right)$ on the graph of f?
   (b) If $x = -2$, what is $f(x)$? What point is on the graph of f?
   (c) If $f(x) = 2$, what is x? What point is on the graph of f?

12. Is the function $f(x) = \dfrac{x^2}{2x + 1}$ even, odd, or neither?

13. Approximate the local maximum values and local minimum values of $f(x) = x^3 - 5x + 1$ on $[-4, 4]$. Determine where the function is increasing and where it is decreasing.

14. If $f(x) = 3x + 5$ and $g(x) = 2x + 1$:
   (a) Solve $f(x) = g(x)$.
   (b) Solve $f(x) > g(x)$.

15. Consider the graph below of the function f.
   (a) Find the domain and the range of f.
   (b) Find the intercepts.
   (c) Is the graph of f symmetric with respect to the x-axis, the y-axis, or the origin?
   (d) Find $f(2)$.
   (e) For what value(s) of x is $f(x) = 3$?
   (f) Solve $f(x) < 0$.
   (g) Graph $y = f(x) + 2$.
   (h) Graph $y = f(-x)$.
   (i) Graph $y = 2f(x)$.
   (j) Is f even, odd, or neither?
   (k) Find the interval(s) on which f is increasing.

Chapter Projects

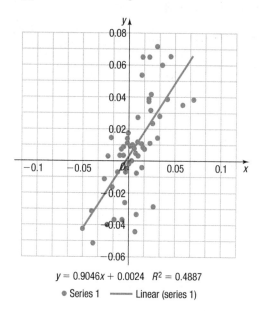

((•)) Internet-based Project

I. The Beta of a Stock You want to invest in the stock market but are not sure which stock to purchase. Information is the key to making an informed investment decision. One piece of information that many stock analysts use is the beta of the stock. Go to Wikipedia (*http://en.wikipedia.org/wiki/Beta_(finance)*) and research what beta measures and what it represents.

1. **Approximating the beta of a stock.** Choose a well-known company such as Google or Coca-Cola. Go to a website such as Yahoo! Finance (*http://finance.yahoo.com/*) and find the weekly closing price of the company's stock for the past year. Then find the closing price of the Standard & Poor's 500 (S&P500) for the same time period.

   To get the historical prices in Yahoo! Finance, select Historical Data from the menu. Choose the appropriate time period. Select Weekly and Apply. Finally, select Download Data, and Open with Microsoft Excel. Repeat this for the S&P500, and copy the data into the same spreadsheet. Finally, rearrange the data in chronological order. Be sure to expand the selection to sort all the data. Now, using the adjusted close price, compute the percentage change in price for each week, using the formula

$$\% \text{ change} = \frac{P_1 - P_0}{P_0}.$$

   For example, if week 1 price is in cell D1 and week 2 price is in cell D2, then $\% \text{ change} = \frac{D2 - D1}{D1}$. Repeat this for the S&P500 data.

2. **Using Excel to draw a scatter plot.** Treat the percentage change in the S&P500 as the independent variable and the percentage change in the stock you chose as the dependent variable. The easiest way to draw a scatter plot in Excel is to place the two columns of data next to each other (for example, have the percentage change in the S&P500 in column F and the percentage change in the stock you chose in column G). Then highlight the data and select the Scatter Plot icon under Insert. Comment on the type of relation that appears to exist between the two variables.

3. **Finding beta.** To find beta requires that we find the line of best fit using least-squares regression. The easiest approach is to click inside the scatter plot. Select the Chart Elements icon (+). Check the box for Trendline, select the arrow to the right, and choose More Options. Select Linear and check the box for Display Equation on chart. The line of best fit appears on the scatter plot. See below.

$y = 0.9046x + 0.0024 \quad R^2 = 0.4887$

● Series 1 ——— Linear (series 1)

The line of best fit for this data is $y = 0.9046x + 0.0024$. You may click on Chart Title or either axis title and insert the appropriate names. The beta is the slope of the line of best fit, 0.9046. We interpret this by saying, "If the S&P500 increases by 1%, then this stock will increase by 0.9%, on average." Find the beta of your stock and provide an interpretation. NOTE: Another way to use Excel to find the line of best fit requires using the Data Analysis Tool Pack under add-ins.

The following projects are available on the Instructor's Resource Center (IRC):

II. Cannons A battery commander uses the weight of a missile, its initial velocity, and the position of its gun to determine where the missile will travel.

III. First and Second Differences Finite differences provide a numerical method that is used to estimate the graph of an unknown function.

IV. CBL Experiment Computer simulation is used to study the physical properties of a bouncing ball.

5 Polynomial and Rational Functions

Day Length

Day length is the length of time each day from the moment the upper limb of the sun's disk appears above the horizon during sunrise to the moment when the upper limb disappears below the horizon during sunset. The length of a day depends on the day of the year as well as the latitude of the location. Latitude gives the location of a point on Earth north or south of the equator. In the Internet Project at the end of this chapter, we use information from the chapter to investigate the relation between day length and latitude for a specific day of the year.

—*See the Internet-based Chapter Project I*—

Outline

⬅ A Look Back

In Chapter 3, we began our discussion of functions. We defined domain, range, and independent and dependent variables, found the value of a function, and graphed functions. We continued our study of functions by listing properties of functions, such as being even or odd, and created a library of functions, naming and graphing key functions and listing their properties.

In Chapter 4, we discussed linear functions and quadratic functions, which belong to the class of *polynomial functions*.

A Look Ahead ➡

In this chapter, we look at two general classes of functions, polynomial functions and rational functions, and examine their properties. Polynomial functions are arguably the simplest functions in algebra. They are often used to approximate other, more complicated functions. Rational functions are ratios of polynomial functions.

5.1 Polynomial Functions

PREPARING FOR THIS SECTION *Before getting started, review the following:*

- Polynomials (Chapter R, Section R.4, pp. 39–47)
- Obtain Information from or about the Graph of a Function (Section 3.2, pp. 220–223)

- Graphing Techniques: Transformations (Section 3.5, pp. 254–263)
- Intercepts (Section 2.2, pp. 160–161)
- Library of Functions (Section 3.4, pp. 242–246)

> Now Work the 'Are You Prepared?' problems on page 343.

OBJECTIVES 1 Identify Polynomial Functions and Their Degree (p. 331)
2 Graph Polynomial Functions Using Transformations (p. 335)
3 Identify the Real Zeros of a Polynomial Function and Their Multiplicity (p. 336)

1 Identify Polynomial Functions and Their Degree

In Chapter 4, we studied the linear function $f(x) = mx + b$, which can be written as

$$f(x) = a_1x + a_0$$

and the quadratic function $f(x) = ax^2 + bx + c, a \neq 0$, which can be written as

$$f(x) = a_2x^2 + a_1x + a_0 \qquad a_2 \neq 0$$

Both of these functions are examples of a *polynomial function*.

> **DEFINITION Polynomial Function**
>
> A **polynomial function** in one variable is a function of the form
>
> $$f(x) = a_nx^n + a_{n-1}x^{n-1} + \cdots + a_1x + a_0 \qquad \textbf{(1)}$$
>
> where $a_n, a_{n-1}, \ldots, a_1, a_0$ are constants, called the **coefficients** of the polynomial, $n \geq 0$ is an integer, and x is a variable. If $a_n \neq 0$, it is called the **leading coefficient**, and n is the **degree** of the polynomial.
>
> The domain of a polynomial function is the set of all real numbers.

In Words

A polynomial function is a sum of monomials.

The monomials that make up a polynomial function are called its **terms**. If $a_n \neq 0$, a_nx^n is called the **leading term**; a_0 is called the **constant term**. If all of the coefficients are 0, the polynomial is called the **zero polynomial**, which has no degree.

Polynomial functions are usually written in **standard form**, beginning with the nonzero term of highest degree and continuing with terms in descending order according to degree. If a power of x is missing, it is because its coefficient is zero.

Polynomial functions are among the simplest in algebra. They are easy to evaluate: only addition and repeated multiplication are required. Because of this, they are often used to approximate other, more complicated functions. In this section, we investigate properties of this important class of functions.

EXAMPLE 1 **Identifying Polynomial Functions**

Determine which of the following are polynomial functions. For those that are, state the degree; for those that are not, state why not. Write each polynomial function in standard form, and then identify the leading term and the constant term.

(a) $p(x) = 5x^3 - \dfrac{1}{4}x^2 - 9$ (b) $f(x) = x + 2 - 3x^4$ (c) $g(x) = \sqrt{x}$

(d) $h(x) = \dfrac{x^2 - 2}{x^3 - 1}$ (e) $G(x) = 8$ (f) $H(x) = -2x^3(x - 1)^2$

Solution
(a) p is a polynomial function of degree 3, and it is in standard form. The leading term is $5x^3$, and the constant term is -9.

(b) f is a polynomial function of degree 4. Its standard form is $f(x) = -3x^4 + x + 2$. The leading term is $-3x^4$, and the constant term is 2.

(c) g is not a polynomial function because $g(x) = \sqrt{x} = x^{\frac{1}{2}}$, so the variable x is raised to the $\frac{1}{2}$ power, which is not a nonnegative integer.

(d) h is not a polynomial function. It is the ratio of two distinct polynomials, and the polynomial in the denominator is of positive degree.

(e) G is a nonzero constant polynomial function, so it is of degree 0. The polynomial is in standard form. The leading term and constant term are both 8.

(f) $H(x) = -2x^3(x - 1)^2 = -2x^3(x^2 - 2x + 1) = -2x^5 + 4x^4 - 2x^3$. So, H is a polynomial function of degree 5. The leading term is $-2x^5$. Since no constant term is shown, the constant term is 0.

Do you see a way to find the degree of H, in part (f), without multiplying it out?

Now Work PROBLEMS 15 AND 19

We have discussed in detail polynomial functions of degrees 0, 1, and 2. See Table 1 for a summary of properties of the graphs of these polynomial functions.

Table 1

Degree	Form	Name	Graph
No degree	$f(x) = 0$	Zero function	The x-axis
0	$f(x) = a_0, \quad a_0 \neq 0$	Constant function	Horizontal line with y-intercept a_0
1	$f(x) = a_1x + a_0, \quad a_1 \neq 0$	Linear function	Nonvertical, nonhorizontal line with slope a_1 and y-intercept a_0
2	$f(x) = a_2x^2 + a_1x + a_0, \quad a_2 \neq 0$	Quadratic function	Parabola: graph is concave up if $a_2 > 0$; graph is concave down if $a_2 < 0$

If you take a course in calculus, you will learn that the graph of every polynomial function is both smooth and continuous. By **smooth**, we mean that the graph contains no sharp corners or cusps; by **continuous**, we mean that the graph has no gaps or holes and can be drawn without lifting your pencil from the paper. See Figures 1(a) and (b).

Figure 1

(a) Graph of a polynomial function: smooth, continuous

(b) Cannot be the graph of a polynomial function

Power Functions

Polynomial functions of degree n, $n > 0$, that have only one term are called *power functions*.

In Words
A power function is defined by a single monomial.

DEFINITION Power Function

A **power function of degree n** is a monomial function of the form

$$f(x) = ax^n \tag{2}$$

where a is a real number, $a \neq 0$, and $n > 0$ is an integer.

Examples of power functions are

$$f(x) = 3x \qquad f(x) = -5x^2 \qquad f(x) = 8x^3 \qquad f(x) = -\frac{1}{2}x^4$$

degree 1   degree 2   degree 3   degree 4

The graph of a power function of degree 1, $f(x) = ax$, is a line with slope a that passes through the origin. The graph of a power function of degree 2, $f(x) = ax^2$, is a parabola with vertex at the origin. The parabola is concave up if $a > 0$ and is concave down if $a < 0$.

If we know how to graph a power function of the form $f(x) = x^n$, a compression, a stretch, or, perhaps, a reflection about the x-axis gives us the graph of $g(x) = ax^n$. Consequently, we concentrate on graphing power functions of the form $f(x) = x^n$.

We begin with power functions of even degree of the form $f(x) = x^n, n \geq 2$ and n even. The domain of f is the set of all real numbers, and the range is the set of nonnegative real numbers. Such a power function is an even function. (Do you see why?) Its graph is symmetric with respect to the y-axis. The graph of a power function $f(x) = x^n$, n even, always contains the origin $(0,0)$ and the points $(-1,1)$ and $(1,1)$.

If $n = 2$, the graph is the familiar parabola $y = x^2$ that is concave up, with vertex at the origin. If $n \geq 4$, the graph of $f(x) = x^n$, n even, will be closer to the x-axis than the parabola $y = x^2$ if $-1 < x < 1, x \neq 0$, and farther from the x-axis than the parabola $y = x^2$ if $x < -1$ or if $x > 1$. Figure 2(a) illustrates this conclusion. Figure 2(b) shows the graphs of $y = x^4$ and $y = x^8$ for further comparison.

Figure 2   (a)   (b)

RECALL Infinity, ∞, is not a real number, but a symbol used to indicate unboundedness in the positive direction. Similarly, $-\infty$ is a symbol used to indicate unboundedness in the negative direction.

Figure 2 shows that as n increases, the graph of $f(x) = x^n, n \geq 2$ and n even, tends to flatten out near the origin and is steeper when x is far from 0. For large n, it may appear that the graph coincides with the x-axis near the origin, but it does not; the graph actually touches the x-axis only at the origin. Also, for large n, it may appear that for $x < -1$ or for $x > 1$ the graph is vertical, but it is not; it is only increasing very rapidly in these intervals. If the graphs were enlarged many times, these distinctions would be clear.

Figure 2 also shows that as x becomes unbounded in the negative direction, $x \to -\infty$, or as x becomes unbounded in the positive direction, $x \to \infty$, the power function $f(x) = x^n$, n even, becomes unbounded in the positive direction, $f(x) \to \infty$. See Table 2.

Table 2

	$x = -10{,}000$	$x = -100{,}000$	$x \to -\infty$	$x = 10{,}000$	$x = 100{,}000$	$x \to \infty$
$f(x) = x^2$	10^8	10^{10}	∞	10^8	10^{10}	∞
$f(x) = x^4$	10^{16}	10^{20}	∞	10^{16}	10^{20}	∞
$f(x) = x^6$	10^{24}	10^{30}	∞	10^{24}	10^{30}	∞

We define the behavior of the graph of a function for large values of x, either positive or negative, as its **end behavior**.

Seeing the Concept

Graph $Y_1 = x^4$, $Y_2 = x^8$, and $Y_3 = x^{12}$ using the viewing rectangle $-2 \leq x \leq 2$, $-4 \leq y \leq 16$. Then graph each again using the viewing rectangle $-1 \leq x \leq 1$, $0 \leq y \leq 1$. See Figure 3. TRACE along one of the graphs to confirm that for x close to 0 the graph is above the x-axis and that for $x > 0$ the graph is increasing.

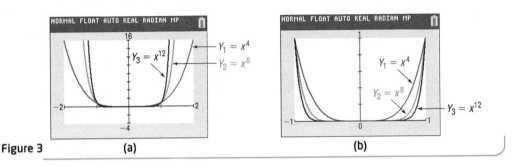

Figure 3
(a)
(b)

Properties of Power Functions, $f(x) = x^n$, n Is a Positive Even Integer

- f is an even function, so its graph is symmetric with respect to the y-axis.
- The domain is the set of all real numbers. The range is the set of nonnegative real numbers.
- The graph always contains the points $(-1, 1)$, $(0, 0)$, and $(1, 1)$.
- As the exponent n increases in magnitude, the graph is steeper when $x < -1$ or $x > 1$; but for x near the origin, the graph tends to flatten out and lie closer to the x-axis.
- End behavior:
    - As $x \rightarrow -\infty$, $f(x) = x^n \rightarrow \infty$.
    - As $x \rightarrow \infty$, $f(x) = x^n \rightarrow \infty$.

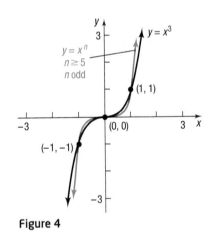

Figure 4

Now consider the power functions $f(x) = x^n$, $n \geq 3$, and n odd. The domain and the range of f are the set of real numbers. Such a power function is an odd function. (Do you see why?) Its graph is symmetric with respect to the origin. The graph of a power function $f(x) = x^n$, n odd, always contains the origin $(0, 0)$ and the points $(-1, -1)$ and $(1, 1)$.

The power function $f(x) = x^3$ is the cube function. Its graph is shown in Figure 4. If $n \geq 5$, the graph of $f(x) = x^n$, n odd, will be closer to the x-axis than that of $y = x^3$ if $-1 < x < 1$ and farther from the x-axis than that of $y = x^3$ if $x < -1$ or if $x > 1$. Figure 4 illustrates this fact. Figure 5 shows the graphs of $y = x^5$ and $y = x^9$ for further comparison.

It appears that each graph in Figure 5 coincides with the x-axis near the origin, but it does not. Each graph actually crosses the x-axis at the origin. Also, it appears that as x increases, the graphs become vertical, but they do not; each graph is just increasing very rapidly.

Figures 4 and 5 show the end behavior of the power function $f(x) = x^n$, n odd. As x becomes unbounded in the negative direction, $x \rightarrow -\infty$, the power function becomes unbounded in the negative direction, that is $f(x) \rightarrow -\infty$. But, as x becomes unbounded in the positive direction, $x \rightarrow \infty$, the power function becomes unbounded in the positive direction, that is, $f(x) \rightarrow \infty$.

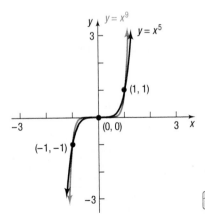

Figure 5

Seeing the Concept

Graph $Y_1 = x^3$, $Y_2 = x^7$, and $Y_3 = x^{11}$ using the viewing rectangle $-2 \leq x \leq 2$, $-16 \leq y \leq 16$. Then graph each again using the viewing rectangle $-1 \leq x \leq 1$, $-1 \leq y \leq 1$. See Figure 6. TRACE along one of the graphs to confirm that the graph is increasing and crosses the x-axis at the origin.

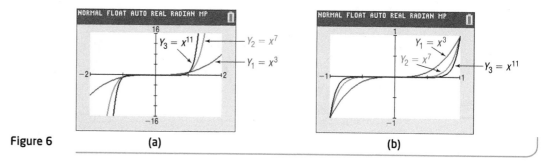

Figure 6   **(a)**   **(b)**

To summarize:

Properties of Power Functions, $f(x) = x^n$, n Is a Positive Odd Integer

- f is an odd function, so its graph is symmetric with respect to the origin.
- The domain and the range are the set of all real numbers.
- The graph always contains the points $(-1, -1)$, $(0, 0)$, and $(1, 1)$.
- As the exponent n increases in magnitude, the graph is steeper when $x < -1$ or $x > 1$; but for x near the origin, the graph tends to flatten out and lie closer to the x-axis.
- End behavior:
  - As $x \to -\infty$, $f(x) = x^n \to -\infty$.
  - As $x \to \infty$, $f(x) = x^n \to \infty$.

---

2 Graph Polynomial Functions Using Transformations

The methods of shifting, compression, stretching, and reflection studied in Section 3.5, when used with the facts just presented, enable us to graph polynomial functions that are transformations of power functions.

EXAMPLE 2   **Graphing a Polynomial Function Using Transformations**

Graph: $f(x) = 1 - x^5$

Solution   It is helpful to rewrite f as $f(x) = -x^5 + 1$. Figure 7 shows the required steps.

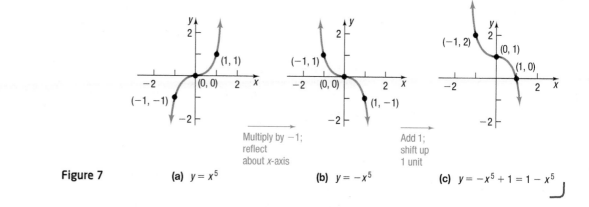

Figure 7   **(a)** $y = x^5$ **(b)** $y = -x^5$ **(c)** $y = -x^5 + 1 = 1 - x^5$

EXAMPLE 3   **Graphing a Polynomial Function Using Transformations**

Graph: $f(x) = \dfrac{1}{2}(x - 1)^4$

Solution   Figure 8 shows the required steps.

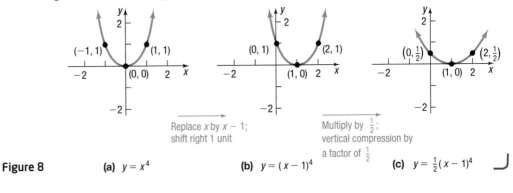

Figure 8

(a) $y = x^4$

Replace x by $x - 1$;
shift right 1 unit

(b) $y = (x - 1)^4$

Multiply by $\frac{1}{2}$;
vertical compression by
a factor of $\frac{1}{2}$

(c) $y = \frac{1}{2}(x - 1)^4$

━━━━━**Now Work** PROBLEMS 27 AND 33

3 Identify the Real Zeros of a Polynomial Function and Their Multiplicity

Figure 9 shows the graph of a polynomial function with four x-intercepts. Notice that at the x-intercepts, the graph must either cross the x-axis or touch the x-axis. Consequently, between consecutive x-intercepts the graph is either above the x-axis or below the x-axis.

Figure 9 Graph of a polynomial function

If a polynomial function f is factored completely, it is easy to locate the x-intercepts of the graph by solving the equation $f(x) = 0$ using the Zero-Product Property. For example, if $f(x) = (x - 1)^2(x + 3)$, then the solutions of the equation

$$f(x) = (x - 1)^2(x + 3) = 0$$

are 1 and -3. That is, $f(1) = 0$ and $f(-3) = 0$.

DEFINITION Real Zero

If f is a function and r is a real number for which $f(r) = 0$, then r is called a **real zero** of f.

As a consequence of this definition, the following statements are equivalent.

- r is a real zero of a polynomial function f.
- r is an x-intercept of the graph of f.
- $x - r$ is a factor of f.
- r is a real solution to the equation $f(x) = 0$.

So the real zeros of a polynomial function are the x-intercepts of its graph, and they are found by solving the equation $f(x) = 0$.

EXAMPLE 4 **Finding a Polynomial Function from Its Real Zeros**

(a) Find a polynomial function of degree 3 whose real zeros are $-3, 2$, and 5.

(b) Use a graphing utility to graph the polynomial found in part (a) to verify your result.

Solution (a) If r is a real zero of a polynomial function f, then $x - r$ is a factor of f. This means that $x - (-3) = x + 3, x - 2$, and $x - 5$ are factors of f. As a result, any polynomial function of the form

$$f(x) = a(x + 3)(x - 2)(x - 5)$$

where a is a nonzero real number, qualifies. The value of a causes a stretch, compression, or reflection, but it does not affect the x-intercepts of the graph. Do you know why?

(b) We choose to graph f with $a = 1$. Then

$$f(x) = (x + 3)(x - 2)(x - 5)$$

Figure 10
$f(x) = (x + 3)(x - 2)(x - 5)$

Figure 10 shows the graph of f. Notice that the x-intercepts are $-3, 2$, and 5.

Seeing the Concept

Graph the function found in Example 4 for $a = 2$ and $a = -1$. Does the value of a affect the real zeros of f? How does the value of a affect the graph of f?

Now Work PROBLEM 41

If, when f is factored, the factor $x - r$ occurs more than once, r is called a **repeated**, or **multiple**, **real zero of** f. More precisely, we have the following definition.

DEFINITION **Real Zero of Multiplicity m**

If $(x - r)^m$ is a factor of a polynomial f and $(x - r)^{m+1}$ is not a factor of f, then r is called a **real zero of multiplicity m of** f.*

EXAMPLE 5 **Identifying Real Zeros and Their Multiplicities**

For the polynomial

$$f(x) = 5x^2(x + 2)\left(x - \frac{1}{2}\right)^4$$

In Words
The multiplicity of a real zero is the number of times its corresponding factor occurs.

- 0 is a real zero of multiplicity 2 because the exponent on the factor x is 2.
- -2 is a real zero of multiplicity 1 because the exponent on the factor $x + 2$ is 1.
- $\dfrac{1}{2}$ is a real zero of multiplicity 4 because the exponent on the factor $x - \dfrac{1}{2}$ is 4.

Now Work PROBLEM 59(a)

Suppose that it is possible to completely factor a polynomial function and, as a result, locate all the x-intercepts of its graph (the real zeros of the function). These x-intercepts then divide the x-axis into open intervals and, on each interval, the graph of the polynomial is either above or below the x-axis over the entire interval. Let's look at an example.

*Some texts use the terms **multiple root** and **root of multiplicity m**.

EXAMPLE 6

Graphing a Polynomial Function Using Its *x*-Intercepts

Consider the polynomial function: $f(x) = (x + 1)^2(x - 2)$

(a) Find the *x*- and *y*-intercepts of the graph of *f*.

(b) Use the *x*-intercepts to find the intervals on which the graph of *f* is above the *x*-axis and the intervals on which the graph of *f* is below the *x*-axis.

(c) Locate other points on the graph, and connect the points with a smooth, continuous curve.

Solution

(a) The *y*-intercept is $f(0) = (0 + 1)^2(0 - 2) = -2$.
The *x*-intercepts satisfy the equation

$$f(x) = (x + 1)^2(x - 2) = 0$$

from which we find

$$(x + 1)^2 = 0 \quad \text{or} \quad x - 2 = 0$$
$$x = -1 \quad \text{or} \quad x = 2$$

The *x*-intercepts are -1 and 2.

(b) The two *x*-intercepts divide the *x*-axis into three intervals:

$$(-\infty, -1) \quad (-1, 2) \quad (2, \infty)$$

Since the graph of *f* crosses or touches the *x*-axis only at $x = -1$ and $x = 2$, it follows that the graph of *f* is either above the *x*-axis $[f(x) > 0]$ or below the *x*-axis $[f(x) < 0]$ on each of these three intervals. To see where the graph lies, we need only pick a number in each interval, evaluate *f* at the number, and see whether the value is positive (above the *x*-axis) or negative (below the *x*-axis). See Table 3.

(c) In constructing Table 3, we obtained three additional points on the graph: $(-2, -4)$, $(1, -4)$ and $(3, 16)$. Figure 11 shows these points, the intercepts, and a smooth, continuous curve (the graph of *f*) connecting them.

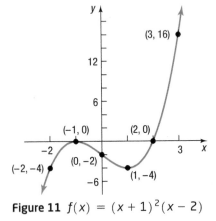

Figure 11 $f(x) = (x + 1)^2(x - 2)$

Table 3

		-1		2	
Interval		$(-\infty, -1)$	$(-1, 2)$	$(2, \infty)$	
Number chosen		-2	1	3	
Value of f		$f(-2) = -4$	$f(1) = -4$	$f(3) = 16$	
Location of graph		Below x-axis	Below x-axis	Above x-axis	
Point on graph		$(-2, -4)$	$(1, -4)$	$(3, 16)$	

Look again at Table 3. Since the graph of $f(x) = (x + 1)^2(x - 2)$ is below the *x*-axis on both sides of -1, the graph of *f* *touches* the *x*-axis at $x = -1$, a *real zero of multiplicity 2*. Since the graph of *f* is below the *x*-axis for $x < 2$ and above the *x*-axis for $x > 2$, the graph of *f* *crosses* the *x*-axis at $x = 2$, a *real zero of multiplicity 1*.

SUMMARY

If *r* Is a Real Zero of Even Multiplicity

- Numerically: The sign of $f(x)$ does not change from one side to the other side of *r*.
- Graphically: The graph of *f* **touches** the *x*-axis at *r*.

If *r* Is a Real Zero of Odd Multiplicity

- Numerically: The sign of $f(x)$ changes from one side to the other side of *r*.
- Graphically: The graph of *f* **crosses** the *x*-axis at *r*.

Now Work PROBLEM 59(b)

Turning Points

Look again at Figure 11. We cannot be sure how low the graph actually goes between $x = -1$ and $x = 2$. But we do know that somewhere in the interval $(-1, 2)$ the graph of f must change direction (from decreasing to increasing). The points at which a graph of a function changes direction are called **turning points**.* Each turning point results in either a **local maximum** or a **local minimum** of f. The following theorem from calculus tells us the maximum number of turning points that the graph of a polynomial function can have.

Need To Review?
Local maximum and local minimum are discussed in Section 3.3, pp. 232–233.

THEOREM Turning Points

- If f is a polynomial function of degree n, then the graph of f has at most $n - 1$ turning points.
- If the graph of a polynomial function f has $n - 1$ turning points, then the degree of f is at least n.

Based on the first bullet of the theorem, a polynomial function of degree 5 will have at most $5 - 1 = 4$ turning points. Based on the second bullet of the theorem, if the graph of a polynomial function has three turning points, then the degree of the function must be at least 4.

Exploration

A graphing utility can be used to locate the turning points of a graph. Graph $Y_1 = (x + 1)^2(x - 2)$. Use MINIMUM to find the location of the turning point for $0 \le x \le 2$. See Figure 12.

Figure 12 $Y_1 = (x + 1)^2(x - 2)$

——Now Work PROBLEM 59(c)

EXAMPLE 7

Identifying the Graph of a Polynomial Function

Which graphs in Figure 13 could be the graph of a polynomial function? For those that could, list the real zeros and state the least degree the polynomial can have. For those that could not, say why not.

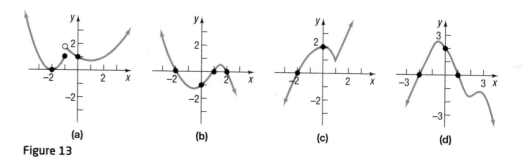

(a)   (b)   (c)   (d)

Figure 13

---

*Graphing utilities can be used to approximate turning points. For most polynomials, calculus is needed to find the exact turning points.

Solution

(a) The graph in Figure 13(a) cannot be the graph of a polynomial function because there is a gap at $x = -1$. The graph of a polynomial function is always continuous—no gaps or holes. (Refer back to Figure 1.)

(b) The graph in Figure 13(b) could be the graph of a polynomial function because the graph is smooth and continuous. It has three real zeros: -2, 1, and 2. Since the graph has two turning points, the degree of the polynomial function must be at least 3.

(c) The graph in Figure 13(c) cannot be the graph of a polynomial function because it has a corner at $x = 1$. The graph of a polynomial function is smooth.

(d) The graph in Figure 13(d) could be the graph of a polynomial function. It has two real zeros: -2 and 1. Since the graph has three turning points, the degree of the polynomial function is at least 4.

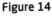 **Now Work** PROBLEM 71

End Behavior

Take one last look at Figure 11. For large values of x, either positive or negative, the graph of $f(x) = (x + 1)^2(x - 2)$ resembles the graph of $y = x^3$. To see why, write f in the form

$$f(x) = (x + 1)^2(x - 2) = x^3 - 3x - 2 = x^3\left(1 - \frac{3}{x^2} - \frac{2}{x^3}\right)$$

For large values of x, either positive or negative, the terms $\dfrac{3}{x^2}$ and $\dfrac{2}{x^3}$ are close to 0. So for large values of x,

$$f(x) = x^3 - 3x - 2 = x^3\left(1 - \frac{3}{x^2} - \frac{2}{x^3}\right) \approx x^3$$

> **THEOREM End Behavior of the Graph of a Polynomial Function**
>
> The end behavior of the graph of the polynomial function
>
> $$f(x) = a_n x^n + a_{n-1} x^{n-1} + \cdots + a_1 x + a_0 \quad a_n \neq 0$$
>
> is the same as that of the graph of the power function
>
> $$y = a_n x^n$$

In Words

The end behavior of the graph of a polynomial function resembles that of its leading term.

For example, the end behavior of the graph of the polynomial function

$$f(x) = -2x^3 + 5x^2 + x - 4$$

is the same as that of the graph of the power function $y = -2x^3$. We can see that the graphs "behave" similarly by considering Table 4 and Figure 14.

Table 4

x	$f(x)$	$y = -2x^3$
10	$-1{,}494$	$-2{,}000$
100	$-1{,}949{,}904$	$-2{,}000{,}000$
500	$-248{,}749{,}504$	$-250{,}000{,}000$
1,000	$-1{,}994{,}999{,}004$	$-2{,}000{,}000{,}000$

Figure 14

In calculus, **limits** are used to convey the idea of end behavior. There the symbolism $\lim\limits_{x\to\infty} f(x) = -\infty$, read "the limit as x approaches infinity of $f(x)$ equals negative infinity," means that $f(x) \to -\infty$ as $x \to \infty$.

Look back at Figures 2 and 4. Based on the preceding theorem and the discussion on power functions, the end behavior of a polynomial function can only be of four types. See Figure 15.

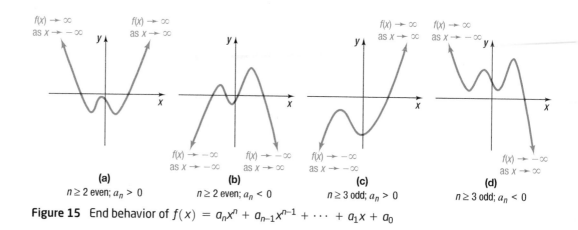

Figure 15   End behavior of $f(x) = a_n x^n + a_{n-1} x^{n-1} + \cdots + a_1 x + a_0$

─── **Now Work** PROBLEM 59(d)

EXAMPLE 8

Identifying the Graph of a Polynomial Function

Which graph in Figure 16 could be the graph of

$$f(x) = x^4 + ax^3 + bx^2 - 5x - 6$$

where $a > 0, b > 0$?

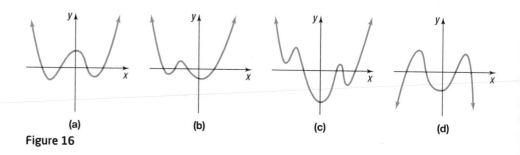

Figure 16

Solution   The y-intercept of f is $f(0) = -6$. We can eliminate the graph in Figure 16(a), whose y-intercept is positive.

We now look at end behavior. For large values of x, the graph of f will behave like the graph of $y = x^4$. This eliminates the graph in Figure 16(d), whose end behavior is like the graph of $y = -x^4$.

We are not able to solve $f(x) = 0$ to find the x-intercepts of f, so we move on to investigate the turning points of each graph. Since f is of degree 4, the graph of f has at most 3 turning points. We eliminate the graph in Figure 16(c) because that graph has 5 turning points.

Only the graph in Figure 16(b) could be the graph of

$$f(x) = x^4 + ax^3 + bx^2 - 5x - 6$$

where $a > 0, b > 0$.

| EXAMPLE 9 | **Finding a Polynomial Function from a Graph** |

Find a polynomial function whose graph is shown in Figure 17 (use the smallest degree possible).

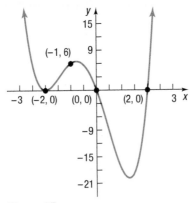

Figure 17

Solution

The x-intercepts are -2, 0, and 2. Therefore, the polynomial must have the factors $(x + 2)$, x, and $(x - 2)$, respectively. There are three turning points, so the degree of the polynomial must be at least 4. The graph touches the x-axis at $x = -2$, so -2 has an even multiplicity. The graph crosses the x-axis at $x = 0$ and $x = 2$, so 0 and 2 have odd multiplicities. Using the smallest degree possible (1 for odd multiplicity and 2 for even multiplicity), we write

$$f(x) = ax(x + 2)^2(x - 2)$$

All that remains is to find the leading coefficient, a. From Figure 17, the point $(-1, 6)$ is on the graph of f.

$$6 = a(-1)(-1 + 2)^2(-1 - 2) \quad f(-1) = 6$$
$$6 = 3a$$
$$2 = a$$

The polynomial function $f(x) = 2x(x + 2)^2(x - 2)$ has the graph in Figure 17.

✓ **Check:** Graph $Y_1 = 2x(x + 2)^2(x - 2)$ using a graphing utility to verify this result.

Now Work PROBLEMS 75 AND 79

SUMMARY

Graph of a Polynomial Function $f(x) = a_n x^n + a_{n-1} x^{n-1} + \cdots + a_1 x + a_0 \quad a_n \neq 0$

- The domain of a polynomial function is the set of all real numbers.
- Degree of the polynomial function f: n
- y-intercept: $f(0) = a_0$
- Graph is smooth and continuous.
- Maximum number of turning points: $n - 1$
- At a real zero of even multiplicity: The graph of f touches the x-axis.
- At a real zero of odd multiplicity: The graph of f crosses the x-axis.
- Between real zeros, the graph of f is either above or below the x-axis.
- End behavior: For large $|x|$, the graph of f resembles the graph of $y = a_n x^n$.

5.1 Assess Your Understanding

'Are You Prepared?' *Answers are given at the end of these exercises. If you get a wrong answer, read the pages listed in* red.

1. The intercepts of the graph of $9x^2 + 4y = 36$ are _____.
   (pp. 160–161)

2. Is the expression $4x^3 - 3.6x^2 - \sqrt{2}$ a polynomial? If so, what is its degree? (pp. 39–47)

3. To graph $y = x^2 - 4$, you would shift the graph of $y = x^2$ _____ a distance of _____ units. (pp. 254–263)

4. **True or False** The x-intercepts of the graph of a function $y = f(x)$ are the real solutions of the equation $f(x) = 0$. (pp. 220–222)

5. **Multiple Choice** The cube function $f(x) = x^3$ is _____.
   (a) even **(b)** odd **(c)** neither

   The graph of the cube function _____.
   (a) has no symmetry
   (b) is symmetric about the y-axis
   (c) is symmetric about the origin
   (d) is symmetric about the line $y = x$
   (pp. 242–246)

Concepts and Vocabulary

6. The graph of every polynomial function is both _____ and _____.

7. **Multiple Choice** If r is a real zero of even multiplicity of a polynomial function f, then the graph of f _____ the x-axis at r.
   (a) crosses **(b)** touches

8. The graphs of power functions of the form $f(x) = x^n$, where n is an even integer, always contain the points _____, _____, and _____.

9. If r is a real solution of the equation $f(x) = 0$, list three equivalent statements regarding f and r.

10. The points at which a graph changes direction (from increasing to decreasing or decreasing to increasing) are called _____ _____.

11. The graph of the function $f(x) = 3x^4 - x^3 + 5x^2 - 2x - 7$ resembles the graph of _____ for large values of $|x|$.

12. If $f(x) = -2x^5 + x^3 - 5x^2 + 7$, then $f(x) \rightarrow$ _____ as $x \rightarrow -\infty$, and $f(x) \rightarrow$ _____ as $x \rightarrow \infty$.

13. **Multiple Choice** The _____ of a real zero is the number of times its corresponding factor occurs.
    (a) degree **(b)** multiplicity **(c)** turning point **(d)** limit

14. **Multiple Choice** The graph of $y = 5x^6 - 3x^4 + 2x - 9$ has at most how many turning points?
    (a) -9 **(b)** 14 **(c)** 6 **(d)** 5

Skill Building

In Problems 15–26, determine which functions are polynomial functions. For those that are, state the degree. For those that are not, state why not. Write each polynomial in standard form. Then identify the leading term and the constant term.

15. $f(x) = 4x + x^3$

16. $f(x) = 5x^2 + 4x^4$

17. $g(x) = \dfrac{2 + 3x^2}{5}$

18. $h(x) = 3 - \dfrac{1}{2}x$

19. $f(x) = 1 - \dfrac{1}{x}$

20. $f(x) = x(x - 1)$

21. $g(x) = x^{2/3} - x^{1/3} + 2$

22. $h(x) = \sqrt{x}(\sqrt{x} - 1)$

23. $F(x) = 5x^4 - \pi x^3 + \dfrac{1}{2}$

24. $F(x) = \dfrac{x^2 - 5}{x^3}$

25. $G(x) = 2(x - 1)^2(x^2 + 1)$

26. $G(x) = -3x^2(x + 2)^3$

In Problems 27–40, use transformations of the graph of $y = x^4$ or $y = x^5$ to graph each function.

27. $f(x) = (x + 1)^4$

28. $f(x) = (x - 2)^5$

29. $f(x) = x^5 - 3$

30. $f(x) = x^4 + 2$

31. $f(x) = \dfrac{1}{2}x^4$

32. $f(x) = 3x^5$

33. $f(x) = -x^5$

34. $f(x) = -x^4$

35. $f(x) = (x - 1)^5 + 2$

36. $f(x) = (x + 2)^4 - 3$

37. $f(x) = 2(x + 1)^4 + 1$

38. $f(x) = \dfrac{1}{2}(x - 1)^5 - 2$

39. $f(x) = 4 - (x - 2)^5$

40. $f(x) = 3 - (x + 2)^4$

In Problems 41–48, find a polynomial function whose real zeros and degree are given. Answers will vary depending on the choice of the leading coefficient.

41. Zeros: $-1, 1, 3$; degree 3

42. Zeros: $-2, 2, 3$; degree 3

43. Zeros: $-5, 0, 6$; degree 3

44. Zeros: $-4, 0, 2$; degree 3

45. Zeros: $-5, -2, 3, 5$; degree 4

46. Zeros: $-3, -1, 2, 5$; degree 4

47. Zeros: -1, multiplicity 1; 3, multiplicity 2; degree 3

48. Zeros: -2, multiplicity 2; 4, multiplicity 1; degree 3

In Problems 49–58, find a polynomial function with the given real zeros whose graph contains the given point.

49. Zeros: $-2, 3, 5$
Degree 3
Point: $(2, 36)$

50. Zeros: $-2, 0, 2$
Degree 3
Point: $(-4, 16)$

51. Zeros: $-2, 0, 1, 3$
Degree 4
Point: $\left(-\dfrac{1}{2}, -63\right)$

52. Zeros: $-5, -1, 2, 6$
Degree 4
Point: $\left(\dfrac{5}{2}, 15\right)$

53. Zeros: $-3, 1, 4$
Degree 3
y-intercept: 36

54. Zeros: $-4, -1, 2$
Degree 3
y-intercept: 16

55. Zeros: -1 (multiplicity 2), 1 (multiplicity 2)
Degree 4
Point: $(-2, 45)$

56. Zeros: 0 (multiplicity 1), -1 (multiplicity 2), 3 (multiplicity 2)
Degree 5
Point: $(1, -48)$

57. Zeros: -5 (multiplicity 2), 2 (multiplicity 1), 4 (multiplicity 1); degree 4; contains the point $(3, 128)$

58. Zeros: -4 (multiplicity 1), 0 (multiplicity 3), 2 (multiplicity 1); degree 5; contains the point $(-2, 64)$

In Problems 59–70, for each polynomial function:
  (a) List each real zero and its multiplicity.
  (b) Determine whether the graph crosses or touches the x-axis at each x-intercept.
  (c) Determine the maximum number of turning points on the graph.
  (d) Determine the end behavior; that is, find the power function that the graph of f resembles for large values of $|x|$.

59. $f(x) = 3(x - 7)(x + 3)^2$

60. $f(x) = 4(x + 4)(x + 3)^3$

61. $f(x) = 7(x^2 + 4)^2(x - 5)^3$

62. $f(x) = 2(x - 3)(x^2 + 4)^3$

63. $f(x) = -2\left(x + \dfrac{1}{2}\right)^2(x + 4)^3$

64. $f(x) = \left(x - \dfrac{1}{3}\right)^2(x - 1)^3$

65. $f(x) = (x - 5)^3(x + 4)^2$

66. $f(x) = (x + \sqrt{3})^2(x - 2)^4$

67. $f(x) = \dfrac{1}{2}(2x^2 + 9)^2(x^2 + 7)$

68. $f(x) = -2(x^2 + 3)^3$

69. $f(x) = -2x^2(x^2 - 2)$

70. $f(x) = 4x(x^2 - 3)$

In Problems 71–74, identify which of the graphs could be the graph of a polynomial function. For those that could, list the real zeros and state the least degree the polynomial can have. For those that could not, say why not.

71. **72.** **73.** **74.**

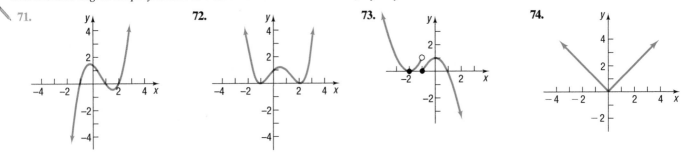

In Problems 75–78, find a polynomial function that might have the given graph. (More than one answer may be possible.)

75. **76.** **77.** **78.**

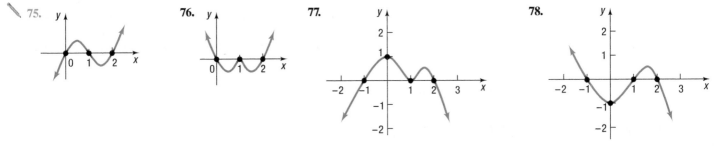

In Problems 79–82, write a polynomial function whose graph is shown (use the smallest degree possible).

79.

80.

(3, 8)

81.

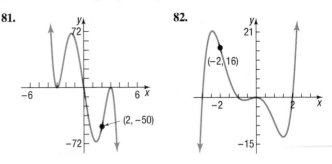

(2, −50)

82.

(−2, 16)

83. *Mixed Practice* $G(x) = (x + 3)^2(x - 2)$
  (a) Identify the x-intercepts of the graph of G.
  (b) What are the x-intercepts of the graph
     of $y = G(x + 3)$?

84. *Mixed Practice* $h(x) = (x + 2)(x - 4)^3$
  (a) Identify the x-intercepts of the graph of h.
  (b) What are the x-intercepts of the graph
     of $y = h(x - 2)$?

85. *Challenge Problem* Find the real zeros of

$$f(x) = 3(x^2 - 1)(x^2 + 4x + 3)^2$$

and their multiplicity.

86. *Challenge Problem* Determine the power function that resembles the end behavior of

$$g(x) = -4x^2(4 - 5x)^2(2x - 3)\left(\frac{1}{2}x + 1\right)^3$$

Explaining Concepts: Discussion and Writing

87. Can the graph of a polynomial function have no y-intercept? Can it have no x-intercepts? Explain.

88. The illustration shows the graph of a polynomial function.

  (a) Is the degree of the polynomial even or odd?
  (b) Is the leading coefficient positive or negative?
  (c) Is the function even, odd, or neither?
  (d) Why is x^2 necessarily a factor of the polynomial?
  (e) What is the minimum degree of the polynomial?
  (f) Formulate five different polynomials whose graphs could look like the one shown. Compare yours to those of other students. What similarities do you see? What differences?

89. Which of the following statements are true regarding the graph of the polynomial function $f(x) = x^3 + bx^2 + cx + d$? (Give reasons for your conclusions.)
  (a) It intersects the y-axis in one and only one point.
  (b) It intersects the x-axis in at most three points.
  (c) It intersects the x-axis at least once.
  (d) For $|x|$ very large, it behaves like the graph of $y = x^3$.
  (e) It is symmetric with respect to the origin.
  (f) It passes through the origin.

90. The graph of a polynomial function is always smooth and continuous. Name a function studied earlier that is smooth but not continuous. Name one that is continuous but not smooth.

91. Make up two polynomial functions, not of the same degree, with the following characteristics: crosses the x-axis at -2, touches the x-axis at 1, and is above the x-axis between -2 and 1. Give your polynomials to a fellow classmate and ask for a written critique.

92. Make up a polynomial function that has the following characteristics: crosses the x-axis at -1 and 4, touches the x-axis at 0 and 2, and is above the x-axis between 0 and 2. Give your polynomial to a fellow classmate and ask for a written critique.

93. Write a few paragraphs that provide a general strategy for graphing a polynomial function. Be sure to mention the following: degree, intercepts, end behavior, and turning points.

94. Design a polynomial function with the following characteristics: degree 6; four distinct real zeros, one of multiplicity 3; y-intercept 3; behaves like $y = -5x^6$ for large values of $|x|$. Is this polynomial unique? Compare your polynomial with those of other students. What terms will be the same as everyone else's? Add some more characteristics, such as symmetry or naming the real zeros. How does this modify the polynomial?

Retain Your Knowledge

Problems 95–104 are based on material learned earlier in the course. The purpose of these problems is to keep the material fresh in your mind so that you are better prepared for the final exam.

95. Find an equation of the line that contains the point $(2, -3)$ and is perpendicular to the line $5x - 2y = 6$.

96. Find the domain of the function $h(x) = \dfrac{x - 3}{x + 5}$.

97. Use the quadratic formula to find the real zeros of the function $f(x) = 4x^2 + 8x - 3$.

98. Solve: $|5x - 3| = 7$

99. Determine whether the function $f(x) = -3x + 2$ is increasing, decreasing, or constant.

100. Find the function that is finally graphed if the graph of $f(x) = x^2$ is shifted left 2 units and up 5 units.

101. The function $f(x) = 2x^3 - 3x^2 + 4$ is increasing where its derivative $f'(x) = 6x^2 - 6x > 0$. Where is f increasing?

102. Find the difference quotient of $f(x) = -\dfrac{1}{3}x + 2$.

103. The midpoint of a line segment is $(3, -5)$ and one endpoint is $(-2, 4)$. Find the other endpoint.

104. Find the quotient and remainder if $4x^3 - 7x^2 + 5$ is divided by $x^2 - 1$.

'Are You Prepared?' Answers

1. $(-2, 0), (2, 0), (0, 9)$ **2.** Yes; 3 **3.** Down; 4 **4.** True **5.** b; c

5.2 Graphing Polynomial Functions; Models

PREPARING FOR THIS SECTION *Before getting started, review the following:*

- Local Maxima and Local Minima (Section 3.3, pp. 232–233)
- Using a Graphing Utility to Approximate Local Maxima and Local Minima (Section 3.3, p. 235)
- Intercepts (Section 2.2, pp. 160–161)
- Build Quadratic Models from Data (Section 4.4, pp. 316–317)
- Polynomials (Chapter R, Section R.4, pp. 39–47)

Now Work the 'Are You Prepared?' problems on page 351.

OBJECTIVES 1 Graph a Polynomial Function (p. 346)
   2 Graph a Polynomial Function Using a Graphing Utility (p. 348)
   3 Build Cubic Models from Data (p. 350)

1 Graph a Polynomial Function

EXAMPLE 1 **Graphing a Polynomial Function**

Graph the polynomial function $f(x) = (2x + 1)(x - 3)^2$.

Step-by-Step Solution

Expand the polynomial:

Step 1 Determine the end behavior of the graph of the function.

$$f(x) = (2x + 1)(x - 3)^2$$
$$= (2x + 1)(x^2 - 6x + 9) \quad \text{Multiply out } (x - 3)^2.$$
$$= 2x^3 - 12x^2 + 18x + x^2 - 6x + 9 \quad \text{Multiply.}$$
$$= 2x^3 - 11x^2 + 12x + 9 \quad \text{Combine like terms.}$$

The polynomial function f is of degree 3. The graph of f resembles the graph of $y = 2x^3$ for large values of $|x|$.

Step 2 Find the *x*- and *y*-intercepts of the graph of the function.

The *y*-intercept is $f(0) = 9$. To find the *x*-intercepts, solve $f(x) = 0$.

$$f(x) = 0$$
$$(2x + 1)(x - 3)^2 = 0$$
$$2x + 1 = 0 \quad \text{or} \quad (x - 3)^2 = 0$$
$$x = -\frac{1}{2} \quad \text{or} \quad x = 3$$

The *x*-intercepts are $-\frac{1}{2}$ and 3.

Step 3 Determine the real zeros of the function and their multiplicity. Use this information to determine whether the graph crosses or touches the *x*-axis at each *x*-intercept.

The real zeros of *f* are $-\frac{1}{2}$ and 3. The zero $-\frac{1}{2}$ is a real zero of multiplicity 1, so the graph of *f* crosses the *x*-axis at $x = -\frac{1}{2}$. The zero 3 is a real zero of multiplicity 2, so the graph of *f* touches the *x*-axis at $x = 3$.

Step 4 Determine the maximum number of turning points on the graph of the function.

Because the polynomial function is of degree 3 (Step 1), the graph of the function has at most $3 - 1 = 2$ turning points.

Step 5 Put all the information from Steps 1 through 4 together to obtain the graph of *f*. To help establish the *y*-axis scale, find additional points on the graph on each side of any *x*-intercept.

Figure 18(a) illustrates the information obtained from Steps 1 through 4. We evaluate *f* at −1, 1, and 4 to help establish the scale on the *y*-axis.

We find that $f(-1) = -16$, $f(1) = 12$, and $f(4) = 9$, so we plot the points $(-1, -16)$, $(1, 12)$, and $(4, 9)$. The graph of *f* is given in Figure 18(b).

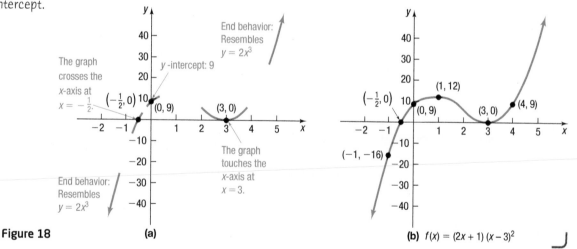

Figure 18   **(a)**   **(b)** $f(x) = (2x + 1)(x - 3)^2$

SUMMARY

Steps for Graphing a Polynomial Function

STEP 1: Determine the end behavior of the graph of the function.

STEP 2: Find the *x*- and *y*-intercepts of the graph of the function.

STEP 3: Determine the real zeros of the function and their multiplicity. Use this information to determine whether the graph crosses or touches the *x*-axis at each *x*-intercept.

STEP 4: Determine the maximum number of turning points on the graph of the function.

STEP 5: Use the information in Steps 1 through 4 to draw a complete graph of the function. To help establish the *y*-axis scale, find additional points on the graph on each side of any *x*-intercept.

EXAMPLE 2   **Graphing a Polynomial Function**

Graph the polynomial function $f(x) = (x - 2)(x + 3)^2(x - 5)$.

Solution   **STEP 1:** $f(x) = (x^2 - 7x + 10)(x + 3)^2$ Multiply $(x - 2)(x - 5)$.

$= (x^2 - 7x + 10)(x^2 + 6x + 9)$ Multiply $(x + 3)^2$.

$= x^4 - x^3 - 23x^2 - 3x + 90$ Multiply.

The polynomial function is of degree 4. The graph of f behaves like $y = x^4$ for large values of $|x|$. So, $f(x) \to \infty$ as $x \to -\infty$ and as $x \to \infty$.

STEP 2: The y-intercept is $f(0) = 90$. The x-intercepts are found by solving $f(x) = 0$. Using the Zero-Product Property, solve $(x - 2) = 0$, $(x + 3) = 0$, and $(x - 5) = 0$. The x-intercepts are $2, -3$, and 5.

STEP 3: The real zeros of the function are $2, -3$, and 5. The zeros 2 and 5 each have multiplicity 1, so the graph of f crosses the x-axis at 2 and 5. The zero -3 has multiplicity 2, so the graph of f touches the x-axis at -3.

STEP 4: The graph of f has at most $n - 1 = 4 - 1 = 3$ turning points.

STEP 5: To help establish the y-scale, evaluate f at $-5, -1, 3$, and 6.

$$f(-5) = 280 \qquad f(-1) = 72 \qquad f(3) = -72 \qquad f(6) = 324$$

Plot the additional points $(-5, 280)$, $(-1, 72)$, $(3, -72)$, and $(6, 324)$. Then connect the points with a smooth curve as shown in Figure 19.

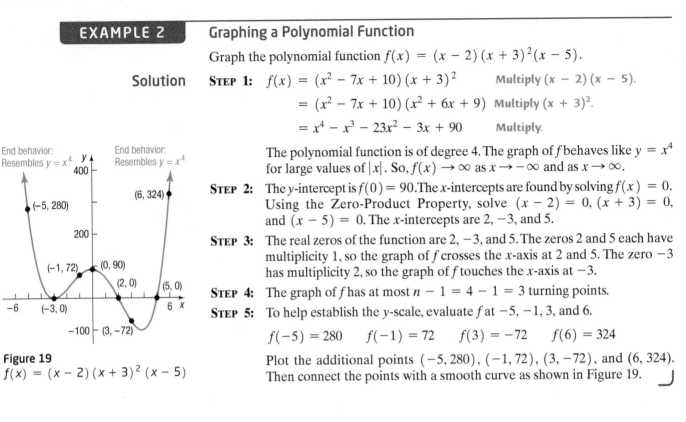

End behavior:
Resembles $y = x^4$

End behavior:
Resembles $y = x^4$

(6, 324)

(−5, 280)

(−1, 72)   (0, 90)

(2, 0)   (5, 0)

(−3, 0)

(3, −72)

Figure 19
$f(x) = (x - 2)(x + 3)^2(x - 5)$

──── **Now Work** PROBLEM 5

2 Graph a Polynomial Function Using a Graphing Utility

For polynomial functions that have noninteger coefficients and for polynomials that are not easily factored, we use a graphing utility early in the analysis. This is because the information that can be obtained from algebraic analysis is limited.

EXAMPLE 3   **Graphing a Polynomial Function Using a Graphing Utility**

Analyze the graph of the polynomial function

$$f(x) = x^3 + 2.48x^2 - 4.3155x + 2.484406$$

Step-by-Step Solution

Step 1 Determine the end behavior of the graph of the function.

The polynomial function f is of degree 3. The graph of f behaves like $y = x^3$ for large values of $|x|$.

Step 2 Graph the function using a graphing utility.

See Figure 20 for the graph of f using a TI-84 Plus C.

NORMAL FLOAT AUTO REAL RADIAN MP

Figure 20 $Y_1 = x^3 + 2.48x^2 - 4.3155x + 2.484406$

Step 3 Use the graphing utility to approximate the x- and y-intercepts of the graph.

The y-intercept is $f(0) = 2.484406$.

Since it is not readily apparent how to factor f, use a graphing utility's ZERO (or ROOT or SOLVE) feature and determine the x-intercept is -3.79, rounded to two decimal places.

Step 4 Use the graphing utility to create a TABLE to find points on the graph around each x-intercept.

Table 5 below shows values of x on each side of the x-intercept using a TI-84 Plus C. The points $(-4, -4.57)$ and $(-2, 13.04)$, rounded to two decimal places, are on the graph.

Table 5

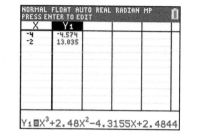

Step 5 Approximate the turning points of the graph.

From the graph of f shown in Figure 20, we see that f has two turning points. Using MAXIMUM reveals one turning point is at $(-2.28, 13.36)$, rounded to two decimal places. Using MINIMUM shows that the other turning point is at $(0.63, 1)$, rounded to two decimal places.

Step 6 Use the information in Steps 1 through 5 to draw a complete graph of the function by hand.

Figure 21 shows a graph of f drawn by hand using the information in Steps 1 through 5.

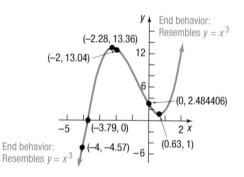

Figure 21

Step 7 From the graph, find the range of the polynomial function.

The range of f is the set of all real numbers.

Step 8 Use the graph to determine where the function is increasing and where it is decreasing.

Based on the graph, f is increasing on the intervals $(-\infty, -2.28]$ and $[0.63, \infty)$. Also, f is decreasing on the interval $[-2.28, 0.63]$.

SUMMARY

Steps for Using a Graphing Utility to Analyze the Graph of a Polynomial Function

STEP 1: Determine the end behavior of the graph of the function.
STEP 2: Graph the function using a graphing utility.
STEP 3: Use the graphing utility to approximate the x- and y-intercepts of the graph.
STEP 4: Use the graphing utility to create a TABLE to find points on the graph around each x-intercept.
STEP 5: Approximate the turning points of the graph.
STEP 6: Use the information in Steps 1 through 5 to draw a complete graph of the function by hand.
STEP 7: From the graph, find the range of the polynomial function.
STEP 8: Use the graph to determine where the function is increasing and where it is decreasing.

Now Work PROBLEM 23

3 Build Cubic Models from Data

In Section 4.2 we found the line of best fit from data, and in Section 4.4 we found the quadratic function of best fit. It is also possible to find other polynomial functions of best fit. However, most statisticians do not recommend finding polynomials of best fit of degree higher than 3.

Data that follow a cubic relation should look like Figure 22(a) or (b).

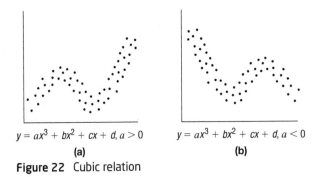

$y = ax^3 + bx^2 + cx + d, a > 0$ $y = ax^3 + bx^2 + cx + d, a < 0$

(a)   (b)

Figure 22 Cubic relation

EXAMPLE 4

A Cubic Function of Best Fit

The data in Table 6 represent the weekly cost C (in thousands of dollars) of printing x thousand textbooks.

Table 6

Number of Textbooks, x (thousands)	Cost, C ($1000s)
0	100
5	128.1
10	144
13	153.5
17	161.2
18	162.6
20	166.3
23	178.9
25	190.2
27	221.8

(a) Draw a scatter plot of the data using x as the independent variable and C as the dependent variable. Comment on the type of relation that may exist between the two variables x and C.

(b) Using a graphing utility, find the cubic function of best fit $C = C(x)$ that models the relation between number of textbooks and cost.

(c) Graph the cubic function of best fit on your scatter plot.

(d) Use the function found in part (b) to predict the cost of printing 22 thousand textbooks per week.

Solution

(a) Figure 23 shows the scatter plot. A cubic relation may exist between the two variables.

(b) Upon executing the CUBIC REGression program on a TI-84 Plus C, we obtain the results shown in Figure 24. The output shows the equation $y = ax^3 + bx^2 + cx + d$. The cubic function of best fit to the data is $C(x) = 0.0155x^3 - 0.5951x^2 + 9.1502x + 98.4327$.

(c) Figure 25 shows the graph of the cubic function of best fit on the scatter plot on a TI-84 Plus C, and Figure 26 shows the result using Desmos.

Figure 23   **Figure 24**   **Figure 25**

(d) Evaluate the function $C(x)$ at $x = 22$.

$$C(22) = 0.0155 \cdot 22^3 - 0.5951 \cdot 22^2 + 9.1502 \cdot 22 + 98.4327 \approx 176.8$$

The model predicts that the cost of printing 22 thousand textbooks in a week will be 176.8 thousand dollars—that is, $176,800.

Figure 26

Now Work PROBLEM 43

5.2 Assess Your Understanding

'Are You Prepared?' *Answers are given at the end of these exercises. If you get a wrong answer, read the pages listed in* red.

1. What are the intercepts of $y = 5x + 10$? (pp. 160–161)

2. Determine the leading term of $3 + 2x - 7x^3$. (p. 40)

3. Use a graphing utility to approximate (rounded to two decimal places) any local maximum values and local minimum values of $f(x) = x^3 - 2x^2 - 4x + 5$, for $-3 \le x \le 3$. (p. 235)

4. Use a graphing utility to find the quadratic function of best fit for the data below. (pp. 316–317)

x	2	2.5	3	3.5	4
y	3.08	3.42	3.65	3.82	3.6

Skill Building

In Problems 5–22, graph each polynomial function by following Steps 1 through 5 on page 347.

5. $f(x) = x^2(x - 3)$

6. $f(x) = x(x + 2)^2$

7. $f(x) = (x + 4)^2(1 - x)$

8. $f(x) = (x - 1)(x + 3)^2$

9. $f(x) = -2(x + 2)(x - 2)^3$

10. $f(x) = -\dfrac{1}{2}(x + 4)(x - 1)^3$

11. $f(x) = (x + 1)(x - 2)(x + 4)$

12. $f(x) = (x - 1)(x + 4)(x - 3)$

13. $f(x) = x(1 - x)(2 - x)$

14. $f(x) = (3 - x)(2 + x)(x + 1)$

15. $f(x) = (x + 1)^2(x - 2)^2$

16. $f(x) = (x - 4)^2(x + 2)^2$

17. $f(x) = -2(x - 1)^2(x^2 - 16)$

18. $f(x) = (x + 1)^3(x - 3)$

19. $f(x) = 5x(x^2 - 4)(x + 3)$

20. $f(x) = (x - 2)^2(x + 2)(x + 4)$

21. $f(x) = x^2(x - 2)(x^2 + 3)$

22. $f(x) = x^2(x^2 + 1)(x + 4)$

In Problems 23–30, use a graphing utility to graph each polynomial function f. Follow Steps 1 through 8 on page 349.

23. $f(x) = x^3 + 0.2x^2 - 1.5876x - 0.31752$

24. $f(x) = x^3 - 0.8x^2 - 4.6656x + 3.73248$

25. $f(x) = x^3 + 2.56x^2 - 3.31x + 0.89$

26. $f(x) = x^3 - 2.91x^2 - 7.668x - 3.8151$

27. $f(x) = x^4 - 2.5x^2 + 0.5625$

28. $f(x) = x^4 - 18.5x^2 + 50.2619$

29. $f(x) = 2x^4 - \pi x^3 + \sqrt{5}x - 4$

30. $f(x) = -1.2x^4 + 0.5x^2 - \sqrt{3}x + 2$

In Problems 31–42, graph each polynomial function f by following Steps 1 through 5 on page 347.

31. $f(x) = 4x - x^3$

32. $f(x) = x - x^3$

33. $f(x) = x^3 + x^2 - 12x$

34. $f(x) = x^3 + 2x^2 - 8x$

35. $f(x) = 2x^4 + 12x^3 - 8x^2 - 48x$

36. $f(x) = 4x^3 + 10x^2 - 4x - 10$

37. $f(x) = -x^5 - x^4 + x^3 + x^2$

38. $f(x) = -x^5 + 5x^4 + 4x^3 - 20x^2$

39. $f(x) = 15x^5 + 80x^4 + 80x^3$

40. $f(x) = 3x^6 + 6x^5 - 12x^4 - 24x^3$

41. $f(x) = \dfrac{1}{5}x^3 - \dfrac{4}{5}x^2 - 5x + 20$

42. $f(x) = \dfrac{3}{2}x^3 - \dfrac{15}{4}x^2 - 6x + 15$

Applications and Extensions

43. Hurricanes In 2012, Hurricane Sandy struck the East Coast of the United States, killing 147 people and causing an estimated $75 billion in damage. With a gale diameter of about 1000 miles, it was the largest ever to form over the Atlantic Basin. The accompanying data represent the number of major hurricane strikes in the Atlantic Basin (category 3, 4, or 5) each decade from 1921 to 2010.

Decade, x	Major Hurricanes Striking Atlantic Basin, H
1921–1930, 1	17
1931–1940, 2	16
1941–1950, 3	29
1951–1960, 4	33
1961–1970, 5	27
1971–1980, 6	16
1981–1990, 7	16
1991–2000, 8	27
2001–2010, 9	33

Source: National Oceanic & Atmospheric Administration

(a) Draw a scatter plot of the data. Comment on the type of relation that may exist between the two variables.

(b) Use a graphing utility to find the cubic function of best fit that models the relation between decade and number of major hurricanes.

(c) Use the model found in part (b) to predict the number of major hurricanes that struck the Atlantic Basin between 1961 and 1970.

(d) With a graphing utility, draw a scatter plot of the data and then graph the cubic function of best fit on the scatter plot.

(e) Concern has risen about the increase in the number and intensity of hurricanes, but some scientists believe this is just a natural fluctuation that could last another decade or two. Use your model to predict the number of major hurricanes that will strike the Atlantic Basin between 2011 and 2020. Is your result reasonable? What does this result suggest about the reliability of using a model to predict an event outside the domain of the data?

44. Poverty Rates The data (top, right) represent the percentage of families with children in the United States whose income is below the poverty level.

(a) With a graphing utility, draw a scatter plot of the data. Comment on the type of relation that appears to exist between the two variables.

(b) Decide on a function of best fit to these data (linear, quadratic, or cubic), and use this function to predict the percentage of U.S. families with children that were below the poverty level in 2016 ($t = 13$). Compare your prediction to the actual value of 15.0.

(c) Draw the function of best fit on the scatter plot drawn in part (a).

Year, t	Percent below Poverty Level, p	Year, t	Percent below Poverty Level, p
2004, 1	14.8	2010, 7	18.5
2005, 2	14.5	2011, 8	18.5
2006, 3	14.6	2012, 9	18.4
2007, 4	15.0	2013, 10	18.1
2008, 5	15.7	2014, 11	17.6
2009, 6	17.1	2015, 12	16.3

Source: U.S. Census Bureau

45. Temperature The following data represent the temperature T (°Fahrenheit) in Kansas City, Missouri, x hours after midnight on March 18, 2018.

Hours after Midnight, x	Temperature (°F), T
3	42.1
6	41.3
9	41.0
12	43.1
15	48.9
18	50.0
21	45.0
24	44.1

Source: The Weather Underground

(a) Draw a scatter plot of the data. Comment on the type of relation that may exist between the two variables.

(b) Find the average rate of change in temperature from 9 AM to 12 noon.

(c) What is the average rate of change in temperature from 3 PM to 9 PM?

(d) Decide on a function of best fit to these data (linear, quadratic, or cubic) and use this function to predict the temperature at 5 PM.

(e) With a graphing utility, draw a scatter plot of the data and then graph the function of best fit on the scatter plot.

(f) Interpret the y-intercept.

46. Future Value of Money Suppose that you make deposits of $500 at the beginning of every year into an Individual Retirement Account (IRA), earning interest r (expressed as a decimal). At the beginning of the first year, the value of the account will be $500; at the beginning of the second year, the value of the account, will be

$$\$500 + \$500r \quad + \quad \$500 = 500r + 1000$$

(a) Verify that the value of the account at the beginning of the third year is $T(r) = 500r^2 + 1500r + 1500$.

(b) The account value at the beginning of the fourth year is $F(r) = 500r^3 + 2000r^2 + 3000r + 2000$. If the annual rate of interest is 5% = 0.05, what will be the value of the account at the beginning of the fourth year?

47. *Challenge Problem* **A Geometric Series** In calculus, you will learn that certain functions can be approximated by polynomial functions. We will explore one such function now.

   (a) Using a graphing utility, create a table of values with

$$Y_1 = f(x) = \frac{1}{1-x} \text{ and}$$
$$Y_2 = g_2(x) = 1 + x + x^2 + x^3$$

     for $-1 < x < 1$ and $\Delta\text{Tbl} = 0.1$.

   (b) Using a graphing utility, create a table of values with

$$Y_1 = f(x) = \frac{1}{1-x} \text{ and}$$
$$Y_2 = g_3(x) = 1 + x + x^2 + x^3 + x^4$$

     for $-1 < x < 1$ and $\Delta\text{Tbl} = 0.1$.

   (c) Using a graphing utility, create a table of values with

$$Y_1 = f(x) = \frac{1}{1-x} \text{ and}$$
$$Y_2 = g_4(x) = 1 + x + x^2 + x^3 + x^4 + x^5$$

     for $-1 < x < 1$ and $\Delta\text{Tbl} = 0.1$.

   (d) What do you notice about the values of the function as more terms are added to the polynomial? Are there some values of x for which the approximations are better?

48. *Challenge Problem* **Tennis Anyone?** Assume that the probability of winning a point on serve or return is treated as constant throughout the match. Further suppose that x is the probability that the better player in a match wins a set.

   (a) The probability P_3 that the better player wins a best-of-three match is $P_3(x) = x^2\,[1 + 2(1 - x)]$. Suppose the probability that the better player wins a set is 0.6. What is the probability that this player wins a best-of-three match?

   (b) The probability P_5 that the better player wins a best-of-five match is

$$P_5(x) = x^3\,[1 + 3(1 - x) + 6(1 - x)^2]$$

Suppose the probability that the better player wins a set is 0.6. What is the probability that this player wins a best-of-five match?

   (c) The difference between the probability of winning and losing is known as the *win advantage*. For a best-of-*n* match, the win advantage is

$$P_n - (1 - P_n) = 2P_n - 1$$

The *edge, E,* is defined as the difference in win advantage between a best-of-five and best-of-three match. That is,

$$E = (2P_5 - 1) - (2P_3 - 1) = 2(P_5 - P_3)$$

Edge as a function of win probability of a set x is

$$E(x) = 2x^2\left\{x\left[1 + 3(1 - x) + 6(1 - x)^2\right] - \left(1 + 2(1 - x)\right)\right\}$$

Graph $E = E(x)$ for $0.5 \le x \le 1$.

   (d) Find the probability of winning a set x that maximizes the edge. What is the maximum edge?

   (e) Explain the meaning of $E(0.5)$.

   (f) Explain the meaning of $E(1)$.

   Source: Stephanie Kovalchik, "Grand Slams Are Short-Changing Women's Tennis," Significance, *October 2015.*

49. *Challenge Problem* Suppose $f(x) = -ax^2(x - b)(x + c)^2$, where $0 < a < b < c$.

   (a) Graph f.

   (b) In what interval(s) is there a local maximum value?

   (c) Which numbers yield a local minimum value?

   (d) Where is $f(x) < 0$?

   (e) Where is $f(-x - 4) < 0$?

   (f) Is f increasing, decreasing, or neither on $(-\infty, -c]$?

Retain Your Knowledge

Problems 50–59 are based on material learned earlier in the course. The purpose of these problems is to keep the material fresh in your mind so that you are better prepared for the final exam.

50. Solve $2|3x - 1| + 4 > 10$.

51. Determine the function that is graphed if the graph of $f(x) = \sqrt{x}$ is reflected about the x-axis and then vertically compressed by a factor of $\frac{1}{3}$.

52. Find the vertex of the graph of $f(x) = -2x^2 + 7x - 3$.

53. The strain on a solid object varies directly with the external tension force acting on the solid and inversely with the cross-sectional area. If a certain steel rod has a strain of 150 when the force is 1.47×10^4 N and the cross-sectional area is 4.9×10^{-4} m^2, find the strain for a similar rod with a cross-sectional area of 8.75×10^{-3} m^2 and a tension force of 2.45×10^5 N.

54. Given $f(x) = 2x^3 - 7x + 1$, find $f\left(-\frac{1}{2}\right)$.

55. Find the domain of $f(x) = -9\sqrt{x - 4} + 1$.

56. Find the average rate of change of $f(x) = x^2 + 4x - 3$ from -2 to 1.

57. Find the center and radius of the circle

$$x^2 + 4x + y^2 - 2y = 11$$

58. Determine if the function $g(x) = \dfrac{\sqrt[3]{x}}{x^3 - x}$ is even, odd, or neither.

59. How long will it take $\$5000$ to grow to $\$7500$ at a simple interest rate of 8%?

'Are You Prepared?' Answers

1. $(0, 10)$, $(-2, 0)$
2. $-7x^3$
3. Local maximum value 6.48 at $x = -0.67$; local minimum value -3 at $x = 2$
4. $y = -0.337x^2 + 2.311x - 0.216$

5.3 Properties of Rational Functions

PREPARING FOR THIS SECTION *Before getting started, review the following:*

- Rational Expressions (Chapter R, Section R.7, pp. 61–69)
- Polynomial Division (Chapter R, Section R.4, pp. 44–46)
- Graph of $f(x) = \dfrac{1}{x}$ (Section 2.2, Example 13, p. 165)
- Graphing Techniques: Transformations (Section 3.5, pp. 254–263)

Now Work the 'Are You Prepared?' problems on page 362.

> **OBJECTIVES**   1 Find the Domain of a Rational Function (p. 354)
>
> 2 Find the Vertical Asymptotes of a Rational Function (p. 357)
>
> 3 Find a Horizontal or an Oblique Asymptote of a Rational Function (p. 359)

Ratios of integers are called *rational numbers*. Similarly, ratios of polynomial functions are called *rational functions*. Examples of rational functions are

$$R(x) = \frac{x^2 - 4}{x^2 + x + 1} \qquad F(x) = \frac{x^3}{x^2 - 4} \qquad G(x) = \frac{3x^2}{x^4 - 1}$$

> **DEFINITION Rational Function**
>
> A **rational function** is a function of the form
>
> $$R(x) = \frac{p(x)}{q(x)}$$
>
> where p and q are polynomial functions and q is not the zero polynomial. The domain of R is the set of all real numbers, except those for which the denominator q is 0.

1 Find the Domain of a Rational Function

EXAMPLE 1

Finding the Domain of a Rational Function

(a) The domain of $R(x) = \dfrac{2x^2 - 4}{x + 5}$ is the set of all real numbers x except -5; that is, the domain is $\{x \mid x \neq -5\}$.

(b) The domain of $R(x) = \dfrac{1}{x^2 - 4} = \dfrac{1}{(x + 2)(x - 2)}$ is the set of all real numbers x except -2 and 2; that is, the domain is $\{x \mid x \neq -2, x \neq 2\}$.

(c) The domain of $R(x) = \dfrac{x^3}{x^2 + 1}$ is the set of all real numbers.

(d) The domain of $R(x) = \dfrac{x^2 - 1}{x - 1}$ is the set of all real numbers x except 1; that is, the domain is $\{x \mid x \neq 1\}$.

Although $\dfrac{x^2 - 1}{x - 1}$ simplifies to $x + 1$, it is important to observe that the functions

$$R(x) = \frac{x^2 - 1}{x - 1} \quad \text{and} \quad f(x) = x + 1$$

are not equal, since the domain of R is $\{x \mid x \neq 1\}$ and the domain of f is the set of all real numbers.

Now Work PROBLEM 17

If $R(x) = \dfrac{p(x)}{q(x)}$ is a rational function, and if p and q have no common factors,

WARNING The domain of a rational function must be found *before* writing the function in lowest terms. ∎

then the rational function R is said to be in **lowest terms**. For a rational function $R(x) = \dfrac{p(x)}{q(x)}$ in lowest terms, the real zeros, if any, of the numerator, which are also in the domain of R, are the x-intercepts of the graph of R. The real zeros of the denominator of R [that is, the numbers x, if any, for which $q(x) = 0$], although not in the domain of R, also play a major role in the graph of R.

We have already discussed the properties of the rational function $y = \dfrac{1}{x}$. (Refer to Example 13, page 165). The next rational function that we analyze is $H(x) = \dfrac{1}{x^2}$.

EXAMPLE 2

Graphing $H(x) = \dfrac{1}{x^2}$

Analyze the graph of $H(x) = \dfrac{1}{x^2}$.

Solution

The domain of $H(x) = \dfrac{1}{x^2}$ is the set of all real numbers x except 0. The graph has no y-intercept, because x cannot equal 0. The graph has no x-intercept because the equation $H(x) = 0$ has no solution. Therefore, the graph of H does not cross or touch either of the coordinate axes. Because

$$H(-x) = \frac{1}{(-x)^2} = \frac{1}{x^2} = H(x)$$

H is an even function, so its graph is symmetric with respect to the y-axis.

Table 7 shows values of $H(x) = \dfrac{1}{x^2}$ for selected positive numbers x. (We use symmetry to obtain values of H when $x < 0$.) From the first three rows of Table 7, we see that as the values of x approach (get closer to) 0, the values of $H(x)$ become unbounded in the positive direction. That is, as $x \to 0$, $H(x) \to \infty$. (In calculus we use limit notation, $\lim\limits_{x \to 0} H(x) = \infty$, to convey this).

Look at the last four rows of Table 7. As $x \to \infty$, then $H(x) \to 0$, so we have the end behavior of the graph. Figure 27 shows the graph. Notice the use of red dashed lines to convey these ideas.

Table 7

x	$H(x) = \dfrac{1}{x^2}$
$\dfrac{1}{2}$	4
$\dfrac{1}{100}$	10,000
$\dfrac{1}{10,000}$	100,000,000
1	1
2	$\dfrac{1}{4}$
100	$\dfrac{1}{10,000}$
10,000	$\dfrac{1}{100,000,000}$

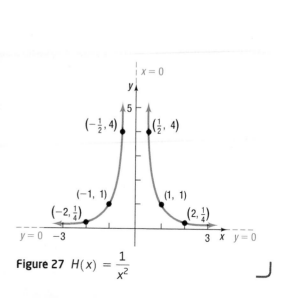

Figure 27 $H(x) = \dfrac{1}{x^2}$

EXAMPLE 3	**Using Transformations to Graph a Rational Function**

Graph the rational function: $R(x) = \dfrac{1}{(x-2)^2} + 1$

Solution

The domain of R is the set of all real numbers except $x = 2$. To graph R, start with the graph of $y = \dfrac{1}{x^2}$. See Figure 28 for the steps.

Figure 28 (a) $y = \dfrac{1}{x^2}$ (b) $y = \dfrac{1}{(x-2)^2}$ (c) $y = \dfrac{1}{(x-2)^2} + 1$

Replace x by $x - 2$;
shift right 2 units.

Add 1;
shift up 1 unit.

Now Work PROBLEMS 35(a) AND 35(b)

Asymptotes

Let's investigate the roles of the vertical line $x = 2$ and the horizontal line $y = 1$ in Figure 28(c).

First, we look at the end behavior of $R(x) = \dfrac{1}{(x-2)^2} + 1$. Table 8(a) shows the values of R at $x = 10, 100, 1000,$ and $10,000$. Note that as x becomes unbounded in the positive direction, the values of R approach 1. That is, as $x \to \infty$, $R(x) \to 1$. From Table 8(b) we see that as x becomes unbounded in the negative direction, the values of R also approach 1. That is, as $x \to -\infty$, $R(x) \to 1$.

Although $x = 2$ is not in the domain of R, the behavior of the graph of R near $x = 2$ is important. Table 8(c) shows the values of R at $x = 1.5, 1.9, 1.99, 1.999,$ and 1.9999. We see that as x approaches 2 for $x < 2$, denoted $x \to 2^-$, the values of R increase without bound. That is, as $x \to 2^-$ (from the left), $R(x) \to \infty$. From Table 8(d), we see that as x approaches 2 for $x > 2$, denoted $x \to 2^+$, the values of R also increase without bound. That is, as $x \to 2^+$ (from the right), $R(x) \to \infty$.

Table 8

x	R(x)
10	1.0156
100	1.0001
1000	1.000001
10,000	1.00000001

(a)

x	R(x)
−10	1.0069
−100	1.0001
−1000	1.000001
−10,000	1.00000001

(b)

x	R(x)
1.5	5
1.9	101
1.99	10,001
1.999	1,000,001
1.9999	100,000,001

(c)

x	R(x)
2.5	5
2.1	101
2.01	10,001
2.001	1,000,001
2.0001	100,000,001

(d)

The vertical line $x = 2$ and the horizontal line $y = 1$ are called *asymptotes* of the graph of R.

> **DEFINITION** Horizontal and Vertical Asymptotes
>
> Let R denote a function.
>
> If, as $x \to -\infty$ or as $x \to \infty$, the values of $R(x)$ approach some fixed number L, then the line $y = L$ is a **horizontal asymptote** of the graph of R. [Refer to Figures 29(a) and (b).]
>
> If, as x approaches some number c, the values $|R(x)| \to \infty$ [that is, $R(x) \to -\infty$ or $R(x) \to \infty$], then the line $x = c$ is a **vertical asymptote** of the graph of R. [Refer to Figures 29(c) and (d).]

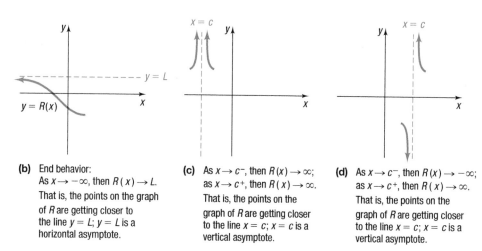

(a) End behavior:
As $x \to \infty$, then $R(x) \to L$.
That is, the points on the graph of R are getting closer to the line $y = L$; $y = L$ is a horizontal asymptote.

(b) End behavior:
As $x \to -\infty$, then $R(x) \to L$.
That is, the points on the graph of R are getting closer to the line $y = L$; $y = L$ is a horizontal asymptote.

(c) As $x \to c^-$, then $R(x) \to \infty$; as $x \to c^+$, then $R(x) \to \infty$.
That is, the points on the graph of R are getting closer to the line $x = c$; $x = c$ is a vertical asymptote.

(d) As $x \to c^-$, then $R(x) \to -\infty$; as $x \to c^+$, then $R(x) \to \infty$.
That is, the points on the graph of R are getting closer to the line $x = c$; $x = c$ is a vertical asymptote.

Figure 29

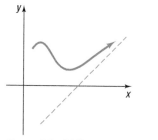

Figure 30 Oblique asymptote

A horizontal asymptote, when it occurs, describes the **end behavior** of the graph as $x \to \infty$ or as $x \to -\infty$. **The graph of a function may intersect a horizontal asymptote.**

A vertical asymptote, when it occurs, describes the behavior of the graph when x is close to some number c. **The graph of a rational function never intersects a vertical asymptote.**

There is a third possibility. If, as $x \to -\infty$ or as $x \to \infty$, the value of a rational function $R(x)$ approaches a linear expression $ax + b$, $a \neq 0$, then the line $y = ax + b$, $a \neq 0$, is an **oblique (or slant) asymptote** of R. Figure 30 shows an oblique asymptote. An oblique asymptote, when it occurs, describes the end behavior of the graph. **The graph of a function may intersect an oblique asymptote.**

Now Work PROBLEMS 27 AND 35(c)

2 Find the Vertical Asymptotes of a Rational Function

The vertical asymptotes of a rational function $R(x) = \dfrac{p(x)}{q(x)}$, in lowest terms, are located at the real zeros of the denominator $q(x)$. Suppose that r is a real zero of q, so $x - r$ is a factor of q. As $x \to r$, the values $x - r \to 0$, causing the ratio to become unbounded; that is, $|R(x)| \to \infty$. Based on the definition, we conclude that the line $x = r$ is a vertical asymptote.

WARNING If a rational function is not in lowest terms, this theorem may result in an incorrect listing of vertical asymptotes. ■

> **THEOREM** Locating Vertical Asymptotes
>
> The graph of a rational function $R(x) = \dfrac{p(x)}{q(x)}$, *in lowest terms*, has a vertical asymptote $x = r$ if r is a real zero of the *denominator* q. That is, if $x - r$ is a factor of the denominator q of the rational function R, in lowest terms, the graph of R has a vertical asymptote $x = r$.

EXAMPLE 4 Finding Vertical Asymptotes

Find the vertical asymptotes, if any, of the graph of each rational function.

(a) $F(x) = \dfrac{x + 3}{x - 1}$ (b) $R(x) = \dfrac{x}{x^2 - 4}$

(c) $H(x) = \dfrac{x^2}{x^2 + 1}$ (d) $G(x) = \dfrac{x^2 - 9}{x^2 + 4x - 21}$

Solution

(a) $F(x) = \dfrac{x + 3}{x - 1}$ is in lowest terms, and the only real zero of the denominator is 1. The line $x = 1$ is the vertical asymptote of the graph of F.

(b) $R(x) = \dfrac{x}{x^2 - 4}$ is in lowest terms, and the real zeros of the denominator $x^2 - 4$ are -2 and 2. The lines $x = -2$ and $x = 2$ are the vertical asymptotes of the graph of R.

(c) $H(x) = \dfrac{x^2}{x^2 + 1}$ is in lowest terms, and the denominator has no real zeros. The graph of H has no vertical asymptotes.

(d) Factor the numerator and denominator of $G(x) = \dfrac{x^2 - 9}{x^2 + 4x - 21}$ to determine whether G is in lowest terms.

$$G(x) = \frac{x^2 - 9}{x^2 + 4x - 21} = \frac{(x + 3)(x - 3)}{(x + 7)(x - 3)} = \frac{x + 3}{x + 7} \qquad x \neq 3$$

The only real zero of the denominator of G in lowest terms is -7. The line $x = -7$ is the only vertical asymptote of the graph of G.

WARNING When identifying a vertical asymptote, as in the solution to Example 4(a), write the equation of the vertical asymptote as $x = 1$. Do not say that the vertical asymptote is 1. ∎

As Example 4 illustrates, rational functions can have no vertical asymptotes, one vertical asymptote, or more than one vertical asymptote.

─**Now Work** PROBLEMS **45, 47,** AND **49** (FIND THE VERTICAL ASYMPTOTES, IF ANY.)

Multiplicity and Vertical Asymptotes

Recall from Figure 15 in Section 5.1 that the end behavior of a polynomial function is always one of four types. For polynomials of odd degree, the ends of the graph go in opposite directions (one up and one down), whereas for polynomials of even degree, the ends go in the same direction (both up or both down).

For a rational function in lowest terms, the multiplicities of the real zeros in the denominator can be used in a similar fashion to determine the behavior of the graph around each vertical asymptote. Consider the following four functions, each with a single vertical asymptote, $x = 2$.

$$R_1(x) = \frac{1}{x - 2} \qquad R_2(x) = -\frac{1}{x - 2} \qquad R_3(x) = \frac{1}{(x - 2)^2} \qquad R_4(x) = -\frac{1}{(x - 2)^2}$$

Figure 31 shows the graphs of each function. The graphs of R_1 and R_2 are transformations of the graph of $y = \dfrac{1}{x}$, and the graphs of R_3 and R_4 are transformations of the graph of $y = \dfrac{1}{x^2}$.

Based on Figure 31, we can make the following conclusions:

- If the multiplicity of the real zero that gives rise to a vertical asymptote is odd, the graph approaches ∞ on one side of the vertical asymptote and approaches $-\infty$ on the other side.
- If the multiplicity of the real zero that gives rise to the vertical asymptote is even, the graph approaches either ∞ or $-\infty$ on both sides of the vertical asymptote.

These results are true in general and will be helpful when graphing rational functions in the next section.

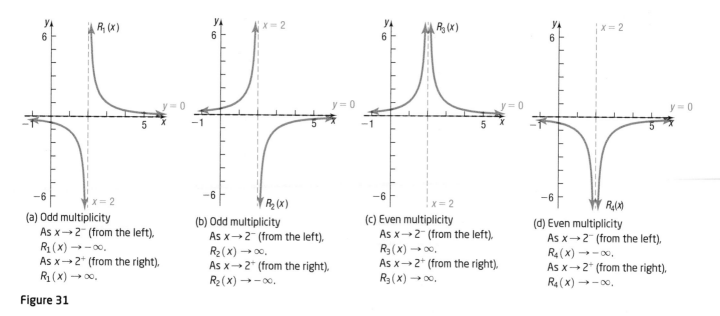

(a) Odd multiplicity
As $x \rightarrow 2^{-}$ (from the left),
$R_1(x) \rightarrow -\infty$.
As $x \rightarrow 2^{+}$ (from the right),
$R_1(x) \rightarrow \infty$.

(b) Odd multiplicity
As $x \rightarrow 2^{-}$ (from the left),
$R_2(x) \rightarrow \infty$.
As $x \rightarrow 2^{+}$ (from the right),
$R_2(x) \rightarrow -\infty$.

(c) Even multiplicity
As $x \rightarrow 2^{-}$ (from the left),
$R_3(x) \rightarrow \infty$.
As $x \rightarrow 2^{+}$ (from the right),
$R_3(x) \rightarrow \infty$.

(d) Even multiplicity
As $x \rightarrow 2^{-}$ (from the left),
$R_4(x) \rightarrow -\infty$.
As $x \rightarrow 2^{+}$ (from the right),
$R_4(x) \rightarrow -\infty$.

Figure 31

3 Find a Horizontal or an Oblique Asymptote of a Rational Function

To find horizontal or oblique asymptotes, we need to know how the function behaves as $x \rightarrow -\infty$ or as $x \rightarrow \infty$. That is, we need to determine the end behavior of the function. This can be done by examining the degrees of the numerator and denominator, and the respective power functions that each resembles. For example, consider the rational function

$$R(x) = \frac{3x - 2}{5x^2 - 7x + 1}$$

The degree of the numerator, 1, is less than the degree of the denominator, 2. When $|x|$ is very large, the numerator of R can be approximated by the power function $y = 3x$, and the denominator can be approximated by the power function $y = 5x^2$. This means

$$R(x) = \underbrace{\frac{3x - 2}{5x^2 - 7x + 1}}_{\text{For } |x| \text{ very large}} \approx \frac{3x}{5x^2} = \underbrace{\frac{3}{5x} \rightarrow 0}_{\text{As } x \rightarrow -\infty \text{ or } x \rightarrow \infty}$$

which shows that the line $y = 0$ is a horizontal asymptote. This result is true for all rational functions that are **proper** (that is, the degree of the numerator is less than the degree of the denominator). If a rational function is **improper** (that is, if the degree of the numerator is greater than or equal to the degree of the denominator), there could be a horizontal asymptote, an oblique asymptote, or neither. The following summary details how to find horizontal or oblique asymptotes.

Finding a Horizontal or an Oblique Asymptote of a Rational Function

Consider the rational function

$$R(x) = \frac{p(x)}{q(x)} = \frac{a_n x^n + a_{n-1} x^{n-1} + \cdots + a_1 x + a_0}{b_m x^m + b_{m-1} x^{m-1} + \cdots + b_1 x + b_0}$$

in which the degree of the numerator is n and the degree of the denominator is m.

- If $n < m$ (the degree of the numerator is less than the degree of the denominator), the line $y = 0$ is a horizontal asymptote.

- If $n = m$ (the degree of the numerator equals the degree of the denominator), the line $y = \dfrac{a_n}{b_m}$ is a horizontal asymptote. (That is, the horizontal asymptote equals the ratio of the leading coefficients.)

- If $n = m + 1$ (the degree of the numerator is one more than the degree of the denominator), the line $y = ax + b$ is an oblique asymptote, which is the quotient found using polynomial division.

- If $n \geq m + 2$ (the degree of the numerator is two or more greater than the degree of the denominator), there are no horizontal or oblique asymptotes, and the end behavior of the graph resembles the power function $y = \dfrac{a_n}{b_m} x^{n-m}$.

Note: A rational function never has both a horizontal asymptote and an oblique asymptote. A rational function may have neither a horizontal nor an oblique asymptote.

EXAMPLE 5   **Finding a Horizontal Asymptote**

Find the horizontal asymptote, if one exists, of the graph of

$$R(x) = \frac{4x^3 - 5x + 2}{7x^5 + 2x^4 - 3x}$$

Solution   Since the degree of the numerator, 3, is less than the degree of the denominator, 5, the rational function R is proper. The line $y = 0$ is a horizontal asymptote of the graph of R.

EXAMPLE 6   **Finding a Horizontal or an Oblique Asymptote**

Find the horizontal or oblique asymptote, if one exists, of the graph of

$$H(x) = \frac{3x^4 - x^2}{x^3 - x^2 + 1}$$

Solution   Since the degree of the numerator, 4, is exactly one greater than the degree of the denominator, 3, the rational function H has an oblique asymptote. Find the asymptote by using polynomial division.

$$
\begin{array}{r}
3x + 3 \\
x^3 - x^2 + 1 \overline{)3x^4 \qquad\quad - x^2 \qquad\quad} \\
\underline{3x^4 - 3x^3 \qquad\quad + 3x} \\
3x^3 - x^2 - 3x \\
\underline{3x^3 - 3x^2 \qquad + 3} \\
2x^2 - 3x - 3
\end{array}
$$

As a result,

$$H(x) = \frac{3x^4 - x^2}{x^3 - x^2 + 1} = 3x + 3 + \frac{2x^2 - 3x - 3}{x^3 - x^2 + 1}$$

As $x \to -\infty$ or as $x \to \infty$,

$$\frac{2x^2 - 3x - 3}{x^3 - x^2 + 1} \approx \frac{2x^2}{x^3} = \frac{2}{x} \to 0$$

As $x \to -\infty$ or as $x \to \infty$, we have $H(x) \to 3x + 3$. The graph of the rational function H has an oblique asymptote $y = 3x + 3$. Put another way, as $x \to \pm\infty$, the graph of H resembles the graph of $y = 3x + 3$.

Figure 32 $H(x) = \dfrac{3x^4 - x^2}{x^3 - x^2 + 1}$

Figure 32 shows the graph of $H(x) = \dfrac{3x^4 - x^2}{x^3 - x^2 + 1}$.

EXAMPLE 7

Finding a Horizontal or an Oblique Asymptote

Find the horizontal or oblique asymptote, if one exists, of the graph of

$$R(x) = \frac{8x^2 - x + 2}{4x^2 - 1}$$

Solution

Since the degree of the numerator, 2, equals the degree of the denominator, 2, the rational function R has a horizontal asymptote equal to the ratio of the leading coefficients.

$$y = \frac{a_n}{b_m} = \frac{8}{4} = 2$$

To see why the horizontal asymptote equals the ratio of the leading coefficients, investigate the behavior of R as $x \to -\infty$ or as $x \to \infty$. When $|x|$ is very large, the numerator of R can be approximated by the power function $y = 8x^2$, and the denominator can be approximated by the power function $y = 4x^2$. This means that as $x \to -\infty$ or as $x \to \infty$,

$$R(x) = \frac{8x^2 - x + 2}{4x^2 - 1} \approx \frac{8x^2}{4x^2} = \frac{8}{4} = 2$$

The graph of the rational function R has a horizontal asymptote $y = 2$. The graph of R resembles the graph of $y = 2$ as $x \to \pm\infty$.

EXAMPLE 8

Finding a Horizontal or an Oblique Asymptote

Find the horizontal or oblique asymptote, if one exists, of the graph of

$$G(x) = \frac{2x^5 - x^3 + 2}{x^3 - 1}$$

Solution

Since the degree of the numerator, 5, is greater than the degree of the denominator, 3, by more than one, the rational function G has no horizontal or oblique asymptote. The end behavior of the graph resembles the power function $y = 2x^{5-3} = 2x^2$.

To see why this is the case, investigate the behavior of G as $x \to -\infty$ or as $x \to \infty$. When $|x|$ is very large, the numerator of G can be approximated by the power function $y = 2x^5$, and the denominator can be approximated by the power function $y = x^3$. This means as $x \to -\infty$ or as $x \to \infty$,

$$G(x) = \frac{2x^5 - x^3 + 2}{x^3 - 1} \approx \frac{2x^5}{x^3} = 2x^{5-3} = 2x^2$$

Since this is not linear, the graph of G has no horizontal or oblique asymptote.

Figure 33 shows the graph of $G(x) = \frac{2x^5 - x^3 + 2}{x^3 - 1}$.

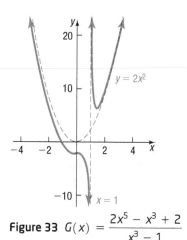

Figure 33 $G(x) = \dfrac{2x^5 - x^3 + 2}{x^3 - 1}$

Now Work PROBLEMS 45, 47, AND 49 (FIND THE HORIZONTAL OR OBLIQUE ASYMPTOTE, IF ONE EXISTS.)

5.3 Assess Your Understanding

'Are You Prepared?' *Answers are given at the end of these exercises. If you get a wrong answer, read the pages listed in* red.

1. **True or False** The quotient of two polynomial expressions is a rational expression. (pp. 61–69)

2. What are the quotient and remainder when $3x^4 - x^2$ is divided by $x^3 - x^2 + 1$. (pp. 44–46)

3. Graph $y = \dfrac{1}{x}$. (p. 165)

4. Graph $y = 2(x + 1)^2 - 3$ using transformations. (pp. 254–263)

Concepts and Vocabulary

5. **True or False** The domain of every rational function is the set of all real numbers.

6. If, as $x \to -\infty$ or as $x \to \infty$, the values of $R(x)$ approach some fixed number L, then the line $y = L$ is a _____ _____ of the graph of R.

7. If, as x approaches some number c, the values of $|R(x)| \to \infty$, then the line $x = c$ is a _____ _____ of the graph of R.

8. For a rational function R, if the degree of the numerator is less than the degree of the denominator, then R is _____.

9. **True or False** The graph of a rational function may intersect a horizontal asymptote.

10. **True or False** The graph of a rational function may intersect a vertical asymptote.

11. If a rational function is proper, then _____ is a horizontal asymptote.

12. **True or False** If the degree of the numerator of a rational function equals the degree of the denominator, then the rational function has a horizontal asymptote.

13. **Multiple Choice** If $R(x) = \dfrac{p(x)}{q(x)}$ is a rational function and if p and q have no common factors, then R is _____.
    (a) improper
    (b) proper
    (c) undefined
    (d) in lowest terms

14. **Multiple Choice** Which type of asymptote, when it occurs, describes the behavior of a graph when x is close to some number?
    (a) vertical   (b) horizontal   (c) oblique   (d) all of these

Skill Building

In Problems 15–26, find the domain of each rational function.

15. $R(x) = \dfrac{4x}{x - 7}$

16. $R(x) = \dfrac{5x^2}{3 + x}$

17. $H(x) = \dfrac{-4x^2}{(x - 2)(x + 4)}$

18. $G(x) = \dfrac{6}{(x + 3)(4 - x)}$

19. $F(x) = \dfrac{3x(x - 1)}{2x^2 - 5x - 12}$

20. $Q(x) = \dfrac{-x(1 - x)}{3x^2 + 5x - 2}$

21. $R(x) = \dfrac{x}{x^3 - 64}$

22. $R(x) = \dfrac{x}{x^4 - 1}$

23. $H(x) = \dfrac{3x^2 + x}{x^2 + 9}$

24. $G(x) = \dfrac{x - 3}{x^4 + 1}$

25. $R(x) = \dfrac{3(x^2 - x - 6)}{5(x^2 - 4)}$

26. $F(x) = \dfrac{-2(x^2 - 4)}{3(x^2 + 4x + 4)}$

In Problems 27–32, use the graph shown to find
(a) The domain and range of each function
(d) Vertical asymptotes, if any

(b) The intercepts, if any
(e) Oblique asymptotes, if any

(c) Horizontal asymptotes, if any

27.

28.

29.

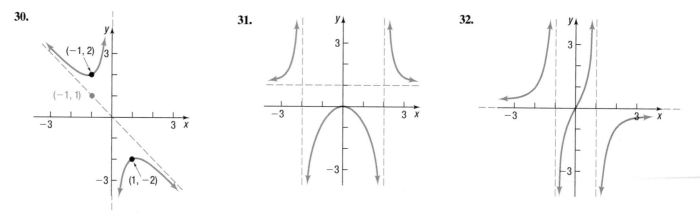

In Problems 33–44, (a) graph the rational function using transformations, (b) use the final graph to find the domain and range, and (c) use the final graph to list any vertical, horizontal, or oblique asymptotes.

33. $F(x) = 2 + \dfrac{1}{x}$

34. $Q(x) = 3 + \dfrac{1}{x^2}$

35. $R(x) = \dfrac{1}{(x-1)^2}$

36. $R(x) = \dfrac{3}{x}$

37. $H(x) = \dfrac{-2}{x+1}$

38. $G(x) = \dfrac{2}{(x+2)^2}$

39. $R(x) = \dfrac{-1}{x^2 + 4x + 4}$

40. $R(x) = \dfrac{1}{x-1} + 1$

41. $G(x) = 1 + \dfrac{2}{(x-3)^2}$

42. $F(x) = 2 - \dfrac{1}{x+1}$

43. $R(x) = \dfrac{x^2 - 4}{x^2}$

44. $R(x) = \dfrac{x-4}{x}$

In Problems 45–56, find the vertical, horizontal, and oblique asymptotes, if any, of each rational function.

45. $R(x) = \dfrac{3x}{x+4}$

46. $R(x) = \dfrac{3x+5}{x-6}$

47. $H(x) = \dfrac{x^3 - 8}{x^2 - 5x + 6}$

48. $G(x) = \dfrac{x^3 + 1}{x^2 - 5x - 14}$

49. $T(x) = \dfrac{x^3}{x^4 - 1}$

50. $P(x) = \dfrac{4x^2}{x^3 - 1}$

51. $Q(x) = \dfrac{2x^2 - 5x - 12}{3x^2 - 11x - 4}$

52. $F(x) = \dfrac{x^2 + 6x + 5}{2x^2 + 7x + 5}$

53. $R(x) = \dfrac{6x^2 + 19x - 7}{3x - 1}$

54. $R(x) = \dfrac{8x^2 + 26x - 7}{4x - 1}$

55. $G(x) = \dfrac{x^4 - 1}{x^2 - x}$

56. $F(x) = \dfrac{x^4 - 16}{x^2 - 2x}$

Applications and Extensions

57. Gravity In physics, it is established that the acceleration due to gravity, g (in meters/sec^2), at a height h meters above sea level is given by

$$g(h) = \frac{3.99 \times 10^{14}}{(6.374 \times 10^6 + h)^2}$$

where 6.374×10^6 is the radius of Earth in meters.
(a) What is the acceleration due to gravity at sea level?
(b) The Willis Tower in Chicago, Illinois, is 443 meters tall. What is the acceleration due to gravity at the top of the Willis Tower?
(c) The peak of Mount Everest is 8848 meters above sea level. What is the acceleration due to gravity on the peak of Mount Everest?
(d) Find the horizontal asymptote of $g(h)$.
(e) Solve $g(h) = 0$. How do you interpret your answer?

58. Population Model A rare species of insect was discovered in the Amazon Rain Forest. To protect the species, environmentalists declared the insect endangered and transplanted the insect into a protected area. The population P of the insect t months after being transplanted is

$$P(t) = \frac{50(1 + 0.5t)}{2 + 0.01t}$$

(a) How many insects were discovered? In other words, what was the population when $t = 0$?
(b) What will the population be after 5 years?
(c) Determine the horizontal asymptote of $P(t)$. What is the largest population that the protected area can sustain?

59. Resistance in Parallel Circuits From Ohm's Law for circuits, it follows that the total resistance R_{tot} of two components hooked in parallel is given by the equation

$$R_{tot} = \frac{R_1 R_2}{R_1 + R_2}$$

where R_1 and R_2 are the individual resistances.
(a) Let $R_1 = 10$ ohms, and graph R_{tot} as a function of R_2.
(b) Find and interpret any asymptotes of the graph obtained in part (a).
(c) If $R_2 = 2\sqrt{R_1}$, what value of R_1 will yield an R_{tot} of 17 ohms?

60. Challenge Problem Newton's Method In calculus you will learn that if

$$p(x) = a_n x^n + a_{n-1}x^{n-1} + \cdots + a_1 x + a_0$$

is a polynomial function, then the *derivative* of $p(x)$ is

$$p'(x) = na_n x^{n-1} + (n-1)a_{n-1}x^{n-2} + \cdots + 2a_2 x + a_1$$

Newton's Method is an efficient method for approximating the x-intercepts (or real zeros) of a function, such as $p(x)$. The following steps outline Newton's Method.

STEP 1: Select an initial value x_0 that is somewhat close to the x-intercept being sought.

STEP 2: Find values for x using the relation

$$x_{n+1} = x_n - \frac{p(x_n)}{p'(x_n)} \quad n = 0, 1, 2, \ldots$$

until you get two consecutive values x_n and x_{n+1} that agree to whatever decimal place accuracy you desire.

STEP 3: The approximate zero will be x_{n+1}.

Consider the polynomial $p(x) = x^3 - 7x - 40$.

(a) Evaluate $p(3)$ and $p(5)$.

(b) What might we conclude about a zero of p?

(c) Use Newton's Method to approximate an x-intercept, r, $3 < r < 5$, of $p(x)$ to four decimal places.

61. Challenge Problem The **standard form** of the rational function $R(x) = \dfrac{mx + b}{cx + d}, c \neq 0$, is $R(x) = a\left(\dfrac{1}{x - h}\right) + k$.

To write a rational function in standard form requires polynomial division.

(a) Write the rational function $R(x) = \dfrac{2x + 3}{x - 1}$ in standard form by writing R in the form

$$\text{Quotient} + \frac{\text{remainder}}{\text{divisor}}$$

(b) Graph R using transformations.

(c) Find the vertical asymptote and the horizontal asymptote of R.

62. Challenge Problem Repeat Problem 61 for the rational function $R(x) = \dfrac{-6x + 16}{2x - 7}$.

63. Challenge Problem Make up a rational function that has $y = 2x + 1$ as an oblique asymptote.

Explaining Concepts: Discussion and Writing

64. If the graph of a rational function R has the horizontal asymptote $y = 2$, the degree of the numerator of R equals the degree of the denominator of R. Explain why.

65. The graph of a rational function cannot have both a horizontal and an oblique asymptote. Explain why.

66. If the graph of a rational function R has the vertical asymptote $x = 4$, the factor $x - 4$ must be present in the denominator of R. Explain why.

Retain Your Knowledge

Problems 67–76 are based on material learned earlier in the course. The purpose of these problems is to keep the material fresh in your mind so that you are better prepared for the final exam.

67. Find the equation of a vertical line passing through the point $(5, -3)$.

68. Solve: $\dfrac{2}{5}(3x - 7) + 1 = \dfrac{x}{4} - 2$

69. Is the graph of the equation $2x^3 - xy^2 = 4$ symmetric with respect to the x-axis, the y-axis, the origin, or none of these?

70. What are the points of intersection of the graphs of the functions $f(x) = -3x + 2$ and $g(x) = x^2 - 2x - 4$?

71. Find the intercepts of the graph of $f(x) = \dfrac{x - 6}{x + 2}$.

72. Use a graphing utility to find the local maximum of

$$f(x) = x^3 + 4x^2 - 3x + 1$$

73. Where is $f(x) = 5x^2 - 13x - 6 < 0$?

74. Determine whether the function $f(x) = \dfrac{\sqrt[3]{x}}{x^2 + 6}$ is even, odd, or neither.

75. Simplify: $\dfrac{3}{x^2 - 9} - \dfrac{2}{x + 3}$

76. Solve: $3 - (2x + 4) > 5x + 13$

'Are You Prepared?' Answers

1. True   **2.** Quotient: $3x + 3$; remainder: $2x^2 - 3x - 3$ **3.**

4.

5.4 The Graph of a Rational Function

PREPARING FOR THIS SECTION *Before getting started, review the following:*

- Intercepts Section 2.2, pp. 160–161

Now Work the 'Are You Prepared?' problem on page 376.

OBJECTIVES 1 Graph a Rational Function (p. 365)
                2 Solve Applied Problems Involving Rational Functions (p. 375)

1 Graph a Rational Function

We commented earlier that calculus provides the tools required to graph a polynomial function accurately. The same holds true for rational functions. However, we can gather together quite a bit of information about their graphs to get an idea of the general shape and position of the graph.

EXAMPLE 1

Graphing a Rational Function

Graph the rational function $R(x) = \dfrac{x-1}{x^2-4}$.

Step-by-Step Solution

Step 1 Factor the numerator and denominator of R. Find the domain of the rational function.

$$R(x) = \frac{x-1}{x^2-4} = \frac{x-1}{(x+2)(x-2)}$$

The domain of R is $\{x \mid x \neq -2, x \neq 2\}$.

Step 2 Write R in lowest terms.

Because there are no common factors between the numerator and denominator, $R(x) = \dfrac{x-1}{x^2-4}$ is in lowest terms.

Step 3 Find and plot the intercepts of the graph. Use multiplicity to determine the behavior of the graph of R at each x-intercept.

Since 0 is in the domain of R, the y-intercept is $R(0) = \dfrac{1}{4}$. Plot the point $\left(0, \dfrac{1}{4}\right)$.

The x-intercepts are the real zeros of the numerator of R written in lowest terms. Solving $x - 1 = 0$, we find that the only real zero of the numerator is 1. So the only x-intercept of the graph of R is 1. Plot the point $(1, 0)$. The multiplicity of 1 is odd, so the graph crosses the x-axis at $x = 1$.

Step 4 Find the vertical asymptotes. Graph each vertical asymptote using a dashed line. Determine the behavior of the graph on either side of each vertical asymptote.

The vertical asymptotes are the zeros of the denominator with the rational function in lowest terms. With R written in lowest terms, the graph of R has two vertical asymptotes: the lines $x = -2$ and $x = 2$. The multiplicities of the zeros that give rise to the vertical asymptotes are both odd. Therefore, the graph approaches ∞ on one side of each vertical asymptote and approaches $-\infty$ on the other side.

Step 5 Find the horizontal or oblique asymptote, if one exists. Graph the asymptote using a dashed line. Find points, if any, where the graph of R intersects the asymptote. Plot the points.

Because the degree of the numerator is less than the degree of the denominator, R is proper and the line $y = 0$ (the x-axis) is a horizontal asymptote of the graph. The graph of R intersects the horizontal asymptote at the x-intercept(s) of R. That is, the graph of R intersects the horizontal asymptote at $(1, 0)$.

Step 6 Use the real zeros of the numerator and denominator of R to divide the x-axis into intervals. Determine where the graph of R is above or below the x-axis by choosing a number in each interval and evaluating R. Plot the points found.

The real zero of the numerator, 1, and the real zeros of the denominator, -2 and 2, divide the x-axis into four intervals:

$$(-\infty, -2) \qquad (-2, 1) \qquad (1, 2) \qquad (2, \infty)$$

Now construct Table 9.

Table 9

	$(-\infty, -2)$	$(-2, 1)$	$(1, 2)$	$(2, \infty)$
Interval	$(-\infty, -2)$	$(-2, 1)$	$(1, 2)$	$(2, \infty)$
Number chosen	-3	-1	$\dfrac{3}{2}$	3
Value of R	$R(-3) = -0.8$	$R(-1) = \dfrac{2}{3}$	$R\left(\dfrac{3}{2}\right) = -\dfrac{2}{7}$	$R(3) = 0.4$
Location of graph	Below x-axis	Above x-axis	Below x-axis	Above x-axis
Point on graph	$(-3, -0.8)$	$\left(-1, \dfrac{2}{3}\right)$	$\left(\dfrac{3}{2}, -\dfrac{2}{7}\right)$	$(3, 0.4)$

Figure 34(a) shows the asymptotes, the points from Table 9, the y-intercept, and the x-intercept.

Step 7 Use the results obtained in Steps 1 through 6 to graph R.

- The graph crosses the x-axis at $x = 1$, changing from above the x-axis for $x < 1$ to below it for $x > 1$. Indicate this on the graph. See Figure 34(b).
- Since $y = 0$ (the x-axis) is a horizontal asymptote and the graph lies below the x-axis for $x < -2$, we can sketch a portion of the graph by placing a small arrow to the far left and under the x-axis.
- Since the line $x = -2$ is a vertical asymptote and the graph lies below the x-axis for $x < -2$, we place an arrow well below the x-axis and approaching the line $x = -2$ from the left. (As $x \to -2$ from the left, $R(x) \to -\infty$.)
- Since the graph approaches $-\infty$ on one side of $x = -2$, and -2 is a zero of odd multiplicity, the graph will approach ∞ on the other side of $x = -2$. That is, as $x \to -2$ from the right, $R(x) \to \infty$. Similarly, as $x \to 2$ from the left, $R(x) \to -\infty$ and as $x \to 2$ from the right, $R(x) \to \infty$.
- End behavior: As $x \to -\infty$, $R(x) \to 0$; and as $x \to \infty$, $R(x) \to 0$.

Figure 34(b) illustrates these conclusions and Figure 34(c) shows the graph of R.

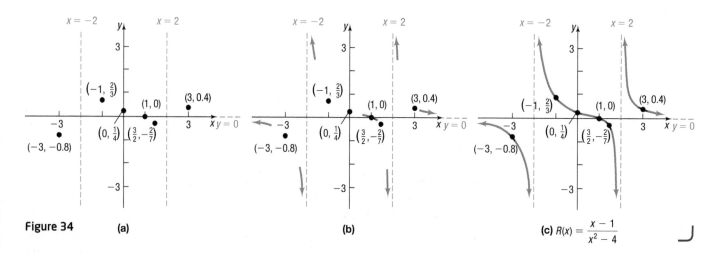

Figure 34   **(a)**   **(b)**   **(c)** $R(x) = \dfrac{x-1}{x^2-4}$

⊞**Exploration**

Graph the rational function: $R(x) = \dfrac{x-1}{x^2-4}$

Result The analysis in Example 1 helps us to set the viewing rectangle to obtain a complete graph. Figure 35(a) shows the graph of $R(x) = \dfrac{x-1}{x^2-4}$ in connected mode, and Figure 35(b) shows it in dot mode. Notice in Figure 35(a) that the graph has vertical lines at $x = -2$ and $x = 2$. This is due to the fact that when a graphing utility is in connected mode, some will connect the dots between consecutive pixels, and vertical lines may occur. We know that the graph of R does not cross the lines $x = -2$ and $x = 2$, since R is not defined at $x = -2$ or $x = 2$. So, when graphing rational functions, use dot mode if extraneous vertical lines are present in connected mode. Other graphing utilities may not have extraneous vertical lines in connected mode. See Figure 35(c) and (d).

Connected mode with extraneous vertical lines	Dot mode	Connected mode without extraneous vertical lines	
Figure 35 **(a)** TI-84 Plus	**(b)** TI-84 Plus	**(c)** TI-84 Plus C	**(d)** Desmos

SUMMARY

Steps for Graphing a Rational Function R

Step 1: Factor the numerator and denominator of R. Find the domain of the rational function.

Step 2: Write R in lowest terms.

Step 3: Find and plot the intercepts of the graph. Use multiplicity to determine the behavior of the graph of R at each x-intercept.

Step 4: Find the vertical asymptotes. Graph each vertical asymptote using a dashed line. Determine the behavior of the graph of R on either side of each vertical asymptote.

Step 5: Find the horizontal or oblique asymptote, if one exists. Graph the asymptote using a dashed line. Find points, if any, where the graph of R intersects the asymptote. Plot the points.

Step 6: Use the real zeros of the numerator and denominator of R to divide the x-axis into intervals. Determine where the graph of R is above or below the x-axis by choosing a number in each interval and evaluating R. Plot the points found.

Step 7: Use the results obtained in Steps 1 through 6 to graph R.

━━━━━**Now Work** PROBLEM 7

EXAMPLE 2

Graphing a Rational Function

Graph the rational function: $R(x) = \dfrac{x^2-1}{x}$

Solution **STEP 1:** $R(x) = \dfrac{(x+1)(x-1)}{x}$. The domain of R is $\{x | x \neq 0\}$.

STEP 2: R is in lowest terms.

STEP 3: Because x cannot equal 0, there is no y-intercept. The graph has two x-intercepts, -1 and 1, each with odd multiplicity. Plot the points $(-1, 0)$ and $(1, 0)$. The graph crosses the x-axis at both points.

STEP 4: The real zero of the denominator with R in lowest terms is 0, so the graph of R has the line $x = 0$ (the y-axis) as a vertical asymptote. Graph $x = 0$ using a dashed line. The multiplicity of 0 is odd, so the graph approaches ∞ on one side of the asymptote $x = 0$ and $-\infty$ on the other side.

NOTE Because the denominator of the rational function is a monomial, we can also find the oblique asymptote as follows:

$$\frac{x^2 - 1}{x} = \frac{x^2}{x} - \frac{1}{x} = x - \frac{1}{x}$$

Since $\frac{1}{x} \to 0$ as $x \to \infty$, $y = x$ is the oblique asymptote. ∎

STEP 5: Since the degree of the numerator, 2, is one greater than the degree of the denominator, 1, the graph of R has an oblique asymptote. To find the oblique asymptote, use polynomial division.

$$\begin{array}{r} x \\ x\overline{\smash{)}x^2 - 1} \\ \underline{x^2 } \\ -1 \end{array}$$

The quotient is x, so the line $y = x$ is an oblique asymptote of the graph. Graph $y = x$ using a dashed line.

To determine whether the graph of R intersects the asymptote $y = x$, solve the equation $R(x) = x$.

$$R(x) = \frac{x^2 - 1}{x} = x$$

$$x^2 - 1 = x^2$$

$$-1 = 0 \quad \textbf{Impossible}$$

The equation $\frac{x^2 - 1}{x} = x$ has no solution, so the graph of R does not intersect the line $y = x$.

STEP 6: The real zeros of the numerator are -1 and 1; the real zero of the denominator is 0. Use these numbers to divide the x-axis into four intervals:

$$(-\infty, -1) \qquad (-1, 0) \qquad (0, 1) \qquad (1, \infty)$$

Now construct Table 10. Plot the points from Table 10. You should now have Figure 36(a).

Table 10

Interval	$(-\infty, -1)$	$(-1, 0)$	$(0, 1)$	$(1, \infty)$
Number chosen	-2	$-\dfrac{1}{2}$	$\dfrac{1}{2}$	2
Value of R	$R(-2) = -\dfrac{3}{2}$	$R\left(-\dfrac{1}{2}\right) = \dfrac{3}{2}$	$R\left(\dfrac{1}{2}\right) = -\dfrac{3}{2}$	$R(2) = \dfrac{3}{2}$
Location of graph	Below x-axis	Above x-axis	Below x-axis	Above x-axis
Point on graph	$\left(-2, -\dfrac{3}{2}\right)$	$\left(-\dfrac{1}{2}, \dfrac{3}{2}\right)$	$\left(\dfrac{1}{2}, -\dfrac{3}{2}\right)$	$\left(2, \dfrac{3}{2}\right)$

STEP 7: To graph the function, begin by recalling that the graph crosses the x-axis at both x-intercepts (Step 3).

Since the graph of R is below the x-axis for $x < -1$ and is above the x-axis for $x > 1$, and since the graph of R does not intersect the oblique asymptote $y = x$ (Step 5), the graph of R approaches the line $y = x$, as shown in Figure 36(b).

Since the graph of R is above the x-axis for $-1 < x < 0$, the graph of R approaches ∞ as R approaches the vertical asymptote $x = 0$ from the left. Since the graph of R is below the x-axis for $0 < x < 1$, the graph of R approaches $-\infty$ as R approaches the vertical asymptote $x = 0$ from the right. See Figure 36(b).

NOTE Notice that R in Example 2 is an odd function. Do you see the symmetry about the origin in the graph of R in Figure 36(c)? ∎

The complete graph is given in Figure 36(c).

Figure 36 (a)　(b)　(c) $R(x) = \dfrac{x^2-1}{x}$

- **Now Work** PROBLEM 15

EXAMPLE 3 **Graphing a Rational Function**

Graph the rational function: $R(x) = \dfrac{x^4 + 1}{x^2}$

Solution

STEP 1: R is completely factored. The domain of R is $\{x \mid x \neq 0\}$.

STEP 2: R is in lowest terms.

STEP 3: There is no y-intercept. Since $x^4 + 1 = 0$ has no real solutions, there are no x-intercepts.

STEP 4: R is in lowest terms, so $x = 0$ (the y-axis) is a vertical asymptote of R. Graph the line $x = 0$ using dashes. The multiplicity of 0 is even, so the graph approaches either ∞ or $-\infty$ on both sides of the asymptote.

STEP 5: Since the degree of the numerator, 4, is two more than the degree of the denominator, 2, the rational function does not have a horizontal or oblique asymptote. Find the end behavior of R. As $|x| \to \infty$,

$$R(x) = \frac{x^4 + 1}{x^2} \approx \frac{x^4}{x^2} = x^2$$

The graph of R approaches the graph of $y = x^2$ as $x \to -\infty$ and as $x \to \infty$. The graph of R does not intersect $y = x^2$. Do you know why? Graph $y = x^2$ using dashes.

STEP 6: The numerator has no real zeros, and the denominator has one real zero at 0. Divide the x-axis into the two intervals $(-\infty, 0)$ and $(0, \infty)$. Construct Table 11.

Table 11

	$(-\infty, 0)$	$(0, \infty)$
Interval	$(-\infty, 0)$	$(0, \infty)$
Number chosen	-1	1
Value of R	$R(-1) = 2$	$R(1) = 2$
Location of graph	Above x-axis	Above x-axis
Point on graph	$(-1, 2)$	$(1, 2)$

STEP 7: Since the graph of R is above the x-axis and does not intersect $y = x^2$, place arrows on the graph of R above $y = x^2$ as shown in Figure 37. Also, since the graph of R lies completely above the x-axis, and the multiplicity of the zero that gives rise to the vertical asymptote, $x = 0$, is even, the graph of R approaches ∞ from both the left and the right of $x = 0$. See Figure 37.

NOTE Notice that R in Example 3 is an even function. Do you see the symmetry about the y-axis in the graph of R? ∎

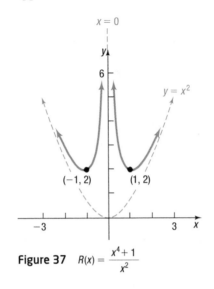

Figure 37 $R(x) = \dfrac{x^4 + 1}{x^2}$

 Now Work PROBLEM 13

EXAMPLE 4

Graphing a Rational Function

Graph the rational function: $R(x) = \dfrac{3x^2 - 3x}{x^2 + x - 12}$

Solution

STEP 1: Factor R.

$$R(x) = \frac{3x(x - 1)}{(x + 4)(x - 3)}$$

The domain of R is $\{x \mid x \neq -4, x \neq 3\}$.

STEP 2: R is in lowest terms.

STEP 3: The y-intercept is $R(0) = 0$. Plot the point $(0, 0)$. Since the real solutions of the equation $3x(x - 1) = 0$ are $x = 0$ and $x = 1$, the graph has two x-intercepts, 0 and 1, each with odd multiplicity. Plot the points $(0, 0)$ and $(1, 0)$; the graph crosses the x-axis at both points.

STEP 4: R is in lowest terms. The real solutions of the equation $(x + 4)(x - 3) = 0$ are $x = -4$ and $x = 3$, so the graph of R has two vertical asymptotes, the lines $x = -4$ and $x = 3$. Graph these lines using dashes. The multiplicities that give rise to the vertical asymptotes are both odd, so the graph approaches ∞ on one side of each vertical asymptote and $-\infty$ on the other side.

STEP 5: Since the degree of the numerator equals the degree of the denominator, the graph has a horizontal asymptote. To find it, form the quotient of the leading coefficient of the numerator, 3, and the leading coefficient of the denominator, 1. The graph of R has the horizontal asymptote $y = 3$.

To find out whether the graph of R intersects the asymptote, solve the equation $R(x) = 3$.

$$R(x) = \frac{3x^2 - 3x}{x^2 + x - 12} = 3$$
$$3x^2 - 3x = 3x^2 + 3x - 36$$
$$-6x = -36$$
$$x = 6$$

The graph intersects the line $y = 3$ at $x = 6$, so $(6, 3)$ is a point on the graph of R. Plot the point $(6, 3)$ and graph the line $y = 3$ using dashes.

STEP 6: The real zeros of the numerator, 0 and 1, and the real zeros of the denominator, -4 and 3, divide the x-axis into five intervals:

$$(-\infty, -4) \quad (-4, 0) \quad (0, 1) \quad (1, 3) \quad (3, \infty)$$

Construct Table 12. Plot the points from Table 12.

STEP 7: We analyze the graph from left to right. We know that $y = 3$ is a horizontal asymptote as $x \to -\infty$. Since the graph of R is above the x-axis for $x < -4$ and intersects the line $y = 3$ at $(6, 3)$, as x approaches $-\infty$ the graph of R approaches the horizontal asymptote $y = 3$ from above. The graph of R approaches ∞ as x approaches -4 from the left and approaches $-\infty$ as x approaches -4 from the right.

The graph crosses the x-axis at $x = 0$, changing from being below the x-axis to being above. The graph also crosses the x-axis at $x = 1$, changing from being above the x-axis to being below.

Next, the graph of R approaches $-\infty$ as x approaches 3 from the left and approaches ∞ as x approaches 3 from the right. See Figure 38(a).

We do not know whether the graph of R crosses or touches the line $y = 3$ at $(6, 3)$. To see whether the graph crosses or touches the line $y = 3$, plot an additional point to the right of $(6, 3)$. We use $x = 7$ to find $R(7) = \dfrac{63}{22} < 3$. The graph crosses $y = 3$ at $x = 6$. Because $(6, 3)$ is the only point where the graph of R intersects the asymptote $y = 3$, the graph approaches the line $y = 3$ from below as $x \to \infty$.

See Figure 38(b).

Table 12

	$(-\infty, -4)$	$(-4, 0)$	$(0, 1)$	$(1, 3)$	$(3, \infty)$
Interval	$(-\infty, -4)$	$(-4, 0)$	$(0, 1)$	$(1, 3)$	$(3, \infty)$
Number chosen	-5	-2	$\dfrac{1}{2}$	2	4
Value of R	$R(-5) = 11.25$	$R(-2) = -1.8$	$R\left(\dfrac{1}{2}\right) = \dfrac{1}{15}$	$R(2) = -1$	$R(4) = 4.5$
Location of graph	Above x-axis	Below x-axis	Above x-axis	Below x-axis	Above x-axis
Point on graph	$(-5, 11.25)$	$(-2, -1.8)$	$\left(\dfrac{1}{2}, \dfrac{1}{15}\right)$	$(2, -1)$	$(4, 4.5)$

(a)

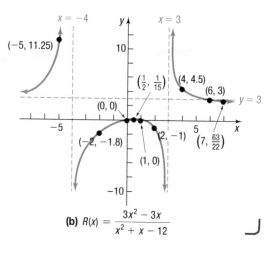

(b) $R(x) = \dfrac{3x^2 - 3x}{x^2 + x - 12}$

Figure 38

> **Exploration**
>
> Graph the rational function: $R(x) = \dfrac{3x^2 - 3x}{x^2 + x - 12}$
>
> **Result** Figure 39(a) shows the graph on a TI-84 Plus C. The graph does not clearly display the behavior of the function between the two x-intercepts, 0 and 1. Nor does it clearly display the fact that the graph crosses the horizontal asymptote at (6, 3). To see these parts better, graph R for $-1 \le x \le 2$ [Figure 39(b)] and for $4 \le x \le 60$ [Figure 39(c)].

Figure 39   (a)   (b)   (c)

> The new graphs show the expected behavior. Furthermore, we observe two turning points, one between 0 and 1 and the other to the right of 6. Rounded to two decimal places, these turning points are (0.52, 0.07) and (11.48, 2.75).

Now Work PROBLEM 31

EXAMPLE 5   **Graphing a Rational Function with a Hole**

Graph the rational function: $R(x) = \dfrac{2x^2 - 5x + 2}{x^2 - 4}$

Solution   **STEP 1:** Factor R and obtain

$$R(x) = \frac{(2x - 1)(x - 2)}{(x + 2)(x - 2)}$$

The domain of R is $\{x \mid x \ne -2, x \ne 2\}$.

STEP 2: In lowest terms,

$$R(x) = \frac{2x - 1}{x + 2} \qquad x \ne -2, x \ne 2$$

STEP 3: The y-intercept is $R(0) = -\dfrac{1}{2}$. Plot the point $\left(0, -\dfrac{1}{2}\right)$. The graph has one x-intercept, $\dfrac{1}{2}$, with odd multiplicity. Plot the point $\left(\dfrac{1}{2}, 0\right)$. The graph will cross the x-axis at $x = \dfrac{1}{2}$.

STEP 4: Since $x + 2$ is the only factor of the denominator of $R(x)$ in lowest terms, the graph has one vertical asymptote, $x = -2$. Graph the line $x = -2$ using dashes. The multiplicity of -2 is odd, so the graph approaches ∞ on one side of the vertical asymptote and $-\infty$ on the other side.

Since 2 is not in the domain of R, and $x = 2$ is not a vertical asymptote of the graph of R, the graph of R has a **hole** at the point $\left(2, \dfrac{3}{4}\right)$. Draw an open circle at the point $\left(2, \dfrac{3}{4}\right)$ as shown in Figure 40(a) on page 374.

NOTE The coordinates of the hole are obtained by evaluating R in lowest terms at $x = 2$. R in lowest terms is $\dfrac{2x - 1}{x + 2}$, which, at $x = 2$, is $\dfrac{2 \cdot 2 - 1}{2 + 2} = \dfrac{3}{4}$. ∎

STEP 5: Since the degree of the numerator equals the degree of the denominator, the graph has a horizontal asymptote. To find it, form the quotient of the leading coefficient of the numerator, 2, and the leading coefficient of the denominator, 1. The graph of R has the horizontal asymptote $y = 2$. Graph the line $y = 2$ using dashes.

To find out whether the graph of R intersects the horizontal asymptote $y = 2$, solve the equation $R(x) = 2$.

$$R(x) = \frac{2x - 1}{x + 2} = 2$$
$$2x - 1 = 2(x + 2)$$
$$2x - 1 = 2x + 4$$
$$-1 = 4 \qquad \text{Impossible}$$

The graph does not intersect the line $y = 2$. See Figure 40(a).

STEP 6: The real zeros of the numerator and denominator, $-2, \frac{1}{2}$, and 2, divide the x-axis into four intervals:

$$(-\infty, -2) \qquad \left(-2, \frac{1}{2}\right) \qquad \left(\frac{1}{2}, 2\right) \qquad (2, \infty)$$

Construct Table 13. Plot the points in Table 13.

Table 13

	$(-\infty, -2)$	$\left(-2, \dfrac{1}{2}\right)$	$\left(\dfrac{1}{2}, 2\right)$	$(2, \infty)$
Interval	$(-\infty, -2)$	$\left(-2, \dfrac{1}{2}\right)$	$\left(\dfrac{1}{2}, 2\right)$	$(2, \infty)$
Number chosen	-3	-1	1	3
Value of R	$R(-3) = 7$	$R(-1) = -3$	$R(1) = \dfrac{1}{3}$	$R(3) = 1$
Location of graph	Above x-axis	Below x-axis	Above x-axis	Above x-axis
Point on graph	$(-3, 7)$	$(-1, -3)$	$\left(1, \dfrac{1}{3}\right)$	$(3, 1)$

STEP 7: From Table 13 we know that the graph of R is above the x-axis for $x < -2$. From Step 5 we know that the graph of R does not intersect the asymptote $y = 2$. Therefore, the graph of R approaches $y = 2$ from above as $x \to -\infty$ and approaches ∞ as x approaches -2 from the left.

Since the graph of R is below the x-axis for $-2 < x < \frac{1}{2}$, the graph of R approaches $-\infty$ as x approaches -2 from the right. Finally, since the graph of R is above the x-axis for $x > \frac{1}{2}$ and does not intersect the horizontal asymptote $y = 2$, the graph of R approaches $y = 2$ from below as $x \to \infty$. The graph crosses the x-axis at $x = \frac{1}{2}$, changing from being below the x-axis to being above. See Figure 40(a).

See Figure 40(b) for the complete graph.

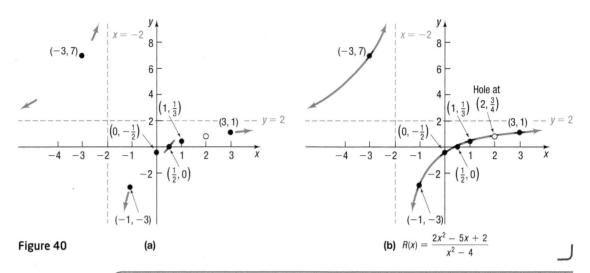

Figure 40

(a)

(b) $R(x) = \dfrac{2x^2 - 5x + 2}{x^2 - 4}$

Exploration

Graph $R(x) = \dfrac{2x^2 - 5x + 2}{x^2 - 4}$. Do you see the hole at $\left(2, \dfrac{3}{4}\right)$? TRACE along the graph. Did you obtain an ERROR at $x = 2$? Are you convinced that an algebraic analysis of a rational function is required in order to accurately interpret the graph obtained with a graphing utility?

As Example 5 shows, **the zeros of the denominator of a rational function give rise to either vertical asymptotes or holes on the graph.**

Now Work PROBLEM 33

EXAMPLE 6

Constructing a Rational Function from Its Graph

Find a rational function that might have the graph shown in Figure 41.

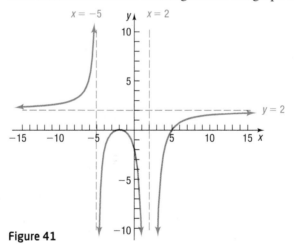

Figure 41

Solution The numerator of a rational function $R(x) = \dfrac{p(x)}{q(x)}$ in lowest terms determines the x-intercepts of its graph. The graph shown in Figure 41 has x-intercepts -2 (even multiplicity; graph touches the x-axis) and 5 (odd multiplicity; graph crosses the x-axis). So one possibility for the numerator is $p(x) = (x + 2)^2(x - 5)$.

The denominator of a rational function in lowest terms determines the vertical asymptotes of its graph. The vertical asymptotes of the graph are $x = -5$ and $x = 2$. Since $R(x)$ approaches ∞ to the left of $x = -5$ and $R(x)$ approaches $-\infty$ to the right of $x = -5$, we know that $(x + 5)$ is a factor of odd multiplicity in $q(x)$. Also, $R(x)$ approaches $-\infty$ on both sides of $x = 2$, so $(x - 2)$ is a factor of even multiplicity in $q(x)$. A possibility for the denominator is $q(x) = (x + 5)(x - 2)^2$.

So far we have $R(x) = \dfrac{(x + 2)^2(x - 5)}{(x + 5)(x - 2)^2}$.

Figure 42

The horizontal asymptote of the graph given in Figure 41 is $y = 2$, so we know that the degree of the numerator must equal the degree of the denominator, which it does, and that the quotient of leading coefficients must be $\dfrac{2}{1}$. This leads to

$$R(x) = \frac{2(x + 2)^2(x - 5)}{(x + 5)(x - 2)^2}$$

✓ **Check:** Figure 42 shows the graph of R on a TI-84 Plus C. Since Figure 42 looks similar to Figure 41, we have found a rational function R for the graph in Figure 41.

Now Work PROBLEM 51

2 Solve Applied Problems Involving Rational Functions

EXAMPLE 7

Finding the Least Cost of a Can

Reynolds Metal Company manufactures aluminum cans in the shape of a cylinder with a capacity of 500 cubic centimeters $\left(\dfrac{1}{2} \text{ liter}\right)$. The top and bottom of the can are made of a special aluminum alloy that costs 0.05¢ per square centimeter. The sides of the can are made of material that costs 0.02¢ per square centimeter.

(a) Express the cost of material for the can as a function of the radius r of the can.
(b) Use a graphing utility to graph the function $C = C(r)$.
(c) What value of r will result in the least cost?
(d) What is this least cost?

Solution

Figure 43

(a) Figure 43 illustrates the components of a can in the shape of a right circular cylinder. Notice that the material required to produce a cylindrical can of height h and radius r consists of a rectangle of area $2\pi rh$ and two circles, each of area πr^2. The total cost C (in cents) of manufacturing the can is

$$C = \text{Cost of the top and bottom} + \text{Cost of the side}$$

$$= \underbrace{2 \cdot \pi r^2}_{\substack{\text{Total area} \\ \text{of top and} \\ \text{bottom}}} \cdot \underbrace{0.05}_{\substack{\text{Cost/unit} \\ \text{area}}} + \underbrace{2\pi rh}_{\substack{\text{Total} \\ \text{area of} \\ \text{side}}} \cdot \underbrace{0.02}_{\substack{\text{Cost/unit} \\ \text{area}}}$$

$$= 0.10\pi r^2 + 0.04\pi rh$$

There is an additional restriction that the height h and radius r must be chosen so that the volume V of the can is 500 cubic centimeters. Since $V = \pi r^2 h$, we have

$$500 = \pi r^2 h \quad \text{so} \quad h = \frac{500}{\pi r^2}$$

Figure 44

Substituting $\dfrac{500}{\pi r^2}$ for h, we find that the cost C, in cents, as a function of the radius r is

$$C(r) = 0.10\pi r^2 + 0.04\pi r \cdot \frac{500}{\pi r^2} = 0.10\pi r^2 + \frac{20}{r} = \frac{0.10\pi r^3 + 20}{r}$$

(b) See Figure 44 for the graph of $C = C(r)$.
(c) Using the MINIMUM command, the cost is least for a radius of about 3.17 centimeters.
(d) The least cost is $C(3.17) \approx 9.47$¢.

Now Work PROBLEM 63

5.4 Assess Your Understanding

'Are You Prepared?' *The answer is given at the end of these exercises. If you get a wrong answer, read the pages listed in* red.

1. Find the intercepts of the graph of the equation $y = \dfrac{x^2 - 1}{x^2 - 4}$. (pp. 160–161)

Concepts and Vocabulary

2. **True or False** The graph of every rational function has at least one asymptote.

3. **Multiple Choice** Which type of asymptote will never intersect the graph of a rational function?
   (a) horizontal   (b) oblique
   (c) vertical   (d) all of these

4. **True or False** The graph of a rational function sometimes has a hole.

5. $R(x) = \dfrac{x(x-2)^2}{x-2}$
   (a) Find the domain of R.
   (b) Find the x-intercepts of R.

6. **Multiple Choice** Identify the y-intercept of the graph of
   $$R(x) = \dfrac{6(x-1)}{(x+1)(x+2)}.$$
   (a) −3   (b) −2   (c) −1   (d) 1

Skill Building

In Problems 7–50, follow Steps 1 through 7 on page 367 to graph each function.

7. $R(x) = \dfrac{x+1}{x(x+4)}$

8. $R(x) = \dfrac{x}{(x-1)(x+2)}$

9. $R(x) = \dfrac{3x+3}{2x+4}$

10. $R(x) = \dfrac{2x+4}{x-1}$

11. $R(x) = \dfrac{3}{x^2-4}$

12. $R(x) = \dfrac{6}{x^2-x-6}$

13. $P(x) = \dfrac{x^4+x^2+1}{x^2-1}$

14. $Q(x) = \dfrac{x^4-1}{x^2-4}$

15. $H(x) = \dfrac{x^3-1}{x^2-9}$

16. $G(x) = \dfrac{x^3+1}{x^2+2x}$

17. $R(x) = \dfrac{x^2}{x^2+x-6}$

18. $R(x) = \dfrac{x^2+x-12}{x^2-4}$

19. $G(x) = \dfrac{x}{x^2-4}$

20. $G(x) = \dfrac{3x}{x^2-1}$

21. $R(x) = \dfrac{3}{(x-1)(x^2-4)}$

22. $R(x) = \dfrac{-4}{(x+1)(x^2-9)}$

23. $H(x) = \dfrac{x^2-1}{x^4-16}$

24. $H(x) = \dfrac{x^2+4}{x^4-1}$

25. $F(x) = \dfrac{x^2-3x-4}{x+2}$

26. $F(x) = \dfrac{x^2+3x+2}{x-1}$

27. $R(x) = \dfrac{x^2+x-12}{x-4}$

28. $R(x) = \dfrac{x^2-x-12}{x+5}$

29. $F(x) = \dfrac{x^2+x-12}{x+2}$

30. $G(x) = \dfrac{x^2-x-12}{x+1}$

31. $R(x) = \dfrac{x(x-1)^2}{(x+3)^3}$

32. $R(x) = \dfrac{(x-1)(x+2)(x-3)}{x(x-4)^2}$

33. $R(x) = \dfrac{x^2+x-12}{x^2-x-6}$

34. $R(x) = \dfrac{x^2+3x-10}{x^2+8x+15}$

35. $R(x) = \dfrac{6x^2-7x-3}{2x^2-7x+6}$

36. $R(x) = \dfrac{8x^2+26x+15}{2x^2-x-15}$

37. $R(x) = \dfrac{x^2+5x+6}{x+3}$

38. $R(x) = \dfrac{x^2+x-30}{x+6}$

39. $H(x) = \dfrac{3x-6}{4-x^2}$

40. $H(x) = \dfrac{2-2x}{x^2-1}$

41. $F(x) = \dfrac{x^2-5x+4}{x^2-2x+1}$

42. $F(x) = \dfrac{x^2-2x-15}{x^2+6x+9}$

43. $G(x) = \dfrac{x}{(x+2)^2}$

44. $G(x) = \dfrac{2-x}{(x-1)^2}$

45. $f(x) = x + \dfrac{1}{x}$

46. $f(x) = 2x + \dfrac{9}{x}$

47. $f(x) = x^2 + \dfrac{1}{x}$

48. $f(x) = 2x^2 + \dfrac{16}{x}$

49. $f(x) = x + \dfrac{1}{x^3}$

50. $f(x) = 2x + \dfrac{9}{x^3}$

In Problems 51–54, find a rational function that might have the given graph. (More than one answer might be possible.)

51.

52.

53.

54.

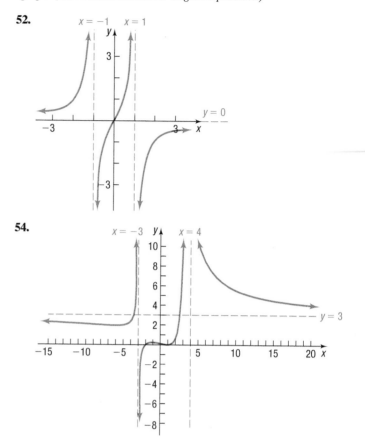

Applications and Extensions

55. Probability At a fundraiser, each person in attendance is given a ball marked with a different number from 1 through x. All the balls are then placed in an urn, and a ball is chosen at random from the urn. The probability that a particular ball is selected is $\dfrac{1}{x}$. So the probability that a particular ball is not chosen is $1 - \dfrac{1}{x}$. Graph $P(x) = 1 - \dfrac{1}{x}$ using transformations. Comment on the probability a particular ball is not chosen as x increases.

56. Waiting in Line Suppose two employees at a fast-food restaurant can serve customers at the rate of 6 customers per minute. Further suppose that customers are arriving at the restaurant at the rate of x customers per minute. The average time T, in minutes, spent waiting in line and having your order taken and filled is given by the function
$$T(x) = -\frac{1}{x - 6}, \text{ where } 0 < x < 6. \text{ Graph this function}$$
using transformations.

57. Drug Concentration The concentration C of a certain drug in a patient's bloodstream t hours after injection is given by
$$C(t) = \frac{t}{2t^2 + 1}$$

    (a) Find the horizontal asymptote of $C(t)$. What happens to the concentration of the drug as t increases?

    (b) Using a graphing utility, graph $C = C(t)$.

    (c) Determine the time at which the concentration is highest.

58. Drug Concentration The concentration C of a certain drug in a patient's bloodstream t minutes after injection is given by
$$C(t) = \frac{50t}{t^2 + 25}$$

    (a) Find the horizontal asymptote of $C(t)$. What happens to the concentration of the drug as t increases?

    (b) Using a graphing utility, graph $C = C(t)$.

    (c) Determine the time at which the concentration is highest.

59. Minimum Cost A rectangular area adjacent to a river is to be fenced in; no fence is needed on the river side. The enclosed area is to be 1000 square feet. Fencing for the side parallel to the river is $5 per linear foot, and fencing for the other two sides is $8 per linear foot; the four corner posts are $25 apiece. Let x be the length of one of the sides perpendicular to the river.

    (a) Write a function $C(x)$ that describes the cost of the project.

    (b) What is the domain of C?

    (c) Use a graphing utility to graph $C = C(x)$.

    (d) Find the dimensions of the cheapest enclosure.

60. Doppler Effect The Doppler effect (named after Christian Doppler) is the change in the pitch (frequency) of the sound from a source (*s*) as heard by an observer (*o*) when one or both are in motion. If we assume both the source and the observer are moving in the same direction, the relationship is

$$f' = f_a \left(\frac{v - v_o}{v - v_s} \right)$$

where f' = perceived pitch by the observer

f_a = actual pitch of the source

v = speed of sound in air (assume 772.4 mph)

v_o = speed of the observer

v_s = speed of the source

Suppose that you are traveling down a road at 45 mph and you hear an ambulance (with siren) coming toward you from the rear. The actual pitch of the siren is 600 hertz (Hz).

(a) Write a function $f'(v_s)$ that describes this scenario.

(b) If $f' = 620$ Hz, find the speed of the ambulance.

(c) Use a graphing utility to graph the function.

(d) Verify your answer from part (b).

Source: www.acs.psu.edu/drussell/

61. Minimizing Surface Area United Parcel Service has contracted you to design a closed box with a square base that has a volume of 10,000 cubic inches. See the illustration.

(a) Express the surface area *S* of the box as a function of *x*.

(b) Using a graphing utility, graph the function found in part (a).

(c) What is the minimum amount of cardboard that can be used to construct the box?

(d) What are the dimensions of the box that minimize the surface area?

(e) Why might UPS be interested in designing a box that minimizes the surface area?

62. Minimizing Surface Area United Parcel Service has contracted you to design an open box with a square base that has a volume of 5000 cubic inches. See the illustration.

(a) Express the surface area *S* of the box as a function of *x*.

(b) Using a graphing utility, graph the function found in part (a).

(c) What is the minimum amount of cardboard that can be used to construct the box?

(d) What are the dimensions of the box that minimize the surface area?

(e) Why might UPS be interested in designing a box that minimizes the surface area?

63. Cost of a Can A can in the shape of a right circular cylinder is required to have a volume of 500 cubic centimeters. The top and bottom are made of material that costs 6¢ per square centimeter, while the sides are made of material that costs 4¢ per square centimeter.

(a) Express the total cost *C* of the material as a function of the radius *r* of the cylinder. (Refer to Figure 43.)

(b) Graph $C = C(r)$. For what value of *r* is the cost *C* a minimum?

64. Material Needed to Make a Drum A steel drum in the shape of a right circular cylinder is required to have a volume of 100 cubic feet.

(a) Express the amount *A* of material required to make the drum as a function of the radius *r* of the cylinder.

(b) How much material is required if the drum's radius is 3 feet?

(c) How much material is required if the drum's radius is 4 feet?

(d) How much material is required if the drum's radius is 5 feet?

(e) Graph $A = A(r)$. For what value of *r* is *A* smallest?

65. Tennis Anyone? To win a game in tennis, a player must win four points. If both players have won three points, the play continues until a player is ahead by two points to win the game. The model

$$P(x) = \frac{x^4(-8x^3 + 28x^2 - 34x + 15)}{2x^2 - 2x + 1}$$

represents the probability *P* of a player winning a game in which the player is serving the game and *x* is the probability of winning a point on serve. The player serving is the first to put the ball in play.

Source: Chris Gray, "Game, set and stats," Significance, February 2015.

(a) What is the probability that a player who is serving will win the game if the probability of the player winning a point on serve is 0.64?

(b) Find and interpret $P(0.62)$.

(c) Solve $P(x) = 0.9$.

(d) Graph $P = P(x)$ for $0 \le x \le 1$. Describe what happens to *P* as *x* approaches 1.

66. Texting Speed A study of a new keyboard layout for smartphones found that the average number of words users could text per minute could be approximated by

$$N(t) = \frac{32(t + 2)}{t + 5}$$

where *t* is the number of days of practie with the keyboard.

(a) What was the average number of words users could text with the new layout at the beginning of the study?

(b) What was the average number of words users could text after using the layout for 1 week?

(c) Find and interpret the horizontal asymptote of *N*.

67. *Challenge Problem* **Removing a Discontinuity** In Example 5, we graphed the rational function $R(x) = \dfrac{2x^2 - 5x + 2}{x^2 - 4}$ and found that the graph has a hole at the point $\left(2, \dfrac{3}{4}\right)$.

Therefore, the graph of R is discontinuous at $\left(2, \dfrac{3}{4}\right)$. We can remove this discontinuity by defining the rational function R using the following piecewise-defined function:

$$R(x) = \begin{cases} \dfrac{2x^2 - 5x + 2}{x^2 - 4} & \text{if } x \neq 2 \\ \dfrac{3}{4} & \text{if } x = 2 \end{cases}$$

(a) Redefine R from Problem 33 so that the discontinuity at $x = 3$ is removed.

(b) Redefine R from Problem 35 so that the discontinuity at $x = \dfrac{3}{2}$ is removed.

68. *Challenge Problem* **Removing a Discontinuity** Refer to Problem 67.

(a) Redefine R from Problem 34 so that the discontinuity at $x = -5$ is removed.

(b) Redefine R from Problem 36 so that the discontinuity at $x = -\dfrac{5}{2}$ is removed.

Discussion and Writing

69. Graph each of the following functions:

$$y = \frac{x^2 - 1}{x - 1} \quad y = \frac{x^3 - 1}{x - 1} \quad y = \frac{x^4 - 1}{x - 1} \quad y = \frac{x^5 - 1}{x - 1}$$

Is $x = 1$ a vertical asymptote? Why? What happens for $x = 1$? What do you conjecture about the graph of $y = \dfrac{x^n - 1}{x - 1}, n \geq 1$ an integer, for $x = 1$?

70. Graph each of the following functions:

$$y = \frac{x^2}{x - 1} \quad y = \frac{x^4}{x - 1} \quad y = \frac{x^6}{x - 1} \quad y = \frac{x^8}{x - 1}$$

What similarities do you see? What differences?

71. Create a rational function that has the following characteristics: crosses the x-axis at 3; touches the x-axis at −2; one vertical asymptote, $x = 1$; and one horizontal asymptote, $y = 2$. Give your rational function to a fellow classmate and ask for a written critique of your rational function.

72. Create a rational function that has the following characteristics: crosses the x-axis at 2; touches the x-axis at −1; one vertical asymptote at $x = -5$ and another at $x = 6$; and one horizontal asymptote, $y = 3$. Compare your function to a fellow classmate's. How do they differ? What are their similarities?

73. Write a few paragraphs that provide a general strategy for graphing a rational function. Be sure to mention the following: proper, improper, intercepts, and asymptotes.

74. Create a rational function with the following characteristics: three real zeros, one of multiplicity 2; y-intercept 1; vertical asymptotes, $x = -2$ and $x = 3$; oblique asymptote, $y = 2x + 1$. Is this rational function unique? Compare your function with those of other students. What will be the same as everyone else's? Add some more characteristics, such as symmetry or naming the real zeros. How does this modify the rational function?

75. Explain the circumstances under which the graph of a rational function has a hole.

Retain Your Knowledge

Problems 76–85 are based on material learned earlier in the course. The purpose of these problems is to keep the material fresh in your mind so that you are better prepared for the final exam.

76. Subtract: $(4x^3 - 7x + 1) - (5x^2 - 9x + 3)$

77. Solve: $\dfrac{3x}{3x + 1} = \dfrac{x - 2}{x + 5}$

78. Find the absolute maximum of $f(x) = -\dfrac{2}{3}x^2 + 6x - 5$.

79. Find the vertex of the graph of $f(x) = 3x^2 - 12x + 7$.

80. Find the function whose graph is the same as the graph of $y = |x|$ but shifted down 4 units.

81. Find $g(3)$ where
$$g(x) = \begin{cases} 3x^2 - 7x & \text{if } x < 0 \\ 5x - 9 & \text{if } x \geq 0 \end{cases}$$

82. Given $f(x) = x^2 + 3x - 2$, find $f(x - 2)$.

83. Determine whether the lines $y = 3x - 2$ and $2x + 6y = 7$ are parallel, perpendicular, or neither.

84. Solve: $x - \sqrt{x + 7} = 5$

85. Solve: $2\left[\dfrac{-x^2}{\sqrt{4 - x^2}} + \sqrt{4 - x^2}\right] = 0$

'Are You Prepared?' Answer

1. $\left(0, \dfrac{1}{4}\right), (1, 0), (-1, 0)$

5.5 Polynomial and Rational Inequalities

PREPARING FOR THIS SECTION *Before getting started, review the following:*

- Solving Linear Inequalities (Section 1.5, pp. 123–124)
- Solving Quadratic Inequalities (Section 4.5, pp. 321–323)

Now Work the 'Are You Prepared?' problems on page 384.

> **OBJECTIVES** 1 Solve Polynomial Inequalities (p. 380)
>   2 Solve Rational Inequalities (p. 382)

1 Solve Polynomial Inequalities

In this section we solve inequalities that involve polynomials of degree 3 and higher, along with inequalities that involve rational functions. To help understand the algebraic procedure for solving such inequalities, we use the information obtained in the previous four sections about the graphs of polynomial and rational functions. The approach follows the same methodology that we used to solve inequalities involving quadratic functions.

EXAMPLE 1

Solving a Polynomial Inequality Using A Graph

Solve $(x + 3)(x - 1)^2 > 0$ by graphing $f(x) = (x + 3)(x - 1)^2$.

Solution

Graph $f(x) = (x + 3)(x - 1)^2$ and determine the intervals of x for which the graph is above the x-axis. The function is postive on these intervals. Using Steps 1 through 5 on page 347, we obtain the graph shown in Figure 45.

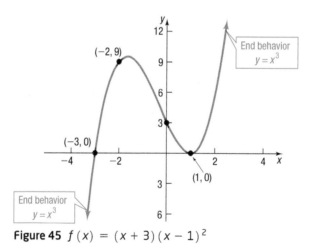

Figure 45 $f(x) = (x + 3)(x - 1)^2$

From the graph, we can see that $f(x) > 0$ for $-3 < x < 1$ or for $x > 1$. The solution set is $\{x \mid -3 < x < 1 \text{ or } x > 1\}$ or, using interval notation, $(-3, 1) \cup (1, \infty)$.

Now Work PROBLEM 9

The results of Example 1 lead to the following approach to solving polynomial and rational inequalities algebraically. Suppose that the polynomial or rational inequality is in one of the forms

$$f(x) < 0 \qquad f(x) > 0 \qquad f(x) \leq 0 \qquad f(x) \geq 0$$

Locate the real zeros of f if f is a polynomial function, and locate the real zeros of the numerator and the denominator if f is a rational function. Use these zeros to divide the real number line into intervals because on each interval, the graph of f is either above the x-axis $[f(x) > 0]$ or below the x-axis $[f(x) < 0]$. This enables us to identify the solution of the inequality.

| EXAMPLE 2 | Solving a Polynomial Inequality Algebraically |

Solve the inequality $x^4 > x$ algebraically, and graph the solution set.

Step-by-Step Solution

Step 1 Write the inequality so that a polynomial function f is on the left side and zero is on the right side.

Rearrange the inequality so that 0 is on the right side.

$$x^4 > x$$

$$x^4 - x > 0 \quad \text{Subtract } x \text{ from both sides of the inequality.}$$

This inequality is equivalent to the one we are solving.

Step 2 Determine the real zeros (x-intercepts of the graph) of f.

Find the real zeros of $f(x) = x^4 - x$ by solving $x^4 - x = 0$.

$$x^4 - x = 0$$

$$x(x^3 - 1) = 0 \quad \text{Factor out } x.$$

$$x(x - 1)(x^2 + x + 1) = 0 \quad \text{Factor the difference of two cubes.}$$

$$x = 0 \quad \text{or} \quad x - 1 = 0 \quad \text{or} \quad x^2 + x + 1 = 0 \quad \text{Use the Zero-Product Property.}$$

$$x = 0 \quad \text{or} \quad x = 1$$

The equation $x^2 + x + 1 = 0$ has no real solutions. Do you see why?

Step 3 Use the real zeros found in Step 2 to divide the real number line into intervals.

Use the real zeros to separate the real number line into three intervals:

$$(-\infty, 0) \qquad (0, 1) \qquad (1, \infty)$$

Step 4 Select a number in each interval, evaluate f at the number, and determine whether the value of f is positive or negative. If the value of f is positive, all values of f in the interval are positive. If the value of f is negative, all values of f in the interval are negative.

Select a test number in each interval found in Step 3 and evaluate $f(x) = x^4 - x$ at each number to determine whether the value of f is positive or negative. See Table 14.

Table 14

	$(-\infty, 0)$	$(0, 1)$	$(1, \infty)$
Interval	$(-\infty, 0)$	$(0, 1)$	$(1, \infty)$
Number chosen	-1	$\dfrac{1}{2}$	2
Value of f	$f(-1) = 2$	$f\left(\dfrac{1}{2}\right) = -\dfrac{7}{16}$	$f(2) = 14$
Conclusion	Positive	Negative	Positive

NOTE If the inequality is not strict (that is, if it is $\leq$ or $\geq$), include the solutions of $f(x) = 0$ in the solution set. ∎

Conclude that $f(x) > 0$ for all numbers x for which $x < 0$ or $x > 1$. The solution set of the inequality $x^4 > x$ is $\{x | x < 0 \text{ or } x > 1\}$ or, using interval notation, $(-\infty, 0) \cup (1, \infty)$.

Figure 46 shows the graph of the solution set. ⌐

Figure 46

The Role of Multiplicity in Solving Polynomial Inequalities

In Example 2, we used the number -1 and found that f is positive for all $x < 0$. Because the "cut point" of 0 is a zero of odd multiplicity (x is a factor to the first power), the sign of f changes on either side of 0, so for $0 < x < 1$, f is negative. Similarly, f is positive for $x > 1$, since the multiplicity of the zero 1 is odd.

━━━ **Now Work** PROBLEM 21

2 Solve Rational Inequalities

Just as we used a graphical approach to help understand the algebraic procedure for solving inequalities involving polynomials, we use a graphical approach to motivate the algebraic procedure for solving inequalities involving rational expressions.

EXAMPLE 3 **Solving a Rational Inequality Using a Graph**

Solve $\dfrac{x-1}{x^2-4} \geq 0$ by graphing $R(x) = \dfrac{x-1}{x^2-4}$.

Solution Graph $R(x) = \dfrac{x-1}{x^2-4}$ and determine the intervals of x for which the graph is above or on the x-axis. The function R is nonnegative on these intervals. We graphed $R(x) = \dfrac{x-1}{x^2-4}$ in Example 1, Section 5.4 (pp. 365–366). We reproduce the graph in Figure 47.

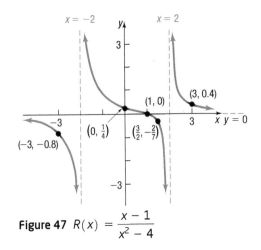

Figure 47 $R(x) = \dfrac{x-1}{x^2-4}$

From the graph, we can see that $R(x) \geq 0$ for $-2 < x \leq 1$ or $x > 2$. The solution set is $\{x \mid -2 < x \leq 1 \text{ or } x > 2\}$ or, using interval notation, $(-2, 1] \cup (2, \infty)$.

Now Work PROBLEM 15

To solve a rational inequality algebraically, we follow the same approach that we used to solve a polynomial inequality algebraically. However, we must also identify the real zeros of the denominator of the rational function because the sign of a rational function may change on either side of a vertical asymptote. Convince yourself of this by looking at Figure 47. Notice that the function values are negative for $x < -2$ and are positive for $x > -2$ (but less than 1).

EXAMPLE 4 **Solving a Rational Inequality Algebraically**

Solve the inequality $\dfrac{3x^2+13x+9}{(x+2)^2} \leq 3$ algebraically, and graph the solution set.

Step-by-Step Solution

Step 1 Write the inequality so that a rational function f is on the left side and zero is on the right side.

Rearrange the inequality so that 0 is on the right side.

$$\frac{3x^2 + 13x + 9}{(x + 2)^2} \leq 3$$

$$\frac{3x^2 + 13x + 9}{x^2 + 4x + 4} - 3 \leq 0 \qquad \text{Subtract 3 from both sides of the inequality; Expand } (x + 2)^2.$$

$$\frac{3x^2 + 13x + 9}{x^2 + 4x + 4} - 3 \cdot \frac{x^2 + 4x + 4}{x^2 + 4x + 4} \leq 0 \qquad \text{Multiply 3 by } \frac{x^2 + 4x + 4}{x^2 + 4x + 4}.$$

$$\frac{3x^2 + 13x + 9 - 3x^2 - 12x - 12}{x^2 + 4x + 4} \leq 0 \qquad \text{Write as a single quotient.}$$

$$\frac{x - 3}{(x + 2)^2} \leq 0 \qquad \text{Combine like terms.}$$

Step 2 Determine the real zeros (x-intercepts of the graph) of f and the real numbers for which f is undefined.

The real zero of $f(x) = \dfrac{x - 3}{(x + 2)^2}$ is 3. Also, f is undefined for $x = -2$.

Step 3 Use the real zeros and undefined values found in Step 2 to divide the real number line into intervals.

Use the real zero and the undefined value to divide the real number line into three intervals:

$$(-\infty, -2) \quad (-2, 3) \quad (3, \infty)$$

Step 4 Select a number in each interval, evaluate f at the number, and determine whether the value of f is positive or negative. If the value of f is positive, all values of f in the interval are positive. If the value of f is negative, all values of f in the interval are negative.

NOTE If the inequality is not strict ($\leq$ or $\geq$), include the solutions of $f(x) = 0$ in the solution set. ∎

Select a test number in each interval from Step 3, and evaluate f at each number to determine whether the value of f is positive or negative. See Table 15.

Table 15

	$(-\infty, -2)$	$(-2, 3)$	$(3, \infty)$
Interval	$(-\infty, -2)$	$(-2, 3)$	$(3, \infty)$
Number chosen	-3	0	4
Value of f	$f(-3) = -6$	$f(0) = -\dfrac{3}{4}$	$f(4) = \dfrac{1}{36}$
Conclusion	Negative	Negative	Positive

We conclude that $f(x) \leq 0$ for all numbers for which $x < -2$ or $-2 < x \leq 3$. Notice that we do not include -2 in the solution because -2 is not in the domain of f.

The solution set of the inequality $\dfrac{3x^2 + 13x + 9}{(x + 2)^2} \leq 3$ is $\{x \mid x < -2 \text{ or } -2 < x \leq 3\}$ or, using interval notation, $(-\infty, -2) \cup (-2, 3]$. Figure 48 shows the graph of the solution set.

Figure 48

The Role of Multiplicity in Solving Rational Inequalities

In Example 4, we used the number -3 and found that $f(x)$ is negative for all $x < -2$. Because the "cut point" of -2 is a zero of even multiplicity, we know the sign of $f(x)$ does not change on either side of -2, so for $-2 < x < 3$, $f(x)$ is negative. Because the "cut point" of 3 is a zero of odd multiplicity, the sign of $f(x)$ changes on either side of 3, so for $x > 3$, $f(x)$ is positive. Therefore, the solution set of $\dfrac{3x^2 + 13x + 9}{(x + 2)^2} \leq 3$ is $\{x \mid x < -2 \text{ or } -2 < x \leq 3\}$ or, using interval notation, $(-\infty, -2) \cup (-2, 3]$.

━━━**Now Work** PROBLEMS 35 AND 41

SUMMARY

> **Steps for Solving Polynomial and Rational Inequalities Algebraically**
>
> **STEP 1:** Write the inequality so that a polynomial or rational function f is on the left side and zero is on the right side in one of the following forms:
>
> $$f(x) > 0 \qquad f(x) \geq 0 \qquad f(x) < 0 \qquad f(x) \leq 0$$
>
> For rational functions, be sure that the left side is written as a single quotient. Find the domain of f.
>
> **STEP 2:** Determine the real numbers at which $f(x) = 0$ and, if the function is rational, the real numbers at which the function f is undefined.
>
> **STEP 3:** Use the numbers found in Step 2 to divide the real number line into intervals.
>
> **STEP 4:** Select a number in each interval and evaluate f at the number.
>
> - If the value of f is positive, then $f(x) > 0$ for all numbers x in the interval.
> - If the value of f is negative, then $f(x) < 0$ for all numbers x in the interval.
> - If the inequality is not strict ($\geq$ or $\leq$), include the solutions of $f(x) = 0$ that are in the domain of f in the solution set. Be careful to exclude values of x where f is undefined.

5.5 Assess Your Understanding

'Are You Prepared?' *Answers are given at the end of these exercises. If you get a wrong answer, read the pages listed in* red.

1. Solve the inequality $3 - 4x > 5$. Graph the solution set. (pp. 123–124)

2. Solve the inequality $x^2 - 5x \leq 24$. Graph the solution set. (pp. 321–323)

Concepts and Vocabulary

3. *Multiple Choice* Which of the following could be a test number for the interval $-2 < x < 5$?

   (a) -3 (b) -2 (c) 4   (d) 7

4. *True or False* The graph of $f(x) = \dfrac{x}{x - 3}$ is above the x-axis for $x < 0$ or $x > 3$, so the solution set of the inequality $\dfrac{x}{x - 3} \geq 0$ is $\{x \mid x \leq 0 \text{ or } x \geq 3\}$.

Skill Building

In Problems 5–8, use the graph of the function f to solve the inequality.

5. **(a)** $f(x) > 0$
   (b) $f(x) \leq 0$

6. **(a)** $f(x) < 0$
   (b) $f(x) \geq 0$

7. **(a)** $f(x) < 0$
   (b) $f(x) \geq 0$

8. **(a)** $f(x) > 0$
   (b) $f(x) \leq 0$

In Problems 9–14, solve the inequality by using the graph of the function.

[**Hint:** The graphs were drawn in Problems 5–10 of Section 5.2.]

9. Solve $f(x) < 0$, where $f(x) = x^2(x - 3)$.

10. Solve $f(x) \leq 0$, where $f(x) = x(x + 2)^2$.

11. Solve $f(x) \geq 0$, where $f(x) = (x + 4)^2(1 - x)$.

12. Solve $f(x) > 0$, where $f(x) = (x - 1)(x + 3)^2$.

13. Solve $f(x) \leq 0$, where $f(x) = -2(x + 2)(x - 2)^3$.

14. Solve $f(x) < 0$, where $f(x) = -\dfrac{1}{2}(x + 4)(x - 1)^3$.

In Problems 15–18, solve the inequality by using the graph of the function.

[**Hint:** The graphs were drawn in Problems 7–10 of Section 5.4.]

15. Solve $R(x) > 0$, where $R(x) = \dfrac{x + 1}{x(x + 4)}$.

16. Solve $R(x) < 0$, where $R(x) = \dfrac{x}{(x - 1)(x + 2)}$.

17. Solve $R(x) \leq 0$, where $R(x) = \dfrac{3x + 3}{2x + 4}$.

18. Solve $R(x) \geq 0$, where $R(x) = \dfrac{2x + 4}{x - 1}$.

In Problems 19–54, solve each inequality algebraically.

19. $(x - 4)^2(x + 6) < 0$

20. $(x - 5)(x + 2)^2 > 0$

21. $x^3 - 4x^2 > 0$

22. $x^3 + 8x^2 < 0$

23. $2x^3 > -8x^2$

24. $3x^3 < -15x^2$

25. $(x + 2)(x - 4)(x - 6) \leq 0$

26. $(x + 1)(x + 2)(x + 3) \leq 0$

27. $x^3 - 4x^2 - 12x > 0$

28. $x^3 + 2x^2 - 3x > 0$

29. $x^4 > x^2$

30. $x^4 < 9x^2$

31. $x^4 > 1$

32. $x^3 > 1$

33. $3(x^2 - 2) < 2(x - 1)^2 + x^2$

34. $(x - 3)(x + 2) < x^2 + 3x + 5$

35. $\dfrac{x + 1}{x - 1} > 0$

36. $\dfrac{x - 3}{x + 1} > 0$

37. $\dfrac{(x - 2)(x + 2)}{x} \leq 0$

38. $\dfrac{(x - 3)(x + 2)}{x - 1} \leq 0$

39. $\dfrac{(x - 3)^2}{x^2 - 4} \geq 0$

40. $\dfrac{(x + 5)^2}{x^2 - 4} \geq 0$

41. $\dfrac{x + 4}{x - 2} \leq 1$

42. $\dfrac{x + 2}{x - 4} \geq 1$

43. $\dfrac{3x - 5}{x + 2} \leq 2$

44. $\dfrac{x - 4}{2x + 4} \geq 1$

45. $\dfrac{x + 1}{x - 3} \leq 2$

46. $\dfrac{x - 1}{x + 2} \geq -2$

47. $\dfrac{1}{x - 2} < \dfrac{2}{3x - 9}$

48. $\dfrac{5}{x - 3} > \dfrac{3}{x + 1}$

49. $\dfrac{x^2(3 + x)(x + 4)}{(x + 5)(x - 1)} \geq 0$

50. $\dfrac{x(x^2 + 1)(x - 2)}{(x - 1)(x + 1)} \geq 0$

51. $\dfrac{(3 - x)^3(2x + 1)}{x^3 - 1} < 0$

52. $\dfrac{(2 - x)^3(3x - 2)}{x^3 + 1} < 0$

53. $6x - 5 < \dfrac{6}{x}$

54. $x + \dfrac{12}{x} < 7$

Mixed Practice *In Problems 55–58, (a) graph each function by hand, and (b) solve* $f(x) \geq 0$.

55. $f(x) = \dfrac{x^2 + 5x - 6}{x^2 - 4x + 4}$

56. $f(x) = \dfrac{2x^2 + 9x + 9}{x^2 - 4}$

57. $f(x) = \dfrac{(x + 4)(x^2 - 2x - 3)}{x^2 - x - 6}$

58. $f(x) = \dfrac{(x - 1)(x^2 - 5x + 4)}{x^2 + x - 20}$

Applications and Extensions

59. For what positive numbers is the cube of the number greater than four times its square?

60. For what positive numbers is the cube of the number less than the number?

61. What is the domain of the function $f(x) = \sqrt{x^4 - 16}$?

62. What is the domain of the function $f(x) = \sqrt{x^3 - 3x^2}$?

63. What is the domain of the function $f(x) = \sqrt{\dfrac{x - 2}{x + 4}}$?

64. What is the domain of the function $f(x) = \sqrt{\dfrac{x - 1}{x + 4}}$?

In Problems 65–68, determine where the graph of f is below the graph of g by solving the inequality $f(x) \leq g(x)$. Graph f and g together.

65. $f(x) = x^4 - 1$
    $g(x) = -2x^2 + 2$

66. $f(x) = x^4 - 1$
    $g(x) = x - 1$

67. $f(x) = x^4 - 4$
    $g(x) = 3x^2$

68. $f(x) = x^4$
    $g(x) = 2 - x^2$

69. Where is the graph of $R(x) = \dfrac{x^4 - 16}{x^2 - 9}$ above the x-axis?

70. Where is the graph of $R(x) = \dfrac{x^3 - 8}{x^2 - 25}$ above the x-axis?

71. Average Cost Suppose that the daily cost C of manufacturing bicycles is given by $C(x) = 80x + 5000$. Then the average daily cost $\overline{C}$ is given by $\overline{C}(x) = \dfrac{80x + 5000}{x}$. How many bicycles must be produced each day for the average cost to be no more than $100?

72. Average Cost See Problem 71. Suppose that the government imposes a $1000-per-day tax on the bicycle manufacturer so that the daily cost C of manufacturing x bicycles is now given by $C(x) = 80x + 6000$. Now the average daily cost $\overline{C}$ is given by $\overline{C}(x) = \dfrac{80x + 6000}{x}$. How many bicycles must be produced each day for the average cost to be no more than $100?

73. Challenge Problem Bungee Jumping Originating on Pentecost Island in the Pacific, the practice of a person jumping from a high place harnessed to a flexible attachment was introduced to Western culture in 1979 by the Oxford University Dangerous Sport Club. One important parameter to know before attempting a bungee jump is the amount the cord will stretch at the bottom of the fall. The stiffness of the cord is related to the amount of stretch by the equation

$$K = \frac{2W(S + L)}{S^2}$$

where $W = $ weight of the jumper (pounds)

$K = $ cord's stiffness (pounds per foot)

$L = $ free length of the cord (feet)

$S = $ stretch (feet)

(a) A 150-pound person plans to jump off a ledge attached to a cord of length 42 feet. If the stiffness of the cord is no less than 16 pounds per foot, how much will the cord stretch?

(b) If safety requirements will not permit the jumper to get any closer than 3 feet to the ground, what is the minimum height required for the ledge in part (a)?

Source: American Institute of Physics, Physics News Update, No. 150, November 5, 1993.

74. Challenge Problem Gravitational Force According to Newton's Law of Universal Gravitation, the attractive force F between two bodies is given by

$$F = G\frac{m_1 m_2}{r^2}$$

where $m_1, m_2 = $ the masses of the two bodies

$r = $ distance between the two bodies

$G = $ gravitational constant $= 6.6742 \times 10^{-11}$ newtons $\cdot$ meter$^2 \cdot$ kilogram^{-2}

Suppose an object is traveling directly from Earth to the moon. The mass of Earth is 5.9742×10^{24} kilograms, the mass of the moon is 7.349×10^{22} kilograms, and the mean distance from Earth to the moon is 384,400 kilometers. For an object between Earth and the moon, how far from Earth is the force on the object due to the moon greater than the force on the object due to Earth?

Source: www.solarviews.com; en.wikipedia.org

Explaining Concepts: Discussion and Writing

75. The inequality $x^4 + 1 < -5$ has no solution. Explain why.

76. A student attempted to solve the inequality $\dfrac{x + 4}{x - 3} \le 0$ by multiplying both sides of the inequality by $x - 3$ to get $x + 4 \le 0$. This led to a solution of $\{x \mid x \le -4\}$. Is the student correct? Explain.

77. Write a rational inequality whose solution set is $\{x \mid -3 < x \le 5\}$.

78. Make up an inequality that has no solution. Make up one that has exactly one solution.

Retain Your Knowledge

Problems 79–88 are based on material learned earlier in the course. The purpose of these problems is to keep the material fresh in your mind so that you are better prepared for the final exam.

79. Solve: $9 - 2x \le 4x + 1$

80. Factor completely: $6x^4y^4 + 3x^3y^5 - 18x^2y^6$

81. Suppose y varies directly with $\sqrt{x}$. Write a general formula to describe the variation if $y = 2$ when $x = 9$.

82. If $f(x) = \sqrt{3x - 1}$ and $g(x) = \sqrt{3x + 1}$, find $(f \cdot g)(x)$ and state its domain.

83. If $f(x) = 4x + 3$, find $f\left(\dfrac{x - 3}{4}\right)$.

84. Solve $\omega = \dfrac{1}{\sqrt{LC}}$ for C.

85. Determine whether the graph of

$$(x^2 + y^2 - 2x)^2 = 9(x^2 + y^2)$$

is symmetric with respect to the x-axis, y-axis, origin, or none of these.

86. Approximate the turning points of $f(x) = x^3 - 2x^2 + 4$. Round answers to two decimal places.

87. Solve: $5x^2 - 3 = 2x^2 + 11x + 1$

88. What are the quotient and remainder when $8x^2 - 4x + 5$ is divided by $4x + 1$?

'Are You Prepared?' Answers

1. $\left\{x \mid x < -\dfrac{1}{2}\right\}$ or $\left(-\infty, -\dfrac{1}{2}\right)$

2. $\{x \mid -3 \le x \le 8\}$ or $[-3, 8]$

5.6 The Real Zeros of a Polynomial Function

PREPARING FOR THIS SECTION *Before getting started, review the following:*

- Evaluating Functions (Section 3.1, pp. 207–209)
- Factoring Polynomials (Chapter R, Section R.5, pp. 49–55)
- Synthetic Division (Chapter R, Section R.6, pp. 57–60)
- Polynomial Division (Chapter R, Section R.4, pp. 44–46)
- Solve a Quadratic Equation (Section 1.2, pp. 92–99)

Now Work the Are You Prepared?' problems on page 398.

OBJECTIVES 1 Use the Remainder and Factor Theorems (p. 387)

    2 Use Descartes' Rule of Signs to Determine the Number of Positive and the Number of Negative Real Zeros of a Polynomial Function (p. 390)

    3 Use the Rational Zeros Theorem to List the Potential Rational Zeros of a Polynomial Function (p. 391)

    4 Find the Real Zeros of a Polynomial Function (p. 392)

    5 Solve Polynomial Equations (p. 394)

    6 Use the Theorem for Bounds on Zeros (p. 395)

    7 Use the Intermediate Value Theorem (p. 396)

In Section 5.1, we were able to identify the real zeros of a polynomial function because either the polynomial function was in factored form or it could be easily factored. But how do we find the real zeros of a polynomial function if it is not factored or cannot be easily factored?

Recall that if r is a real zero of a polynomial function f, then $f(r) = 0$, r is an x-intercept of the graph of f, $x - r$ is a factor of f, and r is a solution of the equation $f(x) = 0$. For example, if $x - 4$ is a factor of f, then 4 is a real zero of f, and 4 is a solution to the equation $f(x) = 0$. For polynomial functions, we have seen the importance of the real zeros for graphing. In most cases, however, the real zeros of a polynomial function are difficult to find using algebraic methods. No nice formulas like the quadratic formula are available to help us find zeros for polynomials of degree 3 or higher. Formulas do exist for solving any third- or fourth-degree polynomial equation, but they are somewhat complicated. No general formulas exist for polynomial equations of degree 5 or higher. Refer to the Historical Feature at the end of this section for more information.

1 Use the Remainder and Factor Theorems

When one polynomial (the dividend) is divided by another (the divisor), a quotient polynomial and a remainder are obtained. The remainder is either the zero polynomial or a polynomial whose degree is less than the degree of the divisor. To check, verify that

$$(\text{Quotient})(\text{Divisor}) + \text{Remainder} = \text{Dividend}$$

This checking routine is the basis for a famous theorem called the **division algorithm*** **for polynomials**, which we now state without proof.

---

*A systematic process in which certain steps are repeated a finite number of times is called an **algorithm**. For example, polynomial division is an algorithm.

THEOREM Division Algorithm for Polynomials

If $f(x)$ and $g(x)$ denote polynomial functions and if $g(x)$ is a polynomial whose degree is greater than zero, then there are unique polynomial functions $q(x)$ and $r(x)$ for which

$$\underset{\uparrow}{\frac{f(x)}{g(x)}} = \underset{\uparrow}{q(x)} + \underset{\uparrow}{\frac{r(x)}{g(x)}} \quad \text{or} \quad \underset{\uparrow}{f(x)} = \underset{\uparrow}{q(x)}\underset{\uparrow}{g(x)} + \underset{\uparrow}{r(x)} \qquad \textbf{(1)}$$

dividend quotient divisor remainder

where $r(x)$ is either the zero polynomial or a polynomial of degree less than that of $g(x)$.

In equation (1), $f(x)$ is the **dividend**, $g(x)$ is the **divisor**, $q(x)$ is the **quotient**, and $r(x)$ is the **remainder**.

If the divisor $g(x)$ is a first-degree polynomial of the form

$$g(x) = x - c \qquad c \text{ a real number}$$

then the remainder $r(x)$ is either the zero polynomial or a polynomial of degree 0. As a result, for such divisors, the remainder is some number, say R, and we may write

$$f(x) = (x - c)q(x) + R \qquad \textbf{(2)}$$

This equation is an identity in x and is true for all real numbers x. Suppose that $x = c$. Then equation (2) becomes

$$f(c) = (c - c)q(c) + R$$
$$f(c) = R$$

Substitute $f(c)$ for R in equation (2) to obtain

$$f(x) = (x - c)q(x) + f(c) \qquad \textbf{(3)}$$

which proves the *Remainder Theorem*.

REMAINDER THEOREM

Suppose f is a polynomial function. If $f(x)$ is divided by $x - c$, then the remainder is $f(c)$.

EXAMPLE 1 Using the Remainder Theorem

Find the remainder when $f(x) = x^3 - 4x^2 - 5$ is divided by

(a) $x - 3$ \qquad (b) $x + 2$

Solution

(a) Either polynomial division or synthetic division could be used, but it is easier to use the Remainder Theorem, which states that the remainder is $f(3)$.

$$f(3) = 3^3 - 4 \cdot 3^2 - 5 = 27 - 36 - 5 = -14$$

The remainder is -14.

(b) To find the remainder when $f(x)$ is divided by $x + 2 = x - (-2)$, find $f(-2)$.

$$f(-2) = (-2)^3 - 4(-2)^2 - 5 = -8 - 16 - 5 = -29$$

The remainder is -29.

COMMENT A graphing utility provides another way to find the value of a function using the eVALUEate feature. Consult your manual for details. Then check the results of Example 1. ∎

Compare the method used in Example 1(a) with the method used in Example 1 of Chapter R, Section R.6. Which method do you prefer?

An important and useful consequence of the Remainder Theorem is the *Factor Theorem.*

FACTOR THEOREM

Suppose f is a polynomial function. Then $x - c$ is a factor of $f(x)$ if and only if $f(c) = 0$.

The Factor Theorem actually consists of two separate statements:

1. If $f(c) = 0$, then $x - c$ is a factor of $f(x)$.
2. If $x - c$ is a factor of $f(x)$, then $f(c) = 0$.

The proof requires two parts.

Proof

1. Suppose that $f(c) = 0$. Then, by equation (3), we have
$$f(x) = (x - c)q(x)$$
for some polynomial $q(x)$. That is, $x - c$ is a factor of $f(x)$.

2. Suppose that $x - c$ is a factor of $f(x)$. Then there is a polynomial function q for which
$$f(x) = (x - c)q(x)$$
Replacing x by c, we find that
$$f(c) = (c - c)q(c) = 0 \cdot q(c) = 0$$
This completes the proof. ∎

One use of the Factor Theorem is to determine whether a polynomial has a particular factor.

EXAMPLE 2

Using the Factor Theorem

Use the Factor Theorem to determine whether the function
$$f(x) = 2x^3 - x^2 + 2x - 3$$

has the factor

(a) $x - 1$　　　　(b) $x + 2$

Solution　The Factor Theorem states that if $f(c) = 0$, then $x - c$ is a factor.

(a) Because $x - 1$ is of the form $x - c$ with $c = 1$, find the value of $f(1)$.
$$f(1) = 2 \cdot 1^3 - 1^2 + 2 \cdot 1 - 3 = 2 - 1 + 2 - 3 = 0$$

By the Factor Theorem, $x - 1$ is a factor of $f(x)$.

(b) To test the factor $x + 2$, first write it in the form $x - c$. Since $x + 2 = x - (-2)$, find the value of $f(-2)$. Using synthetic division,

$$
\begin{array}{r|rrr}
-2 & 2 & -1 & 2 & -3 \\
   &   & -4 & 10 & -24 \\
\hline
   & 2 & -5 & 12 & -27
\end{array}
$$

Because $f(-2) = -27 \neq 0$, conclude from the Factor Theorem that $x - (-2) = x + 2$ is not a factor of $f(x)$.

━━━━**Now Work** PROBLEM **11**

From Example 2(a), $x - 1$ is a factor of f. To write f in factored form, divide f by $(x - 1)$.

$$
\begin{array}{r}
1)\overline{2 \quad -1 \quad 2 \quad -3} \\
2 \quad 1 \quad 3 \\
\hline
2 \quad 1 \quad 3 \quad 0
\end{array}
$$

The quotient is $q(x) = 2x^2 + x + 3$ with a remainder of 0, as expected. Write f in factored form as

$$f(x) = 2x^3 - x^2 + 2x - 3 = (x - 1)(2x^2 + x + 3)$$

But how many real zeros can a polynomial function have? In counting the zeros of a polynomial, count each zero as many times as its multiplicity.

THEOREM Number of Real Zeros

A polynomial function cannot have more real zeros than its degree.

Proof The proof is based on the Factor Theorem. If r is a real zero of a polynomial function f, then $f(r) = 0$, and $x - r$ is a factor of $f(x)$. Each real zero corresponds to a factor of degree 1. Because f cannot have more first-degree factors than its degree, the result follows. ∎

2 Use Descartes' Rule of Signs to Determine the Number of Positive and the Number of Negative Real Zeros of a Polynomial Function

Descartes' Rule of Signs provides information about the number and location of the real zeros of a polynomial function written in standard form (omitting terms with a 0 coefficient). It uses the number of variations in the sign of the coefficients of $f(x)$ and $f(-x)$.

For example, the following polynomial function has two variations in the signs of the coefficients.

$$f(x) = -3x^7 + 4x^4 + 3x^2 - 2x - 1$$
$$\underbrace{}_{-\text{ to }+} \quad \underbrace{}_{+\text{ to }-}$$

Replacing x by $-x$ gives

$$f(-x) = -3(-x)^7 + 4(-x)^4 + 3(-x)^2 - 2(-x) - 1$$
$$= 3x^7 + 4x^4 + 3x^2 + 2x - 1$$
$$\underbrace{}_{+\text{ to }-}$$

which has one variation in sign.

THEOREM Descartes' Rule of Signs

Suppose f is a polynomial function written in standard form.

- The number of positive real zeros of f either equals the number of variations in the sign of the nonzero coefficients of $f(x)$ or else equals that number less an even integer.
- The number of negative real zeros of f either equals the number of variations in the sign of the nonzero coefficients of $f(-x)$ or else equals that number less an even integer.

We do not prove Descartes' Rule of Signs. Let's see how it is used.

EXAMPLE 3	Using the Number of Real Zeros Theorem and Descartes' Rule of Signs

Discuss the real zeros of $f(x) = 3x^7 - 4x^4 + 3x^3 + 2x^2 - x - 3$.

Solution Because the polynomial is of degree 7, by the Number of Real Zeros Theorem there are at most seven real zeros. Since there are three variations in the sign of the nonzero coefficients of $f(x)$, by Descartes' Rule of Signs we expect either three positive real zeros or one positive real zero.

To continue, look at $f(-x)$.

$$f(-x) = -3x^7 - 4x^4 - 3x^3 + 2x^2 + x - 3$$

There are two variations in sign, so we expect either two negative real zeros or no negative real zeros. Equivalently, we now know that the graph of f has either three positive x-intercepts or one positive x-intercept and two negative x-intercepts or no negative x-intercepts.

╼ **Now Work** PROBLEM 21

3 Use the Rational Zeros Theorem to List the Potential Rational Zeros of a Polynomial Function

The next result, called the *Rational Zeros Theorem*, provides information about the rational zeros of a polynomial *with integer coefficients*.

> **THEOREM** **Rational Zeros Theorem**
>
> Let f be a polynomial function of degree 1 or higher of the form
>
> $$f(x) = a_n x^n + a_{n-1} x^{n-1} + \cdots + a_1 x + a_0 \qquad a_n \neq 0 \quad a_0 \neq 0$$
>
> where each coefficient is an integer. If $\dfrac{p}{q}$, in lowest terms, is a rational zero of f, then p must be a factor of a_0, and q must be a factor of a_n.

EXAMPLE 4	Listing Potential Rational Zeros

List the potential rational zeros of

$$f(x) = 2x^3 + 11x^2 - 7x - 6$$

Solution Because f has integer coefficients, the Rational Zeros Theorem may be used. First, list all the integers p that are factors of the constant term $a_0 = -6$ and all the integers q that are factors of the leading coefficient $a_3 = 2$.

$$p: \quad \pm 1, \pm 2, \pm 3, \pm 6 \quad \text{Factors of } -6$$
$$q: \quad \pm 1, \pm 2 \qquad\qquad \text{Factors of 2}$$

Now form all possible ratios $\dfrac{p}{q}$.

$$\frac{p}{q}: \quad \pm\frac{1}{1}, \pm\frac{2}{1}, \pm\frac{3}{1}, \pm\frac{6}{1}, \pm\frac{1}{2}, \pm\frac{2}{2}, \pm\frac{3}{2}, \pm\frac{6}{2}$$

which simplify to

$$\frac{p}{q}: \quad \pm 1, \pm 2, \pm 3, \pm 6, \pm\frac{1}{2}, \pm\frac{3}{2}$$

If f has a rational zero, it will be found in this list of 12 possibilities.

╼ **Now Work** PROBLEM 33

Be sure that you understand what the Rational Zeros Theorem says: For a polynomial with integer coefficients, *if* there is a rational zero, it is one of those listed. It may be the case that the function does not have any rational zeros.

Polynomial division, synthetic division, or substitution can be used to test each potential rational zero to determine whether it is indeed a zero. To make the work easier, integers are usually tested first.

4 Find the Real Zeros of a Polynomial Function

EXAMPLE 5   Finding the Real Zeros of a Polynomial Function

Find the real zeros of the polynomial function $f(x) = 2x^3 + 11x^2 - 7x - 6$. Write f in factored form.

Step-by-Step Solution

Step 1 Use the degree of the polynomial to determine the maximum number of zeros.

Since f is a polynomial of degree 3, there are at most three real zeros.

Step 2 Use Descartes' Rule of Signs to determine the possible number of positive real zeros and negative real zeros.

By Descartes' Rule of Signs, there is one positive real zero. Also, because

$$f(-x) = -2x^3 + 11x^2 + 7x - 6$$

there are two negative real zeros or no negative real zeros.

Step 3 If the polynomial has integer coefficients, use the Rational Zeros Theorem to identify the rational numbers that are potential zeros. Use substitution, synthetic division, or polynomial division to determine whether each potential rational zero is a zero. If it is, factor the polynomial function. Repeat Step 3 until all the rational zeros of the polynomial function have been identified, or until the function can be factored.

List the potential rational zeros obtained in Example 4:

$$\pm 1, \ \pm 2, \ \pm 3, \ \pm 6, \ \pm \frac{1}{2}, \ \pm \frac{3}{2}$$

From the list, test 1 first using synthetic divison.

$$\begin{array}{r} 1{\overline{)}}\ \ 2 \quad 11 \quad -7 \quad -6 \\ \ \ \ \ \ \ \ \ \ 2 \quad 13 \quad \ \ 6 \\ \hline \ \ \ \ \ 2 \quad 13 \quad \ \ 6 \quad \ \ 0 \end{array}$$

Since $f(1) = 0$, 1 is a zero of f and $x - 1$ is a factor of f. So,

$$f(x) = 2x^3 + 11x^2 - 7x - 6 = (x - 1)(2x^2 + 13x + 6)$$

Now any solution of the equation $2x^2 + 13x + 6 = 0$ is also a zero of f. The equation $2x^2 + 13x + 6 = 0$ is called a **depressed equation** of f. Because any solution to the equation $2x^2 + 13x + 6 = 0$ is a zero of f, work with the depressed equation to find the remaining zeros of f.

The depressed equation $2x^2 + 13x + 6 = 0$ can be factored.

$$2x^2 + 13x + 6 = (2x + 1)(x + 6) = 0$$

$$2x + 1 = 0 \quad \text{or} \quad x + 6 = 0$$

$$x = -\frac{1}{2} \quad \text{or} \quad x = -6$$

The zeros of f are -6, $-\frac{1}{2}$, and 1.

The factored form of f is:

$$f(x) = 2x^3 + 11x^2 - 7x - 6 = (x - 1)(2x^2 + 13x + 6)$$
$$= (x - 1)(2x + 1)(x + 6)$$

Notice in Example 5 that all three zeros of f are in the list of potential rational zeros and agree with what was expected from Descartes' Rule of Signs.

SUMMARY

Steps for Finding the Real Zeros of a Polynomial Function

STEP 1: Use the degree of the polynomial to determine the maximum number of real zeros.

STEP 2: Use Descartes' Rule of Signs to determine the possible number of positive real zeros and negative real zeros.

STEP 3:
- If the polynomial has integer coefficients, use the Rational Zeros Theorem to identify those rational numbers that potentially could be zeros.
- Use substitution, synthetic division, or polynomial division to test each potential rational zero. Each time a zero is obtained, a factor is found. Repeat Step 3 on the depressed equation.
- When searching for the zeros, use factoring techniques already known when possible.

EXAMPLE 6 Finding the Real Zeros of a Polynomial Function

Find the real zeros of $f(x) = x^5 - 7x^4 + 19x^3 - 37x^2 + 60x - 36$. Write f in factored form.

Solution **STEP 1:** Because f is a polynomial of degree 5, there are at most five real zeros.

STEP 2: By Descartes' Rule of Signs, there are five, three, or one positive real zeros. There are no negative real zeros because

$$f(-x) = -x^5 - 7x^4 - 19x^3 - 37x^2 - 60x - 36$$

has no sign variation.

STEP 3: Because the leading coefficient $a_5 = 1$ and there are no negative real zeros, the potential rational zeros are limited to the positive integers 1, 2, 3, 4, 6, 9, 12, 18, and 36 (the positive factors of the constant term, 36). Test the potential rational zero 1 first, using synthetic division.

$$
\begin{array}{r|rrrrrr}
1) & 1 & -7 & 19 & -37 & 60 & -36 \\
  &   & 1 & -6 & 13 & -24 & 36 \\
\hline
  & 1 & -6 & 13 & -24 & 36 & 0
\end{array}
$$

The remainder is $f(1) = 0$, so 1 is a zero and $x - 1$ is a factor of f. Use the entries in the bottom row of the synthetic division to begin factoring f.

$$f(x) = x^5 - 7x^4 + 19x^3 - 37x^2 + 60x - 36$$
$$= (x - 1)(x^4 - 6x^3 + 13x^2 - 24x + 36)$$

Continue the process using the depressed equation:

$$q_1(x) = x^4 - 6x^3 + 13x^2 - 24x + 36 = 0$$

REPEAT STEP 3: The potential rational zeros of q_1 are still 1, 2, 3, 4, 6, 9, 12, 18, and 36. Test 1 again, since it may be a repeated zero of f.

$$
\begin{array}{r|rrrrr}
1) & 1 & -6 & 13 & -24 & 36 \\
  &   & 1 & -5 & 8 & -16 \\
\hline
  & 1 & -5 & 8 & -16 & 20
\end{array}
$$

Since the remainder is 20, 1 is not a repeated zero. Try 2 next.

$$
\begin{array}{r|rrrrr}
2) & 1 & -6 & 13 & -24 & 36 \\
  &   & 2 & -8 & 10 & -38 \\
\hline
  & 1 & -4 & 5 & -14 & 8
\end{array}
$$

Since the remainder is 8, 2 is not a zero. Try 3 next.

$$
\begin{array}{r|rrrrr}
3) & 1 & -6 & 13 & -24 & 36 \\
  &   & 3 & -9 & 12 & -36 \\
\hline
  & 1 & -3 & 4 & -12 & 0
\end{array}
$$

The remainder is $f(3) = 0$, so 3 is a zero and $x - 3$ is a factor of f. Use the bottom row of the synthetic division to continue the factoring of f.

$$f(x) = x^5 - 7x^4 + 19x^3 - 37x^2 + 60x - 36$$
$$= (x - 1)(x - 3)(x^3 - 3x^2 + 4x - 12)$$

The remaining zeros satisfy the new depressed equation

$$q_2(x) = x^3 - 3x^2 + 4x - 12 = 0$$

Notice that $q_2(x)$ can be factored by grouping. Alternatively, Step 3 could be repeated to again check the potential rational zero 3. The potential rational zeros 1 and 2 would no longer be checked because they have already been eliminated.

Now

$$x^3 - 3x^2 + 4x - 12 = 0$$
$$x^2(x - 3) + 4(x - 3) = 0$$
$$(x^2 + 4)(x - 3) = 0$$
$$x^2 + 4 = 0 \quad \text{or} \quad x - 3 = 0$$
$$x = 3$$

Since $x^2 + 4 = 0$ has no real solutions, the real zeros of f are 1 and 3, with 3 being a repeated zero of multiplicity 2. The factored form of f is

$$f(x) = x^5 - 7x^4 + 19x^3 - 37x^2 + 60x - 36$$
$$= (x - 1)(x - 3)^2(x^2 + 4)$$

—**Now Work** PROBLEM 45

5 Solve Polynomial Equations

| EXAMPLE 7 | Solving a Polynomial Equation |

Find the real solutions of the equation: $x^5 - 7x^4 + 19x^3 - 37x^2 + 60x - 36 = 0$

Solution   The real solutions of this equation are the real zeros of the polynomial function

$$f(x) = x^5 - 7x^4 + 19x^3 - 37x^2 + 60x - 36$$

Using the result of Example 6, the real zeros of f are 1 and 3. The real solutions of the equation $x^5 - 7x^4 + 19x^3 - 37x^2 + 60x - 36 = 0$ are 1 and 3.

—**Now Work** PROBLEM 57

In Example 6, the quadratic factor $x^2 + 4$ that appears in the factored form of f is called *irreducible*, because the polynomial $x^2 + 4$ cannot be factored over the real numbers. In general, a quadratic factor $ax^2 + bx + c$ is **irreducible** if it cannot be factored over the real numbers—that is, if it is prime over the real numbers.

Refer to Examples 5 and 6. The polynomial function of Example 5 has three real zeros, and its factored form contains three linear factors. The polynomial function of Example 6 has two distinct real zeros, and its factored form contains two distinct linear factors and one irreducible quadratic factor.

> **THEOREM**
>
> Every polynomial function with real coefficients can be uniquely factored into a product of linear factors and/or irreducible quadratic factors.

We prove this result in Section 5.7, and in fact, we shall draw several additional conclusions about the zeros of a polynomial function. One conclusion is worth noting now. If a polynomial with real coefficients is of odd degree, it must have at least one linear factor. (Do you see why? Consider the end behavior of polynomial functions of odd degree.) This means that it must have at least one real zero.

┌─ **THEOREM**

A polynomial function with real coefficients of odd degree has at least one real zero.
─┘

6 Use the Theorem for Bounds on Zeros

The work involved in finding the zeros of a polynomial function can be reduced somewhat if upper and lower bounds to the zeros can be found. A number M is an **upper bound** to the zeros of a polynomial f if no zero of f is greater than M. The number m is a **lower bound** if no zero of f is less than m. Accordingly, if m is a lower bound and M is an upper bound to the zeros of a polynomial function f, then

$$m \leq \text{any zero of } f \leq M$$

For polynomials with integer coefficients, knowing the values of a lower bound m and an upper bound M may enable you to eliminate some potential rational zeros— that is, any zeros outside the interval $[m, M]$.

> **COMMENT** The bounds on the real zeros of a polynomial provide good choices for setting Xmin and Xmax of the viewing rectangle. With these choices, all the x-intercepts of the graph can be seen. ■

┌─ **THEOREM** Bounds on Zeros

Let f denote a polynomial function whose leading coefficient is positive.
- If $M > 0$ is a real number and if the third row in the process of synthetic division of f by $x - M$ contains only numbers that are positive or zero, then M is an upper bound to the real zeros of f.
- If $m < 0$ is a real number and if the third row in the process of synthetic division of f by $x - m$ contains numbers that alternate positive (or 0) and negative (or 0), then m is a lower bound to the real zeros of f.
─┘

NOTE When finding a lower bound, a 0 can be treated as either positive or negative, but not both. For example, the numbers 3, 0, 5 would be considered to alternate sign, whereas 3, 0, −5 would not. ■

Proof (Outline) We give only an outline of the proof of the first part of the theorem. Suppose that M is a positive real number, and the third row in the process of synthetic division of the polynomial f by $x - M$ contains only numbers that are positive or 0. Then there are a quotient q and a remainder R for which

$$f(x) = (x - M)q(x) + R$$

where the coefficients of $q(x)$ are positive or 0 and the remainder $R \geq 0$. Then, for any $x > M$, we must have $x - M > 0$, $q(x) > 0$, and $R \geq 0$, so that $f(x) > 0$. That is, there is no zero of f larger than M. The proof of the second part follows similar reasoning. ■

In finding bounds, it is preferable to find the smallest upper bound and largest lower bound. This will require repeated synthetic division until a desired pattern is observed. For simplicity, we consider only potential rational zeros that are integers. If a bound is not found using these values, continue checking positive and/or negative integers until you find both an upper and a lower bound.

EXAMPLE 8 **Finding Upper and Lower Bounds of Zeros**

For the polynomial function $f(x) = 2x^3 + 11x^2 - 7x - 6$, use the Bounds on Zeros Theorem to find integer upper and lower bounds to the zeros of f.

Solution From Example 4, the potential rational zeros of f are $\pm 1, \pm 2, \pm 3, \pm 6, \pm \dfrac{1}{2}, \pm \dfrac{3}{2}$.

To find an upper bound, start with the smallest positive integer that is a potential rational zero, which is 1. Continue checking 2, 3, and 6 (and then subsequent positive integers), if necessary, until an upper bound is found. To find a lower bound, start with the largest negative integer that is a potential rational zero, which is −1. Continue checking −2, −3, and −6 (and then subsequent negative integers), if necessary, until a lower bound is found. Table 16 summarizes the results of doing repeated synthetic divisions by showing only the third row of each division. For example, the first row of the table shows the result of dividing $f(x)$ by $x - 1$.

$$
\begin{array}{r|rrr}
1) & 2 & 11 & -7 & -6 \\
  &   & 2 & 13 & 6 \\
\hline
  & 2 & 13 & 6 & 0
\end{array}
$$

Table 16   Synthetic Division Summary

	r	Coefficients of $q(x)$			Remainder	
Upper bound	1	2	13	6	0	All nonnegative
	−1	2	9	−16	10	
	−2	2	7	−21	36	
	−3	2	5	−22	60	
	−6	2	−1	−1	0	
Lower bound	−7	2	−3	14	−104	Alternating signs

NOTE Keep track of any zeros that are found when looking for bounds. ∎

For $r = 1$, the third row of synthetic division contains only numbers that are positive or 0, so we know there are no real zeros greater than 1. Since the third row of synthetic division for $r = -7$ results in alternating positive (or 0) and negative (or 0) values, we know that -7 is a lower bound. There are no real zeros less than -7. Notice that in looking for bounds, two zeros were discovered. These zeros are 1 and -6. ⌐

Now Work PROBLEM 69

If the leading coefficient of f is negative, the upper and lower bounds can still be found by first multiplying the polynomial by -1. Since $-f(x) = (-1)f(x)$, the zeros of $-f(x)$ are the same as the zeros of $f(x)$.

7 Use the Intermediate Value Theorem

The next result, called the *Intermediate Value Theorem*, is based on the fact that the graph of a polynomial function is continuous; that is, it contains no "holes" or "gaps."

THEOREM   Intermediate Value Theorem

Let f denote a polynomial function. If $a < b$ and if $f(a)$ and $f(b)$ are of opposite sign, there is at least one real zero of f between a and b.

Although the proof of the Intermediate Value Theorem requires advanced methods in calculus, it is easy to "see" why the result is true. Look at Figure 49.

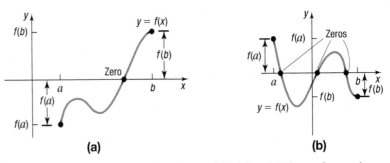

Figure 49 If f is a polynomial function and if $f(a)$ and $f(b)$ are of opposite sign, then there is at least one real zero between a and b.

EXAMPLE 9

Using the Intermediate Value Theorem to Locate a Real Zero

Show that $f(x) = x^5 - x^3 - 1$ has a real zero between 1 and 2.

Solution   Evaluate f at 1 and at 2.

$$f(1) = -1 \quad \text{and} \quad f(2) = 23$$

Because $f(1) < 0$ and $f(2) > 0$, it follows from the Intermediate Value Theorem that the polynomial function f has at least one real zero between 1 and 2.

Now Work PROBLEM 79

Let's look at the polynomial function f of Example 9 more closely. From Descartes' Rule of Signs, f has exactly one positive real zero. From the Rational Zeros Theorem, 1 is the only potential positive rational zero. Since $f(1) \neq 0$, the zero between 1 and 2 is irrational. The Intermediate Value Theorem can be used to approximate it.

Steps for Approximating the Real Zeros of a Polynomial Function

STEP 1: Find two consecutive integers a and $a + 1$ for which f has a real zero between them.

STEP 2: Divide the interval $[a, a + 1]$ into 10 equal subintervals.

STEP 3: Evaluate f at the endpoints of each subinterval until the sign of f changes. Then by the Intermediate Value Theorem, this interval contains a real zero.

STEP 4: Now divide the new interval into 10 equal subintervals and repeat Step 3.

STEP 5: Continue with Steps 3 and 4 until the desired accuracy is achieved.

Note: If at Step 3 the value of f equals 0, the process ends since that value is a zero.

EXAMPLE 10   **Approximating a Real Zero of a Polynomial Function**

Find the positive real zero of $f(x) = x^5 - x^3 - 1$ correct to two decimal places.

Solution   From Example 9 we know that the positive real zero is between 1 and 2. Divide the interval $[1, 2]$ into 10 equal subintervals: $[1, 1.1]$, $[1.1, 1.2]$, $[1.2, 1.3]$, $[1.3, 1.4]$, $[1.4, 1.5]$, $[1.5, 1.6]$, $[1.6, 1.7]$, $[1.7, 1.8]$, $[1.8, 1.9]$, $[1.9, 2]$. Now find the value of $f(x) = x^5 - x^3 - 1$ at each endpoint until the Intermediate Value Theorem applies.

$$f(1.0) = -1 \quad f(1.1) = -0.72049 \quad f(1.2) = -0.23968 \quad f(1.3) = 0.51593$$

We can stop here and conclude that the zero is between 1.2 and 1.3. Now divide the interval $[1.2, 1.3]$ into 10 equal subintervals and evaluate f at each endpoint.

$$f(1.20) = -0.23968 \quad f(1.21) \approx -0.1778185 \quad f(1.22) \approx -0.1131398$$

$$f(1.23) \approx -0.0455613 \quad f(1.24) \approx 0.025001$$

The zero lies between 1.23 and 1.24, and so, correct to two decimal places, the zero is 1.23.

Figure 50 TI-84 Plus C

 COMMENT The TABLE feature of a graphing calculator makes the computations in the solution to Example 10 a lot easier. ■

Exploration

We examine the polynomial function f given in Example 10. The Theorem on Bounds of Zeros tells us that every real zero is between -1 and 2. Graphing f using $-1 \leq x \leq 2$ (see Figure 50), we see that f has exactly one x-intercept. Using ZERO or ROOT, we find this zero to be 1.23 correct to two decimal places.

Now Work PROBLEM 91

There are many other numerical techniques for approximating a real zero of a polynomial. The one outlined in Example 10 (a variation of the *bisection method*) has the advantages that it always works, it can be programmed on a computer, and each time it is used, another decimal place of accuracy is achieved. See Problem 118 for the bisection method, which places the zero in a succession of intervals, with each new interval being half the length of the preceding one.

Historical Feature

ormulas for the solution of third- and fourth-degree polynomial equations exist, and while not very practical, they do have an interesting history.

In the 1500s in Italy, mathematical contests were a popular pastime, and people who possessed methods for solving problems kept them secret. (Solutions that were published were already common knowledge.) Niccolo of Brescia (1499–1557), commonly referred to as Tartaglia ("the stammerer"), had the secret for solving cubic (third-degree) equations, which gave him a decided advantage in the contests. Girolamo Cardano (1501–1576) learned that Tartaglia had the secret, and, being interested in cubics, he requested it from Tartaglia. The reluctant Tartaglia hesitated for some time, but finally, swearing Cardano to secrecy with midnight oaths by candlelight, told him the secret. Cardano then published the solution in his book

Ars Magna (1545), giving Tartaglia the credit but rather compromising the secrecy. Tartaglia exploded into bitter recriminations, and each wrote pamphlets that reflected on the other's mathematics, moral character, and ancestry.

The quartic (fourth-degree) equation was solved by Cardano's student Lodovico Ferrari, and this solution also was included, with credit and this time with permission, in the *Ars Magna*.

Attempts were made to solve the fifth-degree equation in similar ways, all of which failed. In the early 1800s, P. Ruffini, Niels Abel, and Evariste Galois all found ways to show that it is not possible to solve fifth-degree equations by formula, but the proofs required the introduction of new methods. Galois's methods eventually developed into a large part of modern algebra.

Historical Problems

Problems 1–8 develop the Tartaglia-Cardano solution of the cubic equation and show why it is not altogether practical.

1. Show that the general cubic equation $y^3 + by^2 + cy + d = 0$ can be transformed into an equation of the form

   $x^3 + px + q = 0$ by using the substitution $y = x - \dfrac{b}{3}$.

2. In the equation $x^3 + px + q = 0$, replace x by $H + K$. Let $3HK = -p$, and show that $H^3 + K^3 = -q$.

3. Based on Problem 2, we have the two equations

   $$3HK = -p \quad \text{and} \quad H^3 + K^3 = -q$$

   Solve for K in $3HK = -p$ and substitute into $H^3 + K^3 = -q$. Then show that

   $$H = \sqrt[3]{\dfrac{-q}{2} + \sqrt{\dfrac{q^2}{4} + \dfrac{p^3}{27}}}$$

   [**Hint:** Look for an equation that is quadratic in form.]

4. Use the solution for H from Problem 3 and the equation $H^3 + K^3 = -q$ to show that

   $$K = \sqrt[3]{\dfrac{-q}{2} - \sqrt{\dfrac{q^2}{4} + \dfrac{p^3}{27}}}$$

5. Use the results from Problems 2 to 4 to show that the solution of $x^3 + px + q = 0$ is

   $$x = \sqrt[3]{\dfrac{-q}{2} + \sqrt{\dfrac{q^2}{4} + \dfrac{p^3}{27}}} + \sqrt[3]{\dfrac{-q}{2} - \sqrt{\dfrac{q^2}{4} + \dfrac{p^3}{27}}}$$

6. Use the result of Problem 5 to solve the equation $x^3 - 6x - 9 = 0$.

7. Use a calculator and the result of Problem 5 to solve the equation $x^3 + 3x - 14 = 0$.

8. Use the methods of this section to solve the equation $x^3 + 3x - 14 = 0$.

5.6 Assess Your Understanding

'Are You Prepared?' *Answers are given at the end of these exercises. If you get a wrong answer, read the pages listed in* red.

1. Find $f(-1)$ if $f(x) = 2x^2 - x$. (pp. 207–209)

2. Factor the expression $6x^2 + x - 2$. (pp. 49–55)

3. Find the quotient and remainder when $3x^4 - 5x^3 + 7x - 4$ is divided by $x - 3$. (pp. 44–46 or 57–60)

4. Solve $x^2 = 3 - x$. (pp. 92–99)

Concepts and Vocabulary

5. *Multiple Choice* If $f(x) = q(x)g(x) + r(x)$, the function $r(x)$ is called the _____.

   (a) remainder   (b) dividend   (c) quotient (d) divisor

6. When a polynomial function f is divided by $x - c$, the remainder is _____.

7. *Multiple Choice* Given $f(x) = 3x^4 - 2x^3 + 7x - 2$, how many sign changes are there in the coefficients of $f(-x)$?

   (a) 0   (b) 1   (c) 2   (d) 3

8. *True or False* Every polynomial function of degree 3 with real coefficients has exactly three real zeros.

9. If f is a polynomial function and $x - 4$ is a factor of f, then $f(4) = $ _____.

10. *True or False* If f is a polynomial function of degree 4 and if $f(2) = 5$, then

    $$\dfrac{f(x)}{x - 2} = p(x) + \dfrac{5}{x - 2}$$

    where $p(x)$ is a polynomial of degree 3.

Skill Building

In Problems 11–20, use the Remainder Theorem to find the remainder when f(x) is divided by x − c. Then use the Factor Theorem to determine whether x − c is a factor of f(x).

11. $f(x) = 4x^3 - 3x^2 - 8x + 4; x - 2$

12. $f(x) = -4x^3 + 5x^2 + 8; x + 3$

13. $f(x) = 5x^4 - 20x^3 + x - 4; x - 2$

14. $f(x) = 4x^4 - 15x^2 - 4; x - 2$

15. $f(x) = 2x^6 + 129x^3 + 64; x + 4$

16. $f(x) = 2x^6 - 18x^4 + x^2 - 9; x + 3$

17. $f(x) = 4x^6 - 64x^4 + x^2 - 15; x + 4$

18. $f(x) = x^6 - 16x^4 + x^2 - 16; x + 4$

19. $f(x) = 2x^4 - x^3 + 2x - 1; x - \dfrac{1}{2}$

20. $f(x) = 3x^4 + x^3 - 3x + 1; x + \dfrac{1}{3}$

In Problems 21–32, determine the maximum number of real zeros that each polynomial function may have. Then use Descartes' Rule of Signs to determine how many positive and how many negative real zeros each polynomial function may have. Do not attempt to find the zeros.

21. $f(x) = -4x^7 + x^3 - x^2 + 2$

22. $f(x) = 5x^4 + 2x^2 - 6x - 5$

23. $f(x) = 8x^6 - 7x^2 - x + 5$

24. $f(x) = -3x^5 + 4x^4 + 2$

25. $f(x) = -2x^3 + 5x^2 - x - 7$

26. $f(x) = -x^3 - x^2 + x + 1$

27. $f(x) = -x^4 + x^2 - 1$

28. $f(x) = x^4 + 5x^3 - 2$

29. $f(x) = x^5 + x^4 + x^2 + x + 1$

30. $f(x) = x^5 - x^4 + x^3 - x^2 + x - 1$

31. $f(x) = x^6 - 1$

32. $f(x) = x^6 + 1$

In Problems 33–44, list the potential rational zeros of each polynomial function. Do not attempt to find the zeros.

33. $f(x) = 3x^4 - 3x^3 + x^2 - x + 1$

34. $f(x) = x^5 - x^4 + 2x^2 + 3$

35. $f(x) = x^5 - 2x^2 + 8x - 5$

36. $f(x) = 2x^5 - x^4 - x^2 + 1$

37. $f(x) = -9x^3 - x^2 + x + 3$

38. $f(x) = 6x^4 - x^2 + 2$

39. $f(x) = 6x^4 - x^2 + 9$

40. $f(x) = -4x^3 + x^2 + x + 6$

41. $f(x) = 2x^5 - x^3 + 2x^2 + 12$

42. $f(x) = 3x^5 - x^2 + 2x + 18$

43. $f(x) = 6x^4 + 2x^3 - x^2 + 20$

44. $f(x) = -6x^3 - x^2 + x + 10$

In Problems 45–56, use the Rational Zeros Theorem to find all the real zeros of each polynomial function. Use the zeros to factor f over the real numbers.

45. $f(x) = x^3 + 2x^2 - 5x - 6$

46. $f(x) = x^3 + 8x^2 + 11x - 20$

47. $f(x) = 2x^3 - x^2 + 2x - 1$

48. $f(x) = 2x^3 + x^2 + 2x + 1$

49. $f(x) = 2x^3 - 4x^2 - 10x + 20$

50. $f(x) = 3x^3 + 6x^2 - 15x - 30$

51. $f(x) = 2x^4 + x^3 - 7x^2 - 3x + 3$

52. $f(x) = 2x^4 - x^3 - 5x^2 + 2x + 2$

53. $f(x) = x^4 + x^3 - 3x^2 - x + 2$

54. $f(x) = x^4 - x^3 - 6x^2 + 4x + 8$

55. $f(x) = 4x^4 + 5x^3 + 9x^2 + 10x + 2$

56. $f(x) = 3x^4 + 4x^3 + 7x^2 + 8x + 2$

In Problems 57–68, solve each equation in the real number system.

57. $x^4 - x^3 + 2x^2 - 4x - 8 = 0$

58. $2x^3 + 3x^2 + 2x + 3 = 0$

59. $3x^3 + 4x^2 - 7x + 2 = 0$

60. $2x^3 - 3x^2 - 3x - 5 = 0$

61. $3x^3 - x^2 - 15x + 5 = 0$

62. $2x^3 - 11x^2 + 10x + 8 = 0$

63. $x^4 + 4x^3 + 2x^2 - x + 6 = 0$

64. $x^4 - 2x^3 + 10x^2 - 18x + 9 = 0$

65. $x^3 - \dfrac{2}{3}x^2 + \dfrac{8}{3}x + 1 = 0$

66. $x^3 + \dfrac{3}{2}x^2 + 3x - 2 = 0$

67. $2x^4 - 19x^3 + 57x^2 - 64x + 20 = 0$

68. $2x^4 + x^3 - 24x^2 + 20x + 16 = 0$

In Problems 69–78, find bounds on the real zeros of each polynomial function.

69. $f(x) = x^4 - 3x^2 - 4$

70. $f(x) = x^4 - 5x^2 - 36$

71. $f(x) = x^4 + x^3 - x - 1$

72. $f(x) = x^4 - x^3 + x - 1$

73. $f(x) = 3x^4 + 3x^3 - x^2 - 12x - 12$

74. $f(x) = 3x^4 - 3x^3 - 5x^2 + 27x - 36$

75. $f(x) = 4x^5 - x^4 + 2x^3 - 2x^2 + x - 1$

76. $f(x) = 4x^5 + x^4 + x^3 + x^2 - 2x - 2$

77. $f(x) = -x^4 + 3x^3 - 4x^2 - 2x + 9$

78. $f(x) = -4x^5 + 5x^3 + 9x^2 + 3x - 12$

In Problems 79–84, use the Intermediate Value Theorem to show that each polynomial function has a real zero in the given interval.

79. $f(x) = 8x^4 - 2x^2 + 5x - 1;\ [0, 1]$

80. $f(x) = x^4 + 8x^3 - x^2 + 2;\ [-1, 0]$

81. $f(x) = 2x^3 + 6x^2 - 8x + 2;\ [-5, -4]$

82. $f(x) = 3x^3 - 10x + 9;\ [-3, -2]$

83. $f(x) = x^5 - x^4 + 7x^3 - 7x^2 - 18x + 18;\ [1.4, 1.5]$

84. $f(x) = x^5 - 3x^4 - 2x^3 + 6x^2 + x + 2;\ [1.7, 1.8]$

In Problems 85–88, each equation has a solution r in the interval indicated. Use the method of Example 10 to approximate this solution correct to two decimal places.

85. $8x^4 - 2x^2 + 5x - 1 = 0;\ 0 \le r \le 1$

86. $x^4 + 8x^3 - x^2 + 2 = 0;\ -1 \le r \le 0$

87. $2x^3 + 6x^2 - 8x + 2 = 0;\ -5 \le r \le -4$

88. $3x^3 - 10x + 9 = 0;\ -3 \le r \le -2$

In Problems 89–92, each polynomial function has exactly one positive real zero. Use the method of Example 10 to approximate the zero correct to two decimal places.

89. $f(x) = x^3 + x^2 + x - 4$

90. $f(x) = 2x^4 + x^2 - 1$

91. $f(x) = 2x^4 - 3x^3 - 4x^2 - 8$

92. $f(x) = 3x^3 - 2x^2 - 20$

Mixed Practice *In Problems 93–104, graph each polynomial function.*

93. $f(x) = x^3 + 2x^2 - 5x - 6$

94. $f(x) = x^3 + 8x^2 + 11x - 20$

95. $f(x) = 2x^3 - x^2 + 2x - 1$

96. $f(x) = 2x^3 + x^2 + 2x + 1$

97. $f(x) = x^4 + x^2 - 2$

98. $f(x) = x^4 - 3x^2 - 4$

99. $f(x) = 4x^4 + 7x^2 - 2$

100. $f(x) = 4x^4 + 15x^2 - 4$

101. $f(x) = x^4 + x^3 - 3x^2 - x + 2$

102. $f(x) = x^4 - x^3 - 6x^2 + 4x + 8$

103. $f(x) = 4x^5 - 8x^4 - x + 2$

104. $f(x) = 4x^5 + 12x^4 - x - 3$

Applications and Extensions

105. Suppose that $f(x) = 3x^3 + 16x^2 + 3x - 10$. Find the zeros of $f(x + 3)$.

106. Suppose that $f(x) = 4x^3 - 11x^2 - 26x + 24$. Find the zeros of $f(x - 2)$.

107. Find k so that $x - 2$ is a factor of
$$f(x) = x^3 - kx^2 + kx + 2$$

108. Find k so that $x + 2$ is a factor of
$$f(x) = x^4 - kx^3 + kx^2 + 1$$

109. What is the remainder when $f(x) = 2x^{20} - 8x^{10} + x - 2$ is divided by $x - 1$?

110. What is the remainder when $f(x) = -3x^{17} + x^9 - x^5 + 2x$ is divided by $x + 1$?

111. Use the Factor Theorem to prove that $x - c$ is a factor of $x^n - c^n$ for any positive integer n.

112. Use the Factor Theorem to prove that $x + c$ is a factor of $x^n + c^n$ if $n \ge 1$ is an odd integer.

113. One solution of the equation $x^3 - 8x^2 + 16x - 3 = 0$ is 3. Find the sum of the remaining solutions.

114. One solution of the equation $x^3 + 5x^2 + 5x - 2 = 0$ is -2. Find the sum of the remaining solutions.

115. **Geometry** What is the length of the edge of a cube if, after a slice 1-inch thick is cut from one side, the volume remaining is 294 cubic inches?

116. **Geometry** What is the length of the edge of a cube if its volume is doubled by an increase of 6 centimeters in one edge, an increase of 12 centimeters in a second edge, and a decrease of 4 centimeters in the third edge?

117. Let $f(x)$ be a polynomial function whose coefficients are integers. Suppose that r is a real zero of f and that the leading coefficient of f is 1. Use the Rational Zeros Theorem to show that r is either an integer or an irrational number.

118. **Bisection Method for Approximating Real Zeros of a Polynomial Function** We begin with two consecutive integers, a and $a + 1$, for which $f(a)$ and $f(a + 1)$ are of opposite sign. Evaluate f at the midpoint m_1 of a and $a + 1$. If $f(m_1) = 0$, then m_1 is the zero of f, and we are finished. Otherwise, $f(m_1)$ is of opposite sign to either $f(a)$ or $f(a + 1)$. Suppose that it is $f(a)$ and $f(m_1)$ that are of opposite sign. Now evaluate f at the midpoint m_2 of a and m_1. Repeat this process until the desired degree of accuracy is obtained. Note that each iteration places the zero in an interval whose length is half that of the previous interval. Use the bisection method to approximate the zero of $f(x) = 8x^4 - 2x^2 + 5x - 1$ in the interval $[0, 1]$ correct to three decimal places.

[**Hint:** The process ends when both endpoints agree to the desired number of decimal places.]

119. Challenge Problem Suppose f is a polynomial function. If $f(-2) = 7$ and $f(6) = -1$, then the Intermediate Value Theorem guarantees which of the following? Justify your answer.

  (a) $f(0) = 0$
  (b) $f(c) = 3$ for at least one number c between -2 and 6.
  (c) $f(c) = 0$ for at least one number between -1 and 7.
  (d) $-1 \le f(x) \le 7$ for all numbers in the closed interval $[-2, 6]$.

120. Challenge Problem Prove the Rational Zeros Theorem.
  [Hint: Let $\dfrac{p}{q}$, where p and q have no common factors

except 1 and -1, be a zero of the polynomial function

$$f(x) = a_n x^n + a_{n-1} x^{n-1} + \cdots + a_1 x + a_0$$

whose coefficients are all integers. Show that

$$a_n p^n + a_{n-1} p^{n-1} q + \cdots + a_1 p q^{n-1} + a_0 q^n = 0$$

Now, show that p must be a factor of a_0, and that q must be a factor of a_n.]

121. Challenge Problem Use the Intermediate Value Theorem to show that the functions $y = x^3$ and $y = 1 - x^2$ intersect somewhere between $x = 0$ and $x = 1$.

Explaining Concepts: Discussion and Writing

122. Is $\dfrac{1}{3}$ a zero of $f(x) = 2x^3 + 3x^2 - 6x + 7$? Explain.

123. Is $\dfrac{1}{3}$ a zero of $f(x) = 4x^3 - 5x^2 - 3x + 1$? Explain.

124. Is $\dfrac{3}{5}$ a zero of $f(x) = 2x^6 - 5x^4 + x^3 - x + 1$? Explain.

125. Is $\dfrac{2}{3}$ a zero of $f(x) = x^7 + 6x^5 - x^4 + x + 2$? Explain.

Retain Your Knowledge

Problems 126–135 are based on material learned earlier in the course. The purpose of these problems is to keep the material fresh in your mind so that you are better prepared for the final exam.

126. Write $f(x) = -3x^2 + 30x - 4$ in the form
$$f(x) = a(x - h)^2 + k$$

127. Express the inequality $3 \le x < 8$ using interval notation.

128. Solve $2x - 5y = 3$ for y.

129. Solve $\dfrac{1}{3}x^2 - 2x + 9 = 0$.

For Problems 130–135, use the graph on the right.

130. On which interval(s) is f increasing?
131. On which interval(s) is f decreasing?
132. What are the zeros of f, if any?
133. What are the intercepts of the graph of f?
134. What are the turning points?
135. What are the absolute extrema, if any?

'Are You Prepared?' Answers

  1. 3   **2.** $(3x + 2)(2x - 1)$ **3.** Quotient: $3x^3 + 4x^2 + 12x + 43$; Remainder: 125   **4.** $\dfrac{-1 - \sqrt{13}}{2}, \dfrac{-1 + \sqrt{13}}{2}$

5.7 Complex Zeros; Fundamental Theorem of Algebra

PREPARING FOR THIS SECTION *Before getting started, review the following:*

- Complex Numbers (Section 1.3, pp. 104–109)
- Complex Solutions of a Quadratic Equation (Section 1.3, pp. 109–111)

  Now Work the 'Are You Prepared?' problems on page 406.

OBJECTIVES 1 Use the Conjugate Pairs Theorem (p. 403)
        2 Find a Polynomial Function with Specified Zeros (p. 404)
        3 Find the Complex Zeros of a Polynomial Function (p. 405)

In Section 1.2, we found the real solutions of a quadratic equation. That is, we found the real zeros of a polynomial function of degree 2. Then, in Section 1.3 we found the complex solutions of a quadratic equation. That is, we found the complex zeros of a polynomial function of degree 2.

Need to Review?

Complex numbers and the complex solutions of a quadratic equation are discussed in Section 1.3, pp. 104–111.

In Section 5.6, we found the real zeros of polynomial functions of degree 3 or higher. In this section we will find the *complex zeros* of polynomial functions of degree 3 or higher.

DEFINITION Complex Zeros

A variable in the complex number system is referred to as a **complex variable**. A **complex polynomial function** f of degree n is a function of the form

$$f(x) = a_n x^n + a_{n-1} x^{n-1} + \cdots + a_1 x + a_0 \qquad (1)$$

where $a_n, a_{n-1}, \ldots, a_1, a_0$ are complex numbers, $a_n \neq 0$, n is a nonnegative integer, and x is a complex variable. As before, a_n is called the **leading coefficient** of f. A complex number r is called a **complex zero** of f if $f(r) = 0$.

In most of our work, the coefficients in (1) are real numbers.

We have learned that some quadratic equations have no real solutions, but that in the complex number system every quadratic equation has at least one solution, either real or complex. The next result, proved by Karl Friedrich Gauss (1777–1855) when he was 22 years old,* gives an extension to complex polynomials. In fact, this result is so important and useful that it has become known as the *Fundamental Theorem of Algebra*.

Fundamental Theorem of Algebra

Every complex polynomial function f of degree $n \geq 1$ has at least one complex zero.

We shall not prove this result, as the proof is beyond the scope of this text. However, using the Fundamental Theorem of Algebra and the Factor Theorem, we can prove the following result:

THEOREM

Every complex polynomial function f of degree $n \geq 1$ can be factored into n linear factors (not necessarily distinct) of the form

$$f(x) = a_n (x - r_1)(x - r_2) \cdot \cdots \cdot (x - r_n) \qquad (2)$$

where $a_n, r_1, r_2, \ldots, r_n$ are complex numbers. That is, every complex polynomial function of degree $n \geq 1$ has exactly n complex zeros, some of which may repeat.

Proof Let $f(x) = a_n x^n + a_{n-1} x^{n-1} + \cdots + a_1 x + a_0$.

By the Fundamental Theorem of Algebra, f has at least one zero, say r_1. Then, by the Factor Theorem, $x - r_1$ is a factor, and

$$f(x) = (x - r_1) q_1(x)$$

where $q_1(x)$ is a complex polynomial of degree $n - 1$ whose leading coefficient is a_n. Repeating this argument n times, we arrive at

$$f(x) = (x - r_1)(x - r_2) \cdot \cdots \cdot (x - r_n) q_n(x)$$

where $q_n(x)$ is a complex polynomial of degree $n - n = 0$ whose leading coefficient is a_n. That is, $q_n(x) = a_n x^0 = a_n$, and so

$$f(x) = a_n (x - r_1)(x - r_2) \cdot \cdots \cdot (x - r_n)$$

We conclude that every complex polynomial function f of degree $n \geq 1$ has exactly n (not necessarily distinct) zeros. ∎

*In all, Gauss gave four different proofs of this theorem, the first one in 1799 being the subject of his doctoral dissertation.

1 Use the Conjugate Pairs Theorem

The Fundamental Theorem of Algebra can be used to obtain valuable information about the complex zeros of polynomial functions whose coefficients are real numbers.

Conjugate Pairs Theorem

Suppose f is a polynomial function whose coefficients are real numbers. If $r = a + bi$ is a zero of f, the complex conjugate $\bar{r} = a - bi$ is also a zero of f.

In other words, for polynomial functions whose coefficients are real numbers, the complex zeros occur in conjugate pairs. This result should not be all that surprising since the complex zeros of a quadratic function occurred in conjugate pairs.

Proof Let

$$f(x) = a_n x^n + a_{n-1}x^{n-1} + \cdots + a_1 x + a_0$$

where $a_n, a_{n-1}, \ldots, a_1, a_0$ are real numbers and $a_n \neq 0$. If $r = a + bi$ is a zero of f, then $f(r) = f(a + bi) = 0$, so

$$a_n r^n + a_{n-1}r^{n-1} + \cdots + a_1 r + a_0 = 0$$

Take the conjugate of both sides to get

$$\overline{a_n r^n + a_{n-1}r^{n-1} + \cdots + a_1 r + a_0} = \bar{0}$$

$$\overline{a_n r^n} + \overline{a_{n-1}r^{n-1}} + \cdots + \overline{a_1 r} + \overline{a_0} = \bar{0} \qquad \begin{array}{l}\text{The conjugate of a sum equals the sum}\\ \text{of the conjugates (see Section 1.3).}\end{array}$$

$$\overline{a_n}(\bar{r})^n + \overline{a_{n-1}}(\bar{r})^{n-1} + \cdots + \overline{a_1}\bar{r} + \overline{a_0} = \bar{0} \qquad \begin{array}{l}\text{The conjugate of a product equals}\\ \text{the product of the conjugates.}\end{array}$$

$$a_n(\bar{r})^n + a_{n-1}(\bar{r})^{n-1} + \cdots + a_1\bar{r} + a_0 = 0 \qquad \begin{array}{l}\text{The conjugate of a real number}\\ \text{equals the real number.}\end{array}$$

This last equation states that $f(\bar{r}) = 0$; that is, $\bar{r} = a - bi$ is a zero of f. ∎

The importance of this result is that once we know a complex number, say $3 + 4i$, is a zero of a polynomial function with real coefficients, then we know that its conjugate $3 - 4i$ is also a zero. This result has an important corollary.

Corollary

A polynomial function f of odd degree with real coefficients has at least one real zero.

Proof Because complex zeros occur as conjugate pairs in a polynomial function with real coefficients, there will always be an even number of zeros that are not real numbers. Consequently, since f is of odd degree, one of its zeros must be a real number. ∎

For example, the polynomial function $f(x) = x^5 - 3x^4 + 4x^3 - 5$ has at least one zero that is a real number, since f is of degree 5 (odd) and has real coefficients.

EXAMPLE 1 | Using the Conjugate Pairs Theorem

A polynomial function f of degree 5 whose coefficients are real numbers has the zeros $1, 5i,$ and $1 + i$. Find the remaining two zeros.

Solution Since f has coefficients that are real numbers, complex zeros appear as conjugate pairs. It follows that $-5i$, the conjugate of $5i$, and $1 - i$, the conjugate of $1 + i$, are the two remaining zeros.

━━━━━ **Now Work** PROBLEM 9

2 Find a Polynomial Function with Specified Zeros

EXAMPLE 2 | Finding a Polynomial Function Whose Zeros Are Given

Find a polynomial function f of degree 4 whose coefficients are real numbers that has the zeros $1, 1,$ and $-4 + i$.

Solution Since $-4 + i$ is a zero, by the Conjugate Pairs Theorem, $-4 - i$ is also a zero of f. From the Factor Theorem, if $f(c) = 0$, then $x - c$ is a factor of $f(x)$. So f can now be written as

$$f(x) = a(x - 1)(x - 1)[x - (-4 + i)][x - (-4 - i)]$$

where a is any nonzero real number. Then

$$
\begin{aligned}
f(x) &= a(x - 1)(x - 1)[x - (-4 + i)][x - (-4 - i)] \\
&= a(x^2 - 2x + 1)[(x + 4) - i][(x + 4) + i] \\
&= a(x^2 - 2x + 1)[(x + 4)^2 - i^2] \\
&= a(x^2 - 2x + 1)[x^2 + 8x + 16 - (-1)] \qquad i^2 = -1 \\
&= a(x^2 - 2x + 1)(x^2 + 8x + 17) \\
&= a(x^4 + 8x^3 + 17x^2 - 2x^3 - 16x^2 - 34x + x^2 + 8x + 17) \\
&= a(x^4 + 6x^3 + 2x^2 - 26x + 17)
\end{aligned}
$$

NORMAL FLOAT AUTO REAL RADIAN MP

Figure 51
$f(x) = x^4 + 6x^3 + 2x^2 - 26x + 17$

Exploration

Graph the function f found in Example 2 for $a = 1$. Does the value of a affect the zeros of f? How does the value of a affect the graph of f? What information about f is sufficient to uniquely determine a?

Result An analysis of the polynomial function f tells us what to expect:

- At most three turning points.
- For large $|x|$, the graph resembles the graph of $y = x^4$.
- A repeated real zero at 1, so the graph will touch the x-axis at 1.
- The only x-intercept is 1; the y-intercept is 17.

Figure 51 shows the complete graph on a TI-84 Plus C. (Do you see why? The graph has exactly three turning points.) The value of a causes a stretch or compression; a reflection also occurs if $a < 0$. The zeros are not affected.

If any point other than an x-intercept on the graph of f is known, then a can be determined. For example, if $(2, 3)$ is on the graph, then $f(2) = 3 = a(37)$, so $a = 3/37$. Why won't an x-intercept work?

We can now prove the theorem stated in Section 5.6.

> **THEOREM**
>
> Every polynomial function with real coefficients can be uniquely factored over the real numbers into a product of linear factors and/or irreducible quadratic factors.

Proof Every complex polynomial function f of degree n has exactly n zeros and can be factored into a product of n linear factors. If its coefficients are real, the zeros that are complex numbers always occur in conjugate pairs. As a result, if $r = a + bi$ is a complex zero, then so is $\bar{r} = a - bi$. Consequently, when the linear factors $x - r$ and $x - \bar{r}$ of f are multiplied, we have

$$(x - r)(x - \bar{r}) = x^2 - (r + \bar{r})x + r\bar{r} = x^2 - 2ax + a^2 + b^2$$

This second-degree polynomial has real coefficients and is irreducible (over the real numbers). So, the factors of f are either linear or irreducible quadratic factors. ∎

- **Now Work** PROBLEM 19

3 Find the Complex Zeros of a Polynomial Function

The steps for finding the complex zeros of a polynomial function are the same as those for finding the real zeros.

EXAMPLE 3 | **Finding the Complex Zeros of a Polynomial Function**

Find the complex zeros of the polynomial function

$$f(x) = 3x^4 + 5x^3 + 25x^2 + 45x - 18$$

Write f in factored form.

Solution **STEP 1:** The degree of f is 4, so f has four complex zeros.

STEP 2: Since the coefficients of f are real numbers, Descartes' Rule of Signs can be used to obtain information about the real zeros. For this polynomial function, there is one positive real zero. There are three negative real zeros or one negative real zero, because

$$f(-x) = 3x^4 - 5x^3 + 25x^2 - 45x - 18$$

has three variations in sign.

STEP 3: Since the coefficients of f are integers, the Rational Zeros Theorem can be used to obtain information about the potential rational zeros of f. The potential rational zeros are

$$\pm \frac{1}{3}, \pm \frac{2}{3}, \pm 1, \pm 2, \pm 3, \pm 6, \pm 9, \pm 18$$

Table 17 summarizes some results of synthetic division.

Table 17

r	Coefficients of $q(x)$				Remainder	
1	3	8	33	78	60	1 is not a zero.
−1	3	2	23	22	−40	−1 is not a zero.
2	3	11	47	139	260	2 is not a zero.
−2	3	−1	27	−9	⓪	−2 is a zero.

Since $f(-2) = 0$, then -2 is a zero and $x + 2$ is a factor of f. The depressed equation is

$$3x^3 - x^2 + 27x - 9 = 0$$

REPEAT STEP 3: The depressed equation can be factored by grouping.

$$3x^3 - x^2 + 27x - 9 = 0$$
$$x^2(3x - 1) + 9(3x - 1) = 0 \quad \text{Factor } x^2 \text{ from } 3x^3 - x^2 \text{ and 9 from } 27x - 9.$$
$$(x^2 + 9)(3x - 1) = 0 \quad \text{Factor out the common factor } 3x - 1.$$
$$x^2 + 9 = 0 \quad \text{or} \quad 3x - 1 = 0 \quad \text{Use the Zero-Product Property.}$$
$$x^2 = -9 \quad \text{or} \quad x = \frac{1}{3}$$
$$x = -3i, \quad x = 3i \quad \text{or} \quad x = \frac{1}{3} \quad \text{Use the Square Root Method.}$$

The four complex zeros of f are $-3i, 3i, -2,$ and $\frac{1}{3}$.

The factored form of f is

$$f(x) = 3x^4 + 5x^3 + 25x^2 + 45x - 18$$
$$= 3(x + 3i)(x - 3i)(x + 2)\left(x - \frac{1}{3}\right)$$
$$= 3(x^2 + 9)(x + 2)\left(x - \frac{1}{3}\right)$$

⬛

──**Now Work** PROBLEM 35

5.7 Assess Your Understanding

'Are You Prepared?' *Answers are given at the end of these exercises. If you get a wrong answer, read the pages listed in* red.

1. Find the sum and the product of the complex numbers $3 - 2i$ and $-3 + 5i$. (pp. 104–109)

2. Solve $x^2 + 2x + 2 = 0$ in the complex number system. (pp. 109–111)

3. What is the conjugate of $-3 + 4i$? (p. 106)

4. Given $z = 5 + 2i$, find the product $z \cdot \bar{z}$. (p. 107)

Concepts and Vocabulary

5. Every polynomial function of odd degree with real coefficients will have at least _____ real zero(s).

6. If $3 + 4i$ is a zero of a polynomial function of degree 5 with real coefficients, then so is _____ .

7. **True or False** A polynomial function of degree n with real coefficients has exactly n complex zeros. At most n of them are real zeros.

8. **True or False** A polynomial function of degree 4 with real coefficients could have $-3, 2 + i, 2 - i,$ and $-3 + 5i$ as its zeros.

Skill Building

In Problems 9–18, information is given about a polynomial function f whose coefficients are real numbers. Find the remaining zeros of f.

9. Degree 3; zeros: $3, 4 - i$

10. Degree 3; zeros: $4, 3 + i$

11. Degree 4; zeros: $i, 3 + i$

12. Degree 4; zeros: $1, 2, 2 + i$

13. Degree 5; zeros: $1, i, 5i$

14. Degree 5; zeros: $0, 1, 2, i$

15. Degree 4; zeros: $i, 7, -7$

16. Degree 4; zeros: $2 - i, -i$

17. Degree 6; zeros: $2, 4 + 9i, -7 - 2i, 0$

18. Degree 6; zeros: $i, 3 - 2i, -2 + i$

In Problems 19–24, find a polynomial function f with real coefficients having the given degree and zeros. Answers will vary depending on the choice of leading coefficient.

19. Degree 4; zeros: $3 + 2i$; 4, multiplicity 2

20. Degree 4; zeros: i, $1 + 2i$

21. Degree 5; zeros: 2; $-i$; $1 + i$

22. Degree 6; zeros: i, $4 - i$; $2 + i$

23. Degree 4; zeros: 3, multiplicity 2; $-i$

24. Degree 5; zeros: 1, multiplicity 3; $1 + i$

In Problems 25–32, use the given zero to find the remaining zeros of each polynomial function.

25. $f(x) = x^3 - 5x^2 + 9x - 45$; zero: $3i$

26. $g(x) = x^3 + 3x^2 + 25x + 75$; zero: $-5i$

27. $f(x) = 4x^4 + 7x^3 + 62x^2 + 112x - 32$; zero: $-4i$

28. $h(x) = 3x^4 + 5x^3 + 25x^2 + 45x - 18$; zero: $3i$

29. $h(x) = x^4 - 7x^3 + 23x^2 - 15x - 522$; zero: $2 - 5i$

30. $f(x) = x^4 - 7x^3 + 14x^2 - 38x - 60$; zero: $1 + 3i$

31. $h(x) = 3x^5 + 2x^4 - 9x^3 - 6x^2 - 84x - 56$; zero: $-2i$

32. $g(x) = 2x^5 - 3x^4 - 5x^3 - 15x^2 - 207x + 108$; zero: $3i$

In Problems 33–42, find the complex zeros of each polynomial function. Write f in factored form.

33. $f(x) = x^3 - 1$

34. $f(x) = x^4 - 1$

35. $f(x) = x^3 - 8x^2 + 25x - 26$

36. $f(x) = x^3 + 13x^2 + 57x + 85$

37. $f(x) = x^4 + 5x^2 + 4$

38. $f(x) = x^4 + 13x^2 + 36$

39. $f(x) = x^4 + 2x^3 + 22x^2 + 50x - 75$

40. $f(x) = x^4 + 3x^3 - 19x^2 + 27x - 252$

41. $f(x) = 3x^4 - x^3 - 9x^2 + 159x - 52$

42. $f(x) = 2x^4 + x^3 - 35x^2 - 113x + 65$

43. Challenge Problem Suppose $f(x) = 2x^3 - 14x^2 + bx - 3$ with $f(2) = 0$ and $g(x) = x^3 + cx^2 - 8x + 30$, with the zero $x = 3 - i$, where b and c are real numbers. Find $(f \cdot g)(1)$.[†]

44. Challenge Problem Let f be the polynomial function of degree 4 with real coefficients, leading coefficient 1, and zeros $x = 3 + i, 2, -2$. Let g be the polynomial function of degree 4 with intercept $(0, -4)$ and zeros $x = i, 2i$. Find $(f + g)(1)$.[†]

45. Challenge Problem **The complex zeros of $f(x) = x^4 + 1$** For the function $f(x) = x^4 + 1$:

(a) Factor f into the product of two irreducible quadratics.
(b) Find the zeros of f by finding the zeros of each irreducible quadratic.

Explaining Concepts: Discussion and Writing

In Problems 46 and 47, explain why the facts given are contradictory.

46. f is a polynomial function of degree 3 whose coefficients are real numbers; its zeros are $2, i$, and $3 + i$.

47. f is a polynomial function of degree 3 whose coefficients are real numbers; its zeros are $4 + i, 4 - i$, and $2 + i$.

48. f is a polynomial function of degree 4 whose coefficients are real numbers; two of its zeros are -3 and $4 - i$. Explain why one of the remaining zeros must be a real number. Write down one of the missing zeros.

49. f is a polynomial function of degree 4 whose coefficients are real numbers; three of its zeros are $2, 1 + 2i$, and $1 - 2i$. Explain why the remaining zero must be a real number.

50. For the polynomial function $f(x) = x^2 + 2ix - 10$:
(a) Verify that $3 - i$ is a zero of f.
(b) Verify that $3 + i$ is not a zero of f.
(c) Explain why these results do not contradict the Conjugate Pairs Theorem.

[†]Courtesy of the Joliet Junior College Mathematics Department

Retain Your Knowledge

Problems 51–60 are based on material learned earlier in the course. The purpose of these problems is to keep the material fresh in your mind so that you are better prepared for the final exam.

51. Draw a scatter plot for the given data.

x	-1	1	2	5	8	10
y	-4	0	3	1	5	7

52. Given $f(x) = \sqrt{3 - x}$, find x so that $f(x) = 5$.

53. Multiply: $(2x - 5)(3x^2 + x - 4)$

54. Find the area and circumference of a circle with a diameter of 6 feet.

55. If $f(x) = \dfrac{x + 1}{x}$ and $g(x) = 3x - 2$, find $\left(\dfrac{g}{f}\right)(x)$ and state its domain.

56. Solve $x = \sqrt{y + 3} - 5$ for y.

57. Find the domain of $g(x) = \dfrac{x - \sqrt{x}}{x + 2}$.

58. Find the intercepts of the graph of the equation $3x + y^2 = 12$.

59. Rationalize the numerator: $\dfrac{\sqrt{x} - 3}{x + 7}$

60. Find the difference quotient of $f(x) = x^3 + 8$.

'Are You Prepared?' Answers

1. Sum: $3i$; product: $1 + 21i$ **2.** $-1 - i, \ -1 + i$ **3.** $-3 - 4i$ **4.** 29

Chapter Review

Things to Know

Power function (pp. 332–335)

$f(x) = x^n, n \geq 2$ even

Domain: all real numbers; Range nonnegative real numbers
Contains the points $(-1, 1), (0, 0), (1, 1)$
Even function
Decreasing on $(-\infty, 0]$, increasing on $[0, \infty)$

$f(x) = x^n, n \geq 3$ odd

Domain: all real numbers; Range all real numbers
Contains the points $(-1, -1), (0, 0), (1, 1)$
Odd function
Increasing on $(-\infty, \infty)$

Polynomial function (pp. 331–332, 335–342)

$f(x) = a_n x^n + a_{n-1}x^{n-1}$
$\quad + \cdots + a_1 x + a_0, a_n \neq 0$

Domain: all real numbers
At most $n - 1$ turning points
End behavior: Behaves like $y = a_n x^n$ for large $|x|$

Real zeros of a polynomial function f (p. 336)

Real numbers for which $f(x) = 0$; the real zeros of f are the x-intercepts of the graph of f.

Multiplicity (p. 337)

If $(x - r)^m$ is a factor of a polynomial f and $(x - r)^{m+1}$ is not a factor of f, then r is called a **real zero of multiplicity m of f**.

Rational function (pp. 354–361)

$R(x) = \dfrac{p(x)}{q(x)}$

p, q are polynomial functions and q is not the zero polynomial.

Domain: $\{x \mid q(x) \neq 0\}$
Vertical asymptotes: With $R(x)$ in lowest terms, if $q(r) = 0$ for some real number, then $x = r$ is a vertical asymptote.
Horizontal or oblique asymptote: See the summary on pages 359–360.

Remainder Theorem (p. 388)

If a polynomial function $f(x)$ is divided by $x - c$, then the remainder is $f(c)$.

Factor Theorem (p. 389)

$x - c$ is a factor of a polynomial function $f(x)$ if and only if $f(c) = 0$.

Descartes' Rule of Signs (p. 390)

Let f denote a polynomial function written in standard form.
- The number of positive zeros of f either equals the number of variations in the sign of the nonzero coefficients of $f(x)$ or else equals that number less an even integer.
- The number of negative real zeros of f either equals the number of variations in the sign of the nonzero coefficients of $f(-x)$ or else equals that number less some even integer.

Rational Zeros Theorem (p. 391)

Let f be a polynomial function of degree 1 or higher of the form
$$f(x) = a_n x^n + a_{n-1}x^{n-1} + \cdots + a_1 x + a_0 \quad a_n \neq 0, a_0 \neq 0$$
where each coefficient is an integer. If $\frac{p}{q}$, in lowest terms, is a rational zero of f, then p must be a factor of a_0, and q must be a factor of a_n.

Intermediate Value Theorem (p. 396)

Let f be a polynomial function. If $a < b$ and $f(a)$ and $f(b)$ are of opposite sign, then there is at least one real zero of f between a and b.

Fundamental Theorem of Algebra (p. 402)

Every complex polynomial function f of degree $n \geq 1$ has at least one complex zero.

Conjugate Pairs Theorem (p. 403)

Let f be a polynomial function whose coefficients are real numbers. If $r = a + bi$ is a zero of f, then its complex conjugate $\bar{r} = a - bi$ is also a zero of f.

Objectives

Section	You should be able to . . .	Example(s)	Review Exercises
5.1	1 Identify polynomial functions and their degree (p. 331)	1	1–4
	2 Graph polynomial functions using transformations (p. 335)	2, 3	5–7
	3 Identify the real zeros of a polynomial function and their multiplicity (p. 336)	4–9	8–11
5.2	1 Graph a polynomial function (p. 346)	1, 2	8–11
	2 Graph a polynomial function using a graphing utility (p. 348)	3	49
	3 Build cubic models from data (p. 350)	4	51
5.3	1 Find the domain of a rational function (p. 354)	1–3	12–14
	2 Find the vertical asymptotes of a rational function (p. 357)	4	12–14
	3 Find a horizontal or an oblique asymptote of a rational function (p. 359)	5–8	12–14
5.4	1 Graph a rational function (p. 365)	1–6	15–20
	2 Solve applied problems involving rational functions (p. 375)	7	50
5.5	1 Solve polynomial inequalities (p. 380)	1, 2	21, 22
	2 Solve rational inequalities (p. 382)	3, 4	23–25
5.6	1 Use the Remainder and Factor Theorems (p. 387)	1, 2	26–28
	2 Use Descartes' Rule of Signs to determine the number of positive and the number of negative real zeros of a polynomial function (p. 390)	3	29, 30
	3 Use the Rational Zeros Theorem to list the potential rational zeros of a polynomial function (p. 391)	4	31–34
	4 Find the real zeros of a polynomial function (p. 392)	5, 6	32–34
	5 Solve polynomial equations (p. 394)	7	35, 36
	6 Use the Theorem for Bounds on Zeros (p. 395)	8	37, 38
	7 Use the Intermediate Value Theorem (p. 396)	9, 10	39–42
5.7	1 Use the Conjugate Pairs Theorem (p. 403)	1	43, 44
	2 Find a polynomial function with specified zeros (p. 404)	2	43, 44
	3 Find the complex zeros of a polynomial function (p. 405)	3	45–48

Review Exercises

In Problems 1–4, determine whether the function is a polynomial function, a rational function, or neither. For those that are polynomial functions, state the degree. For those that are not polynomial functions, tell why not.

1. $f(x) = 4x^5 - 3x^2 + 5x - 2$ **2.** $f(x) = \dfrac{3x^5}{2x + 1}$ **3.** $f(x) = 3x^2 + 5x^{1/2} - 1$ **4.** $f(x) = 3$

In Problems 5–7, graph each function using transformations (shifting, compressing, stretching, and reflecting). Show all the stages.

5. $f(x) = (x + 2)^3$ **6.** $f(x) = -(x - 1)^4$ **7.** $f(x) = (x - 1)^4 + 2$

In Problems 8–11, graph each polynomial function by following Steps 1 through 5 on page 347.

8. $f(x) = x(x + 2)(x + 4)$ **9.** $f(x) = (x - 2)^2(x + 4)$

10. $f(x) = -2x^3 + 4x^2$ **11.** $f(x) = (x - 1)^2(x + 3)(x + 1)$

In Problems 12–14, find the domain of each rational function. Find any horizontal, vertical, or oblique asymptotes.

12. $R(x) = \dfrac{x + 2}{x^2 - 9}$ **13.** $R(x) = \dfrac{x^2 + 4}{x - 2}$ **14.** $R(x) = \dfrac{x^2 + 3x + 2}{(x + 2)^2}$

In Problems 15–20, graph each rational function following the seven steps given on page 367.

15. $R(x) = \dfrac{2x - 6}{x}$ **16.** $H(x) = \dfrac{x + 2}{x(x - 2)}$ **17.** $R(x) = \dfrac{x^2 + x - 6}{x^2 - x - 6}$

18. $F(x) = \dfrac{x^3}{x^2 - 4}$ **19.** $R(x) = \dfrac{2x^4}{(x - 1)^2}$ **20.** $G(x) = \dfrac{x^2 - 4}{x^2 - x - 2}$

In Problems 21–25, solve each inequality. Graph the solution set.

21. $x^3 + x^2 < 4x + 4$ **22.** $x^3 + 4x^2 \geq x + 4$ **23.** $\dfrac{2x - 6}{1 - x} < 2$ **24.** $\dfrac{(x - 2)(x - 1)}{x - 3} \geq 0$ **25.** $\dfrac{x^2 - 8x + 12}{x^2 - 16} > 0$

In Problems 26 and 27, find the remainder R when f(x) is divided by g(x). Is g a factor of f?

26. $f(x) = 8x^3 - 3x^2 + x + 4$; $g(x) = x - 1$ **27.** $f(x) = x^4 - 2x^3 + 15x - 2$; $g(x) = x + 2$

28. Find the value of $f(x) = 12x^6 - 8x^4 + 1$ at $x = 4$.

In Problems 29 and 30, use Descartes' Rule of Signs to determine how many positive and negative real zeros each polynomial function may have. Do not attempt to find the zeros.

29. $f(x) = 12x^8 - x^7 + 8x^4 - 2x^3 + x + 3$

30. $f(x) = -6x^5 + x^4 + 5x^3 + x + 1$

31. List all the potential rational zeros of $f(x) = 12x^8 - x^7 + 6x^4 - x^3 + x - 3$.

In Problems 32–34, use the Rational Zeros Theorem to find all the real zeros of each polynomial function. Use the zeros to factor f over the real numbers.

32. $f(x) = x^3 - 3x^2 - 6x + 8$ **33.** $f(x) = 4x^3 + 4x^2 - 7x + 2$ **34.** $f(x) = x^4 - 4x^3 + 9x^2 - 20x + 20$

In Problems 35 and 36, solve each equation in the real number system.

35. $2x^4 + 2x^3 - 11x^2 + x - 6 = 0$ **36.** $2x^4 + 7x^3 + x^2 - 7x - 3 = 0$

In Problems 37 and 38, find bounds on the real zeros of each polynomial function.

37. $f(x) = x^3 - x^2 - 4x + 2$ **38.** $f(x) = 2x^3 - 7x^2 - 10x + 35$

In Problems 39 and 40, use the Intermediate Value Theorem to show that each polynomial function has a real zero in the given interval.

39. $f(x) = 3x^3 - x - 1$; $[0, 1]$ **40.** $f(x) = 8x^4 - 4x^3 - 2x - 1$; $[0, 1]$

In Problems 41 and 42, each polynomial function has exactly one positive zero. Approximate the zero correct to two decimal places.

41. $f(x) = x^3 - x - 2$ **42.** $f(x) = 8x^4 - 4x^3 - 2x - 1$

In Problems 43 and 44, information is given about a complex polynomial f whose coefficients are real numbers. Find the remaining zeros of f. Then find a polynomial function with real coefficients that has the zeros.

43. Degree 3; zeros: $4 + i, 6$ **44.** Degree 4; zeros: $i, 1 + i$

In Problems 45–48, find the complex zeros of each polynomial function f. Write f in factored form.

45. $f(x) = x^3 - 3x^2 - 6x + 8$ **46.** $f(x) = 4x^3 + 4x^2 - 7x + 2$

47. $f(x) = x^4 - 4x^3 + 9x^2 - 20x + 20$ **48.** $f(x) = 2x^4 + 2x^3 - 11x^2 + x - 6$

49. Graph the polynomial function

$$f(x) = x^3 - 2.37x^2 - 4.68x + 6.93$$

by following Steps 1 through 8 on page 349.

50. Making a Can A can in the shape of a right circular cylinder is required to have a volume of 250 cubic centimeters.
  (a) Express the amount A of material needed to make the can as a function of the radius r of the cylinder.
  (b) How much material is required if the can is of radius 3 centimeters?
  (c) How much material is required if the can is of radius 5 centimeters?
  (d) Graph $A = A(r)$. For what value of r is A smallest?

51. Housing Prices The data in the table on the right represent the January median new-home prices in the United States for the years shown.
  (a) With a graphing utility, draw a scatter plot of the data. Comment on the type of relation that appears to exist between the two variables.

(b) Decide on the function of best fit to these data (linear, quadratic, or cubic), and use this function to predict the median new-home price in the United States for January 2022 ($t = 10$).

(c) Draw the function of best fit on the scatter plot obtained in part (a).

Year, t	Median Price, P ($1000s)
2004, 1	209.5
2006, 2	244.9
2008, 3	232.4
2010, 4	218.2
2012, 5	221.7
2014, 6	269.8
2016, 7	288.4
2018, 8	324.9

Chapter Test

CHAPTER
Test Prep
VIDEOS

The Chapter Test Prep Videos include step-by-step solutions to all chapter test exercises. These videos are available in MyLab™ Math, or on this text's YouTube Channel. Refer to the Preface for a link to the YouTube channel.

1. Graph $f(x) = (x - 3)^4 - 2$ using transformations.

2. For the polynomial function $g(x) = 2x^3 + 5x^2 - 28x - 15$,
  (a) Determine the maximum number of real zeros that the function may have.
  (b) List the potential rational zeros.
  (c) Determine the real zeros of g. Factor g over the reals.
  (d) Find the x- and y-intercepts of the graph of g.
  (e) Determine whether the graph crosses or touches the x-axis at each x-intercept.
  (f) Find the power function that the graph of g resembles for large values of $|x|$.
  (g) Put all the information together to obtain the graph of g.

3. Find the complex zeros of $f(x) = x^3 - 4x^2 + 25x - 100$.

4. Solve $3x^3 + 2x - 1 = 8x^2 - 4$ in the complex number system.

In Problems 5 and 6, find the domain of each function. Find any horizontal, vertical, or oblique asymptotes.

5. $g(x) = \dfrac{2x^2 - 14x + 24}{x^2 + 6x - 40}$

6. $r(x) = \dfrac{x^2 + 2x - 3}{x + 1}$

7. Graph the function in Problem 6. Label all intercepts, vertical asymptotes, horizontal asymptotes, and oblique asymptotes.

In Problems 8 and 9, write a function that meets the given conditions.

8. Fourth-degree polynomial with real coefficients; zeros: $-2, 0, 3 + i$.

9. Rational function; asymptotes: $y = 2, x = 4$; domain: $\{x | x \neq 4, x \neq 9\}$

10. Use the Intermediate Value Theorem to show that the function $f(x) = -2x^2 - 3x + 8$ has at least one real zero on the interval $[0, 4]$.

11. Solve: $\dfrac{x + 2}{x - 3} < 2$

12. Solve: $x^3 + 7x^2 \leq 2x^2 - 6x$

Cumulative Review

1. Find the distance between the points $P = (1, 3)$ and $Q = (-4, 2)$.

2. Solve the inequality $x^2 \geq x$ and graph the solution set.

3. Solve the inequality $x^2 - 3x < 4$ and graph the solution set.

4. Find a linear function with slope -3 that contains the point $(-1, 4)$. Graph the function.

5. Find the equation of the line parallel to the line $y = 2x + 1$ and containing the point $(3, 5)$. Express your answer in slope-intercept form, and graph the line.

6. Graph the equation $y = x^3$.

7. Does the relation $\{(3, 6), (1, 3), (2, 5), (3, 8)\}$ represent a function? Why or why not?

8. Solve the equation $x^3 - 6x^2 + 8x = 0$.

9. Solve the inequality $3x + 2 \leq 5x - 1$ and graph the solution set.

10. Find the center and the radius of the circle
$$x^2 + 4x + y^2 - 2y - 4 = 0$$
Graph the circle.

11. For the equation $y = x^3 - 9x$, determine the intercepts and test for symmetry.

12. Find an equation of the line perpendicular to $3x - 2y = 7$ that contains the point $(1, 5)$.

13. Is the following the graph of a function? Why or why not?

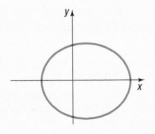

14. For the function $f(x) = x^2 + 5x - 2$, find
   (a) $f(3)$
   (b) $f(-x)$
   (c) $-f(x)$
   (d) $f(3x)$
   (e) $\dfrac{f(x + h) - f(x)}{h}$ $h \neq 0$

15. Given the function
$$f(x) = \frac{x + 5}{x - 1}$$
   (a) What is the domain of f?
   (b) Is the point $(2, 6)$ on the graph of f?
   (c) If $x = 3$, what is $f(x)$? What point is on the graph of f?
   (d) If $f(x) = 9$, what is x? What point is on the graph of f?
   (e) Is f a polynomial or a rational function?

16. Graph the function $f(x) = -3x + 7$.

17. Graph $f(x) = 2x^2 - 4x + 1$ by determining whether its graph is concave up or concave down and by finding its vertex, axis of symmetry, y-intercept, and x-intercepts, if any.

18. Find the average rate of change of $f(x) = x^2 + 3x + 1$ from 1 to 2. Use this result to find the equation of the secant line containing $(1, f(1))$ and $(2, f(2))$.

19. In parts (a) to (f), use the following graph.

   (a) Find the intercepts.
   (b) Based on the graph, tell whether the graph is symmetric with respect to the x-axis, the y-axis, and/or the origin.
   (c) Based on the graph, tell whether the function is even, odd, or neither.
   (d) List the intervals on which f is increasing. List the intervals on which f is decreasing.
   (e) List the numbers, if any, at which f has a local maximum. What are the local maximum values?
   (f) List the numbers, if any, at which f has a local minimum. What are the local minimum values?

20. Determine algebraically whether the function
$$f(x) = \frac{5x}{x^2 - 9}$$
is even, odd, or neither.

21. For the function $f(x) = \begin{cases} 2x + 1 & \text{if } -3 < x < 2 \\ -3x + 4 & \text{if } x \geq 2 \end{cases}$
   (a) Find the domain of f.
   (b) Locate any intercepts.
   (c) Graph the function.
   (d) Based on the graph, find the range.

22. Graph the function $f(x) = -3(x + 1)^2 + 5$ using transformations.

23. Suppose that $f(x) = x^2 - 5x + 1$ and $g(x) = -4x - 7$.
   (a) Find $f + g$ and state its domain.
   (b) Find $\dfrac{f}{g}$ and state its domain.

24. **Demand Equation** The price p (in dollars) and the quantity x sold of a certain product obey the demand equation
$$p = -\frac{1}{10}x + 150, \quad 0 \leq x \leq 1500$$
   (a) Express the revenue R as a function of x.
   (b) What is the revenue if 100 units are sold?
   (c) What quantity x maximizes revenue? What is the maximum revenue?
   (d) What price should the company charge to maximize revenue?

Chapter Projects

Format Trendline panel

Format Trendline ▾ ✕

TRENDLINE OPTIONS ▾

◁ TRENDLINE OPTIONS

○ Exponential

⊙ Linear

○ Logarithmic

○ Polynomial   Order 2

○ Power

○ Moving Average   Period 2

Trendline Name

⊙ Automatic   Linear (Series1)

○ Custom

Forecast

Forward   0.0   period

Backward   0.0   period

☐ Set Intercept   0.0

☑ Display Equation on chart

☐ Display R-squared value on chart

◀   ▶

Figure 52

Internet-based Project

I. Length of Day Go to http://en.wikipedia.org/wiki/Latitude and read about latitude through the subhead "Preliminaries." Now go to http://www.orchidculture.com/COD/daylength.html.

1. For a particular day of the year, record in a table the length of day for the equator (0°N), 5°N, 10°N, . . . , 60°N. Enter the data into an Excel spreadsheet, TI-graphing calculator, or some other spreadsheet capable of finding linear, quadratic, and cubic functions of best fit.

2. Draw a scatter diagram of the data with latitude as the independent variable and length of day as the dependent variable using Excel, a TI-graphing calculator, or some other spreadsheet. The Chapter 4 project describes how to draw a scatter diagram in Excel.

3. Determine the linear function of best fit. Graph the linear function of best fit on the scatter diagram. To do this in Excel, right click on any data point in the scatter diagram. Now click the Add Trendline . . . menu. Under Trendline Options, select the Linear radio button and select Display Equation on Chart. See Figure 52. Move the Trendline Options window off to the side and you will see the linear function of best fit displayed on the scatter diagram. Do you think the function accurately describes the relation between latitude and length of day?

4. Determine the quadratic function of best fit. Graph the quadratic function of best fit on the scatter diagram. To do this in Excel, right-click on any data point in the scatter diagram. Now click the Add Trendline . . . menu. Under Trendline Options, select the Polynomial radio button with Order set to 2. Select Display Equation on chart. Move the Trendline Options window off to the side and you will see the quadratic function of best fit displayed on the scatter diagram. Do you think the function accurately describes the relation between latitude and length of day?

5. Determine the cubic function of best fit. Graph the cubic function of best fit on the scatter diagram. To do this in Excel right-click on any data point in the scatter diagram. Now click the Add Trendline . . . menu. Under Trendline Options, select the Polynomial radio button with Order set to 3. Select Display Equation on chart. Move the Trendline Options window off to the side and you will see the cubic function of best fit displayed on the scatter diagram. Do you think the function accurately describes the relation between latitude and length of day?

6. Which of the three models seems to fit the data best? Explain your reasoning.

7. Use your model to predict the hours of daylight on the day you selected for Chicago (41.85 degrees north latitude). Go to the Old Farmer's Almanac or another website to determine the hours of daylight in Chicago for the day you selected. How do the two compare?

Citation: Excel © 2018 Microsoft Corporation. Used with permission from Microsoft.

The following project is available at the Instructor's Resource Center (IRC):

II. Theory of Equations The coefficients of a polynomial function can be found if its zeros are known, which is an advantage of using polynomials in modeling.

6

Exponential and Logarithmic Functions

Depreciation of Cars

You are ready to buy that first new car. You know that cars lose value over time due to depreciation and that different cars have different rates of depreciation. So you will research the depreciation rates for the cars you are thinking of buying. After all, for cars that sell for about the same price, the lower the depreciation rate, the more the car will be worth each year.

—See the Internet-based Chapter Project I—

Outline

⬅ A Look Back

Until now, our study of functions has concentrated on polynomial and rational functions. These functions belong to the class of **algebraic functions**—that is, functions that can be expressed in terms of sums, differences, products, quotients, powers, or roots of polynomials. Functions that are not algebraic are termed **transcendental**. That is, they transcend, or go beyond, algebraic functions.

A Look Ahead ➡

In this chapter, we study two transcendental functions: the exponential function and the logarithmic function. These functions occur frequently in a wide variety of applications, such as biology, chemistry, economics, and psychology.

The chapter begins with a discussion of composite, one-to-one, and inverse functions—concepts that are needed to explain the relationship between exponential and logarithmic functions.

6.1 Composite Functions

PREPARING FOR THIS SECTION *Before getting started, review the following:*

- Find the Value of a Function (Section 3.1, pp. 207–209)
- Domain of a Function (Section 3.1, pp. 211–213)

Now Work the 'Are You Prepared?' problems on page 419.

OBJECTIVES 1 Form a Composite Function (p. 415)
2 Find the Domain of a Composite Function (p. 416)

1 Form a Composite Function

Suppose that an oil tanker is leaking oil and you want to determine the area of the circular oil patch around the ship. See Figure 1. It is determined that the radius of the circular patch of oil around the ship is increasing at a rate of 3 feet per minute. Then the radius r of the oil patch at any time t, in minutes, is given by $r(t) = 3t$. So after 20 minutes, the radius of the oil patch is $r(20) = 3(20) = 60$ feet.

The area A of a circle is a function of the radius r given by $A(r) = \pi r^2$. The area of the circular patch of oil after 20 minutes is $A(60) = \pi(60)^2 = 3600\pi$ square feet. Note that $60 = r(20)$, so $A(60) = A(r(20))$. The argument of the function A is the output of the function r.

In general, the area of the oil patch can be expressed as a function of time t by evaluating $A(r(t))$ and obtaining $A(r(t)) = A(3t) = \pi(3t)^2 = 9\pi t^2$. The function $A(r(t))$ is a special type of function called a *composite function*.

As another example, consider the function $y = (2x + 3)^2$. Let $y = f(u) = u^2$ and $u = g(x) = 2x + 3$. Then by a substitution process, the original function is obtained as follows: $y = f(u) = f(g(x)) = (2x + 3)^2$.

In general, suppose that f and g are two functions and that x is a number in the domain of g. Evaluating g at x yields $g(x)$. If $g(x)$ is in the domain of f, then evaluating f at $g(x)$ yields the expression $f(g(x))$. The correspondence from x to $f(g(x))$ is called a *composite function* $f \circ g$.

Figure 1

DEFINITION Composite Function

Given two functions f and g, the **composite function**, denoted by $f \circ g$ (read as "f composed with g"), is defined by

$$(f \circ g)(x) = f(g(x))$$

The domain of $f \circ g$ is the set of all numbers x in the domain of g for which $g(x)$ is in the domain of f.

Look carefully at Figure 2. Only those values of x in the domain of g for which $g(x)$ is in the domain of f can be in the domain of $f \circ g$. The reason is if $g(x)$ is not in the domain of f, then $f(g(x))$ is not defined. Because of this, the domain of $f \circ g$ is a subset of the domain of g; the range of $f \circ g$ is a subset of the range of f.

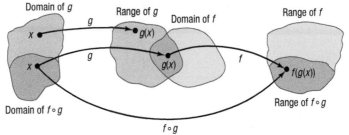

Figure 2

Figure 3 provides a second illustration of the definition. Here x is the input to the function g, yielding $g(x)$. Then $g(x)$ is the input to the function f, yielding $f(g(x))$. Note that the "inside" function g in $f(g(x))$ is "processed" first.

Figure 3

EXAMPLE 1

Evaluating a Composite Function

Suppose that $f(x) = 2x^2 - 3$ and $g(x) = 4x$. Find:

(a) $(f \circ g)(1)$ (b) $(g \circ f)(1)$ (c) $(f \circ f)(-2)$ (d) $(g \circ g)(-1)$

Solution

(a) $(f \circ g)(1) = f(g(1)) = f(4) = 2 \cdot 4^2 - 3 = 29$

$$g(x) = 4x \quad f(x) = 2x^2 - 3$$
$$g(1) = 4$$

(b) $(g \circ f)(1) = g(f(1)) = g(-1) = 4 \cdot (-1) = -4$

$$f(x) = 2x^2 - 3 \quad g(x) = 4x$$
$$f(1) = -1$$

(c) $(f \circ f)(-2) = f(f(-2)) = f(5) = 2 \cdot 5^2 - 3 = 47$

$$f(-2) = 2(-2)^2 - 3 = 5$$

(d) $(g \circ g)(-1) = g(g(-1)) = g(-4) = 4 \cdot (-4) = -16$

$$g(-1) = -4$$

COMMENT Graphing utilities can evaluate composite functions.* Using a TI-84 Plus C graphing calculator, let $Y_1 = f(x) = 2x^2 - 3$ and $Y_2 = g(x) = 4x$, and find $(f \circ g)(1)$ as shown in Figure 4(a). Using Desmos, find $(f \circ g)(1)$ as shown in Figure 4(b). Note that these give the result obtained in Example 1(a). ∎

NORMAL FLOAT AUTO REAL RADIAN MP

$Y_1(Y_2(1))$

          29

(a) TI-84 Plus C

1. $f(x) = 2x^2 - 3$
2. $g(x) = 4x$
3. $f(g(1))$ = 29

(b) Desmos

Figure 4

-**Now Work** PROBLEM 13

2 Find the Domain of a Composite Function

EXAMPLE 2

Finding a Composite Function and Its Domain

Suppose that $f(x) = x^2 + 3x - 1$ and $g(x) = 2x + 3$.

Find: (a) $f \circ g$ (b) $g \circ f$

Then find the domain of each composite function.

Solution

The domain of f and the domain of g are the set of all real numbers.

(a) $(f \circ g)(x) = f(g(x)) = f(2x + 3) = (2x + 3)^2 + 3(2x + 3) - 1$

$$g(x) = 2x + 3 \quad f(x) = x^2 + 3x - 1$$

$$= 4x^2 + 12x + 9 + 6x + 9 - 1 = 4x^2 + 18x + 17$$

Because the domains of both f and g are the set of all real numbers, the domain of $f \circ g$ is the set of all real numbers.

*Consult your user's manual for the appropriate keystrokes.

(b) $(g \circ f)(x) = g(f(x)) = g(x^2 + 3x - 1) = 2(x^2 + 3x - 1) + 3$

$$f(x) = x^2 + 3x - 1 \quad g(x) = 2x + 3$$

$$= 2x^2 + 6x - 2 + 3 = 2x^2 + 6x + 1$$

Because the domains of both f and g are the set of all real numbers, the domain of $g \circ f$ is the set of all real numbers.

Example 2 illustrates that, in general, $f \circ g \neq g \circ f$. Sometimes $f \circ g$ does equal $g \circ f$, as we shall see in Example 5.

Look back at Figure 2 on page 415.

In determining the domain of the composite function $(f \circ g)(x) = f(g(x))$, keep the following two thoughts in mind about the input x.

- Any x not in the domain of g must be excluded.
- Any x for which $g(x)$ is not in the domain of f must be excluded.

EXAMPLE 3 **Finding the Domain of** $f \circ g$

Find the domain of $f \circ g$ if $f(x) = \dfrac{1}{x + 2}$ and $g(x) = \dfrac{4}{x - 1}$.

Solution For $(f \circ g)(x) = f(g(x))$, first note that the domain of g is $\{x \mid x \neq 1\}$, so 1 is excluded from the domain of $f \circ g$. Next note that the domain of f is $\{x \mid x \neq -2\}$, which means that $g(x)$ cannot equal -2. Solve the equation $g(x) = -2$ to determine what additional value(s) of x to exclude.

$$\frac{4}{x - 1} = -2 \qquad\qquad g(x) = -2$$

$$4 = -2(x - 1) \quad \text{Multiply both sides by } x - 1.$$

$$4 = -2x + 2$$

$$2x = -2$$

$$x = -1$$

Also exclude -1 from the domain of $f \circ g$.

The domain of $f \circ g$ is $\{x \mid x \neq -1, x \neq 1\}$.

✓ **Check:** For $x = 1$, $g(x) = \dfrac{4}{x - 1}$ is not defined, so $(f \circ g)(1) = f(g(1))$ is not defined.

For $x = -1$, $g(-1) = -2$, and $(f \circ g)(-1) = f(g(-1)) = f(-2)$ is not defined.

EXAMPLE 4 **Finding a Composite Function and Its Domain**

Suppose $f(x) = \dfrac{1}{x + 2}$ and $g(x) = \dfrac{4}{x - 1}$.

Find: (a) $f \circ g$ (b) $f \circ f$

Then find the domain of each composite function.

Solution The domain of f is $\{x \mid x \neq -2\}$ and the domain of g is $\{x \mid x \neq 1\}$.

(a) $(f \circ g)(x) = f(g(x)) = f\left(\dfrac{4}{x - 1}\right) = \dfrac{1}{\dfrac{4}{x - 1} + 2} = \dfrac{x - 1}{4 + 2(x - 1)} = \dfrac{x - 1}{2x + 2} = \dfrac{x - 1}{2(x + 1)}$

$$g(x) = \frac{4}{x - 1} \quad f(x) = \frac{1}{x + 2} \qquad \text{Multiply by } \frac{x - 1}{x - 1}.$$

(continued)

In Example 3, the domain of $f \circ g$ was found to be $\{x \mid x \neq -1, x \neq 1\}$.

The domain of $f \circ g$ also can be found by first looking at the domain of g: $\{x \mid x \neq 1\}$. Exclude 1 from the domain of $f \circ g$ as a result. Then look at $f \circ g$ and note that x cannot equal -1, because $x = -1$ results in division by 0. So exclude -1 from the domain of $f \circ g$. Therefore, the domain of $f \circ g$ is $\{x \mid x \neq -1, x \neq 1\}$.

(b) $(f \circ f)(x) = f(f(x)) = f\left(\dfrac{1}{x+2}\right) = \dfrac{1}{\dfrac{1}{x+2}+2} = \dfrac{x+2}{1+2(x+2)} = \dfrac{x+2}{2x+5}$

$$\uparrow \qquad\qquad\qquad \uparrow \qquad\qquad\qquad \uparrow$$

$$f(x) = \dfrac{1}{x+2} \qquad f(x) = \dfrac{1}{x+2} \qquad \text{Multiply by } \dfrac{x+2}{x+2}.$$

The domain of $f \circ f$ consists of all values of x in the domain of f, $\{x \mid x \neq -2\}$, for which $f(x) = \dfrac{1}{x+2} \neq -2$. To find other numbers x to exclude, solve the equation

$$\dfrac{1}{x+2} = -2$$

$$1 = -2(x+2)$$

$$1 = -2x - 4$$

$$2x = -5$$

$$x = -\dfrac{5}{2}$$

So $f(x) \neq -2$ if $x \neq -\dfrac{5}{2}$. The domain of $f \circ f$ is $\left\{x \mid x \neq -\dfrac{5}{2}, x \neq -2\right\}$.

The domain of $f \circ f$ also can be found by noting that -2 is not in the domain of f and so is not in the domain of $f \circ f$. Then, looking at $f \circ f$, note that x cannot equal $-\dfrac{5}{2}$. Do you see why? Therefore, the domain of $f \circ f$ is $\left\{x \mid x \neq -\dfrac{5}{2}, x \neq -2\right\}$.

Now Work PROBLEMS 27 AND 29

EXAMPLE 5

Showing That Two Composite Functions Are Equal

If $f(x) = 3x - 4$ and $g(x) = \dfrac{x+4}{3}$, show that

$$(f \circ g)(x) = (g \circ f)(x) = x$$

for every x in the domain of $f \circ g$ and $g \circ f$.

Solution

$$(f \circ g)(x) = f(g(x)) = f\left(\dfrac{x+4}{3}\right) = 3\left(\dfrac{x+4}{3}\right) - 4 = x + 4 - 4 = x$$

$$\uparrow \qquad\qquad\qquad \uparrow$$

$$g(x) = \dfrac{x+4}{3} \qquad f(x) = 3x - 4$$

$$(g \circ f)(x) = g(f(x)) = g(3x - 4) = \dfrac{(3x-4)+4}{3} = \dfrac{3x}{3} = x$$

$$\uparrow \qquad\qquad\qquad \uparrow$$

$$f(x) = 3x - 4 \qquad g(x) = \dfrac{x+4}{3}$$

We conclude that $(f \circ g)(x) = (g \circ f)(x) = x$, x any real number.

Seeing the Concept

Using a graphing calculator, let

$Y_1 = f(x) = 3x - 4$

$Y_2 = g(x) = \dfrac{1}{3}(x+4)$

$Y_3 = f \circ g, Y_4 = g \circ f$

Using the viewing window $-3 \leq x \leq 3$, $-2 \leq y \leq 2$, graph only Y_3 and Y_4. What do you see? TRACE to verify that $Y_3 = Y_4$.

In Section 6.2, we shall see that there is an important relationship between functions f and g for which $(f \circ g)(x) = (g \circ f)(x) = x$.

─── **Now Work** PROBLEM 39

Calculus Application

Some techniques in calculus require the ability to determine the components of a composite function. For example, the function $H(x) = \sqrt{x + 1}$ is the composition of the functions f and g, where $f(x) = \sqrt{x}$ and $g(x) = x + 1$, because $H(x) = (f \circ g)(x) = f(g(x)) = f(x + 1) = \sqrt{x + 1}$.

EXAMPLE 6 **Finding the Components of a Composite Function**

Find functions f and g so that $f \circ g = H$ when $H(x) = (x^2 + 1)^{50}$.

Solution The function H raises the expression $x^2 + 1$ to the power 50. A natural way to decompose H is to raise the function $g(x) = x^2 + 1$ to the power 50. Let $f(x) = x^{50}$ and $g(x) = x^2 + 1$. Then

$$f(g(x)) = f(x^2 + 1)$$
$$= (x^2 + 1)^{50}$$

$$(f \circ g)(x) = f(g(x)) = f(x^2 + 1) = (x^2 + 1)^{50} = H(x)$$

$$H(x) = (x^2 + 1)^{50}$$

See Figure 5.

Figure 5

Other functions f and g may be found for which $f \circ g = H$ in Example 6. For instance, if $f(x) = x^2$ and $g(x) = (x^2 + 1)^{25}$, then

$$(f \circ g)(x) = f(g(x)) = f((x^2 + 1)^{25}) = [(x^2 + 1)^{25}]^2 = (x^2 + 1)^{50}$$

Although the functions f and g found as a solution to Example 6 are not unique, there is usually a "natural" selection for f and g that comes to mind first.

EXAMPLE 7 **Finding the Components of a Composite Function**

Find functions f and g so that $f \circ g = H$ when $H(x) = \dfrac{1}{x + 1}$.

Solution Here H is the reciprocal of $g(x) = x + 1$. Let $f(x) = \dfrac{1}{x}$ and $g(x) = x + 1$. Then

$$(f \circ g)(x) = f(g(x)) = f(x + 1) = \frac{1}{x + 1} = H(x)$$

─── **Now Work** PROBLEM 47

6.1 Assess Your Understanding

'Are You Prepared?' *Answers are given at the end of these exercises. If you get a wrong answer, read the pages listed in* red.

1. Find $f(3)$ if $f(x) = -4x^2 + 5x$. (pp. 207–209)

2. Find $f(3x)$ if $f(x) = 4 - 2x^2$. (pp. 207–209)

3. Find the domain of the function $f(x) = \dfrac{x^2 - 1}{x^2 - 25}$. (pp. 211–213)

Concepts and Vocabulary

4. Given two functions f and g, the _____, _____, denoted $f \circ g$, is defined by $(f \circ g)(x) =$ _____

5. *True or False* If $f(x) = x^2$ and $g(x) = \sqrt{x + 9}$, then $(f \circ g)(4) = 5$.

6. *Multiple Choice* If $f(x) = \sqrt{x + 2}$ and $g(x) = \dfrac{3}{x}$, then $(f \circ g)(x)$ equals

(a) $\dfrac{3}{\sqrt{x + 2}}$ 　　　　　(b) $\dfrac{3}{\sqrt{x}} + 2$

(c) $\sqrt{\dfrac{3}{x} + 2}$ 　　　　　(d) $\sqrt{\dfrac{3}{x + 2}}$

7. *Multiple Choice* If $H = f \circ g$ and $H(x) = \sqrt{25 - 4x^2}$, which of the following cannot be the component functions f and g?

(a) $f(x) = \sqrt{25 - x^2}; g(x) = 4x$

(b) $f(x) = \sqrt{x}; g(x) = 25 - 4x^2$

(c) $f(x) = \sqrt{25 - x}; g(x) = 4x^2$

(d) $f(x) = \sqrt{25 - 4x}; g(x) = x^2$

8. *True or False* The domain of the composite function $(f \circ g)(x)$ is the same as the domain of $g(x)$.

Skill Building

In Problems 9 and 10, evaluate each expression using the values given in the table.

9.

x	−3	−2	−1	0	1	2	3
f(x)	−7	−5	−3	−1	3	5	7
g(x)	8	3	0	−1	0	3	8

(a) $(f \circ g)(1)$ 　　　　(b) $(f \circ g)(-1)$
(c) $(g \circ f)(-1)$ 　　　　(d) $(g \circ f)(0)$
(e) $(g \circ g)(-2)$ 　　　　(f) $(f \circ f)(-1)$

10.

x	−3	−2	−1	0	1	2	3
f(x)	11	9	7	5	3	1	−1
g(x)	−8	−3	0	1	0	−3	−8

(a) $(f \circ g)(1)$ 　　　　(b) $(f \circ g)(2)$
(c) $(g \circ f)(2)$ 　　　　(d) $(g \circ f)(3)$
(e) $(g \circ g)(1)$ 　　　　(f) $(f \circ f)(3)$

In Problems 11 and 12, evaluate each expression using the graphs of $y = f(x)$ and $y = g(x)$ shown in the figure.

11. (a) $(g \circ f)(-1)$ 　　　　**(b)** $(g \circ f)(0)$
　　(c) $(f \circ g)(-1)$ 　　　　**(d)** $(f \circ g)(4)$

12. (a) $(g \circ f)(1)$ 　　　　**(b)** $(g \circ f)(5)$
　　(c) $(f \circ g)(0)$ 　　　　**(d)** $(f \circ g)(2)$

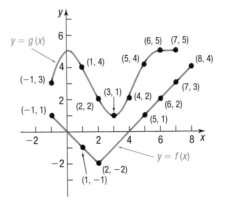

In Problems 13–22, for the given functions f and g, find:

　(a) $(f \circ g)(4)$ 　　(b) $(g \circ f)(2)$ 　　(c) $(f \circ f)(1)$ 　　(d) $(g \circ g)(0)$

13. $f(x) = 2x$; 　$g(x) = 3x^2 + 1$

14. $f(x) = 3x + 2$; 　$g(x) = 2x^2 - 1$

15. $f(x) = 8x^2 - 3$; 　$g(x) = 3 - \dfrac{1}{2}x^2$

16. $f(x) = 2x^2$; 　$g(x) = 1 - 3x^2$

17. $f(x) = \sqrt{x}$; 　$g(x) = 5x$

18. $f(x) = \sqrt{x + 1}$; 　$g(x) = 3x$

19. $f(x) = |x|$; 　$g(x) = \dfrac{1}{x^2 + 9}$

20. $f(x) = |x - 2|$; 　$g(x) = \dfrac{3}{x^2 + 2}$

21. $f(x) = \dfrac{3}{x + 1}$; 　$g(x) = \sqrt[3]{x}$

22. $f(x) = x^{3/2}$; 　$g(x) = \dfrac{2}{x + 1}$

In Problems 23–38, for the given functions f and g, find:

　(a) $f \circ g$ 　　(b) $g \circ f$ 　　(c) $f \circ f$ 　　(d) $g \circ g$

State the domain of each composite function.

23. $f(x) = 2x + 3$; 　$g(x) = 4x$

24. $f(x) = -x$; 　$g(x) = 2x - 4$

25. $f(x) = 3x - 1$; 　$g(x) = x^2$

26. $f(x) = x + 1$; 　$g(x) = x^2 + 4$

27. $f(x) = x^2$; $g(x) = x^2 + 4$

28. $f(x) = x^2 + 1$; $g(x) = 2x^2 + 3$

29. $f(x) = \dfrac{3}{x - 1}$; $g(x) = \dfrac{2}{x}$

30. $f(x) = \dfrac{1}{x + 3}$; $g(x) = -\dfrac{2}{x}$

31. $f(x) = \dfrac{x}{x - 1}$; $g(x) = -\dfrac{4}{x}$

32. $f(x) = \dfrac{x}{x + 3}$; $g(x) = \dfrac{2}{x}$

33. $f(x) = \sqrt{x}$; $g(x) = 2x + 5$

34. $f(x) = \sqrt{x - 2}$; $g(x) = 1 - 2x$

35. $f(x) = x^2 + 7$; $g(x) = \sqrt{x - 7}$

36. $f(x) = x^2 + 4$; $g(x) = \sqrt{x - 2}$

37. $f(x) = \dfrac{x - 5}{x + 1}$; $g(x) = \dfrac{x + 2}{x - 3}$

38. $f(x) = \dfrac{2x - 1}{x - 2}$; $g(x) = \dfrac{x + 4}{2x - 5}$

In Problems 39–46, show that $(f \circ g)(x) = (g \circ f)(x) = x$.

39. $f(x) = 2x$; $g(x) = \dfrac{1}{2}x$

40. $f(x) = 4x$; $g(x) = \dfrac{1}{4}x$

41. $f(x) = x^3$; $g(x) = \sqrt[3]{x}$

42. $f(x) = x + 5$; $g(x) = x - 5$

43. $f(x) = 9x - 6$; $g(x) = \dfrac{1}{9}(x + 6)$

44. $f(x) = 4 - 3x$; $g(x) = \dfrac{1}{3}(4 - x)$

45. $f(x) = ax + b$; $g(x) = \dfrac{1}{a}(x - b)$ $a \neq 0$

46. $f(x) = \dfrac{1}{x}$; $g(x) = \dfrac{1}{x}$

In Problems 47–52, find functions f and g so that $f \circ g = H$.

47. $H(x) = (2x + 3)^4$

48. $H(x) = (1 + x^2)^3$

49. $H(x) = \sqrt{x^2 + 1}$

50. $H(x) = \sqrt{1 - x^2}$

51. $H(x) = |2x + 1|$

52. $H(x) = |2x^2 + 3|$

Applications and Extensions

53. If $f(x) = 2x^3 - 3x^2 + 4x - 1$ and $g(x) = 2$, find $(f \circ g)(x)$ and $(g \circ f)(x)$.

54. If $f(x) = \dfrac{x + 1}{x - 1}$, find $(f \circ f)(x)$.

55. If $f(x) = 2x^2 + 5$ and $g(x) = 3x + a$, find a so that the y-intercept of $f \circ g$ is 23.

56. If $f(x) = 3x^2 - 7$ and $g(x) = 2x + a$, find a so that the y-intercept of $f \circ g$ is 68.

In Problems 57 and 58, use the functions f and g to find:

(a) $f \circ g$ (b) $g \circ f$
(c) *the domain of* $f \circ g$ *and of* $g \circ f$
(d) *the conditions for which* $f \circ g = g \circ f$

57. $f(x) = ax + b$ $g(x) = cx + d$

58. $f(x) = \dfrac{ax + b}{cx + d}$ $g(x) = mx$

59. **Surface Area of a Balloon** The surface area S (in square meters) of a hot-air balloon is given by

$$S(r) = 4\pi r^2$$

where r is the radius of the balloon (in meters). If the radius r is increasing with time t (in seconds) according to the formula $r(t) = \dfrac{2}{3}t^3, t \geq 0$, find the surface area S of the balloon as a function of the time t.

60. **Volume of a Balloon** The volume V (in cubic meters) of the hot-air balloon described in Problem 59 is given by $V(r) = \dfrac{4}{3}\pi r^3$. If the radius r is the same function of t as in Problem 59, find the volume V as a function of the time t.

61. **Automobile Production** The number N of cars produced at a certain factory in one day after t hours of operation is

given by $N(t) = 100t - 5t^2, 0 \leq t \leq 10$. If the cost C (in dollars) of producing N cars is $C(N) = 15{,}000 + 8000N$, find the cost C as a function of the time t of operation of the factory.

62. **Environmental Concerns** The spread of oil leaking from a tanker is in the shape of a circle. If the radius r (in feet) of the spread after t hours is $r(t) = 200\sqrt{t}$, find the area A of the oil slick as a function of the time t.

63. **Production Cost** The price p, in dollars, of a certain product and the quantity x sold follow the demand equation

$$p = -\dfrac{1}{4}x + 100 \quad 0 \leq x \leq 400$$

Suppose that the cost C, in dollars, of producing x units is

$$C = \dfrac{\sqrt{x}}{25} + 600$$

Assuming that all items produced are sold, find the cost C as a function of the price p.

[**Hint:** Solve for x in the demand equation and then form the composite function.]

64. **Cost of a Commodity** The price p, in dollars, of a certain commodity and the quantity x sold follow the demand equation

$$p = -\dfrac{1}{5}x + 200 \quad 0 \leq x \leq 1000$$

Suppose that the cost C, in dollars, of producing x units is

$$C = \dfrac{\sqrt{x}}{10} + 400$$

Assuming that all items produced are sold, find the cost C as a function of the price p.

65. Volume of a Cylinder The volume V of a right circular cylinder of height h and radius r is $V = \pi r^2 h$. If the height is twice the radius, express the volume V as a function of r.

66. Volume of a Cone The volume V of a right circular cone is $V = \frac{1}{3}\pi r^2 h$. If the height is twice the radius, express the volume V as a function of r.

67. Foreign Exchange Traders often buy foreign currency in the hope of making money when the currency's value changes. For example, on April 10, 2018, one U.S. dollar could purchase 0.8101 euro, and one euro could purchase 132.317 yen. Let $f(x)$ represent the number of euros you can buy with x dollars, and let $g(x)$ represent the number of yen you can buy with x euros.
  (a) Find a function that relates dollars to euros.
  (b) Find a function that relates euros to yen.
  (c) Use the results of parts (a) and (b) to find a function that relates dollars to yen. That is, find $(g \circ f)(x)$.
  (d) What is $(g \circ f)(1000)$?

68. Temperature Conversion The function $C(F) = \frac{5}{9}(F - 32)$ converts a temperature in degrees Fahrenheit, F, to a temperature in degrees Celsius, C. The function $K(C) = C + 273$, converts a temperature in degrees Celsius to a temperature in kelvins, K.
  (a) Find a function that converts a temperature in degrees Fahrenheit to a temperature in kelvins.
  (b) Determine 80 degrees Fahrenheit in kelvins.

69. Discounts The manufacturer of a computer is offering two discounts on last year's model computer. The first discount is a $200 rebate and the second discount is 20% off the regular price, p.
  (a) Write a function f that represents the sale price if only the rebate applies.

  (b) Write a function g that represents the sale price if only the 20% discount applies.
  (c) Find $f \circ g$ and $g \circ f$. What does each of these functions represent? Which combination of discounts represents a better deal for the consumer? Why?

70. Taxes Suppose that you work for $15 per hour. Write a function that represents gross salary G as a function of hours worked h. Your employer is required to withhold taxes (federal income tax, Social Security, Medicare) from your paycheck. Suppose your employer withholds 20% of your income for taxes. Write a function that represents net salary N as a function of gross salary G. Find and interpret $N \circ G$.

71. Suppose $f(x) = x^3 + x^2 - 16x - 16$ and $g(x) = x^2 - 4$. Find the zeros of $(f \circ g)(x)$.

72. Suppose $f(x) = 2x^3 - 3x^2 - 8x + 12$ and $g(x) = x + 5$. Find the zeros of $(f \circ g)(x)$.

73. Let $f(x) = ax + b$ and $g(x) = bx + a$, where a and b are integers. If $f(1) = 8$ and $f(g(20)) - g(f(20)) = -14$, find the product of a and b.*

74. *Challenge Problem* If f and g are odd functions, show that the composite function $f \circ g$ is also odd.

75. *Challenge Problem* If f is an odd function and g is an even function, show that the composite functions $f \circ g$ and $g \circ f$ are both even.

76. *Challenge Problem* If $f(x) = x^2 + 5x + c, g(x) = ax + b$, and $(f \circ g)(x) = 4x^2 + 22x + 31$, find a, b, and c.

77. *Challenge Problem* Given three functions f, g, and h, define $(f \circ g \circ h)(x) = f[g(h(x))]$. Find $(f \circ g \circ h)(2)$ if $f(x) = 6x - 7, g(x) = \frac{1}{x}$, and $h(x) = \sqrt{x + 7}$.

78. *Challenge Problem* If $f(x) = \frac{1}{x + 4}, g(x) = \frac{x}{x - 2}$, and $h(x) = \sqrt{x + 3}$, find the domain of $(f \circ g \circ h)(x)$.

* Courtesy of the Joliet Junior College Mathematics Department

Retain Your Knowledge

Problems 79–88 are based on material learned earlier in the course. The purpose of these problems is to keep the material fresh in your mind so that you are better prepared for the final exam.

79. Given $f(x) = 3x + 8$ and $g(x) = x - 5$, find
$$(f + g)(x), (f - g)(x), (f \cdot g)(x), \text{ and } \left(\frac{f}{g}\right)(x)$$
State the domain of each.

80. Find the real zeros of $f(x) = 2x - 5\sqrt{x} + 2$.

81. Find the domain of $R(x) = \frac{x^2 + 6x + 5}{x - 3}$. Find any horizontal, vertical, or oblique asymptotes.

82. For the quadratic function $f(x) = -\frac{1}{3}x^2 + 2x + 5$, find the vertex and the axis of symmetry, and determine whether the graph is concave up or concave down.

83. Solve: $x^2 - 6x - 7 \leq 0$

84. If a right triangle has hypotenuse $c = 2$ and leg $a = 1$, find the length of the other leg b.

85. Find the points of intersection of the graphs of the functions $f(x) = x^2 + 3x + 7$ and $g(x) = -2x + 3$.

86. Find the distance between the points $(-3, 8)$ and $(2, -7)$.

87. Simplify: $\dfrac{\dfrac{3}{x + 1} - \dfrac{3}{c + 1}}{x - c}$

88. Solve: $-x^3(9 - x^2)^{-1/2} + 2x(9 - x^2)^{1/2} = 0$

'Are You Prepared?' Answers

1. -21 **2.** $4 - 18x^2$ **3.** $\{x \mid x \neq -5, x \neq 5\}$

6.2 One-to-One Functions; Inverse Functions

PREPARING FOR THIS SECTION *Before getting started, review the following:*

- Functions (Section 3.1, pp. 203–215)
- Increasing/Decreasing Functions (Section 3.3, pp. 231–232)

- Rational Expressions (Chapter R, Section R.7, pp. 61–69)
- Properties of Rational Functions (Section 5.3, pp. 354–361)

Now Work the 'Are You Prepared?' problems on page 431.

OBJECTIVES 1 Determine Whether a Function Is One-to-One (p. 423)
　　　　　　　 2 Obtain the Graph of the Inverse Function from the Graph of a One-to-One Function (p. 426)
　　　　　　　 3 Verify an Inverse Function (p. 427)
　　　　　　　 4 Find the Inverse of a Function Defined by an Equation (p. 428)

1 Determine Whether a Function Is One-to-One

Section 3.1 presented five ways to represent a function: (1) verbally, (2) with a mapping, (3) as a set of ordered pairs, (4) with a graph, and (5) with an equation. For example, Figures 6 and 7 illustrate two different functions represented as mappings. The function in Figure 6 shows the correspondence between states and their populations (in millions). The function in Figure 7 shows a correspondence between animals and life expectancies (in years).

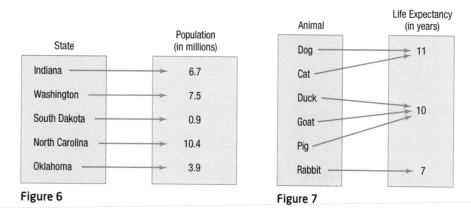

Figure 6   Figure 7

Suppose several people are asked to name a state that has a population of 0.9 million based on the function in Figure 6. Everyone will respond "South Dakota." Now, if the same people are asked to name an animal whose life expectancy is 11 years based on the function in Figure 7, some may respond "dog," while others may respond "cat." What is the difference between the functions in Figures 6 and 7? In Figure 6, no two elements in the domain correspond to the same element in the range. In Figure 7, this is not the case: Different elements in the domain correspond to the same element in the range. Functions such as the one in Figure 6 are given a special name.

DEFINITION One-to-One

A function is **one-to-one** if any two different inputs in the domain correspond to two different outputs in the range. That is, if x_1 and x_2 are two different inputs of a function f, then f is one-to-one if $f(x_1) \neq f(x_2)$.

In Words

A function is not one-to-one if two different inputs correspond to the same output.

Put another way, a function f is one-to-one if no y in the range is the image of more than one x in the domain. A function is not one-to-one if any two (or more) different elements in the domain correspond to the same element in the range. So the function in Figure 7 is not one-to-one because two different elements in the domain, *dog* and *cat*, both correspond to 11 (and also because three different elements in the domain correspond to 10).

Figure 8 illustrates the distinction among one-to-one functions, functions that are not one-to-one, and relations that are not functions.

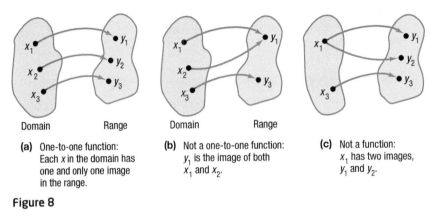

(a) One-to-one function: Each *x* in the domain has one and only one image in the range.

(b) Not a one-to-one function: y_1 is the image of both x_1 and x_2.

(c) Not a function: x_1 has two images, y_1 and y_2.

Figure 8

EXAMPLE 1

Determining Whether a Function Is One-to-One

Determine whether the following functions are one-to-one.

(a) For the following function, the domain represents the ages of five males, and the range represents their HDL (good) cholesterol scores (mg/dL).

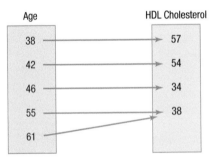

(b) $\{ (-2, 6), (-1, 3), (0, 2), (1, 5), (2, 8) \}$

Solution

(a) The function is not one-to-one because there are two different inputs, 55 and 61, that correspond to the same output, 38.

(b) The function is one-to-one because every distinct input corresponds to a different output.

━━━━ **Now Work** PROBLEMS **13** AND **17**

For functions defined by an equation $y = f(x)$ and for which the graph of f is known, there is a simple test, called the **horizontal-line test**, to determine whether f is one-to-one.

> **THEOREM Horizontal-line Test**
>
> If every horizontal line intersects the graph of a function f in at most one point, then f is one-to-one.

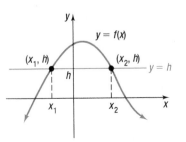

Figure 9
$f(x_1) = f(x_2) = h$ and $x_1 \neq x_2$; f is not a one-to-one function.

The reason why this test works can be seen in Figure 9, where the horizontal line $y = h$ intersects the graph at two distinct points, (x_1, h) and (x_2, h). Since h is the image of both x_1 and x_2 and $x_1 \neq x_2$, f is not one-to-one. Based on Figure 9, we can state the horizontal-line test in another way: If the graph of any horizontal line intersects the graph of a function f at more than one point, then f is not one-to-one.

EXAMPLE 2   **Using the Horizontal-line Test**

For each function, use its graph to determine whether the function is one-to-one.

(a) $f(x) = x^2$ (b) $g(x) = x^3$

Solution   (a) Figure 10(a) illustrates the horizontal-line test for $f(x) = x^2$. The horizontal line $y = 1$ intersects the graph of f twice, at $(1, 1)$ and at $(-1, 1)$, so f is not one-to-one.

(b) Figure 10(b) illustrates the horizontal-line test for $g(x) = x^3$. Because every horizontal line intersects the graph of g exactly once, the function g is one-to-one.

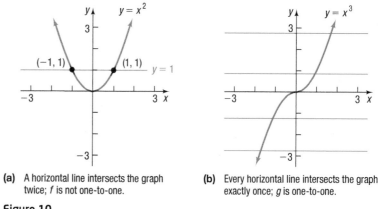

(a) A horizontal line intersects the graph twice; f is not one-to-one.

(b) Every horizontal line intersects the graph exactly once; g is one-to-one.

Figure 10

Now Work PROBLEM 21

Look more closely at the one-to-one function $g(x) = x^3$. This function is an increasing function. Because an increasing (or decreasing) function always has different y-values for unequal x-values, it follows that a function that is increasing (or decreasing) over its domain is also a one-to-one function.

> **THEOREM**
>
> • A function that is increasing on an interval I is a one-to-one function on I.
> • A function that is decreasing on an interval I is a one-to-one function on I.

One-to-one functions $y = f(x)$ have an important property. Corresponding to each x in the domain of f, there is exactly one y in the range because f is a function. And corresponding to each y in the range of f, there is exactly one x in the domain because f is one-to-one. The function represented by the correspondence from the range back to the domain is called the *inverse function of f*.

In Words

Suppose that f is a one-to-one function so that the input 5 corresponds to the output 10. For the inverse function f^{-1}, the input 10 will correspond to the output 5.

> **DEFINITION Inverse Function**
>
> Suppose $y = f(x)$ is a one-to-one function. The correspondence from the range of f to the domain of f is called the **inverse function of f**. The symbol f^{-1} is used to denote the inverse function of f. In other words, if $y = f(x)$ is a one-to-one function, then f has an inverse function f^{-1} and $x = f^{-1}(y)$.

2 Obtain the Graph of the Inverse Function from the Graph of a One-to-One Function

Suppose (a, b) is a point on the graph of a one-to-one function f defined by $y = f(x)$. Then $b = f(a)$. This means that $a = f^{-1}(b)$, so (b, a) is a point on the graph of the inverse function f^{-1}. The relationship between the point (a, b) on f and the point (b, a) on f^{-1} is shown in Figure 11. The line segment with endpoints (a, b) and (b, a) is perpendicular to the line $y = x$ and is bisected by the line $y = x$. (Do you see why?) It follows that the point (b, a) on f^{-1} is the reflection about the line $y = x$ of the point (a, b) on f.

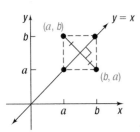

Figure 11

THEOREM

The graph of a one-to-one function f and the graph of its inverse function f^{-1} are symmetric with respect to the line $y = x$.

Figure 12 illustrates the theorem. Once the graph of f is known, the graph of f^{-1} can be obtained by reflecting the graph of f about the line $y = x$.

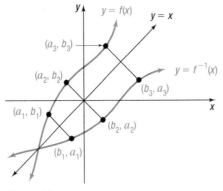

Figure 12

EXAMPLE 3

Graphing an Inverse Function

The graph in Figure 13(a) shows a one-to-one function $y = f(x)$. Draw the graph of its inverse.

Solution

First add the graph of $y = x$ to Figure 13(a). Since the points $(-2, -1)$, $(-1, 0)$, and $(2, 1)$ are on the graph of f, the points $(-1, -2)$, $(0, -1)$, and $(1, 2)$ must be on the graph of f^{-1}. Keeping in mind that the graph of f^{-1} is the reflection about the line $y = x$ of the graph of f, graph f^{-1}. See Figure 13(b).

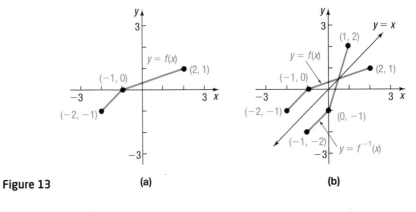

Figure 13   (a)   (b)

Now Work PROBLEM 27

3 Verify an Inverse Function

Domain of f   **Range of f**

x

f

f^{-1}

Range of f⁻¹   **Domain of f⁻¹**

Figure 14

WARNING Be careful! f^{-1} is a symbol for the inverse function of f. The -1 used in f^{-1} is not an exponent. That is, f^{-1} does not mean the reciprocal of f; $f^{-1}(x)$ is not equal to $\dfrac{1}{f(x)}$. ∎

Suppose f is a one-to-one function. Then f has an inverse function f^{-1}. Figure 14 shows the relationship between the domain and range of f and the domain and range of f^{-1}. As Figure 14 illustrates,

> • Domain of f = Range of f^{-1} • Range of f = Domain of f^{-1}

Look again at Figure 14 to visualize the relationship. Starting with x, applying f, and then applying f^{-1} gets x back again. Starting with x, applying f^{-1}, and then applying f gets the number x back again. To put it simply, what f does, f^{-1} undoes, and vice versa. See the illustration that follows.

$$\boxed{\text{Input } x \text{ from domain of } f} \xrightarrow{\text{Apply } f} \boxed{f(x)} \xrightarrow{\text{Apply } f^{-1}} \boxed{f^{-1}(f(x)) = x}$$

$$\boxed{\text{Input } x \text{ from domain of } f^{-1}} \xrightarrow{\text{Apply } f^{-1}} \boxed{f^{-1}(x)} \xrightarrow{\text{Apply } f} \boxed{f(f^{-1}(x)) = x}$$

In other words,

> • $f^{-1}(f(x)) = x$ where x is in the domain of f
>
> • $f(f^{-1}(x)) = x$ where x is in the domain of f^{-1}

Consider the function $f(x) = 2x$, which multiplies the argument x by 2. The inverse function f^{-1} undoes whatever f does. So the inverse function of f is $f^{-1}(x) = \dfrac{1}{2}x$, which divides the argument by 2. For example, $f(3) = 2 \cdot 3 = 6$ and $f^{-1}(6) = \dfrac{1}{2} \cdot 6 = 3$, so f^{-1} undoes f. This is verified by showing that $f^{-1}(f(x)) = f^{-1}(2x) = \dfrac{1}{2} \cdot 2x = x$ and $f(f^{-1}(x)) = f\left(\dfrac{1}{2}x\right) = 2 \cdot \dfrac{1}{2}x = x$.

See Figure 15.

x   f   f⁻¹ $f(x) = 2x$

$f^{-1}(2x) = \frac{1}{2} \cdot 2x = x$

Figure 15

EXAMPLE 4   **Verifying Inverse Functions**

(a) Verify that the inverse of $g(x) = x^3$ is $g^{-1}(x) = \sqrt[3]{x}$.

(b) Verify that the inverse of $f(x) = 2x + 3$ is $f^{-1}(x) = \dfrac{1}{2}(x - 3)$.

Solution   (a) $g^{-1}(g(x)) = g^{-1}(x^3) = \sqrt[3]{x^3} = x$ for all x in the domain of g

$\quad\quad\quad g(g^{-1}(x)) = g(\sqrt[3]{x}) = (\sqrt[3]{x})^3 = x$ for all x in the domain of g^{-1}

(b) $f^{-1}(f(x)) = f^{-1}(2x + 3) = \dfrac{1}{2}[(2x + 3) - 3] = \dfrac{1}{2} \cdot 2x = x$ for all x in the domain of f

$\quad\quad f(f^{-1}(x)) = f\left(\dfrac{1}{2}(x - 3)\right) = 2 \cdot \dfrac{1}{2}(x - 3) + 3 = x$ for all x in the domain of f^{-1}

| EXAMPLE 5 | Verifying Inverse Functions |

Verify that the inverse of $f(x) = \dfrac{1}{x-1}$ is $f^{-1}(x) = \dfrac{1}{x} + 1$. For what values of x is $f^{-1}(f(x)) = x$? For what values of x is $f(f^{-1}(x)) = x$?

Solution The domain of f is $\{x \mid x \neq 1\}$ and the domain of f^{-1} is $\{x \mid x \neq 0\}$. Now

$$f^{-1}(f(x)) = f^{-1}\left(\frac{1}{x-1}\right) = \frac{1}{\dfrac{1}{x-1}} + 1 = x - 1 + 1 = x \quad \text{provided } x \neq 1$$

$$f(f^{-1}(x)) = f\left(\frac{1}{x} + 1\right) = \frac{1}{\left(\dfrac{1}{x} + 1\right) - 1} = \frac{1}{\dfrac{1}{x}} = x \quad \text{provided } x \neq 0$$

⌐

━━━ **Now Work** PROBLEMS 33 AND 37

4 Find the Inverse of a Function Defined by an Equation

The fact that the graphs of a one-to-one function f and its inverse function f^{-1} are symmetric with respect to the line $y = x$ tells us more. It says that we can obtain f^{-1} by interchanging the roles of x and y in f. If f is defined by the equation

$$y = f(x)$$

then f^{-1} is defined by the equation

┌─────────────────────┐
│ **Need to Review?** │
│ Implicit functions are │
│ discussed in Section 3.1, │
│ p. 210. │
└─────────────────────┘

$$x = f(y)$$

The equation $x = f(y)$ defines f^{-1} *implicitly*. If we can solve this equation for y, we will have the *explicit* form of f^{-1}, that is,

$$y = f^{-1}(x)$$

Let's use this procedure to find the inverse of $f(x) = 2x + 3$. Because f is a linear function and is increasing, f is one-to-one and so has an inverse function.

| EXAMPLE 6 | Finding the Inverse of a Function Defined by an Equation |

Find the inverse of $f(x) = 2x + 3$. Graph f and f^{-1} on the same coordinate axes.

Step-by-Step Solution

Replace $f(x)$ with y in $f(x) = 2x + 3$ and obtain $y = 2x + 3$. Now interchange the variables x and y to obtain

Step 1 Replace $f(x)$ with y. Then interchange the variables x and y. This equation defines the inverse function f^{-1} implicitly.

$$x = 2y + 3$$

This equation defines the inverse function f^{-1} implicitly.

Step 2 If possible, solve the implicit equation for y in terms of x to obtain the explicit form of f^{-1}, $y = f^{-1}(x)$.

To find the explicit form of the inverse, solve $x = 2y + 3$ for y.

$$x = 2y + 3$$

$$2y + 3 = x$$

$$2y = x - 3 \qquad \text{Subtract 3 from both sides.}$$

$$y = \frac{1}{2}(x - 3) \qquad \text{Multiply both sides by } \frac{1}{2}.$$

The explicit form of the inverse function f^{-1} is

$$f^{-1}(x) = \frac{1}{2}(x - 3)$$

Step 3 Check the result by showing that $f^{-1}(f(x)) = x$ and $f(f^{-1}(x)) = x$.

We verified that f and f^{-1} are inverses in Example 4(b).

The graphs of $f(x) = 2x + 3$ and its inverse $f^{-1}(x) = \dfrac{1}{2}(x - 3)$ are shown in Figure 16. Note the symmetry of the graphs with respect to the line $y = x$.

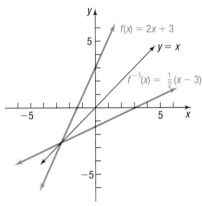

Figure 16

Procedure for Finding the Inverse of a One-to-One Function

STEP 1: In $y = f(x)$, interchange the variables x and y to obtain

$$x = f(y)$$

This equation defines the inverse function f^{-1} implicitly.

STEP 2: If possible, solve the implicit equation for y in terms of x to obtain the explicit form of f^{-1}:

$$y = f^{-1}(x)$$

STEP 3: Check the result by showing that

$$f^{-1}(f(x)) = x \quad \text{and} \quad f(f^{-1}(x)) = x$$

EXAMPLE 7

Finding the Inverse of a Function Defined by an Equation

The function

$$f(x) = \frac{2x + 1}{x - 1} \qquad x \neq 1$$

is one-to-one. Find its inverse function and check the result.

Solution **STEP 1:** Replace $f(x)$ with y and interchange the variables x and y in

$$y = \frac{2x + 1}{x - 1}$$

to obtain

$$x = \frac{2y + 1}{y - 1}$$

STEP 2: Solve for y.

$$x = \frac{2y + 1}{y - 1}$$

$x(y - 1) = 2y + 1$ Multiply both sides by $y - 1$.

$xy - x = 2y + 1$ Use the Distributive Property.

$xy - 2y = x + 1$ Subtract $2y$ from both sides; add x to both sides.

$(x - 2)y = x + 1$ Factor.

$y = \dfrac{x + 1}{x - 2}$ Divide by $x - 2$.

The inverse function is

$$f^{-1}(x) = \frac{x + 1}{x - 2} \qquad x \neq 2 \qquad \text{Replace } y \text{ by } f^{-1}(x).$$

STEP 3: ✓ Check:

$$f^{-1}(f(x)) = f^{-1}\left(\frac{2x+1}{x-1}\right) = \frac{\dfrac{2x+1}{x-1}+1}{\dfrac{2x+1}{x-1}-2} = \frac{2x+1+x-1}{2x+1-2(x-1)} = \frac{3x}{3} = x, \ x \neq 1$$

$$f(f^{-1}(x)) = f\left(\frac{x+1}{x-2}\right) = \frac{2\cdot\dfrac{x+1}{x-2}+1}{\dfrac{x+1}{x-2}-1} = \frac{2(x+1)+x-2}{x+1-(x-2)} = \frac{3x}{3} = x, \ x \neq 2$$

Exploration

In Example 7, we found that if $f(x) = \dfrac{2x+1}{x-1}$, then $f^{-1}(x) = \dfrac{x+1}{x-2}$. Compare the vertical and horizontal asymptotes of f and f^{-1}.

Result The vertical asymptote of f is $x = 1$, and the horizontal asymptote is $y = 2$. The vertical asymptote of f^{-1} is $x = 2$, and the horizontal asymptote is $y = 1$.

- **Now Work** PROBLEMS 45 AND 59

If a function is not one-to-one, it has no inverse function. Sometimes, though, an appropriate restriction on the domain of such a function yields a new function that is one-to-one. Then the function defined on the restricted domain has an inverse function.

EXAMPLE 8

Finding the Inverse of a Domain-restricted Function

Find the inverse of $y = f(x) = x^2$ if $x \geq 0$. Graph f and f^{-1}.

Solution

The function $y = x^2$ is not one-to-one. [Refer to Example 2(a).] However, restricting the domain of this function to $x \geq 0$, as indicated, results in a new function that is increasing and therefore is one-to-one. Consequently, the function defined by $y = f(x) = x^2, x \geq 0$, has an inverse function, f^{-1}.

Follow the steps to find f^{-1}.

STEP 1: In the equation $y = x^2, x \geq 0$, interchange the variables x and y. The result is

$$x = y^2 \quad y \geq 0$$

This equation defines the inverse function implicitly.

STEP 2: Solve for y to get the explicit form of the inverse. Because $y \geq 0$, only one solution for y is obtained: $y = \sqrt{x}$. So $f^{-1}(x) = \sqrt{x}$.

STEP 3: ✓ Check: $f^{-1}(f(x)) = f^{-1}(x^2) = \sqrt{x^2} = |x| = x$ because $x \geq 0$

$$f(f^{-1}(x)) = f(\sqrt{x}) = (\sqrt{x})^2 = x$$

Figure 17

Figure 17 illustrates the graphs of $f(x) = x^2, x \geq 0$, and $f^{-1}(x) = \sqrt{x}$.

SUMMARY

- If a function f is one-to-one, then it has an inverse function f^{-1}.
- Domain of f = Range of f^{-1}; Range of f = Domain of f^{-1}.
- To verify that f^{-1} is the inverse of f, show that $f^{-1}(f(x)) = x$ for every x in the domain of f and that $f(f^{-1}(x)) = x$ for every x in the domain of f^{-1}.
- The graphs of f and f^{-1} are symmetric with respect to the line $y = x$.

6.2 Assess Your Understanding

'Are You Prepared?' *Answers are given at the end of these exercises. If you get a wrong answer, read the pages listed in* red.

1. Suppose $f(x) = 4x - 2$. Find $f\left(\dfrac{x + 2}{4}\right)$ (pp. 207–209)

2. Where is the function $f(x) = x^2$ increasing? Where is it decreasing? (pp. 231–232)

3. What is the domain of $f(x) = \dfrac{x + 5}{x^2 + 3x - 18}$? (pp. 211–213)

4. Simplify: $\dfrac{\dfrac{1}{x} + 1}{\dfrac{1}{x^2} - 1}$ (pp. 67–69)

Concepts and Vocabulary

5. If x_1 and x_2 are any two different inputs of a function f, then f is one-to-one if _____.

6. If every horizontal line intersects the graph of a function f at no more than one point, then f is a(n) _____ function.

7. If f is a one-to-one function and $f(3) = 8$, then $f^{-1}(8) =$ ____.

8. If f^{-1} is the inverse of a function f, then the graphs of f and f^{-1} are symmetric with respect to the line _____.

9. If the domain of a one-to-one function f is $[4, \infty)$, then the range of its inverse function f^{-1} is _____.

10. **True or False** If f and g are inverse functions, then the domain of f is the same as the range of g.

11. **Multiple Choice** If $(-2, 3)$ is a point on the graph of a one-to-one function f, which of the following points is on the graph of f^{-1}?

   (a) $(3, -2)$ **(b)** $(2, -3)$ **(c)** $(-3, 2)$ **(d)** $(-2, -3)$

12. **Multiple Choice** Suppose f is a one-to-one function with a domain of $\{x | x \neq 3\}$ and a range of $\left\{y \,\middle|\, y \neq \dfrac{2}{3}\right\}$. Which of the following is the domain of f^{-1}?

   (a) $\{x | x \neq 3\}$ **(b)** All real numbers

   (c) $\left\{x \,\middle|\, x \neq \dfrac{2}{3}, x \neq 3\right\}$ **(d)** $\left\{x \,\middle|\, x \neq \dfrac{2}{3}\right\}$

Skill Building

In Problems 13–20, determine whether the function is one-to-one.

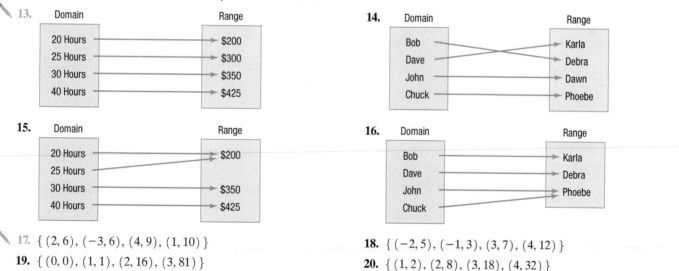

17. $\{(2, 6), (-3, 6), (4, 9), (1, 10)\}$

18. $\{(-2, 5), (-1, 3), (3, 7), (4, 12)\}$

19. $\{(0, 0), (1, 1), (2, 16), (3, 81)\}$

20. $\{(1, 2), (2, 8), (3, 18), (4, 32)\}$

In Problems 21–26, the graph of a function f is given. Use the horizontal-line test to determine whether f is one-to-one.

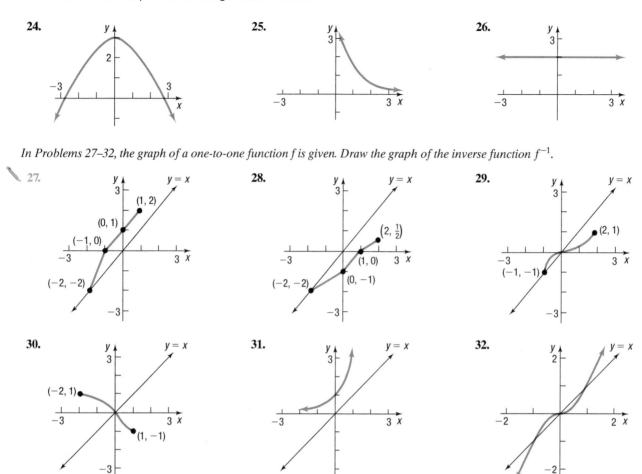

In Problems 27–32, the graph of a one-to-one function f is given. Draw the graph of the inverse function f^{-1}.

In Problems 33–42, verify that the functions f and g are inverses of each other by showing that $f(g(x)) = x$ and $g(f(x)) = x$. Give any values of x that need to be excluded from the domain of f and the domain of g.

33. $f(x) = 3x + 4$; $g(x) = \dfrac{1}{3}(x - 4)$

34. $f(x) = 3 - 2x$; $g(x) = -\dfrac{1}{2}(x - 3)$

35. $f(x) = 4x - 8$; $g(x) = \dfrac{x}{4} + 2$

36. $f(x) = 2x + 6$; $g(x) = \dfrac{1}{2}x - 3$

37. $f(x) = x^3 - 8$; $g(x) = \sqrt[3]{x + 8}$

38. $f(x) = (x - 2)^2, x \geq 2$; $g(x) = \sqrt{x} + 2$

39. $f(x) = \dfrac{1}{x}$; $g(x) = \dfrac{1}{x}$

40. $f(x) = x$; $g(x) = x$

41. $f(x) = \dfrac{2x + 3}{x + 4}$; $g(x) = \dfrac{4x - 3}{2 - x}$

42. $f(x) = \dfrac{x - 5}{2x + 3}$; $g(x) = \dfrac{3x + 5}{1 - 2x}$

In Problems 43–54, the function f is one-to-one. (a) Find its inverse function f^{-1} and check your answer. (b) Find the domain and the range of f and f^{-1}. (c) Graph f, f^{-1}, and $y = x$ on the same coordinate axes.

43. $f(x) = 3x$

44. $f(x) = -4x$

45. $f(x) = 4x + 2$

46. $f(x) = 1 - 3x$

47. $f(x) = x^3 - 1$

48. $f(x) = x^3 + 1$

49. $f(x) = x^2 + 4, \ x \geq 0$

50. $f(x) = x^2 + 9, \ x \geq 0$

51. $f(x) = \dfrac{4}{x}$

52. $f(x) = -\dfrac{3}{x}$

53. $f(x) = \dfrac{1}{x - 2}$

54. $f(x) = \dfrac{4}{x + 2}$

In Problems 55–72, the function f is one-to-one. (a) Find its inverse function f^{-1} and check your answer. (b) Find the domain and the range of f and f^{-1}.

55. $f(x) = \dfrac{2}{3+x}$

56. $f(x) = \dfrac{4}{2-x}$

57. $f(x) = \dfrac{3x}{x+2}$

58. $f(x) = -\dfrac{2x}{x-1}$

59. $f(x) = \dfrac{2x}{3x-1}$

60. $f(x) = -\dfrac{3x+1}{x}$

61. $f(x) = \dfrac{3x+4}{2x-3}$

62. $f(x) = \dfrac{2x-3}{x+4}$

63. $f(x) = \dfrac{2x+3}{x+2}$

64. $f(x) = \dfrac{-3x-4}{x-2}$

65. $f(x) = \dfrac{x^2-4}{2x^2}, \quad x > 0$

66. $f(x) = \dfrac{x^2+3}{3x^2}, \quad x > 0$

67. $f(x) = x^{\frac{2}{3}} - 4, \quad x \geq 0$

68. $f(x) = x^{\frac{3}{2}} + 5$

69. $f(x) = \sqrt[3]{x^5 - 2}$

70. $f(x) = \sqrt[5]{x^3 + 13}$

71. $f(x) = \dfrac{1}{9}(x-1)^2 + 2, \quad x \geq 1$

72. $f(x) = 2\sqrt{x+3} - 5$

Applications and Extensions

73. Use the graph of $y = f(x)$ given in Problem 27 to evaluate the following:

   (a) $f(-1)$ **(b)** $f(1)$ **(c)** $f^{-1}(1)$ **(d)** $f^{-1}(2)$

74. Use the graph of $y = f(x)$ given in Problem 28 to evaluate the following:

   (a) $f(2)$ **(b)** $f(1)$ **(c)** $f^{-1}(0)$ **(d)** $f^{-1}(-1)$

75. If $f(7) = 13$ and f is one-to-one, what is $f^{-1}(13)$?

76. If $g(-5) = 3$ and g is one-to-one, what is $g^{-1}(3)$?

77. The domain of a one-to-one function f is $[5, \infty)$, and its range is $[-2, \infty)$. State the domain and the range of f^{-1}.

78. The domain of a one-to-one function f is $[0, \infty)$, and its range is $[5, \infty)$. State the domain and the range of f^{-1}.

79. The domain of a one-to-one function g is $(-\infty, 0]$, and its range is $[0, \infty)$. State the domain and the range of g^{-1}.

80. The domain of a one-to-one function g is $[0, 15]$, and its range is $(0, 8)$. State the domain and the range of g^{-1}.

81. A function $y = f(x)$ is increasing on the interval $[0, 5]$. What conclusions can you draw about the graph of $y = f^{-1}(x)$?

82. A function $y = f(x)$ is decreasing on the interval $[0, 5]$. What conclusions can you draw about the graph of $y = f^{-1}(x)$?

83. Find the inverse of the linear function

$$f(x) = mx + b, \quad m \neq 0$$

84. Find the inverse of the function

$$f(x) = \sqrt{r^2 - x^2}, \quad 0 \leq x \leq r$$

85. A function f has an inverse function f^{-1}. If the graph of f lies in quadrant I, in which quadrant does the graph of f^{-1} lie?

86. A function f has an inverse function f^{-1}. If the graph of f lies in quadrant II, in which quadrant does the graph of f^{-1} lie?

87. The function $f(x) = |x|$ is not one-to-one. Find a suitable restriction on the domain of f so that the new function that results is one-to-one. Then find the inverse of the new function.

88. The function $f(x) = x^4$ is not one-to-one. Find a suitable restriction on the domain of f so that the new function that results is one-to-one. Then find the inverse of the new function.

In applications, the symbols used for the independent and dependent variables are often based on common usage. So, rather than using $y = f(x)$ to represent a function, an applied problem might use $C = C(q)$ to represent the cost C of manufacturing q units of a good. Because of this, the inverse notation f^{-1} used in a pure mathematics problem is not used when finding inverses of applied problems. Rather, the inverse of a function such as $C = C(q)$ will be $q = q(C)$. So $C = C(q)$ is a function that represents the cost C as a function of the number q of units manufactured, and $q = q(C)$ is a function that represents the number q as a function of the cost C. Problems 89–96 illustrate this idea.

89. Vehicle Stopping Distance Taking into account reaction time, the distance d (in feet) that a car requires to come to a complete stop while traveling r miles per hour is given by the function

$$d(r) = 6.97r - 90.39$$

   (a) Express the speed r at which the car is traveling as a function of the distance d required to come to a complete stop.

   (b) Verify that $r = r(d)$ is the inverse of $d = d(r)$ by showing that $r(d(r)) = r$ and $d(r(d)) = d$.

   (c) Predict the speed that a car was traveling if the distance required to stop was 300 feet.

90. Height and Head Circumference The head circumference C of a child is related to the height H of the child (both in inches) through the function

$$H(C) = 2.15C - 10.53$$

   (a) Express the head circumference C as a function of height H.

   (b) Verify that $C = C(H)$ is the inverse of $H = H(C)$ by showing that $H(C(H)) = H$ and $C(H(C)) = C$.

   (c) Predict the head circumference of a child who is 26 inches tall.

91. Ideal Body Weight One model for the ideal body weight W for men (in kilograms) as a function of height h (in inches) is given by the function

$$W(h) = 50 + 2.3(h - 60)$$

   (a) What is the ideal weight of a 6-foot male?

   (b) Express the height h as a function of weight W.

   (c) Verify that $h = h(W)$ is the inverse of $W = W(h)$ by showing that $h(W(h)) = h$ and $W(h(W)) = W$.

   (d) What is the height of a male who is at his ideal weight of 80 kilograms?

92. Temperature Conversion The function $F(C) = \dfrac{9}{5}C + 32$ converts a temperature from C degrees Celsius to F degrees Fahrenheit.

(a) Express the temperature in degrees Celsius C as a function of the temperature in degrees Fahrenheit F.

(b) Verify that $C = C(F)$ is the inverse of $F = F(C)$ by showing that $C(F(C)) = C$ and $F(C(F)) = F$.

(c) What is the temperature in degrees Celsius if it is 70 degrees Fahrenheit?

93. Income Taxes The function

$$T(g) = 4453.05 + 0.22(g - 38{,}700)$$

represents the 2018 federal income tax T (in dollars) due for a "single" filer whose modified adjusted gross income is g dollars, where $38{,}700 < g \le 82{,}500$.

(a) What is the domain of the function T?

(b) Given that the tax due T is an increasing linear function of modified adjusted gross income g, find the range of the function T.

(c) Find adjusted gross income g as a function of federal income tax T. What are the domain and the range of this function?

94. Income Taxes The function

$$T(g) = 1905 + 0.12(g - 19{,}050)$$

represents the 2018 federal income tax T (in dollars) due for a "married filing jointly" filer whose modified adjusted gross income is g dollars, where $19{,}050 < g \le 77{,}400$.

(a) What is the domain of the function T?

(b) Given that the tax due T is an increasing linear function of modified adjusted gross income g, find the range of the function T.

(c) Find adjusted gross income g as a function of federal income tax T. What are the domain and the range of this function?

95. Gravity on Earth If a rock falls from a height of 100 meters above Earth, the height H (in meters) after t seconds is approximately

$$H(t) = 100 - 4.9t^2$$

(a) In general, quadratic functions are not one-to-one. However, the function H is one-to-one. Why?

(b) Find the inverse of H and verify your result.

(c) How long will it take a rock to fall 80 meters?

96. Period of a Pendulum The period T (in seconds) of a simple pendulum as a function of its length l (in feet) is given by

$$T(l) = 2\pi\sqrt{\dfrac{l}{32.2}}$$

(a) Express the length l as a function of the period T.

(b) How long is a pendulum whose period is 3 seconds?

97. Challenge Problem Given

$$f(x) = \dfrac{ax + b}{cx + d}$$

find $f^{-1}(x)$. If $c \ne 0$, under what conditions on $a, b, c,$ and d is $f = f^{-1}$?

98. Challenge Problem If $h(x) = (f \circ g)(x)$, find h^{-1} in terms of f^{-1} and g^{-1}.

99. Challenge Problem For $f(x) = \begin{cases} 2x + 3, & x < 0 \\ 3x + 4, & x \ge 0 \end{cases}$

(a) Find the domain and range of f.

(b) Find f^{-1}.

(c) Find the domain and range of f^{-1}.

Explaining Concepts: Discussion and Writing

100. Can a one-to-one function and its inverse be equal? What must be true about the graph of f for this to happen? Give some examples to support your conclusion.

101. Draw the graph of a one-to-one function that contains the points $(-2, -3)$, $(0, 0)$, and $(1, 5)$. Now draw the graph of its inverse. Compare your graph to those of other students. Discuss any similarities. What differences do you see?

102. Give an example of a function whose domain is the set of real numbers and that is neither increasing nor decreasing on its domain, but is one-to-one.
[**Hint:** Use a piecewise-defined function.]

103. Is every odd function one-to-one? Explain.

104. Suppose that $C(g)$ represents the cost C, in dollars, of manufacturing g cars. Explain what $C^{-1}(800{,}000)$ represents.

105. Explain why the horizontal-line test can be used to identify one-to-one functions from a graph.

106. Explain why a function must be one-to-one in order to have an inverse that is a function. Use the function $y = x^2$ to support your explanation.

Retain Your Knowledge

Problems 107–116 are based on material learned earlier in the course. The purpose of these problems is to keep the material fresh in your mind so that you are better prepared for the final exam.

107. If $f(x) = 3x^2 - 7x$, find $f(x + h) - f(x)$.

108. Find the zeros of the quadratic function

$$f(x) = 3x^2 + 5x + 1$$

What are the x-intercepts, if any, of the graph of the function? Find the vertex. Is it a maximum or minimum? Is the graph concave up or concave down?

109. Use the techniques of shifting, compressing or stretching, and reflections to graph $f(x) = -|x + 2| + 3$.

110. Find the domain of $R(x) = \dfrac{6x^2 - 11x - 2}{2x^2 - x - 6}$. Find any horizontal, vertical, or oblique asymptotes.

111. Find an equation of a circle with center $(-3, 5)$ and radius 7.

112. Find an equation of the line that contains the point $(-4, 1)$ and is perpendicular to the line $3x - 6y = 5$. Write the equation in slope-intercept form.

113. Is the function $f(x) = \dfrac{3x}{5x^3 + 7x}$ even, odd, or neither?

114. Solve for D: $2x + 2yD = xD + y$

115. Find the average rate of change of $f(x) = -3x^2 + 2x + 1$ from 2 to 4.

116. Find the difference quotient of f: $f(x) = \sqrt{2x + 3}$

'Are You Prepared?' Answers

1. x **2.** Increasing on $[0, \infty)$; decreasing on $(-\infty, 0]$ **3.** $\{x \mid x \neq -6, x \neq 3\}$ **4.** $\dfrac{x}{1 - x}, x \neq 0, x \neq -1$

6.3 Exponential Functions

PREPARING FOR THIS SECTION *Before getting started, review the following:*

- Exponents (Chapter R, Section R.2, pp. 21–23, and Section R.8, pp. 76–77)
- Graphing Techniques: Transformations (Section 3.5, pp. 254–263)
- Solving Equations (Section 1.1, pp. 82–87 and Section 1.2, pp. 92–99)

- Average Rate of Change (Section 3.3, pp. 235–237)
- Quadratic Functions (Section 4.3, pp. 299–308)
- Linear Functions (Section 4.1, pp. 281–284)
- Horizontal Asymptotes (Section 5.3, pp. 359–361)

Now Work the 'Are You Prepared?' problems on page 446.

OBJECTIVES 1 Evaluate Exponential Functions (p. 435)
2 Graph Exponential Functions (p. 439)
3 Define the Number e (p. 442)
4 Solve Exponential Equations (p. 444)

1 Evaluate Exponential Functions

Chapter R, Section R.8 gives a definition for raising a real number a to a rational power. That discussion provides meaning to expressions of the form

$$a^r$$

where the base a is a positive real number and the exponent r is a rational number. But what is the meaning of a^x, where the base a is a positive real number and the exponent x is an irrational number? Although a rigorous definition requires methods discussed in calculus, the basis for the definition is easy to follow: Select a rational number r that is formed by truncating (removing) all but a finite number of digits from the irrational number x. Then it is reasonable to expect that

$$a^x \approx a^r$$

For example, take the irrational number $\pi = 3.14159\ldots$. Then an approximation to a^π is

$$a^\pi \approx a^{3.14}$$

where the digits of π after the hundredths position are truncated. A better approximation is

$$a^\pi \approx a^{3.14159}$$

where the digits after the hundred-thousandths position are truncated. Continuing in this way, we can obtain approximations to a^π to any desired degree of accuracy.

Most calculators have an $\boxed{x^y}$ key or a caret key $\boxed{\wedge}$ for working with exponents. To evaluate expressions of the form a^x, enter the base a, then press the $\boxed{x^y}$ key (or the $\boxed{\wedge}$ key), enter the exponent x, and press $\boxed{=}$ (or $\boxed{\text{ENTER}}$).

EXAMPLE 1

Using a Calculator to Evaluate Powers of 2

Use a calculator to evaluate:

(a) $2^{1.4}$ (b) $2^{1.41}$ (c) $2^{1.414}$ (d) $2^{1.4142}$ (e) $2^{\sqrt{2}}$

Solution

(a) $2^{1.4} \approx 2.639015822$ (b) $2^{1.41} \approx 2.657371628$

(c) $2^{1.414} \approx 2.66474965$ (d) $2^{1.4142} \approx 2.665119089$

(e) $2^{\sqrt{2}} \approx 2.665144143$

Now Work PROBLEM 17

It can be shown that the laws for rational exponents hold for real exponents.

THEOREM Laws of Exponents

If s, t, a, and b are real numbers with $a > 0$ and $b > 0$, then

$$
\begin{array}{lll}
\bullet\ a^s \cdot a^t = a^{s+t} & \bullet\ (a^s)^t = a^{st} & \bullet\ (ab)^s = a^s \cdot b^s \\[2mm]
\bullet\ 1^s = 1 & \bullet\ a^{-s} = \dfrac{1}{a^s} = \left(\dfrac{1}{a}\right)^s & \bullet\ a^0 = 1
\end{array}
\tag{1}
$$

Introduction to Exponential Growth

Suppose a function f has the following two properties:

- The value of f doubles with every 1-unit increase in the independent variable x.
- The value of f at $x = 0$ is 5, so $f(0) = 5$.

Table 1 shows values of the function f for $x = 0, 1, 2, 3$, and 4.

Let's find an equation $y = f(x)$ that describes this function f. The key fact is that the value of f doubles for every 1-unit increase in x.

Table 1

x	f(x)
0	5
1	10
2	20
3	40
4	80

$f(0) = 5$

$f(1) = 2f(0) = 2 \cdot 5 = 5 \cdot 2^1$ Double the value of f at 0 to get the value at 1.

$f(2) = 2f(1) = 2(5 \cdot 2) = 5 \cdot 2^2$ Double the value of f at 1 to get the value at 2.

$f(3) = 2f(2) = 2(5 \cdot 2^2) = 5 \cdot 2^3$

$f(4) = 2f(3) = 2(5 \cdot 2^3) = 5 \cdot 2^4$

The pattern leads to

$$f(x) = 2f(x-1) = 2(5 \cdot 2^{x-1}) = 5 \cdot 2^x$$

DEFINITION Exponential Function

An **exponential function** is a function of the form

$$f(x) = Ca^x$$

where a is a positive real number $(a > 0)$, $a \neq 1$, and $C \neq 0$ is a real number. The domain of f is the set of all real numbers. The base a is the **growth factor**, and, because $f(0) = Ca^0 = C$, C is called the **initial value**.

In the definition of an exponential function, the base $a = 1$ is excluded because this function is simply the constant function $f(x) = C \cdot 1^x = C$. Bases that are negative are also excluded; otherwise, many values of x would have to be excluded from the domain, such as $x = \dfrac{1}{2}$ and $x = \dfrac{3}{4}$. [Recall that $(-2)^{1/2} = \sqrt{-2}$, $(-3)^{3/4} = \sqrt[4]{(-3)^3} = \sqrt[4]{-27}$, and so on, are not defined in the set of real numbers.]

Transformations (vertical shifts, horizontal shifts, reflections, and so on) of a function of the form $f(x) = Ca^x$ are also exponential functions. Examples of such exponential functions are

$$f(x) = 2^x \qquad F(x) = \left(\frac{1}{3}\right)^x + 5 \qquad G(x) = 2 \cdot 3^{x-3}$$

For each function, note that the base of the exponential expression is a constant and the exponent contains a variable.

In the function $f(x) = 5 \cdot 2^x$, notice that the ratio of consecutive outputs is constant for 1-unit increases in the input. This ratio equals the constant 2, the base of the exponential function. In other words,

$$\frac{f(1)}{f(0)} = \frac{5 \cdot 2^1}{5} = 2 \quad \frac{f(2)}{f(1)} = \frac{5 \cdot 2^2}{5 \cdot 2^1} = 2 \quad \frac{f(3)}{f(2)} = \frac{5 \cdot 2^3}{5 \cdot 2^2} = 2 \quad \text{and so on}$$

This leads to the following result.

WARNING It is important to distinguish a power function, $g(x) = ax^n$, $n \geq 2$ an integer, from an exponential function, $f(x) = C \cdot a^x$, $a \neq 1$, $a > 0$. In a power function, the base is a variable and the exponent is a constant. In an exponential function, the base is a constant and the exponent is a variable. ∎

THEOREM

For an exponential function $f(x) = Ca^x$, $a > 0$, $a \neq 1$, and $C \neq 0$, if x is any real number, then

$$\frac{f(x+1)}{f(x)} = a \quad \text{or} \quad f(x+1) = af(x)$$

In Words
For 1-unit changes in the input x of an exponential function $f(x) = C \cdot a^x$, the ratio of consecutive outputs is the constant a.

Proof

$$\frac{f(x+1)}{f(x)} = \frac{Ca^{x+1}}{Ca^x} = a^{x+1-x} = a^1 = a \qquad ∎$$

EXAMPLE 2   Identifying Linear or Exponential Functions

Determine whether the given function is linear, exponential, or neither. For those that are linear, find a linear function that models the data. For those that are exponential, find an exponential function that models the data.

(a)

x	y
−1	5
0	2
1	−1
2	−4
3	−7

(b)

x	y
−1	32
0	16
1	8
2	4
3	2

(c)

x	y
−1	2
0	4
1	7
2	11
3	16

Solution   For each function, compute the average rate of change of y with respect to x and the ratio of consecutive outputs. If the average rate of change is constant, then the function is linear. If the ratio of consecutive outputs is constant, then the function is exponential.

(continued)

Table 2 (a)

x	y	Average Rate of Change	Ratio of Consecutive Outputs
−1	5		
		$\dfrac{\Delta y}{\Delta x} = \dfrac{2-5}{0-(-1)} = -3$	$\dfrac{2}{5}$
0	2		
		$\dfrac{-1-2}{1-0} = -3$	$\dfrac{-1}{2} = -\dfrac{1}{2}$
1	−1		
		$\dfrac{-4-(-1)}{2-1} = -3$	$\dfrac{-4}{-1} = 4$
2	−4		
		$\dfrac{-7-(-4)}{3-2} = -3$	$\dfrac{-7}{-4} = \dfrac{7}{4}$
3	−7		

(b)

x	y	Average Rate of Change	Ratio of Consecutive Outputs
−1	32		
		$\dfrac{\Delta y}{\Delta x} = \dfrac{16-32}{0-(-1)} = -16$	$\dfrac{16}{32} = \dfrac{1}{2}$
0	16		
		−8	$\dfrac{8}{16} = \dfrac{1}{2}$
1	8		
		−4	$\dfrac{4}{8} = \dfrac{1}{2}$
2	4		
		−2	$\dfrac{2}{4} = \dfrac{1}{2}$
3	2		

(c)

x	y	Average Rate of Change	Ratio of Consecutive Outputs
−1	2		
		$\dfrac{\Delta y}{\Delta x} = \dfrac{4-2}{0-(-1)} = 2$	2
0	4		
		3	$\dfrac{7}{4}$
1	7		
		4	$\dfrac{11}{7}$
2	11		
		5	$\dfrac{16}{11}$
3	16		

(a) See Table 2(a). The average rate of change for every 1-unit increase in x is -3. Therefore, the function is a linear function. In a linear function the average rate of change is the slope m, so $m = -3$. The y-intercept b is the value of the function at $x = 0$, so $b = 2$. The linear function that models the data is $f(x) = mx + b = -3x + 2$.

(b) See Table 2(b). For this function, the average rate of change is not constant. So the function is not a linear function. The ratio of consecutive outputs for a 1-unit increase in the inputs is a constant, $\dfrac{1}{2}$. Because the ratio of consecutive outputs is constant, the function is an exponential function with growth factor $a = \dfrac{1}{2}$. The initial value C of the exponential function is $C = 16$, the value of the function at 0. Therefore, the exponential function that models the data is $g(x) = Ca^x = 16 \cdot \left(\dfrac{1}{2}\right)^x$.

(c) See Table 2(c). For this function, neither the average rate of change nor the ratio of two consecutive outputs is constant. Because the average rate of change is not constant, the function is not a linear function. Because the ratio of consecutive outputs is not a constant, the function is not an exponential function.

> Now Work PROBLEM 29

2 Graph Exponential Functions

If we know how to graph an exponential function of the form $f(x) = a^x$, then we can use transformations (shifting, stretching, and so on) to obtain the graph of any exponential function.

First, let's graph the exponential function $f(x) = 2^x$.

EXAMPLE 3 **Graphing an Exponential Function**

Graph the exponential function: $f(x) = 2^x$

Solution The domain of $f(x) = 2^x$ is the set of all real numbers. Begin by locating some points on the graph of $f(x) = 2^x$, as listed in Table 3.

Because $2^x > 0$ for all x, the graph has no x-intercepts and lies above the x-axis for all x. The y-intercept is 1.

Table 3 suggests that as $x \to -\infty$, the value of f approaches 0. Therefore, the x-axis $(y = 0)$ is a horizontal asymptote of the graph of f as $x \to -\infty$. This provides the end behavior for x large and negative.

To determine the end behavior for x large and positive, look again at Table 3. As $x \to \infty$, $f(x) = 2^x$ grows very quickly, causing the graph of $f(x) = 2^x$ to rise very rapidly.

Using all this information, plot some of the points from Table 3 and connect them with a smooth, continuous curve, as shown in Figure 18. From the graph, we conclude that the range of f is $(0, \infty)$. We also conclude that f is an increasing function, and so f is one-to-one.

Table 3

x	$f(x) = 2^x$
-10	$2^{-10} \approx 0.00098$
-3	$2^{-3} = \dfrac{1}{8}$
-2	$2^{-2} = \dfrac{1}{4}$
-1	$2^{-1} = \dfrac{1}{2}$
0	$2^0 = 1$
1	$2^1 = 2$
2	$2^2 = 4$
3	$2^3 = 8$
10	$2^{10} = 1024$

NOTE Recall $a^{-n} = \dfrac{1}{a^n}$. ∎

Figure 18 $f(x) = 2^x$

Graphs that look like the one in Figure 18 occur very frequently in a variety of situations. For example, the graph in Figure 19 on the next page shows the annual revenue of Amazon, Inc. from 2000 to 2017. One might conclude from this graph that Amazon's annual revenue is growing *exponentially*.

Later in this chapter, we discuss other situations that exhibit exponential growth. For now, we continue to explore properties of exponential functions.

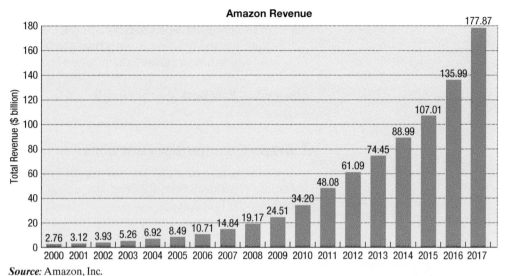

Amazon Revenue

Source: Amazon, Inc.

Figure 19

Figure 20

The graph of $f(x) = 2^x$ in Figure 18 is typical of all exponential functions of the form $f(x) = a^x$ with $a > 1$. Such functions are increasing functions and so are one-to-one. Their graphs lie above the x-axis, contain the point $(0, 1)$, and rise rapidly as $x \to \infty$. As $x \to -\infty$, the x-axis $(y = 0)$ is a horizontal asymptote. There are no vertical asymptotes. Finally, the graphs are smooth and continuous with no corners or gaps.

Figure 20 illustrates the graphs of two other exponential functions whose bases are larger than 1. Notice that the larger the base, the steeper the graph is when $x > 0$, and when $x < 0$, the larger the base, the closer the graph is to the x-axis.

Seeing the Concept

Graph $Y_1 = 2^x$ and compare what you see to Figure 18. Clear the screen, graph $Y_1 = 3^x$ and $Y_2 = 6^x$, and compare what you see to Figure 20. Clear the screen and graph $Y_1 = 10^x$ and $Y_2 = 100^x$.

Properties of the Exponential Function $f(x) = a^x$, $a > 1$

- The domain is the set of all real numbers, or $(-\infty, \infty)$ using interval notation; the range is the set of positive real numbers, or $(0, \infty)$ using interval notation.

- There are no x-intercepts; the y-intercept is 1.

- The x-axis $(y = 0)$ is a horizontal asymptote of the graph of f as $x \to -\infty$.

- $f(x) = a^x$, $a > 1$, is an increasing function and is one-to-one.

- The graph of f contains the points $\left(-1, \dfrac{1}{a}\right)$, $(0, 1)$ and $(1, a)$.

- The graph of f is smooth and continuous, with no corners or gaps. See Figure 21.

Figure 21 $f(x) = a^x$, $a > 1$

Now consider $f(x) = a^x$ when $0 < a < 1$.

EXAMPLE 4   **Graphing an Exponential Function**

Graph the exponential function: $f(x) = \left(\dfrac{1}{2}\right)^x$

Solution

The domain of $f(x) = \left(\dfrac{1}{2}\right)^x$ is all real numbers. As before, locate some points on the graph as shown in Table 4. Because $\left(\dfrac{1}{2}\right)^x > 0$ for all x, the range of f is the interval $(0, \infty)$. The graph lies above the x-axis and has no x-intercepts. The y-intercept is 1. As $x \to -\infty$, $f(x) = \left(\dfrac{1}{2}\right)^x$ grows very quickly. As $x \to \infty$, the values of $f(x)$ approach 0. The x-axis $(y = 0)$ is a horizontal asymptote of the graph of f as $x \to \infty$. The function f is a decreasing function and so is one-to-one. Figure 22 shows the graph of f.

Table 4

x	$f(x) = \left(\dfrac{1}{2}\right)^x$
-10	$\left(\dfrac{1}{2}\right)^{-10} = 2^{10} = 1024$
-3	$\left(\dfrac{1}{2}\right)^{-3} = 2^3 = 8$
-2	$\left(\dfrac{1}{2}\right)^{-2} = 2^2 = 4$
-1	$\left(\dfrac{1}{2}\right)^{-1} = 2^1 = 2$
0	$\left(\dfrac{1}{2}\right)^{0} = 1$
1	$\left(\dfrac{1}{2}\right)^{1} = \dfrac{1}{2}$
2	$\left(\dfrac{1}{2}\right)^{2} = \dfrac{1}{4}$
3	$\left(\dfrac{1}{2}\right)^{3} = \dfrac{1}{8}$
10	$\left(\dfrac{1}{2}\right)^{10} \approx 0.00098$

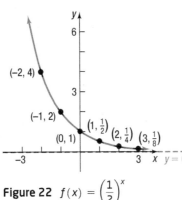

Figure 22 $f(x) = \left(\dfrac{1}{2}\right)^x$

The graph of $y = \left(\dfrac{1}{2}\right)^x$ also can be obtained from the graph of $y = 2^x$ using transformations. The graph of $y = \left(\dfrac{1}{2}\right)^x = 2^{-x}$ is a reflection about the y-axis of the graph of $y = 2^x$ (replace x by $-x$). See Figures 23(a) and (b).

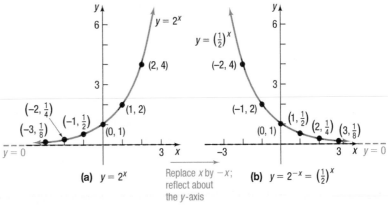

Figure 23

The graph of $f(x) = \left(\dfrac{1}{2}\right)^x$ in Figure 22 is typical of all exponential functions of the form $f(x) = a^x$ with $0 < a < 1$. Such functions are decreasing and one-to-one. Their graphs lie above the x-axis and contain the point $(0, 1)$. The graphs rise rapidly as $x \to -\infty$. As $x \to \infty$, the x-axis $(y = 0)$ is a horizontal asymptote. There are no vertical asymptotes. Finally, the graphs are smooth and continuous, with no corners or gaps.

Figure 24 illustrates the graphs of two other exponential functions whose bases are between 0 and 1. Notice that the smaller base results in a graph that is steeper when $x < 0$. When $x > 0$, the graph of the equation with the smaller base is closer to the x-axis.

Need to Review?

Reflections about the y-axis are discussed in Section 3.5, p. 261.

Seeing the Concept

Using a graphing utility, simultaneously graph:

(a) $Y_1 = 3^x$, $Y_2 = \left(\dfrac{1}{3}\right)^x$

(b) $Y_1 = 6^x$, $Y_2 = \left(\dfrac{1}{6}\right)^x$

Conclude that the graph of $Y_2 = \left(\dfrac{1}{a}\right)^x$, for $a > 0$, is the reflection about the y-axis of the graph of $Y_1 = a^x$.

Figure 24

Figure 25 $f(x) = a^x, 0 < a < 1$

Properties of the Exponential Function $f(x) = a^x, 0 < a < 1$

- The domain is the set of all real numbers, or $(-\infty, \infty)$ using interval notation; the range is the set of positive real numbers, or $(0, \infty)$ using interval notation.
- There are no x-intercepts; the y-intercept is 1.
- The x-axis ($y = 0$) is a horizontal asymptote of the graph of f as $x \to \infty$.
- $f(x) = a^x, 0 < a < 1$, is a decreasing function and is one-to-one.
- The graph of f contains the points $\left(-1, \dfrac{1}{a}\right)$, $(0, 1)$, and $(1, a)$.
- The graph of f is smooth and continuous, with no corners or gaps. See Figure 25.

EXAMPLE 5 **Graphing an Exponential Function Using Transformations**

Graph $f(x) = 2^{-x} - 3$ and determine the domain, range, horizontal asymptote, and y-intercept of f.

Solution Begin with the graph of $y = 2^x$. Figure 26 shows the steps.

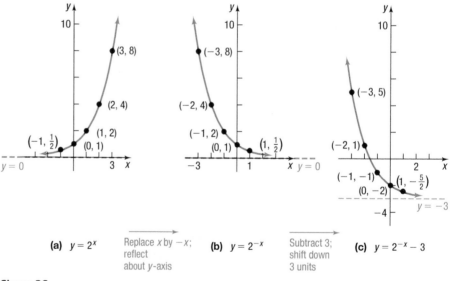

(a) $y = 2^x$ Replace x by $-x$; reflect about y-axis   **(b)** $y = 2^{-x}$ Subtract 3; shift down 3 units   **(c)** $y = 2^{-x} - 3$

Figure 26

As Figure 26(c) illustrates, the domain of $f(x) = 2^{-x} - 3$ is the interval $(-\infty, \infty)$ and the range is the interval $(-3, \infty)$. The horizontal asymptote of the graph of f is the line $y = -3$. The y-intercept is $f(0) = 2^0 - 3 = 1 - 3 = -2$.

 Now Work PROBLEM 45

---

3 Define the Number e

Many problems that occur in nature require the use of an exponential function whose base is a certain irrational number, symbolized by the letter e.

One way of arriving at this important number e is given next.

Historical Feature

The number *e* is called *Euler's* number in honor of the Swiss mathematician Leonard Euler (1707–1783). ■

> **DEFINITION** Number e
>
> The **number e** is defined as the number that the expression
>
> $$\left(1 + \frac{1}{n}\right)^n \tag{2}$$
>
> approaches as $n \to \infty$. In calculus, this is expressed, using limit notation, as
>
> $$e = \lim_{n \to \infty}\left(1 + \frac{1}{n}\right)^n$$

Table 5 illustrates what happens to the defining expression (2) as *n* takes on increasingly large values. The last number in the right column in the table approximates *e* correct to nine decimal places. That is, $e = 2.718281828\ldots$. Remember, the three dots indicate that the decimal places continue. Because these decimal places continue but do not repeat, *e* is an irrational number. The number *e* is often expressed as a decimal rounded to a specific number of places. For example, $e \approx 2.71828$ is rounded to five decimal places.

Table 5

n	$\dfrac{1}{n}$	$1 + \dfrac{1}{n}$	$\left(1 + \dfrac{1}{n}\right)^n$
1	1	2	2
2	0.5	1.5	2.25
5	0.2	1.2	2.48832
10	0.1	1.1	2.59374246
100	0.01	1.01	2.704813829
1,000	0.001	1.001	2.716923932
10,000	0.0001	1.0001	2.718145927
100,000	0.00001	1.00001	2.718268237
1,000,000	0.000001	1.000001	2.718280469
10,000,000,000	10^{-10}	$1 + 10^{-10}$	2.718281828

Table 6

x	e^x
-2	$e^{-2} \approx 0.14$
-1	$e^{-1} \approx 0.37$
0	$e^0 = 1$
1	$e^1 \approx 2.72$
2	$e^2 \approx 7.39$

The exponential function $f(x) = e^x$, whose base is the number *e*, occurs with such frequency in applications that it is usually referred to as *the* exponential function. Most calculators have the key $\boxed{e^x}$ or $\boxed{\exp(x)}$, which may be used to approximate the exponential function for a given value of *x*.*

Use your calculator to approximate e^x for $x = -2, x = -1, x = 0, x = 1,$ and $x = 2$. See Table 6. The graph of the exponential function $f(x) = e^x$ is given in Figure 27. Since $2 < e < 3$, the graph of $y = e^x$ lies between the graphs of $y = 2^x$ and $y = 3^x$. Do you see why? (Refer to Figures 18 and 20.)

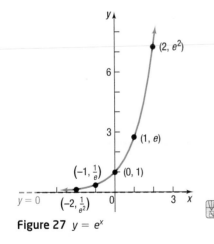

Figure 27 $y = e^x$

Seeing the Concept

Graph $Y_1 = e^x$ and compare what you see to Figure 27. Use eVALUEate or TABLE to verify the points on the graph shown in Figure 27. Now graph $Y_2 = 2^x$ and $Y_3 = 3^x$ on the same screen as $Y_1 = e^x$. Notice that the graph of $Y_1 = e^x$ lies between these two graphs.

EXAMPLE 6

Graphing an Exponential Function Using Transformations

Graph $f(x) = -e^{x-3}$ and determine the domain, range, horizontal asymptote, and *y*-intercept of *f*.

---

*If your calculator does not have one of these keys, refer to your owner's manual.

Solution   Begin with the graph of $y = e^x$. Figure 28 shows the steps.

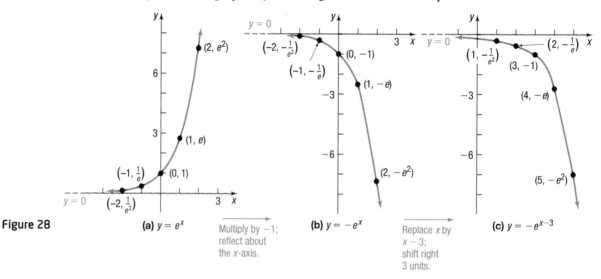

Figure 28

(a) $y = e^x$

Multiply by -1;
reflect about
the x-axis.

(b) $y = -e^x$

Replace x by
$x - 3$;
shift right
3 units.

(c) $y = -e^{x-3}$

As Figure 28(c) illustrates, the domain of $f(x) = -e^{x-3}$ is the interval $(-\infty, \infty)$, and the range is the interval $(-\infty, 0)$. The horizontal asymptote is the line $y = 0$. The y-intercept is $f(0) = -e^{0-3} = -e^{-3} \approx -0.05$.

➤ **Now Work** PROBLEM 57

4 Solve Exponential Equations

Equations that involve terms of the form a^x, where $a > 0$ and $a \neq 1$, are referred to as **exponential equations**. Such equations can sometimes be solved by using the Laws of Exponents and property (3):

In Words
When two exponential expressions with the same base are equal, then their exponents are equal.

$$\text{If} \quad a^u = a^v, \quad \text{then} \quad u = v. \tag{3}$$

Property (3) is a consequence of the fact that exponential functions are one-to-one. To use property (3), each side of the equality must be written with the same base.

EXAMPLE 7   **Solving Exponential Equations**

Solve each exponential equation.

(a) $3^{x+1} = 81$ (b) $4^{2x-1} = 8^{x+3}$

Solution   (a) Since $81 = 3^4$, write the equation as

$$3^{x+1} = 3^4$$

Now the expressions on both sides of the equation have the same base, 3. Set the exponents equal to each other to obtain

$$x + 1 = 4$$
$$x = 3$$

The solution set is $\{3\}$.

(b)
$$4^{2x-1} = 8^{x+3}$$
$$(2^2)^{(2x-1)} = (2^3)^{(x+3)} \qquad 4 = 2^2; 8 = 2^3$$
$$2^{2(2x-1)} = 2^{3(x+3)} \qquad (a^r)^s = a^{rs}$$
$$2(2x - 1) = 3(x + 3) \qquad \text{If } a^u = a^v, \text{then } u = v, \text{(property (3))}.$$
$$4x - 2 = 3x + 9$$
$$x = 11$$

The solution set is $\{11\}$.

➤ **Now Work** PROBLEMS 67 AND 77

EXAMPLE 8	**Solving an Exponential Equation**

Solve: $e^{-x^2} = (e^x)^2 \cdot \dfrac{1}{e^3}$

Solution   Use the Laws of Exponents first to get a single expression with the base e on the right side.

$$(e^x)^2 \cdot \frac{1}{e^3} = e^{2x} \cdot e^{-3} = e^{2x-3}$$

Then,

$$e^{-x^2} = e^{2x-3}$$

$$-x^2 = 2x - 3 \qquad \text{Use property (3).}$$

$$x^2 + 2x - 3 = 0 \qquad \text{Place the quadratic equation in standard form.}$$

$$(x + 3)(x - 1) = 0 \qquad \text{Factor.}$$

$$x = -3 \quad \text{or} \quad x = 1 \qquad \text{Use the Zero-Product Property.}$$

The solution set is $\{-3, 1\}$.

-**Now Work** PROBLEM 83

EXAMPLE 9	**Exponential Probability**

Between 9:00 PM and 10:00 PM, cars arrive at Burger King's drive-thru at the rate of 12 cars per hour (0.2 car per minute). The following formula from probability theory can be used to determine the probability that a car will arrive within t minutes of 9:00 PM.

$$F(t) = 1 - e^{-0.2t}$$

(a) Determine the probability that a car will arrive within 5 minutes of 9 PM (that is, before 9:05 PM).

(b) Determine the probability that a car will arrive within 30 minutes of 9 PM (before 9:30 PM).

(c) Graph F.

(d) What does F approach as t increases without bound in the positive direction?

Solution   (a) The probability that a car will arrive within 5 minutes is found by evaluating $F(t)$ at $t = 5$.

$$F(5) = 1 - e^{-0.2(5)} \approx 0.63212$$
$$\uparrow$$
$$\text{Use a calculator.}$$

There is a 63% probability that a car will arrive within 5 minutes.

(b) The probability that a car will arrive within 30 minutes is found by evaluating $F(t)$ at $t = 30$.

$$F(30) = 1 - e^{-0.2(30)} \approx 0.9975$$
$$\uparrow$$
$$\text{Use a calculator.}$$

There is a 99.75% probability that a car will arrive within 30 minutes.

(c) See Figure 29 for the graph of F.

(d) As time passes, the probability that a car will arrive increases. The value that F approaches can be found by letting $t \to \infty$. Since $e^{-0.2t} = \dfrac{1}{e^{0.2t}}$, it follows that $e^{-0.2t} \to 0$ as $t \to \infty$. Therefore, $F(t) = 1 - e^{-0.2t} \to 1$ as $t \to \infty$. The algebraic analysis is supported by Figure 29.

NORMAL FLOAT AUTO REAL RADIAN MP

Figure 29 $F(t) = 1 - e^{-0.2t}$

-**Now Work** PROBLEM 115

SUMMARY

Properties of the Exponential Function

$f(x) = a^x, \quad a > 1$

- Domain: the interval $(-\infty, \infty)$; range: the interval $(0, \infty)$
- x-intercepts: none; y-intercept: 1
- Horizontal asymptote: x-axis $(y = 0)$ as $x \to -\infty$
- Increasing; one-to-one; smooth; continuous
- See Figure 21 for a typical graph.

$f(x) = a^x, \quad 0 < a < 1$

- Domain: the interval $(-\infty, \infty)$; range: the interval $(0, \infty)$
- x-intercepts: none; y-intercept: 1
- Horizontal asymptote: x-axis $(y = 0)$ as $x \to \infty$
- Decreasing; one-to-one; smooth; continuous
- See Figure 25 for a typical graph.

If $a^u = a^v$, then $u = v$.

6.3 Assess Your Understanding

'Are You Prepared?' *Answers are given at the end of these exercises. If you get a wrong answer, read the pages listed in* red.

1. $4^3 = $ _____ ; $8^{2/3} = $ _____ ; $3^{-2} = $ _____ . (pp. 21–23 and pp. 76–77)

2. Solve: $x^2 + 3x = 4$ (pp. 93–99)

3. *True or False* To graph $y = (x - 2)^3$, shift the graph of $y = x^3$ to the left 2 units. (pp. 254–263)

4. Find the average rate of change of $f(x) = 3x - 5$ from $x = 0$ to $x = 4$. (pp. 235–267)

5. *True or False* The graph of the function $f(x) = \dfrac{2x}{x - 3}$ has $y = 2$ as a horizontal asymptote. (pp. 359–361)

6. If $f(x) = -3x + 10$, then the graph of f is a _____ with slope _____ and y-intercept _____ . (pp. 281–284)

7. Where is the function $f(x) = x^2 - 4x + 3$ increasing? Where is it decreasing? (pp. 302–305)

Concepts and Vocabulary

8. A(n) _____ _____ is a function of the form $f(x) = Ca^x$, where $a > 0, a \neq 1$, and $C \neq 0$ are real numbers. The base a is the _____ _____ and C is the _____ _____ .

9. For an exponential function $f(x) = Ca^x$, $\dfrac{f(x + 1)}{f(x)} = $ ___ .

10. *True or False* The domain of the function $f(x) = a^x$, where $a > 0$ and $a \neq 1$, is the set of all real numbers.

11. *True or False* The function $f(x) = e^x$ is increasing and is one-to-one.

12. The graph of every exponential function $f(x) = a^x$, where $a > 0$ and $a \neq 1$, contains the three points: _____ , _____ , and _____ .

13. If $3^x = 3^4$, then $x = $ ___

14. *True or False* The graphs of $y = 3^x$ and $y = \left(\dfrac{1}{3}\right)^x$ are identical.

15. *Multiple Choice* Which exponential function is increasing?

(a) $f(x) = 0.5^x$
(b) $f(x) = \left(\dfrac{5}{2}\right)^x$
(c) $f(x) = \left(\dfrac{2}{3}\right)^x$
(d) $f(x) = 0.9^x$

16. *Multiple Choice* The range of the function $f(x) = a^x$, where $a > 0$ and $a \neq 1$, is the interval

(a) $(-\infty, \infty)$ (b) $(-\infty, 0)$ (c) $(0, \infty)$ (d) $[0, \infty)$

Skill Building

In Problems 17–28, approximate each number using a calculator. Express your answer rounded to three decimal places.

17. (a) $2^{3.14}$ (b) $2^{3.141}$ (c) $2^{3.1415}$ (d) 2^π

18. (a) $2^{2.7}$ (b) $2^{2.71}$ (c) $2^{2.718}$ (d) 2^e

19. (a) $3.1^{2.7}$ (b) $3.14^{2.71}$ (c) $3.141^{2.718}$ (d) π^e

20. (a) $2.7^{3.1}$ (b) $2.71^{3.14}$ (c) $2.718^{3.141}$ (d) e^π

21. $(1 + 0.04)^6$

22. $\left(1 + \dfrac{0.09}{12}\right)^{24}$

23. $8.4\left(\dfrac{1}{3}\right)^{2.9}$

24. $158\left(\dfrac{5}{6}\right)^{8.63}$

25. $e^{1.2}$

26. $e^{-1.3}$

27. $125e^{0.026(7)}$

28. $83.6e^{-0.157(9.5)}$

In Problems 29–36, determine whether the given function is linear, exponential, or neither. For those that are linear functions, find a linear function that models the data; for those that are exponential, find an exponential function that models the data.

29.

x	f(x)
−1	3
0	6
1	12
2	18
3	30

30.

x	g(x)
−1	2
0	5
1	8
2	11
3	14

31.

x	H(x)
−1	$\frac{1}{4}$
0	1
1	4
2	16
3	64

32.

x	F(x)
−1	$\frac{2}{3}$
0	1
1	$\frac{3}{2}$
2	$\frac{9}{4}$
3	$\frac{27}{8}$

33.

x	f(x)
−1	$\frac{3}{2}$
0	3
1	6
2	12
3	24

34.

x	g(x)
−1	6
0	1
1	0
2	3
3	10

35.

x	H(x)
−1	2
0	4
1	6
2	8
3	10

36.

x	F(x)
−1	$\frac{1}{2}$
0	$\frac{1}{4}$
1	$\frac{1}{8}$
2	$\frac{1}{16}$
3	$\frac{1}{32}$

In Problems 37–44, the graph of an exponential function is given. Match each graph to one of the following functions.

(A) $y = 3^x$ (B) $y = 3^{-x}$ (C) $y = -3^x$ (D) $y = -3^{-x}$

(E) $y = 3^x - 1$ (F) $y = 3^{x-1}$ (G) $y = 3^{1-x}$ (H) $y = 1 - 3^x$

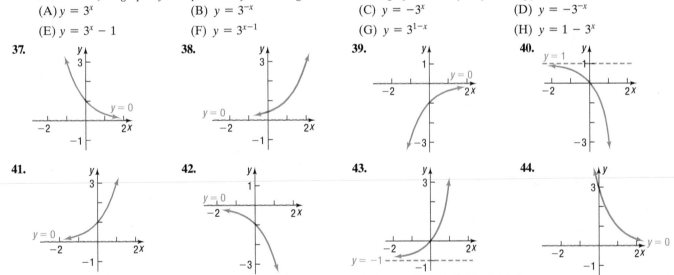

37.   **38.**   **39.**   **40.**

41.   **42.**   **43.**   **44.**

In Problems 45–56, use transformations to graph each function. Determine the domain, range, horizontal asymptote, and y-intercept of each function.

45. $f(x) = 2^x + 1$ **46.** $f(x) = 3^x - 2$ **47.** $f(x) = 3^{x-1}$ **48.** $f(x) = 2^{x+2}$

49. $f(x) = 3 \cdot \left(\frac{1}{2}\right)^x$ **50.** $f(x) = 4 \cdot \left(\frac{1}{3}\right)^x$ **51.** $f(x) = 3^{-x} - 2$ **52.** $f(x) = -3^x + 1$

53. $f(x) = 2 + 4^{x-1}$ **54.** $f(x) = 1 - 2^{x+3}$ **55.** $f(x) = 2 + 3^{x/2}$ **56.** $f(x) = 1 - 2^{-x/3}$

In Problems 57–64, begin with the graph of $y = e^x$ (Figure 27) and use transformations to graph each function. Determine the domain, range, horizontal asymptote, and y-intercept of each function.

57. $f(x) = e^{-x}$ **58.** $f(x) = -e^x$ **59.** $f(x) = e^{x+2}$ **60.** $f(x) = e^x - 1$

61. $f(x) = 5 - e^{-x}$ **62.** $f(x) = 9 - 3e^{-x}$ **63.** $f(x) = 2 - e^{-x/2}$ **64.** $f(x) = 7 - 3e^{2x}$

In Problems 65–84, solve each equation.

65. $6^x = 6^5$

66. $5^x = 5^{-6}$

67. $2^{-x} = 16$

68. $3^{-x} = 81$

69. $\left(\dfrac{1}{5}\right)^x = \dfrac{1}{25}$

70. $\left(\dfrac{1}{4}\right)^x = \dfrac{1}{64}$

71. $3^{2x-5} = 9$

72. $5^{x+3} = \dfrac{1}{5}$

73. $3^{x^3} = 9^x$

74. $4^{x^2} = 2^x$

75. $8^{-x+11} = 16^{2x}$

76. $9^{-x+15} = 27^x$

77. $3^{x^2-7} = 27^{2x}$

78. $5^{x^2+8} = 125^{2x}$

79. $4^x \cdot 2^{x^2} = 16^2$

80. $9^{2x} \cdot 27^{x^2} = 3^{-1}$

81. $e^{2x} = e^{5x+12}$

82. $e^{3x} = e^{2-x}$

83. $e^{x^2} = e^{3x} \cdot \dfrac{1}{e^2}$

84. $\left(e^4\right)^x \cdot e^{x^2} = e^{12}$

85. If $4^x = 7$, what does 4^{-2x} equal?

86. If $2^x = 3$, what does 4^{-x} equal?

87. If $3^{-x} = 2$, what does 3^{2x} equal?

88. If $5^{-x} = 3$, what does 5^{3x} equal?

89. If $9^x = 25$, what does 3^x equal?

90. If $2^{-3x} = \dfrac{1}{1000}$, what does 2^x equal?

In Problems 91–94, determine the exponential function whose graph is given.

91.

92.

93.

94.

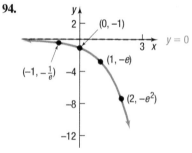

95. Find an exponential function whose graph has the horizontal asymptote $y = 2$ and contains the points $(0, 3)$ and $(1, 5)$.

96. Find an exponential function whose graph has the horizontal asymptote $y = -3$ and contains the points $(0, -2)$ and $(-2, 1)$.

97. Suppose that $f(x) = 2^x$.
  (a) What is $f(4)$? What point is on the graph of f?
  (b) If $f(x) = \dfrac{1}{16}$, what is x? What point is on the graph of f?

98. Suppose that $f(x) = 3^x$.
  (a) What is $f(4)$? What point is on the graph of f?
  (b) If $f(x) = \dfrac{1}{9}$, what is x? What point is on the graph of f?

99. Suppose that $g(x) = 4^x + 2$.
  (a) What is $g(-1)$? What point is on the graph of g?
  (b) If $g(x) = 66$, what is x? What point is on the graph of g?

100. Suppose that $g(x) = 5^x - 3$.
  (a) What is $g(-1)$? What point is on the graph of g?
  (b) If $g(x) = 122$, what is x? What point is on the graph of g?

101. Suppose that $H(x) = \left(\dfrac{1}{2}\right)^x - 4$.
  (a) What is $H(-6)$? What point is on the graph of H?
  (b) If $H(x) = 12$, what is x? What point is on the graph of H?
  (c) Find the zero of H.

102. Suppose that $F(x) = \left(\dfrac{1}{3}\right)^x - 3$.
  (a) What is $F(-5)$? What point is on the graph of F?
  (b) If $F(x) = 24$, what is x? What point is on the graph of F?
  (c) Find the zero of F.

Mixed Practice *In Problems 103–106, graph each function. Based on the graph, state the domain and the range, and find any intercepts.*

103. $f(x) = \begin{cases} e^{-x} & \text{if } x < 0 \\ e^x & \text{if } x \geq 0 \end{cases}$

104. $f(x) = \begin{cases} e^x & \text{if } x < 0 \\ e^{-x} & \text{if } x \geq 0 \end{cases}$

105. $f(x) = \begin{cases} -e^x & \text{if } x < 0 \\ -e^{-x} & \text{if } x \geq 0 \end{cases}$

106. $f(x) = \begin{cases} -e^{-x} & \text{if } x < 0 \\ -e^x & \text{if } x \geq 0 \end{cases}$

Applications and Extensions

107. Optics If a single pane of glass obliterates 3% of the light passing through it, the percent p of light that passes through n successive panes is given approximately by the function

$$p(n) = 100 \cdot 0.97^n$$

(a) What percent of light will pass through 10 panes?
(b) What percent of light will pass through 25 panes?
(c) Explain the meaning of the base 0.97 in this problem.

108. Atmospheric Pressure The atmospheric pressure p on a balloon or airplane decreases with increasing height. This pressure, measured in millimeters of mercury, is related to the height h (in kilometers) above sea level by the function

$$p(h) = 760e^{-0.145h}$$

(a) Find the atmospheric pressure at a height of 2 km (over a mile).
(b) What is it at a height of 10 kilometers (over 30,000 feet)?

109. Depreciation The price p, in dollars, of a Honda Civic EX-L sedan that is x years old is modeled by

$$p(x) = 22{,}265 \cdot 0.90^x$$

(a) How much should a 3-year-old Civic EX-L sedan cost?
(b) How much should a 9-year-old Civic EX-L sedan cost?
(c) Explain the meaning of the base 0.90 in this problem.

110. Healing of Wounds The normal healing of wounds can be modeled by an exponential function. If A_0 represents the original area of the wound and if A equals the area of the wound, then the function

$$A(n) = A_0 e^{-0.35n}$$

describes the area of a wound after n days following an injury when no infection is present to retard the healing. Suppose that a wound initially had an area of 100 square millimeters.
(a) If healing is taking place, how large will the area of the wound be after 3 days?
(b) How large will it be after 10 days?

111. Advanced-Stage Pancreatic Cancer The percentage of patients P who have survived t years after initial diagnosis of advanced-stage pancreatic cancer is modeled by the function

$$P(t) = 100 \cdot 0.3^t$$

Source: Cancer Treatment Centers of America
(a) According to the model, what percent of patients survive 1 year after initial diagnosis?
(b) What percent of patients survive 2 years after initial diagnosis?
(c) Explain the meaning of the base 0.3 in the context of this problem.

112. Endangered Species In a protected environment, the population P of a certain endangered species recovers over time t (in years) according to the model

$$P(t) = 30 \cdot 1.149^x$$

(a) What is the size of the initial population of the species?
(b) According to the model, what will be the population of the species in 5 years?

(c) According to the model, what will be the population of the species in 10 years?
(d) According to the model, what will be the population of the species in 15 years?
(e) What is happening to the population every 5 years?

113. Drug Medication The function

$$D(h) = 5e^{-0.4h}$$

can be used to find the number of milligrams D of a certain drug that is in a patient's bloodstream h hours after the drug has been administered. How many milligrams will be present after 1 hour? After 6 hours?

114. Spreading of Rumors A model for the number N of people in a college community who have heard a certain rumor is

$$N = P(1 - e^{-0.15d})$$

where P is the total population of the community and d is the number of days that have elapsed since the rumor began. In a community of 1000 students, how many students will have heard the rumor after 3 days?

115. Exponential Probability Between 12:00 PM and 1:00 PM, cars arrive at Citibank's drive-thru at the rate of 6 cars per hour (0.1 car per minute). The following formula from probability can be used to determine the probability that a car arrives within t minutes of 12:00 PM.

$$F(t) = 1 - e^{-0.1t}$$

(a) Determine the probability that a car arrives within 10 minutes of 12:00 PM (that is, before 12:10 PM).
(b) Determine the probability that a car arrives within 40 minutes of 12:00 PM (before 12:40 PM).
(c) What does F approach as t becomes unbounded in the positive direction?
(d) Graph F using a graphing utility.
(e) Using INTERSECT, determine how many minutes are needed for the probability to reach 50%.

116. Exponential Probability Between 5:00 PM and 6:00 PM, cars arrive at Jiffy Lube at the rate of 9 cars per hour (0.15 car per minute). This formula from probability can be used to determine the probability that a car arrives within t minutes of 5:00 PM:

$$F(t) = 1 - e^{-0.15t}$$

(a) Determine the probability that a car arrives within 15 minutes of 5:00 PM (that is, before 5:15 PM).
(b) Determine the probability that a car arrives within 30 minutes of 5:00 PM (before 5:30 PM).
(c) What does F approach as t becomes unbounded in the positive direction?
(d) Graph F.
(e) Using INTERSECT, determine how many minutes are needed for the probability to reach 60%.

117. Poisson Probability Between 5:00 PM and 6:00 PM, cars arrive at a McDonald's drive-thru at the rate of 20 cars per hour. The following formula from probability can be used to determine the probability that x cars arrive between 5:00 PM and 6:00 PM.

$$P(x) = \frac{20^x e^{-20}}{x!}$$

(continued)

where

$$x! = x \cdot (x - 1) \cdot (x - 2) \cdot \cdots \cdot 3 \cdot 2 \cdot 1$$

(a) Determine the probability that $x = 15$ cars arrive between 5:00 PM and 6:00 PM.

(b) Determine the probability that $x = 20$ cars arrive between 5:00 PM and 6:00 PM.

118. Poisson Probability People enter a line for the *Demon Roller Coaster* at the rate of 4 per minute. The following formula from probability can be used to determine the probability that x people arrive within the next minute.

$$P(x) = \frac{4^x e^{-4}}{x!}$$

where

$$x! = x \cdot (x - 1) \cdot (x - 2) \cdot \cdots \cdot 3 \cdot 2 \cdot 1$$

(a) Determine the probability that $x = 5$ people arrive within the next minute.

(b) Determine the probability that $x = 8$ people arrive within the next minute.

119. Relative Humidity The relative humidity is the ratio (expressed as a percent) of the amount of water vapor in the air to the maximum amount that the air can hold at a specific temperature. The relative humidity, R, is found using the following formula:

$$R = 10^{\left(\frac{4221}{T + 459.4} - \frac{4221}{D + 459.4} + 2\right)}$$

where T is the air temperature (in °F) and D is the dew point temperature (in °F).

(a) Determine the relative humidity if the air temperature is 50° Fahrenheit and the dew point temperature is 41° Fahrenheit.

(b) Determine the relative humidity if the air temperature is 68° Fahrenheit and the dew point temperature is 59° Fahrenheit.

(c) What is the relative humidity if the air temperature and the dew point temperature are the same?

120. Learning Curve Suppose that a student has 500 vocabulary words to learn. If the student learns 15 words after 5 minutes, the function

$$L(t) = 500(1 - e^{-0.0061t})$$

approximates the number of words L that the student will have learned after t minutes.

(a) How many words will the student have learned after 30 minutes?

(b) How many words will the student have learned after 60 minutes?

121. Current in an RL Circuit The equation governing the amount of current I (in amperes) after time t (in seconds) in a single RL circuit consisting of a resistance R (in ohms), an inductance L (in henrys), and an electromotive force E (in volts) is

$$I = \frac{E}{R} \left[1 - e^{-(R/L)t} \right]$$

(a) If $E = 120$ volts, $R = 10$ ohms, and $L = 5$ henrys, how much current I_1 is flowing after 0.3 second? After 0.5 second? After 1 second?

(b) What is the maximum current?

(c) Graph this function $I = I_1(t)$, measuring I along the y-axis and t along the x-axis.

(d) If $E = 120$ volts, $R = 5$ ohms, and $L = 10$ henrys, how much current I_2 is flowing after 0.3 second? After 0.5 second? After 1 second?

(e) What is the maximum current?

(f) Graph the function $I = I_2(t)$ on the same coordinate axes as $I_1(t)$.

122. Current in an RC Circuit The equation governing the amount of current I (in amperes) after time t (in microseconds) in a single RC circuit consisting of a resistance R (in ohms), a capacitance C (in microfarads), and an electromotive force E (in volts) is

$$I = \frac{E}{R} e^{-t/(RC)}$$

(a) If $E = 120$ volts, $R = 2000$ ohms, and $C = 1.0$ microfarad, how much current I_1 is flowing initially $(t = 0)$? After 1000 microseconds? After 3000 microseconds?

(b) What is the maximum current?

(c) Graph the function $I = I_1(t)$, measuring I along the y-axis and t along the x-axis.

(d) If $E = 120$ volts, $R = 1000$ ohms, and $C = 2.0$ microfarads, how much current I_2 is flowing initially? After 1000 microseconds? After 3000 microseconds?

(e) What is the maximum current?

(f) Graph the function $I = I_2(t)$ on the same coordinate axes as $I_1(t)$.

123. If f is an exponential function of the form $f(x) = Ca^x$ with growth factor 3, and if $f(6) = 12$, what is $f(7)$?

124. Another Formula for e Use a calculator to compute the values of

$$2 + \frac{1}{2!} + \frac{1}{3!} + \cdots + \frac{1}{n!}$$

for $n = 4, 6, 8,$ and 10. Compare each result with e.

[**Hint:** $1! = 1, 2! = 2 \cdot 1, 3! = 3 \cdot 2 \cdot 1,$
$$n! = n(n - 1) \cdot \cdots \cdot (3)(2)(1).]$$

125. Another Formula for e Use a calculator to compute the various values of the expression. Compare the values to e.

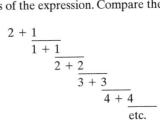

126. Difference Quotient If $f(x) = a^x$, show that

$$\frac{f(x+h) - f(x)}{h} = a^x \cdot \frac{a^h - 1}{h} \quad h \neq 0$$

127. If $f(x) = a^x$, show that $f(A + B) = f(A) \cdot f(B)$.

128. If $f(x) = a^x$, show that $f(-x) = \dfrac{1}{f(x)}$.

129. If $f(x) = a^x$, show that $f(\alpha x) = [f(x)]^\alpha$.

Problems 130 and 131 define two other transcendental functions.

130. The **hyperbolic sine function**, designated by $\sinh x$, is defined as

$$\sinh x = \frac{1}{2}(e^x - e^{-x})$$

(a) Show that $f(x) = \sinh x$ is an odd function.

(b) Graph $f(x) = \sinh x$.

131. The **hyperbolic cosine function**, designated by $\cosh x$, is defined as

$$\cosh x = \frac{1}{2}(e^x + e^{-x})$$

(a) Show that $f(x) = \cosh x$ is an even function.

(b) Graph $f(x) = \cosh x$.

(c) Refer to Problem 130. Show that, for every x,

$$(\cosh x)^2 - (\sinh x)^2 = 1$$

132. Historical Problem Pierre de Fermat (1601–1665) conjectured that the function

$$f(x) = 2^{(2^x)} + 1$$

for $x = 1, 2, 3, \ldots$, would always have a value equal to a prime number. But Leonhard Euler (1707–1783) showed that this formula fails for $x = 5$. Use a calculator to determine the prime numbers produced by f for $x = 1, 2, 3, 4$. Then show that $f(5) = 641 \times 6{,}700{,}417$, which is not prime.

133. *Challenge Problem* Solve: $3^{2x-1} - 4 \cdot 3^x + 9 = 0$

134. *Challenge Problem* Solve: $2^{\frac{2}{3}x+1} - 3 \cdot 2^{\frac{1}{3}x} - 20 = 0$

Explaining Concepts: Discussion and Writing

135. The bacteria in a 4-liter container double every minute. After 60 minutes the container is full. How long did it take to fill half the container?

136. Explain in your own words what the number e is. Provide at least two applications that use this number.

137. Do you think that there is a power function that increases more rapidly than an exponential function? Explain.

138. As the base a of an exponential function $f(x) = a^x$, where $a > 1$, increases, what happens to its graph for $x > 0$? What happens to its graph for $x < 0$?

139. The graphs of $y = a^{-x}$ and $y = \left(\dfrac{1}{a}\right)^x$ are identical. Why?

Retain Your Knowledge

Problems 140–149 are based on material learned earlier in the course. The purpose of these problems is to keep the material fresh in your mind so that you are better prepared for the final exam.

140. Solve the inequality: $x^3 + 5x^2 \leq 4x + 20$.

141. Solve the inequality: $\dfrac{x+1}{x-2} \geq 1$.

142. Find the equation of the quadratic function f that has its vertex at $(3, 5)$ and contains the point $(2, 3)$.

143. Suppose $f(x) = x^2 + 2x - 3$.

(a) Graph f by determining whether its graph is concave up or concave down and by finding its vertex, axis of symmetry, y-intercept, and x-intercepts, if any.

(b) Find the domain and range of f.

(c) Determine where f is increasing and where it is decreasing.

144. Solve: $13x - (5x - 6) = 2x - (8x - 27)$

145. Find an equation for the circle with center $(0, 0)$ and radius $r = 1$.

146. Solve: $x - 16\sqrt{x} + 48 = 0$

147. If \$12,000 is invested at 3.5% simple interest for 2.5 years, how much interest is earned?

148. Determine where the graph of $f(x) = x^4 - 5x^2 - 6$ is below the graph of $g(x) = 2x^2 + 12$ by solving the inequality $f(x) \leq g(x)$.

149. Find the difference quotient of $f(x) = 2x^2 - 7x$.

'Are You Prepared?' Answers

1. $64; 4; \dfrac{1}{9}$ **2.** $\{-4, 1\}$ **3.** False   **4.** 3   **5.** True   **6.** line; $-3; 10$ **7.** $[2, \infty); (-\infty, 2]$

6.4 Logarithmic Functions

PREPARING FOR THIS SECTION *Before getting started, review the following:*

- Solving Inequalities (Section 1.5, pp. 123–126)
- Quadratic Inequalities (Section 4.5, pp. 321–323)
- Polynomial and Rational Inequalities (Section 5.5, pp. 380–384)

Now Work the 'Are You Prepared?' problems on page 460.

OBJECTIVES
1. Change Exponential Statements to Logarithmic Statements and Logarithmic Statements to Exponential Statements (p. 452)
2. Evaluate Logarithmic Expressions (p. 453)
3. Determine the Domain of a Logarithmic Function (p. 453)
4. Graph Logarithmic Functions (p. 454)
5. Solve Logarithmic Equations (p. 458)

Recall that a one-to-one function $y = f(x)$ has an inverse function that is defined implicitly by the equation $x = f(y)$. In particular, the exponential function $y = f(x) = a^x$, where $a > 0$ and $a \neq 1$, is one-to-one, so it has an inverse function that is defined implicitly by the equation

$$x = a^y \qquad a > 0 \qquad a \neq 1$$

This inverse function is so important that it is given a name, the *logarithmic function*.

> **DEFINITION Logarithmic Function with Base *a***
>
> The **logarithmic function with base *a***, where $a > 0$ and $a \neq 1$, is denoted by $y = \log_a x$ (read as "*y* is the logarithm with base *a* of *x*") and is defined by
>
> $$y = \log_a x \quad \text{if and only if} \quad x = a^y$$
>
> The domain of the logarithmic function $y = \log_a x$ is $x > 0$.

In Words
When you need to evaluate $\log_a x$, think to yourself "*a* raised to what power gives me *x*?"

As this definition illustrates, **a logarithm is a name for a certain exponent.** So $\log_a x$ equals the exponent to which *a* must be raised to obtain *x*.

EXAMPLE 1 **Relating Logarithms to Exponents**

(a) If $y = \log_3 x$, then $x = 3^y$. For example, the logarithmic statement $4 = \log_3 81$ is equivalent to the exponential statement $81 = 3^4$.

(b) If $y = \log_5 x$, then $x = 5^y$. For example, $-1 = \log_5\left(\dfrac{1}{5}\right)$ is equivalent to $\dfrac{1}{5} = 5^{-1}$.

1 Change Exponential Statements to Logarithmic Statements and Logarithmic Statements to Exponential Statements

The definition of a logarithm can be used to convert from exponential form to logarithmic form, and vice versa, as illustrated in the next two examples.

EXAMPLE 2 **Changing Exponential Statements to Logarithmic Statements**

Change each exponential statement to an equivalent statement involving a logarithm.

(a) $1.2^3 = m$ (b) $e^b = 9$ (c) $a^4 = 24$

Solution Use the fact that $y = \log_a x$ and $x = a^y$, where $a > 0$ and $a \neq 1$, are equivalent.

(a) If $1.2^3 = m$, then $3 = \log_{1.2} m$.   (b) If $e^b = 9$, then $b = \log_e 9$.

(c) If $a^4 = 24$, then $4 = \log_a 24$.

Now Work PROBLEM 11

| EXAMPLE 3 | **Changing Logarithmic Statements to Exponential Statements** |

Change each logarithmic statement to an equivalent statement involving an exponent.

(a) $\log_a 4 = 5$ (b) $\log_e b = -3$ (c) $\log_3 5 = c$

Solution
(a) If $\log_a 4 = 5$, then $a^5 = 4$.
(b) If $\log_e b = -3$, then $e^{-3} = b$.
(c) If $\log_3 5 = c$, then $3^c = 5$.

➤ **Now Work** PROBLEM 19

2 Evaluate Logarithmic Expressions

To find the exact value of a logarithm, write the logarithm in exponential notation using the fact that $y = \log_a x$ is equivalent to $a^y = x$, and use the fact that if $a^u = a^v$, then $u = v$.

| EXAMPLE 4 | **Finding the Exact Value of a Logarithmic Expression** |

Find the exact value of:

(a) $\log_2 16$ (b) $\log_3 \dfrac{1}{27}$

Solution
(a) To find $\log_2 16$, think "2 raised to what power equals 16?" Then,

$$y = \log_2 16$$
$$2^y = 16 \quad \text{Change to exponential form.}$$
$$2^y = 2^4 \quad 16 = 2^4$$
$$y = 4 \quad \text{Equate exponents.}$$

Therefore, $\log_2 16 = 4$.

(b) To find $\log_3 \dfrac{1}{27}$, think "3 raised to what power equals $\dfrac{1}{27}$?" Then,

$$y = \log_3 \frac{1}{27}$$
$$3^y = \frac{1}{27} \quad \text{Change to exponential form.}$$
$$3^y = 3^{-3} \quad \frac{1}{27} = \frac{1}{3^3} = 3^{-3}$$
$$y = -3 \quad \text{Equate exponents.}$$

Therefore, $\log_3 \dfrac{1}{27} = -3$.

➤ **Now Work** PROBLEM 27

3 Determine the Domain of a Logarithmic Function

The logarithmic function $y = \log_a x$ has been defined as the inverse of the exponential function $y = a^x$. That is, if $f(x) = a^x$, then $f^{-1}(x) = \log_a x$. Based on the discussion in Section 6.2 on inverse functions, for a function f and its inverse f^{-1},

$$\text{Domain of } f^{-1} = \text{Range of } f \quad \text{and} \quad \text{Range of } f^{-1} = \text{Domain of } f$$

Consequently, it follows that

- Domain of the logarithmic function = Range of the exponential function = $(0, \infty)$
- Range of the logarithmic function = Domain of the exponential function = $(-\infty, \infty)$

The next box summarizes some properties of the logarithmic function.

> $y = \log_a x$ if and only if $x = a^y$
> Domain: $0 < x < \infty$ Range: $-\infty < y < \infty$

The domain of a logarithmic function consists of the *positive* real numbers, so the argument of a logarithmic function must be greater than zero.

EXAMPLE 5

Finding the Domain of a Logarithmic Function

Find the domain of each logarithmic function.

(a) $F(x) = \log_2(x + 3)$ (b) $g(x) = \log_5\left(\dfrac{1 + x}{1 - x}\right)$ (c) $h(x) = \log_{1/2}|x|$

Solution

(a) The domain of F consists of all x for which $x + 3 > 0$, that is, $x > -3$. Using interval notation, the domain of F is $(-3, \infty)$.

(b) The domain of g is restricted to

$$\frac{1 + x}{1 - x} > 0$$

Solve this inequality to find that the domain of g consists of all x between -1 and 1, that is, $-1 < x < 1$, or, using interval notation, $(-1, 1)$.

(c) Since $|x| > 0$, provided that $x \neq 0$, the domain of h consists of all real numbers except zero, or, using interval notation, $(-\infty, 0) \cup (0, \infty)$.

Now Work PROBLEMS 41 AND 47

4 Graph Logarithmic Functions

Because exponential functions and logarithmic functions are inverses of each other, the graph of the logarithmic function $y = \log_a x$ is the reflection about the line $y = x$ of the graph of the exponential function $y = a^x$, as shown in Figures 30(a) and (b).

For example, to graph $y = \log_2 x$, graph $y = 2^x$ and reflect it about the line $y = x$. See Figure 31. To graph $y = \log_{1/3} x$, graph $y = \left(\dfrac{1}{3}\right)^x$ and reflect it about the line $y = x$. See Figure 32.

(a) $0 < a < 1$

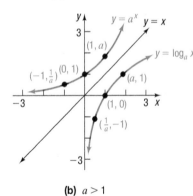

(b) $a > 1$

Figure 30

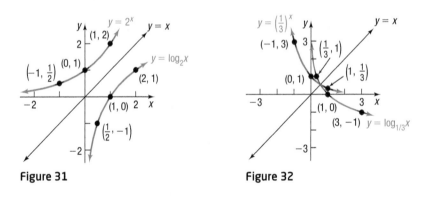

Figure 31

Figure 32

Now Work PROBLEM 61

The graphs of $y = \log_a x$ in Figures 30(a) and (b) lead to the following properties.

> **Properties of the Logarithmic Function $f(x) = \log_a x; a > 0, a \neq 1$**
> - The domain is the set of positive real numbers, or $(0, \infty)$ using interval notation; the range is the set of all real numbers, or $(-\infty, \infty)$ using interval notation.
> - The x-intercept of the graph is 1. There is no y-intercept.
> - The y-axis ($x = 0$) is a vertical asymptote of the graph of f.
> - A logarithmic function is decreasing if $0 < a < 1$ and is increasing if $a > 1$.
> - The graph of f contains the points $(1, 0)$, $(a, 1)$, and $\left(\dfrac{1}{a}, -1\right)$.
> - The graph is smooth and continuous, with no corners or gaps.

In Words
$y = \log_e x$ is written $y = \ln x$.

If the base of a logarithmic function is the number e, the result is the **natural logarithm function**. This function occurs so frequently in applications that it is given a special symbol, **ln** (from the Latin, *logarithmus naturalis*). That is,

$$y = \ln x \quad \text{if and only if} \quad x = e^y \qquad \text{(1)}$$

Because $y = \ln x$ and the exponential function $y = e^x$ are inverse functions, the graph of $y = \ln x$ can be obtained by reflecting the graph of $y = e^x$ about the line $y = x$. See Figure 33.

Using a calculator with an $\boxed{\ln}$ key, we can obtain other points on the graph of $f(x) = \ln x$. See Table 7.

Seeing the Concept
Graph $Y_1 = e^x$ and $Y_2 = \ln x$ on the same square screen. Use eVALUEate to verify the points on the graph given in Figure 33. Do you see the symmetry of the two graphs with respect to the line $y = x$?

Table 7

x	ln x
$\dfrac{1}{2}$	−0.69
2	0.69
3	1.10

Figure 33

EXAMPLE 6

Graphing a Logarithmic Function and Its Inverse

(a) Find the domain of the logarithmic function $f(x) = -\ln(x - 2)$.
(b) Graph f.
(c) From the graph, determine the range and vertical asymptote of f.
(d) Find f^{-1}, the inverse of f.
(e) Find the domain and the range of f^{-1}.
(f) Graph f^{-1}.

Solution

(a) The domain of f consists of all x for which $x - 2 > 0$, or equivalently, $x > 2$. The domain of f is $\{x \mid x > 2\}$, or $(2, \infty)$ in interval notation.

(b) To obtain the graph of $y = -\ln(x - 2)$, begin with the graph of $y = \ln x$ and use transformations. See Figure 34.

Figure 34   **(a)** $y = \ln x$

Multiply by −1; reflect about the *x*-axis.   **(b)** $y = -\ln x$

Replace *x* by *x* − 2; shift right 2 units.   **(c)** $y = -\ln(x - 2)$

(continued)

(c) The range of $f(x) = -\ln(x - 2)$ is the set of all real numbers. The vertical asymptote is $x = 2$. [Do you see why? The original asymptote $(x = 0)$ is shifted to the right 2 units.]

(d) To find f^{-1}, begin with $y = -\ln(x - 2)$. The inverse function is defined implicitly by the equation

$$x = -\ln(y - 2)$$

Now solve for y.

$$-x = \ln(y - 2) \qquad \text{Isolate the logarithm.}$$
$$e^{-x} = y - 2 \qquad \text{Change to exponential form.}$$
$$y = e^{-x} + 2 \qquad \text{Solve for } y.$$

The inverse of f is $f^{-1}(x) = e^{-x} + 2$.

(e) The domain of f^{-1} equals the range of f, which is the set of all real numbers, from part (c). The range of f^{-1} is the domain of f, which is $(2, \infty)$ in interval notation.

(f) To graph f^{-1}, use the graph of f in Figure 34(c) and reflect it about the line $y = x$. See Figure 35. We could also graph $f^{-1}(x) = e^{-x} + 2$ using transformations.

Figure 35

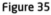 **Now Work** PROBLEM 73

If the base of a logarithmic function is the number 10, the result is the **common logarithm function**. If the base a of the logarithmic function is not indicated, it is understood to be 10. That is,

$$y = \log x \quad \text{if and only if} \quad x = 10^y$$

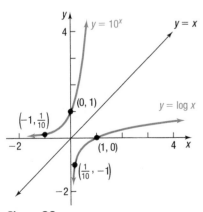

Figure 36

Because $y = \log x$ and the exponential function $y = 10^x$ are inverse functions, the graph of $y = \log x$ can be obtained by reflecting the graph of $y = 10^x$ about the line $y = x$. See Figure 36.

EXAMPLE 7

Graphing a Logarithmic Function and Its Inverse

(a) Find the domain of the logarithmic function $f(x) = 3 \log(x - 1)$.
(b) Graph f.
(c) From the graph, determine the range and vertical asymptote of f.
(d) Find f^{-1}, the inverse of f.
(e) Find the domain and the range of f^{-1}.
(f) Graph f^{-1}.

Solution

(a) The domain of f consists of all x for which $x - 1 > 0$, or equivalently, $x > 1$. The domain of f is $\{x \mid x > 1\}$, or $(1, \infty)$ in interval notation.
(b) To obtain the graph of $y = 3 \log(x - 1)$, begin with the graph of $y = \log x$ and use transformations. See Figure 37.

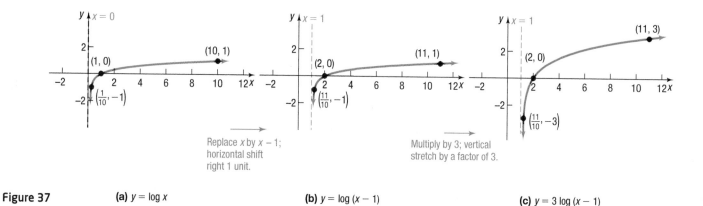

Figure 37   **(a)** $y = \log x$ **(b)** $y = \log(x - 1)$ **(c)** $y = 3 \log(x - 1)$

(c) The range of $f(x) = 3 \log(x - 1)$ is the set of all real numbers. The vertical asymptote is $x = 1$.
(d) Begin with $y = 3 \log(x - 1)$. The inverse function is defined implicitly by the equation

$$x = 3 \log(y - 1)$$

Solve for y.

$$\frac{x}{3} = \log(y - 1) \quad \text{Isolate the logarithm.}$$

$$10^{x/3} = y - 1 \quad \text{Change to exponential form.}$$

$$y = 10^{x/3} + 1 \quad \text{Solve for y.}$$

The inverse of f is $f^{-1}(x) = 10^{x/3} + 1$.

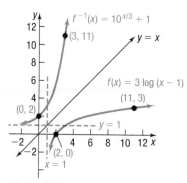

Figure 38

(e) The domain of f^{-1} is the range of f, which is the set of all real numbers, from part (c). The range of f^{-1} is the domain of f, which is $(1, \infty)$ in interval notation.
(f) To graph f^{-1}, use the graph of f in Figure 37(c) and reflect it about the line $y = x$. See Figure 38.

We could also graph $f^{-1}(x) = 10^{x/3} + 1$ using transformations.

 Now Work PROBLEM 81

5 Solve Logarithmic Equations

Equations that contain logarithms are called **logarithmic equations**. Care must be taken when solving logarithmic equations algebraically. In the expression $\log_a M$, remember that a and M are positive and $a \neq 1$. Be sure to check each apparent solution in the original equation and discard any that are extraneous.

Some logarithmic equations can be solved by changing the logarithmic equation to exponential form using the fact that $y = \log_a x$ means $a^y = x$.

EXAMPLE 8

Solving Logarithmic Equations

Solve:

(a) $\log_3(4x - 7) = 2$ (a) $\log_x 64 = 2$

Solution   (a) To solve, change the logarithmic equation to exponential form.

$$\log_3(4x - 7) = 2$$
$$4x - 7 = 3^2 \quad \text{Change to exponential form.}$$
$$4x - 7 = 9$$
$$4x = 16$$
$$x = 4$$

✓ Check: $\log_3(4x - 7) = \log_3(4 \cdot 4 - 7) = \log_3 9 = 2 \quad 3^2 = 9$
The solution set is $\{4\}$.

(b) To solve, change the logarithmic equation to exponential form.

$$\log_x 64 = 2$$
$$x^2 = 64 \quad \text{Change to exponential form.}$$
$$x = \pm\sqrt{64} = \pm 8 \quad \text{Use the Square Root Method.}$$

Because the base of a logarithm must be positive, discard -8. Check the potential solution 8.

✓ Check: $\log_8 64 = 2 \quad 8^2 = 64$

The solution set is $\{8\}$.

EXAMPLE 9

Using Logarithms to Solve an Exponential Equation

Solve: $e^{2x} = 5$

Solution   To solve, change the exponential equation to logarithmic form.

$$e^{2x} = 5$$
$$\ln 5 = 2x \quad \text{Change to logarithmic form.}$$
$$x = \frac{\ln 5}{2} \quad \text{Exact solution}$$
$$\approx 0.805 \quad \text{Approximate solution}$$

The solution set is $\left\{\dfrac{\ln 5}{2}\right\}$.

Now Work PROBLEMS 89 AND 101

EXAMPLE 10   Alcohol and Driving

Blood alcohol concentration (BAC) is a measure of the amount of alcohol in a person's bloodstream. A BAC of 0.04% means that a person has 4 parts alcohol per 10,000 parts blood in the body. Relative risk is defined as the likelihood of one event occurring divided by the likelihood of a second event occurring. For example, if an individual with a BAC of 0.02% is 1.4 times as likely to have a car accident as an individual who has not been drinking, the relative risk of an accident with a BAC of 0.02% is 1.4. Recent medical research suggests that the relative risk R of having an accident while driving a car can be modeled by an equation of the form

$$R = e^{kx}$$

where x is the percent concentration of alcohol in the bloodstream and k is a constant.

(a) Research indicates that the relative risk of a person having an accident with a BAC of 0.02% is 1.4. Find the constant k in the equation.

(b) Using this value of k, what is the relative risk if the BAC is 0.17%?

(c) Using this same value of k, what BAC corresponds to a relative risk of 100?

(d) If the law asserts that anyone with a relative risk of 4 or more should not have driving privileges, at what concentration of alcohol in the bloodstream should a driver be arrested and charged with DUI (driving under the influence)?

Solution   (a) For a blood alcohol concentration of 0.02% and a relative risk R of 1.4, let $x = 0.02$ and solve for k.

$$R = e^{kx}$$

$$1.4 = e^{k(0.02)} \qquad R = 1.4; x = 0.02$$

$$0.02k = \ln 1.4 \qquad \text{Change to a logarithmic statement.}$$

$$k = \frac{\ln 1.4}{0.02} \approx 16.82 \quad \text{Solve for } k.$$

(b) A BAC of 0.17% means $x = 0.17$. Use $k = 16.82$ in the equation to find the relative risk R:

$$R = e^{kx} = e^{(16.82)(0.17)} \approx 17.5$$

For a blood alcohol concentration of 0.17%, the relative risk R of an accident is about 17.5. That is, a person with a BAC of 0.17% is 17.5 times as likely to have a car accident as a person with no alcohol in the bloodstream.

(c) A relative risk of 100 means $R = 100$. Use $k = 16.82$ in the equation $R = e^{kx}$. The blood alcohol concentration x obeys

$$100 = e^{16.82x} \qquad R = e^{kx}, R = 100, k = 16.82$$

$$16.82x = \ln 100 \qquad \text{Change to a logarithmic statement.}$$

$$x = \frac{\ln 100}{16.82} \approx 0.27 \quad \text{Solve for } x.$$

NOTE A BAC of 0.30% results in a loss of consciousness in most people. ∎

For a blood alcohol concentration of 0.27%, the relative risk R of an accident is 100.

(d) A relative risk of 4 means $R = 4$. Use $k = 16.82$ in the equation $R = e^{kx}$. The blood alcohol concentration x obeys

$$4 = e^{16.82x}$$

$$16.82x = \ln 4$$

$$x = \frac{\ln 4}{16.82} \approx 0.082$$

NOTE In most states, the blood alcohol content at which a DUI citation is given is 0.08%. ∎

A driver with a BAC of 0.082% or more should be arrested and charged with DUI.

SUMMARY

Properties of the Logarithmic Function

$f(x) = \log_a x, \quad a > 1$
$(y = \log_a x \text{ if and only if } x = a^y)$

- Domain: the interval $(0, \infty)$; Range: the interval $(-\infty, \infty)$
- x-intercept: 1; y-intercept: none;
  vertical asymptote: $x = 0$ (y-axis); increasing; one-to-one
- The graph of f is smooth and continuous. It contains the
  points $\left(\dfrac{1}{a}, -1\right)$, $(1, 0)$, and $(a, 1)$.
- See Figure 39(a) for a typical graph.

$f(x) = \log_a x, \quad 0 < a < 1$
$(y = \log_a x \text{ if and only if } x = a^y)$

- Domain: the interval $(0, \infty)$; Range: the interval $(-\infty, \infty)$
- x-intercept: 1; y-intercept: none;
  vertical asymptote: $x = 0$ (y-axis); decreasing; one-to-one
- The graph of f is smooth and continuous. It contains the
  points $(a, 1)$, $(1, 0)$, and $\left(\dfrac{1}{a}, -1\right)$.
- See Figure 39(b) for a typical graph.

Figure 39 **(a)** $a > 1$ **(b)** $0 < a < 1$

6.4 Assess Your Understanding

'Are You Prepared?' *Answers are given at the end of these exercises. If you get a wrong answer, read the pages listed in red.*

1. Solve each inequality:
   (a) $3x - 7 \leq 8 - 2x$ (pp. 123–126)
   (b) $x^2 - x - 6 > 0$ (pp. 321–323)

2. Solve the inequality: $\dfrac{x - 1}{x + 4} > 0$ (pp. 382–384)

Concepts and Vocabulary

3. The domain of the logarithmic function $f(x) = \log_a x$ is _____.

4. The graph of every logarithmic function $f(x) = \log_a x$, where $a > 0$ and $a \neq 1$, contains the three points: _____, _____, and _____.

5. If the graph of a logarithmic function $f(x) = \log_a x$, where $a > 0$ and $a \neq 1$, is increasing, then its base is larger than _____.

6. **True or False** If $y = \log_a x$, then $y = a^x$.

7. **True or False** The graph of $f(x) = \log_a x$, where $a > 0$ and $a \neq 1$, has an x-intercept equal to 1 and no y-intercept.

8. **Multiple Choice** Select the answer that completes the statement: $y = \ln x$ if and only if _____.
   (a) $x = e^y$ **(b)** $y = e^x$ **(c)** $x = 10^y$ **(d)** $y = 10^x$

9. **Multiple Choice** The domain of $f(x) = \log_3(x + 2)$ is
   (a) $(-\infty, \infty)$ **(b)** $(2, \infty)$ **(c)** $(-2, \infty)$ **(d)** $(0, \infty)$

10. **Multiple Choice** $\log_3 81$ equals
    (a) 9 **(b)** 4 **(c)** 2 **(d)** 3

Skill Building

In Problems 11–18, change each exponential statement to an equivalent statement involving a logarithm.

11. $9 = 3^2$　　　　　　**12.** $16 = 4^2$　　　　　　**13.** $a^2 = 1.6$　　　　　　**14.** $a^3 = 2.1$

15. $2^x = 7.2$　　　　　　**16.** $3^x = 4.6$　　　　　　**17.** $e^x = 8$　　　　　　**18.** $e^{2.2} = M$

In Problems 19–26, change each logarithmic statement to an equivalent statement involving an exponent.

19. $\log_2 8 = 3$　　　　**20.** $\log_3\left(\dfrac{1}{9}\right) = -2$　　　　**21.** $\log_a 3 = 6$　　　　**22.** $\log_b 4 = 2$

23. $\log_3 2 = x$　　　　**24.** $\log_2 6 = x$　　　　**25.** $\ln 4 = x$　　　　**26.** $\ln x = 4$

In Problems 27–38, find the exact value of each logarithm without using a calculator.

27. $\log_2 1$　　　　　　**28.** $\log_8 8$　　　　　　**29.** $\log_7 49$　　　　　　**30.** $\log_3\left(\dfrac{1}{9}\right)$

31. $\log_{1/5} 125$　　　　**32.** $\log_{1/3} 9$　　　　**33.** $\log \sqrt{10}$　　　　**34.** $\log \sqrt[3]{100}$

35. $\log_{\sqrt{2}} 4$　　　　**36.** $\log_{\sqrt{3}} 9$　　　　**37.** $\ln \sqrt{e}$　　　　**38.** $\ln e^3$

In Problems 39–50, find the domain of each function.

39. $f(x) = \ln(x - 3)$　　　　**40.** $g(x) = \ln(x - 1)$　　　　**41.** $F(x) = \log_2 x^2$

42. $H(x) = \log_5 x^3$　　　　**43.** $f(x) = 3 - 2\log_4\left(\dfrac{x}{2} - 5\right)$　　　　**44.** $g(x) = 8 + 5\ln(2x + 3)$

45. $f(x) = \ln\left(\dfrac{1}{x + 1}\right)$　　　　**46.** $g(x) = \ln\left(\dfrac{1}{x - 5}\right)$　　　　**47.** $g(x) = \log_5\left(\dfrac{x + 1}{x}\right)$

48. $h(x) = \log_3\left(\dfrac{x}{x - 1}\right)$　　　　**49.** $f(x) = \sqrt{\ln x}$　　　　**50.** $g(x) = \dfrac{1}{\ln x}$

In Problems 51–58, use a calculator to evaluate each expression. Round your answer to three decimal places.

51. $\ln \dfrac{5}{3}$　　　**52.** $\dfrac{\ln 5}{3}$　　　**53.** $\dfrac{\ln \dfrac{10}{3}}{0.04}$　　　**54.** $\dfrac{\ln \dfrac{2}{3}}{-0.1}$

55. $\dfrac{\ln 4 + \ln 2}{\log 4 + \log 2}$　　　**56.** $\dfrac{\log 15 + \log 20}{\ln 15 + \ln 20}$　　　**57.** $\dfrac{2\ln 5 + \log 50}{\log 4 - \ln 2}$　　　**58.** $\dfrac{3\log 80 - \ln 5}{\log 5 + \ln 20}$

59. Find a so that the graph of $f(x) = \log_a x$ contains the point $(2, 2)$.

60. Find a so that the graph of $f(x) = \log_a x$ contains the point $\left(\dfrac{1}{2}, -4\right)$.

In Problems 61–64, graph each function and its inverse on the same set of axes.

61. $f(x) = 3^x; f^{-1}(x) = \log_3 x$　　　　　　**62.** $f(x) = 4^x; f^{-1}(x) = \log_4 x$

63. $f(x) = \left(\dfrac{1}{2}\right)^x; f^{-1}(x) = \log_{1/2} x$　　　　　　**64.** $f(x) = \left(\dfrac{1}{3}\right)^x; f^{-1}(x) = \log_{1/3} x$

In Problems 65–72, the graph of a logarithmic function is given. Match each graph to one of the following functions:

(A) $y = \log_3 x$　　　(B) $y = \log_3(-x)$　　　(C) $y = -\log_3 x$　　　(D) $y = -\log_3(-x)$

(E) $y = \log_3 x - 1$　　　(F) $y = \log_3(x - 1)$　　　(G) $y = \log_3(1 - x)$　　　(H) $y = 1 - \log_3 x$

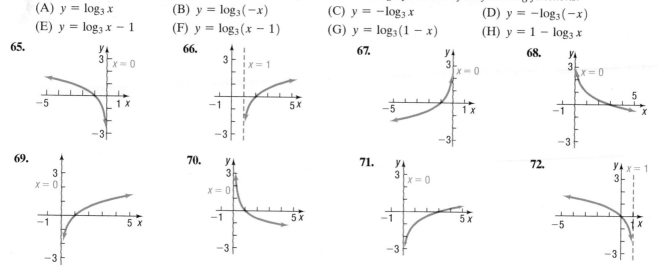

In Problems 73–88, use the given function f.
(a) *Find the domain of f.*   (b) *Graph f.*   (c) *From the graph, determine the range and any asymptotes of f.*
(d) *Find* f^{-1}, *the inverse of f.*   (e) *Find the domain and the range of* f^{-1}.   (f) *Graph* f^{-1}.

73. $f(x) = \ln(x + 4)$ 74. $f(x) = \ln(x - 3)$ 75. $f(x) = 2 + \ln x$ 76. $f(x) = -\ln(-x)$

77. $f(x) = \ln(2x) - 3$ 78. $f(x) = -2\ln(x + 1)$ 79. $f(x) = \log(x - 4) + 2$ 80. $f(x) = \frac{1}{2}\log x - 5$

81. $f(x) = \frac{1}{2}\log(2x)$ 82. $f(x) = \log(-2x)$ 83. $f(x) = 3 + \log_3(x + 2)$ 84. $f(x) = 2 - \log_3(x + 1)$

85. $f(x) = e^{x+2} - 3$ 86. $f(x) = 3e^x + 2$ 87. $f(x) = 2^{x/3} + 4$ 88. $f(x) = -3^{x+1}$

In Problems 89–112, solve each equation.

89. $\log_3 x = 2$ 90. $\log_5 x = 3$ 91. $\log_2(3x + 4) = 5$ 92. $\log_3(3x - 2) = 2$

93. $\log_x 16 = 2$ 94. $\log_x\left(\frac{1}{8}\right) = 3$ 95. $\ln e^x = 5$ 96. $\ln e^{-2x} = 8$

97. $\log_4 64 = x$ 98. $\log_5 625 = x$ 99. $\log_3 243 = 2x + 1$ 100. $\log_6 36 = 5x + 3$

101. $e^{3x} = 10$ 102. $e^{-2x} = \frac{1}{3}$ 103. $e^{2x+5} = 8$ 104. $e^{-2x+1} = 13$

105. $\log_7(x^2 + 4) = 2$ 106. $\log_5(x^2 + x + 4) = 2$ 107. $\log_2 8^x = -6$ 108. $\log_3 3^x = -1$

109. $5e^{0.2x} = 7$ 110. $8 \cdot 10^{2x-7} = 3$ 111. $2 \cdot 10^{2-x} = 5$ 112. $4e^{x+1} = 5$

113. Suppose that $G(x) = \log_3(2x + 1) - 2$.

    (a) What is the domain of G?
    (b) What is $G(40)$? What point is on the graph of G?
    (c) If $G(x) = 3$, what is x? What point is on the graph of G?
    (d) What is the zero of G?

114. Suppose that $F(x) = \log_2(x + 1) - 3$.

    (a) What is the domain of F?
    (b) What is $F(7)$? What point is on the graph of F?
    (c) If $F(x) = -1$, what is x? What point is on the graph of F?
    (d) What is the zero of F?

Mixed Practice In Problems 115–118, graph each function. Based on the graph, state the domain and the range, and find any intercepts.

115. $f(x) = \begin{cases} \ln(-x) & \text{if } x < 0 \\ \ln x & \text{if } x > 0 \end{cases}$

116. $f(x) = \begin{cases} \ln(-x) & \text{if } x \le -1 \\ -\ln(-x) & \text{if } -1 < x < 0 \end{cases}$

117. $f(x) = \begin{cases} -\ln x & \text{if } 0 < x < 1 \\ \ln x & \text{if } x \ge 1 \end{cases}$

118. $f(x) = \begin{cases} \ln x & \text{if } 0 < x < 1 \\ -\ln x & \text{if } x \ge 1 \end{cases}$

Applications and Extensions

119. **Chemistry** The pH of a chemical solution is given by the formula

$$\text{pH} = -\log_{10}[H^+]$$

where $[H^+]$ is the concentration of hydrogen ions in moles per liter. Values of pH range from 0 (acidic) to 14 (alkaline).
    (a) What is the pH of a solution for which $[H^+]$ is 0.1?
    (b) What is the pH of a solution for which $[H^+]$ is 0.01?
    (c) What is the pH of a solution for which $[H^+]$ is 0.001?
    (d) What happens to pH as the hydrogen ion concentration decreases?
    (e) Determine the hydrogen ion concentration of an orange (pH = 3.5).
    (f) Determine the hydrogen ion concentration of human blood (pH = 7.4).

120. **Diversity Index** Shannon's diversity index is a measure of the diversity of a population. The diversity index is given by the formula

$$H = -(p_1 \log p_1 + p_2 \log p_2 + \cdots + p_n \log p_n)$$

where p_1 is the proportion of the population that is species 1, p_2 is the proportion of the population that is species 2, and so on. In this problem, the population is people in the United States and the species is race.
    (a) According to the U.S. Census Bureau, the distribution of race in the United States in 2015 was:

Race	Proportion
White	0.617
Black or African American	0.124
American Indian and Alaska Native	0.007
Asian	0.053
Native Hawaiian and Other Pacific Islander	0.002
Hispanic	0.177
Two or More Races	0.020

Source: U.S. Census Bureau

Compute the diversity index of the United States in 2015.
    (b) The largest value of the diversity index is given by $H_{max} = \log(S)$, where S is the number of categories of race. Compute H_{max}.
    (c) The **evenness ratio** is given by $E_H = \dfrac{H}{H_{max}}$, where $0 \le E_H \le 1$. If $E_H = 1$, there is complete evenness. Compute the evenness ratio for the United States.
    (d) Obtain the distribution of race for the United States in 2010 from the Census Bureau. Compute Shannon's diversity index. Is the United States becoming more diverse? Why?

121. Atmospheric Pressure The atmospheric pressure p on an object decreases with increasing height. This pressure, measured in millimeters of mercury, is related to the height h (in kilometers) above sea level by the function

$$p(h) = 760e^{-0.145h}$$

(a) Find the height of an aircraft if the atmospheric pressure is 320 millimeters of mercury.
(b) Find the height of a mountain if the atmospheric pressure is 667 millimeters of mercury.

122. Healing of Wounds The normal healing of wounds can be modeled by an exponential function. If A_0 represents the original area of the wound, and if A equals the area of the wound, then the function

$$A(n) = A_0e^{-0.35n}$$

describes the area of a wound after n days following an injury when no infection is present to retard the healing. Suppose that a wound initially had an area of 100 square millimeters.
(a) If healing is taking place, after how many days will the wound be one-half its original size?
(b) How long before the wound is 10% of its original size?

123. Exponential Probability Between 12:00 PM and 1:00 PM, cars arrive at Citibank's drive-thru at the rate of 6 cars per hour (0.1 car per minute). The following formula from probability can be used to determine the probability that a car will arrive within t minutes of 12:00 PM.

$$F(t) = 1 - e^{-0.1t}$$

(a) Determine how many minutes are needed for the probability to reach 50%.
(b) Determine how many minutes are needed for the probability to reach 80%.
(c) Is it possible for the probability to equal 100%? Explain.

124. Exponential Probability Between 5:00 PM and 6:00 PM, cars arrive at Jiffy Lube at the rate of 9 cars per hour (0.15 car per minute). The following formula from probability can be used to determine the probability that a car will arrive within t minutes of 5:00 PM.

$$F(t) = 1 - e^{-0.15t}$$

(a) Determine how many minutes are needed for the probability to reach 50%.
(b) Determine how many minutes are needed for the probability to reach 80%.

125. Drug Medication The function

$$D(h) = 5e^{-0.4h}$$

can be used to find the number of milligrams D of a certain drug that is in a patient's bloodstream h hours after the drug was administered. When the number of milligrams reaches 2, the drug is to be administered again. What is the time between injections?

126. Spreading of Rumors A model for the number N of people in a college community who have heard a certain rumor is

$$N(d) = P(1 - e^{-0.15d})$$

where P is the total population of the community and d is the number of days that have elapsed since the rumor began. In a community of 1000 students, how many days will elapse before 450 students have heard the rumor?

127. Current in an RL Circuit The equation governing the amount of current I (in amperes) after time t (in seconds) in a simple RL circuit consisting of a resistance R (in ohms), an inductance L (in henrys), and an electromotive force E (in volts) is

$$I = \frac{E}{R}\left[1 - e^{-(R/L)t}\right]$$

If $E = 12$ volts, $R = 10$ ohms, and $L = 5$ henrys, how long does it take to obtain a current of 0.5 ampere? Of 1.0 ampere? Graph the equation.

128. Learning Curve Psychologists sometimes use the function

$$L(t) = A(1 - e^{-kt})$$

to measure the amount L learned at time t. Here A represents the amount to be learned, and the number k measures the rate of learning. Suppose that a student has an amount A of 200 vocabulary words to learn. A psychologist determines that the student has learned 20 vocabulary words after 5 minutes.
(a) Determine the rate of learning k.
(b) Approximately how many words will the student have learned after 10 minutes?
(c) After 15 minutes?
(d) How long does it take for the student to learn 180 words?

Loudness of Sound Problems 129–132 use the following discussion: The **loudness** $L(x)$, measured in decibels (dB), of a sound of intensity x, measured in watts per square meter, is defined as $L(x) = 10 \log \frac{x}{I_0}$, where $I_0 = 10^{-12}$ watt per square meter is the least intense sound that a human ear can detect. Determine the loudness, in decibels, of each of the following sounds.

129. Normal conversation: intensity of $x = 10^{-7}$ watt per square meter.

130. Amplified rock music: intensity of 10^{-1} watt per square meter.

131. Heavy city traffic: intensity of $x = 10^{-3}$ watt per square meter.

132. Diesel truck traveling 40 miles per hour 50 feet away: intensity 10 times that of a passenger car traveling 50 miles per hour 50 feet away, whose loudness is 70 decibels.

The Richter Scale Problems 133 and 134 use the following discussion: The **Richter scale** is one way of converting seismographic readings into numbers that provide an easy reference for measuring the magnitude M of an earthquake. All earthquakes are compared to a **zero-level earthquake** whose seismographic reading measures 0.001 millimeter at a distance of 100 kilometers from the epicenter. An earthquake whose seismographic reading measures x millimeters has **magnitude** M(x), given by

$$M(x) = \log\left(\frac{x}{x_0}\right)$$

where $x_0 = 10^{-3}$ is the reading of a zero-level earthquake the same distance from its epicenter. In Problems 133 and 134, determine the magnitude of each earthquake.

133. Magnitude of an Earthquake Mexico City in 1985: seismographic reading of 125,892 millimeters 100 kilometers from the center

134. Magnitude of an Earthquake San Francisco in 1906: seismographic reading of 50,119 millimeters 100 kilometers from the center

135. Alcohol and Driving The concentration of alcohol in a person's bloodstream is measurable. Suppose that the relative risk R of having an accident while driving a car can be modeled by an equation of the form

$$R = e^{kx}$$

where x is the percent concentration of alcohol in the bloodstream and k is a constant.

(a) Suppose that a concentration of alcohol in the bloodstream of 0.03 percent results in a relative risk of an accident of 1.4. Find the constant k in the equation.

(b) Using this value of k, what is the relative risk if the concentration is 0.17 percent?

(c) Using the same value of k, what concentration of alcohol corresponds to a relative risk of 100?

(d) If the law asserts that anyone with a relative risk of having an accident of 5 or more should not have driving privileges, at what concentration of alcohol in the bloodstream should a driver be arrested and charged with a DUI?

(e) Compare this situation with that of Example 10. If you were a lawmaker, which situation would you support? Give your reasons.

136. The Marriage Problem There is an infamous problem from mathematics that attempts to quantify the number of potential mates one should date before choosing one's "true love." The function

$$L(x) = -x \ln x$$

represents the probability of finding the ideal mate after rejecting the first x proportion of potential mates. For example, if you reject the first 20% = 0.20 of individuals you date, the probability of finding the ideal mate is $L(0.2) \approx 0.322$. So, if you want the probability of finding the ideal mate to be greater than 0.332 and you are only willing to date up to 20 individuals, you should reject the first 0.2(20) = 4 individuals before attempting to decide on the ideal mate. Presumably, you are using those first 4 individuals to help you decide which traits you value in a mate.

(a) Determine and interpret L(0.1).

(b) Determine and interpret L(0.6).

(c) What is the domain of L?

(d) Graph L = L(x) over the domain.

(e) Judging on the basis of the approach suggested by the model, what is the value of x that maximizes L? What is the highest probability of finding the ideal mate?

137. Challenge Problem Solve: $\log_6(\log_2 x) = 1$

138. Challenge Problem Solve: $\log_2[\log_4(\log_3 x)] = 0$

139. Challenge Problem Solve: $\log_3 9^{2x+3} = x^2 + 1$

Explaining Concepts: Discussion and Writing

140. Is there any function of the form $y = x^\alpha, 0 < \alpha < 1$, that increases more slowly than a logarithmic function whose base is greater than 1? Explain.

141. In the definition of the logarithmic function, the base a is not allowed to equal 1. Why?

142. **Critical Thinking** In buying a new car, one consideration might be how well the price of the car holds up over time. Different makes of cars have different depreciation rates. One way to compute a depreciation rate for a car is given here. Suppose that the current prices of a certain automobile are as shown in the table.

	Age in Years					
New	**1**	**2**	**3**	**4**	**5**
$38,000	$36,600	$32,400	$28,750	$25,400	$21,200

Use the formula New = Old(e^{Rt}) to find R, the annual depreciation rate, for a specific time t. When might be the best time to trade in the car? Consult the NADA ("blue") book and compare two like models that you are interested in. Which has the better depreciation rate?

Retain Your Knowledge

Problems 143–152 are based on material learned earlier in the course. The purpose of these problems is to keep the material fresh in your mind so that you are better prepared for the final exam.

143. Find the real zeros of $g(x) = 4x^4 - 37x^2 + 9$. What are the x-intercepts of the graph of g?

144. Find the average rate of change of $f(x) = 9^x$ from $\frac{1}{2}$ to 1.

145. Use the Intermediate Value Theorem to show that the function $f(x) = 4x^3 - 2x^2 - 7$ has a real zero in the interval $[1, 2]$.

146. A complex polynomial function f of degree 4 with real coefficients has the zeros $-1, 2$, and $3 - i$. Find the remaining zero(s) of f. Then find a polynomial function that has the zeros.

147. Solve: $2x^2 - 7x - 1 = 0$

148. Find an equation of the line that contains the points $(0, 1)$ and $(8, -4)$. Write the equation in general form.

149. Solve: $|2x + 17| = 45$

150. Forensic Science The relationship between the height H of an adult female and the length x of her tibia, in centimeters, is estimated by the linear model $H(x) = 2.90x + 61.53$. If incomplete skeletal remains of an adult female include a tibia measuring 30.9 centimeters, estimate the height of the female. Round to the nearest tenth.

151. For $f(x) = x^3$, find $\dfrac{f(x) - f(2)}{x - 2}$.

152. Factor completely:
$$(x + 5)^4 \cdot 7(x - 3)^6 + (x - 3)^7 \cdot 4(x + 5)^3$$

'Are You Prepared?' Answers

1. (a) $x \le 3$ **(b)** $x < -2$ or $x > 3$ **2.** $x < -4$ or $x > 1$

6.5 Properties of Logarithms

OBJECTIVES   **1** Work with the Properties of Logarithms (p. 465)
   2 Write a Logarithmic Expression as a Sum or Difference of Logarithms (p. 467)
   3 Write a Logarithmic Expression as a Single Logarithm (p. 468)
   4 Evaluate Logarithms Whose Base Is Neither 10 Nor e (p. 469)

1 Work with the Properties of Logarithms

Logarithms have some very useful properties that can be derived directly from the definition and the laws of exponents.

EXAMPLE 1   **Establishing Properties of Logarithms**

(a) Show that $\log_a 1 = 0$.   (b) Show that $\log_a a = 1$.

Solution   (a) This fact was established when we graphed $y = \log_a x$ (see Figure 30 on page 454). To show the result algebraically, let $y = \log_a 1$. Then

$$y = \log_a 1$$
$$a^y = 1 \qquad \text{Change to exponential form.}$$
$$a^y = a^0 \qquad a^0 = 1 \text{ since } a > 0, a \neq 1$$
$$y = 0 \qquad \text{Equate exponents.}$$
$$\log_a 1 = 0 \qquad y = \log_a 1$$

(b) Let $y = \log_a a$. Then

$$y = \log_a a$$
$$a^y = a \qquad \text{Change to exponential form.}$$
$$a^y = a^1 \qquad a = a^1$$
$$y = 1 \qquad \text{Equate exponents.}$$
$$\log_a a = 1 \qquad y = \log_a a$$

To summarize:

$$\log_a 1 = 0 \qquad \log_a a = 1$$

THEOREM **Properties of Logarithms**

In these properties, M and a are positive real numbers, $a \neq 1$, and r is any real number.

- The number $\log_a M$ is the exponent to which a must be raised to obtain M. That is,

$$a^{\log_a M} = M \tag{1}$$

- The logarithm with base a of a raised to a power equals that power. That is,

$$\log_a a^r = r \tag{2}$$

The proof uses the fact that $y = a^x$ and $y = \log_a x$ are inverse functions.

Proof of Property (1) For inverse functions,

$$f(f^{-1}(x)) = x \quad \text{for all } x \text{ in the domain of } f^{-1}$$

Use $f(x) = a^x$ and $f^{-1}(x) = \log_a x$ to find

$$f(f^{-1}(x)) = a^{\log_a x} = x \quad \text{for } x > 0$$

Now let $x = M$ to obtain $a^{\log_a M} = M$, where $M > 0$. ∎

Proof of Property (2) For inverse functions,

$$f^{-1}(f(x)) = x \quad \text{for all } x \text{ in the domain of } f$$

Use $f(x) = a^x$ and $f^{-1}(x) = \log_a x$ to find

$$f^{-1}(f(x)) = \log_a a^x = x \quad \text{for all real numbers } x$$

Now let $x = r$ to obtain $\log_a a^r = r$, where r is any real number. ∎

EXAMPLE 2

Using Properties of Logarithms (1) and (2)

(a) $2^{\log_2 \pi} = \pi$ (b) $\log_{0.2} 0.2^{-\sqrt{2}} = -\sqrt{2}$ (c) $\ln e^{kt} = kt$

━━━━▶ **Now Work** PROBLEM 15

Other useful properties of logarithms are given next.

THEOREM **Properties of Logarithms**

In these properties, M, N, and a are positive real numbers, $a \neq 1$, and r is any real number.

The Log of a Product Equals the Sum of the Logs

$$\log_a(MN) = \log_a M + \log_a N \tag{3}$$

The Log of a Quotient Equals the Difference of the Logs

$$\log_a\left(\frac{M}{N}\right) = \log_a M - \log_a N \tag{4}$$

The Log of a Power Equals the Product of the Power and the Log

$$\log_a M^r = r \log_a M \tag{5}$$

$$a^r = e^{r \ln a} \tag{6}$$

We prove properties (3), (5), and (6) and leave the proof of property (4) as an exercise (see Problem 109).

Proof of Property (3) Let $A = \log_a M$ and let $B = \log_a N$. These expressions are equivalent to the exponential expressions

$$a^A = M \quad \text{and} \quad a^B = N$$

Now

$$
\begin{aligned}
\log_a(MN) = \log_a(a^A a^B) &= \log_a a^{A+B} &&\text{Law of Exponents} \\
&= A + B &&\text{Property (2) of logarithms} \\
&= \log_a M + \log_a N
\end{aligned}
$$

$\blacksquare$

Proof of Property (5) Let $A = \log_a M$. This expression is equivalent to

$$a^A = M$$

Now

$$
\begin{aligned}
\log_a M^r = \log_a(a^A)^r &= \log_a a^{rA} &&\text{Law of Exponents} \\
&= rA &&\text{Property (2) of logarithms} \\
&= r\log_a M
\end{aligned}
$$

$\blacksquare$

Proof of Property (6) Property (1), with $a = e$, gives

$$e^{\ln M} = M$$

Now let $M = a^r$ and use property (5).

$$e^{\ln a^r} = e^{r \ln a} = a^r$$

$\blacksquare$

━━━ **Now Work** PROBLEM 19

2 Write a Logarithmic Expression as a Sum or Difference of Logarithms

Logarithms can be used to transform products into sums, quotients into differences, and powers into factors. Such transformations are useful in certain calculus problems.

EXAMPLE 3

Writing a Logarithmic Expression as a Sum of Logarithms

Write $\log_a(x\sqrt{x^2 + 1})$, $x > 0$, as a sum of logarithms. Express all powers as factors.

Solution

$$
\begin{aligned}
\log_a(x\sqrt{x^2 + 1}) &= \log_a x + \log_a \sqrt{x^2 + 1} &&\log_a(M \cdot N) = \log_a M + \log_a N \\
&= \log_a x + \log_a(x^2 + 1)^{1/2} \\
&= \log_a x + \frac{1}{2}\log_a(x^2 + 1) &&\log_a M^r = r\log_a M
\end{aligned}
$$

EXAMPLE 4

Writing a Logarithmic Expression as a Difference of Logarithms

Write

$$\ln \frac{x^2}{(x - 1)^3} \qquad x > 1$$

as a difference of logarithms. Express all powers as factors.

Solution

$$\ln \frac{x^2}{(x - 1)^3} = \ln x^2 - \ln(x - 1)^3 = 2\ln x - 3\ln(x - 1)$$

$$\log_a\left(\frac{M}{N}\right) = \log_a M - \log_a N \qquad \log_a M^r = r\log_a M$$

EXAMPLE 5

Writing a Logarithmic Expression as a Sum and Difference of Logarithms

Write

$$\log_a \frac{\sqrt{x^2+1}}{x^3(x+1)^4} \quad x > 0$$

as a sum and difference of logarithms. Express all powers as factors.

Solution

$$\log_a \frac{\sqrt{x^2+1}}{x^3(x+1)^4} = \log_a \sqrt{x^2+1} - \log_a[x^3(x+1)^4] \qquad \text{Property (4)}$$

$$= \log_a \sqrt{x^2+1} - [\log_a x^3 + \log_a(x+1)^4] \qquad \text{Property (3)}$$

$$= \log_a(x^2+1)^{1/2} - \log_a x^3 - \log_a(x+1)^4$$

$$= \frac{1}{2}\log_a(x^2+1) - 3\log_a x - 4\log_a(x+1) \qquad \text{Property (5)}$$

WARNING In using properties (3) through (5), be careful about the values that the variable may assume. For example, the domain of the variable for $\log_a x$ is $x > 0$ and for $\log_a(x-1)$ is $x > 1$. If these expressions are added, the domain is $x > 1$. That is,
$\log_a x + \log_a(x-1) = \log_a[x(x-1)]$
is true only for $x > 1$. ∎

━━━ **Now Work** PROBLEM 51

3 Write a Logarithmic Expression as a Single Logarithm

Another use of properties (3) through (5) is to write sums and/or differences of logarithms with the same base as a single logarithm. This skill is needed to solve certain logarithmic equations discussed in the next section.

EXAMPLE 6

Writing Expressions as a Single Logarithm

Write each of the following as a single logarithm.

(a) $\log_a 7 + 4\log_a 3$ (b) $\dfrac{2}{3}\ln 8 - \ln(5^2 - 1)$

(c) $\log_a x + \log_a 9 + \log_a(x^2 + 1) - \log_a 5$

Solution

(a) $\log_a 7 + 4\log_a 3 = \log_a 7 + \log_a 3^4 \qquad r\log_a M = \log_a M^r$

$$= \log_a 7 + \log_a 81$$

$$= \log_a(7 \cdot 81) \qquad \log_a M + \log_a N = \log_a(M \cdot N)$$

$$= \log_a 567$$

(b) $\dfrac{2}{3}\ln 8 - \ln(5^2 - 1) = \ln 8^{2/3} - \ln(25 - 1) \qquad r\log_a M = \log_a M^r$

$$= \ln 4 - \ln 24 \qquad\qquad 8^{2/3} = \left(\sqrt[3]{8}\right)^2 = 2^2 = 4$$

$$= \ln\left(\frac{4}{24}\right) \qquad\qquad \log_a M - \log_a N = \log_a\left(\frac{M}{N}\right)$$

$$= \ln\left(\frac{1}{6}\right)$$

$$= \ln 1 - \ln 6$$

$$= -\ln 6 \qquad\qquad \ln 1 = 0$$

(c) $\log_a x + \log_a 9 + \log_a(x^2 + 1) - \log_a 5 = \log_a(9x) + \log_a(x^2 + 1) - \log_a 5$

$$= \log_a[9x(x^2 + 1)] - \log_a 5$$

$$= \log_a\left[\frac{9x(x^2 + 1)}{5}\right]$$

WARNING A common error that some students make is to express the logarithm of a sum as the sum of logarithms.

$$\log_a(M + N) \quad \text{is not equal to} \quad \log_a M + \log_a N$$

Correct statement $\log_a(MN) = \log_a M + \log_a N$ **Property (3)**

Another common error is to express the difference of logarithms as the quotient of logarithms.

$$\log_a M - \log_a N \quad \text{is not equal to} \quad \frac{\log_a M}{\log_a N}$$

Correct statement $\log_a M - \log_a N = \log_a\left(\dfrac{M}{N}\right)$ **Property (4)**

A third common error is to express a logarithm raised to a power as the product of the power times the logarithm.

$$(\log_a M)^r \quad \text{is not equal to} \quad r\log_a M$$

Correct statement $\log_a M^r = r\log_a M$ **Property (5)** ∎

───── **Now Work** PROBLEMS 57 AND 63

Two other important properties of logarithms are consequences of the fact that the logarithmic function $y = \log_a x$ is a one-to-one function.

THEOREM **Properties of Logarithms**

In these properties, M, N, and a are positive real numbers, $a \neq 1$.

- If $M = N$, then $\log_a M = \log_a N$. **(7)**
- If $\log_a M = \log_a N$, then $M = N$. **(8)**

Property (7) is used as follows: Starting with the equation $M = N$, "take the logarithm of both sides" to obtain $\log_a M = \log_a N$.

Properties (7) and (8) are useful for solving *exponential and logarithmic equations*, a topic discussed in the next section.

4 Evaluate Logarithms Whose Base Is Neither 10 Nor e

Logarithms with base 10—common logarithms—were used to facilitate arithmetic computations before the widespread use of calculators. (See the Historical Feature at the end of this section.) Natural logarithms—that is, logarithms whose base is the number *e*—remain very important because they arise frequently in the study of natural phenomena.

Common logarithms are usually abbreviated by writing **log**, with the base understood to be 10, just as natural logarithms are abbreviated by **ln**, with the base understood to be *e*.

Most calculators have both ⌈log⌉ and ⌈ln⌉ keys to calculate the common logarithm and the natural logarithm of a number, respectively. Let's look at an example to see how to approximate logarithms having a base other than 10 or *e*.

EXAMPLE 7 **Approximating a Logarithm Whose Base Is Neither 10 Nor e**

Approximate $\log_2 7$. Round the answer to four decimal places.

Solution Remember, evaluating $\log_2 7$ means answering the question "2 raised to what exponent equals 7?" Let $y = \log_2 7$. Then $2^y = 7$. Because $2^2 = 4$ and $2^3 = 8$, the value of $\log_2 7$ is between 2 and 3.

$$2^y = 7$$
$$\ln 2^y = \ln 7 \qquad \text{Property (7)}$$
$$y \ln 2 = \ln 7 \qquad \text{Property (5)}$$
$$y = \frac{\ln 7}{\ln 2} \qquad \text{Exact value}$$
$$y \approx 2.8074 \qquad \text{Approximate value rounded to four decimal places}$$

Example 7 shows how to approximate a logarithm whose base is 2 by changing to logarithms involving the base e. In general, the *Change-of-Base Formula* is used.

THEOREM Change-of-Base Formula

If $a \neq 1$, $b \neq 1$, and M are positive real numbers, then

$$\log_a M = \frac{\log_b M}{\log_b a} \qquad \qquad (9)$$

Proof Let $y = \log_a M$. Then

$$a^y = M$$
$$\log_b a^y = \log_b M \qquad \text{Property (7)}$$
$$y \log_b a = \log_b M \qquad \text{Property (5)}$$
$$y = \frac{\log_b M}{\log_b a} \qquad \text{Solve for y.}$$
$$\log_a M = \frac{\log_b M}{\log_b a} \qquad y = \log_a M \qquad \blacksquare$$

TECHNOLOGY NOTE Some calculators have features for evaluating logarithms with bases other than 10 or e. For example, the TI-84 Plus C has the logBASE function (under Math > Math > A: logBASE). Consult the user's manual for your calculator. ∎

Because most calculators have keys for $\boxed{\log}$ and $\boxed{\ln}$, in practice, the Change-of-Base Formula uses either $b = 10$ or $b = e$. That is,

$$\log_a M = \frac{\log M}{\log a} \quad \text{and} \quad \log_a M = \frac{\ln M}{\ln a}$$

EXAMPLE 8 **Using the Change-of-Base Formula**

Approximate:

(a) $\log_5 89$

(b) $\log_{\sqrt{2}} \sqrt{5}$

Round answers to four decimal places.

Solution (a) $\log_5 89 = \dfrac{\log 89}{\log 5} \approx 2.7889$

or

$\log_5 89 = \dfrac{\ln 89}{\ln 5} \approx 2.7889$

(b) $\log_{\sqrt{2}} \sqrt{5} = \dfrac{\log \sqrt{5}}{\log \sqrt{2}} \approx 2.3219$

or

$\log_{\sqrt{2}} \sqrt{5} = \dfrac{\ln \sqrt{5}}{\ln \sqrt{2}} \approx 2.3219$

Now Work PROBLEMS 23 AND 71

⊻ **COMMENT** We can use the Change-of-Base Formula to graph logarithmic functions when the base is different from e or 10. For example, to graph $y = \log_2 x$, graph either $y = \dfrac{\ln x}{\ln 2}$ or $y = \dfrac{\log x}{\log 2}$. ∎

━━━━ **Now Work** PROBLEM 79

SUMMARY

Properties of Logarithms

In the list that follows, $a, b, M, N,$ and r are real numbers. Also, $a > 0, a \neq 1, b > 0, b \neq 1, M > 0,$ and $N > 0$.

Definition	$y = \log_a x$ if and only if $x = a^y$

Properties of logarithms

- $\log_a 1 = 0$
- $\log_a a = 1$
- $\log_a M^r = r \log_a M$
- $a^{\log_a M} = M$
- $\log_a a^r = r$
- $a^r = e^{r \ln a}$

- $\log_a (MN) = \log_a M + \log_a N$
- $\log_a\left(\dfrac{M}{N}\right) = \log_a M - \log_a N$
- If $M = N$, then $\log_a M = \log_a N$.
- If $\log_a M = \log_a N$, then $M = N$.

Change-of-Base Formula $\qquad \log_a M = \dfrac{\log_b M}{\log_b a}$

Historical Feature

John Napier (1550–1617)

ogarithms were invented about 1590 by John Napier (1550-1617) and Joost Bürgi (1552-1632), working independently. Napier, whose work had the greater influence, was a Scottish lord, a secretive man whose neighbors were inclined to believe him to be in league with the devil. His approach to logarithms was very different from ours; it was based on the relationship between arithmetic and geometric sequences, discussed in a later chapter, and not on the inverse function relationship of logarithms to exponential functions (described in Section 6.4). Napier's tables, published in 1614, listed what would now be called *natural logarithms* of sines and were rather difficult to use. A London professor, Henry Briggs, became interested in the tables and visited Napier. In their conversations, they developed the idea of common logarithms, which were published in 1617. The importance of this tool for calculation was immediately recognized, and by 1650 common logarithms were being printed as far away as China. They remained an important calculation tool until the advent of inexpensive handheld calculators in about 1972, which has decreased their calculational–but not their theoretical–importance.

A side effect of the invention of logarithms was the popularization of the decimal system of notation for real numbers.

6.5 Assess Your Understanding

Concepts and Vocabulary

1. $\log_a 1 = $ _____
2. $a^{\log_a M} = $ _____
3. $\log_a a^r = $ _____
4. $\log_a (MN) = $ _____ + _____
5. $\log_a\left(\dfrac{M}{N}\right) = $ _____ − _____
6. $\log_a M^r = $ _____
7. If $\log_8 M = \dfrac{\log_5 7}{\log_5 8}$, then $M = $ _____ .
8. *True or False* $\ln(x + 3) - \ln(2x) = \dfrac{\ln(x + 3)}{\ln(2x)}$

9. *True or False* $\log_2(3x^4) = 4 \log_2(3x)$
10. *True or False* $\log\left(\dfrac{2}{3}\right) = \dfrac{\log 2}{\log 3}$
11. *Multiple Choice* Choose the expression equivalent to 2^x.
    (a) e^{2x}　(b) $e^{x \ln 2}$　(c) $e^{\log_2 x}$　(d) $e^{2 \ln x}$
12. *Multiple Choice* Writing $\log_a x - \log_a y + 2 \log_a z$ as a single logarithm results in which of the following?
    (a) $\log_a (x - y + 2z)$　(b) $\log_a\left(\dfrac{xz^2}{y}\right)$
    (c) $\log_a\left(\dfrac{2xz}{y}\right)$　(d) $\log_a\left(\dfrac{x}{yz^2}\right)$

Skill Building

In Problems 13–28, use properties of logarithms to find the exact value of each expression. Do not use a calculator.

13. $\log_7 7^{29}$

14. $\log_2 2^{-13}$

15. $\ln e^{-4}$

16. $\ln e^{\sqrt{2}}$

17. $9^{\log_9 13}$

18. $e^{\ln 8}$

19. $\log_8 2 + \log_8 4$

20. $\log_6 9 + \log_6 4$

21. $\log_5 35 - \log_5 7$

22. $\log_8 16 - \log_8 2$

23. $\log_2 6 \cdot \log_6 8$

24. $\log_3 8 \cdot \log_8 9$

25. $4^{\log_4 6 - \log_4 5}$

26. $5^{\log_5 6 + \log_5 7}$

27. $e^{\log_{e^2} 16}$

28. $e^{\log_{e^2} 9}$

In Problems 29–36, suppose that $\ln 2 = a$ and $\ln 3 = b$. Use properties of logarithms to write each logarithm in terms of a and b.

29. $\ln 6$

30. $\ln \dfrac{2}{3}$

31. $\ln 1.5$

32. $\ln 0.5$

33. $\ln 8$

34. $\ln 27$

35. $\ln \sqrt[5]{6}$

36. $\ln \sqrt[4]{\dfrac{2}{3}}$

In Problems 37–56, write each expression as a sum and/or difference of logarithms. Express powers as factors.

37. $\log_6 36x$

38. $\log_3 \dfrac{x}{9}$

39. $\log_5 y^6$

40. $\log_7 x^5$

41. $\ln(ex)$

42. $\ln \dfrac{e}{x}$

43. $\ln \dfrac{x}{e^x}$

44. $\ln(xe^x)$

45. $\log_a(u^2 v^3)$ $\;u > 0, v > 0$

46. $\log_2 \left(\dfrac{a}{b^2}\right)$ $\;a > 0, b > 0$

47. $\ln(x^2 \sqrt{1 - x})$ $\;0 < x < 1$

48. $\ln(x\sqrt{1 + x^2})$ $\;x > 0$

49. $\log_2 \left(\dfrac{x^3}{x - 3}\right)$ $\;x > 3$

50. $\log_5 \left(\dfrac{\sqrt[3]{x^2 + 1}}{x^2 - 1}\right)$ $\;x > 1$

51. $\log\left[\dfrac{x(x + 2)}{(x + 3)^2}\right]$ $\;x > 0$

52. $\log\left[\dfrac{x^3 \sqrt{x + 1}}{(x - 2)^2}\right]$ $\;x > 2$

53. $\ln\left[\dfrac{x^2 - x - 2}{(x + 4)^2}\right]^{1/3}$ $\;x > 2$

54. $\ln\left[\dfrac{(x - 4)^2}{x^2 - 1}\right]^{2/3}$ $\;x > 4$

55. $\ln \dfrac{5x\sqrt{1 + 3x}}{(x - 4)^3}$ $\;x > 4$

56. $\ln\left[\dfrac{5x^2 \sqrt[3]{1 - x}}{4(x + 1)^2}\right]$ $\;0 < x < 1$

In Problems 57–70, write each expression as a single logarithm.

57. $3 \log_5 u + 4 \log_5 v$

58. $2 \log_3 u - \log_3 v$

59. $\log_3 \sqrt{x} - \log_3 x^3$

60. $\log_2 \left(\dfrac{1}{x}\right) + \log_2 \left(\dfrac{1}{x^2}\right)$

61. $\log_4(x^2 - 1) - 5 \log_4(x + 1)$

62. $\log(x^2 + 3x + 2) - 2 \log(x + 1)$

63. $\ln\left(\dfrac{x}{x - 1}\right) + \ln\left(\dfrac{x + 1}{x}\right) - \ln(x^2 - 1)$

64. $\log\left(\dfrac{x^2 + 2x - 3}{x^2 - 4}\right) - \log\left(\dfrac{x^2 + 7x + 6}{x + 2}\right)$

65. $8 \log_2 \sqrt{3x - 2} - \log_2 \left(\dfrac{4}{x}\right) + \log_2 4$

66. $21 \log_3 \sqrt[3]{x} + \log_3(9x^2) - \log_3 9$

67. $2 \log_a(5x^3) - \dfrac{1}{2} \log_a(2x + 3)$

68. $\dfrac{1}{3} \log(x^3 + 1) + \dfrac{1}{2} \log(x^2 + 1)$

69. $2 \log_2(x + 1) - \log_2(x + 3) - \log_2(x - 1)$

70. $3 \log_5(3x + 1) - 2 \log_5(2x - 1) - \log_5 x$

In Problems 71–78, use the Change-of-Base Formula and a calculator to evaluate each logarithm. Round your answer to three decimal places.

71. $\log_3 21$

72. $\log_5 18$

73. $\log_{1/3} 71$

74. $\log_{1/2} 15$

75. $\log_{\sqrt{2}} 7$

76. $\log_{\sqrt{5}} 8$

77. $\log_\pi e$

78. $\log_\pi \sqrt{2}$

In Problems 79–84, graph each function using a graphing utility and the Change-of-Base Formula.

79. $y = \log_4 x$

80. $y = \log_5 x$

81. $y = \log_2(x + 2)$

82. $y = \log_4(x - 3)$

83. $y = \log_{x-1}(x + 1)$

84. $y = \log_{x+2}(x - 2)$

85. Mixed Practice If $f(x) = \ln x$, $g(x) = e^x$, and $h(x) = x^2$, find:

(a) $(f \circ g)(x)$. What is the domain of $f \circ g$?

(b) $(g \circ f)(x)$. What is the domain of $g \circ f$?

(c) $(f \circ g)(5)$

(d) $(f \circ h)(x)$. What is the domain of $f \circ h$?

(e) $(f \circ h)(e)$

86. Mixed Practice If $f(x) = \log_2 x$, $g(x) = 2^x$, and $h(x) = 4x$, find:

(a) $(f \circ g)(x)$. What is the domain of $f \circ g$?

(b) $(g \circ f)(x)$. What is the domain of $g \circ f$?

(c) $(f \circ g)(3)$

(d) $(f \circ h)(x)$. What is the domain of $f \circ h$?

(e) $(f \circ h)(8)$

Applications and Extensions

In Problems 87–96, express y as a function of x. The constant C is a positive number.

87. $\ln y = \ln x + \ln C$

88. $\ln y = \ln(x + C)$

89. $\ln y = \ln x + \ln(x + 1) + \ln C$

90. $\ln y = 2 \ln x - \ln(x + 1) + \ln C$

91. $\ln y = 3x + \ln C$

92. $\ln y = -2x + \ln C$

93. $\ln(y - 3) = -4x + \ln C$

94. $\ln(y + 4) = 5x + \ln C$

95. $3 \ln y = \frac{1}{2} \ln(2x + 1) - \frac{1}{3} \ln(x + 4) + \ln C$

96. $2 \ln y = -\frac{1}{2} \ln x + \frac{1}{3} \ln(x^2 + 1) + \ln C$

97. Find the value of $\log_2 3 \cdot \log_3 4 \cdot \log_4 5 \cdot \log_5 6 \cdot \log_6 7 \cdot \log_7 8$.

98. Find the value of $\log_2 4 \cdot \log_4 6 \cdot \log_6 8$.

99. Find the value of $\log_2 3 \cdot \log_3 4 \cdot \cdots \cdot \log_n(n + 1) \cdot \log_{n+1} 2$.

100. Find the value of $\log_2 2 \cdot \log_2 4 \cdot \log_2 8 \cdot \cdots \cdot \log_2 2^n$.

101. Show that $\log_a(x + \sqrt{x^2 - 1}) + \log_a(x - \sqrt{x^2 - 1}) = 0$.

102. Show that $\log_a(\sqrt{x} + \sqrt{x - 1}) + \log_a(\sqrt{x} - \sqrt{x - 1}) = 0$.

103. Show that $\ln(1 + e^{2x}) = 2x + \ln(1 + e^{-2x})$.

104. Difference Quotient If $f(x) = \log_a x$, show that $\dfrac{f(x + h) - f(x)}{h} = \log_a\left(1 + \dfrac{h}{x}\right)^{1/h}$, $h \neq 0$.

105. If $f(x) = \log_a x$, show that $-f(x) = \log_{1/a} x$.

106. If $f(x) = \log_a x$, show that $f(AB) = f(A) + f(B)$.

107. If $f(x) = \log_a x$, show that $f\left(\dfrac{1}{x}\right) = -f(x)$.

108. If $f(x) = \log_a x$, show that $f(x^\alpha) = \alpha f(x)$.

109. Show that $\log_a\left(\dfrac{M}{N}\right) = \log_a M - \log_a N$, where a, M, and N are positive real numbers and $a \neq 1$.

110. Show that $\log_a\left(\dfrac{1}{N}\right) = -\log_a N$, where a and N are positive real numbers and $a \neq 1$.

111. Challenge Problem Show that $\log_a b = \dfrac{1}{\log_b a}$, where a and b are positive real numbers, $a \neq 1$, and $b \neq 1$.

112. Challenge Problem Show that $\log_{\sqrt{a}} m = \log_a m^2$, where a and m are positive real numbers and $a \neq 1$.

113. Challenge Problem Show that $\log_{a^n} b^m = \dfrac{m}{n} \log_a b$, where a, b, m, and n are positive real numbers, $a \neq 1$, and $b \neq 1$.

114. Challenge Problem Find n:
$$\log_2 3 \cdot \log_3 4 \cdot \log_4 5 \cdot \ldots \cdot \log_n(n + 1) = 10$$

Explaining Concepts: Discussion and Writing

115. Graph $Y_1 = \log(x^2)$ and $Y_2 = 2 \log(x)$ using a graphing utility. Are they equivalent? What might account for any differences in the two functions?

116. Write an example that illustrates why $(\log_a x)^r \neq r \log_a x$.

117. Write an example that illustrates why
$$\log_2(x + y) \neq \log_2 x + \log_2 y$$

118. Does $3^{\log_3(-5)} = -5$? Why or why not?

⌐ Retain Your Knowledge

Problems 119–128 are based on material learned earlier in the course. The purpose of these problems is to keep the material fresh in your mind so that you are better prepared for the final exam.

119. Use a graphing utility to solve $x^3 - 3x^2 - 4x + 8 = 0$. Round answers to two decimal places.

120. Without solving, determine the character of the solution of the quadratic equation $4x^2 - 28x + 49 = 0$ in the complex number system.

121. Find the real zeros of
$$f(x) = 5x^5 + 44x^4 + 116x^3 + 95x^2 - 4x - 4$$

122. Graph $f(x) = \sqrt{2 - x}$ using the techniques of shifting, compressing or stretching, and reflecting. State the domain and the range of f.

123. Find the domain of $f(x) = 2\sqrt{3 - 5x} - 4$.

124. Solve: $4|x + 1| - 9 < 23$

125. Find the vertex of $f(x) = -\dfrac{1}{2}x^2 + 4x + 5$, and determine if the graph is concave up or concave down.

126. Find the center and radius of the circle
$$x^2 - 10x + y^2 + 4y = 35$$

127. Find the average rate of change $f(x) = x^3$ from -1 to 3.

128. Is the function $f(x) = \dfrac{5x^2 - 3x^4}{\sqrt[3]{x}}$ even, odd, or neither?

6.6 Logarithmic and Exponential Equations

PREPARING FOR THIS SECTION *Before getting started, review the following:*

- Using a Graphing Utility to Solve Equations (Appendix, Section 4, pp. A6–A8)
- Solving Quadratic Equations (Section 1.2, pp. 92–99)
- Solving Equations Quadratic in Form (Section 1.4, pp. 114–116)

Now Work the 'Are You Prepared?' problems on page 478.

OBJECTIVES 1 Solve Logarithmic Equations (p. 474)
2 Solve Exponential Equations (p. 476)
3 Solve Logarithmic and Exponential Equations Using a Graphing Utility (p. 477)

1 Solve Logarithmic Equations

In Section 6.4, we solved logarithmic equations by changing a logarithmic expression to an exponential expression. That is, we used the definition of a logarithm:

$$y = \log_a x \quad \text{if and only if} \quad x = a^y \qquad a > 0 \quad a \neq 1$$

For example, to solve the equation $\log_2(1 - 2x) = 3$, write the logarithmic equation as an equivalent exponential equation $1 - 2x = 2^3$ and solve for x.

$$\log_2(1 - 2x) = 3$$
$$1 - 2x = 2^3 \qquad \text{Change to exponential form.}$$
$$-2x = 7 \qquad \text{Simplify.}$$
$$x = -\frac{7}{2} \qquad \text{Solve.}$$

You should check this solution for yourself.

For most logarithmic equations, some manipulation of the equation (usually using properties of logarithms) is required to obtain a solution. Also, to avoid extraneous solutions with logarithmic equations, determine the domain of the variable first.

Let's begin with an example of a logarithmic equation that requires using the fact that a logarithmic function is a one-to-one function:

$$\text{If } \log_a M = \log_a N, \text{ then } M = N \qquad M, N, \text{ and } a \text{ are positive and } a \neq 1$$

EXAMPLE 1 **Solving a Logarithmic Equation**

Solve: $2 \log_5 x = \log_5 9$

Solution The domain of the variable in this equation is $x > 0$. Note that each logarithm has the same base, 5. Now use properties of logarithms to solve the equation.

$$2 \log_5 x = \log_5 9$$
$$\log_5 x^2 = \log_5 9 \qquad r \log_a M = \log_a M^r$$
$$x^2 = 9 \qquad \text{If } \log_a M = \log_a N, \text{ then } M = N.$$
$$x = 3 \quad \text{or} \quad x = -3$$

Since the domain of the variable is $x > 0$, -3 is extraneous and is discarded.

✓ Check: $2 \log_5 3 \overset{?}{=} \log_5 9$

$\qquad\qquad \log_5 3^2 \overset{?}{=} \log_5 9 \quad r \log_a M = \log_a M^r$

$\qquad\qquad\quad \log_5 9 = \log_5 9$

The solution set is $\{3\}$.

⌐────── **Now Work** PROBLEM 17

Often one or more properties of logarithms are needed to rewrite the equation as a single logarithm. In the next example, the log of a product property is used.

EXAMPLE 2

Solving a Logarithmic Equation

Solve: $\log_5 (x + 6) + \log_5 (x + 2) = 1$

Solution

The domain of the variable requires that $x + 6 > 0$ and $x + 2 > 0$, so $x > -6$ and $x > -2$. This means any solution must satisfy $x > -2$. To obtain an exact solution, first express the left side as a single logarithm. Then change the equation to an equivalent exponential equation.

$$\log_5 (x + 6) + \log_5 (x + 2) = 1$$

$$\log_5 [(x + 6)(x + 2)] = 1 \qquad \log_a M + \log_a N = \log_a (MN)$$

$$(x + 6)(x + 2) = 5^1 = 5 \qquad \text{Change to exponential form.}$$

$$x^2 + 8x + 12 = 5 \qquad \text{Multiply out.}$$

$$x^2 + 8x + 7 = 0 \qquad \text{Place the quadratic equation in standard form.}$$

$$(x + 7)(x + 1) = 0 \qquad \text{Factor.}$$

$$x = -7 \quad \text{or} \quad x = -1 \qquad \text{Use the Zero-Product Property.}$$

WARNING A negative solution is not automatically extraneous. You must determine whether the potential solution causes the argument of any logarithmic expression in the equation to be negative or 0. ∎

Only $x = -1$ satisfies the restriction that $x > -2$, so $x = -7$ is extraneous. The solution set is $\{-1\}$, which you should check.

⌐────── **Now Work** PROBLEM 25

EXAMPLE 3

Solving a Logarithmic Equation

Solve: $\ln x = \ln (x + 6) - \ln (x - 4)$

Solution

The domain of the variable requires that $x > 0$, $x + 6 > 0$, and $x - 4 > 0$. As a result, the domain of the variable here is $x > 4$. Begin the solution using the log of a difference property.

$$\ln x = \ln (x + 6) - \ln (x - 4)$$

$$\ln x = \ln\left(\frac{x + 6}{x - 4}\right) \qquad \ln M - \ln N = \ln\left(\frac{M}{N}\right)$$

$$x = \frac{x + 6}{x - 4} \qquad \text{If } \ln M = \ln N, \text{ then } M = N.$$

$$x(x - 4) = x + 6 \qquad \text{Multiply both sides by } x - 4.$$

$$x^2 - 4x = x + 6 \qquad \text{Distribute.}$$

$$x^2 - 5x - 6 = 0 \qquad \text{Place the quadratic equation in standard form.}$$

$$(x - 6)(x + 1) = 0 \qquad \text{Factor.}$$

$$x = 6 \quad \text{or} \quad x = -1 \qquad \text{Use the Zero-Product Property.}$$

Because the domain of the variable is $x > 4$, discard -1 as extraneous. The solution set is $\{6\}$, which you should check.

WARNING In using properties of logarithms to solve logarithmic equations, avoid using the property $\log_a x^r = r \log_a x$, when r is even. The reason can be seen in this example:

Solve: $\log_3 x^2 = 4$

Solution: The domain of the variable x is all real numbers except 0.

(a) $\log_3 x^2 = 4$

$\quad x^2 = 3^4 = 81$ Change to exponential form.

$\quad x = -9 \text{ or } x = 9$

(b) $\log_3 x^2 = 4$ $\log_a x^r = r \log_a x$

$\quad 2 \log_3 x = 4$ Domain of variable is $x > 0$.

$\quad\quad \log_3 x = 2$

$\quad\quad\quad x = 9$

Both -9 and 9 are solutions of $\log_3 x^2 = 4$ (as you can verify). The solution in part (b) does not find the solution -9 because the domain of the variable was further restricted due to the application of the property $\log_a x^r = r \log_a x$. ∎

━━━━━ **Now Work** PROBLEM 35

2 Solve Exponential Equations

In Sections 6.3 and 6.4, we solved exponential equations algebraically by expressing each side of the equation using the same base. That is, we used the one-to-one property of the exponential function:

$$\text{If } a^u = a^v, \text{ then } u = v \quad a > 0 \quad a \neq 1$$

For example, to solve the exponential equation $4^{2x+1} = 16$, notice that $16 = 4^2$ and use the one-to-one property to obtain the equation $2x + 1 = 2$, from which we find $x = \dfrac{1}{2}$.

Not all exponential equations can be expressed so that each side of the equation has the same base. For such equations, other techniques often can be used to obtain exact solutions.

EXAMPLE 4

Solving Exponential Equations

Solve: (a) $2^x = 5$ (b) $8 \cdot 3^x = 5$

Solution

(a) Because 5 cannot be written as an integer power of 2 ($2^2 = 4$ and $2^3 = 8$), write the exponential equation as the equivalent logarithmic equation.

$$2^x = 5$$

$$x = \log_2 5 = \frac{\ln 5}{\ln 2}$$

$$\underset{\uparrow}{}$$

Change-of-Base Formula

Alternatively, the equation $2^x = 5$ can be solved by taking the natural logarithm (or common logarithm) of each side.

$$2^x = 5$$

$$\ln 2^x = \ln 5 \quad \text{If } M = N, \text{ then } \ln M = \ln N.$$

$$x \ln 2 = \ln 5 \quad \ln M^r = r \ln M$$

$$x = \frac{\ln 5}{\ln 2} \quad \text{Exact solution}$$

$$\approx 2.322 \quad \text{Approximate solution}$$

The solution set is $\left\{ \dfrac{\ln 5}{\ln 2} \right\}$.

(b) $8 \cdot 3^x = 5$

$$3^x = \frac{5}{8} \quad\quad\quad \text{Solve for } 3^x.$$

$$x = \log_3\left(\frac{5}{8}\right) = \frac{\ln\left(\frac{5}{8}\right)}{\ln 3} \quad \text{Exact solution}$$

$$\approx -0.428 \quad \text{Approximate solution}$$

The solution set is $\left\{ \dfrac{\ln\left(\frac{5}{8}\right)}{\ln 3} \right\}$.

➤ **Now Work** PROBLEM 47

EXAMPLE 5 — Solving an Exponential Equation

Solve: $5^{x-2} = 3^{3x+2}$

Solution

In this equation, the bases are 3 and 5, and we cannot express one as a power of the other as we did in Section 6.3 (pp. 444–445). Instead begin by taking the natural logarithm of both sides and then use the fact that $\ln M^r = r \ln M$. This results in an equation we know how to solve.

$$5^{x-2} = 3^{3x+2}$$

$$\ln 5^{x-2} = \ln 3^{3x+2} \qquad \text{If } M = N, \ln M = \ln N.$$

$$(x-2)\ln 5 = (3x+2)\ln 3 \qquad \ln M^r = r \ln M$$

$$(\ln 5)x - 2\ln 5 = (3\ln 3)x + 2\ln 3 \qquad \text{Distribute.}$$

$$(\ln 5)x - (3\ln 3)x = 2\ln 3 + 2\ln 5 \qquad \text{Place terms involving } x \text{ on the left.}$$

$$(\ln 5 - 3\ln 3)x = 2(\ln 3 + \ln 5) \qquad \text{Factor.}$$

$$x = \frac{2(\ln 3 + \ln 5)}{\ln 5 - 3\ln 3} \qquad \text{Exact solution}$$

$$\approx -3.212 \qquad \text{Approximate solution}$$

NOTE Because of the properties of logarithms, exact solutions involving logarithms often can be expressed in multiple ways. For example, the solution to $5^{x-2} = 3^{3x+2}$ from Example 5 can be expressed equivalently as $\dfrac{2\ln 15}{\ln 5 - \ln 27}$ or as $\dfrac{\ln 225}{\ln\left(\frac{5}{27}\right)}$, among others. Do you see why? ∎

The solution set is $\left\{ \dfrac{2(\ln 3 + \ln 5)}{\ln 5 - 3\ln 3} \right\}$.

➤ **Now Work** PROBLEM 57

EXAMPLE 6 — Solving an Exponential Equation That Is Quadratic in Form

Solve: $4^x - 2^x - 12 = 0$

Solution

Note that $4^x = (2^2)^x = 2^{2x} = (2^x)^2$, so the equation is quadratic in form and can be written as

$$(2^x)^2 - 2^x - 12 = 0 \qquad \text{Let } u = 2^x; \text{ then } u^2 - u - 12 = 0.$$

Now factor as usual.

$$(2^x - 4)(2^x + 3) = 0 \qquad (u - 4)(u + 3) = 0$$

$$2^x - 4 = 0 \quad \text{or} \quad 2^x + 3 = 0 \qquad u - 4 = 0 \quad \text{or} \quad u + 3 = 0$$

$$2^x = 4 \qquad\qquad 2^x = -3 \qquad u = 2^x = 4 \qquad u = 2^x = -3$$

The equation on the left has the solution $x = 2$; the equation on the right has no solution, since $2^x > 0$ for all x. The solution set is $\{2\}$.

➤ **Now Work** PROBLEM 65

---

3 Solve Logarithmic and Exponential Equations Using a Graphing Utility

The algebraic techniques introduced in this section to obtain exact solutions can be used to solve only certain types of logarithmic and exponential equations. Solutions for other types are generally studied in calculus, using numerical methods. For such types, we can use a graphing utility to approximate the solution.

EXAMPLE 7	Solving Equations Using a Graphing Utility

Solve: $x + e^x = 2$
Express the solution(s) rounded to two decimal places.

Solution The solution is found using a TI-84 Plus C by graphing $Y_1 = x + e^x$ and $Y_2 = 2$ as shown in Figure 40(a). Note that because Y_1 is an increasing function (do you know why?), there is only one point of intersection for Y_1 and Y_2. Using the INTERSECT command reveals that the solution is 0.44, rounded to two decimal places. Figure 40(b) shows the solution using Desmos.

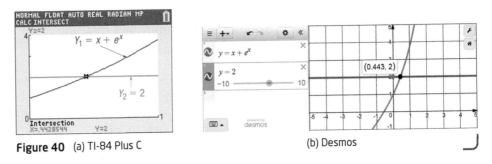

Figure 40 (a) TI-84 Plus C   (b) Desmos

- **Now Work** PROBLEM 75

6.6 Assess Your Understanding

'Are You Prepared?' *Answers are given at the end of these exercises. If you get a wrong answer, read the pages listed in* red.

1. Solve $x^2 - 7x - 30 = 0$. (pp. 92–99)

2. Solve $(x + 3)^2 - 4(x + 3) + 3 = 0$. (pp. 114–116)

3. Approximate the solution(s) to $x^3 = x^2 - 5$ using a graphing utility. (pp. A6–A8)

4. Approximate the solution(s) to $x^3 - 2x + 2 = 0$ using a graphing utility. (pp. A6–A8)

Skill Building

In Problems 5–44, solve each logarithmic equation. Express irrational solutions in exact form.

5. $\log_4 x = 2$

6. $\log(x + 6) = 1$

7. $\log_2(5x) = 4$

8. $\log_3(3x - 1) = 2$

9. $\log_4(x + 4) = \log_4 15$

10. $\log_5(2x + 3) = \log_5 3$

11. $\log_4|x| = 3$

12. $\log_2|x - 7| = 4$

13. $\log_5|2x - 1| = \log_5 13$

14. $\log_9|3x + 4| = \log_9|5x - 12|$

15. $\dfrac{1}{2}\log_7 x = 3\log_7 2$

16. $-2\log_4 x = \log_4 9$

17. $3\log_2 x = -\log_2 27$

18. $2\log_5 x = 3\log_5 4$

19. $2\log_6(x - 5) + \log_6 9 = 2$

20. $2\log_3(x + 4) - \log_3 9 = 2$

21. $\log x + \log(x + 15) = 2$

22. $\log x + \log(x - 21) = 2$

23. $\log(7x + 6) = 1 + \log(x - 1)$

24. $\log(2x) - \log(x - 3) = 1$

25. $\log_2(x + 7) + \log_2(x + 8) = 1$

26. $\log_6(x + 4) + \log_6(x + 3) = 1$

27. $\log_8(x + 6) = 1 - \log_8(x + 4)$

28. $\log_5(x + 3) = 1 - \log_5(x - 1)$

29. $\ln x + \ln(x + 2) = 4$

30. $\ln(x + 1) - \ln x = 2$

31. $\log_9(x + 8) + \log_9(x + 7) = 2$

32. $\log_2(x + 1) + \log_2(x + 7) = 3$

33. $\log_{1/3}(x^2 + x) - \log_{1/3}(x^2 - x) = -1$

34. $\log_4(x^2 - 9) - \log_4(x + 3) = 3$

35. $\log_a(x - 1) - \log_a(x + 6) = \log_a(x - 2) - \log_a(x + 3)$

36. $\log_a x + \log_a(x - 2) = \log_a(x + 4)$

37. $2\log_5(x - 3) - \log_5 8 = \log_5 2$

38. $\log_3 x - 2\log_3 5 = \log_3(x + 1) - 2\log_3 10$

39. $2\log_6(x + 2) = 3\log_6 2 + \log_6 4$

40. $3(\log_7 x - \log_7 2) = 2\log_7 4$

41. $2\log_{13}(x + 2) = \log_{13}(4x + 7)$

42. $\log(x - 1) = \dfrac{1}{3}\log 2$

43. $(\log_3 x)^2 - 3\log_3 x = 10$

44. $\ln x - 3\sqrt{\ln x} + 2 = 0$

In Problems 45–72, solve each exponential equation. Express irrational solutions in exact form.

45. $2^{x-5} = 8$

46. $5^{-x} = 25$

47. $2^x = 10$

48. $3^x = 14$

49. $8^{-x} = 1.2$

50. $2^{-x} = 1.5$

51. $5(2^{3x}) = 8$

52. $0.3(4^{0.2x}) = 0.2$

53. $3^{1-2x} = 4^x$

54. $2^{x+1} = 5^{1-2x}$

55. $\left(\dfrac{3}{5}\right)^x = 7^{1-x}$

56. $\left(\dfrac{4}{3}\right)^{1-x} = 5^x$

57. $1.2^x = (0.5)^{-x}$

58. $0.3^{1+x} = 1.7^{2x-1}$

59. $\pi^{1-x} = e^x$

60. $e^{x+3} = \pi^x$

61. $2^{2x} + 2^x - 12 = 0$

62. $3^{2x} + 3^x - 2 = 0$

63. $3^{2x} + 3^{x+1} - 4 = 0$

64. $2^{2x} + 2^{x+2} - 12 = 0$

65. $16^x + 4^{x+1} - 3 = 0$

66. $9^x - 3^{x+1} + 1 = 0$

67. $25^x - 8 \cdot 5^x = -16$

68. $36^x - 6 \cdot 6^x = -9$

69. $3 \cdot 4^x + 4 \cdot 2^x + 8 = 0$

70. $2 \cdot 49^x + 11 \cdot 7^x + 5 = 0$

71. $4^x - 10 \cdot 4^{-x} = 3$

72. $3^x - 14 \cdot 3^{-x} = 5$

In Problems 73–86, use a graphing utility to solve each equation. Express your answer rounded to two decimal places.

73. $\log_5(x + 1) - \log_4(x - 2) = 1$

74. $\log_2(x - 1) - \log_6(x + 2) = 2$

75. $e^x = -x$

76. $e^{2x} = x + 2$

77. $e^x = x^2$

78. $e^x = x^3$

79. $\ln x = -x$

80. $\ln(2x) = -x + 2$

81. $\ln x = x^3 - 1$

82. $\ln x = -x^2$

83. $e^x + \ln x = 4$

84. $e^x - \ln x = 4$

85. $e^{-x} = \ln x$

86. $e^{-x} = -\ln x$

Applications and Extensions

87. $f(x) = \log_2(x + 3)$ and $g(x) = \log_2(3x + 1)$.

(a) Solve $f(x) = 3$. What point is on the graph of f?

(b) Solve $g(x) = 4$. What point is on the graph of g?

(c) Solve $f(x) = g(x)$. Do the graphs of f and g intersect? If so, where?

(d) Solve $(f + g)(x) = 7$.

(e) Solve $(f - g)(x) = 2$.

88. $f(x) = \log_3(x + 5)$ and $g(x) = \log_3(x - 1)$.

(a) Solve $f(x) = 2$. What point is on the graph of f?

(b) Solve $g(x) = 3$. What point is on the graph of g?

(c) Solve $f(x) = g(x)$. Do the graphs of f and g intersect? If so, where?

(d) Solve $(f + g)(x) = 3$.

(e) Solve $(f - g)(x) = 2$.

89. (a) If $f(x) = 3^{x+1}$ and $g(x) = 2^{x+2}$, graph f and g on the same Cartesian plane.

(b) Find the point(s) of intersection of the graphs of f and g by solving $f(x) = g(x)$. Round answers to three decimal places. Label any intersection points on the graph drawn in part (a).

(c) Based on the graph, solve $f(x) > g(x)$.

90. (a) If $f(x) = 5^{x-1}$ and $g(x) = 2^{x+1}$, graph f and g on the same Cartesian plane.

(b) Find the point(s) of intersection of the graphs of f and g by solving $f(x) = g(x)$. Label any intersection points on the graph drawn in part (a).

(c) Based on the graph, solve $f(x) > g(x)$.

91. (a) Graph $f(x) = 3^x$ and $g(x) = 10$ on the same Cartesian plane.

(b) Shade the region bounded by the y-axis, $f(x) = 3^x$, and $g(x) = 10$ on the graph drawn in part (a).

(c) Solve $f(x) = g(x)$ and label the point of intersection on the graph drawn in part (a).

92. (a) Graph $f(x) = 2^x$ and $g(x) = 12$ on the same Cartesian plane.

(b) Shade the region bounded by the y-axis, $f(x) = 2^x$, and $g(x) = 12$ on the graph drawn in part (a).

(c) Solve $f(x) = g(x)$ and label the point of intersection on the graph drawn in part (a).

93. (a) Graph $f(x) = 2^{x+1}$ and $g(x) = 2^{-x+2}$ on the same Cartesian plane.

(b) Shade the region bounded by the y-axis, $f(x) = 2^{x+1}$, and $g(x) = 2^{-x+2}$ on the graph drawn in part (a).

(c) Solve $f(x) = g(x)$ and label the point of intersection on the graph drawn in part (a).

94. (a) Graph $f(x) = 3^{-x+1}$ and $g(x) = 3^{x-2}$ on the same Cartesian plane.

(b) Shade the region bounded by the y-axis, $f(x) = 3^{-x+1}$, and $g(x) = 3^{x-2}$ on the graph drawn in part (a).

(c) Solve $f(x) = g(x)$ and label the point of intersection on the graph drawn in part (a).

95. (a) Graph $f(x) = 2^x - 4$.

(b) Find the zero of f.

(c) Based on the graph, solve $f(x) < 0$.

96. (a) Graph $g(x) = 3^x - 9$.

(b) Find the zero of g.

(c) Based on the graph, solve $g(x) > 0$.

97. A Population Model The resident population of the United States in 2018 was 327 million people and was growing at a rate of 0.7% per year. Assuming that this growth rate continues, the model $P(t) = 327(1.007)^{t-2018}$ represents the population P (in millions of people) in year t.

(a) According to this model, when will the population of the United States be 415 million people?

(b) According to this model, when will the population of the United States be 470 million people?

Source: U.S. Census Bureau

98. A Population Model The population of the world in 2018 was 7.63 billion people and was growing at a rate of 1.1% per year. Assuming that this growth rate continues, the model $P(t) = 7.63(1.011)^{t-2018}$ represents the population P (in billions of people) in year t.

(a) According to this model, when will the population of the world be 9 billion people?

(b) According to this model, when will the population of the world be 12.5 billion people?

Source: U.S. Census Bureau

99. Depreciation The value V of a Chevy Cruze LT that is t years old can be modeled by $V(t) = 19{,}200(0.82)^t$.
  (a) According to the model, when will the car be worth $12,000?
  (b) According to the model, when will the car be worth $9000?
  (c) According to the model, when will the car be worth $3000?

Source: Kelley Blue Book

100. Depreciation The value V of a Honda Civic LX that is t years old can be modeled by $V(t) = 19{,}705(0.848)^t$.
  (a) According to the model, when will the car be worth $14,000?
  (b) According to the model, when will the car be worth $10,000?

  (c) According to the model, when will the car be worth $7500?

Source: Kelley Blue Book

Challenge Problems In Problems 101–105, solve each equation. Express irrational solutions in exact form.

101. $\log_2(x + 1) - \log_4 x = 1$

102. $\left(\sqrt[3]{2}\right)^{2-x} = 2^{x^2}$

103. $\log_2 x^{\log_2 x} = 4$

104. $\ln x^2 = (\ln x)^2$

105. $\sqrt{\log x} = 2 \log \sqrt{3}$

Explaining Concepts: Discussion and Writing

106. Fill in the reason for each step in the following two solutions.

Solve: $\log_3(x - 1)^2 = 2$

Solution A

$\log_3(x - 1)^2 = 2$

$(x - 1)^2 = 3^2 = 9$ _____

$(x - 1) = \pm 3$ _____

$x - 1 = -3$ or $x - 1 = 3$ _____

$x = -2$ or $x = 4$ _____

Solution B

$\log_3(x - 1)^2 = 2$

$2 \log_3(x - 1) = 2$ _____

$\log_3(x - 1) = 1$ _____

$x - 1 = 3^1 = 3$ _____

$x = 4$ _____

Both solutions given in Solution A check. Explain what caused the solution $x = -2$ to be lost in Solution B.

Retain Your Knowledge

Problems 107–116, are based on material learned earlier in the course. The purpose of these problems is to keep the material fresh in your mind so that you are better prepared for the final exam.

107. Solve: $4x^3 + 3x^2 - 25x + 6 = 0$

108. Determine whether the function is one-to-one:
$$\{(0, -4), (2, -2), (4, 0), (6, 2)\}$$

109. For $f(x) = \dfrac{x}{x - 2}$ and $g(x) = \dfrac{x + 5}{x - 3}$, find $f \circ g$.
Then find the domain of $f \circ g$.

110. Find the domain of $f(x) = \sqrt{x + 3} + \sqrt{x - 1}$.

111. Solve: $x - \sqrt{x + 7} = 5$

112. If y is inversely proportion to the square of x and $y = 2.16$ when $x = 5$, find y when $x = 3$.

113. If $f(x) = \dfrac{x}{x - 2}$ and $g(x) = \dfrac{5}{x + 2}$, find $(f + g)(x)$.

114. Find the distance between the center of the circle
$$(x - 2)^2 + (y + 3)^2 = 25$$
and the vertex of the parabola $y = -2(x - 6)^2 + 9$.

115. Find the average rate of change $f(x) = \log_2 x$ from 4 to 16.

116. Rationalize the numerator: $\dfrac{\sqrt{x + 6} - \sqrt{x}}{6}$

'Are You Prepared?' Answers

1. $\{-3, 10\}$ **2.** $\{-2, 0\}$ **3.** $\{-1.43\}$ **4.** $\{-1.77\}$

6.7 Financial Models

PREPARING FOR THIS SECTION *Before getting started, review the following:*

- Simple Interest (Section 1.7, pp. 136–137)

Now Work the 'Are You Prepared?' problems on page 487.

OBJECTIVES 1 Determine the Future Value of a Lump Sum of Money (p. 481)
2 Calculate Effective Rates of Return (p. 484)
3 Determine the Present Value of a Lump Sum of Money (p. 485)
4 Determine the Rate of Interest or the Time Required to Double a Lump Sum of Money (p. 486)

1 Determine the Future Value of a Lump Sum of Money

Interest is money paid for the use of money. The total amount borrowed (whether by an individual from a bank in the form of a loan or by a bank from an individual in the form of a savings account) is called the **principal**. The **rate of interest**, expressed as a percent, is the amount charged for the use of the principal for a given period of time, usually on a yearly (that is, per annum) basis.

THEOREM Simple Interest Formula

If a principal of P dollars is borrowed for a period of t years at a per annum interest rate r, expressed as a decimal, the interest I charged is

$$I = Prt \qquad\qquad (1)$$

Interest charged according to formula (1) is called **simple interest**.

In problems involving interest, the term **payment period** is defined as follows.

Annually: Once per year   **Monthly:** 12 times per year
Semiannually: Twice per year   **Daily:** 365 times per year[*]
Quarterly: Four times per year

When the interest due at the end of a payment period is added to the principal so that the interest computed at the end of the next payment period is based on this new principal amount (old principal + interest), the interest is said to have been **compounded**. **Compound interest** is interest paid on the principal and on previously earned interest.

EXAMPLE 1 **Computing Compound Interest**

A credit union pays interest of 2% per annum compounded quarterly on a certain savings plan. If $1000 is deposited in the plan and the interest is left to accumulate, how much is in the account after 1 year?

Solution Use the simple interest formula, $I = Prt$. The principal P is $1000 and the rate of interest is 2% $= 0.02$. After the first quarter of a year, the time t is $\frac{1}{4}$ year, so the interest earned is

$$I = Prt = \$1000 \cdot 0.02 \cdot \frac{1}{4} = \$5$$

[*]Some banks use a 360-day "year." Why do you think they do?   *(continued)*

The new principal is $P + I = \$1000 + \$5 = \$1005$. At the end of the second quarter, the interest on this principal is

$$I = \$1005 \cdot 0.02 \cdot \frac{1}{4} = \$5.03$$

At the end of the third quarter, the interest on the new principal of $\$1005 + \$5.03 = \$1010.03$ is

$$I = \$1010.03 \cdot 0.02 \cdot \frac{1}{4} = \$5.05$$

Finally, after the fourth quarter, the interest is

$$I = \$1015.08 \cdot 0.02 \cdot \frac{1}{4} = \$5.08$$

After 1 year the account contains $\$1015.08 + \$5.08 = \$1020.16$.

The pattern of the calculations performed in Example 1 leads to a general formula for compound interest. For this purpose, let P represent the principal to be invested at a per annum interest rate r that is compounded n times per year, so the time of each compounding period is $\frac{1}{n}$ year. (For computing purposes, r is expressed as a decimal.) The interest earned after each compounding period is given by formula (1).

$$\text{Interest} = \text{principal} \cdot \text{rate} \cdot \text{time} = P \cdot r \cdot \frac{1}{n} = P \cdot \frac{r}{n}$$

The amount A after one compounding period is

$$A = P + P \cdot \frac{r}{n} = P \cdot \left(1 + \frac{r}{n}\right)$$

After two compounding periods, the amount A, based on the new principal $P \cdot \left(1 + \frac{r}{n}\right)$, is

$$A = \underbrace{P \cdot \left(1 + \frac{r}{n}\right)}_{\substack{\text{New} \\ \text{principal}}} + \underbrace{P \cdot \left(1 + \frac{r}{n}\right) \cdot \frac{r}{n}}_{\substack{\text{Interest on} \\ \text{new principal}}} \underset{\uparrow}{=} \underset{\substack{\text{Factor out } P \cdot \left(1 + \frac{r}{n}\right)}}{P \cdot \left(1 + \frac{r}{n}\right) \cdot \left(1 + \frac{r}{n}\right)} = P \cdot \left(1 + \frac{r}{n}\right)^2$$

After three compounding periods, the amount A is

$$A = P \cdot \left(1 + \frac{r}{n}\right)^2 + P \cdot \left(1 + \frac{r}{n}\right)^2 \cdot \frac{r}{n} = P \cdot \left(1 + \frac{r}{n}\right)^2 \cdot \left(1 + \frac{r}{n}\right) = P \cdot \left(1 + \frac{r}{n}\right)^3$$

Continuing this way, after n compounding periods (1 year), the amount A is

$$A = P \cdot \left(1 + \frac{r}{n}\right)^n$$

Because t years will contain $n \cdot t$ compounding periods, the amount after t years is

$$A = P \cdot \left(1 + \frac{r}{n}\right)^{nt}$$

THEOREM **Compound Interest Formula**

The amount A after t years due to a principal P invested at an annual interest rate r, expressed as a decimal, compounded n times per year is

$$A = P \cdot \left(1 + \frac{r}{n}\right)^{nt} \tag{2}$$

In equation (2), the amount A is typically referred to as the **accumulated value** of the account, and P is called the **present value**.

Exploration

To observe the effects of compounding interest monthly on an initial deposit of $1, graph $Y_1 = \left(1 + \dfrac{r}{12}\right)^{12x}$ with $r = 0.06$ and $r = 0.12$ for $0 \le x \le 30$. What is the future value of $1 in 30 years when the interest rate per annum is $r = 0.06$ (6%)? What is the future value of $1 in 30 years when the interest rate per annum is $r = 0.12$ (12%)? Does doubling the interest rate double the future value?

For example, to rework Example 1, use $P = \$1000$, $r = 0.02$, $n = 4$ (quarterly compounding), and $t = 1$ year to obtain

$$A = P \cdot \left(1 + \frac{r}{n}\right)^{nt} = 1000\left(1 + \frac{0.02}{4}\right)^{4 \cdot 1} = \$1020.15$$

The result obtained here differs slightly from that obtained in Example 1 because of rounding.

Now Work PROBLEM 7

EXAMPLE 2

Comparing Investments Using Different Compounding Periods

Investing $1000 at an annual rate of 10% compounded annually, semiannually, quarterly, monthly, and daily will yield the following amounts after 1 year:

Annual compounding ($n = 1$):
$$A = P \cdot (1 + r)$$
$$= \$1000(1 + 0.10) = \$1100.00$$

Semiannual compounding ($n = 2$):
$$A = P \cdot \left(1 + \frac{r}{2}\right)^2$$
$$= \$1000(1 + 0.05)^2 = \$1102.50$$

Quarterly compounding ($n = 4$):
$$A = P \cdot \left(1 + \frac{r}{4}\right)^4$$
$$= \$1000(1 + 0.025)^4 = \$1103.81$$

Monthly compounding ($n = 12$):
$$A = P \cdot \left(1 + \frac{r}{12}\right)^{12}$$
$$= \$1000\left(1 + \frac{0.10}{12}\right)^{12} = \$1104.71$$

Daily compounding ($n = 365$):
$$A = P \cdot \left(1 + \frac{r}{365}\right)^{365}$$
$$= \$1000\left(1 + \frac{0.10}{365}\right)^{365} = \$1105.16$$

From Example 2, note that the effect of compounding more frequently is that the amount after 1 year is higher. This leads to the following question: What would happen to the amount after 1 year if the number of times that the interest is compounded were increased without bound?

Let's find the answer. Suppose that P is the principal, r is the per annum interest rate, and n is the number of times that the interest is compounded each year. The amount A after 1 year is

$$A = P \cdot \left(1 + \frac{r}{n}\right)^n$$

Rewrite this expression as follows:

$$A = P \cdot \left(1 + \frac{r}{n}\right)^n = P \cdot \left(1 + \frac{1}{\frac{n}{r}}\right)^n = P \cdot \left[\left(1 + \frac{1}{\frac{n}{r}}\right)^{n/r}\right]^r \underset{\underset{h = \frac{n}{r}}{\uparrow}}{=} P \cdot \left[\left(1 + \frac{1}{h}\right)^h\right]^r \quad \textbf{(3)}$$

Need to Review?
The number *e* is defined on page 443.

Now suppose that the number *n* of times that the interest is compounded per year gets larger and larger; that is, suppose that $n \rightarrow \infty$. Then $h = \dfrac{n}{r} \rightarrow \infty$, and the expression in brackets in equation (3) equals *e*. That is, $A \rightarrow Pe^r$.

Table 8 compares $\left(1 + \dfrac{r}{n}\right)^n$, for large values of *n*, to e^r for $r = 0.05$, $r = 0.10$, $r = 0.15$, and $r = 1$. As *n* becomes larger, the closer $\left(1 + \dfrac{r}{n}\right)^n$ gets to e^r. No matter how frequent the compounding, the amount after 1 year has the upper bound Pe^r.

Table 8

	$\left(1 + \frac{r}{n}\right)^n$			
	n = 100	*n* = 1000	*n* = 10,000	e^r
r = 0.05	1.0512580	1.0512698	1.051271	1.0512711
r = 0.10	1.1051157	1.1051654	1.1051704	1.1051709
r = 0.15	1.1617037	1.1618212	1.1618329	1.1618342
r = 1	2.7048138	2.7169239	2.7181459	2.7182818

When interest is compounded so that the amount after 1 year is Pe^r, the interest is said to be **compounded continuously**.

> **THEOREM Continuous Compounding**
>
> The amount *A* after *t* years due to a principal *P* invested at an annual interest rate *r* compounded continuously is
>
> $$A = Pe^{rt} \qquad \qquad (4)$$

EXAMPLE 3 **Using Continuous Compounding**

The amount *A* that results from investing a principal *P* of $1000 at an annual rate *r* of 10% compounded continuously for a time *t* of 1 year is

$$A = \$1000e^{0.10} = \$1000 \cdot 1.10517 = \$1105.17$$

→ **Now Work** PROBLEM 13

2 Calculate Effective Rates of Return

Suppose that you have $1000 to invest and a bank offers to pay you 3 percent annual interest compounded monthly. What simple interest rate is needed to earn an equal amount after one year? To answer this question, first determine the value after one year of the $1000 investment that earns 3 percent compounded monthly.

$$A = \$1000\left(1 + \frac{0.03}{12}\right)^{12} \quad \text{Use } A = P\left(1 + \frac{r}{n}\right)^n \text{ with } P = \$1000, r = 0.03, n = 12.$$
$$= \$1030.42$$

So the interest earned is $30.42. Using $I = Prt$ with $t = 1$, $I = \$30.42$, and $P = \$1000$, the annual simple interest rate is $0.03042 = 3.042\%$. This interest rate is known as the *effective rate of interest*.

The **effective rate of interest** is the annual simple interest rate that would yield the same amount as compounding n times per year, or continuously, after one year.

> **THEOREM Effective Rate of Interest**
>
> The effective rate of interest r_E of an investment earning an annual interest rate r is given by
>
> - Compounding n times per year: $r_E = \left(1 + \dfrac{r}{n}\right)^n - 1$
> - Continuous compounding: $r_E = e^r - 1$

EXAMPLE 4   **Computing the Effective Rate of Interest—Which Is the Best Deal?**

Suppose you want to buy a 5-year certificate of deposit (CD). You visit three banks to determine their CD rates. American Express offers you 2.15% annual interest compounded monthly, and First Internet Bank offers you 2.20% compounded quarterly. Discover offers 2.12% compounded daily. Determine which bank is offering the best deal.

Solution   The bank that offers the best deal is the one with the highest effective interest rate.

American Express	**First Internet Bank**	**Discover**
$r_E = \left(1 + \dfrac{0.0215}{12}\right)^{12} - 1$	$r_E = \left(1 + \dfrac{0.022}{4}\right)^{4} - 1$	$r_E = \left(1 + \dfrac{0.0212}{365}\right)^{365} - 1$
$\approx 1.02171 - 1$	$\approx 1.02218 - 1$	$\approx 1.02143 - 1$
$= 0.02171$	$= 0.02218$	$= 0.02143$
$= 2.171\%$	$= 2.218\%$	$= 2.143\%$

The effective rate of interest is highest for First Internet Bank, so First Internet Bank is offering the best deal.

———— **Now Work** PROBLEM 23

3 Determine the Present Value of a Lump Sum of Money

When people in finance speak of the "time value of money," they are usually referring to the *present value* of money. The **present value** of A dollars to be received at a future date is the principal that you would need to invest now so that it will grow to A dollars in the specified time period. The present value of money to be received at a future date is always less than the amount to be received, since the amount to be received will equal the present value (money invested now) *plus* the interest accrued over the time period.

The compound interest formula (2) is used to develop a formula for present value. If P is the present value of A dollars to be received after t years at a per annum interest rate r compounded n times per year, then, by formula (2),

$$A = P \cdot \left(1 + \frac{r}{n}\right)^{nt}$$

To solve for P, divide both sides by $\left(1 + \dfrac{r}{n}\right)^{nt}$. The result is

$$\frac{A}{\left(1 + \dfrac{r}{n}\right)^{nt}} = P \quad \text{or} \quad P = A \cdot \left(1 + \frac{r}{n}\right)^{-nt}$$

THEOREM Present Value Formulas

The present value P of A dollars to be received after t years, assuming a per annum interest rate r compounded n times per year, is

$$P = A \cdot \left(1 + \frac{r}{n}\right)^{-nt} \tag{5}$$

If the interest is compounded continuously, then

$$P = Ae^{-rt} \tag{6}$$

To derive (6), solve $A = Pe^{rt}$ for P.

EXAMPLE 5 **Computing the Value of a Zero-Coupon Bond**

A zero-coupon (noninterest-bearing) bond can be redeemed in 10 years for $1000. How much should you be willing to pay for it now if you want a return of

(a) 8% compounded monthly? (b) 7% compounded continuously?

Solution (a) To find the present value of $1000, use formula (5) with $A = \$1000$, $n = 12$, $r = 0.08$, and $t = 10$.

$$P = A \cdot \left(1 + \frac{r}{n}\right)^{-nt} = \$1000\left(1 + \frac{0.08}{12}\right)^{-12 \cdot 10} = \$450.52$$

For a return of 8% compounded monthly, pay $450.52 for the bond.

(b) Here use formula (6) with $A = \$1000$, $r = 0.07$, and $t = 10$.

$$P = Ae^{-rt} = \$1000e^{-0.07 \cdot 10} = \$496.59$$

For a return of 7% compounded continuously, pay $496.59 for the bond.

━━━━ **Now Work** PROBLEM 15

4 Determine the Rate of Interest or the Time Required to Double a Lump Sum of Money

---

EXAMPLE 6 **Determining the Rate of Interest Required to Double an Investment**

What rate of interest compounded annually is needed in order to double an investment in 5 years?

Solution If P is the principal and P is to double, then the amount A will be $2P$. Use the compound interest formula with $n = 1$ and $t = 5$ to find r.

$$A = P \cdot \left(1 + \frac{r}{n}\right)^{nt}$$

$$2P = P \cdot (1 + r)^5 \qquad\qquad A = 2P, n = 1, t = 5$$

$$2 = (1 + r)^5 \qquad\qquad \text{Divide both sides by } P.$$

$$1 + r = \sqrt[5]{2} \qquad\qquad \text{Take the fifth root of both sides.}$$

$$r = \sqrt[5]{2} - 1 \approx 1.148698 - 1 = 0.148698$$

The annual rate of interest needed to double the principal in 5 years is 14.87%.

━━━━ **Now Work** PROBLEM 31

| EXAMPLE 7 | Determining the Time Required to Double or Triple an Investment |

(a) How long will it take for an investment to double in value if it earns 5% compounded continuously?

(b) How long will it take to triple at this rate?

Solution

(a) If P is the initial investment and P is to double, then the amount A will be $2P$. Use formula (4) for continuously compounded interest with $r = 0.05$.

$$A = Pe^{rt}$$
$$2P = Pe^{0.05t} \qquad \text{A = 2P, r = 0.05}$$
$$2 = e^{0.05t} \qquad \text{Cancel the P's.}$$
$$0.05t = \ln 2 \qquad \text{Rewrite as a logarithm.}$$
$$t = \frac{\ln 2}{0.05} \approx 13.86 \qquad \text{Solve for t.}$$

It will take about 14 years to double the investment.

(b) To triple the investment, let $A = 3P$ in formula (4).

$$A = Pe^{rt}$$
$$3P = Pe^{0.05t} \qquad \text{A = 3P, r = 0.05}$$
$$3 = e^{0.05t} \qquad \text{Cancel the P's.}$$
$$0.05t = \ln 3 \qquad \text{Rewrite as a logarithm.}$$
$$t = \frac{\ln 3}{0.05} \approx 21.97 \qquad \text{Solve for t.}$$

It will take about 22 years to triple the investment.

Now Work PROBLEM 35

6.7 Assess Your Understanding

'Are You Prepared?' *Answers are given at the end of these exercises. If you get a wrong answer, read the pages listed in* red.

1. What is the interest due when $500 is borrowed for 6 months at a simple interest rate of 6% per annum? (pp. 136–137)

2. If you borrow $5000 and, after 9 months, pay off the loan in the amount of $5500, what per annum rate of interest was charged? (pp. 136–137)

Concepts and Vocabulary

3. The total amount borrowed (whether by an individual from a bank in the form of a loan or by a bank from an individual in the form of a savings account) is called the _____ .

4. If a principal of P dollars is borrowed for a period of t years at a per annum interest rate r, expressed as a decimal, the interest I charged is _____ = _____ . Interest charged according to this formula is called _____ .

5. The _____ _____ ____ _____ is the annual simple interest rate that would yield the same amount as compounding n times per year, or continuously, after 1 year.

6. *Multiple Choice* The principal that must be invested now so that it will grow to a given amount in a specified time period is called the _____ .
   (a) present value   (b) future value
   (c) interest   (d) effective rate

Skill Building

In Problems 7–14, find the amount that results from each investment.

7. $100 invested at 4% compounded quarterly after a period of 2 years

8. $50 invested at 6% compounded monthly after a period of 3 years

9. $900 invested at 3% compounded semiannually after a period of $2\frac{1}{2}$ years

10. $300 invested at 12% compounded monthly after a period of $1\frac{1}{2}$ years

11. $1200 invested at 5% compounded daily after a period of 3 years

12. $700 invested at 6% compounded daily after a period of 2 years

13. $1000 invested at 11% compounded continuously after a period of 2 years

14. $400 invested at 7% compounded continuously after a period of 3 years

In Problems 15–22, find the principal needed now to get each amount; that is, find the present value.

15. To get $100 after 2 years at 6% compounded monthly

16. To get $75 after 3 years at 8% compounded quarterly

17. To get $1500 after $2\frac{1}{2}$ years at 1.5% compounded daily

18. To get $800 after $3\frac{1}{2}$ years at 7% compounded monthly

19. To get $750 after 2 years at 2.5% compounded quarterly

20. To get $300 after 4 years at 3% compounded daily

21. To get $120 after $3\frac{1}{4}$ years at 5% compounded continuously

22. To get $800 after $2\frac{1}{2}$ years at 8% compounded continuously

In Problems 23–26, find the effective rate of interest.

23. For 5% compounded quarterly

24. For 6% compounded monthly

25. For 4% compounded continuously

26. For 6% compounded continuously

In Problems 27–30, determine the rate that represents the better deal.

27. 6% compounded quarterly or $6\frac{1}{4}$% compounded annually

28. 9% compounded quarterly or $9\frac{1}{4}$% compounded annually

29. 9% compounded monthly or 8.8% compounded daily

30. 8% compounded semiannually or 7.9% compounded daily

31. What rate of interest compounded annually is required to double an investment in 3 years?

32. What rate of interest compounded annually is required to double an investment in 6 years?

33. What rate of interest compounded annually is required to triple an investment in 5 years?

34. What rate of interest compounded annually is required to triple an investment in 10 years?

35. (a) How long does it take for an investment to double in value if it is invested at 8% compounded monthly?
    (b) How long does it take if the interest is compounded continuously?

36. (a) How long does it take for an investment to triple in value if it is invested at 6% compounded monthly?
    (b) How long does it take if the interest is compounded continuously?

37. What rate of interest compounded quarterly will yield an effective interest rate of 7%?

38. What rate of interest compounded continuously will yield an effective interest rate of 6%?

Applications and Extensions

39. **Time Required to Reach a Goal** If Tanisha has $100 to invest at 4% per annum compounded monthly, how long will it be before she has $150? If the compounding is continuous, how long will it be?

40. **Time Required to Reach a Goal** If Angela has $100 to invest at 2.5% per annum compounded monthly, how long will it be before she has $175? If the compounding is continuous, how long will it be?

41. **Time Required to Reach a Goal** How many years will it take for an initial investment of $10,000 to grow to $25,000? Assume a rate of interest of 6% compounded continuously.

42. **Time Required to Reach a Goal** How many years will it take for an initial investment of $25,000 to grow to $80,000? Assume a rate of interest of 7% compounded continuously.

43. **Price Appreciation of Homes** What will a $90,000 condominium cost 5 years from now if the price appreciation for condos over that period averages 3% compounded annually?

44. **Credit Card Interest** A department store charges 1.25% per month on the unpaid balance for customers with charge accounts (interest is compounded monthly). A customer charges $200 and does not pay her bill for 6 months. What is the bill at that time?

45. **Saving for a Car** Jerome will be buying a used car for $15,000 in 3 years. How much money should he ask his parents for now so that, if he invests it at 5% compounded continuously, he will have enough to buy the car?

46. **Paying off a Loan** John requires $3000 in 6 months to pay off a loan that has no prepayment privileges. If he has the $3000 now, how much of it should he save in an account paying 3% compounded monthly so that in 6 months he will have exactly $3000?

47. **Return on a Stock** George contemplates the purchase of 100 shares of a stock selling for $15 per share. The stock pays no dividends. The history of the stock indicates that it should grow at an annual rate of 15% per year. How much should the 100 shares of stock be worth in 5 years?

48. **Return on an Investment** A business purchased for $650,000 in 2010 is sold in 2013 for $850,000. What is the annual rate of return for this investment?

49. **Comparing Savings Plans** Jim places $1000 in a bank account that pays 5.6% compounded continuously. After 1 year, will he have enough money to buy a computer system that costs $1060? If another bank will pay Jim 5.9% compounded monthly, is this a better deal?

50. **Savings Plans** On January 1, Kim places $1000 in a certificate of deposit that pays 6.8% compounded continuously and matures in 3 months. Then Kim places the $1000 and the interest in a passbook account that pays 5.25% compounded monthly. How much does Kim have in the passbook account on May 1?

51. Comparing IRA Investments Will invests $2000 of the money in his IRA in a bond trust that pays 9% interest compounded semiannually. His friend Henry invests $2000 in his IRA in a certificate of deposit that pays $8\frac{1}{2}$% compounded continuously. Who has more money after 20 years, Will or Henry?

52. Comparing Two Alternatives Suppose that April has access to an investment that will pay 10% interest compounded continuously. Which is better: to be given $1000 now so that she can take advantage of this investment opportunity or to be given $1325 after 3 years?

53. College Costs The average annual cost of college at 4-year private colleges was $34,740 in the 2017–2018 academic year. This was a 3.6% increase from the previous year.
   (a) If the cost of college increases by 3.6% each year, what will be the average cost of college at a 4-year private college for the 2037–2038 academic year?
   (b) College savings plans, such as a 529 plan, allow individuals to put money aside now to help pay for college later. If one such plan offers a rate of 2% compounded continuously, how much should be put in a college savings plan in 2019 to pay for 1 year of the cost of college at a 4-year private college for an incoming freshman in 2037?

   Source: The College Board

54. Analyzing Interest Rates on a Mortgage Colleen and Bill have just purchased a house for $650,000, with the seller holding a second mortgage of $100,000. They promise to pay the seller $100,000 plus all accrued interest 5 years from now. The seller offers them three interest options on the second mortgage:
   (a) Simple interest at 6% per annum
   (b) 5.5% interest compounded monthly
   (c) 5.25% interest compounded continuously

   Which option is best? That is, which results in paying the least interest on the loan?

55. Comparing Bank Accounts Two bank accounts are opened at the same time. The first has a principal of $1000 in an account earning 5% compounded monthly. The second has a principal of $2000 in an account earning 4% interest compounded annually. Determine the number of years, to the nearest tenth, at which the account balances will be equal.

56. Per Capita Federal Debt In 2018, the federal debt was about $21 trillion. In 2018, the U.S. population was about 327 million. Assuming that the federal debt is increasing about 5.5% per year and the U.S. population is increasing about 0.7% per year, determine the per capita debt (total debt divided by population) in 2030.

Problems 57–62 require the following discussion. **Inflation** *is a term used to describe the erosion of the purchasing power of money. For example, if the annual inflation rate is 3%, then $1000 worth of purchasing power now will have only $970 worth of purchasing power in 1 year because 3% of the original $1000 $(0.03 \times 1000 = 30)$ has been eroded due to inflation. In general, if the rate of inflation averages r% per annum over n years, the amount A that $P will purchase after n years is*

$$A = P \cdot (1 - r)^n$$

where r is expressed as a decimal.

57. Inflation If the inflation rate averages 3%, what will be the purchasing power of $1000 in 2 years?

58. Inflation If the inflation rate averages 2%, what will be the purchasing power of $1000 in 3 years?

59. Inflation If the purchasing power of $1000 is only $950 after 2 years, what was the average inflation rate?

60. Inflation If the purchasing power of $1000 is only $930 after 2 years, what was the average inflation rate?

61. Inflation If the average inflation rate is 2%, how long is it until purchasing power is cut in half?

62. Inflation If the average inflation rate is 4%, how long is it until purchasing power is cut in half?

Problems 63–66 involve zero-coupon bonds. A **zero-coupon bond** *is a bond that is sold now at a discount and will pay its face value at the time when it matures; no interest payments are made.*

63. Zero-Coupon Bonds A zero-coupon bond can be redeemed in 20 years for $10,000. How much should you be willing to pay for it now if you want a return of:
   (a) 5% compounded monthly?
   (b) 5% compounded continuously?

64. Zero-Coupon Bonds A child's grandparents are considering buying an $80,000 face-value, zero-coupon bond at her birth so that she will have enough money for her college education 17 years later. If they want a rate of return of 6% compounded annually, what should they pay for the bond?

65. Zero-Coupon Bonds How much should a $10,000 face-value, zero-coupon bond, maturing in 10 years, be sold for now if its rate of return is to be 4.5% compounded annually?

66. Zero-Coupon Bonds If Pat pays $15,334.65 for a $25,000 face-value, zero-coupon bond that matures in 8 years, what is his annual rate of return?

67. Time to Double or Triple an Investment The formula

$$t = \frac{\ln m}{n \ln\left(1 + \frac{r}{n}\right)}$$

can be used to find the number of years t required to multiply an investment m times when r is the per annum interest rate compounded n times a year.
   (a) How many years will it take to double the value of an IRA that compounds annually at the rate of 6%?
   (b) How many years will it take to triple the value of a savings account that compounds quarterly at an annual rate of 5%?
   (c) Give a derivation of this formula.

68. Time to Reach an Investment Goal The formula

$$t = \frac{\ln A - \ln P}{r}$$

can be used to find the number of years t required for an investment P to grow to a value A when compounded continuously at an annual rate r.

(a) How long will it take to increase an initial investment of $1000 to $4500 at an annual rate of 5.75%?

(b) What annual rate is required to increase the value of a $2000 IRA to $30,000 in 35 years?

(c) Give a derivation of this formula.

---

*Problems 69–72 require the following discussion. The **consumer price index (CPI)** indicates the relative change in price over time for a fixed basket of goods and services. It is a cost-of-living index that helps measure the effect of inflation on the cost of goods and services. The CPI uses the base period 1982–1984 for comparison (the CPI for this period is 100). The CPI for March 2018 was 249.55. This means that $100 in the period 1982–1984 had the same purchasing power as $249.55 in March 2018. In general, if the rate of inflation averages r% per annum over n years, then the CPI index after n years is*

$$\text{CPI} = \text{CPI}_0 \left(1 + \frac{r}{100} \right)^n$$

where CPI_0 is the CPI index at the beginning of the n-year period.
Source: *U.S. Bureau of Labor Statistics*

69. Consumer Price Index

(a) The CPI was 229.39 for 2012 and 243.80 for 2017. Assuming that annual inflation remained constant for this time period, determine the average annual inflation rate.

(b) Using the inflation rate from part (a), in what year will the CPI reach 300?

70. Consumer Price Index If the current CPI is 234.2 and the average annual inflation rate is 2.8%, what will be the CPI in 5 years?

71. Consumer Price Index If the average annual inflation rate is 3.1%, how long will it take for the CPI index to double? (A doubling of the CPI index means purchasing power is cut in half.)

72. Consumer Price Index The base period for the CPI changed in 1998. Under the previous weight and item structure, the CPI for 1995 was 456.5. If the average annual inflation rate was 5.57%, what year was used as the base period for the CPI?

Explaining Concepts: Discussion and Writing

73. Explain in your own words what the term *compound interest* means. What does *continuous compounding* mean?

74. Explain in your own words the meaning of *present value*.

75. Critical Thinking You have just contracted to buy a house and will seek financing in the amount of $100,000. You go to several banks. Bank 1 will lend you $100,000 at the rate of 4.125% amortized over 30 years with a loan origination fee of 0.45%. Bank 2 will lend you $100,000 at the rate of 3.375% amortized over 15 years with a loan origination fee of 0.95%. Bank 3 will lend you $100,000 at the rate of 4.25% amortized over 30 years with no loan origination fee. Bank 4 will lend you $100,000 at the rate of 3.625% amortized over 15 years with no loan origination fee. Which loan would you take? Why? Be

sure to have sound reasons for your choice. Use the information in the table to assist you. If the amount of the monthly payment does not matter to you, which loan would you take? Again, have sound reasons for your choice. Compare your final decision with others in the class. Discuss.

	Monthly Payment	Loan Origination Fee
Bank 1	$485	$450
Bank 2	$709	$950
Bank 3	$492	$0
Bank 4	$721	$0

Retain Your Knowledge

Problems 76–85 are based on material learned earlier in the course. The purpose of these problems is to keep the material fresh in your mind so that you are better prepared for the final exam.

76. Find the remainder R when $f(x) = 6x^3 + 3x^2 + 2x - 11$ is divided by $g(x) = x - 1$. Is g a factor of f?

77. The function $f(x) = \dfrac{x}{x-2}$ is one-to-one. Find f^{-1}.

78. Find the real zeros of
$$f(x) = x^5 - x^4 - 15x^3 - 21x^2 - 16x - 20$$
Then write f in factored form.

79. Solve: $\log_2(x+3) = 2 \log_2(x-3)$

80. Factor completely: $2x^4 + 6x^3 - 50x^2 - 150x$

81. If $f(x) = 5x^2 + 4x - 8$ and $g(x) = 3x - 1$, find $(f \circ g)(x)$.

82. Find the domain and range of $f(x) = -2x^2 - 8x + 1$.

83. For $f(x) = \dfrac{2x^2 - 5x - 4}{x - 7}$, find all vertical asymptotes, horizontal asymptotes, and oblique asymptotes, if any.

84. If $f(x) = x^2 - 4x - 3$, find an equation of the secant line containing the points $(3, f(3))$ and $(5, f(5))$.

85. Find the difference quotient for $f(x) = 3x - 5$.

'Are You Prepared?' Answers

1. $15   **2.** $13\frac{1}{3}\%$

6.8 Exponential Growth and Decay Models; Newton's Law; Logistic Growth and Decay Models

OBJECTIVES 1 Model Populations That Obey the Law of Uninhibited Growth (p. 491)
2 Model Populations That Obey the Law of Uninhibited Decay (p. 493)
3 Use Newton's Law of Cooling (p. 494)
4 Use Logistic Models (p. 496)

1 Model Populations That Obey the Law of Uninhibited Growth

Many natural phenomena follow the law that an amount A varies with time t according to the function

$$A(t) = A_0 e^{kt} \tag{1}$$

Here A_0 is the original amount ($t = 0$) and $k \neq 0$ is a constant.

If $k > 0$, then equation (1) states that the amount A is increasing over time; if $k < 0$, the amount A is decreasing over time. In either case, when an amount A varies over time according to equation (1), it is said to follow the **exponential law**, or the **law of uninhibited growth** ($k > 0$) **or decay** ($k < 0$). See Figure 41.

For example, in Section 6.7, continuously compounded interest was shown to follow the law of uninhibited growth. In this section we shall look at some additional phenomena that follow the exponential law.

Cell division is the growth process of many living organisms, such as amoebas, plants, and human skin cells. Based on an ideal situation in which no cells die and no by-products are produced, the number of cells present at a given time follows the law of uninhibited growth. Actually, however, after enough time has passed, growth at an exponential rate will cease as a consequence of factors such as lack of living space and dwindling food supply. The law of uninhibited growth accurately models only the early stages of the cell division process.

The cell division process begins with a culture containing N_0 cells. Each cell in the culture grows for a certain period of time and then divides into two identical cells. Assume that the time needed for each cell to divide in two is constant and does not change as the number of cells increases. These new cells then grow, and eventually each divides in two, and so on.

(a) $A(t) = A_0 e^{kt}, k > 0$
Exponential growth

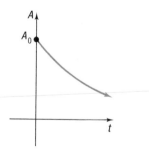

(b) $A(t) = A_0 e^{kt}, k < 0$
Exponential decay
Figure 41

Uninhibited Growth of Cells

A model that gives the number N of cells in a culture after a time t has passed (in the early stages of growth) is

$$N(t) = N_0 e^{kt} \qquad k > 0 \tag{2}$$

where N_0 is the initial number of cells and k is a positive constant that represents the growth rate of the cells.

Using formula (2) to model the growth of cells employs a function that yields positive real numbers, even though the number of cells being counted must be an integer. This is a common practice in many applications.

EXAMPLE 1 Bacterial Growth

A colony of bacteria that grows according to the law of uninhibited growth is modeled by the function $N(t) = 100e^{0.045t}$, where N is measured in grams and t is measured in days.

(a) Determine the initial amount of bacteria.
(b) What is the growth rate of the bacteria?
(c) What is the population after 5 days?
(d) How long will it take for the population to reach 140 grams?
(e) What is the doubling time for the population?

Solution
(a) The initial amount of bacteria, N_0, is obtained when $t = 0$, so

$$N_0 = N(0) = 100e^{0.045 \cdot 0} = 100 \text{ grams}$$

(b) Compare $N(t) = 100e^{0.045t}$ to $N(t) = N_0 e^{kt}$. The value of k, 0.045, indicates a growth rate of 4.5%.

(c) The population after 5 days is $N(5) = 100e^{0.045 \cdot 5} \approx 125.2$ grams.

(d) To find how long it takes for the population to reach 140 grams, solve the equation $N(t) = 140$.

$$100e^{0.045t} = 140$$

$$e^{0.045t} = 1.4 \qquad \text{Divide both sides of the equation by 100.}$$

$$0.045t = \ln 1.4 \qquad \text{Rewrite as a logarithm.}$$

$$t = \frac{\ln 1.4}{0.045} \qquad \text{Divide both sides of the equation by 0.045.}$$

$$\approx 7.5 \text{ days}$$

The population reaches 140 grams in about 7.5 days.

(e) The population doubles when $N(t) = 200$ grams, so the doubling time is found by solving the equation $200 = 100e^{0.045t}$ for t.

$$200 = 100e^{0.045t}$$

$$2 = e^{0.045t} \qquad \text{Divide both sides of the equation by 100.}$$

$$\ln 2 = 0.045t \qquad \text{Rewrite as a logarithm.}$$

$$t = \frac{\ln 2}{0.045} \qquad \text{Divide both sides of the equation by 0.045.}$$

$$\approx 15.4 \text{ days}$$

The population doubles approximately every 15.4 days.

Now Work PROBLEM 1

EXAMPLE 2 Bacterial Growth

A colony of bacteria increases according to the law of uninhibited growth.

(a) If N is the number of cells and t is the time in hours, express N as a function of t.
(b) If the number of bacteria doubles in 3 hours, find the function that gives the number of cells in the culture.
(c) How long will it take for the size of the colony to triple?
(d) How long will it take for the population to double a second time (that is, to increase four times)?

Solution
(a) Using formula (2), the number N of cells at time t is

$$N(t) = N_0 e^{kt}$$

where N_0 is the initial number of bacteria present and k is a positive number.

(b) To find the growth rate k, note that the number of cells doubles in 3 hours, so

$$N(3) = 2N_0$$

Since $N(3) = N_0 e^{k \cdot 3}$,

$$N_0 e^{k \cdot 3} = 2N_0$$

$$e^{3k} = 2 \qquad\qquad \text{Divide both sides by } N_0.$$

$$3k = \ln 2 \qquad\qquad \text{Write the exponential equation as a logarithm.}$$

$$k = \frac{1}{3} \ln 2 \approx 0.23105$$

The function that models this growth process is therefore

$$N(t) = N_0 \, e^{0.23105t}$$

(c) The time t needed for the size of the colony to triple requires that $N = 3N_0$. Substitute $3N_0$ for N to get

$$3N_0 = N_0 \, e^{0.23105t}$$

$$3 = e^{0.23105t}$$

$$0.23105t = \ln 3$$

$$t = \frac{\ln 3}{0.23105} \approx 4.755 \text{ hours}$$

It will take about 4.755 hours, or 4 hours and 45 minutes, for the size of the colony to triple.

(d) If a population doubles in 3 hours, it will double a second time in 3 more hours, for a total time of 6 hours.

2 Model Populations That Obey the Law of Uninhibited Decay

Radioactive materials follow the law of uninhibited decay.

Uninhibited Radioactive Decay

The amount A of a radioactive material present at time t is given by

$$A(t) = A_0 e^{kt} \qquad k < 0 \tag{3}$$

where A_0 is the original amount of radioactive material and k is a negative number that represents the rate of decay.

All radioactive substances have a specific **half-life**, which is the time required for half of the radioactive substance to decay. **Carbon dating** uses the fact that all living organisms contain two kinds of carbon, carbon-12 (a stable carbon) and carbon-14 (a radioactive carbon with a half-life of 5730 years). While an organism is living, the ratio of carbon-12 to carbon-14 is constant. But when an organism dies, the original amount of carbon-12 present remains unchanged, whereas the amount of carbon-14 begins to decrease. This change in the amount of carbon-14 present relative to the amount of carbon-12 present makes it possible to calculate when the organism died.

EXAMPLE 3

Estimating the Age of Ancient Tools

Traces of burned wood along with ancient stone tools in an archeological dig in Chile were found to contain approximately 1.67% of the original amount of carbon-14. If the half-life of carbon-14 is 5730 years, approximately when was the tree cut and burned?

Solution   Using formula (3), the amount A of carbon-14 present at time t is

$$A(t) = A_0 e^{kt}$$

where A_0 is the original amount of carbon-14 present and k is a negative number. We first seek the number k. To find it, we use the fact that after 5730 years, half of the original amount of carbon-14 remains, so $A(5730) = \dfrac{1}{2}A_0$. Then

$$\frac{1}{2}A_0 = A_0 e^{k \cdot 5730}$$

$$\frac{1}{2} = e^{5730k} \qquad\qquad \textsf{Divide both sides of the equation by } A_0.$$

$$5730k = \ln\frac{1}{2} \qquad\qquad \textsf{Rewrite as a logarithm.}$$

$$k = \frac{1}{5730}\ln\frac{1}{2} \approx -0.000120968$$

Formula (3) therefore becomes

$$A(t) = A_0\, e^{-0.000120968t}$$

If the amount A of carbon-14 now present is 1.67% of the original amount, it follows that

$$0.0167A_0 = A_0\, e^{-0.000120968t}$$

$$0.0167 = e^{-0.000120968t} \qquad\qquad \textsf{Divide both sides of the equation by } A_0.$$

$$-0.000120968t = \ln 0.0167 \qquad\qquad \textsf{Rewrite as a logarithm.}$$

$$t = \frac{\ln 0.0167}{-0.000120968} \approx 33{,}830 \text{ years}$$

The tree was cut and burned about 33,830 years ago. Some archeologists use this conclusion to argue that humans lived in the Americas nearly 34,000 years ago, much earlier than is generally accepted.

Now Work PROBLEM 3

3 Use Newton's Law of Cooling

Newton's Law of Cooling[*] states that the temperature of a heated object decreases exponentially over time toward the temperature of the surrounding medium.

Newton's Law of Cooling

The temperature u of a heated object at a given time t can be modeled by the following function:

$$u(t) = T + (u_0 - T)e^{kt} \qquad k < 0 \qquad\qquad \textbf{(4)}$$

where T is the constant temperature of the surrounding medium, u_0 is the initial temperature of the heated object, and k is a negative constant.

EXAMPLE 4   **Using Newton's Law of Cooling**

An object is heated to 100°C (degrees Celsius) and is then allowed to cool in a room whose air temperature is 30°C.

(a) If the temperature of the object is 80°C after 5 minutes, when will its temperature be 50°C?

(b) Determine the elapsed time before the temperature of the object is 35°C.

(c) What do you notice about the temperature as time passes?

[*]Named after Sir Isaac Newton (1643–1727), one of the cofounders of calculus.

Solution   (a) Using formula (4) with $T = 30$ and $u_0 = 100$, the temperature $u(t)$ (in degrees Celsius) of the object at time t (in minutes) is

$$u(t) = 30 + (100 - 30)e^{kt} = 30 + 70e^{kt} \tag{5}$$

where k is a negative constant. To find k, use the fact that $u = 80$ when $t = 5$. Then

$$u(t) = 30 + 70e^{kt}$$
$$80 = 30 + 70e^{k \cdot 5} \qquad\qquad u(5) = 80$$
$$50 = 70e^{5k} \qquad\qquad \text{Simplify.}$$
$$e^{5k} = \frac{50}{70} \qquad\qquad \text{Solve for } e^{5k}.$$
$$5k = \ln\frac{5}{7} \qquad\qquad \text{Rewrite as a logarithm.}$$
$$k = \frac{1}{5}\ln\frac{5}{7} \approx -0.0673 \quad \text{Solve for } k.$$

Formula (5) therefore becomes

$$u(t) = 30 + 70e^{-0.0673t} \tag{6}$$

To find t when $u = 50°C$, solve the equation

$$50 = 30 + 70e^{-0.0673t}$$
$$20 = 70e^{-0.0673t} \qquad\qquad \text{Simplify.}$$
$$e^{-0.0673t} = \frac{20}{70}$$
$$-0.0673t = \ln\frac{2}{7} \qquad\qquad \text{Rewrite as a logarithm.}$$
$$t = \frac{\ln\dfrac{2}{7}}{-0.0673} \approx 18.6 \text{ minutes} \quad \text{Solve for } t.$$

The temperature of the object will be 50°C after about 18.6 minutes, or 18 minutes, 36 seconds.

(b) Use equation (6) to find t when $u = 35°C$.

$$35 = 30 + 70e^{-0.0673t}$$
$$5 = 70e^{-0.0673t} \qquad\qquad \text{Simplify.}$$
$$e^{-0.0673t} = \frac{5}{70}$$
$$-0.0673t = \ln\frac{5}{70} \qquad\qquad \text{Rewrite as a logarithm.}$$
$$t = \frac{\ln\dfrac{5}{70}}{-0.0673} \approx 39.2 \text{ minutes} \quad \text{Solve for } t.$$

The object will reach a temperature of 35°C after about 39.2 minutes.

(c) Look at equation (6). As t increases, the exponent $-0.0673t$ becomes unbounded in the negative direction. As a result, the value of $e^{-0.0673t}$ approaches zero, so the value of u, the temperature of the object, approaches 30°C, the air temperature of the room.

━━━ **Now Work** PROBLEM 13

4 Use Logistic Models

The exponential growth model $A(t) = A_0 e^{kt}$, $k > 0$, assumes uninhibited growth, meaning that the value of the function grows without limit. Recall that cell division could be modeled using this function, assuming that no cells die and no by-products are produced. However, cell division eventually is limited by factors such as living space and food supply. The **logistic model**, given next, can describe situations where the growth or decay of the dependent variable is limited.

Logistic Model

In a logistic model, the population P after time t is given by the function

$$P(t) = \frac{c}{1 + ae^{-bt}} \tag{7}$$

where a, b, and c are constants with $a > 0$ and $c > 0$. The model is a growth model if $b > 0$; the model is a decay model if $b < 0$.

The number c is called the **carrying capacity** (for growth models) because the value $P(t)$ approaches c as t approaches infinity; that is, $\lim\limits_{t \to \infty} P(t) = c$. The number $|b|$ is the growth rate for $b > 0$ and the decay rate for $b < 0$. Figure 42(a) shows the graph of a typical logistic growth function, and Figure 42(b) shows the graph of a typical logistic decay function.

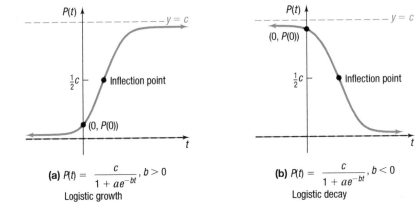

Figure 42

(a) $P(t) = \dfrac{c}{1 + ae^{-bt}}$, $b > 0$
Logistic growth

(b) $P(t) = \dfrac{c}{1 + ae^{-bt}}$, $b < 0$
Logistic decay

Based on the figures, the following properties of logistic functions emerge.

Properties of the Logistic Model, Equation (7)

- The domain is the set of all real numbers. The range is the interval $(0, c)$ where c is the carrying capacity.
- There are no x-intercepts; the y-intercept is $P(0)$.
- There are two horizontal asymptotes: $y = 0$ and $y = c$.
- $P(t)$ is an increasing function if $b > 0$ and a decreasing function if $b < 0$.
- There is an **inflection point** where $P(t)$ equals $\dfrac{1}{2}$ of the carrying capacity.

  The inflection point is the point on the graph where the graph changes from being concave up to being concave down for growth functions, and the point where the graph changes from being concave down to being concave up for decay functions.
- The graph is smooth and continuous, with no corners or gaps.

EXAMPLE 5   **Fruit Fly Population**

Fruit flies are placed in a half-pint milk bottle with a banana (for food) and yeast plants (for food and to provide a stimulus to lay eggs). Suppose that the fruit fly population after t days is given by

$$P(t) = \frac{230}{1 + 56.5e^{-0.37t}}$$

(a) State the carrying capacity and the growth rate.
(b) Determine the initial population.
(c) What is the population after 5 days?
(d) How long does it take for the population to reach 180?
(e) Use a graphing utility to determine how long it takes for the population to reach one-half of the carrying capacity.

Solution   (a) As $t \to \infty$, $e^{-0.37t} \to 0$ and $P(t) \to \frac{230}{1}$. The carrying capacity of the half-pint bottle is 230 fruit flies. The growth rate is $|b| = |0.37| = 37\%$ per day.

(b) To find the initial number of fruit flies in the half-pint bottle, evaluate $P(0)$.

$$P(0) = \frac{230}{1 + 56.5e^{-0.37 \cdot 0}} = \frac{230}{1 + 56.5} = 4$$

So, initially, there were 4 fruit flies in the half-pint bottle.

(c) After 5 days the number of fruit flies in the half-pint bottle is

$$P(5) = \frac{230}{1 + 56.5e^{-0.37 \cdot 5}} \approx 23 \text{ fruit flies}$$

After 5 days, there are approximately 23 fruit flies in the bottle.

(d) To determine when the population of fruit flies will be 180, solve the equation

$$P(t) = 180$$

$$\frac{230}{1 + 56.5e^{-0.37t}} = 180$$

$$230 = 180(1 + 56.5e^{-0.37t})$$

$0.2778 \approx 56.5e^{-0.37t}$ Divide both sides by 180.

$0.2778 \approx 56.5e^{-0.37t}$ Subtract 1 from both sides.

$0.0049 \approx e^{-0.37t}$ Divide both sides by 56.5.

$\ln(0.0049) \approx -0.37t$ Rewrite as a logarithmic expression.

$t \approx 14.4 \text{ days}$ Divide both sides by -0.37.

It will take approximately 14.4 days (14 days, 10 hours) for the population to reach 180 fruit flies.

(e) One-half of the carrying capacity is 115 fruit flies. Solve $P(t) = 115$ by graphing $Y_1 = \frac{230}{1 + 56.5e^{-0.37t}}$ and $Y_2 = 115$, and using INTERSECT. See Figure 43. The population will reach one-half of the carrying capacity in about 10.9 days (10 days, 22 hours).

$Y_1 = \dfrac{230}{1 + 56.5e^{-0.37x}}$

Figure 43 TI-84 Plus C

     Look at Figure 43. Notice the point where the graph reaches 115 fruit flies (one-half of the carrying capacity): The graph changes from being concave up to being concave down. Using the language of calculus, we say the graph changes from increasing at an increasing rate to increasing at a decreasing rate. For any logistic growth function, when the population reaches one-half the carrying capacity, the population growth starts to slow down.

━━━━▶**Now Work** PROBLEM 23

Exploration

On the same viewing rectangle, graph

$$Y_1 = \frac{500}{1 + 24e^{-0.03t}} \quad \text{and} \quad Y_2 = \frac{500}{1 + 24e^{-0.08t}}$$

What effect does the growth rate $|b|$ have on the logistic growth function?

EXAMPLE 6

Wood Products

The EFISCEN wood product model classifies wood products according to their life-span. There are four classifications: short (1 year), medium short (4 years), medium long (16 years), and long (50 years). Based on data obtained from the European Forest Institute, the percentage of remaining wood products after t years for wood products with long life spans (such as those used in the building industry) is given by

$$P(t) = \frac{100.3952}{1 + 0.0316e^{0.0581t}}$$

(a) What is the decay rate?
(b) What is the percentage of remaining wood products after 10 years?
(c) How long does it take for the percentage of remaining wood products to reach 50%?
(d) Explain why the numerator given in the model is reasonable.

Solution

(a) The decay rate is $|b| = |-0.0581| = 5.81\%$ per year.
(b) Evaluate $P(10)$.

$$P(10) = \frac{100.3952}{1 + 0.0316e^{0.0581 \cdot 10}} \approx 95.0$$

So 95% of long-life-span wood products remain after 10 years.
(c) Solve the equation $P(t) = 50$.

$$\frac{100.3952}{1 + 0.0316e^{0.0581t}} = 50$$

$$100.3952 = 50(1 + 0.0316e^{0.0581t})$$

$2.0079 \approx 1 + 0.0316e^{0.0581t}$ Divide both sides by 50.

$1.0079 \approx 0.0316e^{0.0581t}$ Subtract 1 from both sides.

$31.8956 \approx e^{0.0581t}$ Divide both sides by 0.0316.

$\ln(31.8956) \approx 0.0581t$ Rewrite as a logarithmic expression.

$t \approx 59.6$ years   Divide both sides by 0.0581.

It will take approximately 59.6 years for the percentage of long-life-span wood products remaining to reach 50%.

(d) The numerator of 100.3952 is reasonable because the maximum percentage of wood products remaining that is possible is 100%.

6.8 Assess Your Understanding

Applications and Extensions

1. **Growth of an Insect Population** The size P of a certain insect population at time t (in days) obeys the law of uninhibited growth $P(t) = 500e^{0.02t}$.
   (a) Determine the number of insects at $t = 0$ days.
   (b) What is the growth rate of the insect population?
   (c) What is the population after 10 days?
   (d) When will the insect population reach 800?
   (e) When will the insect population double?

2. **Growth of Bacteria** The number N of bacteria present in a culture at time t (in hours) obeys the law of uninhibited growth $N(t) = 1000e^{0.01t}$.
   (a) Determine the number of bacteria at $t = 0$ hours.
   (b) What is the growth rate of the bacteria?
   (c) What is the population after 4 hours?
   (d) When will the number of bacteria reach 1700?
   (e) When will the number of bacteria double?

3. **Radioactive Decay** Strontium-90 is a radioactive material that decays according to the function $A(t) = A_0 e^{-0.0244t}$, where A_0 is the initial amount present and A is the amount present at time t (in years). Assume that a scientist has a sample of 500 grams of strontium-90.
   (a) What is the decay rate of strontium-90?
   (b) How much strontium-90 is left after 10 years?
   (c) When will 400 grams of strontium-90 be left?
   (d) What is the half-life of strontium-90?

4. **Radioactive Decay** Iodine-131 is a radioactive material that decays according to the function $A(t) = A_0 e^{-0.087t}$, where A_0 is the initial amount present and A is the amount present at time t (in days). Assume that a scientist has a sample of 100 grams of iodine-131.
   (a) What is the decay rate of iodine-131?
   (b) How much iodine-131 is left after 9 days?
   (c) When will 70 grams of iodine-131 be left?
   (d) What is the half-life of iodine-131?

5. **Growth of a Colony of Mosquitoes** The population of a colony of mosquitoes obeys the law of uninhibited growth.
   (a) If N is the population of the colony and t is the time in days, express N as a function of t.
   (b) If there are 1000 mosquitoes initially and there are 1800 after 1 day, what is the size of the colony after 3 days?
   (c) How long is it until there are 10,000 mosquitoes?

6. **Bacterial Growth** A culture of bacteria obeys the law of uninhibited growth.
   (a) If N is the number of bacteria in the culture and t is the time in hours, express N as a function of t.
   (b) If 500 bacteria are present initially and there are 800 after 1 hour, how many will be present in the culture after 5 hours?
   (c) How long is it until there are 20,000 bacteria?

7. **Population Growth** The population of a southern city is growing according to the exponential law.
   (a) If N is the population of the city and t is the time in years, express N as a function of t.
   (b) If the population doubled in size over an 18-month period and the current population is 10,000, what will the population be 2 years from now?

8. **Population Decline** The population of a midwestern city is declining according to the exponential law.
   (a) If N is the population of the city and t is the time in years, express N as a function of t.
   (b) If the population decreased from 900,000 to 800,000 from 2016 to 2018, what will the population be in 2020?

9. **Radioactive Decay** The half-life of radium is 1690 years. If 10 grams is present now, how much will be present in 50 years?

10. **Radioactive Decay** The half-life of radioactive potassium is 1.3 billion years. If 10 grams is present now, how much will be present in 100 years? In 1000 years?

11. **Estimating the Age of a Tree** A piece of charcoal is found to contain 30% of the carbon-14 that it originally had. When did the tree die from which the charcoal came? Use 5730 years as the half-life of carbon-14.

12. **Estimating Age** A fossilized leaf contains 70% of its normal amount of carbon-14. How old is the fossil?

13. **Cooling Time of a Pizza** A pizza baked at 450°F is removed from the oven at 5:00 PM and placed in a room that is a constant 70°F. After 5 minutes, the pizza is at 300°F.
    (a) At what time can you begin eating the pizza if you want its temperature to be 135°F?
    (b) Determine the time that needs to elapse before the temperature of the pizza is 160°F.
    (c) What do you notice about the temperature as time passes?

14. **Newton's Law of Cooling** A thermometer reading 72°F is placed in a refrigerator where the temperature is a constant 38°F.
    (a) If the thermometer reads 60°F after 2 minutes, what will it read after 7 minutes?
    (b) How long will it take before the thermometer reads 39°F?
    (c) Determine the time that must elapse before the thermometer reads 45°F.
    (d) What do you notice about the temperature as time passes?

15. **Newton's Law of Heating** A thermometer reading 8°C is brought into a room with a constant temperature of 35°C. If the thermometer reads 15°C after 3 minutes, what will it read after being in the room for 5 minutes? For 10 minutes?
    [**Hint:** You need to construct a formula similar to equation (4).]

16. **Warming Time of a Beer Stein** A beer stein has a temperature of 28°F. It is placed in a room with a constant temperature of 70°F. After 10 minutes, the temperature of the stein has risen to 35°F. What will the temperature of the stein be after 30 minutes? How long will it take the stein to reach a temperature of 45°F? (See the hint given for Problem 15.)

17. **Decomposition of Chlorine in a Pool** Under certain water conditions, the free chlorine (hypochlorous acid, HOCl) in a swimming pool decomposes according to the law of uninhibited decay. After shocking his pool, Ben tested the water and found the amount of free chlorine to be 2.5 parts per million (ppm). Twenty-four hours later, Ben tested the water again and found the amount of free chlorine to be 2.2 ppm. What will be the reading after 3 days (that is, 72 hours)? When the chlorine level reaches 1.0 ppm, Ben must shock the pool again. How long can Ben go before he must shock the pool again?

18. **Decomposition of Dinitrogen Pentoxide** At 45°C, dinitrogen pentoxide (N_2O_5) decomposes into nitrous dioxide (NO_2) and oxygen (O_2) according to the law of uninhibited decay. An initial amount of 0.25 M N_2O_5 (M is a measure of concentration known as molarity) decomposes to 0.15 M N_2O_5 in 17 minutes. What concentration of N_2O_5 will remain after 30 minutes? How long will it take until only 0.01 M N_2O_5 remains?

19. **Decomposition of Sucrose** Reacting with water in an acidic solution at 35°C, sucrose ($C_{12}H_{22}O_{11}$) decomposes into glucose ($C_6H_{12}O_6$) and fructose ($C_6H_{12}O_6$)* according to the law of uninhibited decay. An initial concentration of 0.40 M of sucrose decomposes to 0.36 M sucrose in 30 minutes. What concentration of sucrose will remain after 2 hours? How long will it take until only 0.10 M sucrose remains?

20. **Decomposition of Salt in Water** Salt (NaCl) decomposes in water into sodium (Na^+) and chloride (Cl^-) ions according to the law of uninhibited decay. If the initial amount of salt is 25 kilograms and, after 10 hours, 15 kilograms of salt is left, how much salt is left after 1 day? How long does it take until $\frac{1}{2}$ kilogram of salt is left?

21. **Radioactivity from Chernobyl** After the release of radioactive material into the atmosphere from a nuclear power plant at Chernobyl (Ukraine) in 1986, the hay in Austria was contaminated by iodine 131 (half-life 8 days). If it is safe to feed the hay to cows when 10% of the iodine 131 remains, how long did the farmers need to wait to use this hay?

22. **Word Users** According to a survey by Olsten Staffing Services, the percentage of companies reporting usage of Microsoft Word t years since 1984 is given by

$$P(t) = \frac{99.744}{1 + 3.014e^{-0.799t}}$$

(a) What is the growth rate in the percentage of Microsoft Word users?
(b) Use a graphing utility to graph $P = P(t)$.
(c) What was the percentage of Microsoft Word users in 1990?
(d) During what year did the percentage of Microsoft Word users reach 90%?
(e) Explain why the numerator given in the model is reasonable. What does it imply?

23. **Home Computers** The logistic model

$$P(t) = \frac{95.4993}{1 + 0.0405e^{0.1968t}}$$

represents the percentage of households that do not own a personal computer t years since 1984.

(a) Evaluate and interpret $P(0)$.
(b) Use a graphing utility to graph $P = P(t)$.
(c) What percentage of households did not own a personal computer in 1995?
(d) In what year did the percentage of households that do not own a personal computer reach 10%?

Source: U.S. Department of Commerce

24. **Farmers** The logistic model

$$W(t) = \frac{14,656,248}{1 + 0.059e^{0.057t}}$$

represents the number of farm workers in the United States t years after 1910.

(a) Evaluate and interpret $W(0)$.
(b) Use a graphing utility to graph $W = W(t)$.
(c) How many farm workers were there in the United States in 2010?
(d) When did the number of farm workers in the United States reach 10,000,000?
(e) According to this model, what happens to the number of farm workers in the United States as t approaches ∞? Based on this result, do you think that it is reasonable to use this model to predict the number of farm workers in the United States in 2060? Why?

Source: U.S. Department of Agriculture

25. **Birthdays** The logistic model

$$P(n) = \frac{113.3198}{1 + 0.115e^{0.0912n}}$$

models the probability that, in a room of n people, no two people share the same birthday.

(a) Use a graphing utility to graph $P = P(n)$.
(b) In a room of $n = 15$ people, what is the probability that no two share the same birthday?
(c) How many people must be in a room before the probability that no two people share the same birthday falls below 10%?
(d) What happens to the probability as n increases? Explain what this result means.

26. **Population of an Endangered Species** Environmentalists often capture an endangered species and transport the species to a controlled environment where the species can produce offspring and regenerate its population. Suppose that six American bald eagles are captured, transported to Montana, and set free. Based on experience, the environmentalists expect the population to grow according to the model

$$P(t) = \frac{500}{1 + 83.33e^{-0.162t}}$$

where t is measured in years.

*Author's Note: Surprisingly, the chemical formulas for glucose and fructose are the same: This is not a typo.

(a) Determine the carrying capacity of the environment.
(b) What is the growth rate of the bald eagle?
(c) What is the population after 3 years?
(d) When will the population be 300 eagles?
(e) How long does it take for the population to reach one-half of the carrying capacity?

27. Invasive Species A habitat can be altered by invasive species that crowd out or replace native species. The logistic model

$$P(t) = \frac{431}{1 + 7.91e^{-0.017t}}$$

represents the number of invasive species present in the Great Lakes t years after 1900.
(a) Evaluate and interpret $P(0)$.
(b) What is the growth rate of invasive species?
(c) Use a graphing utility to graph $P = P(t)$.
(d) How many invasive species were present in the Great Lakes in 2000?
(e) In what year was the number of invasive species 175?

Source: NOAA

28. Social Networking The logistic model

$$P(t) = \frac{86.1}{1 + 2.12e^{-0.361t}}$$

gives the percentage of Americans who have a social media profile, where t represents the number of years after 2008.
(a) Evaluate and interpret $P(0)$.
(b) What is the growth rate?
(c) Use a graphing utility to graph $P = P(t)$.
(d) During 2017, what percentage of Americans had a social media profile?
(e) In what year did 69.3% of Americans have a social media profile?

Source: Statista, 2018

Problems 29 and 30 use the following discussion: Uninhibited growth can be modeled by exponential functions other than $A(t) = A_0 e^{kt}$. For example, if an initial population P_0 requires n units of time to double, then the function $P(t) = P_0 \cdot 2^{t/n}$ models the size of the population at time t. Likewise, a population requiring n units of time to triple can be modeled by $P(t) = P_0 \cdot 3^{t/n}$.

29. Growth of an Insect Population An insect population grows exponentially.
(a) If the population triples in 20 days, and 50 insects are present initially, write an exponential function of the form $P(t) = P_0 \cdot 3^{t/n}$ that models the population.
(b) What will the population be in 47 days?
(c) When will the population reach 700?
(d) Express the model from part (a) in the form $A(t) = A_0 e^{kt}$.

30. Growth of a Human Population The population of a town is growing exponentially.
(a) If its population doubled in size over an 8-year period and the current population is 25,000, write an exponential function of the form $P(t) = P_0 \cdot 2^{t/n}$ that models the population.
(b) What will the population be in 3 years?
(c) When will the population reach 80,000?
(d) Express the model from part (a) in the form $A(t) = A_0 e^{kt}$.

Retain Your Knowledge

Problems 31–40 are based on material learned earlier in the course. The purpose of these problems is to keep the material fresh in your mind so that you are better prepared for the final exam.

31. Find the linear function f whose graph contains the points $(4, 1)$ and $(8, -5)$.

32. Determine whether the graphs of the linear functions $f(x) = 5x - 1$ and $g(x) = \frac{1}{5}x + 1$ are parallel, perpendicular, or neither.

33. Write the logarithmic expression $\ln\left(\frac{x^2 \sqrt{y}}{z}\right)$ as the sum and/or difference of logarithms. Express powers as factors.

34. Find the domain of $f(x) = \frac{x + 3}{x^2 + 2x - 8}$.

35. If $f(x) = \frac{2x - 3}{x - 4}$ and $g(x) = \frac{3x + 1}{x - 3}$, find $(g - f)(x)$.

36. Find the x-intercept(s) and y-intercept(s) of the graph of $f(x) = 2x^2 - 5x + 1$.

37. Solve: $\frac{x + 1}{x} - \frac{x}{x + 1} = 2$

38. For the data provided, use a graphing utility to find the line of best fit. What is the correlation coefficient?

x	-4	-2	0	2	4	6
y	9	5	4	2	-1	-2

39. Use a graphing utility to graph $f(x) = x^4 - 3x^2 + 2x - 1$ over the interval $[-3, 3]$. Then, approximate any local maximum values and local minimum values, and determine where f is increasing and where f is decreasing. Round answers to two decimal places.

40. Write $\frac{10x}{3(2x + 3)^{2/3}} + 5(2x + 3)^{1/3}$ as a single quotient in which only positive exponents appear.

6.9 Building Exponential, Logarithmic, and Logistic Models from Data

PREPARING FOR THIS SECTION *Before getting started, review the following:*

- Building Linear Models from Data (Section 4.2, pp. 291–295)
- Building Cubic Models from Data (Section 5.2, pp. 350–351)
- Building Quadratic Models from Data (Section 4.4, pp. 316–317)

OBJECTIVES 1 Build an Exponential Model from Data (p. 502)
2 Build a Logarithmic Model from Data (p. 504)
3 Build a Logistic Model from Data (p. 504)

In Section 4.2 we discussed how to find the linear function of best fit $(y = ax + b)$, in Section 4.4 we discussed how to find the quadratic function of best fit $(y = ax^2 + bx + c)$, and in Section 5.2 we discussed how to find the cubic function of best fit $(y = ax^3 + bx^2 + cx + d)$.

In this section we discuss how to use a graphing utility to find equations of best fit that describe the relation between two variables when the relation is thought to be exponential $(y = ab^x)$, logarithmic $(y = a + b \ln x)$, or logistic $\left(y = \dfrac{c}{1 + ae^{-bx}} \right)$.

As before, we draw a scatter plot of the data to help to determine the appropriate model to use.

Figure 44 shows scatter plots that are typically observed for the three models.

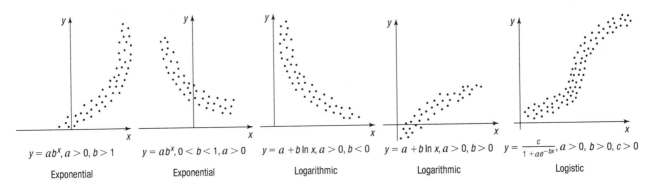

$y = ab^x, a > 0, b > 1$	$y = ab^x, 0 < b < 1, a > 0$	$y = a + b \ln x, a > 0, b < 0$	$y = a + b \ln x, a > 0, b > 0$	$y = \dfrac{c}{1 + ae^{-bx}}, a > 0, b > 0, c > 0$
Exponential	Exponential	Logarithmic	Logarithmic	Logistic

Figure 44

Most graphing utilities have REGression options that fit data to a specific type of curve. Once the data have been entered and a scatter plot obtained, the type of curve that you want to fit to the data is selected. Then that REGression option is used to obtain the curve of *best fit* of the type selected.

The correlation coefficient r will appear only if the model can be written as a linear expression. As it turns out, r will appear for the linear, power, exponential, and logarithmic models, since these models can be written as a linear expression. Remember, the closer $|r|$ is to 1, the better the fit.

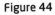 1 Build an Exponential Model from Data

We saw in Section 6.7 that money earning compound interest grows exponentially, and we saw in Section 6.8 that growth and decay models also can behave exponentially. The next example shows how data can lead to an exponential model.

EXAMPLE 1 Fitting an Exponential Function to Data

Mariah deposited $20,000 into a well-diversified mutual fund 6 years ago. The data in Table 9 represent the value of the account each year for the last 7 years.

Table 9

Year, x	Account Value, y
0	20,000
1	21,516
2	23,355
3	24,885
4	27,484
5	30,053
6	32,622

(a) Using a graphing utility, draw a scatter plot with year as the independent variable.

(b) Using a graphing utility, build an exponential model from the data.

(c) Express the function found in part (b) in the form $A = A_0 e^{kt}$.

(d) Graph the exponential function found in part (b) or (c) on the scatter plot.

(e) Using the solution to part (b) or (c), predict the value of the account after 10 years.

(f) Interpret the value of k found in part (c).

Solution

(a) Enter the data into the graphing utility and draw the scatter plot as shown in Figure 45 on a TI-84 Plus C.

Figure 45 TI-84 Plus C

(b) A graphing utility fits the data in Table 9 to an exponential model of the form $y = ab^x$ using the EXPonential REGression option. Figure 46 shows that $y = ab^x = 19,820.43\,(1.085568)^x$ on a TI-84 Plus C. Notice that $|r| = 0.999$, which is close to 1, indicating a good fit.

(c) To express $y = ab^x$ in the form $A = A_0 e^{kt}$, where $x = t$ and $y = A$, proceed as follows:

$$ab^x = A_0 e^{kt}$$

If $x = t = 0$, then $a = A_0$. This leads to

$$b^x = e^{kt}$$
$$b^x = (e^k)^t$$
$$b = e^k \qquad x = t$$

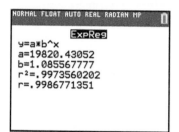

Figure 46 Exponential model using a TI-84 Plus C

Because $y = ab^x = 19,820.43\,(1.085568)^x$, this means that $a = 19,820.43$ and $b = 1.085568$.

$$a = A_0 = 19,820.43 \quad \text{and} \quad b = e^k = 1.085568$$

To find k, rewrite $e^k = 1.085568$ as a logarithm to obtain

$$k = \ln\,(1.085568) \approx 0.08210$$

As a result, $A = A_0 e^{kt} = 19,820.43 e^{0.08210t}$.

(d) See Figure 47 for the graph of the exponential function of best fit on a TI-84 Plus C. Figure 48 shows the exponential model using Desmos.

Figure 47

Figure 48 Exponential model using Desmos*

(e) Let $t = 10$ in the function found in part (c). The predicted value of the account after 10 years is

$$A = A_0 e^{kt} = 19,820.43 e^{0.08210(10)} \approx \$45,047$$

(f) The value of $k = 0.08210 = 8.210\%$ represents the annual growth rate of the account. It represents the rate of interest earned, assuming the account is growing continuously.

———— **Now Work** PROBLEM 1

*For this result in Desmos to agree precisely with the result of a TI-84 Plus C, the "Log Mode" option must be selected. Consult the help feature in Desmos for more information about this option.

2 Build a Logarithmic Model from Data

Some relations between variables follow a logarithmic model.

EXAMPLE 2 | **Fitting a Logarithmic Function to Data**

Jodi, a meteorologist, is interested in finding a function that explains the relation between the height of a weather balloon (in kilometers) and the atmospheric pressure (measured in millimeters of mercury) on the balloon. She collects the data shown in Table 10.

Table 10

Atmospheric Pressure, p	Height, h
760	0
740	0.184
725	0.328
700	0.565
650	1.079
630	1.291
600	1.634
580	1.862
550	2.235

(a) Using a graphing utility, draw a scatter plot of the data with atmospheric pressure as the independent variable.

(b) It is known that the relation between atmospheric pressure and height follows a logarithmic model. Using a graphing utility, build a logarithmic model from the data.

(c) Graph the logarithmic function found in part (b) on the scatter plot.

(d) Use the function found in part (b) to predict the height of the weather balloon if the atmospheric pressure is 560 millimeters of mercury.

Solution

(a) Enter the data into the graphing utility, and draw the scatter plot. See Figure 49.

(b) A graphing utility fits the data in Table 10 to a logarithmic function of the form $y = a + b \ln x$ by using the LOGarithm REGression option. Figure 50 shows the result on a TI-84 Plus C. The logarithmic model from the data is

$$h(p) = 45.7863 - 6.9025 \ln p$$

where h is the height of the weather balloon and p is the atmospheric pressure. Notice that $|r|$ is close to 1, indicating a good fit.

(c) Figure 51 shows the graph of $h(p) = 45.7863 - 6.9025 \ln p$ on the scatter plot. Figure 52 shows the logarithmic model using Desmos.

Figure 49 TI-84 Plus C

Figure 50 Logarithmic model using a TI-84 Plus C

Figure 51

Figure 52 Logarithmic model using Desmos

(d) Using the function found in part (b), Jodi predicts the height of the weather balloon when the atmospheric pressure is 560 to be

$$h(560) = 45.7863 - 6.9025 \ln 560$$
$$\approx 2.108 \text{ kilometers}$$

Now Work PROBLEM 5

3 Build a Logistic Model from Data

Logistic growth models can be used to model situations for which the value of the dependent variable is limited. Many real-world situations conform to this scenario. For example, the population of the human race is limited by the availability of natural resources, such as food and shelter. When the value of the dependent variable is limited, a logistic growth model is often appropriate.

EXAMPLE 3   **Fitting a Logistic Function to Data**

The data in Table 11 represent the amount of yeast biomass in a culture after t hours.

Table 11

Time (hours)	Yeast Biomass	Time (hours)	Yeast Biomass	Time (hours)	Yeast Biomass
0	9.6	7	257.3	14	640.8
1	18.3	8	350.7	15	651.1
2	29.0	9	441.0	16	655.9
3	47.2	10	513.3	17	659.6
4	71.1	11	559.7	18	661.8
5	119.1	12	594.8		
6	174.6	13	629.4		

Source: Tor Carlson (Über Geschwindigkeit und Grösse der Hefevermehrung in Würze, Biochemische Zeitschrift, Bd. 57, pp. 313–334, 1913)

(a) Using a graphing utility, draw a scatter plot of the data with time as the independent variable.

(b) Using a graphing utility, build a logistic model from the data.

(c) Using a graphing utility, graph the function found in part (b) on the scatter plot.

(d) What is the predicted carrying capacity of the culture?

(e) Use the function found in part (b) to predict the population of the culture at $t = 19$ hours.

Solution   (a) See Figure 53 for a scatter plot of the data on a TI-84 Plus C.

(b) A graphing utility fits the data in Table 11 to a logistic growth model of the form $y = \dfrac{c}{1 + ae^{-bx}}$ by using the LOGISTIC regression option. Figure 54 shows the result on a TI-84 Plus C. The logistic model from the data is

$$y = \frac{663.0}{1 + 71.6e^{-0.5470x}}$$

where y is the amount of yeast biomass in the culture and x is the time.

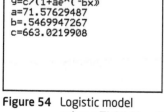

Figure 53   TI-84 Plus C

(c) See Figure 55 for the graph of the logistic model on a TI-84 Plus C. Figure 56 shows the logistic model using Desmos.

Figure 54   Logistic model using a TI-84 Plus C

Figure 55

Figure 56   Logistic model using Demos

(d) Based on the logistic growth model found in part (b), the carrying capacity of the culture is 663.

(e) Using the logistic growth model found in part (b), the predicted amount of yeast biomass at $t = 19$ hours is

$$y = \frac{663.0}{1 + 71.6e^{-0.5470 \cdot 19}} \approx 661.5$$

Now Work PROBLEM 7

6.9 Assess Your Understanding

Applications and Extensions

1. **Biology** A strain of E-coli Beu 397-recA441 is placed into a nutrient broth at 30° Celsius and allowed to grow. The following data are collected. Theory states that the number of bacteria in the petri dish will initially grow according to the law of uninhibited growth. The population is measured using an optical device in which the amount of light that passes through the petri dish is measured.

Time (hours), x	Population, y
0	0.09
2.5	0.18
3.5	0.26
4.5	0.35
6	0.50

Source: Dr. Polly Lavery, Joliet Junior College

(a) Draw a scatter plot treating time as the independent variable.
(b) Using a graphing utility, build an exponential model from the data.
(c) Express the function found in part (b) in the form $N(t) = N_0 e^{kt}$.
(d) Graph the exponential function found in part (b) or (c) on the scatter plot.
(e) Use the exponential function from part (b) or (c) to predict the population at $x = 7$ hours.
(f) Use the exponential function from part (b) or (c) to predict when the population will reach 0.75.

2. **Tesla, Inc. Revenue** The data in the table below represent annual revenue of Tesla, Inc. from 2010 to 2017.

Year	Revenue ($ Billion)
2010 ($x = 0$)	0.12
2011 ($x = 1$)	0.20
2012 ($x = 2$)	0.41
2013 ($x = 3$)	2.01
2014 ($x = 4$)	3.20
2015 ($x = 5$)	4.05
2016 ($x = 6$)	7.00
2017 ($x = 7$)	11.76

Source: Tesla, Inc.

(a) Using a graphing utility, draw a scatter plot of the data using 0 for 2010, 1 for 2011, and so on, as the independent variable.
(b) Using a graphing utility, build an exponential model from the data.
(c) Express the function found in part (b) in the form $A(t) = A_0 e^{kt}$.

(d) Graph the exponential function found in part (b) or (c) on the scatter plot.
(e) Use the exponential function from part (b) or (c) to predict Tesla's revenue in 2019.
(f) Interpret the meaning of k in the function found in part (c).

3. **Advanced-Stage Breast Cancer** The data in the table below represent the percentage of patients who have survived after diagnosis of advanced-stage breast cancer at 6-month intervals of time.

Time after Diagnosis (years)	Percentage Surviving
0.5	95.7
1	83.6
1.5	74.0
2	58.6
2.5	47.4
3	41.9
3.5	33.6

Source: Cancer Treatment Centers of America

(a) Using a graphing utility, draw a scatter plot of the data with time after diagnosis as the independent variable.
(b) Using a graphing utility, build an exponential model from the data.
(c) Express the function found in part (b) in the form $A(t) = A_0 e^{kt}$.
(d) Graph the exponential function found in part (b) or (c) on the scatter plot.
(e) Use the model to predict the percentage of patients diagnosed with advanced-stage cancer who survive for 4 years after initial diagnosis?
(f) Interpret the meaning of k in the function found in part (c).

4. **Chemistry** A chemist has a 100-gram sample of a radioactive material. He records the amount of radioactive material every week for 7 weeks and obtains the following data:

Week	Weight (in grams)
0	100.0
1	88.3
2	75.9
3	69.4
4	59.1
5	51.8
6	45.5

(a) Using a graphing utility, draw a scatter plot with week as the independent variable.
(b) Using a graphing utility, build an exponential model from the data.
(c) Express the function found in part (b) in the form $A(t) = A_0 e^{kt}$.
(d) Graph the exponential function found in part (b) or (c) on the scatter plot.
(e) From the result found in part (b), determine the half-life of the radioactive material.
(f) How much radioactive material will be left after 50 weeks?
(g) When will there be 20 grams of radioactive material?

5. **Milk Production** The data in the table below represent the number of dairy farms (in thousands) and the amount of milk produced (in billions of pounds) in the United States for various years.

Year	Dairy Farms (thousands)	Milk Produced (billion pounds)
1980	334	128
1985	269	143
1990	193	148
1995	140	155
2000	105	167
2005	78	177
2010	63	193
2015	44	209

Source: National Agricultural Statistics Services

(a) Using a graphing utility, draw a scatter plot of the data with the number of dairy farms as the independent variable.
(b) Using a graphing utility, build a logarithmic model from the data.
(c) Graph the logarithmic function found in part (b) on the scatter plot.
(d) In 2008, there were 67 thousand dairy farms in the United States. Use the function in part (b) to predict the amount of milk produced in 2008.
(e) The actual amount of milk produced in 2008 was 190 billion pounds. How does your prediction in part (d) compare to this?

6. **Social Networking** The data in the table below represent the percent of U.S. citizens aged 12 and older who have a profile on at least one social network.

Year	Percent on a Social Networking Site
2008 ($x = 8$)	24
2009 ($x = 9$)	34
2010 ($x = 10$)	48
2011 ($x = 11$)	52
2012 ($x = 12$)	56
2013 ($x = 13$)	62
2014 ($x = 14$)	67
2015 ($x = 15$)	73
2016 ($x = 16$)	78
2017 ($x = 17$)	81

Source: Statista.com

(a) Using a graphing utility, draw a scatter plot of the data using 8 for 2008, 9 for 2009, and so on, as the independent variable, and percent on social networking site as the dependent variable.
(b) Using a graphing utility, build a logarithmic model from the data.
(c) Graph the logarithmic function found in part (b) on the scatter plot.
(d) Use the model to predict the percent of U.S. citizens on social networking sites in 2019.
(e) Use the model to predict the year in which 98% of U.S. citizens will be on social networking sites.

7. **Population Model** The following data represent the population of the United States. An ecologist is interested in building a model that describes the population of the United States.

Year	Population
1900	76,212,168
1910	92,228,496
1920	106,021,537
1930	123,202,624
1940	132,164,569
1950	151,325,798
1960	179,323,175
1970	203,302,031
1980	226,542,203
1990	248,709,873
2000	281,421,906
2010	308,745,538

Source: U.S. Census Bureau

(a) Using a graphing utility, draw a scatter plot of the data using years since 1900 as the independent variable and population as the dependent variable.
(b) Using a graphing utility, build a logistic model from the data.
(c) Using a graphing utility, graph the function found in part (b) on the scatter plot.
(d) Based on the function found in part (b), what is the carrying capacity of the United States?
(e) Use the function found in part (b) to predict the population of the United States in 2012.
(f) When will the United States population be 350,000,000?
(g) Compare actual U.S. Census figures to the predictions found in parts (e) and (f). Discuss any differences.

8. **Population Model** The data that follow on the next page represent world population. An ecologist is interested in building a model that describes the world population.
(a) Using a graphing utility, draw a scatter plot of the data using years since 2000 as the independent variable and population as the dependent variable.
(b) Using a graphing utility, build a logistic model from the data.
(c) Using a graphing utility, graph the function found in part (b) on the scatter plot.
(d) Based on the function found in part (b), what is the carrying capacity of the world?

(continued)

(e) Use the function found in part (b) to predict the population of the world in 2025.

(f) When will world population be 10 billion?

Year	Population (billions)	Year	Population (billions)
2001	6.22	2010	6.96
2002	6.30	2011	7.04
2003	6.38	2012	7.13
2004	6.46	2013	7.21
2005	6.54	2014	7.30
2006	6.62	2015	7.38
2007	6.71	2016	7.47
2008	6.79	2017	7.55
2009	6.87	2018	7.63

Source: worldometers.info

9. *Mixed Practice* **Online Advertising Revenue** The data in the table below represent the U.S. online advertising revenues for the years 2005–2016.

Year	U.S. Online Advertising Revenue ($ billions)
2005 ($x = 0$)	12.5
2006 ($x = 1$)	16.9
2007 ($x = 2$)	21.2
2008 ($x = 3$)	23.4
2009 ($x = 4$)	22.7
2010 ($x = 5$)	26.0
2011 ($x = 6$)	31.7
2012 ($x = 7$)	36.6
2013 ($x = 8$)	42.8
2014 ($x = 9$)	49.5
2015 ($x = 10$)	59.6
2016 ($x = 11$)	72.5

Source: marketingcharts.com

(a) Using a graphing utility, draw a scatter plot of the data using 0 for 2005, 1 for 2006, and so on as the independent variable, and online advertising revenue as the dependent variable.

(b) Based on the scatter plot drawn in part (a), decide what model (linear, quadratic, cubic, exponential, logarithmic, or logistic) that you think best describes the relation between year and revenue.

(c) Using a graphing utitlity, find the model of best fit.

(d) Using a graphing utility, graph the function found in part (c) on the scatter plot drawn in part (a).

(e) Use your model to predict the online advertising revenue in 2021.

10. *Mixed Practice* **Age versus Total Cholesterol** The data on top of the next column represent the age and average total cholesterol for adult males at various ages.

Age	Total Cholesterol
27	189
40	205
50	215
60	210
70	210
80	194

(a) Using a graphing utility, draw a scatter plot of the data using age, x, as the independent variable and total cholesterol, y, as the dependent variable.

(b) Based on the scatter plot drawn in part (a), decide on a model (linear, quadratic, cubic, exponential, logarithmic, or logistic) that you think best describes the relation between age and total cholesterol. Be sure to justify your choice of model.

(c) Using a graphing utility, find the model of best fit.

(d) Using a graphing utility, graph the function found in part (c) on the scatter plot drawn in part (a).

(e) Use your model to predict the total cholesterol of a 35-year-old male.

11. *Mixed Practice* **Golfing** The data below represent the expected percentage of putts that will be made by professional golfers on the PGA Tour, depending on distance. For example, it is expected that 99.3% of 2-foot putts will be made.

Distance (feet)	Expected Percentage	Distance (feet)	Expected Percentage
2	99.3	14	25.0
3	94.8	15	22.0
4	85.8	16	20.0
5	74.7	17	19.0
6	64.7	18	17.0
7	55.6	19	16.0
8	48.5	20	14.0
9	43.4	21	13.0
10	38.3	22	12.0
11	34.2	23	11.0
12	30.1	24	11.0
13	27.0	25	10.0

Source: TheSandTrap.com

(a) Using a graphing utility, draw a scatter plot of the data with distance as the independent variable.

(b) Based on the scatter plot drawn in part (a), decide on a model (linear, quadratic, cubic, exponential, logarithmic, or logistic) that you think best describes the relation between distance and expected percentage. Be sure to justify your choice of model.

(c) Using a graphing utility, find the model of best fit.

(d) Graph the function found in part (c) on the scatter plot.

(e) Use the function found in part (c) to predict what percentage of 30-foot putts will be made.

Retain Your Knowledge

Problems 12–21 are based on material learned earlier in the course. The purpose of these problems is to keep the material fresh in your mind so that you are better prepared for the final exam.

12. Construct a polynomial function that might have the graph shown. (More than one answer is possible.)

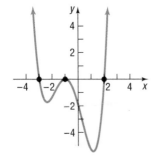

13. Use the Pythagorean Theorem to find the exact length of the unlabeled side in the given right triangle.

14. Graph the equation $(x - 3)^2 + y^2 = 25$.

15. Find the midpoint of the line segment with endpoints $(-7, 5)$ and $(1, -9)$.

16. Find the function whose graph is the shape of $y = \sqrt{x}$, but shifted to the right 4 units and reflected about the x-axis.

17. Solve: $3x^2 - 4x - 5 = 0$

18. Solve: $\dfrac{x + 1}{x^2 - 25} \geq 0$

19. Use the Remainder Theorem to find the remainder when $f(x) = 3x^5 - 7x^4 - 27x^3 + 67x^2 - 36$ is divided by $x + 3$. Is $x + 3$ a factor of $f(x)$?

20. Find the average rate of change of $f(x) = 8^x$ from $\dfrac{1}{3}$ to $\dfrac{2}{3}$.

21. Use the Intermediate Value Theorem to show that $f(x) = -x^4 + 2x^3 - 5x + 1$ has a zero in the interval $[-2, -1]$. Then, approximate the zero correct to two decimal places.

Chapter Review

Things to Know

Composite function (p. 415)

$(f \circ g)(x) = f(g(x))$; The domain of $f \circ g$ is the set of all numbers x in the domain of g for which $g(x)$ is in the domain of f.

One-to-one function f (p. 423)

A function for which any two different inputs in the domain correspond to two different outputs in the range

For any choice of elements x_1, x_2 in the domain of f, if $x_1 \neq x_2$, then $f(x_1) \neq f(x_2)$.

Horizontal-line test (p. 424)

If every horizontal line intersects the graph of a function f in at most one point, f is one-to-one.

Inverse function f^{-1} of f (pp. 425, 430)

For a one-to-one function $y = f(x)$, the correspondence from the range of f to the domain of f. Domain of f = range of f^{-1}; range of f = domain of f^{-1}

$f^{-1}(f(x)) = x$ for all x in the domain of f

$f(f^{-1}(x)) = x$ for all x in the domain of f^{-1}

The graphs of f and f^{-1} are symmetric with respect to the line $y = x$.

Properties of the exponential function (pp. 440, 442, 446)

$f(x) = Ca^x, \quad a > 1, C > 0$

Domain: the interval $(-\infty, \infty)$

Range: the interval $(0, \infty)$

x-intercepts: none; y-intercept: C

Horizontal asymptote: x-axis $(y = 0)$ as $x \to -\infty$

Increasing; one-to-one; smooth; continuous

See Figure 21 for a typical graph.

$f(x) = Ca^x, \quad 0 < a < 1, C > 0$

Domain: the interval $(-\infty, \infty)$

Range: the interval $(0, \infty)$

x-intercepts: none; y-intercept: C

Horizontal asymptote: x-axis $(y = 0)$ as $x \to \infty$

Decreasing; one-to-one; smooth; continuous

See Figure 25 for a typical graph.

Number e (p. 443)	Number approached by the expression $\left(1 + \dfrac{1}{n}\right)^n$ as $n \to \infty$
Property of exponents (p. 444)	If $a^u = a^v$, then $u = v$.
Natural logarithm (p. 455)	$y = \ln x$ if and only if $x = e^y$.

Properties of the logarithmic function (p. 460)

$f(x) = \log_a x, \quad a > 1$

$(y = \log_a x \text{ if and only if } x = a^y)$

Domain: the interval $(0, \infty)$
Range: the interval $(-\infty, \infty)$
x-intercept: 1; y-intercept: none
Vertical asymptote: $x = 0$ (y-axis)
Increasing; one-to-one; smooth; continuous
See Figure 39(a) for a typical graph.

$f(x) = \log_a x, \quad 0 < a < 1$

$(y = \log_a x \text{ if and only if } x = a^y)$

Domain: the interval $(0, \infty)$
Range: the interval $(-\infty, \infty)$
x-intercept: 1; y-intercept: none
Vertical asymptote: $x = 0$ (y-axis)
Decreasing; one-to-one; smooth; continuous
See Figure 39(b) for a typical graph.

Properties of logarithms (pp. 465–466, 469)

$\log_a 1 = 0 \qquad \log_a a = 1 \qquad a^{\log_a M} = M \qquad \log_a a^r = r \qquad a^r = e^{r \ln a}$

$\log_a(MN) = \log_a M + \log_a N \qquad \log_a\left(\dfrac{M}{N}\right) = \log_a M - \log_a N$

$\log_a M^r = r \log_a M$

If $M = N$, then $\log_a M = \log_a N$.

If $\log_a M = \log_a N$, then $M = N$.

Formulas

Change-of-Base Formula (p. 470)

$\log_a M = \dfrac{\log_b M}{\log_b a} \qquad a \neq 1, b \neq 1, \text{ and } M \text{ are positive real numbers}$

Compound Interest Formula (p. 482)

$A = P \cdot \left(1 + \dfrac{r}{n}\right)^{nt}$

Continuous compounding (p. 484)

$A = Pe^{rt}$

Effective rate of interest (p. 485)

Compounding n times per year: $r_E = \left(1 + \dfrac{r}{n}\right)^n - 1$

Continuous compounding: $r_E = e^r - 1$

Present Value Formulas (p. 486)

$P = A \cdot \left(1 + \dfrac{r}{n}\right)^{-nt} \quad \text{or} \quad P = Ae^{-rt}$

Uninhibited growth or decay (pp. 491, 493)

$A(t) = A_0 e^{kt} \quad k \neq 0; \text{growth}, k > 0; \text{decay}, k < 0$

Newton's Law of Cooling (p. 494)

$u(t) = T + (u_0 - T)e^{kt} \quad k < 0$

Logistic model (p. 496)

$P(t) = \dfrac{c}{1 + ae^{-bt}} \quad a > 0, c > 0, b \neq 0$

Objectives

Section	You should be able to . . .	Example(s)	Review Exercises
6.1	1 Form a composite function (p. 415)	1, 2, 4, 5	1–6
	2 Find the domain of a composite function (p. 416)	2–4	4–6
6.2	1 Determine whether a function is one-to-one (p. 423)	1, 2	7(a), 8
	2 Obtain the graph of the inverse function from the graph of a one-to-one function (p. 426)	3	8
	3 Verify an inverse function (p. 427)	4, 5	9, 10
	4 Find the inverse of a function defined by an equation (p. 428)	6, 7, 8	11–14
6.3	1 Evaluate exponential functions (p. 435)	1	15(a), (c), 48(a)
	2 Graph exponential functions (p. 439)	3–6	32–34, (a)–(c), 35(d)–(f)
	3 Define the number e (p. 442)	p. 443	
	4 Solve exponential equations (p. 444)	7, 8	36, 37, 40, 42, 48(b)

Section	You should be able to . . .	Example(s)	Review Exercises
6.4	**1** Change exponential statements to logarithmic statements and logarithmic statements to exponential statements (p. 452)	2, 3	16, 17
	2 Evaluate logarithmic expressions (p. 453)	4	15(b), (d), 20, 47(b), 49(a), 50
	3 Determine the domain of a logarithmic function (p. 453)	5	18, 19, 35(a)
	4 Graph logarithmic functions (p. 454)	6, 7	32–34, (d)–(f), 35(b)–(c), 47(a)
	5 Solve logarithmic equations (p. 458)	8, 9	38, 47(c), 49(b)
6.5	**1** Work with the properties of logarithms (p. 465)	1, 2	21, 22
	2 Write a logarithmic expression as a sum or difference of logarithms (p. 467)	3–5	23–26
	3 Write a logarithmic expression as a single logarithm (p. 468)	6	27–29
	4 Evaluate logarithms whose base is neither 10 nor e (p. 469)	7, 8	30, 31
6.6	**1** Solve logarithmic equations (p. 474)	1–3	38, 41, 44
	2 Solve exponential equations (p. 476)	4–6	39, 43, 45, 46
	3 Solve logarithmic and exponential equations using a graphing utility (p. 477)	7	36–46
6.7	**1** Determine the future value of a lump sum of money (p. 481)	1–3	51
	2 Calculate effective rates of return (p. 484)	4	51
	3 Determine the present value of a lump sum of money (p. 485)	5	52
	4 Determine the rate of interest or the time required to double a lump sum of money (p. 486)	6, 7	51
6.8	**1** Model populations that obey the law of uninhibited growth (p. 491)	1, 2	55
	2 Model populations that obey the law of uninhibited decay (p. 493)	3	53, 56
	3 Use Newton's Law of Cooling (p. 494)	4	54
	4 Use logistic models (p. 496)	5, 6	57
6.9	**1** Build an exponential model from data (p. 502)	1	58
	2 Build a logarithmic model from data (p. 504)	2	59
	3 Build a logistic model from data (p. 504)	3	60

Review Exercises

In Problems 1–3, for each pair of functions f and g, find:

(a) $(f \circ g)(2)$ (b) $(g \circ f)(-2)$ (c) $(f \circ f)(4)$ (b) $(g \circ g)(-1)$

1. $f(x) = 3x - 5$; $g(x) = 1 - 2x^2$ **2.** $f(x) = \sqrt{x + 2}$; $g(x) = 2x^2 + 1$ **3.** $f(x) = e^x$; $g(x) = 3x - 2$

In Problems 4–6, find $f \circ g$, $g \circ f$, $f \circ f$, and $g \circ g$ for each pair of functions. State the domain of each composite function.

4. $f(x) = 2 - x$; $g(x) = 3x + 1$ **5.** $f(x) = \sqrt{3x}$; $g(x) = 1 + x + x^2$ **6.** $f(x) = \dfrac{x + 1}{x - 1}$; $g(x) = \dfrac{1}{x}$

7. (a) Verify that the function below is one-to-one.   **(b)** Find its inverse.
  $\{(1, 2), (3, 5), (5, 8), (6, 10)\}$

8. The graph of a function f is given below. State why f is one-to-one. Then draw the graph of the inverse function f^{-1}.

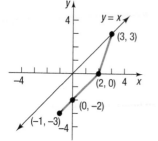

In Problems 9 and 10, verify that the functions f and g are inverses of each other by showing that $f(g(x)) = x$ and $g(f(x)) = x$. Give any values of x that need to be excluded from the domain of f and the domain of g.

9. $f(x) = 5x - 10$; $g(x) = \dfrac{1}{5}x + 2$ **10.** $f(x) = \dfrac{x - 4}{x}$; $g(x) = \dfrac{4}{1 - x}$

In Problems 11–14, each function is one-to-one. Find the inverse of each function and check your answer. Find the domain and range of f and f^{-1}.

11. $f(x) = \dfrac{2x + 3}{5x - 2}$
12. $f(x) = \dfrac{1}{x - 1}$
13. $f(x) = \sqrt{x - 2}$
14. $f(x) = x^{1/3} + 1$

15. Given $f(x) = 3^x$ and $g(x) = \log_3 x$, evaluate each of the following.

(a) $f(4)$
(b) $g(9)$
(c) $f(-2)$
(d) $g\!\left(\dfrac{1}{27}\right)$

16. Change $5^2 = z$ to an equivalent statement involving a logarithm.

17. Change $\log_5 u = 13$ to an equivalent statement involving an exponent.

In Problems 18 and 19, find the domain of each logarithmic function.

18. $f(x) = \log(3x - 2)$
19. $H(x) = \log_2(x^2 - 3x + 2)$

In Problems 20–22, find the exact value of each expression. Do not use a calculator.

20. $\log_2\!\left(\dfrac{1}{8}\right)$
21. $\ln e^{\sqrt{2}}$
22. $2^{\log_2 0.4}$

In Problems 23–26, write each expression as the sum and/or difference of logarithms. Express powers as factors.

23. $\log_3\!\left(\dfrac{uv^2}{w}\right)$ $\quad u > 0, v > 0, w > 0$
24. $\log_2\!\left(a^2 \sqrt{b}\right)^4$ $\quad a > 0, \quad b > 0$

25. $\log\!\left(x^2 \sqrt{x^3 + 1}\right)$ $\quad x > 0$
26. $\ln\!\left(\dfrac{2x + 3}{x^2 - 3x + 2}\right)^2$ $\quad x > 2$

In Problems 27–29, write each expression as a single logarithm.

27. $3 \log_4 x^2 + \dfrac{1}{2} \log_4 \sqrt{x}$
28. $\ln\!\left(\dfrac{x - 1}{x}\right) + \ln\!\left(\dfrac{x}{x + 1}\right) - \ln(x^2 - 1)$

29. $\dfrac{1}{2} \ln(x^2 + 1) - 4 \ln\dfrac{1}{2} - \dfrac{1}{2}\left[\ln(x - 4) + \ln x\right]$

30. Use the Change-of-Base Formula and a calculator to evaluate $\log_4 19$. Round your answer to three decimal places.

31. Graph $y = \log_3 x$ using a graphing utility and the Change-of-Base Formula.

In Problems 32–35, for each function f:

(a) Find the domain of f.
(b) Graph f.
(c) From the graph, determine the range and any asymptotes of f.
(d) Find f^{-1}, the inverse function of f.
(e) Find the domain and the range of f^{-1}.
(f) Graph f^{-1}.

32. $f(x) = 2^{x-3}$
33. $f(x) = 1 + 3^{-x}$
34. $f(x) = 3e^{x-2}$
35. $f(x) = \dfrac{1}{2} \ln(x + 3)$

In Problems 36–46, solve each equation. Express irrational solutions in exact form.

36. $8^{6+3x} = 4$
37. $3^{x^2+x} = \sqrt{3}$
38. $\log_x 64 = -3$
39. $5^x = 3^{x+2}$
40. $25^{2x} = 5^{x^2-12}$
41. $\log_3 \sqrt{x - 2} = 2$
42. $8 = 4^{x^2} \cdot 2^{5x}$
43. $2^x \cdot 5 = 10^x$
44. $\log_6(x + 3) + \log_6(x + 4) = 1$
45. $e^{1-x} = 5$
46. $9^x + 4 \cdot 3^x - 3 = 0$

---

47. Suppose that $f(x) = \log_2(x - 2) + 1$.

(a) Graph f.
(b) What is $f(6)$? What point is on the graph of f?
(c) Solve $f(x) = 4$. What point is on the graph of f?
(d) Based on the graph drawn in part (a), solve $f(x) > 0$.
(e) Find $f^{-1}(x)$. Graph f^{-1} on the same Cartesian plane as f.

48. Amplifying Sound An amplifier's power output P (in watts) is related to its decibel voltage gain d by the formula
$$P = 25e^{0.1d}$$

(a) Find the power output for a decibel voltage gain of 4 decibels.
(b) For a power output of 50 watts, what is the decibel voltage gain?

49. Limiting Magnitude of a Telescope A telescope is limited in its usefulness by the brightness of the star that it is aimed at and by the diameter of its lens. One measure of a star's brightness is its *magnitude;* the dimmer the star, the larger its magnitude. A formula for the limiting magnitude L of a telescope—that is, the magnitude of the dimmest star that it can be used to view—is given by
$$L = 9 + 5.1 \log d$$
where d is the diameter (in inches) of the lens.

(a) What is the limiting magnitude of a 3.5-inch telescope?
(b) What diameter is required to view a star of magnitude 14?

50. **Salvage Value** The number of years n for a piece of machinery to depreciate to a known salvage value can be found using the formula

$$n = \frac{\log s - \log i}{\log (1 - d)}$$

where s is the salvage value of the machinery, i is its initial value, and d is the annual rate of depreciation.
  (a) How many years will it take for a piece of machinery to decline in value from \$90,000 to \$10,000 if the annual rate of depreciation is 0.20 (20%)?
  (b) How many years will it take for a piece of machinery to lose half of its value if the annual rate of depreciation is 15%?

51. **Funding a College Education** A child's grandparents purchase a \$10,000 bond fund that matures in 18 years to be used for her college education. The bond fund pays 4% interest compounded semiannually. How much will the bond fund be worth at maturity? What is the effective rate of interest? How long will it take the bond to double in value under these terms?

52. **Funding a College Education** A child's grandparents wish to purchase a bond that matures in 18 years to be used for her college education. The bond pays 4% interest compounded semiannually. How much should they pay so that the bond will be worth \$85,000 at maturity?

53. **Estimating the Date When a Prehistoric Man Died** The bones of a prehistoric man found in the desert of New Mexico contain approximately 5% of the original amount of carbon-14. If the half-life of carbon-14 is 5730 years, approximately how long ago did the man die?

54. **Temperature of a Skillet** A skillet is removed from an oven where the temperature is 450°F and placed in a room where the temperature is 70°F. After 5 minutes, the temperature of the skillet is 400°F. How long will it be until its temperature is 150°F?

55. **World Population** The annual growth rate of the world's population in 2018 was $k = 1.1\% = 0.011$. The population of the world in 2018 was 7,632,819,325. Letting $t = 0$ represent 2018, use the uninhibited growth model to predict the world's population in the year 2024.

  Source: worldometers.info

56. **Radioactive Decay** The half-life of radioactive cobalt is 5.27 years. If 100 grams of radioactive cobalt is present now, how much will be present in 20 years? In 40 years?

57. **Logistic Growth** The logistic growth model

$$P(t) = \frac{0.8}{1 + 1.67e^{-0.16t}}$$

represents the proportion of new cars with a global positioning system (GPS). Let $t = 0$ represent 2006, $t = 1$ represent 2007, and so on.
  (a) What proportion of new cars in 2006 had a GPS?
  (b) Determine the maximum proportion of new cars that have a GPS.
  (c) Using a graphing utility, graph $P = P(t)$.
  (d) When will 75% of new cars have a GPS?

58. **Rising Tuition** The following data represent the average in-state tuition and fees (in 2017 dollars) at public four-year colleges and universities in the United States from the academic year 1990–91 to the academic year 2017–18.

Academic Year	Tuition and Fees (2017 dollars)
1990–91 ($x = 0$)	3580
1995–96 ($x = 5$)	4510
2000–01 ($x = 10$)	4970
2005–06 ($x = 15$)	6880
2009–10 ($x = 19$)	8040
2013–14 ($x = 23$)	9310
2017–18 ($x = 27$)	9970

  Source: The College Board

  (a) Using a graphing utility, draw a scatter plot with academic year as the independent variable.
  (b) Using a graphing utility, build an exponential model from the data.
  (c) Express the function found in part (b) in the form $A(t) = A_0 e^{kt}$.
  (d) Graph the exponential function found in part (b) or (c) on the scatter plot.
  (e) Predict the academic year when the average tuition will reach \$16,000.

59. **Wind Chill Factor** The data represent the wind speed (mph) and the wind chill factor at an air temperature of 15°F.

Wind Speed (mph)	Wind Chill Factor (°F)
5	7
10	3
15	0
20	−2
25	−4
30	−5
35	−7

  Source: U.S. National Weather Service

  (a) Using a graphing utility, draw a scatter plot with wind speed as the independent variable.
  (b) Using a graphing utility, build a logarithmic model from the data.
  (c) Using a graphing utility, draw the logarithmic function found in part (b) on the scatter plot.
  (d) Use the function found in part (b) to predict the wind chill factor if the air temperature is 15°F and the wind speed is 23 mph.

60. **Spreading of a Disease** Jack and Diane live in a small town of 50 people. Unfortunately, both Jack and Diane have a cold. Those who come in contact with someone who has this cold will themselves catch the cold. The data that follow on the next page represent the number of people in the small town who have caught the cold after t days.

(continued)

Days, *t*	Number of People with Cold, *C*
0	2
1	4
2	8
3	14
4	22
5	30
6	37
7	42
8	44

(a) Using a graphing utility, draw a scatter plot of the data. Comment on the type of relation that appears to exist between the number of days that have passed and the number of people with a cold.

(b) Using a graphing utility, build a logistic model from the data.

(c) Graph the function found in part (b) on the scatter plot.

(d) According to the function found in part (b), what is the maximum number of people who will catch the cold? In reality, what is the maximum number of people who could catch the cold?

(e) Sometime between the second and third day, 10 people in the town had a cold. According to the model found in part (b), when did 10 people have a cold?

(f) How long will it take for 46 people to catch the cold?

Chapter Test

CHAPTER
Test Prep
VIDEOS

The Chapter Test Prep Videos include step-by-step solutions to all chapter test exercises. These videos are available in MyLab™ Math, or on this text's YouTube Channel. Refer to the Preface for a link to the YouTube channel.

1. If $f(x) = \dfrac{x+2}{x-2}$ and $g(x) = 2x + 5$, find:
   (a) $f \circ g$ and state its domain
   (b) $(g \circ f)(-2)$
   (c) $(f \circ g)(-2)$

2. Determine whether the function is one-to-one.
   (a) $y = 4x^2 + 3$
   (b) $y = \sqrt{x+3} - 5$

3. Find the inverse of $f(x) = \dfrac{2}{3x-5}$ and check your answer. State the domain and the range of f and f^{-1}.

4. If the point $(3, -5)$ is on the graph of a one-to-one function f, what point must be on the graph of f^{-1}?

In Problems 5–7, solve each equation.

5. $3^x = 243$ 6. $\log_b 16 = 2$ 7. $\log_5 x = 4$

In Problems 8–11, evaluate each expression. without using a calculator.

8. $\log_6 \dfrac{1}{36}$ 9. $\log 10,000$

10. $8^{\log_2 5}$ 11. $\ln e^7$

In Problems 12 and 13, for each function f:
   (a) Find the domain of f.
   (b) Graph f.
   (c) From the graph, of f, find the range and any asymptotes.
   (d) Find f^{-1}, the inverse of f.
   (e) Find the domain and the range of f^{-1}.
   (f) Graph f^{-1}.

12. $f(x) = 4^{x+1} - 2$ 13. $f(x) = 1 - \log_5(x - 2)$

In Problems 14–19, solve each equation. Express irrational solutions in exact form.

14. $5^{x+2} = 125$ 15. $\log(x+9) = 2$

16. $8 - 2e^{-x} = 4$ 17. $\log(x^2 + 3) = \log(x + 6)$

18. $7^{x+3} = e^x$ 19. $\log_2(x-4) + \log_2(x+4) = 3$

20. Write $\log_2\left(\dfrac{4x^3}{x^2 - 3x - 18}\right)$ as the sum and/or difference of logarithms. Express powers as factors.

21. A 50-mg sample of a radioactive substance decays to 34 mg after 30 days. How long will it take for there to be 2 mg remaining?

22. (a) If \$1000 is invested at 5% compounded monthly, how much is there after 8 months?
   (b) If you want to have \$1000 in 9 months, how much do you need to place in a savings account now that pays 5% compounded quarterly?
   (c) How long does it take to double your money if you can invest it at 6% compounded annually?

23. The decibel level, D, of sound is given by the equation

$$D = 10 \log\left(\frac{I}{I_0}\right)$$

where I is the intensity of the sound and $I_0 = 10^{-12}$ watt per square meter.

   (a) If the shout of a single person measures 80 decibels, how loud would the sound be if two people shout at the same time? That is, how loud would the sound be if the intensity doubled?
   (b) The pain threshold for sound is 125 decibels. If the Athens Olympic Stadium 2004 (Olympiako Stadio Athinas 'Spyros Louis') can seat 74,400 people, how many people in the crowd need to shout at the same time for the resulting sound level to meet or exceed the pain threshold? (Ignore any possible sound dampening.)

Cumulative Review

1. Is the following graph the graph of a function? If it is, is the function one-to-one?

2. For the function $f(x) = 2x^2 - 3x + 1$, find:
   (a) $f(3)$ **(b)** $f(-x)$ **(c)** $f(x + h)$

3. Determine which points are on the graph of $x^2 + y^2 = 1$.

   (a) $\left(\dfrac{1}{2}, \dfrac{1}{2}\right)$ **(b)** $\left(\dfrac{1}{2}, \dfrac{\sqrt{3}}{2}\right)$

4. Solve the equation $3(x - 2) = 4(x + 5)$.

5. Graph the line $2x - 4y = 16$.

6. **(a)** Graph the quadratic function $f(x) = -x^2 + 2x - 3$ by determining whether its graph is concave up or concave down and by finding its vertex, axis of symmetry, y-intercept, and x-intercept(s), if any.
   (b) Solve $f(x) \le 0$.

7. Determine the quadratic function whose graph is given in the figure.

8. Graph $f(x) = 3(x + 1)^3 - 2$ using transformations.

9. Given that $f(x) = x^2 + 2$ and $g(x) = \dfrac{2}{x - 3}$, find $(f \circ g)(x)$ and state its domain. What is $(f \circ g)(5)$?

10. For the polynomial function
$$f(x) = 3x^4 - 15x^3 - 12x^2 + 60x$$
   (a) Determine the end behavior of the graph.
   (b) Find the x- and y-intercepts of the graph.
   (c) Find the real zeros and their multiplicity, and determine if the graph crosses or touches the x-axis at each intercept.
   (d) Determine the maximum number of turning points on the graph.
   (e) Graph the function.

11. For the function $g(x) = 3^x + 2$:
   (a) Graph g using transformations. State the domain, range, and horizontal asymptote of the graph of g.
   (b) Determine the inverse of g. State the domain, range, and vertical asymptote of the graph of g^{-1}.
   (c) On the same coordinate axes as g, graph g^{-1}.

12. Solve the equation: $4^{x-3} = 8^{2x}$

13. Solve the equation: $\log_3(x + 1) + \log_3(2x - 3) = \log_9 9$

14. Suppose that $f(x) = \log_3(x + 2)$. Solve:
   (a) $f(x) = 0$
   (b) $f(x) > 0$
   (c) $f(x) = 3$

15. **Data Analysis** The following data represent the percent of all drivers by age who have been stopped by the police for any reason within the past year. The median age represents the midpoint of the upper and lower limit for the age range.

Age Range	Median Age, x	Percent Stopped, y
16–19	17.5	18.2
20–29	24.5	16.8
30–39	34.5	11.3
40–49	44.5	9.4
50–59	54.5	7.7
≥ 60	69.5	3.8

   (a) Using a graphing utility, draw a scatter plot of the data treating median age, x, as the independent variable.
   (b) Determine a model that best describes the relation between median age and percent stopped. You may choose from among linear, quadratic, cubic, exponential, logarithmic, and logistic models.
   (c) Provide a justification for the model that you selected in part (b).

Chapter Projects

Internet-based Project

I. Depreciation of Cars Kelley Blue Book is a guide that provides the current retail price of cars. You can access the Kelley Blue Book online at *www.kbb.com.*

1. Identify three cars that you are considering purchasing, and find the Kelley Blue Book value of the cars for 0 (brand new), 1, 2, 3, 4, and 5 years of age. Online, the value of the car can be found by selecting Price New/Used. Enter the year, make, and model of the new or used car you are selecting. To be consistent, assume the cars will be driven 12,000 miles per year, so a 1-year-old car will have 12,000 miles, a 2-year-old car will have 24,000 miles, and so on. Choose the same options for each year, and select Buy from a Private Party when choosing a price type. Finally, determine the suggested retail price for cars that are in Excellent, Good, and Fair shape. You should have a total of 16 observations (1 for a brand new car, 3 for a 1-year-old car, 3 for a 2-year-old car, and so on).

2. Draw a scatter plot of the data with age as the independent variable and value as the dependent variable using Excel, a TI-graphing calculator, or some other spreadsheet. The Chapter 4 project describes how to draw a scatter plot in Excel.

3. Determine the exponential function of best fit. Graph the exponential function of best fit on the scatter plot. To do this in Excel, right click on any data point in the scatter plot. Now select Add Trendline. Select the Exponential radio button and select Display Equation on Chart. See

Figure 57. Move the Trendline Options window off to the side, if necessary, and you will see the exponential function of best fit displayed on the scatter plot. Do you think the function accurately describes the relation between age of the car and suggested retail price?

Format Trendline ▼ ✕

Trendline Options ▼

◢ **Trendline Options**

- ◉ Exponential
- ○ Linear
- ○ Logarithmic
- ○ Polynomial   Order 2
- ○ Power
- ○ Moving Average   Period 2

Trendline Name
- ◉ Automatic   Expon. (Series1)
- ○ Custom

Forecast
- Forward   0.0   periods
- Backward   0.0   periods
- ☐ Set Intercept   0.0
- ☑ Display Equation on chart
- ☐ Display R-squared value on chart

Figure 57

4. The exponential function of best fit is of the form $y = Ce^{rx}$, where y is the suggested retail value of the car and x is the age of the car (in years). What does the value of C represent? What does the value of r represent? What is the depreciation rate for each car that you are considering?

5. Write a report detailing which car you would purchase based on the depreciation rate you found for each car.

Citation: Excel © 2018 Microsoft Corporation. Used with permission from Microsoft.

The following projects are available on the Instructor's Resource Center (IRC):

II. Hot Coffee A fast-food restaurant wants a special container to hold coffee. The restaurant wishes the container to quickly cool the coffee from 200° to 130°F and keep the liquid between 110° and 130°F as long as possible. The restaurant has three containers to select from. Which one should be purchased?

III. Project at Motorola ***Thermal Fatigue of Solder Connections*** Product reliability is a major concern of a manufacturer. Here a logarithmic transformation is used to simplify the analysis of a cell phone's ability to withstand temperature change.

Analytic Geometry

The Orbit of Comet Hale-Bopp

The orbits of Comet Hale-Bopp and Earth can be modeled using *ellipses*, the subject of Section 7.3. The Internet-based Project at the end of this chapter explores the possibility of Comet Hale-Bopp colliding with Earth.

—See the Internet-based Chapter Project I—

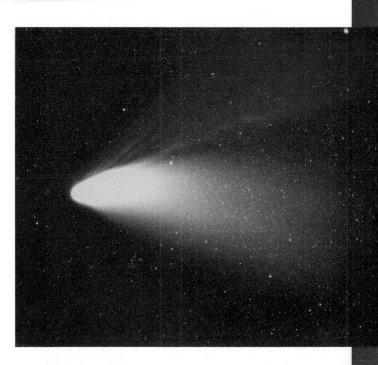

⟵ A Look Back

In Chapter 2, we introduced rectangular coordinates and showed how geometry problems can be solved algebraically. We defined a circle geometrically and then used the distance formula and rectangular coordinates to obtain an equation for a circle.

A Look Ahead ⟶

In this chapter, geometric definitions are given for the *conics*, and the distance formula and rectangular coordinates are used to obtain their equations.

Historically, Apollonius (200 BC) was among the first to study conics and discover some of their interesting properties. Today, conics are still studied because of their many uses. *Paraboloids of revolution* (parabolas rotated about their axes of symmetry) are used as signal collectors (the satellite dishes used with radar and dish TV, for example), as solar energy collectors, and as reflectors (telescopes, light projection, and so on). The planets circle the Sun in approximately *elliptical* orbits. Elliptical surfaces are used to reflect signals such as light and sound from one place to another. A third conic, the *hyperbola*, is used to determine the location of ships or sound sources, such as lightning strikes.

The Greeks used Euclidean geometry to study conics. However, we use the more powerful methods of analytic geometry, which uses both algebra and geometry, for our study of conics.

Outline

7.1 Conics

1 Know the Names of the Conics

The word *conic* derives from the word *cone*, which is a geometric figure that can be constructed in the following way: Let *a* and *g* be two distinct lines that intersect at a point *V*. Keep the line *a* fixed. Now rotate the line *g* about *a*, while maintaining the same angle between *a* and *g*. The collection of points swept out (generated) by the line *g* is called a **right circular cone**. See Figure 1. The fixed line *a* is called the **axis** of the cone; the point *V* is its **vertex**; the lines that pass through *V* and make the same angle with *a* as *g* are **generators** of the cone. Each generator is a line that lies entirely on the cone. The cone consists of two parts, called **nappes**, that intersect at the vertex.

Conics, an abbreviation for **conic sections**, are curves that result from the intersection of a right circular cone and a plane. We discuss only conics formed where the plane does not contain the vertex. These conics are **circles** when the plane is perpendicular to the axis of the cone and intersects each generator; **ellipses** when the plane is tilted slightly so that it intersects each generator, but intersects only one nappe of the cone; **parabolas** when the plane is tilted farther so that it is parallel to one (and only one) generator and intersects only one nappe of the cone; and **hyperbolas** when the plane intersects both nappes. See Figure 2.

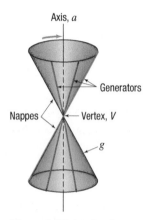

Figure 1 Right circular cone

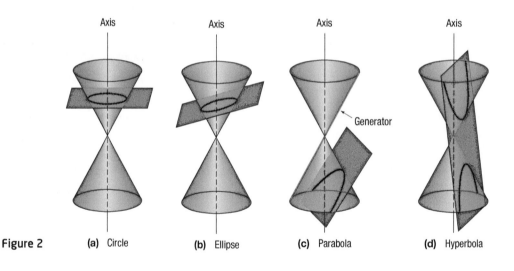

Figure 2   (a) Circle   (b) Ellipse   (c) Parabola   (d) Hyperbola

If the plane contains the vertex, the intersection of the plane and the cone is a point, a line, or a pair of intersecting lines. These are called **degenerate conics**.

Conic sections are used in modeling many applications. For example, parabolas are used in describing searchlights and telescopes (see Figures 14 and 15 on page 524). Ellipses are used to model the orbits of planets and whispering chambers (see pages 534–535). And hyperbolas are used to locate lightning strikes and model nuclear cooling towers (see Problems 76 and 77 in Section 7.4).

7.2 The Parabola

PREPARING FOR THIS SECTION *Before getting started, review the following:*

- Distance Formula (Section 2.1, p. 151)
- Symmetry (Section 2.2, pp. 161–164)
- Square Root Method (Section 1.2, pp. 94–95)
- Completing the Square (Section R.5, pp. 55–56)

- Graphing Techniques: Transformations (Section 3.5, pp. 254–263)
- Quadratic Functions and Their Properties (Section 4.3, pp. 299–308)

Now Work the 'Are You Prepared?' problems on page 525.

OBJECTIVES 1 Analyze Parabolas with Vertex at the Origin (p. 519)
2 Analyze Parabolas with Vertex at (h, k) (p. 522)
3 Solve Applied Problems Involving Parabolas (p. 524)

In Section 4.3, we learned that the graph of a quadratic function is a parabola. In this section, we give a geometric definition of a parabola and use it to obtain an equation.

> **DEFINITION Parabola**
>
> A **parabola** is the collection of all points P in a plane that are the same distance d from a fixed point F as they are from a fixed line D. The point F is called the **focus** of the parabola, and the line D is its **directrix**. As a result, a parabola is the set of points P for which
>
> $$d(F, P) = d(P, D) \qquad (1)$$

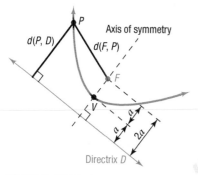

Figure 3 Parabola

Figure 3 shows a parabola (in blue). The line through the focus F and perpendicular to the directrix D is the **axis of symmetry** of the parabola. The point of intersection of the parabola with its axis of symmetry is the **vertex** V.

Because the vertex V lies on the parabola, it must satisfy equation (1): $d(F, V) = d(V, D)$. The vertex is midway between the focus and the directrix. We let a equal the distance $d(F, V)$ from F to V. To derive an equation for a parabola, we use a rectangular system of coordinates positioned so that the vertex V, focus F, and directrix D of the parabola are conveniently located.

1 Analyze Parabolas with Vertex at the Origin

If we locate the vertex V at the origin $(0, 0)$, we can conveniently position the focus F on either the x-axis or the y-axis. First, consider the case where the focus F is on the positive x-axis, as shown in Figure 4. Because the distance from F to V is a, the coordinates of F will be $(a, 0)$ with $a > 0$. Similarly, because the distance from V to the directrix D is also a, and because D is perpendicular to the x-axis (since the x-axis is the axis of symmetry), the equation of the directrix D is $x = -a$.

Now, if $P = (x, y)$ is any point on the parabola, then P satisfies equation (1):

$$d(F, P) = d(P, D)$$

So we have

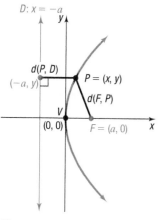

Figure 4

$$\sqrt{(x - a)^2 + (y - 0)^2} = |x + a| \qquad \text{Use the Distance Formula.}$$
$$(x - a)^2 + y^2 = (x + a)^2 \qquad \text{Square both sides.}$$
$$x^2 - 2ax + a^2 + y^2 = x^2 + 2ax + a^2 \qquad \text{Multiply out.}$$
$$y^2 = 4ax \qquad \text{Simplify.}$$

THEOREM **Equation of a Parabola: Vertex at (0, 0), Focus at (a, 0), $a > 0$**

The equation of a parabola with vertex at $(0, 0)$, focus at $(a, 0)$, and directrix $x = -a, a > 0$, is

$$y^2 = 4ax \qquad (2)$$

Recall that a is the distance from the vertex to the focus of a parabola. When graphing the parabola $y^2 = 4ax$, it is helpful to determine the "opening" by finding the points that lie directly above and below the focus $(a, 0)$. Do this by substituting $x = a$ in $y^2 = 4ax$, so $y^2 = 4a \cdot a = 4a^2$, or $y = \pm 2a$. The line segment joining the two points, $(a, 2a)$ and $(a, -2a)$, is called the **latus rectum**; its length is $4a$.

EXAMPLE 1

Finding the Equation of a Parabola and Graphing It

Find an equation of the parabola with vertex at $(0, 0)$ and focus at $(3, 0)$. Graph the equation.

Solution

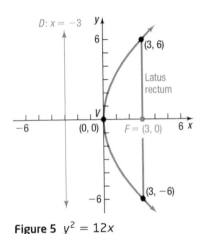

Figure 5 $y^2 = 12x$

The distance from the vertex $(0, 0)$ to the focus $(3, 0)$ is $a = 3$. Then, the equation of the parabola is

$$y^2 = 4ax \qquad \textbf{Equation (2)}$$
$$y^2 = 12x \qquad a = 3$$

To graph the parabola, find the two points that determine the latus rectum by substituting $x = 3$. Then

$$y^2 = 12x = 12 \cdot 3 = 36 \qquad x = 3$$
$$y = \pm 6 \qquad \textbf{Solve for y.}$$

The points $(3, 6)$ and $(3, -6)$ determine the latus rectum. These points help graph the parabola because they determine the "opening." See Figure 5.

 Now Work PROBLEM 21

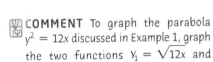

COMMENT To graph the parabola $y^2 = 12x$ discussed in Example 1, graph the two functions $Y_1 = \sqrt{12x}$ and $Y_2 = -\sqrt{12x}$. Do this and compare what you see with Figure 5. ∎

By reversing the steps used to obtain equation (2), it follows that the graph of an equation of the form $y^2 = 4ax$ is a parabola; its vertex is at $(0, 0)$, its focus is at $(a, 0)$, its directrix is the line $x = -a$, and its axis of symmetry is the x-axis.

For the remainder of this section, the direction "**Analyze the equation**" means to find the vertex, focus, and directrix of the parabola and graph it.

EXAMPLE 2

Analyzing the Equation of a Parabola

Analyze the equation $y^2 = 8x$.

Solution The equation $y^2 = 8x$ is of the form $y^2 = 4ax$, where $4a = 8$, so $a = 2$. Consequently, the graph of the equation is a parabola with vertex at $(0, 0)$ and focus on the positive x-axis at $(a, 0) = (2, 0)$. The directrix is the vertical line $a = 2$ units to the left of the y-axis. That is, $x = -2$. The two points that determine the latus rectum are obtained by substituting $x = 2$ in the equation $y^2 = 8x$. Then $y^2 = 16$, so $y = \pm 4$. The points $(2, -4)$ and $(2, 4)$ determine the latus rectum. See Figure 6 for the graph.

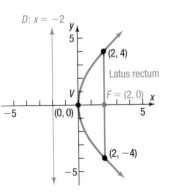

Figure 6 $y^2 = 8x$

Recall that we obtained equation (2) after placing the focus on the positive *x*-axis. Placing the focus on the negative *x*-axis, positive *y*-axis, or negative *y*-axis results in a different form of the equation for the parabola. The four forms of the equation of a parabola with vertex at $(0, 0)$ and focus on a coordinate axis a distance *a* from $(0, 0)$ are given in Table 1, and their graphs are given in Figure 7. Notice that each graph is symmetric with respect to its axis of symmetry.

Table 1

Equations of a Parabola: Vertex at (0, 0); Focus on an Axis; $a > 0$				
Vertex	**Focus**	**Directrix**	**Equation**	**Description**
$(0, 0)$	$(a, 0)$	$x = -a$	$y^2 = 4ax$	Axis of symmetry is the *x*-axis, the parabola opens right
$(0, 0)$	$(-a, 0)$	$x = a$	$y^2 = -4ax$	Axis of symmetry is the *x*-axis, the parabola opens left
$(0, 0)$	$(0, a)$	$y = -a$	$x^2 = 4ay$	Axis of symmetry is the *y*-axis, the parabola opens up (is concave up)
$(0, 0)$	$(0, -a)$	$y = a$	$x^2 = -4ay$	Axis of symmetry is the *y*-axis, the parabola opens down (is concave down)

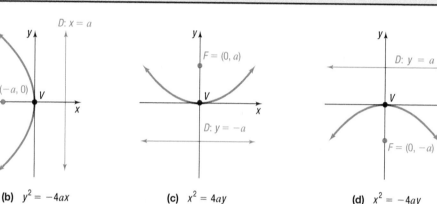

Figure 7   (a) $y^2 = 4ax$ (b) $y^2 = -4ax$ (c) $x^2 = 4ay$ (d) $x^2 = -4ay$

EXAMPLE 3   **Analyzing the Equation of a Parabola**

Analyze the equation $x^2 = -12y$.

Solution   The equation $x^2 = -12y$ is of the form $x^2 = -4ay$, with $a = 3$. Consequently, the graph of the equation is a parabola with vertex at $(0, 0)$, focus at $(0, -3)$, and directrix the line $y = 3$. The parabola opens down (is concave down), and its axis of symmetry is the *y*-axis. To obtain the points defining the latus rectum, let $y = -3$. Then $x^2 = 36$, so $x = \pm 6$. The points $(-6, -3)$ and $(6, -3)$ determine the latus rectum. See Figure 8 for the graph.

━━━ **Now Work** PROBLEM 41

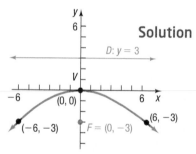

Figure 8 $x^2 = -12y$

EXAMPLE 4   **Finding the Equation of a Parabola**

Find the equation of the parabola with focus at $(0, 4)$ and directrix the line $y = -4$. Graph the equation.

Solution   A parabola whose focus is at $(0, 4)$ and whose directrix is the horizontal line $y = -4$ has its vertex at $(0, 0)$. (Do you see why? The vertex is midway between the focus and the directrix.) Since the focus is on the positive *y*-axis at $(0, 4)$, the equation of the parabola is of the form $x^2 = 4ay$, with $a = 4$. That is,

$$x^2 = 4ay = 4 \cdot 4y = 16y$$
$$\uparrow$$
$$a = 4$$

Substituting $y = 4$ in the equation $x^2 = 16y$ yields $x^2 = 64$, so $x = \pm 8$. The points $(8, 4)$ and $(-8, 4)$ determine the latus rectum. Figure 9 shows the graph of $x^2 = 16y$.

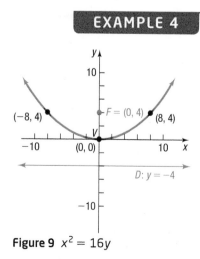

Figure 9 $x^2 = 16y$

EXAMPLE 5	Finding the Equation of a Parabola

Find the equation of a parabola with vertex at $(0, 0)$ if its axis of symmetry is the x-axis and its graph contains the point $\left(-\frac{1}{2}, 2\right)$. Find its focus and directrix, and graph the equation.

Solution

The vertex is at the origin, the axis of symmetry is the x-axis, and the graph contains a point in the second quadrant, so the parabola opens to the left. From Table 1, note that the form of the equation is

$$y^2 = -4ax$$

Because the point $\left(-\frac{1}{2}, 2\right)$ is on the parabola, the coordinates $x = -\frac{1}{2}, y = 2$ satisfy $y^2 = -4ax$. Substituting $x = -\frac{1}{2}$ and $y = 2$ into the equation leads to

$$2^2 = -4a\left(-\frac{1}{2}\right) \qquad x = -\frac{1}{2}, y = 2$$

$$a = 2 \qquad\qquad \text{Solve for } a.$$

The equation of the parabola is

$$y^2 = -4 \cdot 2x = -8x$$

The focus is $(-2, 0)$ and the directrix is the line $x = 2$. Substituting $x = -2$ in the equation $y^2 = -8x$ gives $y = \pm 4$. The points $(-2, 4)$ and $(-2, -4)$ determine the latus rectum. See Figure 10.

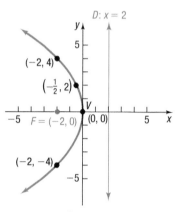

Figure 10 $y^2 = -8x$

 Now Work PROBLEM 29

2 Analyze Parabolas with Vertex at (h, k)

If a parabola with vertex at the origin and axis of symmetry along a coordinate axis is shifted horizontally h units and then vertically k units, the result is a parabola with vertex at (h, k) and axis of symmetry parallel to a coordinate axis. The equations of such parabolas have the same forms as those in Table 1, but with x replaced by $x - h$ (the horizontal shift) and y replaced by $y - k$ (the vertical shift). Table 2 gives the forms of the equations of such parabolas. Figures 11(a)–(d) on page 523 illustrate the graphs for $h > 0, k > 0$.

NOTE Rather than memorizing Table 2, use transformations (shift horizontally h units, vertically k units) and the fact that a is the distance from the vertex to the focus to determine the parabola. ∎

Table 2

Equations of a Parabola: Vertex at (h, k); Axis of Symmetry Parallel to a Coordinate Axis; $a > 0$				
Vertex	Focus	Directrix	Equation	Description
(h, k)	$(h + a, k)$	$x = h - a$	$(y - k)^2 = 4a(x - h)$	Axis of symmetry is parallel to the x-axis, the parabola opens right
(h, k)	$(h - a, k)$	$x = h + a$	$(y - k)^2 = -4a(x - h)$	Axis of symmetry is parallel to the x-axis, the parabola opens left
(h, k)	$(h, k + a)$	$y = k - a$	$(x - h)^2 = 4a(y - k)$	Axis of symmetry is parallel to the y-axis, the parabola opens up (is concave up)
(h, k)	$(h, k - a)$	$y = k + a$	$(x - h)^2 = -4a(y - k)$	Axis of symmetry is parallel to the y-axis, the parabola opens down (is concave down)

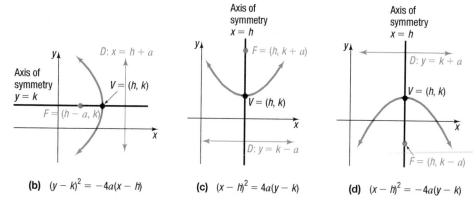

(a) $(y - k)^2 = 4a(x - h)$ **(b)** $(y - k)^2 = -4a(x - h)$ **(c)** $(x - h)^2 = 4a(y - k)$ **(d)** $(x - h)^2 = -4a(y - k)$

Figure 11

| EXAMPLE 6 | Finding the Equation of a Parabola, Vertex Not at the Origin |

Find an equation of the parabola with vertex at $(-2, 3)$ and focus at $(0, 3)$. Graph the equation.

Solution

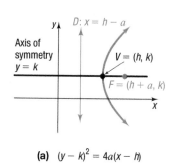

Figure 12 $(y - 3)^2 = 8(x + 2)$

The vertex $(-2, 3)$ and focus $(0, 3)$ both lie on the horizontal line $y = 3$ (the axis of symmetry). The distance a from the vertex $(-2, 3)$ to the focus $(0, 3)$ is $a = 2$. Because the focus lies to the right of the vertex, the parabola opens to the right. Consequently, the form of the equation is

$$(y - k)^2 = 4a(x - h)$$

where $(h, k) = (-2, 3)$ and $a = 2$. Therefore, the equation is

$$(y - 3)^2 = 4 \cdot 2[x - (-2)]$$
$$(y - 3)^2 = 8(x + 2)$$

Since the vertex is midway between the focus and the directrix, the line $x = -4$ is the directrix of the parabola. To find the points that define the latus rectum, substitute $x = 0$ in the equation $(y - 3)^2 = 8(x + 2)$. Then $y - 3 = \pm 4$, so $y = -1$ or $y = 7$. The points $(0, -1)$ and $(0, 7)$ determine the latus rectum. See Figure 12.

Now Work PROBLEM 31

Polynomial equations involving two variables define parabolas whenever they are quadratic in one variable and linear in the other.

| EXAMPLE 7 | Analyzing the Equation of a Parabola |

Analyze the equation $x^2 + 4x - 4y = 0$.

Solution

To analyze the equation $x^2 + 4x - 4y = 0$, complete the square involving the variable x.

$$x^2 + 4x - 4y = 0$$
$$x^2 + 4x = 4y \qquad \text{Isolate the terms involving } x \text{ on the left side.}$$
$$x^2 + 4x + 4 = 4y + 4 \qquad \text{Complete the square on the left side.}$$
$$(x + 2)^2 = 4(y + 1) \qquad \text{Factor.}$$

The equation is of the form $(x - h)^2 = 4a(y - k)$, with $h = -2, k = -1$, and $a = 1$. The graph is a parabola with vertex at $(h, k) = (-2, -1)$ that opens up (is concave up). The focus is at $(-2, 0)$, and the directrix is the line $y = -2$. See Figure 13.

Figure 13 $x^2 + 4x - 4y = 0$

Now Work PROBLEM 49

3 Solve Applied Problems Involving Parabolas

Parabolas occur in many applications. For example, as discussed in Section 4.4, suspension bridges have cables in the shape of a parabola. Parabolas also have a reflecting property that is used in applications.

A parabola that is rotated about its axis of symmetry generates a surface called a **paraboloid of revolution**. If a light (or any other emitting source) is placed at the focus of the parabola, all the rays emanating from the light will reflect off the paraboloid of revolution in lines parallel to the axis of symmetry of the parabola. This principle is used in the design of searchlights, flashlights, certain automobile headlights, and other such devices. See Figure 14.

Conversely, suppose that rays of light (or other signals) emanate from a distant source so that they are essentially parallel. When these rays strike the surface of a parabolic mirror whose axis of symmetry is parallel to these rays, they are reflected to a single point at the focus. This principle is used in the design of some solar energy devices, satellite dishes, and the mirrors used in some types of telescopes. See Figure 15.

Figure 14 Searchlight   **Figure 15** Telescope

EXAMPLE 8	Satellite Dish

A satellite dish is shaped like a paraboloid of revolution. The signals that emanate from a satellite strike the surface of the dish and are reflected to a single point, where the receiver is located. If the dish is 8 feet across at its opening and 3 feet deep at its center, at what position should the receiver be placed? That is, where is the focus?

Solution   Figure 16(a) shows the satellite dish. On a rectangular coordinate system, draw the parabola used to form the dish so that the vertex of the parabola is at the origin and its focus is on the positive y-axis. See Figure 16(b).

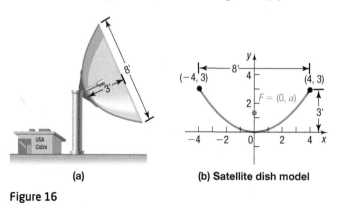

(a)   (b) Satellite dish model

Figure 16

The form of the equation of the parabola is

$$x^2 = 4ay$$

and its focus is at $(0, a)$. Since $(4, 3)$ is a point on the graph, this gives

$$4^2 = 4a \cdot 3 \qquad x^2 = 4ay; x = 4, y = 3$$

$$a = \frac{4}{3} \qquad \text{Solve for } a.$$

The receiver should be located $\frac{4}{3}$ feet (1 foot, 4 inches) from the base of the dish, along its axis of symmetry.

───── **Now Work** PROBLEM 69

7.2 Assess Your Understanding

'Are You Prepared?' *Answers are given at the end of these exercises. If you get a wrong answer, read the pages listed in* red.

1. The formula for the distance d from $P_1 = (x_1, y_1)$ to $P_2 = (x_2, y_2)$ is $d =$ _____. (p. 151)

2. To complete the square of $x^2 - 4x$, add _____. (p. 55)

3. Use the Square Root Method to find the real solutions of $(x + 4)^2 = 9$. (pp. 94–95)

4. The point that is symmetric with respect to the x-axis to the point $(-2, 5)$ is _____. (pp. 161–164)

5. To graph $y = (x - 3)^2 + 1$, shift the graph of $y = x^2$ to the right _____ units and then _____ 1 unit. (pp. 254–258)

6. The graph of $y = (x - 3)^2 - 5$ has vertex _____ and axis of symmetry _____. (pp. 300–302)

Concepts and Vocabulary

7. A(n) _____ is the collection of all points in a plane that are the same distance from a fixed point as they are from a fixed line. The line through the focus and perpendicular to the directrix is called the _____ of the parabola.

8. For the parabola $y^2 = 4ax$, the line segment joining the two points $(a, 2a)$ and $(a, -2a)$ is called the _____ _____.

Answer Problems 9–12 using the figure.

9. *Multiple Choice* If $a > 0$, the equation of the parabola is of the form
 (a) $(y - k)^2 = 4a(x - h)$ **(b)** $(y - k)^2 = -4a(x - h)$
 (c) $(x - h)^2 = 4a(y - k)$ **(d)** $(x - h)^2 = -4a(y - k)$

10. The coordinates of the vertex are _____.

11. *Multiple Choice* If $a = 4$, then the coordinates of the focus are _____.
 (a) $(-1, 2)$ **(b)** $(3, -2)$ **(c)** $(7, 2)$ **(d)** $(3, 6)$

12. *True or False* If $a = 4$, then the equation of the directrix is $x = 3$.

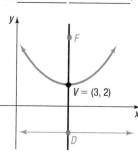

Skill Building

In Problems 13–20, the graph of a parabola is given. Match each graph to its equation.

(A) $y^2 = 4x$
(B) $x^2 = 4y$
(C) $y^2 = -4x$
(D) $x^2 = -4y$
(E) $(y - 1)^2 = 4(x - 1)$
(F) $(x + 1)^2 = 4(y + 1)$
(G) $(y - 1)^2 = -4(x - 1)$
(H) $(x + 1)^2 = -4(y + 1)$

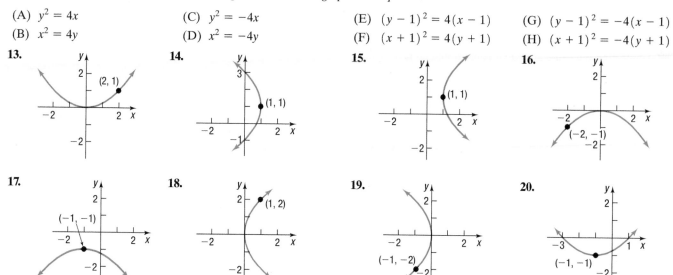

In Problems 21–38, find the equation of the parabola described. Find the two points that define the latus rectum, and graph the equation.

21. Focus at $(4, 0)$ vertex at $(0, 0)$

22. Focus at $(0, 2)$; vertex at $(0, 0)$

23. Focus at $(0, -3)$; vertex at $(0, 0)$

24. Focus at $(-4, 0)$; vertex at $(0, 0)$

25. Focus at $(-2, 0)$; directrix the line $x = 2$

26. Focus at $(0, -1)$; directrix the line $y = 1$

27. Directrix the line $y = -\dfrac{1}{2}$; vertex at $(0, 0)$

28. Directrix the line $x = -\dfrac{1}{2}$; vertex at $(0, 0)$

29. Vertex at $(0, 0)$; axis of symmetry the y-axis; containing the point $(2, 3)$

30. Vertex at $(0, 0)$; axis of symmetry the x-axis; containing the point $(2, 3)$

31. Vertex at $(2, -3)$; focus at $(2, -5)$

32. Vertex at $(4, -2)$; focus at $(6, -2)$

33. Vertex at $(-1, -2)$; focus at $(0, -2)$

34. Vertex at $(3, 0)$; focus at $(3, -2)$

35. Focus at $(-3, 4)$; directrix the line $y = 2$

36. Focus at $(2, 4)$; directrix the line $x = -4$

37. Focus at $(-3, -2)$; directrix the line $x = 1$

38. Focus at $(-4, 4)$; directrix the line $y = -2$

In Problems 39–56, find the vertex, focus, and directrix of each parabola. Graph the equation.

39. $x^2 = 4y$

40. $y^2 = 8x$

41. $y^2 = -16x$

42. $x^2 = -4y$

43. $(y - 2)^2 = 8(x + 1)$

44. $(x + 4)^2 = 16(y + 2)$

45. $(x - 3)^2 = -(y + 1)$

46. $(y + 1)^2 = -4(x - 2)$

47. $(y + 3)^2 = 8(x - 2)$

48. $(x - 2)^2 = 4(y - 3)$

49. $y^2 - 4y + 4x + 4 = 0$

50. $x^2 + 6x - 4y + 1 = 0$

51. $x^2 + 8x = 4y - 8$

52. $y^2 - 2y = 8x - 1$

53. $y^2 + 2y - x = 0$

54. $x^2 - 4x = 2y$

55. $x^2 - 4x = y + 4$

56. $y^2 + 12y = -x + 1$

In Problems 57–64, write an equation for each parabola.

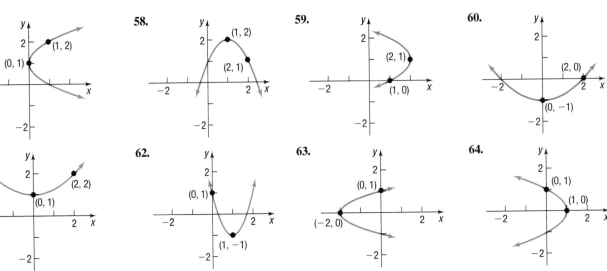

57. (0, 1), (1, 2)

58. (1, 2), (2, 1)

59. (2, 1), (1, 0)

60. (2, 0), (0, −1)

61. (2, 2), (0, 1)

62. (0, 1), (1, −1)

63. (0, 1), (−2, 0)

64. (0, 1), (1, 0)

Applications and Extensions

65. Suspension Bridge The cables of a suspension bridge are in the shape of a parabola, as shown in the figure. The towers supporting the cable are 600 feet apart and 80 feet high. If the cables touch the road surface midway between the towers, what is the height of the cable from the road at a point 150 feet from the center of the bridge?

66. Suspension Bridge The cables of a suspension bridge are in the shape of a parabola. The towers supporting the cable are 400 feet apart and 100 feet high. If the cables are at a height of 10 feet midway between the towers, what is the height of the cable at a point 50 feet from the center of the bridge?

67. Parabolic Arch Bridge A bridge is built in the shape of a parabolic arch. The arch has a span of 120 feet and a maximum height of 25 feet above the water. See the figure. Choose a suitable rectangular coordinate system and find the height of the arch at distances of 10, 30, and 50 feet from the center.

68. Parabolic Arch Bridge A bridge is to be built in the shape of a parabolic arch and is to have a span of 100 feet. The height of the arch a distance of 40 feet from the center is to be 10 feet. Find the height of the arch at its center.

69. Satellite Dish A satellite dish is shaped like a paraboloid of revolution. The signals that emanate from a satellite strike the surface of the dish and are reflected to a single point, where the receiver is located. If the dish is 10 feet across at its opening and 4 feet deep at its center, at what position should the receiver be placed?

70. Constructing a TV Dish A cable TV receiving dish is in the shape of a paraboloid of revolution. Find the location of the receiver, which is placed at the focus, if the dish is 6 feet across at its opening and 2 feet deep.

71. Constructing a Flashlight The reflector of a flashlight is in the shape of a paraboloid of revolution. Its diameter is 4 inches and its depth is 1 inch. How far from the vertex should the light bulb be placed so that the rays will be reflected parallel to the axis?

72. Constructing a Headlight A sealed-beam headlight is in the shape of a paraboloid of revolution. The bulb, which is placed at the focus, is 1 inch from the vertex. If the depth is to be 2 inches, what is the diameter of the headlight at its opening?

73. Searchlight A searchlight is shaped like a paraboloid of revolution. If the light source is located 2 feet from the base along the axis of symmetry and the opening is 5 feet across, how deep should the searchlight be?

74. Searchlight A searchlight is shaped like a paraboloid of revolution. If the light source is located 2 feet from the base along the axis of symmetry and the depth of the searchlight is 4 feet, what should the width of the opening be?

75. Solar Heat A mirror is shaped like a paraboloid of revolution and will be used to concentrate the rays of the sun at its focus, creating a heat source. See the figure. If the mirror is 20 inches across at its opening and is 6 inches deep, where will the heat source be concentrated?

Sun's rays

20"

6"

76. Reflecting Telescope A reflecting telescope contains a mirror shaped like a paraboloid of revolution. If the mirror is 4 inches across at its opening and is 3 inches deep, where will the collected light be concentrated?

77. Gateway Arch The Gateway Arch in St. Louis is often mistaken to be parabolic in shape. In fact, it is a *catenary*, which has a more complicated formula than a parabola. The Arch is 630 feet high and 630 feet wide at its base.
 (a) Find the equation of a parabola with the same dimensions. Let x equal the horizontal distance from the center of the arch.
 (b) The table below gives the height of the Arch at various widths; find the corresponding heights for the parabola found in (a).

Width (ft)	Height (ft)
567	100
478	312.5
308	525

 (c) Do the data support the notion that the Arch is in the shape of a parabola?

Source: gatewayarch.com

78. Show that an equation of the form
$$Ax^2 + Ey = 0 \qquad A \neq 0, E \neq 0$$
is the equation of a parabola with vertex at $(0, 0)$ and axis of symmetry the y-axis. Find its focus and directrix.

79. Show that an equation of the form
$$Cy^2 + Dx = 0 \qquad C \neq 0, D \neq 0$$
is the equation of a parabola with vertex at $(0, 0)$ and axis of symmetry the x-axis. Find its focus and directrix.

80. *Challenge Problem* Show that the graph of an equation of the form
$$Ax^2 + Dx + Ey + F = 0 \qquad A \neq 0$$
 (a) Is a parabola if $E \neq 0$.
 (b) Is a vertical line if $E = 0$ and $D^2 - 4AF = 0$.
 (c) Is two vertical lines if $E = 0$ and $D^2 - 4AF > 0$.
 (d) Contains no points if $E = 0$ and $D^2 - 4AF < 0$.

81. *Challenge Problem* Show that the graph of an equation of the form
$$Cy^2 + Dx + Ey + F = 0 \qquad C \neq 0$$
 (a) Is a parabola if $D \neq 0$.
 (b) Is a horizontal line if $D = 0$ and $E^2 - 4CF = 0$.
 (c) Is two horizontal lines if $D = 0$ and $E^2 - 4CF > 0$.
 (d) Contains no points if $D = 0$ and $E^2 - 4CF < 0$.

82. *Challenge Problem* Let A be either endpoint of the latus rectum of the parabola $y^2 - 2y - 8x + 1 = 0$, and let V be the vertex. Find the exact distance from A to V.[†]

[†]Courtesy of the Joliet Junior College Mathematics Department

Retain Your Knowledge

Problems 83–89 are based on material learned earlier in the course. The purpose of these problems is to keep the material fresh in your mind so that you are better prepared for the final exam.

83. For $x = 9y^2 - 36$, list the intercepts and test for symmetry.

84. Solve: $4^{x+1} = 8^{x-1}$

85. Find the exact distance between the points $\left(-3, \dfrac{1}{2}\right)$ and $\left(\dfrac{2}{3}, -5\right)$.

86. Find the standard form of the equation of a circle with radius $\sqrt{6}$ and center $(-12, 7)$.

87. In 1978, Congress created a gas guzzler tax on vehicles with a fuel economy of less than 22.5 miles per gallon (mpg). Today, a car getting 20 mpg has a tax of $1700 and a car getting 15 mpg has a tax of $4500. If the tax decreases exponentially as fuel economy increases, determine the tax on a vehicle getting 13 mpg to the nearest $100.

$\triangle$**88.** Given $f(x) = \ln(x + 3)$, find the average rate of change of f from 1 to 5.

89. Solve: $|x^2 - 5x| - 2 = 4$

'Are You Prepared?' Answers

1. $\sqrt{(x_2 - x_1)^2 + (y_2 - y_1)^2}$ **2.** 4   **3.** $\{-7, -1\}$ **4.** $(-2, -5)$ **5.** 3; up   **6.** $(3, -5); x = 3$

7.3 The Ellipse

PREPARING FOR THIS SECTION *Before getting started, review the following:*

- Distance Formula (Section 2.1, p. 151)
- Completing the Square (Section R.5, pp. 55–56)
- Intercepts (Section 2.2, pp. 160–161)
- Symmetry (Section 2.2, pp. 161–164)

- Circles (Section 2.4, pp. 185–188)
- Graphing Techniques: Transformations (Section 3.5, pp. 254–263)

Now Work the 'Are You Prepared?' problems on page 535.

OBJECTIVES   **1** Analyze Ellipses with Center at the Origin (p. 528)
                2 Analyze Ellipses with Center at (h, k) (p. 532)
                3 Solve Applied Problems Involving Ellipses (p. 534)

DEFINITION **Ellipse**

An **ellipse** is the collection of all points in a plane the sum of whose distances from two fixed points, called the **foci**, is a constant.

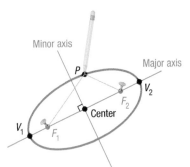

Figure 17 Ellipse

The definition contains within it a physical means for drawing an ellipse. Find a piece of string (the length of the string is the constant referred to in the definition). Then take two thumbtacks (the foci) and stick them into a piece of cardboard so that the distance between them is less than the length of the string. Now attach the ends of the string to the thumbtacks and, using the point of a pencil, pull the string taut. See Figure 17. Keeping the string taut, rotate the pencil around the two thumbtacks. The pencil traces out an ellipse, as shown in Figure 17.

In Figure 17, the foci are labeled F_1 and F_2. The line containing the foci is called the **major axis**. The midpoint of the line segment joining the foci is the **center** of the ellipse. The line through the center and perpendicular to the major axis is the **minor axis**.

The two points of intersection of the ellipse and the major axis are the **vertices**, V_1 and V_2, of the ellipse. The distance from one vertex to the other is the **length of the major axis**. The ellipse is symmetric with respect to its major axis, with respect to its minor axis, and with respect to its center.

1 Analyze Ellipses with Center at the Origin

With these ideas in mind, we are ready to find the equation of an ellipse in a rectangular coordinate system. First, place the center of the ellipse at the origin. Second, position the ellipse so that its major axis coincides with a coordinate axis, say the x-axis, as shown in Figure 18. If c is the distance from the center to a focus, one focus will be at $F_1 = (-c, 0)$ and the other at $F_2 = (c, 0)$.

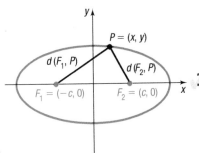

Figure 18

As we shall see, it is convenient to let $2a$ denote the constant distance referred to in the definition. Then, if $P = (x, y)$ is any point on the ellipse,

$$d(F_1, P) + d(F_2, P) = 2a \qquad \text{The sum of the distances from } P \text{ to the foci equals a constant, } 2a.$$

$$\sqrt{(x + c)^2 + y^2} + \sqrt{(x - c)^2 + y^2} = 2a \qquad \text{Use the Distance Formula.}$$

$$\sqrt{(x + c)^2 + y^2} = 2a - \sqrt{(x - c)^2 + y^2} \qquad \text{Isolate one radical.}$$

$$(x + c)^2 + y^2 = 4a^2 - 4a\sqrt{(x - c)^2 + y^2} \qquad \text{Square both sides.}$$
$$+ (x - c)^2 + y^2$$

$$x^2 + 2cx + c^2 + y^2 = 4a^2 - 4a\sqrt{(x - c)^2 + y^2} \qquad \text{Multiply out.}$$
$$+ x^2 - 2cx + c^2 + y^2$$

$$4cx - 4a^2 = -4a\sqrt{(x - c)^2 + y^2} \qquad \text{Simplify; isolate the radical.}$$

$$cx - a^2 = -a\sqrt{(x - c)^2 + y^2} \qquad \text{Divide both sides by 4.}$$

$$(cx - a^2)^2 = a^2[(x - c)^2 + y^2] \qquad \text{Square both sides again.}$$

$$c^2x^2 - 2a^2cx + a^4 = a^2(x^2 - 2cx + c^2 + y^2) \qquad \text{Multiply out.}$$

$$(c^2 - a^2)x^2 - a^2y^2 = a^2c^2 - a^4 \qquad \text{Rearrange the terms.}$$

$$(a^2 - c^2)x^2 + a^2y^2 = a^2(a^2 - c^2) \qquad \text{Multiply both sides by } -1; \quad \textbf{(1)}$$
$$\text{factor out } a^2 \text{ on the right side.}$$

To obtain points on the ellipse that are not on the major axis, we must have $a > c$. To see why, look again at Figure 18. Then

$$d(F_1, P) + d(F_2, P) > d(F_1, F_2) \qquad \begin{array}{l}\text{The sum of the lengths of any two sides of a} \\ \text{triangle is greater than the length of the third side.}\end{array}$$

$$2a > 2c \qquad d(F_1, P) + d(F_2, P) = 2a; d(F_1, F_2) = 2c$$

$$a > c$$

Because $a > c > 0$, this means $a^2 > c^2$, so $a^2 - c^2 > 0$. Let $b^2 = a^2 - c^2$, $b > 0$. Then $a > b$ and equation (1) can be written as

$$b^2x^2 + a^2y^2 = a^2b^2$$

$$\frac{x^2}{a^2} + \frac{y^2}{b^2} = 1 \qquad \text{Divide both sides by } a^2b^2.$$

The graph of the equation has symmetry with respect to the x-axis, the y-axis, and the origin.

Because the major axis is the x-axis, the vertices lie on the x-axis. So the vertices satisfy the equation $\dfrac{x^2}{a^2} = 1$, the solutions of which are $x = \pm a$. Consequently, the vertices of the ellipse are $V_1 = (-a, 0)$ and $V_2 = (a, 0)$. The y-intercepts of the ellipse, found by substituting $x = 0$ in the equation, have coordinates $(0, -b)$ and $(0, b)$. The four intercepts, $(a, 0)$, $(-a, 0)$, $(0, b)$, and $(0, -b)$, are used to graph the ellipse.

> **THEOREM Equation of an Ellipse: Center at (0, 0); Major Axis along the x-Axis**
>
> An equation of the ellipse with center at $(0, 0)$, foci at $(-c, 0)$ and $(c, 0)$, and vertices at $(-a, 0)$ and $(a, 0)$ is
>
> $$\frac{x^2}{a^2} + \frac{y^2}{b^2} = 1 \qquad \text{where } a > b > 0 \text{ and } b^2 = a^2 - c^2 \qquad \textbf{(2)}$$
>
> The major axis is the x-axis. See Figure 19.

Figure 19 $\dfrac{x^2}{a^2} + \dfrac{y^2}{b^2} = 1, a > b > 0$

Notice in Figure 19 the points $(0,0)$, $(c,0)$, and $(0,b)$ form a right triangle. Because $b^2 = a^2 - c^2$ (or $b^2 + c^2 = a^2$), the distance from the focus at $(c,0)$ to the point $(0,b)$ is a.

This can be seen another way. Look at the two right triangles in Figure 19. They are congruent. Do you see why? Because the sum of the distances from the foci to a point on the ellipse is $2a$, it follows that the distance from $(c,0)$ to $(0,b)$ is a.

EXAMPLE 1   Finding an Equation of an Ellipse

Find an equation of the ellipse with center at the origin, one focus at $(3,0)$, and a vertex at $(-4,0)$. Graph the equation.

Solution   The ellipse has its center at the origin, and since the given focus and vertex lie on the x-axis, the major axis is the x-axis. The distance from the center, $(0,0)$, to one of the foci, $(3,0)$, is $c = 3$. The distance from the center, $(0,0)$, to one of the vertices, $(-4,0)$, is $a = 4$. From equation (2), it follows that

$$b^2 = a^2 - c^2 = 16 - 9 = 7$$

so an equation of the ellipse is

$$\frac{x^2}{16} + \frac{y^2}{7} = 1$$

Figure 20 shows the graph.

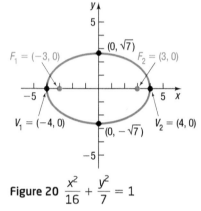

Figure 20 $\dfrac{x^2}{16} + \dfrac{y^2}{7} = 1$

In Figure 20, the intercepts of the equation are used to graph the ellipse. Following this practice makes it easier to obtain an accurate graph of an ellipse.

—**Now Work** PROBLEM 27

COMMENT The intercepts of the ellipse also provide information about how to set the viewing rectangle for graphing an ellipse. To graph the ellipse

$$\frac{x^2}{16} + \frac{y^2}{7} = 1$$

set the viewing rectangle using a square screen that includes the intercepts, perhaps $-4.8 \le x \le 4.8$, $-3 \le y \le 3$. Then solve the equation for y:

$$\frac{x^2}{16} + \frac{y^2}{7} = 1$$

$$\frac{y^2}{7} = 1 - \frac{x^2}{16} \qquad \text{Subtract } \frac{x^2}{16} \text{ from both sides.}$$

$$y^2 = 7\left(1 - \frac{x^2}{16}\right) \qquad \text{Multiply both sides by 7.}$$

$$y = \pm\sqrt{7\left(1 - \frac{x^2}{16}\right)} \qquad \text{Use the Square Root Method.}$$

Now graph the two functions

$$Y_1 = \sqrt{7\left(1 - \frac{x^2}{16}\right)} \text{ and } Y_2 = -\sqrt{7\left(1 - \frac{x^2}{16}\right)}$$

Figure 21 shows the result on a TI-84 Plus C.

$Y_1 = \sqrt{7\left(1 - \frac{x^2}{16}\right)}$

$Y_2 = -\sqrt{7\left(1 - \frac{x^2}{16}\right)}$

Figure 21

For the remainder of this section, the direction "**Analyze the equation**" means to find the center, major axis, foci, and vertices of the ellipse and graph it.

EXAMPLE 2

Analyzing the Equation of an Ellipse

Analyze the equation $\dfrac{x^2}{25} + \dfrac{y^2}{9} = 1$.

Solution

The equation is of the form of equation (2), with $a^2 = 25$ and $b^2 = 9$. The equation is that of an ellipse with center at $(0, 0)$ and major axis along the x-axis. The vertices are at $(\pm a, 0) = (\pm 5, 0)$. Because $b^2 = a^2 - c^2$, this means

$$c^2 = a^2 - b^2 = 25 - 9 = 16$$

The foci are at $(\pm c, 0) = (\pm 4, 0)$. The y-intercepts are $(0, \pm b) = (0, \pm 3)$. Figure 22 shows the graph.

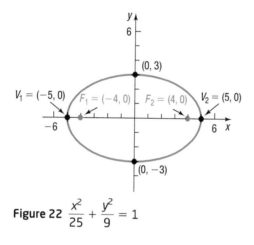

Figure 22 $\dfrac{x^2}{25} + \dfrac{y^2}{9} = 1$

— **Now Work** PROBLEM 17

If the major axis of an ellipse with center at $(0, 0)$ lies on the y-axis, the foci are at $(0, -c)$ and $(0, c)$. Using the same steps as before, the definition of an ellipse leads to the following result.

> **THEOREM Equation of an Ellipse: Center at (0, 0); Major Axis along the y-Axis**
>
> An equation of the ellipse with center at $(0, 0)$, foci at $(0, -c)$ and $(0, c)$, and vertices at $(0, -a)$ and $(0, a)$ is
>
> $$\dfrac{x^2}{b^2} + \dfrac{y^2}{a^2} = 1 \qquad \text{where } a > b > 0 \text{ and } b^2 = a^2 - c^2 \tag{3}$$
>
> The major axis is the y-axis.

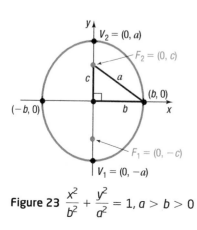

Figure 23 $\dfrac{x^2}{b^2} + \dfrac{y^2}{a^2} = 1, a > b > 0$

Figure 23 illustrates the graph of such an ellipse. Again, notice the right triangle formed by the points at $(0, 0)$, $(b, 0)$, and $(0, c)$, so that $a^2 = b^2 + c^2$ (or $b^2 = a^2 - c^2$).

Look closely at equations (2) and (3). Although they may look alike, there is a difference! In equation (2), the larger number, a^2, is in the denominator of the x^2-term, so the major axis of the ellipse is along the x-axis. In equation (3), the larger number, a^2, is in the denominator of the y^2-term, so the major axis is along the y-axis.

EXAMPLE 3

Analyzing the Equation of an Ellipse

Analyze the equation $9x^2 + y^2 = 9$.

Solution

To put the equation in proper form, divide both sides by 9.

$$x^2 + \frac{y^2}{9} = 1$$

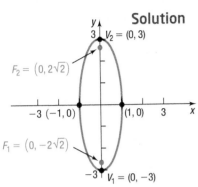

The larger denominator, 9, is in the y^2-term so, based on equation (3), this is the equation of an ellipse with center at the origin and major axis along the y-axis. Also, $a^2 = 9, b^2 = 1$, and $c^2 = a^2 - b^2 = 9 - 1 = 8$. The vertices are at $(0, \pm a) = (0, \pm 3)$, and the foci are at $(0, \pm c) = (0, \pm 2\sqrt{2})$. The x-intercepts are at $(\pm b, 0) = (\pm 1, 0)$. Figure 24 shows the graph.

Figure 24 $9x^2 + y^2 = 9$

━━ **Now Work** PROBLEM 21

EXAMPLE 4

Finding an Equation of an Ellipse

Find an equation of the ellipse having one focus at $(0, 2)$ and vertices at $(0, -3)$ and $(0, 3)$. Graph the equation.

Solution

Plot the given focus and vertices, and note that the major axis is the y-axis. Because the vertices are at $(0, -3)$ and $(0, 3)$, the center of the ellipse is at their midpoint, the origin. The distance from the center, $(0, 0)$, to the given focus, $(0, 2)$, is $c = 2$. The distance from the center, $(0, 0)$, to one of the vertices, $(0, 3)$, is $a = 3$. So $b^2 = a^2 - c^2 = 9 - 4 = 5$. The form of the equation of the ellipse is given by equation (3).

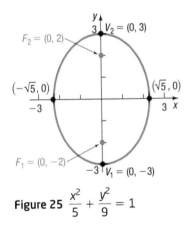

$$\frac{x^2}{b^2} + \frac{y^2}{a^2} = 1$$

$$\frac{x^2}{5} + \frac{y^2}{9} = 1$$

Figure 25 shows the graph.

Figure 25 $\frac{x^2}{5} + \frac{y^2}{9} = 1$

━━ **Now Work** PROBLEM 29

A circle may be considered a special kind of ellipse. To see why, let $a = b$ in equation (2) or (3). Then

$$\frac{x^2}{a^2} + \frac{y^2}{a^2} = 1$$

$$x^2 + y^2 = a^2$$

This is the equation of a circle with center at the origin and radius a. The value of c is

$$c^2 = a^2 - b^2 = 0$$
$$\uparrow$$
$$a = b$$

This indicates that the closer the two foci of an ellipse are to the center, the more the ellipse will look like a circle.

2 Analyze Ellipses with Center at (h, k)

If an ellipse with center at the origin and major axis coinciding with a coordinate axis is shifted horizontally h units and then vertically k units, the result is an ellipse with center at (h, k) and major axis parallel to a coordinate axis. The equations of such ellipses have the same forms as those given in equations (2) and (3), except that x is replaced by $x - h$ (the horizontal shift) and y is replaced by $y - k$ (the vertical shift). Table 3 (on the next page) gives the forms of the equations of such ellipses, and Figure 26 shows their graphs.

Table 3

NOTE Rather than memorizing Table 3, use transformations (shift horizontally *h* units, vertically *k* units), along with the facts that *a* represents the distance from the center to the vertices, *c* represents the distance from the center to the foci, and $b^2 = a^2 - c^2$ (or $c^2 = a^2 - b^2$). ∎

Equations of an Ellipse: Center at (h, k); Major Axis Parallel to a Coordinate Axis				
Center	Major Axis	Foci	Vertices	Equation
(h, k)	Parallel to the *x*-axis	$(h + c, k)$	$(h + a, k)$	$\dfrac{(x - h)^2}{a^2} + \dfrac{(y - k)^2}{b^2} = 1$
		$(h - c, k)$	$(h - a, k)$	$a > b > 0$ and $b^2 = a^2 - c^2$
(h, k)	Parallel to the *y*-axis	$(h, k + c)$	$(h, k + a)$	$\dfrac{(x - h)^2}{b^2} + \dfrac{(y - k)^2}{a^2} = 1$
		$(h, k - c)$	$(h, k - a)$	$a > b > 0$ and $b^2 = a^2 - c^2$

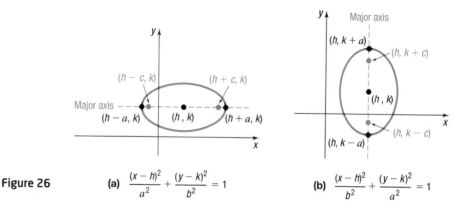

Figure 26

(a) $\dfrac{(x - h)^2}{a^2} + \dfrac{(y - k)^2}{b^2} = 1$

(b) $\dfrac{(x - h)^2}{b^2} + \dfrac{(y - k)^2}{a^2} = 1$

EXAMPLE 5

Finding an Equation of an Ellipse, Center Not at the Origin

Find an equation of the ellipse with center at $(2, -3)$, one focus at $(3, -3)$, and one vertex at $(5, -3)$. Graph the equation.

Solution

The center is at $(h, k) = (2, -3)$, so $h = 2$ and $k = -3$. Note that the center, focus, and vertex all lie on the line $y = -3$. Therefore, the major axis is parallel to the *x*-axis. The distance from the center $(2, -3)$ to a focus $(3, -3)$ is $c = 1$; the distance from the center $(2, -3)$ to a vertex $(5, -3)$ is $a = 3$. Then $b^2 = a^2 - c^2 = 9 - 1 = 8$. The form of the equation is

$$\frac{(x - h)^2}{a^2} + \frac{(y - k)^2}{b^2} = 1$$

$$\frac{(x - 2)^2}{9} + \frac{(y + 3)^2}{8} = 1 \quad h = 2, k = -3, a^2 = 9, b^2 = 8$$

The major axis is parallel to the *x*-axis, so the vertices are $a = 3$ units left and right of the center $(2, -3)$. Therefore, the vertices are

$$V_1 = (2 - 3, -3) = (-1, -3) \quad \text{and} \quad V_2 = (2 + 3, -3) = (5, -3)$$

Note that the vertex $(5, -3)$ agrees with the information given in the problem.

Since $c = 1$ and the major axis is parallel to the *x*-axis, the foci are 1 unit left and right of the center. Therefore, the foci are

$$F_1 = (2 - 1, -3) = (1, -3) \quad \text{and} \quad F_2 = (2 + 1, -3) = (3, -3)$$

Finally, use the value of $b = 2\sqrt{2}$ to find the two points above and below the center.

$$\left(2, -3 - 2\sqrt{2}\right) \quad \text{and} \quad \left(2, -3 + 2\sqrt{2}\right)$$

Use these two points and the vertices to obtain the graph. See Figure 27.

Figure 27 $\dfrac{(x - 2)^2}{9} + \dfrac{(y - 3)^2}{8} = 1$

──── **Now Work** PROBLEM 55

EXAMPLE 6

Analyzing the Equation of an Ellipse

Analyze the equation $4x^2 + y^2 - 8x + 4y + 4 = 0$.

Solution

Complete the squares in x and in y.

$$4x^2 + y^2 - 8x + 4y + 4 = 0$$

$$4x^2 - 8x + y^2 + 4y = -4 \qquad \text{Group like variables; place the constant on the right side.}$$

$$4(x^2 - 2x) + (y^2 + 4y) = -4 \qquad \text{Factor out 4 from the first two terms.}$$

$$4(x^2 - 2x + 1) + (y^2 + 4y + 4) = -4 + 4 + 4 \qquad \text{Complete each square.}$$

$$4(x - 1)^2 + (y + 2)^2 = 4 \qquad \text{Factor.}$$

$$(x - 1)^2 + \frac{(y + 2)^2}{4} = 1 \qquad \text{Divide both sides by 4.}$$

This is the equation of an ellipse with center at $(1, -2)$ and major axis parallel to the y-axis. Since $a^2 = 4$ and $b^2 = 1$, we have $c^2 = a^2 - b^2 = 4 - 1 = 3$. The vertices are at $(h, k \pm a) = (1, -2 \pm 2)$, or $(1, -4)$ and $(1, 0)$. The foci are at $(h, k \pm c) = (1, -2 \pm \sqrt{3})$, or $(1, -2 - \sqrt{3})$ and $(1, -2 + \sqrt{3})$. Figure 28 shows the graph.

Figure 28
$4x^2 + y^2 - 8x + 4y + 4 = 0$

Now Work PROBLEM 47

3 Solve Applied Problems Involving Ellipses

Ellipses are found in many applications in science and engineering. For example, the orbits of the planets around the Sun are elliptical, with the Sun's position at a focus. See Figure 29.

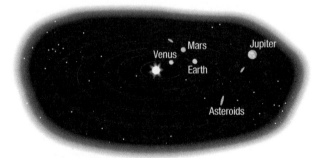

Figure 29 Elliptical orbits

Stone and concrete bridges are often shaped as semielliptical arches. Elliptical gears are used in machinery when a variable rate of motion is required.

Ellipses also have an interesting reflection property. If a source of light (or sound) is placed at one focus, the waves transmitted by the source will reflect off the ellipse and concentrate at the other focus. This is the principle behind *whispering galleries*, which are rooms designed with elliptical ceilings. A person standing at one focus of the ellipse can whisper and be heard by a person standing at the other focus, because all the sound waves that reach the ceiling are reflected to the other person.

EXAMPLE 7

A Whispering Gallery

The whispering gallery in the Museum of Science and Industry in Chicago can be modeled by the top half of a three-dimensional ellipse, which is called an **ellipsoid**. The gallery is 47.3 feet long. The distance from the center of the room to the foci is 20.3 feet. Find an equation that describes the shape of the room. How high is the room at its center?

Source: Chicago Museum of Science and Industry Web site; www.msichicago.org

Solution Set up a rectangular coordinate system so that the center of the ellipse is at the origin and the major axis is along the x-axis. The equation of the ellipse is

$$\frac{x^2}{a^2} + \frac{y^2}{b^2} = 1$$

Since the length of the room is 47.3 feet, the distance from the center of the room to each vertex (the end of the room) will be $\frac{47.3}{2} = 23.65$ feet; so $a = 23.65$ feet. The distance from the center of the room to each focus is $c = 20.3$ feet. See Figure 30.
Because $b^2 = a^2 - c^2$, this means that $b^2 = 23.65^2 - 20.3^2 = 147.2325$. An equation that describes the shape of the room is given by

$$\frac{x^2}{23.65^2} + \frac{y^2}{147.2325} = 1$$

The height of the room at its center is $b = \sqrt{147.2325} \approx 12.1$ feet.

Figure 30 Whispering gallery model

━ **Now Work** PROBLEM 71

7.3 Assess Your Understanding

'Are You Prepared?' *Answers are given at the end of these exercises. If you get a wrong answer, read the pages listed in* red.

1. The distance d from $P_1 = (2, -5)$ to $P_2 = (4, -2)$ is
$d =$ _____. (p. 151)

2. To complete the square of $x^2 - 3x$, add _____. (pp. 55–56)

3. Find the intercepts of the equation $y^2 = 16 - 4x^2$. (pp. 160–161)

4. The point symmetric with respect to the y-axis to the point $(-2, 5)$ is _____. (pp. 161–164)

5. To graph $y = (x + 1)^2 - 4$, shift the graph of $y = x^2$ to the _____ unit(s) and then _____ unit(s).
(left/right) (up/down)
(pp. 254–260)

6. The standard equation of a circle with center at $(2, -3)$ and radius 1 is _____. (pp. 185–188)

Concepts and Vocabulary

7. A(n) _____ is the collection of all points in a plane the sum of whose distances from two fixed points is a constant.

8. *Multiple Choice* For an ellipse, the foci lie on a line called the _____.
(a) minor axis   (b) major axis
(c) directrix   (d) latus rectum

9. For the ellipse $\frac{x^2}{4} + \frac{y^2}{25} = 1$, the vertices are the points _____ and _____.

10. For the ellipse $\frac{x^2}{25} + \frac{y^2}{9} = 1$, the value of a is _____, the value of b is _____, and the major axis is the _____-axis.

11. If the center of an ellipse is $(2, -3)$, the major axis is parallel to the x-axis, and the distance from the center of the ellipse to a vertex is $a = 4$ units, then the coordinates of the vertices are _____ and _____.

12. *Multiple Choice* If the foci of an ellipse are $(-4, 4)$ and $(6, 4)$, then the coordinates of the center of the ellipse are _____.
(a) $(1, 4)$ (b) $(4, 1)$
(c) $(1, 0)$ (d) $(5, 4)$

Skill Building

In Problems 13–16, the graph of an ellipse is given. Match each graph to its equation.

(A) $\frac{x^2}{4} + y^2 = 1$ (B) $x^2 + \frac{y^2}{4} = 1$ (C) $\frac{x^2}{16} + \frac{y^2}{4} = 1$ (D) $\frac{x^2}{4} + \frac{y^2}{16} = 1$

13.

14.

15.

16.

In Problems 17–26, analyze each equation. That is, find the center, vertices, and foci of each ellipse and graph it.

17. $\dfrac{x^2}{25} + \dfrac{y^2}{4} = 1$ **18.** $\dfrac{x^2}{9} + \dfrac{y^2}{4} = 1$ **19.** $\dfrac{x^2}{9} + \dfrac{y^2}{25} = 1$ **20.** $x^2 + \dfrac{y^2}{16} = 1$ **21.** $4x^2 + y^2 = 16$

22. $x^2 + 9y^2 = 18$ **23.** $4y^2 + x^2 = 8$ **24.** $4y^2 + 9x^2 = 36$ **25.** $x^2 + y^2 = 16$ **26.** $x^2 + y^2 = 4$

In Problems 27–38, find an equation for each ellipse. Graph the equation.

27. Center at $(0,0)$; focus at $(3,0)$; vertex at $(5,0)$

28. Center at $(0,0)$; focus at $(-1,0)$; vertex at $(3,0)$

29. Center at $(0,0)$; focus at $(0,-4)$; vertex at $(0,5)$

30. Center at $(0,0)$; focus at $(0,1)$; vertex at $(0,-2)$

31. Foci at $(\pm 2,0)$; length of the major axis is 6

32. Foci at $(0,\pm 2)$; length of the major axis is 8

33. Focus at $(-4,0)$; vertices at $(\pm 5,0)$

34. Focus at $(0,-4)$; vertices at $(0,\pm 8)$

35. Foci at $(0,\pm 3)$; x-intercepts are ± 2

36. Vertices at $(\pm 4,0)$; y-intercepts are ± 1

37. Center at $(0,0)$; vertex at $(0,4)$; $b = 1$

38. Vertices at $(\pm 5,0)$; $c = 2$

In Problems 39–42, find an equation for each ellipse.

39.

40.

41.

42.

In Problems 43–54, analyze each equation; that is, find the center, foci, and vertices of each ellipse. Graph each equation.

43. $\dfrac{(x-3)^2}{4} + \dfrac{(y+1)^2}{9} = 1$ **44.** $\dfrac{(x+4)^2}{9} + \dfrac{(y+2)^2}{4} = 1$ **45.** $(x+5)^2 + 4(y-4)^2 = 16$

46. $9(x-3)^2 + (y+2)^2 = 18$ **47.** $x^2 + 4x + 4y^2 - 8y + 4 = 0$ **48.** $x^2 + 3y^2 - 12y + 9 = 0$

49. $2x^2 + 3y^2 - 8x + 6y + 5 = 0$ **50.** $4x^2 + 3y^2 + 8x - 6y = 5$ **51.** $9x^2 + 4y^2 - 18x + 16y - 11 = 0$

52. $x^2 + 9y^2 + 6x - 18y + 9 = 0$ **53.** $4x^2 + y^2 + 4y = 0$ **54.** $9x^2 + y^2 - 18x = 0$

In Problems 55–64, find an equation for each ellipse. Graph the equation.

55. Center at $(2,-2)$; vertex at $(7,-2)$; focus at $(4,-2)$

56. Center at $(-3,1)$; vertex at $(-3,3)$; focus at $(-3,0)$

57. Vertices at $(4,3)$ and $(4,9)$; focus at $(4,8)$

58. Foci at $(1,2)$ and $(-3,2)$; vertex at $(-4,2)$

59. Foci at $(5,1)$ and $(-1,1)$; length of the major axis is 8

60. Vertices at $(2,5)$ and $(2,-1)$; $c = 2$

61. Center at $(1,2)$; focus at $(4,2)$; contains the point $(1,3)$

62. Center at $(1,2)$; focus at $(1,4)$; contains the point $(2,2)$

63. Center at $(1,2)$; vertex at $(4,2)$; contains the point $(1,5)$

64. Center at $(1,2)$; vertex at $(1,4)$; contains the point $\left(1 + \sqrt{3}, 3\right)$

In Problems 65–68, graph each function. Be sure to label all the intercepts. **[Hint:** *Notice that each function is half an ellipse.*]

65. $f(x) = \sqrt{16 - 4x^2}$ **66.** $f(x) = \sqrt{9 - 9x^2}$ **67.** $f(x) = -\sqrt{64 - 16x^2}$ **68.** $f(x) = -\sqrt{4 - 4x^2}$

Applications and Extensions

69. Semielliptical Arch Bridge An arch in the shape of the upper half of an ellipse is used to support a bridge that is to span a river 20 meters wide. The center of the arch is 6 meters above the center of the river. See the figure. Find an equation for the ellipse in which the x-axis coincides with the water level and the y-axis passes through the center of the arch.

70. Semielliptical Arch Bridge The arch of a bridge is a semiellipse with a horizontal major axis. The span is 30 feet,

and the top of the arch is 10 feet above the major axis. The roadway is horizontal and is 2 feet above the top of the arch. Find the vertical distance from the roadway to the arch at 5-foot intervals along the roadway.

71. Whispering Gallery A hall 100 feet in length is to be designed as a whispering gallery. If the foci are located 25 feet from the center, how high will the ceiling be at the center?

72. Whispering Gallery Jim, standing at one focus of a whispering gallery, is 6 feet from the nearest wall. His friend is standing at the other focus, 100 feet away. What is the length of this whispering gallery? How high is its elliptical ceiling at the center?

73. Semielliptical Arch Bridge A bridge is built in the shape of a semielliptical arch. The arch has a span of 120 feet and a maximum height of 25 feet. Choose a suitable rectangular coordinate system and find the height of the arch at distances of 10, 30, and 50 feet from the center.

74. Semielliptical Arch Bridge A bridge is to be built in the shape of a semielliptical arch and is to have a span of 100 feet. The height of the arch, at a distance of 40 feet from the center, is to be 10 feet. Find the height of the arch at its center.

75. Racetrack Design Consult the figure. A racetrack is in the shape of an ellipse 100 feet long and 50 feet wide. What is the width 10 feet from a vertex?

76. Semielliptical Arch Bridge An arch for a bridge over a highway is in the form of half an ellipse. The top of the arch is 20 feet above the ground level (the major axis). The highway has four lanes, each 12 feet wide; a center safety strip 8 feet wide; and two side strips, each 4 feet wide. What should the span of the arch be (the length of its major axis) if the height 28 feet from the center is to be 13 feet?

77. Installing a Vent Pipe A homeowner is putting in a fireplace that has a 4-inch-radius vent pipe. He needs to cut an elliptical hole in his roof to accommodate the pipe. If the pitch of his roof is $\frac{5}{4}$ (a rise of 5, run of 4), what are the dimensions of the hole?
Source: www.doe.virginia.gov

78. Volume of a Football A football is in the shape of a **prolate spheroid**, which is simply a solid obtained by rotating an ellipse $\left(\dfrac{x^2}{a^2} + \dfrac{y^2}{b^2} = 1\right)$ about its major axis. An inflated NFL football averages 11.125 inches in length and 28.25 inches in center circumference. If the volume of a prolate spheroid is $\frac{4}{3}\pi ab^2$, how much air does the football contain? (Neglect material thickness.)

Source: nfl.com

*In Problems 79–83, use the fact that the orbit of a planet about the Sun is an ellipse, with the Sun at one focus. The **aphelion** of a planet is its greatest distance from the Sun, and the **perihelion** is its shortest distance. The **mean distance** of a planet from the Sun is the length of the semimajor axis of the elliptical orbit. See the figure.*

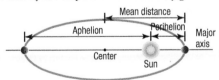

79. Earth The mean distance of Earth from the Sun is 93 million miles. If the aphelion of Earth is 94.5 million miles, what is the perihelion? Find an equation for the orbit of Earth around the Sun.

80. Mars The mean distance of Mars from the Sun is 142 million miles. If the perihelion of Mars is 128.5 million miles, what is the aphelion? Find an equation for the orbit of Mars about the Sun.

81. Jupiter The aphelion of Jupiter is 507 million miles. If the distance from the center of its elliptical orbit to the Sun is 23.2 million miles, what is the perihelion? What is the mean distance? Find an equation for the orbit of Jupiter around the Sun.

82. Pluto The perihelion of Pluto is 4551 million miles, and the distance from the center of its elliptical orbit to the Sun is 897.5 million miles. Find the aphelion of Pluto. What is the mean distance of Pluto from the Sun? Find an equation for the orbit of Pluto about the Sun.

83. Elliptical Orbit A planet orbits a star in an elliptical orbit with the star located at one focus. The perihelion of the planet is 5 million miles. The **eccentricity** e of a conic section is $e = \dfrac{c}{a}$. If the eccentricity of the orbit is 0.75, find the aphelion of the planet.[†]

84. A rectangle is inscribed in an ellipse with major axis of length 14 meters and minor axis of length 4 meters. Find the maximum area of a rectangle inscribed in the ellipse. Round your answer to two decimal places.[†]

85. Lithotripsy Extracorporeal shock wave lithotripsy is a procedure that uses shockwaves to fragment kidney stones without the need for surgery. Using an elliptical reflector, a shock wave generator is placed at one focus and the kidney stone is positioned at the other focus. Shock waves pass through a water column and disintegrate the stone, allowing it to be passed out of the body naturally. If the equation of the ellipse formed by the reflector is $\dfrac{x^2}{324} + \dfrac{y^2}{100} = 1$, how far from the kidney stone does the shock wave generator need to be placed? (Units are in centimeters.)

86. Elliptical Trainer The pedals of an elliptical exercise machine travel an elliptical path as the user is exercising. If the stride length (length of the major axis) for one machine is 20 inches and the maximum vertical pedal displacement (length of the minor axis) is 9 inches, find the equation of the pedal path, assuming it is centered at the origin.

87. Challenge Problem For the ellipse, $x^2 + 5y^2 = 20$, let V be the vertex with the smaller x-coordinate and let B be the endpoint on the minor axis with the larger y-coordinate. Find the y-coordinate of the point M that is on the line $x + 5 = 0$ and is equidistant from V and B.

88. Challenge Problem Consider the circle $(x - 2)^2 + y^2 = 1$ and the ellipse with vertices at $(2, 0)$ and $(6, 0)$ and one focus at $(4 + \sqrt{3}, 0)$. Find the points of intersection of the circle and the ellipse.[†]

89. Challenge Problem Show that an equation of the form
$$Ax^2 + Cy^2 + F = 0 \qquad A \neq 0, C \neq 0, F \neq 0$$
where A and C are of the same sign and F is of opposite sign,
(a) is the equation of an ellipse with center at $(0, 0)$ if $A \neq C$.
(b) is the equation of a circle with center $(0, 0)$ if $A = C$.

90. Challenge Problem Show that the graph of an equation of the form
$$Ax^2 + Cy^2 + Dx + Ey + F = 0 \qquad A \neq 0, C \neq 0$$
where A and C are of the same sign,
(a) is an ellipse if $\dfrac{D^2}{4A} + \dfrac{E^2}{4C} - F$ is the same sign as A.
(b) is a point if $\dfrac{D^2}{4A} + \dfrac{E^2}{4C} - F = 0$.
(c) contains no points if $\dfrac{D^2}{4A} + \dfrac{E^2}{4C} - F$ is of opposite sign to A.

[†]Courtesy of the Joliet Junior College Mathematics Department

Discussion and Writing

91. The **eccentricity** e of an ellipse is defined as the number $\dfrac{c}{a}$, where a is the distance of a vertex from the center and c is the distance of a focus from the center. Because $a > c$, it follows that $e < 1$. Write a brief paragraph about the general shape of each of the following ellipses. Be sure to justify your conclusions.

(a) Eccentricity close to 0   **(b)** Eccentricity $= 0.5$ **(c)** Eccentricity close to 1

Retain Your Knowledge

Problems 92–99 are based on material learned earlier in the course. The purpose of these problems is to keep the material fresh in your mind so that you are better prepared for the final exam.

92. Find the zeros of the quadratic function $f(x) = (x - 5)^2 - 12$. What are the x-intercepts, if any, of the graph of the function?

93. Find the domain of the rational function $f(x) = \dfrac{2x - 3}{x - 5}$. Find any horizontal, vertical, or oblique asymptotes.

94. For the right triangle below, find c and B.

95. What value does $R(x) = \dfrac{3x^2 + 14x + 8}{x^2 + x - 12}$ approach as $x \to -4$?

96. Solve $e^{-2x+1} = 8$ rounded to four decimal places.

97. Find the difference quotient of $f(x) = 2x^2 - 7x$ as $h \to 0$.

98. Solve: $\log_3\left(\dfrac{x}{2} - 1\right) = 4$

99. Solve: $(x + 3)^2 = 20$

'Are You Prepared?' Answers

1. $\sqrt{13}$ **2.** $\dfrac{9}{4}$ **3.** $(-2, 0), (2, 0), (0, -4), (0, 4)$ **4.** $(2, 5)$ **5.** left; 1; down; 4   **6.** $(x - 2)^2 + (y + 3)^2 = 1$

7.4 The Hyperbola

PREPARING FOR THIS SECTION *Before getting started, review the following:*

- Distance Formula (Section 2.1, p. 151)
- Completing the Square (Section R.5, pp. 55–56)
- Intercepts (Section 2.2, pp. 160–161)
- Symmetry (Section 2.2, pp. 161–164)
- Asymptotes (Section 5.3, pp. 356–361)
- Graphing Techniques: Transformations (Section 3.5, pp. 254–263)
- Square Root Method (Section 1.2, pp. 94–95)

Now Work the 'Are You Prepared?' problems on page 548.

OBJECTIVES 1 Analyze Hyperbolas with Center at the Origin (p. 539)
2 Find the Asymptotes of a Hyperbola (p. 543)
3 Analyze Hyperbolas with Center at (h, k) (p. 545)
4 Solve Applied Problems Involving Hyperbolas (p. 546)

DEFINITION

A **hyperbola** is the collection of all points in a plane the difference of whose distances from two fixed points, called the **foci**, is a constant.

Figure 31 Hyperbola

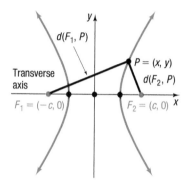

Figure 32
$d(F_1, P) - d(F_2, P) = \pm 2a$

Figure 31 illustrates a hyperbola with foci F_1 and F_2. The line containing the foci is called the **transverse axis**. The midpoint of the line segment joining the foci is the **center** of the hyperbola. The line through the center and perpendicular to the transverse axis is the **conjugate axis**. The hyperbola consists of two separate curves, called **branches**, that are symmetric with respect to the transverse axis, conjugate axis, and center. The two points of intersection of the hyperbola and the transverse axis are the **vertices**, V_1 and V_2, of the hyperbola.

1 Analyze Hyperbolas with Center at the Origin

With these ideas in mind, we are now ready to find the equation of a hyperbola in the rectangular coordinate system. First, place the center at the origin. Next, position the hyperbola so that its transverse axis coincides with a coordinate axis. Suppose that the transverse axis coincides with the x-axis, as shown in Figure 32.

If c is the distance from the center to a focus, one focus will be at $F_1 = (-c, 0)$ and the other at $F_2 = (c, 0)$. Now we let the constant difference of the distances from any point $P = (x, y)$ on the hyperbola to the foci F_1 and F_2 be denoted by $\pm 2a$, where $a > 0$. (If P is on the right branch, the $+$ sign is used; if P is on the left branch, the $-$ sign is used.) The coordinates of P must satisfy the equation

$d(F_1, P) - d(F_2, P) = \pm 2a$	**Difference of the distances from P to the foci equals $\pm 2a$.**
$\sqrt{(x + c)^2 + y^2} - \sqrt{(x - c)^2 + y^2} = \pm 2a$	**Use the Distance Formula.**
$\sqrt{(x + c)^2 + y^2} = \pm 2a + \sqrt{(x - c)^2 + y^2}$	**Isolate one radical.**
$(x + c)^2 + y^2 = 4a^2 \pm 4a\sqrt{(x - c)^2 + y^2}$ $+ (x - c)^2 + y^2$	**Square both sides.**
$x^2 + 2cx + c^2 + y^2 = 4a^2 \pm 4a\sqrt{(x - c)^2 + y^2}$ $+ x^2 - 2cx + c^2 + y^2$	**Multiply.**
$4cx - 4a^2 = \pm 4a\sqrt{(x - c)^2 + y^2}$	**Simplify; isolate the radical.**
$cx - a^2 = \pm a\sqrt{(x - c)^2 + y^2}$	**Divide both sides by 4.**
$(cx - a^2)^2 = a^2[(x - c)^2 + y^2]$	**Square both sides.**
$c^2x^2 - 2ca^2x + a^4 = a^2(x^2 - 2cx + c^2 + y^2)$	**Multiply.**
$c^2x^2 + a^4 = a^2x^2 + a^2c^2 + a^2y^2$	**Distribute and simplify.**
$(c^2 - a^2)x^2 - a^2y^2 = a^2c^2 - a^4$	**Rearrange terms.**
$(c^2 - a^2)x^2 - a^2y^2 = a^2(c^2 - a^2)$	**Factor a^2 on the right side.** (1)

To obtain points on the hyperbola off the x-axis, we must have $a < c$. To see why, look again at Figure 32.

$d(F_1, P) < d(F_2, P) + d(F_1, F_2)$	**Use triangle F_1PF_2.**
$d(F_1, P) - d(F_2, P) < d(F_1, F_2)$	**P is on the right branch, so**
$2a < 2c$	$d(F_1, P) - d(F_2, P) = 2a;$
$a < c$	$d(F_1, F_2) = 2c.$

Since $a < c$, we also have $a^2 < c^2$, so $c^2 - a^2 > 0$. Let $b^2 = c^2 - a^2, b > 0$. Then equation (1) can be written as

$$b^2x^2 - a^2y^2 = a^2b^2$$
$$\frac{x^2}{a^2} - \frac{y^2}{b^2} = 1 \qquad \textbf{Divide both sides by } a^2b^2.$$

To find the vertices of the hyperbola, substitute $y = 0$ in the equation. The vertices satisfy the equation $\dfrac{x^2}{a^2} = 1$, the solutions of which are $x = \pm a$. Consequently, the vertices of the hyperbola are $V_1 = (-a, 0)$ and $V_2 = (a, 0)$. Notice that the distance from the center $(0, 0)$ to either vertex is a.

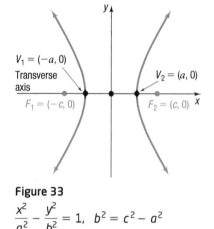

Figure 33
$$\frac{x^2}{a^2} - \frac{y^2}{b^2} = 1, \quad b^2 = c^2 - a^2$$

> **THEOREM** **Equation of a Hyperbola: Center at (0, 0); Transverse Axis along the x-Axis**
>
> An equation of the hyperbola with center at $(0, 0)$, foci at $(-c, 0)$ and $(c, 0)$, and vertices at $(-a, 0)$ and $(a, 0)$ is
>
> $$\frac{x^2}{a^2} - \frac{y^2}{b^2} = 1 \qquad \text{where } b^2 = c^2 - a^2 \tag{2}$$
>
> The transverse axis is the x-axis.

See Figure 33. The hyperbola defined by equation (2) is symmetric with respect to the x-axis, y-axis, and origin. To find the y-intercepts, if any, let $x = 0$ in equation (2). This results in the equation $\dfrac{y^2}{b^2} = -1$, which has no real solution, so the hyperbola defined by equation (2) has no y-intercepts. In fact, since $\dfrac{x^2}{a^2} - 1 = \dfrac{y^2}{b^2} \geq 0$, it follows that $\dfrac{x^2}{a^2} \geq 1$. There are no points on the graph for $-a < x < a$.

EXAMPLE 1   **Finding and Graphing an Equation of a Hyperbola**

Find an equation of the hyperbola with center at the origin, one focus at $(3, 0)$ and one vertex at $(-2, 0)$. Graph the equation.

Solution   The hyperbola has its center at the origin. Plot the center, focus, and vertex. Since they all lie on the x-axis, the transverse axis coincides with the x-axis. One focus is at $(c, 0) = (3, 0)$, so $c = 3$. One vertex is at $(-a, 0) = (-2, 0)$, so $a = 2$. From equation (2), it follows that $b^2 = c^2 - a^2 = 9 - 4 = 5$, so an equation of the hyperbola is

$$\frac{x^2}{4} - \frac{y^2}{5} = 1$$

To graph a hyperbola, it is helpful to locate and plot other points on the graph. For example, to find the points above and below the foci, let $x = \pm 3$. Then

$$\frac{x^2}{4} - \frac{y^2}{5} = 1$$

$$\frac{(\pm 3)^2}{4} - \frac{y^2}{5} = 1 \qquad x = \pm 3$$

$$\frac{9}{4} - \frac{y^2}{5} = 1$$

$$\frac{y^2}{5} = \frac{5}{4}$$

$$y^2 = \frac{25}{4}$$

$$y = \pm \frac{5}{2}$$

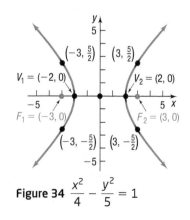

Figure 34 $\dfrac{x^2}{4} - \dfrac{y^2}{5} = 1$

The points above and below the foci are $\left(\pm 3, \dfrac{5}{2}\right)$ and $\left(\pm 3, -\dfrac{5}{2}\right)$. These points determine the "opening" of the hyperbola. See Figure 34.

🖩 **COMMENT** To graph the hyperbola $\dfrac{x^2}{4} - \dfrac{y^2}{5} = 1$, graph the two functions $Y_1 = \sqrt{5}\sqrt{\dfrac{x^2}{4} - 1}$ and $Y_2 = -\sqrt{5}\sqrt{\dfrac{x^2}{4} - 1}$. Do this and compare the result with Figure 34. ∎

✏️ **Now Work** PROBLEM 19

For the next two examples, the direction "**Analyze the equation**" means to find the center, transverse axis, vertices, and foci of the hyperbola and graph it.

EXAMPLE 2

Analyzing the Equation of a Hyperbola

Analyze the equation $\dfrac{x^2}{16} - \dfrac{y^2}{4} = 1$.

Solution

The equation is of the form of equation (2), with $a^2 = 16$ and $b^2 = 4$. The graph of the equation is a hyperbola with center at $(0, 0)$ and transverse axis along the x-axis. Also, $c^2 = a^2 + b^2 = 16 + 4 = 20$. The vertices are at $(\pm a, 0) = (\pm 4, 0)$, and the foci are at $(\pm c, 0) = (\pm 2\sqrt{5}, 0)$.

To locate the points on the graph above and below the foci, let $x = \pm 2\sqrt{5}$ in the equation. Then

$$\frac{x^2}{16} - \frac{y^2}{4} = 1$$

$$\frac{(\pm 2\sqrt{5})^2}{16} - \frac{y^2}{4} = 1 \qquad x = \pm 2\sqrt{5}$$

$$\frac{20}{16} - \frac{y^2}{4} = 1$$

$$\frac{5}{4} - \frac{y^2}{4} = 1$$

$$\frac{y^2}{4} = \frac{1}{4}$$

$$y = \pm 1$$

The points above and below the foci are $(\pm 2\sqrt{5}, 1)$ and $(\pm 2\sqrt{5}, -1)$. See Figure 35.

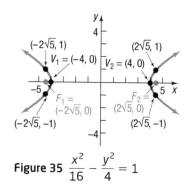

Figure 35 $\dfrac{x^2}{16} - \dfrac{y^2}{4} = 1$

THEOREM Equation of a Hyperbola: Center at (0, 0); Transverse Axis along the y-Axis

An equation of the hyperbola with center at $(0, 0)$, foci at $(0, -c)$ and $(0, c)$, and vertices at $(0, -a)$ and $(0, a)$ is

$$\frac{y^2}{a^2} - \frac{x^2}{b^2} = 1 \qquad \text{where } b^2 = c^2 - a^2 \tag{3}$$

The transverse axis is the y-axis.

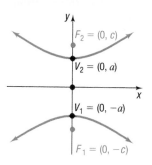

Figure 36
$\dfrac{y^2}{a^2} - \dfrac{x^2}{b^2} = 1, b^2 = c^2 - a^2$

Figure 36 shows the graph of a typical hyperbola defined by equation (3). Let's compare equations (2) and (3).

An equation of the form of equation (2), $\dfrac{x^2}{a^2} - \dfrac{y^2}{b^2} = 1$, is the equation of a hyperbola with center at the origin, foci on the x-axis at $(-c, 0)$ and $(c, 0)$, where $b^2 = c^2 - a^2$, and transverse axis along the x-axis.

An equation of the form of equation (3), $\dfrac{y^2}{a^2} - \dfrac{x^2}{b^2} = 1$, is the equation of a hyperbola with center at the origin, foci on the y-axis at $(0, -c)$ and $(0, c)$, where $b^2 = c^2 - a^2$, and transverse axis along the y-axis.

Notice the difference in the forms of equations (2) and (3). When the y^2-term is subtracted from the x^2-term, the transverse axis is along the x-axis. When the x^2-term is subtracted from the y^2-term, the transverse axis is along the y-axis.

EXAMPLE 3   Analyzing the Equation of a Hyperbola

Analyze the equation $4y^2 - x^2 = 4$.

Solution   To put the equation in proper form, divide both sides by 4:

$$y^2 - \frac{x^2}{4} = 1$$

Since the x^2-term is subtracted from the y^2-term, the equation is that of a hyperbola with center at the origin and transverse axis along the y-axis. Also, comparing the equation to equation (3), note that $a^2 = 1$, $b^2 = 4$, and $c^2 = a^2 + b^2 = 5$. The vertices are at $(0, \pm a) = (0, \pm 1)$, and the foci are at $(0, \pm c) = (0, \pm\sqrt{5})$.

To locate points on the graph to the left and right of the foci, let $y = \pm\sqrt{5}$ in the equation. Then

$$4y^2 - x^2 = 4$$
$$4(\pm\sqrt{5})^2 - x^2 = 4 \qquad y = \pm\sqrt{5}$$
$$20 - x^2 = 4$$
$$x^2 = 16$$
$$x = \pm 4$$

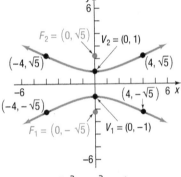

Figure 37 $4y^2 - x^2 = 4$

Four other points on the graph are $(\pm 4, \sqrt{5})$ and $(\pm 4, -\sqrt{5})$. See Figure 37.

EXAMPLE 4   Finding an Equation of a Hyperbola

Find an equation of the hyperbola that has one vertex at $(0, 2)$ and foci at $(0, -3)$ and $(0, 3)$. Graph the equation.

Solution   Since the foci are at $(0, -3)$ and $(0, 3)$, the center of the hyperbola, which is at their midpoint, is the origin. Also, the transverse axis is along the y-axis. This information tells us that $c = 3$, $a = 2$, and $b^2 = c^2 - a^2 = 9 - 4 = 5$. The form of the equation of the hyperbola is given by equation (3):

$$\frac{y^2}{a^2} - \frac{x^2}{b^2} = 1$$

$$\frac{y^2}{4} - \frac{x^2}{5} = 1$$

Let $y = \pm 3$ to obtain points on the graph on either side of each focus. See Figure 38.

Figure 38 $\dfrac{y^2}{4} - \dfrac{x^2}{5} = 1$

Now Work PROBLEMS 15 AND 21

Look at the equations of the hyperbolas in Examples 2 and 4. For the hyperbola in Example 2, $a^2 = 16$ and $b^2 = 4$, so $a > b$; for the hyperbola in Example 4, $a^2 = 4$ and $b^2 = 5$, so $a < b$. This indicates that for hyperbolas, there are no requirements involving the relative sizes of a and b. Contrast this situation to the case of an ellipse, in which the relative sizes of a and b dictate which axis is the major axis. Hyperbolas have another feature to distinguish them from ellipses and parabolas: hyperbolas have asymptotes.

2 Find the Asymptotes of a Hyperbola

Recall from Section 5.3 that a horizontal or oblique asymptote of a graph is a line with the property that the distance from the line to points on the graph approaches 0 as $x \to -\infty$ or as $x \to \infty$. Asymptotes provide information about the end behavior of the graph of a hyperbola.

> **THEOREM** Asymptotes of a Hyperbola; Transverse Axis along the *x*-Axis
>
> The hyperbola $\dfrac{x^2}{a^2} - \dfrac{y^2}{b^2} = 1$ has the two oblique asymptotes
>
> $$y = \frac{b}{a}x \quad \text{and} \quad y = -\frac{b}{a}x \qquad (4)$$

Proof Begin by solving for y in the equation of the hyperbola.

$$\frac{x^2}{a^2} - \frac{y^2}{b^2} = 1$$

$$\frac{y^2}{b^2} = \frac{x^2}{a^2} - 1$$

$$y^2 = b^2\left(\frac{x^2}{a^2} - 1\right)$$

Since $x \neq 0$, the right side can be factored and rewritten as

$$y^2 = \frac{b^2 x^2}{a^2}\left(1 - \frac{a^2}{x^2}\right)$$

$$y = \pm\frac{bx}{a}\sqrt{1 - \frac{a^2}{x^2}}$$

Now, as $x \to -\infty$ or as $x \to \infty$, the term $\dfrac{a^2}{x^2}$ approaches 0, so the expression under the radical approaches 1. So as $x \to -\infty$ or as $x \to \infty$, the value of y approaches $\pm\dfrac{bx}{a}$; that is, the graph of the hyperbola approaches the lines

$$y = -\frac{b}{a}x \quad \text{and} \quad y = \frac{b}{a}x$$

These lines are oblique asymptotes of the hyperbola. ∎

The asymptotes of a hyperbola are not part of the hyperbola, but they serve as a guide for graphing the hyperbola. For example, suppose that we want to graph the equation

$$\frac{x^2}{a^2} - \frac{y^2}{b^2} = 1$$

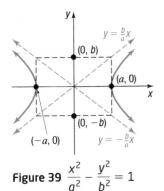

Figure 39 $\dfrac{x^2}{a^2} - \dfrac{y^2}{b^2} = 1$

Begin by plotting the vertices $(-a, 0)$ and $(a, 0)$. Then plot the points $(0, -b)$ and $(0, b)$ and use these four points to construct a rectangle, as shown in Figure 39. The diagonals of this rectangle have slopes $\dfrac{b}{a}$ and $-\dfrac{b}{a}$, and their extensions are the asymptotes of the hyperbola, $y = \dfrac{b}{a}x$ and $y = -\dfrac{b}{a}x$. If we graph the asymptotes, we can use them to establish the "opening" of the hyperbola and avoid plotting other points.

THEOREM Asymptotes of a Hyperbola; Transverse Axis along the y-Axis

The hyperbola $\dfrac{y^2}{a^2} - \dfrac{x^2}{b^2} = 1$ has the two oblique asymptotes

$$y = \frac{a}{b}x \quad \text{and} \quad y = -\frac{a}{b}x \qquad\qquad (5)$$

You are asked to prove this result in Problem 86.

For the remainder of this section, the direction "**Analyze the equation**" means to find the center, transverse axis, vertices, foci, and asymptotes of the hyperbola and graph it.

EXAMPLE 5   Analyzing the Equation of a Hyperbola

Analyze the equation $\dfrac{y^2}{4} - x^2 = 1$.

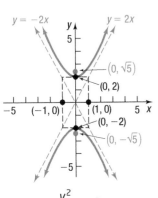

Figure 40 $\dfrac{y^2}{4} - x^2 = 1$

Solution   Since the x^2-term is subtracted from the y^2-term, the equation is of the form of equation (3) and is a hyperbola with center at the origin and transverse axis along the y-axis. Comparing the equation to equation (3), note that $a^2 = 4$, $b^2 = 1$, and $c^2 = a^2 + b^2 = 5$. The vertices are at $(0, \pm a) = (0, \pm 2)$, and the foci are at $(0, \pm c) = (0, \pm \sqrt{5})$. Using equation (5) with $a = 2$ and $b = 1$, the asymptotes are the lines $y = \dfrac{a}{b}x = 2x$ and $y = -\dfrac{a}{b}x = -2x$. Form the rectangle containing the points $(0, \pm a) = (0, \pm 2)$ and $(\pm b, 0) = (\pm 1, 0)$. The extensions of the diagonals of the rectangle are the asymptotes. Now graph the asymptotes and the hyperbola. See Figure 40.

EXAMPLE 6   Analyzing the Equation of a Hyperbola

Analyze the equation $9x^2 - 4y^2 = 36$.

Solution   Divide both sides of the equation by 36 to put the equation in proper form.

$$\frac{x^2}{4} - \frac{y^2}{9} = 1$$

The center of the hyperbola is the origin. Since the y^2-term is subtracted from the x^2-term, the transverse axis is along the x-axis, and the vertices and foci will lie on the x-axis. Using equation (2), note that $a^2 = 4$, $b^2 = 9$, and $c^2 = a^2 + b^2 = 13$. The vertices are $a = 2$ units left and right of the center at $(\pm a, 0) = (\pm 2, 0)$, the foci

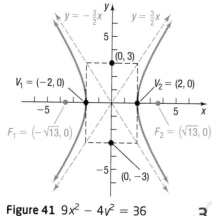

Figure 41 $9x^2 - 4y^2 = 36$

are $c = \sqrt{13}$ units left and right of the center at $(\pm c, 0) = (\pm\sqrt{13}, 0)$, and the asymptotes have the equations

$$y = \frac{b}{a}x = \frac{3}{2}x \quad \text{and} \quad y = -\frac{b}{a}x = -\frac{3}{2}x$$

To graph the hyperbola, form the rectangle containing the points $(\pm a, 0)$ and $(0, \pm b)$, that is, $(-2, 0)$, $(2, 0)$, $(0, -3)$, and $(0, 3)$. The extensions of the diagonals of the rectangle are the asymptotes. See Figure 41 for the graph.

──── **Now Work** PROBLEM 31

3 Analyze Hyperbolas with Center at (h, k)

If a hyperbola with center at the origin and transverse axis coinciding with a coordinate axis is shifted horizontally h units and then vertically k units, the result is a hyperbola with center at (h, k) and transverse axis parallel to a coordinate axis. The equations of such hyperbolas have the same forms as those given in equations (2) and (3), except that x is replaced by $x - h$ (the horizontal shift) and y is replaced by $y - k$ (the vertical shift). Table 4 gives the forms of the equations of such hyperbolas. See Figure 42 for typical graphs.

Table 4

Equations of a Hyperbola: Center at (h, k); Transverse Axis Parallel to a Coordinate Axis					
Center	**Transverse Axis**	**Foci**	**Vertices**	**Equation**	**Asymptotes**
(h, k)	Parallel to the x-axis	$(h \pm c, k)$	$(h \pm a, k)$	$\dfrac{(x-h)^2}{a^2} - \dfrac{(y-k)^2}{b^2} = 1, \; b^2 = c^2 - a^2$	$y - k = \pm\dfrac{b}{a}(x - h)$
(h, k)	Parallel to the y-axis	$(h, k \pm c)$	$(h, k \pm a)$	$\dfrac{(y-k)^2}{a^2} - \dfrac{(x-h)^2}{b^2} = 1, \; b^2 = c^2 - a^2$	$y - k = \pm\dfrac{a}{b}(x - h)$

NOTE Rather than memorize Table 4, use transformations (shift horizontally h units, vertically k units), along with the facts that a represents the distance from the center to the vertices, c represents the distance from the center to the foci, and $b^2 = c^2 - a^2$. ∎

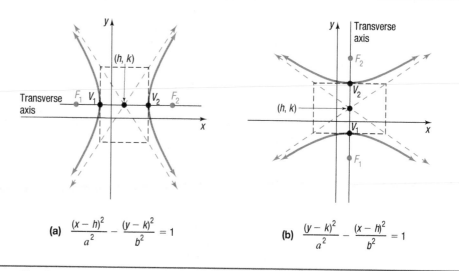

Figure 42

(a) $\dfrac{(x-h)^2}{a^2} - \dfrac{(y-k)^2}{b^2} = 1$

(b) $\dfrac{(y-k)^2}{a^2} - \dfrac{(x-h)^2}{b^2} = 1$

EXAMPLE 7

Finding an Equation of a Hyperbola, Center Not at the Origin

Find an equation for the hyperbola with center at $(1, -2)$, one focus at $(4, -2)$, and one vertex at $(3, -2)$. Graph the equation.

Solution   The center is at $(h, k) = (1, -2)$, so $h = 1$ and $k = -2$. Since the center, focus, and vertex all lie on the line $y = -2$, the transverse axis is parallel to the x-axis. The distance from the center $(1, -2)$ to the focus $(4, -2)$ is $c = 3$; the distance from the center $(1, -2)$ to the vertex $(3, -2)$ is $a = 2$. Then $b^2 = c^2 - a^2 = 9 - 4 = 5$.

(continued)

The equation of the hyperbola is

$$\frac{(x - h)^2}{a^2} - \frac{(y - k)^2}{b^2} = 1$$

$$\frac{(x - 1)^2}{4} - \frac{(y + 2)^2}{5} = 1$$

Form the rectangle containing the vertices: $(-1, -2)$, $(3, -2)$, and the points $(h, k \pm b)$: $(1, -2 - \sqrt{5})$, $(1, -2 + \sqrt{5})$. Extend the diagonals of the rectangle to obtain the asymptotes. See Figure 43.

━━━━ **Now Work** PROBLEM 41

Figure 43 $\dfrac{(x - 1)^2}{4} - \dfrac{(y + 2)^2}{5} = 1$

EXAMPLE 8

Analyzing the Equation of a Hyperbola

Analyze the equation $-x^2 + 4y^2 - 2x - 16y + 11 = 0$.

Solution

Complete the squares in x and in y.

$$-x^2 + 4y^2 - 2x - 16y + 11 = 0$$

$-(x^2 + 2x) + 4(y^2 - 4y) = -11$ Group terms.

$-(x^2 + 2x + 1) + 4(y^2 - 4y + 4) = -11 - 1 + 16$ Complete each square.

$-(x + 1)^2 + 4(y - 2)^2 = 4$ Factor.

$$(y - 2)^2 - \frac{(x + 1)^2}{4} = 1$$ Divide both sides by 4.

This is the equation of a hyperbola with center at $(-1, 2)$ and transverse axis parallel to the y-axis. Also, $a^2 = 1$ and $b^2 = 4$, so $c^2 = a^2 + b^2 = 5$. Since the transverse axis is parallel to the y-axis, the vertices and foci are located a and c units above and below the center, respectively. The vertices are at $(h, k \pm a) = (-1, 2 \pm 1)$, or $(-1, 1)$ and $(-1, 3)$. The foci are at $(h, k \pm c) = (-1, 2 \pm \sqrt{5})$. The asymptotes are $y - 2 = \dfrac{1}{2}(x + 1)$ and $y - 2 = -\dfrac{1}{2}(x + 1)$. Figure 44 shows the graph.

Figure 44
$-x^2 + 4y^2 - 2x - 16y + 11 = 0$

━━━━ **Now Work** PROBLEM 55

4 Solve Applied Problems Involving Hyperbolas

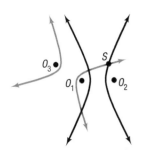

Look at Figure 45. Suppose that three microphones are located at points O_1, O_2, and O_3 (the foci of the two hyperbolas). In addition, suppose that a gun is fired at S and the microphone at O_1 records the gunshot 1 second after the microphone at O_2. Because sound travels at about 1100 feet per second, we conclude that the microphone at O_1 is 1100 feet farther from the gunshot than O_2. This situation is modeled by placing S on a branch of a hyperbola with foci at O_1 and O_2. (Do you see why? The difference of the distances from S to O_1 and from S to O_2 is the constant 1100.) If the third microphone at O_3 records the gunshot 2 seconds after O_1, then S lies on a branch of a second hyperbola with foci at O_1 and O_3. In this case, the constant difference will be 2200. The intersection of the two hyperbolas identifies the location of S.

Figure 45

EXAMPLE 9

Lightning Strikes

Suppose that two people standing 1 mile apart both see a flash of lightning. After a period of time, the person standing at point A hears the thunder. One second later, the person standing at point B hears the thunder. If the person at B is due west of the person at A and the lightning strike is known to occur due north of the person standing at A, where did the lightning strike occur?

Solution See Figure 46, where the point (x, y) represents the location of the lightning strike. Sound travels at 1100 feet per second, so the person at point A is 1100 feet closer to the lightning strike than the person at point B. Since the difference of the distance from (x, y) to B and the distance from (x, y) to A is the constant 1100, the point (x, y) lies on a hyperbola whose foci are at A and B.

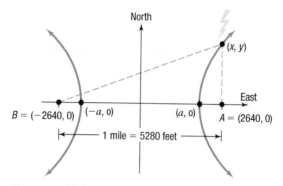

Figure 46 Lightning strike model

An equation of the hyperbola is

$$\frac{x^2}{a^2} - \frac{y^2}{b^2} = 1$$

where $2a = 1100$, or $a = 550$.

Because the distance between the two people is 1 mile (5280 feet) and each person is at a focus of the hyperbola, then

$$2c = 5280$$
$$c = \frac{5280}{2} = 2640$$

Since $b^2 = c^2 - a^2 = 2640^2 - 550^2 = 6{,}667{,}100$, the equation of the hyperbola that describes the location of the lightning strike is

$$\frac{x^2}{550^2} - \frac{y^2}{6{,}667{,}100} = 1$$

Refer to Figure 46. Since the lightning strike occurred due north of the individual at the point $A = (2640, 0)$, let $x = 2640$ and solve the resulting equation.

$$\frac{2640^2}{550^2} - \frac{y^2}{6{,}667{,}100} = 1 \qquad x = 2640$$

$$-\frac{y^2}{6{,}667{,}100} = -22.04 \qquad \text{Subtract } \frac{2640^2}{550^2} \text{ from both sides.}$$

$$y^2 = 146{,}942{,}884 \qquad \text{Multiply both sides by } -6{,}667{,}100.$$

$$y = 12{,}122 \qquad y > 0 \text{ since the lightning strike occurred in quadrant I.}$$

The lightning strike occurred 12,122 feet north of the person standing at point A.

✓**Check:** The difference between the distance from $(2640, 12122)$ to the person at the point $B = (-2640, 0)$ and the distance from $(2640, 12122)$ to the person at the point $A = (2640, 0)$ should be 1100. Using the distance formula, the difference of the distances is

$$\sqrt{[2640 - (-2640)]^2 + (12{,}122 - 0)^2} - \sqrt{(2640 - 2640)^2 + (12{,}122 - 0)^2} = 1100$$

as required.

◢

Now Work PROBLEM 75

7.4 Assess Your Understanding

'Are You Prepared?' *Answers are given at the end of these exercises. If you get a wrong answer, read the pages listed in red.*

1. The distance d from $P_1 = (3, -4)$ to $P_2 = (-2, 1)$ is $d = $ _____ . (p. 151)

2. To complete the square of $x^2 + 5x$, add _____ . (pp. 55–56)

3. Find the intercepts of the equation $y^2 = 9 + 4x^2$. (pp. 160–161)

4. *True or False* The equation $y^2 = 9 + x^2$ is symmetric with respect to the x-axis, the y-axis, and the origin. (pp. 161–164)

5. To graph $y = (x - 5)^3 - 4$, shift the graph of $y = x^3$ to the _____ _____ unit(s) and then _____ _____ unit(s).
(left/right)   (up/down)
(pp. 254–258)

6. Find the vertical asymptotes, if any, and the horizontal or oblique asymptote, if any, of $y = \dfrac{x^2 - 9}{x^2 - 4}$. (pp. 356–361)

Concepts and Vocabulary

7. A(n) _____ is the collection of points in a plane, the difference of whose distances from two fixed points is a constant.

8. For a hyperbola, the foci lie on a line called the _____ _____ .

Answer Problems 9–11 using the figure to the right.

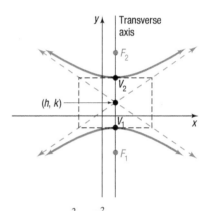

9. *Multiple Choice* The equation of the hyperbola is of the form

(a) $\dfrac{(x - h)^2}{a^2} - \dfrac{(y - k)^2}{b^2} = 1$ **(b)** $\dfrac{(y - k)^2}{a^2} - \dfrac{(x - h)^2}{b^2} = 1$

(c) $\dfrac{(x - h)^2}{a^2} + \dfrac{(y - k)^2}{b^2} = 1$ **(d)** $\dfrac{(x - h)^2}{b^2} + \dfrac{(y - k)^2}{a^2} = 1$

10. If the center of the hyperbola is $(2, 1)$ and $a = 3$, then the coordinates of the vertices are _____ and _____ .

11. If the center of the hyperbola is $(2, 1)$ and $c = 5$, then the coordinates of the foci are _____ and _____ .

12. *Multiple Choice* In a hyperbola, if $a = 3$ and $c = 5$, then $b = $ _____ .
(a) 1   **(b)** 2   **(c)** 4   **(d)** 8

13. For the hyperbola $\dfrac{x^2}{4} - \dfrac{y^2}{9} = 1$, the value of a is _____ , the value of b is _____ , and the transverse axis is the _____ -axis.

14. For the hyperbola $\dfrac{y^2}{16} - \dfrac{x^2}{81} = 1$, the asymptotes are _____ and _____ .

Skill Building

In Problems 15–18, the graph of a hyperbola is given. Match each graph to its equation.

(A) $\dfrac{x^2}{4} - y^2 = 1$ **(B)** $x^2 - \dfrac{y^2}{4} = 1$ **(C)** $\dfrac{y^2}{4} - x^2 = 1$ **(D)** $y^2 - \dfrac{x^2}{4} = 1$

15.   **16.**   **17.**   **18.**

In Problems 19–28, find an equation for the hyperbola described. Graph the equation.

19. Center at $(0, 0)$; focus at $(3, 0)$; vertex at $(1, 0)$

20. Center at $(0, 0)$; focus at $(0, 5)$; vertex at $(0, 3)$

21. Center at $(0, 0)$; focus at $(0, -6)$; vertex at $(0, 4)$

22. Center at $(0, 0)$; focus at $(-3, 0)$; vertex at $(2, 0)$

23. Foci at $(-5, 0)$ and $(5, 0)$; vertex at $(3, 0)$

24. Focus at $(0, 6)$; vertices at $(0, -2)$ and $(0, 2)$

25. Vertices at $(0, -6)$ and $(0, 6)$; asymptote the line $y = 2x$

26. Vertices at $(-4, 0)$ and $(4, 0)$; asymptote the line $y = 2x$

27. Foci at $(-4, 0)$ and $(4, 0)$; asymptote the line $y = -x$

28. Foci at $(0, -2)$ and $(0, 2)$; asymptote the line $y = -x$

In Problems 29–36, find the center, transverse axis, vertices, foci, and asymptotes. Graph each equation.

29. $\dfrac{x^2}{25} - \dfrac{y^2}{9} = 1$

30. $\dfrac{y^2}{16} - \dfrac{x^2}{4} = 1$

31. $4x^2 - y^2 = 16$

32. $4y^2 - x^2 = 16$

33. $y^2 - 9x^2 = 9$

34. $x^2 - y^2 = 4$

35. $y^2 - x^2 = 25$

36. $2x^2 - y^2 = 4$

In Problems 37–40, find an equation for each hyperbola.

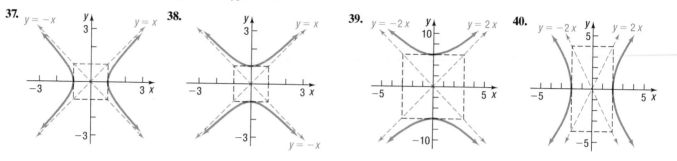

37. $y = -x$, $y = x$

38. $y = x$, $y = -x$

39. $y = -2x$, $y = 2x$

40. $y = -2x$, $y = 2x$

In Problems 41–48, find an equation for the hyperbola described. Graph the equation.

41. Center at $(4, -1)$; focus at $(7, -1)$; vertex at $(6, -1)$

42. Center at $(-3, 1)$; focus at $(-3, 6)$; vertex at $(-3, 4)$

43. Center at $(-3, -4)$; focus at $(-3, -8)$; vertex at $(-3, -2)$

44. Center at $(1, 4)$; focus at $(-2, 4)$; vertex at $(0, 4)$

45. Foci at $(3, 7)$ and $(7, 7)$; vertex at $(6, 7)$

46. Focus at $(-4, 0)$; vertices at $(-4, 4)$ and $(-4, 2)$

47. Vertices at $(-1, -1)$ and $(3, -1)$;
  asymptote the line $y + 1 = \dfrac{3}{2}(x - 1)$

48. Vertices at $(1, -3)$ and $(1, 1)$;
  asymptote the line $y + 1 = \dfrac{3}{2}(x - 1)$

In Problems 49–62, find the center, transverse axis, vertices, foci, and asymptotes. Graph each equation.

49. $\dfrac{(x - 2)^2}{4} - \dfrac{(y + 3)^2}{9} = 1$

50. $\dfrac{(y + 3)^2}{4} - \dfrac{(x - 2)^2}{9} = 1$

51. $(y - 2)^2 - 4(x + 2)^2 = 4$

52. $(x + 4)^2 - 9(y - 3)^2 = 9$

53. $(x + 1)^2 - (y + 2)^2 = 4$

54. $(y - 3)^2 - (x + 2)^2 = 4$

55. $x^2 - y^2 - 2x - 2y - 1 = 0$

56. $y^2 - x^2 - 4y + 4x - 1 = 0$

57. $y^2 - 4x^2 - 4y - 8x - 4 = 0$

58. $2x^2 - y^2 + 4x + 4y - 4 = 0$

59. $4x^2 - y^2 - 24x - 4y + 16 = 0$

60. $2y^2 - x^2 + 2x + 8y + 3 = 0$

61. $y^2 - 4x^2 - 16x - 2y - 19 = 0$

62. $x^2 - 3y^2 + 8x - 6y + 4 = 0$

In Problems 63–66, graph each function. Be sure to label any intercepts. [**Hint:** Notice that each function is half a hyperbola.]

63. $f(x) = \sqrt{16 + 4x^2}$

64. $f(x) = -\sqrt{9 + 9x^2}$

65. $f(x) = -\sqrt{-25 + x^2}$

66. $f(x) = \sqrt{-1 + x^2}$

Mixed Practice In Problems 67–74, analyze each equation.

67. $\dfrac{(x - 3)^2}{4} - \dfrac{y^2}{25} = 1$

68. $\dfrac{(y + 2)^2}{16} - \dfrac{(x - 2)^2}{4} = 1$

69. $x^2 = 16(y - 3)$

70. $y^2 = -12(x + 1)$

71. $25x^2 + 9y^2 - 250x + 400 = 0$

72. $x^2 + 36y^2 - 2x + 288y + 541 = 0$

73. $x^2 - 6x - 8y - 31 = 0$

74. $9x^2 - y^2 - 18x - 8y - 88 = 0$

Applications and Extensions

75. Fireworks Display Suppose that two people standing 2 miles apart both see the burst from a fireworks display. After a period of time the first person, standing at point A, hears the burst. One second later the second person, standing at point B, hears the burst. If the person at point B is due west of the person at point A, and if the display is known to occur due north of the person at point A, where did the fireworks display occur?

76. Lightning Strikes Suppose that two people standing 1 mile apart both see a flash of lightning. After a period of time the first person, standing at point A, hears the thunder. Two seconds later the second person, standing at point B, hears the thunder. If the person at point B is due west of the person at point A, and if the lightning strike is known to occur due north of the person standing at point A, where did the lightning strike occur?

77. Nuclear Power Plant Some nuclear power plants utilize "natural draft" cooling towers in the shape of a **hyperboloid**, a solid obtained by rotating a hyperbola about its conjugate axis. Suppose that such a cooling tower has a base diameter of 400 feet, and the diameter at its narrowest point, 360 feet above the ground, is 200 feet. If the diameter at the top of the tower is 300 feet, how tall is the tower?

Source: Bay Area Air Quality Management District

78. An Explosion Two recording devices are set 2400 feet apart, with the device at point A to the west of the device at point B. At a point between the devices 300 feet from point B, a small amount of explosive is detonated. The recording devices record the time until the sound reaches each. How far directly north of point B should a second explosion be done so that the measured time difference recorded by the devices is the same as that for the first detonation?

79. Rutherford's Experiment In May 1911, Ernest Rutherford published a paper in *Philosophical Magazine*. In this article, he described the motion of alpha particles as they are shot at a piece of gold foil 0.00004 cm thick. Before conducting this experiment, Rutherford expected that the alpha particles would shoot through the foil just as a bullet would shoot through snow. Instead, a small fraction of the alpha particles bounced off the foil. This led to the conclusion that the nucleus of an atom is dense, while the remainder of the atom is sparse. Only the density of the nucleus could cause the alpha particles to deviate from their path. The figure shows a diagram from Rutherford's paper that indicates that the deflected alpha particles follow the path of one branch of a hyperbola.

(a) Find an equation of the asymptotes under this scenario.
(b) If the vertex of the path of the alpha particles is 10 cm from the center of the hyperbola, find a model that describes the path of the particle.

80. Hyperbolic Mirrors Hyperbolas have interesting reflective properties that make them useful for lenses and mirrors. For example, if a ray of light strikes a convex hyperbolic mirror on a line that would (theoretically) pass through its rear focus, it is reflected through the front focus. This property, and that of the parabola, were used to develop the *Cassegrain* telescope in 1672. The focus of the parabolic mirror and the rear focus of the hyperbolic mirror are the same point. The rays are collected by the parabolic mirror, then reflected toward the (common) focus, and thus are reflected by the hyperbolic mirror through the opening to its front focus, where the eyepiece is located. If the equation of the hyperbola is $\dfrac{y^2}{9} - \dfrac{x^2}{16} = 1$ and the focal length (distance from the vertex to the focus) of the parabola is 6, find the equation of the parabola.

Source: www.enchantedlearning.com

81. Lamp Shadow The light from a lamp creates a shadow on a wall with a hyperbolic border. Find the equation of the border if the distance between the vertices is 18 inches and the foci are 4 inches from the vertices. Assume the center of the hyperbola is at the origin.

82. Sonic Boom Aircraft such as fighter jets routinely go supersonic (faster than the speed of sound). An aircraft moving faster than the speed of sound produces a cone-shaped shock wave that "booms" as it trails the vehicle. The wave intersects the ground in the shape of one half of a hyperbola and the area over which the "boom" is audible is called the "boom carpet." If an aircraft creates a shock wave that intersects the ground in the shape of the hyperbola $\dfrac{x^2}{484} - \dfrac{y^2}{100} = 1$ (units in miles), how wide is the "boom carpet" 32 miles behind the aircraft?

83. The **eccentricity** e of a hyperbola is defined as the number $\dfrac{c}{a}$, where a is the distance of a vertex from the center and c is the distance of a focus from the center. Because $c > a$, it follows that $e > 1$. Describe the general shape of a hyperbola whose eccentricity is close to 1. What is the shape if e is very large?

84. A hyperbola for which $a = b$ is called an **equilateral hyperbola**. Find the eccentricity e of an equilateral hyperbola.

[**Note:** The eccentricity of a hyperbola is defined in Problem 83.]

85. Two hyperbolas that have the same set of asymptotes are called **conjugate**. Show that the hyperbolas

$$\frac{x^2}{4} - y^2 = 1 \quad \text{and} \quad y^2 - \frac{x^2}{4} = 1$$

are conjugate. Graph each hyperbola on the same set of coordinate axes.

86. Prove that the hyperbola

$$\frac{y^2}{a^2} - \frac{x^2}{b^2} = 1$$

has the two oblique asymptotes

$$y = \frac{a}{b}x \quad \text{and} \quad y = -\frac{a}{b}x$$

87. *Challenge Problem* Show that the graph of an equation of the form

$$Ax^2 + Cy^2 + F = 0 \qquad A \neq 0, C \neq 0, F \neq 0$$

where A and C are opposite in sign, is a hyperbola with center at $(0, 0)$.

88. *Challenge Problem* Show that the graph of an equation of the form

$$Ax^2 + Cy^2 + Dx + Ey + F = 0 \qquad A \neq 0, C \neq 0$$

where A and C are opposite in sign,

(a) is a hyperbola if $\dfrac{D^2}{4A} + \dfrac{E^2}{4C} - F \neq 0$.

(b) is two intersecting lines if $\dfrac{D^2}{4A} + \dfrac{E^2}{4C} - F = 0$.

Retain Your Knowledge

Problems 89–96 are based on material learned earlier in the course. The purpose of these problems is to keep the material fresh in your mind so that you are better prepared for the final exam.

89. Solve the equation $\log_5 x + \log_5(x - 4) = 1$.

90. The graph of $f(x) = \sqrt{x^2 + 25} + \dfrac{1}{2}(8 - x)$ has an absolute minimum when $\dfrac{x}{\sqrt{x^2 + 25}} - \dfrac{1}{2} = 0$. What is the minimum value rounded to two decimal places?

91. If $g(x) = \sqrt{x - 7} + 2$, find $g^{-1}(3)$.

92. Find the horizontal asymptote for the graph of
$$f(x) = 4e^{x+1} - 5$$

93. What is the inverse function for $f(x) = 3e^{x-1} + 4$?

94. Find the area of the region enclosed by the graphs of $y = \sqrt{9 - x^2}$ and $y = x + 3$.

95. Solve $(2x + 3)^2 + x^2 = 5x(2 + x) + 1$.

96. Find the midpoint of the line segment connecting the points $(3, -8)$ and $(-2, 5)$.

'Are You Prepared?' Answers

1. $5\sqrt{2}$ **2.** $\dfrac{25}{4}$ **3.** $(0, -3), (0, 3)$ **4.** True   **5.** right; 5; down; 4   **6.** Vertical: $x = -2, x = 2$; horizontal: $y = 1$

Chapter Review

Things to Know

Equations

Parabola (pp. 519–523)	See Tables 1 and 2 (pp. 521 and 522).
Ellipse (pp. 528–534)	See Table 3 (p. 533).
Hyperbola (pp. 538–546)	See Table 4 (p. 545).

Definitions

Parabola (p. 519)	Set of points P in a plane for which $d(F, P) = d(P, D)$, where F is the focus and D is the directrix
Ellipse (p. 528)	Set of points P in a plane the sum of whose distances from two fixed points (the foci) is a constant
Hyperbola (p. 538)	Set of points P in a plane the difference of whose distances from two fixed points (the foci) is a constant

Objectives

Section		You should be able to . . .	Example(s)	Review Exercises
7.1	1	Know the names of the conics (p. 518)		1–16
7.2	1	Analyze parabolas with vertex at the origin (p. 519)	1–5	1, 11
	2	Analyze parabolas with vertex at (h, k) (p. 522)	6, 7	4, 6, 9, 14
	3	Solve applied problems involving parabolas (p. 524)	8	19
7.3	1	Analyze ellipses with center at the origin (p. 528)	1–4	3, 13
	2	Analyze ellipses with center at (h, k) (p. 532)	5, 6	8, 10, 16
	3	Solve applied problems involving ellipses (p. 534)	7	20
7.4	1	Analyze hyperbolas with center at the origin (p. 539)	1–4	2, 5, 12
	2	Find the asymptotes of a hyperbola (p. 543)	5, 6	2, 5, 7
	3	Analyze hyperbolas with center at (h, k) (p. 545)	7, 8	7, 15, 17, 18
	4	Solve applied problems involving hyperbolas (p. 546)	9	21

Review Exercises

In Problems 1–10, identify each equation. If it is a parabola, give its vertex, focus, and directrix; if it is an ellipse, give its center, vertices, and foci; if it is a hyperbola, give its center, vertices, foci, and asymptotes.

1. $y^2 = -16x$

2. $\dfrac{x^2}{25} - y^2 = 1$

3. $\dfrac{y^2}{25} + \dfrac{x^2}{16} = 1$

4. $x^2 + 4y = 4$

5. $4x^2 - y^2 = 8$

6. $x^2 - 4x = 2y$

7. $y^2 - 4y - 4x^2 + 8x = 4$

8. $4x^2 + 9y^2 - 16x - 18y = 11$

9. $4x^2 - 16x + 16y + 32 = 0$

10. $9x^2 + 4y^2 - 18x + 8y = 23$

In Problems 11–18, find an equation of the conic described. Graph the equation.

11. Parabola; focus at $(-2, 0)$; directrix the line $x = 2$

12. Hyperbola; center at $(0, 0)$; focus at $(0, 4)$; vertex at $(0, -2)$

13. Ellipse; foci at $(-3, 0)$ and $(3, 0)$; vertex at $(4, 0)$

14. Parabola; vertex at $(2, -3)$; focus at $(2, -4)$

15. Hyperbola; center at $(-2, -3)$; focus at $(-4, -3)$; vertex at $(-3, -3)$

16. Ellipse; foci at $(-4, 2)$ and $(-4, 8)$; vertex at $(-4, 10)$

17. Center at $(-1, 2)$; $a = 3$; $c = 4$; transverse axis parallel to the x-axis

18. Vertices at $(0, 1)$ and $(6, 1)$; asymptote the line $3y + 2x = 9$

19. Searchlight A searchlight is shaped like a paraboloid of revolution. If a light source is located 1 foot from the vertex along the axis of symmetry and the opening is 2 feet across, how deep should the mirror be in order to reflect the light rays parallel to the axis of symmetry?

20. Semielliptical Arch Bridge A bridge is built in the shape of a semielliptical arch. The bridge has a span of 60 feet and a maximum height of 20 feet. Find the height of the arch at distances of 5, 10, and 20 feet from the center.

21. Calibrating Instruments In a test of their recording devices, a team of seismologists positioned two devices 2000 feet apart, with the device at point A to the west of the device at point B. At a point between the devices and 200 feet from point B, a small amount of explosive was detonated and a note made of the time at which the sound reached each device. A second explosion is to be carried out at a point directly north of point B. How far north should the site of the second explosion be chosen so that the measured time difference recorded by the devices for the second detonation is the same as that recorded for the first detonation?

CHAPTER **Test Prep** **VIDEOS** The Chapter Test Prep Videos include step-by-step solutions to all chapter test exercises. These videos are available in MyLab™ Math, or on this text's YouTube Channel. Refer to the Preface for a link to the YouTube channel.

Chapter Test

In Problems 1–3, identify each equation. If it is a parabola, give its vertex, focus, and directrix; if an ellipse, give its center, vertices, and foci; if a hyperbola, give its center, vertices, foci, and asymptotes.

1. $\dfrac{(x+1)^2}{4} - \dfrac{y^2}{9} = 1$

2. $8y = (x-1)^2 - 4$

3. $2x^2 + 3y^2 + 4x - 6y = 13$

In Problems 4–6, find an equation of the conic described; graph the equation.

4. Parabola: focus $(-1, 4.5)$, vertex $(-1, 3)$

5. Ellipse: center $(0, 0)$, vertex $(0, -4)$, focus $(0, 3)$

6. Hyperbola: center $(2, 2)$, vertex $(2, 4)$, contains the point $\left(2 + \sqrt{10}, 5\right)$

7. A parabolic reflector (paraboloid of revolution) is used by TV crews at football games to pick up the referee's announcements, quarterback signals, and so on. A microphone is placed at the focus of the parabola. If a certain reflector is 4 feet wide and 1.5 feet deep, where should the microphone be placed?

Cumulative Review

1. For $f(x) = -3x^2 + 5x - 2$, find
$$\frac{f(x+h) - f(x)}{h} \qquad h \neq 0$$

2. In the complex number system, solve the equation
$$9x^4 + 33x^3 - 71x^2 - 57x - 10 = 0$$

3. For what numbers x is $6 - x \geq x^2$?

4. (a) Find the domain and range of $y = 3^x + 2$.
   (b) Find the inverse of $y = 3^x + 2$ and state its domain and range.

5. $f(x) = \log_4 (x - 2)$
   (a) Solve $f(x) = 2$.
   (b) Solve $f(x) \leq 2$.

6. Find an equation for each of the following graphs.

(a) Line:

(b) Circle:

(c) Ellipse:

(d) Parabola:

(e) Hyperbola:

(f) Exponential:

Chapter Projects

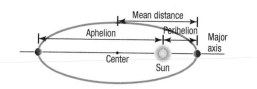

1. Research the history of Comet Hale-Bopp on the Internet. In particular, determine the aphelion and perihelion. Often these values are given in terms of astronomical units. What is an astronomical unit? What is it equivalent to in miles? In kilometers? What is the orbital period of Comet Hale-Bopp? When will it next be visible from Earth? How close does it come to Earth?

2. Find a model for the orbit of Comet Hale-Bopp around the Sun. Use the x-axis as the major axis.

3. Comet Hale-Bopp has an orbit that is roughly perpendicular to that of Earth. Find a model for the orbit of Earth using the y-axis as the major axis.

4. Use a graphing utility or some other graphing technology to graph the paths of the orbits. Based on the graphs, do the paths of the orbits intersect? Does this mean that Comet Hale-Bopp will collide with Earth?

(((•))) Internet-based Project

I. Comet Hale-Bopp The orbits of planets and some comets about the Sun are ellipses, with the Sun at one focus. The **aphelion** of a planet is its greatest distance from the Sun, and the **perihelion** is its shortest distance. The **mean distance** of a planet from the Sun is the length of the semimajor axis of the elliptical orbit. See the figure.

The following projects can be found at the Instructor's Resource Center (IRC):

II. The Orbits of Neptune and Pluto The astronomical body known as Pluto and the planet Neptune travel around the Sun in elliptical orbits. Pluto, at times, comes closer to the Sun than Neptune, the outermost planet. This project examines and analyzes the two orbits.

III. Project at Motorola ***Distorted Deployable Space Reflector Antennas*** An engineer designs an antenna that will deploy in space to collect sunlight.

IV. Constructing a Bridge over the East River The size of ships using a river and fluctuations in water height due to tides or flooding must be considered when designing a bridge that will cross a major waterway.

Systems of Equations and Inequalities

Economic Outcomes

Annual Earnings of Young Adults

For both males and females, earnings increase with education: full-time workers with at least a bachelor's degree have higher median earnings than those with less education. Males and females who dropped out of high school earned 19% and 25% less, respectively, than male and female high school completers.

The median earnings of young adults who had at least a bachelor's degree declined in the 1970s relative to their counterparts who were high school completers, before increasing between 1980 and 2016. Males with a bachelor's degree or higher had earnings 19% higher than male high school completers in 1980 and had earnings 83% higher in 2016. Among females, those with at least a bachelor's degree had earnings 34% higher than female high school completers in 1980, compared with earnings 111% higher in 2016.

Source: U. S. Census Bureau

—See Chapter Project I—

⬅ A Look Back

In Chapters 1, 4, 5, and 6 we solved various kinds of equations and inequalities involving a single variable.

A Look Ahead ➡

In this chapter we take up the problem of solving equations and inequalities containing two or more variables. There are various ways to solve such problems.

The *method of substitution* for solving equations in several variables dates back to ancient times.

The *method of elimination*, although it had existed for centuries, was put into systematic order by Karl Friedrich Gauss (1777–1855) and by Camille Jordan (1838-1922).

The theory of *matrices* was developed in 1857 by Arthur Cayley (1821–1895), although only later were matrices used as we use them in this chapter. Matrices are useful in almost all areas of mathematics.

The method of *determinants* was invented by Takakazu Seki Kôwa (1642–1708) in 1683 in Japan and by Gottfried Wilhelm von Leibniz (1646–1716) in 1693 in Germany. *Cramer's Rule* is named after Gabriel Cramer (1704–1752) of Switzerland, who popularized the use of determinants for solving linear systems.

Section 8.5, on *partial fraction decomposition*, is an application of systems of equations and is used in integral calculus.

Section 8.8 introduces *linear programming*, a modern application of linear inequalities. This topic is particularly useful for students interested in operations research.

Outline

8.1 Systems of Linear Equations: Substitution and Elimination

PREPARING FOR THIS SECTION *Before getting started, review the following:*

- Linear Equations (Section 1.1, pp. 84–85)
- Lines (Section 2.3, pp. 169–179)

Now Work the 'Are You Prepared?' problems on page 566.

OBJECTIVES 1 Solve Systems of Equations by Substitution (p. 559)
2 Solve Systems of Equations by Elimination (p. 559)
3 Identify Inconsistent Systems of Equations Containing Two Variables (p. 561)
4 Express the Solution of a System of Dependent Equations Containing Two Variables (p. 561)
5 Solve Systems of Three Equations Containing Three Variables (p. 562)
6 Identify Inconsistent Systems of Equations Containing Three Variables (p. 564)
7 Express the Solution of a System of Dependent Equations Containing Three Variables (p. 564)

EXAMPLE 1 Movie Theater Ticket Sales

A movie theater sells tickets for $10.00 each, with seniors receiving a discount of $2.00. One evening the theater had $4630 in revenue. If x represents the number of tickets sold at $10.00 and y the number of tickets sold at the discounted price of $8.00, write an equation that relates these variables.

Solution Each nondiscounted ticket costs $10.00, so x tickets bring in $10x$ dollars. Similarly, y discounted tickets bring in $8y$ dollars. Because the total revenue is $4630, we must have

$$10x + 8y = 4630$$

In Example 1, suppose that we also know that 525 tickets were sold that evening. Then we have another equation relating the variables x and y:

$$x + y = 525$$

The two equations

$$\begin{cases} 10x + 8y = 4630 \\ x + y = 525 \end{cases}$$

form a *system* of equations.

In general, a **system of equations** is a collection of two or more equations, each containing one or more variables. Example 2 gives some illustrations of systems of equations.

EXAMPLE 2 Examples of Systems of Equations

(a) $\begin{cases} 2x + y = 5 & (1) \\ -4x + 6y = -2 & (2) \end{cases}$ **Two equations containing two variables, x and y**

(b) $\begin{cases} x + y^2 = 5 & (1) \\ 2x + y = 4 & (2) \end{cases}$ **Two equations containing two variables, x and y**

(c) $\begin{cases} x + y + z = 6 & \text{(1)} \\ 3x - 2y + 4z = 9 & \text{(2)} \\ x - y - z = 0 & \text{(3)} \end{cases}$ **Three equations containing three variables, x, y, and z**

(d) $\begin{cases} x + y + z = 5 & \text{(1)} \\ x - y = 2 & \text{(2)} \end{cases}$ **Two equations containing three variables, x, y, and z**

(e) $\begin{cases} x + y + z = 6 & \text{(1)} \\ 2x + 2z = 4 & \text{(2)} \\ y + z = 2 & \text{(3)} \\ x = 4 & \text{(4)} \end{cases}$ **Four equations containing three variables, x, y, and z**

We use a brace to remind us that we are dealing with a system of equations, and we number each equation in the system for convenient reference.

A **solution** of a system of equations consists of values for the variables that are solutions of each equation of the system. To **solve** a system of equations means to find all solutions of the system.

For example, $x = 2$, $y = 1$ is a solution of the system in Example 2(a), because

$$\begin{cases} 2x + y = 5 & \text{(1)} \\ -4x + 6y = -2 & \text{(2)} \end{cases} \qquad \begin{cases} 2\cdot 2 + 1 = 4 + 1 = 5 \\ -4\cdot 2 + 6\cdot 1 = -8 + 6 = -2 \end{cases}$$

This solution may also be written as the ordered pair $(2, 1)$.

A solution of the system in Example 2(b) is $x = 1$, $y = 2$, because

$$\begin{cases} x + y^2 = 5 & \text{(1)} \\ 2x + y = 4 & \text{(2)} \end{cases} \qquad \begin{cases} 1 + 2^2 = 1 + 4 = 5 \\ 2\cdot 1 + 2 = 2 + 2 = 4 \end{cases}$$

Another solution of the system in Example 2(b) is $x = \dfrac{11}{4}$, $y = -\dfrac{3}{2}$, which you can check for yourself.

A solution of the system in Example 2(c) is $x = 3$, $y = 2$, $z = 1$, because

$$\begin{cases} x + y + z = 6 & \text{(1)} \\ 3x - 2y + 4z = 9 & \text{(2)} \\ x - y - z = 0 & \text{(3)} \end{cases} \qquad \begin{cases} 3 + 2 + 1 = 6 \\ 3\cdot 3 - 2\cdot 2 + 4\cdot 1 = 9 - 4 + 4 = 9 \\ 3 - 2 - 1 = 0 \end{cases}$$

This solution may also be written as the ordered triplet $(3, 2, 1)$.

Note that $x = 3$, $y = 3$, $z = 0$ is not a solution of the system in Example 2(c).

$$\begin{cases} x + y + z = 6 & \text{(1)} \\ 3x - 2y + 4z = 9 & \text{(2)} \\ x - y - z = 0 & \text{(3)} \end{cases} \qquad \begin{cases} 3 + 3 + 0 = 6 \\ 3\cdot 3 - 2\cdot 3 + 4\cdot 0 = 3 \neq 9 \\ 3 - 3 - 0 = 0 \end{cases}$$

Although $x = 3$, $y = 3$, and $z = 0$ satisfy equations (1) and (3), they do not satisfy equation (2). Any solution of the system must satisfy *each* equation of the system.

Now Work PROBLEM 11

When a system of equations has at least one solution, it is said to be **consistent**. When a system of equations has no solution, it is called **inconsistent**.

An equation in n variables is said to be **linear** if it is equivalent to an equation of the form

$$a_1 x_1 + a_2 x_2 + \cdots + a_n x_n = b$$

where $x_1, x_2, \ldots, x_n$ are n distinct variables, $a_1, a_2, \ldots, a_n, b$ are constants, and at least one of the a's is not 0.

Some examples of linear equations are

$$2x + 3y = 2 \qquad 5x - 2y + 3z = 10 \qquad 8x + 8y - 2z + 5w = 0$$

If each equation in a system of equations is linear, we have a **system of linear equations**. The systems in Examples 2(a), (c), (d), and (e) are linear, but the system in Example 2(b) is nonlinear. In this chapter we solve linear systems in Sections 8.1 to 8.3. Nonlinear systems are discussed in Section 8.6.

We begin by discussing a system of two linear equations containing two variables. The problem of solving such a system can be viewed as a geometry problem. The graph of each equation in such a system is a line. So a system of two linear equations containing two variables represents a pair of lines. The lines may intersect, be parallel, or be **coincident** (that is, identical).

- If the lines intersect, the system of equations has one solution, given by the point of intersection. The system is **consistent** and the equations are **independent**. See Figure 1(a).

- If the lines are parallel, the system of equations has no solution, because the lines never intersect. The system is **inconsistent**. See Figure 1(b).

- If the lines are coincident (the lines lie on top of each other), the system of equations has infinitely many solutions, represented by all of the points on the line. The system is **consistent**, and the equations are **dependent**. See Figure 1(c).

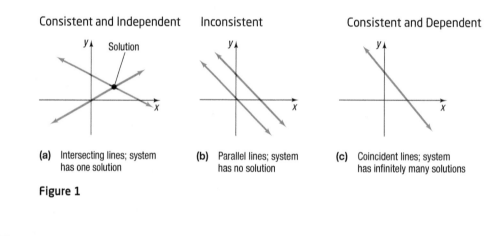

Consistent and Independent Inconsistent Consistent and Dependent

(a) Intersecting lines; system has one solution

(b) Parallel lines; system has no solution

(c) Coincident lines; system has infinitely many solutions

Figure 1

EXAMPLE 3 **Graphing a System of Linear Equations**

Graph the system of linear equations: $\begin{cases} 2x + y = -1 & \text{(1)} \\ -4x + 6y = 42 & \text{(2)} \end{cases}$

Solution First, solve each equation for y. That is, write each equation in slope-intercept form. Equation (1) in slope-intercept form is $y = -2x - 1$, which has slope -2 and y-intercept -1. Equation (2) in slope-intercept form is $y = \dfrac{2}{3}x + 7$, which has slope $\dfrac{2}{3}$ and y-intercept 7. Figure 2 shows their graphs.

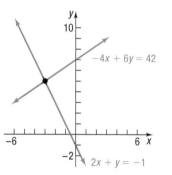

Figure 2

From the graph in Figure 2, we see that the lines intersect, so the system given in Example 3 is consistent. The graph can also be used to approximate the solution. For this system, the solution appears to be the point $(-3, 5)$.

Most of the time we use algebraic methods to obtain exact solutions. A number of methods are available for solving systems of linear equations algebraically. In this section, we introduce two methods: *substitution* and *elimination*. We illustrate the **method of substitution** by solving the system given in Example 3.

1 Solve Systems of Equations by Substitution

EXAMPLE 4	**Solving a System of Linear Equations by Substitution**

Solve: $\begin{cases} 2x + y = -1 & (1) \\ -4x + 6y = 42 & (2) \end{cases}$

Step-by-Step Solution

Step 1: Pick one of the equations, and solve for one variable in terms of the remaining variable(s).

Solve equation (1) for y.

$$2x + y = -1 \qquad \text{Equation (1)}$$
$$y = -2x - 1$$

Step 2: Substitute the result into the remaining equation(s).

Substitute $-2x - 1$ for y in equation (2). The result is an equation containing just the variable x, which we can solve.

$$-4x + 6y = 42 \qquad \text{Equation (2)}$$
$$-4x + 6(-2x - 1) = 42 \qquad \text{Substitute } -2x - 1 \text{ for y in (2)}.$$

Step 3: If one equation in one variable results, solve this equation. Otherwise, repeat Steps 1 and 2 until a single equation with one variable remains.

$$-4x - 12x - 6 = 42 \qquad \text{Distribute.}$$
$$-16x - 6 = 42 \qquad \text{Combine like terms.}$$
$$-16x = 48 \qquad \text{Add 6 to both sides.}$$
$$x = -3 \qquad \text{Solve for x.}$$

Step 4: Find the values of the remaining variables by back-substitution.

Because we know that $x = -3$, we can find the value of y by **back-substitution**, that is, by substituting -3 for x in one of the original equations. Equation (1) seems easier to work with, so we will back-substitute into equation (1).

$$2x + y = -1 \qquad \text{Equation (1)}$$
$$2(-3) + y = -1 \qquad \text{Substitute } -3 \text{ for x.}$$
$$-6 + y = -1 \qquad \text{Simplify.}$$
$$y = 5 \qquad \text{Solve for y.}$$

Step 5: Check the solution found.

We have $x = -3$ and $y = 5$. Verify that both equations are satisfied (true) for these values.

$$\begin{cases} 2x + y = -1 & 2(-3) + 5 = -6 + 5 = -1 \\ -4x + 6y = 42 & -4(-3) + 6 \cdot 5 = 12 + 30 = 42 \end{cases}$$

The solution of the system is $x = -3$ and $y = 5$. The solution can also be written as the ordered pair $(-3, 5)$.

━━━━━ **Now Use Substitution to Work** PROBLEM 21

2 Solve Systems of Equations by Elimination

A second method for solving a system of linear equations is the *method of elimination*. This method is usually preferred over substitution if substitution leads to fractions or if the system contains more than two variables. Elimination also provides the motivation for solving systems using matrices (the subject of Section 8.2).

In Words

When using elimination, get the coefficient of one variable to be opposite that of the other.

The idea behind the **method of elimination** is to replace the original system of equations by an equivalent system so that adding two of the equations eliminates a variable. The rules for obtaining equivalent equations are the same as those studied earlier. We may also interchange any two equations of the system and/or replace any equation in the system by the sum (or difference) of that equation and a nonzero multiple of any other equation in the system.

Rules for Obtaining an Equivalent System of Equations

- Interchange any two equations of the system.
- Multiply (or divide) both sides of an equation by the same nonzero constant.
- Replace any equation in the system by the sum (or difference) of that equation and a nonzero multiple of any other equation in the system.

An example will give you the idea. As you work through the example, pay particular attention to the pattern being followed.

EXAMPLE 5

Solving a System of Linear Equations by Elimination

Solve: $\begin{cases} 2x + 3y = 1 & (1) \\ -x + y = -3 & (2) \end{cases}$

Step-by-Step Solution

Step 1: Multiply both sides of one or both equations by a nonzero constant so that the coefficients of one of the variables are additive inverses.

Multiply equation (2) by 2 so that the coefficients of x in the two equations are additive inverses.

$\begin{cases} 2x + 3y = 1 & (1) \\ -x + y = -3 & (2) \end{cases}$

$\begin{cases} 2x + 3y = 1 & (1) \\ 2(-x + y) = 2(-3) & (2) \quad \textbf{Multiply by 2.} \end{cases}$

$\begin{cases} 2x + 3y = 1 & (1) \\ -2x + 2y = -6 & (2) \end{cases}$

Step 2: Add the equations to eliminate the variable. Solve the resulting equation.

$\begin{cases} 2x + 3y = 1 & (1) \\ -2x + 2y = -6 & (2) \end{cases}$

$\phantom{-2x + 2{}}5y = -5 \qquad \text{Add equations (1) and (2).}$

$\phantom{-2x + 2{}}y = -1 \qquad \text{Solve for } y.$

Step 3: Back-substitute the value of the variable found in Step 2 into one of the *original* equations to find the value of the remaining variable.

Back-substitute $y = -1$ into equation (1) and solve for x.

$2x + 3y = 1 \quad \text{Equation (1)}$

$2x + 3(-1) = 1 \quad \text{Substitute } y = -1.$

$2x - 3 = 1 \quad \text{Simplify.}$

$2x = 4 \quad \text{Add 3 to both sides.}$

$x = 2 \quad \text{Solve for } x.$

Step 4: Check the solution found.

The check is left to you.

The solution of the system is $x = 2$ and $y = -1$. The solution also can be written as the ordered pair $(2, -1)$.

Now Use Elimination to Work PROBLEM 21

EXAMPLE 6

Movie Theater Ticket Sales

A movie theater sells tickets for $10.00 each, with seniors receiving a discount of $2.00. One evening the theater sold 525 tickets and had revenue of $4630. How many of each type of ticket were sold?

Solution

If x represents the number of tickets sold at $10.00 and y the number of tickets sold at the discounted price of $8.00, then the given information results in the system of equations

$$\begin{cases} 10x + 8y = 4630 & (1) \\ x + y = 525 & (2) \end{cases}$$

Using the method of elimination, first multiply equation (2) by -8, and then add the equations.

$$\begin{cases} 10x + 8y = 4630 \\ \underline{-8x - 8y = -4200} \quad \text{Multiply equation (2) by } -8. \\ 2x = 430 \quad \text{Add the equations.} \\ x = 215 \end{cases}$$

Since $x + y = 525$, then $y = 525 - x = 525 - 215 = 310$. So 215 nondiscounted tickets and 310 senior discount tickets were sold.

3 Identify Inconsistent Systems of Equations Containing Two Variables

The previous examples dealt with consistent systems of equations that had a single solution. The next two examples deal with two other possibilities that may occur, the first being a system that has no solution.

EXAMPLE 7

Identifying an Inconsistent System of Linear Equations

Solve: $\begin{cases} 2x + y = 5 & (1) \\ 4x + 2y = 8 & (2) \end{cases}$

Solution

We choose to use the method of substitution and solve equation (1) for y.

$$2x + y = 5 \qquad (1)$$
$$y = -2x + 5 \quad \text{Subtract } 2x \text{ from both sides.}$$

Now substitute $-2x + 5$ for y in equation (2) and solve for x.

$$4x + 2y = 8 \qquad (2)$$
$$4x + 2(-2x + 5) = 8 \quad \text{Substitute } y = -2x + 5.$$
$$4x - 4x + 10 = 8 \quad \text{Multiply out.}$$
$$10 = 8 \quad \text{Simplify.}$$

This statement is false. We conclude that the system has no solution and is inconsistent.

$y = -2x + 4$

$y = -2x + 5$

Figure 3

Figure 3 illustrates the pair of lines whose equations form the system in Example 7. Notice that the graphs of the two equations are lines, each with slope -2; one has a y-intercept of 5, the other a y-intercept of 4. The lines are parallel and have no point of intersection. This geometric statement is equivalent to the algebraic statement that the system has no solution.

4 Express the Solution of a System of Dependent Equations Containing Two Variables

EXAMPLE 8

Solving a System of Dependent Equations

Solve: $\begin{cases} 2x + y = 4 & (1) \\ -6x - 3y = -12 & (2) \end{cases}$

Solution We choose to use the method of elimination.

$$\begin{cases} 2x + y = 4 & (1) \\ -6x - 3y = -12 & (2) \end{cases}$$

$$\begin{cases} 6x + 3y = 12 & (1) \quad \text{Multiply equation (1) by 3.} \\ -6x - 3y = -12 & (2) \end{cases}$$
$$\overline{0 = 0 \quad \text{Add equations (1) and (2).}}$$

The statement $0 = 0$ means the original system is equivalent to a system containing one equation, $2x + y = 4$. So the equations of the system are dependent. Furthermore, any values of x and y that satisfy $2x + y = 4$ are solutions.

For example, $x = 2$, $y = 0$; $x = 0$, $y = 4$; $x = -1$, $y = 6$; $x = 4$, $y = -4$ are solutions. There are, in fact, infinitely many values of x and y for which $2x + y = 4$, so the original system has infinitely many solutions. We write the solution of the original system either as

$$y = -2x + 4, \text{ where } x \text{ can be any real number}$$

or as

$$x = -\frac{1}{2}y + 2, \text{ where } y \text{ can be any real number}$$

The solution can also be expressed using set notation:

$$\{(x, y) \mid y = -2x + 4, x \text{ any real number}\} \text{ or}$$
$$\left\{(x, y) \mid x = -\frac{1}{2}y + 2, y \text{ any real number}\right\}.$$

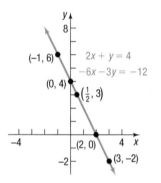

(−1, 6) $2x + y = 4$
     $-6x - 3y = -12$
(0, 4)
    $(\frac{1}{2}, 3)$
(2, 0)   (3, −2)

Figure 4 $y = -2x + 4$

Figure 4 illustrates the system given in Example 8. Notice that the graphs of the two equations are lines, each with slope -2 and each with y-intercept 4. The lines are coincident, and the solutions of the system, such as $(-1, 6)$, $(0, 4)$, and $(2, 0)$, are points on the line. Notice also that equation (2) in the original system is -3 times equation (1), indicating that the two equations are dependent.

—**Now Work** PROBLEMS 27 AND 31

5 Solve Systems of Three Equations Containing Three Variables

Just like a system of two linear equations containing two variables, a system of three linear equations containing three variables has

- Exactly one solution (a consistent system with independent equations)
- No solution (an inconsistent system)
- Infinitely many solutions (a consistent system with dependent equations)

The problem of solving a system of three linear equations containing three variables can be viewed as a geometry problem. The graph of each equation in such a system is a plane in space. A system of three linear equations containing three variables represents three planes in space. Figure 5 illustrates some of the possibilities.

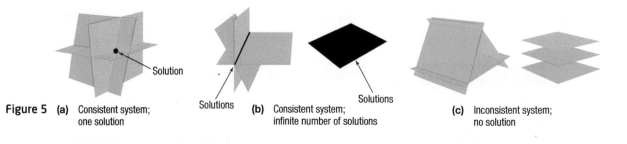

Figure 5 **(a)** Consistent system; one solution   **(b)** Consistent system; infinite number of solutions   **(c)** Inconsistent system; no solution

Recall that a **solution** to a system of equations consists of values for the variables that are solutions of each equation of the system.

For example, $x = 3$, $y = -1$, $z = -5$ is a solution to the system of equations

$$\begin{cases} x + y + z = -3 & (1) \\ 2x - 3y + 6z = -21 & (2) \\ -3x + 5y = -14 & (3) \end{cases} \quad \begin{aligned} & 3 + (-1) + (-5) = -3 \\ & 2 \cdot 3 - 3(-1) + 6(-5) = 6 + 3 - 30 = -21 \\ & -3 \cdot 3 + 5(-1) = -9 - 5 = -14 \end{aligned}$$

because these values of the variables are solutions of each equation.

Typically, when solving a system of three linear equations containing three variables, we use the method of elimination. Recall that the idea behind the method of elimination is to form equivalent equations so that adding two of the equations eliminates a variable.

EXAMPLE 9   **Solving a System of Three Linear Equations with Three Variables**

Use the method of elimination to solve the system of equations.

$$\begin{cases} x + y - z = -1 & (1) \\ 4x - 3y + 2z = 16 & (2) \\ 2x - 2y - 3z = 5 & (3) \end{cases}$$

Solution   For a system of three equations, attempt to eliminate one variable at a time, using pairs of equations, until an equation with a single variable remains. Our strategy for solving this system will be to use equation (1) to eliminate the variable x from equations (2) and (3). We can then treat the new equations (2) and (3) as a system with two variables. Alternatively, we could use equation (1) to eliminate either y or z from equations (2) and (3). Try one of these approaches for yourself.

We begin by multiplying both sides of equation (1) by -4 and adding the result to equation (2). (Do you see why? The coefficients of x are now opposites of one another.) We also multiply equation (1) by -2 and add the result to equation (3). Notice that these two procedures result in the elimination of the variable x from equations (2) and (3).

$$\begin{cases} x + y - z = -1 & (1) \\ 4x - 3y + 2z = 16 & (2) \end{cases} \quad \text{Multiply by } -4.$$

$$\begin{cases} -4x - 4y + 4z = 4 & (1) \\ \underline{4x - 3y + 2z = 16} & (2) \\ -7y + 6z = 20 & \text{Add.} \end{cases}$$

$$\begin{cases} x + y - z = -1 & (1) \\ 2x - 2y - 3z = 5 & (3) \end{cases} \quad \text{Multiply by } -2.$$

$$\begin{cases} -2x - 2y + 2z = 2 & (1) \\ \underline{2x - 2y - 3z = 5} & (3) \\ -4y - z = 7 & \text{Add.} \end{cases}$$

$$\begin{cases} x + y - z = -1 & (1) \\ -7y + 6z = 20 & (2) \\ -4y - z = 7 & (3) \end{cases}$$

Now concentrate on the new equations (2) and (3), treating them as a system of two equations containing two variables. It is easier to eliminate z. Multiply equation (3) by 6, and add equations (2) and (3).

$$\begin{aligned} -7y + 6z &= 20 & (2) \\ -4y - z &= 7 & (3) \end{aligned} \quad \text{Multiply by 6.}$$

$$\begin{aligned} -7y + 6z &= 20 & (2) \\ \underline{-24y - 6z = 42} & (3) \\ -31y &= 62 & \text{Add.} \end{aligned}$$

$$\begin{cases} x + y - z = -1 & (1) \\ -7y + 6z = 20 & (2) \\ -31y = 62 & (3) \end{cases}$$

Now solve the new equation (3) for y by dividing both sides of the equation by -31.

$$\begin{cases} x + y - z = -1 & (1) \\ -7y + 6z = 20 & (2) \\ y = -2 & (3) \end{cases}$$

Back-substitute $y = -2$ in equation (2) and solve for z.

$$\begin{aligned} -7y + 6z &= 20 & (2) \\ -7(-2) + 6z &= 20 & \text{Substitute } y = -2. \\ 6z &= 6 & \text{Subtract 14 from both sides of the equation.} \\ z &= 1 & \text{Solve for } z. \end{aligned}$$

(continued)

Finally, back-substitute $y = -2$ and $z = 1$ in equation (1) and solve for x.

$$
\begin{array}{rll}
x + y - z = -1 & \text{(1)} \\
x + (-2) - 1 = -1 & \text{Substitute } y = -2 \text{ and } z = 1. \\
x - 3 = -1 & \text{Simplify.} \\
x = 2 & \text{Solve for } x.
\end{array}
$$

The solution of the original system is $x = 2, y = -2, z = 1$ or, using an ordered triplet, $(2, -2, 1)$. You should check this solution. ⏌

Look back over the solution to Example 9. Note the pattern of eliminating one of the variables from two of the equations, followed by solving the resulting system of two equations and two variables. Although the variables to eliminate is your choice, the method is the same for all systems.

━━━ **Now Work** PROBLEM 45

6 Identify Inconsistent Systems of Equations Containing Three Variables

EXAMPLE 10 **Identifying an Inconsistent System of Linear Equations**

$$
\text{Solve:} \quad
\begin{cases}
2x + y - z = -2 & \text{(1)} \\
x + 2y - z = -9 & \text{(2)} \\
x - 4y + z = 1 & \text{(3)}
\end{cases}
$$

Solution Our strategy is the same as in Example 9. However, in this system, it seems easiest to eliminate the variable z first. Do you see why?

Multiply equation (1) by -1, and add the result to equation (2). Also, add equations (2) and (3).

$$
\begin{array}{ll}
2x + y - z = -2 & \text{(1)} \quad \text{Multiply by } -1. \\
x + 2y - z = -9 & \text{(2)}
\end{array}
$$

$$
\begin{array}{ll}
-2x - y + z = 2 & \text{(1)} \\
\underline{x + 2y - z = -9} & \text{(2)} \\
-x + y = -7 & \text{Add.}
\end{array}
$$

$$
\begin{array}{ll}
x + 2y - z = -9 & \text{(2)} \\
\underline{x - 4y + z = 1} & \text{(3)} \\
2x - 2y = -8 & \text{Add.}
\end{array}
$$

$$
\begin{cases}
2x + y - z = -2 & \text{(1)} \\
-x + y = -7 & \text{(2)} \\
2x - 2y = -8 & \text{(3)}
\end{cases}
$$

Now concentrate on the new equations (2) and (3), treating them as a system of two equations containing two variables. Multiply equation (2) by 2, and add the result to equation (3).

$$
\begin{array}{ll}
-x + y = -7 & \text{(2)} \quad \text{Multiply by 2.} \\
2x - 2y = -8 & \text{(3)}
\end{array}
$$

$$
\begin{array}{ll}
-2x + 2y = -14 & \text{(2)} \\
\underline{2x - 2y = -8} & \text{(3)} \\
0 = -22 & \text{Add.}
\end{array}
$$

$$
\begin{cases}
2x + y - z = -2 & \text{(1)} \\
-x + y = -7 & \text{(2)} \\
0 = -22 & \text{(3)}
\end{cases}
$$

Equation (3) has no solution, so the system is inconsistent. ⏌

7 Express the Solution of a System of Dependent Equations Containing Three Variables

EXAMPLE 11 **Solving a System of Dependent Equations**

$$
\text{Solve:} \quad
\begin{cases}
x - 2y - z = 8 & \text{(1)} \\
2x - 3y + z = 23 & \text{(2)} \\
4x - 5y + 5z = 53 & \text{(3)}
\end{cases}
$$

Solution   Our plan is to eliminate x from equations (2) and (3). Multiply equation (1) by -2, and add the result to equation (2). Also, multiply equation (1) by -4, and add the result to equation (3).

$$\begin{array}{rl} x - 2y - z = 8 & (1) \\ 2x - 3y + z = 23 & (2) \end{array} \quad \text{Multiply by } -2.$$

$$\begin{array}{rl} -2x + 4y + 2z = -16 & (1) \\ \underline{2x - 3y + z = 23} & (2) \\ y + 3z = 7 & \text{Add.} \end{array}$$

$$\begin{array}{rl} x - 2y - z = 8 & (1) \\ 4x - 5y + 5z = 53 & (3) \end{array} \quad \text{Multiply by } -4.$$

$$\begin{array}{rl} -4x + 8y + 4z = -32 & (1) \\ \underline{4x - 5y + 5z = 53} & (3) \\ 3y + 9z = 21 & \text{Add.} \end{array}$$

$$\begin{cases} x - 2y - z = 8 & (1) \\ y + 3z = 7 & (2) \\ 3y + 9z = 21 & (3) \end{cases}$$

Treat equations (2) and (3) as a system of two equations containing two variables, and eliminate the variable y by multiplying equation (2) by -3 and adding the result to equation (3).

$$\begin{array}{rl} y + 3z = 7 & (2) \\ 3y + 9z = 21 & (3) \end{array} \quad \text{Multiply by } -3.$$

$$\begin{array}{rl} -3y - 9z = -21 \\ \underline{3y + 9z = 21} \quad \text{Add.} \\ 0 = 0 \end{array}$$

$$\begin{cases} x - 2y - z = 8 & (1) \\ y + 3z = 7 & (2) \\ 0 = 0 & (3) \end{cases}$$

The original system is equivalent to a system containing two equations, so the equations are dependent and the system has infinitely many solutions. If we solve equation (2) for y, we can express y in terms of z as $y = -3z + 7$. Substitute this expression into equation (1) to determine x in terms of z.

$$\begin{array}{ll} x - 2y - z = 8 & \text{Equation (1)} \\ x - 2(-3z + 7) - z = 8 & \text{Substitute } y = -3z + 7. \\ x + 6z - 14 - z = 8 & \text{Multiply out.} \\ x + 5z = 22 & \text{Combine like terms.} \\ x = -5z + 22 & \text{Solve for } x. \end{array}$$

We will write the solution to the system as

$$\begin{cases} x = -5z + 22 \\ y = -3z + 7 \end{cases} \quad \text{where } z \text{ can be any real number.}$$

This way of writing the solution makes it easier to find specific solutions. To find specific solutions, choose any value of z and use the equations $x = -5z + 22$ and $y = -3z + 7$ to determine x and y. For example, if $z = 0$, then $x = 22$ and $y = 7$, and if $z = 1$, then $x = 17$ and $y = 4$.

Using ordered triplets, the solution is

$$\{(x, y, z) \mid x = -5z + 22, y = -3z + 7, z \text{ any real number}\}$$

━━━━▶**Now Work** PROBLEM 47

Two distinct points in the Cartesian plane determine a unique line. Given three noncollinear points, we can find the unique quadratic function whose graph contains these three points.

EXAMPLE 12   **Curve Fitting**

Find real numbers a, b, and c so that the graph of the quadratic function $y = ax^2 + bx + c$ contains the points $(-1, -4)$, $(1, 6)$, and $(3, 0)$.

Solution   The three points must satisfy the equation $y = ax^2 + bx + c$.

For the point $(-1, -4)$ we have: $-4 = a(-1)^2 + b(-1) + c$ $-4 = a - b + c$
For the point $(1, 6)$ we have: $6 = a \cdot 1^2 + b \cdot 1 + c$ $6 = a + b + c$
For the point $(3, 0)$ we have: $0 = a \cdot 3^2 + b \cdot 3 + c$ $0 = 9a + 3b + c$

(continued)

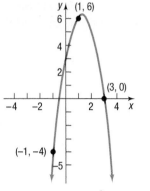

Figure 6 $y = -2x^2 + 5x + 3$

Determine a, b, and c so that each equation is satisfied. That is, solve the system of three equations containing three variables:

$$\begin{cases} a - b + c = -4 & (1) \\ a + b + c = 6 & (2) \\ 9a + 3b + c = 0 & (3) \end{cases}$$

Solving this system of equations, we obtain $a = -2$, $b = 5$, and $c = 3$. So the quadratic function whose graph contains the points $(-1, -4)$, $(1, 6)$, and $(3, 0)$ is

$$y = -2x^2 + 5x + 3 \quad y = ax^2 + bx + c, \quad a = -2, b = 5, c = 3$$

Figure 6 shows the graph of the function, along with the three points.

━━━━━ **Now Work** PROBLEM 73

8.1 Assess Your Understanding

'Are You Prepared?' *Answers are given at the end of these exercises. If you get a wrong answer, read the pages listed in red.*

1. Solve the equation: $3x + 4 = 8 - x$. (pp. 84–85)

2. **(a)** Graph the line: $3x + 4y = 12$.
   (b) What is the slope of a line parallel to this line? (pp. 169–179)

Concepts and Vocabulary

3. *True or False* If a system of equations has no solution, it is said to be dependent.

4. If a system of equations has one solution, the system is _____ and the equations are _____.

5. If the only solution to a system of two linear equations containing two variables is $x = 3$, $y = -2$, then the graphs of the lines in the system intersect at the point _____.

6. If the lines that make up a system of two linear equations are coincident, then the system is _____ and the equations are _____.

7. *Multiple Choice* If a system of two linear equations in two variables is inconsistent, then the graphs of the lines in the system are _____.
   (a) intersecting   **(b)** parallel
   (c) coincident   **(d)** perpendicular

8. *Multiple Choice* If a system of dependent equations containing three variables has the general solution $\{(x, y, z) \mid x = -z + 4, y = -2z + 5, z \text{ is any real number}\}$ then _____ is one of the infinite number of solutions of the system.
   (a) $(1, -1, 3)$ **(b)** $(0, 4, 5)$ **(c)** $(4, -3, 0)$ **(d)** $(-1, 5, 7)$

Skill Building

In Problems 9–18, verify that the values of the variables listed are solutions of the system of equations.

9. $\begin{cases} 2x - y = 5 \\ 5x + 2y = 8 \end{cases}$
$x = 2, y = -1; (2, -1)$

10. $\begin{cases} 3x + 2y = 2 \\ x - 7y = -30 \end{cases}$
$x = -2, y = 4; (-2, 4)$

11. $\begin{cases} 3x - 4y = 4 \\ \frac{1}{2}x - 3y = -\frac{1}{2} \end{cases}$
$x = 2, y = \frac{1}{2}; \left(2, \frac{1}{2}\right)$

12. $\begin{cases} 2x + \frac{1}{2}y = 0 \\ 3x - 4y = -\frac{19}{2} \end{cases}$
$x = -\frac{1}{2}, y = 2; \left(-\frac{1}{2}, 2\right)$

13. $\begin{cases} x - y = 3 \\ \frac{1}{2}x + y = 3 \end{cases}$
$x = 4, y = 1; (4, 1)$

14. $\begin{cases} x - y = 3 \\ -3x + y = 1 \end{cases}$
$x = -2, y = -5; (-2, -5)$

15. $\begin{cases} 3x + 3y + 2z = 4 \\ x - y - z = 0 \\ 2y - 3z = -8 \end{cases}$
$x = 1, y = -1, z = 2;$
$(1, -1, 2)$

16. $\begin{cases} 4x - z = 7 \\ 8x + 5y - z = 0 \\ -x - y + 5z = 6 \end{cases}$
$x = 2, y = -3, z = 1;$
$(2, -3, 1)$

17. $\begin{cases} 3x + 3y + 2z = 4 \\ x - 3y + z = 10 \\ 5x - 2y - 3z = 8 \end{cases}$
$x = 2, y = -2, z = 2; (2, -2, 2)$

18. $\begin{cases} 4x - 5z = 6 \\ 5y - z = -17 \\ -x - 6y + 5z = 24 \end{cases}$
$x = 4, y = -3, z = 2; (4, -3, 2)$

In Problems 19–56, solve each system of equations. If the system has no solution, state that it is inconsistent. For Problems 19–30, graph the lines of the system.

19. $\begin{cases} x + y = 8 \\ x - y = 4 \end{cases}$

20. $\begin{cases} x + 2y = -7 \\ x + y = -3 \end{cases}$

21. $\begin{cases} 5x - y = 21 \\ 2x + 3y = -12 \end{cases}$

22. $\begin{cases} x + 3y = 5 \\ 2x - 3y = -8 \end{cases}$

23. $\begin{cases} 3x = 24 \\ x + 2y = 0 \end{cases}$

24. $\begin{cases} 4x + 5y = -3 \\ -2y = -8 \end{cases}$

25. $\begin{cases} 3x - 6y = 2 \\ 5x + 4y = 1 \end{cases}$

26. $\begin{cases} 2x + 4y = \dfrac{2}{3} \\ 3x - 5y = -10 \end{cases}$

27. $\begin{cases} 2x + y = 1 \\ 4x + 2y = 3 \end{cases}$

28. $\begin{cases} x - y = 5 \\ -3x + 3y = 2 \end{cases}$

29. $\begin{cases} 2x - y = 0 \\ 4x + 2y = 12 \end{cases}$

30. $\begin{cases} 3x + 3y = -1 \\ 4x + y = \dfrac{8}{3} \end{cases}$

31. $\begin{cases} x + 2y = 4 \\ 2x + 4y = 8 \end{cases}$

32. $\begin{cases} 3x - y = 7 \\ 9x - 3y = 21 \end{cases}$

33. $\begin{cases} 2x - 3y = -1 \\ 10x + y = 11 \end{cases}$

34. $\begin{cases} 3x - 2y = 0 \\ 5x + 10y = 4 \end{cases}$

35. $\begin{cases} 2x + 3y = 6 \\ x - y = \dfrac{1}{2} \end{cases}$

36. $\begin{cases} \dfrac{1}{2}x + y = -2 \\ x - 2y = 8 \end{cases}$

37. $\begin{cases} \dfrac{1}{2}x + \dfrac{1}{3}y = 3 \\ \dfrac{1}{4}x - \dfrac{2}{3}y = -1 \end{cases}$

38. $\begin{cases} \dfrac{1}{3}x - \dfrac{3}{2}y = -5 \\ \dfrac{3}{4}x + \dfrac{1}{3}y = 11 \end{cases}$

39. $\begin{cases} 3x - 6y = 7 \\ 5x - 2y = 5 \end{cases}$

40. $\begin{cases} 2x - y = -1 \\ x + \dfrac{1}{2}y = \dfrac{3}{2} \end{cases}$

41. $\begin{cases} \dfrac{1}{x} + \dfrac{1}{y} = 8 \\ \dfrac{3}{x} - \dfrac{5}{y} = 0 \end{cases}$

42. $\begin{cases} \dfrac{4}{x} - \dfrac{3}{y} = 0 \\ \dfrac{6}{x} + \dfrac{3}{2y} = 2 \end{cases}$

$\left[\text{\textbf{Hint:} Let } u = \dfrac{1}{x} \text{ and } v = \dfrac{1}{y}, \text{ and solve for } u \text{ and } v. \text{ Then } x = \dfrac{1}{u} \text{ and } y = \dfrac{1}{v}.\right]$

43. $\begin{cases} x + y = 9 \\ 2x - z = 13 \\ 3y + 2z = 7 \end{cases}$

44. $\begin{cases} 2x + y = -4 \\ -2y + 4z = 0 \\ 3x - 2z = -11 \end{cases}$

45. $\begin{cases} x - 2y + 3z = 7 \\ 2x + y + z = 4 \\ -3x + 2y - 2z = -10 \end{cases}$

46. $\begin{cases} 2x + y - 3z = 0 \\ -2x + 2y + z = -7 \\ 3x - 4y - 3z = 7 \end{cases}$

47. $\begin{cases} x - y - z = 1 \\ 2x + 3y + z = 2 \\ 3x + 2y = 0 \end{cases}$

48. $\begin{cases} 2x - 3y - z = 0 \\ -x + 2y + z = 5 \\ 3x - 4y - z = 1 \end{cases}$

49. $\begin{cases} x - y - z = 1 \\ -x + 2y - 3z = -4 \\ 3x - 2y - 7z = 0 \end{cases}$

50. $\begin{cases} 2x - 3y - z = 0 \\ 3x + 2y + 2z = 2 \\ x + 5y + 3z = 2 \end{cases}$

51. $\begin{cases} 2x - 2y + 3z = 6 \\ 4x - 3y + 2z = 0 \\ -2x + 3y - 7z = 1 \end{cases}$

52. $\begin{cases} 3x - 2y + 2z = 6 \\ 7x - 3y + 2z = -1 \\ 2x - 3y + 4z = 0 \end{cases}$

53. $\begin{cases} x + y - z = 6 \\ 3x - 2y + z = -5 \\ x + 3y - 2z = 14 \end{cases}$

54. $\begin{cases} x - y + z = -4 \\ 2x - 3y + 4z = -15 \\ 5x + y - 2z = 12 \end{cases}$

55. $\begin{cases} x + 2y - z = -3 \\ 2x - 4y + z = -7 \\ -2x + 2y - 3z = 4 \end{cases}$

56. $\begin{cases} x + 4y - 3z = -8 \\ 3x - y + 3z = 12 \\ x + y + 6z = 1 \end{cases}$

Applications and Extensions

57. The perimeter of a rectangular floor is 90 feet. Find the dimensions of the floor if the length is twice the width.

58. The length of fence required to enclose a rectangular field is 3000 meters. What are the dimensions of the field if it is known that the difference between its length and width is 50 meters?

59. Orbital Launches In 2017 there was a total of 469 commercial and noncommercial orbital launches worldwide. In addition, the number of noncommercial orbital launches was 31 more than half the number of commercial orbital launches. Determine the number of commercial and noncommercial orbital launches in 2017.
Source: Federal Aviation Administration

60. Movie Theater Tickets A movie theater charges $9.00 for adults and $7.00 for senior citizens. On a day when 325 people paid for admission, the total receipts were $2495. How many who paid were adults? How many were seniors?

61. Mixing Nuts A store sells cashews for $5.00 per pound and peanuts for $1.50 per pound. The manager decides to mix 30 pounds of peanuts with some cashews and sell the mixture for $3.00 per pound. How many pounds of cashews should be mixed with the peanuts so that the mixture will produce the same revenue as selling the nuts separately?

62. Mixing a Solution A chemist wants to make 14 liters of a 40% acid solution. She has solutions that are 30% acid and 65% acid. How much of each must she mix?

63. Presale Order A wireless store owner takes presale orders for a new smartphone and tablet. He gets 340 preorders for the smartphone and 250 preorders for the tablet. The combined value of the preorders is $486,000. If the price of a smartphone and tablet together is $1665, how much does each device cost?

64. Financial Planning A recently retired couple needs $12,000 per year to supplement their Social Security. They have $300,000 to invest to obtain this income. They have decided on two investment options: AA bonds yielding 5% per annum and a Bank Certificate yielding 2.5%.

  (a) How much should be invested in each to realize exactly $12,000?

  (b) If, after 2 years, the couple requires $14,000 per year in income, how should they reallocate their investment to achieve the new amount?

65. Computing Wind Speed With a tail wind, a small Piper aircraft can fly 600 miles in 3 hours. Against this same wind, the Piper can fly the same distance in 4 hours. Find the average wind speed and the average airspeed of the Piper.

3 hours   4 hours

600 mi.

66. Computing Wind Speed The average airspeed of a single-engine aircraft is 150 miles per hour. If the aircraft flew the same distance in 2 hours with the wind as it flew in 3 hours against the wind, what was the wind speed?

67. Restaurant Management A restaurant manager wants to purchase 200 sets of dishes. One design costs $25 per set, and another costs $45 per set. If she has only $7400 to spend, how many sets of each design should she order?

68. Cost of Fast Food One group of people purchased 10 hot dogs and 5 soft drinks at a cost of $35.00. A second bought 7 hot dogs and 4 soft drinks at a cost of $25.25. What is the cost of a single hot dog? A single soft drink?

We paid $35.00.
How much is one hot dog?
How much is one soda?

We paid $25.25.
How much is one hot dog?
How much is one soda?

69. Computing a Refund The grocery store we use does not mark prices on its goods. My wife went to this store, bought three 1-pound packages of bacon and two cartons of eggs, and paid a total of $13.45. Not knowing that she went to the store, I also went to the same store, purchased two 1-pound packages of bacon and three cartons of eggs, and paid a total of $11.45. Now we want to return two 1-pound packages of bacon and two cartons of eggs. How much will be refunded?

70. Finding the Current of a Stream Pamela requires 3 hours to swim 15 miles downstream on the Illinois River. The return trip upstream takes 5 hours. Find Pamela's average speed in still water. How fast is the current? (Assume that Pamela's speed is the same in each direction.)

71. Pharmacy A doctor's prescription calls for a daily intake containing 40 milligrams (mg) of vitamin C and 30 mg of vitamin D. Your pharmacy stocks two liquids that can be used: One contains 20% vitamin C and 30% vitamin D, the other 40% vitamin C and 20% vitamin D. How many milligrams of each compound should be mixed to fill the prescription?

72. Pharmacy A doctor's prescription calls for the creation of pills that contain 12 units of vitamin B_{12} and 12 units of vitamin E. Your pharmacy stocks two powders that can be used to make these pills: One contains 20% vitamin B_{12} and 30% vitamin E, the other 40% vitamin B_{12} and 20% vitamin E. How many units of each powder should be mixed in each pill?

73. Curve Fitting Find real numbers a, b, and c so that the graph of the function $y = ax^2 + bx + c$ contains the points $(-1, 4)$, $(2, 3)$, and $(0, 1)$.

74. Curve Fitting Find real numbers a, b, and c so that the graph of the function $y = ax^2 + bx + c$ contains the points $(-1, -2)$, $(1, -4)$, and $(2, 4)$.

75. IS–LM Model in Economics In economics, the IS curve is a linear equation that represents all combinations of income Y and interest rates r that maintain an equilibrium in the market for goods in the economy. The LM curve is a linear equation that represents all combinations of income Y and interest rates r that maintain an equilibrium in the market for money in the economy. In an economy, suppose that the equilibrium level of income (in millions of dollars) and interest rates satisfy the system of equations

$$\begin{cases} 0.06Y - 5000r = 240 \\ 0.06Y + 6000r = 900 \end{cases}$$

Find the equilibrium level of income and interest rates.

76. IS–LM Model in Economics In economics, the IS curve is a linear equation that represents all combinations of income Y and interest rates r that maintain an equilibrium in the market for goods in the economy. The LM curve is a linear equation that represents all combinations of income Y and interest rates r that maintain an equilibrium in the market for money in the economy. In an economy, suppose that the equilibrium level of income (in millions of dollars) and interest rates satisfy the system of equations

$$\begin{cases} 0.05Y - 1000r = 10 \\ 0.05Y + 800r = 100 \end{cases}$$

Find the equilibrium level of income and interest rates.

77. Electricity: Kirchhoff's Rules An application of Kirchhoff's Rules to the circuit shown results in the following system of equations:

$$\begin{cases} I_2 = I_1 + I_3 \\ 5 - 3I_1 - 5I_2 = 0 \\ 10 - 5I_2 - 7I_3 = 0 \end{cases}$$

Find the currents I_1, I_2, and I_3.

Source: *Physics for Scientists & Engineers*, 9th ed., by Serway. © 2013 Cengage Learning.

78. Electricity: Kirchhoff's Rules An application of Kirchhoff's Rules to the circuit shown below results in the following system of equations:

$$\begin{cases} I_3 = I_1 + I_2 \\ 8 = 4I_3 + 6I_2 \\ 8I_1 = 4 + 6I_2 \end{cases}$$

Find the currents I_1, I_2, and I_3.

Source: *Physics for Scientists & Engineers*, 9th ed., by Serway. © 2013 Cengage Learning.

79. Theater Revenues A Broadway theater has 500 seats, divided into orchestra, main, and balcony seating. Orchestra seats sell for $150, main seats for $135, and balcony seats for $110. If all the seats are sold, the gross revenue to the theater is $64,250. If all the main and balcony seats are sold, but only half the orchestra seats are sold, the gross revenue is $56,750. How many of each kind of seat are there?

80. Theater Revenues A movie theater charges $11.00 for adults, $6.50 for children, and $9.00 for senior citizens. One day the theater sold 405 tickets and collected $3315 in receipts. Twice as many children's tickets were sold as adult tickets. How many adults, children, and senior citizens went to the theater that day?

81. Nutrition A dietitian wishes a patient to have a meal that has 66 grams (g) of protein, 94.5 g of carbohydrates, and 910 milligrams (mg) of calcium. The hospital food service tells the dietitian that the dinner for today is chicken, corn,

and 2% milk. Each serving of chicken has 30 g of protein, 35 g of carbohydrates, and 200 mg of calcium. Each serving of corn has 3 g of protein, 16 g of carbohydrates, and 10 mg of calcium. Each glass of 2% milk has 9 g of protein, 13 g of carbohydrates, and 300 mg of calcium. How many servings of each food should the dietitian provide for the patient?

82. Investments Kelly has $20,000 to invest. As her financial planner, you recommend that she diversify into three investments: Treasury bills that yield 5% simple interest, Treasury bonds that yield 7% simple interest, and corporate bonds that yield 10% simple interest. Kelly wishes to earn $1390 per year in income. Also, Kelly wants her investment in Treasury bills to be $3000 more than her investment in corporate bonds. How much money should Kelly place in each investment?

83. Prices of Fast Food One group of customers bought 8 deluxe hamburgers, 6 orders of large fries, and 6 large colas for $26.10. A second group ordered 10 deluxe hamburgers, 6 large fries, and 8 large colas and paid $31.60. Is there sufficient information to determine the price of each food item? If not, construct a table showing the various possibilities. Assume that the hamburgers cost between $1.75 and $2.25, the fries between $0.75 and $1.00, and the colas between $0.60 and $0.90.

84. Prices of Fast Food Use the information given in Problem 83. Suppose that a third group purchased 3 deluxe hamburgers, 2 large fries, and 4 large colas for $10.95. Now is there sufficient information to determine the price of each food item? If so, determine each price.

85. Painting a House Three painters (Beth, Dan, and Edie), working together, can paint the exterior of a home in 10 hours (h). Dan and Edie together have painted a similar house in 15 h. One day, all three worked on this same kind of house for 4 h, after which Edie left. Beth and Dan required 8 more hours to finish. Assuming no gain or loss in efficiency, how long should it take each person to complete such a job alone?

86. Challenge Problem Solve for x and y, assuming $a \neq 0$ and $b \neq 0$.

$$\begin{cases} ax + by = a + b \\ abx - b^2y = b^2 - ab \end{cases}$$

87. Challenge Problem Solve for x, y, and z, assuming $a \neq 0$, $b \neq 0$, and $c \neq 0$.

$$\begin{cases} ax + by + cz = a + b + c \\ a^2x + b^2y + c^2z = ac + ab + bc \\ abx + bcy = bc + ac \end{cases}$$

Explaining Concepts: Discussion and Writing

88. Make up a system of three linear equations containing three variables that has:
    (a) No solution
    (b) Exactly one solution
    (c) Infinitely many solutions

    Give the three systems to a friend to solve and critique.

89. Write a brief paragraph outlining your strategy for solving a system of two linear equations containing two variables.

90. Do you prefer the method of substitution or the method of elimination for solving a system of two linear equations containing two variables? Give your reasons.

Retain Your Knowledge

Problems 91–100 are based on material learned earlier in the course. The purpose of these problems is to keep the material fresh in your mind so that you are better prepared for the final exam.

91. Graph $f(x) = -3^{1-x} + 2$.

92. Factor each of the following:
    (a) $4(2x - 3)^3 \cdot 2 \cdot (x^3 + 5)^2 + 2(x^3 + 5) \cdot 3x^2 \cdot (2x - 3)^4$
    (b) $\frac{1}{2}(3x - 5)^{-\frac{1}{2}} \cdot 3 \cdot (x + 3)^{-\frac{1}{2}} - \frac{1}{2}(x + 3)^{-\frac{3}{2}}(3x - 5)^{\frac{1}{2}}$

93. Solve: $|2x - 3| + 7 < 12$

94. The exponential function $f(x) = 5^{x-2} + 3$ is one-to-one. Find $f^{-1}(x)$.

95. If $A = \{2, 4, 6, \dots, 30\}$ and $B = \{3, 6, 9, \dots, 30\}$, find $A \cap B$.

96. Find an equation of an ellipse if the center is at the origin, the length of the major axis is 20 along the x-axis, and the length of the minor axis is 12.

97. Determine whether $f(x) = -3x^2 - 24x + 10$ has a maximum or minimum value, and then find the value.

98. Find the principal needed now to get $5000 after 18 months at 4% interest compounded monthly.

99. Find the average rate of change of $f(x) = 2x^3 + 5x - 7$ from $x = 0$ to $x = 4$.

100. Find the domain of $f(x) = \sqrt{-x^2 + 2x + 8}$.

'Are You Prepared?' Answers

1. $\{1\}$

2. (a)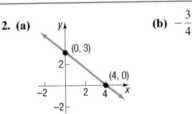

   (b) $-\dfrac{3}{4}$

8.2 Systems of Linear Equations: Matrices

OBJECTIVES 1 Write the Augmented Matrix of a System of Linear Equations (p. 571)
2 Write the System of Equations from the Augmented Matrix (p. 572)
3 Perform Row Operations on a Matrix (p. 572)
4 Solve a System of Linear Equations Using Matrices (p. 573)

The systematic approach of the method of elimination for solving a system of linear equations provides another method of solution that involves a simplified notation. Consider the following system of linear equations:

$$\begin{cases} x + 4y = 14 \\ 3x - 2y = 0 \end{cases}$$

If we choose not to write the variables, we can represent this system as

$$\begin{bmatrix} 1 & 4 & | & 14 \\ 3 & -2 & | & 0 \end{bmatrix}$$

where it is understood that the first column represents the coefficients of the variable x, the second column the coefficients of y, and the third column the constants on the right side of the equal signs. The vertical bar serves as a reminder of the equal signs. The large square brackets are used to denote a *matrix* in algebra.

> **DEFINITION Matrix**
>
> A **matrix** is defined as a rectangular array of numbers:
>
> $$\begin{array}{c} & \text{Column 1} & \text{Column 2} & & \text{Column } j & & \text{Column } n \\ \text{Row 1} \\ \text{Row 2} \\ \\ \text{Row } i \\ \\ \text{Row } m \end{array} \begin{bmatrix} a_{11} & a_{12} & \cdots & a_{1j} & \cdots & a_{1n} \\ a_{21} & a_{22} & \cdots & a_{2j} & \cdots & a_{2n} \\ \vdots & \vdots & & \vdots & & \vdots \\ a_{i1} & a_{i2} & \cdots & a_{ij} & \cdots & a_{in} \\ \vdots & \vdots & & \vdots & & \vdots \\ a_{m1} & a_{m2} & \cdots & a_{mj} & \cdots & a_{mn} \end{bmatrix} \quad \textbf{(1)}$$

Each number a_{ij} in the matrix has two indexes: the **row index** i and the **column index** j. The matrix shown in display (1) has m rows and n columns. The numbers a_{ij} are usually referred to as the **entries** of the matrix. For example, a_{23} refers to the entry in the second row, third column.

1 Write the Augmented Matrix of a System of Linear Equations

In Words

To augment means to increase or expand. An augmented matrix broadens the idea of matrices to systems of linear equations.

Now we will use matrix notation to represent a system of linear equations. The matrix used to represent a system of linear equations is called an **augmented matrix**. In writing the augmented matrix of a system, the variables of each equation must be on the left side of the equal sign and the constants on the right side. A variable that does not appear in an equation has a coefficient of 0.

EXAMPLE 1 Writing the Augmented Matrix of a System of Linear Equations

Write the augmented matrix of each system of equations.

(a) $\begin{cases} 3x - 4y = -6 & (1) \\ 2x - 3y = -5 & (2) \end{cases}$ (b) $\begin{cases} 2x - y + z = 0 & (1) \\ x + z - 1 = 0 & (2) \\ x + 2y - 8 = 0 & (3) \end{cases}$

Solution

(a) The augmented matrix is

$$\left[\begin{array}{cc|c} 3 & -4 & -6 \\ 2 & -3 & -5 \end{array} \right]$$

(b) Care must be taken that the system be written so that the coefficients of all variables are present (if any variable is missing, its coefficient is 0). Also, all constants must be to the right of the equal sign. We need to rearrange the given system to put it into the required form.

$$\begin{cases} 2x - y + z = 0 & (1) \\ x + z - 1 = 0 & (2) \\ x + 2y - 8 = 0 & (3) \end{cases}$$

$$\begin{cases} 2x - y + z = 0 & (1) \\ x + 0 \cdot y + z = 1 & (2) \\ x + 2y + 0 \cdot z = 8 & (3) \end{cases}$$

CAUTION Be sure variables and constants are lined up correctly before writing the augmented matrix. ∎

The augmented matrix is

$$\left[\begin{array}{ccc|c} 2 & -1 & 1 & 0 \\ 1 & 0 & 1 & 1 \\ 1 & 2 & 0 & 8 \end{array} \right]$$

If we do not include the constants to the right of the equal sign (that is, to the right of the vertical bar in the augmented matrix of a system of equations), the resulting matrix is called the **coefficient matrix** of the system. For the systems discussed in Example 1, the coefficient matrices are

$$\begin{bmatrix} 3 & -4 \\ 2 & -3 \end{bmatrix} \text{ and } \begin{bmatrix} 2 & -1 & 1 \\ 1 & 0 & 1 \\ 1 & 2 & 0 \end{bmatrix}$$

Now Work PROBLEM 9

2 Write the System of Equations from the Augmented Matrix

EXAMPLE 2

Writing the System of Linear Equations from the Augmented Matrix

Write the system of linear equations that corresponds to each augmented matrix.

(a) $\begin{bmatrix} 5 & 2 & | & 13 \\ -3 & 1 & | & -10 \end{bmatrix}$ (b) $\begin{bmatrix} 3 & -1 & -1 & | & 7 \\ 2 & 0 & 2 & | & 8 \\ 0 & 1 & 1 & | & 0 \end{bmatrix}$

Solution

(a) The augmented matrix has two rows and so represents a system of two equations. The two columns to the left of the vertical bar indicate that the system has two variables. If x and y are used to denote these variables, the system of equations is

$$\begin{cases} 5x + 2y = 13 & (1) \\ -3x + y = -10 & (2) \end{cases}$$

(b) Since the augmented matrix has three rows, it represents a system of three equations. Since there are three columns to the left of the vertical bar, the system contains three variables. If $x, y,$ and z are the three variables, the system of equations is

$$\begin{cases} 3x - y - z = 7 & (1) \\ 2x + 2z = 8 & (2) \\ y + z = 0 & (3) \end{cases}$$

3 Perform Row Operations on a Matrix

Row operations on a matrix are used to solve systems of equations when the system is written as an augmented matrix. There are three basic row operations.

Row Operations

- Interchange any two rows.
- Replace a row by a nonzero multiple of that row.
- Replace a row by the sum of that row and a constant nonzero multiple of some other row.

These three row operations correspond to the three rules given earlier for obtaining an equivalent system of equations. When a row operation is performed on a matrix, the resulting matrix represents a system of equations equivalent to the system represented by the original matrix.

For example, consider the augmented matrix

$$\begin{bmatrix} 1 & 2 & | & 3 \\ 4 & -1 & | & 2 \end{bmatrix}$$

Suppose we want to use a row operation that results in a matrix whose entry in row 2, column 1 is 0. The row operation to use is

Multiply each entry in row 1 by -4, and add the result to the corresponding entries in row 2. **(2)**

If we use R_2 to represent the new entries in row 2 and r_1 and r_2 to represent the original entries in rows 1 and 2, respectively, we can represent the row operation in statement (2) by

$$R_2 = -4r_1 + r_2$$

Then

$$\begin{bmatrix} 1 & 2 & | & 3 \\ 4 & -1 & | & 2 \end{bmatrix} \underset{\underset{R_2 = -4r_1 + r_2}{\uparrow}}{\rightarrow} \begin{bmatrix} 1 & 2 & | & 3 \\ -4 \cdot 1 + 4 & -4 \cdot 2 + (-1) & | & -4 \cdot 3 + 2 \end{bmatrix} = \begin{bmatrix} 1 & 2 & | & 3 \\ 0 & -9 & | & -10 \end{bmatrix}$$

We now have the entry 0 in row 2, column 1.

EXAMPLE 3   **Using a Row Operation on an Augmented Matrix**

Use the row operation $R_2 = -3r_1 + r_2$ on the augmented matrix

$$\begin{bmatrix} 1 & -2 & | & 2 \\ 3 & -5 & | & 9 \end{bmatrix}$$

Solution   The row operation $R_2 = -3r_1 + r_2$ replaces the entries in row 2 by the entries obtained after multiplying each entry in row 1 by -3 and adding the result to the corresponding entries in row 2.

$$\begin{bmatrix} 1 & -2 & | & 2 \\ 3 & -5 & | & 9 \end{bmatrix} \underset{\underset{R_2 = -3r_1 + r_2}{\uparrow}}{\rightarrow} \begin{bmatrix} 1 & -2 & | & 2 \\ -3 \cdot 1 + 3 & -3 \cdot (-2) + (-5) & | & -3 \cdot 2 + 9 \end{bmatrix} = \begin{bmatrix} 1 & -2 & | & 2 \\ 0 & 1 & | & 3 \end{bmatrix}$$

Now Work PROBLEM 19

EXAMPLE 4   **Finding a Row Operation**

Find a row operation that results in the augmented matrix

$$\begin{bmatrix} 1 & -2 & | & 2 \\ 0 & 1 & | & 3 \end{bmatrix}$$

having 0 in row 1, column 2.

Solution   We want 0 in row 1, column 2. Because the entry in row 2, column 2 is 1, multiply row 2 by 2 and add the result to row 1. That is, use the row operation $R_1 = 2r_2 + r_1$.

$$\begin{bmatrix} 1 & -2 & | & 2 \\ 0 & 1 & | & 3 \end{bmatrix} \underset{\underset{R_1 = 2r_2 + r_1}{\uparrow}}{\rightarrow} \begin{bmatrix} 2 \cdot 0 + 1 & 2 \cdot 1 + (-2) & | & 2 \cdot 3 + 2 \\ 0 & 1 & | & 3 \end{bmatrix} = \begin{bmatrix} 1 & 0 & | & 8 \\ 0 & 1 & | & 3 \end{bmatrix}$$

A word about notation: The row operation $R_1 = 2r_2 + r_1$ changes the entries in row 1. We change the entries in row 1 by multiplying the entries in some other row by a nonzero number and adding the results to the original entries of row 1.

4 Solve a System of Linear Equations Using Matrices

To solve a system of linear equations using matrices, use row operations on the augmented matrix of the system to obtain a matrix that is in *row echelon form*.

DEFINITION Row Echelon Form

A matrix is in **row echelon form** when the following conditions are met:

- The entry in row 1, column 1 is a 1, and only 0's appear below it.
- The first nonzero entry in each row after the first row is a 1, only 0's appear below it, and the 1 appears to the right of the first nonzero entry in any row above.
- Any rows that contain all 0's to the left of the vertical bar appear at the bottom.

For example, for a system of three equations containing three variables, x, y, and z, with a unique solution, the augmented matrix is in row echelon form if it is of the form

$$\left[\begin{array}{ccc|c} 1 & a & b & d \\ 0 & 1 & c & e \\ 0 & 0 & 1 & f \end{array}\right]$$

where a, b, c, d, e, and f are real numbers. The last row of this augmented matrix states that $z = f$. We then determine the value of y using back-substitution with $z = f$, since row 2 represents the equation $y + cz = e$. Finally, x is determined using back-substitution again.

Two advantages of solving a system of equations by writing the augmented matrix in row echelon form are the following:

- The process is algorithmic; that is, it consists of repetitive steps that can be programmed on a computer.
- The process works on any system of linear equations, no matter how many equations or variables are present.

The next example shows how to solve a system of linear equations by writing its augmented matrix in row echelon form.

EXAMPLE 5

Solving a System of Linear Equations Using Matrices (Row Echelon Form)

Solve: $\begin{cases} 2x + 2y & = 6 \quad (1) \\ x + y + z = 1 \quad (2) \\ 3x + 4y - z = 13 \quad (3) \end{cases}$

Step-by-Step Solution

Step 1: Write the augmented matrix that represents the system.

Write the augmented matrix of the system.

$$\left[\begin{array}{ccc|c} 2 & 2 & 0 & 6 \\ 1 & 1 & 1 & 1 \\ 3 & 4 & -1 & 13 \end{array}\right]$$

Step 2: Use row operations to obtain 1 in row 1, column 1.

To get 1 in row 1, column 1, interchange rows 1 and 2. [Note that this is equivalent to interchanging equations (1) and (2) of the system.]

$$\left[\begin{array}{ccc|c} 1 & 1 & 1 & 1 \\ 2 & 2 & 0 & 6 \\ 3 & 4 & -1 & 13 \end{array}\right]$$

Step 3: Use row operations that leave row 1 unchanged, but change the entries in column 1 below row 1 to 0's.

Next, we want 0 in row 2, column 1 and 0 in row 3, column 1. Use the row operations $R_2 = -2r_1 + r_2$ and $R_3 = -3r_1 + r_3$. Note that row 1 is unchanged using these row operations.

$$\left[\begin{array}{ccc|c} 1 & 1 & 1 & 1 \\ 2 & 2 & 0 & 6 \\ 3 & 4 & -1 & 13 \end{array}\right] \rightarrow \left[\begin{array}{ccc|c} 1 & 1 & 1 & 1 \\ 0 & 0 & -2 & 4 \\ 0 & 1 & -4 & 10 \end{array}\right]$$

$$\begin{array}{l} R_2 = -2r_1 + r_2 \\ R_3 = -3r_1 + r_3 \end{array}$$

Step 4: Use row operations to obtain 1 in row 2, column 2, and 0's below it.

We want the entry in row 2, column 2 to be 1. We also want to have 0 below the 1 in row 2, column 2. Interchanging rows 2 and 3 will accomplish both goals.

$$\left[\begin{array}{ccc|c} 1 & 1 & 1 & 1 \\ 0 & 0 & -2 & 4 \\ 0 & 1 & -4 & 10 \end{array}\right] \rightarrow \left[\begin{array}{ccc|c} 1 & 1 & 1 & 1 \\ 0 & 1 & -4 & 10 \\ 0 & 0 & -2 & 4 \end{array}\right]$$

Step 5: Repeat Step 4 to obtain 1 in row 3, column 3.

To obtain 1 in row 3, column 3, use the row operation $R_3 = -\dfrac{1}{2}r_3$. The result is

$$
\begin{bmatrix}
1 & 1 & 1 & | & 1 \\
0 & 1 & -4 & | & 10 \\
0 & 0 & -2 & | & 4
\end{bmatrix}
\rightarrow
\begin{bmatrix}
1 & 1 & 1 & | & 1 \\
0 & 1 & -4 & | & 10 \\
0 & 0 & 1 & | & -2
\end{bmatrix}
$$

$$R_3 = -\frac{1}{2}r_3$$

Step 6: The matrix on the right in Step 5 is the row echelon form of the augmented matrix. Use back-substitution to solve the original system.

The third row of the augmented matrix represents the equation $z = -2$. Using $z = -2$, back-substitute into the equation $y - 4z = 10$ (row 2) and obtain

$$y - 4z = 10$$
$$y - 4(-2) = 10 \quad z = -2$$
$$y = 2 \quad \text{Solve for } y.$$

Finally, back-substitute $y = 2$ and $z = -2$ into the equation $x + y + z = 1$ (row 1) and obtain

$$x + y + z = 1$$
$$x + 2 + (-2) = 1 \quad y = 2, z = -2$$
$$x = 1 \quad \text{Solve for } x.$$

The solution of the system is $x = 1, y = 2, z = -2$ or, using an ordered triplet, $(1, 2, -2)$.

Matrix Method for Solving a System of Linear Equations (Row Echelon Form)

STEP 1: Write the augmented matrix that represents the system.

STEP 2: Use row operations to obtain 1 in row 1, column 1.

STEP 3: Use row operations that leave row 1 unchanged, but change the entries in column 1 below row 1 to 0's.

STEP 4: Use row operations to obtain 1 in row 2, column 2, but leave the entries in columns to the left unchanged. If it is impossible to place 1 in row 2, column 2, place 1 in row 2, column 3. Once the 1 is in place, use row operations to obtain 0's below it. (Place any rows that contain only 0's on the left side of the vertical bar, at the bottom of the matrix.)

STEP 5: Now repeat Step 4 to obtain 1 in the next row, but one column to the right. Continue until the bottom row or the vertical bar is reached.

STEP 6: The matrix that results is the row echelon form of the augmented matrix. Analyze the system of equations corresponding to it to solve the original system.

EXAMPLE 6

Solving a System of Linear Equations Using Matrices (Row Echelon Form)

Solve: $\begin{cases} x - y + z = 8 & \text{(1)} \\ 2x + 3y - z = -2 & \text{(2)} \\ 3x - 2y - 9z = 9 & \text{(3)} \end{cases}$

Solution

STEP 1: The augmented matrix of the system is

$$
\begin{bmatrix}
1 & -1 & 1 & | & 8 \\
2 & 3 & -1 & | & -2 \\
3 & -2 & -9 & | & 9
\end{bmatrix}
$$

(continued)

STEP 2: Because 1 is already in row 1, column 1, go to Step 3.

STEP 3: Perform the row operations $R_2 = -2r_1 + r_2$ and $R_3 = -3r_1 + r_3$. Each of these leaves row 1 unchanged, while causing 0's in column 1 of the other rows.

$$\left[\begin{array}{rrr|r} 1 & -1 & 1 & 8 \\ 2 & 3 & -1 & -2 \\ 3 & -2 & -9 & 9 \end{array}\right] \underset{\underset{\displaystyle R_3 = -3r_1 + r_3}{\displaystyle R_2 = -2r_1 + r_2}}{\longrightarrow} \left[\begin{array}{rrr|r} 1 & -1 & 1 & 8 \\ 0 & 5 & -3 & -18 \\ 0 & 1 & -12 & -15 \end{array}\right]$$

STEP 4: The easiest way to obtain the entry 1 in row 2, column 2 without altering column 1 is to interchange rows 2 and 3 (another way would be to multiply row 2 by $\dfrac{1}{5}$, but this introduces fractions).

$$\left[\begin{array}{rrr|r} 1 & -1 & 1 & 8 \\ 0 & 1 & -12 & -15 \\ 0 & 5 & -3 & -18 \end{array}\right]$$

To obtain 0 under the 1 in row 2, column 2, use the row operation $R_3 = -5r_2 + r_3$.

$$\left[\begin{array}{rrr|r} 1 & -1 & 1 & 8 \\ 0 & 1 & -12 & -15 \\ 0 & 5 & -3 & -18 \end{array}\right] \underset{\displaystyle R_3 = -5r_2 + r_3}{\longrightarrow} \left[\begin{array}{rrr|r} 1 & -1 & 1 & 8 \\ 0 & 1 & -12 & -15 \\ 0 & 0 & 57 & 57 \end{array}\right]$$

STEP 5: Continuing, obtain 1 in row 3, column 3 by using $R_3 = \dfrac{1}{57}r_3$.

$$\left[\begin{array}{rrr|r} 1 & -1 & 1 & 8 \\ 0 & 1 & -12 & -15 \\ 0 & 0 & 57 & 57 \end{array}\right] \underset{\displaystyle R_3 = \frac{1}{57}r_3}{\longrightarrow} \left[\begin{array}{rrr|r} 1 & -1 & 1 & 8 \\ 0 & 1 & -12 & -15 \\ 0 & 0 & 1 & 1 \end{array}\right]$$

STEP 6: The matrix on the right is the row echelon form of the augmented matrix. The system of equations represented by the matrix in row echelon form is

$$\begin{cases} x - y + z = 8 & \text{(1)} \\ \quad\; y - 12z = -15 & \text{(2)} \\ \qquad\qquad z = 1 & \text{(3)} \end{cases}$$

Using $z = 1$, back-substitute to get

$$\begin{cases} x - y + 1 = 8 & \text{(1)} \\ \quad\; y - 12 \cdot 1 = -15 & \text{(2)} \end{cases} \underset{\text{Simplify.}}{\longrightarrow} \begin{cases} x - y = 7 & \text{(1)} \\ \quad\; y = -3 & \text{(2)} \end{cases}$$

Using $y = -3$, back-substitute into $x - y = 7$ to get $x = 4$. The solution of the system is $x = 4$, $y = -3$, $z = 1$ or, using an ordered triplet, $(4, -3, 1)$. ⌡

Sometimes it is advantageous to write a matrix in **reduced row echelon form**. In this form, row operations are used to obtain entries that are 0 above (as well as below) the leading 1 in a row. For example, the row echelon form obtained in the solution to Example 6 is

$$\left[\begin{array}{rrr|r} 1 & -1 & 1 & 8 \\ 0 & 1 & -12 & -15 \\ 0 & 0 & 1 & 1 \end{array}\right]$$

To write this matrix in reduced row echelon form, proceed as follows:

$$\begin{bmatrix} 1 & -1 & 1 & | & 8 \\ 0 & 1 & -12 & | & -15 \\ 0 & 0 & 1 & | & 1 \end{bmatrix} \rightarrow \begin{bmatrix} 1 & 0 & -11 & | & -7 \\ 0 & 1 & -12 & | & -15 \\ 0 & 0 & 1 & | & 1 \end{bmatrix} \rightarrow \begin{bmatrix} 1 & 0 & 0 & | & 4 \\ 0 & 1 & 0 & | & -3 \\ 0 & 0 & 1 & | & 1 \end{bmatrix}$$

$$\underset{R_1 = r_2 + r_1}{\uparrow} \qquad \underset{\substack{R_1 = 11r_3 + r_1 \\ R_2 = 12r_3 + r_2}}{\uparrow}$$

The matrix is now written in reduced row echelon form. The advantage of writing the matrix in this form is that the solution to the system, $x = 4$, $y = -3$, $z = 1$, can be read from the matrix without the need to back-substitute. Another advantage will be seen in Section 8.4, where the inverse of a matrix is discussed. The method used to write a matrix in reduced row echelon form is called **Gauss-Jordan elimination**.

-**Now Work** PROBLEMS 39 AND 49

The matrix method for solving a system of linear equations also identifies systems that have infinitely many solutions and systems that are inconsistent.

EXAMPLE 7 **Solving a Dependent System of Linear Equations Using Matrices**

Solve: $\begin{cases} 6x - y - z = 4 & (1) \\ -12x + 2y + 2z = -8 & (2) \\ 5x + y - z = 3 & (3) \end{cases}$

Solution Start with the augmented matrix of the system and use row operations to obtain 1 in row 1, column 1 with 0's below.

$$\begin{bmatrix} 6 & -1 & -1 & | & 4 \\ -12 & 2 & 2 & | & -8 \\ 5 & 1 & -1 & | & 3 \end{bmatrix} \rightarrow \begin{bmatrix} 1 & -2 & 0 & | & 1 \\ -12 & 2 & 2 & | & -8 \\ 5 & 1 & -1 & | & 3 \end{bmatrix} \rightarrow \begin{bmatrix} 1 & -2 & 0 & | & 1 \\ 0 & -22 & 2 & | & 4 \\ 0 & 11 & -1 & | & -2 \end{bmatrix}$$

$$\underset{R_1 = -1r_3 + r_1}{\uparrow} \qquad \underset{\substack{R_2 = 12r_1 + r_2 \\ R_3 = -5r_1 + r_3}}{\uparrow}$$

We can obtain 1 in row 2, column 2 without altering column 1 by using either $R_2 = -\dfrac{1}{22}r_2$ or $R_2 = \dfrac{23}{11}r_3 + r_2$. We use the first of these here.

$$\begin{bmatrix} 1 & -2 & 0 & | & 1 \\ 0 & -22 & 2 & | & 4 \\ 0 & 11 & -1 & | & -2 \end{bmatrix} \rightarrow \begin{bmatrix} 1 & -2 & 0 & | & 1 \\ 0 & 1 & -\dfrac{1}{11} & | & -\dfrac{2}{11} \\ 0 & 11 & -1 & | & -2 \end{bmatrix} \rightarrow \begin{bmatrix} 1 & -2 & 0 & | & 1 \\ 0 & 1 & -\dfrac{1}{11} & | & -\dfrac{2}{11} \\ 0 & 0 & 0 & | & 0 \end{bmatrix}$$

$$\underset{R_2 = -\frac{1}{22}r_2}{\uparrow} \qquad \underset{R_3 = -11r_2 + r_3}{\uparrow}$$

This matrix is in row echelon form. Because the bottom row consists entirely of 0's, the system actually consists of only two equations.

$$\begin{cases} x - 2y = 1 & (1) \\ y - \dfrac{1}{11}z = -\dfrac{2}{11} & (2) \end{cases}$$

To make it easier to write some of the solutions, we express both x and y in terms of z.

From the second equation, $y = \dfrac{1}{11}z - \dfrac{2}{11}$. Now back-substitute this solution for y into the first equation to get

$$x = 2y + 1 = 2\left(\dfrac{1}{11}z - \dfrac{2}{11}\right) + 1 = \dfrac{2}{11}z + \dfrac{7}{11}$$

(continued)

The original system is equivalent to the system

$$\begin{cases} x = \dfrac{2}{11}z + \dfrac{7}{11} \quad \text{(1)} \\ y = \dfrac{1}{11}z - \dfrac{2}{11} \quad \text{(2)} \end{cases} \quad \text{where } z \text{ can be any real number}$$

Let's look at the situation. The original system of three equations is equivalent to a system containing two equations. This means that any values of x, y, z that satisfy both

$$x = \dfrac{2}{11}z + \dfrac{7}{11} \quad \text{and} \quad y = \dfrac{1}{11}z - \dfrac{2}{11}$$

are solutions. For example, $z = 0, x = \dfrac{7}{11}, y = -\dfrac{2}{11}; z = 1, x = \dfrac{9}{11}, y = -\dfrac{1}{11};$ and $z = -1, x = \dfrac{5}{11}, y = -\dfrac{3}{11}$ are three of the solutions of the original system. There are, in fact, infinitely many values of $x, y,$ and z for which the two equations are satisfied. That is, the original system has infinitely many solutions. We write the solution of the original system as

$$\begin{cases} x = \dfrac{2}{11}z + \dfrac{7}{11} \\ y = \dfrac{1}{11}z - \dfrac{2}{11} \end{cases} \quad \text{where } z \text{ can be any real number}$$

or, using ordered triplets, as

$$\left\{ (x, y, z) \,\middle|\, x = \dfrac{2}{11}z + \dfrac{7}{11}, y = \dfrac{1}{11}z - \dfrac{2}{11}, z \text{ any real number} \right\}$$

Now Work PROBLEM 55

EXAMPLE 8

Identifying an Inconsistent System of Linear Equations Using Matrices

Solve: $\begin{cases} x + y + z = 6 \\ 2x - y - z = 3 \\ x + 2y + 2z = 0 \end{cases}$

Solution Begin with the augmented matrix, and use row operations to write the matrix in row echelon form.

$$\begin{bmatrix} 1 & 1 & 1 & | & 6 \\ 2 & -1 & -1 & | & 3 \\ 1 & 2 & 2 & | & 0 \end{bmatrix} \rightarrow \begin{bmatrix} 1 & 1 & 1 & | & 6 \\ 0 & -3 & -3 & | & -9 \\ 0 & 1 & 1 & | & -6 \end{bmatrix} \rightarrow \begin{bmatrix} 1 & 1 & 1 & | & 6 \\ 0 & 1 & 1 & | & -6 \\ 0 & -3 & -3 & | & -9 \end{bmatrix} \rightarrow \begin{bmatrix} 1 & 1 & 1 & | & 6 \\ 0 & 1 & 1 & | & -6 \\ 0 & 0 & 0 & | & -27 \end{bmatrix}$$

$$\begin{array}{c} \uparrow \\ R_2 = -2r_1 + r_2 \\ R_3 = -1r_1 + r_3 \end{array} \qquad \begin{array}{c} \uparrow \\ \text{Interchange rows 2 and 3.} \end{array} \qquad \begin{array}{c} \uparrow \\ R_3 = 3r_2 + r_3 \end{array}$$

The bottom row is equivalent to the equation

$$0x + 0y + 0z = -27$$

which has no solution. The original system is inconsistent.

Now Work PROBLEM 29

The matrix method is especially effective for systems of equations for which the number of equations and the number of variables are unequal. Here, too, such a system is either inconsistent or consistent. If it is consistent, it will have either exactly one solution or infinitely many solutions.

EXAMPLE 9

Solving a System of Linear Equations Using Matrices

Solve:
$$\begin{cases} x - 2y + z = 0 & (1) \\ 2x + 2y - 3z = -3 & (2) \\ y - z = -1 & (3) \\ -x + 4y + 2z = 13 & (4) \end{cases}$$

Solution

This is a system of four equations containing three variables. Begin with the augmented matrix, and use row operations to write the matrix in row echelon form.

$$\begin{bmatrix} 1 & -2 & 1 & | & 0 \\ 2 & 2 & -3 & | & -3 \\ 0 & 1 & -1 & | & -1 \\ -1 & 4 & 2 & | & 13 \end{bmatrix} \rightarrow \begin{bmatrix} 1 & -2 & 1 & | & 0 \\ 0 & 6 & -5 & | & -3 \\ 0 & 1 & -1 & | & -1 \\ 0 & 2 & 3 & | & 13 \end{bmatrix} \rightarrow \begin{bmatrix} 1 & -2 & 1 & | & 0 \\ 0 & 1 & -1 & | & -1 \\ 0 & 6 & -5 & | & -3 \\ 0 & 2 & 3 & | & 13 \end{bmatrix}$$

$$\begin{array}{c} \uparrow \\ R_2 = -2r_1 + r_2 \\ R_4 = r_1 + r_4 \end{array} \qquad\qquad \begin{array}{c} \uparrow \\ \text{Interchange rows 2 and 3.} \end{array}$$

$$\rightarrow \begin{bmatrix} 1 & -2 & 1 & | & 0 \\ 0 & 1 & -1 & | & -1 \\ 0 & 0 & 1 & | & 3 \\ 0 & 0 & 5 & | & 15 \end{bmatrix} \rightarrow \begin{bmatrix} 1 & -2 & 1 & | & 0 \\ 0 & 1 & -1 & | & -1 \\ 0 & 0 & 1 & | & 3 \\ 0 & 0 & 0 & | & 0 \end{bmatrix}$$

$$\begin{array}{c} \uparrow \\ R_3 = -6r_2 + r_3 \\ R_4 = -2r_2 + r_4 \end{array} \qquad\qquad \begin{array}{c} \uparrow \\ R_4 = -5r_3 + r_4 \end{array}$$

Since the matrix is in row echelon form, we could now back-substitute $z = 3$ to find x and y. Or we can continue and obtain the reduced row echelon form.

$$\begin{bmatrix} 1 & -2 & 1 & | & 0 \\ 0 & 1 & -1 & | & -1 \\ 0 & 0 & 1 & | & 3 \\ 0 & 0 & 0 & | & 0 \end{bmatrix} \rightarrow \begin{bmatrix} 1 & 0 & -1 & | & -2 \\ 0 & 1 & -1 & | & -1 \\ 0 & 0 & 1 & | & 3 \\ 0 & 0 & 0 & | & 0 \end{bmatrix} \rightarrow \begin{bmatrix} 1 & 0 & 0 & | & 1 \\ 0 & 1 & 0 & | & 2 \\ 0 & 0 & 1 & | & 3 \\ 0 & 0 & 0 & | & 0 \end{bmatrix}$$

$$\begin{array}{c} \uparrow \\ R_1 = 2r_2 + r_1 \end{array} \qquad\qquad \begin{array}{c} \uparrow \\ R_1 = r_3 + r_1 \\ R_2 = r_3 + r_2 \end{array}$$

The matrix is now in reduced row echelon form, and we can see that the solution is $x = 1$, $y = 2$, $z = 3$ or, using an ordered triplet, $(1, 2, 3)$.

Now Work PROBLEM 71

EXAMPLE 10

Financial Planning

Adam and Michelle require an additional $25,000 in annual income (beyond their pension benefits). They are rather risk averse and have narrowed their investment choices down to Treasury notes that yield 3%, Treasury bonds that yield 5%, and corporate bonds that yield 6%. They have $600,000 to invest and want the amount invested in Treasury notes to equal the total amount invested in Treasury bonds and corporate bonds. How much should they place in each investment?

Solution

Let $n, b,$ and c represent the amounts invested in Treasury notes, Treasury bonds, and corporate bonds, respectively. There is a total of $600,000 to invest, which means that the sum of the amounts invested in Treasury notes, Treasury bonds, and corporate bonds should equal $600,000. The first equation is

$$n + b + c = 600{,}000 \quad (1)$$

(continued)

If $100,000 is invested in Treasury notes, the income is $0.03 \cdot \$100,000 = \$3000$. In general, if n dollars are invested in Treasury notes, the income is $0.03n$. Since the total income is to be $25,000, the second equation is

$$0.03n + 0.05b + 0.06c = 25,000 \quad (2)$$

The amount invested in Treasury notes must equal the sum of the amounts invested in Treasury bonds and corporate bonds, so the third equation is

$$n = b + c \quad \text{or} \quad n - b - c = 0 \quad (3)$$

We have the following system of equations:

$$\begin{cases} n + b + c = 600,000 & (1) \\ 0.03n + 0.05b + 0.06c = 25,000 & (2) \\ n - b - c = 0 & (3) \end{cases}$$

Begin with the augmented matrix, and use row operations to write the matrix in row echelon form.

$$\begin{bmatrix} 1 & 1 & 1 & | & 600,000 \\ 0.03 & 0.05 & 0.06 & | & 25,000 \\ 1 & -1 & -1 & | & 0 \end{bmatrix} \rightarrow \begin{bmatrix} 1 & 1 & 1 & | & 600,000 \\ 0 & 0.02 & 0.03 & | & 7000 \\ 0 & -2 & -2 & | & -600,000 \end{bmatrix}$$

$$R_2 = -0.03r_1 + r_2$$
$$R_3 = -r_1 + r_3$$

$$\rightarrow \begin{bmatrix} 1 & 1 & 1 & | & 600,000 \\ 0 & 1 & 1.5 & | & 350,000 \\ 0 & -2 & -2 & | & -600,000 \end{bmatrix} \rightarrow \begin{bmatrix} 1 & 1 & 1 & | & 600,000 \\ 0 & 1 & 1.5 & | & 350,000 \\ 0 & 0 & 1 & | & 100,000 \end{bmatrix}$$

$$R_2 = \frac{1}{0.02}r_2$$
$$R_3 = 2r_2 + r_3$$

The matrix is now in row echelon form. The final matrix represents the system

$$\begin{cases} n + b + c = 600,000 & (1) \\ b + 1.5c = 350,000 & (2) \\ c = 100,000 & (3) \end{cases}$$

COMMENT Most graphing utilities have the capability to put an augmented matrix into row echelon form (ref) and also reduced row echelon form (rref). See the Appendix, Section 7, for a discussion. ∎

From equation (3), we determine that Adam and Michelle should invest $100,000 in corporate bonds. Back-substitute $100,000 into equation (2) to find that $b = 200,000$, so Adam and Michelle should invest $200,000 in Treasury bonds. Back-substitute these values into equation (1) and find that $n = \$300,000$, so $300,000 should be invested in Treasury notes.

8.2 Assess Your Understanding

Concepts and Vocabulary

1. An m by n rectangular array of numbers is called a(n) _____ .

2. The matrix used to represent a system of linear equations is called a(n) _____ matrix.

3. The notation a_{35} refers to the entry in the _____ row and _____ column of a matrix.

4. *Multiple Choice* Which matrix is in reduced row echelon form?

(a) $\begin{bmatrix} 1 & 2 & | & 9 \\ 3 & -1 & | & -1 \end{bmatrix}$

(b) $\begin{bmatrix} 1 & 0 & | & 1 \\ 0 & 1 & | & 4 \end{bmatrix}$

(c) $\begin{bmatrix} 1 & 2 & | & 9 \\ 0 & 0 & | & 28 \end{bmatrix}$

(d) $\begin{bmatrix} 1 & 2 & | & 9 \\ 0 & 1 & | & 4 \end{bmatrix}$

5. *True or False* The matrix $\begin{bmatrix} 1 & 3 & | & -2 \\ 0 & 1 & | & 5 \\ 0 & 0 & | & 0 \end{bmatrix}$ is in row echelon form.

6. *Multiple Choice* Which statement describes the system of equations represented by $\begin{bmatrix} 1 & 5 & -2 & | & 3 \\ 0 & 1 & 3 & | & -2 \\ 0 & 0 & 0 & | & 5 \end{bmatrix}$?

(a) The system has one solution.
(b) The system has infinitely many solutions.
(c) The system has no solution.
(d) The number of solutions cannot be determined.

Skill Building

In Problems 7–18, write the augmented matrix of the given system of equations.

7. $\begin{cases} x - 5y = 5 \\ 4x + 3y = 6 \end{cases}$

8. $\begin{cases} 3x + 4y = 7 \\ 4x - 2y = 5 \end{cases}$

9. $\begin{cases} 2x + 3y - 6 = 0 \\ 4x - 6y + 2 = 0 \end{cases}$

10. $\begin{cases} 9x - y = 0 \\ 3x - y - 4 = 0 \end{cases}$

11. $\begin{cases} 0.01x - 0.03y = 0.06 \\ 0.13x + 0.10y = 0.20 \end{cases}$

12. $\begin{cases} \dfrac{4}{3}x - \dfrac{3}{2}y = \dfrac{3}{4} \\ -\dfrac{1}{4}x + \dfrac{1}{3}y = \dfrac{2}{3} \end{cases}$

13. $\begin{cases} x - y + z = 10 \\ 3x + 3y = 5 \\ x + y + 2z = 2 \end{cases}$

14. $\begin{cases} 5x - y - z = 0 \\ x + y = 5 \\ 2x - 3z = 2 \end{cases}$

15. $\begin{cases} x + y - z = 2 \\ 3x - 2y = 2 \\ 5x + 3y - z = 1 \end{cases}$

16. $\begin{cases} 2x + 3y - 4z = 0 \\ x - 5z + 2 = 0 \\ x + 2y - 3z = -2 \end{cases}$

17. $\begin{cases} x - y - z = 10 \\ 2x + y + 2z = -1 \\ -3x + 4y = 5 \\ 4x - 5y + z = 0 \end{cases}$

18. $\begin{cases} x - y + 2z - w = 5 \\ x + 3y - 4z + 2w = 2 \\ 3x - y - 5z - w = -1 \end{cases}$

In Problems 19–26, write the system of equations corresponding to each augmented matrix. Then perform the indicated row operation(s) on the given augmented matrix.

19. $\left[\begin{array}{cc|c} 1 & -3 & -2 \\ 2 & -5 & 5 \end{array}\right]$ $R_2 = -2r_1 + r_2$

20. $\left[\begin{array}{cc|c} 1 & -3 & -3 \\ 2 & -5 & -4 \end{array}\right]$ $R_2 = -2r_1 + r_2$

21. $\left[\begin{array}{ccc|c} 1 & -3 & 4 & 3 \\ 3 & -5 & 6 & 6 \\ -5 & 3 & 4 & 6 \end{array}\right]$ $\begin{array}{l} R_2 = -3r_1 + r_2 \\ R_3 = 5r_1 + r_3 \end{array}$

22. $\left[\begin{array}{ccc|c} 1 & -3 & 3 & -5 \\ -4 & -5 & -3 & -5 \\ -3 & -2 & 4 & 6 \end{array}\right]$ $\begin{array}{l} R_2 = 4r_1 + r_2 \\ R_3 = 3r_1 + r_3 \end{array}$

23. $\left[\begin{array}{ccc|c} 1 & -3 & 2 & -6 \\ 2 & -5 & 3 & -4 \\ -3 & -6 & 4 & 6 \end{array}\right]$ $\begin{array}{l} R_2 = -2r_1 + r_2 \\ R_3 = 3r_1 + r_3 \end{array}$

24. $\left[\begin{array}{ccc|c} 1 & -3 & -4 & -6 \\ 6 & -5 & 6 & -6 \\ -1 & 1 & 4 & 6 \end{array}\right]$ $\begin{array}{l} R_2 = -6r_1 + r_2 \\ R_3 = r_1 + r_3 \end{array}$

25. $\left[\begin{array}{ccc|c} 5 & -3 & 1 & -2 \\ 2 & -5 & 6 & -2 \\ -4 & 1 & 4 & 6 \end{array}\right]$ $\begin{array}{l} R_1 = -2r_2 + r_1 \\ R_3 = 2r_2 + r_3 \end{array}$

26. $\left[\begin{array}{ccc|c} 4 & -3 & -1 & 2 \\ 3 & -5 & 2 & 6 \\ -3 & -6 & 4 & 6 \end{array}\right]$ $\begin{array}{l} R_1 = -r_2 + r_1 \\ R_3 = r_2 + r_3 \end{array}$

In Problems 27–38, the reduced row echelon form of a system of linear equations is given. Write the system of equations corresponding to the given matrix. Use x, y; or x, y, z; or x_1, x_2, x_3, x_4 as variables. Determine whether the system is consistent or inconsistent. If it is consistent, give the solution.

27. $\left[\begin{array}{cc|c} 1 & 0 & 5 \\ 0 & 1 & -1 \end{array}\right]$

28. $\left[\begin{array}{cc|c} 1 & 0 & -4 \\ 0 & 1 & 0 \end{array}\right]$

29. $\left[\begin{array}{ccc|c} 1 & 0 & 0 & 1 \\ 0 & 1 & 0 & 2 \\ 0 & 0 & 0 & 3 \end{array}\right]$

30. $\left[\begin{array}{ccc|c} 1 & 0 & 0 & 0 \\ 0 & 1 & 0 & 0 \\ 0 & 0 & 0 & 2 \end{array}\right]$

31. $\left[\begin{array}{ccc|c} 1 & 0 & 2 & -1 \\ 0 & 1 & -4 & -2 \\ 0 & 0 & 0 & 0 \end{array}\right]$

32. $\left[\begin{array}{ccc|c} 1 & 0 & 4 & 4 \\ 0 & 1 & 3 & 2 \\ 0 & 0 & 0 & 0 \end{array}\right]$

33. $\left[\begin{array}{cccc|c} 1 & 0 & 0 & 0 & 1 \\ 0 & 1 & 0 & 1 & 2 \\ 0 & 0 & 1 & 2 & 3 \end{array}\right]$

34. $\left[\begin{array}{cccc|c} 1 & 0 & 0 & 0 & 1 \\ 0 & 1 & 0 & 2 & 2 \\ 0 & 0 & 1 & 3 & 0 \end{array}\right]$

35. $\left[\begin{array}{cccc|c} 1 & 0 & 0 & 4 & 2 \\ 0 & 1 & 1 & 3 & 3 \\ 0 & 0 & 0 & 0 & 0 \end{array}\right]$

36. $\left[\begin{array}{cccc|c} 1 & 0 & 0 & 0 & 1 \\ 0 & 1 & 0 & 0 & 2 \\ 0 & 0 & 1 & 2 & 3 \end{array}\right]$

37. $\left[\begin{array}{cccc|c} 1 & 0 & 0 & 1 & -2 \\ 0 & 1 & 0 & 2 & 2 \\ 0 & 0 & 1 & -1 & 0 \\ 0 & 0 & 0 & 0 & 0 \end{array}\right]$

38. $\left[\begin{array}{cccc|c} 1 & 0 & 0 & 0 & 1 \\ 0 & 1 & 0 & 0 & 2 \\ 0 & 0 & 1 & 0 & 3 \\ 0 & 0 & 0 & 1 & 0 \end{array}\right]$

In Problems 39–74, solve each system of equations using matrices (row operations). If the system has no solution, say that it is inconsistent.

39. $\begin{cases} x + y = 8 \\ x - y = 4 \end{cases}$

40. $\begin{cases} x + 2y = 5 \\ x + y = 3 \end{cases}$

41. $\begin{cases} 3x - 6y = -4 \\ 5x + 4y = 5 \end{cases}$

42. $\begin{cases} 3x + 3y = 3 \\ 4x + 2y = \dfrac{8}{3} \end{cases}$

43. $\begin{cases} x + 2y = 4 \\ 2x + 4y = 8 \end{cases}$

44. $\begin{cases} 3x - y = 7 \\ 9x - 3y = 21 \end{cases}$

45. $\begin{cases} 2x + 3y = 6 \\ x - y = \dfrac{1}{2} \end{cases}$

46. $\begin{cases} \dfrac{1}{2}x + y = -2 \\ x - 2y = 8 \end{cases}$

47. $\begin{cases} 3x - 5y = 3 \\ 15x + 5y = 21 \end{cases}$

48. $\begin{cases} 2x - y = -1 \\ x + \dfrac{1}{2}y = \dfrac{3}{2} \end{cases}$

49. $\begin{cases} x - y = 6 \\ 2x - 3z = 16 \\ 2y + z = 4 \end{cases}$

50. $\begin{cases} 2x + y = -4 \\ -2y + 4z = 0 \\ 3x - 2z = -11 \end{cases}$

51. $\begin{cases} x - 4y + 2z = -9 \\ 3x + y + z = 4 \\ -2x + 3y - 3z = 7 \end{cases}$

52. $\begin{cases} 2x + y - 3z = 0 \\ -2x + 2y + z = -7 \\ 3x - 4y - 3z = 7 \end{cases}$

53. $\begin{cases} 2x - 2y - 2z = 2 \\ 2x + 3y + z = 2 \\ 3x + 2y = 0 \end{cases}$

54. $\begin{cases} 2x - 3y - z = 0 \\ -x + 2y + z = 5 \\ 3x - 4y - z = 1 \end{cases}$

55. $\begin{cases} -x + y + z = -1 \\ -x + 2y - 3z = -4 \\ 3x - 2y - 7z = 0 \end{cases}$

56. $\begin{cases} 2x - 3y - z = 0 \\ 3x + 2y + 2z = 2 \\ x + 5y + 3z = 2 \end{cases}$

57. $\begin{cases} 2x - 2y + 3z = 6 \\ 4x - 3y + 2z = 0 \\ -2x + 3y - 7z = 1 \end{cases}$

58. $\begin{cases} 3x - 2y + 2z = 6 \\ 7x - 3y + 2z = -1 \\ 2x - 3y + 4z = 0 \end{cases}$

59. $\begin{cases} x + y - z = 6 \\ 3x - 2y + z = -5 \\ x + 3y - 2z = 14 \end{cases}$

60. $\begin{cases} x - y + z = -4 \\ 2x - 3y + 4z = -15 \\ 5x + y - 2z = 12 \end{cases}$

61. $\begin{cases} x + 2y - z = -3 \\ 2x - 4y + z = -7 \\ -2x + 2y - 3z = 4 \end{cases}$

62. $\begin{cases} x + 4y - 3z = -8 \\ 3x - y + 3z = 12 \\ x + y + 6z = 1 \end{cases}$

63. $\begin{cases} 3x + y - z = \dfrac{2}{3} \\ 2x - y + z = 1 \\ 4x + 2y = \dfrac{8}{3} \end{cases}$

64. $\begin{cases} x + y = 1 \\ 2x - y + z = 1 \\ x + 2y + z = \dfrac{8}{3} \end{cases}$

65. $\begin{cases} x + y + z + w = 4 \\ 2x - y + z = 0 \\ 3x + 2y + z - w = 6 \\ x - 2y - 2z + 2w = -1 \end{cases}$

66. $\begin{cases} x + y + z + w = 4 \\ -x + 2y + z = 0 \\ 2x + 3y + z - w = 6 \\ -2x + y - 2z + 2w = -1 \end{cases}$

67. $\begin{cases} x + 2y + z = 1 \\ 2x - y + 2z = 2 \\ 3x + y + 3z = 3 \end{cases}$

68. $\begin{cases} x + 2y - z = 3 \\ 2x - y + 2z = 6 \\ x - 3y + 3z = 4 \end{cases}$

69. $\begin{cases} x - y + z = 5 \\ 3x + 2y - 2z = 0 \end{cases}$

70. $\begin{cases} 2x + y - z = 4 \\ -x + y + 3z = 1 \end{cases}$

71. $\begin{cases} 2x + 3y - z = 3 \\ x - y - z = 0 \\ -x + y + z = 0 \\ x + y + 3z = 5 \end{cases}$

72. $\begin{cases} x - 3y + z = 1 \\ 2x - y - 4z = 0 \\ x - 3y + 2z = 1 \\ x - 2y = 5 \end{cases}$

73. $\begin{cases} 4x + y + z - w = 4 \\ x - y + 2z + 3w = 3 \end{cases}$

74. $\begin{cases} -4x + y = 5 \\ 2x - y + z - w = 5 \\ z + w = 4 \end{cases}$

Applications and Extensions

75. Curve Fitting Find the function $y = ax^2 + bx + c$ whose graph contains the points $(1, 2)$, $(-2, -7)$, and $(2, -3)$.

76. Curve Fitting Find the function $y = ax^2 + bx + c$ whose graph contains the points $(1, -1)$, $(3, -1)$, and $(-2, 14)$.

77. Curve Fitting Find the function $f(x) = ax^3 + bx^2 + cx + d$ for which $f(-3) = -112, f(-1) = -2, f(1) = 4$, and $f(2) = 13$.

78. Curve Fitting Find the function $f(x) = ax^3 + bx^2 + cx + d$ for which $f(-2) = -10, f(-1) = 3, f(1) = 5$, and $f(3) = 15$.

79. Nutrition A dietitian at Palos Community Hospital wants a patient to have a meal that has 78 grams (g) of protein, 59 g of carbohydrates, and 75 milligrams (mg) of vitamin A. The hospital food service tells the dietitian that the dinner for today is salmon steak, baked eggs, and acorn squash. Each serving of salmon steak has 30 g of protein, 20 g of carbohydrates, and 2 mg of vitamin A. Each serving of baked eggs contains 15 g of protein, 2 g of carbohydrates, and 20 mg of vitamin A. Each serving of acorn squash contains 3 g of protein, 25 g of carbohydrates, and 32 mg of vitamin A. How many servings of each food should the dietitian provide for the patient?

80. Nutrition A dietitian at General Hospital wants a patient to have a meal that has 47 grams (g) of protein, 58 g of carbohydrates, and 630 milligrams (mg) of calcium. The hospital food service tells the dietitian that the dinner for today is pork chops, corn on the cob, and 2% milk. Each serving of pork chops has 23 g of protein, 0 g of carbohydrates, and 10 mg of calcium. Each serving of corn on the cob contains 3 g of protein, 16 g of carbohydrates, and 10 mg of calcium. Each glass of 2% milk contains 9 g of protein, 13 g of carbohydrates, and 300 mg of calcium. How many servings of each food should the dietitian provide for the patient?

81. Financial Planning Carletta has $10,000 to invest. As her financial consultant, you recommend that she invest in Treasury bills that yield 6%, Treasury bonds that yield 7%, and corporate bonds that yield 8%. Carletta wants to have an annual income of $680, and the amount invested in corporate bonds must be half that invested in Treasury bills. Find the amount in each investment.

82. Landscaping A landscape company is hired to plant trees in three new subdivisions. The company charges the developer for each tree planted, an hourly rate to plant the trees, and a fixed delivery charge. In one subdivision it took 166 labor hours to plant 250 trees for a cost of $7520. In a second subdivision it took 124 labor hours to plant 200 trees for a cost of $5945. In the final subdivision it took 200 labor hours to plant 300 trees for a cost of $8985. Determine the cost for each tree, the hourly labor charge, and the fixed delivery charge.

Source: www.bx.org

83. Production To manufacture an automobile requires painting, drying, and polishing. Epsilon Motor Company produces three types of cars: the Delta, the Beta, and the Sigma. Each Delta requires 10 hours (h) for painting, 3 h for drying, and 2 h for polishing. A Beta requires 16 h for painting, 5 h for drying, and 3 h for polishing, and a Sigma requires 8 h for painting, 2 h for drying, and 1 h for polishing. If the company has 240 h for painting, 69 h for drying, and 41 h for polishing per month, how many of each type of car are produced?

84. Production A Florida juice company completes the preparation of its products by sterilizing, filling, and labeling bottles. Each case of orange juice requires 9 minutes (min) for sterilizing, 6 min for filling, and 1 min for labeling. Each case of grapefruit juice requires 10 min for sterilizing, 4 min for filling, and 2 min for labeling. Each case of tomato juice requires 12 min for sterilizing, 4 min for filling, and 1 min for labeling. If the company runs the sterilizing machine for 398 min, the filling machine for 164 min, and the labeling machine for 58 min, how many cases of each type of juice are prepared?

85. Electricity: Kirchhoff's Rules An application of Kirchhoff's Rules to the circuit shown results in the following system of equations:

$$\begin{cases} I_1 + I_2 + I_4 = I_3 \\ -200I_1 - 40 + 80I_2 = 0 \\ 360 - 20I_3 - 80I_2 + 40 = 0 \\ -80 + 70I_4 - 200I_1 = 0 \end{cases}$$

Find the currents I_1, I_2, I_3, and I_4.

Source: Based on Raymond A. Serway and John W. Jewett, Jr. *Physics for Scientists and Engineers with Modern Physics*, *9th ed.* (Boston: Brooks/Cole, Cengage Learning, 2014) Prob. 30, pp. 860–861.

86. Electricity: Kirchhoff's Rules An application of Kirchhoff's Rules to the circuit shown results in the following system of equations:

$$\begin{cases} I_1 = I_3 + I_2 \\ 24 - 6I_1 - 3I_3 = 0 \\ 12 + 24 - 6I_1 - 6I_2 = 0 \end{cases}$$

Find the currents I_1, I_2, and I_3.

Source: Ibid., Prob. 36, p. 861.

87. Financial Planning Three retired couples each require an additional annual income of $2000 per year. As their financial consultant, you recommend that they invest some money in Treasury bills that yield 7%, some money in corporate bonds that yield 9%, and some money in "junk bonds" that yield 11%. Prepare a table for each couple showing the various ways that their goals can be achieved:
  (a) If the first couple has $20,000 to invest.
  (b) If the second couple has $25,000 to invest.
  (c) If the third couple has $30,000 to invest.
  (d) What advice would you give each couple regarding the amount to invest and the choices available?

[**Hint**: Higher yields generally carry more risk.]

88. Financial Planning A young couple has $25,000 to invest. As their financial consultant, you recommend that they invest some money in Treasury bills that yield 7%, some money in corporate bonds that yield 9%, and some money in junk bonds that yield 11%. Prepare a table showing the various ways that this couple can achieve the following goals:
  (a) $1500 per year in income
  (b) $2000 per year in income
  (c) $2500 per year in income
  (d) What advice would you give this couple regarding the income that they require and the choices available?

[**Hint**: Higher yields generally carry more risk.]

89. Pharmacy A doctor's prescription calls for a daily intake of a supplement containing 40 milligrams (mg) of vitamin C and 30 mg of vitamin D. Your pharmacy stocks three supplements that can be used: one contains 20% vitamin C and 30% vitamin D; a second, 40% vitamin C and 20% vitamin D; and a third, 30% vitamin C and 50% vitamin D. Create a table showing the possible combinations that could be used to fill the prescription.

90. Pharmacy A doctor's prescription calls for the creation of pills that contain 12 units of vitamin B_{12} and 12 units of vitamin E. Your pharmacy stocks three powders that can be used to make these pills: one contains 20% vitamin B_{12} and 30% vitamin E; a second, 40% vitamin B_{12} and 20% vitamin E; and a third, 30% vitamin B_{12} and 40% vitamin E. Create a table showing the possible combinations of these powders that could be mixed in each pill. Hint: 10 units of the first powder contains $10 \cdot 0.2 = 2$ units of vitamin B_{12}.

Explaining Concepts: Discussion and Writing

91. Write a brief paragraph or two outlining your strategy for solving a system of linear equations using matrices.

92. When solving a system of linear equations using matrices, do you prefer to place the augmented matrix in row echelon form or in reduced row echelon form? Give reasons for your choice.

93. Make up a system of three linear equations containing three variables that have:

(a) No solution

(b) Exactly one solution

(c) Infinitely many solutions

Give the three systems to a friend to solve and critique.

Retain Your Knowledge

Problems 94–101 are based on material learned earlier in the course. The purpose of these problems is to keep the material fresh in your mind so that you are better prepared for the final exam.

94. Solve: $x^2 - 3x < 6 + 2x$

95. Graph: $f(x) = \dfrac{2x^2 - x - 1}{x^2 + 2x + 1}$

96. State the domain of $f(x) = -e^{x+5} - 6$.

97. Simplify: $\left(\dfrac{18x^4y^5}{27x^3y^9}\right)^3$

98. Find an equation of the hyperbola with vertices $(4, 1)$ and $(4, 9)$ and foci $(4, 0)$ and $(4, 10)$.

99. What is the amount that results if $2700 is invested at 3.6% compounded monthly for 3 years?

△ 100. Find the average rate of change of $f(x) = \sqrt{5 - x}$ from $x = -4$ to $x = 1$.

△ 101. Find the difference quotient for $f(x) = -\dfrac{1}{x^2}$. Express the answer as a single fraction.

8.3 Systems of Linear Equations: Determinants

OBJECTIVES 1 Evaluate 2 by 2 Determinants (p. 584)

2 Use Cramer's Rule to Solve a System of Two Equations Containing Two Variables (p. 585)

3 Evaluate 3 by 3 Determinants (p. 587)

4 Use Cramer's Rule to Solve a System of Three Equations Containing Three Variables (p. 589)

5 Know Properties of Determinants (p. 591)

The previous section described a method of using matrices to solve a system of linear equations. This section describes yet another method for solving systems of linear equations; however, it can be used only when the number of equations equals the number of variables. This method, called *Cramer's Rule*, is based on the concept of a *determinant*. Although the method works for all systems where the number of equations equals the number of variables, it is most often used for systems of two equations containing two variables or three equations containing three variables.

1 Evaluate 2 by 2 Determinants

DEFINITION **2 by 2 Determinant**

If a, b, c, and d are four real numbers, the symbol

$$D = \begin{vmatrix} a & b \\ c & d \end{vmatrix}$$

is called a **2 by 2 determinant**. Its value is the number $ad - bc$; that is,

$$D = \begin{vmatrix} a & b \\ c & d \end{vmatrix} = ad - bc \tag{1}$$

The following illustration may be helpful for remembering the value of a 2 by 2 determinant:

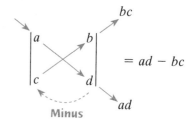

| EXAMPLE 1 | Evaluating a 2 by 2 Determinant |

Evaluate: $\begin{vmatrix} 3 & -2 \\ 6 & 1 \end{vmatrix}$

Solution

$$\begin{vmatrix} 3 & -2 \\ 6 & 1 \end{vmatrix} = 3 \cdot 1 - 6(-2) = 3 - (-12) = 15$$

- **Now Work** PROBLEM 7

2 Use Cramer's Rule to Solve a System of Two Equations Containing Two Variables

Let's see the role that a 2 by 2 determinant plays in the solution of a system of two equations containing two variables. Consider the system

$$\begin{cases} ax + by = s & \text{(1)} \\ cx + dy = t & \text{(2)} \end{cases} \qquad \textbf{(2)}$$

We use the method of elimination to solve this system.

Provided that $d \neq 0$ and $b \neq 0$, this system is equivalent to the system

$$\begin{cases} adx + bdy = sd & \text{(1)} \quad \text{Multiply by } d. \\ bcx + bdy = tb & \text{(2)} \quad \text{Multiply by } b. \end{cases}$$

Subtract the second equation from the first equation and obtain

$$\begin{cases} (ad - bc)x + 0 \cdot y = sd - tb & \text{(1)} \\ \qquad\quad bcx + bdy = tb & \text{(2)} \end{cases}$$

Now the first equation can be rewritten using determinant notation.

$$\begin{vmatrix} a & b \\ c & d \end{vmatrix} x = \begin{vmatrix} s & b \\ t & d \end{vmatrix}$$

If $D = \begin{vmatrix} a & b \\ c & d \end{vmatrix} = ad - bc \neq 0$, solve for x to get

$$x = \dfrac{\begin{vmatrix} s & b \\ t & d \end{vmatrix}}{\begin{vmatrix} a & b \\ c & d \end{vmatrix}} = \dfrac{\begin{vmatrix} s & b \\ t & d \end{vmatrix}}{D} \qquad \textbf{(3)}$$

Return now to the original system (2). Provided that $a \neq 0$ and $c \neq 0$, the system is equivalent to

$$\begin{cases} acx + bcy = cs & \text{(1)} \quad \text{Multiply by } c. \\ acx + ady = at & \text{(2)} \quad \text{Multiply by } a. \end{cases}$$

Subtract the first equation from the second equation and obtain

$$\begin{cases} acx + \quad bcy \quad = \quad cs \quad (1) \\ 0 \cdot x + (ad - bc)y = at - cs \quad (2) \end{cases}$$

The second equation can now be rewritten using determinant notation.

$$\begin{vmatrix} a & b \\ c & d \end{vmatrix} y = \begin{vmatrix} a & s \\ c & t \end{vmatrix}$$

If $D = \begin{vmatrix} a & b \\ c & d \end{vmatrix} = ad - bc \neq 0$, solve for y to get

$$y = \frac{\begin{vmatrix} a & s \\ c & t \end{vmatrix}}{\begin{vmatrix} a & b \\ c & d \end{vmatrix}} = \frac{\begin{vmatrix} a & s \\ c & t \end{vmatrix}}{D} \tag{4}$$

Equations (3) and (4) lead to the following result, called **Cramer's Rule**.

THEOREM Cramer's Rule for Two Equations Containing Two Variables

The solution to the system of equations

$$\begin{cases} ax + by = s \quad (1) \\ cx + dy = t \quad (2) \end{cases} \tag{5}$$

is given by

$$x = \frac{\begin{vmatrix} s & b \\ t & d \end{vmatrix}}{\begin{vmatrix} a & b \\ c & d \end{vmatrix}} \qquad y = \frac{\begin{vmatrix} a & s \\ c & t \end{vmatrix}}{\begin{vmatrix} a & b \\ c & d \end{vmatrix}} \tag{6}$$

provided that

$$D = \begin{vmatrix} a & b \\ c & d \end{vmatrix} = ad - bc \neq 0$$

In the derivation given for Cramer's Rule, we assumed that none of the numbers $a, b, c,$ and d was 0. In Problem 65 you are asked to complete the proof under the less stringent condition that $D = ad - bc \neq 0$.

Now look carefully at the pattern in Cramer's Rule. The denominator in the solution (6) is the determinant of the coefficients of the variables.

$$\begin{cases} ax + by = s \\ cx + dy = t \end{cases} \qquad D = \begin{vmatrix} a & b \\ c & d \end{vmatrix}$$

In the solution for x, the numerator is the determinant, denoted by D_x, formed by replacing the entries in the first column (the coefficients of x) of D by the constants on the right side of the equal sign.

$$D_x = \begin{vmatrix} s & b \\ t & d \end{vmatrix}$$

In the solution for y, the numerator is the determinant, denoted by D_y, formed by replacing the entries in the second column (the coefficients of y) of D by the constants on the right side of the equal sign.

$$D_y = \begin{vmatrix} a & s \\ c & t \end{vmatrix}$$

Cramer's Rule then states that if $D \neq 0$,

$$x = \frac{D_x}{D} \qquad y = \frac{D_y}{D} \tag{7}$$

EXAMPLE 2

Solving a System of Linear Equations Using Determinants

Use Cramer's Rule, if applicable, to solve the system

$$\begin{cases} 3x - 2y = 4 & (1) \\ 6x + y = 13 & (2) \end{cases}$$

Solution

The determinant D of the coefficients of the variables is

$$D = \begin{vmatrix} 3 & -2 \\ 6 & 1 \end{vmatrix} = 3 \cdot 1 - 6(-2) = 15$$

Because $D \neq 0$, Cramer's Rule (7) can be used.

$$x = \frac{D_x}{D} = \frac{\begin{vmatrix} 4 & -2 \\ 13 & 1 \end{vmatrix}}{15} \qquad\qquad y = \frac{D_y}{D} = \frac{\begin{vmatrix} 3 & 4 \\ 6 & 13 \end{vmatrix}}{15}$$

$$= \frac{4 \cdot 1 - 13 \cdot (-2)}{15} \qquad\qquad = \frac{3 \cdot 13 - 6 \cdot 4}{15}$$

$$= \frac{30}{15} \qquad\qquad\qquad\qquad = \frac{15}{15}$$

$$= 2 \qquad\qquad\qquad\qquad\quad = 1$$

The solution is $x = 2$, $y = 1$, or, using an ordered pair, $(2, 1)$.

If the determinant D of the coefficients of the variables equals 0 (so that Cramer's Rule cannot be used), then the system either is consistent with dependent equations or is inconsistent. To determine whether the system has no solution or infinitely many solutions, solve the system using the methods of Section 8.1 or Section 8.2.

— **Now Work** PROBLEM 15

3 Evaluate 3 by 3 Determinants

To use Cramer's Rule to solve a system of three equations containing three variables, we need to define a 3 by 3 determinant.

A **3 by 3 determinant** is symbolized by

$$\begin{vmatrix} a_{11} & a_{12} & a_{13} \\ a_{21} & a_{22} & a_{23} \\ a_{31} & a_{32} & a_{33} \end{vmatrix} \tag{8}$$

in which $a_{11}, a_{12}, \ldots, a_{33}$ are real numbers.

As with matrices, we use a double subscript to identify an entry by indicating its row and column numbers. For example, the entry a_{23} is in row 2, column 3.

The value of a 3 by 3 determinant may be defined in terms of 2 by 2 determinants by the following formula:

$$\begin{vmatrix} a_{11} & a_{12} & a_{13} \\ a_{21} & a_{22} & a_{23} \\ a_{31} & a_{32} & a_{33} \end{vmatrix} = a_{11} \overset{\text{Minus}}{\begin{vmatrix} a_{22} & a_{23} \\ a_{32} & a_{33} \end{vmatrix}} \overset{\downarrow}{-} a_{12} \begin{vmatrix} a_{21} & a_{23} \\ a_{31} & a_{33} \end{vmatrix} \overset{\text{Plus}}{\underset{\downarrow}{+}} a_{13} \begin{vmatrix} a_{21} & a_{22} \\ a_{31} & a_{32} \end{vmatrix} \tag{9}$$

$$\uparrow \qquad\qquad \uparrow \qquad\qquad \uparrow$$

| 2 by 2 determinant left after removing the row and column containing a_{11} | 2 by 2 determinant left after removing the row and column containing a_{12} | 2 by 2 determinant left after removing the row and column containing a_{13} |

The 2 by 2 determinants in formula (9) are called **minors** of the 3 by 3 determinant. For an n by n determinant, the **minor** M_{ij} of entry a_{ij} is the $(n-1)$ by $(n-1)$ determinant that results from removing the ith row and the jth column.

EXAMPLE 3   **Finding Minors of a 3 by 3 Determinant**

For the determinant $A = \begin{vmatrix} 2 & -1 & 3 \\ -2 & 5 & 1 \\ 0 & 6 & -9 \end{vmatrix}$, find: (a) M_{12} (b) M_{23}

Solution   (a) M_{12} is the determinant that results from removing the first row and the second column from A.

$$A = \begin{vmatrix} 2 & \cancel{-1} & 3 \\ -2 & 5 & 1 \\ 0 & 6 & -9 \end{vmatrix} \qquad M_{12} = \begin{vmatrix} -2 & 1 \\ 0 & -9 \end{vmatrix} = (-2)(-9) - 0 \cdot 1 = 18$$

(b) M_{23} is the determinant that results from removing the second row and the third column from A.

$$A = \begin{vmatrix} 2 & -1 & 3 \\ -2 & 5 & 1 \\ 0 & 6 & -9 \end{vmatrix} \qquad M_{23} = \begin{vmatrix} 2 & -1 \\ 0 & 6 \end{vmatrix} = 2 \cdot 6 - 0(-1) = 12$$

Referring to formula (9), note that each element a_{ij} in the first row of the determinant is multiplied by its minor, but sometimes this term is added and other times subtracted. To determine whether to add or subtract a term, we must consider the *cofactor*.

DEFINITION Cofactor

For an n by n determinant A, the **cofactor** of entry a_{ij}, denoted by A_{ij}, is given by

$$A_{ij} = (-1)^{i+j} M_{ij}$$

where M_{ij} is the minor of entry a_{ij}.

The exponent of $(-1)^{i+j}$ is the sum of the row and column of the entry a_{ij}, so if $i + j$ is even, $(-1)^{i+j} = 1$, and if $i + j$ is odd, $(-1)^{i+j} = -1$.

To find the value of a determinant, multiply each entry in any row or column by its cofactor and sum the results. This process is referred to as **expanding across a row or column**. For example, the value of the 3 by 3 determinant in formula (9) was found by expanding across row 1.

Expanding down column 2 gives

$$\begin{vmatrix} a_{11} & a_{12} & a_{13} \\ a_{21} & a_{22} & a_{23} \\ a_{31} & a_{32} & a_{33} \end{vmatrix} = (-1)^{1+2} a_{12} \begin{vmatrix} a_{21} & a_{23} \\ a_{31} & a_{33} \end{vmatrix} + (-1)^{2+2} a_{22} \begin{vmatrix} a_{11} & a_{13} \\ a_{31} & a_{33} \end{vmatrix} + (-1)^{3+2} a_{32} \begin{vmatrix} a_{11} & a_{13} \\ a_{21} & a_{23} \end{vmatrix}$$

↑ Expand down column 2.

$$= -1 \cdot a_{12} \begin{vmatrix} a_{21} & a_{23} \\ a_{31} & a_{33} \end{vmatrix} + 1 \cdot a_{22} \begin{vmatrix} a_{11} & a_{13} \\ a_{31} & a_{33} \end{vmatrix} + (-1) \cdot a_{32} \begin{vmatrix} a_{11} & a_{13} \\ a_{21} & a_{23} \end{vmatrix}$$

Expanding across row 3 gives

$$\begin{vmatrix} a_{11} & a_{12} & a_{13} \\ a_{21} & a_{22} & a_{23} \\ a_{31} & a_{32} & a_{33} \end{vmatrix} = (-1)^{3+1} a_{31} \begin{vmatrix} a_{12} & a_{13} \\ a_{22} & a_{23} \end{vmatrix} + (-1)^{3+2} a_{32} \begin{vmatrix} a_{11} & a_{13} \\ a_{21} & a_{23} \end{vmatrix} + (-1)^{3+3} a_{33} \begin{vmatrix} a_{11} & a_{12} \\ a_{21} & a_{22} \end{vmatrix}$$

↑ Expand across row 3.

$$= 1 \cdot a_{31} \begin{vmatrix} a_{12} & a_{13} \\ a_{22} & a_{23} \end{vmatrix} + (-1) \cdot a_{32} \begin{vmatrix} a_{11} & a_{13} \\ a_{21} & a_{23} \end{vmatrix} + 1 \cdot a_{33} \begin{vmatrix} a_{11} & a_{12} \\ a_{21} & a_{22} \end{vmatrix}$$

Observe that the signs, $(-1)^{i+j}$, associated with the cofactors alternate between positive and negative. For example, for a 3 by 3 determinant, the signs follow the pattern

$$\begin{bmatrix} 1 & -1 & 1 \\ -1 & 1 & -1 \\ 1 & -1 & 1 \end{bmatrix} \tag{10}$$

It can be shown that the value of a determinant does not depend on the choice of the row or column used in the expansion. So, expanding across a row or column that has an entry equal to 0 reduces the amount of work needed to compute the value of the determinant.

EXAMPLE 4   **Evaluating a 3 by 3 Determinant**

Find the value of the 3 by 3 determinant: $\begin{vmatrix} 3 & 0 & -4 \\ 4 & 6 & 2 \\ 8 & -2 & 3 \end{vmatrix}$

Solution   Because 0 is in row 1, column 2, it is easiest to expand across row 1 or down column 2. We choose to expand across row 1. For the signs of the cofactors, we use "$1, -1, 1$" from row 1 of the 3 by 3 determinant in (10).

$$\begin{vmatrix} 3 & 0 & -4 \\ 4 & 6 & 2 \\ 8 & -2 & 3 \end{vmatrix} = 1 \cdot 3 \cdot \begin{vmatrix} 6 & 2 \\ -2 & 3 \end{vmatrix} + (-1) \cdot 0 \cdot \begin{vmatrix} 4 & 2 \\ 8 & 3 \end{vmatrix} + 1 \cdot (-4) \cdot \begin{vmatrix} 4 & 6 \\ 8 & -2 \end{vmatrix}$$

$$= 3[18 - (-4)] + 0 + (-4)(-8 - 48)$$

$$= 3 \cdot 22 + (-4)(-56)$$

$$= 66 + 224 = 290$$

— **Now Work** PROBLEM 11

---

4 Use Cramer's Rule to Solve a System of Three Equations Containing Three Variables

Consider the following system of three equations containing three variables.

$$\begin{cases} a_{11}x + a_{12}y + a_{13}z = c_1 \\ a_{21}x + a_{22}y + a_{23}z = c_2 \\ a_{31}x + a_{32}y + a_{33}z = c_3 \end{cases} \tag{11}$$

It can be shown that if the determinant D of the coefficients of the variables is not 0, that is, if

$$D = \begin{vmatrix} a_{11} & a_{12} & a_{13} \\ a_{21} & a_{22} & a_{23} \\ a_{31} & a_{32} & a_{33} \end{vmatrix} \neq 0$$

the system in (11) has a unique solution.

> **THEOREM** **Cramer's Rule for Three Equations Containing Three Variables**
>
> If $D \neq 0$, the solution of the system in (11) is
>
> $$x = \frac{D_x}{D} \qquad y = \frac{D_y}{D} \qquad z = \frac{D_z}{D}$$
>
> where
>
> $$D_x = \begin{vmatrix} c_1 & a_{12} & a_{13} \\ c_2 & a_{22} & a_{23} \\ c_3 & a_{32} & a_{33} \end{vmatrix} \qquad D_y = \begin{vmatrix} a_{11} & c_1 & a_{13} \\ a_{21} & c_2 & a_{23} \\ a_{31} & c_3 & a_{33} \end{vmatrix} \qquad D_z = \begin{vmatrix} a_{11} & a_{12} & c_1 \\ a_{21} & a_{22} & c_2 \\ a_{31} & a_{32} & c_3 \end{vmatrix}$$

Notice the similarity between this pattern and the pattern observed earlier for a system of two equations containing two variables.

EXAMPLE 5

Using Cramer's Rule

Use Cramer's Rule, if applicable, to solve the following system:

$$\begin{cases} 2x + y - z = 3 & (1) \\ -x + 2y + 4z = -3 & (2) \\ x - 2y - 3z = 4 & (3) \end{cases}$$

Solution The value of the determinant D of the coefficients of the variables is

$$D = \begin{vmatrix} 2 & 1 & -1 \\ -1 & 2 & 4 \\ 1 & -2 & -3 \end{vmatrix} = (-1)^{1+1} \cdot 2 \cdot \begin{vmatrix} 2 & 4 \\ -2 & -3 \end{vmatrix} + (-1)^{1+2} \cdot 1 \cdot \begin{vmatrix} -1 & 4 \\ 1 & -3 \end{vmatrix} + (-1)^{1+3}(-1) \begin{vmatrix} -1 & 2 \\ 1 & -2 \end{vmatrix}$$

$$= 2 \cdot 2 - 1(-1) + (-1) \cdot 0 = 4 + 1 = 5$$

Because $D \neq 0$, proceed to find the values of D_x, D_y, and D_z. To find D_x, replace the coefficients of x in D with the constants and then evaluate the determinant.

$$D_x = \begin{vmatrix} 3 & 1 & -1 \\ -3 & 2 & 4 \\ 4 & -2 & -3 \end{vmatrix} = (-1)^{1+1} \cdot 3 \cdot \begin{vmatrix} 2 & 4 \\ -2 & -3 \end{vmatrix} + (-1)^{1+2} \cdot 1 \cdot \begin{vmatrix} -3 & 4 \\ 4 & -3 \end{vmatrix} + (-1)^{1+3}(-1) \begin{vmatrix} -3 & 2 \\ 4 & -2 \end{vmatrix}$$

$$= 3 \cdot 2 - 1(-7) + (-1)(-2) = 15$$

$$D_y = \begin{vmatrix} 2 & 3 & -1 \\ -1 & -3 & 4 \\ 1 & 4 & -3 \end{vmatrix} = (-1)^{1+1} \cdot 2 \cdot \begin{vmatrix} -3 & 4 \\ 4 & -3 \end{vmatrix} + (-1)^{1+2} \cdot 3 \cdot \begin{vmatrix} -1 & 4 \\ 1 & -3 \end{vmatrix} + (-1)^{1+3}(-1) \begin{vmatrix} -1 & -3 \\ 1 & 4 \end{vmatrix}$$

$$= 2(-7) - 3(-1) + (-1)(-1) = -10$$

$$D_z = \begin{vmatrix} 2 & 1 & 3 \\ -1 & 2 & -3 \\ 1 & -2 & 4 \end{vmatrix} = (-1)^{1+1} \cdot 2 \cdot \begin{vmatrix} 2 & -3 \\ -2 & 4 \end{vmatrix} + (-1)^{1+2} \cdot 1 \cdot \begin{vmatrix} -1 & -3 \\ 1 & 4 \end{vmatrix} + (-1)^{1+3} \cdot 3 \cdot \begin{vmatrix} -1 & 2 \\ 1 & -2 \end{vmatrix}$$

$$= 2 \cdot 2 - 1(-1) + 3 \cdot 0 = 5$$

As a result,

$$x = \frac{D_x}{D} = \frac{15}{5} = 3 \qquad y = \frac{D_y}{D} = \frac{-10}{5} = -2 \qquad z = \frac{D_z}{D} = \frac{5}{5} = 1$$

The solution is $x = 3$, $y = -2$, $z = 1$ or, using an ordered triplet, $(3, -2, 1)$.

Cramer's Rule cannot be used when the determinant of the coefficients of the variables, D, is 0. But can anything be learned about the system other than it is not a consistent and independent system if $D = 0$? The answer is yes!

Cramer's Rule with Inconsistent or Dependent Systems

- If $D = 0$ and at least one of the determinants D_x, D_y, or D_z is different from 0, then the system is inconsistent and the solution set is $\varnothing$, or $\{\ \}$.
- If $D = 0$ and all the determinants D_x, D_y, and D_z equal 0, then the system is consistent and dependent, so there are infinitely many solutions. The system must be solved using row reduction techniques.

──────**Now Work** PROBLEM 33

5 Know Properties of Determinants

Determinants have several properties that are sometimes helpful for obtaining their value. We list some of them here.

THEOREM

The value of a determinant changes sign if any two rows (or any two columns) are interchanged.

(12)

Proof for 2 by 2 Determinants

$$\begin{vmatrix} a & b \\ c & d \end{vmatrix} = ad - bc \quad \text{and} \quad \begin{vmatrix} c & d \\ a & b \end{vmatrix} = bc - ad = -(ad - bc) \qquad ■$$

EXAMPLE 6 **Demonstrating Theorem (12)**

$$\begin{vmatrix} 3 & 4 \\ 1 & 2 \end{vmatrix} = 6 - 4 = 2 \qquad \begin{vmatrix} 1 & 2 \\ 3 & 4 \end{vmatrix} = 4 - 6 = -2$$

THEOREM

If all the entries in any row (or any column) equal 0, the value of the determinant is 0.

(13)

Proof Expand across the row (or down the column) containing the 0's. ■

THEOREM

If any two rows (or any two columns) of a determinant have corresponding entries that are equal, the value of the determinant is 0.

(14)

In Problem 68, you are asked to prove the theorem for a 3 by 3 determinant in which the entries in column 1 equal the entries in column 3.

EXAMPLE 7   Demonstrating Theorem (14)

$$\begin{vmatrix} 1 & 2 & 3 \\ 1 & 2 & 3 \\ 4 & 5 & 6 \end{vmatrix} = (-1)^{1+1} \cdot 1 \cdot \begin{vmatrix} 2 & 3 \\ 5 & 6 \end{vmatrix} + (-1)^{1+2} \cdot 2 \cdot \begin{vmatrix} 1 & 3 \\ 4 & 6 \end{vmatrix} + (-1)^{1+3} \cdot 3 \cdot \begin{vmatrix} 1 & 2 \\ 4 & 5 \end{vmatrix}$$

$$= 1(-3) - 2(-6) + 3(-3) = -3 + 12 - 9 = 0$$

THEOREM

If any row (or any column) of a determinant is multiplied by a nonzero number k, the value of the determinant is also changed by a factor of k.   **(15)**

In Problem 67, you are asked to prove the theorem for a 3 by 3 determinant using row 2.

EXAMPLE 8   Demonstrating Theorem (15)

$$\begin{vmatrix} 1 & 2 \\ 4 & 6 \end{vmatrix} = 6 - 8 = -2$$

$$\begin{vmatrix} k & 2k \\ 4 & 6 \end{vmatrix} = 6k - 8k = -2k = k(-2) = k \begin{vmatrix} 1 & 2 \\ 4 & 6 \end{vmatrix}$$

THEOREM

If the entries of any row (or any column) of a determinant are multiplied by a nonzero number k and the result is added to the corresponding entries of another row (or column), the value of the determinant remains unchanged.   **(16)**

In Problem 69, you are asked to prove the theorem for a 3 by 3 determinant using rows 1 and 2.

EXAMPLE 9   Demonstrating Theorem (16)

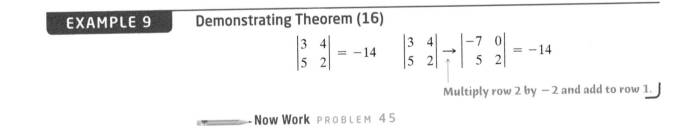

Now Work PROBLEM 45

8.3 Assess Your Understanding

Concepts and Vocabulary

1. $D = \begin{vmatrix} a & b \\ c & d \end{vmatrix} = $ _____ .

2. Using Cramer's Rule, the value of x that satisfies the system of

equations $\begin{cases} 2x + 3y = 5 \\ x - 4y = -3 \end{cases}$ is $x = \dfrac{}{\begin{vmatrix} 2 & 3 \\ 1 & -4 \end{vmatrix}}$.

3. *True or False* A determinant can never equal 0.

4. *True or False* When using Cramer's Rule, if $D = 0$, then the system of linear equations is inconsistent.

5. *True or False* If any row (or any column) of a determinant is multiplied by a nonzero number k, the value of the determinant remains unchanged.

6. *Multiple Choice* If any two rows of a determinant are interchanged, its value:
(a) changes sign   (b) becomes zero   (c) remains the same
(d) no longer relates to the original value

Skill Building

In Problems 7–14, find the value of each determinant.

7. $\begin{vmatrix} 6 & 4 \\ -1 & 3 \end{vmatrix}$

8. $\begin{vmatrix} 8 & -3 \\ 4 & 2 \end{vmatrix}$

9. $\begin{vmatrix} -3 & -1 \\ 4 & 2 \end{vmatrix}$

10. $\begin{vmatrix} -4 & 2 \\ -5 & 3 \end{vmatrix}$

11. $\begin{vmatrix} 3 & 4 & 2 \\ 1 & -1 & 5 \\ 1 & 2 & -2 \end{vmatrix}$

12. $\begin{vmatrix} 1 & 3 & -2 \\ 6 & 1 & -5 \\ 8 & 2 & 3 \end{vmatrix}$

13. $\begin{vmatrix} 4 & -1 & 2 \\ 6 & -1 & 0 \\ 1 & -3 & 4 \end{vmatrix}$

14. $\begin{vmatrix} 3 & -9 & 4 \\ 1 & 4 & 0 \\ 8 & -3 & 1 \end{vmatrix}$

In Problems 15–42, solve each system of equations using Cramer's Rule if it is applicable. If Cramer's Rule is not applicable, write, "Not applicable".

15. $\begin{cases} x + y = 8 \\ x - y = 4 \end{cases}$

16. $\begin{cases} x + 2y = 5 \\ x - y = 3 \end{cases}$

17. $\begin{cases} 5x - y = 13 \\ 2x + 3y = 12 \end{cases}$

18. $\begin{cases} x + 3y = 5 \\ 2x - 3y = -8 \end{cases}$

19. $\begin{cases} 3x = 24 \\ x + 2y = 0 \end{cases}$

20. $\begin{cases} 4x + 5y = -3 \\ -2y = -4 \end{cases}$

21. $\begin{cases} 4x - 6y = -42 \\ 7x + 4y = -1 \end{cases}$

22. $\begin{cases} 2x + 4y = 16 \\ 3x - 5y = -9 \end{cases}$

23. $\begin{cases} 3x - 2y = 4 \\ 6x - 4y = 0 \end{cases}$

24. $\begin{cases} -x + 2y = 5 \\ 4x - 8y = 6 \end{cases}$

25. $\begin{cases} 2x - 4y = -2 \\ 3x + 2y = 3 \end{cases}$

26. $\begin{cases} 3x + 3y = 3 \\ 4x + 2y = \dfrac{8}{3} \end{cases}$

27. $\begin{cases} 2x - 3y = -1 \\ 10x + 10y = 5 \end{cases}$

28. $\begin{cases} 3x - 2y = 0 \\ 5x + 10y = 4 \end{cases}$

29. $\begin{cases} 2x + 3y = 6 \\ x - y = \dfrac{1}{2} \end{cases}$

30. $\begin{cases} \dfrac{1}{2}x + y = -2 \\ x - 2y = 8 \end{cases}$

31. $\begin{cases} 3x - 5y = 3 \\ 15x + 5y = 21 \end{cases}$

32. $\begin{cases} 2x - y = -1 \\ x + \dfrac{1}{2}y = \dfrac{3}{2} \end{cases}$

33. $\begin{cases} x + y - z = 6 \\ 3x - 2y + z = -5 \\ x + 3y - 2z = 14 \end{cases}$

34. $\begin{cases} x - y + z = -4 \\ 2x - 3y + 4z = -15 \\ 5x + y - 2z = 12 \end{cases}$

35. $\begin{cases} x + 3y - z = -2 \\ 2x - 6y + z = -5 \\ -3x + 3y - 2z = 5 \end{cases}$

36. $\begin{cases} x + 4y - 3z = -8 \\ 3x - y + 3z = 12 \\ x + y + 6z = 1 \end{cases}$

37. $\begin{cases} x - 2y + 3z = 1 \\ 3x + y - 2z = 0 \\ 2x - 4y + 6z = 2 \end{cases}$

38. $\begin{cases} x - y + 2z = 5 \\ 3x + 2y = 4 \\ -2x + 2y - 4z = -10 \end{cases}$

39. $\begin{cases} x + 2y - z = 0 \\ 2x - 4y + z = 0 \\ -2x + 2y - 3z = 0 \end{cases}$

40. $\begin{cases} x + 4y - 3z = 0 \\ 3x - y + 3z = 0 \\ x + y + 6z = 0 \end{cases}$

41. $\begin{cases} x - 2y + 3z = 0 \\ 3x + y - 2z = 0 \\ 2x - 4y + 6z = 0 \end{cases}$

42. $\begin{cases} x - y + 2z = 0 \\ 3x + 2y = 0 \\ -2x + 2y - 4z = 0 \end{cases}$

In Problems 43–50, use properties of determinants to find the value of each determinant if it is known that

$$\begin{vmatrix} x & y & z \\ u & v & w \\ 1 & 2 & 3 \end{vmatrix} = 4$$

43. $\begin{vmatrix} 1 & 2 & 3 \\ u & v & w \\ x & y & z \end{vmatrix}$

44. $\begin{vmatrix} x & y & z \\ u & v & w \\ 2 & 4 & 6 \end{vmatrix}$

45. $\begin{vmatrix} x & y & z \\ -3 & -6 & -9 \\ u & v & w \end{vmatrix}$

46. $\begin{vmatrix} 1 & 2 & 3 \\ x - u & y - v & z - w \\ u & v & w \end{vmatrix}$

47. $\begin{vmatrix} 1 & 2 & 3 \\ x - 3 & y - 6 & z - 9 \\ 2u & 2v & 2w \end{vmatrix}$

48. $\begin{vmatrix} x & y & z - x \\ u & v & w - u \\ 1 & 2 & 2 \end{vmatrix}$

49. $\begin{vmatrix} 1 & 2 & 3 \\ 2x & 2y & 2z \\ u - 1 & v - 2 & w - 3 \end{vmatrix}$

50. $\begin{vmatrix} x + 3 & y + 6 & z + 9 \\ 3u - 1 & 3v - 2 & 3w - 3 \\ 1 & 2 & 3 \end{vmatrix}$

Applications and Extensions

In Problems 51–56, solve for x.

51. $\begin{vmatrix} x & x \\ 4 & 3 \end{vmatrix} = 5$

52. $\begin{vmatrix} x & 1 \\ 3 & x \end{vmatrix} = -2$

53. $\begin{vmatrix} x & 1 & 1 \\ 4 & 3 & 2 \\ -1 & 2 & 5 \end{vmatrix} = 2$

54. $\begin{vmatrix} 3 & 2 & 4 \\ 1 & x & 5 \\ 0 & 1 & -2 \end{vmatrix} = 0$

55. $\begin{vmatrix} x & 2 & 3 \\ 1 & x & 0 \\ 6 & 1 & -2 \end{vmatrix} = 7$

56. $\begin{vmatrix} x & 1 & 2 \\ 1 & x & 3 \\ 0 & 1 & 2 \end{vmatrix} = -4x$

57. Geometry: Equation of a Line An equation of the line containing the two points (x_1, y_1) and (x_2, y_2) may be expressed as the determinant

$$\begin{vmatrix} x & y & 1 \\ x_1 & y_1 & 1 \\ x_2 & y_2 & 1 \end{vmatrix} = 0$$

Prove this result by expanding the determinant and comparing the result to the two-point form of the equation of a line.

58. Geometry: Collinear Points Using the result obtained in Problem 57, show that three distinct points (x_1, y_1), (x_2, y_2), and (x_3, y_3) are collinear (lie on the same line) if and only if

$$\begin{vmatrix} x_1 & y_1 & 1 \\ x_2 & y_2 & 1 \\ x_3 & y_3 & 1 \end{vmatrix} = 0$$

59. Geometry: Area of a Triangle A triangle has vertices (x_1, y_1), (x_2, y_2), and (x_3, y_3). The area of the triangle is given by the absolute value of D, where $D = \dfrac{1}{2}\begin{vmatrix} x_1 & x_2 & x_3 \\ y_1 & y_2 & y_3 \\ 1 & 1 & 1 \end{vmatrix}$.

Use this formula to find the area of a triangle with vertices $(2, 3)$, $(5, 2)$, and $(6, 5)$.

60. Geometry: Area of a Polygon The formula from Problem 59 can be used to find the area of a polygon. To do so, divide the polygon into non-overlapping triangular regions and find the sum of the areas. Use this approach to find the area of the given polygon.

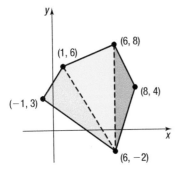

61. Geometry: Area of a Polygon Another approach for finding the area of a polygon by using determinants is to use the formula

$$A = \frac{1}{2}\left(\begin{vmatrix} x_1 & y_1 \\ x_2 & y_2 \end{vmatrix} + \begin{vmatrix} x_2 & y_2 \\ x_3 & y_3 \end{vmatrix} + \begin{vmatrix} x_3 & y_3 \\ x_4 & y_4 \end{vmatrix} + \cdots + \begin{vmatrix} x_n & y_n \\ x_1 & y_1 \end{vmatrix} \right)$$

where (x_1, y_1), (x_2, y_2), ..., (x_n, y_n) are the *n* corner points in counterclockwise order. Use this formula to compute the area of the polygon from Problem 60 again. Which method do you prefer?

62. Geometry: Volume of a Tetrahedron A tetrahedron (triangular pyramid) has vertices (x_1, y_1, z_1), (x_2, y_2, z_2), (x_3, y_3, z_3), and (x_4, y_4, z_4). The volume of the tetrahedron is given by the absolute value of D, where $D = \dfrac{1}{6}\begin{vmatrix} x_1 & y_1 & z_1 & 1 \\ x_2 & y_2 & z_2 & 1 \\ x_3 & y_3 & z_3 & 1 \\ x_4 & y_4 & z_4 & 1 \end{vmatrix}$.

Use this formula to find the volume of the tetrahedron with vertices $(0, 0, 8)$, $(2, 8, 0)$, $(10, 4, 4)$, and $(4, 10, 6)$.

63. Geometry: Equation of a Circle An equation of the circle containing the distinct points (x_1, y_1), (x_2, y_2), and (x_3, y_3) can be found using the following equation.

$$\begin{vmatrix} 1 & 1 & 1 & 1 \\ x & x_1 & x_2 & x_3 \\ y & y_1 & y_2 & y_3 \\ x^2 + y^2 & x_1^2 + y_1^2 & x_2^2 + y_2^2 & x_3^2 + y_3^2 \end{vmatrix} = 0$$

Find the standard equation of the circle containing the points $(7, -5)$, $(3, 3)$, and $(6, 2)$.

64. Show that $\begin{vmatrix} x^2 & x & 1 \\ y^2 & y & 1 \\ z^2 & z & 1 \end{vmatrix} = (y - z)(x - y)(x - z)$.

65. Complete the proof of Cramer's Rule for two equations containing two variables.

[**Hint**: In system (5), page 586, if $a = 0$, then $b \neq 0$ and $c \neq 0$, since $D = -bc \neq 0$. Now show that equation (6) provides a solution of the system when $a = 0$. Then three cases remain: $b = 0$, $c = 0$, and $d = 0$.]

66. *Challenge Problem* Interchange columns 1 and 3 of a 3 by 3 determinant. Show that the value of the new determinant is -1 times the value of the original determinant.

67. *Challenge Problem* Multiply each entry in row 2 of a 3 by 3 determinant by the number $k, k \neq 0$. Show that the value of the new determinant is k times the value of the original determinant.

68. *Challenge Problem* Prove that a 3 by 3 determinant in which the entries in column 1 equal those in column 3 has the value 0.

69. *Challenge Problem* If row 2 of a 3 by 3 determinant is multiplied by $k, k \neq 0$, and the result is added to the entries in row 1, prove that there is no change in the value of the determinant.

Retain Your Knowledge

Problems 70–77 are based on material learned earlier in the course. The purpose of these problems is to keep the material fresh in your mind so that you are better prepared for the final exam.

70. Radioactive Decay The half-life of uranium-232 is 68.9 years. If 25 grams is present now, how much will be present in 40 years?

71. List the potential rational zeros of the polynomial function $P(x) = 2x^3 - 5x^2 + x - 10$.

72. Graph $f(x) = (x + 1)^2 - 4$ using transformations (shifting, compressing, stretching, and/or reflecting).

73. The function $f(x) = 3 + \log_5(x - 1)$ is one-to-one. Find f^{-1}.

74. Find the distance between the vertices of $f(x) = 2x^2 - 12x + 20$ and $g(x) = -3x^2 - 30x - 77$.

75. Expand: $(2x - 5)^3$

76. Rationalize the numerator: $\dfrac{\sqrt{x + 7} - 10}{x}$

77. Find an equation of the line perpendicular to $f(x) = -\dfrac{2}{5}x + 7$ where $x = 10$.

8.4 Matrix Algebra

OBJECTIVES
1. Find the Sum and Difference of Two Matrices (p. 596)
2. Find Scalar Multiples of a Matrix (p. 598)
3. Find the Product of Two Matrices (p. 599)
4. Find the Inverse of a Matrix (p. 603)
5. Solve a System of Linear Equations Using an Inverse Matrix (p. 607)

Section 8.2 defined a matrix as a rectangular array of real numbers and used an augmented matrix to represent a system of linear equations. There is, however, a branch of mathematics, called **linear algebra**, in which an algebra of matrices is defined. This section is a survey of how this **matrix algebra** is developed.

Before getting started, recall the definition of a matrix.

DEFINITION Matrix

A **matrix** is defined as a rectangular array of numbers:

$$
\begin{array}{c}
\text{Row 1} \\
\text{Row 2} \\
\vdots \\
\text{Row } i \\
\vdots \\
\text{Row } m
\end{array}
\begin{bmatrix}
a_{11} & a_{12} & \cdots & a_{1j} & \cdots & a_{1n} \\
a_{21} & a_{22} & \cdots & a_{2j} & \cdots & a_{2n} \\
\vdots & \vdots & & \vdots & & \vdots \\
a_{i1} & a_{i2} & \cdots & a_{ij} & \cdots & a_{in} \\
\vdots & \vdots & & \vdots & & \vdots \\
a_{m1} & a_{m2} & \cdots & a_{mj} & \cdots & a_{mn}
\end{bmatrix}
$$

with column headings Column 1, Column 2, Column j, Column n.

In Words

In a matrix, the rows go across, and the columns go down.

Each number a_{ij} of the matrix has two indexes: the **row index** i and the **column index** j. The matrix shown here has m rows and n columns. The numbers a_{ij} are usually referred to as the **entries** of the matrix. For example, a_{23} refers to the entry in the second row, third column.

EXAMPLE 1

Arranging Data in a Matrix

In a survey of 900 people, the following information was obtained:

200 males	Thought federal defense spending was too high
150 males	Thought federal defense spending was too low
45 males	Had no opinion
315 females	Thought federal defense spending was too high
125 females	Thought federal defense spending was too low
65 females	Had no opinion

We can arrange these data in a rectangular array as follows:

	Too High	Too Low	No Opinion
Male	200	150	45
Female	315	125	65

or as the matrix

$$\begin{bmatrix} 200 & 150 & 45 \\ 315 & 125 & 65 \end{bmatrix}$$

This matrix has two rows (representing male and female) and three columns (representing "too high," "too low," and "no opinion").

The matrix developed in Example 1 has 2 rows and 3 columns. In general, a matrix with m rows and n columns is called an **m by n matrix**. An m by n matrix contains $m \cdot n$ entries. The matrix developed in Example 1 is a 2 by 3 matrix and contains $2 \cdot 3 = 6$ entries.

If an m by n matrix has the same number of rows as columns, that is, if $m = n$, then the matrix is a **square matrix**.

In Words

For an m by n matrix, the number of rows is listed first and the number of columns second.

EXAMPLE 2

Examples of Matrices

NOTE Matrices and determinants are different. A matrix is a rectangular array of numbers or expressions, nothing more. A determinant has a value, usually a real number. ∎

(a) $\begin{bmatrix} 5 & 0 \\ -6 & 1 \end{bmatrix}$ A 2 by 2 square matrix   (b) $\begin{bmatrix} 1 & 0 & 3 \end{bmatrix}$ A 1 by 3 matrix

(c) $\begin{bmatrix} 6 & -2 & 4 \\ 4 & 3 & 5 \\ 8 & 0 & 1 \end{bmatrix}$ A 3 by 3 square matrix

1 Find the Sum and Difference of Two Matrices

We begin our discussion of matrix algebra by defining equal matrices and then defining the operations of addition and subtraction. It is important to note that these definitions require both matrices to have the same number of rows *and* the same number of columns as a condition for equality and for addition and subtraction.

Matrices usually are represented by capital letters, such as A, B, and C.

DEFINITION Equal Matrices

Two matrices A and B are **equal**, written as

$$A = B$$

provided that A and B have the same number of rows and the same number of columns and each entry a_{ij} in A is equal to the corresponding entry b_{ij} in B.

For example,

$$\begin{bmatrix} 2 & 1 \\ 0.5 & -1 \end{bmatrix} = \begin{bmatrix} \sqrt{4} & 1 \\ \frac{1}{2} & -1 \end{bmatrix} \text{ and } \begin{bmatrix} 3 & 2 & 1 \\ 0 & 1 & -2 \end{bmatrix} = \begin{bmatrix} \sqrt{9} & \sqrt{4} & 1 \\ 0 & 1 & \sqrt[3]{-8} \end{bmatrix}$$

$$\begin{bmatrix} 4 & 1 \\ 6 & 1 \end{bmatrix} \neq \begin{bmatrix} 4 & 0 \\ 6 & 1 \end{bmatrix} \qquad \text{Because the entries in row 1, column 2 are not equal}$$

$$\begin{bmatrix} 4 & 1 & 2 \\ 6 & 1 & 2 \end{bmatrix} \neq \begin{bmatrix} 4 & 1 & 2 & 3 \\ 6 & 1 & 2 & 4 \end{bmatrix} \qquad \text{Because the matrix on the left has 3 columns and the matrix on the right has 4 columns}$$

Suppose that A and B represent two m by n matrices. The **sum**, $A + B$, is defined as the m by n matrix formed by adding the corresponding entries a_{ij} of A and b_{ij} of B. The **difference**, $A - B$, is defined as the m by n matrix formed by subtracting the entries b_{ij} in B from the corresponding entries a_{ij} in A. Addition and subtraction of matrices are defined only for matrices having the same number m of rows and the same number n of columns. For example, a 2 by 3 matrix and a 2 by 4 matrix cannot be added or subtracted.

EXAMPLE 3

Adding and Subtracting Matrices

Suppose that

$$A = \begin{bmatrix} 2 & 4 & 8 & -3 \\ 0 & 1 & 2 & 3 \end{bmatrix} \text{ and } B = \begin{bmatrix} -3 & 4 & 0 & 1 \\ 6 & 8 & 2 & 0 \end{bmatrix}$$

Find: (a) $A + B$ (b) $A - B$

Solution   Both A and B are 2 by 4 matrices, so they can be added and subtracted.

(a) $A + B = \begin{bmatrix} 2 & 4 & 8 & -3 \\ 0 & 1 & 2 & 3 \end{bmatrix} + \begin{bmatrix} -3 & 4 & 0 & 1 \\ 6 & 8 & 2 & 0 \end{bmatrix}$

$$= \begin{bmatrix} 2 + (-3) & 4 + 4 & 8 + 0 & -3 + 1 \\ 0 + 6 & 1 + 8 & 2 + 2 & 3 + 0 \end{bmatrix} \qquad \text{Add corresponding entries.}$$

$$= \begin{bmatrix} -1 & 8 & 8 & -2 \\ 6 & 9 & 4 & 3 \end{bmatrix}$$

(b) $A - B = \begin{bmatrix} 2 & 4 & 8 & -3 \\ 0 & 1 & 2 & 3 \end{bmatrix} - \begin{bmatrix} -3 & 4 & 0 & 1 \\ 6 & 8 & 2 & 0 \end{bmatrix}$

$$= \begin{bmatrix} 2 - (-3) & 4 - 4 & 8 - 0 & -3 - 1 \\ 0 - 6 & 1 - 8 & 2 - 2 & 3 - 0 \end{bmatrix} \qquad \text{Subtract corresponding entries.}$$

$$= \begin{bmatrix} 5 & 0 & 8 & -4 \\ -6 & -7 & 0 & 3 \end{bmatrix}$$

NORMAL FLOAT AUTO REAL RADIAN MP

[A]+[B]

$\begin{bmatrix} -1 & 8 & 8 & -2 \\ 6 & 9 & 4 & 3 \end{bmatrix}$

[A]-[B]

$\begin{bmatrix} 5 & 0 & 8 & -4 \\ -6 & -7 & 0 & 3 \end{bmatrix}$

Figure 7 Matrix addition and subtraction on a TI-84 Plus C

Seeing the Concept

Graphing utilities can make the sometimes tedious process of matrix algebra easy. In fact, most graphing calculators can handle matrices as large as 9 by 9, some even larger ones. Enter the matrices from Example 3 into a graphing utility. Name them A and B. Figure 7 shows the results of adding and subtracting A and B on a TI-84 Plus C.

Now Work PROBLEM 9

Many of the algebraic properties of sums of real numbers are also true for sums of matrices. Suppose that A, B, and C are m by n matrices. Then matrix addition is **commutative**. That is,

Commutative Property of Matrix Addition

$$A + B = B + A$$

Matrix addition is also **associative**. That is,

Associative Property of Matrix Addition

$$(A + B) + C = A + (B + C)$$

Although we do not prove these results, the proofs, as the following example illustrates, are based on the commutative and associative properties for real numbers.

EXAMPLE 4   **Demonstrating the Commutative Property of Matrix Addition**

$$\begin{bmatrix} 2 & 3 & -1 \\ 4 & 0 & 7 \end{bmatrix} + \begin{bmatrix} -1 & 2 & 1 \\ 5 & -3 & 4 \end{bmatrix} = \begin{bmatrix} 2 + (-1) & 3 + 2 & -1 + 1 \\ 4 + 5 & 0 + (-3) & 7 + 4 \end{bmatrix}$$

$$= \begin{bmatrix} -1 + 2 & 2 + 3 & 1 + (-1) \\ 5 + 4 & -3 + 0 & 4 + 7 \end{bmatrix}$$

$$= \begin{bmatrix} -1 & 2 & 1 \\ 5 & -3 & 4 \end{bmatrix} + \begin{bmatrix} 2 & 3 & -1 \\ 4 & 0 & 7 \end{bmatrix}$$

A matrix whose entries are all equal to 0 is called a **zero matrix**. Each of the following matrices is a zero matrix.

$$\begin{bmatrix} 0 & 0 \\ 0 & 0 \end{bmatrix} \quad \begin{matrix} \text{2 by 2 square} \\ \text{zero matrix} \end{matrix} \qquad \begin{bmatrix} 0 & 0 & 0 \\ 0 & 0 & 0 \end{bmatrix} \quad \begin{matrix} \text{2 by 3 zero} \\ \text{matrix} \end{matrix} \qquad \begin{bmatrix} 0 & 0 & 0 \end{bmatrix} \quad \begin{matrix} \text{1 by 3 zero} \\ \text{matrix} \end{matrix}$$

Zero matrices have properties similar to the real number 0. If A is an m by n matrix and 0 is the m by n zero matrix, then

$$A + 0 = 0 + A = A$$

In other words, a zero matrix is the additive identity in matrix algebra.

2 Find Scalar Multiples of a Matrix

We can also multiply a matrix by a real number. If k is a real number and A is an m by n matrix, the matrix kA is the m by n matrix formed by multiplying each entry a_{ij} in A by k. The number k is sometimes referred to as a **scalar**, and the matrix kA is called a **scalar multiple** of A.

EXAMPLE 5   **Operations Using Matrices**

Suppose

$$A = \begin{bmatrix} 3 & 1 & 5 \\ -2 & 0 & 6 \end{bmatrix} \qquad B = \begin{bmatrix} 4 & 1 & 0 \\ 8 & 1 & -3 \end{bmatrix} \qquad C = \begin{bmatrix} 9 & 0 \\ -3 & 6 \end{bmatrix}$$

Find: (a) $4A$ (b) $\dfrac{1}{3}C$ (c) $3A - 2B$

Solution   (a) $4A = 4\begin{bmatrix} 3 & 1 & 5 \\ -2 & 0 & 6 \end{bmatrix} = \begin{bmatrix} 4\cdot3 & 4\cdot1 & 4\cdot5 \\ 4(-2) & 4\cdot0 & 4\cdot6 \end{bmatrix} = \begin{bmatrix} 12 & 4 & 20 \\ -8 & 0 & 24 \end{bmatrix}$

(b) $\dfrac{1}{3}C = \dfrac{1}{3}\begin{bmatrix} 9 & 0 \\ -3 & 6 \end{bmatrix} = \begin{bmatrix} \dfrac{1}{3}\cdot9 & \dfrac{1}{3}\cdot0 \\ \dfrac{1}{3}(-3) & \dfrac{1}{3}\cdot6 \end{bmatrix} = \begin{bmatrix} 3 & 0 \\ -1 & 2 \end{bmatrix}$

(c) $3A - 2B = 3\begin{bmatrix} 3 & 1 & 5 \\ -2 & 0 & 6 \end{bmatrix} - 2\begin{bmatrix} 4 & 1 & 0 \\ 8 & 1 & -3 \end{bmatrix}$

$= \begin{bmatrix} 3\cdot3 & 3\cdot1 & 3\cdot5 \\ 3(-2) & 3\cdot0 & 3\cdot6 \end{bmatrix} - \begin{bmatrix} 2\cdot4 & 2\cdot1 & 2\cdot0 \\ 2\cdot8 & 2\cdot1 & 2(-3) \end{bmatrix}$

$= \begin{bmatrix} 9 & 3 & 15 \\ -6 & 0 & 18 \end{bmatrix} - \begin{bmatrix} 8 & 2 & 0 \\ 16 & 2 & -6 \end{bmatrix}$

$= \begin{bmatrix} 9-8 & 3-2 & 15-0 \\ -6-16 & 0-2 & 18-(-6) \end{bmatrix}$

$= \begin{bmatrix} 1 & 1 & 15 \\ -22 & -2 & 24 \end{bmatrix}$

Now Work PROBLEM 13

Some of the algebraic properties of scalar multiplication are listed next.

Properties of Scalar Multiplication

Suppose h and k are real numbers, and A and B are m by n matrices. Then

- $k(hA) = (kh)A$
- $(k + h)A = kA + hA$
- $k(A + B) = kA + kB$

3 Find the Product of Two Matrices

Unlike the straightforward definition for adding two matrices, the definition for multiplying two matrices is not what might be expected. In preparation for the definition, we need the following definitions:

DEFINITION Product of a Row Vector and a Column Vector

A **row vector** R is a 1 by n matrix

$$R = [r_1 \quad r_2 \quad \cdots \quad r_n]$$

A **column vector** C is an n by 1 matrix

$$C = \begin{bmatrix} c_1 \\ c_2 \\ \vdots \\ c_n \end{bmatrix}$$

The **product** RC of R times C is defined as the number

$$RC = [r_1 \quad r_2 \quad \cdots \quad r_n]\begin{bmatrix} c_1 \\ c_2 \\ \vdots \\ c_n \end{bmatrix} = r_1c_1 + r_2c_2 + \cdots + r_nc_n$$

Note that a row vector and a column vector can be multiplied if and only if they both contain the same number of entries.

EXAMPLE 6 The Product of a Row Vector and a Column Vector

If $R = \begin{bmatrix} 3 & -5 & 2 \end{bmatrix}$ and $C = \begin{bmatrix} 3 \\ 4 \\ -5 \end{bmatrix}$, then

$$RC = \begin{bmatrix} 3 & -5 & 2 \end{bmatrix} \begin{bmatrix} 3 \\ 4 \\ -5 \end{bmatrix} = 3 \cdot 3 + (-5)4 + 2(-5) = 9 - 20 - 10 = -21$$

EXAMPLE 7 Using Matrices to Compute Revenue

A clothing store sells men's shirts for \$40, silk ties for \$20, and wool suits for \$400. Last month, the store sold 100 shirts, 200 ties, and 50 suits. What was the total revenue from these sales?

Solution Set up a row vector R to represent the prices of these three items and a column vector C to represent the corresponding number of items sold. Then

$$
\begin{array}{cc}
\begin{array}{c} \text{Prices} \\ \text{Shirts \ Ties \ Suits} \\ R = \begin{bmatrix} 40 & 20 & 400 \end{bmatrix} \end{array}
&
\begin{array}{c} \text{Number} \\ \text{sold} \\ C = \begin{bmatrix} 100 \\ 200 \\ 50 \end{bmatrix} \begin{array}{l} \text{Shirts} \\ \text{Ties} \\ \text{Suits} \end{array} \end{array}
\end{array}
$$

The total revenue from these sales equals the product RC. That is,

$$RC = \begin{bmatrix} 40 & 20 & 400 \end{bmatrix} \begin{bmatrix} 100 \\ 200 \\ 50 \end{bmatrix}$$

$$= \underbrace{40 \cdot 100}_{\text{Shirt revenue}} + \underbrace{20 \cdot 200}_{\text{Tie revenue}} + \underbrace{400 \cdot 50}_{\text{Suit revenue}} = \underbrace{\$28{,}000}_{\text{Total revenue}}$$

The definition for multiplying two matrices is based on the definition of a row vector times a column vector.

DEFINITION Matrix Multiplication

Let A denote an m by r matrix and B denote an r by n matrix. The **product** AB is defined as the m by n matrix whose entry in row i, column j is the product of the ith row of A and the jth column of B.

The definition of the product AB of two matrices A and B, in this order, requires that the number of columns of A equals the number of rows of B; otherwise, the product is not defined.

In Words
To find the product AB, the number of columns in the left matrix A must equal the number of rows in the right matrix B.

An example will help clarify the definition.

EXAMPLE 8   **Multiplying Two Matrices**

Find the product AB if

$$A = \begin{bmatrix} 2 & 4 & -1 \\ 5 & 8 & 0 \end{bmatrix} \quad \text{and} \quad B = \begin{bmatrix} 2 & 5 & 1 & 4 \\ 4 & 8 & 0 & 6 \\ -3 & 1 & -2 & -1 \end{bmatrix}$$

Solution   First, observe that A is 2 by 3 and B is 3 by 4. The number of columns in A equals the number of rows in B, so the product AB is defined and will be a 2 by 4 matrix.

Suppose we want the entry in row 2, column 3 of AB. It equals the product of the row vector from row 2 of A and the column vector from column 3 of B.

$$\underset{\text{Row 2 of } A}{[5 \quad 8 \quad 0]} \overset{\text{Column 3 of } B}{\begin{bmatrix} 1 \\ 0 \\ -2 \end{bmatrix}} = 5 \cdot 1 + 8 \cdot 0 + 0(-2) = 5$$

So far, we have

$$AB = \begin{bmatrix} \underline{} & \underline{} & \overset{\text{Column 3}}{\underset{5}{\downarrow}} & \underline{} \\ \underline{} & \underline{} & & \underline{} \end{bmatrix} \leftarrow \text{Row 2}$$

Now, to find the entry in row 1, column 4 of AB, find the product of row 1 of A and column 4 of B.

$$\underset{\text{Row 1 of } A}{[2 \quad 4 \quad -1]} \overset{\text{Column 4 of } B}{\begin{bmatrix} 4 \\ 6 \\ -1 \end{bmatrix}} = 2 \cdot 4 + 4 \cdot 6 + (-1)(-1) = 33$$

Continuing in this fashion, we find AB.

$$AB = \begin{bmatrix} 2 & 4 & -1 \\ 5 & 8 & 0 \end{bmatrix} \begin{bmatrix} 2 & 5 & 1 & 4 \\ 4 & 8 & 0 & 6 \\ -3 & 1 & -2 & -1 \end{bmatrix}$$

$$= \begin{bmatrix} \begin{array}{l} \text{Row 1 of } A \\ \text{times} \\ \text{column 1 of } B \end{array} & \begin{array}{l} \text{Row 1 of } A \\ \text{times} \\ \text{column 2 of } B \end{array} & \begin{array}{l} \text{Row 1 of } A \\ \text{times} \\ \text{column 3 of } B \end{array} & \begin{array}{l} \text{Row 1 of } A \\ \text{times} \\ \text{column 4 of } B \end{array} \\ \begin{array}{l} \text{Row 2 of } A \\ \text{times} \\ \text{column 1 of } B \end{array} & \begin{array}{l} \text{Row 2 of } A \\ \text{times} \\ \text{column 2 of } B \end{array} & \begin{array}{l} \text{Row 2 of } A \\ \text{times} \\ \text{column 3 of } B \end{array} & \begin{array}{l} \text{Row 2 of } A \\ \text{times} \\ \text{column 4 of } B \end{array} \end{bmatrix}$$

$$= \begin{bmatrix} 2 \cdot 2 + 4 \cdot 4 + (-1)(-3) & 2 \cdot 5 + 4 \cdot 8 + (-1)1 & 2 \cdot 1 + 4 \cdot 0 + (-1)(-2) & 33 \, (\text{from earlier}) \\ 5 \cdot 2 + 8 \cdot 4 + 0(-3) & 5 \cdot 5 + 8 \cdot 8 + 0 \cdot 1 & 5 \, (\text{from earlier}) & 5 \cdot 4 + 8 \cdot 6 + 0(-1) \end{bmatrix}$$

$$= \begin{bmatrix} 23 & 41 & 4 & 33 \\ 42 & 89 & 5 & 68 \end{bmatrix}$$

 Check: Enter the matrices A and B. Then find AB. (See what happens if you try to find BA.)

Notice that the product AB in Example 8 is a 2 by 4 matrix, as we expected. Also notice that, for the matrices given in Example 8, the product BA is not defined because B is 3 by 4 and A is 2 by 3.

━━━➤**Now Work** PROBLEM 27

EXAMPLE 9   **Multiplying Two Matrices**

If

$$A = \begin{bmatrix} 2 & 1 & 3 \\ 1 & -1 & 0 \end{bmatrix} \quad \text{and} \quad B = \begin{bmatrix} 1 & 0 \\ 2 & 1 \\ 3 & 2 \end{bmatrix}$$

find: (a) AB (b) BA

Solution   (a) $AB = \begin{bmatrix} 2 & 1 & 3 \\ 1 & -1 & 0 \end{bmatrix}\begin{bmatrix} 1 & 0 \\ 2 & 1 \\ 3 & 2 \end{bmatrix} = \begin{bmatrix} 13 & 7 \\ -1 & -1 \end{bmatrix}$

   2 by 3$\qquad$3 by 2$\qquad$2 by 2

  (b) $BA = \begin{bmatrix} 1 & 0 \\ 2 & 1 \\ 3 & 2 \end{bmatrix}\begin{bmatrix} 2 & 1 & 3 \\ 1 & -1 & 0 \end{bmatrix} = \begin{bmatrix} 2 & 1 & 3 \\ 5 & 1 & 6 \\ 8 & 1 & 9 \end{bmatrix}$

   3 by 2$\qquad$2 by 3$\qquad$3 by 3

Notice in Example 9 that AB is 2 by 2 and BA is 3 by 3. It is possible for both AB and BA to be defined and yet be unequal. In fact, even if A and B are both n by n matrices so that AB and BA are both defined and n by n, AB and BA will usually be unequal.

EXAMPLE 10   **Multiplying Two Square Matrices**

If

$$A = \begin{bmatrix} 2 & 1 \\ 0 & 4 \end{bmatrix} \quad \text{and} \quad B = \begin{bmatrix} -3 & 1 \\ 1 & 2 \end{bmatrix}$$

find: (a) AB (b) BA

Solution   (a) $AB = \begin{bmatrix} 2 & 1 \\ 0 & 4 \end{bmatrix}\begin{bmatrix} -3 & 1 \\ 1 & 2 \end{bmatrix} = \begin{bmatrix} -5 & 4 \\ 4 & 8 \end{bmatrix}$

  (b) $BA = \begin{bmatrix} -3 & 1 \\ 1 & 2 \end{bmatrix}\begin{bmatrix} 2 & 1 \\ 0 & 4 \end{bmatrix} = \begin{bmatrix} -6 & 1 \\ 2 & 9 \end{bmatrix}$

Examples 9 and 10 prove that, unlike real number multiplication, matrix multiplication is not commutative.

THEOREM

Matrix multiplication is not commutative.

Now Work PROBLEMS **15** AND **17**

Next, consider two of the properties of real numbers that *are* shared by matrices. Assuming that each product and each sum is defined, the following are true:

Associative Property of Matrix Multiplication

$$A(BC) = (AB)C$$

Distributive Property

$$A(B + C) = AB + AC$$

For an n by n square matrix, the entries located in row i, column i, $1 \leq i \leq n$, are called the **diagonal entries** or **the main diagonal**. The n by n square matrix whose diagonal entries are 1's, and all other entries are 0's, is called the **identity matrix I_n**. For example,

$$I_2 = \begin{bmatrix} 1 & 0 \\ 0 & 1 \end{bmatrix} \qquad I_3 = \begin{bmatrix} 1 & 0 & 0 \\ 0 & 1 & 0 \\ 0 & 0 & 1 \end{bmatrix}$$

and so on.

EXAMPLE 11

Multiplication with an Identity Matrix

Let

$$A = \begin{bmatrix} -1 & 2 & 0 \\ 0 & 1 & 3 \end{bmatrix} \quad \text{and} \quad B = \begin{bmatrix} 3 & 2 \\ 4 & 6 \\ 5 & 2 \end{bmatrix}$$

Find: (a) AI_3 (b) I_2A (c) BI_2

Solution   (a) $AI_3 = \begin{bmatrix} -1 & 2 & 0 \\ 0 & 1 & 3 \end{bmatrix} \begin{bmatrix} 1 & 0 & 0 \\ 0 & 1 & 0 \\ 0 & 0 & 1 \end{bmatrix} = \begin{bmatrix} -1 & 2 & 0 \\ 0 & 1 & 3 \end{bmatrix} = A$

           (b) $I_2A = \begin{bmatrix} 1 & 0 \\ 0 & 1 \end{bmatrix} \begin{bmatrix} -1 & 2 & 0 \\ 0 & 1 & 3 \end{bmatrix} = \begin{bmatrix} -1 & 2 & 0 \\ 0 & 1 & 3 \end{bmatrix} = A$

           (c) $BI_2 = \begin{bmatrix} 3 & 2 \\ 4 & 6 \\ 5 & 2 \end{bmatrix} \begin{bmatrix} 1 & 0 \\ 0 & 1 \end{bmatrix} = \begin{bmatrix} 3 & 2 \\ 4 & 6 \\ 5 & 2 \end{bmatrix} = B$

Example 11 demonstrates the following property:

Identity Property

- If A is an m by n matrix, then

$$I_mA = A \text{ and } AI_n = A$$

- If A is an n by n square matrix, then

$$AI_n = I_nA = A$$

An identity matrix has properties similar to those of the real number 1. In other words, the identity matrix is a multiplicative identity in matrix algebra.

4 Find the Inverse of a Matrix

DEFINITION Inverse of a Matrix

Let A be a square n by n matrix. If there exists an n by n matrix A^{-1} (read as "A inverse") for which

$$AA^{-1} = A^{-1}A = I_n$$

then A^{-1} is called the **inverse** of the matrix A.

Not every square matrix has an inverse. When a matrix A has an inverse A^{-1}, then A is said to be **nonsingular**. If a matrix A has no inverse, it is called **singular**.

EXAMPLE 12   **Multiplying a Matrix by Its Inverse**

Show that the inverse of

$$A = \begin{bmatrix} 3 & 1 \\ 2 & 1 \end{bmatrix} \quad \text{is} \quad A^{-1} = \begin{bmatrix} 1 & -1 \\ -2 & 3 \end{bmatrix}$$

Solution   We need to show that $AA^{-1} = A^{-1}A = I_2$.

$$AA^{-1} = \begin{bmatrix} 3 & 1 \\ 2 & 1 \end{bmatrix}\begin{bmatrix} 1 & -1 \\ -2 & 3 \end{bmatrix} = \begin{bmatrix} 1 & 0 \\ 0 & 1 \end{bmatrix} = I_2$$

$$A^{-1}A = \begin{bmatrix} 1 & -1 \\ -2 & 3 \end{bmatrix}\begin{bmatrix} 3 & 1 \\ 2 & 1 \end{bmatrix} = \begin{bmatrix} 1 & 0 \\ 0 & 1 \end{bmatrix} = I_2$$

The following shows one way to find the inverse of

$$A = \begin{bmatrix} 3 & 1 \\ 2 & 1 \end{bmatrix}$$

Suppose that A^{-1} is given by

$$A^{-1} = \begin{bmatrix} x & y \\ z & w \end{bmatrix} \tag{1}$$

where x, y, z, and w are four variables. Based on the definition of an inverse, if A has an inverse, then

$$AA^{-1} = I_2$$

$$\begin{bmatrix} 3 & 1 \\ 2 & 1 \end{bmatrix}\begin{bmatrix} x & y \\ z & w \end{bmatrix} = \begin{bmatrix} 1 & 0 \\ 0 & 1 \end{bmatrix}$$

$$\begin{bmatrix} 3x + z & 3y + w \\ 2x + z & 2y + w \end{bmatrix} = \begin{bmatrix} 1 & 0 \\ 0 & 1 \end{bmatrix}$$

Because corresponding entries must be equal, it follows that this matrix equation is equivalent to two systems of linear equations.

$$\begin{cases} 3x + z = 1 \\ 2x + z = 0 \end{cases} \qquad \begin{cases} 3y + w = 0 \\ 2y + w = 1 \end{cases}$$

The augmented matrix of each system is

$$\begin{bmatrix} 3 & 1 & | & 1 \\ 2 & 1 & | & 0 \end{bmatrix} \qquad \begin{bmatrix} 3 & 1 & | & 0 \\ 2 & 1 & | & 1 \end{bmatrix} \tag{2}$$

The usual procedure would be to transform each augmented matrix into reduced row echelon form. Notice, though, that the left sides of the augmented matrices are equal, so the same row operations (see Section 8.2) can be used to reduce both matrices. It is more efficient to combine the two augmented matrices (2) into a single matrix, as shown next.

$$\begin{bmatrix} 3 & 1 & | & 1 & 0 \\ 2 & 1 & | & 0 & 1 \end{bmatrix}$$

Now, use row operations to transform the matrix into reduced row echelon form.

$$\begin{bmatrix} 3 & 1 & | & 1 & 0 \\ 2 & 1 & | & 0 & 1 \end{bmatrix} \rightarrow \begin{bmatrix} 1 & 0 & | & 1 & -1 \\ 2 & 1 & | & 0 & 1 \end{bmatrix}$$

$$\uparrow$$
$$R_1 = -1r_2 + r_1$$

$$\rightarrow \begin{bmatrix} 1 & 0 & | & 1 & -1 \\ 0 & 1 & | & -2 & 3 \end{bmatrix} \tag{3}$$

$$\uparrow$$
$$R_2 = -2r_1 + r_2$$

Matrix (3) is in reduced row echelon form.

Now reverse the earlier step of combining the two augmented matrices in (2), and write the single matrix (3) as two augmented matrices.

$$\begin{bmatrix} 1 & 0 & | & 1 \\ 0 & 1 & | & -2 \end{bmatrix} \text{ and } \begin{bmatrix} 1 & 0 & | & -1 \\ 0 & 1 & | & 3 \end{bmatrix}$$

The conclusion from these matrices is that $x = 1, z = -2$, and $y = -1, w = 3$. Substituting these values into matrix (1) results in

$$A^{-1} = \begin{bmatrix} 1 & -1 \\ -2 & 3 \end{bmatrix}$$

Notice in the augmented matrix (3) that the 2 by 2 matrix to the right of the vertical bar is the inverse of A. Also notice that the identity matrix I_2 appears to the left of the vertical bar. In general, using row operations to transform a nonsingular square matrix A, augmented by an identity matrix of the same dimensions, into reduced row echelon form results in the inverse matrix A^{-1}.

In Words

If A is nonsingular, begin with the matrix $[A|I_n]$, and after transforming it into reduced row echelon form, you end up with the matrix $[I_n|A^{-1}]$.

Steps for Finding the Inverse of an *n* by *n* Nonsingular Matrix *A*

STEP 1: Form the augmented matrix $[A|I_n]$.

STEP 2: Transform the matrix $[A|I_n]$ into reduced row echelon form.

STEP 3: The reduced row echelon form of $[A|I_n]$ contains the identity matrix I_n on the left of the vertical bar; the n by n matrix on the right of the vertical bar is the inverse of A.

EXAMPLE 13 **Finding the Inverse of a Matrix**

The matrix

$$A = \begin{bmatrix} 1 & 1 & 0 \\ -1 & 3 & 4 \\ 0 & 4 & 3 \end{bmatrix}$$

is nonsingular. Find its inverse.

Solution First, form the matrix

$$[A|I_3] = \begin{bmatrix} 1 & 1 & 0 & | & 1 & 0 & 0 \\ -1 & 3 & 4 & | & 0 & 1 & 0 \\ 0 & 4 & 3 & | & 0 & 0 & 1 \end{bmatrix}$$

Next, use row operations to transform $[A|I_3]$ into reduced row echelon form.

$$\begin{bmatrix} 1 & 1 & 0 & | & 1 & 0 & 0 \\ -1 & 3 & 4 & | & 0 & 1 & 0 \\ 0 & 4 & 3 & | & 0 & 0 & 1 \end{bmatrix} \rightarrow \begin{bmatrix} 1 & 1 & 0 & | & 1 & 0 & 0 \\ 0 & 4 & 4 & | & 1 & 1 & 0 \\ 0 & 4 & 3 & | & 0 & 0 & 1 \end{bmatrix} \rightarrow \begin{bmatrix} 1 & 1 & 0 & | & 1 & 0 & 0 \\ 0 & 1 & 1 & | & \frac{1}{4} & \frac{1}{4} & 0 \\ 0 & 4 & 3 & | & 0 & 0 & 1 \end{bmatrix}$$
$$\qquad\qquad \uparrow \qquad\qquad\qquad\qquad \uparrow$$
$$\qquad\quad R_2 = r_1 + r_2 \qquad\qquad\qquad R_2 = \frac{1}{4}r_2$$

$$\rightarrow \begin{bmatrix} 1 & 0 & -1 & | & \frac{3}{4} & -\frac{1}{4} & 0 \\ 0 & 1 & 1 & | & \frac{1}{4} & \frac{1}{4} & 0 \\ 0 & 0 & -1 & | & -1 & -1 & 1 \end{bmatrix} \rightarrow \begin{bmatrix} 1 & 0 & -1 & | & \frac{3}{4} & -\frac{1}{4} & 0 \\ 0 & 1 & 1 & | & \frac{1}{4} & \frac{1}{4} & 0 \\ 0 & 0 & 1 & | & 1 & 1 & -1 \end{bmatrix} \rightarrow \begin{bmatrix} 1 & 0 & 0 & | & \frac{7}{4} & \frac{3}{4} & -1 \\ 0 & 1 & 0 & | & -\frac{3}{4} & -\frac{3}{4} & 1 \\ 0 & 0 & 1 & | & 1 & 1 & -1 \end{bmatrix}$$
$$\uparrow \qquad\qquad\qquad\qquad\qquad \uparrow \qquad\qquad\qquad\qquad\qquad \uparrow$$
$$R_1 = -1r_2 + r_1 \qquad\qquad R_3 = -1r_3 \qquad\qquad\qquad R_1 = r_3 + r_1$$
$$R_3 = -4r_2 + r_3 \qquad\qquad\qquad\qquad\qquad\qquad\qquad R_2 = -1r_3 + r_2$$

(continued)

The matrix $[A|I_3]$ is now in reduced row echelon form, and the identity matrix I_3 is on the left of the vertical bar. The inverse of A is

$$A^{-1} = \begin{bmatrix} \dfrac{7}{4} & \dfrac{3}{4} & -1 \\ -\dfrac{3}{4} & -\dfrac{3}{4} & 1 \\ 1 & 1 & -1 \end{bmatrix}$$

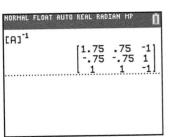

Figure 8 Inverse matrix on a TI-84 Plus C

You should verify that this is the correct inverse by showing that

$$AA^{-1} = A^{-1}A = I_3$$

✓ **Check:** Enter the matrix A into a graphing utility. Figure 8 shows A^{-1} on a TI-84 Plus C.

── **Now Work** PROBLEM 37

If transforming the matrix $[A|I_n]$ into reduced row echelon form does not result in the identity matrix I_n to the left of the vertical bar, then A is singular and has no inverse.

EXAMPLE 14 **Showing That a Matrix Has No Inverse**

Show that the matrix $A = \begin{bmatrix} 4 & 6 \\ 2 & 3 \end{bmatrix}$ has no inverse.

Solution Begin by writing the matrix $[A|I_2]$.

$$[A|I_2] = \begin{bmatrix} 4 & 6 & | & 1 & 0 \\ 2 & 3 & | & 0 & 1 \end{bmatrix}$$

Then use row operations to transform $[A|I_2]$ into reduced row echelon form.

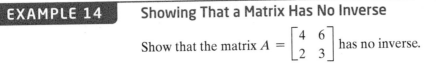

$$[A|I_2] = \begin{bmatrix} 4 & 6 & | & 1 & 0 \\ 2 & 3 & | & 0 & 1 \end{bmatrix} \rightarrow \begin{bmatrix} 1 & \dfrac{3}{2} & | & \dfrac{1}{4} & 0 \\ 2 & 3 & | & 0 & 1 \end{bmatrix} \rightarrow \begin{bmatrix} 1 & \dfrac{3}{2} & | & \dfrac{1}{4} & 0 \\ 0 & 0 & | & -\dfrac{1}{2} & 1 \end{bmatrix}$$
$$\uparrow \qquad\qquad\qquad \uparrow$$
$$R_1 = \dfrac{1}{4}r_1 \qquad\qquad R_2 = -2r_1 + r_2$$

The matrix $[A|I_2]$ is sufficiently reduced to see that the identity matrix cannot appear to the left of the vertical bar, so A is singular and has no inverse.

It can be shown that if the determinant of a matrix is 0, the matrix is singular. For example, the determinant of matrix A from Example 14 is

$$\begin{vmatrix} 4 & 6 \\ 2 & 3 \end{vmatrix} = 4 \cdot 3 - 6 \cdot 2 = 0$$

── **Now Work** PROBLEM 65

5 Solve a System of Linear Equations Using an Inverse Matrix

Inverse matrices can be used to solve systems of equations in which the number of equations is the same as the number of variables.

EXAMPLE 15

Using the Inverse Matrix to Solve a System of Linear Equations

Solve the system of equations: $\begin{cases} x + y = 3 \\ -x + 3y + 4z = -3 \\ 4y + 3z = 2 \end{cases}$

Solution   Let

$$A = \begin{bmatrix} 1 & 1 & 0 \\ -1 & 3 & 4 \\ 0 & 4 & 3 \end{bmatrix} \qquad X = \begin{bmatrix} x \\ y \\ z \end{bmatrix} \qquad B = \begin{bmatrix} 3 \\ -3 \\ 2 \end{bmatrix}$$

Then the original system of equations can be written compactly as the matrix equation

$$AX = B \tag{4}$$

From Example 13, the matrix A has the inverse A^{-1}. Multiply both sides of equation (4) by A^{-1}.

$$
\begin{array}{ll}
AX = B & \\
A^{-1}(AX) = A^{-1}B & \text{Multiply both sides by } A^{-1}. \\
(A^{-1}A)X = A^{-1}B & \text{Associative Property of matrix multiplication} \\
I_3 X = A^{-1}B & \text{Definition of an inverse matrix} \\
X = A^{-1}B & \text{Property of the identity matrix} \tag{5}
\end{array}
$$

Now use (5) to find $X = \begin{bmatrix} x \\ y \\ z \end{bmatrix}$.

$$X = \begin{bmatrix} x \\ y \\ z \end{bmatrix} = A^{-1}B = \underset{\substack{\uparrow \\ \text{Example 13}}}{\begin{bmatrix} \dfrac{7}{4} & \dfrac{3}{4} & -1 \\[2mm] -\dfrac{3}{4} & -\dfrac{3}{4} & 1 \\[2mm] 1 & 1 & -1 \end{bmatrix}} \begin{bmatrix} 3 \\ -3 \\ 2 \end{bmatrix} = \begin{bmatrix} 1 \\ 2 \\ -2 \end{bmatrix}$$

The solution is $x = 1, y = 2, z = -2$ or, using an ordered triplet, $(1, 2, -2)$.

The method used in Example 15 to solve a system of equations is particularly useful when it is necessary to solve several systems of equations in which the constants appearing to the right of the equal signs change, but the coefficients of the variables on the left side remain the same. See Problems 45–64 for some illustrations.

Be careful; this method can be used only if the inverse exists. If the matrix of the coefficients is singular, a different method must be used and the system is either inconsistent or dependent.

Now Work PROBLEM 49

Historical Feature

Arthur Cayley (1821–1895)

Matrices were invented in 1857 by Arthur Cayley (1821–1895) as a way of efficiently computing the result of substituting one linear system into another (see Historical Problem 3). The resulting system had incredible richness, in the sense that a wide variety of mathematical systems could be mimicked by the matrices. Cayley and his friend James J. Sylvester (1814–1897) spent much of the rest of their lives elaborating the theory. The torch was then passed to Georg Frobenius (1849–1917), whose deep investigations established a central place for matrices in modern mathematics. In 1924, rather to the surprise of physicists, it was found that matrices (with complex numbers in them) were exactly the right tool for describing the behavior of atomic systems. Today, matrices are used in a wide variety of applications.

Historical Problems

1. **Matrices and Complex Numbers** Frobenius emphasized in his research how matrices could be used to mimic other mathematical systems. Here, we mimic the behavior of complex numbers using matrices. Mathematicians call such a relationship an *isomorphism*.

$$\text{Complex number} \longleftrightarrow \text{Matrix}$$

$$a + bi \longleftrightarrow \begin{bmatrix} a & b \\ -b & a \end{bmatrix}$$

Note that the complex number can be read off the top line of the matrix. Then

$$2 + 3i \longleftrightarrow \begin{bmatrix} 2 & 3 \\ -3 & 2 \end{bmatrix} \text{ and } \begin{bmatrix} 4 & -2 \\ 2 & 4 \end{bmatrix} \longleftrightarrow 4 - 2i$$

(a) Find the matrices corresponding to $2 - 5i$ and $1 + 3i$.

(b) Multiply the two matrices.

(c) Find the corresponding complex number for the matrix found in part (b).

(d) Multiply $2 - 5i$ and $1 + 3i$. The result should be the same as that found in part (c).

The process also works for addition and subtraction. Try it for yourself.

2. Compute $(a + bi)(a - bi)$ using matrices. Interpret the result.

3. **Cayley's Definition of Matrix Multiplication** Cayley devised matrix multiplication to simplify the following problem:

$$\begin{cases} u = ar + bs \\ v = cr + ds \end{cases} \qquad \begin{cases} x = ku + lv \\ y = mu + nv \end{cases}$$

(a) Find x and y in terms of r and s by substituting u and v from the first system of equations into the second system of equations.

(b) Use the result of part (a) to find the 2 by 2 matrix A in

$$\begin{bmatrix} x \\ y \end{bmatrix} = A \begin{bmatrix} r \\ s \end{bmatrix}$$

(c) Now look at the following way to do it. Write the equations in matrix form.

$$\begin{bmatrix} u \\ v \end{bmatrix} = \begin{bmatrix} a & b \\ c & d \end{bmatrix} \begin{bmatrix} r \\ s \end{bmatrix} \qquad \begin{bmatrix} x \\ y \end{bmatrix} = \begin{bmatrix} k & l \\ m & n \end{bmatrix} \begin{bmatrix} u \\ v \end{bmatrix}$$

So

$$\begin{bmatrix} x \\ y \end{bmatrix} = \begin{bmatrix} k & l \\ m & n \end{bmatrix} \begin{bmatrix} a & b \\ c & d \end{bmatrix} \begin{bmatrix} r \\ s \end{bmatrix}$$

Do you see how Cayley defined matrix multiplication?

8.4 Assess Your Understanding

Concepts and Vocabulary

1. A matrix that has the same number of rows as columns is called a(n) _____ matrix.

2. *True or False* Matrix addition is commutative.

3. *True or False* If A and B are square matrices, then $AB = BA$.

4. Suppose that A is a square n by n matrix that is nonsingular. The matrix B for which $AB = BA = I_n$ is the _____ of the matrix A.

5. *True or False* The identity matrix has properties similar to those of the real number 1.

6. If $AX = B$ represents a matrix equation where A is a nonsingular matrix, then we can solve the equation using $X =$ _____.

7. *Multiple Choice* To find the product AB of two matrices A and B, which statement must be true?
   (a) The number of columns in A must equal the number of rows in B.
   (b) The number of rows in A must equal the number of columns in B.
   (c) A and B must have the same number of rows and the same number of columns.
   (d) A and B must both be square matrices.

8. *Multiple Choice* A matrix that has no inverse is called a(n):
   (a) zero matrix
   (b) nonsingular matrix
   (c) identity matrix
   (d) singular matrix

Skill Building

In Problems 9–26, use the following matrices. Determine whether the given expression is defined. If it is defined, express the result as a single matrix; if it is not, write "not defined".

$$A = \begin{bmatrix} 0 & 3 & -5 \\ 1 & 2 & 6 \end{bmatrix} \quad B = \begin{bmatrix} 4 & 1 & 0 \\ -2 & 3 & -2 \end{bmatrix} \quad C = \begin{bmatrix} 4 & 1 \\ 6 & 2 \\ -2 & 3 \end{bmatrix}$$

9. $A + B$

10. $A - B$

11. $4A$

12. $-3B$

13. $3A - 2B$

14. $2A + 4B$

15. AC

16. BC

17. CA

18. CB

19. AB

20. BA

21. $C(A + B)$

22. $(A + B)C$

23. $AC - 3I_2$

24. $CA + 5I_3$

25. $CA - CB$

26. $AC + BC$

In Problems 27–34, determine whether the product is defined. If it is defined, find the product; if it is not write "not defined."

27. $\begin{bmatrix} 2 & -2 \\ 1 & 0 \end{bmatrix}\begin{bmatrix} 2 & 1 & 4 & 6 \\ 3 & -1 & 3 & 2 \end{bmatrix}$

28. $\begin{bmatrix} 4 & 1 \\ 2 & 1 \end{bmatrix}\begin{bmatrix} -6 & 6 & 1 & 0 \\ 2 & 5 & 4 & -1 \end{bmatrix}$

29. $\begin{bmatrix} 1 & 2 & 3 \\ 0 & -1 & 4 \end{bmatrix}\begin{bmatrix} 1 & 2 \\ -1 & 0 \\ 2 & 4 \end{bmatrix}$

30. $\begin{bmatrix} 1 & -1 \\ -3 & 2 \\ 0 & 5 \end{bmatrix}\begin{bmatrix} 2 & 8 & -1 \\ 3 & 6 & 0 \end{bmatrix}$

31. $\begin{bmatrix} -4 \\ 2 \end{bmatrix}\begin{bmatrix} 1 & 0 \\ 3 & -1 \end{bmatrix}$

32. $\begin{bmatrix} 2 & -1 \\ 5 & 8 \\ -6 & 0 \end{bmatrix}\begin{bmatrix} 6 & 4 & 2 \\ -3 & 5 & -1 \\ 9 & 0 & 7 \end{bmatrix}$

33. $\begin{bmatrix} 1 & 0 & 1 \\ 2 & 4 & 1 \\ 3 & 6 & 1 \end{bmatrix}\begin{bmatrix} 1 & 3 \\ 6 & 2 \\ 8 & -1 \end{bmatrix}$

34. $\begin{bmatrix} 4 & -2 & 3 \\ 0 & 1 & 2 \\ -1 & 0 & 1 \end{bmatrix}\begin{bmatrix} 2 & 6 \\ 1 & -1 \\ 0 & 2 \end{bmatrix}$

In Problems 35–44, each matrix is nonsingular. Find the inverse of each matrix.

35. $\begin{bmatrix} 2 & 1 \\ 1 & 1 \end{bmatrix}$

36. $\begin{bmatrix} 3 & -1 \\ -2 & 1 \end{bmatrix}$

37. $\begin{bmatrix} 6 & 5 \\ 2 & 2 \end{bmatrix}$

38. $\begin{bmatrix} -4 & 1 \\ 6 & -2 \end{bmatrix}$

39. $\begin{bmatrix} 2 & 1 \\ a & a \end{bmatrix} a \neq 0$

40. $\begin{bmatrix} b & 3 \\ b & 2 \end{bmatrix} b \neq 0$

41. $\begin{bmatrix} 1 & -1 & 1 \\ 0 & -2 & 1 \\ -2 & -3 & 0 \end{bmatrix}$

42. $\begin{bmatrix} 1 & 0 & 2 \\ -1 & 2 & 3 \\ 1 & -1 & 0 \end{bmatrix}$

43. $\begin{bmatrix} 1 & 1 & 1 \\ 3 & 2 & -1 \\ 3 & 1 & 2 \end{bmatrix}$

44. $\begin{bmatrix} 3 & 3 & 1 \\ 1 & 2 & 1 \\ 2 & -1 & 1 \end{bmatrix}$

In Problems 45–64, use the inverses found in Problems 35–44 to solve each system of equations.

45. $\begin{cases} 2x + y = -1 \\ x + y = 3 \end{cases}$

46. $\begin{cases} 3x - y = 8 \\ -2x + y = 4 \end{cases}$

47. $\begin{cases} 2x + y = 0 \\ x + y = 5 \end{cases}$

48. $\begin{cases} 3x - y = 4 \\ -2x + y = 5 \end{cases}$

49. $\begin{cases} 6x + 5y = 7 \\ 2x + 2y = 2 \end{cases}$

50. $\begin{cases} -4x + y = 0 \\ 6x - 2y = 14 \end{cases}$

51. $\begin{cases} 6x + 5y = 13 \\ 2x + 2y = 5 \end{cases}$

52. $\begin{cases} -4x + y = 5 \\ 6x - 2y = -9 \end{cases}$

53. $\begin{cases} 2x + y = -3 \\ ax + ay = -a \end{cases} a \neq 0$

54. $\begin{cases} bx + 3y = 2b + 3 \\ bx + 2y = 2b + 2 \end{cases} b \neq 0$

55. $\begin{cases} 2x + y = \dfrac{7}{a} \\ ax + ay = 5 \end{cases} a \neq 0$

56. $\begin{cases} bx + 3y = 14 \\ bx + 2y = 10 \end{cases} b \neq 0$

57. $\begin{cases} x - y + z = 4 \\ -2y + z = 1 \\ -2x - 3y = -4 \end{cases}$

58. $\begin{cases} x + 2z = 6 \\ -x + 2y + 3z = -5 \\ x - y = 6 \end{cases}$

59. $\begin{cases} x - y + z = 2 \\ -2y + z = 2 \\ -2x - 3y = \dfrac{1}{2} \end{cases}$

60. $\begin{cases} x + 2z = 2 \\ -x + 2y + 3z = -\dfrac{3}{2} \\ x - y = 2 \end{cases}$

61. $\begin{cases} x + y + z = 9 \\ 3x + 2y - z = 8 \\ 3x + y + 2z = 1 \end{cases}$

62. $\begin{cases} 3x + 3y + z = 8 \\ x + 2y + z = 5 \\ 2x - y + z = 4 \end{cases}$

63. $\begin{cases} x + y + z = 2 \\ 3x + 2y - z = \dfrac{7}{3} \\ 3x + y + 2z = \dfrac{10}{3} \end{cases}$

64. $\begin{cases} 3x + 3y + z = 1 \\ x + 2y + z = 0 \\ 2x - y + z = 4 \end{cases}$

In Problems 65–70 show that each matrix has no inverse.

65. $\begin{bmatrix} 4 & 2 \\ 2 & 1 \end{bmatrix}$

66. $\begin{bmatrix} -3 & \dfrac{1}{2} \\ 6 & -1 \end{bmatrix}$

67. $\begin{bmatrix} 15 & 3 \\ 10 & 2 \end{bmatrix}$

68. $\begin{bmatrix} -3 & 0 \\ 4 & 0 \end{bmatrix}$

69. $\begin{bmatrix} -3 & 1 & -1 \\ 1 & -4 & -7 \\ 1 & 2 & 5 \end{bmatrix}$

70. $\begin{bmatrix} 1 & 1 & -3 \\ 2 & -4 & 1 \\ -5 & 7 & 1 \end{bmatrix}$

In Problems 71–74, use a graphing utility to find the inverse, if it exists, of each matrix. Round answers to two decimal places.

71. $\begin{bmatrix} 25 & 61 & -12 \\ 18 & -12 & 7 \\ 3 & 4 & -1 \end{bmatrix}$

72. $\begin{bmatrix} 18 & -3 & 4 \\ 6 & -20 & 14 \\ 10 & 25 & -15 \end{bmatrix}$

73. $\begin{bmatrix} 44 & 21 & 18 & 6 \\ -2 & 10 & 15 & 5 \\ 21 & 12 & -12 & 4 \\ -8 & -16 & 4 & 9 \end{bmatrix}$

74. $\begin{bmatrix} 16 & 22 & -3 & 5 \\ 21 & -17 & 4 & 8 \\ 2 & 8 & 27 & 20 \\ 5 & 15 & -3 & -10 \end{bmatrix}$

In Problems 75–78, use the inverse matrix found in Problem 71 to solve the following systems of equations. Round answers to two decimal places.

75. $\begin{cases} 25x + 61y - 12z = 10 \\ 18x - 12y + 7y = -9 \\ 3x + 4y - z = 12 \end{cases}$

76. $\begin{cases} 25x + 61y - 12z = 15 \\ 18x - 12y + 7z = -3 \\ 3x + 4y - z = 12 \end{cases}$

77. $\begin{cases} 25x + 61y - 12z = 21 \\ 18x - 12y + 7z = 7 \\ 3x + 4y - z = -2 \end{cases}$

78. $\begin{cases} 25x + 61y - 12z = 25 \\ 18x - 12y + 7z = 10 \\ 3x + 4y - z = -4 \end{cases}$

Mixed Practice *In Problems 79–86, solve each system of equations using any method you wish.*

79. $\begin{cases} 2x + 3y = 11 \\ 5x + 7y = 24 \end{cases}$

80. $\begin{cases} 2x + 8y = -8 \\ x + 7y = -13 \end{cases}$

81. $\begin{cases} x - 2y + 4z = 2 \\ -3x + 5y - 2z = 17 \\ 4x - 3y = -22 \end{cases}$

82. $\begin{cases} 2x + 3y - z = -2 \\ 4x + 3z = 6 \\ 6y - 2z = 2 \end{cases}$

83. $\begin{cases} 5x - y + 4z = 2 \\ -x + 5y - 4z = 3 \\ 7x + 13y - 4z = 17 \end{cases}$

84. $\begin{cases} 3x + 2y - z = 2 \\ 2x + y + 6z = -7 \\ 2x + 2y - 14z = 17 \end{cases}$

85. $\begin{cases} 2x - 3y + z = 4 \\ -3x + 2y - z = -3 \\ -5y + z = 6 \end{cases}$

86. $\begin{cases} -4x + 3y + 2z = 6 \\ 3x + y - z = -2 \\ x + 9y + z = 6 \end{cases}$

Applications and Extensions

87. College Tuition Nikki and Joe take classes at a community college, LCCC, and a local university, SIUE. The number of credit hours taken and the cost per credit hour (2018–2019 academic year, tuition and approximate fees) are as follows:

	LCCC	SIUE
Nikki	6	9
Joe	3	12

Cost per Credit Hour	
LCCC	$148.00
SIUE	$404.40

(a) Write a matrix A for the credit hours taken by each student and a matrix B for the cost per credit hour.
(b) Compute AB and interpret the results.

Sources: lc.edu, siue.edu

88. School Loan Interest Jamal and Stephanie both have school loans issued from the same two banks. The amounts borrowed and the monthly interest rates are given next (interest is compounded monthly).

	Lender 1	Lender 2
Jamal	$4000	$3000
Stephanie	$2500	$3800

	Monthly Interest Rate
Lender 1	0.011 (1.1%)
Lender 2	0.006 (0.6%)

(a) Write a matrix A for the amounts borrowed by each student and a matrix B for the monthly interest rates.
(b) Compute AB and interpret the result.
(c) Let $C = \begin{bmatrix} 1 \\ 1 \end{bmatrix}$. Compute $A(C + B)$ and interpret the result.

89. Computing the Cost of Production The Acme Steel Company is a producer of stainless steel and aluminum containers. On a certain day, the following stainless steel containers were manufactured: 500 with 10-gallon (gal) capacity, 350 with 5-gal capacity, and 400 with 1-gal capacity. On the same day, the following aluminum containers were manufactured: 700 with 10-gal capacity, 500 with 5-gal capacity, and 850 with 1-gal capacity.

(a) Find a 2 by 3 matrix representing these data. Find a 3 by 2 matrix to represent the same data.
(b) If the amount of material used in the 10-gal containers is 15 pounds (lb), the amount used in the 5-gal containers is 8 lb, and the amount used in the 1-gal containers is 3 lb, find a 3 by 1 matrix representing the amount of material used.
(c) Multiply the 2 by 3 matrix found in part (a) and the 3 by 1 matrix found in part (b) to get a 2 by 1 matrix showing the day's usage of material.
(d) If stainless steel costs Acme $0.10 per pound and aluminum costs $0.05 per pound, find a 1 by 2 matrix representing cost.
(e) Multiply the matrices found in parts (c) and (d) to find the total cost of the day's production.

90. Computing Profit Rizza's Used Cars has two locations, one in the city and the other in the suburbs. In January, the city location sold 400 subcompacts, 250 intermediate-size cars, and 50 SUVs; in February, it sold 350 subcompacts, 100 intermediates, and 30 SUVs. At the suburban location in January, 450 subcompacts, 200 intermediates, and 140 SUVs were sold. In February, the suburban location sold 350 subcompacts, 300 intermediates, and 100 SUVs.

(a) Find 2 by 3 matrices that summarize the sales data for each location for January and February (one matrix for each month).
(b) Use matrix addition to obtain total sales for the 2-month period.
(c) The profit on each kind of car is $100 per subcompact, $150 per intermediate, and $200 per SUV. Find a 3 by 1 matrix representing this profit.
(d) Multiply the matrices found in parts (b) and (c) to get a 2 by 1 matrix showing the profit at each location.

91. Cryptography One method of encryption is to use a matrix to encrypt the message and then use the corresponding inverse matrix to decode the message. The encrypted matrix, E, is obtained by multiplying the message matrix, M, by a key matrix, K. The original message can be retrieved by multiplying the encrypted matrix by the inverse of the key matrix. That is, $E = M \cdot K$ and $M = E \cdot K^{-1}$.

(a) The key matrix $K = \begin{bmatrix} 2 & 1 & 1 \\ 1 & 1 & 0 \\ 1 & 1 & 1 \end{bmatrix}$. Find its inverse, K^{-1}.

[**Note**: This key matrix is known as the Q_2^3 Fibonacci encryption matrix.]

(b) Use the result from part (a) to decode the encrypted

matrix $E = \begin{bmatrix} 47 & 34 & 33 \\ 44 & 36 & 27 \\ 47 & 41 & 20 \end{bmatrix}$.

(c) Each entry in the result for part (b) represents the position of a letter in the English alphabet ($A = 1$, $B = 2$, $C = 3$, and so on). What is the original message?

Source: *goldenmuseum.com*

92. Economic Mobility The income of a child (low, medium, or high) generally depends on the income of the child's parents. The matrix P, given by

Parent's Income

$$P = \begin{array}{c} \\ \\ \\ \end{array} \begin{array}{ccc} L & M & H \\ \begin{bmatrix} 0.4 & 0.2 & 0.1 \\ 0.5 & 0.6 & 0.5 \\ 0.1 & 0.2 & 0.4 \end{bmatrix} & \begin{array}{c} L \\ M \\ H \end{array} \end{array} \text{ Child's income}$$

is called a *left stochastic transition matrix*. For example, the entry $P_{21} = 0.5$ means that 50% of the children of low-income parents will transition to the medium level of income. The diagonal entry P_{ii} represents the percent of children who remain in the same income level as their parents. Assuming that the transition matrix is valid from one generation to the next, compute and interpret P^2.

Source: *Understanding Mobility in America, April 2006*

93. Solve for matrix X:

$$\begin{bmatrix} 3 & 2 \\ -1 & 5 \end{bmatrix} X = \begin{bmatrix} 24 & -13 & 1 \\ 26 & -7 & -40 \end{bmatrix}$$

*Use the following discussion for Problems 94 and 95. In graph theory, an **adjacency matrix**, A, is a way of representing which nodes (or vertices) are connected. For a simple directed graph, each entry, a_{ij}, is either 1 (if a direct path exists from node i to node j) or 0 (if no direct path exists from node i to node j). For example, consider the following graph and corresponding adjacency matrix.*

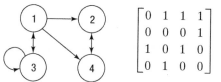

$$\begin{bmatrix} 0 & 1 & 1 & 1 \\ 0 & 0 & 0 & 1 \\ 1 & 0 & 1 & 0 \\ 0 & 1 & 0 & 0 \end{bmatrix}$$

The entry a_{14} is 1 because a direct path exists from node 1 to node 4. However, the entry a_{41} is 0 because no path exists from node 4 to node 1. The entry a_{33} is 1 because a direct path exists from node 3 to itself. The matrix $B_k = A + A^2 + \cdots + A^k$ indicates the number of ways to get from node i to node j within k moves (steps).

94. Website Map A content map can be used to show how different pages on a website are connected. For example, the following content map shows the relationship among the five pages of a certain website with links between pages represented by arrows.

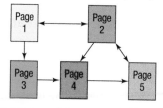

The content map can be represented by a 5 by 5 adjacency matrix where each entry, a_{ij}, is either 1 (if a link exists from page i to page j) or 0 (if no link exists from page i to page j).

(a) Write the 5 by 5 adjacency matrix that represents the given content map.

(b) Explain the significance of the entries on the main diagonal in your result from part (a).

(c) Find and interpret A^2.

95. Three-Click Rule An unofficial, and often contested, guideline for website design is to make all website content available to a user within three clicks. The webpage adjacency matrix for a certain website is given by

$$A = \begin{bmatrix} 0 & 1 & 1 & 0 & 0 \\ 1 & 0 & 0 & 1 & 1 \\ 1 & 0 & 0 & 1 & 0 \\ 0 & 0 & 1 & 0 & 1 \\ 0 & 1 & 0 & 0 & 0 \end{bmatrix}$$

(a) Find B_3. Does this website satisfy the Three-Click Rule?

(b) Which page can be reached the most number of ways from page 1 within three clicks?

96. Computer Graphics: Translating An important aspect of computer graphics is the ability to transform the coordinates of points within a graphic. For transformation purposes, a point (x, y) is represented as the column matrix $X = \begin{bmatrix} x \\ y \\ 1 \end{bmatrix}$.

To translate a point (x, y) horizontally h units and vertically k units, we use the translation matrix $S = \begin{bmatrix} 1 & 0 & h \\ 0 & 1 & k \\ 0 & 0 & 1 \end{bmatrix}$ and compute the matrix product SX. The translation is to the right for $h > 0$ and to the left for $h < 0$. Likewise, the translation is up for $k > 0$ and down for $k < 0$. The transformed coordinates are the first two entries in the resulting column matrix.

(a) Write the translation matrix needed to translate a point 3 units to the left and 5 units up.

(b) Find and interpret S^{-1}.

97. Computer Graphics: Rotating Besides translating a point, it is also important in computer graphics to be able to rotate a point. This is achieved by multiplying a point's column matrix (see Problem 96) by an appropriate rotation matrix R to form the matrix product RX. For example, to rotate a point 60°, the rotation matrix is $R = \begin{bmatrix} \dfrac{1}{2} & -\dfrac{\sqrt{3}}{2} & 0 \\ \dfrac{\sqrt{3}}{2} & \dfrac{1}{2} & 0 \\ 0 & 0 & 1 \end{bmatrix}$.

(a) Write the coordinates of the point $(6, 4)$ after it has been rotated 60°.

(b) Find and interpret R^{-1}.

98. Challenge Problem If $A = \begin{bmatrix} a & b \\ c & d \end{bmatrix}$ and $B = \begin{bmatrix} 1 & 1 \\ -1 & 1 \end{bmatrix}$, find a, b, c, d so that $AB = BA$.

99. Challenge Problem If $A = \begin{bmatrix} a & b \\ b & a \end{bmatrix}$, find a and b so that $A^2 + A = 0$.

Explaining Concepts: Discussion and Writing

100. Create a situation different from any found in the text that can be represented by a matrix.

101. Explain why the number of columns in matrix A must equal the number of rows in matrix B to find the product AB.

102. If $a, b,$ and $c \neq 0$ are real numbers with $ac = bc$, then $a = b$. Does this same property hold for matrices? In other words, if $A, B,$ and C are matrices and $AC = BC$, must $A = B$?

103. What is the solution of the system of equations $AX = 0$ if A^{-1} exists? Discuss the solution of $AX = 0$ if A^{-1} does not exist.

Retain Your Knowledge

Problems 104–111 are based on material learned earlier in the course. The purpose of these problems is to keep the material fresh in your mind so that you are better prepared for the final exam.

104. Find a polynomial with minimum degree and leading coefficient 1 that has zeros $x = 3$ (multiplicity 2), $x = 0$ (multiplicity 3), and $x = -2$ (multiplicity 1).

105. Solve: $\dfrac{5x}{x + 2} = \dfrac{x}{x - 2}$

106. Solve: $4^{x+3} = 8^{x+4}$

107. Add: $\dfrac{x + 1}{x - 3} + \dfrac{4}{x + 3}$

108. Find the domain of $f(x) = \dfrac{\sqrt{10 - 2x}}{x + 3}$.

109. Factor completely: $3x^4 + 12x^3 - 108x^2 - 432x$

110. Find the area of the region enclosed by the graphs of $y = \sqrt{4 - x^2}$, $y = x - 2$, and $y = -x - 2$.

111. If $f(x) = \dfrac{8}{x^2}$, find an equation of the secant line containing the points $(1, f(1))$ and $(4, f(4))$.

8.5 Partial Fraction Decomposition

PREPARING FOR THIS SECTION *Before getting started, review the following:*

- Identity (Section 1.1, p. 82)
- Properties of Rational Functions (Section 5.3, pp. 354–361)

- Reducing a Rational Expression to Lowest Terms (Section R.7, pp. 61–62)
- Complex Zeros; Fundamental Theorem of Algebra (Section 5.7, pp. 401–406)

Now Work the 'Are You Prepared?' problems on page 619.

OBJECTIVES 1 Decompose $\dfrac{P}{Q}$ Where Q Has Only Nonrepeated Linear Factors (p. 614)

2 Decompose $\dfrac{P}{Q}$ Where Q Has Repeated Linear Factors (p. 615)

3 Decompose $\dfrac{P}{Q}$ Where Q Has a Nonrepeated Irreducible Quadratic Factor (p. 617)

4 Decompose $\dfrac{P}{Q}$ Where Q Has a Repeated Irreducible Quadratic Factor (p. 618)

Consider the problem of adding two rational expressions:

$$\frac{3}{x + 4} \quad \text{and} \quad \frac{2}{x - 3}$$

The sum is

$$\frac{3}{x + 4} + \frac{2}{x - 3} = \frac{3(x - 3) + 2(x + 4)}{(x + 4)(x - 3)} = \frac{5x - 1}{x^2 + x - 12}$$

The reverse procedure, starting with the rational expression $\dfrac{5x - 1}{x^2 + x - 12}$ and

writing it as the sum (or difference) of the two simpler fractions $\dfrac{3}{x + 4}$ and $\dfrac{2}{x - 3}$, is

referred to as **partial fraction decomposition**, and the two simpler fractions are called **partial fractions**. Decomposing a rational expression into a sum of partial fractions is important in solving certain types of calculus problems. This section presents a systematic way to decompose rational expressions.

Recall that a rational expression is the ratio of two polynomials, say P and $Q \neq 0$.

Recall also that a rational expression $\dfrac{P}{Q}$ is called **proper** if the degree of the polynomial in the numerator is less than the degree of the polynomial in the denominator. Otherwise, the rational expression is called **improper**.

Also, we assume P and Q are in lowest terms. That is, P and Q have no common factors.

EXAMPLE 1　**Identifying Proper and Improper Rational Expressions**

Determine whether the rational expression is proper or improper. If the expression is improper, rewrite it as the sum of a polynomial and a proper rational expression.

(a) $\dfrac{x + 5}{x^2 + 3x + 2}$ 　　　　　(b) $\dfrac{6x^2 - x + 5}{3x - 2}$

Solution　(a) The numerator, $x + 5$, is a polynomial of degree 1, and the denominator, $x^2 + 3x + 2$, is a polynomial of degree 2. Since the degree of the numerator is less than the degree of the denominator, the rational expression is proper.

(b) The numerator, $6x^2 - x + 5$, is a polynomial of degree 2, and the denominator, $3x - 2$, is a polynomial of degree 1. Since the degree of the numerator is greater than the degree of the denominator, the rational expression is improper. We use long division to rewrite this expression as the sum of a polynomial and a proper rational expression:

$$
\begin{array}{r}
2x + 1 \\
3x - 2 \overline{) 6x^2 - x + 5} \\
\underline{6x^2 - 4x } \\
3x + 5 \\
\underline{3x - 2} \\
7
\end{array}
$$

So, $\dfrac{6x^2 - x + 5}{3x - 2} = 2x + 1 + \dfrac{7}{3x - 2}$.

Now Work　PROBLEMS 5 AND 13

Because by using long division, every improper rational expression can be written as the sum of a polynomial and a proper rational expression, we restrict the discussion that follows to proper rational expressions.

The partial fraction decomposition of the rational expression $\dfrac{P}{Q}$, in lowest terms, depends on the factors of the denominator Q. Recall from Section 5.7 that any polynomial with real coefficients can be factored over the real numbers into a product of linear and/or irreducible quadratic factors.

This means that the denominator Q of the rational expression $\dfrac{P}{Q}$ contains only factors of one or both of the following types:

- *Linear factors* of the form $x - a$, where a is a real number.
- *Irreducible quadratic factors* of the form $ax^2 + bx + c$, where a, b, and c are real numbers, $a \neq 0$, and $b^2 - 4ac < 0$. The negative discriminant guarantees that $ax^2 + bx + c$ cannot be written as the product of two linear factors with real coefficients.

As it turns out, there are four cases to be examined. We begin with the case for which Q has only nonrepeated linear factors. Throughout we assume the rational expression $\dfrac{P}{Q}$ is in lowest terms.

1 Decompose $\dfrac{P}{Q}$ Where Q Has Only Nonrepeated Linear Factors

Case 1: Q has only nonrepeated linear factors.

Under the assumption that Q has only nonrepeated linear factors, the polynomial Q has the form

$$Q(x) = (x - a_1)(x - a_2) \cdots \cdots (x - a_n)$$

where no two of the numbers $a_1, a_2, \ldots, a_n$ are equal. In this case, the partial fraction decomposition of $\dfrac{P}{Q}$ is of the form

$$\frac{P(x)}{Q(x)} = \frac{A_1}{x - a_1} + \frac{A_2}{x - a_2} + \cdots + \frac{A_n}{x - a_n} \tag{1}$$

where the numbers $A_1, A_2, \ldots, A_n$ are to be determined.

EXAMPLE 2 Decomposing $\dfrac{P}{Q}$ Where Q Has Only Nonrepeated Linear Factors

Find the partial fraction decomposition of $\dfrac{x}{x^2 - 5x + 6}$.

Solution First, factor the denominator,

$$x^2 - 5x + 6 = (x - 2)(x - 3)$$

and notice that the denominator contains only nonrepeated linear factors. Then decompose the rational expression according to equation (1):

$$\frac{x}{x^2 - 5x + 6} = \frac{A}{x - 2} + \frac{B}{x - 3} \tag{2}$$

where A and B are numbers to be determined. To find A and B, clear the fractions by multiplying both sides by $(x - 2)(x - 3) = x^2 - 5x + 6$. The result is

$$x = A(x - 3) + B(x - 2) \tag{3}$$

$$x = (A + B)x + (-3A - 2B)$$

This equation is an identity in x. Equate the coefficients of like powers of x to get

$$\begin{cases} 1 = \quad A + B & \text{Equate the coefficients of } x\text{: } 1x = (A + B)x. \\ 0 = -3A - 2B & \text{Equate the constants: } 0 = -3A - 2B. \end{cases}$$

This system of two equations containing two variables, A and B, can be solved using whatever method you wish. The solution is

$$A = -2 \qquad B = 3$$

From equation (2), the partial fraction decomposition is

$$\frac{x}{x^2 - 5x + 6} = \frac{-2}{x - 2} + \frac{3}{x - 3}$$

✔ **Check:** The decomposition can be checked by adding the rational expressions.

$$\frac{-2}{x - 2} + \frac{3}{x - 3} = \frac{-2(x - 3) + 3(x - 2)}{(x - 2)(x - 3)} = \frac{x}{(x - 2)(x - 3)}$$

$$= \frac{x}{x^2 - 5x + 6}$$

The numbers to be found in the partial fraction decomposition can sometimes be found more easily by using suitable choices for x in the identity obtained after fractions have been cleared. In Example 2, the identity after clearing fractions is equation (3):

$$x = A(x - 3) + B(x - 2)$$

Let $x = 2$ in this expression, and the term containing B drops out, leaving $2 = A(-1)$, or $A = -2$. Similarly, let $x = 3$, and the term containing A drops out, leaving $3 = B$. As before, $A = -2$ and $B = 3$.

━━➤ **Now Work** PROBLEM 17

2 Decompose $\dfrac{P}{Q}$ Where Q Has Repeated Linear Factors

Case 2: Q has repeated linear factors.

If the polynomial Q has a repeated linear factor, say $(x - a)^n, n \geq 2$ an integer, then, in the partial fraction decomposition of $\dfrac{P}{Q}$, allow for the terms

$$\frac{A_1}{x - a} + \frac{A_2}{(x - a)^2} + \cdots + \frac{A_n}{(x - a)^n}$$

where the numbers $A_1, A_2, \ldots, A_n$ are to be determined.

EXAMPLE 3

Decomposing $\dfrac{P}{Q}$ Where Q Has Repeated Linear Factors

Find the partial fraction decomposition of $\dfrac{x + 2}{x^3 - 2x^2 + x}$.

Solution First, factor the denominator,

$$x^3 - 2x^2 + x = x(x^2 - 2x + 1) = x(x - 1)^2$$

and notice that the denominator has the nonrepeated linear factor x and the repeated linear factor $(x - 1)^2$. By Case 1, the term $\dfrac{A}{x}$ is in the decomposition; and by Case 2, the terms $\dfrac{B}{x - 1} + \dfrac{C}{(x - 1)^2}$ are in the decomposition.

(continued)

Now write

$$\frac{x+2}{x^3-2x^2+x}=\frac{A}{x}+\frac{B}{x-1}+\frac{C}{(x-1)^2} \tag{4}$$

Again, clear fractions by multiplying both sides by $x^3-2x^2+x=x(x-1)^2$. The result is the identity

$$x+2=A(x-1)^2+Bx(x-1)+Cx \tag{5}$$

Let $x=0$ in this expression and the terms containing B and C drop out, leaving $2=A(-1)^2$, or $A=2$. Similarly, let $x=1$, and the terms containing A and B drop out, leaving $3=C$. Then equation (5) becomes

$$x+2=2(x-1)^2+Bx(x-1)+3x$$

Let $x=2$ (any number other than 0 or 1 will work as well). The result is

$$4=2\cdot1^2+B\cdot2\cdot1+3\cdot2$$
$$4=2+2B+6$$
$$2B=-4$$
$$B=-2$$

Therefore, $A=2$, $B=-2$, and $C=3$.

From equation (4), the partial fraction decomposition is

$$\frac{x+2}{x^3-2x^2+x}=\frac{2}{x}+\frac{-2}{x-1}+\frac{3}{(x-1)^2}$$

EXAMPLE 4   **Decomposing $\dfrac{P}{Q}$ Where Q Has Repeated Linear Factors**

Find the partial fraction decomposition of $\dfrac{x^3-8}{x^2(x-1)^3}$.

Solution   The denominator contains the repeated linear factors x^2 and $(x-1)^3$. The partial fraction decomposition has the form

$$\frac{x^3-8}{x^2(x-1)^3}=\frac{A}{x}+\frac{B}{x^2}+\frac{C}{x-1}+\frac{D}{(x-1)^2}+\frac{E}{(x-1)^3} \tag{6}$$

As before, clear fractions and obtain the identity

$$x^3-8=Ax(x-1)^3+B(x-1)^3+Cx^2(x-1)^2+Dx^2(x-1)+Ex^2 \tag{7}$$

Let $x=0$. (Do you see why this choice was made?) Then

$$-8=B(-1)$$
$$B=8$$

Let $x=1$ in equation (7). Then

$$-7=E$$

Use $B=8$ and $E=-7$ in equation (7), and collect like terms.

$$x^3-8=Ax(x-1)^3+8(x-1)^3+Cx^2(x-1)^2+Dx^2(x-1)-7x^2$$
$$x^3-8-8(x^3-3x^2+3x-1)+7x^2=Ax(x-1)^3+Cx^2(x-1)^2+Dx^2(x-1)$$
$$-7x^3+31x^2-24x=x(x-1)[A(x-1)^2+Cx(x-1)+Dx]$$
$$x(x-1)(-7x+24)=x(x-1)[A(x-1)^2+Cx(x-1)+Dx]$$
$$-7x+24=A(x-1)^2+Cx(x-1)+Dx \tag{8}$$

Now work with equation (8). Let $x = 0$. Then

$$24 = A$$

Let $x = 1$ in equation (8). Then

$$17 = D$$

Use $A = 24$ and $D = 17$ in equation (8).

$$-7x + 24 = 24(x - 1)^2 + Cx(x - 1) + 17x$$

Let $x = 2$ and simplify.

$$-14 + 24 = 24 + C(2) + 34$$
$$-48 = 2C$$
$$-24 = C$$

The numbers A, B, C, D, and E are now known. So, from equation (6),

$$\frac{x^3 - 8}{x^2(x - 1)^3} = \frac{24}{x} + \frac{8}{x^2} + \frac{-24}{x - 1} + \frac{17}{(x - 1)^2} + \frac{-7}{(x - 1)^3}$$

➤ **Now Work Example 4 by solving the system of five equations containing five variables that results by expanding equation (7).**

➤ **Now Work** PROBLEM 23

The final two cases involve irreducible quadratic factors. A quadratic factor is irreducible if it cannot be factored into linear factors with real coefficients. A quadratic expression $ax^2 + bx + c$ is irreducible whenever $b^2 - 4ac < 0$. For example, $x^2 + x + 1$ and $x^2 + 4$ are irreducible.

3 Decompose $\dfrac{P}{Q}$ Where Q Has a Nonrepeated Irreducible Quadratic Factor

Case 3: Q contains a nonrepeated irreducible quadratic factor.

Suppose Q contains a nonrepeated irreducible quadratic factor of the form $ax^2 + bx + c$. Then, in the partial fraction decomposition of $\dfrac{P}{Q}$, allow for the term

$$\frac{Ax + B}{ax^2 + bx + c}$$

where the numbers A and B are to be determined.

EXAMPLE 5

Decomposing $\dfrac{P}{Q}$ Where Q Has a Nonrepeated Irreducible Quadratic Factor

Find the partial fraction decomposition of $\dfrac{3x - 5}{x^3 - 1}$.

Solution Factor the denominator,

$$x^3 - 1 = (x - 1)(x^2 + x + 1)$$

Notice the nonrepeated linear factor $x - 1$ and the nonrepeated irreducible quadratic factor $x^2 + x + 1$. Allow for the term $\dfrac{A}{x - 1}$ by Case 1, and allow for the term $\dfrac{Bx + C}{x^2 + x + 1}$ by Case 3. Then

$$\frac{3x - 5}{x^3 - 1} = \frac{A}{x - 1} + \frac{Bx + C}{x^2 + x + 1} \tag{9}$$

Multiply both sides of equation (9) by $x^3 - 1 = (x - 1)(x^2 + x + 1)$ to obtain the identity

$$3x - 5 = A(x^2 + x + 1) + (Bx + C)(x - 1) \tag{10}$$

Expand the identity in (10) to obtain

$$3x - 5 = (A + B)x^2 + (A - B + C)x + (A - C)$$

This identity leads to the system of equations

$$\begin{cases} A + B & = & 0 & (1) \\ A - B + C & = & 3 & (2) \\ A & - C & = & -5 & (3) \end{cases}$$

The solution of this system is $A = -\dfrac{2}{3}, B = \dfrac{2}{3}, C = \dfrac{13}{3}$. Then, from equation (9),

$$\frac{3x - 5}{x^3 - 1} = \frac{-\dfrac{2}{3}}{x - 1} + \frac{\dfrac{2}{3}x + \dfrac{13}{3}}{x^2 + x + 1}$$

> **Now Work Example 5 using equation (10) and assigning values to x.**

> **Now Work** PROBLEM 25

4 Decompose $\dfrac{P}{Q}$ Where Q Has a Repeated Irreducible Quadratic Factor

Case 4: Q contains a repeated irreducible quadratic factor.

Suppose the polynomial Q contains a repeated irreducible quadratic factor $(ax^2 + bx + c)^n, n \geq 2, n$ an integer, and $b^2 - 4ac < 0$. Then, in the partial fraction decomposition of $\dfrac{P}{Q}$, allow for the terms

$$\frac{A_1 x + B_1}{ax^2 + bx + c} + \frac{A_2 x + B_2}{(ax^2 + bx + c)^2} + \cdots + \frac{A_n x + B_n}{(ax^2 + bx + c)^n}$$

where the numbers $A_1, B_1, A_2, B_2, \ldots, A_n, B_n$ are to be determined.

EXAMPLE 6 **Decomposing $\dfrac{P}{Q}$ Where Q Has a Repeated Irreducible Quadratic Factor**

Find the partial fraction decomposition of $\dfrac{x^3 + x^2}{(x^2 + 4)^2}$.

Solution The denominator contains the repeated irreducible quadratic factor $(x^2 + 4)^2$, so by Case 4,

$$\frac{x^3 + x^2}{(x^2 + 4)^2} = \frac{Ax + B}{x^2 + 4} + \frac{Cx + D}{(x^2 + 4)^2} \qquad \textbf{(11)}$$

Clear fractions to obtain

$$x^3 + x^2 = (Ax + B)(x^2 + 4) + Cx + D$$

Collecting like terms yields the identity

$$x^3 + x^2 = Ax^3 + Bx^2 + (4A + C)x + 4B + D$$

Equating coefficients results in the system

$$\begin{cases} A = 1 \\ B = 1 \\ 4A + C = 0 \\ 4B + D = 0 \end{cases}$$

The solution is $A = 1$, $B = 1$, $C = -4$, $D = -4$. From equation (11),

$$\frac{x^3 + x^2}{(x^2 + 4)^2} = \frac{x + 1}{x^2 + 4} + \frac{-4x - 4}{(x^2 + 4)^2}$$

Now Work PROBLEM 39

8.5 Assess Your Understanding

'Are You Prepared?' *Answers are given at the end of these exercises. If you get a wrong answer, read the pages listed in red.*

1. **True or False** The equation $(x - 1)^2 - 1 = x(x - 2)$ is an example of an identity. (p. 82)

2. **True or False** The rational expression $\dfrac{5x^2 - 1}{x^3 + 1}$ is proper. (p. 359)

3. Reduce to lowest terms: $\dfrac{3x - 12}{x^2 - 16}$ (pp. 61–62)

4. **True or False** Every polynomial with real numbers as coefficients can be factored into products of linear and/or irreducible quadratic factors. (p. 405)

Skill Building

In Problems 5–16, determine whether the given rational expression is proper or improper. If the expression is improper, rewrite it as the sum of a polynomial and a proper rational expression.

5. $\dfrac{x}{x^2 - 1}$

6. $\dfrac{5x + 2}{x^3 - 1}$

7. $\dfrac{x^2 + 5}{x^2 - 4}$

8. $\dfrac{3x^2 - 2}{x^2 - 1}$

9. $\dfrac{x^3 + x^2 - 12x + 9}{x^2 + 2x - 15}$

10. $\dfrac{6x^3 - 5x^2 - 7x - 3}{2x - 5}$

11. $\dfrac{5x^2 - 7x - 6}{x + x^3}$

12. $\dfrac{x^3 + 12x^2 - 9x}{9x^2 - x^4}$

13. $\dfrac{5x^3 + 2x - 1}{x^2 - 4}$

14. $\dfrac{3x^4 + x^2 - 2}{x^3 + 8}$

15. $\dfrac{x(x - 1)}{(x + 4)(x - 3)}$

16. $\dfrac{2x(x^2 + 4)}{x^2 + 1}$

In Problems 17–50, find the partial fraction decomposition of each rational expression.

17. $\dfrac{4}{x(x - 1)}$

18. $\dfrac{3x}{(x + 2)(x - 1)}$

19. $\dfrac{1}{x(x^2 + 1)}$

20. $\dfrac{1}{(x + 1)(x^2 + 4)}$

21. $\dfrac{x}{(x - 1)(x - 2)}$

22. $\dfrac{3x}{(x + 2)(x - 4)}$

23. $\dfrac{x^2}{(x - 1)^2(x + 1)}$

24. $\dfrac{x + 1}{x^2(x - 2)}$

25. $\dfrac{1}{x^3 - 8}$

26. $\dfrac{2x + 4}{x^3 - 1}$

27. $\dfrac{x^2}{(x - 1)^2(x + 1)^2}$

28. $\dfrac{x + 1}{x^2(x - 2)^2}$

29. $\dfrac{x - 3}{(x + 2)(x + 1)^2}$

30. $\dfrac{x^2 + x}{(x + 2)(x - 1)^2}$

31. $\dfrac{x + 4}{x^2(x^2 + 4)}$

32. $\dfrac{10x^2 + 2x}{(x - 1)^2(x^2 + 2)}$

33. $\dfrac{x^2 + 2x + 3}{(x + 1)(x^2 + 2x + 4)}$

34. $\dfrac{x^2 - 11x - 18}{x(x^2 + 3x + 3)}$

35. $\dfrac{x}{(3x - 2)(2x + 1)}$

36. $\dfrac{1}{(2x + 3)(4x - 1)}$

37. $\dfrac{x}{x^2 + 2x - 3}$

38. $\dfrac{x^2 - x - 8}{(x + 1)(x^2 + 5x + 6)}$

39. $\dfrac{x^2 + 2x + 3}{(x^2 + 4)^2}$

40. $\dfrac{x^3 + 1}{(x^2 + 16)^2}$

41. $\dfrac{7x + 3}{x^3 - 2x^2 - 3x}$

42. $\dfrac{x^3 + 1}{x^5 - x^4}$

43. $\dfrac{x^2}{x^3 - 4x^2 + 5x - 2}$

44. $\dfrac{x^2 + 1}{x^3 + x^2 - 5x + 3}$

45. $\dfrac{x^3}{(x^2 + 16)^3}$

46. $\dfrac{x^2}{(x^2 + 4)^3}$

47. $\dfrac{4}{2x^2 - 5x - 3}$

48. $\dfrac{4x}{2x^2 + 3x - 2}$

49. $\dfrac{2x + 3}{x^4 - 9x^2}$

50. $\dfrac{x^2 + 9}{x^4 - 2x^2 - 8}$

Mixed Practice In Problems 51–58, use the division algorithm to rewrite each improper rational expression as the sum of a polynomial and a proper rational expression. Find the partial fraction decomposition of the proper rational expression. Finally, express the improper rational expression as the sum of a polynomial and the partial fraction decomposition.

51. $\dfrac{x^3 + x^2 - 3}{x^2 + 3x - 4}$

52. $\dfrac{x^3 - 3x^2 + 1}{x^2 + 5x + 6}$

53. $\dfrac{x^3}{x^2 + 1}$

54. $\dfrac{x^3 + x}{x^2 + 4}$

55. $\dfrac{x^4 - 5x^2 + x - 4}{x^2 + 4x + 4}$

56. $\dfrac{x^4 + x^3 - x + 2}{x^2 - 2x + 1}$

57. $\dfrac{x^5 + x^4 - x^2 + 2}{x^4 - 2x^2 + 1}$

58. $\dfrac{x^5 - x^3 + x^2 + 1}{x^4 + 6x^2 + 9}$

59. *Challenge Problem* Use a substitution and partial fraction decomposition to express $\dfrac{3e^x}{e^{2x} + e^x - 2}$ in terms of e^x.

60. *Challenge Problem* Use a substitution and partial fraction decomposition to express $\dfrac{2}{x - \sqrt[3]{x}}$ in terms of $\sqrt[3]{x}$.

Retain Your Knowledge

Problems 61–68 are based on material learned earlier in the course. The purpose of these problems is to keep the material fresh in your mind so that you are better prepared for the final exam.

61. **Credit Card Balance** Nick has a credit card balance of $4200. If the credit card company charges 18% interest compound daily, and Nick does not make any payments on the account, how long will it take for his balance to double? Round to two decimal places.

62. If $f(x) = x + 4$ and $g(x) = x^2 - 3x$, find $(g \circ f)(-3)$.

63. Determine whether $f(x) = -\dfrac{3x}{x^2 - 10}$ is even, odd, or neither.

64. The function $f(x) = 8^{x-3} - 4$ is one-to-one. Find f^{-1}.

65. Find an equation for the hyperbola with vertices $(0, -5)$ and $(0, 5)$, and a focus at $(0, 13)$.

66. Solve: $\dfrac{5}{2}x - 1 \geq x + \dfrac{4}{5}$

67. Solve for D: $2x - 4xD - 4y + 2yD = D$

68. The **normal line** is the line that is perpendicular to the tangent line at the point of tangency. If $y = -2x + 2$ is the tangent line to $f(x) = \dfrac{1}{3}x^2 - 4x + 5$, find an equation of the normal line to f at the point of tangency.

'Are You Prepared?' Answers

1. True   **2.** True   **3.** $\dfrac{3}{x + 4}$ **4.** True

8.6 Systems of Nonlinear Equations

PREPARING FOR THIS SECTION *Before getting started, review the following:*

- Lines (Section 2.3, pp. 169–179)
- Circles (Section 2.4, pp. 185–188)
- Parabolas (Section 7.2, pp. 519–523)
- Ellipses (Section 7.3, pp. 528–534)
- Hyperbolas (Section 7.4, pp. 538–546)

Now Work the 'Are You Prepared?' problems on page 626.

OBJECTIVES 1 Solve a System of Nonlinear Equations Using Substitution (p. 621)
2 Solve a System of Nonlinear Equations Using Elimination (p. 622)

In Section 8.1, we observed that the solution to a system of linear equations could be found geometrically by determining the point(s) of intersection (if any) of the equations in the system. Similarly, in solving systems of nonlinear equations, the solution(s) also represent(s) the point(s) of intersection (if any) of the graphs of the equations.

There is no general method for solving a system of nonlinear equations. Sometimes substitution is best; other times elimination is best; and sometimes neither of these methods works. Experience and a certain degree of imagination are your allies here.

Before we begin, two comments are in order.

- If the system contains two variables and if the equations in the system are easy to graph, then graph them. By graphing each equation in the system, you can get an idea of how many solutions a system has and approximate their location.

- Extraneous solutions can creep in when solving nonlinear systems, so it is imperative to check all apparent solutions.

1 Solve a System of Nonlinear Equations Using Substitution

EXAMPLE 1 Solving a System of Nonlinear Equations Using Substitution

Solve the following system of equations:

$$\begin{cases} 3x - y = -2 & (1) \\ 2x^2 - y = 0 & (2) \end{cases}$$

Solution First, notice that the system contains two variables and that we know how to graph each equation. Equation (1) is the line $y = 3x + 2$, and equation (2) is the parabola $y = 2x^2$. See Figure 9. The system apparently has two solutions.

To use substitution to solve the system, we choose to solve equation (1) for y.

$$3x - y = -2 \qquad \text{Equation (1)}$$
$$y = 3x + 2$$

Substitute this expression for y in equation (2). The result is an equation containing just the variable x, which we can solve.

$$2x^2 - y = 0 \qquad \text{Equation (2)}$$
$$2x^2 - (3x + 2) = 0 \qquad \text{Substitute } 3x + 2 \text{ for } y.$$
$$2x^2 - 3x - 2 = 0 \qquad \text{Simplify.}$$
$$(2x + 1)(x - 2) = 0 \qquad \text{Factor.}$$
$$2x + 1 = 0 \quad \text{or} \quad x - 2 = 0 \qquad \text{Use the Zero-Product Property.}$$
$$x = -\frac{1}{2} \quad \text{or} \qquad x = 2$$

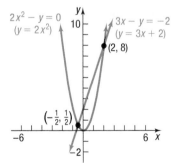

Figure 9

(continued)

Use these values for x in $y = 3x + 2$ to find

$$y = 3\left(-\frac{1}{2}\right) + 2 = \frac{1}{2} \quad \text{or} \quad y = 3 \cdot 2 + 2 = 8$$

The apparent solutions are $x = -\frac{1}{2}, y = \frac{1}{2}$ and $x = 2, y = 8$.

✓ **Check:** For $x = -\frac{1}{2}, y = \frac{1}{2}$,

$$\begin{cases} 3\left(-\dfrac{1}{2}\right) - \dfrac{1}{2} = -\dfrac{3}{2} - \dfrac{1}{2} = -2 & \text{(1)} \\ 2\left(-\dfrac{1}{2}\right)^2 - \dfrac{1}{2} = 2 \cdot \dfrac{1}{4} - \dfrac{1}{2} = 0 & \text{(2)} \end{cases}$$

For $x = 2, y = 8$,

$$\begin{cases} 3 \cdot 2 \;-\; 8 = 6 - 8 = -2 & \text{(1)} \\ 2 \cdot 2^2 - 8 = 2 \cdot 4 - 8 = 0 & \text{(2)} \end{cases}$$

Each solution checks. The graphs of the two equations intersect at the points $\left(-\dfrac{1}{2}, \dfrac{1}{2}\right)$ and $(2, 8)$, as shown in Figure 9 on the previous page. ⌐

━━━ **Now Work** PROBLEM 15 USING SUBSTITUTION

2 Solve a System of Nonlinear Equations Using Elimination

EXAMPLE 2

Solving a System of Nonlinear Equations Using Elimination

Solve: $\begin{cases} x^2 + y^2 = 13 & \text{(1)} \quad \text{A circle} \\ x^2 - y = 7 & \text{(2)} \quad \text{A parabola} \end{cases}$

Solution

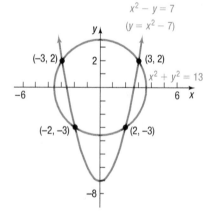

First graph each equation, as shown in Figure 10. Based on the graph, we expect four solutions. Notice that subtracting equation (2) from equation (1) eliminates the variable x.

$$\begin{cases} x^2 + y^2 = 13 \\ \underline{x^2 - y = 7} \\ y^2 + y = 6 \quad \text{Subtract.} \end{cases}$$

This quadratic equation in y can be solved by factoring.

$$y^2 + y - 6 = 0$$
$$(y + 3)(y - 2) = 0$$
$$y = -3 \quad \text{or} \quad y = 2$$

Use these values for y in equation (2) to find x.

- If $y = 2$, then $x^2 = y + 7 = 9$, so $x = 3$ or -3.
- If $y = -3$, then $x^2 = y + 7 = 4$, so $x = 2$ or -2.

There are four solutions: $x = 3, y = 2$; $x = -3, y = 2$; $x = 2, y = -3$; and $x = -2, y = -3$.

You should verify that these four solutions also satisfy equation (1), so all four are solutions of the system. The four points, $(3, 2)$, $(-3, 2)$, $(2, -3)$, and $(-2, -3)$, are the points of intersection of the graphs. Look again at Figure 10. ⌐

Figure 10

━━━ **Now Work** PROBLEM 13 USING ELIMINATION

EXAMPLE 3

Solving a System of Nonlinear Equations

Solve: $\begin{cases} x^2 - y^2 = 4 & \text{(1)} \quad \text{A hyperbola} \\ \quad\;\; y = x^2 & \text{(2)} \quad \text{A parabola} \end{cases}$

Solution

Either substitution or elimination can be used here. To use substitution, replace x^2 by y in equation (1).

$$x^2 - y^2 = 4 \quad \text{Equation (1)}$$
$$y - y^2 = 4 \quad y = x^2$$
$$y^2 - y + 4 = 0 \quad \text{Place in standard form.}$$

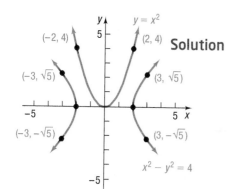

This is a quadratic equation. Its discriminant is $(-1)^2 - 4 \cdot 1 \cdot 4 = 1 - 16 = -15 < 0$. The equation has no real solutions, so the system is inconsistent. The graphs of these two equations do not intersect. See Figure 11.

Figure 11

EXAMPLE 4

Solving a System of Nonlinear Equations Using Elimination

Solve: $\begin{cases} x^2 + x + y^2 - 3y + 2 = 0 & \text{(1)} \\ x + 1 + \dfrac{y^2 - y}{x} = 0 & \text{(2)} \end{cases}$

Solution

First, multiply equation (2) by x to clear the denominator. The result is an equivalent system because x cannot be 0. [Look at equation (2) to see why.]

$$\begin{cases} x^2 + x + y^2 - 3y + 2 = 0 & \text{(1)} \\ x^2 + x + y^2 - y = 0 \quad x \neq 0 & \text{(2)} \end{cases}$$

Now subtract equation (2) from equation (1) to eliminate x. The result is

$$-2y + 2 = 0$$
$$y = 1 \quad \text{Solve for y.}$$

To find x, back-substitute $y = 1$ in equation (1).

$$x^2 + x + y^2 - 3y + 2 = 0 \quad \text{Equation (1)}$$
$$x^2 + x + 1 - 3 + 2 = 0 \quad \text{Substitute 1 for y.}$$
$$x^2 + x = 0 \quad \text{Simplify.}$$
$$x(x + 1) = 0 \quad \text{Factor.}$$
$$x = 0 \quad \text{or} \quad x = -1 \quad \text{Use the Zero-Product Property.}$$

Because x cannot be 0, the value $x = 0$ is extraneous, so discard it.

✓ **Check:** Check $x = -1, y = 1$:

$$\begin{cases} (-1)^2 + (-1) + 1^2 - 3 \cdot 1 + 2 = 1 - 1 + 1 - 3 + 2 = 0 & \text{(1)} \\ -1 + 1 + \dfrac{1^2 - 1}{-1} = 0 + \dfrac{0}{-1} = 0 & \text{(2)} \end{cases}$$

The solution is $x = -1, y = 1$. The point of intersection of the graphs of the equations is $(-1, 1)$.

In Problem 55 you are asked to graph the equations given in Example 4. Be sure to show holes in the graph of equation (2) for $x = 0$.

Now Work PROBLEMS 29 AND 49

EXAMPLE 5

Solving a System of Nonlinear Equations

Solve: $\begin{cases} 3xy - 2y^2 = -2 & \text{(1)} \\ 9x^2 + 4y^2 = 10 & \text{(2)} \end{cases}$

Solution   Multiply equation (1) by 2, and add the result to equation (2), to eliminate the y^2 terms.

$$\begin{cases} 6xy - 4y^2 = -4 & \text{(1)} \\ 9x^2 + 4y^2 = 10 & \text{(2)} \end{cases}$$

$$\begin{aligned} 9x^2 + 6xy &= 6 \quad \text{Add.} \\ 3x^2 + 2xy &= 2 \quad \text{Divide both sides by 3.} \end{aligned}$$

Since $x \neq 0$ (do you see why?), solve $3x^2 + 2xy = 2$ for y.

$$y = \frac{2 - 3x^2}{2x} \qquad x \neq 0 \quad \text{(3)}$$

Now substitute for y in equation (2) of the system.

$$9x^2 + 4y^2 = 10 \qquad \text{Equation (2)}$$

$$9x^2 + 4\left(\frac{2 - 3x^2}{2x}\right)^2 = 10 \qquad \text{Substitute } y = \frac{2 - 3x^2}{2x}.$$

$$9x^2 + \frac{4 - 12x^2 + 9x^4}{x^2} = 10 \qquad \text{Expand and simplify.}$$

$$9x^4 + 4 - 12x^2 + 9x^4 = 10x^2 \qquad \text{Multiply both sides by } x^2.$$

$$18x^4 - 22x^2 + 4 = 0 \qquad \text{Subtract } 10x^2 \text{ from both sides.}$$

$$9x^4 - 11x^2 + 2 = 0 \qquad \text{Divide both sides by 2.}$$

This quadratic equation (in x^2) can be factored:

$$(9x^2 - 2)(x^2 - 1) = 0$$

$$9x^2 - 2 = 0 \qquad \text{or} \quad x^2 - 1 = 0$$

$$x^2 = \frac{2}{9} \qquad \text{or} \qquad x^2 = 1$$

$$x = \pm\sqrt{\frac{2}{9}} = \pm\frac{\sqrt{2}}{3} \quad \text{or} \qquad x = \pm 1$$

To find y, use equation (3).

- If $x = \dfrac{\sqrt{2}}{3}$: $y = \dfrac{2 - 3x^2}{2x} = \dfrac{2 - \dfrac{2}{3}}{2 \cdot \dfrac{\sqrt{2}}{3}} = \dfrac{4}{2\sqrt{2}} = \sqrt{2}$

- If $x = -\dfrac{\sqrt{2}}{3}$: $y = \dfrac{2 - 3x^2}{2x} = \dfrac{2 - \dfrac{2}{3}}{2\left(-\dfrac{\sqrt{2}}{3}\right)} = \dfrac{4}{-2\sqrt{2}} = -\sqrt{2}$

- If $x = 1$: $y = \dfrac{2 - 3x^2}{2x} = \dfrac{2 - 3 \cdot 1^2}{2} = -\dfrac{1}{2}$

- If $x = -1$: $y = \dfrac{2 - 3x^2}{2x} = \dfrac{2 - 3(-1)^2}{-2} = \dfrac{1}{2}$

The system has four solutions: $\left(\dfrac{\sqrt{2}}{3}, \sqrt{2}\right), \left(-\dfrac{\sqrt{2}}{3}, -\sqrt{2}\right), \left(1, -\dfrac{1}{2}\right), \left(-1, \dfrac{1}{2}\right)$. Check them for yourself.

━━━━━➤ **Now Work** PROBLEM 47

The next example illustrates an imaginative solution to a system of nonlinear equations.

EXAMPLE 6

Running a Long-Distance Race

In a 50-mile race, the winner crosses the finish line 1 mile ahead of the second-place runner and 4 miles ahead of the third-place runner. Assuming that each runner maintains a constant speed throughout the race, by how many miles does the second-place runner beat the third-place runner?

Solution

Let v_1, v_2, and v_3 denote the speeds of the first-, second-, and third-place runners, respectively. Let t_1 and t_2 denote the times (in hours) required for the first-place runner and the second-place runner to finish the race. Then the following system of equations results:

$$\begin{cases} 50 = v_1 t_1 & (1) & \text{First-place runner goes 50 miles in } t_1 \text{ hours.} \\ 49 = v_2 t_1 & (2) & \text{Second-place runner goes 49 miles in } t_1 \text{ hours.} \\ 46 = v_3 t_1 & (3) & \text{Third-place runner goes 46 miles in } t_1 \text{ hours.} \\ 50 = v_2 t_2 & (4) & \text{Second-place runner goes 50 miles in } t_2 \text{ hours.} \end{cases}$$

We want the distance d of the third-place runner from the finish at time t_2. At time t_2, the third-place runner has gone a distance of $v_3 t_2$ miles, so the distance d remaining is $50 - v_3 t_2$. Now

$$d = 50 - v_3 t_2$$

$$= 50 - v_3 t_1 \cdot \frac{t_2}{t_1} \quad \text{Multiply and divide by } t_1.$$

$$= 50 - 46 \cdot \frac{\dfrac{50}{v_2}}{\dfrac{50}{v_1}} \quad \left\{ \begin{array}{l} \text{From equation (3), } v_3 t_1 = 46 \\[2mm] \text{From equation (4), } t_2 = \dfrac{50}{v_2} \\[2mm] \text{From equation (1), } t_1 = \dfrac{50}{v_1} \end{array} \right.$$

$$= 50 - 46 \cdot \frac{v_1}{v_2}$$

$$= 50 - 46 \cdot \frac{50}{49} \quad \text{From the quotient of equations (1) and (2)}$$

$$\approx 3.06 \text{ miles}$$

Historical Feature

I n the beginning of this section, it was stated that imagination and experience are important in solving systems of nonlinear equations. Indeed, these kinds of problems lead into some of the deepest and most difficult parts of modern mathematics. Look again at the graphs in Examples 1 and 2 of this section (Figures 9 and 10). Example 1 has two solutions, and Example 2 has four solutions. We might conjecture that the number of solutions is equal to the product of the degrees of the equations involved. This conjecture was made

by Étienne Bézout (1730–1783), but working out the details took about 150 years. It turns out that arriving at the correct number of intersections requires counting not only the complex number intersections, but also those intersections that, in a certain sense, lie at infinity. For example, a parabola and a line lying on the axis of the parabola intersect at the vertex and at infinity. This topic is part of the study of algebraic geometry.

Historical Problems

A papyrus dating back to 1950 BC contains the following problem: "A given surface area of 100 units of area shall be represented as the sum of two squares whose sides are to each other as 1 is to $\dfrac{3}{4}$."

Solve for the sides by solving the system of equations

$$\begin{cases} x^2 + y^2 = 100 \\ \quad x = \dfrac{3}{4}y \end{cases}$$

8.6 Assess Your Understanding

'Are You Prepared?' *Answers are given at the end of these exercises. If you get a wrong answer, read the pages listed in* red.

1. Graph the equation: $y = 3x + 2$ (pp. 169–179)
2. Graph the equation: $y + 4 = x^2$ (pp. 519–523)
3. Graph the equation: $y^2 = x^2 - 1$ (pp. 538–545)
4. Graph the equation: $x^2 + 4y^2 = 4$ (pp. 528–532)

Skill Building

In Problems 5–24, graph each equation of the system. Then solve the system to find the points of intersection.

5. $\begin{cases} y = x^2 + 1 \\ y = x + 1 \end{cases}$

6. $\begin{cases} y = x^2 + 1 \\ y = 4x + 1 \end{cases}$

7. $\begin{cases} y = \sqrt{36 - x^2} \\ y = 8 - x \end{cases}$

8. $\begin{cases} y = \sqrt{4 - x^2} \\ y = 2x + 4 \end{cases}$

9. $\begin{cases} y = \sqrt{x} \\ y = 2 - x \end{cases}$

10. $\begin{cases} y = \sqrt{x} \\ y = 6 - x \end{cases}$

11. $\begin{cases} x = 2y \\ x = y^2 - 2y \end{cases}$

12. $\begin{cases} y = x - 1 \\ y = x^2 - 6x + 9 \end{cases}$

13. $\begin{cases} x^2 + y^2 = 4 \\ x^2 + 2x + y^2 = 0 \end{cases}$

14. $\begin{cases} x^2 + y^2 = 8 \\ x^2 + y^2 + 4y = 0 \end{cases}$

15. $\begin{cases} y = 3x - 5 \\ x^2 + y^2 = 5 \end{cases}$

16. $\begin{cases} x^2 + y^2 = 10 \\ y = x + 2 \end{cases}$

17. $\begin{cases} x^2 + y^2 = 4 \\ y^2 - x = 4 \end{cases}$

18. $\begin{cases} x^2 + y^2 = 16 \\ x^2 - 2y = 8 \end{cases}$

19. $\begin{cases} xy = 4 \\ x^2 + y^2 = 8 \end{cases}$

20. $\begin{cases} x^2 = y \\ xy = 1 \end{cases}$

21. $\begin{cases} x^2 + y^2 = 4 \\ y = x^2 - 9 \end{cases}$

22. $\begin{cases} xy = 1 \\ y = 2x + 1 \end{cases}$

23. $\begin{cases} y = x^2 - 4 \\ y = 6x - 13 \end{cases}$

24. $\begin{cases} x^2 + y^2 = 10 \\ xy = 3 \end{cases}$

In Problems 25–54, solve each system. Use any method you wish.

25. $\begin{cases} 2x^2 + y^2 = 18 \\ \quad xy = 4 \end{cases}$

26. $\begin{cases} x^2 - y^2 = 21 \\ x + y = 7 \end{cases}$

27. $\begin{cases} y = 3x + 2 \\ 3x^2 + y^2 = 4 \end{cases}$

28. $\begin{cases} x^2 - 4y^2 = 16 \\ 2y - x = 2 \end{cases}$

29. $\begin{cases} x + y + 1 = 0 \\ x^2 + y^2 + 6y - x = -5 \end{cases}$

30. $\begin{cases} 2x^2 - xy + y^2 = 8 \\ \quad xy = 4 \end{cases}$

31. $\begin{cases} 9x^2 - 8xy + 4y^2 = 70 \\ 3x + 2y = 10 \end{cases}$

32. $\begin{cases} 2y^2 - 3xy + 6y + 2x + 4 = 0 \\ 2x - 3y + 4 = 0 \end{cases}$

33. $\begin{cases} x^2 - 4y^2 + 7 = 0 \\ 3x^2 + y^2 = 31 \end{cases}$

34. $\begin{cases} 3x^2 - 2y^2 + 5 = 0 \\ 2x^2 - y^2 + 2 = 0 \end{cases}$

35. $\begin{cases} 7x^2 - 3y^2 + 5 = 0 \\ 3x^2 + 5y^2 = 12 \end{cases}$

36. $\begin{cases} x^2 - 3y^2 + 1 = 0 \\ 2x^2 - 7y^2 + 5 = 0 \end{cases}$

37. $\begin{cases} x^2 + 2xy = 10 \\ 3x^2 - xy = 2 \end{cases}$

38. $\begin{cases} 5xy + 13y^2 + 36 = 0 \\ xy + 7y^2 = 6 \end{cases}$

39. $\begin{cases} 2x^2 + y^2 = 2 \\ x^2 - 2y^2 + 8 = 0 \end{cases}$

40. $\begin{cases} y^2 - x^2 + 4 = 0 \\ 2x^2 + 3y^2 = 6 \end{cases}$

41. $\begin{cases} x^2 + 2y^2 = 16 \\ 4x^2 - y^2 = 24 \end{cases}$

42. $\begin{cases} 4x^2 + 3y^2 = 4 \\ 2x^2 - 6y^2 = -3 \end{cases}$

43. $\begin{cases} \dfrac{5}{x^2} - \dfrac{2}{y^2} + 3 = 0 \\ \dfrac{3}{x^2} + \dfrac{1}{y^2} = 7 \end{cases}$

44. $\begin{cases} \dfrac{2}{x^2} - \dfrac{3}{y^2} + 1 = 0 \\ \dfrac{6}{x^2} - \dfrac{7}{y^2} + 2 = 0 \end{cases}$

45. $\begin{cases} \dfrac{1}{x^4} + \dfrac{6}{y^4} = 6 \\ \dfrac{2}{x^4} - \dfrac{2}{y^4} = 19 \end{cases}$

46. $\begin{cases} \dfrac{1}{x^4} - \dfrac{1}{y^4} = 1 \\ \dfrac{1}{x^4} + \dfrac{1}{y^4} = 4 \end{cases}$

47. $\begin{cases} x^2 - 3xy + 2y^2 = 0 \\ x^2 + xy = 6 \end{cases}$

48. $\begin{cases} x^2 - xy - 2y^2 = 0 \\ xy + x + 6 = 0 \end{cases}$

49. $\begin{cases} y^2 + y + x^2 - x - 2 = 0 \\ y + 1 + \dfrac{x - 2}{y} = 0 \end{cases}$

50. $\begin{cases} x^3 - 2x^2 + y^2 + 3y - 4 = 0 \\ x - 2 + \dfrac{y^2 - y}{x^2} = 0 \end{cases}$

51. $\begin{cases} \log_x y = 3 \\ \log_x(4y) = 5 \end{cases}$

52. $\begin{cases} \log_x(2y) = 3 \\ \log_x(4y) = 2 \end{cases}$

53. $\begin{cases} \ln x = 4 \ln y \\ \log_3 x = 2 + 2 \log_3 y \end{cases}$

54. $\begin{cases} \ln x = 5 \ln y \\ \log_2 x = 3 + 2 \log_2 y \end{cases}$

55. Graph the equations given in Example 4.

56. Graph the equations given in Problem 49.

In Problems 57–64, use a graphing utility to solve each system of equations. Express the solution(s) rounded to two decimal places.

57. $\begin{cases} y = x^{2/3} \\ y = e^{-x} \end{cases}$

58. $\begin{cases} y = x^{3/2} \\ y = e^{-x} \end{cases}$

59. $\begin{cases} x^2 + y^3 = 2 \\ x^3 y = 4 \end{cases}$

60. $\begin{cases} x^3 + y^2 = 2 \\ x^2 y = 4 \end{cases}$

61. $\begin{cases} x^4 + y^4 = 12 \\ xy^2 = 2 \end{cases}$

62. $\begin{cases} x^4 + y^4 = 6 \\ xy = 1 \end{cases}$

63. $\begin{cases} xy = 2 \\ y = \ln x \end{cases}$

64. $\begin{cases} x^2 + y^2 = 4 \\ y = \ln x \end{cases}$

Mixed Practice *In Problems 65–70, graph each equation and find the point(s) of intersection, if any.*

65. The line $x + 2y = 0$ and
the circle $(x - 1)^2 + (y - 1)^2 = 5$

66. The line $x + 2y + 6 = 0$ and
the circle $(x + 1)^2 + (y + 1)^2 = 5$

67. The circle $(x - 1)^2 + (y + 2)^2 = 4$ and
the parabola $y^2 + 4y - x + 1 = 0$

68. The circle $(x + 2)^2 + (y - 1)^2 = 4$ and
the parabola $y^2 - 2y - x - 5 = 0$

69. $y = \dfrac{4}{x - 3}$ and the circle $x^2 - 6x + y^2 + 1 = 0$

70. $y = \dfrac{4}{x + 2}$ and the circle $x^2 + 4x + y^2 - 4 = 0$

Applications and Extensions

71. The difference of two numbers is 2 and the sum of their squares is 10. Find the numbers.

72. The sum of two numbers is 7 and the difference of their squares is 21. Find the numbers.

73. The product of two numbers is 4 and the sum of their squares is 8. Find the numbers.

74. The product of two numbers is 10 and the difference of their squares is 21. Find the numbers.

75. The difference of two numbers is the same as their product, and the sum of their reciprocals is 5. Find the numbers.

76. The sum of two numbers is the same as their product, and the difference of their reciprocals is 3. Find the numbers.

77. The ratio of a to b is $\dfrac{2}{3}$. The sum of a and b is 10. What is the ratio of $a + b$ to $b - a$?

78. The ratio of a to b is 4:3. The sum of a and b is 14. What is the ratio of $a - b$ to $a + b$?

79. Geometry The perimeter of a rectangle is 16 inches and its area is 15 square inches. What are its dimensions?

80. Geometry An area of 52 square feet is to be enclosed by two squares whose sides are in the ratio of 2:3. Find the sides of the squares.

81. Geometry Two circles have circumferences that add up to 12π centimeters and areas that add up to 20π square centimeters. Find the radius of each circle.

82. Geometry The altitude of an isosceles triangle drawn to its base is 3 centimeters, and its perimeter is 18 centimeters. Find the length of its base.

83. The Tortoise and the Hare In a 21-meter race between a tortoise and a hare, the tortoise leaves 9 minutes before the hare. The hare, by running at an average speed of 0.5 meter per hour faster than the tortoise, crosses the finish line 3 minutes before the tortoise. What are the average speeds of the tortoise and the hare?

84. Running a Race In a 1-mile race, the winner crosses the finish line 10 feet ahead of the second-place runner and 20 feet ahead of the third-place runner. Assuming that each runner maintains a constant speed throughout the race, by how many feet does the second-place runner beat the third-place runner?

85. Constructing a Box A rectangular piece of cardboard, whose area is 216 square centimeters, is made into an open box by cutting a 2-centimeter square from each corner and turning up the sides. See the figure. If the box is to have a volume of 224 cubic centimeters, what size cardboard should you start with?

86. Constructing a Cylindrical Tube A rectangular piece of cardboard, whose area is 216 square centimeters, is made into a cylindrical tube by joining together two sides of the rectangle. See the figure. If the tube is to have a volume of 224 cubic centimeters, what size cardboard should you start with?

87. Fencing A farmer has 300 feet of fence available to enclose a 4500-square-foot region in the shape of adjoining squares, with sides of length x and y. See the figure. Find x and y.

88. Bending Wire A wire 60 feet long is cut into two pieces. Is it possible to bend one piece into the shape of a square and the other into the shape of a circle so that the total area enclosed by the two pieces is 100 square feet? If this is possible, find the length of the side of the square and the radius of the circle.

89. Descartes' Method of Equal Roots Descartes' method for finding tangent lines depends on the idea that, for many graphs, the tangent line at a given point is the *unique* line that intersects the graph at that point only. We use his method to find an equation of the tangent line to the parabola $y = x^2$ at the point $(2, 4)$. See the figure.

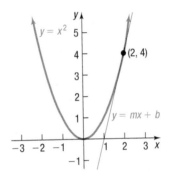

First, an equation of the tangent line can be written as $y = mx + b$. Using the fact that the point $(2, 4)$ is on the line, we can solve for b in terms of m and get the equation $y = mx + (4 - 2m)$. Now we want $(2, 4)$ to be the *unique* solution to the system

$$\begin{cases} y = x^2 \\ y = mx + 4 - 2m \end{cases}$$

From this system, we get $x^2 - mx + (2m - 4) = 0$. Using the quadratic formula, we get

$$x = \frac{m \pm \sqrt{m^2 - 4(2m - 4)}}{2}$$

To obtain a unique solution for x, the two roots must be equal; in other words, the discriminant $m^2 - 4(2m - 4)$ must be 0. Complete the work to get m, and write an equation of the tangent line.

In Problems 90–96, use Descartes' method from Problem 89 to find an equation of the tangent line to each graph at the given point.

90. $x^2 + y^2 = 10$; at $(1, 3)$

91. $y = x^2 + 2$; at $(1, 3)$

92. $x^2 + y = 5$; at $(-2, 1)$

93. $2x^2 + 3y^2 = 14$; at $(1, 2)$

94. $3x^2 + y^2 = 7$; at $(-1, 2)$

95. $x^2 - y^2 = 3$; at $(2, 1)$

96. $2y^2 - x^2 = 14$; at $(2, 3)$

97. If r_1 and r_2 are two solutions of a quadratic equation $ax^2 + bx + c = 0$, it can be shown that

$$r_1 + r_2 = -\frac{b}{a} \quad \text{and} \quad r_1 r_2 = \frac{c}{a}$$

Solve this system of equations for r_1 and r_2.

98. Challenge Problem Solve for x and y in terms of $a \neq 0$ and $b \neq 0$:

$$\begin{cases} \dfrac{x^2}{a^2} + \dfrac{y^2}{b^2} = \dfrac{a^2 + b^2}{a^2 b^2} \\ \dfrac{x}{a} + \dfrac{y}{b} = \dfrac{a + b}{ab} \end{cases}$$

99. Challenge Problem Geometry Find formulas for the length l and width w of a rectangle in terms of its area A and perimeter P.

100. Challenge Problem Geometry Find formulas for the base b and one of the equal sides l of an isosceles triangle in terms of its altitude h and perimeter P.

Explaining Concepts: Discussion and Writing

101. A circle and a line intersect at most twice. A circle and a parabola intersect at most four times. Deduce that a circle and the graph of a polynomial of degree 3 intersect at most six times. What do you conjecture about a polynomial of degree 4? What about a polynomial of degree n? Can you explain your conclusions using an algebraic argument?

102. Suppose you are the manager of a sheet metal shop. A customer asks you to manufacture 10,000 boxes, each box being open on top. The boxes are required to have a square base and a 9-cubic-foot capacity. You construct the boxes by cutting out a square from each corner of a square piece of sheet metal and folding along the edges.

(a) Find the dimensions of the square to be cut if the area of the square piece of sheet metal is 100 square feet.

(b) Could you make the box using a smaller piece of sheet metal? Make a list of the dimensions of the box for various pieces of sheet metal.

Retain Your Knowledge

Problems 103–110 are based on material learned earlier in the course. The purpose of these problems is to keep the material fresh in your mind so that you are better prepared for the final exam.

103. Solve: $7x^2 = 8 - 6x$

104. Find an equation of the line with slope $-\dfrac{2}{5}$ that contains the point $(10, -7)$.

105. Find an equation of the circle with center at $(-3, 4)$ and radius 10.

106. Solve: $x^2 < 4x + 21$

107. Find the function that is finally graphed after $y = \sqrt{25 - x^2}$ is reflected about the x-axis and shifted right 4 units.

108. If $f(x) = 2x^2 - 8x + 7$, find $f(x - 3)$.

109. Find the difference quotient of $f(x) = \dfrac{3x}{x - 8}$. Simplify the answer.

110. Simplify: $\dfrac{(2x - 5)^9 \cdot 3 - 3x \cdot 9(2x - 5)^8 \cdot 2}{[(2x - 5)^9]^2}$

'Are You Prepared?' Answers

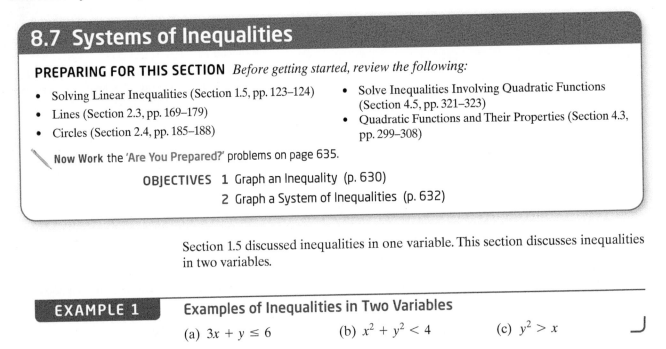

8.7 Systems of Inequalities

PREPARING FOR THIS SECTION *Before getting started, review the following:*

- Solving Linear Inequalities (Section 1.5, pp. 123–124)
- Lines (Section 2.3, pp. 169–179)
- Circles (Section 2.4, pp. 185–188)

- Solve Inequalities Involving Quadratic Functions (Section 4.5, pp. 321–323)
- Quadratic Functions and Their Properties (Section 4.3, pp. 299–308)

Now Work the 'Are You Prepared?' problems on page 635.

OBJECTIVES 1 Graph an Inequality (p. 630)
2 Graph a System of Inequalities (p. 632)

Section 1.5 discussed inequalities in one variable. This section discusses inequalities in two variables.

EXAMPLE 1

Examples of Inequalities in Two Variables

(a) $3x + y \leq 6$ 　　　(b) $x^2 + y^2 < 4$ 　　　(c) $y^2 > x$

---

1 Graph an Inequality

An inequality in two variables x and y is **satisfied** by an ordered pair (a, b) if, when x is replaced by a and y by b, a true statement results. The **graph of an inequality in two variables** x and y consists of all points (x, y) whose coordinates satisfy the inequality.

EXAMPLE 2

Graphing an Inequality

Graph the linear inequality: $3x + y \leq 6$

Solution Begin by graphing the equation

$$3x + y = 6$$

formed by replacing (for now) the $\leq$ symbol with an $=$ sign. The graph of the equation is a line. See Figure 12(a). This line is part of the graph of the inequality because the inequality is nonstrict, so the line is drawn as a solid line. (Do you see why? We are seeking points for which $3x + y$ is less than *or equal to* 6.)

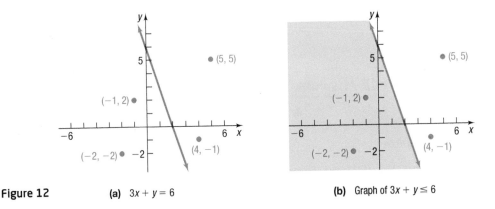

Figure 12　　　**(a)** $3x + y = 6$ 　　　　　　　　**(b)** Graph of $3x + y \leq 6$

Now test a few randomly selected points to see whether they belong to the graph of the inequality.

	$3x + y \le 6$	**Conclusion**
$(4, -1)$	$3 \cdot 4 + (-1) = 11 > 6$	Does not belong to the graph
$(5, 5)$	$3 \cdot 5 + 5 = 20 > 6$	Does not belong to the graph
$(-1, 2)$	$3(-1) + 2 = -1 \le 6$	Belongs to the graph
$(-2, -2)$	$3(-2) + (-2) = -8 \le 6$	Belongs to the graph

Look again at Figure 12(a). Notice that the two points that belong to the graph both lie on the same side of the line, and the two points that do not belong to the graph lie on the opposite side. As it turns out, all the points that satisfy the inequality will lie on one side of the line or on the line itself. All the points that do not satisfy the inequality will lie on the other side. The graph of $3x + y \le 6$ consists of all points that lie on the line or on the same side of the line as $(-1, 2)$ and $(-2, -2)$. This graph is shown as the shaded region in Figure 12(b).

━━━ **Now Work** PROBLEM 15

The graph of any inequality in two variables may be obtained similarly. The steps to follow are given next.

RECALL The strict inequalities are $<$ and $>$. The nonstrict inequalities are $\le$ and $\ge$. ∎

Steps for Graphing an Inequality

STEP 1: Replace the inequality symbol by an equal sign, and graph the resulting equation. If the inequality is strict, use dashes; if it is nonstrict, use a solid mark. This graph separates the xy-plane into two or more regions.

STEP 2: In each region, select a test point P.
- If the coordinates of P satisfy the inequality, so do all the points in that region. Indicate this by shading the region.
- If the coordinates of P do not satisfy the inequality, no point in that region satisfies the inequality.

EXAMPLE 3

Graphing an Inequality

Graph: $x^2 + y^2 \le 4$

Solution

STEP 1: Graph the equation $x^2 + y^2 = 4$, a circle of radius 2, with center at the origin. A solid circle is used because the inequality is not strict.

STEP 2: Use two test points, one inside the circle, the other outside.

Inside $(0, 0)$: $x^2 + y^2 = 0^2 + 0^2 = 0 \le 4$ **Belongs to the graph**

Outside $(4, 0)$: $x^2 + y^2 = 4^2 + 0^2 = 16 > 4$ **Does not belong to the graph**

All the points inside and on the circle satisfy the inequality. See Figure 13.

Figure 13 $x^2 + y^2 \le 4$

━━━ **Now Work** PROBLEM 17

Linear Inequalities

A linear inequality is an inequality in one of the forms

$$Ax + By < C \qquad Ax + By > C \qquad Ax + By \le C \qquad Ax + By \ge C$$

where A and B are not both zero.

The graph of the corresponding equation of a linear inequality is a line that separates the xy-plane into two regions called **half-planes**. See Figure 14.

As shown, $Ax + By = C$ is the equation of the boundary line, and it divides the plane into two half-planes: one for which $Ax + By < C$ and the other for which $Ax + By > C$. Because of this, for linear inequalities, only one test point is required.

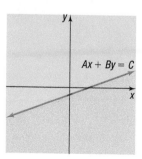

Figure 14

EXAMPLE 4 | **Graphing Linear Inequalities**

Graph: (a) $y < 2$ (b) $y \geq 2x$

Solution

(a) Points on the horizontal line $y = 2$ are not part of the graph of the inequality, so the graph is shown as a dashed line. Since $(0, 0)$ satisfies the inequality, the graph consists of the half-plane below the line $y = 2$. See Figure 15.

(b) Points on the line $y = 2x$ are part of the graph of the inequality, so the graph is shown as a solid line. Use $(3, 0)$ as a test point. It does not satisfy the inequality $[0 < 2 \cdot 3]$. Points in the half-plane opposite the side containing $(3, 0)$ satisfy the inequality. See Figure 16.

COMMENT A graphing utility can be used to graph inequalities. To see how, read Section 6 in the Appendix. ∎

Figure 15 $y < 2$

Figure 16 $y \geq 2x$

Now Work PROBLEM 13

2 Graph a System of Inequalities

The **graph of a system of inequalities** in two variables x and y is the set of all points (x, y) that simultaneously satisfy *each* inequality in the system. The graph of a system of inequalities can be obtained by graphing each inequality individually and then determining where, if at all, they intersect.

EXAMPLE 5 | **Graphing a System of Linear Inequalities**

Graph the system: $\begin{cases} x + y \geq 2 \\ 2x - y \leq 4 \end{cases}$

Solution

Begin by graphing the lines $x + y = 2$ and $2x - y = 4$ using a solid line since both inequalities are nonstrict. Use the test point $(0, 0)$ on each inequality. For example, $(0, 0)$ does not satisfy $x + y \geq 2$, so shade above the line $x + y = 2$. See Figure 17(a). But $(0, 0)$ does satisfy $2x - y \leq 4$, so shade above the line $2x - y = 4$. See Figure 17(b). The intersection of the shaded regions (in purple) gives the result presented in Figure 17(c).

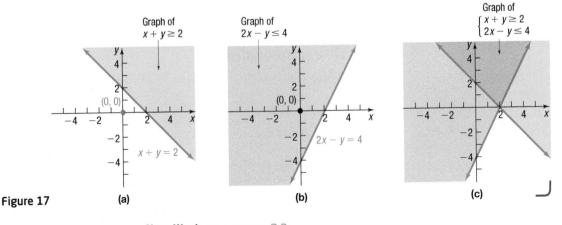

Figure 17   (a)   (b)   (c)

Now Work PROBLEM 23

EXAMPLE 6   **Graphing a System of Linear Inequalities**

Graph the system: $\begin{cases} x + y \leq 2 \\ x + y \geq 0 \end{cases}$

Solution   See Figure 18. The overlapping purple-shaded region between the two boundary lines is the graph of the system.

Figure 18

Now Work PROBLEM 29

EXAMPLE 7   **Graphing a System of Linear Inequalities**

Graph the systems:

(a) $\begin{cases} 2x - y \geq 0 \\ 2x - y \geq 2 \end{cases}$

(b) $\begin{cases} x + 2y \leq 2 \\ x + 2y \geq 6 \end{cases}$

Solution   (a) See Figure 19. The overlapping purple-shaded region is the graph of the system. Note that the graph of the system is identical to the graph of the single inequality $2x - y \geq 2$.

(b) See Figure 20. Here, because no overlapping region results, there are no points in the xy-plane that simultaneously satisfy each inequality. The system has no solution.

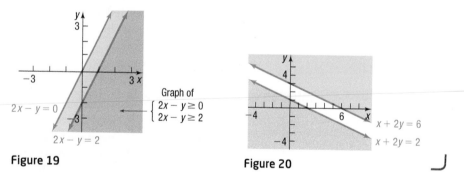

Figure 19

Figure 20

EXAMPLE 8   **Graphing a System of Nonlinear Inequalities**

Graph the region below the graph of $x + y = 2$ and above the graph of $y = x^2 - 4$ by graphing the system

$$\begin{cases} y \geq x^2 - 4 \\ x + y \leq 2 \end{cases}$$

Label all points of intersection.

Solution   Figure 21 on the next page shows the graph of the region above the graph of the parabola $y = x^2 - 4$ and below the graph of the line $x + y = 2$. The points of intersection are found by solving the system of equations

$$\begin{cases} y = x^2 - 4 \\ x + y = 2 \end{cases}$$

(continued)

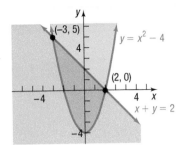

Figure 21

Use substitution to find

$$x + (x^2 - 4) = 2$$
$$x^2 + x - 6 = 0$$
$$(x + 3)(x - 2) = 0$$
$$x = -3 \quad \text{or} \quad x = 2$$

The two points of intersection are $(-3, 5)$ and $(2, 0)$.

 Now Work PROBLEM 37

EXAMPLE 9   Graphing a System of Four Linear Inequalities

Graph the system: $\begin{cases} x + y \geq 3 \\ 2x + y \geq 4 \\ x \geq 0 \\ y \geq 0 \end{cases}$

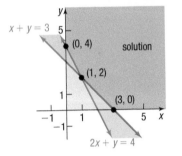

Figure 22

Solution   See Figure 22. The two inequalities $x \geq 0$ and $y \geq 0$ require the graph of the system to be in quadrant I. Concentrate on the remaining two inequalities. The intersection of the graphs of these two inequalities and quadrant I is shown in dark purple.

EXAMPLE 10   Financial Planning

A retired couple can invest up to $25,000. As their financial adviser, you recommend that they place at least $15,000 in Treasury bills yielding 2% and at most $5000 in corporate bonds yielding 3%.

(a) Using x to denote the amount of money invested in Treasury bills and y to denote the amount invested in corporate bonds, write a system of linear inequalities that describes the possible amounts of each investment. Assume that x and y are in thousands of dollars.

(b) Graph the system.

Solution   (a) The system of linear inequalities is

$\begin{cases} x \geq 0 & \text{x and y are nonnegative variables since they represent} \\ y \geq 0 & \text{money invested, in thousands of dollars.} \\ x + y \leq 25 & \text{The total of the two investments, $x + y$, cannot exceed \$25,000.} \\ x \geq 15 & \text{At least \$15,000 in Treasury bills} \\ y \leq 5 & \text{At most \$5000 in corporate bonds} \end{cases}$

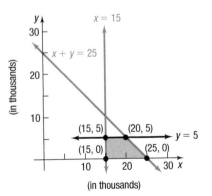

Figure 23

(b) See the shaded region in Figure 23. Note that the inequalities $x \geq 0$ and $y \geq 0$ require that the graph of the system be in quadrant I.

The graph of the system of linear inequalities in Figure 23 is **bounded**, because it can be contained within some circle of sufficiently large radius. A graph that cannot be contained in any circle is **unbounded**. For example, the graph of the system of linear inequalities in Figure 22 is unbounded, since it extends indefinitely in the positive x and positive y directions.

Notice in Figures 22 and 23 that those points that belong to the graph and are also points of intersection of boundary lines have been plotted. Such points are referred to as **vertices** or **corner points** of the graph. The system graphed in Figure 22 has three corner points: $(0, 4)$, $(1, 2)$, and $(3, 0)$. The system graphed in Figure 23 has four corner points: $(15, 0)$, $(25, 0)$, $(20, 5)$, and $(15, 5)$.

These ideas are used in the next section in developing a method for solving linear programming problems, an important application of linear inequalities.

━━━━━**Now Work** PROBLEM 45

8.7 Assess Your Understanding

'Are You Prepared?' *Answers are given at the end of these exercises. If you get a wrong answer, read the pages listed in* red.

1. Solve the inequality: $3x + 4 < 8 - x$ (pp. 123–124)
2. Graph the equation: $3x - 2y = 6$ (pp. 176–177)
3. Graph the equation: $x^2 + y^2 = 9$ (pp. 185–188)
4. Graph the equation: $y = x^2 + 4$ (pp. 300–305)

5. **True or False** The lines $2x + y = 4$ and $4x + 2y = 0$ are parallel. (pp. 177–178)
6. Solve the inequality: $x^2 - 4 \le 5$ (pp. 321–323)

Concepts and Vocabulary

7. When graphing an inequality in two variables, use _____ if the inequality is strict; if the inequality is nonstrict, use a _____ mark.
8. The graph of a linear equation is a line that separates the xy-plane into two regions called _____.

9. **True or False** The graph of a system of inequalities must have an overlapping region.
10. **Multiple Choice** If the graph of a system of inequalities cannot be contained in any circle, then the graph is:
    (a) bounded   (b) unbounded
    (c) decomposed   (d) composed

Skill Building

In Problems 11–22, graph each inequality.

11. $x \ge 0$
12. $y \ge 0$
13. $x \ge 4$
14. $y \le 2$
15. $2x + y \ge 6$
16. $3x + 2y \le 6$
17. $x^2 + y^2 > 1$
18. $x^2 + y^2 \le 9$
19. $y \le x^2 - 1$
20. $y > x^2 + 2$
21. $xy \ge 4$
22. $xy \le 1$

In Problems 23–34, graph each system of linear inequalities.

23. $\begin{cases} x + y \le 2 \\ 2x + y \ge 4 \end{cases}$
24. $\begin{cases} 3x - y \ge 6 \\ x + 2y \le 2 \end{cases}$
25. $\begin{cases} 2x - y \le 4 \\ 3x + 2y \ge -6 \end{cases}$
26. $\begin{cases} 4x - 5y \le 0 \\ 2x - y \ge 2 \end{cases}$

27. $\begin{cases} 2x - 3y \le 0 \\ 3x + 2y \le 6 \end{cases}$
28. $\begin{cases} 4x - y \ge 2 \\ x + 2y \ge 2 \end{cases}$
29. $\begin{cases} x - 2y \le 6 \\ 2x - 4y \ge 0 \end{cases}$
30. $\begin{cases} x + 4y \le 8 \\ x + 4y \ge 4 \end{cases}$

31. $\begin{cases} 2x + y \ge -2 \\ 2x + y \ge 2 \end{cases}$
32. $\begin{cases} x - 4y \le 4 \\ x - 4y \ge 0 \end{cases}$
33. $\begin{cases} 2x + 3y \ge 6 \\ 2x + 3y \le 0 \end{cases}$
34. $\begin{cases} 2x + y \ge 0 \\ 2x + y \ge 2 \end{cases}$

In Problems 35–42, graph each system of inequalities.

35. $\begin{cases} x^2 + y^2 \le 9 \\ x + y \ge 3 \end{cases}$
36. $\begin{cases} x^2 + y^2 \ge 9 \\ x + y \le 3 \end{cases}$
37. $\begin{cases} y \ge x^2 - 4 \\ y \le x - 2 \end{cases}$
38. $\begin{cases} y^2 \le x \\ y \ge x \end{cases}$

39. $\begin{cases} x^2 + y^2 \le 16 \\ y \ge x^2 - 4 \end{cases}$
40. $\begin{cases} x^2 + y^2 \le 25 \\ y \le x^2 - 5 \end{cases}$
41. $\begin{cases} xy \ge 4 \\ y \ge x^2 + 1 \end{cases}$
42. $\begin{cases} y + x^2 \le 1 \\ y \ge x^2 - 1 \end{cases}$

In Problems 43–52, graph each system of linear inequalities. State whether the graph is bounded or unbounded, and label the corner points.

43. $\begin{cases} x \ge 0 \\ y \ge 0 \\ 2x + y \le 6 \\ x + 2y \le 6 \end{cases}$
44. $\begin{cases} x \ge 0 \\ y \ge 0 \\ x + y \ge 4 \\ 2x + 3y \ge 6 \end{cases}$
45. $\begin{cases} x \ge 0 \\ y \ge 0 \\ x + y \ge 2 \\ 2x + y \ge 4 \end{cases}$
46. $\begin{cases} x \ge 0 \\ y \ge 0 \\ 3x + y \le 6 \\ 2x + y \le 2 \end{cases}$

47. $\begin{cases} x \ge 0 \\ y \ge 0 \\ x + y \ge 2 \\ 2x + 3y \le 12 \\ 3x + y \le 12 \end{cases}$
48. $\begin{cases} x \ge 0 \\ y \ge 0 \\ x + y \ge 1 \\ x + y \le 7 \\ 2x + y \le 10 \end{cases}$
49. $\begin{cases} x \ge 0 \\ y \ge 0 \\ x + y \ge 2 \\ x + y \le 8 \\ 2x + y \le 10 \end{cases}$
50. $\begin{cases} x \ge 0 \\ y \ge 0 \\ x + y \ge 2 \\ x + y \le 8 \\ x + 2y \ge 1 \end{cases}$

51. $\begin{cases} x \geq 0 \\ y \geq 0 \\ x + 2y \geq 1 \\ x + 2y \leq 10 \end{cases}$

52. $\begin{cases} x \geq 0 \\ y \geq 0 \\ x + 2y \geq 1 \\ x + 2y \leq 10 \\ x + y \geq 2 \\ x + y \leq 8 \end{cases}$

In Problems 53–56, write a system of linear inequalities for the given graph.

53.

54.

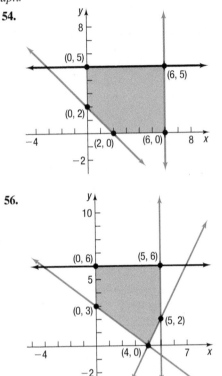

55.

56.

Applications and Extensions

57. Financial Planning A retired couple has up to $50,000 to invest. As their financial adviser, you recommend that they place at least $35,000 in Treasury bills yielding 1% and at most $10,000 in corporate bonds yielding 3%.

  (a) Using x to denote the amount of money invested in Treasury bills and y to denote the amount invested in corporate bonds, write a system of linear inequalities that describes the possible amounts of each investment.

  (b) Graph the system and label the corner points.

58. Manufacturing Trucks Mike's Toy Truck Company manufactures two models of toy trucks, a standard model and a deluxe model. Each standard model requires 2 hours (h) for painting and 3 h for detail work; each deluxe model requires 3 h for painting and 4 h for detail work. Two painters and three detail workers are employed by the company, and each works 40 h per week.

  (a) Using x to denote the number of standard-model trucks and y to denote the number of deluxe-model trucks, write a system of linear inequalities that describes the possible numbers of each model of truck that can be manufactured in a week.

  (b) Graph the system and label the corner points.

59. Blending Coffee Bill's Coffee House, a store that specializes in coffee, has available 75 pounds (lb) of A grade coffee and 120 lb of B grade coffee. These will be blended into 1-lb packages as follows: an economy blend that contains 4 ounces (oz) of A grade coffee and 12 oz of B grade coffee, and a superior blend that contains 8 oz of A grade coffee and 8 oz of B grade coffee.

  (a) Using x to denote the number of packages of the economy blend and y to denote the number of packages of the superior blend, write a system of linear inequalities that describes the possible numbers of packages of each kind of blend.

  (b) Graph the system and label the corner points.

60. Mixed Nuts Nola's Nuts, a store that specializes in selling nuts, has available 90 pounds (lb) of cashews and 120 lb of peanuts. These are to be mixed in 12-ounce (oz) packages as follows: a lower-priced package containing 8 oz of peanuts and 4 oz of cashews, and a quality package containing 6 oz of peanuts and 6 oz of cashews.
  (a) Use x to denote the number of lower-priced packages, and use y to denote the number of quality packages. Write a system of linear inequalities that describes the possible numbers of each kind of package.
  (b) Graph the system and label the corner points.

61. Transporting Goods A small truck can carry no more than 1600 pounds (lb) of cargo and no more than 150 cubic

feet (ft^3) of cargo. A printer weighs 20 lb and occupies 3 ft^3 of space. A microwave oven weighs 30 lb and occupies 2 ft^3 of space.
  (a) Using x to represent the number of microwave ovens and y to represent the number of printers, write a system of linear inequalities that describes the number of ovens and printers that can be hauled by the truck.
  (b) Graph the system and label the corner points.

62. Challenge Problem Graph the system of inequalities.
$$\begin{cases} |x| + |y| \le 4 \\ |y| \le |x^2 - 3| \end{cases}$$

Retain Your Knowledge

Problems 63–70 are based on material learned earlier in the course. The purpose of these problems is to keep the material fresh in your mind so that you are better prepared for the final exam.

63. Solve $2(x + 1)^2 + 8 = 0$ in the complex number system.

64. Use the Intermediate Value Theorem to show that $f(x) = 6x^2 + 5x - 6$ has a real zero on the interval $[-1, 2]$.

65. Solve: $x - 2 \le -4x + 3 \le x + 18$

66. If \$7500 is invested in an account paying 3.25% interest compounded daily, how much money will be in the account after 5 years?

67. The horsepower P needed to propel a boat through water is directly proportional to the cube of the boat's speed s. If a boat needs 150 horsepower to travel 12 miles per hour, what horsepower does it need to travel 6 miles per hour?

68. Change $y = \log_5 x$ to an equivalent statement involving an exponent.

69. Given $f(x) = \dfrac{2}{x - 5}$ and $g(x) = \sqrt{x + 2}$, find the domain of $(f \circ g)(x)$.

70. Consider the functions $f(x) = x^3 - 7x^2 - 5x + 4$ and $f'(x) = 3x^2 - 14x - 5$. Given that f is increasing where $f'(x) > 0$ and f is decreasing where $f'(x) < 0$, find where f is increasing and where f is decreasing.

'Are You Prepared?' Answers

1. $\{x | x < 1\}$ or $(-\infty, 1)$ **2.**   **3.**   **4.**

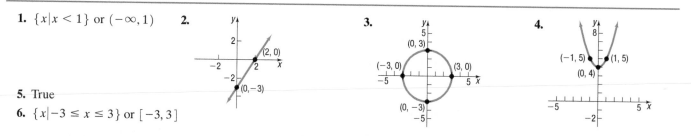

5. True

6. $\{x | -3 \le x \le 3\}$ or $[-3, 3]$

8.8 Linear Programming

> **OBJECTIVES** **1** Set Up a Linear Programming Problem (p. 638)
>   **2** Solve a Linear Programming Problem (p. 638)

Historically, linear programming evolved as a technique for solving problems involving resource allocation of goods and materials for the U.S. Air Force during World War II. Today, linear programming techniques are used to solve a wide variety of problems, such as optimizing airline scheduling. Although most practical linear programming problems involve systems of several hundred linear inequalities containing several hundred variables, we limit our discussion to problems containing only two variables, because we can solve such problems using graphing techniques.*

*The **simplex method** is a way to solve linear programming problems involving many inequalities and variables. Developed by George Dantzig in 1946, it is particularly well suited for computerization. In 1984, Narendra Karmarkar of Bell Laboratories discovered a way to improve the simplex method.

1 Set Up a Linear Programming Problem

Let's begin by returning to Example 10 from Section 8.7.

EXAMPLE 1

Financial Planning

A retired couple has up to $25,000 to invest. As their financial adviser, you recommend that they place at least $15,000 in Treasury bills yielding 2% and at most $5000 in corporate bonds yielding 3%. Develop a model that can be used to determine how much money they should place in each investment so that income is maximized.

Solution   The problem is typical of a *linear programming problem*. The problem requires that a certain linear function, the income, be maximized. If I represents income, x the amount invested in Treasury bills at 2%, and y the amount invested in corporate bonds at 3%, then

$$I = 0.02x + 0.03y$$

Assume, as before, that I, x, and y are in thousands of dollars.

The linear function $I = 0.02x + 0.03y$ is called the **objective function**. Further, the problem requires that the maximum income be achieved under certain conditions, or **constraints**, each of which is a linear inequality involving the variables. (See Example 10 in Section 8.7.) The linear programming problem is modeled as

$$\text{Maximize} \qquad I = 0.02x + 0.03y$$

subject to the constraints

$$\begin{cases} x \geq 0 \\ y \geq 0 \\ x + y \leq 25 \\ x \geq 15 \\ y \leq 5 \end{cases}$$

In general, every linear programming problem has two components:

- A linear objective function that is to be maximized or minimized
- A collection of linear inequalities that must be satisfied simultaneously

---

DEFINITION   Linear Programming Problem

A **linear programming problem** in two variables x and y consists of maximizing (or minimizing) a linear objective function

$$z = Ax + By \quad A \text{ and } B \text{ are real numbers, not both } 0$$

subject to certain constraints, or conditions, expressible as linear inequalities in x and y.

---

2 Solve a Linear Programming Problem

To maximize (or minimize) the quantity $z = Ax + By$, we need to identify points (x, y) that make the expression for z the largest (or smallest) possible. But not all points (x, y) are eligible; only those that also satisfy each linear inequality (constraint) can be used. Each point (x, y) that satisfies the system of linear inequalities (the constraints) is a **feasible point**. Linear programming problems seek the feasible point(s) that maximizes (or minimizes) the objective function.

Look again at the linear programming problem in Example 1.

EXAMPLE 2 | Analyzing a Linear Programming Problem

Consider the linear programming problem

$$\text{Maximize} \quad I = 0.02x + 0.03y$$

subject to the constraints

$$\begin{cases} x \geq 0 \\ y \geq 0 \\ x + y \leq 25 \\ x \geq 15 \\ y \leq 5 \end{cases}$$

Graph the constraints. Then graph the objective function for $I = 0, 0.3, 0.45, 0.55,$ and 0.6.

Solution   Figure 24 shows the graph of the constraints. We superimpose on this graph the graph of the objective function for the given values of I.

For $I = 0$, the objective function is the line $0 = 0.02x + 0.03y$.
For $I = 0.3$, the objective function is the line $0.3 = 0.02x + 0.03y$.
For $I = 0.45$, the objective function is the line $0.45 = 0.02x + 0.03y$.
For $I = 0.55$, the objective function is the line $0.55 = 0.02x + 0.03y$.
For $I = 0.6$, the objective function is the line $0.6 = 0.02x + 0.03y$.

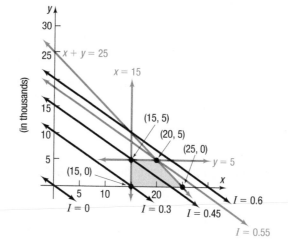

Figure 24

DEFINITION Solution to a Linear Programming Problem

A **solution** to a linear programming problem consists of a feasible point that maximizes (or minimizes) the objective function, together with the corresponding value of the objective function.

RECALL: The graph of a system of linear inequalities is bounded if it can be enclosed by a circle of sufficiently large radius. ∎

One condition for a linear programming problem in two variables to have a solution is that the graph of the feasible points be bounded.

If none of the feasible points maximizes (or minimizes) the objective function or if there are no feasible points, the linear programming problem has no solution.

Consider the linear programming problem posed in Example 2, and look again at Figure 24. The feasible points are the points that lie in the shaded region. For example, $(20, 3)$ is a feasible point, as are $(15, 5)$, $(20, 5)$, $(18, 4)$, and so on. To find the solution of the problem requires finding a feasible point (x, y) that makes $I = 0.02x + 0.03y$ as large as possible.

Notice that as I increases in value from $I = 0$ to $I = 0.3$ to $I = 0.45$ to $I = 0.55$ to $I = 0.6$, the result is a collection of parallel lines. Further, notice that the largest value of I that can be obtained using feasible points is $I = 0.55$, which corresponds to the line $0.55 = 0.02x + 0.03y$. Any larger value of I results in a line that does not pass through any feasible points. Finally, notice that the feasible point that yields $I = 0.55$ is the point $(20, 5)$, a corner point. These observations form the basis of the following results, which are stated without proof.

> **THEOREM** **Location of the Solution of a Linear Programming Problem**
>
> - If a linear programming problem has a solution, it is located at a corner point of the graph of the feasible points.
> - If a linear programming problem has multiple solutions, at least one of them is located at a corner point of the graph of the feasible points.
> - In either case, the corresponding value of the objective function is unique.

We do not consider linear programming problems that have no solution. As a result, we can outline the procedure for solving a linear programming problem as follows:

> **Steps for Solving a Linear Programming Problem**
>
> **STEP 1:** Assign symbols for the variables in the problem, and write an expression for the quantity to be maximized (or minimized). This expression is the objective function.
>
> **STEP 2:** Write all the constraints as a system of linear inequalities.
>
> **STEP 3:** Graph the system (the set of feasible points) and find the corner points.
>
> **STEP 4:** Evaluate the objective function at each corner point. The largest (or smallest) of these is the solution.

EXAMPLE 3 **Solving a Minimum Linear Programming Problem**

Minimize the objective function

$$z = 2x + 3y$$

subject to the constraints

$$y \leq 5 \qquad x \leq 6 \qquad x + y \geq 2 \qquad x \geq 0 \qquad y \geq 0$$

Solution **STEP 1:** We want to minimize the objective function $z = 2x + 3y$.

STEP 2: The constraints form the system of linear inequalities

$$\begin{cases} y \leq 5 \\ x \leq 6 \\ x + y \geq 2 \\ x \geq 0 \\ y \geq 0 \end{cases}$$

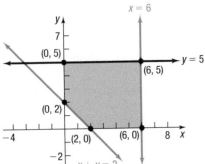

Figure 25

STEP 3: The graph of this system (the set of feasible points) is shown as the shaded region in Figure 25. The corner points are labeled.

STEP 4: Table 1 lists the corner points and the corresponding values of the objective function. From the table, the minimum value of z is 4, and it occurs at the point $(2, 0)$.

Table 1

Corner Point (x, y)	Value of the Objective Function $z = 2x + 3y$
(0, 2)	$z = 2 \cdot 0 + 3 \cdot 2 = 6$
(0, 5)	$z = 2 \cdot 0 + 3 \cdot 5 = 15$
(6, 5)	$z = 2 \cdot 6 + 3 \cdot 5 = 27$
(6, 0)	$z = 2 \cdot 6 + 3 \cdot 0 = 12$
(2, 0)	$z = 2 \cdot 2 + 3 \cdot 0 = 4$

Now Work PROBLEMS 5 AND 11

EXAMPLE 4   **Maximizing Profit**

At the end of every month, after filling orders for its regular customers, a coffee company has some pure Colombian coffee and some special-blend coffee remaining. The practice of the company has been to package a mixture of the two coffees into 1-pound (lb) packages as follows: a low-grade mixture containing 4 ounces (oz) of Colombian coffee and 12 oz of special-blend coffee, and a high-grade mixture containing 8 oz of Colombian and 8 oz of special-blend coffee. A profit of $1.25 per package is made on the low-grade mixture, whereas a profit of $1.75 per package is made on the high-grade mixture. This month, 120 lb of special-blend coffee and 100 lb of pure Colombian coffee remain. How many packages of each mixture should be prepared to achieve a maximum profit? Assume that all packages prepared can be sold.

Solution   **STEP 1:** Begin by assigning symbols for the two variables.

x = Number of packages of the low-grade mixture

y = Number of packages of the high-grade mixture

The goal is to maximize the profit subject to constraints on x and y. If P denotes the profit, then the objective function is

$$P = \$1.25x + \$1.75y$$

STEP 2: Because x and y represent numbers of packages, the only meaningful values for x and y are nonnegative integers. This yields the two constraints

$$x \geq 0 \quad y \geq 0 \quad \text{Nonnegative constraints}$$

There is only so much of each type of coffee available. For example, the total amount of Colombian coffee used in the two mixtures cannot exceed 100 lb, or 1600 oz. Because 4 oz are used in each low-grade package and 8 oz are used in each high-grade package, this leads to the constraint

$$4x + 8y \leq 1600 \quad \text{Colombian coffee constraint}$$

Similarly, the supply of 120 lb, or 1920 oz, of special-blend coffee leads to the constraint

$$12x + 8y \leq 1920 \quad \text{Special-blend coffee constraint}$$

The linear programming problem is

$$\text{Maximize} \quad P = 1.25x + 1.75y$$

subject to the constraints

$$x \geq 0 \quad y \geq 0 \quad 4x + 8y \leq 1600 \quad 12x + 8y \leq 1920$$

(continued)

STEP 3: The graph of the constraints (the feasible points) is shown in Figure 26 with the corner points labeled.

STEP 4: Table 2 shows the value of P at each corner point. From the table, the maximum profit of $365 is achieved with 40 packages of the low-grade mixture and 180 packages of the high-grade mixture.

Figure 26

Table 2

Corner Point (x, y)	Value of Profit $P = 1.25x + 1.75y$
(0, 0)	$P = 0$
(0, 200)	$P = 1.25 \cdot 0 + 1.75 \cdot 200 = \$350$
(40, 180)	$P = 1.25 \cdot 40 + 1.75 \cdot 180 = \$365$
(160, 0)	$P = 1.25 \cdot 160 + 1.75 \cdot 0 = \$200$

 Now Work PROBLEM 19

8.8 Assess Your Understanding

Concepts and Vocabulary

1. A linear programming problem requires that a linear expression, called the _____ _____, be maximized or minimized.

2. *True or False* If a linear programming problem has a solution, it is located at a corner point of the graph of the feasible points.

Skill Building

In Problems 3–8, find the maximum and minimum value of the given objective function of a linear programming problem. The figure illustrates the graph of the feasible points.

3. $z = x + y$

4. $z = 2x + 3y$

5. $z = x + 10y$

6. $z = 10x + y$

7. $z = 5x + 7y$

8. $z = 7x + 5y$

In Problems 9–18, solve each linear programming problem.

9. Maximize $z = 2x + y$ subject to the constraints $x \geq 0$, $y \geq 0$, $x + y \leq 6$, $x + y \geq 1$

10. Maximize $z = x + 3y$ subject to the constraints $x \geq 0$, $y \geq 0$, $x + y \geq 3$, $x \leq 5$, $y \leq 7$

11. Minimize $z = 2x + 5y$ subject to the constraints $x \geq 0$, $y \geq 0$, $x + y \geq 2$, $x \leq 5$, $y \leq 3$

12. Minimize $z = 3x + 4y$ subject to the constraints $x \geq 0$, $y \geq 0$, $2x + 3y \geq 6$, $x + y \leq 8$

13. Maximize $z = 3x + 5y$ subject to the constraints $x \geq 0$, $y \geq 0$, $x + y \geq 2$, $2x + 3y \leq 12$, $3x + 2y \leq 12$

14. Maximize $z = 5x + 3y$ subject to the constraints $x \geq 0$, $y \geq 0$, $x + y \geq 2$, $x + y \leq 8$, $2x + y \leq 10$

15. Minimize $z = 5x + 4y$ subject to the constraints $x \geq 0$, $y \geq 0$, $x + y \geq 2$, $2x + 3y \leq 12$, $3x + y \leq 12$

16. Minimize $z = 2x + 3y$ subject to the constraints $x \geq 0$, $y \geq 0$, $x + y \geq 3$, $x + y \leq 9$, $x + 3y \geq 6$

17. Maximize $z = 5x + 2y$ subject to the constraints $x \geq 0$, $y \geq 0$, $x + y \leq 10$, $2x + y \geq 10$, $x + 2y \geq 10$

18. Maximize $z = 2x + 4y$ subject to the constraints $x \geq 0$, $y \geq 0$, $2x + y \geq 4$, $x + y \leq 9$

Applications and Extensions

19. Maximizing Profit A manufacturer of skis produces two types: downhill and cross-country. Use the following table to determine how many of each kind of ski should be produced to achieve a maximum profit. What is the maximum profit? What would the maximum profit be if the time available for manufacturing were increased to 48 hours?

	Downhill	Cross-country	Time Available
Manufacturing time per ski	2 hours	1 hour	40 hours
Finishing time per ski	1 hour	1 hour	32 hours
Profit per ski	$70	$50	

20. Farm Management A farmer has 70 acres of land available for planting either soybeans or wheat. The cost of preparing the soil, the workdays required, and the expected profit per acre planted for each type of crop are given in the following table.

	Soybeans	Wheat
Preparation cost per acre	$60	$30
Workdays required per acre	3	4
Profit per acre	$180	$100

The farmer cannot spend more than $1800 in preparation costs and cannot use a total of more than 120 workdays. How many acres of each crop should be planted to maximize the profit? What is the maximum profit? What is the maximum profit if the farmer is willing to spend no more than $2400 on preparation?

21. Banquet Seating A banquet hall offers two types of tables for rent: 6-person rectangular tables at a cost of $28 each and 10-person round tables at a cost of $52 each. Kathleen would like to rent the hall for a wedding banquet and needs tables for 250 people. The hall can have a maximum of 35 tables, and the hall has only 15 rectangular tables available. How many of each type of table should be rented to minimize cost and what is the minimum cost?

Source: facilities.princeton.edu

22. Spring Break The student activities department of a community college plans to rent buses and vans for a spring-break trip. Each bus has 40 regular seats and 1 special seat designed to accommodate travelers with disabilities. Each van has 8 regular seats and 3 special seats. The rental cost is $350 for each van and $975 for each bus. If 320 regular and 36 special seats are required for the trip, how many vehicles of each type should be rented to minimize cost?

Source: www.busrates.com

23. Return on Investment An investment broker is instructed by her client to invest up to $20,000, some in a junk bond yielding 9% per annum and some in Treasury bills yielding 7% per annum. The client wants to invest at least $8000 in T-bills and no more than $12,000 in the junk bond.

(a) How much should the broker recommend that the client place in each investment to maximize income if the client insists that the amount invested in T-bills must equal or exceed the amount placed in the junk bond?

(b) How much should the broker recommend that the client place in each investment to maximize income if the client insists that the amount invested in T-bills must not exceed the amount placed in the junk bond?

24. Production Scheduling In a factory, machine 1 produces 8-inch (in.) pliers at the rate of 60 units per hour (h) and 6-in. pliers at the rate of 70 units/h. Machine 2 produces 8-in. pliers at the rate of 40 units/h and 6-in. pliers at the rate of 20 units/h. It costs $50/h to operate machine 1, and machine 2 costs $30/h to operate. The production schedule requires that at least 240 units of 8-in. pliers and at least 140 units of 6-in. pliers be produced during each 10-h day. Which combination of machines will cost the least money to operate?

25. Managing a Meat Market A meat market combines ground beef and ground pork in a single package for meat loaf. The ground beef is 75% lean (75% beef, 25% fat) and costs the market $2.25 per pound (lb). The ground pork is 60% lean and costs the market $1.35/lb. The meat loaf must be at least 70% lean. If the market wants to make at least 180 lb of meat loaf by using at least 50 lb of its available pork, but no more than 200 lb of its available ground beef, how much ground beef should be mixed with ground pork so that the cost is minimized?

26. Ice Cream The Mom and Pop Ice Cream Company makes two kinds of chocolate ice cream: regular and premium. The properties of 1 gallon (gal) of each type are shown in the table:

	Regular	Premium
Flavoring	24 oz	20 oz
Milk-fat products	12 oz	20 oz
Shipping weight	5 lbs	6 lbs
Profit	$0.75	$0.90

In addition, current commitments require the company to make at least 1 gal of premium for every 4 gal of regular. Each day, the company has available 725 pounds (lb) of flavoring and 425 lb of milk-fat products. If the company can ship no more than 3000 lb of product per day, how many gallons of each type should be produced daily to maximize profit?

Source: www.scitoys.com/ingredients/ice_cream.html

27. **Maximizing Profit on Ice Skates** A factory manufactures two kinds of ice skates: racing skates and figure skates. The racing skates require 6 work-hours in the fabrication department, whereas the figure skates require 4 work-hours there. The racing skates require 1 work-hour in the finishing department, whereas the figure skates require 2 work-hours there. The fabricating department has available at most 120 work-hours per day, and the finishing department has no more than 40 work-hours per day available. If the profit on each racing skate is $10 and the profit on each figure skate is $12, how many of each should be manufactured each day to maximize profit? (Assume that all skates made are sold.)

28. **Financial Planning** A retired couple have up to $50,000 to place in fixed-income securities. Their financial adviser suggests two securities to them: one is an AAA bond that yields 8% per annum; the other is a certificate of deposit (CD) that yields 4%. After careful consideration of the alternatives, the couple decide to place at most $20,000 in the AAA bond and at least $15,000 in the CD. They also instruct the financial adviser to place at least as much in the CD as in the AAA bond. How should the financial adviser proceed to maximize the return on their investment?

29. **Product Design** An entrepreneur is having a design group produce at least six samples of a new kind of fastener that he wants to market. It costs $9.00 to produce each metal fastener and $4.00 to produce each plastic fastener. He wants to have at least two of each version of the fastener and needs to have all the samples 24 hours (h) from now. It takes 4 h to produce each metal sample and 2 h to produce each plastic sample. To minimize the cost of the samples, how many of each kind should the entrepreneur order? What will be the cost of the samples?

30. **Animal Nutrition** Kevin's dog Amadeus likes two kinds of canned dog food. Gourmet Dog costs $1.40 per can and has 20 units of a vitamin complex; the calorie content is 75 calories. Chow Hound costs $1.12 per can and has 35 units of vitamins and 50 calories. Kevin likes Amadeus to have at least 1175 units of vitamins a month and at least 2375 calories during the same time period. Kevin has space to store only 60 cans of dog food at a time. How much of each kind of dog food should Kevin buy each month to minimize his cost?

31. **Airline Revenue** An airline has two classes of service: first class and coach. Management's experience has been that each aircraft should have at least 8 but no more than 16 first-class seats and at least 80 but no more than 120 coach seats.

   (a) If management decides that the ratio of first class to coach seats should never exceed 1:12, with how many of each type of seat should an aircraft be configured to maximize revenue?

   (b) If management decides that the ratio of first class to coach seats should never exceed 1:8, with how many of each type of seat should an aircraft be configured to maximize revenue?

   (c) If you were management, what would you do?

   [**Hint:** Assume that the airline charges C for a coach seat and F for a first-class seat; $C > 0$, $F > C$.]

32. **Challenge Problem** Maximize $z = 10x + 4y$ subject to the constraints $x \geq 0$, $y \geq 0$, $4x - y \geq -9$, $x - 2y \geq -25$, $x + 2y \leq 31$, $x + y \leq 19$, $4x + y \leq 43$, $5x - y \leq 38$, $x - 2y \leq 4$.

Explaining Concepts: Discussion and Writing

33. Explain in your own words what a linear programming problem is and how it can be solved.

Retain Your Knowledge

Problems 34–43 are based on material learned earlier in the course. The purpose of these problems is to keep the material fresh in your mind so that you are better prepared for the final exam.

34. Solve: $2m^{2/5} - m^{1/5} = 1$

35. Find the complex zeros of $f(x) = x^4 - 12x^2 - 64$.

36. **Radioactive Decay** The half-life of titanium-44 is 63 years. How long will it take 200 grams to decay to 75 grams? Round to one decimal place.

37. Find the equation of the line that is parallel to $y = 3x + 11$ and passes through the point $(-2, 1)$.

38. The sum of two numbers is 16. If the larger number is 4 less than 3 times the smaller number, find the two numbers.

39. What amount must be invested at 4% interest compounded daily to have $15,000 in 3 years?

40. Find an equation of the ellipse that has vertices $(0, \pm 5)$ and foci $(0, \pm 3)$.

41. If $f(x) = \dfrac{2x - 7}{5x + 1}$, find $f^{-1}(x)$.

42. Factor completely: $30x^2(x - 7)^{3/2} + 15x^3(x - 7)^{1/2}$

43. Consider the functions $f(x) = \dfrac{1}{3}x^3 - 3x^2 + 5x + 7$ and $f''(x) = 2x - 6$. Given that f is concave up where $f''(x) > 0$ and f is concave down where $f''(x) < 0$, find where f is concave up and where f is concave down.

Chapter Review

Things to Know

Systems of equations (pp. 556–558)

Systems with no solutions are inconsistent.   Systems with a solution are consistent.

Consistent systems of linear equations have either a unique solution (independent) or an infinite number of solutions (dependent).

Matrix (p. 571)

Rectangular array of numbers, called entries

Augmented matrix (p. 571)

Row operations (p. 572)

Row echelon form (p. 573)

Reduced row echelon form (p. 576)

Determinants and Cramer's Rule (pp. 584–592)

Matrix Algebra (pp. 595–607)

m by n matrix (p. 596)   Matrix with m rows and n columns

Identity matrix I_n (p. 603)   An n by n square matrix whose diagonal entries are 1's, while all other entries are 0's

Inverse of a matrix (p. 603) A^{-1} is the inverse of A if $AA^{-1} = A^{-1}A = I_n$.

Nonsingular matrix (p. 604)   A square matrix that has an inverse

Linear programming problem (p. 638)

Maximize (or minimize) a linear objective function, $z = Ax + By$, subject to certain conditions, or constraints, expressible as linear inequalities in x and y. A feasible point (x, y) is a point that satisfies the constraints (linear inequalities) of a linear programming problem.

Location of the solution of a linear programming problem (p. 640)

If a linear programming problem has a solution, it is located at a corner point of the graph of the feasible points. If a linear programming problem has multiple solutions, at least one of them is located at a corner point of the graph of the feasible points. In either case, the corresponding value of the objective function is unique.

Objectives

Section	You should be able to . . .	Example(s)	Review Exercises
8.1	1 Solve systems of equations by substitution (p. 559)	4	1–7, 56, 59
	2 Solve systems of equations by elimination (p. 559)	5, 6	1–7, 56, 59
	3 Identify inconsistent systems of equations containing two variables (p. 561)	7	5, 54
	4 Express the solution of a system of dependent equations containing two variables (p. 561)	8	7, 53
	5 Solve systems of three equations containing three variables (p. 562)	9, 12	8–10, 55, 57, 60
	6 Identify inconsistent systems of equations containing three variables (p. 564)	10	10
	7 Express the solution of a system of dependent equations containing three variables (p. 564)	11	9
8.2	1 Write the augmented matrix of a system of linear equations (p. 571)	1	20–25
	2 Write the system of equations from the augmented matrix (p. 572)	2	11, 12
	3 Perform row operations on a matrix (p. 572)	3, 4	20–25
	4 Solve a system of linear equations using matrices (p. 573)	5–10	20–25
8.3	1 Evaluate 2 by 2 determinants (p. 584)	1	26
	2 Use Cramer's Rule to solve a system of two equations containing two variables (p. 585)	2	29, 30
	3 Evaluate 3 by 3 determinants (p. 587)	4	27, 28
	4 Use Cramer's Rule to solve a system of three equations containing three variables (p. 589)	5	31
	5 Know properties of determinants (p. 591)	6–9	32, 33

(continued)

Section	You should be able to . . .	Example(s)	Review Exercises
8.4	1 Find the sum and difference of two matrices (p. 596)	3, 4	13
	2 Find scalar multiples of a matrix (p. 598)	5	14
	3 Find the product of two matrices (p. 599)	6–11	15, 16
	4 Find the inverse of a matrix (p. 603)	12–14	17–19
	5 Solve a system of linear equations using an inverse matrix (p. 607)	15	20–25
8.5	1 Decompose $\dfrac{P}{Q}$ where Q has only nonrepeated linear factors (p. 614)	2	34
	2 Decompose $\dfrac{P}{Q}$ where Q has repeated linear factors (p. 615)	3, 4	35
	3 Decompose $\dfrac{P}{Q}$ where Q has a nonrepeated irreducible quadratic factor (p. 617)	5	36, 38
	4 Decompose $\dfrac{P}{Q}$ where Q has a repeated irreducible quadratic factor (p. 618)	6	37
8.6	1 Solve a system of nonlinear equations using substitution (p. 621)	1, 3	39–43
	2 Solve a system of nonlinear equations using elimination (p. 622)	2, 4, 5	39–43
8.7	1 Graph an inequality (p. 630)	2–4	44, 45
	2 Graph a system of inequalities (p. 632)	5–10	46–50, 58
8.8	1 Set up a linear programming problem (p. 638)	1	61
	2 Solve a linear programming problem (p. 638)	2–4	51, 52, 61

Review Exercises

In Problems 1–10, solve each system of equations using the method of substitution or the method of elimination. If the system has no solution, state that it is inconsistent.

1. $\begin{cases} 2x - y = 5 \\ 5x + 2y = 8 \end{cases}$

2. $\begin{cases} 3x - 4y = 4 \\ x - 3y = \dfrac{1}{2} \end{cases}$

3. $\begin{cases} x - 2y - 4 = 0 \\ 3x + 2y - 4 = 0 \end{cases}$

4. $\begin{cases} y = 2x - 5 \\ x = 3y + 4 \end{cases}$

5. $\begin{cases} x - 3y + 4 = 0 \\ \dfrac{1}{2}x - \dfrac{3}{2}y + \dfrac{4}{3} = 0 \end{cases}$

6. $\begin{cases} 2x + 3y - 13 = 0 \\ 3x - 2y = 0 \end{cases}$

7. $\begin{cases} 2x + 5y = 10 \\ 4x + 10y = 20 \end{cases}$

8. $\begin{cases} x + 2y - z = 6 \\ 2x - y + 3z = -13 \\ 3x - 2y + 3z = -16 \end{cases}$

9. $\begin{cases} 2x - 4y + z = -15 \\ x + 2y - 4z = 27 \\ 5x - 6y - 2z = -3 \end{cases}$

10. $\begin{cases} x - 4y + 3z = 15 \\ -3x + y - 5z = -5 \\ -7x - 5y - 9z = 10 \end{cases}$

In Problems 11 and 12, write the system of equations that corresponds to the given augmented matrix.

11. $\begin{bmatrix} 3 & 2 & | & 8 \\ 1 & 4 & | & -1 \end{bmatrix}$

12. $\begin{bmatrix} 1 & 2 & 5 & | & -2 \\ 5 & 0 & -3 & | & 8 \\ 2 & -1 & 0 & | & 0 \end{bmatrix}$

In Problems 13–16, use the following matrices to compute each expression.

$$A = \begin{bmatrix} 1 & 0 \\ 2 & 4 \\ -1 & 2 \end{bmatrix} \quad B = \begin{bmatrix} 4 & -3 & 0 \\ 1 & 1 & -2 \end{bmatrix} \quad C = \begin{bmatrix} 3 & -4 \\ 1 & 5 \\ 5 & 2 \end{bmatrix}$$

13. $A + C$

14. $6A$

15. AB

16. BC

In Problems 17–19, find the inverse, if there is one, of each matrix. If there is no inverse, state that the matrix is singular.

17. $\begin{bmatrix} 4 & 6 \\ 1 & 3 \end{bmatrix}$

18. $\begin{bmatrix} 1 & 3 & 3 \\ 1 & 2 & 1 \\ 1 & -1 & 2 \end{bmatrix}$

19. $\begin{bmatrix} 4 & -8 \\ -1 & 2 \end{bmatrix}$

In Problems 20–25, solve each system of equations using matrices. If the system has no solution, state that it is inconsistent.

20. $\begin{cases} 3x - 2y = 1 \\ 10x + 10y = 5 \end{cases}$

21. $\begin{cases} 5x - 6y - 3z = 6 \\ 4x - 7y - 2z = -3 \\ 3x + y - 7z = 1 \end{cases}$

22. $\begin{cases} 2x + y + z = 5 \\ 4x - y - 3z = 1 \\ 8x + y - z = 5 \end{cases}$

23. $\begin{cases} x - 2z = 1 \\ 2x + 3y = -3 \\ 4x - 3y - 4z = 3 \end{cases}$

24. $\begin{cases} x - y + z = 0 \\ x - y - 5z - 6 = 0 \\ 2x - 2y + z - 1 = 0 \end{cases}$

25. $\begin{cases} x - y - z - t = 1 \\ 2x + y - z + 2t = 3 \\ x - 2y - 2z - 3t = 0 \\ 3x - 4y + z + 5t = -3 \end{cases}$

In Problems 26–28, find the value of each determinant.

26. $\begin{vmatrix} 3 & 4 \\ 1 & 3 \end{vmatrix}$

27. $\begin{vmatrix} 1 & 4 & 0 \\ -1 & 2 & 6 \\ 4 & 1 & 3 \end{vmatrix}$

28. $\begin{vmatrix} 2 & 1 & -3 \\ 5 & 0 & 1 \\ 2 & 6 & 0 \end{vmatrix}$

In Problems 29–31, use Cramer's Rule, if applicable, to solve each system.

29. $\begin{cases} x - 2y = 4 \\ 3x + 2y = 4 \end{cases}$

30. $\begin{cases} 2x + 3y - 13 = 0 \\ 3x - 2y = 0 \end{cases}$

31. $\begin{cases} x + 2y - z = 6 \\ 2x - y + 3z = -13 \\ 3x - 2y + 3z = -16 \end{cases}$

In Problems 32 and 33, use properties of determinants to find the value of each determinant if it is known that $\begin{vmatrix} x & y \\ a & b \end{vmatrix} = 8$.

32. $\begin{vmatrix} 2x & y \\ 2a & b \end{vmatrix}$

33. $\begin{vmatrix} y & x \\ b & a \end{vmatrix}$

In Problems 34–38, find the partial fraction decomposition of each rational expression.

34. $\dfrac{6}{x(x-4)}$

35. $\dfrac{x-4}{x^2(x-1)}$

36. $\dfrac{x}{(x^2+9)(x+1)}$

37. $\dfrac{x^3}{(x^2+4)^2}$

38. $\dfrac{x^2}{(x^2+1)(x^2-1)}$

In Problems 39–43, solve each system of equations.

39. $\begin{cases} 2x + y + 3 = 0 \\ x^2 + y^2 = 5 \end{cases}$

40. $\begin{cases} 2xy + y^2 = 10 \\ 3y^2 - xy = 2 \end{cases}$

41. $\begin{cases} x^2 + y^2 = 6y \\ x^2 = 3y \end{cases}$

42. $\begin{cases} 3x^2 + 4xy + 5y^2 = 8 \\ x^2 + 3xy + 2y^2 = 0 \end{cases}$

43. $\begin{cases} x^2 - 3x + y^2 + y = -2 \\ \dfrac{x^2 - x}{y} + y + 1 = 0 \end{cases}$

In Problems 44 and 45 graph each inequality.

44. $3x + 4y \le 12$

45. $y \le x^2$

In Problems 46–48, graph each system of inequalities. State whether the graph is bounded or unbounded, and label the corner points.

46. $\begin{cases} -2x + y \le 2 \\ x + y \ge 2 \end{cases}$

47. $\begin{cases} x \ge 0 \\ y \ge 0 \\ x + y \le 4 \\ 2x + 3y \le 6 \end{cases}$

48. $\begin{cases} x \ge 0 \\ y \ge 0 \\ 2x + y \le 8 \\ x + 2y \ge 2 \end{cases}$

In Problems 49 and 50, graph each system of inequalities.

49. $\begin{cases} x^2 + y^2 \le 16 \\ x + y \ge 2 \end{cases}$

50. $\begin{cases} y \le x^2 \\ xy \le 4 \end{cases}$

In Problems 51 and 52, solve each linear programming problem.

51. Maximize $z = 3x + 4y$ subject to the constraints $x \ge 0, y \ge 0, 3x + 2y \ge 6, x + y \le 8$

52. Minimize $z = 3x + 5y$ subject to the constraints $x \ge 0, y \ge 0, x + y \ge 1, 3x + 2y \le 12, x + 3y \le 12$

53. Find A so that the system of equations has infinitely many solutions.

$$\begin{cases} 2x + 5y = 5 \\ 4x + 10y = A \end{cases}$$

54. Find A so that the system in Problem 53 is inconsistent.

55. Curve Fitting Find the quadratic function $y = ax^2 + bx + c$ that passes through the three points $(0, 1), (1, 0),$ and $(-2, 1)$.

56. Blending Coffee A coffee distributor is blending a new coffee that will cost $6.90 per pound. It will consist of a blend of $6.00-per-pound coffee and $9.00-per-pound coffee. What amounts of each type of coffee should be mixed to achieve the desired blend?

[**Hint**: Assume that the weight of the blended coffee is 100 pounds.]

$6.00/lb $6.90/lb $9.00/lb

57. Cookie Orders A cookie company makes three kinds of cookies (oatmeal raisin, chocolate chip, and shortbread) packaged in small, medium, and large boxes. The small box contains 1 dozen oatmeal raisin and 1 dozen chocolate chip; the medium box has 2 dozen oatmeal raisin, 1 dozen chocolate chip, and 1 dozen shortbread; the large box contains 2 dozen oatmeal raisin, 2 dozen chocolate chip, and 3 dozen shortbread. If you require exactly 15 dozen oatmeal raisin, 10 dozen chocolate chip, and 11 dozen shortbread, how many of each size box should you buy?

58. Mixed Nuts A store that specializes in selling nuts has available 72 pounds (lb) of cashews and 120 lb of peanuts. These are to be mixed in 12-ounce (oz) packages as follows: a lower-priced package containing 8 oz of peanuts and 4 oz of cashews, and a quality package containing 6 oz of peanuts and 6 oz of cashews.

(a) Use x to denote the number of lower-priced packages, and use y to denote the number of quality packages. Write a system of linear inequalities that describes the possible numbers of each kind of package.

(b) Graph the system and label the corner points.

59. Determining the Speed of the Current of the Aguarico River On a recent trip to the Cuyabeno Wildlife Reserve in the Amazon region of Ecuador, Mike took a 100-kilometer trip by speedboat down the Aguarico River from Chiritza to the Flotel Orellana. As Mike watched the Amazon unfold, he wondered how fast the speedboat was going and how fast the current of the white-water Aguarico River was. Mike timed the trip downstream at 2.5 hours and the return trip at 3 hours. What were the two speeds?

60. Constant Rate Jobs If Bruce and Bryce work together for 1 hour and 20 minutes, they will finish a certain job. If Bryce and Marty work together for 1 hour and 36 minutes, the same job can be finished. If Marty and Bruce work together, they can complete this job in 2 hours and 40 minutes. How long would it take each of them, working alone, to finish the job?

61. Minimizing Production Cost A factory produces gasoline engines and diesel engines. Each week the factory is obligated to deliver at least 20 gasoline engines and at least 15 diesel engines. Due to physical limitations, however, the factory cannot make more than 60 gasoline engines or more than 40 diesel engines in any given week. Finally, to prevent layoffs, a total of at least 50 engines must be produced. If gasoline engines cost $450 each to produce and diesel engines cost $550 each to produce, how many of each should be produced per week to minimize the cost? What is the excess capacity of the factory? That is, how many of each kind of engine are being produced in excess of the number that the factory is obligated to deliver?

62. Describe four ways of solving a system of three linear equations containing three variables. Which method do you prefer? Why?

Chapter Test

CHAPTER
Test Prep
VIDEOS

The Chapter Test Prep Videos include step-by-step solutions to all chapter test exercises. These videos are available in MyLab™ Math, or on this text's YouTube Channel. Refer to the Preface for a link to the YouTube channel.

In Problems 1–4, solve each system of equations using the method of substitution or the method of elimination. If the system has no solution, state that it is inconsistent.

1. $\begin{cases} -2x + y = -7 \\ 4x + 3y = 9 \end{cases}$

2. $\begin{cases} \dfrac{1}{3}x - 2y = 1 \\ 5x - 30y = 18 \end{cases}$

3. $\begin{cases} x - y + 2z = 5 \\ 3x + 4y - z = -2 \\ 5x + 2y + 3z = 8 \end{cases}$

4. $\begin{cases} 3x + 2y - 8z = -3 \\ -x - \dfrac{2}{3}y + z = 1 \\ 6x - 3y + 15z = 8 \end{cases}$

5. Write the augmented matrix corresponding to the system of equations:

$$\begin{cases} 4x - 5y + z = 0 \\ -2x - y + 6 = -19 \\ x + 5y - 5z = 10 \end{cases}$$

6. Write the system of equations corresponding to the augmented matrix:

$$\begin{bmatrix} 3 & 2 & 4 & | & -6 \\ 1 & 0 & 8 & | & 2 \\ -2 & 1 & 3 & | & -11 \end{bmatrix}$$

In Problems 7–10, use the matrices below to compute each expression.

$$A = \begin{bmatrix} 1 & -1 \\ 0 & -4 \\ 3 & 2 \end{bmatrix} \quad B = \begin{bmatrix} 1 & -2 & 5 \\ 0 & 3 & 1 \end{bmatrix} \quad C = \begin{bmatrix} 4 & 6 \\ 1 & -3 \\ -1 & 8 \end{bmatrix}$$

7. $2A + C$ **8.** $A - 3C$ **9.** CB **10.** BA

In Problems 11 and 12, find the inverse of each nonsingular matrix.

11. $A = \begin{bmatrix} 3 & 2 \\ 5 & 4 \end{bmatrix}$ **12.** $B = \begin{bmatrix} 1 & -1 & 1 \\ 2 & 5 & -1 \\ 2 & 3 & 0 \end{bmatrix}$

In Problems 13–16, solve each system of equations using matrices. If the system has no solution, state that it is inconsistent.

13. $\begin{cases} 6x + 3y = 12 \\ 2x - y = -2 \end{cases}$ **14.** $\begin{cases} x + \dfrac{1}{4}y = 7 \\ 8x + 2y = 56 \end{cases}$

15. $\begin{cases} x + 2y + 4z = -3 \\ 2x + 7y + 15z = -12 \\ 4x + 7y + 13z = -10 \end{cases}$ **16.** $\begin{cases} 2x + 2y - 3z = 5 \\ x - y + 2z = 8 \\ 3x + 5y - 8z = -2 \end{cases}$

In Problems 17 and 18, find the value of each determinant.

17. $\begin{vmatrix} -2 & 5 \\ 3 & 7 \end{vmatrix}$ **18.** $\begin{vmatrix} 2 & -4 & 6 \\ 1 & 4 & 0 \\ -1 & 2 & -4 \end{vmatrix}$

In Problems 19 and 20, use Cramer's Rule, if possible, to solve each system.

19. $\begin{cases} 4x + 3y = -23 \\ 3x - 5y = 19 \end{cases}$ **20.** $\begin{cases} 4x - 3y + 2z = 15 \\ -2x + y - 3z = -15 \\ 5x - 5y + 2z = 18 \end{cases}$

In Problems 21 and 22, solve each system of equations.

21. $\begin{cases} 3x^2 + y^2 = 12 \\ y^2 = 9x \end{cases}$ **22.** $\begin{cases} 2y^2 - 3x^2 = 5 \\ y - x = 1 \end{cases}$

23. Graph the system of inequalities: $\begin{cases} x^2 + y^2 \leq 100 \\ 4x - 3y \geq 0 \end{cases}$

In Problems 24 and 25, find the partial fraction decomposition of each rational expression.

24. $\dfrac{3x + 7}{(x + 3)^2}$ **25.** $\dfrac{4x^2 - 3}{x(x^2 + 3)^2}$

26. Graph the system of inequalities. State whether the graph is bounded or unbounded, and label all corner points.

$$\begin{cases} x \geq 0 \\ y \geq 0 \\ x + 2y \geq 8 \\ 2x - 3y \geq 2 \end{cases}$$

27. Maximize $z = 5x + 8y$ subject to the constraints $x \geq 0$, $2x + y \leq 8$, and $x - 3y \leq -3$.

28. Megan went clothes shopping and bought 2 pairs of jeans, 2 camisoles, and 4 T-shirts for $90.00. At the same store, Paige bought one pair of jeans and 3 T-shirts for $42.50, while Kara bought 1 pair of jeans, 3 camisoles, and 2 T-shirts for $62.00. Determine the price of each clothing item.

Cumulative Review

In Problems 1–6, solve each equation.

1. $2x^2 - x = 0$ **2.** $\sqrt{3x + 1} = 4$ **3.** $2x^3 - 3x^2 - 8x - 3 = 0$

4. $3^x = 9^{x+1}$ **5.** $\log_3(x - 1) + \log_3(2x + 1) = 2$ **6.** $3^x = e$

7. Determine whether the function $g(x) = \dfrac{2x^3}{x^4 + 1}$ is even, odd, or neither. Is the graph of g symmetric with respect to the x-axis, y-axis, or origin?

8. Find the center and radius of the circle

$$x^2 + y^2 - 2x + 4y - 11 = 0$$

Graph the circle.

9. Graph $f(x) = 3^{x-2} + 1$ using transformations. What are the domain, range, and horizontal asymptote of f?

10. The function $f(x) = \dfrac{5}{x + 2}$ is one-to-one. Find f^{-1}. Find the domain and the range of f and the domain and the range of f^{-1}.

11. Graph each equation.

(a) $y = 3x + 6$ (b) $x^2 + y^2 = 4$

(c) $y = x^3$ (d) $y = \dfrac{1}{x}$

(e) $y = \sqrt{x}$ (f) $y = e^x$

(g) $y = \ln x$ (h) $2x^2 + 5y^2 = 1$

(i) $x^2 - 3y^2 = 1$ (j) $x^2 - 2x - 4y + 1 = 0$

12. $f(x) = x^3 - 3x + 5$

(a) Using a graphing utility, graph f and approximate the zero(s) of f.

(b) Using a graphing utility, approximate the local maxima and the local minima.

(c) Determine the intervals on which f is increasing.

Chapter Projects

I. Markov Chains A **Markov chain** (or process) is one in which future outcomes are determined by a current state. Future outcomes are based on probabilities. The probability of moving to a certain state depends only on the state previously occupied and does not vary with time. An example of a Markov chain is the maximum education achieved by children based on the highest educational level attained by their parents, where the states are (1) earned college degree, (2) high school diploma only, (3) elementary school only. If p_{ij} is the probability of moving from state i to state j, the **transition matrix** is the m by m matrix

$$P = \begin{bmatrix} p_{11} & p_{12} & \cdots & p_{1m} \\ \vdots & \vdots & & \vdots \\ p_{m1} & p_{m2} & \cdots & p_{mm} \end{bmatrix}$$

The table represents the probabilities for the highest educational level of children based on the highest educational level of their parents. For example, the table shows that the probability p_{21} is 40% that parents with a high-school education (row 2) will have children with a college education (column 1).

Highest Educational Level of Parents	Maximum Education That Children Achieve		
	College	High School	Elementary
College	80%	18%	2%
High school	40%	50%	10%
Elementary	20%	60%	20%

1. Convert the percentages to decimals.
2. What is the transition matrix?
3. Sum across the rows. What do you notice? Why do you think that you obtained this result?

4. If P is the transition matrix of a Markov chain, the (i, j)th entry of P^n (nth power of P) gives the probability of passing from state i to state j in n stages. What is the probability that the grandchild of a college graduate is a college graduate?

5. What is the probability that the grandchild of a high school graduate finishes college?

6. The row vector $v^{(0)} = \begin{bmatrix} 0.342 & 0.554 & 0.104 \end{bmatrix}$ represents the proportion of the U.S. population 25 years or older that has college, high school, and elementary school, respectively, as the highest educational level in 2017.* In a Markov chain the probability distribution $v^{(k)}$ after k stages is $v^{(k)} = v^{(0)}P^k$, where P^k is the kth power of the transition matrix. What will be the distribution of highest educational attainment of the grandchildren of the current population?

7. Calculate P^3, P^4, P^5, ... Continue until the matrix does not change. This is called the long-run or steady-state distribution. What is the long-run distribution of highest educational attainment of the population?

__Source__: U.S. Census Bureau.

The following projects are available at the Instructor's Resource Center (IRC).

II. Project at Motorola: ***Error Control Coding*** The high-powered engineering needed to ensure that wireless communications are transmitted correctly is analyzed using matrices to control coding errors.

III. Using Matrices to Find the Line of Best Fit Have you wondered how our calculators get a line of best fit? See how to find the line by solving a matrix equation.

IV. CBL Experiment Simulate two people walking toward each other at a constant rate. Then solve the resulting system of equations to determine when and where they will meet.

Sequences; Induction; the Binomial Theorem

World Population Projected to Reach 9.8 Billion by 2050

The current world population of 7.6 billion is expected to reach 8.6 billion by 2030, 9.8 billion in 2050, and 11.2 billion in 2100, according to a United Nations DESA report titled "World Population Prospects: The 2017 Revision."

Most of the projected increase in the world's population can be attributed to a short list of high-fertility countries, mainly in Africa and countries that already have large populations. During 2017–2050, half of the world's population growth is expected to be concentrated in nine countries: India, Nigeria, Democratic Republic of the Congo, Pakistan, Ethiopia, United Republic of Tanzania, United States of America, Uganda, and Indonesia (listed in order of the size of their contribution to the total population growth). Among the ten countries with the largest populations, Nigeria is growing the most rapidly. Currently ranked seventh, Nigeria is projected to become the third largest shortly before 2050.

China and India remain the two countries with the largest populations. With more than 1 billion people each, they represent 19% and 18% of the world's population, respectively. By 2024, the population of India is expected to surpass that of China.

Future population growth is highly dependent on the path of future fertility. Relatively small changes in the fertility rate, when projected over decades, can generate large differences in total population. In recent years, the fertility rate has declined in virtually all areas of the world, even in Africa, where fertility levels remain the highest of all major areas. Europe has been an exception to this trend in recent years, with the mean total fertility rate increasing from 1.4 births per woman in 2000 to 1.6 in 2015.

Source*: Adapted from United Nations Department of Economic and Social Affairs, June 21, 2017, New York (https://www.un.org.development/desa/en/news/population/world-population-prospects-2017.html)*

—*See the Internet-based Chapter Project I—*

← A Look Back, A Look Ahead →

This chapter is divided into three independent parts:
Sections 9.1–9.3, Section 9.4, and Section 9.5.

In Chapter 3, we defined a function and its domain, which was usually some set of real numbers. In Sections 9.1–9.3, we discuss a sequence, which is a function whose domain is the set of positive integers.

Throughout this text, where it seemed appropriate, we gave proofs of the results. In Section 9.4, a technique for proving theorems involving natural numbers is discussed.

In Chapter R, Review, Section R.4, there are formulas for expanding $(x + a)^2$ and $(x + a)^3$. In Section 9.5, we discuss the Binomial Theorem, a formula for the expansion of $(x + a)^n$, where n is any positive integer.

The topics introduced in this chapter are covered in more detail in courses titled *Discrete Mathematics*. Applications of these topics are found in the fields of computer science, engineering, business and economics, the social sciences, and the physical and biological sciences.

Outline

9.1 Sequences

PREPARING FOR THIS SECTION *Before getting started, review the following:*

- Functions (Section 3.1, pp. 203–210)

Now Work the 'Are You Prepared?' problems on page 658.

OBJECTIVES 1 List the First Several Terms of a Sequence (p. 652)
   2 List the Terms of a Sequence Defined by a Recursive Formula (p. 655)
   3 Use Summation Notation (p. 656)
   4 Find the Sum of a Sequence (p. 657)

When you hear the word *sequence* as it is used in the phrase "a sequence of events," you probably think of a collection of events, one of which happens first, another second, and so on. In mathematics, the word *sequence* also refers to outcomes that are first, second, and so on.

> **DEFINITION** **Sequence**
>
> A **sequence** is a function whose domain is the set of positive integers and whose range is a subset of the real numbers.

In a sequence, the inputs are $1, 2, 3, \ldots$. Because a sequence is a function, it has a graph. Figure 1(a) shows the graph of the function $f(x) = \dfrac{1}{x}, x > 0$. If all the points on this graph were removed except those whose x-coordinates are positive integers—that is, if all points were removed except $(1, 1), \left(2, \dfrac{1}{2}\right), \left(3, \dfrac{1}{3}\right)$, and so on—the remaining points would be the graph of the sequence $f(n) = \dfrac{1}{n}$, as shown in Figure 1(b). Note that n is used to represent the independent variable in a sequence. This serves to remind us that n is a positive integer.

Figure 1

(a) $f(x) = \frac{1}{x}, x > 0$

(b) $f(n) = \frac{1}{n}, n$ a positive integer

1 List the First Several Terms of a Sequence

A sequence is usually represented by listing its values in order. For example, the sequence whose graph is given in Figure 1(b) might be represented as

$$f(1), f(2), f(3), f(4), \ldots \quad \text{or} \quad 1, \frac{1}{2}, \frac{1}{3}, \frac{1}{4}, \ldots$$

The list never ends, as the ellipsis indicates. The numbers in this ordered list are called the **terms** of the sequence.

In dealing with sequences, subscripted letters are used such as a_1 to represent the first term, a_2 for the second term, a_3 for the third term, and so on.

For the sequence $f(n) = \dfrac{1}{n}$, this means

$$\underbrace{a_1 = f(1) = 1,}_{\text{first term}} \quad \underbrace{a_2 = f(2) = \frac{1}{2},}_{\text{second term}} \quad \underbrace{a_3 = f(3) = \frac{1}{3},}_{\text{third term}} \quad \underbrace{a_4 = f(4) = \frac{1}{4},}_{\text{fourth term}} \ldots, \quad \underbrace{a_n = f(n) = \frac{1}{n},}_{\text{nth term}} \ldots$$

In other words, the traditional function notation $f(n)$ is typically not used for sequences. For the sequence $f(n) = \dfrac{1}{n}$, we have a rule for the nth term, which is $a_n = \dfrac{1}{n}$, so it is easy to find any term of the sequence.

When a formula for the nth term (sometimes called the **general term**) of a sequence is known, the entire sequence can be represented by placing braces around the formula for the nth term.

For example, the sequence whose nth term is $b_n = \left(\dfrac{1}{2}\right)^n$ can be represented by

$$\{b_n\} = \left\{\left(\frac{1}{2}\right)^n\right\}$$

or by listing the terms

$$b_1 = \frac{1}{2}, \quad b_2 = \frac{1}{4}, \quad b_3 = \frac{1}{8}, \quad \dots, \quad b_n = \left(\frac{1}{2}\right)^n, \quad \dots$$

EXAMPLE 1   **Listing the First Several Terms of a Sequence**

List the first six terms of the sequence $\{a_n\}$ and graph it.

$$\{a_n\} = \left\{\frac{n-1}{n}\right\}$$

Solution   The first six terms of the sequence are

$$a_1 = \frac{1-1}{1} = 0, \quad a_2 = \frac{2-1}{2} = \frac{1}{2}, \quad a_3 = \frac{3-1}{3} = \frac{2}{3}, \quad a_4 = \frac{3}{4}, \quad a_5 = \frac{4}{5}, \quad a_6 = \frac{5}{6}$$

See Figure 2 for the graph.

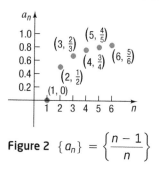

Figure 2 $\{a_n\} = \left\{\dfrac{n-1}{n}\right\}$

COMMENT Graphing utilities can be used to list the terms of a sequence and graph them. Figure 3 shows the sequence $\left\{\dfrac{n-1}{n}\right\}$ generated on a TI-84 Plus C graphing calculator. The first six terms of the sequence are shown on the viewing window. Figure 4 shows a graph of the sequence after pressing Y = in SEQuence mode and entering the formula for the sequence. Note that the first term of the sequence is barely visible since it lies on the x-axis. TRACEing the graph will enable you to see the terms of the sequence. The TABLE feature can also be used to generate the terms of the sequence. See Table 1.

Table 1

Figure 3

Figure 4

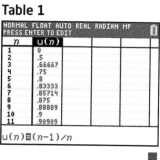

Now Work PROBLEM 17

EXAMPLE 2   **Listing the First Several Terms of a Sequence**

List the first six terms of the sequence $\{b_n\}$ and graph it.

$$\{b_n\} = \left\{(-1)^{n+1}\left(\frac{2}{n}\right)\right\}$$

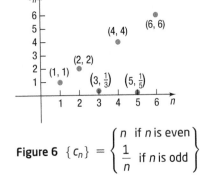

Figure 5 $\{b_n\} = \left\{(-1)^{n+1}\left(\dfrac{2}{n}\right)\right\}$

Solution The first six terms of the sequence are

$$b_1 = (-1)^{1+1}\left(\frac{2}{1}\right) = 2, \quad b_2 = (-1)^{2+1}\left(\frac{2}{2}\right) = -1, \quad b_3 = (-1)^{3+1}\left(\frac{2}{3}\right) = \frac{2}{3},$$

$$b_4 = -\frac{1}{2}, \qquad\qquad\qquad b_5 = \frac{2}{5}, \qquad\qquad\qquad b_6 = -\frac{1}{3}$$

See Figure 5 for the graph.

Note that in the sequence $\{b_n\}$ in Example 2, the signs of the terms *alternate*. This occurs when we use factors such as $(-1)^{n+1}$, which equals 1 if n is odd and -1 if n is even, or $(-1)^n$, which equals -1 if n is odd and 1 if n is even.

EXAMPLE 3 **Listing the First Several Terms of a Sequence**

List the first six terms of the sequence $\{c_n\}$ and graph it.

$$\{c_n\} = \begin{cases} n & \text{if } n \text{ is even} \\ \dfrac{1}{n} & \text{if } n \text{ is odd} \end{cases}$$

Solution The first six terms of the sequence are

$$c_1 = \frac{1}{1} = 1, \quad c_2 = 2, \quad c_3 = \frac{1}{3}, \quad c_4 = 4, \quad c_5 = \frac{1}{5}, \quad c_6 = 6$$

See Figure 6 for the graph.

- **Now Work** PROBLEM 19

Figure 6 $\{c_n\} = \begin{cases} n & \text{if } n \text{ is even} \\ \dfrac{1}{n} & \text{if } n \text{ is odd} \end{cases}$

The formula that generates the terms of a sequence is not unique. For example, the terms of the sequence in Example 3 could also be found using

$$\{d_n\} = \left\{n^{(-1)^n}\right\}$$

Sometimes a sequence is indicated by an observed pattern in the first few terms that makes it possible to infer the makeup of the nth term. In the examples that follow, enough terms of the sequence are given so that a natural choice for the nth term is suggested.

EXAMPLE 4 **Determining a Sequence from a Pattern**

(a) $e, \dfrac{e^2}{2}, \dfrac{e^3}{3}, \dfrac{e^4}{4}, \ldots$ $\qquad$ $a_n = \dfrac{e^n}{n}$

(b) $1, \dfrac{1}{3}, \dfrac{1}{9}, \dfrac{1}{27}, \ldots$ $\qquad$ $b_n = \dfrac{1}{3^{n-1}}$

(c) $1, 3, 5, 7, \ldots$ $\qquad$ $c_n = 2n - 1$

(d) $1, 4, 9, 16, 25, \ldots$ $\qquad$ $d_n = n^2$

(e) $1, -\dfrac{1}{2}, \dfrac{1}{3}, -\dfrac{1}{4}, \dfrac{1}{5}, \ldots$ $\qquad$ $e_n = (-1)^{n-1}\left(\dfrac{1}{n}\right)$

- **Now Work** PROBLEM 27

The Factorial Symbol

Some sequences in mathematics involve a special product called a *factorial*.

> **DEFINITION** **Factorial Symbol**
>
> If $n \geq 0$ is an integer, the **factorial symbol** $n!$ is defined as follows:
>
> - $0! = 1$ $\quad$ - $1! = 1$
> - $n! = n(n-1) \cdot \cdots \cdot 3 \cdot 2 \cdot 1$ $\quad$ if $n \geq 2$

Table 2

n	$n!$
0	1
1	1
2	2
3	6
4	24
5	120
6	720

For example, $2! = 2 \cdot 1 = 2$, $3! = 3 \cdot 2 \cdot 1 = 6$, $4! = 4 \cdot 3 \cdot 2 \cdot 1 = 24$, and so on. Table 2 lists the values of $n!$ for $0 \le n \le 6$.

Because

$$n! = n\,\underbrace{(n-1)\,(n-2)\cdot\,\cdots\,\cdot 3\cdot 2\cdot 1}_{(n-1)!}$$

the formula

$$n! = n(n-1)!$$

is used to find successive factorials. For example, because $6! = 720$,

$$7! = 7\cdot 6! = 7\cdot 720 = 5040$$

and

$$8! = 8\cdot 7! = 8\cdot 5040 = 40{,}320$$

Now Work PROBLEM 11

Exploration

Use your calculator's factorial key to see how fast factorials increase in value. Find the value of 69!. What happens when you try to find 70!? In fact, 70! is larger than 10^{100} (a googol), which is the largest number most calculators can display.

---

2 List the Terms of a Sequence Defined by a Recursive Formula

A second way of defining a sequence is to assign a value to the first (or the first few) term(s) and specify the nth term by a formula or equation that involves one or more of the terms preceding it. Such sequences are said to be defined **recursively**, and the rule or formula is called a **recursive formula**.

EXAMPLE 5 **Listing the Terms of a Recursively Defined Sequence**

List the first five terms of the recursively defined sequence

$$s_1 = 1 \qquad s_n = ns_{n-1}$$

Solution The first term is given as $s_1 = 1$. To get the second term, use $n = 2$ in the formula $s_n = ns_{n-1}$ to get $s_2 = 2s_1 = 2\cdot 1 = 2$. To get the third term, use $n = 3$ in the formula to get $s_3 = 3s_2 = 3\cdot 2 = 6$. To get a new term requires knowing the value of the preceding term. The first five terms are

$$s_1 = 1$$
$$s_2 = 2\cdot 1 = 2$$
$$s_3 = 3\cdot 2 = 6$$
$$s_4 = 4\cdot 6 = 24$$
$$s_5 = 5\cdot 24 = 120$$

Do you recognize this sequence? $s_n = n!$

EXAMPLE 6 **Listing the Terms of a Recursively Defined Sequence**

List the first five terms of the recursively defined sequence

$$u_1 = 1 \qquad u_2 = 1 \qquad u_n = u_{n-2} + u_{n-1}$$

Solution The first two terms are given. Finding each successive term requires knowing the previous two terms. That is,

$$u_1 = 1$$
$$u_2 = 1$$
$$u_3 = u_1 + u_2 = 1 + 1 = 2$$
$$u_4 = u_2 + u_3 = 1 + 2 = 3$$
$$u_5 = u_3 + u_4 = 2 + 3 = 5$$

The sequence given in Example 6 is called the **Fibonacci sequence**, and the terms of the sequence are called **Fibonacci numbers**. These numbers appear in a wide variety of applications (see Problems 85–88).

Now Work PROBLEMS 35 AND 43

3 Use Summation Notation

It is often important to find the sum of the first n terms of a sequence $\{a_n\}$, namely

$$a_1 + a_2 + a_3 + \cdots + a_n$$

Rather than writing down all these terms, we use **summation notation** to express the sum more concisely:

$$a_1 + a_2 + a_3 + \cdots + a_n = \sum_{k=1}^{n} a_k$$

The symbol Σ (the Greek letter sigma, which is an S in our alphabet) is simply an instruction to sum, or add up, the terms. The integer k is called the **index** of the sum; it tells where to start the sum and where to end it. The expression

$$\sum_{k=1}^{n} a_k$$

is an instruction to add the terms a_k of the sequence $\{a_n\}$ starting with $k = 1$ and ending with $k = n$. The expression is read as "the sum of a_k from $k = 1$ to $k = n$."

EXAMPLE 7 **Expanding Summation Notation**

Expand each sum.

(a) $\displaystyle\sum_{k=1}^{n} \frac{1}{k}$ (b) $\displaystyle\sum_{k=1}^{n} k!$

Solution (a) $\displaystyle\sum_{k=1}^{n} \frac{1}{k} = 1 + \frac{1}{2} + \frac{1}{3} + \cdots + \frac{1}{n}$ (b) $\displaystyle\sum_{k=1}^{n} k! = 1! + 2! + \cdots + n!$

Now Work PROBLEM 51

EXAMPLE 8 **Expressing a Sum Using Summation Notation**

Express each sum using summation notation.

(a) $1^2 + 2^2 + 3^2 + \cdots + 9^2$ (b) $1 + \frac{1}{2} + \frac{1}{4} + \frac{1}{8} + \cdots + \frac{1}{2^{n-1}}$

Solution (a) The sum $1^2 + 2^2 + 3^2 + \cdots + 9^2$ has 9 terms, each of the form k^2, starting at $k = 1$ and ending at $k = 9$:

$$1^2 + 2^2 + 3^2 + \cdots + 9^2 = \sum_{k=1}^{9} k^2$$

(b) The sum

$$1 + \frac{1}{2} + \frac{1}{4} + \frac{1}{8} + \cdots + \frac{1}{2^{n-1}}$$

has n terms, each of the form $\dfrac{1}{2^{k-1}}$, starting at $k = 1$ and ending at $k = n$:

$$1 + \frac{1}{2} + \frac{1}{4} + \frac{1}{8} + \cdots + \frac{1}{2^{n-1}} = \sum_{k=1}^{n} \frac{1}{2^{k-1}}$$

Now Work PROBLEM 61

The index of summation need not begin at 1, nor end at n; for example, the sum in Example 8(b) could also be expressed as

$$\sum_{k=0}^{n-1} \frac{1}{2^k} = 1 + \frac{1}{2} + \frac{1}{4} + \cdots + \frac{1}{2^{n-1}}$$

Letters other than k are also used as the index. For example,

$$\sum_{j=1}^{n} j! \quad \text{and} \quad \sum_{i=1}^{n} i!$$

both represent the same sum given in Example 7(b).

4 Find the Sum of a Sequence

The following theorem lists some properties of summation notation. These properties are useful for adding the terms of a sequence.

THEOREM **Summation Notation Properties**

If $\{a_n\}$ and $\{b_n\}$ are two sequences and c is a real number, then

$$\sum_{k=1}^{n} (ca_k) = ca_1 + ca_2 + \cdots + ca_n = c(a_1 + a_2 + \cdots + a_n) = c\sum_{k=1}^{n} a_k \quad \textbf{(1)}$$

$$\sum_{k=1}^{n} (a_k + b_k) = \sum_{k=1}^{n} a_k + \sum_{k=1}^{n} b_k \quad \textbf{(2)}$$

$$\sum_{k=1}^{n} (a_k - b_k) = \sum_{k=1}^{n} a_k - \sum_{k=1}^{n} b_k \quad \textbf{(3)}$$

$$\sum_{k=j+1}^{n} a_k = \sum_{k=1}^{n} a_k - \sum_{k=1}^{j} a_k \quad \text{where } 0 < j < n \quad \textbf{(4)}$$

The proof of property (1) follows from the distributive property of real numbers. The proofs of properties (2) and (3) are based on the commutative and associative properties of real numbers. Property (4) states that the sum from $j + 1$ to n equals the sum from 1 to n minus the sum from 1 to j. This property is helpful when the index of summation begins at a number larger than 1.

The next theorem provides some formulas for finding the sum of certain sequences.

THEOREM **Formulas for Sums of the First n Terms of a Sequence**

$$\sum_{k=1}^{n} c = \underbrace{c + c + \cdots + c}_{n \text{ terms}} = cn \quad c \text{ is a real number} \quad \textbf{(5)}$$

$$\sum_{k=1}^{n} k = 1 + 2 + 3 + \cdots + n = \frac{n(n+1)}{2} \quad \textbf{(6)}$$

$$\sum_{k=1}^{n} k^2 = 1^2 + 2^2 + 3^2 + \cdots + n^2 = \frac{n(n+1)(2n+1)}{6} \quad \textbf{(7)}$$

$$\sum_{k=1}^{n} k^3 = 1^3 + 2^3 + 3^3 + \cdots + n^3 = \left[\frac{n(n+1)}{2}\right]^2 \quad \textbf{(8)}$$

The proof of formula (5) follows from the definition of summation notation. You are asked to prove formula (6) in Problem 94. The proofs of formulas (7) and (8) require mathematical induction, which is discussed in Section 9.4.

Notice the difference between formulas (5) and (6). In (5) the constant c is being summed from 1 to n, while in (6) the index of summation k is being summed from 1 to n.

| EXAMPLE 9 | **Finding the Sum of a Sequence** |

Find the sum of each sequence.

(a) $\displaystyle\sum_{k=1}^{5} (3k)$
(b) $\displaystyle\sum_{k=1}^{10} (k^3 + 1)$

(c) $\displaystyle\sum_{k=1}^{24} (k^2 - 7k + 2)$
(d) $\displaystyle\sum_{k=6}^{20} (4k^2)$

Solution

(a) $\displaystyle\sum_{k=1}^{5} (3k) = 3 \sum_{k=1}^{5} k$ **Property (1)**

$\qquad\qquad\quad = 3 \cdot \dfrac{5(5+1)}{2}$ $\displaystyle\sum_{k=1}^{n} k = \dfrac{n(n+1)}{2}$

$\qquad\qquad\quad = 3 \cdot 15$

$\qquad\qquad\quad = 45$

(b) $\displaystyle\sum_{k=1}^{10} (k^3 + 1) = \sum_{k=1}^{10} k^3 + \sum_{k=1}^{10} 1$ **Property (2)**

$\qquad\qquad\quad = \left(\dfrac{10(10+1)}{2}\right)^2 + 1 \cdot 10$ $\displaystyle\sum_{k=1}^{n} k^3 = \left[\dfrac{n(n+1)}{2}\right]^2 ; \sum_{k=1}^{n} 1 = 1 \cdot n$

$\qquad\qquad\quad = 3025 + 10$

$\qquad\qquad\quad = 3035$

(c) $\displaystyle\sum_{k=1}^{24} (k^2 - 7k + 2) = \sum_{k=1}^{24} k^2 - \sum_{k=1}^{24} (7k) + \sum_{k=1}^{24} 2$ **Properties (2) and (3)**

$\qquad\qquad\quad = \displaystyle\sum_{k=1}^{24} k^2 - 7 \sum_{k=1}^{24} k + \sum_{k=1}^{24} 2$ **Property (1)**

$\qquad\qquad\quad = \dfrac{24(24+1)(2 \cdot 24 + 1)}{6} - 7 \cdot \dfrac{24(24+1)}{2} + 2 \cdot 24$ **Formulas (7), (6), (5)**

$\qquad\qquad\quad = 4900 - 2100 + 48$

$\qquad\qquad\quad = 2848$

(d) Notice that the index of summation starts at 6. Use property (4) as follows:

$$\sum_{k=6}^{20} (4k^2) = 4 \sum_{k=6}^{20} k^2 = 4 \left[\sum_{k=1}^{20} k^2 - \sum_{k=1}^{5} k^2 \right] = 4 \left[\dfrac{20 \cdot 21 \cdot 41}{6} - \dfrac{5 \cdot 6 \cdot 11}{6} \right]$$

$\qquad$↑   ↑   ↑

$\qquad$**Property (1)**   **Property (4)**   **Formula (7)**

$$= 4[2870 - 55] = 11{,}260$$

➤ **Now Work** PROBLEM 73

9.1 Assess Your Understanding

'Are You Prepared?' *Answers are given at the end of these exercises. If you get a wrong answer, read the pages listed in red.*

1. For the function $f(x) = \dfrac{x-1}{x}$, find $f(2)$ and $f(3)$. (pp. 207–210)

2. *True or False* A function is a relation between two sets D and R so that each element x in the first set D is related to exactly one element y in the second set R. (pp. 205–207)

Concepts and Vocabulary

3. $A(n)$ _____ is a function whose domain is the set of positive integers.

4. *True or False* The notation a_5 represents the fifth term of a sequence.

5. *True or False* If $n \geq 2$ is an integer, then
$$n! = n(n-1) \cdots 3 \cdot 2 \cdot 1$$

6. *Multiple Choice* The sequence $a_1 = 5$, $a_n = 3a_{n-1}$ is an example of a(n) _____ sequence.
- **(a)** alternating
- **(b)** recursive
- **(c)** Fibonacci
- **(d)** summation

7. The notation
$$a_1 + a_2 + a_3 + \cdots + a_n = \sum_{k=1}^{n} a_k$$
is an example of _____ notation.

8. *Multiple Choice* $\displaystyle\sum_{k=1}^{n} k = 1 + 2 + 3 + \cdots + n = $ _____.
- **(a)** $n!$
- **(b)** $\dfrac{n(n+1)}{2}$
- **(c)** nk
- **(d)** $\dfrac{n(n+1)(2n+1)}{6}$

Skill Building

In Problems 9–14, evaluate each factorial expression.

9. $10!$

10. $9!$

11. $\dfrac{9!}{6!}$

12. $\dfrac{12!}{10!}$

13. $\dfrac{4! \, 11!}{7!}$

14. $\dfrac{5! \, 8!}{3!}$

In Problems 15–26, list the first five terms of each sequence.

15. $\{s_n\} = \{n\}$

16. $\{s_n\} = \{n^2 + 1\}$

17. $\{a_n\} = \left\{\dfrac{n}{n+2}\right\}$

18. $\{b_n\} = \left\{\dfrac{2n+1}{2n}\right\}$

19. $\{c_n\} = \{(-1)^{n+1} n^2\}$

20. $\{d_n\} = \left\{(-1)^{n-1}\left(\dfrac{n}{2n-1}\right)\right\}$

21. $\{s_n\} = \left\{\dfrac{3^n}{2^n + 3}\right\}$

22. $\{s_n\} = \left\{\left(\dfrac{4}{3}\right)^n\right\}$

23. $\{t_n\} = \left\{\dfrac{(-1)^n}{(n+1)(n+2)}\right\}$

24. $\{a_n\} = \left\{\dfrac{3^n}{n}\right\}$

25. $\{b_n\} = \left\{\dfrac{n}{e^n}\right\}$

26. $\{c_n\} = \left\{\dfrac{n^2}{2^n}\right\}$

In Problems 27–34, the given pattern continues. Write down the nth term of a sequence $\{a_n\}$ suggested by the pattern.

27. $\dfrac{1}{2}, \dfrac{2}{3}, \dfrac{3}{4}, \dfrac{4}{5}, \cdots$

28. $\dfrac{1}{1 \cdot 2}, \dfrac{1}{2 \cdot 3}, \dfrac{1}{3 \cdot 4}, \dfrac{1}{4 \cdot 5}, \cdots$

29. $1, \dfrac{1}{2}, \dfrac{1}{4}, \dfrac{1}{8}, \cdots$

30. $\dfrac{2}{3}, \dfrac{4}{9}, \dfrac{8}{27}, \dfrac{16}{81}, \cdots$

31. $1, -1, 1, -1, 1, -1, \ldots$

32. $1, \dfrac{1}{2}, 3, \dfrac{1}{4}, 5, \dfrac{1}{6}, 7, \dfrac{1}{8}, \ldots$

33. $1, -2, 3, -4, 5, -6, \ldots$

34. $2, -4, 6, -8, 10, \ldots$

In Problems 35–48, a sequence is defined recursively. List the first five terms.

35. $a_1 = 2$; $a_n = 3 + a_{n-1}$

36. $a_1 = 3$; $a_n = 4 - a_{n-1}$

37. $a_1 = -2$; $a_n = n + a_{n-1}$

38. $a_1 = 1$; $a_n = n - a_{n-1}$

39. $a_1 = 4$; $a_n = 3a_{n-1}$

40. $a_1 = 2$; $a_n = -a_{n-1}$

41. $a_1 = 3$; $a_n = \dfrac{a_{n-1}}{n}$

42. $a_1 = -2$; $a_n = n + 3a_{n-1}$

43. $a_1 = 1$; $a_2 = 2$; $a_n = a_{n-1} \cdot a_{n-2}$

44. $a_1 = -1$; $a_2 = 1$; $a_n = a_{n-2} + n a_{n-1}$

45. $a_1 = A$; $a_n = a_{n-1} + d$

46. $a_1 = A$; $a_n = r a_{n-1}$, $r \neq 0$

47. $a_1 = \sqrt{2}$; $a_n = \sqrt{2 + a_{n-1}}$

48. $a_1 = \sqrt{2}$; $a_n = \sqrt{\dfrac{a_{n-1}}{2}}$

In Problems 49–58, expand each sum.

49. $\displaystyle\sum_{k=1}^{n} (k+2)$

50. $\displaystyle\sum_{k=1}^{n} (2k+1)$

51. $\displaystyle\sum_{k=1}^{n} \dfrac{k^2}{2}$

52. $\displaystyle\sum_{k=1}^{n} (k+1)^2$

53. $\displaystyle\sum_{k=0}^{n} \dfrac{1}{3^k}$

54. $\displaystyle\sum_{k=0}^{n} \left(\dfrac{3}{2}\right)^k$

55. $\displaystyle\sum_{k=0}^{n-1} \dfrac{1}{3^{k+1}}$

56. $\displaystyle\sum_{k=0}^{n-1} (2k+1)$

57. $\displaystyle\sum_{k=2}^{n} (-1)^k \ln k$

58. $\displaystyle\sum_{k=3}^{n} (-1)^{k+1} 2^k$

In Problems 59–68, express each sum using summation notation.

59. $1 + 2 + 3 + \cdots + 20$

60. $1^3 + 2^3 + 3^3 + \cdots + 8^3$

61. $\dfrac{1}{2} + \dfrac{2}{3} + \dfrac{3}{4} + \cdots + \dfrac{13}{13+1}$

62. $1 + 3 + 5 + 7 + \cdots + [2(12) - 1]$

63. $1 - \dfrac{1}{3} + \dfrac{1}{9} - \dfrac{1}{27} + \cdots + (-1)^6 \left(\dfrac{1}{3^6}\right)$

64. $\dfrac{2}{3} - \dfrac{4}{9} + \dfrac{8}{27} - \cdots + (-1)^{12} \left(\dfrac{2}{3}\right)^{11}$

65. $3 + \dfrac{3^2}{2} + \dfrac{3^3}{3} + \cdots + \dfrac{3^n}{n}$

66. $\dfrac{1}{e} + \dfrac{2}{e^2} + \dfrac{3}{e^3} + \cdots + \dfrac{n}{e^n}$

67. $a + (a+d) + (a+2d) + \cdots + (a+nd)$

68. $a + ar + ar^2 + \cdots + ar^{n-1}$

In Problems 69–80, find the sum of each sequence.

69. $\sum_{k=1}^{40} 5$

70. $\sum_{k=1}^{50} 8$

71. $\sum_{k=1}^{40} k$

72. $\sum_{k=1}^{24} (-k)$

73. $\sum_{k=1}^{20} (5k + 3)$

74. $\sum_{k=1}^{26} (3k - 7)$

75. $\sum_{k=1}^{16} (k^2 + 4)$

76. $\sum_{k=0}^{14} (k^2 - 4)$

77. $\sum_{k=10}^{60} (2k)$

78. $\sum_{k=8}^{40} (-3k)$

79. $\sum_{k=5}^{20} k^3$

80. $\sum_{k=4}^{24} k^3$

Applications and Extensions

81. Credit Card Debt John has a balance of $3000 on his Discover card, which charges 1% interest per month on any unpaid balance from the previous month. John can afford to pay $100 toward the balance each month. His balance each month after making a $100 payment is given by the recursively defined sequence

$$B_0 = \$3000 \qquad B_n = 1.01B_{n-1} - 100$$

Determine John's balance after making the first payment. That is, determine B_1.

82. Trout Population A pond currently contains 2000 trout. A fish hatchery decides to add 20 trout each month. It is also known that the trout population is growing at a rate of 3% per month. The size of the population after n months is given by the recursively defined sequence

$$p_0 = 2000 \qquad p_n = 1.03p_{n-1} + 20$$

How many trout are in the pond after 2 months? That is, what is p_2?

83. Car Loans Phil bought a car by taking out a loan for $18,500 at 0.5% interest per month. Phil's normal monthly payment is $434.47 per month, but he decides that he can afford to pay $100 extra toward the balance each month. His balance each month is given by the recursively defined sequence

$$B_0 = 18,500 \qquad B_n = 1.005B_{n-1} - 534.47$$

Determine Phil's balance after making the first payment. That is, determine B_1.

84. Environmental Control The Environmental Protection Agency (EPA) determines that Maple Lake has 250 tons of pollutant as a result of industrial waste and that 10% of the pollutant present is neutralized by solar oxidation every year. The EPA imposes new pollution control laws that result in 15 tons of new pollutant entering the lake each year. The amount of pollutant in the lake after n years is given by the recursively defined sequence

$$p_0 = 250 \qquad p_n = 0.9p_{n-1} + 15$$

Determine the amount of pollutant in the lake after 2 years. That is, determine p_2.

85. Growth of a Rabbit Colony A colony of rabbits begins with one pair of mature rabbits, which produces a pair of offspring (one male, one female) each month. Assume that all rabbits mature in 1 month and produce a pair of offspring (one male, one female) after 2 months. If no rabbits ever die, how many pairs of mature rabbits are there after 7 months? See illustration, top right.

[**Hint:** A Fibonacci sequence models this colony. Do you see why?]

86. Fibonacci Sequence Let

$$u_n = \frac{\left(1 + \sqrt{5}\right)^n - \left(1 - \sqrt{5}\right)^n}{2^n \sqrt{5}}$$

define the nth term of a sequence.
(a) Show that $u_1 = 1$ and $u_2 = 1$.
(b) Show that $u_{n+2} = u_{n+1} + u_n$.
(c) Draw the conclusion that $\{u_n\}$ is a Fibonacci sequence.

87. The Pascal Triangle The triangular array shown, called the Pascal triangle, is partitioned using diagonal lines as shown. Find the sum of the numbers in each diagonal row. Do you recognize this sequence?

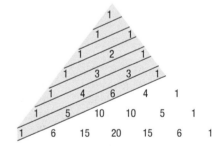

88. Fibonacci Sequence Use the result of Problem 86 to do the following problems.
(a) List the first 11 terms of the Fibonacci sequence.
(b) List the first 10 terms of the ratio $\dfrac{u_{n+1}}{u_n}$.
(c) As n gets large, what number does the ratio approach? This number is referred to as the **golden ratio**. Rectangles whose sides are in this ratio were considered pleasing to the eye by the Greeks. For example, the façade of the Parthenon was constructed using the golden ratio.
(d) Write down the first 10 terms of the ratio $\dfrac{u_n}{u_{n+1}}$.
(e) As n gets large, what number does the ratio approach? This number is referred to as the **conjugate golden ratio**. This ratio is believed to have been used in the construction of the Great Pyramid in Egypt. The ratio equals the sum of the areas of the four face triangles divided by the total surface area of the Great Pyramid.

89. Approximating $f(x) = e^x$ In calculus, it can be shown that

$$f(x) = e^x = \sum_{k=0}^{\infty} \frac{x^k}{k!}$$

We can approximate the value of $f(x) = e^x$ for any x using the following sum

$$f(x) = e^x \approx \sum_{k=0}^{n} \frac{x^k}{k!}$$

for some n.
(a) Approximate $f(1.3)$ with $n = 4$.
(b) Approximate $f(1.3)$ with $n = 7$.
(c) Use a calculator to approximate $f(1.3)$.
(d) Using trial and error, along with a graphing utility's SEQuence mode, determine the value of n required to approximate $f(1.3)$ correct to eight decimal places.

90. Approximating $f(x) = e^x$ Refer to Problem 89.
(a) Approximate $f(-2.4)$ with $n = 3$.
(b) Approximate $f(-2.4)$ with $n = 6$.
(c) Use a calculator to approximate $f(-2.4)$.
(d) Using trial and error, along with a graphing utility's SEQuence mode, determine the value of n required to approximate $f(-2.4)$ correct to eight decimal places.

91. Bode's Law In 1772, Johann Bode published the following formula for predicting the mean distances, in astronomical units (AU), of the planets from the sun:

$$a_1 = 0.4 \qquad a_n = 0.4 + 0.3 \cdot 2^{n-2}$$

where $n \geq 2$ is the number of the planet from the sun.
(a) Determine the first eight terms of the sequence.
(b) At the time of Bode's publication, the known planets were Mercury (0.39 AU), Venus (0.72 AU), Earth (1 AU), Mars (1.52 AU), Jupiter (5.20 AU), and Saturn (9.54 AU).
How do the actual distances compare to the terms of the sequence?
(c) The planet Uranus was discovered in 1781, and the asteroid Ceres was discovered in 1801. The mean orbital distances from the sun to Uranus and Ceres* are 19.2 AU and 2.77 AU, respectively. How well do these values fit within the sequence?
(d) Determine the ninth and tenth terms of Bode's sequence.
(e) The planets Neptune and Pluto* were discovered in 1846 and 1930, respectively. Their mean orbital distances from the sun are 30.07 AU and 39.44 AU, respectively. How do these actual distances compare to the terms of the sequence?
(f) On July 29, 2005, NASA announced the discovery of a dwarf planet* ($n = 11$), which has been named Eris. Use Bode's Law to predict the mean orbital distance of Eris from the sun. Its actual mean distance is not yet known, but Eris is currently about 97 astronomical units from the sun.

Source: NASA

92. Droste Effect The *Droste Effect*, named after the image on boxes of Droste cocoa powder, refers to an image that contains within it a smaller version of the image, which in turn contains an even smaller version, and so on. If each version of the image is $\frac{1}{5}$ the height of the previous version, the height of the nth version is given by $a_n = \frac{1}{5}a_{n-1}$. Suppose a Droste image on a package has a height of 4 inches. How tall would the image be in the 6th version?

* Ceres, Haumea, Makemake, Pluto, and Eris are referred to as dwarf planets.

93. Reflections in a Mirror A highly reflective mirror reflects 95% of the light that falls on it. In a light box having walls made of the mirror, the light reflects back-and-forth between the mirrors.
(a) If the original intensity of the light is I_0 before it falls on a mirror, write the nth term of the sequence that describes the intensity of the light after n reflections.
(b) How many reflections are needed to reduce the light intensity by at least 98%?

94. Show that

$$1 + 2 + \cdots + (n-1) + n = \frac{n(n+1)}{2}$$

[**Hint:** Let

$$S = 1 + 2 + \cdots + (n-1) + n$$
$$S = n + (n-1) + (n-2) + \cdots + 1$$

Add these equations. Then

$$2S = \underbrace{[1+n] + [2+(n-1)] + \cdots + [n+1]}_{n \text{ terms in bracket}}$$

Now complete the derivation.]

Computing Square Roots *A method for approximating $\sqrt{p}$ can be traced back to the Babylonians. The formula is given by the recursively defined sequence*

$$a_0 = k \qquad a_n = \frac{1}{2}\left(a_{n-1} + \frac{p}{a_{n-1}}\right)$$

where k is an initial guess as to the value of the square root. Use this recursive formula to approximate the following square roots by finding a_5. Compare this result to the value provided by your calculator.

95. $\sqrt{5}$ **96.** $\sqrt{8}$ **97.** $\sqrt{21}$ **98.** $\sqrt{89}$

99. Triangular Numbers A **triangular number** is a term of the sequence

$$u_1 = 1 \qquad u_{n+1} = u_n + (n+1)$$

List the first seven triangular numbers.

100. *Challenge Problem* For the sequence given in Problem 99, show that

$$u_{n+1} = \frac{(n+1)(n+2)}{2}.$$

101. *Challenge Problem* For the sequence given in Problem 99, show that

$$u_{n+1} + u_n = (n+1)^2$$

102. *Challenge Problem* If the terms of a sequence have the property that $\dfrac{a_1}{a_2} = \dfrac{a_2}{a_3} = \cdots = \dfrac{a_{n-1}}{a_n}$, show that $\dfrac{a_1{}^n}{a_2{}^n} = \dfrac{a_1}{a_{n+1}}$.

[**Hint**: Let r equal the common ratio so $\dfrac{a_1}{a_2} = \dfrac{a_2}{a_3} = \cdots = \dfrac{a_{n-1}}{a_n} = r.$]

Explaining Concepts: Discussion and Writing

103. Investigate various applications that lead to a Fibonacci sequence, such as in art, architecture, or financial markets. Write an essay on these applications.

104. Write a paragraph that explains why the numbers found in Problem 99 are called triangular.

Retain Your Knowledge

Problems 105–111 are based on material learned earlier in the course. The purpose of these problems is to keep the material fresh in your mind so that you are better prepared for the final exam.

105. If \$2500 is invested at 3% compounded monthly, find the amount that results after a period of 2 years.

106. Multiply: $(3x - 2)^3$

107. Find an equation of the parabola with vertex $(-3, 4)$ and focus $(1, 4)$.

108. Find the horizontal asymptote, if one exists, of
$$f(x) = \frac{9x}{3x^2 - 2x - 1}$$

109. If $f(x) = 5x^2 - 2x + 9$ and $f(a + 1) = 16$, find the possible values for a.

△ 110. In calculus, the critical numbers for a function are numbers in the domain of f where $f'(x) = 0$ or $f'(x)$ is undefined. Find the critical numbers for $f(x) = \dfrac{x^2 - 3x + 18}{x - 2}$ if $f'(x) = \dfrac{x^2 - 4x - 12}{(x - 2)^2}$.

△ 111. Find the difference quotient of
$$f(x) = 2x^3 - 5$$

'Are You Prepared?' Answers

1. $f(2) = \dfrac{1}{2}; f(3) = \dfrac{2}{3}$ 2. True

9.2 Arithmetic Sequences

OBJECTIVES 1 Determine Whether a Sequence Is Arithmetic (p. 662)
 2 Find a Formula for an Arithmetic Sequence (p. 663)
 3 Find the Sum of an Arithmetic Sequence (p. 665)

1 Determine Whether a Sequence Is Arithmetic

When the difference between successive terms of a sequence is always the same number, the sequence is called **arithmetic**.

DEFINITION Arithmetic Sequence

An **arithmetic sequence*** is defined recursively as $a_1 = a$, $a_n - a_{n-1} = d$, or as

$$a_1 = a \qquad a_n = a_{n-1} + d \qquad \qquad (1)$$

where $a_1 = a$ and d are real numbers. The number a is the first term, and the number d is called the **common difference**.

*Sometimes called an **arithmetic progression**

EXAMPLE 1 — **Determining Whether a Sequence Is Arithmetic**

The sequence

$$4, \ 6, \ 8, \ 10, \ldots$$

is arithmetic since the difference of successive terms is 2. The first term is $a_1 = 4$, and the common difference is $d = 2$.

EXAMPLE 2 — **Determining Whether a Sequence Is Arithmetic**

Show that the following sequence is arithmetic. Find the first term and the common difference.

$$\{s_n\} = \{3n + 5\}$$

Solution   The first term is $s_1 = 3 \cdot 1 + 5 = 8$. The nth term and the $(n - 1)$st term of the sequence $\{s_n\}$ are

$$s_n = 3n + 5 \quad \text{and} \quad s_{n-1} = 3(n - 1) + 5 = 3n + 2$$

Their difference d is

$$d = s_n - s_{n-1} = (3n + 5) - (3n + 2) = 5 - 2 = 3$$

Since the difference of any two successive terms is 3, $\{s_n\}$ is an arithmetic sequence. The common difference is $d = 3$.

EXAMPLE 3 — **Determining Whether a Sequence Is Arithmetic**

Show that the sequence $\{t_n\} = \{4 - n\}$ is arithmetic. Find the first term and the common difference.

Solution   The first term is $t_1 = 4 - 1 = 3$. The nth term and the $(n - 1)$st term are

$$t_n = 4 - n \quad \text{and} \quad t_{n-1} = 4 - (n - 1) = 5 - n$$

Their difference d is

$$d = t_n - t_{n-1} = (4 - n) - (5 - n) = 4 - 5 = -1$$

Since the difference of any two successive terms is -1, $\{t_n\}$ is an arithmetic sequence. The common difference is $d = -1$.

- **Now Work** PROBLEM 9

2 Find a Formula for an Arithmetic Sequence

Suppose that a is the first term of an arithmetic sequence whose common difference is d. We seek a formula for the nth term, a_n. To see the pattern, consider the first few terms.

$$
\begin{aligned}
a_1 &= a \\
a_2 &= a_1 + d = a_1 + 1 \cdot d \\
a_3 &= a_2 + d = (a_1 + d) + d = a_1 + 2 \cdot d \\
a_4 &= a_3 + d = (a_1 + 2 \cdot d) + d = a_1 + 3 \cdot d \\
a_5 &= a_4 + d = (a_1 + 3 \cdot d) + d = a_1 + 4 \cdot d \\
&\vdots \\
a_n &= a_{n-1} + d = [a_1 + (n - 2)d] + d = a_1 + (n - 1)d
\end{aligned}
$$

The terms of an arithmetic sequence with first term a_1 and common difference d follow the pattern

$$a_1, \ a_1 + d, \ a_1 + 2d, \ a_1 + 3d, \ldots$$

> **THEOREM** *n*th Term of an Arithmetic Sequence
>
> For an arithmetic sequence $\{a_n\}$ whose first term is a_1 and whose common difference is d, the nth term is determined by the formula
>
> $$a_n = a_1 + (n-1)d \qquad n \text{ a positive integer} \qquad (2)$$

EXAMPLE 4 **Finding a Particular Term of an Arithmetic Sequence**

Find the 41st term of the arithmetic sequence: $2, 6, 10, 14, 18, \ldots$

Solution The first term of the arithmetic sequence is $a_1 = 2$, and the common difference is $d = 4$. By formula (2), the nth term is

$$a_n = 2 + (n-1)4 = 4n - 2 \qquad a_n = a_1 + (n-1)d$$

The 41st term is

$$a_{41} = 4 \cdot 41 - 2 = 164 - 2 = 162$$

> **Now Work** PROBLEM 25

EXAMPLE 5 **Finding a Recursive Formula for an Arithmetic Sequence**

The 8th term of an arithmetic sequence is 75, and the 20th term is 39.

(a) Find the first term and the common difference.
(b) Find a recursive formula for the sequence.
(c) What is the nth term of the sequence?

Solution (a) The nth term of an arithmetic sequence is $a_n = a_1 + (n-1)d$. As a result,

$$\begin{cases} a_8 = a_1 + 7d = 75 \\ a_{20} = a_1 + 19d = 39 \end{cases}$$

This is a system of two linear equations containing two variables, a_1 and d, which can be solved by elimination. Subtracting the second equation from the first gives

$$-12d = 36$$
$$d = -3$$

Exploration

Graph the recursive formula from Example 5, $a_1 = 96$, $a_n = a_{n-1} - 3$, using a graphing utility. Conclude that the graph of the recursive formula behaves like the graph of a linear function. How is d, the common difference, related to m, the slope of a line?

With $d = -3$, use $a_1 + 7d = 75$ to find that $a_1 = 75 - 7d = 75 - 7(-3) = 96$. The first term is $a_1 = 96$, and the common difference is $d = -3$.

(b) Using formula (1), a recursive formula for this sequence is

$$a_1 = 96 \qquad a_n = a_{n-1} - 3$$

(c) Using formula (2), the nth term of the sequence $\{a_n\}$ is

$$a_n = a_1 + (n-1)d = 96 + (n-1)(-3) = 99 - 3n$$

> **Now Work** PROBLEMS 17 AND 31

3 Find the Sum of an Arithmetic Sequence

The next theorem gives two formulas for finding the sum of the first n terms of an arithmetic sequence.

> **THEOREM Sum of the First n Terms of an Arithmetic Sequence**
>
> Suppose $\{a_n\}$ is an arithmetic sequence with first term a_1 and common difference d. The sum S_n of the first n terms of $\{a_n\}$ may be found in two ways:
>
> $$S_n = a_1 + a_2 + a_3 + \cdots + a_n$$
>
> $$= \frac{n}{2}[2a_1 + (n-1)d] \qquad (3)$$
>
> $$= \frac{n}{2}(a_1 + a_n) \qquad (4)$$

Proof

$$S_n = a_1 + a_2 + a_3 + \cdots + a_n \qquad \text{Sum of first } n \text{ terms}$$

$$= a_1 + (a_1 + d) + (a_1 + 2d) + \cdots + [a_1 + (n-1)d] \qquad \text{Formula (2)}$$

$$= \underbrace{(a_1 + a_1 + \cdots + a_1)}_{n \text{ terms}} + [d + 2d + \cdots + (n-1)d] \qquad \text{Rearrange terms.}$$

$$= na_1 + d[1 + 2 + \cdots + (n-1)]$$

$$= na_1 + d \cdot \frac{(n-1)n}{2} \qquad \sum_{k=1}^{n-1} k = \frac{(n-1)n}{2}$$

$$= na_1 + \frac{n}{2}(n-1)d$$

$$= \frac{n}{2}[2a_1 + (n-1)d] \qquad \text{Factor out } \frac{n}{2}; \text{ this is formula (3).}$$

$$= \frac{n}{2}[a_1 + a_1 + (n-1)d]$$

$$= \frac{n}{2}(a_1 + a_n) \qquad a_n = a_1 + (n-1)d; \text{ this is formula (4).} \qquad \blacksquare$$

There are two ways to find the sum of the first n terms of an arithmetic sequence. Formula (3) uses the first term and common difference, and formula (4) uses the first term and the nth term. Use whichever form is easier.

EXAMPLE 6

Finding the Sum of an Arithmetic Sequence

Find the sum S_n of the first n terms of the arithmetic sequence

$$8 + 11 + 14 + 17 + \cdots$$

Solution

The sequence is an arithmetic sequence with first term $a_1 = 8$ and common difference $d = 11 - 8 = 3$. To find the sum of the first n terms, use formula (3).

$$S_n = \frac{n}{2}[2 \cdot 8 + (n-1) \cdot 3] = \frac{n}{2}(3n + 13)$$

$$\underset{\uparrow}{} $$

$$S_n = \frac{n}{2}[2a_1 + (n-1)d]$$

Now Work PROBLEM 39

| EXAMPLE 7 | **Finding the Sum of an Arithmetic Sequence** |

Find the sum: $60 + 64 + 68 + 72 + \cdots + 120$

Solution This is the sum S_n of an arithmetic sequence $\{a_n\}$ whose first term is $a_1 = 60$ and whose common difference is $d = 4$. The nth term is $a_n = 120$. Use formula (2) to find n.

$$a_n = a_1 + (n - 1)d \qquad \text{Formula (2)}$$
$$120 = 60 + (n - 1) \cdot 4 \qquad a_n = 120, a_1 = 60, d = 4$$
$$60 = 4(n - 1) \qquad \text{Simplify.}$$
$$15 = n - 1 \qquad \text{Simplify.}$$
$$n = 16 \qquad \text{Solve for } n.$$

Now use formula (4) to find the sum S_{16}.

$$60 + 64 + 68 + \cdots + 120 = S_{16} \underset{\uparrow}{=} \frac{16}{2}(60 + 120) = 1440$$

$$S_n = \frac{n}{2}(a_1 + a_n)$$

➤ **Now Work** PROBLEM 43

| EXAMPLE 8 | **Creating a Floor Design** |

A ceramic tile floor is designed in the shape of a trapezoid 20 feet wide at the base and 10 feet wide at the top. See Figure 7. The tiles, which measure 12 inches by 12 inches, are to be placed so that each successive row contains one fewer tile than the preceding row. How many tiles will be required?

Solution The bottom row requires 20 tiles and the top row, 10 tiles. Since each successive row requires one fewer tile, the total number of tiles required is

$$S = 20 + 19 + 18 + \cdots + 11 + 10$$

This is the sum of an arithmetic sequence; the common difference is -1. The number of terms to be added is $n = 11$, with the first term $a_1 = 20$ and the last term $a_{11} = 10$. The sum S is

$$S = \frac{n}{2}(a_1 + a_{11}) = \frac{11}{2}(20 + 10) = 165$$

In all, 165 tiles will be required.

Figure 7

9.2 Assess Your Understanding

Concepts and Vocabulary

1. In a(n) _____ sequence, the difference between successive terms is a constant.

2. *True or False* For an arithmetic sequence $\{a_n\}$ whose first term is a_1 and whose common difference is d, the nth term is determined by the formula $a_n = a_1 + nd$.

3. If the 5th term of an arithmetic sequence is 12 and the common difference is 5, then the 6th term of the sequence is ____.

4. *True or False* The sum S_n of the first n terms of an arithmetic sequence $\{a_n\}$ whose first term is a_1 is found using the formula $S_n = \frac{n}{2}(a_1 + a_n)$.

5. *Multiple Choice* An arithmetic sequence can always be expressed as a(n) _____ sequence.

(a) Fibonacci   (b) alternating

(c) increasing   (d) recursive

6. *Multiple Choice* If $a_n = -2n + 7$ is the nth term of an arithmetic sequence, the first term is _____.

(a) -2 (b) 0

(c) 5   (d) 7

Skill Building

In Problems 7–16, show that each sequence is arithmetic. Find the common difference, and list the first four terms.

7. $\{s_n\} = \{n + 4\}$ **8.** $\{s_n\} = \{n - 5\}$ **9.** $\{a_n\} = \{2n - 5\}$ **10.** $\{b_n\} = \{3n + 1\}$

11. $\{c_n\} = \{6 - 2n\}$ **12.** $\{a_n\} = \{4 - 2n\}$ **13.** $\{t_n\} = \left\{\dfrac{1}{2} - \dfrac{1}{3}n\right\}$ **14.** $\{t_n\} = \left\{\dfrac{2}{3} + \dfrac{n}{4}\right\}$

15. $\{s_n\} = \{\ln 3^n\}$ **16.** $\{s_n\} = \{e^{\ln n}\}$

In Problems 17–24, find the nth term of the arithmetic sequence $\{a_n\}$ whose first term a_1 and common difference d are given. What is the 51st term?

17. $a_1 = 2;\ \ d = 3$ **18.** $a_1 = -2;\ \ d = 4$ **19.** $a_1 = 8;\ \ d = -7$ **20.** $a_1 = 6;\ \ d = -2$

21. $a_1 = 0;\ \ d = \dfrac{1}{2}$ **22.** $a_1 = 1;\ \ d = -\dfrac{1}{3}$ **23.** $a_1 = \sqrt{2};\ \ d = \sqrt{2}$ **24.** $a_1 = 0;\ \ d = \pi$

In Problems 25–30, find the indicated term in each arithmetic sequence.

25. 100th term of $2, 4, 6, \ldots$ **26.** 80th term of $-1, 1, 3, \ldots$ **27.** 90th term of $3, -3, -9, \ldots$

28. 80th term of $5, 0, -5, \ldots$ **29.** 80th term of $2, \dfrac{5}{2}, 3, \dfrac{7}{2}, \ldots$ **30.** 70th term of $2\sqrt{5}, 4\sqrt{5}, 6\sqrt{5}, \ldots$

In Problems 31–38, find the first term and the common difference of the arithmetic sequence described. Find a recursive formula for the sequence. Find a formula for the nth term.

31. 8th term is 8; 20th term is 44   **32.** 4th term is 3; 20th term is 35   **33.** 9th term is -5; 15th term is 31

34. 8th term is 4; 18th term is -96 **35.** 15th term is 0; 40th term is -50 **36.** 5th term is -2; 13th term is 30

37. 14th term is -1; 18th term is -9 **38.** 12th term is 4; 18th term is 28

In Problems 39–56, find each sum.

39. $1 + 3 + 5 + \cdots + (2n - 1)$ **40.** $2 + 4 + 6 + \cdots + 2n$ **41.** $7 + 12 + 17 + \cdots + (2 + 5n)$

42. $-1 + 3 + 7 + \cdots + (4n - 5)$ **43.** $2 + 4 + 6 + \cdots + 70$ **44.** $1 + 3 + 5 + \cdots + 59$

45. $-9 - 5 - 1 + \cdots + 39$ **46.** $2 + 5 + 8 + \cdots + 41$ **47.** $93 + 89 + 85 + \cdots - 287$

48. $7 + 1 - 5 - 11 - \cdots - 299$ **49.** $4 + 4.5 + 5 + 5.5 + \cdots + 100$ **50.** $8 + 8\dfrac{1}{4} + 8\dfrac{1}{2} + 8\dfrac{3}{4} + 9 + \cdots + 50$

51. $\displaystyle\sum_{n=1}^{80}(4n - 9)$ **52.** $\displaystyle\sum_{n=1}^{90}(3 - 2n)$ **53.** $\displaystyle\sum_{n=1}^{100}\left(6 - \dfrac{1}{2}n\right)$ **54.** $\displaystyle\sum_{n=1}^{80}\left(\dfrac{1}{3}n + \dfrac{1}{2}\right)$

55. The sum of the first 120 terms of the sequence
$$14, 16, 18, 20, \ldots$$

56. The sum of the first 46 terms of the sequence
$$2, -1, -4, -7, \ldots$$

Applications and Extensions

57. Find x so that $x + 3, 2x + 1,$ and $5x + 2$ are consecutive terms of an arithmetic sequence.

58. Find x so that $2x, 3x + 2,$ and $5x + 3$ are consecutive terms of an arithmetic sequence.

59. How many terms must be added in an arithmetic sequence whose first term is 11 and whose common difference is 3 to obtain a sum of 1092?

60. How many terms must be added in an arithmetic sequence whose first term is 78 and whose common difference is -4 to obtain a sum of 702?

61. Drury Lane Theater The Drury Lane Theater has 25 seats in the first row and 30 rows in all. Each successive row contains one additional seat. How many seats are in the theater?

62. Seats in an Amphitheater An outdoor amphitheater has 35 seats in the first row, 37 in the second row, 39 in the third

row, and so on. There are 27 rows altogether. How many can the amphitheater seat?

63. Football Stadium The corner section of a football stadium has 15 seats in the first row and 40 rows in all. Each successive row contains two additional seats. How many seats are in this section?

64. Constructing a Brick Staircase A brick staircase has a total of 30 steps. The bottom step requires 100 bricks. Each successive step requires two fewer bricks than the prior step.
 (a) How many bricks are required for the top step?
 (b) How many bricks are required to build the staircase?

65. Salary If you take a job with a starting salary of $35,000 per year and a guaranteed raise of $1400 per year, how many years will it be before your aggregate salary is $280,000?

 [**Hint**: Remember that your aggregate salary after 2 years is $35,000 + ($35,000 + $1400).]

66. Stadium Construction How many rows are in the corner section of a stadium containing 2040 seats if the first row has 10 seats and each successive row has 4 additional seats?

67. Creating a Mosaic A mosaic is designed in the shape of an equilateral triangle, 20 feet on each side. Each tile in the mosaic is in the shape of an equilateral triangle, 12 inches to a side. The tiles are to alternate in color as shown in the illustration. How many tiles of each color will be required?

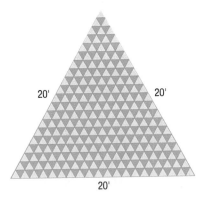

68. Old Faithful Old Faithful is a geyser in Yellowstone National Park named for its regular eruption pattern. Past data indicates that the average time between eruptions is 1h 35m.

(a) Suppose rangers log the first eruption on a given day at 12:57 am. Using $a_1 = 57$, write a prediction formula for the sequence of eruption times that day in terms of the number of minutes after midnight.

(b) At what time of day (e.g., 7:15 am) is the 10th eruption expected to occur?

(c) At what time of day is the last eruption expected to occur?

69. Cooling Air As a parcel of air rises (for example, as it is pushed over a mountain), it cools at the *dry adiabatic lapse rate* of 5.5°F per 1000 feet until it reaches its dew point. If the ground temperature is 67°F, write a formula for the sequence of temperatures, $\{T_n\}$, of a parcel of air that has risen n thousand feet. What is the temperature of a parcel of air if it has risen 5000 feet?

 Source: National Aeronautics and Space Administration

70. Citrus Ladders Ladders used by fruit pickers are typically tapered with a wide bottom for stability and a narrow top for ease of picking. If the bottom rung of such a ladder is 49 inches wide and the top rung is 24 inches wide, how many rungs does the ladder have if each rung is 2.5 inches shorter than the one below it? How much material would be needed to make the rungs for the ladder described?

 Source: www.stokesladders.com

71. Challenge Problem If $\{a_n\}$ is an arithmetic sequence with 100 terms where $a_1 = 2$ and $a_2 = 9$, and $\{b_n\}$ is an arithmetic sequence with 100 terms where $b_1 = 5$ and $b_2 = 11$, how many terms are the same in each sequence?

72. Challenge Problem Suppose $\{a_n\}$ is an arithmetic sequence. If S_n is the sum of the first n terms of $\{a_n\}$, and $\dfrac{S_{2n}}{S_n}$ is a positive constant for all n, find an expression for the nth term, a_n, in terms of only n and the common difference, d.

Explaining Concepts: Discussion and Writing

73. Make up an arithmetic sequence. Give it to a friend and ask for its 20th term.

74. Describe the similarities and differences between arithmetic sequences and linear functions.

Retain Your Knowledge

Problems 75–82 are based on material learned earlier in the course. The purpose of these problems is to keep the material fresh in your mind so that you are better prepared for the final exam.

75. If a credit card charges 15.3% interest compounded monthly, find the effective rate of interest.

76. Determine whether $x + 7$ is a factor of
$$x^4 + 5x^3 - 19x^2 - 29x + 42$$

77. Analyze and graph the equation: $25x^2 + 4y^2 = 100$

78. Find the inverse of the matrix $\begin{bmatrix} 2 & 0 \\ 3 & -1 \end{bmatrix}$, if it exists; otherwise, state that the matrix is singular.

79. Find the partial fraction decomposition of $\dfrac{3x}{x^3 - 1}$.

80. If g is a function with domain $[-4, 10]$, what is the domain of the function $2g(x - 1)$?

81. Find the real zeros of
$$h(x) = \frac{(x^4 + 1) \cdot 2x - (x^2 - 1) \cdot 4x^3}{(x^4 + 1)}$$

82. Solve: $(x + 3)^2 = (x + 3)(x - 5) + 7$

9.3 Geometric Sequences; Geometric Series

PREPARING FOR THIS SECTION *Before getting started, review the following:*

- Compound Interest (Section 6.7, pp. 481–487)

Now Work the 'Are You Prepared?' problems on page 677.

> OBJECTIVES 1 Determine Whether a Sequence Is Geometric (p. 669)
> 2 Find a Formula for a Geometric Sequence (p. 670)
> 3 Find the Sum of a Geometric Sequence (p. 671)
> 4 Determine Whether a Geometric Series Converges or Diverges (p. 672)
> 5 Solve Annuity Problems (p. 675)

1 Determine Whether a Sequence Is Geometric

When the ratio of successive terms of a sequence is always the same nonzero number, the sequence is called **geometric**.

> **DEFINITION Geometric Sequence**
>
> A **geometric sequence*** is defined recursively as $a_1 = a$, $\dfrac{a_n}{a_{n-1}} = r$, or as
>
> $$a_1 = a \qquad a_n = r a_{n-1} \tag{1}$$
>
> where $a_1 = a$ and $r \neq 0$ are real numbers. The number a_1 is the first term, and the nonzero number r is called the **common ratio**.

EXAMPLE 1 **Determining Whether a Sequence Is Geometric**

The sequence

$$2,\ 6,\ 18,\ 54,\ 162,\ \dots$$

is geometric because the ratio of successive terms is 3; $\left(\dfrac{6}{2} = \dfrac{18}{6} = \dfrac{54}{18} = \cdots = 3 \right)$. The first term is $a_1 = 2$, and the common ratio is 3.

EXAMPLE 2 **Determining Whether a Sequence Is Geometric**

Show that the following sequence is geometric.

$$\{s_n\} = \{2^{-n}\}$$

Find the first term and the common ratio.

Solution The first term of the sequence is $s_1 = 2^{-1} = \dfrac{1}{2}$. The nth term and the $(n-1)$st term of the sequence $\{s_n\}$ are

$$s_n = 2^{-n} \quad \text{and} \quad s_{n-1} = 2^{-(n-1)}$$

Their ratio is

$$\frac{s_n}{s_{n-1}} = \frac{2^{-n}}{2^{-(n-1)}} = 2^{-n+(n-1)} = 2^{-1} = \frac{1}{2}$$

Because the ratio of successive terms is the nonzero constant $\dfrac{1}{2}$, the sequence $\{s_n\}$ is geometric and the common ratio is $\dfrac{1}{2}$.

---

*Sometimes called a **geometric progression**.

EXAMPLE 3

Determining Whether a Sequence Is Geometric

Show that the following sequence is geometric.

$$\{t_n\} = \{3 \cdot 4^n\}$$

Find the first term and the common ratio.

Solution

The first term is $t_1 = 3 \cdot 4^1 = 12$. The nth term and the $(n-1)$st term are

$$t_n = 3 \cdot 4^n \quad \text{and} \quad t_{n-1} = 3 \cdot 4^{n-1}$$

Their ratio is

$$\frac{t_n}{t_{n-1}} = \frac{3 \cdot 4^n}{3 \cdot 4^{n-1}} = 4^{n-(n-1)} = 4$$

The sequence, $\{t_n\}$, is a geometric sequence with common ratio 4.

━━━━━ **Now Work** PROBLEM 11

2 Find a Formula for a Geometric Sequence

Suppose that a_1 is the first term of a geometric sequence with common ratio $r \neq 0$. We seek a formula for the nth term, a_n. To see the pattern, consider the first few terms:

$$a_1 = a_1 \cdot 1 = a_1 r^0$$

$$a_2 = r a_1 = a_1 r^1$$

$$a_3 = r a_2 = r(a_1 r) = a_1 r^2$$

$$a_4 = r a_3 = r(a_1 r^2) = a_1 r^3$$

$$a_5 = r a_4 = r(a_1 r^3) = a_1 r^4$$

$$\vdots$$

$$a_n = r a_{n-1} = r(a_1 r^{n-2}) = a_1 r^{n-1}$$

The terms of a geometric sequence with first term a_1 and common ratio r follow the pattern

$$a_1, \quad a_1 r, \quad a_1 r^2, \quad a_1 r^3, \quad \ldots$$

THEOREM *n*th Term of a Geometric Sequence

For a geometric sequence $\{a_n\}$ whose first term is a_1 and whose common ratio is r, the nth term is determined by the formula

$$a_n = a_1 r^{n-1} \qquad r \neq 0 \tag{2}$$

EXAMPLE 4

Finding a Particular Term of a Geometric Sequence

(a) Find the nth term of the geometric sequence: $10, 9, \dfrac{81}{10}, \dfrac{729}{100}, \ldots$

(b) Find the 9th term of the sequence.

(c) Find a recursive formula for the sequence.

Solution

(a) The first term of the geometric sequence is $a_1 = 10$. The common ratio $r = \dfrac{a_n}{a_{n-1}}$ is the ratio of any two consecutive terms. So, $r = \dfrac{a_2}{a_1} = \dfrac{9}{10}$. Then, by formula (2), the nth term of the geometric sequence is

$$a_n = 10\left(\frac{9}{10}\right)^{n-1} \qquad a_n = a_1 r^{n-1}; a_1 = 10, r = \frac{9}{10}$$

(b) The 9th term is

$$a_9 = 10\left(\frac{9}{10}\right)^{9-1} = 10\left(\frac{9}{10}\right)^{8} = 4.3046721$$

(c) The first term in the sequence is 10, and the common ratio is $r = \dfrac{9}{10}$. Using formula (1), the recursive formula is $a_1 = 10, \ a_n = \dfrac{9}{10} a_{n-1}$.

Now Work PROBLEMS 19, 27, AND 35

Exploration

Use a graphing utility to find the ninth term of the sequence in Example 4. Use it to find the 20th and 50th terms. Now use a graphing utility to graph the recursive formula found in Example 4(c). Conclude that the graph of the recursive formula behaves like the graph of an exponential function. How is r, the common ratio, related to a, the base of the exponential function $y = a^x$?

---

3 Find the Sum of a Geometric Sequence

> **THEOREM Sum of the First n Terms of a Geometric Sequence**
>
> Let $\{a_n\}$ be a geometric sequence with first term a_1 and common ratio r, where $r \neq 0, r \neq 1$. The sum S_n of the first n terms of $\{a_n\}$ is
>
> $$S_n = a_1 + a_1 r + a_1 r^2 + \cdots + a_1 r^{n-1} = \sum_{k=1}^{n} a_1 r^{k-1}$$
>
> $$= a_1 \cdot \frac{1-r^n}{1-r} \qquad r \neq 0, 1 \tag{3}$$

Proof The sum S_n of the first n terms of $\{a_n\} = \{a_1 r^{n-1}\}$ is

$$S_n = a_1 + a_1 r + \cdots + a_1 r^{n-1} \tag{4}$$

Multiply both sides by r to obtain

$$rS_n = a_1 r + a_1 r^2 + \cdots + a_1 r^n \tag{5}$$

Now, subtract (5) from (4). The result is

$$S_n - rS_n = a_1 - a_1 r^n$$
$$(1-r)S_n = a_1(1-r^n)$$

Since $r \neq 1$, solve for S_n.

$$S_n = a_1 \cdot \frac{1-r^n}{1-r}$$

■

EXAMPLE 5 **Finding the Sum of the First n Terms of a Geometric Sequence**

Find the sum S_n of the first n terms of the sequence $\left\{\left(\dfrac{1}{2}\right)^n\right\}$; that is, find

$$\sum_{k=1}^{n}\left(\frac{1}{2}\right)^k = \frac{1}{2} + \frac{1}{4} + \frac{1}{8} + \cdots + \left(\frac{1}{2}\right)^n$$

Solution The sequence $\left\{\left(\dfrac{1}{2}\right)^n\right\}$ is a geometric sequence with $a_1 = \dfrac{1}{2}$ and $r = \dfrac{1}{2}$. Use formula (3) to get

$$S_n = \sum_{k=1}^{n}\left(\frac{1}{2}\right)^k = \frac{1}{2} + \frac{1}{4} + \frac{1}{8} + \cdots + \left(\frac{1}{2}\right)^n = \sum_{k=1}^{n} \frac{1}{2}\left(\frac{1}{2}\right)^{k-1}$$

$$= \frac{1}{2} \cdot \frac{1 - \left(\dfrac{1}{2}\right)^n}{1 - \dfrac{1}{2}} \qquad \text{Formula (3); } a_1 = \frac{1}{2}, r = \frac{1}{2}$$

$$= \frac{1}{2} \cdot \frac{1 - \left(\dfrac{1}{2}\right)^n}{\dfrac{1}{2}}$$

$$= 1 - \left(\frac{1}{2}\right)^n$$

- Now Work PROBLEM 41

EXAMPLE 6

Using a Graphing Utility to Find the Sum of a Geometric Sequence

Use a graphing utility to find the sum of the first 15 terms of the sequence $\left\{\left(\dfrac{1}{3}\right)^n\right\}$; that is, find

$$\sum_{k=1}^{15}\left(\frac{1}{3}\right)^k = \frac{1}{3} + \frac{1}{9} + \frac{1}{27} + \cdots + \left(\frac{1}{3}\right)^{15}$$

Solution Figure 8 shows the result using a TI-84 Plus C graphing calculator. The sum of the first 15 terms of the sequence $\left\{\left(\dfrac{1}{3}\right)^n\right\}$ is approximately 0.4999999652.

- Now Work PROBLEM 47

```
NORMAL FLOAT AUTO REAL RADIAN MP
sum(seq((1/3)ⁿ,n,1,15,1)
                 .4999999652
```

Figure 8

4 Determine Whether a Geometric Series Converges or Diverges

DEFINITION Infinite Geometric Series

An infinite sum of the form

$$a_1 + a_1 r + a_1 r^2 + \cdots + a_1 r^{n-1} + \cdots$$

with first term a_1 and common ratio r, is called an **infinite geometric series** and is denoted by

$$\sum_{k=1}^{\infty} a_1 r^{k-1}$$

The sum S_n of the first n terms of a geometric series is

$$S_n = a_1 + a_1 \cdot r + a_1 \cdot r^2 + \cdots + a_1 \cdot r^{n-1} \qquad \textbf{(6)}$$

If this finite sum S_n approaches a number L as $n \to \infty$, then the infinite geometric series $\sum_{k=1}^{\infty} a_1 r^{k-1}$ **converges** to L and L is called the **sum of the infinite geometric series**. The sum is written as

$$L = \sum_{k=1}^{\infty} a_1 r^{k-1}$$

A series that does not converge is called a **divergent series**.

THEOREM **Convergence of an Infinite Geometric Series**

If $|r| < 1$, the infinite geometric series $\sum_{k=1}^{\infty} a_1 r^{k-1}$ converges. Its sum is

$$\sum_{k=1}^{\infty} a_1 r^{k-1} = \frac{a_1}{1-r} \tag{7}$$

Intuitive Proof

- If $r = 0$, then $S_n = a_1 + 0 + \cdots + 0 = a_1$, so formula (7) is true for $r = 0$.
- If $r \neq 0$ and $|r| < 1$, then, based on formula (3),

$$S_n = a_1 \cdot \frac{1 - r^n}{1 - r} = \frac{a_1}{1-r} - \frac{a_1 r^n}{1-r} \tag{8}$$

Since $|r| < 1$, it follows that $|r^n|$ approaches 0 as $n \to \infty$. Then, in formula (8), the term $\dfrac{a_1 r^n}{1-r}$ approaches 0, so the sum S_n approaches $\dfrac{a_1}{1-r}$ as $n \to \infty$. ∎

EXAMPLE 7 **Determining Whether a Geometric Series Converges or Diverges**

Determine whether the geometric series

$$\sum_{k=1}^{\infty} 2\left(\frac{2}{3}\right)^{k-1} = 2 + \frac{4}{3} + \frac{8}{9} + \cdots$$

converges or diverges. If it converges, find its sum.

Solution Comparing $\sum_{k=1}^{\infty} 2\left(\dfrac{2}{3}\right)^{k-1}$ to $\sum_{k=1}^{\infty} a_1 r^{k-1}$, the first term is $a_1 = 2$ and the common ratio is $r = \dfrac{2}{3}$. Since $|r| < 1$, the series converges. Use formula (7) to find its sum:

$$\sum_{k=1}^{\infty} 2\left(\frac{2}{3}\right)^{k-1} = 2 + \frac{4}{3} + \frac{8}{9} + \cdots = \frac{2}{1 - \dfrac{2}{3}} = 6$$

⬤══ **Now Work** PROBLEM 53

EXAMPLE 8 **Repeating Decimals**

Show that the repeating decimal $0.999\ldots$ equals 1.

Solution The decimal $0.999\ldots = 0.9 + 0.09 + 0.009 + \cdots = \dfrac{9}{10} + \dfrac{9}{100} + \dfrac{9}{1000} + \cdots$ is an infinite geometric series. Write it in the form $\sum_{k=1}^{\infty} a_1 r^{k-1}$ and use formula (7).

$$0.999\ldots = \frac{9}{10} + \frac{9}{100} + \frac{9}{1000} + \cdots = \sum_{k=1}^{\infty} \frac{9}{10^k} = \sum_{k=1}^{\infty} \frac{9}{10 \cdot 10^{k-1}} = \sum_{k=1}^{\infty} \frac{9}{10}\left(\frac{1}{10}\right)^{k-1}$$

(continued)

Compare this series to $\sum_{k=1}^{\infty} a_1 r^{k-1}$ and note that $a_1 = \dfrac{9}{10}$ and $r = \dfrac{1}{10}$. Since $|r| < 1$, the series converges and its sum is

$$0.999\ldots = \dfrac{\dfrac{9}{10}}{1 - \dfrac{1}{10}} = \dfrac{\dfrac{9}{10}}{\dfrac{9}{10}} = 1$$

The repeating decimal $0.999\ldots$ equals 1.

EXAMPLE 9

Pendulum Swings

Initially, a pendulum swings through an arc of length 18 inches. See Figure 9. On each successive swing, the length of the arc is 0.98 of the previous length.

(a) What is the length of the arc of the 10th swing?
(b) On which swing is the length of the arc first less than 12 inches?
(c) After 15 swings, what total distance has the pendulum swung?
(d) When it stops, what total distance has the pendulum swung?

Solution

(a) The length of the first swing is 18 inches.
The length of the second swing is $0.98 \cdot 18$ inches.
The length of the third swing is $0.98 \cdot 0.98 \cdot 18 = 0.98^2 \cdot 18$ inches.
The length of the arc of the 10th swing is
$$(0.98)^9 \cdot 18 \approx 15.007 \text{ inches}$$

(b) The length of the arc of the nth swing is $(0.98)^{n-1} \cdot 18$. For the length of the arc to be exactly 12 inches requires that

$$(0.98)^{n-1} \cdot 18 = 12$$

$$(0.98)^{n-1} = \dfrac{12}{18} = \dfrac{2}{3} \qquad \text{Divide both sides by 18.}$$

$$n - 1 = \log_{0.98}\left(\dfrac{2}{3}\right) \qquad \text{Express as a logarithm.}$$

$$n = 1 + \dfrac{\ln\left(\dfrac{2}{3}\right)}{\ln 0.98} \approx 1 + 20.07 = 21.07 \qquad \text{Solve for } n; \text{ use the Change of Base Formula.}$$

The length of the arc of the pendulum exceeds 12 inches on the 21st swing and is first less than 12 inches on the 22nd swing.

(c) After 15 swings, the total distance swung is
$$L = \underset{\text{1st}}{18} + \underset{\text{2nd}}{0.98 \cdot 18} + \underset{\text{3rd}}{(0.98)^2 \cdot 18} + \underset{\text{4th}}{(0.98)^3 \cdot 18} + \cdots + \underset{\text{15th}}{(0.98)^{14} \cdot 18}$$

This is the sum of a geometric sequence. The common ratio is 0.98; the first term is 18. The sum has 15 terms, so
$$L = 18 \cdot \dfrac{1 - 0.98^{15}}{1 - 0.98} \approx 18 \cdot 13.07 \approx 235.3 \text{ inches}$$

The pendulum has swung approximately 235.3 inches after 15 swings.

(d) When the pendulum stops, it has swung the total distance
$$T = 18 + 0.98 \cdot 18 + (0.98)^2 \cdot 18 + (0.98)^3 \cdot 18 + \cdots$$

This is the sum of an infinite geometric series. The common ratio is $r = 0.98$; the first term is $a_1 = 18$. Since $|r| < 1$, the series converges. Its sum is
$$T = \dfrac{a_1}{1 - r} = \dfrac{18}{1 - 0.98} = 900$$

The pendulum has swung a total of 900 inches when it finally stops.

Figure 9

18"

----Now Work PROBLEM 87

5 Solve Annuity Problems

Section 6.7 developed the compound interest formula, which gives the future value when a fixed amount of money is deposited in an account that pays interest compounded periodically. Often, though, money is invested in small amounts at periodic intervals. An **annuity** is a sequence of equal periodic deposits. The periodic deposits may be made annually, quarterly, monthly, or daily.

When deposits are made at the same time that the interest is credited, the annuity is called **ordinary**. We discuss only ordinary annuities here. The **amount of an annuity** is the sum of all deposits made plus all interest paid.

Suppose that the interest rate that an account earns is i percent per payment period (expressed as a decimal). For example, if an account pays 12% compounded monthly (12 times a year), then $i = \dfrac{0.12}{12} = 0.01$. If an account pays 8% compounded quarterly (4 times a year), then $i = \dfrac{0.08}{4} = 0.02$.

To develop a formula for the amount of an annuity, suppose that P is deposited each payment period for n payment periods in an account that earns i percent per payment period. When the last deposit is made at the nth payment period, the first deposit of P has earned interest compounded for $n - 1$ payment periods, the second deposit of P has earned interest compounded for $n - 2$ payment periods, and so on. Table 3 shows the value of each deposit after n deposits have been made.

Table 3

Deposit	1	2	3	$\cdots$	$n-1$	n
Amount	$P(1+i)^{n-1}$	$P(1+i)^{n-2}$	$P(1+i)^{n-3}$	$\cdots$	$P(1+i)$	P

The amount A of the annuity is the sum of the amounts shown in Table 3; that is,

$$A = P \cdot (1+i)^{n-1} + P \cdot (1+i)^{n-2} + \cdots + P \cdot (1+i) + P$$
$$= P[1 + (1+i) + \cdots + (1+i)^{n-1}]$$

The expression in brackets is the sum of a geometric sequence with n terms and a common ratio of $(1+i)$. As a result,

$$A = P[1 + (1+i) + \cdots + (1+i)^{n-2} + (1+i)^{n-1}]$$
$$= P\frac{1 - (1+i)^n}{1 - (1+i)} = P\frac{1 - (1+i)^n}{-i} = P\frac{(1+i)^n - 1}{i}$$

The following theorem has been proved:

THEOREM Amount of an Annuity

Suppose that P is the deposit in dollars made at the end of each payment period for an annuity paying i percent interest per payment period. The amount A of the annuity after n deposits is

$$A = P\frac{(1+i)^n - 1}{i} \tag{9}$$

NOTE In formula (9), remember that when the nth deposit is made, the first deposit has earned interest for $n - 1$ compounding periods and the nth deposit has earned no interest. ∎

EXAMPLE 10   **Determining the Amount of an Annuity**

To save for retirement, Brett decides to place $4000 into an individual retirement account (IRA) each year for the next 30 years. What will the value of the IRA be when Brett makes his 30th deposit? Assume that the rate of return of the IRA is 7% per annum compounded annually. (This is the historical rate of return in the stock market.)

Solution

This is an ordinary annuity with $n = 30$ annual deposits of $P = \$4000$. The rate of interest per payment period is $i = \dfrac{0.07}{1} = 0.07$. The amount A of the annuity after 30 deposits is

$$A = \$4000 \frac{(1 + 0.07)^{30} - 1}{0.07} \approx \$4000 \cdot 94.46078632 \approx \$377{,}843.15$$

EXAMPLE 11 **Determining the Amount of an Annuity**

To save for her daughter's college education, Miranda decides to put $100 aside every month in a credit union account paying 2% interest compounded monthly. She begins this savings program when her daughter is 3 years old. How much will she have saved by the time she makes the 180th deposit? How old is her daughter at this time?

Solution

This is an annuity with $P = \$100$, $n = 180$, and $i = \dfrac{0.02}{12}$. The amount A of the annuity after 180 deposits is

$$A = \$100 \frac{\left(1 + \dfrac{0.02}{12}\right)^{180} - 1}{\dfrac{0.02}{12}} \approx \$100 \cdot 209.71306 \approx \$20{,}971.31$$

Because there are 12 deposits per year, when the 180th deposit is made $\dfrac{180}{12} = 15$ years have passed, and Miranda's daughter is 18 years old.

➤ **Now Work** PROBLEM 91

Historical Feature

Fibonacci

Sequences are among the oldest objects of mathematical investigation, having been studied for over 3500 years. After the initial steps, however, little progress was made until about 1600.

Arithmetic and geometric sequences appear in the Rhind papyrus, a mathematical text containing 85 problems copied around 1650 BC by the Egyptian scribe Ahmes from an earlier work (see Historical Problem 1). Fibonacci (AD 1220) wrote about problems similar to those found in the Rhind papyrus, leading one to suspect that Fibonacci may have had material available that is now lost. This material would have been in the non-Euclidean Greek tradition of Heron (about AD 75) and Diophantus (about AD 250). One problem, again modified slightly, is still with us in the familiar puzzle rhyme "As I was going to St. Ives ..." (see Historical Problem 2).

The Rhind papyrus indicates that the Egyptians knew how to add up the terms of an arithmetic or geometric sequence, as did the Babylonians. The rule for summing up a geometric sequence is found in Euclid's *Elements* (Book IX, 35, 36), where, like all Euclid's algebra, it is presented in a geometric form.

Investigations of other kinds of sequences began in the 1500s, when algebra became sufficiently developed to handle the more complicated problems. The development of calculus in the 1600s added a powerful new tool, especially for finding the sum of an infinite series, and the subject continues to flourish today.

Historical Problems

1. *Arithmetic sequence problem from the Rhind papyrus (statement modified slightly for clarity)* One hundred loaves of bread are to be divided among five people so that the amounts that they receive form an arithmetic sequence. The first two together receive one-seventh of what the last three receive. How many loaves does each receive?

   [*Partial answer:* First person receives $1\dfrac{2}{3}$ loaves.]

2. The following old English children's rhyme resembles one of the Rhind papyrus problems.

   As I was going to St. Ives
   I met a man with seven wives

   Each wife had seven sacks
   Each sack had seven cats
   Each cat had seven kits [kittens]
   Kits, cats, sacks, wives
   How many were going to St. Ives?

   (a) Assuming that the speaker and the cat fanciers met by traveling in opposite directions, what is the answer?

   (b) How many kittens are being transported?

   (c) Kits, cats, sacks, wives; how many?

9.3 Assess Your Understanding

'Are You Prepared?' *Answers are given at the end of these exercises. If you get a wrong answer, read the pages listed in* red.

1. If $1000 is invested at 4% per annum compounded semiannually, how much is in the account after 2 years? (pp. 481–484)

2. How much do you need to invest now at 5% per annum compounded monthly so that in 1 year you will have $10,000? (pp. 485–486)

Concepts and Vocabulary

3. In a(n) _____ sequence, the ratio of successive terms is a constant.

4. If $|r| < 1$, the sum of the geometric series $\sum_{k=1}^{\infty} ar^{k-1}$ is _____.

5. *Multiple Choice* If a series does not converge, it is called a(n) _____ series.
(a) arithmetic   (b) divergent   (c) geometric   (d) recursive

6. *True or False* A geometric sequence may be defined recursively.

7. *True or False* In a geometric sequence, the common ratio is always a positive number.

8. *True or False* For a geometric sequence with first term a_1 and common ratio r, where $r \neq 0, r \neq 1$, the sum of the first n terms is $S_n = a_1 \cdot \dfrac{1 - r^n}{1 - r}$.

Skill Building

In Problems 9–18, show that each sequence is geometric. Then find the common ratio and list the first four terms.

9. $\{s_n\} = \{4^n\}$ **10.** $\{s_n\} = \{(-5)^n\}$ **11.** $\{a_n\} = \left\{-3\left(\dfrac{1}{2}\right)^n\right\}$ **12.** $\{b_n\} = \left\{\left(\dfrac{5}{2}\right)^n\right\}$ **13.** $\{c_n\} = \left\{\dfrac{2^{n-1}}{4}\right\}$

14. $\{d_n\} = \left\{\dfrac{3^n}{9}\right\}$ **15.** $\{e_n\} = \{7^{n/4}\}$ **16.** $\{f_n\} = \{3^{2n}\}$ **17.** $\{t_n\} = \left\{\dfrac{3^{n-1}}{2^n}\right\}$ **18.** $\{u_n\} = \left\{\dfrac{2^n}{3^{n-1}}\right\}$

In Problems 19–26, find the fifth term and the nth term of the geometric sequence whose first term a_1 and common ratio r are given.

19. $a_1 = 2$; $r = 3$ **20.** $a_1 = -2$; $r = 4$ **21.** $a_1 = 5$; $r = -1$ **22.** $a_1 = 6$; $r = -2$

23. $a_1 = 0$; $r = \dfrac{1}{7}$ **24.** $a_1 = 1$; $r = -\dfrac{1}{3}$ **25.** $a_1 = \sqrt{3}$; $r = \sqrt{3}$ **26.** $a_1 = 0$; $r = \dfrac{1}{\pi}$

In Problems 27–32, find the indicated term of each geometric sequence.

27. 7th term of $1, \dfrac{1}{2}, \dfrac{1}{4}, \ldots$ **28.** 8th term of $1, 3, 9, \ldots$ **29.** 15th term of $1, -1, 1, \ldots$

30. 10th term of $-1, 2, -4, \ldots$ **31.** 8th term of $0.4, 0.04, 0.004, \ldots$ **32.** 7th term of $0.1, 1.0, 10.0, \ldots$

In Problems 33–40, find the nth term a_n of each geometric sequence. When given, r is the common ratio.

33. $6, 18, 54, 162, \ldots$ **34.** $5, 10, 20, 40, \ldots$ **35.** $-3, 1, -\dfrac{1}{3}, \dfrac{1}{9}, \ldots$ **36.** $4, 1, \dfrac{1}{4}, \dfrac{1}{16}, \ldots$

37. $a_6 = 243$; $r = -3$ **38.** $a_2 = 7$; $r = \dfrac{1}{3}$ **39.** $a_2 = 7$; $a_4 = 1575$ **40.** $a_3 = \dfrac{1}{3}$; $a_6 = \dfrac{1}{81}$

In Problems 41–46, find each sum.

41. $\dfrac{1}{4} + \dfrac{2}{4} + \dfrac{2^2}{4} + \dfrac{2^3}{4} + \cdots + \dfrac{2^{n-1}}{4}$ **42.** $\dfrac{3}{9} + \dfrac{3^2}{9} + \dfrac{3^3}{9} + \cdots + \dfrac{3^n}{9}$ **43.** $\sum_{k=1}^{n} \left(\dfrac{2}{3}\right)^k$

44. $\sum_{k=1}^{n} 4 \cdot 3^{k-1}$ **45.** $-1 - 2 - 4 - 8 - \cdots - (2^{n-1})$ **46.** $2 + \dfrac{6}{5} + \dfrac{18}{25} + \cdots + 2\left(\dfrac{3}{5}\right)^{n-1}$

For Problems 47–52, use a graphing utility to find the sum of each geometric sequence.

47. $\dfrac{1}{4} + \dfrac{2}{4} + \dfrac{2^2}{4} + \dfrac{2^3}{4} + \cdots + \dfrac{2^{14}}{4}$ **48.** $\dfrac{3}{9} + \dfrac{3^2}{9} + \dfrac{3^3}{9} + \cdots + \dfrac{3^{15}}{9}$ **49.** $\sum_{n=1}^{15} \left(\dfrac{2}{3}\right)^n$

50. $\sum_{n=1}^{15} 4 \cdot 3^{n-1}$ **51.** $-1 - 2 - 4 - 8 - \cdots - 2^{14}$ **52.** $2 + \dfrac{6}{5} + \dfrac{18}{25} + \cdots + 2\left(\dfrac{3}{5}\right)^{15}$

In Problems 53–68, determine whether each infinite geometric series converges or diverges. If it converges, find its sum.

53. $1 + \dfrac{1}{3} + \dfrac{1}{9} + \cdots$

54. $2 + \dfrac{4}{3} + \dfrac{8}{9} + \cdots$

55. $8 + 4 + 2 + \cdots$

56. $6 + 2 + \dfrac{2}{3} + \cdots$

57. $2 - \dfrac{1}{2} + \dfrac{1}{8} - \dfrac{1}{32} + \cdots$

58. $1 - \dfrac{3}{4} + \dfrac{9}{16} - \dfrac{27}{64} + \cdots$

59. $8 + 12 + 18 + 27 + \cdots$

60. $9 + 12 + 16 + \dfrac{64}{3} + \cdots$

61. $\displaystyle\sum_{k=1}^{\infty} 5\left(\dfrac{1}{4}\right)^{k-1}$

62. $\displaystyle\sum_{k=1}^{\infty} 8\left(\dfrac{1}{3}\right)^{k-1}$

63. $\displaystyle\sum_{k=1}^{\infty} \dfrac{1}{2} \cdot 3^{k-1}$

64. $\displaystyle\sum_{k=1}^{\infty} 3\left(\dfrac{3}{2}\right)^{k-1}$

65. $\displaystyle\sum_{k=1}^{\infty} 6\left(-\dfrac{2}{3}\right)^{k-1}$

66. $\displaystyle\sum_{k=1}^{\infty} 4\left(-\dfrac{1}{2}\right)^{k-1}$

67. $\displaystyle\sum_{k=1}^{\infty} 3\left(\dfrac{2}{3}\right)^{k}$

68. $\displaystyle\sum_{k=1}^{\infty} 2\left(\dfrac{3}{4}\right)^{k}$

Mixed Practice *In Problems 69–82, determine whether the given sequence is arithmetic, geometric, or neither. If the sequence is arithmetic, find the common difference; if it is geometric, find the common ratio. If the sequence is arithmetic or geometric, find the sum of the first 50 terms.*

69. $\{n + 2\}$

70. $\{2n - 5\}$

71. $\{4n^2\}$

72. $\{5n^2 + 1\}$

73. $\left\{3 - \dfrac{2}{3}n\right\}$

74. $\left\{8 - \dfrac{3}{4}n\right\}$

75. $1, 3, 6, 10, \ldots$

76. $2, 4, 6, 8, \ldots$

77. $\left\{\left(\dfrac{2}{3}\right)^n\right\}$

78. $\left\{\left(\dfrac{5}{4}\right)^n\right\}$

79. $-1, 2, -4, 8, \ldots$

80. $1, 1, 2, 3, 5, 8, \ldots$

81. $\{3^{n/2}\}$

82. $\{(-1)^n\}$

Applications and Extensions

83. Find x so that x, $x + 2$, and $x + 3$ are consecutive terms of a geometric sequence.

84. Find x so that $x - 1$, x, and $x + 2$ are consecutive terms of a geometric sequence.

85. Salary Increases If you have been hired at an annual salary of $42,000 and expect to receive annual increases of 3%, what will your salary be when you begin your fifth year?

86. Equipment Depreciation A new piece of equipment cost a company $15,000. Each year, for tax purposes, the company depreciates the value by 15%. What value should the company give the equipment after 5 years?

87. Pendulum Swings Initially, a pendulum swings through an arc of length 2 feet. On each successive swing, the length of the arc is 0.9 of the previous length.
  (a) What is the length of the arc of the 10th swing?
  (b) On which swing is the length of the arc first less than 1 foot?
  (c) After 15 swings, what total length has the pendulum swung?
  (d) When it stops, what total length has the pendulum swung?

88. Bouncing Balls A ball is dropped from a height of 30 feet. Each time it strikes the ground, it bounces up to 0.8 of the previous height.
  (a) What height does the ball bounce up to after it strikes the ground for the third time?
  (b) How high does it bounce after it strikes the ground for the nth time?

30'   24' 19.2'

  (c) How many times does the ball need to strike the ground before its bounce is less than 6 inches?
  (d) What total vertical distance does the ball travel before it stops bouncing?

89. Retirement Christine contributes $100 each month to her 401(k). What will be the value of Christine's 401(k) after the 360th deposit (30 years) if the per annum rate of return is assumed to be 8% compounded monthly?

90. Saving for a Home Jolene wants to purchase a new home. Suppose that she invests $400 per month into a mutual fund. If the per annum rate of return of the mutual fund is assumed to be 6% compounded monthly, how much will Jolene have for a down payment after the 36th deposit (3 years)?

91. Tax-Sheltered Annuity Don contributes $500 at the end of each quarter to a tax-sheltered annuity (TSA). What will the value of the TSA be after the 80th deposit (20 years) if the per annum rate of return is assumed to be 5% compounded quarterly?

92. Retirement Ray contributes $1000 to an individual retirement account (IRA) semiannually. What will be the value of the IRA be when Ray makes his 30th deposit (after 15 years) if the per annum rate of return is assumed to be 7% compounded semiannually?

93. Sinking Fund Scott and Alice want to purchase a vacation home in 10 years and need $50,000 for a down payment. How much should they place in a savings account each month if the per annum rate of return is assumed to be 3.5% compounded monthly?

94. Sinking Fund For a child born in 2018, the cost of a 4-year college education at a public university is projected to be $185,000. Assuming a 4.75% per annum rate of return compounded monthly, how much must be contributed to a college fund every month to have $185,000 in 18 years when the child begins college?

95. Grains of Wheat on a Chess Board In an old fable, a commoner who had saved the king's life was told he could ask the king for any just reward. Being a shrewd man, the commoner said, "A simple wish, sire. Place one grain of wheat on the first square of a chessboard, two grains on the second square, four grains on the third square, continuing until you have filled the board. This is all I seek." Compute the total number of grains needed to do this to see why the request, seemingly simple, could not be granted. (A chessboard consists of $8 \times 8 = 64$ squares.)

96. Shading Squares Look at the figure. What fraction of the square is eventually shaded if the indicated shading process continues indefinitely?

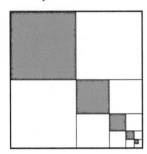

97. Multiplier Suppose that, throughout the U.S. economy, individuals spend 90% of every additional dollar that they earn. Economists would say that an individual's **marginal propensity to consume** is 0.90. For example, if Jane earns an additional dollar, she will spend $0.9(1) = \$0.90$ of it. The individual who earns \$0.90 (from Jane) will spend 90% of it, or \$0.81. This process of spending continues and results in an infinite geometric series as follows:

$$1, 0.90, 0.90^2, 0.90^3, 0.90^4, \ldots$$

The sum of this infinite geometric series is called the **multiplier**. What is the multiplier if individuals spend 90% of every additional dollar that they earn?

98. Multiplier Refer to Problem 97. Suppose that the marginal propensity to consume throughout the U.S. economy is 0.95. What is the multiplier for the U.S. economy?

99. Stock Price One method of pricing a stock is to discount the stream of future dividends of the stock. Suppose that a stock pays $\$P$ per year in dividends, and historically, the dividend has been increased i% per year. If you desire an annual rate of return of r%, this method of pricing a stock states that the price that you should pay is the present value of an infinite stream of payments:

$$\text{Price} = P + P \cdot \frac{1+i}{1+r} + P \cdot \left(\frac{1+i}{1+r}\right)^2 + P \cdot \left(\frac{1+i}{1+r}\right)^3 + \cdots$$

The price of the stock is the sum of an infinite geometric series. Suppose that a stock pays an annual dividend of \$4.00, and historically, the dividend has been increased 3% per year. You desire an annual rate of return of 9%. What is the most you should pay for the stock?

100. Stock Price Refer to Problem 99. Suppose that a stock pays an annual dividend of \$2.50, and historically, the dividend has increased 4% per year. You desire an annual rate of return of 11%. What is the most that you should pay for the stock?

101. A Rich Man's Promise A rich man promises to give you \$1000 on September 1. Each day thereafter he will give you $\frac{9}{10}$ of what he gave you the previous day. What is the first date on which the amount you receive is less than 1¢? How much have you received when this happens?

102. Seating Revenue A special section in the end zone of a football stadium has 2 seats in the first row and 14 rows total. Each successive row has 2 seats more than the row before. In this particular section, the first seat is sold for 1 cent, and each following seat sells for 5% more than the previous seat. Find the total revenue generated if every seat in the section is sold. Round only the final answer, and state the final answer in dollars rounded to two decimal places. (JJC)[†]

103. *Challenge Problem* Suppose x, y, z are consecutive terms in a geometric sequence. If $x + y + z = 103$ and $x^2 + y^2 + z^2 = 6901$, find the value of y. [**Hint:** Let r be the common ratio so $y = xr$ and $z = yr = xr^2$.]

104. *Challenge Problem* Koch's Snowflake The area inside the fractal known as the Koch snowflake can be described as the sum of the areas of infinitely many equilateral triangles, as pictured below.

For all but the center (largest) triangle, a triangle in the Koch snowflake is $\frac{1}{9}$ the area of the next largest triangle in the fractal. Suppose the area of the largest triangle has area of 2 square meters.

(a) Show that the area of the Koch snowflake is given by the series

$$A = 2 + 2 \cdot 3\left(\frac{1}{9}\right) + 2 \cdot 12\left(\frac{1}{9}\right)^2 + 2 \cdot 48\left(\frac{1}{9}\right)^3 + 2 \cdot 192\left(\frac{1}{9}\right)^4 + \cdots$$

(b) Find the exact area of the Koch snowflake by finding the sum of the series.

[†]Courtesy of the Joliet Junior College Mathematics Department.

Explaining Concepts: Discussion and Writing

105. Critical Thinking You are interviewing for a job and receive two offers for a five-year contract:

> A: $40,000 to start, with guaranteed annual increases of 6% for the first 5 years
>
> B: $44,000 to start, with guaranteed annual increases of 3% for the first 5 years

Which offer is better if your goal is to be making as much as possible after 5 years? Which is better if your goal is to make as much money as possible over the contract (5 years)?

106. Critical Thinking Which of the following choices, A or B, results in more money?

> A: To receive $1000 on day 1, $999 on day 2, $998 on day 3, with the process to end after 1000 days
>
> B: To receive $1 on day 1, $2 on day 2, $4 on day 3, for 19 days

107. Critical Thinking You have just signed a 7-year professional football league contract with a beginning salary of $2,000,000 per year. Management gives you the following options with regard to your salary over the 7 years.

1. A bonus of $100,000 each year
2. An annual increase of 4.5% per year beginning after 1 year
3. An annual increase of $95,000 per year beginning after 1 year

Which option provides the most money over the 7-year period? Which the least? Which would you choose? Why?

108. Critical Thinking Suppose you were offered a job in which you would work 8 hours per day for 5 workdays per week for 1 month at hard manual labor. Your pay the first day would be 1 penny. On the second day your pay would be two pennies; the third day 4 pennies. Your pay would double on each successive workday. There are 22 workdays in the month. There will be no sick days. If you miss a day of work, there is no pay or pay increase. How much do you get paid if you work all 22 days? How much do you get paid for the 22nd workday? What risks do you run if you take this job offer? Would you take the job?

109. Can a sequence be both arithmetic and geometric? Give reasons for your answer.

110. Make up a geometric sequence. Give it to a friend and ask for its 20th term.

111. Make up two infinite geometric series, one that has a sum and one that does not. Give them to a friend and ask for the sum of each series.

112. Describe the similarities and differences between geometric sequences and exponential functions.

Retain Your Knowledge

Problems 113–120 are based on material learned earlier in the course. The purpose of these problems is to keep the material fresh in your mind so that you are better prepared for the final exam.

113. Use the Change-of-Base Formula and a calculator to evaluate $\log_7 62$. Round the answer to three decimal places.

114. Find the equation of the hyperbola with vertices at $(-2, 0)$ and $(2, 0)$, and a focus at $(4, 0)$.

115. Find the value of the determinant: $\begin{vmatrix} 3 & 1 & 0 \\ 0 & -2 & 6 \\ 4 & -1 & -2 \end{vmatrix}$

116. Write the factored form of the polynomial function of smallest degree that touches the x-axis at $x = 4$, crosses the x-axis at $x = -2$ and $x = 1$, and has a y-intercept of 4.

117. Given $s(t) = -16t^2 + 3t$, find the difference quotient $\dfrac{s(t) - s(1)}{t - 1}$.

118. Find the function g whose graph is the graph of $y = \sqrt{x}$ but is stretched vertically by a factor of 7 and shifted left 5 units.

119. Factor completely: $x^4 - 29x^2 + 100$

120. Use a graphing utility to determine the interval(s) where $g(x) = x^3 + 2x^2 - 3.59x - 2.9$ is decreasing.

'Are You Prepared?' Answers

1. $1082.43   **2.** $9513.28

9.4 Mathematical Induction

OBJECTIVE 1 Prove Statements Using Mathematical Induction (p. 681)

1 Prove Statements Using Mathematical Induction

Mathematical induction is a method for proving that statements involving natural numbers are true for all natural numbers.[*]

For example, the statement "the sum of the first n positive odd integers equals n^2," that is,

$$1 + 3 + 5 + \cdots + (2n - 1) = n^2 \tag{1}$$

can be proved for all natural numbers n by using mathematical induction.

Before stating the method of mathematical induction, let's try to gain a sense of the power of the method. We use the statement in equation (1) for this purpose by restating it for various values of $n = 1, 2, 3, \ldots$.

$n = 1$	The sum of the first positive odd integer is 1^2; $1 = 1^2$.
$n = 2$	The sum of the first 2 positive odd integers is 2^2; $1 + 3 = 4 = 2^2$.
$n = 3$	The sum of the first 3 positive odd integers is 3^2; $1 + 3 + 5 = 9 = 3^2$.
$n = 4$	The sum of the first 4 positive odd integers is 4^2; $1 + 3 + 5 + 7 = 16 = 4^2$.

Although from this pattern we might conjecture that statement (1) is true for any natural number n, can we really be sure that it does not fail for some choice of n? The method of proof by mathematical induction allows us to prove that the statement is true for all n.

THEOREM **The Principle of Mathematical Induction**

Suppose that the following two conditions are satisfied with regard to a statement about natural numbers:

CONDITION I: The statement is true for the natural number 1.

CONDITION II: If the statement is true for some natural number k, and it can be shown to be true for the next natural number $k + 1$,

then the statement is true for all natural numbers.

Figure 10

The following physical interpretation illustrates why the principle works. Think of a collection of natural numbers obeying a statement as a collection of infinitely many dominoes. See Figure 10.

Now, suppose that two facts are given:

1.   The first domino is pushed over.
2.   If one domino falls over, say the kth domino, so will the next one, the $(k + 1)$st domino.

Is it safe to conclude that *all* the dominoes fall over? The answer is yes, because if the first one falls (Condition I), the second one does also (by Condition II); and if the second one falls, so does the third (by Condition II); and so on.

EXAMPLE 1   **Using Mathematical Induction**

Show that the following statement is true for all natural numbers n.

$$1 + 3 + 5 + \cdots + (2n - 1) = n^2 \tag{2}$$

---

[*]Recall that the natural numbers are the numbers $1, 2, 3, 4, \ldots$. In other words, the terms *natural numbers* and *positive integers* are synonymous.

Solution First show that statement (2) holds for $n = 1$. Because $1 = 1^2$, statement (2) is true for $n = 1$. Condition I holds.

Next, show that Condition II holds. From statement (2), assume that

$$1 + 3 + 5 + \cdots + (2k - 1) = k^2 \tag{3}$$

is true for some natural number k.

Now show that, based on equation (3), statement (2) holds for $k + 1$. Look at the sum of the first $k + 1$ positive odd integers to determine whether this sum equals $(k + 1)^2$.

$$1 + 3 + 5 + \cdots + (2k - 1) + [2(k + 1) - 1] = \underbrace{[1 + 3 + 5 + \cdots + (2k - 1)]}_{= k^2 \text{ by equation (3)}} + (2k + 1)$$

$$= k^2 + (2k + 1)$$
$$= k^2 + 2k + 1 = (k + 1)^2$$

Conditions I and II are satisfied; by the Principle of Mathematical Induction, statement (2) is true for all natural numbers n.

EXAMPLE 2   **Using Mathematical Induction**

Show that the following statement is true for all natural numbers n.

$$2^n > n$$

Solution First, show that the statement $2^n > n$ holds when $n = 1$. Because $2^1 = 2 > 1$, the inequality is true for $n = 1$. Condition I holds.

Next, assume the statement holds for some natural number k; that is, $2^k > k$. Now show that the statement holds for $k + 1$; that is, show that $2^{k+1} > k + 1$.

$$2^{k+1} = 2 \cdot 2^k \underset{\underset{2^k > k}{\uparrow}}{>} 2 \cdot k = k + k \underset{\underset{k \geq 1}{\uparrow}}{\geq} k + 1$$

If $2^k > k$, then $2^{k+1} > k + 1$, so Condition II of the Principle of Mathematical Induction is satisfied. The statement $2^n > n$ is true for all natural numbers n.

EXAMPLE 3   **Using Mathematical Induction**

Show that the following formula is true for all natural numbers n.

$$1 + 2 + 3 + \cdots + n = \frac{n(n + 1)}{2} \tag{4}$$

Solution First, show that formula (4) is true when $n = 1$. Because

$$\frac{1(1 + 1)}{2} = \frac{1 \cdot 2}{2} = 1$$

Condition I of the Principle of Mathematical Induction holds.

Next, assume that formula (4) holds for some natural number k, and determine whether the formula then holds for the next natural number, $k + 1$.

Assume that

$$1 + 2 + 3 + \cdots + k = \frac{k(k + 1)}{2} \quad \text{for some } k \tag{5}$$

Now show that

$$1 + 2 + 3 + \cdots + k + (k + 1) = \frac{(k + 1)[(k + 1) + 1]}{2} = \frac{(k + 1)(k + 2)}{2}$$

as follows:

$$1 + 2 + 3 + \cdots + k + (k + 1) = \underbrace{[1 + 2 + 3 + \cdots + k]}_{\frac{k(k+1)}{2} \text{ by equation (5)}} + (k + 1)$$

$$= \frac{k(k + 1)}{2} + (k + 1)$$

$$= \frac{k^2 + k + 2k + 2}{2}$$

$$= \frac{k^2 + 3k + 2}{2} = \frac{(k + 1)(k + 2)}{2}$$

Condition II also holds. As a result, formula (4) is true for all natural numbers n.

— **Now Work** PROBLEM 1

EXAMPLE 4   **Using Mathematical Induction**

Show that $3^n - 1$ is divisible by 2 for all natural numbers n.

Solution   First, show that the statement is true when $n = 1$. Because $3^1 - 1 = 3 - 1 = 2$ is divisible by 2, the statement is true when $n = 1$. Condition I is satisfied.

Next, assume that the statement holds for some natural number k, and determine whether the statement holds for the next natural number, $k + 1$.

Assume that $3^k - 1$ is divisible by 2 for some k. Now show that $3^{k+1} - 1$ is divisible by 2.

$$3^{k+1} - 1 = 3^{k+1} - 3^k + 3^k - 1 \qquad \text{Subtract and add } 3^k.$$

$$= 3^k(3 - 1) + (3^k - 1) = 3^k \cdot 2 + (3^k - 1)$$

Because $3^k \cdot 2$ is divisible by 2 and $3^k - 1$ is divisible by 2, it follows that $3^k \cdot 2 + (3^k - 1) = 3^{k+1} - 1$ is divisible by 2. Condition II is also satisfied. As a result, the statement "$3^n - 1$ is divisible by 2" is true for all natural numbers n.

— **Now Work** PROBLEM 19

WARNING The conclusion that a statement involving natural numbers is true for all natural numbers is made only after *both* Conditions I and II of the Principle of Mathematical Induction have been satisfied. Problem 28 demonstrates a statement for which only Condition I holds, and the statement is *not* true for all natural numbers. Problem 29 demonstrates a statement for which only Condition II holds, and the statement is *not* true for any natural number. ∎

9.4 Assess Your Understanding

Skill Building

In Problems 1–22, use the Principle of Mathematical Induction to show that the given statement is true for all natural numbers n.

1. $2 + 4 + 6 + \cdots + 2n = n(n + 1)$

2. $1 + 5 + 9 + \cdots + (4n - 3) = n(2n - 1)$

3. $3 + 4 + 5 + \cdots + (n + 2) = \frac{1}{2} n(n + 5)$

4. $3 + 5 + 7 + \cdots + (2n + 1) = n(n + 2)$

5. $2 + 5 + 8 + \cdots + (3n - 1) = \frac{1}{2} n(3n + 1)$

6. $1 + 4 + 7 + \cdots + (3n - 2) = \frac{1}{2} n(3n - 1)$

7. $1 + 2 + 2^2 + \cdots + 2^{n-1} = 2^n - 1$

8. $1 + 3 + 3^2 + \cdots + 3^{n-1} = \frac{1}{2}(3^n - 1)$

9. $1 + 4 + 4^2 + \cdots + 4^{n-1} = \frac{1}{3}(4^n - 1)$

10. $1 + 5 + 5^2 + \cdots + 5^{n-1} = \frac{1}{4}(5^n - 1)$

11. $\dfrac{1}{1 \cdot 2} + \dfrac{1}{2 \cdot 3} + \dfrac{1}{3 \cdot 4} + \cdots + \dfrac{1}{n(n + 1)} = \dfrac{n}{n + 1}$

12. $\dfrac{1}{1 \cdot 3} + \dfrac{1}{3 \cdot 5} + \dfrac{1}{5 \cdot 7} + \cdots + \dfrac{1}{(2n - 1)(2n + 1)} = \dfrac{n}{2n + 1}$

13. $1^2 + 2^2 + 3^2 + \cdots + n^2 = \dfrac{1}{6}n(n+1)(2n+1)$

14. $1^3 + 2^3 + 3^3 + \cdots + n^3 = \dfrac{1}{4}n^2(n+1)^2$

15. $4 + 3 + 2 + \cdots + (5-n) = \dfrac{1}{2}n(9-n)$

16. $-2 - 3 - 4 - \cdots - (n+1) = -\dfrac{1}{2}n(n+3)$

17. $1\cdot 2 + 2\cdot 3 + 3\cdot 4 + \cdots + n(n+1) = \dfrac{1}{3}n(n+1)(n+2)$

18. $1\cdot 2 + 3\cdot 4 + 5\cdot 6 + \cdots + (2n-1)(2n) = \dfrac{1}{3}n(n+1)(4n-1)$

19. $n^2 + n$ is divisible by 2.

20. $n^3 + 2n$ is divisible by 3.

21. $n^2 - n + 2$ is divisible by 2.

22. $n(n+1)(n+2)$ is divisible by 6.

Applications and Extensions

In Problems 23–27, prove each statement.

23. If $x > 1$, then $x^n > 1$.

24. If $0 < x < 1$, then $0 < x^n < 1$.

25. $a - b$ is a factor of $a^n - b^n$.

  [**Hint:** $a^{k+1} - b^{k+1} = a(a^k - b^k) + b^k(a - b)$]

26. $a + b$ is a factor of $a^{2n+1} + b^{2n+1}$.

27. $(1+a)^n \geq 1 + na$, for $a > 0$

28. Show that the statement "$n^2 - n + 41$ is a prime number" is true for $n = 1$ but is not true for $n = 41$.

29. Show that the formula

$$2 + 4 + 6 + \cdots + 2n = n^2 + n + 2$$

obeys Condition II of the Principle of Mathematical Induction. That is, show that if the formula is true for some k, it is also true for $k + 1$. Then show that the formula is false for $n = 1$ (or for any other choice of n).

30. Use mathematical induction to prove that if $r \neq 1$, then

$$a + ar + ar^2 + \cdots + ar^{n-1} = a\,\dfrac{1 - r^n}{1 - r}$$

31. Use mathematical induction to prove that

$a + (a + d) + (a + 2d)$

$+ \cdots + [a + (n-1)d] = na + d\dfrac{n(n-1)}{2}$

32. **Extended Principle of Mathematical Induction** The Extended Principle of Mathematical Induction states that if Conditions I and II hold, that is,

  (I) A statement is true for a natural number j.
  (II) If the statement is true for some natural number $k \geq j$, then it is also true for the next natural number $k + 1$.

  then the statement is true for all natural numbers $\geq j$. Use the Extended Principle of Mathematical Induction to show

that the number of diagonals in a convex polygon of n sides is $\dfrac{1}{2}n(n-3)$.

  [**Hint:** Begin by showing that the result is true when $n = 4$ (Condition I).]

33. **Geometry** Use the Extended Principle of Mathematical Induction to show that the sum of the interior angles of a convex polygon of n sides equals $(n-2)\cdot 180°$.

34. ***Challenge Problem*** Use the Principle of Mathematical Induction to prove that

$$\begin{bmatrix} 5 & -8 \\ 2 & -3 \end{bmatrix}^n = \begin{bmatrix} 4n+1 & -8n \\ 2n & 1-4n \end{bmatrix}$$

for all natural numbers n.

35. ***Challenge Problem* Paper Creases** If a sheet of paper is folded in half by folding the top edge down to the bottom edge, one crease will result. If the folded paper is folded in the same manner, the result is three creases. With each fold, the number of creases can be defined recursively by $c_1 = 1$, $c_{n+1} = 2c_n + 1$.

  (a) Find the number of creases for $n = 3$ and $n = 4$ folds.
  (b) Use the given information and your results from part (a) to find a formula for the number of creases after n folds, c_n, in terms of the number of folds alone.
  (c) Use the Principle of Mathematical Induction to prove that the formula found in part (b) is correct for all natural numbers.
  (d) Tosa Tengujo is reportedly the world's thinnest paper with a thickness of 0.02 mm. If a piece of this paper could be folded 25 times, how tall would the stack be?

Explaining Concepts: Discussion and Writing

36. How would you explain the Principle of Mathematical Induction to a friend?

Retain Your Knowledge

Problems 37–43 are based on material learned earlier in the course. The purpose of these problems is to keep the material fresh in your mind so that you are better prepared for the final exam.

37. Solve: $\log_2 \sqrt{x+5} = 4$

38. Solve the system: $\begin{cases} 4x + 3y = -7 \\ 2x - 5y = 16 \end{cases}$

39. Solve: $2\sqrt{x-3} + 5 = 8$

40. For $A = \begin{bmatrix} 1 & 2 & -1 \\ 0 & 1 & 4 \end{bmatrix}$ and $B = \begin{bmatrix} 3 & -1 \\ 1 & 0 \\ -2 & 2 \end{bmatrix}$, find $A \cdot B$.

41. Find the partial fraction decomposition of $\dfrac{3x}{x^2 + x - 2}$.

42. Solve: $e^{3x-7} = 4$

43. If $f'(x) = (x^2 - 2x + 1)(3x^2) + (x^3 - 1)(2x - 2)$, find all real numbers x for which $f'(x) = 0$.

9.5 The Binomial Theorem

OBJECTIVES 1 Evaluate $\dbinom{n}{j}$ (p. 685)

2 Use the Binomial Theorem (p. 687)

Formulas have been given for expanding $(x + a)^n$ for $n = 2$ and $n = 3$. The *Binomial Theorem** is a formula for the expansion of $(x + a)^n$ for any positive integer n. If $n = 1, 2, 3,$ and 4, the expansion of $(x + a)^n$ is straightforward.

$(x + a)^1 = x + a$ Two terms, beginning with x^1 and ending with a^1

$(x + a)^2 = x^2 + 2ax + a^2$ Three terms, beginning with x^2 and ending with a^2

$(x + a)^3 = x^3 + 3ax^2 + 3a^2x + a^3$ Four terms, beginning with x^3 and ending with a^3

$(x + a)^4 = x^4 + 4ax^3 + 6a^2x^2 + 4a^3x + a^4$ Five terms, beginning with x^4 and ending with a^4

Notice that each expansion of $(x + a)^n$ begins with x^n and ends with a^n. From left to right, the powers of x are decreasing by 1, while the powers of a are increasing by 1. Also, the number of terms equals $n + 1$. Notice, too, that the degree of each monomial in the expansion equals n. For example, in the expansion of $(x + a)^3$, each monomial $(x^3, 3ax^2, 3a^2x, a^3)$ is of degree 3. As a result, it is reasonable to conjecture that the expansion of $(x + a)^n$ would look like this:

$$(x + a)^n = x^n + \underline{\hspace{0.8cm}} ax^{n-1} + \underline{\hspace{0.8cm}} a^2 x^{n-2} + \cdots + \underline{\hspace{0.8cm}} a^{n-1} x + a^n$$

where the blanks are numbers to be found. This is in fact the case.

Before we can fill in the blanks, we need to introduce the symbol $\dbinom{n}{j}$.

1 Evaluate $\dbinom{n}{j}$

The symbol $\dbinom{n}{j}$, read "n taken j at a time," is defined next.

---
*The name *binomial* is derived from the fact that $x + a$ is a binomial; that is, it contains two terms.

📱 **COMMENT** On a graphing calculator, the symbol $\binom{n}{j}$ may be denoted by the key $\boxed{nCr}$. ∎

DEFINITION $\binom{n}{j}$

If j and n are integers with $0 \leq j \leq n$, the symbol $\binom{n}{j}$ is defined as

$$\binom{n}{j} = \frac{n!}{j!\,(n-j)!} \tag{1}$$

EXAMPLE 1

Evaluating $\binom{n}{j}$

Find:

(a) $\binom{3}{1}$ (b) $\binom{4}{2}$ (c) $\binom{8}{7}$ 📱 (d) $\binom{65}{15}$

Solution

(a) $\binom{3}{1} = \dfrac{3!}{1!\,(3-1)!} = \dfrac{3!}{1!\,2!} = \dfrac{3 \cdot 2 \cdot 1}{1\,(2 \cdot 1)} = \dfrac{6}{2} = 3$

(b) $\binom{4}{2} = \dfrac{4!}{2!\,(4-2)!} = \dfrac{4!}{2!\,2!} = \dfrac{4 \cdot 3 \cdot 2 \cdot 1}{(2 \cdot 1)(2 \cdot 1)} = \dfrac{24}{4} = 6$

(c) $\binom{8}{7} = \dfrac{8!}{7!\,(8-7)!} = \dfrac{8!}{7!\,1!} = \dfrac{8 \cdot 7!}{7! \cdot 1!} = \dfrac{8}{1} = 8$

$$8! = 8 \cdot 7!$$

NORMAL FLOAT AUTO REAL RADIAN MP

$_{65}C_{15}$
                    2.073746998ᴇ14

Figure 11

📱 (d) Figure 11 shows the solution using a TI-84 Plus C graphing calculator. So

$$\binom{65}{15} \approx 2.073746998 \times 10^{14}$$

✏️ **Now Work** PROBLEM 5

THEOREM

Four useful formulas involving the symbol $\binom{n}{j}$ are

• $\binom{n}{0} = 1$ • $\binom{n}{1} = n$ • $\binom{n}{n-1} = n$ • $\binom{n}{n} = 1$

Proof $\binom{n}{0} = \dfrac{n!}{0!\,(n-0)!} = \dfrac{n!}{0!\,n!} = \dfrac{1}{1} = 1$

$\binom{n}{1} = \dfrac{n!}{1!\,(n-1)!} = \dfrac{n!}{(n-1)!} = \dfrac{n\,(n-1)!}{(n-1)!} = n$

You are asked to prove the remaining two formulas in Problem 45. ∎

Suppose that the values of the symbol $\binom{n}{j}$ are arranged in a triangular display, as shown next and in Figure 12.

$$\binom{0}{0}$$

$$\binom{1}{0} \quad \binom{1}{1}$$

$$\binom{2}{0} \quad \binom{2}{1} \quad \binom{2}{2}$$

$$\binom{3}{0} \quad \binom{3}{1} \quad \binom{3}{2} \quad \binom{3}{3}$$

$$\binom{4}{0} \quad \binom{4}{1} \quad \binom{4}{2} \quad \binom{4}{3} \quad \binom{4}{4}$$

$$\binom{5}{0} \quad \binom{5}{1} \quad \binom{5}{2} \quad \binom{5}{3} \quad \binom{5}{4} \quad \binom{5}{5}$$

NOTE The Pascal triangle is symmetric about a line down its center. If we divide the triangle vertically in half, the entries in each row of the left half are the mirror image of the entries in the same row of the right half.

The vertical symmetry of the entries in the Pascal triangle is a result of the fact that

$$\binom{n}{n-j} = \frac{n!}{(n-j)!\,j!}$$

$$= \frac{n!}{j!\,(n-j)!} = \binom{n}{j} \quad \blacksquare$$

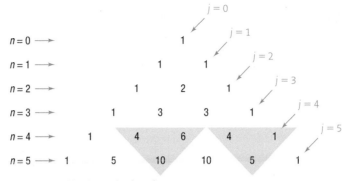

Figure 12 The Pascal triangle

This display is called the **Pascal triangle**, named after Blaise Pascal (1623–1662), a French mathematician.

The Pascal triangle has 1's down the sides. To get any other entry, add the two nearest entries in the row above it. The shaded triangles in Figure 12 illustrate this feature of the Pascal triangle. Based on this feature, the row corresponding to $n = 6$ is found as follows:

$$
\begin{array}{ccccccccccccc}
n = 5 \rightarrow & & 1 & & 5 & & 10 & & 10 & & 5 & & 1 \\
n = 6 \rightarrow & 1 & & 6 & & 15 & & 20 & & 15 & & 6 & & 1
\end{array}
$$

This addition always works (see the theorem on page 689).

Although the Pascal triangle provides an interesting and organized display of the symbol $\binom{n}{j}$, in practice it is not that helpful. For example, if you want the value of $\binom{12}{5}$, you would need 13 rows of the triangle before seeing the answer. It is much faster to use definition (1).

2 Use the Binomial Theorem

> **THEOREM Binomial Theorem**
>
> Let x and a be real numbers. For any positive integer n,
>
> $$(x + a)^n = \binom{n}{0}x^n + \binom{n}{1}ax^{n-1} + \cdots + \binom{n}{j}a^j x^{n-j} + \cdots + \binom{n}{n}a^n$$
>
> $$= \sum_{j=0}^{n}\binom{n}{j}a^j x^{n-j} \tag{2}$$

This is why it was necessary to introduce the symbol $\binom{n}{j}$; the numbers $\dfrac{n!}{j!\,(n-j)!}$ are the numerical coefficients in the expansion of $(x + a)^n$. Because of this, the symbol $\binom{n}{j}$ is called a **binomial coefficient**.

EXAMPLE 2

Expanding a Binomial

Use the Binomial Theorem to expand $(x + 2)^5$.

Solution   In the Binomial Theorem, let $a = 2$ and $n = 5$. Then

$$(x + 2)^5 = \binom{5}{0}x^5 + \binom{5}{1}2x^4 + \binom{5}{2}2^2 x^3 + \binom{5}{3}2^3 x^2 + \binom{5}{4}2^4 x + \binom{5}{5}2^5$$

↑ Use equation (2).

$$= 1 \cdot x^5 + 5 \cdot 2x^4 + 10 \cdot 4x^3 + 10 \cdot 8x^2 + 5 \cdot 16x + 1 \cdot 32$$

↑ Use row $n = 5$ of the Pascal triangle or definition (1) for $\binom{n}{j}$.

$$= x^5 + 10x^4 + 40x^3 + 80x^2 + 80x + 32$$

EXAMPLE 3 **Expanding a Binomial**

Expand $(2y - 3)^4$ using the Binomial Theorem.

Solution First, rewrite the expression $(2y - 3)^4$ as $[2y + (-3)]^4$. Now use the Binomial Theorem with $n = 4$, $x = 2y$, and $a = -3$.

$$[2y + (-3)]^4 = \binom{4}{0}(2y)^4 + \binom{4}{1}(-3)(2y)^3 + \binom{4}{2}(-3)^2(2y)^2$$

$$+ \binom{4}{3}(-3)^3(2y) + \binom{4}{4}(-3)^4$$

$$= 1 \cdot 16y^4 + 4(-3)8y^3 + 6 \cdot 9 \cdot 4y^2 + 4(-27)2y + 1 \cdot 81$$

$\uparrow$

Use row $n = 4$ of the Pascal triangle or definition (1) for $\binom{n}{j}$.

$$= 16y^4 - 96y^3 + 216y^2 - 216y + 81$$

In this expansion, note that the signs alternate because $a = -3 < 0$.

━━━▶ **Now Work** PROBLEM 21

EXAMPLE 4 **Finding a Particular Coefficient in a Binomial Expansion**

Find the coefficient of y^8 in the expansion of $(2y + 3)^{10}$.

Solution Expand $(2y + 3)^{10}$ using the Binomial Theorem.

$$(2y + 3)^{10} = \binom{10}{0}(2y)^{10} + \binom{10}{1}(2y)^9(3)^1 + \binom{10}{2}(2y)^8(3)^2 + \binom{10}{3}(2y)^7(3)^3$$

$$+ \binom{10}{4}(2y)^6(3)^4 + \cdots + \binom{10}{9}(2y)(3)^9 + \binom{10}{10}(3)^{10}$$

From the third term in the expansion, the coefficient of y^8 is

$$\binom{10}{2}(2)^8(3)^2 = \frac{10!}{2! \, 8!} \cdot 2^8 \cdot 9 = \frac{10 \cdot 9 \cdot 8!}{2 \cdot 8!} \cdot 2^8 \cdot 9 = 103,680$$

As this solution demonstrates, the Binomial Theorem can be used to find a particular term in the expansion of $(ax + b)^n$ without writing the entire expansion.

The term containing x^j in the expansion of $(ax + b)^n$ is

$$\binom{n}{n - j} b^{n-j} (ax)^j \qquad (3)$$

Example 4 can be solved by using formula (3) with $n = 10$, $a = 2$, $b = 3$, and $j = 8$. Then the term containing y^8 is

$$\binom{10}{10 - 8} 3^{10-8} (2y)^8 = \binom{10}{2} \cdot 3^2 \cdot 2^8 \cdot y^8 = \frac{10!}{2! \, 8!} \cdot 9 \cdot 2^8 y^8$$

$$= \frac{10 \cdot 9 \cdot 8!}{2 \cdot 8!} \cdot 9 \cdot 2^8 \, y^8 = 103,680 y^8$$

EXAMPLE 5	**Finding a Particular Term in a Binomial Expansion**

Find the 6th term in the expansion of $(x + 2)^9$.

Solution A Expand using the Binomial Theorem until the 6th term is reached.

$$(x + 2)^9 = \binom{9}{0}x^9 + \binom{9}{1}x^8 \cdot 2 + \binom{9}{2}x^7 \cdot 2^2 + \binom{9}{3}x^6 \cdot 2^3 + \binom{9}{4}x^5 \cdot 2^4$$

$$+ \binom{9}{5}x^4 \cdot 2^5 + \cdots$$

The 6th term is

$$\binom{9}{5}x^4 \cdot 2^5 = \frac{9!}{5! \, 4!} \cdot x^4 \cdot 32 = 4032x^4$$

Solution B The 6th term in the expansion of $(x + 2)^9$, which has 10 terms total, contains x^4. (Do you see why?) By formula (3), the 6th term is

$$\binom{9}{9-4}2^{9-4}x^4 = \binom{9}{5}2^5 x^4 = \frac{9!}{5! \, 4!} \cdot 32x^4 = 4032x^4$$

- **Now Work** PROBLEMS 29 AND 35

The following theorem shows that the *triangular addition* feature of the Pascal triangle illustrated in Figure 12 always works.

> **THEOREM** If n and j are integers with $1 \le j \le n$, then
>
> $$\binom{n}{j-1} + \binom{n}{j} = \binom{n+1}{j} \qquad (4)$$

Proof

$$\binom{n}{j-1} + \binom{n}{j} = \frac{n!}{(j-1)![n - (j-1)]!} + \frac{n!}{j!(n-j)!}$$

$$= \frac{n!}{(j-1)!(n-j+1)!} + \frac{n!}{j!(n-j)!}$$

$$= \frac{jn!}{j(j-1)!(n-j+1)!} + \frac{(n-j+1)n!}{j!(n-j+1)(n-j)!} \qquad \text{Multiply the first term by } \frac{j}{j} \text{ and}$$

$$\text{the second term by } \frac{n-j+1}{n-j+1}$$

$$= \frac{jn!}{j!(n-j+1)!} + \frac{(n-j+1)n!}{j!(n-j+1)!} \qquad \text{to make the denominators equal.}$$

$$= \frac{jn! + (n-j+1)n!}{j!(n-j+1)!}$$

$$= \frac{n!(j + n - j + 1)}{j!(n-j+1)!}$$

$$= \frac{n!(n+1)}{j!(n-j+1)!} = \frac{(n+1)!}{j![(n+1)-j]!} = \binom{n+1}{j}$$

Historical Feature

Omar Khayyám (1048–1131)

T he case $n = 2$ of the Binomial Theorem, $(a + b)^2$, was known to Euclid in 300 BC, but the general law seems to have been discovered by the Persian mathematician and astronomer Omar Khayyám (1048–1131), who is also well known as the author of the *Rubáiyát*, a collection of four-line poems making observations on the human condition. Omar Khayyám did not state the Binomial Theorem explicitly, but he claimed to have a method for extracting third, fourth, fifth roots, and so on. A little study shows that one must know the Binomial Theorem to create such a method.

The heart of the Binomial Theorem is the formula for the numerical coefficients, and, as we saw, they can be written in a symmetric triangular form. The Pascal triangle appears first in the books of Yang Hui (about 1270) and Chu Shih-chieh (1303). Pascal's name is attached to the triangle because of the many applications he made of it, especially to counting and probability. In establishing these results, he was one of the earliest users of mathematical induction.

Many people worked on the proof of the Binomial Theorem, which was finally completed for all n (including complex numbers) by Niels Abel (1802–1829).

9.5 Assess Your Understanding

Concepts and Vocabulary

1. The _____ _____ is a triangular display of the binomial coefficients.

2. $\binom{n}{0} = $ ____ and $\binom{n}{1} = $ ____.

3. *True or False* $\binom{n}{j} = \dfrac{j!}{(n-j)!\, n!}$

4. The _____ _____ can be used to expand expressions like $(2x + 3)^6$.

Skill Building

In Problems 5–16, evaluate each expression.

5. $\binom{5}{3}$

6. $\binom{7}{3}$

7. $\binom{7}{5}$

8. $\binom{9}{7}$

9. $\binom{50}{49}$

10. $\binom{100}{98}$

11. $\binom{1000}{1000}$

12. $\binom{1000}{0}$

13. $\binom{55}{23}$

14. $\binom{60}{20}$

15. $\binom{47}{25}$

16. $\binom{37}{19}$

In Problems 17–28, expand each expression using the Binomial Theorem.

17. $(x + 1)^5$

18. $(x - 1)^5$

19. $(x - 2)^6$

20. $(x + 3)^5$

21. $(3x + 1)^4$

22. $(2x + 3)^5$

23. $(x^2 + y^2)^5$

24. $(x^2 - y^2)^6$

25. $(\sqrt{x} + \sqrt{2})^6$

26. $(\sqrt{x} - \sqrt{3})^4$

27. $(ax + by)^5$

28. $(ax - by)^4$

In Problems 29–42, use the Binomial Theorem to find the indicated coefficient or term.

29. The coefficient of x^6 in the expansion of $(x + 3)^{10}$

30. The coefficient of x^3 in the expansion of $(x - 3)^{10}$

31. The coefficient of x^7 in the expansion of $(2x - 1)^{12}$

32. The coefficient of x^3 in the expansion of $(2x + 1)^{12}$

33. The coefficient of x^7 in the expansion of $(2x + 3)^9$

34. The coefficient of x^2 in the expansion of $(2x - 3)^9$

35. The 5th term in the expansion of $(x + 3)^7$

36. The 3rd term in the expansion of $(x - 3)^7$

37. The 3rd term in the expansion of $(3x - 2)^9$

38. The 6th term in the expansion of $(3x + 2)^8$

39. The coefficient of x^0 in the expansion of $\left(x^2 + \dfrac{1}{x}\right)^{12}$

40. The coefficient of x^0 in the expansion of $\left(x - \dfrac{1}{x^2}\right)^9$

41. The coefficient of x^4 in the expansion of $\left(x - \dfrac{2}{\sqrt{x}}\right)^{10}$

42. The coefficient of x^2 in the expansion of $\left(\sqrt{x} + \dfrac{3}{\sqrt{x}}\right)^8$

Applications and Extensions

43. Use the Binomial Theorem to find the numerical value of $(1.001)^5$ correct to five decimal places.

   [**Hint**: $(1.001)^5 = (1 + 10^{-3})^5$]

44. Use the Binomial Theorem to find the numerical value of $(0.998)^6$ correct to five decimal places.

45. Show that $\dbinom{n}{n-1} = n$ and $\dbinom{n}{n} = 1$.

46. Stirling's Formula An approximation for $n!$, when n is large, is given by

$$n! \approx \sqrt{2n\pi}\left(\frac{n}{e}\right)^n\left(1 + \frac{1}{12n-1}\right)$$

Calculate $12!$, $20!$, and $25!$ on your calculator. Then use Stirling's formula to approximate $12!$, $20!$, and $25!$.

47. Challenge Problem If n is a positive integer, show that

$$\binom{n}{0} + \binom{n}{1} + \cdots + \binom{n}{n} = 2^n$$

[**Hint:** $2^n = (1+1)^n$; now use the Binomial Theorem.]

48. Challenge Problem If n is a positive integer, show that

$$\binom{n}{0} - \binom{n}{1} + \binom{n}{2} - \cdots + (-1)^n\binom{n}{n} = 0$$

49. Challenge Problem Find the value of

$$\binom{5}{0}\left(\frac{1}{4}\right)^5 + \binom{5}{1}\left(\frac{1}{4}\right)^4\left(\frac{3}{4}\right) + \binom{5}{2}\left(\frac{1}{4}\right)^3\left(\frac{3}{4}\right)^2$$
$$+ \binom{5}{3}\left(\frac{1}{4}\right)^2\left(\frac{3}{4}\right)^3 + \binom{5}{4}\left(\frac{1}{4}\right)\left(\frac{3}{4}\right)^4 + \binom{5}{5}\left(\frac{3}{4}\right)^5$$

50. Challenge Problem **Pascal Figures** The entries in the Pascal Triangle can, for $n \geq 2$, be used to determine the number of k-sided figures that can be formed using a set of n points on a circle. In general, the first entry in a row indicates the number of n-sided figures that can be formed, the second entry indicates the number of $(n-1)$-sided figures, and so on. For example, if a circle contains 4 points, the row for $n = 4$ in the Pascal Triangle shows the number of possible quadrilaterals (1), the number of triangles (4), and the number of line segments (6) that can be formed using the four points.

   (a) How many hexagons can be formed using 8 points lying on the circumference of a circle?

   (b) How many triangles can be formed using 10 points lying on the circumference of a circle?

   (c) How many dodecagons can be formed using 20 points lying on the circumference of a circle?

51. Challenge Problem Find the coefficient of x^4 in

$$f(x) = (1 - x^2) + (1 - x^2)^2 + \cdots + (1 - x^2)^{10}$$

52. Challenge Problem In the expansion of $[a + (b + c)^2]^8$, find the coefficient of the term containing $a^5b^4c^2$.

Retain Your Knowledge

Problems 53–60 are based on material learned earlier in the course. The purpose of these problems is to keep the material fresh in your mind so that you are better prepared for the final exam.

53. Solve $6^x = 5^{x+1}$. Express the answer both in exact form and as a decimal rounded to three decimal places.

54. Solve the system of equations:

$$\begin{cases} x - y - z = 0 \\ 2x + y + 3z = -1 \\ 4x + 2y - z = 12 \end{cases}$$

55. Graph the system of inequalities. Tell whether the graph is bounded or unbounded, and label the corner points.

$$\begin{cases} x \geq 0 \\ y \geq 0 \\ x + y \leq 6 \\ 2x + y \leq 10 \end{cases}$$

56. If $f(x) = x^2 - 6$ and $g(x) = \sqrt{x + 2}$, find $g(f(x))$ and state its domain.

57. If $y = \dfrac{5}{3}x^3 + 2x + C$ and $y = 5$ when $x = 3$, find the value of C.

58. Simplify: $\dfrac{(x^3 + 1) \cdot \frac{1}{3}x^{-2/3} - x^{1/3}(3x^2)}{(x^3 + 1)^2}$

59. Find the vertical asymptotes, if any, of the graph of

$$f(x) = \frac{3x^2}{(x - 3)(x + 1)}$$

60. If $f(x) = \dfrac{x^2 + 1}{2x + 5}$, find $f(-2)$. What is the corresponding point on the graph of f?

Chapter Review

Things to Know

Sequence (p. 652)	A function whose domain is the set of positive integers and whose range is a subset of the real numbers
Factorials (p. 654)	$0! = 1$, $1! = 1$, $n! = n(n-1) \cdot \cdots \cdot 3 \cdot 2 \cdot 1$ if $n \geq 2$ is an integer
Arithmetic sequence (pp. 662 and 664)	$a_1 = a$, $a_n = a_{n-1} + d$, where $a_1 = a =$ first term, $d =$ common difference $a_n = a_1 + (n-1)d$
Sum of the first n terms of an arithmetic sequence (p. 665)	$S_n = \dfrac{n}{2}[2a_1 + (n-1)d] = \dfrac{n}{2}(a_1 + a_n)$

Geometric sequence (pp. 669 and 670)

$a_1 = a, \quad a_n = ra_{n-1}$, where $a_1 = a =$ first term, $r =$ common ratio

$a_n = a_1 r^{n-1} \quad r \neq 0$

Sum of the first n terms of a geometric sequence (p. 671)

$S_n = a_1 \dfrac{1 - r^n}{1 - r} \quad r \neq 0, 1$

Infinite geometric series (p. 672)

$a_1 + a_1 r + \cdots + a_1 r^{n-1} + \cdots = \displaystyle\sum_{k=1}^{\infty} a_1 r^{k-1}$

Sum of a convergent infinite geometric series (p. 673)

If $|r| < 1$, $\displaystyle\sum_{k=1}^{\infty} a_1 r^{k-1} = \dfrac{a_1}{1 - r}$

Amount of an annuity (p. 675)

$A = P \dfrac{(1 + i)^n - 1}{i}$, where $P =$ the deposit (in dollars) made at the end of each payment period, $i =$ interest rate per payment period (as a decimal), and $A =$ the amount of the annuity after n deposits.

Principle of Mathematical Induction (p. 681)

If the following two conditions are satisfied,

    Condition I: The statement is true for the natural number 1.

    Condition II: If the statement is true for some natural number k, and it can be shown to be true for $k + 1$,

then the statement is true for all natural numbers.

Binomial coefficient (p. 686)

$\dbinom{n}{j} = \dfrac{n!}{j!\,(n - j)!}$

The Pascal triangle (p. 687)

See Figure 12.

Binomial Theorem (p. 687)

$(x + a)^n = \dbinom{n}{0}x^n + \dbinom{n}{1}ax^{n-1} + \cdots + \dbinom{n}{j}a^j x^{n-j} + \cdots + \dbinom{n}{n}a^n = \displaystyle\sum_{j=0}^{n}\dbinom{n}{j}x^{n-j} a^j$

Objectives

Section	You should be able to . . .	Examples	Review Exercises
9.1	1 List the first several terms of a sequence (p. 652)	1–4	1, 2
	2 List the terms of a sequence defined by a recursive formula (p. 655)	5, 6	3, 4
	3 Use summation notation (p. 656)	7, 8	5, 6
	4 Find the sum of a sequence (p. 657)	9	13, 14
9.2	1 Determine whether a sequence is arithmetic (p. 662)	1–3	7–12
	2 Find a formula for an arithmetic sequence (p. 663)	4, 5	17, 19–21, 34(a)
	3 Find the sum of an arithmetic sequence (p. 665)	6–8	7, 10, 14, 34(b), 35
9.3	1 Determine whether a sequence is geometric (p. 669)	1–3	7–12
	2 Find a formula for a geometric sequence (p. 670)	4	11, 18, 36(a)–(c), 38
	3 Find the sum of a geometric sequence (p. 671)	5, 6	9, 11, 15, 16
	4 Determine whether a geometric series converges or diverges (p. 672)	7–9	22–25, 36(d)
	5 Solve annuity problems (p. 675)	10, 11	37
9.4	1 Prove statements using mathematical induction (p. 681)	1–4	26–28
9.5	1 Evaluate $\dbinom{n}{j}$ (p. 685)	1	29
	2 Use the Binomial Theorem (p. 687)	2–5	30–33

Review Exercises

In Problems 1–4, list the first five terms of each sequence.

1. $\{a_n\} = \left\{(-1)^n\left(\dfrac{n+3}{n+2}\right)\right\}$ **2.** $\{c_n\} = \left\{\dfrac{2^n}{n^2}\right\}$

3. $a_1 = 3;\quad a_n = \dfrac{2}{3}a_{n-1}$ **4.** $a_1 = 2;\quad a_n = 2 - a_{n-1}$

5. Expand $\displaystyle\sum_{k=1}^{4}(4k+2)$.

6. Express $1 - \dfrac{1}{2} + \dfrac{1}{3} - \dfrac{1}{4} + \cdots + \dfrac{1}{13}$ using summation notation.

In Problems 7–12, determine whether the given sequence is arithmetic, geometric, or neither. If the sequence is arithmetic, find the common difference and the sum of the first n terms. If the sequence is geometric, find the common ratio and the sum of the first n terms.

7. $\{a_n\} = \{n+5\}$ **8.** $\{c_n\} = \{2n^3\}$ **9.** $\{s_n\} = \{2^{3n}\}$

10. $0, 4, 8, 12, \ldots$ **11.** $3, \dfrac{3}{2}, \dfrac{3}{4}, \dfrac{3}{8}, \dfrac{3}{16}, \ldots$ **12.** $\dfrac{2}{3}, \dfrac{3}{4}, \dfrac{4}{5}, \dfrac{5}{6}, \ldots$

In Problems 13–16, find each sum.

13. $\displaystyle\sum_{k=1}^{30}(k^2+2)$ **14.** $\displaystyle\sum_{k=1}^{40}(-2k+8)$ **15.** $\displaystyle\sum_{k=1}^{7}\left(\dfrac{1}{3}\right)^k$ **16.** $\displaystyle\sum_{k=1}^{10}(-2)^k$

In Problems 17–19, find the indicated term in each sequence. [**Hint:** *Find the general term first.*]

17. 9th term of $3, 7, 11, 15, \ldots$ **18.** 11th term of $1, \dfrac{1}{10}, \dfrac{1}{100}, \ldots$ **19.** 9th term of $\sqrt{2}, 2\sqrt{2}, 3\sqrt{2}, \ldots$

In Problems 20 and 21, find a general formula for each arithmetic sequence.

20. 7th term is 31; 20th term is 96 **21.** 10th term is 0; 18th term is 8

In Problems 22–25, determine whether each infinite geometric series converges or diverges. If it converges, find its sum.

22. $3 + 1 + \dfrac{1}{3} + \dfrac{1}{9} + \cdots$ **23.** $2 - 1 + \dfrac{1}{2} - \dfrac{1}{4} + \cdots$

24. $\dfrac{1}{2} + \dfrac{3}{4} + \dfrac{9}{8} + \cdots$ **25.** $\displaystyle\sum_{k=1}^{\infty} 4\left(\dfrac{1}{2}\right)^{k-1}$

In Problems 26–28, use the Principle of Mathematical Induction to show that the given statement is true for all natural numbers.

26. $3 + 6 + 9 + \cdots + 3n = \dfrac{3n}{2}(n+1)$ **27.** $2 + 6 + 18 + \cdots + 2 \cdot 3^{n-1} = 3^n - 1$

28. $1^2 + 4^2 + 7^2 + \cdots + (3n-2)^2 = \dfrac{1}{2}n(6n^2 - 3n - 1)$ **29.** Evaluate: $\dbinom{5}{2}$

In Problems 30 and 31, expand each expression using the Binomial Theorem.

30. $(x+2)^5$ **31.** $(3x-4)^4$

---

32. Find the coefficient of x^7 in the expansion of $(x+2)^9$.

33. Find the coefficient of x^2 in the expansion of $(2x+1)^7$.

34. Constructing a Brick Staircase A brick staircase has a total of 25 steps. The bottom step requires 80 bricks. Each step thereafter requires three fewer bricks than the prior step.
(a) How many bricks are required for the top step?
(b) How many bricks are required to build the staircase?

35. Creating a Floor Design A mosaic tile floor is designed in the shape of a trapezoid 30 feet wide at the base and 15 feet wide at the top. The tiles, 12 inches by 12 inches, are to be placed so that each successive row contains one fewer tile than the row below. How many tiles will be required?

36. Bouncing Balls A ball is dropped from a height of 20 feet. Each time it strikes the ground, it bounces up to three-quarters of the height of the previous bounce.

(a) What height will the ball bounce up to after it strikes the ground for the 3rd time?
(b) How high will it bounce after it strikes the ground for the nth time?
(c) How many times does the ball need to strike the ground before its bounce is less than 6 inches?
(d) What total distance does the ball travel before it stops bouncing?

37. Retirement Planning Chris gets paid once a month and contributes $350 each pay period into his 401(k). If Chris plans on retiring in 20 years, what will be the value of his 401(k) if the per annum rate of return of the 401(k) is 6.5% compounded monthly?

38. Salary Increases Your friend has just been hired at an annual salary of $50,000. If she expects to receive annual increases of 4%, what will be her salary as she begins her 5th year?

Chapter Test

CHAPTER
Test Prep
VIDEOS

The Chapter Test Prep Videos include step-by-step solutions to all chapter test exercises. These videos are available in MyLab™ Math, or on this text's YouTube Channel. Refer to the Preface for a link to the YouTube channel.

In Problems 1 and 2, list the first five terms of each sequence.

1. $\{s_n\} = \left\{\dfrac{n^2 - 1}{n + 8}\right\}$

2. $a_1 = 4, a_n = 3a_{n-1} + 2$

In Problems 3 and 4, expand each sum. Evaluate each sum.

3. $\displaystyle\sum_{k=1}^{3} (-1)^{k+1}\left(\dfrac{k + 1}{k^2}\right)$

4. $\displaystyle\sum_{k=1}^{4}\left[\left(\dfrac{2}{3}\right)^k - k\right]$

5. Write the following sum using summation notation.

$$-\dfrac{2}{5} + \dfrac{3}{6} - \dfrac{4}{7} + \cdots + \dfrac{11}{14}$$

In Problems 6–11, determine whether the given sequence is arithmetic, geometric, or neither. If the sequence is arithmetic, find the common difference and the sum of the first n terms. If the sequence is geometric, find the common ratio and the sum of the first n terms.

6. $6, 12, 36, 144, \ldots$

7. $\left\{-\dfrac{1}{2} \cdot 4^n\right\}$

8. $-2, -10, -18, -26, \ldots$

9. $\left\{-\dfrac{n}{2} + 7\right\}$

10. $25, 10, 4, \dfrac{8}{5}, \ldots$

11. $\left\{\dfrac{2n - 3}{2n + 1}\right\}$

12. Determine whether the infinite geometric series

$$256 - 64 + 16 - 4 + \cdots$$

converges or diverges. If it converges, find its sum.

13. Expand $(3m + 2)^5$ using the Binomial Theorem.

14. Use the Principle of Mathematical Induction to show that the given statement is true for all natural numbers.

$$\left(1 + \dfrac{1}{1}\right)\left(1 + \dfrac{1}{2}\right)\left(1 + \dfrac{1}{3}\right) \cdots \left(1 + \dfrac{1}{n}\right) = n + 1$$

15. A new car sold for $31,000. If the vehicle loses 15% of its value each year, how much will it be worth after 10 years?

16. A weightlifter begins his routine by benching 100 pounds and increases the weight by 30 pounds for each set. If he does 10 repetitions in each set, what is the total weight lifted after 5 sets?

Cumulative Review

1. Find all the solutions, real and complex, of the equation

$$|x^2| = 9$$

2. (a) Graph the circle $x^2 + y^2 = 100$ and the parabola $y = 3x^2$.

(b) Solve the system of equations: $\begin{cases} x^2 + y^2 = 100 \\ y = 3x^2 \end{cases}$

(c) Where do the circle and the parabola intersect?

3. Solve the equation: $2e^x = 5$

4. Find an equation of the line with slope 5 and x-intercept 2.

5. Find the standard equation of the circle whose center is the point $(-1, 2)$ if $(3, 5)$ is a point on the circle.

6. $f(x) = \dfrac{3x}{x - 2}$ and $g(x) = 2x + 1$

Find:
(a) $(f \circ g)(2)$
(b) $(g \circ f)(4)$
(c) $(f \circ g)(x)$
(d) The domain of $(f \circ g)(x)$
(e) $(g \circ f)(x)$
(f) The domain of $(g \circ f)(x)$
(g) The function g^{-1} and its domain
(h) The function f^{-1} and its domain

7. Find an equation of an ellipse with center at the origin, a focus at $(0, 3)$ and a vertex at $(0, 4)$.

8. Find an equation of a parabola with vertex at $(-1, 2)$ and focus at $(-1, 3)$.

Chapter Projects

((•)) **Internet-based Project**

I. Population Growth The size of the population of the United States essentially depends on its current population, the birth and death rates of the population, and immigration. Let b represent the birth rate of the U.S. population, and let d represent its death rate. Then $r = b - d$ represents the growth rate of the population, where r varies from year to year. The U.S. population after n years can be modeled using the recursive function

$$p_n = (1 + r)p_{n-1} + I$$

where I represents net immigration into the United States.

1. Using data from the CIA World Factbook at *https://www.cia.gov/library/publications/resources/ the-world-factbook/*, determine the birth and death rates in the United States for the most recent year that data

are available. Birth rates and death rates are given as the number of live births per 1000 population. Each must be computed as the number of births (deaths) per individual. For example, in 2017, the birth rate was 12.5 per 1000 and the death rate was 8.2 per 1000, so

$$b = \frac{12.5}{1000} = 0.0125, \text{ while } d = \frac{8.2}{1000} = 0.0082.$$

Next, using data from the Immigration and Naturalization Service at *https://fedstats.sites.usa.gov/*, determine the net immigration into the United States for the same year used to obtain b and d.

2. Determine the value of r, the growth rate of the population.

3. Find a recursive formula for the population of the United States.

4. Use the recursive formula to predict the population of the United States in the following year. In other words, if data are available up to the year 2018, predict the U.S. population in 2019.

5. Does your prediction seem reasonable? Explain.

6. Repeat Problems 1–5 for Uganda using the CIA World Factbook (in 2017, the birth rate was 42.9 per 1000 and the death rate was 10.2 per 1000).

7. Do your results for the United States (a developed country) and Uganda (a developing country) seem in line with the article in the chapter opener? Explain.

8. Do you think the recursive formula found in Problem 3 will be useful in predicting future populations? Why or why not?

The following projects are available at the Instructor's Resource Center (IRC):

II. Project at Motorola **Digital Wireless Communication** Cell phones take speech and change it into digital code using only zeros and ones. See how the code length can be modeled using a mathematical sequence.

III. Economics Economists use the current price of a good and a recursive model to predict future consumer demand and to determine future production.

IV. Standardized Tests Many tests of intelligence, aptitude, and achievement contain questions asking for the terms of a mathematical sequence.

Purchasing a Lottery Ticket

In recent years, the jackpot prizes for the nation's two major multistate lotteries, Mega Millions and Powerball, have climbed to all-time highs. This has happened since Powerball (in October 2015) and Mega Millions (in October 2017) made it more difficult to win their top prizes. The probability of winning the Mega Millions jackpot is now about 1 in 303 million, and the probability for Powerball is about 1 in 292 million.

With such improbable chances of winning the jackpots, one might wonder if there *ever* comes a point when purchasing a lottery ticket is worthwhile. One important consideration in making this determination is the *expected value*. For a game of chance, the **expected value** is a measure of how much a player will win or lose if she or he plays the game a large number of times.

The project at the end of this chapter explores the expected value from playing Mega Millions and Powerball and examines how the expected value is related to the jackpot amount.

—See Chapter Project I—

Outline

← A Look Back

We introduced sets in Chapter R, Review, and have been using them to represent solutions of equations and inequalities and to represent the domain and range of functions.

A Look Ahead →

Here we discuss methods for counting the number of elements in a set and consider the role of sets in probability.

10.1 Counting

PREPARING FOR THIS SECTION *Before getting started, review the following:*

- Sets (Section R.1, pp. 2–3)

Now Work the 'Are You Prepared?' problems on page 701.

OBJECTIVES 1 Find All the Subsets of a Set (p. 697)
2 Count the Number of Elements in a Set (p. 697)
3 Solve Counting Problems Using the Multiplication Principle (p. 699)

Counting plays a major role in many diverse areas, such as probability, statistics, and computer science; counting techniques are a part of a branch of mathematics called **combinatorics**.

1 Find All the Subsets of a Set

We begin by reviewing the ways in which two sets can be compared.

- If two sets A and B have precisely the same elements, we say that A and B are **equal** and write $A = B$.
- If each element of a set A is also an element of a set B, we say that A is a **subset** of B and write $A \subseteq B$.
- If $A \subseteq B$ and $A \neq B$, we say that A is a **proper subset** of B and write $A \subset B$.
- If $A \subseteq B$, every element in set A is also in set B, but B may or may not have additional elements. If $A \subset B$, every element in A is also in B, and B has at least one element not found in A.
- Finally, the empty set, $\varnothing$, is a subset of every set; that is,

$$\varnothing \subseteq A \qquad \text{for any set } A$$

EXAMPLE 1

Finding All the Subsets of a Set

Write down all the subsets of the set $\{a, b, c\}$.

Solution

To organize the work, write down all the subsets with no elements, then those with one element, then those with two elements, and finally those with three elements. This gives all the subsets. Do you see why?

0 Elements	1 Element	2 Elements	3 Elements
$\varnothing$	$\{a\}, \{b\}, \{c\}$	$\{a, b\}, \{b, c\}, \{a, c\}$	$\{a, b, c\}$

Now Work PROBLEM 9

2 Count the Number of Elements in a Set

As you count the number of students in a classroom or the number of pennies in your pocket, what you are really doing is matching, on a one-to-one basis, each object to be counted with the set of counting numbers, $1, 2, 3, \ldots, n$, for some number n. If a set A matched up in this fashion with the set $\{1, 2, \ldots, 25\}$, you would conclude that there are 25 elements in the set A. The notation $n(A) = 25$ is used to indicate that there are 25 elements in the set A.

Because the empty set has no elements, we write

$$n(\varnothing) = 0$$

If the number of elements in a set is a nonnegative integer, the set is **finite**. Otherwise, it is **infinite**. We shall concern ourselves only with finite sets.

In Words
The notation $n(A)$ means "the number of elements in set A."

Look again at Example 1. A set with 3 elements has $2^3 = 8$ subsets. This result can be generalized.

> If A is a set with n elements, then A has 2^n subsets.

For example, the set $\{a, b, c, d, e\}$ has $2^5 = 32$ subsets.

EXAMPLE 2

Analyzing Survey Data

In a survey of 100 college students, 35 were registered in College Algebra, 52 were registered in Computer Science I, and 18 were registered in both courses.

(a) How many students were registered in College Algebra or Computer Science I?

(b) How many were registered in neither course?

Solution

(a) First, let A = set of students in College Algebra

B = set of students in Computer Science I

Then the given information tells us that

$$n(A) = 35 \qquad n(B) = 52 \qquad n(A \cap B) = 18$$

Refer to Figure 1. Since $n(A \cap B) = 18$, the common part of the circles representing set A and set B has 18 elements. In addition, the remaining portion of the circle representing set A will have $35 - 18 = 17$ elements. Similarly, the remaining portion of the circle representing set B has $52 - 18 = 34$ elements. This means that $17 + 18 + 34 = 69$ students were registered in College Algebra or Computer Science I.

(b) Since 100 students were surveyed, it follows that $100 - 69 = 31$ were registered in neither course.

Figure 1

Universal set

A   B
17 18 34

31

 Now Work PROBLEMS 17 AND 27

The solution to Example 2 contains the basis for a general counting formula. If we count the elements in each of two sets A and B, we necessarily count twice any elements that are in both A and B—that is, those elements in $A \cap B$. To count correctly the elements that are in A or B—that is, to find $n(A \cup B)$—subtract those in $A \cap B$ from $n(A) + n(B)$.

> **THEOREM Counting Formula**
>
> If A and B are finite sets,
>
> $$n(A \cup B) = n(A) + n(B) - n(A \cap B) \tag{1}$$

Refer to Example 2. Using formula (1), we have

$$\begin{aligned} n(A \cup B) &= n(A) + n(B) - n(A \cap B) \\ &= 35 + 52 - 18 \\ &= 69 \end{aligned}$$

There are 69 students registered in College Algebra or Computer Science I.

A special case of the counting formula (1) occurs if A and B have no elements in common. In this case, $A \cap B = \varnothing$, so $n(A \cap B) = 0$.

> **THEOREM Addition Principle of Counting**
>
> If two sets A and B have no elements in common, that is,
>
> $$\text{if } A \cap B = \varnothing, \text{ then } n(A \cup B) = n(A) + n(B) \tag{2}$$

Formula (2) can be generalized.

> **THEOREM General Addition Principle of Counting**
>
> If, for n sets $A_1, A_2, \ldots, A_n$, no two have elements in common, then
>
> $$n(A_1 \cup A_2 \cup \cdots \cup A_n) = n(A_1) + n(A_2) + \cdots + n(A_n) \qquad \textbf{(3)}$$

EXAMPLE 3

Counting

Table 1 lists the level of education for all United States residents 25 years of age or older in 2017.

Table 1

Level of Education	Number of U.S. Residents at Least 25 Years Old
Not a high school graduate	22,540,000
High school graduate	62,512,000
Some college, but no degree	35,455,000
Associate's degree	22,310,000
Bachelor's degree	46,262,000
Advanced degree	27,841,000

Source: U.S. Census Bureau

(a) How many U.S. residents 25 years of age or older had an associate's degree or a bachelor's degree?

(b) How many U.S. residents 25 years of age or older had an associate's degree, a bachelor's degree, or an advanced degree?

Solution

Let A represent the set of associate's degree holders, B represent the set of bachelor's degree holders, and C represent the set of advanced degree holders. No two of the sets A, B, and C have elements in common (although the holder of an advanced degree certainly also holds a bachelor's degree, the individual would be part of the set for which the highest degree has been conferred). Then

$$n(A) = 22{,}310{,}000 \quad n(B) = 46{,}262{,}000 \quad n(C) = 27{,}841{,}000$$

(a) Using formula (2),

$$n(A \cup B) = n(A) + n(B) = 22{,}310{,}000 + 46{,}262{,}000 = 68{,}572{,}000$$

There were 68,572,000 U.S. residents 25 years of age or older who had an associate's degree or a bachelor's degree.

(b) Using formula (3),

$$n(A \cup B \cup C) = n(A) + n(B) + n(C)$$
$$= 22{,}310{,}000 + 46{,}262{,}000 + 27{,}841{,}000$$
$$= 96{,}413{,}000$$

There were 96,413,000 U.S. residents 25 years of age or older who had an associate's degree, a bachelor's degree, or an advanced degree.

Now Work PROBLEM 31

3 Solve Counting Problems Using the Multiplication Principle

EXAMPLE 4

Counting the Number of Possible Meals

The fixed-price dinner at Mabenka Restaurant provides the following choices:

Appetizer: soup or salad
Entrée: baked chicken, broiled beef patty, beef liver, or roast beef au jus
Dessert: ice cream or cheesecake

How many different meals can be ordered?

Solution Ordering such a meal requires three separate decisions:

Choose an Appetizer   **Choose an Entrée**   **Choose a Dessert**

2 choices   4 choices   2 choices

Look at the **tree diagram** in Figure 2. Note that for each choice of appetizer, there are 4 choices of entrées. And for each of these $2 \cdot 4 = 8$ choices, there are 2 choices for dessert. A total of

$$2 \cdot 4 \cdot 2 = 16$$

different meals can be ordered.

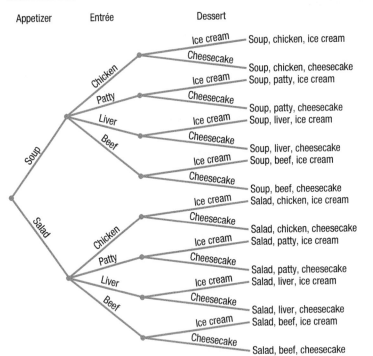

Figure 2

Example 4 demonstrates a general principle of counting.

THEOREM Multiplication Principle of Counting

If a task consists of a sequence of choices in which there are p selections for the first choice, q selections for the second choice, r selections for the third choice, and so on, the task of making these selections can be done in

$$p \cdot q \cdot r \cdot \ldots$$

different ways.

EXAMPLE 5   **Forming Codes**

How many two-symbol code words can be formed if the first symbol is an uppercase letter and the second symbol is a digit?

Solution It sometimes helps to begin by listing some of the possibilities. The code consists of an uppercase letter followed by a digit, so some possibilities are A1, A2, B3, X0, and so on. The task consists of making two selections: The first selection requires choosing an uppercase letter (26 choices), and the second task requires choosing a digit (10 choices). By the Multiplication Principle, there are

$$26 \cdot 10 = 260$$

different code words of the type described.

━━━━━**Now Work** PROBLEM 23

10.1 Assess Your Understanding

'Are You Prepared? *Answers are given at the end of these exercises. If you get a wrong answer, read the pages listed in* red.

1. The _____ of *A* and *B* consists of all elements in either *A* or *B* or both. (pp. 2–3)

2. The _____ of *A* with *B* consists of all elements in both *A* and *B*. (pp. 2–3)

3. *True or False* The intersection of two sets is always a subset of their union. (pp. 2–3)

4. *True or False* If *A* is a set, the complement of *A* is the set of all the elements in the universal set that are not in *A*. (pp. 2–3)

Concepts and Vocabulary

5. If each element of a set *A* is also an element of a set *B*, we say that *A* is a _____ of *B* and write *A* _____ *B*.

6. If the number of elements in a set is a nonnegative integer, we say that the set is _____.

7. *True or False* If a task consists of a sequence of three choices in which there are *p* selections for the first choice, *q* selections for the second choice, and *r* selections for the third choice, then the task of making these selections can be done in $p \cdot q \cdot r$ different ways.

8. *Multiple Choice* The Counting Formula states that if *A* and *B* are finite sets, then $n(A \cup B) = $ _____.

(a) $n(A) + n(B)$ **(b)** $n(A) + n(B) - n(A \cap B)$

(c) $n(A) \cdot n(B)$ **(d)** $n(A) - n(B)$

Skill Building

9. Write down all the subsets of $\{a, b, c, d\}$.

10. Write down all the subsets of $\{a, b, c, d, e\}$.

11. If $n(A) = 15, n(B) = 20$, and $n(A \cap B) = 10$, find $n(A \cup B)$.

12. If $n(A) = 30, n(B) = 40$, and $n(A \cup B) = 45$, find $n(A \cap B)$.

13. If $n(A \cup B) = 50, n(A \cap B) = 10$, and $n(B) = 20$, find $n(A)$.

14. If $n(A \cup B) = 60, n(A \cap B) = 40$, and $n(A) = n(B)$, find $n(A)$.

In Problems 15–22, use the information given in the figure.

15. How many are in set *A*?

16. How many are in set *B*?

17. How many are in *A* or *B*?

18. How many are in *A* and *B*?

19. How many are in *A* but not *C*?

20. How many are not in *A*?

21. How many are in *A* and *B* and *C*?

22. How many are in *A* or *B* or *C*?

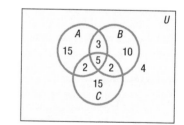

Applications and Extensions

23. Shirts and Ties A man has 5 shirts and 3 ties. How many different shirt-and-tie arrangements can he wear?

24. Blouses and Skirts A woman has 5 blouses and 8 skirts. How many different outfits can she wear?

25. Four-digit Numbers How many four-digit numbers can be formed using the digits 0, 1, 2, 3, 4, 5, 6, 7, 8, and 9 if the first digit cannot be 0? Repeated digits are allowed.

26. Five-digit Numbers How many five-digit numbers can be formed using the digits 0, 1, 2, 3, 4, 5, 6, 7, 8, and 9 if the first digit cannot be 0 or 1? Repeated digits are allowed.

27. Analyzing Survey Data In a consumer survey of 500 people, 200 indicated that they would be buying a major appliance within the next month, 150 indicated that they would buy a car, and 25 said that they would purchase both a major appliance and a car. How many will purchase neither? How many will purchase only a car?

28. Analyzing Survey Data In a student survey, 200 indicated that they would attend Summer Session I, and 150 indicated Summer Session II. If 75 students plan to attend both summer sessions, and 275 indicated that they would attend neither session, how many students participated in the survey?

29. Analyzing Survey Data In a survey of 100 investors in the stock market,

    50 owned shares in IBM
    40 owned shares in AT&T
    45 owned shares in GE
    20 owned shares in both IBM and GE
    15 owned shares in both AT&T and GE
    20 owned shares in both IBM and AT&T
    5 owned shares in all three

  (a) How many of the investors surveyed did not have shares in any of the three companies?

  (b) How many owned just IBM shares?

  (c) How many owned just GE shares?

  (d) How many owned neither IBM nor GE?

  (e) How many owned either IBM or AT&T but no GE?

30. Classifying Blood Types Human blood is classified as either Rh+ or Rh−. Blood is also classified by type: A, if it contains an A antigen but not a B antigen; B, if it contains a B antigen but not an A antigen; AB, if it contains both A and B antigens; and O, if it contains neither antigen. Draw a Venn diagram illustrating the various blood types. Based on this classification, how many different kinds of blood are there?

31. Demographics The following data represent the marital status of males 18 years old and older in the U.S. in 2017.

Marital Status	Number (in millions)
Married	65.3
Widowed	3.3
Divorced	10.9
Separated	2.2
Never married	37.3

Source: Current Population Survey

(a) Determine the number of males 18 years old and older who are widowed or divorced.

(b) Determine the number of males 18 years old and older who are married, divorced, or separated.

32. Demographics The following data represent the marital status of females 18 years old and older in the U.S. in 2017.

Marital Status	Number (in millions)
Married	65.1
Widowed	11.6
Divorced	14.6
Separated	2.8
Never married	32.8

Source: Current Population Survey

(a) Determine the number of females 18 years old and older who are divorced or separated.

(b) Determine the number of females 18 years old and older who are married, widowed, or divorced.

33. Stock Portfolios As a financial planner, you are asked to select one stock each from the following groups: 8 Dow Jones stocks, 15 NASDAQ stocks, and 4 global stocks. How many different portfolios are possible?

Explaining Concepts: Discussion and Writing

34. Make up a problem different from any found in the text that requires the addition principle of counting to solve. Give it to a friend to solve and critique.

35. Investigate the notion of counting as it relates to infinite sets. Write an essay on your findings.

Retain Your Knowledge

Problems 36–44 are based on material learned earlier in the course. The purpose of these problems is to keep the material fresh in your mind so that you are better prepared for the final exam.

36. Graph $(x - 2)^2 + (y + 1)^2 = 9$.

37. Find all the real zeros of the function:
$$f(x) = (x - 2)(x^2 - 3x - 10)$$

38. Solve: $\log_3 x + \log_3 2 = -2$

39. Solve: $x^3 = 72x$

40. Solve the system: $\begin{cases} x - y = 5 \\ x - y^2 = -1 \end{cases}$

41. Multiply: $(2x - 7)(3x^2 - 5x + 4)$

42. Determine whether the infinite series converges or diverges. If it converges, find the sum.
$$4 + \frac{12}{5} + \frac{36}{25} + \frac{108}{125} + \cdots$$

43. If $f''(x) = \frac{2}{3}(x - 2)^{-1/3} + \frac{1}{3}(x - 2)^{-2/3}$, find where

(a) $f''(x) = 0$ (b) $f''(x)$ is undefined

44. Find the partial fraction decomposition: $\dfrac{3x^2 + 15x + 5}{x^3 + 2x^2 + x}$

'Are You Prepared?' Answers

1. union **2.** intersection **3.** True **4.** True

10.2 Permutations and Combinations

PREPARING FOR THIS SECTION *Before getting started, review the following:*

- Factorial (Section 9.1, pp. 654–655)
- Binomial Coefficient (Section 9.5, pp. 685–687)

Now Work the 'Are You Prepared?' problems on page 709.

OBJECTIVES 1 Solve Counting Problems Using Permutations Involving n Distinct Objects (p. 703)

2 Solve Counting Problems Using Combinations (p. 705)

3 Solve Counting Problems Using Permutations Involving n Nondistinct Objects (p. 708)

1 Solve Counting Problems Using Permutations Involving *n* Distinct Objects

> **DEFINITION** Permutation
>
> A **permutation** is an ordered arrangement of *r* objects chosen from *n* objects.

Three types of permutations are discussed:

- The *n* objects are distinct (different), and repetition is allowed in the selection of *r* of them. [Distinct, with repetition]
- The *n* objects are distinct (different), and repetition is not allowed in the selection of *r* of them, where $r \leq n$. [Distinct, without repetition]
- The *n* objects are not distinct, and all of them are used in the arrangement. [Not distinct]

We take up the first two types here and deal with the third type at the end of this section.

The first type of permutation (*n* distinct objects, repetition allowed) is handled using the Multiplication Principle.

EXAMPLE 1 Counting Airport Codes [Permutation: Distinct, with Repetition]

The International Airline Transportation Association (IATA) assigns three-letter codes to represent airport locations. For example, the airport code for Ft. Lauderdale, Florida, is FLL. Notice that repetition is allowed in forming this code. How many airport codes are possible?

Solution An airport code is formed by choosing 3 letters from 26 letters and arranging them in order. In the ordered arrangement, a letter may be repeated. This is an example of a permutation with repetition in which 3 objects are chosen from 26 distinct objects.

The task of counting the number of such arrangements consists of making three selections. Each selection requires choosing a letter of the alphabet (26 choices). By the Multiplication Principle, there are

$$26 \cdot 26 \cdot 26 = 26^3 = 17{,}576$$

possible airport codes.

The solution given to Example 1 can be generalized.

> **THEOREM** Permutations: Distinct Objects with Repetition
>
> The number of ordered arrangements of *r* objects chosen from *n* objects, in which the *n* objects are distinct and repetition is allowed, is n^r.

───── **Now Work** PROBLEM 33

Now let's consider permutations in which the objects are distinct and repetition is not allowed.

EXAMPLE 2 Forming Codes [Permutation: Distinct, without Repetition]

Suppose that a three-letter code is to be formed using any of the 26 uppercase letters of the alphabet, but no letter is to be used more than once. How many different three-letter codes are there?

Solution Some of the possibilities are ABC, ABD, ABZ, ACB, CBA, and so on. The task consists of making three selections. The first selection requires choosing from 26 letters. Since no letter can be used more than once, the second selection requires choosing from 25 letters. The third selection requires choosing from 24 letters. (Do you see why?) By the Multiplication Principle, there are

$$26 \cdot 25 \cdot 24 = 15,600$$

different three-letter codes with no letter repeated.

For the second type of permutation, we introduce the following notation.

> The notation **P(n, r)** represents the number of ordered arrangements of r objects chosen from n distinct objects, where $r \leq n$ and repetition is not allowed.

For example, the question posed in Example 2 asks for the number of ways in which the 26 letters of the alphabet can be arranged, in order, using three nonrepeated letters. The answer is

$$P(26, 3) = 26 \cdot 25 \cdot 24 = 15,600$$

EXAMPLE 3 **Lining People Up**

In how many ways can 5 people be lined up?

Solution The 5 people are distinct. Once a person is in line, that person will not be repeated elsewhere in the line; and, in lining people up, order is important. This is a permutation of 5 objects taken 5 at a time, so 5 people can be lined up in

$$P(5, 5) = \underbrace{5 \cdot 4 \cdot 3 \cdot 2 \cdot 1}_{\textbf{5 factors}} = 120 \text{ ways}$$

-**Now Work** PROBLEM 35

To arrive at a formula for $P(n, r)$, note that the task of obtaining an ordered arrangement of n objects in which only $r \leq n$ of them are used, without repeating any of them, requires making r selections. For the first selection, there are n choices; for the second selection, there are $n - 1$ choices; for the third selection, there are $n - 2$ choices; . . . ; for the rth selection, there are $n - (r - 1)$ choices. By the Multiplication Principle, this means

$$
\begin{array}{cccc}
\text{1st} & \text{2nd} & \text{3rd} & \text{rth} \\
P(n, r) = n \cdot (n - 1) \cdot (n - 2) \cdot & \cdots & \cdot [n - (r - 1)]
\end{array}
$$
$$= n \cdot (n - 1) \cdot (n - 2) \cdot \cdots \cdot (n - r + 1)$$

RECALL
$0! = 1,\ 1! = 1,\ 2! = 2 \cdot 1, \ldots ,$
$n! = n(n-1) \cdot \cdots \cdot 3 \cdot 2 \cdot 1$ ∎

This formula for $P(n, r)$ can be compactly written using factorial notation.

$$P(n, r) = n \cdot (n - 1) \cdot (n - 2) \cdot \cdots \cdot (n - r + 1)$$
$$= n \cdot (n - 1) \cdot (n - 2) \cdot \cdots \cdot (n - r + 1) \cdot \frac{(n - r) \cdot \cdots \cdot 3 \cdot 2 \cdot 1}{(n - r) \cdot \cdots \cdot 3 \cdot 2 \cdot 1} = \frac{n!}{(n - r)!}$$

> **THEOREM** **Permutations of *r* Objects Chosen from *n* Distinct Objects without Repetition**
>
> The number of arrangements of *n* objects using $r \leq n$ of them, in which
>
> - the *n* objects are distinct
> - repetition of objects is not allowed
> - order is important
>
> is given by the formula
>
> $$P(n, r) = \frac{n!}{(n-r)!} \tag{1}$$

EXAMPLE 4   **Computing Permutations**

Evaluate: (a) $P(7, 3)$ (b) $P(6, 1)$ (c) $P(52, 5)$

Solution   Parts (a) and (b) are each worked two ways.

(a) $P(7, 3) = \underbrace{7 \cdot 6 \cdot 5}_{\text{3 factors}} = 210$

or

$$P(7, 3) = \frac{7!}{(7-3)!} = \frac{7!}{4!} = \frac{7 \cdot 6 \cdot 5 \cdot \cancel{4!}}{\cancel{4!}} = 210$$

(b) $P(6, 1) = \underbrace{6}_{\text{1 factor}} = 6$

or

$$P(6, 1) = \frac{6!}{(6-1)!} = \frac{6!}{5!} = \frac{6 \cdot \cancel{5!}}{\cancel{5!}} = 6$$

(c) Figure 3 shows the solution using a TI-84 Plus C graphing calculator. So

$$P(52, 5) = 311{,}875{,}200$$

Figure 3 $P(52, 5)$

—**Now Work** PROBLEM 7

EXAMPLE 5   **The Birthday Problem**

All we know about Shannon, Patrick, and Ryan is that they have different birthdays. If all the possible ways this could occur were listed, how many would there be? Assume that there are 365 days in a year.

Solution   This is an example of a permutation in which 3 birthdays are selected from a possible 365 days, and no birthday may repeat itself. The number of ways this can occur is

$$P(365, 3) = \frac{365!}{(365-3)!} = \frac{365 \cdot 364 \cdot 363 \cdot \cancel{362!}}{\cancel{362!}} = 365 \cdot 364 \cdot 363 = 48{,}228{,}180$$

There are 48,228,180 ways in which three people can all have different birthdays.

—**Now Work** PROBLEM 47

2 Solve Counting Problems Using Combinations

In a permutation, order is important. For example, the arrangements *ABC*, *CAB*, *BAC*, ... are considered different arrangements of the letters *A*, *B*, and *C*. In many situations, though, order is unimportant. For example, in the card game of poker, the order in which the cards are received does not matter; it is the *combination* of the cards that matters.

DEFINITION Combination

A **combination** is an arrangement, without regard to order, of r objects selected from n distinct objects without repetition, where $r \leq n$. The notation $C(n, r)$ represents the number of combinations of n distinct objects taken r at a time.

EXAMPLE 6 **Listing Combinations**

List all the combinations of the 4 objects a, b, c, d taken 2 at a time. What is $C(4, 2)$?

Solution One combination of a, b, c, d taken 2 at a time is

$$ab$$

Exclude ba from the list because order is not important in a combination (this means that we do not distinguish ab from ba). The list of all combinations of a, b, c, d taken 2 at a time is

$$ab, \quad ac, \quad ad, \quad bc, \quad bd, \quad cd$$

so

$$C(4, 2) = 6$$

A formula for $C(n, r)$ can be found by noting that the only difference between a permutation of r objects chosen from n distinct objects without repetition and a combination is that order is disregarded in combinations. To determine $C(n, r)$, eliminate from the formula for $P(n, r)$ the number of permutations that are simply rearrangements of a given set of r objects. This can be determined from the formula for $P(n, r)$ by calculating $P(r, r) = r!$. So, dividing $P(n, r)$ by $r!$ gives the desired formula for $C(n, r)$:

$$C(n, r) = \frac{P(n, r)}{r!} = \frac{\dfrac{n!}{(n - r)!}}{r!} = \frac{n!}{(n - r)! \, r!}$$

Use formula (1).

We have proved the following result:

THEOREM Number of Combinations of n Distinct Objects Taken r at a Time

The number of ways of selecting r objects from n distinct objects, $r \leq n$, in which

- repetition of objects is not allowed
- order is not important

is given by the formula

$$C(n, r) = \frac{n!}{(n - r)! \, r!} \tag{2}$$

Based on formula (2), we discover that the symbol $C(n, r)$ and the symbol $\dbinom{n}{r}$ for the binomial coefficients are, in fact, the same. The Pascal triangle (see Sections 9.5) can be used to find the value of $C(n, r)$. However, because it is more practical and convenient, we will use formula (2) instead.

EXAMPLE 7

Using Formula (2)

Use formula (2) to find the value of each combination.

(a) $C(3,1)$ (b) $C(6,3)$ (c) $C(n,n)$ (d) $C(n,0)$ (e) $C(52,5)$

Solution (a) $C(3,1) = \dfrac{3!}{(3-1)!1!} = \dfrac{3!}{2!1!} = \dfrac{3 \cdot \cancel{2} \cdot 1}{\cancel{2} \cdot 1 \cdot 1} = 3$

(b) $C(6,3) = \dfrac{6!}{(6-3)!3!} = \dfrac{6 \cdot 5 \cdot 4 \cdot \cancel{3!}}{3! \cancel{3!}} = \dfrac{6 \cdot 5 \cdot 4}{\cancel{6}} = 20$

(c) $C(n,n) = \dfrac{n!}{(n-n)!n!} = \dfrac{\cancel{n!}}{0!\cancel{n!}} = \dfrac{1}{1} = 1$

(d) $C(n,0) = \dfrac{n!}{(n-0)!0!} = \dfrac{\cancel{n!}}{\cancel{n!}0!} = \dfrac{1}{1} = 1$

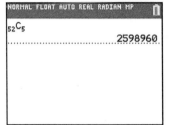

(e) Figure 4 shows the solution using a TI-84 Plus C graphing calculator.

$$C(52,5) = 2{,}598{,}960$$

Figure 4 $C(52,5)$

Now Work PROBLEM 15

EXAMPLE 8

Forming Committees

How many different committees of 3 people can be formed from a group of 7 people?

Solution The 7 people are distinct. More important, though, is the observation that the order of being selected for a committee is not significant. The problem asks for the number of combinations of 7 objects taken 3 at a time.

$$C(7,3) = \dfrac{7!}{4!3!} = \dfrac{7 \cdot 6 \cdot 5 \cdot \cancel{4!}}{\cancel{4!}3!} = \dfrac{7 \cdot \cancel{6} \cdot 5}{\cancel{6}} = 35$$

Thirty-five different committees can be formed.

EXAMPLE 9

Forming Committees

In how many ways can a committee consisting of 2 faculty members and 3 students be formed if 6 faculty members and 10 students are eligible to serve on the committee?

Solution The problem can be separated into two parts: the number of ways in which the faculty members can be chosen, $C(6,2)$, and the number of ways in which the student members can be chosen, $C(10,3)$. By the Multiplication Principle, the committee can be formed in

$$C(6,2) \cdot C(10,3) = \dfrac{6!}{4!2!} \cdot \dfrac{10!}{7!3!} = \dfrac{6 \cdot 5 \cdot \cancel{4!}}{\cancel{4!}2!} \cdot \dfrac{10 \cdot 9 \cdot 8 \cdot \cancel{7!}}{\cancel{7!}3!}$$

$$= \dfrac{30}{2} \cdot \dfrac{720}{6} = 1800 \text{ ways}$$

Now Work PROBLEM 49

3 Solve Counting Problems Using Permutations Involving *n* Nondistinct Objects

EXAMPLE 10 **Forming Different Words**

How many different words (meaningful or not) can be formed using all the letters in the word REARRANGE?

Solution Each word formed will have 9 letters: 3 R's, 2 A's, 2 E's, 1 N, and 1 G. To construct each word, we need to fill in 9 positions with the 9 letters:

$$\overline{1}\ \overline{2}\ \overline{3}\ \overline{4}\ \overline{5}\ \overline{6}\ \overline{7}\ \overline{8}\ \overline{9}$$

The process of forming a word consists of five tasks.

Task 1: Choose the positions for the 3 R's.

Task 2: Choose the positions for the 2 A's.

Task 3: Choose the positions for the 2 E's.

Task 4: Choose the position for the 1 N.

Task 5: Choose the position for the 1 G.

Task 1 can be done in $C(9,3)$ ways. There then remain 6 positions to be filled, so Task 2 can be done in $C(6,2)$ ways. There remain 4 positions to be filled, so Task 3 can be done in $C(4,2)$ ways. There remain 2 positions to be filled, so Task 4 can be done in $C(2,1)$ ways. The last position can be filled in $C(1,1)$ way. Using the Multiplication Principle, the number of possible words that can be formed is

$$C(9,3) \cdot C(6,2) \cdot C(4,2) \cdot C(2,1) \cdot C(1,1) = \frac{9!}{3! \cdot \cancel{6!}} \cdot \frac{\cancel{6!}}{2! \cdot \cancel{4!}} \cdot \frac{\cancel{4!}}{2! \cdot \cancel{2!}} \cdot \frac{\cancel{2!}}{1! \cdot \cancel{1!}} \cdot \frac{\cancel{1!}}{0! \cdot 1!}$$

$$= \frac{9!}{3! \cdot 2! \cdot 2! \cdot 1! \cdot 1!} = 15,120$$

15,120 possible words can be formed.

The form of the expression before the answer to Example 10 is suggestive of a general result. Had all the letters in REARRANGE been different, there would have been $P(9,9) = 9!$ possible words formed. This is the numerator of the answer. The presence of 3 R's, 2 A's, and 2 E's reduces the number of different words, as the entries in the denominator illustrate. This leads to the following result:

> **THEOREM** **Permutations Involving *n* Objects That Are Not Distinct**
>
> The number of permutations of *n* objects of which n_1 are of one kind, n_2 are of a second kind, ..., and n_k are of a *k*th kind is given by
>
> $$\frac{n!}{n_1! \cdot n_2! \cdots \cdots n_k!} \tag{3}$$
>
> where $n = n_1 + n_2 + \cdots + n_k$.

EXAMPLE 11 **Arranging Flags**

How many different vertical arrangements are there of 8 flags if 4 are white, 3 are blue, and 1 is red?

Solution We seek the number of permutations of 8 objects, of which 4 are of one kind, 3 are of a second kind, and 1 is of a third kind. Using formula (3), we find that there are

$$\frac{8!}{4! \cdot 3! \cdot 1!} = \frac{8 \cdot 7 \cdot 6 \cdot 5 \cdot \cancel{4!}}{\cancel{4!} \cdot 3! \cdot 1!} = 280 \text{ different arrangements}$$

━━━━▶ **Now Work** PROBLEM 51

10.2 Assess Your Understanding

'Are You Prepared?' *Answers are given at the end of these exercises. If you get a wrong answer, read the pages listed in red.*

1. $0! =$ _____ ; $1! =$ _____ . (pp. 654–655)

2. *Multiple Choice* The binomial coefficient $\dbinom{6}{4}$ equals (pp. 685–687)

(a) $\dfrac{6!}{4!}$ (b) $\dfrac{6!}{4! \cdot 2!}$ (c) $\dfrac{6!}{2!}$ (d) $\dfrac{(6-4)!}{2!}$

Concepts and Vocabulary

3. A(n) _____ is an ordered arrangement of r objects chosen from n objects.

4. A(n) _____ is an arrangement of r objects chosen from n distinct objects, without repetition and without regard to order.

5. $P(n, r) =$ _____

6. $C(n, r) =$ _____

Skill Building

In Problems 7–14, find the value of each permutation.

7. $P(6, 2)$

8. $P(7, 2)$

9. $P(4, 4)$

10. $P(8, 8)$

11. $P(7, 0)$

12. $P(9, 0)$

13. $P(8, 4)$

14. $P(8, 3)$

In Problems 15–22, use formula (2) *to find the value of each combination.*

15. $C(8, 2)$

16. $C(8, 6)$

17. $C(7, 4)$

18. $C(6, 2)$

19. $C(15, 15)$

20. $C(18, 1)$

21. $C(26, 13)$

22. $C(18, 9)$

Applications and Extensions

23. List all the permutations of 5 objects $a, b, c, d,$ and e choosing 3 at a time without repetition. What is $P(5, 3)$?

24. List all the permutations of 5 objects $a, b, c, d,$ and e choosing 2 at a time without repetition. What is $P(5, 2)$?

25. List all the permutations of 4 objects $1, 2, 3,$ and 4 choosing 3 at a time without repetition. What is $P(4, 3)$?

26. List all the permutations of 6 objects 1, 2, 3, 4, 5, and 6 choosing 3 at a time without repetition. What is $P(6, 3)$?

27. List all the combinations of 5 objects $a, b, c, d,$ and e taken 3 at a time. What is $C(5, 3)$?

28. List all the combinations of 5 objects $a, b, c, d,$ and e taken 2 at a time. What is $C(5, 2)$?

29. List all the combinations of 4 objects $1, 2, 3,$ and 4 taken 3 at a time. What is $C(4, 3)$?

30. List all the combinations of 6 objects $1, 2, 3, 4, 5,$ and 6 taken 3 at a time. What is $C(6, 3)$?

31. Forming Codes How many two-letter codes can be formed using the letters A, B, C, and D? Repeated letters are allowed.

32. Forming Codes How many two-letter codes can be formed using the letters A, B, C, D, and E? Repeated letters are allowed.

33. Forming Numbers How many three-digit numbers can be formed using the digits 0 and 1? Repeated digits are allowed.

34. Forming Numbers How many three-digit numbers can be formed using the digits $0, 1, 2, 3, 4, 5, 6, 7, 8,$ and 9? Repeated digits are allowed.

35. Lining People Up In how many ways can 4 people be lined up?

36. Stacking Boxes In how many ways can 5 different boxes be stacked?

37. Forming Codes How many different three-letter codes are there if only the letters $A, B, C, D,$ and E can be used and no letter can be used more than once?

38. Forming Codes How many different four-letter codes are there if only the letters $A, B, C, D, E,$ and F can be used and no letter can be used more than once?

39. Stocks on the NYSE Companies whose stocks are listed on the New York Stock Exchange (NYSE) have their company name represented by 1, 2, or 3 letters (repetition of letters is allowed). What is the maximum number of companies that can be listed on the NYSE?

40. Stocks on the NASDAQ Companies whose stocks are listed on the NASDAQ stock exchange have their company name represented by either 4 or 5 letters (repetition of letters is allowed). What is the maximum number of companies that can be listed on the NASDAQ?

41. Establishing Committees In how many ways can a committee of 4 students be formed from a pool of 7 students?

42. Establishing Committees In how many ways can a committee of 3 professors be formed from a department that has 8 professors?

43. Possible Answers on a True/False Test How many arrangements of answers are possible for a true/false test with 10 questions?

44. Possible Answers on a Multiple-choice Test How many arrangements of answers are possible in a multiple-choice test with 5 questions, each of which has 4 possible answers?

45. Arranging Books Five different mathematics books are to be arranged on a student's desk. How many arrangements are possible?

46. Forming License Plate Numbers How many different license plate numbers can be made using 2 letters followed by 4 digits selected from the digits 0 through 9, if:
(a) Letters and digits may be repeated?
(b) Letters may be repeated, but digits may not be repeated?
(c) Neither letters nor digits may be repeated?

47. Birthday Problem In how many ways can 2 people each have different birthdays? Assume that there are 365 days in a year.

48. Birthday Problem In how many ways can 5 people all have different birthdays? Assume that there are 365 days in a year.

49. Forming a Committee A student dance committee is to be formed consisting of 2 boys and 3 girls. If the membership is to be chosen from 4 boys and 8 girls, how many different committees are possible?

50. Forming a Committee The student relations committee of a college consists of 2 administrators, 3 faculty members, and 5 students. Four administrators, 8 faculty members, and 20 students are eligible to serve. How many different committees are possible?

51. Forming Words How many different 9-letter words (meaningful or not) can be formed from the letters in the word ECONOMICS?

52. Forming Words How many different 11-letter words (meaningful or not) can be formed from the letters in the word MATHEMATICS?

53. Selecting Objects An urn contains 7 white balls and 3 red balls. Three balls are selected. In how many ways can the 3 balls be drawn from the total of 10 balls:
(a) If 2 balls are white and 1 is red?
(b) If all 3 balls are white?
(c) If all 3 balls are red?

54. Selecting Objects An urn contains 15 red balls and 10 white balls. Five balls are selected. In how many ways can the 5 balls be drawn from the total of 25 balls:
(a) If all 5 balls are red?
(b) If 3 balls are red and 2 are white?
(c) If at least 4 are red balls?

55. Senate Committees The U.S. Senate has 100 members. Suppose that it is desired to place each senator on exactly 1 of 7 possible committees. The first committee has 22 members, the second has 13, the third has 10, the fourth has 5, the fifth has 16, and the sixth and seventh have 17 apiece. In how many ways can these committees be formed?

56. Football Teams A defensive football squad consists of 25 players. Of these, 10 are linemen, 10 are linebackers, and 5 are safeties. How many different teams of 5 linemen, 3 linebackers, and 3 safeties can be formed?

57. Baseball In the American Baseball League, a designated hitter may be used. How many batting orders is it possible for a manager to use? (There are 9 regular players on a team.)

58. Baseball In the National Baseball League, the pitcher usually bats ninth. If this is the case, how many batting orders is it possible for a manager to use?

59. Baseball Teams A baseball team has 15 members. Four of the players are pitchers, and the remaining 11 members can play any position. How many different teams of 9 players can be formed?

60. World Series In the World Series the American League team (*A*) and the National League team (*N*) play until one team wins four games. If the sequence of winners is designated by letters (for example, *NAAAA* means that the National League team won the first game and the American League won the next four), how many different sequences are possible?

61. Basketball Teams A basketball team has 6 players who play guard (2 of 5 starting positions). How many different teams are possible, assuming that the remaining 3 positions are filled and it is not possible to distinguish a left guard from a right guard?

62. Basketball Teams On a basketball team of 12 players, 2 play only center, 3 play only guard, and the rest play forward (5 players on a team: 2 forwards, 2 guards, and 1 center). How many different teams are possible, assuming that it is not possible to distinguish a left guard from a right guard or a left forward from a right forward?

63. Combination Locks A combination lock displays 50 numbers. To open it, you turn clockwise to the first number of the "combination," then rotate counterclockwise to the second number, and then rotate clockwise to the third number.
(a) How many different lock combinations are there?
(b) Comment on the description of such a lock as a *combination* lock.

64. *Challenge Problem* Passwords Suppose a password must have at least 8 characters, but no more than 12 characters, made up of letters (without distinction for case) and digits. If the password must contain at least one letter and at least one digit, how many passwords are possible?

Explaining Concepts: Discussion and Writing

65. Create a problem different from any found in the text that requires a permutation to solve. Give it to a friend to solve and critique.

66. Create a problem different from any found in the text that requires a combination to solve. Give it to a friend to solve and critique.

67. Explain the difference between a permutation and a combination. Give an example to illustrate your explanation.

Retain Your Knowledge

Problems 68–75 are based on material learned earlier in the course. The purpose of these problems is to keep the material fresh in your mind so that you are better prepared for the final exam.

68. Solve: $\dfrac{5}{x-3} \geq 1$

69. If $f(x) = 2x - 1$ and $g(x) = x^2 + x - 2$, find $(g \circ f)(x)$.

70. Find the 5th term of the geometric sequence with first term $a_1 = 5$ and common ratio $r = -2$.

71. Use the binomial theorem to expand: $(x + 2y)^5$

72. Solve the system: $\begin{cases} 3x + 4y = 5 \\ 5x - 2y = 17 \end{cases}$

73. Multiply, if possible: $\begin{bmatrix} 4 & 2 & 0 \\ -1 & 3 & 1 \end{bmatrix} \begin{bmatrix} 0 & -2 \\ 3 & 1 \\ 5 & 0 \end{bmatrix}$

74. Find the partial fraction decomposition: $\dfrac{5x^2 + 3x + 14}{x^4 + 4x^2 + 4}$

75. Write $\dfrac{6x}{(x-3)^{2/5}} + 10(x-3)^{3/5}$ as a single quotient in which only positive exponents appear.

'Are You Prepared?' Answers

1. $1; 1$ 2. b

10.3 Probability

OBJECTIVES 1 Construct Probability Models (p. 711)
2 Compute Probabilities of Equally Likely Outcomes (p. 714)
3 Find Probabilities of the Union of Two Events (p. 715)
4 Use the Complement Rule to Find Probabilities (p. 716)

Probability is an area of mathematics that deals with experiments that yield random results, yet admit a certain regularity. Such experiments do not always produce the same result or outcome, so the result of any one observation is not predictable. However, the results of the experiment over a long period do produce regular patterns that enable us to make predictions with remarkable accuracy.

EXAMPLE 1   **Tossing a Fair Coin**

If a fair coin is tossed, the outcome is either a head or a tail. On any particular throw, we cannot predict what will happen, but if we toss the coin many times, we observe that the number of times that a head comes up is approximately equal to the number of times that a tail comes up. It seems reasonable, therefore, to assign a probability of $\dfrac{1}{2}$ that a head comes up and a probability of $\dfrac{1}{2}$ that a tail comes up.

1 Construct Probability Models

The discussion in Example 1 constitutes the construction of a **probability model** for the experiment of tossing a fair coin once. A probability model has two components: a sample space and an assignment of probabilities. A **sample space** S is a set whose elements represent all the possibilities that can occur as a result of the experiment. Each element of S is called an **outcome**. To each outcome a number is assigned, called the **probability** of that outcome, which has two properties:

- The probability assigned to each outcome is nonnegative.
- The sum of all the probabilities equals 1.

DEFINITION Probability Model

A **probability model** with the sample space

$$S = \{e_1, e_2, \ldots, e_n\}$$

where $e_1, e_2, \ldots, e_n$ are the possible outcomes and $P(e_1), P(e_2), \ldots, P(e_n)$ are the respective probabilities of these outcomes, requires that

$$P(e_1) \geq 0, P(e_2) \geq 0, \ldots, P(e_n) \geq 0 \tag{1}$$

$$\sum_{i=1}^{n} P(e_i) = P(e_1) + P(e_2) + \cdots + P(e_n) = 1 \tag{2}$$

EXAMPLE 2 **Determining Probability Models**

In a bag of M&Ms,^{TM} the candies are colored red, green, blue, brown, yellow, and orange. A candy is drawn from the bag and the color is recorded. The sample space of this experiment is {red, green, blue, brown, yellow, orange}. Determine which of the following are probability models.

(a)

Outcome	Probability
red	0.3
green	0.15
blue	0
brown	0.15
yellow	0.2
orange	0.2

(b)

Outcome	Probability
red	0.1
green	0.1
blue	0.1
brown	0.4
yellow	0.2
orange	0.3

(c)

Outcome	Probability
red	0.3
green	−0.3
blue	0.2
brown	0.4
yellow	0.2
orange	0.2

(d)

Outcome	Probability
red	0
green	0
blue	0
brown	0
yellow	1
orange	0

Solution

(a) This model is a probability model because all the outcomes have probabilities that are nonnegative, and the sum of the probabilities is 1.

(b) This model is not a probability model because the sum of the probabilities is not 1.

(c) This model is not a probability model because P(green) is less than 0. Remember that all probabilities must be nonnegative.

(d) This model is a probability model because all the outcomes have probabilities that are nonnegative, and the sum of the probabilities is 1. Notice that P(yellow) = 1, meaning that this outcome will occur with 100% certainty each time that the experiment is repeated. This means that the bag of M&Ms^{TM} contains only yellow candies.

Now Work PROBLEM 7

EXAMPLE 3	Constructing a Probability Model

An experiment consists of rolling a fair die once. A die is a cube with each face having 1, 2, 3, 4, 5, or 6 dots on it. See Figure 5. Construct a probability model for this experiment.

Solution A sample space S consists of all the possibilities that can occur. Because rolling the die will result in one of six faces showing, the sample space S consists of

$$S = \{1, 2, 3, 4, 5, 6\}$$

Because the die is fair, one face is no more likely to occur than another. As a result, our assignment of probabilities is

$$P(1) = \frac{1}{6} \qquad P(2) = \frac{1}{6}$$

$$P(3) = \frac{1}{6} \qquad P(4) = \frac{1}{6}$$

$$P(5) = \frac{1}{6} \qquad P(6) = \frac{1}{6}$$

Figure 5 A six-sided die

Now suppose that a die is loaded (weighted) so that the probability assignments are

$$P(1) = 0 \quad P(2) = 0 \quad P(3) = \frac{1}{3} \quad P(4) = \frac{2}{3} \quad P(5) = 0 \quad P(6) = 0$$

This assignment would be made if the die were loaded so that only a 3 or 4 could occur and the 4 was twice as likely as the 3 to occur. This assignment is consistent with the definition, since each assignment is nonnegative, and the sum of all the probability assignments equals 1.

✏️ **Now Work** PROBLEM 23

EXAMPLE 4	Constructing a Probability Model

An experiment consists of tossing a coin. The coin is weighted so that heads (H) is three times as likely to occur as tails (T). Construct a probability model for this experiment.

Solution The sample space S is $S = \{H, T\}$. If x denotes the probability that a tail occurs,

$$P(T) = x \quad \text{and} \quad P(H) = 3x$$

The sum of the probabilities of the possible outcomes must equal 1, so

$$P(T) + P(H) = x + 3x = 1$$

$$4x = 1$$

$$x = \frac{1}{4}$$

Assign the probabilities

$$P(T) = \frac{1}{4} \qquad P(H) = \frac{3}{4}$$

✏️ **Now Work** PROBLEM 27

In Words

$P(S) = 1$ means that one of the outcomes in the sample space must occur in an experiment.

In working with probability models, the term **event** is used to describe a set of possible outcomes of the experiment. An event E is some subset of the sample space S. The **probability of an event** E, $E \neq \varnothing$, denoted by $P(E)$, is defined as the sum of the probabilities of the outcomes in E. We can also think of the probability of an event E as the likelihood that the event E occurs. If $E = \varnothing$, then $P(E) = 0$; if $E = S$, then $P(E) = P(S) = 1$.

2 Compute Probabilities of Equally Likely Outcomes

When the same probability is assigned to each outcome of the sample space, the experiment is said to have **equally likely outcomes**.

> **THEOREM** **Probability for Equally Likely Outcomes**
>
> If an experiment has n equally likely outcomes, and if the number of ways in which an event E can occur is m, then the probability of E is
>
> $$P(E) = \frac{\text{Number of ways that } E \text{ can occur}}{\text{Number of possible outcomes}} = \frac{m}{n} \qquad (3)$$
>
> If S is the sample space of this experiment,
>
> $$P(E) = \frac{n(E)}{n(S)} \qquad (4)$$

EXAMPLE 5 **Calculating Probabilities of Events Involving Equally Likely Outcomes**

Calculate the probability that in a 3-child family there are 2 boys and 1 girl. Assume equally likely outcomes.

Solution Begin by constructing a tree diagram to help in listing the possible outcomes of the experiment. See Figure 6, where B stands for "boy" and G for "girl." The sample space of this experiment is

$$S = \{BBB, BBG, BGB, BGG, GBB, GBG, GGB, GGG\}$$

so $n(S) = 8$.

We wish to know the probability of the event E: "having two boys and one girl." From Figure 6, we conclude that $E = \{BBG, BGB, GBB\}$, so $n(E) = 3$. Since the outcomes are equally likely, the probability of E is

$$P(E) = \frac{n(E)}{n(S)} = \frac{3}{8}$$

1st child 2nd child 3rd child

B — BBB
B
  G — BBG
  B — BGB
  G
    G — BGG
B
G
  B — GBB
  B
    G — GBG
  G
    B — GGB
    G
      G — GGG

Figure 6

 Now Work PROBLEM 37

So far, we have calculated probabilities of single events. Now we compute probabilities of multiple events, which are called **compound probabilities**.

EXAMPLE 6 **Computing Compound Probabilities**

Consider the experiment of rolling a single fair die. Let E represent the event "roll an odd number," and let F represent the event "roll a 1 or 2."

(a) Write the event E and F. What is $n(E \cap F)$?
(b) Write the event E or F. What is $n(E \cup F)$?
(c) Compute $P(E)$. Compute $P(F)$.
(d) Compute $P(E \cap F)$.
(e) Compute $P(E \cup F)$.

Solution The sample space S of the experiment is $\{1, 2, 3, 4, 5, 6\}$, so $n(S) = 6$. Since the die is fair, the outcomes are equally likely. The event E: "roll an odd number" is $\{1, 3, 5\}$, and the event F: "roll a 1 or 2" is $\{1, 2\}$, so $n(E) = 3$ and $n(F) = 2$.

(a) In probability, the word *and* means the intersection of two events. The event E and F is

$$E \cap F = \{1, 3, 5\} \cap \{1, 2\} = \{1\} \qquad n(E \cap F) = 1$$

(b) In probability, the word *or* means the union of the two events. The event E or F is

$$E \cup F = \{1, 3, 5\} \cup \{1, 2\} = \{1, 2, 3, 5\} \qquad n(E \cup F) = 4$$

(c) Use formula (4). Then

$$P(E) = \frac{n(E)}{n(S)} = \frac{3}{6} = \frac{1}{2} \qquad P(F) = \frac{n(F)}{n(S)} = \frac{2}{6} = \frac{1}{3}$$

(d) $P(E \cap F) = \dfrac{n(E \cap F)}{n(S)} = \dfrac{1}{6}$

(e) $P(E \cup F) = \dfrac{n(E \cup F)}{n(S)} = \dfrac{4}{6} = \dfrac{2}{3}$

3 Find Probabilities of the Union of Two Events

The next formula can be used to find the probability of the union of two events.

> **THEOREM**
>
> For any two events E and F,
>
> $$P(E \cup F) = P(E) + P(F) - P(E \cap F) \qquad (5)$$

This result is a consequence of the Counting Formula discussed earlier, in Section 10.1.

For example, formula (5) can be used to find $P(E \cup F)$ in Example 6(e). Then

$$P(E \cup F) = P(E) + P(F) - P(E \cap F) = \frac{1}{2} + \frac{1}{3} - \frac{1}{6} = \frac{3}{6} + \frac{2}{6} - \frac{1}{6} = \frac{4}{6} = \frac{2}{3}$$

as before.

EXAMPLE 7

Computing Probabilities of the Union of Two Events

If $P(E) = 0.2$, $P(F) = 0.3$, and $P(E \cap F) = 0.1$, find the probability of E or F. That is, find $P(E \cup F)$.

Solution   Use formula (5).

$$\text{Probability of } E \text{ or } F = P(E \cup F) = P(E) + P(F) - P(E \cap F)$$
$$= 0.2 + 0.3 - 0.1 = 0.4$$

A Venn diagram can sometimes be used to obtain probabilities. To construct a Venn diagram representing the information in Example 7, draw two sets E and F. Begin with the fact that $P(E \cap F) = 0.1$. See Figure 7(a). Then, since $P(E) = 0.2$ and $P(F) = 0.3$, fill in E with $0.2 - 0.1 = 0.1$ and fill in F with $0.3 - 0.1 = 0.2$. See Figure 7(b). Since $P(S) = 1$, complete the diagram by inserting $1 - (0.1 + 0.1 + 0.2) = 0.6$ outside the circles. See Figure 7(c). Now it is easy to see, for example, that the probability of F but not E is 0.2. Also, the probability of neither E nor F is 0.6.

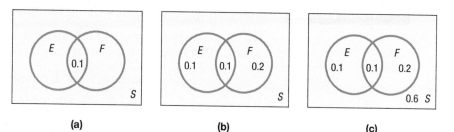

Figure 7   (a)   (b)   (c)

Now Work PROBLEM 45

If events E and F are disjoint so that $E \cap F = \varnothing$, we say they are **mutually exclusive**. In this case, $P(E \cap F) = 0$, and formula (5) takes the following form:

THEOREM **Mutually Exclusive Events**

If E and F are mutually exclusive events, that is, if $E \cap F = \varnothing$, then

$$P(E \cup F) = P(E) + P(F) \qquad \qquad \text{(6)}$$

EXAMPLE 8 **Computing Probabilities of the Union of Two Mutually Exclusive Events**

If $P(E) = 0.4$ and $P(F) = 0.25$, and E and F are mutually exclusive, find $P(E \cup F)$.

Solution Since E and F are mutually exclusive, use formula (6).

$$P(E \cup F) = P(E) + P(F) = 0.4 + 0.25 = 0.65$$

 Now Work PROBLEM 47

4 Use the Complement Rule to Find Probabilities

Recall that if A is a set, the complement of A, denoted $\overline{A}$, is the set of all elements in the universal set U that are not in A. We similarly define the complement of an event.

DEFINITION **Complement of an Event**

Let S denote the sample space of an experiment, and let E denote an event. The **complement of E**, denoted $\overline{E}$, is the set of all outcomes in the sample space S that are not outcomes in the event E.

The complement of an event E—that is, $\overline{E}$—in a sample space S has the following two properties:

$$E \cap \overline{E} = \varnothing \qquad E \cup \overline{E} = S$$

Since E and $\overline{E}$ are mutually exclusive, it follows from (6) that

$$P(E \cup \overline{E}) = P(S) = 1 \qquad P(E) + P(\overline{E}) = 1 \qquad P(\overline{E}) = 1 - P(E)$$

We have the following result:

THEOREM **Computing Probabilities of Complementary Events**

If E represents any event and $\overline{E}$ represents the complement of E, then

$$P(\overline{E}) = 1 - P(E) \qquad \qquad \text{(7)}$$

EXAMPLE 9 **Computing Probabilities Using Complements**

On the local news the weather reporter stated that the probability of rain tomorrow is 40%. What is the probability that it will not rain?

Solution The complement of the event "rain" is "no rain."

$$P(\text{no rain}) = 1 - P(\text{rain}) = 1 - 0.4 = 0.6$$

There is a 60% chance of no rain tomorrow.

 Now Work PROBLEM 51

EXAMPLE 10   **Birthday Problem**

What is the probability that in a group of 10 people, at least 2 people have the same birthday? Assume that there are 365 days in a year and that a person is as likely to be born on one day as another, so all the outcomes are equally likely.

Solution   First determine the number of outcomes in the sample space S. There are 365 possibilities for each person's birthday. Since there are 10 people in the group, there are 365^{10} possibilities for the birthdays. [For one person in the group, there are 365 days on which his or her birthday can fall; for two people, there are $(365)(365) = 365^2$ pairs of days; and, in general, using the Multiplication Principle, for n people there are 365^n possibilities.] So

$$n(S) = 365^{10}$$

We wish to find the probability of the event E: "at least two people have the same birthday." It is difficult to count the elements in this set; it is much easier to count the elements of the complementary event $\overline{E}$: "no two people have the same birthday."

Find $n(\overline{E})$ as follows: Choose one person at random. There are 365 possibilities for his or her birthday. Choose a second person. There are 364 possibilities for this birthday, if no two people are to have the same birthday. Choose a third person. There are 363 possibilities left for this birthday. Finally, arrive at the tenth person. There are 356 possibilities left for this birthday. By the Multiplication Principle, the total number of possibilities is

$$n(\overline{E}) = 365 \cdot 364 \cdot 363 \cdot \,\cdots\, \cdot 356$$

The probability of the event $\overline{E}$ is

$$P(\overline{E}) = \frac{n(\overline{E})}{n(S)} = \frac{365 \cdot 364 \cdot 363 \cdot \,\cdots\, \cdot 356}{365^{10}} \approx 0.883$$

The probability of two or more people in a group of 10 people having the same birthday is then

$$P(E) = 1 - P(\overline{E}) \approx 1 - 0.883 = 0.117$$

The birthday problem can be solved for any group size. The following table gives the probabilities for two or more people having the same birthday for various group sizes. Notice that the probability is greater than $\frac{1}{2}$ for any group of 23 or more people.

					Number of People											
	5	10	15	20	21	22	23	24	25	30	40	50	60	70	80	90
Probability That Two or More Have the Same Birthday	0.027	0.117	0.253	0.411	0.444	0.476	0.507	0.538	0.569	0.706	0.891	0.970	0.994	0.99916	0.99991	0.99999

- **Now Work** PROBLEM 71

Historical Feature

Blaise Pascal (1623–1662)

Set theory, counting, and probability first took form as a systematic theory in an exchange of letters (1654) between Pierre de Fermat (1601–1665) and Blaise Pascal (1623–1662). They discussed the problem of how to divide the stakes in a game that is interrupted before completion, knowing how many points each player needs to win. Fermat solved the problem by listing all possibilities and counting the favorable ones, whereas Pascal made use of the triangle that now bears his name. As mentioned in the text, the entries in Pascal's triangle are equivalent to $C(n, r)$. This recognition of the role of $C(n, r)$ in counting is the foundation of all further developments.

The first book on probability, the work of Christiaan Huygens (1629–1695), appeared in 1657. In it, the notion of mathematical expectation is explored. This allows the calculation of the profit or loss that a gambler might expect, knowing the probabilities involved in the game (see the Historical Problem that follows).   *(continued)*

Although Girolamo Cardano (1501–1576) wrote a treatise on probability, it was not published until 1663 in Cardano's collected works, and this was too late to have had any effect on the early development of the theory.

In 1713, the posthumously published *Ars Conjectandi* of Jakob Bernoulli (1654–1705) gave the theory the form it would have until 1900. Recently, both combinatorics (counting) and probability have undergone rapid development, thanks to the use of computers.

A final comment about notation. The notations $C(n, r)$ and $P(n, r)$ are variants of a form of notation developed in England after 1830.

The notation $\binom{n}{r}$ for $C(n, r)$ goes back to Leonhard Euler (1707–1783) but is now losing ground because it has no clearly related symbolism of the same type for permutations. The set symbols $\cup$ and $\cap$ were introduced by Giuseppe Peano (1858–1932) in 1888 in a slightly different context. The inclusion symbol $\subset$ was introduced by E. Schroeder (1841–1902) about 1890. We owe the treatment of set theory in the text to George Boole (1815–1864), who wrote $A + B$ for $A \cup B$ and AB for $A \cap B$ (statisticians still use AB for $A \cap B$).

Historical Problem

1. *The Problem Discussed by Fermat and Pascal* A game between two equally skilled players, A and B, is interrupted when A needs 2 points to win and B needs 3 points. In what proportion should the stakes be divided?

   (a) *Fermat's solution* List all possible outcomes that can occur as a result of four more plays. Comparing the probabilities for A to win and for B to win then determines how the stakes should be divided.

   (b) *Pascal's solution* Use combinations to determine the number of ways that the 2 points needed for A to win could occur in four plays. Then use combinations to determine the number of ways that the 3 points needed for B to win could occur. This is trickier than it looks, since A can win with 2 points in two plays, in three plays, or in four plays. Compute the probabilities, and compare them with the results in part (a).

10.3 Assess Your Understanding

Concepts and Vocabulary

1. When the same probability is assigned to each outcome of a sample space, the experiment is said to have _____ _____ outcomes.

2. The _____ of an event E is the set of all outcomes in the sample space S that are not outcomes in the event E.

3. *True or False* The probability of an event can never equal 0.

4. *True or False* In a probability model, the sum of all probabilities is 1.

Skill Building

5. In a probability model, which of the following numbers could be the probability of an outcome?

$$0 \quad 0.01 \quad 0.35 \quad -0.4 \quad 1 \quad 1.4$$

6. In a probability model, which of the following numbers could be the probability of an outcome?

$$1.5 \quad \frac{1}{2} \quad \frac{3}{4} \quad \frac{2}{3} \quad 0 \quad -\frac{1}{4}$$

7. Determine whether the following is a probability model.

Outcome	Probability
1	0.2
2	0.3
3	0.1
4	0.4

8. Determine whether the following is a probability model.

Outcome	Probability
Steve	0.4
Bob	0.3
Faye	0.1
Patricia	0.2

9. Determine whether the following is a probability model.

Outcome	Probability
Linda	0.3
Jean	0.2
Grant	0.1
Jim	0.3

10. Determine whether the following is a probability model.

Outcome	Probability
Erica	0.3
Joanne	0.2
Laura	0.1
Donna	0.5
Angela	-0.1

In Problems 11–16, (a) list the sample space S of each experiment and (b) construct a probability model for the experiment.

11. Tossing a fair coin twice

12. Tossing two fair coins once

13. Tossing two fair coins and then a fair die

14. Tossing a fair coin, a fair die, and then a fair coin

15. Tossing three fair coins once

16. Tossing one fair coin three times

In Problems 17–22, use the following spinners to construct a probability model for each experiment.

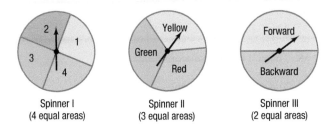

Spinner I
(4 equal areas)

Spinner II
(3 equal areas)

Spinner III
(2 equal areas)

17. Spin spinner I, then spinner II. What is the probability of getting a 2 or a 4, followed by Red?

18. Spin spinner III, then spinner II. What is the probability of getting Forward, followed by Yellow or Green?

19. Spin spinner I, then II, then III. What is the probability of getting a 1, followed by Red or Green, followed by Backward?

20. Spin spinner II, then I, then III. What is the probability of getting Yellow, followed by a 2 or a 4, followed by Forward?

21. Spin spinner I twice, then spinner II. What is the probability of getting a 2, followed by a 2 or a 4, followed by Red or Green?

22. Spin spinner III, then spinner I twice. What is the probability of getting Forward, followed by a 1 or a 3, followed by a 2 or a 4?

In Problems 23–26, consider the experiment of tossing a coin twice. The table lists six possible assignments of probabilities for this experiment. Using this table, answer the following questions.

	Sample Space			
Assignments	**HH**	**HT**	**TH**	**TT**
A	$\frac{1}{4}$	$\frac{1}{4}$	$\frac{1}{4}$	$\frac{1}{4}$
B	0	0	0	1
C	$\frac{3}{16}$	$\frac{5}{16}$	$\frac{5}{16}$	$\frac{3}{16}$
D	$\frac{1}{2}$	$\frac{1}{2}$	$-\frac{1}{2}$	$\frac{1}{2}$
E	$\frac{1}{4}$	$\frac{1}{4}$	$\frac{1}{4}$	$\frac{1}{8}$
F	$\frac{1}{9}$	$\frac{2}{9}$	$\frac{2}{9}$	$\frac{4}{9}$

23. Which of the assignments of probabilities is(are) consistent with the definition of a probability model?

24. Which of the assignments of probabilities should be used if the coin is known to be fair?

25. Which of the assignments of probabilities should be used if the coin is known to always come up tails?

26. Which of the assignments of probabilities should be used if tails is twice as likely as heads to occur?

27. Assigning Probabilities A coin is weighted so that heads is four times as likely as tails to occur. What probability should be assigned to heads? to tails?

28. Assigning Probabilities A coin is weighted so that tails is twice as likely as heads to occur. What probability should be assigned to heads? to tails?

29. Assigning Probabilities A die is weighted so that an odd-numbered face is twice as likely to occur as an even-numbered face. What probability should be assigned to each face?

30. Assigning Probabilities A die is weighted so that a six cannot appear. All the other faces occur with the same probability. What probability should be assigned to each face?

For Problems 31–34, the sample space is $S = \{1, 2, 3, 4, 5, 6, 7, 8, 9, 10\}$. Suppose that the outcomes are equally likely.

31. Compute the probability of the event $E = \{1, 2, 3\}$.

32. Compute the probability of the event $F = \{3, 5, 9, 10\}$.

33. Compute the probability of the event E: "an even number."

34. Compute the probability of the event F: "an odd number."

For Problems 35 and 36, an urn contains 5 white marbles, 10 green marbles, 8 yellow marbles, and 7 black marbles.

35. If one marble is selected, determine the probability that it is white.

36. If one marble is selected, determine the probability that it is black.

In Problems 37–40, assume equally likely outcomes.

37. Determine the probability of having 3 boys in a 3-child family.

38. Determine the probability of having 3 girls in a 3-child family.

39. Determine the probability of having 1 girl and 3 boys in a 4-child family.

40. Determine the probability of having 2 girls and 2 boys in a 4-child family.

For Problems 41–44, two fair dice are rolled.

41. Determine the probability that the sum of the faces is 7.

42. Determine the probability that the sum of the faces is 11.

43. Determine the probability that the sum of the faces is 3.

44. Determine the probability that the sum of the faces is 12.

In Problems 45–48, find the probability of the indicated event if $P(A) = 0.25$ and $P(B) = 0.45$.

45. $P(A \cup B)$ if $P(A \cap B) = 0.15$

46. $P(A \cap B)$ if $P(A \cup B) = 0.6$

47. $P(A \cup B)$ if A, B are mutually exclusive

48. $P(A \cap B)$ if A, B are mutually exclusive

49. If $P(A) = 0.60$, $P(A \cup B) = 0.85$, and $P(A \cap B) = 0.05$, find $P(B)$.

50. If $P(B) = 0.30$, $P(A \cup B) = 0.65$, and $P(A \cap B) = 0.15$, find $P(A)$.

51. **Automobile Theft** According to the Insurance Information Institute, in 2016 there was a 13.3% probability that an automobile theft in the United States would be cleared by arrests. If an automobile theft case from 2016 is randomly selected, what is the probability that it was not cleared by an arrest?

52. **Pet Ownership** According to the American Pet Products Manufacturers Association's *2017–2018 National Pet Owners Survey,* there is a 68% probability that a U.S. household owns a pet. If a U.S. household is randomly selected, what is the probability that it does not own a pet?

53. **Cat Ownership** According to the American Pet Products Manufacturers Association's *2017–2018 National Pet Owners Survey,* there is a 38% probability that a U.S. household owns a cat. If a U.S. household is randomly selected, what is the probability that it does not own a cat?

54. **Doctorate Degrees** According to the National Science Foundation, in 2016 there was a 17.2% probability that a doctoral degree awarded at a U.S. university was awarded in engineering. If a 2016 U.S. doctoral recipient is randomly selected, what is the probability that his or her degree was not in engineering?

55. **Gambling Behavior** According to a 2016 Gallup survey, 26% of U.S. adults visited a casino within the past year. If a U.S. adult is selected at random, what is the probability that he or she has not visited a casino within the past year?

56. **Girl Scout Cookies** According to the Girl Scouts of America, 19% of all Girl Scout cookies sold are Samoas/Caramel deLites. If a box of Girl Scout cookies is selected at random, what is the probability that it does not contain Samoas/Caramel deLites?

For Problems 57–60, a golf ball is selected at random from a container. If the container has 9 white balls, 8 green balls, and 3 orange balls, find the probability of each event.

57. The golf ball is white or green.

58. The golf ball is white or orange.

59. The golf ball is not white.

60. The golf ball is not green.

61. On *The Price Is Right*, there is a game in which a bag is filled with 3 strike chips and 5 numbers. Let's say that the numbers in the bag are 0, 1, 3, 6, and 9. What is the probability of selecting a strike chip or the number 1?

62. Another game on *The Price Is Right* requires the contestant to spin a wheel with the numbers 5, 10, 15, 20, . . . , 100. What is the probability that the contestant spins 100 or 30?

Problems 63–66 are based on a survey of annual incomes in 100 households. The following table gives the data.

Income	$0–24,999	$25,000–49,999	$50,000–74,999	$75,000–99,999	$100,000 or more
Number of households	22	23	17	12	26

63. What is the probability that a household has an annual income of $75,000 or more?

64. What is the probability that a household has an annual income between $25,000 and $74,999, inclusive?

65. What is the probability that a household has an annual income of less than $50,000?

66. What is the probability that a household has an annual income of $50,000 or more?

67. **Surveys** In a survey about the number of TV sets in a house, the following probability table was constructed:

Number of TV sets	0	1	2	3	4 or more
Probability	0.05	0.24	0.33	0.21	0.17

Find the probability of a house having:
(a) 1 or 2 TV sets
(b) 1 or more TV sets
(c) 3 or fewer TV sets
(d) 3 or more TV sets
(e) Fewer than 2 TV sets
(f) Fewer than 1 TV set
(g) 1, 2, or 3 TV sets
(h) 2 or more TV sets

68. **Checkout Lines** Through observation, it has been determined that the probability for a given number of people waiting in line at the "5 items or less" checkout register of a supermarket is as follows:

Number waiting in line	0	1	2	3	4 or more
Probability	0.10	0.15	0.20	0.24	0.31

Find the probability of:
(a) At most 2 people in line
(b) At least 2 people in line
(c) At least 1 person in line

69. In a certain College Algebra class, there are 18 freshmen and 15 sophomores. Of the 18 freshmen, 10 are male, and of the 15 sophomores, 8 are male. Find the probability that a randomly selected student is:

(a) A freshman or female
(b) A sophomore or male

70. The faculty of the mathematics department at Joliet Junior College is composed of 4 females and 9 males. Of the 4 females, 2 are under age 40, and of the 9 males, 3 are under age 40. Find the probability that a randomly selected faculty member is:
  (a) Female or under age 40
  (b) Male or over age 40

✎ **71. Birthday Problem** What is the probability that at least 2 people in a group of 12 people have the same birthday? Assume that there are 365 days in a year.

72. Birthday Problem What is the probability that at least 2 people in a group of 35 people have the same birthday? Assume that there are 365 days in a year.

73. Winning a Lottery Lotto America is a multistate lottery in which 5 red balls from a drum with 52 balls and 1 star ball from a drum with 10 balls are selected. For a $1 ticket, players get one chance at winning the grand prize by matching all 6 numbers. What is the probability of selecting the winning numbers on a $1 play?

74. *Challenge Problem* If 3 six-sided dice are tossed, find the probability that exactly 2 dice have the same reading.

Retain Your Knowledge

Problems 75–83 are based on material learned earlier in the course. The purpose of these problems is to keep the material fresh in your mind so that you are better prepared for the final exam.

75. To graph $g(x) = |x + 2| - 3$, shift the graph of $f(x) = |x|$ _____ units _____ and then _____ units _____ .
  (number) (left/right) (number) (up/down)

76. Solve: $\log_5 (x + 3) = 2$

77. Solve the given system using matrices.
$$\begin{cases} 3x + y + 2z = 1 \\ 2x - 2y + 5z = 5 \\ x + 3y + 2z = -9 \end{cases}$$

78. Evaluate: $\begin{vmatrix} 7 & -6 & 3 \\ -8 & 0 & 5 \\ 6 & -4 & 2 \end{vmatrix}$

79. Simplify: $\sqrt{108} - \sqrt{147} + \sqrt{363}$

80. José drives 60 miles per hour to his friend's house and 40 miles per hour on the way back. What is his average speed?

81. Find the 85th term of the sequence $5, 12, 19, 26, \ldots$.

✎ **82.** Find the area bounded by the graphs of
$$y = \frac{3}{5}x + \frac{12}{5}, \quad y = -x + 4, \quad \text{and} \quad y = -\sqrt{16 - x^2}.$$

✎ **83.** Find the partial fraction decomposition:
$$\frac{7x^2 - 5x + 30}{x^3 - 8}$$

Chapter Review

Things to Know

Counting formula (p. 698)	$n(A \cup B) = n(A) + n(B) - n(A \cap B)$
Addition Principle of Counting (p. 698)	If $A \cap B = \varnothing$, then $n(A \cup B) = n(A) + n(B)$.
Multiplication Principle of Counting (p. 700)	If a task consists of a sequence of choices in which there are p selections for the first choice, q selections for the second choice, and so on, the task of making these selections can be done in $p \cdot q \cdot \ \cdots \ $ different ways.
Permutation (p. 703)	An ordered arrangement of r objects chosen from n objects
Number of permutations: Distinct, with repetition (p. 703)	n^r
	The n objects are distinct (different), and repetition is allowed in the selection of r of them.
Number of permutations: Distinct, without repetition (p. 705)	$P(n, r) = n(n - 1) \cdot \ \cdots \ \cdot [n - (r - 1)] = \dfrac{n!}{(n - r)!}$
	The n objects are distinct (different), and repetition is not allowed in the selection of r of them, where $r \leq n$.
Combination (p. 706)	An arrangement, without regard to order, of r objects selected from n distinct objects, where $r \leq n$

Number of combinations (p. 706)	$C(n, r) = \dfrac{P(n, r)}{r!} = \dfrac{n!}{(n - r)! \, r!}$
Number of permutations: Not distinct (p. 708)	$\dfrac{n!}{n_1! n_2! \cdots n_k!}$
	The number of permutations of n objects of which n_1 are of one kind, n_2 are of a second kind, . . . , and n_k are of a kth kind, where $n = n_1 + n_2 + \cdots + n_k$
Sample space (pp. 711–712)	Set whose elements represent the possible outcomes that can occur as a result of an experiment
Probability (p. 712)	A nonnegative number assigned to each outcome of a sample space; the sum of all the probabilities of the outcomes equals 1.
Probability for equally likely outcomes (p. 714)	$P(E) = \dfrac{n(E)}{n(S)}$
	The same probability is assigned to each outcome.
Probability of the union of two events (p. 715)	$P(E \cup F) = P(E) + P(F) - P(E \cap F)$
Probability of the complement of an event (p. 716)	$P(\overline{E}) = 1 - P(E)$

Objectives

Section	You should be able to . . .	Example(s)	Review Exercises
10.1	1 Find all the subsets of a set (p. 697)	1	1
	2 Count the number of elements in a set (p. 697)	2, 3	2–9
	3 Solve counting problems using the Multiplication Principle (p. 699)	4, 5	12, 13, 17, 18
10.2	1 Solve counting problems using permutations involving n distinct objects (p. 703)	1–5	10, 14, 15, 19, 22(a)
	2 Solve counting problems using combinations (p. 705)	6–9	11, 16, 21
	3 Solve counting problems using permutations involving n nondistinct objects (p. 708)	10, 11	20
10.3	1 Construct probability models (p. 711)	2–4	22(b)
	2 Compute probabilities of equally likely outcomes (p. 714)	5, 6	22(b), 23(a), 24, 25
	3 Find probabilities of the union of two events (p. 715)	7, 8	26
	4 Use the Complement Rule to find probabilities (p. 716)	9, 10	22(c), 23(b)

Review Exercises

1. Write down all the subsets of the set {Dave, Joanne, Erica}.

2. If $n(A) = 8, n(B) = 12$, and $n(A \cap B) = 3$, find $n(A \cup B)$.

3. If $n(A) = 12, n(A \cup B) = 30$, and $n(A \cap B) = 6$, find $n(B)$.

In Problems 4–9, use the information supplied in the figure.

4. How many are in A?

5. How many are in A or B?

6. How many are in A and C?

7. How many are not in B?

8. How many are in neither A nor C?

9. How many are in B but not in C?

In Problems 10 and 11, compute the given expression.

10. $P(8, 3)$ **11.** $C(8, 3)$

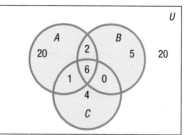

12. **Stocking a Store** A clothing store sells pure wool and polyester-wool suits. Each suit comes in 3 colors and 10 sizes. How many suits are required for a complete assortment?

13. **Baseball** On a given day, the American Baseball League schedules 7 games. How many different outcomes are possible, assuming that each game is played to completion?

14. **Choosing Seats** If 4 people enter a bus that has 9 vacant seats, in how many ways can they be seated?

15. **Choosing a Team** In how many ways can a squad of 4 relay runners be chosen from a track team of 8 runners?

16. **Baseball** In how many ways can 2 teams from 14 teams in the American League be chosen without regard to which team is at home?

17. **Telephone Numbers** Using the digits 0, 1, 2, . . . , 9, how many 7-digit numbers can be formed if the first digit cannot be 0 or 9 and if the last digit is greater than or equal to 2 and less than or equal to 3? Repeated digits are allowed.

18. **License Plate Possibilities** A license plate has 1 letter, excluding O and I, followed by a 4-digit number that cannot have a 0 in the lead position. How many different plates are possible?

19. **Binary Codes** Using the digits 0 and 1, how many different numbers consisting of 8 digits can be formed?

20. **Arranging Flags** How many different vertical arrangements are there of 10 flags if 4 are white, 3 are blue, 2 are green, and 1 is red?

21. **Forming Committees** A group of 9 people is going to be formed into committees of 4, 3, and 2 people. How many committees can be formed if:
    (a) A person can serve on any number of committees?
    (b) No person can serve on more than one committee?

22. **Birthday Problem** For this problem, assume that a year has 365 days.
    (a) In how many ways can 18 people have different birthdays?
    (b) What is the probability that no 2 people in a group of 18 people have the same birthday?
    (c) What is the probability that at least 2 people in a group of 18 people have the same birthday?

23. **Unemployment** According to the U.S. Bureau of Labor Statistics, 3.8% of the U.S. labor force was unemployed in May 2018.
    (a) What is the probability that a randomly selected member of the U.S. labor force was unemployed in May 2018?
    (b) What is the probability that a randomly selected member of the U.S. labor force was not unemployed in May 2018?

24. You have four $1 bills, three $5 bills, and two $10 bills in your wallet. If you pick a bill at random, what is the probability that it will be a $1 bill?

25. Each of the numbers 1, 2, . . . , 100 is written on an index card, and the cards are shuffled. If a card is selected at random, what is the probability that the number on the card is divisible by 5? What is the probability that the card selected either is a 1 or names a prime number?

26. At the Milex tune-up and brake repair shop, the manager has found that a car will require a tune-up with a probability of 0.6, a brake job with a probability of 0.1, and both with a probability of 0.02.
    (a) What is the probability that a car requires either a tune-up or a brake job?
    (b) What is the probability that a car requires a tune-up but not a brake job?
    (c) What is the probability that a car requires neither a tune-up nor a brake job?

Chapter Test

CHAPTER **Test Prep** **VIDEOS** The Chapter Test Prep Videos include step-by-step solutions to all chapter test exercises. These videos are available in MyLab™ Math, or on this text's YouTube Channel. Refer to the Preface for a link to the YouTube channel.

In Problems 1–4, a survey of 70 college freshmen asked whether students planned to take biology, chemistry, or physics during their first year. Use the diagram to answer each question.

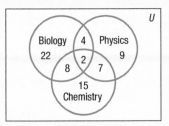

1. How many of the surveyed students plan to take physics during their first year?

2. How many of the surveyed students do not plan to take biology, chemistry, or physics during their first year?

3. How many of the surveyed students plan to take only biology and chemistry during their first year?

4. How many of the surveyed students plan to take physics or chemistry during their first year?

In Problems 5–7, compute the value of the given expression.

5. 7!   6. $P(10, 6)$ 7. $C(11, 5)$

8. M&M's® offers customers the opportunity to create their own color mix of candy. There are 21 colors to choose from, and customers are allowed to select up to 6 different colors. How many different color mixes are possible, assuming that no color is selected more than once and 6 different colors are chosen?

9. How many distinct 8-letter words (meaningful or not) can be formed from the letters in the word REDEEMED?

10. In horse racing, an exacta bet requires the bettor to pick the first two horses in the exact order. If there are 8 horses in a race, in how many ways could you make an exacta bet?

11. On February 20, 2004, the Ohio Bureau of Motor Vehicles unveiled the state's new license plate format. The plate consists of three letters (A–Z) followed by 4 digits (0–9). Assume that all letters and digits may be used, except that the third letter cannot be O, I, or Z. If repetitions are allowed, how many different plates are possible?

12. Kiersten applies for admission to the University of Southern California (USC) and Florida State University (FSU). She estimates that she has a 60% chance of being admitted to USC, a 70% chance of being admitted to FSU, and a 35% chance of being admitted to both universities.
   (a) What is the probability that she will be admitted to either USC or FSU?
   (b) What is the probability that she will not be admitted to FSU?

13. A cooler contains 8 bottles of Pepsi, 5 bottles of Coke, 4 bottles of Mountain Dew, and 3 bottles of IBC.
   (a) What is the probability that a bottle chosen at random is Coke?
   (b) What is the probability that a bottle chosen at random is either Pepsi or IBC?

14. A study on the age distribution of students at a community college yielded the following data:

Age	17 and under	18–20	21–24	25–34	35–64	65 and over
Probability	0.03	???	0.23	0.29	0.25	0.01

What is the probability a randomly selected student at the college is between 18 and 20 years old?

15. In a certain lottery, there are ten balls numbered $1, 2, 3, 4, 5, 6, 7, 8, 9, 10$. Of these, five are drawn in order. If you pick five numbers that match those drawn in the correct order, you win $1,000,000. What is the probability of winning such a lottery?

16. If you roll a die five times, what is the probability that you obtain exactly 2 fours?

Cumulative Review

1. Solve: $3x^2 - 2x = -1$

2. Graph $f(x) = x^2 + 4x - 5$ by determining whether the graph is concave up or concave down and by finding the vertex, axis of symmetry, and intercepts.

3. Graph $f(x) = 2(x + 1)^2 - 4$ using transformations.

4. Solve: $|x - 4| \leq 0.01$

5. Find the complex zeros of

$$f(x) = 5x^4 - 9x^3 - 7x^2 - 31x - 6$$

6. Graph $g(x) = 3^{x-1} + 5$ using transformations. Find the domain, the range, and the horizontal asymptote of g.

7. What is the exact value of $\log_3 9$?

8. Solve: $\log_2(3x - 2) + \log_2 x = 4$

9. Solve the system: $\begin{cases} x - 2y + z = 15 \\ 3x + y - 3z = -8 \\ -2x + 4y - z = -27 \end{cases}$

10. What is the 33rd term in the sequence $-3, 1, 5, 9, \ldots$? What is the sum of the first 20 terms?

Chapter Projects

I. **The Lottery and Expected Profit** When all of the possible outcomes in a probability model are numeric quantities, useful statistics can be computed for such models. The **expected value**, or **mean**, of such a probability model is found by multiplying each possible numeric outcome by its corresponding probability and then adding these products.

For example, Table 2 provides the probability model for rolling a fair six-sided die. The expected value, $E(x)$, is

$$E(x) = 1 \cdot \frac{1}{6} + 2 \cdot \frac{1}{6} + 3 \cdot \frac{1}{6} + 4 \cdot \frac{1}{6} + 5 \cdot \frac{1}{6} + 6 \cdot \frac{1}{6} = 3.5$$

When a fair die is rolled repeatedly, the average of the outcomes will approach 3.5.

Mega Millions is a multistate lottery in which a player selects five different "white" numbers from 1 to 70 and one "gold" number from 1 to 25. The probability model shown in Table 3 lists the possible cash prizes and their corresponding probabilities.

1. Verify that Table 3 is a probability model.

2. To win the jackpot, a player must match all six numbers. Verify the probability given in Table 3 of winning the jackpot.

Table 2

Outcome	Probability
1	$\frac{1}{6}$
2	$\frac{1}{6}$
3	$\frac{1}{6}$
4	$\frac{1}{6}$
5	$\frac{1}{6}$
6	$\frac{1}{6}$

Table 3

Cash Prize	Probability
Jackpot	0.00000000330
$1,000,000	0.00000007932
$10,000	0.00000107411
$500	0.00002577851
$200	0.00006874270
$10	0.00309316646
$4	0.01123595506
$2	0.02702702703
$0	0.95854817351

For questions 3–6, assume a single jackpot winner so that the jackpot does not have to be shared.

3. If the jackpot is $40,000,000, calculate the expected cash prize.
4. If a ticket costs $2, what is the expected financial result from purchasing one ticket? Interpret (give the meaning of) this result.
5. If the jackpot is $250,000,000, what is the expected cash prize? What is the expected financial result from purchasing one $2 ticket? Interpret this result.
6. What amount must the jackpot be so that a profit from one $2 ticket is expected?
7. Research the Powerball lottery, and create a probability model similar to Table 3 for it. Repeat questions 3–6 for Powerball. Based on what you have learned, which lottery would you prefer to play? Justify your decision.

The following projects are available at the Instructor's Resource Center (IRC):

II. Project at Motorola ***Probability of Error in Digital Wireless Communications*** Transmission errors in digital communications can often be detected by adding an extra digit of code to each transmitted signal. Investigate the probability of identifying an erroneous code using this simple coding method.

III. Surveys Polling (or taking a survey) is big business in the United States. Take and analyze a survey; then consider why different pollsters might get different results.

IV. Law of Large Numbers The probability that an event occurs, such as a head in a coin toss, is the proportion of heads you expect in the long run. A simulation is used to show that as a coin is flipped more and more times, the proportion of heads gets close to 0.5.

V. Simulation Electronic simulation of an experiment is often an economical way to investigate a theoretical probability. Develop a theory without leaving your desk.

Graphing Utilities

Outline

1 The Viewing Rectangle

(a) $y = 2x$ on a TI-84 Plus C

(b) $y = 2x$ using Desmos

Figure 1

All graphing utilities (that is, all graphing calculators and all computer software graphing packages) graph equations by plotting points on a screen. The screen itself actually consists of small rectangles called **pixels**. The more pixels the screen has, the better the resolution. Most graphing calculators have 50 to 100 pixels per inch; most smartphones have 300 to 450 pixels per inch. When a point to be plotted lies inside a pixel, the pixel is turned on (lights up). The graph of an equation is a collection of pixels. Figure 1(a) shows how the graph of $y = 2x$ looks on a TI-84 Plus C graphing calculator, and Figure 1(b) shows the same graph using Desmos.com.

The screen of a graphing utility displays the coordinate axes of a rectangular coordinate system. However, the scale must be set on each axis. The smallest and largest values of x and y to be included in the graph must also be set. This is called **setting the viewing rectangle** or **viewing window**. Figure 2 shows a typical viewing window on a TI-84 Plus C.

To select the viewing window, values must be given to the following expressions:

Xmin: the smallest value of x
Xmax: the largest value of x
Xscl: the number of units per tick mark on the x-axis
Ymin: the smallest value of y
Ymax: the largest value of y
Yscl: the number of units per tick mark on the y-axis

Figure 3 illustrates these settings and their relation to the Cartesian coordinate system.

Figure 2 Viewing window on a TI-84 Plus C

Figure 3

If the scale used on each axis is known, the minimum and maximum values of x and y shown on the screen can be determined by counting the tick marks. Look again at Figure 2. For a scale of 1 on each axis, the minimum and maximum values of x are -10 and 10, respectively; the minimum and maximum values of y are

A1

Figure 4

also −10 and 10. If the scale is 2 on each axis, then the minimum and maximum values of x are −20 and 20, respectively; and the minimum and maximum values of y are −20 and 20, respectively.

Conversely, if the minimum and maximum values of x and y are known, the scales can be determined by counting the tick marks displayed. This text follows the practice of showing the minimum and maximum values of x and y in illustrations so that the reader will know how the viewing window was set. See Figure 4. The numbers outside of the viewing window stand for

$$X\text{min} = -3, \quad X\text{max} = 3, \quad X\text{scl} = 1$$
$$Y\text{min} = -4, \quad Y\text{max} = 4, \quad Y\text{scl} = 2$$

EXAMPLE 1 **Finding the Coordinates of a Point Shown on a Graphing Utility Screen**

Find the coordinates of the point shown in Figure 5. Assume that the coordinates are integers.

Figure 5

Solution First note that the viewing window used in Figure 5 is

$$X\text{min} = -3, \quad X\text{max} = 3, \quad X\text{scl} = 1$$
$$Y\text{min} = -4, \quad Y\text{max} = 4, \quad Y\text{scl} = 2$$

The point shown is 2 tick units to the left of the origin on the horizontal axis (scale = 1) and 1 tick up on the vertical axis (scale = 2). The coordinates of the point shown are $(-2, 2)$.

1 Exercises

In Problems 1–4, determine the coordinates of the points shown. Tell in which quadrant each point lies. Assume that the coordinates are integers.

1. **2.**

3. **4.**

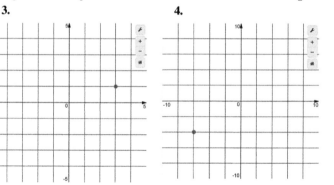

In Problems 5–10, determine the viewing window used.

8. **9.** **10.**

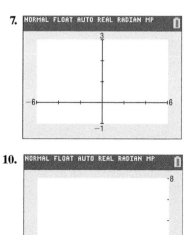

In Problems 11–16, select a setting so that each of the given points will lie within the viewing rectangle.

11. $(-10, 5), (3, -2), (4, -1)$

12. $(5, 0), (6, 8), (-2, -3)$

13. $(40, 20), (-20, -80), (10, 40)$

14. $(-80, 60), (20, -30), (-20, -40)$

15. $(0, 0), (100, 5), (5, 150)$

16. $(0, -1), (100, 50), (-10, 30)$

2 Using a Graphing Utility to Graph Equations

From Examples 2 and 3 in Section 2.2, recall that one way a graph can be obtained is by plotting points in a rectangular coordinate system and connecting them. Graphing utilities perform these same steps when graphing an equation. For example, the TI-84 Plus C determines 265 evenly spaced input values,* starting at Xmin and ending at Xmax; uses the equation to determine the output values; plots these points on the screen; and finally (if in the connected mode) draws a line between consecutive points.

To graph an equation in two variables x and y using a graphing utility often requires that the equation be written explicitly in the form $y = \{expression\ in\ x\}$. If the original equation is not in this form, replace it by equivalent equations until the form $y = \{expression\ in\ x\}$ is obtained.

Steps for Graphing an Equation Using a Graphing Utility

STEP 1: Solve the equation for y in terms of x.

STEP 2: Get into the graphing mode of the graphing utility. The screen will usually display $Y_1 = \quad$, prompting you to enter the expression involving x found in Step 1. (Consult your manual for the correct way to enter the expression; for example, $y = x^2$ might be entered as $x^\wedge 2$ or as $x*x$ or as $x\ x^y\ 2$.)

STEP 3: Select the viewing window. Without prior knowledge about the behavior of the graph of the equation, it is common to select the **standard viewing window**** initially. The viewing window is then adjusted based on the graph that appears. In this text the standard viewing window is

$$X\text{min} = -10 \quad X\text{max} = 10 \quad X\text{scl} = 1$$
$$Y\text{min} = -10 \quad Y\text{max} = 10 \quad Y\text{scl} = 1$$

STEP 4: Graph.

STEP 5: Adjust the viewing window until a complete graph is obtained.

Figure 6 $6x^2 + 3y = 36$

NOTE Some graphing utilities allow input of implicit equations. For example, Figure 6 shows the graph of $6x^2 + 3y = 36$ using Desmos.   ∎

*These input values depend on the values of Xmin and Xmax. For example, if Xmin $= -10$ and Xmax $= 10$, then the first input value will be -10 and the next input value will be $-10 + \dfrac{10 - (-10)}{264} = -9.9242$, and so on.

**Some graphing utilities have a ZOOM-STANDARD feature that automatically sets the viewing window to the standard viewing window and graphs the equation.

| EXAMPLE 1 | Graphing an Equation on a Graphing Utility |

Graph the equation $6x^2 + 3y = 36$.

Solution   **STEP 1:** Solve for y in terms of x.

$$6x^2 + 3y = 36$$
$$3y = -6x^2 + 36 \quad \text{Subtract } 6x^2 \text{ from both sides.}$$
$$y = -2x^2 + 12 \quad \text{Divide both sides by 3 and simplify.}$$

STEP 2: From the $Y_1 =$ screen, enter the expression $-2x^2 + 12$ after the prompt.

STEP 3: Set the viewing window to the standard viewing window.

STEP 4: Graph. The screen should look like Figure 7.

STEP 5: The graph of $y = -2x^2 + 12$ is not complete. The value of Ymax must be increased so that the top portion of the graph is visible. After increasing the value of Ymax to 12, we obtain the graph in Figure 8. The graph is now complete.

Figure 7 $y = -2x^2 + 12$

Figure 8 $y = -2x^2 + 12$

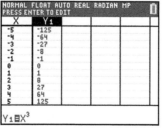

Figure 9 $y = -2x^2 + 12$

Look again at Figure 8. Although a complete graph is shown, the graph might be improved by adjusting the values of Xmin and Xmax. Figure 9 shows the graph of $y = -2x^2 + 12$ using Xmin $= -4$ and Xmax $= 4$. Do you think this is a better choice for the viewing window?

| EXAMPLE 2 | Creating a Table and Graphing an Equation |

Create a table and graph the equation $y = x^3$.

Solution   Most graphing utilities have the capability of creating a table of values for an equation. (Check your manual to see if your graphing utility has this capability.) Table 1 illustrates a table of values for $y = x^3$ on a TI-84 Plus C. See Figure 10 for the graph.

Table1

Figure 10 $y = x^3$

2 Exercises

In Problems 1–16, graph each equation using the following viewing windows:

(a) Xmin $= -5$ Xmax $= 5$ Xscl $= 1$ (b) Xmin $= -10$ Xmax $= 10$ Xscl $= 2$
   Ymin $= -4$ Ymax $= 4$ Yscl $= 1$ Ymin $= -8$ Ymax $= 8$ Yscl $= 2$

1. $y = x + 2$ **2.** $y = x - 2$ **3.** $y = -x + 2$ **4.** $y = -x - 2$

5. $y = 2x + 2$ **6.** $y = 2x - 2$ **7.** $y = -2x + 2$ **8.** $y = -2x - 2$

9. $y = x^2 + 2$ **10.** $y = x^2 - 2$ **11.** $y = -x^2 + 2$ **12.** $y = -x^2 - 2$

13. $3x + 2y = 6$ **14.** $3x - 2y = 6$ **15.** $-3x + 2y = 6$ **16.** $-3x - 2y = 6$

17–32. *For each of the equations in Problems 1–16, create a table,* $-5 \le x \le 5$, *and list points on the graph.*

---

3 Using a Graphing Utility to Locate Intercepts and Check for Symmetry

Value and Zero (or Root)

Most graphing utilities have an eVALUEate feature that, given a value of x, determines the value of y for an equation. This feature is useful for evaluating an equation at $x = 0$ to find the y-intercept. Most graphing utilities also have a ZERO (or ROOT) feature that is used to find the x-intercept(s) of an equation.

Figure 11 $y = x^3 - 8$

NOTE Some graphing utilities automatically identify key points such as intercepts and intersection points. For example, Figure 11 shows the graph of $y = x^3 - 8$ using Desmos where the intercepts are already identified. ∎

EXAMPLE 1   Finding Intercepts Using a Graphing Utility

Use a graphing utility to find the intercepts of the equation $y = x^3 - 8$.

Solution   Figure 12(a) shows the graph of $y = x^3 - 8$ on a TI-84 Plus C graphing calculator.

The eVALUEate feature of a TI-84 Plus C accepts as input a value of x and determines the value of y. Letting $x = 0$, we find that the y-intercept is -8. See Figure 12(b).

The ZERO feature of a TI-84 Plus C is used to find the x-intercept(s). See Figure 12(c). The x-intercept is 2.

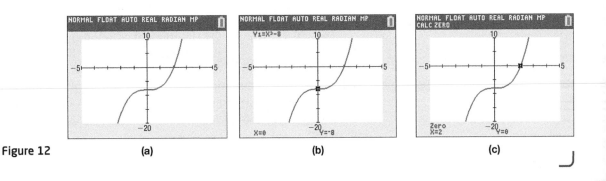

Figure 12   (a)   (b)   (c)

EXAMPLE 2   Graphing the Equation $y = \dfrac{1}{x}$

Graph the equation $y = \dfrac{1}{x}$. Based on the graph, infer information about intercepts and symmetry.

Solution   Figure 13 shows the graph. Infer from the graph that there are no intercepts; also infer that symmetry with respect to the origin is a possibility. The TABLE feature on a graphing utility provides further evidence of symmetry with respect to the origin. Using a TABLE, observe that for any ordered pair (x, y), the ordered pair $(-x, -y)$ is also a point on the graph. See Table 2.

(continued)

Figure 13 $y = \dfrac{1}{x}$

3 Exercises

In Problems 1–6, use ZERO (or ROOT) to approximate the smaller of the two x-intercepts of each equation. Express the answer rounded to two decimal places.

1. $y = x^2 + 4x + 2$

2. $y = x^2 + 4x - 3$

3. $y = 2x^2 + 4x + 1$

4. $y = 3x^2 + 5x + 1$

5. $y = 2x^2 - 3x - 1$

6. $y = 2x^2 - 4x - 1$

*In Problems 7–12, use ZERO (or ROOT) to approximate the **positive** x-intercepts of each equation. Express each answer rounded to two decimal places.*

7. $y = x^3 + 3.2x^2 - 16.83x - 5.31$

8. $y = x^3 + 3.2x^2 - 7.25x - 6.3$

9. $y = x^4 - 1.4x^3 - 33.71x^2 + 23.94x + 292.41$

10. $y = x^4 + 1.2x^3 - 7.46x^2 - 4.692x + 15.2881$

11. $y = x^3 + 19.5x^2 - 1021x + 1000.5$

12. $y = x^3 + 14.2x^2 - 4.8x - 12.4$

4 Using a Graphing Utility to Solve Equations

For many equations, there are no algebraic techniques that lead to a solution. For such equations, a graphing utility is often used to investigate possible solutions. When a graphing utility is used to solve an equation, *approximate* solutions usually are obtained. Unless otherwise stated, this text follows the practice of giving approximate solutions *rounded to two decimal places*.

The ZERO (or ROOT) feature of a graphing utility is often used to find the solutions of an equation when one side of the equation is 0. In using this feature to solve equations, make use of the fact that the x-intercepts (or zeros) of the graph of an equation are found by letting $y = 0$ and solving the equation for x. Solving an equation for x when one side of the equation is 0 is equivalent to finding where the graph of the corresponding equation crosses or touches the x-axis.

EXAMPLE 1 **Using ZERO (or ROOT) to Approximate Solutions of an Equation**

Find the solution(s) of the equation $x^2 - 6x + 7 = 0$. Round answers to two decimal places.

Solution The solutions of the equation $x^2 - 6x + 7 = 0$ are the same as the x-intercepts of the graph of $Y_1 = x^2 - 6x + 7$. Begin by graphing the equation. See Figure 14(a) for the graph using a TI-84 Plus C.

From the graph there appear to be two x-intercepts (solutions to the equation): one between 1 and 2, the other between 4 and 5.

Using the ZERO (or ROOT) feature of the graphing utility, determine that the x-intercepts, and thus the solutions to the equation, are $x = 1.59$ and $x = 4.41$, rounded to two decimal places. See Figures 14(b) and (c).

Figure 14   (a)   (b)   (c)

A second method for solving equations using a graphing utility involves the INTERSECT feature of the graphing utility. This feature is used most effectively when one side of the equation is not 0.

EXAMPLE 2   **Using INTERSECT to Approximate Solutions of an Equation**

Find the solution(s) of the equation $3(x - 2) = 5(x - 1)$.

Solution   Begin by graphing each side of the equation as follows: graph $Y_1 = 3(x - 2)$ and $Y_2 = 5(x - 1)$. See Figure 15(a) for the graph using a TI-84 Plus C.

At the point of intersection of the graphs, the value of the y-coordinate is the same. Conclude that the x-coordinate of the point of intersection represents the solution of the equation. Do you see why? The INTERSECT feature on a graphing utility determines the point of intersection of the graphs. Using this feature, find that the graphs intersect at $(-0.5, -7.5)$. See Figure 15(b). The solution of the equation is therefore $x = -0.5$. Figure 15(c) shows the intersection point using Desmos.

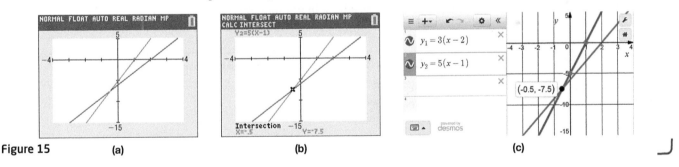

Figure 15   (a)   (b)   (c)

SUMMARY

The following steps can be used for approximating solutions of equations.

Steps for Approximating Solutions of Equations Using ZERO (or ROOT)

STEP 1:   Write the equation in the form $\{expression\ in\ x\} = 0$.

STEP 2:   Graph $Y_1 = \{expression\ in\ x\}$.

Be sure that the graph is complete. That is, be sure that all the intercepts are shown on the screen.

STEP 3:   Use ZERO (or ROOT) to determine each x-intercept of the graph.

Steps for Approximating Solutions of Equations Using INTERSECT

STEP 1:   Graph $Y_1 = \{expression\ in\ x\ on\ the\ left\ side\ of\ the\ equation\}$.

Graph $Y_2 = \{expression\ in\ x\ on\ the\ right\ side\ of\ the\ equation\}$.

STEP 2:   Use INTERSECT to determine each x-coordinate of the point(s) of intersection, if any.

Be sure that the graphs are complete. That is, be sure that all the points of intersection are shown on the screen.

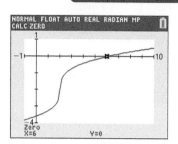

Figure 16 $y = \sqrt[3]{2x - 4} - 2$

EXAMPLE 3

Solving a Radical Equation

Find the real solutions of the equation $\sqrt[3]{2x - 4} - 2 = 0$.

Solution Figure 16 shows the graph of the equation

$$Y_1 = \sqrt[3]{2x - 4} - 2.$$

From the graph, there is one x-intercept near 6. Using ZERO (or ROOT), find that the x-intercept is 6. The only solution is $x = 6$.

---

5 Square Screens

Figure 17 $y = x$ and $y = -x$ using a standard viewing window.

Most graphing utilities have a rectangular screen. Because of this, using the same settings for both x and y results in a distorted view. For example, Figure 17 shows the graphs of $y = x$ and $y = -x$ using a TI-84 Plus C with a standard viewing window.

We expect the lines to intersect at right angles, but they do not. The selections for Xmin, Xmax, Ymin, and Ymax must be adjusted so that a **square screen** results. On the TI-84 Plus C, this is accomplished by setting the ratio of x to y at 8:5.* For example, if

$$X\text{min} = -16 \qquad Y\text{min} = -10$$
$$X\text{max} = 16 \qquad Y\text{max} = 10$$

then the ratio of x to y is

$$\frac{X\text{max} - X\text{min}}{Y\text{max} - Y\text{min}} = \frac{16 - (-16)}{10 - (-10)} = \frac{32}{20} = \frac{8}{5}$$

for a ratio of 8:5, resulting in a square screen.

EXAMPLE 1

Examples of Viewing Rectangles That Result in Square Screens

(a) Xmin $= -8$
 Xmax $= 8$
 Xscl $= 1$
 Ymin $= -5$
 Ymax $= 5$
 Yscl $= 1$

(b) Xmin $= -16$
 Xmax $= 16$
 Xscl $= 1$
 Ymin $= -10$
 Ymax $= 10$
 Yscl $= 1$

(c) Xmin $= -24$
 Xmax $= 24$
 Xscl $= 3$
 Ymin $= -15$
 Ymax $= 15$
 Yscl $= 3$

Figure 18 $y = x$ and $y = -x$ with a square screen.

Figure 18 shows the graphs of $y = x$ and $y = -x$ on a square screen using the viewing rectangle given in part (b). Notice that the lines now intersect at right angles. Compare this illustration to Figure 17.

---

5 Exercises

In Problems 1–4, determine which of the given viewing rectangles result in a square screen.

1. Xmin $= -6$ $\quad$ Xmax $= 6$ $\quad$ Xscl $= 1$
 Ymin $= -2$ $\quad$ Ymax $= 2$ $\quad$ Yscl $= 0.5$

2. Xmin $= -5$ $\quad$ Xmax $= 5$ $\quad$ Xscl $= 1$
 Ymin $= -4$ $\quad$ Ymax $= 4$ $\quad$ Yscl $= 1$

3. Xmin $= 0$ $\quad$ Xmax $= 16$ $\quad$ Xscl $= 4$
 Ymin $= -2$ $\quad$ Ymax $= 8$ $\quad$ Yscl $= 2$

4. Xmin $= -10$ $\quad$ Xmax $= 14$ $\quad$ Xscl $= 2$
 Ymin $= -7$ $\quad$ Ymax $= 8$ $\quad$ Yscl $= 3$

*Some graphing utilities have a built-in function that automatically squares the screen. For example, the TI-84 Plus C has a ZSquare function that does this. Some graphing utilities require a ratio other than 8:5 to square the screen. For example, the HP 48G requires the ratio of x to y to be 2:1 for a square screen. Consult your manual.

5. If $X\text{min} = -4$, $X\text{max} = 12$, and $X\text{scl} = 1$, how should $Y\text{min}$, $Y\text{max}$, and $Y\text{scl}$ be selected so that the viewing rectangle contains the point $(4, 8)$ and the screen is square?

6. If $X\text{min} = -6$, $X\text{max} = 10$, and $X\text{scl} = 2$, how should $Y\text{min}$, $Y\text{max}$, and $Y\text{scl}$ be selected so that the viewing rectangle contains the point $(4, 8)$ and the screen is square?

6 Using a Graphing Utility to Graph Inequalities

EXAMPLE 1	Graphing an Inequality Using a Graphing Utility

Use a graphing utility to graph $3x + y - 6 \leq 0$.

Solution Begin by graphing the equation $3x + y - 6 = 0$ $(Y_1 = -3x + 6)$. See Figure 19.

As with graphing by hand, select test points from each region and determine whether they satisfy the inequality. To test the point $(-1, 2)$, for example, enter $3(-1) + 2 - 6 \leq 0$. See Figure 20(a). The 1 that appears indicates that the statement entered (the inequality) is true. When the point $(5, 5)$ is tested, a 0 appears, indicating that the statement entered is false. Thus, $(-1, 2)$ is a part of the graph of the inequality, and $(5, 5)$ is not. Figure 20(b) shows the graph of the inequality on a TI-84 Plus C.* Figure 20(c) shows the graph using Desmos.

Figure 19

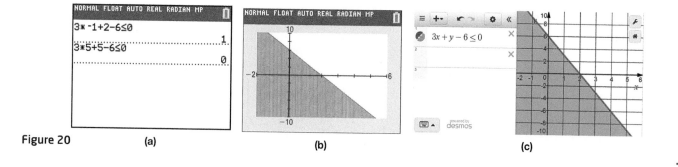

Figure 20 (a) (b) (c)

Steps for Graphing an Inequality Using a Graphing Utility

STEP 1: Replace the inequality symbol by an equal sign, solve the equation for y, and graph the equation.

STEP 2: In each region, select a test point P and determine whether the coordinates of P satisfy the inequality.

- If the test point satisfies the inequality, then so do all the points in the region. Indicate this by using the graphing utility to shade the region.
- If the coordinates of P do not satisfy the inequality, then neither will any of the other points in that region.

*Consult your owner's manual for shading techniques.

7 Using a Graphing Utility to Solve Systems of Linear Equations

Most graphing utilities have the capability to put the augmented matrix of a system of linear equations in row echelon form. The next example, Example 6 from Section 8.2, demonstrates this feature using a TI-84 Plus C graphing calculator.

EXAMPLE 1	Solving a System of Linear Equations Using a Graphing Utility

Solve:
$$\begin{cases} x - y + z = 8 & (1) \\ 2x + 3y - z = -2 & (2) \\ 3x - 2y - 9z = 9 & (3) \end{cases}$$

Solution   The augmented matrix of the system is

$$\begin{bmatrix} 1 & -1 & 1 & | & 8 \\ 2 & 3 & -1 & | & -2 \\ 3 & -2 & -9 & | & 9 \end{bmatrix}$$

Enter this matrix into a graphing utility and name it A. See Figure 21(a). Using the **ref** (row echelon form) command on matrix A, the results shown in Figure 21(b) are obtained. If the entire matrix does not fit on the screen, scroll right to see the rest of it.

Figure 21   (a)   (b)

The system of equations represented by the matrix in row echelon form is

$$\begin{bmatrix} 1 & -\dfrac{2}{3} & -3 & | & 3 \\ 0 & 1 & \dfrac{15}{13} & | & -\dfrac{24}{13} \\ 0 & 0 & 1 & | & 1 \end{bmatrix} \quad \begin{cases} x - \dfrac{2}{3}y - 3z = 3 & (1) \\ y + \dfrac{15}{13}z = -\dfrac{24}{13} & (2) \\ z = 1 & (3) \end{cases}$$

Using $z = 1$, back-substitute to get

$$\begin{cases} x - \dfrac{2}{3}y - 3(1) = 3 & (1) \\ y + \dfrac{15}{13}(1) = -\dfrac{24}{13} & (2) \end{cases} \xrightarrow[\text{Simplify.}]{} \begin{cases} x - \dfrac{2}{3}y = 6 & (1) \\ y = -\dfrac{39}{13} = -3 & (2) \end{cases}$$

Solve the second equation for y to find that $y = -3$. Back-substitute $y = -3$ into $x - \dfrac{2}{3}y = 6$ to find that $x = 4$. The solution of the system is $x = 4$, $y = -3$, $z = 1$.

Figure 22

Notice that the row echelon form of the augmented matrix using the graphing utility differs from the row echelon form obtained in Example 6 of Section 8.2, yet both matrices provide the same solution! This is because the two solutions used different row operations to obtain the row echelon form. In all likelihood, the two solutions parted ways in Step 4 of the algebraic solution, where fractions were avoided by interchanging rows 2 and 3.

Most graphing utilities also have the ability to put a matrix in reduced row echelon form. Figure 22 shows the reduced row echelon form of the augmented matrix from Example 1 using the **rref** command on a TI-84 Plus C graphing calculator. Using this command, note that the solution of the system is still $x = 4$, $y = -3$, $z = 1$.

Answers

CHAPTER R Review

R.1 Assess Your Understanding (page 15)

1. rational **2.** 31 **3.** Distributive **4.** c **5.** a **6.** b **7.** T **8.** F **9.** F **10.** T **11.** $\{1, 2, 3, 4, 5, 6, 7, 8, 9\}$ **13.** $\{4\}$ **15.** $\{1, 3, 4, 6\}$
17. $\{0, 2, 6, 7, 8\}$ **19.** $\{0, 1, 2, 3, 5, 6, 7, 8, 9\}$ **21.** $\{0, 1, 2, 3, 5, 6, 7, 8, 9\}$

23. (a) $\{2, 5\}$ **(b)** $\{-6, 2, 5\}$ **(c)** $\left\{-6, \frac{1}{2}, -1.333\ldots, 2, 5\right\}$ **(d)** $\{\pi\}$ **(e)** $\left\{-6, \frac{1}{2}, -1.333\ldots, \pi, 2, 5\right\}$

25. (a) $\{1\}$ **(b)** $\{0, 1\}$ **(c)** $\left\{0, 1, \frac{1}{2}, \frac{1}{3}, \frac{1}{4}\right\}$ **(d)** None **(e)** $\left\{0, 1, \frac{1}{2}, \frac{1}{3}, \frac{1}{4}\right\}$

27. (a) None **(b)** None **(c)** None **(d)** $\left\{\sqrt{2}, \pi, \sqrt{2} + 1, \pi + \frac{1}{2}\right\}$ **(e)** $\left\{\sqrt{2}, \pi, \sqrt{2} + 1, \pi + \frac{1}{2}\right\}$

29. (a) 18.953 **(b)** 18.952 **31. (a)** 28.653 **(b)** 28.653 **33. (a)** 0.063 **(b)** 0.062 **35. (a)** 9.999 **(b)** 9.998 **37. (a)** 0.429 **(b)** 0.428

39. (a) 34.733 **(b)** 34.733 **41.** $3 + 2 = 5$ **43.** $x + 2 = 3 \cdot 4$ **45.** $3y = 1 + 2$ **47.** $x - 2 = 6$ **49.** $\frac{x}{2} = 6$ **51.** 7 **53.** 6 **55.** 8 **57.** $\frac{13}{3}$

59. -11 **61.** 17 **63.** -4 **65.** 1 **67.** 6 **69.** $\frac{2}{7}$ **71.** $\frac{4}{45}$ **73.** $\frac{23}{20}$ **75.** $\frac{81}{56}$ **77.** $\frac{13}{36}$ **79.** $-\frac{13}{40}$ **81.** $\frac{1}{60}$ **83.** $\frac{15}{22}$ **85.** 1 **87.** $\frac{15}{8}$ **89.** $6x + 24$

91. $x^2 - 4x$ **93.** $\frac{3}{2}x - 1$ **95.** $x^2 + 6x + 8$ **97.** $2x^2 + 11x - 63$ **99.** $x^2 - 10x + 16$ **101.** $k = 4$ **103.** $2x + 3x = (2 + 3)x = 5x$

105. $2(3 \cdot 4) = 2 \cdot 12 = 24; (2 \cdot 3) \cdot (2 \cdot 4) = 6 \cdot 8 = 48$ **107.** No; $2 - 3 \neq 3 - 2$ **109.** No; $\frac{2}{3} \neq \frac{3}{2}$ **111.** Symmetric Property **113.** No; no

R.2 Assess Your Understanding (page 26)

1. variable **2.** origin **3.** strict **4.** base; exponent or power **5.** 1.2345678×10^3 **6.** d **7.** a **8.** b **9.** T **10.** F **11.** F **12.** T
13.
15. $>$ **17.** $>$ **19.** $>$ **21.** $=$ **23.** $<$ **25.** $x > 0$ **27.** $x < 2$ **29.** $x \leq 1$ **31.**
33. **35.** 1 **37.** 2 **39.** 6 **41.** 4 **43.** -28 **45.** $\frac{4}{5}$ **47.** 0 **49.** 1 **51.** 5 **53.** 1 **55.** 22 **57.** 2 **59.** $x = 0$ **61.** $x = 3$ **63.** None
65. $x = 0, x = 1, x = -1$ **67.** $\{x \mid x \neq 5\}$ **69.** $\{x \mid x \neq -4\}$ **71.** 0°C **73.** 25°C **75.** 81 **77.** $\frac{1}{16}$ **79.** $\frac{1}{9}$ **81.** 64 **83.** 10 **85.** 4 **87.** $81x^8$
89. $\frac{x^4}{y^2}$ **91.** $\frac{y}{x}$ **93.** $-\frac{16x^2z}{27y^2}$ **95.** $\frac{9x^6}{4y^2}$ **97.** -4 **99.** 5 **101.** 4 **103.** 2 **105.** $\sqrt{5}$ **107.** $\frac{1}{2}$ **109.** $10; 0$ **111.** 81 **113.** 304,006.671 **115.** 0.004
117. 481.890 **119.** 0.000 **121.** 4.542×10^2 **123.** 1.3×10^{-2} **125.** 3.2155×10^4 **127.** 4.23×10^{-4} **129.** 61,500 **131.** 0.001214 **133.** 110,000,000
135. 0.081 **137.** $A = lw$ **139.** $C = \pi d$ **141.** $A = \frac{\sqrt{3}}{4}x^2$ **143.** $V = \frac{4}{3}\pi r^3$ **145.** $V = x^3$ **147. (a)** \$6000 **(b)** \$8000 **149.** $|x - 4| \geq 6$
151. (a) $2 \leq 5$ **(b)** $6 > 5$ **153. (a)** Yes **(b)** No **155.** 400,000,000 m **157.** 0.0000005 m **159.** 4.03×10^{-2} in. **161.** 5.865696×10^{12} mi
163. No; $\frac{1}{3}$ is larger; 0.000333... **165.** 3.406×10^{20} **167.** No

R.3 Assess Your Understanding (page 36)

1. right; hypotenuse **2.** $A = \frac{1}{2}bh$ **3.** $C = 2\pi r$ **4.** similar **5.** c **6.** b **7.** T **8.** T **9.** F **10.** T **11.** T **12.** F **13.** 13 **15.** 26 **17.** 25
19. Right triangle; 5 **21.** Not a right triangle **23.** Right triangle; 25 **25.** Not a right triangle **27.** 42 in.^{2} **29.** 28 in.^{2} **31.** $A = 25\pi$ m^2; $C = 10\pi$ m
33. $V = 240$ ft^3; $S = 236$ ft^2 **35.** $V = \frac{500}{3}\pi$ cm^3; $S = 100\pi$ cm^2 **37.** $V = 648\pi$ in.^{3}; $S = 306\pi$ in.^{2} **39.** π square units **41.** 2π square units
43. $x = 4$ units; $A = 90°$; $B = 60°$; $C = 30°$ **45.** $x = 67.5$ units; $A = 60°$; $B = 95°$; $C = 25°$ **47.** About 16.8 ft **49.** 64 ft^2
51. $24 + 2\pi \approx 30.28$ ft^2; $16 + 2\pi \approx 22.28$ ft **53.** 160 paces **55.** About 5.477 mi **57.** From 100 ft: 12.2 mi; From 150 ft: 15.0 mi **59.** 10,575 gal
61. No. The area quadruples.

R.4 Assess Your Understanding (page 47)

1. 4; 3 **2.** $x^4 - 16$ **3.** $x^3 - 8$ **4.** a **5.** c **6.** F **7.** T **8.** F **9.** Monomial; variable: x; coefficient: 2; degree: 3
11. Not a monomial; the exponent of the variable is not a nonnegative integer **13.** Monomial; variables: x, y; coefficient: -2; degree: 3
15. Not a monomial; the exponent of one of the variables is not a nonnegative integer **17.** Not a monomial; it has more than one term
19. Yes; 2 **21.** Yes; 0 **23.** No; the exponent of the variable of one of the terms is not a nonnegative integer **25.** Yes; 3
27. No; the polynomial of the denominator has a degree greater than 0 **29.** $4x^2 + 2x + 15$ **31.** $x^3 - 4x^2 + 9x + 7$ **33.** $6x^5 + 5x^4 + 3x^2 + x$
35. $7x^2 - 3x - 11$ **37.** $-2x^3 + 18x^2 - 18$ **39.** $2x^2 - 4x + 6$ **41.** $11y^2 - 35y + 9$ **43.** $x^4 + 2x^3 - 5x^2$ **45.** $-8x^5 - 10x^2$ **47.** $x^3 + 3x^2 - 2x - 4$
49. $x^2 + 6x + 8$ **51.** $2x^2 + 17x + 35$ **53.** $x^2 - 2x - 8$ **55.** $x^2 - 9x + 18$ **57.** $2x^2 - x - 6$ **59.** $-3x^2 + 10x - 8$ **61.** $2x^2 + 17x + 35$
63. $x^2 - xy - 2y^2$ **65.** $-6x^2 - 13xy - 6y^2$ **67.** $x^2 - 49$ **69.** $4x^2 - 9$ **71.** $x^2 + 8x + 16$ **73.** $x^2 - 8x + 16$ **75.** $9x^2 - 16$ **77.** $4x^2 - 12x + 9$
79. $x^2 - y^2$ **81.** $9x^2 - y^2$ **83.** $x^2 + 2xy + y^2$ **85.** $x^2 - 4xy + 4y^2$ **87.** $x^3 - 6x^2 + 12x - 8$ **89.** $8x^3 + 12x^2 + 6x + 1$
91. $4x^2 - 11x + 23$; remainder -45 **93.** $4x - 3$; remainder $x + 1$ **95.** $5x^2 - 13$; remainder $x + 27$ **97.** $2x^2$; remainder $-x^2 + x + 1$
99. $x^2 - 2x + \frac{1}{2}$; remainder $\frac{5}{2}x + \frac{1}{2}$ **101.** $-4x^2 - 3x - 3$; remainder -7 **103.** $x^2 - x - 1$; remainder $2x + 2$
105. $x^2 + ax + a^2$; remainder 0 **107.** $k = 4$ or $k = -4$

R.5 Assess Your Understanding (page 56)

1. $3x(x-2)(x+2)$ **2.** prime   **3.** c   **4.** b   **5.** d   **6.** c   **7.** T   **8.** F   **9.** $3(x+2)$ **11.** $a(x^2+1)$ **13.** $x(x^2+x+1)$ **15.** $2x(x-1)$
17. $3xy(x-2y+4)$ **19.** $(x+1)(x-1)$ **21.** $(2x+1)(2x-1)$ **23.** $(x+4)(x-4)$ **25.** $(5x+2)(5x-2)$ **27.** $(x+1)^2$ **29.** $(x+2)^2$
31. $(x-5)^2$ **33.** $(2x+1)^2$ **35.** $(4x+1)^2$ **37.** $(x-3)(x^2+3x+9)$ **39.** $(x+3)(x^2-3x+9)$ **41.** $(2x+3)(4x^2-6x+9)$ **43.** $(x+2)(x+3)$
45. $(x+6)(x+1)$ **47.** $(x+5)(x+2)$ **49.** $(x-8)(x-2)$ **51.** $(x-8)(x+1)$ **53.** $(x+8)(x-1)$ **55.** $(x+2)(2x+3)$ **57.** $(x-3)(5x+1)$
59. $(3x+4)(2x+7)$ **61.** $(3x+1)(x+1)$ **63.** $(z+1)(2z+7)$ **65.** $(x+2)(5x-4)$ **67.** $(x-2)(5x+4)$ **69.** $(x+4)(5x+2)$

71. $(x+4)(5x-2)$ **73.** $25;(x+5)^2$ **75.** $9;(y-3)^2$ **77.** $\dfrac{1}{16};\left(x-\dfrac{1}{4}\right)^2$ **79.** $(x+6)(x-6)$ **81.** $2(1+2x)(1-2x)$ **83.** $(x+1)(x+10)$

85. $(x-7)(x-3)$ **87.** $4(x^2-2x+8)$ **89.** Prime   **91.** $-(x-5)(x+3)$ **93.** $3(x+2)(x-6)$ **95.** $y^2(y+5)(y+6)$ **97.** $(2x+3)^2$
99. $2(3x+1)(x+1)$ **101.** $(x-3)(x+3)(x^2+9)$ **103.** $(x-1)^2(x^2+x+1)^2$ **105.** $x^5(x-1)(x+1)$ **107.** $(4x+3)^2$ **109.** $-(4x-5)(4x+1)$
111. $(2y-5)(2y-3)$ **113.** $-(3x-1)(3x+1)(x^2+1)$ **115.** $(x+3)(x-6)$ **117.** $(x+2)(x-3)$ **119.** $(3x-5)(9x^2-3x+7)$
121. $(x+5)(3x+11)$ **123.** $(x-1)(x+1)(x+2)$ **125.** $(x-1)(x+1)(x^2-x+1)$ **127.** $2(3x+4)(9x+13)$ **129.** $2x(3x+5)$
131. $5(x+3)(x-2)^2(x+1)$ **133.** $3(4x-3)(4x-1)$ **135.** $6(3x-5)(2x+1)^2(5x-4)$
137. The possibilities are $(x\pm1)(x\pm4)=x^2\pm5x+4$ or $(x\pm2)(x\pm2)=x^2\pm4x+4$, none of which equals x^2+4.

R.6 Assess Your Understanding (page 60)

1. quotient; divisor; remainder   **2.** $-3\overline{)2\ 0\ -5\ 1}$ **3.** d   **4.** a   **5.** T   **6.** T   **7.** x^2-5x-5; remainder 0 **9.** $3x^2+11x+32$; remainder 99
11. $x^4-3x^3+5x^2-15x+46$; remainder -138 **13.** $4x^5+4x^4+x^3+x^2+2x+2$; remainder 7 **15.** $0.1x^2-0.11x+0.321$; remainder -0.3531
17. $x^4+2x^3+4x^2+8x+16$; remainder 0 **19.** No   **21.** Yes   **23.** Yes   **25.** No   **27.** Yes   **29.** -9 **31.** Yes

R.7 Assess Your Understanding (page 70)

1. lowest terms   **2.** least common multiple   **3.** d   **4.** a   **5.** T   **6.** F   **7.** $\dfrac{3}{x-3}$ **9.** $\dfrac{x}{3}$ **11.** $\dfrac{4x}{2x-1}$ **13.** $\dfrac{y+7}{3(y+1)}$ **15.** $\dfrac{x+6}{x-2}$ **17.** $-(x+5)$

19. $\dfrac{3}{5x(x-2)}$ **21.** $\dfrac{2x(x^2+4x+16)}{x+4}$ **23.** $\dfrac{8}{3x}$ **25.** $\dfrac{x-4}{x+8}$ **27.** $\dfrac{4x}{(x-2)(x-3)}$ **29.** $\dfrac{4}{5(x-1)}$ **31.** $-\dfrac{(x-4)^2}{4x}$ **33.** $\dfrac{(x+3)^2}{(x-3)^2}$ **35.** $\dfrac{(x-2)(x+4)}{(x-1)(3x+1)}$

37. $\dfrac{x+5}{2}$ **39.** $\dfrac{(x-2)(x+2)}{2x-3}$ **41.** $\dfrac{4x+3}{x-4}$ **43.** $\dfrac{x+9}{2x-1}$ **45.** $\dfrac{4-x}{x-2}$ **47.** $\dfrac{4(x+4)}{(x-3)(x+1)}$ **49.** $\dfrac{3x^2-2x-3}{(x+1)(x-1)}$ **51.** $\dfrac{-(11x+2)}{(x+2)(x-2)}$

53. $\dfrac{2(x^2-2)}{x(x-2)(x+2)}$ **55.** $(x-2)(x+2)(x+1)$ **57.** $x(x-1)(x+1)$ **59.** $x^3(2x-1)^2$ **61.** $x(x-1)^2(x+1)(x^2+x+1)$

63. $\dfrac{5x}{(x-6)(x-1)(x+4)}$ **65.** $\dfrac{2(2x^2+5x-2)}{(x-2)(x+2)(x+3)}$ **67.** $\dfrac{5x+1}{(x-1)^2(x+1)^2}$ **69.** $\dfrac{-x^2+3x+13}{(x-2)(x+1)(x+4)}$ **71.** $\dfrac{x^3-2x^2+4x+3}{x^2(x+1)(x-1)}$ **73.** $\dfrac{-1}{x(x+h)}$

75. $\dfrac{x+1}{x-1}$ **77.** $\dfrac{(x-1)(x+1)}{2x(2x+1)}$ **79.** $\dfrac{2(5x-1)}{(x-2)(x+1)^2}$ **81.** $\dfrac{-2x(x^2-2)}{(x+2)(x^2-x-3)}$ **83.** $\dfrac{-1}{x-1}$ **85.** $\dfrac{3x-1}{2x+1}$ **87.** $\dfrac{19}{(3x-5)^2}$ **89.** $\dfrac{(x+1)(x-1)}{(x^2+1)^2}$

91. $\dfrac{x(3x+2)}{(3x+1)^2}$ **93.** $-\dfrac{(x+3)(3x-1)}{(x^2+1)^2}$ **95.** $f=\dfrac{R_1\cdot R_2}{(n-1)(R_1+R_2)};\dfrac{2}{15}$ m   **97.** $k=4$

R.8 Assess Your Understanding (page 78)

3. index   **4.** cube root   **5.** b   **6.** d   **7.** c   **8.** c   **9.** T   **10.** F   **11.** 3 **13.** -2 **15.** $2\sqrt{2}$ **17.** $10\sqrt{7}$ **19.** $2\sqrt[3]{4}$ **21.** $-2x\sqrt[3]{x}$ **23.** $3\sqrt[4]{3}$ **25.** x^3y^2
27. x^2y **29.** $8\sqrt{x}$ **31.** $3x^2y^3\sqrt[4]{2x}$ **33.** $5x\sqrt{3x}$ **35.** $15\sqrt[3]{3}$ **37.** $12\sqrt{3}$ **39.** $7\sqrt{2}$ **41.** $6\sqrt{3}$ **43.** $2\sqrt{3}$ **45.** $-\sqrt[3]{2}$ **47.** $x-2\sqrt{x}+1$
49. $(2x-1)\sqrt[3]{2x}$ **51.** $(2x-15)\sqrt{2x}$ **53.** $-(x+5y)\sqrt[3]{2xy}$ **55.** $\dfrac{\sqrt{2}}{2}$ **57.** $-\dfrac{\sqrt{15}}{5}$ **59.** $\dfrac{(5+\sqrt{2})\sqrt{3}}{23}$ **61.** $\dfrac{8\sqrt{5}-19}{41}$ **63.** $5\sqrt{2}+5$ **65.** $\dfrac{5\sqrt[3]{4}}{2}$
67. $\dfrac{2x+h-2\sqrt{x^2+xh}}{h}$ **69.** $\dfrac{5}{\sqrt{11}-1}$ **71.** $-\dfrac{3}{\sqrt{10}+5}$ **73.** $\dfrac{1}{\sqrt{x}+\sqrt{c}}$ **75.** $\dfrac{1}{\sqrt{x-7}+1}$ **77.** 4 **79.** -4 **81.** 1000 **83.** $\dfrac{1}{8}$ **85.** $\dfrac{27\sqrt{2}}{32}$
87. $\dfrac{27\sqrt{2}}{32}$ **89.** $-\dfrac{1}{10}$ **91.** $\dfrac{25}{16}$ **93.** $x^{7/12}$ **95.** xy^2 **97.** $x^{2/3}y$ **99.** $\dfrac{8x^{5/4}}{y^{3/4}}$ **101.** $\dfrac{3x+2}{(1+x)^{1/2}}$ **103.** $\dfrac{x(3x^2+2)}{(x^2+1)^{1/2}}$ **105.** $\dfrac{22x+5}{10\sqrt{x-5}\,\sqrt{4x+3}}$ **107.** $\dfrac{2+x}{2(1+x)^{3/2}}$
109. $\dfrac{4-x}{(x+4)^{3/2}}$ **111.** $\dfrac{1}{x^2(x^2-1)^{1/2}}$ **113.** $\dfrac{1-3x^2}{2\sqrt{x}(1+x^2)^2}$ **115.** $\dfrac{1}{2}(5x+2)(x+1)^{1/2}$ **117.** $2x^{1/2}(3x-4)(x+1)$ **119.** $(x^2+4)^{1/3}(11x^2+12)$
121. $(3x+5)^{1/3}(2x+3)^{1/2}(17x+27)$ **123.** $\dfrac{3(x+2)}{2x^{1/2}}$ **125.** 1.41 **127.** 1.59 **129.** 4.89 **131.** 2.15 **133. (a)** $15{,}660.4$ gal **(b)** 390.7 gal
135. $2\sqrt{2\pi}\approx8.89$ s   **137.** $x^2+(\sqrt{3}-4)x-4\sqrt{3}$; remainder 1

CHAPTER 1 Equations and Inequalities

1.1 Assess Your Understanding (page 90)

4. F   **5.** identity   **6.** linear; first-degree   **7.** F   **8.** T   **9.** b   **10.** d   **11.** $\{3\}$ **13.** $\{-5\}$ **15.** $\left\{\dfrac{3}{2}\right\}$ **17.** $\left\{\dfrac{7}{5}\right\}$ **19.** $\{-2\}$ **21.** $\{3\}$ **23.** $\{-1\}$ **25.** $\{-2\}$

27. $\{-23\}$ **29.** $\{-4\}$ **31.** $\left\{-\dfrac{3}{4}\right\}$ **33.** $\{-20\}$ **35.** $\{2\}$ **37.** $\{-3\}$ **39.** $\left\{\dfrac{29}{10}\right\}$ **41.** $\left\{\dfrac{1}{2}\right\}$ **43.** $\{2\}$ **45.** $\{8\}$ **47.** $\{2\}$ **49.** $\{-1\}$ **51.** $\{4\}$

53. No solution   **55.** No solution   **57.** $\{-6\}$ **59.** $\{41\}$ **61.** $\left\{-\dfrac{20}{39}\right\}$ **63.** $\{-1\}$ **65.** $\left\{-\dfrac{11}{6}\right\}$ **67.** $\{-14\}$ **69.** $\{5.91\}$ **71.** $\{0.41\}$ **73.** $x=\dfrac{b+c}{a}$

75. $x = \dfrac{abc}{a+b}$ **77.** $a = 3$ **79.** $R = \dfrac{R_1 R_2}{R_1 + R_2}$ **81.** $R = \dfrac{mv^2}{F}$ **83.** $r = \dfrac{S-a}{S}$ **85.** $11,500 will be invested in bonds and $8500 in CDs.

87. The regular hourly rate is $17.50.   **89.** Brooke needs a score of 85.   **91.** The original price was $650; purchasing the refurbished phone saves $78.

93. The theater paid $1.20 for the candy.   **95.** There were 2187 adults.   **97.** The length is 19 ft; the width is 11 ft.

99. breakfast: 675 cal; lunch: 550 cal; dinner: 800 cal   **101.** Judy pays $10.80, and Tom $7.20.   **103.** $\left\{ -\dfrac{16}{107} \right\}$

105. To obtain step (7) we divided both sides by $x - 2$, but in step (1) we were told $x = 2$. We divided by zero!

Historical Problems *(page 100)*

1. The area of each shaded square is 9, so the larger square will have area $85 + 4(9) = 121$. The area of the larger square is also given by the expression $(x + 6)^2$ so $(x + 6)^2 = 121$. Taking the positive square root of each side, $x + 6 = 11$ or $x = 5$.

2. Let $z = -6$, so $z^2 + 12z - 85 = -121$. We get the equation $u^2 - 121 = 0$ or $u^2 = 121$. Thus $u = \pm 11$, so $x = \pm 11 - 6$. $x = -17$ or $x = 5$.

3.
$$\left(x + \frac{b}{2a}\right)^2 = \left(\frac{\sqrt{b^2 - 4ac}}{2a}\right)^2$$
$$\left(x + \frac{b}{2a}\right)^2 - \left(\frac{\sqrt{b^2 - 4ac}}{2a}\right)^2 = 0$$
$$\left(x + \frac{b}{2a} - \frac{\sqrt{b^2 - 4ac}}{2a}\right)\left(x + \frac{b}{2a} + \frac{\sqrt{b^2 - 4ac}}{2a}\right) = 0$$
$$\left(x + \frac{b - \sqrt{b^2 - 4ac}}{2a}\right)\left(x + \frac{b + \sqrt{b^2 - 4ac}}{2a}\right) = 0$$
$$x = \frac{-b + \sqrt{b^2 - 4ac}}{2a} \text{ or } x = \frac{-b - \sqrt{b^2 - 4ac}}{2a}$$

1.2 Assess Your Understanding *(page 101)*

6. discriminant; negative   **7.** F   **8.** F   **9.** b   **10.** d   **11.** $\{0, 9\}$ **13.** $\{-5, 5\}$ **15.** $\{-3, 2\}$ **17.** $\left\{ -\dfrac{1}{2}, 3 \right\}$ **19.** $\{-6, 6\}$ **21.** $\{-5, 2\}$ **23.** $\left\{ \dfrac{3}{2} \right\}$

25. $\left\{ -\dfrac{2}{3}, \dfrac{3}{2} \right\}$ **27.** $\left\{ -\dfrac{2}{3}, \dfrac{3}{2} \right\}$ **29.** $\left\{ -\dfrac{3}{4}, 2 \right\}$ **31.** $\{-5, 5\}$ **33.** $\{-1, 3\}$ **35.** $\{-24, 0\}$ **37.** $\{-7, 3\}$ **39.** $\left\{ -\dfrac{1}{4}, \dfrac{3}{4} \right\}$ **41.** $\left\{ \dfrac{-1 - \sqrt{7}}{6}, \dfrac{-1 + \sqrt{7}}{6} \right\}$

43. $\{-1 - \sqrt{14}, -1 + \sqrt{14}\}$ **45.** $\{2 - \sqrt{5}, 2 + \sqrt{5}\}$ **47.** $\left\{ 1, \dfrac{3}{2} \right\}$ **49.** No real solution   **51.** $\left\{ \dfrac{-4 - \sqrt{61}}{9}, \dfrac{-4 + \sqrt{61}}{9} \right\}$ **53.** $\left\{ 0, \dfrac{9}{4} \right\}$

55. $\left\{ \dfrac{1}{3} \right\}$ **57.** $\left\{ -\dfrac{2}{3}, 1 \right\}$ **59.** $\left\{ \dfrac{3 - \sqrt{29}}{10}, \dfrac{3 + \sqrt{29}}{10} \right\}$ **61.** $\left\{ \dfrac{-2 - \sqrt{10}}{2}, \dfrac{-2 + \sqrt{10}}{2} \right\}$ **63.** $\left\{ \dfrac{-1 - \sqrt{17}}{8}, \dfrac{-1 + \sqrt{17}}{8} \right\}$

65. $\left\{ \dfrac{9 - \sqrt{73}}{2}, \dfrac{9 + \sqrt{73}}{2} \right\}$ **67.** $\{0.63, 3.47\}$ **69.** $\{-2.80, 1.07\}$ **71.** $\{-0.85, 1.17\}$ **73.** No real solution   **75.** Repeated real solution

77. Two unequal real solutions   **79.** $\{-\sqrt{5}, \sqrt{5}\}$ **81.** $\left\{ \dfrac{1}{4} \right\}$ **83.** $\left\{ -\dfrac{3}{5}, \dfrac{5}{2} \right\}$ **85.** $\left\{ -\dfrac{1}{2}, \dfrac{2}{3} \right\}$ **87.** $\left\{ \dfrac{-\sqrt{2} + 2}{2}, \dfrac{-\sqrt{2} - 2}{2} \right\}$

89. $\left\{ \dfrac{-1 - \sqrt{17}}{2}, \dfrac{-1 + \sqrt{17}}{2} \right\}$ **91.** $\{5\}$ **93.** 2; 5 meters, 12 meters, 13 meters; 20 meters, 21 meters, 29 meters   **95.** The dimensions are 11 ft by 13 ft.

97. The dimensions are 5 m by 8 m.   **99.** The dimensions should be 4 ft by 4 ft.

101. (a) The ball strikes the ground after 6 sec.   **(b)** The ball passes the top of the building on its way down after 5 sec.

103. The dimensions should be 11.55 cm by 6.55 cm by 3 cm.   **105.** The border will be 2.71 ft wide.   **107.** The border will be 2.56 ft wide.

109. The screen of an iPad Pro in 16:9 format has an area of 71.11 square inches; the screen of the Surface Pro in 3:2 format has an area of 69.83 square inches. The iPad has a larger screen.

111. 35.3 ft   **113.** 29 h   **115.** 23 years old   **117.** 37 consecutive integers must be added.

119. $\dfrac{-b + \sqrt{b^2 - 4ac}}{2a} + \dfrac{-b - \sqrt{b^2 - 4ac}}{2a} = \dfrac{-2b}{2a} = -\dfrac{b}{a}$ **121.** $k = \dfrac{1}{2}$ or $k = -\dfrac{1}{2}$

123. $ax^2 + bx + c = 0, x = \dfrac{-b \pm \sqrt{b^2 - 4ac}}{2a}$; $ax^2 - bx + c = 0, x = \dfrac{b \pm \sqrt{(-b)^2 - 4ac}}{2a} = \dfrac{b \pm \sqrt{b^2 - 4ac}}{2a} = -\dfrac{-b \pm \sqrt{b^2 - 4ac}}{2a}$

125. $\sqrt[4]{2}$ m by $\dfrac{\sqrt[4]{8}}{2}$ m

1.3 Assess Your Understanding *(page 111)*

4. real; imaginary; imaginary unit   **5.** F   **6.** T   **7.** F   **8.** b   **9.** a   **10.** c   **11.** $8 + 5i$ **13.** $-7 + 6i$ **15.** $-6 - 11i$ **17.** $6 - 18i$ **19.** $18 - 21i$

21. $10 - 5i$ **23.** 26   **25.** $\dfrac{6}{5} + \dfrac{8}{5}i$ **27.** $1 - 2i$ **29.** $\dfrac{5}{2} - \dfrac{7}{2}i$ **31.** $-\dfrac{1}{2} + \dfrac{\sqrt{3}}{2}i$ **33.** $2i$ **35.** $-i$ **37.** 1   **39.** -6 **41.** $-10i$ **43.** $-2 + 2i$ **45.** 0

47. 0   **49.** $2i$ **51.** $5i$ **53.** $2\sqrt{3}i$ **55.** $10\sqrt{2}i$ **57.** $5i$ **59.** $\{-2i, 2i\}$ **61.** $\{-4, 4\}$ **63.** $\{3 - 2i, 3 + 2i\}$ **65.** $\{3 - i, 3 + i\}$ **67.** $\left\{ \dfrac{1}{5} - \dfrac{1}{5}i, \dfrac{1}{5} + \dfrac{1}{5}i \right\}$

69. $\left\{ \dfrac{1}{5} - \dfrac{2}{5}i, \dfrac{1}{5} + \dfrac{2}{5}i \right\}$ **71.** $\left\{ -\dfrac{1}{2} - \dfrac{\sqrt{3}}{2}i, -\dfrac{1}{2} + \dfrac{\sqrt{3}}{2}i \right\}$ **73.** $\{4, -2 - 2\sqrt{3}i, -2 + 2\sqrt{3}i\}$ **75.** $\{-2, 2, -2i, 2i\}$ **77.** $\{-3i, -2i, 2i, 3i\}$

79. Two complex solutions that are conjugates of each other   **81.** Two unequal real solutions   **83.** A repeated real solution

85. $2 - 3i$ **87.** 6   **89.** 25   **91.** $2 + 3i$ ohms   **93.** $z + \bar{z} = (a + bi) + (a - bi) = 2a$; $z - \bar{z} = (a + bi) - (a - bi) = 2bi$

95. $\overline{z + w} = \overline{(a + bi) + (c + di)} = \overline{(a + c) + (b + d)i} = (a + c) - (b + d)i = (a - bi) + (c - di) = \bar{z} + \bar{w}$

97. Any complex number of the form $a + ai$ or $a - ai$.   **99.** -10

1.4 Assess Your Understanding *(page 117)*

4. F **5.** quadratic in form **6.** T **7.** a **8.** c **9.** $\{1\}$ **11.** No real solution **13.** $\{-13\}$ **15.** $\{-1\}$ **17.** $\{0, 64\}$ **19.** $\{3\}$ **21.** $\{2\}$ **23.** $\left\{-\dfrac{8}{5}\right\}$ **25.** $\{8\}$

27. $\{2\}$ **29.** $\{18\}$ **31.** $\{1, 5\}$ **33.** $\{1\}$ **35.** $\{5\}$ **37.** $\{2\}$ **39.** $\{-4, 4\}$ **41.** $\{0, 3\}$ **43.** $\{-2, -1, 1, 2\}$ **45.** $\{-1, 1\}$ **47.** $\{-2, 1\}$ **49.** $\{-6, -5\}$

51. $\left\{\dfrac{7}{2}\right\}$ **53.** $\left\{-\dfrac{3}{2}, 2\right\}$ **55.** $\left\{0, \dfrac{1}{16}\right\}$ **57.** $\{16\}$ **59.** $\{1\}$ **61.** $\{1, 16\}$ **63.** $\left\{\sqrt{2}, \sqrt{3}\right\}$ **65.** $\{-4, 1\}$ **67.** $\left\{-2, -\dfrac{1}{2}\right\}$ **69.** $\left\{-\dfrac{3}{2}, \dfrac{1}{3}\right\}$

71. $\left\{-\dfrac{1}{8}, 27\right\}$ **73.** $\left\{-4, -\dfrac{5}{3}\right\}$ **75.** $\{-3, 0, 3\}$ **77.** $\left\{0, \dfrac{3}{4}\right\}$ **79.** $\{-5, 0, 4\}$ **81.** $\{-1, 1\}$ **83.** $\{-4, 3, 4\}$ **85.** $\left\{-2, \dfrac{1}{2}, 2\right\}$ **87.** $\left\{\dfrac{7}{3}\right\}$

89. $\left\{0, \dfrac{5}{2}, 3\right\}$ **91.** $\{0.34, 11.66\}$ **93.** $\{-1.03, 1.03\}$ **95.** $\{-1.85, 0.17\}$ **97.** $\left\{-4, \dfrac{5}{3}\right\}$ **99.** $\left\{-3, -\dfrac{2}{5}, 3\right\}$ **101.** $\{5\}$ **103.** $\left\{-2, -\dfrac{4}{5}\right\}$

105. $\{2\}$ **107.** $\left\{\dfrac{-3 - \sqrt{6}}{3}, \dfrac{-3 + \sqrt{6}}{3}\right\}$ **109.** $\left\{\sqrt{2}, \sqrt{3}\right\}$ **111.** $\left\{\dfrac{3}{2}, 5\right\}$ **113.** The depth of the well is 229.94 ft. **115.** 220.7 ft **117.** $\{11\}$

119. $\left\{-3, -1, \dfrac{1}{2} - \dfrac{\sqrt{3}}{2}i, \dfrac{1}{2} + \dfrac{\sqrt{3}}{2}i, \dfrac{3}{2} - \dfrac{3\sqrt{3}}{2}i, \dfrac{3}{2} + \dfrac{3\sqrt{3}}{2}i\right\}$ **123.** -1 is extraneous.

1.5 Assess Your Understanding *(page 127)*

3. closed interval **4.** Multiplication Properties **5.** T **6.** T **7.** T **8.** F **9.** T **10.** F **11.** d **12.** c **13.** $[0, 2]; 0 \le x \le 2$ **15.** $[2, \infty); x \ge 2$

17. $[0, 3); 0 \le x < 3$ **19. (a)** $6 < 8$ **(b)** $-2 < 0$ **(c)** $9 < 15$ **(d)** $-6 > -10$ **21. (a)** $7 > 0$ **(b)** $-1 > -8$ **(c)** $12 > -9$ **(d)** $-8 < 6$

23. (a) $2x + 4 < 5$ **(b)** $2x - 4 < -3$ **(c)** $6x + 3 < 6$ **(d)** $-4x - 2 > -4$

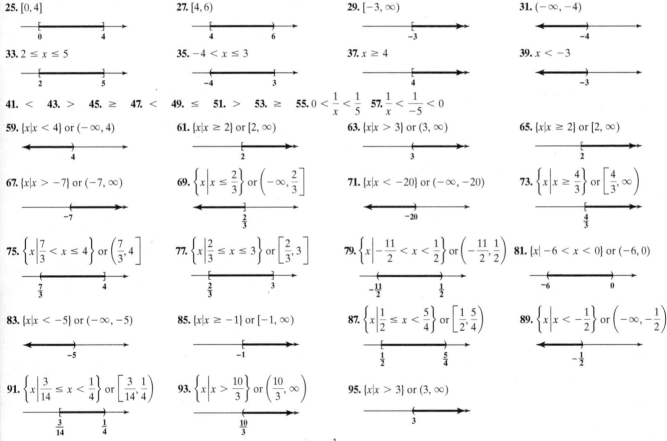

25. $[0, 4]$ **27.** $[4, 6)$ **29.** $[-3, \infty)$ **31.** $(-\infty, -4)$

33. $2 \le x \le 5$ **35.** $-4 < x \le 3$ **37.** $x \ge 4$ **39.** $x < -3$

41. $<$ **43.** $>$ **45.** $\ge$ **47.** $<$ **49.** $\le$ **51.** $>$ **53.** $\ge$ **55.** $0 < \dfrac{1}{x} < \dfrac{1}{5}$ **57.** $\dfrac{1}{x} < \dfrac{1}{-5} < 0$

59. $\{x | x < 4\}$ or $(-\infty, 4)$ **61.** $\{x | x \ge 2\}$ or $[2, \infty)$ **63.** $\{x | x > 3\}$ or $(3, \infty)$ **65.** $\{x | x \ge 2\}$ or $[2, \infty)$

67. $\{x | x > -7\}$ or $(-7, \infty)$ **69.** $\left\{x \middle| x \le \dfrac{2}{3}\right\}$ or $\left(-\infty, \dfrac{2}{3}\right]$ **71.** $\{x | x < -20\}$ or $(-\infty, -20)$ **73.** $\left\{x \middle| x \ge \dfrac{4}{3}\right\}$ or $\left[\dfrac{4}{3}, \infty\right)$

75. $\left\{x \middle| \dfrac{7}{3} < x \le 4\right\}$ or $\left(\dfrac{7}{3}, 4\right]$ **77.** $\left\{x \middle| \dfrac{2}{3} \le x \le 3\right\}$ or $\left[\dfrac{2}{3}, 3\right]$ **79.** $\left\{x \middle| -\dfrac{11}{2} < x < \dfrac{1}{2}\right\}$ or $\left(-\dfrac{11}{2}, \dfrac{1}{2}\right)$ **81.** $\{x | -6 < x < 0\}$ or $(-6, 0)$

83. $\{x | x < -5\}$ or $(-\infty, -5)$ **85.** $\{x | x \ge -1\}$ or $[-1, \infty)$ **87.** $\left\{x \middle| \dfrac{1}{2} \le x < \dfrac{5}{4}\right\}$ or $\left[\dfrac{1}{2}, \dfrac{5}{4}\right)$ **89.** $\left\{x \middle| x < -\dfrac{1}{2}\right\}$ or $\left(-\infty, -\dfrac{1}{2}\right)$

91. $\left\{x \middle| \dfrac{3}{14} \le x < \dfrac{1}{4}\right\}$ or $\left[\dfrac{3}{14}, \dfrac{1}{4}\right)$ **93.** $\left\{x \middle| x > \dfrac{10}{3}\right\}$ or $\left(\dfrac{10}{3}, \infty\right)$ **95.** $\{x | x > 3\}$ or $(3, \infty)$

97. $a = 3, b = 5$ **99.** $a = -12, b = -8$ **101.** $a = 3, b = 11$ **103.** $a = \dfrac{1}{4}, b = 1$ **105.** $a = 4, b = 16$ **107.** $\{x | x \ge -2\}$ **109.** $21 < \text{Age} < 30$

111. (a) Male ≥ 82.2 years **(b)** Female ≥ 85.8 years **(c)** A female could expect to live 3.6 years longer.

113. The agent's commission ranges from $45,000 to $95,000, inclusive. As a percent of selling price, the commission ranges from 5% to 8.6%, inclusive. **115.** The amount withheld varies from $104.32 to $148.32, inclusive. **117.** The usage varied from 1150 kWh to 2050 kWh, inclusive.

119. 5 cookies **121. (a)** You need at least a 74 on the fifth test. **(b)** You need at least a 77 on the fifth test.

123. $\dfrac{a + b}{2} - a = \dfrac{a + b - 2a}{2} = \dfrac{b - a}{2} > 0$; therefore, $a < \dfrac{a + b}{2}$. $b - \dfrac{a + b}{2} = \dfrac{2b - a - b}{2} = \dfrac{b - a}{2} > 0$; therefore, $b > \dfrac{a + b}{2}$.

125. $(\sqrt{ab})^2 - a^2 = ab - a^2 = a(b - a) > 0$; thus $(\sqrt{ab})^2 > a^2$ so $\sqrt{ab} > a$.

$b^2 - (\sqrt{ab})^2 = b^2 - ab = b(b - a) > 0$; thus $b^2 > (\sqrt{ab})^2$ so $b > \sqrt{ab}$.

127. $h - a = \dfrac{2ab}{a + b} - a = \dfrac{ab - a^2}{a + b} = \dfrac{a(b - a)}{a + b} > 0$; so $h > a$. $b - h = b - \dfrac{2ab}{a + b} = \dfrac{b^2 - ab}{a + b} = \dfrac{b(b - a)}{a + b} > 0$; so $h < b$. **129.** $\left(-1, \dfrac{14}{5}\right]$

1.6 Assess Your Understanding *(page 132)*

3. $\{-5, 5\}$ **4.** $\{x|-5 < x < 5\}$ **5.** T **6.** T **7.** d **8.** a **9.** $\{-5, 5\}$ **11.** $\{-4, 1\}$ **13.** $\left\{-1, \frac{3}{2}\right\}$ **15.** $\{-4, 4\}$ **17.** $\{2\}$ **19.** $\left\{-\frac{21}{8}, \frac{21}{8}\right\}$

21. $\left\{-\frac{36}{5}, \frac{24}{5}\right\}$ **23.** No solution **25.** $\left\{-\frac{1}{4}, \frac{1}{4}\right\}$ **27.** $\{-3, 3\}$ **29.** $\{-1, 3\}$ **31.** $\{-2, -1, 0, 1\}$ **33.** $\left\{\frac{13}{11}, 7\right\}$ **35.** $\left\{-\frac{1}{2}, 0\right\}$

37. $\{x|-4 < x < 4\}; (-4, 4)$ **39.** $\{x|x < -6 \text{ or } x > 6\}; (-\infty, -6) \cup (6, \infty)$ **41.** $\{x|1 < x < 3\}; (1, 3)$ **43.** $\left\{t\left|-\frac{2}{3} \le t \le 2\right.\right\}; \left[-\frac{2}{3}, 2\right]$

45. $\left\{x\left|x \le \frac{1}{2} \text{ or } x \ge \frac{5}{2}\right.\right\}; \left(-\infty, \frac{1}{2}\right] \cup \left[\frac{5}{2}, \infty\right)$ **47.** $\left\{x\left|-1 < x < \frac{3}{2}\right.\right\}; \left(-1, \frac{3}{2}\right)$ **49.** $\{x|x < -1 \text{ or } x > 6\}; (-\infty, -1) \cup (6, \infty)$

51. No solution **53.** $\left\{x\left|x < -\frac{3}{2} \text{ or } x > \frac{3}{2}\right.\right\}; \left(-\infty, -\frac{3}{2}\right) \cup \left(\frac{3}{2}, \infty\right)$ **55.** $\{x|-1 \le x \le 6\}; [-1, 6]$

57. No solution **59.** All real numbers; $(-\infty, \infty)$ **61.** $\left\{x\left|-\frac{9}{4} < x < \frac{3}{4}\right.\right\}; \left(-\frac{9}{4}, \frac{3}{4}\right)$ **63.** No solution

65. $\left\{\frac{7}{2}\right\}$ **67.** $\left\{x\left|x < -\frac{13}{3} \text{ or } x > 9\right.\right\}; \left(-\infty, -\frac{13}{3}\right) \cup (9, \infty)$ **69.** All real numbers; $(-\infty, \infty)$

71. $|x - 98.6| \ge 1.5; x \le 97.1 \text{ or } x \ge 100.1$ **73.** $|x - 64| \le 3.9; 60.1 \le x \le 67.9$, between 60.1% and 67.9%, inclusive.
75. $|x - 3| < \frac{1}{2}; \frac{5}{2} < x < \frac{7}{2}$ **77.** $|x + 3| > 2; x < -5 \text{ or } x > -1$ **79.** $a = 2, b = 8$ **81.** $a = -15, b = -7$ **83.** $a = -1, b = -\frac{1}{15}$
85. $b - a = (\sqrt{b} - \sqrt{a})(\sqrt{b} + \sqrt{a})$. Since $\sqrt{b} - \sqrt{a} > 0, \sqrt{a} > 0, \sqrt{b} > 0$, then $b - a > 0$, so $a < b$.
87. $(a + b)^2 = a^2 + 2ab + b^2 \le |a|^2 + 2|a||b| + |b|^2 = (|a| + |b|)^2$; thus, $|a + b| \le |a| + |b|$.
89. $x^2 - a < 0; (x - \sqrt{a})(x + \sqrt{a}) < 0$; therefore, $-\sqrt{a} < x < \sqrt{a}$. **91.** $\{x|-1 < x < 1\}$ **93.** $\{x|x \le -3 \text{ or } x \ge 3\}$ **95.** $\{x|-4 \le x \le 4\}$
97. $\{x|x < -2 \text{ or } x > 2\}$ **99.** $\{-1, 5\}$ **101.** $\frac{1}{9}$

1.7 Assess Your Understanding *(page 140)*

1. mathematical modeling **2.** interest **3.** uniform motion **4.** F **5.** T **6.** a **7.** b **8.** c **9.** $A = \pi r^2; r = $ radius, $A = $ area
11. $A = s^2; A = $ area, $s = $ length of a side **13.** $F = ma; F = $ force, $m = $ mass, $a = $ acceleration **15.** $W = Fd; W = $ work, $F = $ force, $d = $ distance
17. $C = 150x; C = $ total variable cost, $x = $ number of dishwashers **19.** Invest $31,250 in bonds and $18,750 in the CD.
21. $11,600 was loaned out at 8%. **23.** Mix 75 lb of Earl Grey tea with 25 lb of Orange Pekoe tea. **25.** Mix 156 lb of cashews with the almonds.
27. The speed of the current is 2.286 mi/h. **29.** The speed of the current is 5 mi/h. **31.** Karen walked at 3.75 ft/sec.
33. A doubles tennis court is 78 feet long and 36 feet wide. **35.** Working together, it takes 12 min.
37. (a) The dimensions are 10 ft by 5 ft. **(b)** The area is 50 sq ft. **(c)** The dimensions would be 7.5 ft by 7.5 ft. **(d)** The area would be 56.25 sq ft.
39. The defensive back catches up to the tight end at the tight end's 45-yd line. **41.** Add $\frac{2}{3}$ gal of water. **43.** Evaporate 10.67 oz of water.
45. 40 g of 12-karat gold should be mixed with 20 g of pure gold.
47. Mike passes Dan $\frac{1}{3}$ mile from the start, 2 min from the time Mike started to run. **49.** Start the auxiliary pump at 9:45 AM.
51. The tub will fill in 1 hour. **53.** Run: 12 miles; bicycle: 75 miles **55.** Bolt would beat Burke by 18.25 m.
57. It would take Elaine, Brian, and one of their daughters $\frac{12}{7}$ hours to complete the project. **59.** The train is 190.67 feet long.
61. Set the original price at $40. At 50% off, there will be no profit. **63.** The tail wind was 91.47 knots.

Review Exercises *(page 145)*

1. $\{-18\}$ **2.** $\{6\}$ **3.** $\{6\}$ **4.** $\left\{\frac{-7 - 2\sqrt{5}}{2}, \frac{-7 + 2\sqrt{5}}{2}\right\}$ **5.** No real solution **6.** $\{-3, 2\}$ **7.** $\left\{\frac{11}{8}\right\}$ **8.** $\left\{-\frac{27}{13}\right\}$ **9.** $\left\{-2, \frac{3}{2}\right\}$
10. $\left\{\frac{1 - \sqrt{13}}{4}, \frac{1 + \sqrt{13}}{4}\right\}$ **11.** $\{-3, 3\}$ **12.** $\{2\}$ **13.** No real solution **14.** $\{-2, -1, 1, 2\}$ **15.** $\{2\}$ **16.** $\left\{\frac{13}{2}\right\}$ **17.** $\left\{\frac{\sqrt{5}}{2}\right\}$ **18.** $\{5, 41\}$
19. $\left\{\frac{9}{4}\right\}$ **20.** $\left\{-1, \frac{1}{2}\right\}$ **21.** $\left\{\frac{m}{1 - n}, \frac{m}{1 + n}\right\}$ **22.** $\left\{-\frac{9b}{5a}, \frac{2b}{a}\right\}$ **23.** $\left\{-\frac{9}{5}\right\}$ **24.** $\{-5, 2\}$ **25.** $\left\{-\frac{5}{3}, 3\right\}$ **26.** $\left\{0, \frac{3}{2}\right\}$ **27.** $\left\{-\frac{5}{2}, -2, 2\right\}$

28. $\{x|x \geq 14\}$; $[14, \infty)$

29. $\left\{x\left|-\dfrac{31}{2} \leq x \leq \dfrac{33}{2}\right.\right\}$; $\left[-\dfrac{31}{2}, \dfrac{33}{2}\right]$

30. $\{x|-23 < x < -7\}$; $(-23, -7)$

31. $\left\{x\left|-\dfrac{3}{2} < x < -\dfrac{7}{6}\right.\right\}$; $\left(-\dfrac{3}{2}, -\dfrac{7}{6}\right)$

32. $\{x|x \leq -2 \text{ or } x \geq 7\}$; $(-\infty, -2] \cup [7, \infty)$

33. $\left\{x\left|0 \leq x \leq \dfrac{4}{3}\right.\right\}$; $\left[0, \dfrac{4}{3}\right]$

34. $\left\{x\left|x < -1 \text{ or } x > \dfrac{7}{3}\right.\right\}$; $(-\infty, -1) \cup \left(\dfrac{7}{3}, \infty\right)$

35. $4 + 7i$ **36.** $-3 + 2i$ **37.** $\dfrac{9}{10} - \dfrac{3}{10}i$ **38.** -1 **39.** $-46 + 9i$ **40.** $\left\{-\dfrac{1}{2} - \dfrac{\sqrt{3}}{2}i, -\dfrac{1}{2} + \dfrac{\sqrt{3}}{2}i\right\}$ **41.** $\left\{\dfrac{-1 - \sqrt{17}}{4}, \dfrac{-1 + \sqrt{17}}{4}\right\}$

42. $\left\{\dfrac{1}{2} - \dfrac{\sqrt{11}}{2}i, \dfrac{1}{2} + \dfrac{\sqrt{11}}{2}i\right\}$ **43.** $\left\{\dfrac{1}{2} - \dfrac{\sqrt{23}}{2}i, \dfrac{1}{2} + \dfrac{\sqrt{23}}{2}i\right\}$ **44.** $C = 50{,}000 + 95x$ **45.** Bonds: \$50,000; CD: \$20,000

46. The storm is 3300 ft away.   **47.** From 0.5 m to 0.75 m   **48.** The search plane can go as far as 616 miles.
49. The helicopter will reach the life raft in a little less than 1 hr 35 min.   **50. (a)** In 8 sec   **(b)** The height is 896 ft.
51. It takes Clarissa 10 days by herself.   **52.** 12 hr 23 min   **53.** Add 256 oz of water.   **54.** 6 in. by 8 in.
55. Mix 90 cm^3 of 15% HCl with the 60 cm^3 of 40% HCl to obtain 150 cm^3 of 25% HCl.

56. (a) 6.5 in. by 6.5 in.; 12.5 in. by 12.5 in.   **(b)** $8\dfrac{2}{3}$ in. by $4\dfrac{1}{3}$ in.; $14\dfrac{2}{3}$ in. by $10\dfrac{1}{3}$ in.

57. Scott receives \$400,000, Alice receives \$300,000, and Tricia receives \$200,000.   **58.** It would take the older copier 180 min or 3 h.

59. (a) No   **(b)** Todd   **(c)** $\dfrac{1}{4}$ m   **(d)** 5.26 m   **(e)** Yes   **60.** Lenah's interest rate was 4.25%.

Chapter Test *(page 147)*

1. $\left\{\dfrac{5}{2}\right\}$ **2.** $\{-2, 3\}$ **3.** $\{-2, 2\}$ **4.** $\left\{\dfrac{9}{2}\right\}$ **5.** $\{0, 3\}$ **6.** $\left\{-2, -\dfrac{2}{3}, 2\right\}$ **7.** No real solution   **8.** $\left[-\dfrac{2}{3}, \dfrac{16}{3}\right]$

9. $\left(-4, \dfrac{4}{3}\right)$

10. $(-\infty, -1] \cup [6, \infty)$

11. $-\dfrac{3}{5} - \dfrac{1}{5}i$ **12.** $\left\{\dfrac{1}{2} - i, \dfrac{1}{2} + i\right\}$ **13.** Add $6\dfrac{2}{3}$ lb of \$8/lb coffee to get $26\dfrac{2}{3}$ lb of \$5/lb coffee.

CHAPTER 2 Graphs

2.1 Assess Your Understanding *(page 154)*

7. x-coordinate or abscissa; y-coordinate or ordinate   **8.** quadrants   **9.** midpoint   **10.** F   **11.** F   **12.** T   **13.** b   **14.** a

15. (a) Quadrant II   **(b)** x-axis
(c) Quadrant III   **(d)** Quadrant I
(e) y-axis   **(f)** Quadrant IV

17. The points will be on a vertical line that is 2 units to the right of the y-axis.

19. $\sqrt{5}$
21. $\sqrt{10}$
23. $2\sqrt{17}$
25. $\sqrt{130}$
27. $\sqrt{10}$
29. $\sqrt{6.89} \approx 2.62$
31. $\sqrt{a^2 + b^2}$

33. $d(A, B) = \sqrt{13}$
$d(B, C) = \sqrt{13}$
$d(A, C) = \sqrt{26}$
$(\sqrt{13})^2 + (\sqrt{13})^2 = (\sqrt{26})^2$
Area $= \dfrac{13}{2}$ square units

35. $d(A, B) = \sqrt{130}$
$d(B, C) = \sqrt{26}$
$d(A, C) = 2\sqrt{26}$
$(\sqrt{26})^2 + (2\sqrt{26})^2 = (\sqrt{130})^2$
Area $= 26$ square units

37. $d(A, B) = 4$
$d(B, C) = \sqrt{41}$
$d(A, C) = 5$
$4^2 + 5^2 = (\sqrt{41})^2$
Area $= 10$ square units

39. $(4, 0)$ **41.** $\left(\dfrac{7}{2}, 2\right)$
43. $(8, -2)$ **45.** $\left(\dfrac{a}{2}, \dfrac{b}{2}\right)$
47. $(5, 3)$ **49.** $(3, -13), (3, 11)$

51. $(4 + 3\sqrt{3}, 0); (4 - 3\sqrt{3}, 0)$ **53. (a)** $(-1, 1)$ **(b)** $(0, 13)$ **55.** $(1, 2)$ **57.** $\sqrt{17}; 2\sqrt{5}; \sqrt{29}$
59. $d(P_1, P_2) = 6; d(P_2, P_3) = 4; d(P_1, P_3) = 2\sqrt{13}$; right triangle   **61.** $d(P_1, P_2) = 2\sqrt{17}; d(P_2, P_3) = \sqrt{34}; d(P_1, P_3) = \sqrt{34}$; isosceles right triangle
63. $90\sqrt{2} \approx 127.28$ ft   **65. (a)** $(90, 0), (90, 90), (0, 90)$ **(b)** $5\sqrt{2161} \approx 232.43$ ft   **(c)** $30\sqrt{149} \approx 366.20$ ft   **67.** $d = 75t$ mi

69. (a) $(2.65, 1.6)$ **(b)** Approximately 1.285 units   **71.** \$23,307; a slight underestimate   **73.** $\left(\dfrac{s}{2}, \dfrac{s}{2}\right)$

75. Arrange the parallelogram on the coordinate plane so that the vertices are $P_1 = (0, 0)$, $P_2 = (a, 0)$, $P_3 = (a + b, c)$ and $P_4 = (b, c)$.

Then the lengths of the sides are $d(P_1, P_2) = a$, $d(P_2, P_3) = \sqrt{b^2 + c^2}$, $d(P_3, P_4) = a$, and $d(P_1, P_4) = \sqrt{b^2 + c^2}$, and the lengths

of the diagonals are $d(P_1, P_3) = \sqrt{(a + b)^2 + c^2}$ and $d(P_2, P_4) = \sqrt{(a - b)^2 + c^2}$. So, the sum of the squares of the lengths of the sides is

$a^2 + (\sqrt{b^2 + c^2})^2 + a^2 + (\sqrt{b^2 + c^2})^2 = a^2 + b^2 + c^2 + a^2 + b^2 + c^2 = 2a^2 + 2b^2 + 2c^2$. The sum of the squares of the lengths of the diagonals is

$(\sqrt{(a + b)^2 + c^2})^2 + (\sqrt{(a - b)^2 + c^2})^2 = (a + b)^2 + c^2 + (a - b)^2 + c^2 = a^2 + 2ab + b^2 + c^2 + a^2 - 2ab + b^2 + c^2 = 2a^2 + 2b^2 + 2c^2$.

77. $\left\{x \,\middle|\, x \neq \dfrac{5}{2}\right\}$ **78.** -11 **79.** $15x^2 + 29x - 14$ **80.** $\dfrac{x+2}{x-4}$ **81.** $7(x+1)^3 (2x-5)^6 (x-5)$

82. $\left\{-\dfrac{5}{3}, 4\right\}$ **83.** $\{6\}$ **84.** $\left\{-\dfrac{27}{7}, 5\right\}$ **85.** $(-\infty, 2]$ ⟵——● $-3\ -2\ -1\ \ 0\ \ 1\ \ 2\ \ 3\ \ 4$ **86.** $13 - 11i$

2.2 Assess Your Understanding *(page 166)*

3. intercepts **4.** $y = 0$ **5.** y-axis **6.** 4 **7.** $(-3, 4)$ **8.** T **9.** F **10.** F **11.** d **12.** c **13.** $(0, 0)$ is on the graph. **15.** $(0, 3)$ is on the graph.
17. $(0, 2)$ and $(\sqrt{2}, \sqrt{2})$ are on the graph.

19. $(-2, 0), (0, 2)$ **21.** $(-4, 0), (0, 8)$ **23.** $(-1, 0), (1, 0), (0, -1)$ **25.** $(-2, 0), (2, 0), (0, 4)$ **27.** $(3, 0), (0, 2)$

29. $(-2, 0), (2, 0), (0, 9)$ **31.** **33.** **35.** **37.**

39. **41.** (a) $(-1, 0), (1, 0)$ **43.** (a) $\left(-\dfrac{\pi}{2}, 0\right), (0, 1), \left(\dfrac{\pi}{2}, 0\right)$ **45.** (a) $(0, 0)$
(b) Symmetric with respect to the x-axis, the y-axis, and the origin (b) Symmetric with respect to the y-axis (b) Symmetric with respect to the x-axis

47. (a) $(-2, 0), (0, 0), (2, 0)$ **49.** (a) $(x, 0), -2 \le x \le 1$ **51.** (a) No intercepts
(b) Symmetric with respect to the origin (b) No symmetry (b) Symmetric with respect to the origin

53. **55.** **57.** $(-16, 0), (0, -4), (0, 4)$; symmetric with respect to the x-axis
59. $(0, 0)$; symmetric with respect to the origin
61. $(0, 9), (3, 0), (-3, 0)$; symmetric with respect to the y-axis
63. $(-2, 0), (2, 0), (0, -5), (0, 5)$; symmetric with respect to the x-axis, the y-axis, and the origin
65. $(0, -64), (4, 0)$; no symmetry **67.** $(0, -8), (4, 0), (-2, 0)$; no symmetry
69. $(0, 0)$; symmetric with respect to the origin **71.** $(0, 0)$; symmetric with respect to the origin

73. **75.** **77.** $a = -4$ or $a = 1$ **79.** $(-1, -2)$ **81.** 4
83. (a) $(0, 0), (2, 0), (0, 1), (0, -1)$ (b) x-axis symmetry
85. $(0, 0), (-a, 0), (a, 0)$; symmetric with respect to the x-axis, the y-axis, and the origin
94. $\dfrac{1}{2}$ **95.** $3(x - 5)^2$ **96.** $72x^{12}y^{11}$ **97.** $2x^3 - x^2 - 3x - 2$; remainder 0

98. $\{2\}$ **99.** $[-4, 2)$ ⟵—[————)—⟶ $-5\ -4\ -3\ -2\ -1\ \ 0\ \ 1\ \ 2\ \ 3\ \ 4$ **100.** $\left\{4 - 2\sqrt{3}, 4 + 2\sqrt{3}\right\}$ **101.** $\left\{\dfrac{-1 + \sqrt{13}}{6}, \dfrac{-1 - \sqrt{13}}{6}\right\}$ **102.** 24.49 mm **103.** $14i$

2.3 Assess Your Understanding *(page 180)*

1. undefined; 0 **2.** 3; 2 **3.** T **4.** F **5.** T **6.** $m_1 = m_2$; y-intercepts; $m_1 m_2 = -1$ **7.** 2 **8.** $-\dfrac{1}{2}$ **9.** c **10.** d **11.** b **12.** d

13. (a) Slope $= \dfrac{1}{2}$ **15.** (a) Slope $= -\dfrac{1}{3}$ **17.** Slope $= -\dfrac{3}{2}$ **19.** Slope $= -\dfrac{1}{2}$ **21.** Slope $= 0$ **23.** Slope undefined
(b) If x increases by 2 units, y will increase by 1 unit. (b) If x increases by 3 units, y will decrease by 1 unit.

25. **27.** **29.** **31.** **33.** $y - 2 = 3(x - 1)$ **35.** $y - 4 = -\dfrac{3}{4}(x - 2)$
37. $y - 3 = 0$ **39.** $(2, 6); (3, 10); (4, 14)$
41. $(4, -7); (6, -10); (8, -13)$
43. $(-1, -5); (0, -7); (1, -9)$

45. $x - 2y = 0$ or $y = \frac{1}{2}x$ **47.** $x + y = 2$ or $y = -x + 2$ **49.** $2x - y = 3$ or $y = 2x - 3$ **51.** $x + 2y = 5$ or $y = -\frac{1}{2}x + \frac{5}{2}$

53. $3x - y = -9$ or $y = 3x + 9$ **55.** $x - 2y = 1$ or $y = \frac{1}{2}x - \frac{1}{2}$ **57.** $x - 2y = -5$ or $y = \frac{1}{2}x + \frac{5}{2}$ **59.** $3x + y = 3$ or $y = -3x + 3$

61. $x - y = -4$ or $y = x + 4$ **63.** $x = 2$; no slope-intercept form   **65.** $y = 2$ **67.** $2x - y = -4$ or $y = 2x + 4$ **69.** $x - 2y = 0$ or $y = \frac{1}{2}x$

71. $x = 4$; no slope-intercept form   **73.** $2x + y = 0$ or $y = -2x$ **75.** $2x + y = 4$ or $y = -2x + 4$ **77.** $y = 4$

79. Slope $= 2$; y-intercept $= 3$ **81.** Slope $= 2$; y-intercept $= -2$ **83.** Slope $= \frac{1}{2}$; y-intercept $= 2$ **85.** Slope $= -\frac{1}{2}$; y-intercept $= 2$

87. Slope $= \frac{2}{3}$; y-intercept $= -2$ **89.** Slope $= -1$; y-intercept $= 1$ **91.** Slope undefined; no y-intercept   **93.** Slope $= 0$; y-intercept $= 5$

95. Slope $= 1$; y-intercept $= 0$ **97.** Slope $= \frac{3}{2}$; y-intercept $= 0$ **99.** (a) x-intercept: 3; y-intercept: 2   **101.** (a) x-intercept: -10; y-intercept: 8

(b)   (b)

103. (a) x-intercept: 3; y-intercept: $\frac{21}{2}$ **105.** (a) x-intercept: 2; y-intercept: 3   **107.** (a) x-intercept: 5; y-intercept: -2 **109.** $y = 0$
111. Parallel   **113.** Neither
115. $x - y = -2$ or $y = x + 2$

(b)   (b)   (b)   **117.** $x + 3y = 3$ or $y = -\frac{1}{3}x + 1$

119. $P_1 = (-2, 5)$, $P_2 = (1, 3)$, $m_1 = -\frac{2}{3}$; $P_2 = (1, 3)$, $P_3 = (-1, 0)$, $m_2 = \frac{3}{2}$; because $m_1 m_2 = -1$, the lines are perpendicular and the points $(-2, 5)$, $(1, 3)$, and $(-1, 0)$ are the vertices of a right triangle.

121. $P_1 = (-1, 0)$, $P_2 = (2, 3)$, $m = 1$; $P_3 = (1, -2)$, $P_4 = (4, 1)$, $m = 1$; $P_1 = (-1, 0)$, $P_3 = (1, -2)$, $m = -1$; $P_2 = (2, 3)$, $P_4 = (4, 1)$, $m = -1$; opposite sides are parallel, and adjacent sides are perpendicular; the points are the vertices of a rectangle.

123. $C = 0.60x + 39$; \$105.00; \$177.00   **125.** $C = 0.14x + 4252$

127. (a) $C = 0.0889x + 8.01$, $0 \le x \le 1000$

(b)

(c) \$25.79   (d) \$52.46
(e) Each additional kWh used adds \$0.0889 to the bill.

129. $°C = \frac{5}{9}(°F - 32)$; approximately 21.1°C   **131.** (a) $y = -\frac{2}{25}x + 30$ (b) x-intercept: 375; The ramp meets the floor 375 in. (31.25 ft) from the base of the platform.   (c) The ramp does not meet design requirements. It has a run of 31.25 ft.   (d) The only slope possible for the ramp to comply with the requirement is for it to drop 1 in. for every 12-in. run.

133. (a) $A = \frac{1}{5}x + 20,000$ (b) \$80,000
(c) Each additional box sold requires an additional \$0.20 in advertising.

135. $-2, -\frac{2}{3}, \frac{1}{2}$ **137.** $\left(\frac{a+b}{3}, \frac{c}{3}\right)$ **139.** b, c, e, g   **141.** c   **147.** No; no

149. They are the same line.   **151.** Yes, if the y-intercept is 0.   **154.** $x^4 y^{16}$ **155.** 17

156. $\left\{ 3 - 2\sqrt{6}, 3 + 2\sqrt{6} \right\}$ **157.** $\{x | 1 < x < 4\}$ or $(1, 4)$ 〔diagram〕   **158.** $(-\infty, 12]$ **159.** $\frac{9}{64}$ **160.** $-\frac{1}{4}$

161. $2x + 1 + 2\sqrt{x(x + 1)}$ **162.** $\{4, 9\}$

2.4 Assess Your Understanding (*page 188*)

3. F   **4.** radius   **5.** T   **6.** F   **7.** d   **8.** a   **9.** Center $(2, 1)$; radius $= 2$; $(x - 2)^2 + (y - 1)^2 = 4$

11. Center $\left(\frac{5}{2}, 2\right)$; radius $= \frac{3}{2}$; $\left(x - \frac{5}{2}\right)^2 + (y - 2)^2 = \frac{9}{4}$

13. $x^2 + y^2 = 4$;
$x^2 + y^2 - 4 = 0$

15. $x^2 + (y - 2)^2 = 4$;
$x^2 + y^2 - 4y = 0$

17. $(x - 4)^2 + (y + 3)^2 = 25$;
$x^2 + y^2 - 8x + 6y = 0$

19. $(x + 2)^2 + (y - 1)^2 = 16$;
$x^2 + y^2 + 4x - 2y - 11 = 0$

21. $\left(x - \dfrac{1}{2}\right)^2 + y^2 = \dfrac{1}{4}$;
$x^2 + y^2 - x = 0$

23. $(x - 5)^2 + (y + 1)^2 = 13$;
$x^2 + y^2 - 10x + 2y + 13 = 0$

25. (a) $(h, k) = (0, 0); r = 2$

(b)

(c) $(\pm 2, 0); (0, \pm 2)$

27. (a) $(h, k) = (3, 0); r = 2$

(b)

(c) $(1, 0); (5, 0)$

29. (a) $(h, k) = (1, 2); r = 3$

(b)

(c) $(1 \pm \sqrt{5}, 0); (0, 2 \pm 2\sqrt{2})$

31. (a) $(h, k) = (-2, 2); r = 3$

(b)

(c) $\left(-2 \pm \sqrt{5}, 0\right); \left(0, 2 \pm \sqrt{5}\right)$

33. (a) $(h, k) = \left(\dfrac{1}{2}, -1\right); r = \dfrac{1}{2}$

(b)

(c) $(0, -1)$

35. (a) $(h, k) = (3, -2); r = 5$

(b)

(c) $(3 \pm \sqrt{21}, 0); (0, -6), (0, 2)$

37. (a) $(h, k) = (-2, 0); r = 2$

(b)

(c) $(0, 0), (-4, 0)$

39. $x^2 + y^2 = 13$ **41.** $(x + 1)^2 + (y - 3)^2 = 5$ **43.** $(x - 2)^2 + (y + 4)^2 = 64$

45. c **47.** b **49.** $7x + 5y = 18$ **51.** $y = 2$ **53.** 18 units^2 **55.** $x^2 + (y - 139)^2 = 15{,}625$

57. $x^2 + (y - 49.5)^2 = 3660.25$ **59.** $(x - 2)^2 + (y - 3)^2 = 9$ **61.** $(x + 1)^2 + (y - 3)^2 = 1$

63. (a) $x^2 + (mx + b)^2 = r^2$
$(1 + m^2)x^2 + 2mbx + b^2 - r^2 = 0$
The equation has one solution if and
only if the discriminant $= 0$.
$(2mb)^2 - 4(1 + m^2)(b^2 - r^2) = 0$
$-4b^2 + 4r^2 + 4m^2r^2 = 0$
$r^2(1 + m^2) = b^2$

(b) $x = \dfrac{-2mb}{2(1 + m^2)} = \dfrac{-2mb}{2b^2/r^2} = -\dfrac{r^2m}{b}$

$y = m\left(-\dfrac{r^2m}{b}\right) + b = -\dfrac{r^2m^2}{b} + b = \dfrac{-r^2m^2 + b^2}{b} = \dfrac{r^2}{b}$

(c) The slope of the tangent line $= m$.

The slope of the line joining the center to the point of tangency $= \dfrac{r^2/b}{-r^2m/b} = -\dfrac{1}{m}$.

65. $\sqrt{2}x + 4y = 9\sqrt{2}$

67. The center of the circle is $\left(\dfrac{x_1 + x_2}{2}, \dfrac{y_1 + y_2}{2}\right)$ and the radius is $\dfrac{1}{2}\sqrt{(x_1 - x_2)^2 + (y_1 - y_2)^2}$. Then the equation of the

circle is $\left(x - \dfrac{x_1 + x_2}{2}\right)^2 + \left(y - \dfrac{y_1 + y_2}{2}\right)^2 = \dfrac{1}{4}\left[(x_1 - x_2)^2 + (y_1 - y_2)^2\right]$. Expanding, gives

$$x^2 - x(x_1 + x_2) + \dfrac{(x_1 + x_2)^2}{4} + y^2 - y(y_1 + y_2) + \dfrac{(y_1 + y_2)^2}{4} = \dfrac{1}{4}\left[x_1^2 - 2x_1x_2 + x_2^2 + y_1^2 - 2y_1y_2 + y_2^2\right]$$

$$4x^2 - 4x_1x - 4x_2x + x_1^2 + 2x_1x_2 + x_2^2 + 4y^2 - 4y_1y - 4y_2y + y_1^2 + 2y_1y_2 + y_2^2 = x_1^2 - 2x_1x_2 + x_2^2 + y_1^2 - 2y_1y_2 + y_2^2$$

$$4x^2 - 4x_1x - 4x_2x + 4x_1x_2 + 4y^2 - 4y_1y - 4y_2y + 4y_1y_2 = 0$$

$$x^2 - x_1x - x_2x + x_1x_2 + y^2 - y_1y - y_2y + y_1y_2 = 0$$

$$x(x - x_1) - x_2(x - x_1) + y(y - y_1) - y_2(y - y_1) = 0$$

$$(x - x_1)(x - x_2) + (y - y_1)(y - y_2) = 0$$

69. b, c, e, g **73.** $A = 169\pi$ cm^2; $C = 26\pi$ cm **74.** $3x^3 - 8x^2 + 13x - 6$ **75.** $\{1\}$ **76.** 12.32 minutes **77.** 0.0000957

78. $4x^3 - 5x^2 + 7x - 2$; remainder: $5x + 1$ **79.** $3x^2(x^2 + 7)(4x - 5)$ **80.** $3x^2y^5 \sqrt[4]{5x^3}$ **81.** $C = 8\pi\sqrt{2}$ m; $A = 32\pi$ m^2 **82.** 2.8 liters

2.5 Assess Your Understanding *(page 194)*

1. $y = kx$ **2.** F **3.** d **4.** c **5.** $y = \dfrac{1}{5}x$ **7.** $V = \dfrac{4\pi}{3}x^3$ **9.** $y = \dfrac{12}{\sqrt{x}}$ **11.** $z = \dfrac{2}{13}(x^2 + y^2)$ **13.** $M = \dfrac{9d^2}{2\sqrt{x}}$ **15.** $T^2 = \dfrac{8a^3}{d^2}$ **17.** $V = \dfrac{4\pi}{3}r^3$

19. $P = 2(l + w)$ **21.** $F = 6.67 \times 10^{-11}\left(\dfrac{mM}{d^2}\right)$ **23.** $p = 0.00649B$; $941.05 **25.** 144 ft; 2 s **27.** 2.25 **29.** $R = 2.84g$; $29.82

31. (a) $D = \dfrac{429}{p}$ **(b)** 143 bags **33.** 450 cm^3 **35.** 124.76 lb **37.** $V = \pi r^2 h$ **39.** 0.012 foot-candle **41.** $\sqrt[3]{6} \approx 1.82$ in. **43.** 282.2 ft **45.** 384 psi

51. $(3x + 25)(x + 2)(x - 2)$ **52.** $\dfrac{6}{x + 4}$ **53.** $\dfrac{8}{125}$ **54.** $\sqrt{7} + 2$ **55.** 3 **56.** $\{4\}$ **57.** $\{20\}$ **58.** $\left\{-\dfrac{9}{2}, \dfrac{13}{5}\right\}$ **59.** $\left\{\dfrac{3}{2}, 2\right\}$

60. $(-6, \infty)$

Review Exercises *(page 198)*

1. (a) $2\sqrt{5}$ **(b)** $(2, 1)$ **(c)** $\dfrac{1}{2}$ **(d)** For each run of 2, there is a rise of 1. **2. (a)** 5 **(b)** $\left(-\dfrac{1}{2}, 1\right)$ **(c)** $-\dfrac{4}{3}$ **(d)** For each run of 3, there is a rise of -4.
3. (a) 12 **(b)** $(4, 2)$ **(c)** undefined **(d)** no change in x

4.

5. $(-4, 0), (0, 2), (0, 0), (0, -2), (2, 0)$ **6.** $(0, 0)$; symmetric with respect to the x-axis
7. $(\pm 4, 0), (0, \pm 2)$; symmetric with respect to the x-axis, the y-axis, and the origin **8.** $(0, 1)$; symmetric with respect to the y-axis
9. $(0, 0), (\pm 1, 0)$; symmetric with respect to the origin **10.** $(0, 0), (-1, 0), (0, -2)$; no symmetry **11.** $(x + 2)^2 + (y - 3)^2 = 16$
12. $(x + 1)^2 + (y + 2)^2 = 1$

13. Center $(0, 1)$; radius = 2   **14.** Center $(1, -2)$; radius = 3   **15.** Center $(1, -2)$; radius = $\sqrt{5}$ **16.** $2x + y = 5$ or $y = -2x + 5$

17. $x = -3$; no slope-intercept form

18. $x + 5y = -10$ or $y = -\dfrac{1}{5}x - 2$

19. $5x + y = 11$ or $y = -5x + 11$

Intercepts: $\left(-\sqrt{3}, 0\right), \left(\sqrt{3}, 0\right),$ Intercepts: $\left(1 - \sqrt{5}, 0\right), \left(1 + \sqrt{5}, 0\right),$ Intercepts: $(0, 0),$ **20.** $2x - 3y = -19$ or $y = \dfrac{2}{3}x + \dfrac{19}{3}$
$(0, -1), (0, 3)$ $(0, -2 - 2\sqrt{2}), (0, -2 + 2\sqrt{2})$ $(2, 0), (0, -4)$ **21.** $x - y = 7$ or $y = x - 7$

22. Slope = $\dfrac{4}{5}$; y-intercept = 4   **23.** Slope = $\dfrac{3}{2}$; y-intercept = $\dfrac{1}{2}$ **24.** Intercepts: $(6, 0), (0, -4)$ **25.** Intercepts: $(4, 0), (0, 6)$

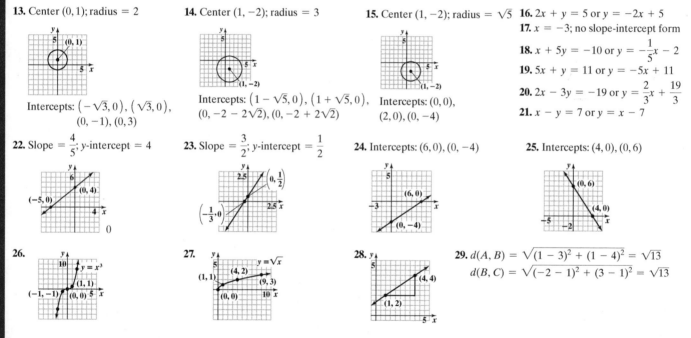

26.   **27.**   **28.**   **29.** $d(A, B) = \sqrt{(1 - 3)^2 + (1 - 4)^2} = \sqrt{13}$
$d(B, C) = \sqrt{(-2 - 1)^2 + (3 - 1)^2} = \sqrt{13}$

30. (a) $d(A, B) = 2\sqrt{5}; d(B, C) = \sqrt{145}; d(A, C) = 5\sqrt{5}; [d(A, B)]^2 + [d(A, C)]^2 = (2\sqrt{5})^2 + (5\sqrt{5})^2 = 20 + 125 = 145 = [d(B, C)]^2$

(b) Slope from A to B is -2; slope from A to C is $\dfrac{1}{2}$. Since $-2 \cdot \dfrac{1}{2} = -1$, the lines are perpendicular.

31. Center $(1, -2)$; Radius = $4\sqrt{2}$; $(x - 1)^2 + (y + 2)^2 = 32$ **32.** Slope from A to B is -1; slope from A to C is -1. **33.** $p = \dfrac{427}{65,000}B$; $1083.92
34. 199.9 lb **35.** 189 Btu

Chapter Test *(page 200)*

1. $d = 2\sqrt{13}$ **2.** $(2, 1)$ **3. (a)** $m = -\dfrac{2}{3}$ **(b)** For every 3-unit change in x, y will change by -2 units.

4.   **5.**   **6.** Intercepts: $(-3, 0),$ **7.** $y = -2x + 2$ **8.** $x^2 + y^2 - 8x + 6y = 0$ **9.** Center: $(-2, 1)$;
$(3, 0), (0, 9)$;   radius: 3
symmetric with
respect to the y-axis

10. Parallel line: $y = -\dfrac{2}{3}x - \dfrac{1}{3}$; perpendicular line: $y = \dfrac{3}{2}x + 3$ **11.** 14.69 ohms

Cumulative Review *(page 200)*

1. $\left\{\dfrac{5}{3}\right\}$ **2.** $\{-3, 4\}$ **3.** $\left\{-\dfrac{1}{2}, 3\right\}$ **4.** $\left\{1 - \sqrt{3}, 1 + \sqrt{3}\right\}$ **5.** No real solution **6.** $\{4\}$ **7.** $\{1, 3\}$ **8.** $\left\{-2 - 2\sqrt{2}, -2 + 2\sqrt{2}\right\}$ **9.** $\{-3i, 3i\}$

10. $\{1 - 2i, 1 + 2i\}$ **11.** $\{x\,|\,x \le 5\}$ or $(-\infty, 5]$; **12.** $\{x\,|\,-5 < x < 1\}$ or $(-5, 1)$;

13. $\{x\,|\,1 \le x \le 3\}$ or $[1, 3]$; **14.** $\{x\,|\,x < -5 \text{ or } x > 1\}$ or $(-\infty, -5) \cup (1, \infty)$;

15. $5\sqrt{2}$; $\left(\dfrac{3}{2}, \dfrac{1}{2}\right)$ **16.** a, b **17.** **18.** $y = -2x + 2$ **19.** $y = -\dfrac{1}{2}x + \dfrac{13}{2}$; $\left(0, \dfrac{13}{2}\right)$ **20.**

CHAPTER 3 Functions and Their Graphs

3.1 Assess Your Understanding *(page 215)*

7. independent; dependent **8.** a **9.** c **10.** F **11.** F **12.** verbally; numerically; graphically; algebraically **13.** F **14.** F
15. difference quotient **16.** explicitly
17. (a) Domain: $\{0, 22, 40, 70, 100\}$ in °C
   Range $\{1.031, 1.121, 1.229, 1.305, 1.411\}$ in kg/m^3

(b)

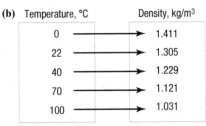

Temperature, °C	Density, kg/m³
0	1.411
22	1.305
40	1.229
70	1.121
100	1.031

(c) $\{(0, 1.411), (22, 1.305), (40, 1.229), (70, 1.121), (100, 1.031)\}$

19. Domain: $\{\text{Elvis, Colleen, Kaleigh, Marissa}\}$; Range: $\{\text{January 8, March 15, September 17}\}$; Function
21. Domain: $\{20, 30, 40\}$; Range: $\{200, 300, 350, 425\}$; Not a function **23.** Domain: $\{-3, 2, 4\}$; Range: $\{6, 9, 10\}$; Not a function
25. Domain: $\{1, 2, 3, 4\}$; Range: $\{3\}$; Function **27.** Domain: $\{-4, 0, 3\}$; Range: $\{1, 3, 5, 6\}$; Not a function **29.** Domain: $\{-1, 0, 2, 4\}$;
   Range: $\{-1, 3, 8\}$; Function **31.** Function **33.** Function **35.** Not a function **37.** Not a function **39.** Function **41.** Not a function
43. (a) -4 **(b)** 1 **(c)** -3 **(d)** $3x^2 - 2x - 4$ **(e)** $-3x^2 - 2x + 4$ **(f)** $3x^2 + 8x + 1$ **(g)** $12x^2 + 4x - 4$ **(h)** $3x^2 + 6xh + 3h^2 + 2x + 2h - 4$
45. (a) 0 **(b)** $\dfrac{1}{2}$ **(c)** $-\dfrac{1}{2}$ **(d)** $-\dfrac{x}{x^2 + 1}$ **(e)** $-\dfrac{x}{x^2 + 1}$ **(f)** $\dfrac{x + 1}{x^2 + 2x + 2}$ **(g)** $\dfrac{2x}{4x^2 + 1}$ **(h)** $\dfrac{x + h}{x^2 + 2xh + h^2 + 1}$ **47. (a)** 4 **(b)** 5 **(c)** 5 **(d)** $|x| + 4$
   (e) $-|x| - 4$ **(f)** $|x + 1| + 4$ **(g)** $2|x| + 4$ **(h)** $|x + h| + 4$ **49. (a)** $-\dfrac{1}{5}$ **(b)** $-\dfrac{3}{2}$ **(c)** $\dfrac{1}{8}$ **(d)** $\dfrac{2x - 1}{3x + 5}$ **(e)** $\dfrac{-2x - 1}{3x - 5}$ **(f)** $\dfrac{2x + 3}{3x - 2}$
   (g) $\dfrac{4x + 1}{6x - 5}$ **(h)** $\dfrac{2x + 2h + 1}{3x + 3h - 5}$ **51.** All real numbers **53.** All real numbers **55.** $\{x\,|\,x \ne -4, x \ne 4\}$ **57.** $\{x\,|\,x \ne 0\}$ **59.** $\{x\,|\,x \ge 4\}$
61. $\{x\,|\,x \ne -2, -1\}$ **63.** $\{x\,|\,x > 4\}$ **65.** $\{t\,|\,t \ge 4, t \ne 7\}$ **67.** All real numbers **69.** $\{t\,|\,t \ne -2, t \ne 7\}$
71. (a) $(f + g)(x) = 5x + 1$; All real numbers **(b)** $(f - g)(x) = x + 7$; All real numbers **(c)** $(f \cdot g)(x) = 6x^2 - x - 12$; All real numbers
   (d) $\left(\dfrac{f}{g}\right)(x) = \dfrac{3x + 4}{2x - 3}$; $\left\{x\,\middle|\,x \ne \dfrac{3}{2}\right\}$ **(e)** 16 **(f)** 11 **(g)** 10 **(h)** -7 **73. (a)** $(f + g)(x) = 2x^2 + x - 1$; All real numbers
   (b) $(f - g)(x) = -2x^2 + x - 1$; All real numbers **(c)** $(f \cdot g)(x) = 2x^3 - 2x^2$; All real numbers **(d)** $\left(\dfrac{f}{g}\right)(x) = \dfrac{x - 1}{2x^2}$; $\{x\,|\,x \ne 0\}$
   (e) 20 **(f)** -29 **(g)** 8 **(h)** 0 **75. (a)** $(f + g)(x) = \sqrt{x} + 3x - 5$; $\{x\,|\,x \ge 0\}$ **(b)** $(f - g)(x) = \sqrt{x} - 3x + 5$; $\{x\,|\,x \ge 0\}$
   (c) $(f \cdot g)(x) = 3x\sqrt{x} - 5\sqrt{x}$; $\{x\,|\,x \ge 0\}$ **(d)** $\left(\dfrac{f}{g}\right)(x) = \dfrac{\sqrt{x}}{3x - 5}$; $\left\{x\,\middle|\,x \ge 0, x \ne \dfrac{5}{3}\right\}$ **(e)** $\sqrt{3} + 4$ **(f)** -5 **(g)** $\sqrt{2}$ **(h)** $-\dfrac{1}{2}$
77. (a) $(f + g)(x) = 1 + \dfrac{2}{x}$; $\{x\,|\,x \ne 0\}$ **(b)** $(f - g)(x) = 1$; $\{x\,|\,x \ne 0\}$ **(c)** $(f \cdot g)(x) = \dfrac{1}{x} + \dfrac{1}{x^2}$; $\{x\,|\,x \ne 0\}$
   (d) $\left(\dfrac{f}{g}\right)(x) = x + 1$; $\{x\,|\,x \ne 0\}$ **(e)** $\dfrac{5}{3}$ **(f)** 1 **(g)** $\dfrac{3}{4}$ **(h)** 2 **79. (a)** $(f + g)(x) = \dfrac{6x + 3}{3x - 2}$; $\left\{x\,\middle|\,x \ne \dfrac{2}{3}\right\}$
   (b) $(f - g)(x) = \dfrac{-2x + 3}{3x - 2}$; $\left\{x\,\middle|\,x \ne \dfrac{2}{3}\right\}$ **(c)** $(f \cdot g)(x) = \dfrac{8x^2 + 12x}{(3x - 2)^2}$; $\left\{x\,\middle|\,x \ne \dfrac{2}{3}\right\}$ **(d)** $\left(\dfrac{f}{g}\right)(x) = \dfrac{2x + 3}{4x}$; $\left\{x\,\middle|\,x \ne 0, x \ne \dfrac{2}{3}\right\}$
   (e) 3 **(f)** $-\dfrac{1}{2}$ **(g)** $\dfrac{7}{2}$ **(h)** $\dfrac{5}{4}$ **81.** $g(x) = 5 - \dfrac{7}{2}x$ **83.** 4 **85.** $2x + h$ **87.** $2x + h - 1$ **89.** $-\dfrac{20}{(4x + 4h - 3)(4x - 3)}$
91. $\dfrac{6}{(x + 3)(x + h + 3)}$ **93.** $\dfrac{1}{\sqrt{x + h - 2} + \sqrt{x - 2}}$ **95.** $-\dfrac{2x + h}{x^2(x + h)^2}$ **97.** $-\dfrac{2x + h}{\sqrt{4 - (x + h)^2} + \sqrt{4 - x^2}}$ **99.** $\{-2, 4\}$ **101.** $A = -\dfrac{7}{2}$
103. $A = -4$ **105.** $A(x) = \dfrac{1}{2}x^2$ **107.** $G(x) = 16x$ **109. (a)** 15.1 m, 14.071 m, 12.944 m **(b)** 1.01 sec, 1.43 sec, 1.75 sec **(c)** 2.02 sec

111. (a) $222   **(b)** $225   **(c)** $220   **(d)** $230   **113.** $R(x) = \dfrac{L(x)}{P(x)}$ **115.** $H(x) = P(x) \cdot I(x)$ **117. (a)** $P(x) = -0.05x^3 + 0.8x^2 + 155x - 500$

(b) $P(15) = 1836.25 **(c)** When 15 hundred smartphones are sold, the profit is $1836.25.   **119. (a)** $D(v) = 0.05v^2 + 2.6v - 15$

(b) 321 feet   **(c)** The car will need 321 feet to stop once the impediment is observed.   **121.** $\dfrac{1}{(x + h)^{2/3} + x^{1/3}(x + h)^{1/3} + x^{2/3}}$

123. $\left\{ x \middle| -2 < x < \dfrac{8}{3} \right\}$ **125.** $H(x) = \dfrac{3x - x^3}{\text{age}}$ **127.** Intercepts: $(-16, 0), (-8, 0)$; x-axis symmetry   **128.** $(4, 32)$ **129.** 7.5 lbs

130. $\{-3, 2, 3\}$ **131.** $a = \dfrac{d - bx}{1 - c}$ or $a = \dfrac{bx - d}{c - 1}$ **132.** $2.5 \text{ kg} \cdot \text{m}^2$ **133.** $-\dfrac{10}{3}$ **134.** $-\dfrac{12x^2 + 40x + 21}{(4x^2 - 7)^2}$ **135.** 7

3.2 Assess Your Understanding *(page 223)*

3. vertical   **4.** 5; -3 **5.** $a = -2$ **6.** F   **7.** F   **8.** T   **9.** c   **10.** a

11. (a) $f(0) = 3; f(-6) = -3$ **(b)** $f(6) = 0; f(11) = 1$ **(c)** Positive   **(d)** Negative   **(e)** $-3, 6$, and 10 **(f)** $-3 < x < 6; 10 < x \le 11$
   (g) $\{x | -6 \le x \le 11\}$ **(h)** $\{y | -3 \le y \le 4\}$ **(i)** $-3, 6, 10$ **(j)** 3   **(k)** 3 times   **(l)** Once   **(m)** $0, 4$ **(n)** $-5, 8$

13. Not a function **(a)** Domain: $\{x | x \le -1 \text{ or } x \ge 1\}$; Range: all real numbers   **(b)** $(-1, 0), (1, 0)$ **(c)** x-axis, y-axis, and origin symmetry

15. Function **(a)** Domain: $\{x | -\pi \le x \le \pi\}$; Range: $\{y | -1 \le y \le 1\}$ **(b)** $\left(-\dfrac{\pi}{2}, 0\right), \left(\dfrac{\pi}{2}, 0\right), (0, 1)$ **(c)** y-axis

17. Not a function **(a)** Domain: $\{x | x \le 0\}$; Range: all real numbers   **(b)** $(0, 0)$ **(c)** x-axis

19. Function **(a)** Domain: $\{x | 0 < x < 3\}$; Range: $\{y | y < 2\}$ **(b)** $(1, 0)$ **(c)** None

21. Function **(a)** Domain: all real numbers; Range: $\{y | y \le 2\}$ **(b)** $(-3, 0), (3, 0), (0, 2)$ **(c)** y-axis

23. Function **(a)** Domain: all real numbers; Range: $\{y | y \ge -3\}$ **(b)** $(1, 0), (3, 0), (0, 9)$ **(c)** None

25. (a) Yes   **(b)** $f(-2) = 8; (-2, 8)$ **(c)** 0 or $-\dfrac{1}{3}; (0, -2), \left(-\dfrac{1}{3}, -2\right)$ **(d)** All real numbers   **(e)** $-1, \dfrac{2}{3}$ **(f)** -2

27. (a) No   **(b)** $f(4) = -3; (4, -3)$ **(c)** $14; (14, 2)$ **(d)** $\{x | x \ne 6\}$ **(e)** -2 **(f)** $-\dfrac{1}{3}$

29. (a) Yes   **(b)** $f(3) = \dfrac{486}{5}; \left(3, \dfrac{486}{5}\right)$ **(c)** $-\dfrac{\sqrt{3}}{3}$ or $\dfrac{\sqrt{3}}{3}; \left(-\dfrac{\sqrt{3}}{3}, 1\right), \left(\dfrac{\sqrt{3}}{3}, 1\right)$ **(d)** All real numbers   **(e)** 0 **(f)** 0

31. (a) 3   **(b)** -2 **(c)** -1 **(d)** 1   **(e)** 2   **(f)** $-\dfrac{1}{3}$

33. (a) Approximately 10.4 ft high
   (b) Approximately 9.9 ft high

(c)

(d) The ball will not go through the hoop; $h(15) \approx 8.4$ ft. If $v = 30$ ft/sec, $h(15) = 10$ ft.

35. (a) About 81.07 ft   **(b)** About 129.59 ft   **(c)** About 26.63 ft; The ball is about 26.63 feet high after it has traveled 500 feet.
   (d) About 528.13 ft

(e)

(f) About 115.07 ft and 413.05 ft
   (g) 275 ft; maximum height shown in the table is 131.8 ft   **(h)** 264 ft

37. (a) $223; $220   **(b)** $\{x | x > 0\}$

(c)

(d)

(e) 600 mi/h

39. (a) $50; It costs $50 if you use 0 gigabytes.   **(b)** $50; It costs $50 if you use 5 gigabytes.   **(c)** $150; It costs $150 if you use 15 gigabytes.   **(d)** $\{g | 0 \le g \le 30\}$. There are at most 30 gigabytes used in a month.
41. 420   **43.** The x-intercepts can number anywhere from 0 to infinitely many. There is at most one y-intercept.
45. (a) III   **(b)** IV   **(c)** I   **(d)** V   **(e)** II

47.

49. (a) 2 hr elapsed during which Kevin was between 0 and 3 mi from home   **(b)** 0.5 hr elapsed during which Kevin was 3 mi from home   **(c)** 0.3 hr elapsed during which Kevin was between 0 and 3 mi from home   **(d)** 0.2 hr elapsed during which Kevin was 0 mi from home   **(e)** 0.9 hr elapsed during which Kevin was between 0 and 2.8 mi from home   **(f)** 0.3 hr elapsed during which Kevin was 2.8 mi from home   **(g)** 1.1 hr elapsed during which Kevin was between 0 and 2.8 mi from home   **(h)** 3 mi   **(i)** Twice   **51.** No points whose x-coordinate is 5 or whose y-coordinate is 0 can be on the graph.   **54.** $-x^2 + 5x - 9$ **55.** $2\sqrt{10}$ **56.** $y = \dfrac{2}{3}x + 8$ **57.** All real numbers or $(-\infty, \infty)$

58. 36   **59.** $\dfrac{1}{\sqrt{x} + \sqrt{6}}$ **60.** 55.8 min   **61.** $\{x | -4 < x \le 1\}$ or $(-4, 1]$ **62.** $5x^2 - 15x + 12$ **63.** $[-3, 10]$

3.3 Assess Your Understanding *(page 237)*

6. increasing   **7.** even; odd   **8.** T   **9.** T   **10.** F   **11.** c   **12.** d   **13.** Yes   **15.** No   **17.** $[-8, -2]; [0, 2]; [5, 7]$ **19.** Yes; 10   **21.** $-2, 2; 6, 10$
23. $f(-8) = -4$ **25. (a)** $(-2, 0), (0, 3), (2, 0)$ **(b)** Domain: $\{x | -4 \le x \le 4\}$ or $[-4, 4]$; Range: $\{y | 0 \le y \le 3\}$ or $[0, 3]$ **(c)** Increasing on $[-2, 0]$ and $[2, 4]$; Decreasing on $[-4, -2]$ and $[0, 2]$ **(d)** Even   **27. (a)** $(0, 1)$ **(b)** Domain: all real numbers; Range: $\{y | y > 0\}$ or $(0, \infty)$
   (c) Increasing on $(-\infty, \infty)$ **(d)** Neither   **29. (a)** $(-\pi, 0), (0, 0), (\pi, 0)$ **(b)** Domain: $\{x | -\pi \le x \le \pi\}$ or $[-\pi, \pi]$; Range: $\{y | -1 \le y \le 1\}$
   or $[-1, 1]$ **(c)** Increasing on $\left[-\dfrac{\pi}{2}, \dfrac{\pi}{2}\right]$; Decreasing on $\left[-\pi, -\dfrac{\pi}{2}\right]$ and $\left[\dfrac{\pi}{2}, \pi\right]$ **(d)** Odd   **31. (a)** $\left(0, \dfrac{1}{2}\right), \left(\dfrac{1}{3}, 0\right), \left(\dfrac{5}{2}, 0\right)$

(b) Domain: $\{x|-3 \le x \le 3\}$ or $[-3, 3]$; Range: $\{y|-1 \le y \le 2\}$ or $[-1, 2]$ **(c)** Increasing on $[2, 3]$; Decreasing on $[-1, 1]$;

Constant on $[-3, -1]$ and $[1, 2]$ **(d)** Neither   **33. (a)** $0; 3$ **(b)** $-2, 2; 0, 0$ **35. (a)** $\dfrac{\pi}{2}; 1$ **(b)** $-\dfrac{\pi}{2}; -1$ **37.** Odd   **39.** Even   **41.** Odd   **43.** Neither

45. Even   **47.** Odd   **49.** Absolute maximum: $f(1) = 4$; absolute minimum: $f(5) = 1$; local maximum: $f(3) = 3$; local minimum: $f(2) = 2$

51. Absolute maximum: $f(3) = 4$; absolute minimum: $f(1) = 1$; local maximum: $f(3) = 4$; local minimum: $f(1) = 1$

53. Absolute maximum: none; absolute minimum: $f(0) = 0$; local maximum: $f(2) = 3$; local minima: $f(0) = 0$ and $f(3) = 2$

55. Absolute maximum: none; absolute minimum: none; local maximum: none; local minimum: none

57.

Increasing: $[-2, -1], [1, 2]$
Decreasing: $[-1, 1]$
Local maximum: $f(-1) = 4$
Local minimum: $f(1) = 0$

59.

Increasing: $[-2, -0.77], [0.77, 2]$
Decreasing: $[-0.77, 0.77]$
Local maximum: $f(-0.77) = 0.19$
Local minimum: $f(0.77) = -0.19$

61.

Increasing: $[-3.77, 1.77]$
Decreasing: $[-6, -3.77], [1.77, 4]$
Local maximum: $f(1.77) = -1.91$
Local minimum: $f(-3.77) = -18.89$

63.

Increasing: $[-1.87, 0], [0.97, 2]$
Decreasing: $[-3, -1.87], [0, 0.97]$
Local maximum: $f(0) = 3$
Local minima: $f(-1.87) = 0.95,$
$f(0.97) = 2.65$

65. (a) -4 **(b)** -8 **(c)** -10 **67. (a)** 15 **(b)** -3 **(c)** 9 **69. (a)** 5 **(b)** $y = 5x - 2$ **71. (a)** -1 **(b)** $y = -x$ **73. (a)** 4 **(b)** $y = 4x - 8$

75. (a) Odd   **(b)** Local maximum value is 54 at $x = -3$.   **77. (a)** Even   **(b)** Local maximum value is 25 at $x = -2$.   **(c)** 50.4 sq. units

79. (a)

(b) 10 riding lawn mowers
(c) $239/mower

81. (a) On average, the population is increasing at a rate of 0.036 g/h from 0 to 2.5 h.   **(b)** On average, from 4.5 to 6 h, the population is increasing at a rate of 0.1 g/h.   **(c)** The average rate of change is increasing over time.

83. (a) 1 **(b)** 0.5 **(c)** 0.1 **(d)** 0.01
(e) 0.001
(f)

(g) They are getting closer to the tangent line at $(0, 0)$.
(h) They are getting closer to 0.

85. (a) 2
(b) $2; 2; 2; 2$
(c) $y = 2x + 5$
(d)

87. (a) $2x + h + 2$ **(b)** $4.5; 4.1; 4.01; 4$
(c) $y = 4.01x - 1.01$
(d)

89. (a) $4x + 2h - 3$ **(b)** $2; 1.2; 1.02; 1$
(c) $y = 1.02x - 1.02$
(d)

91. (a) $-\dfrac{1}{(x+h)x}$

(b) $-\dfrac{2}{3}; -\dfrac{10}{11}; -\dfrac{100}{101}; -1$

(c) $y = -\dfrac{100}{101}x + \dfrac{201}{101}$

(d)

93. $\dfrac{-2 + \sqrt{19}}{3}$

97. At most one

99. Yes; the function $f(x) = 0$ is both even and odd.

101. Not necessarily. It just means $f(5) > f(2)$.   **103.** $6\sqrt{15}$ **104.** $16a^2 - 8ab + b^2$ **105.** $C(x) = 0.80x + 40$ **106.** 8.25 days

107. $(x - 3)^2 + (y + 2)^2 = \dfrac{3}{2}$ **108.** $6x(x^2 - 2)^2 (3x^5 + 1) (8x^5 - 10x^3 + 1)$ **109.** $\left(-\dfrac{7}{10}, -\dfrac{3}{2}\right)$ **110.** $\left\{-5, \dfrac{1}{3}\right\}$ **111.** $\{-2, 1\}$ **112.** $D = \dfrac{x(y^2 - x)}{y(y - x^2)} = \dfrac{xy^2 - x^2}{y^2 - x^2y}$

3.4 Assess Your Understanding *(page 250)*

4. $(-\infty, 0]$ **5.** piecewise-defined   **6.** T   **7.** F   **8.** F   **9.** b   **10.** a   **11.** C   **13.** E   **15.** B   **17.** F

19.

21.

23.

25.

27. (a) -9 **(b)** 4 **(c)** 7 **29. (a)** 0 **(b)** 4 **(c)** 6 **(d)** 26

31. (a) All real numbers
(b) $(0, 1)$
(c)

(d) $\{y \mid y \neq 0\}$; $(-\infty, 0) \cup (0, \infty)$

33. (a) All real numbers
(b) $(0, 3)$
(c)

(d) $\{y \mid y \geq 1\}$; $[1, \infty)$

35. (a) $\{x \mid x \geq -2\}$; $[-2, \infty)$
(b) $(0, 3), (2, 0)$
(c)

(d) $\{y \mid y < 4, y = 5\}$; $(-\infty, 4) \cup \{5\}$

37. (a) All real numbers
(b) $(-1, 0), (0, 0)$
(c)

(d) All real numbers

39. (a) $\{x \mid x \geq -2, x \neq 0\}$; $[-2, 0) \cup (0, \infty)$
(b) No intercepts
(c)

(d) $\{y \mid y > 0\}$; $(0, \infty)$

41. (a) Domain $(0, \infty)$
(b) No intercepts
(c)

(d) Range: $(0, 7]$

43. $f(x) = \begin{cases} -x & \text{if } -1 \leq x \leq 0 \\ \frac{1}{2}x & \text{if } 0 < x \leq 2 \end{cases}$

45. $f(x) = \begin{cases} -x & \text{if } x \leq 0 \\ -x + 2 & \text{if } 0 < x \leq 2 \end{cases}$

47. (a) 3 **(b)** 5 **(c)** -8

49. (a)

(b) $[0, 6]$
(c) Absolute maximum: $f(2) = 6$;
absolute minimum: $f(6) = -2$

51. (a) $\$34.99$
(b) $\$64.99$
(c) $\$184.99$

53. (a) $\$47.07$
(b) $\$119.97$
(c) $C(x) = \begin{cases} 1.18172x + 23.44 & \text{if } 0 \leq x \leq 30 \\ 0.50897x + 43.6225 & \text{if } x > 30 \end{cases}$

(d)

55. $f(x) = \begin{cases} 0.10x & \text{if} & 0 < x \leq 9525 \\ 952.50 + 0.12(x - 9525) & \text{if} & 9525 < x \leq 38{,}700 \\ 4453.50 + 0.22(x - 38{,}700) & \text{if} & 38{,}700 < x \leq 82{,}500 \\ 14{,}089.50 + 0.24(x - 82{,}500) & \text{if} & 82{,}500 < x \leq 157{,}500 \\ 32{,}089.50 + 0.32(x - 157{,}500) & \text{if} & 157{,}500 < x \leq 200{,}000 \\ 45{,}689.50 + 0.35(x - 200{,}000) & \text{if} & 200{,}000 < x \leq 500{,}000 \\ 150{,}689.50 + 0.37(x - 500{,}000) & \text{if} & x > 500{,}000 \end{cases}$

57. (a)

(b) $C(x) = 10 + 0.4x$
(c) $C(x) = 70 + 0.25x$

59. (a) $C(s) = \begin{cases} 9000 & \text{if} & s \leq 659 \\ 8250 & \text{if} & 660 \leq s \leq 679 \\ 5250 & \text{if} & 680 \leq s \leq 699 \\ 3750 & \text{if} & 700 \leq s \leq 719 \\ 2250 & \text{if} & 720 \leq s \leq 739 \\ 1500 & \text{if} & s \geq 740 \end{cases}$

(b) $\$2250$ **(c)** $\$8250$

61. (a) $10°C$ **(b)** $4°C$ **(c)** $-3°C$ **(d)** $-4°C$
(e) The wind chill is equal to the air temperature.
(f) At wind speed greater than 20 m/s, the wind chill factor depends only on the air temperature.

65. $f(x) + g(x) = \begin{cases} -2x + 4 & \text{if } x \leq 0 \\ 3x - 4 & \text{if } 0 < x < 2 \\ x^2 + 6x - 7 & \text{if } x \geq 2 \end{cases}$

63. $C(x) = \begin{cases} 1.00 & \text{if} & 0 < x \leq 1 \\ 1.21 & \text{if} & 1 < x \leq 2 \\ 1.42 & \text{if} & 2 < x \leq 3 \\ 1.63 & \text{if} & 3 < x \leq 4 \\ 1.84 & \text{if} & 4 < x \leq 5 \\ 2.05 & \text{if} & 5 < x \leq 6 \\ 2.26 & \text{if} & 6 < x \leq 7 \\ 2.47 & \text{if} & 7 < x \leq 8 \\ 2.68 & \text{if} & 8 < x \leq 9 \\ 2.89 & \text{if} & 9 < x \leq 10 \\ 3.10 & \text{if} & 10 < x \leq 11 \\ 3.31 & \text{if} & 11 < x \leq 12 \\ 3.52 & \text{if} & 12 < x \leq 13 \end{cases}$

67. Each graph is that of $y = x^2$, but shifted horizontally. If $y = (x - k)^2, k > 0$, the shift is right k units; if $y = (x + k)^2, k > 0$, the shift is left k units.

69. The graph of $y = -f(x)$ is the reflection about the x-axis of the graph of $y = f(x)$. **71.** Yes. The graph of $y = (x - 1)^3 + 2$ is the graph of $y = x^3$ shifted right 1 unit and up 2 units. **73.** They all have the same general shape. All three go through the points $(-1, -1), (0, 0),$ and $(1, 1)$.

As the exponent increases, the steepness of the curve increases (except near $x = 0$). **76.** $\dfrac{x^6}{y^{10}}$ **77.** $(h, k) = (0, 3); r = 5$ **78.** $\{-8\}$

79. CD: $\$22{,}000$; Mutual fund: $\$38{,}000$ **80.** Quotient: $x^2 + x - 2$; Remainder: -2 **81.** $\dfrac{3}{2} + 2i$ **82.** $-2x^7$ **83.** $5t^2\sqrt{1 + 25t^3}$
84. $\{x \mid x \geq -7\}$ or $[-7, \infty)$ **85.** $(3x - 2y)(x^2y + 6)$

3.5 Assess Your Understanding (*page 263*)

1. horizontal; right   **2.** y **3.** F   **4.** T   **5.** d   **6.** a   **7.** B   **9.** H   **11.** I   **13.** L   **15.** F   **17.** G   **19.** $y = (x - 4)^3$ **21.** $y = x^3 + 4$ **23.** $y = -x^3$
25. $y = 5x^3$ **27.** $y = (2x)^3 = 8x^3$ **29.** $y = -(\sqrt{-x} + 2)$ **31.** $y = 3\sqrt{x + 5} + 4$ **33.** c   **35.** c

37.
Domain: $(-\infty, \infty)$;
Range: $[-1, \infty)$

39.
Domain: $[0, \infty)$;
Range: $[0, \infty)$

41.
Domain: $[-2, \infty)$;
Range: $[0, \infty)$

43.
Domain: $(-\infty, \infty)$;
Range: $(-\infty, \infty)$

45.
Domain: $[0, \infty)$;
Range: $[0, \infty)$

47.
Domain: $(-\infty, \infty)$;
Range: $(-\infty, \infty)$

49.
Domain: $(-\infty, \infty)$;
Range: $[-3, \infty)$

51.
Domain: $[2, \infty)$;
Range: $[1, \infty)$

53.
Domain: $(-\infty, 0]$;
Range: $[-2, \infty)$

55.
Domain: $(-\infty, \infty)$;
Range: $(-\infty, \infty)$

57.
Domain: $(-\infty, \infty)$;
Range: $[0, \infty)$

59.
Domain: $(-\infty, 0) \cup (0, \infty)$
Range: $(-\infty, 0) \cup (0, \infty)$

61. (a) $F(x) = f(x) + 3$

(b) $G(x) = f(x + 2)$

(c) $P(x) = -f(x)$

(d) $H(x) = f(x + 1) - 2$

(e) $Q(x) = \dfrac{1}{2} f(x)$

(f) $g(x) = f(-x)$

(g) $h(x) = f(2x)$

63. (a) $F(x) = f(x) + 3$

(b) $G(x) = f(x + 2)$

(c) $P(x) = -f(x)$

(d) $H(x) = f(x + 1) - 2$

(e) $Q(x) = \dfrac{1}{2} f(x)$

(f) $g(x) = f(-x)$

(g) $h(x) = f(2x)$

65. $f(x) = (x + 1)^2 - 1$

67. $f(x) = (x - 4)^2 - 15$

69. $f(x) = 2(x - 3)^2 + 1$

71. $f(x) = -3(x + 2)^2 - 5$

73. (a) -7 and 1 **(b)** -3 and 5
(c) -5 and 3 **(d)** -3 and 5
75. (a) $[-3, 3]$ **(b)** $[4, 10]$
(c) Decreasing on $[-1, 5]$
(d) Decreasing on $[-5, 1]$

77. (a)

(b)

79. (a) $(-2, 2)$
(b) $(3, -5)$
(c) $(-1, 3)$

81. (a) **(b)**   **83. (a)**   **85. (a)** 72°F; 65°F

(b) 9 square units

85. (b) The temperature decreases by 2° to 70°F during the day and 63°F overnight.

(c) The time at which the temperature adjusts between the daytime and overnight settings is moved to 1 hr sooner. It begins warming up at 5:00 AM instead of 6:00 AM, and it begins cooling down at 8:00 PM instead of 9:00 PM.

87.   **89.** $c = -2$ $c = 0$ $c = 3$

91. $[4, 8]$ **93.** The graph of $y = 4f(x)$ is a vertical stretch by a factor of 4. The graph of $y = f(4x)$ is a horizontal compression by a factor of $\frac{1}{4}$.   **95.** $\frac{16}{3}$ sq. units   **97.** The domain of $g(x) = \sqrt{x}$ is $[0, \infty)$. The graph of $g(x - k)$ is the graph of g shifted k units to the right, so the domain of $g(x - k)$ is $[k, \infty)$.   **98.** $m = \frac{3}{5}; b = -6$ **99.** 3.11 mph   **100.** 15.75 gal

101. Intercepts: $(0, -2), (0, 2), (-4, 0)$; x-axis symmetry   **102.** $\{x \mid x \neq 7, x \neq -2\}$ **103.** 7 sec   **104.** $2xy^2\sqrt[3]{2x^2z}$ **105.** $6x + 3h + 2$

106. $(z + 6)(z^2 - 6z + 36)$

3.6 Assess Your Understanding (page 270)

1. (a) $d(x) = \sqrt{x^4 - 15x^2 + 64}$
(b) $d(0) = 8$
(c) $d(1) = \sqrt{50} \approx 7.07$
(d)
(e) d is smallest when $x \approx -2.74$ or $x \approx 2.74$

3. (a) $d(x) = \sqrt{x^2 - x + 1}$
(b)
(c) d is smallest when $x = 0.5$.
(d) $\frac{\sqrt{3}}{2} \approx 0.87$

5. $A(x) = \frac{1}{2}x^4$

7. (a) $A(x) = x(16 - x^2)$
(b) Domain: $\{x \mid 0 < x < 4\}$
(c) The area is largest when $x \approx 2.31$.
(d) The largest area is approximately 24.63 square units.

9. (a) $A(x) = 4x\sqrt{4 - x^2}$ **(b)** $p(x) = 4x + 4\sqrt{4 - x^2}$
(c) A is largest when $x \approx 1.41$.   **(d)** p is largest when $x \approx 1.41$.

11. (a) $A(x) = x^2 + \frac{25 - 20x + 4x^2}{\pi}$
(b) Domain: $\{x \mid 0 < x < 2.5\}$
(c) A is smallest when $x \approx 1.40$ m.

13. (a) $C(x) = x$ **(b)** $A(x) = \frac{x^2}{4\pi}$ **15. (a)** $A(r) = 2r^2$ **(b)** $p(r) = 6r$ **17.** $A(x) = \left(\frac{\pi}{3} - \frac{\sqrt{3}}{4}\right)x^2$

19. (a) $d(t) = \sqrt{2500t^2 - 360t + 13}$ **21.** $V(r) = \frac{\pi H(R - r)r^2}{R}$ **23. (a)** $T(x) = \frac{12 - x}{5} + \frac{\sqrt{x^2 + 4}}{3}$ **25. (a)** $V(x) = x(24 - 2x)^2$
(b) d is smallest when $t \approx 0.07$ hr.
(b) $\{x \mid 0 \leq x \leq 12\}$ **(b)** 972 in.³
(c) 3.09 hr   **(c)** 160 in.³
(d) 3.55 hr   **(d)** V is largest when $x = 4$ in.

27. (a) $C(x) = \frac{9000}{x} + 1.60x + 2910$
(b) The retailer should order 75 drives per order for a minimum yearly cost of $3150.

(e) The largest volume is 1024 in.³

28. $\{0, 3\}$ **29.** 64 mi/h   **30.** $m = -4$ **31.** 5.6   **32.** $y = \frac{u - 1}{u}$ **33.** $\frac{4x + 5}{3x^{2/3}}$ **34.** $\varnothing$ **35.** $(-3, 2)$ **36.** $P = \frac{6.76t^2E}{v^2d^4}$ **37.** 97

Review Exercises *(page 275)*

1. (a) Domain: $\{8, 16, 20, 24\}$; range: $\{\$6.30, \$12.32, \$13.99\}$
(b) $\{(8, \$6.30), (16, \$13.99), (20, \$12.32), (24, \$13.99)\}$

(c)

(d)

2. Domain: $\{-1, 2, 4\}$; range: $\{0, 3\}$; function
3. Domain: $\{2, 4\}$; range: $\{-1, 1, 2\}$; not a function
4. Domain: $[-1, 3]$; range: $[-2, 2]$; not a function
5. Domain: all real numbers; range: $[-3, \infty)$; function
6. (a) 2 **(b)** -2 **(c)** $-\dfrac{3x}{x^2 - 1}$ **(d)** $-\dfrac{3x}{x^2 - 1}$ **(e)** $\dfrac{3(x - 2)}{x^2 - 4x + 3}$ **(f)** $\dfrac{6x}{4x^2 - 1}$
7. (a) 0 **(b)** 0 **(c)** $\sqrt{x^2 - 4}$ **(d)** $-\sqrt{x^2 - 4}$ **(e)** $\sqrt{x^2 - 4x}$ **(f)** $2\sqrt{x^2 - 1}$
8. (a) 0 **(b)** 0 **(c)** $\dfrac{x^2 - 4}{x^2}$ **(d)** $-\dfrac{x^2 - 4}{x^2}$ **(e)** $\dfrac{x(x - 4)}{(x - 2)^2}$ **(f)** $\dfrac{x^2 - 1}{x^2}$
9. $\{x \mid x \neq -3, x \neq 3\}$ **10.** $\{x \mid x \leq 2\}$ **11.** $\{x \mid x \neq 0\}$
12. $\{x \mid x \neq -3, x \neq 1\}$ **13.** $[-1, 2) \cup (2, \infty)$ **14.** $\{x \mid x > -8\}$

15. $(f + g)(x) = 2x + 3$; Domain: all real numbers
$(f - g)(x) = -4x + 1$; Domain: all real numbers
$(f \cdot g)(x) = -3x^2 + 5x + 2$; Domain: all real numbers
$\left(\dfrac{f}{g}\right)(x) = \dfrac{2 - x}{3x + 1}$; Domain: $\left\{x \mid x \neq -\dfrac{1}{3}\right\}$

16. $(f + g)(x) = 3x^2 + 4x + 1$; Domain: all real numbers
$(f - g)(x) = 3x^2 - 2x + 1$; Domain: all real numbers
$(f \cdot g)(x) = 9x^3 + 3x^2 + 3x$; Domain: all real numbers
$\left(\dfrac{f}{g}\right)(x) = \dfrac{3x^2 + x + 1}{3x}$; Domain: $\{x \mid x \neq 0\}$

17. $(f + g)(x) = \dfrac{x^2 + 2x - 1}{x(x - 1)}$; Domain: $\{x \mid x \neq 0, x \neq 1\}$
$(f - g)(x) = \dfrac{x^2 + 1}{x(x - 1)}$; Domain: $\{x \mid x \neq 0, x \neq 1\}$
$(f \cdot g)(x) = \dfrac{x + 1}{x(x - 1)}$; Domain: $\{x \mid x \neq 0, x \neq 1\}$
$\left(\dfrac{f}{g}\right)(x) = \dfrac{x(x + 1)}{x - 1}$; Domain: $\{x \mid x \neq 0, x \neq 1\}$

18. $-4x + 1 - 2h$
19. (a) Domain: $\{x \mid -4 \leq x \leq 3\}$; Range: $\{y \mid -3 \leq y \leq 3\}$
(b) $(0, 0)$ **(c)** -1 **(d)** -4 **(e)** $\{x \mid 0 < x \leq 3\}$
(f)

(g)

(h)

(e) No symmetry
(f) Neither
(g) $(-3, 0), (0, 0), (3, 0)$

21. Odd **22.** Even **23.** Neither **24.** Odd

20. (a) Domain: $\{x \mid x \leq 4\}$ or $(-\infty, 4]$
Range: $\{y \mid y \leq 3\}$ or $(-\infty, 3]$
(b) Increasing on $(-\infty, -2]$ and $[2, 4]$; Decreasing on $[-2, 2]$
(c) Local maximum value is 1 and occurs at $x = -2$.
Local minimum value is -1 and occurs at $x = 2$.
(d) Absolute maximum: $f(4) = 3$
Absolute minimum: none

25.

Local maximum value: 4.04 at $x = -0.91$
Local minimum value: -2.04 at $x = 0.91$
Increasing: $[-3, -0.91], [0.91, 3]$
Decreasing: $[-0.91, 0.91]$

26.

Local maximum value: 1.53 at $x = 0.41$
Local minimum values: 0.54 at $x = -0.34$
and -3.56 at $x = 1.80$
Increasing: $[-0.34, 0.41], [1.80, 3]$
Decreasing: $[-2, -0.34], [0.41, 1.80]$

27. (a) 23 **(b)** 7 **(c)** 47 **28.** -5 **29.** -17
30. $y = -17x + 24$ **31.** No **32.** Yes
33.

34.

35.

Intercepts: $(-4, 0), (4, 0), (0, -4)$
Domain: all real numbers
Range: $\{y \mid y \geq -4\}$ or $[-4, \infty)$

36.

Intercept: $(0, 0)$
Domain: all real numbers
Range: $\{y \mid y \leq 0\}$ or $(-\infty, 0]$

37.

Intercept: $(1, 0)$
Domain: $\{x \mid x \geq 1\}$ or $[1, \infty)$
Range: $\{y \mid y \geq 0\}$ or $[0, \infty)$

38.

Intercepts: $(0, 1), (1, 0)$
Domain: $\{x \mid x \leq 1\}$ or $(-\infty, 1]$
Range: $\{y \mid y \geq 0\}$ or $[0, \infty)$

39.

Intercept: $(0, 3)$
Domain: all real numbers
Range: $\{y \mid y \geq 2\}$ or $[2, \infty)$

40.

Intercepts: $(0, -24)$,
$(-2 - \sqrt[3]{4}, 0)$ or about $(-3.6, 0)$
Domain: all real numbers
Range: all real numbers

41. (a) $\{x \mid x > -2\}$ or $(-2, \infty)$
(b) $(0, 0)$
(c)

(d) $\{y \mid y > -6\}$ or $(-6, \infty)$

42. (a) $\{x \mid x \geq -4\}$ or $[-4, \infty)$
(b) $(0, 1)$
(c)

(d) $\{y \mid -4 \leq y < 0 \text{ or } y > 0\}$
or $[-4, 0) \cup (0, \infty)$

43. $A = 11$ **44. (a)** $A(x) = 2x^2 + \dfrac{40}{x}$ **45. (a)** $A(x) = 10x - x^3$

(b) 42 ft^2 **(c)** 28 ft^2

(b) The largest area that can be enclosed by the rectangle is approximately 12.17 square units.

(d)

A is smallest when $x \approx 2.15$ ft.

Chapter Test *(page 277)*

1. (a) Domain: $\{2, 4, 6, 8\}$; range: $\{5, 6, 7, 8\}$; function   **(b)** Domain: $\{-3, 1, 4\}$; range: $\{-2, 3, 5, 7\}$; not a function   **(c)** Domain: $\{x \mid x \geq -1\}$;

range: all real numbers; not a function   **(d)** Domain: all real numbers; range: $\{y \mid y \geq 2\}$; function   **2.** Domain: $\left\{x \mid x \leq \dfrac{4}{5}\right\}$; $f(-1) = 3$

3. Domain: $\{x \mid x \neq -2\}$; $g(-1) = 1$ **4.** Domain: $\{x \mid x \neq -9, x \neq 4\}$; $h(-1) = \dfrac{1}{8}$ **5. (a)** Domain: $\{x \mid -5 \leq x \leq 5\}$; range: $\{y \mid -3 \leq y \leq 3\}$

(b) $(0, 2)$, $(-2, 0)$, and $(2, 0)$ **(c)** $f(1) = 3$ **(d)** $x = -5$ and $x = 3$ **(e)** $\{x \mid -5 \leq x < -2$ or $2 < x \leq 5\}$ or $[-5, -2) \cup (2, 5]$

6. Local maximum values: $f(-0.85) \approx -0.86$; $f(2.35) \approx 15.55$; local minimum value: $f(0) = -2$; the function is increasing on the intervals
$[-5, -0.85]$ and $[0, 2.35]$ and decreasing on the intervals $[-0.85, 0]$ and $[2.35, 5]$.

7. (a)   **(b)** $(0, -4)$, $(4, 0)$ **8. (a)** 18   **(b)** $y = 18x - 32$ **9. (a)** $(f - g)(x) = 2x^2 - 3x + 3$

(c) $g(-5) = -9$

(d) $g(2) = -2$

(b) $(f \cdot g)(x) = 6x^3 - 4x^2 + 3x - 2$

(c) $f(x + h) - f(x) = 4xh + 2h^2$

10. (a)   **(b)**   **11.** $2x + h - 3$ **12. (a)** $V(x) = \dfrac{x^2}{3} - \dfrac{10x}{3} + \dfrac{\pi x^2}{24}$ **(b)** 3460.29 ft^3 **13.** Even

Cumulative Review *(page 278)*

1. $\{6\}$ **2.** $\left\{0, \dfrac{1}{3}\right\}$ **3.** $\{-1, 9\}$ **4.** $\left\{\dfrac{1}{3}, \dfrac{1}{2}\right\}$ **5.** $\left\{-\dfrac{7}{2}, \dfrac{1}{2}\right\}$ **6.** $\left\{\dfrac{1}{2}\right\}$

7. $\left\{x \mid x < -\dfrac{4}{3}\right\}$; $\left(-\infty, -\dfrac{4}{3}\right)$ **8.** $\{x \mid 1 < x < 4\}$; $(1, 4)$ **9.** $\left\{x \mid x \leq -2 \text{ or } x \geq \dfrac{3}{2}\right\}$; $\left(-\infty, -2\right] \cup \left[\dfrac{3}{2}, \infty\right)$ **10. (a)** distance: $\sqrt{29}$

(b) midpoint: $\left(\dfrac{1}{2}, -4\right)$

(c) slope: $-\dfrac{2}{5}$

11.   **12.**   **13.**   **14.**   **15.** Intercepts: $(0, -3)$, $(-2, 0)$, $(2, 0)$;

symmetry with respect to the y-axis

16. $y = \dfrac{1}{2}x + 5$

17.   **18.**   **19.**

CHAPTER 4   Linear and Quadratic Functions

4.1 Assess Your Understanding *(page 287)*

7. slope; y-intercept   **8.** positive   **9.** T   **10.** F   **11.** a   **12.** d

13. (a) $m = 2$; $b = 3$ **15. (a)** $m = -3$; $b = 4$ **17. (a)** $m = \dfrac{1}{4}$; $b = -3$ **19. (a)** $m = 0$; $b = 4$ **21.** Linear; -3 **23.** Nonlinear

(b)   **(b)**   **(b)**   **(b)**   **25.** Nonlinear   **27.** Linear; 0

(c) 2   **(d)** Increasing   **(c)** -3 **(d)** Decreasing   **(c)** $\dfrac{1}{4}$ **(d)** Increasing   **(c)** 0   **(d)** Constant

29. (a) $\dfrac{1}{4}$ (b) $\left\{x \middle| x > \dfrac{1}{4}\right\}$ or $\left(\dfrac{1}{4}, \infty\right)$
(c) 1 (d) $\{x | x \le 1\}$ or $(-\infty, 1]$
(e)

31. (a) 40 (b) 88 (c) -40 (d) $\{x | x > 40\}$ or $(40, \infty)$ (e) $\{x | x \le 88\}$ or $(-\infty, 88]$
(f) $\{x | -40 < x < 88\}$ or $(-40, 88)$

33. (a) -4 (b) $\{x | x < -4\}$ or $(-\infty, -4)$

35. (a) -6 (b) $\{x | -6 \le x < 5\}$ or $[-6, 5)$

37. (a) \$185 (b) 64 mi (c) 26 mi (d) $\{x | x \ge 0\}$ or $[0, \infty)$

39. (a) 185.3cm (b) 33.6 cm

41. (a) \$24; 600 T-shirts (b) $\$0 \le p < \$24$ (c) The price will increase.

43. (a) $\{x | 9525 < x \le 38,700\}$ or $(9525, 38\,700]$ (b) \$2209.50
(c) The independent variable is adjusted gross income, x.
The dependent variable is the tax bill, T.
(d)

(e) \$27,500

45. (a) $x = 5000$
(b) $x > 5000$

47. (a) $V(x) = -1000x + 3000$
(b) $\{x | 0 \le x \le 3\}$ or $[0, 3]$
(c)

(d) \$1000 (e) After 1 year

49. (a) $C(x) = 90x + 1800$
(b)

(c) \$3060 (d) 22 bicycles

51. (a) $d(w) = 5.5w$ (b) 13.2 cm (c) 3.6 kg

53. $C(R) = \dfrac{5}{9}R - 273.15$ **55.** d, e

57. $b = 0$; yes, $f(x) = b$

59.

60. 6

61. 7

62.

63. $f(x) = (x - 5)^2 - 18$

64. $g(x) = 3\left(x + \dfrac{5}{2}\right)^2 - \dfrac{23}{4}$

65. x-intercepts: $-3\sqrt{2},\ 3\sqrt{2}$;
y-intercept: 8

66. $\{x | x \ge -2,\ x \ne 4\}$

67. $y = -2x + 9$

68.

Local maximum: $f(1) = 4$
Local minimum: $f(4.33) = -14.52$
Increasing: $[-2, 1] \cup [4.33, 8]$
Decreasing: $[1, 4.33]$

4.2 Assess Your Understanding *(page 295)*

3. scatter plot **4.** T **5.** Linear relation, $m > 0$ **7.** Linear relation, $m < 0$ **9.** Nonlinear relation

11. (a)

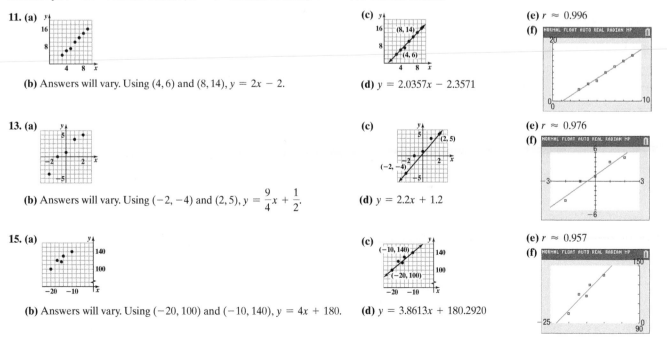

(b) Answers will vary. Using $(4, 6)$ and $(8, 14)$, $y = 2x - 2$.

(c)

(d) $y = 2.0357x - 2.3571$

(e) $r \approx 0.996$
(f)

13. (a)

(b) Answers will vary. Using $(-2, -4)$ and $(2, 5)$, $y = \dfrac{9}{4}x + \dfrac{1}{2}$.

(c)

(d) $y = 2.2x + 1.2$

(e) $r \approx 0.976$
(f)

15. (a)

(b) Answers will vary. Using $(-20, 100)$ and $(-10, 140)$, $y = 4x + 180$.

(c)

(d) $y = 3.8613x + 180.2920$

(e) $r \approx 0.957$
(f)

17. (a)

(b) Linear with positive slope

(c) Answers will vary. Using the points $(39.52, 210)$ and $(66.45, 280)$, $y = 2.599x + 107.288$.

(d)

(e) 269 calories

(f) If the weight of a candy bar is increased by 1 gram, the number of calories will increase by 2.599, on average.

19. (a) The independent variable is the number of hours spent playing video games, and cumulative grade-point average is the dependent variable because we are using number of hours playing video games to predict (or explain) cumulative grade-point average.

(b)

(c) $G(h) = -0.0942h + 3.2763$

(d) If the number of hours playing video games in a week increases by 1 hour, the cumulative grade-point average decreases 0.09, on average.

(e) 2.52

(f) Approximately 9.3 hours

21.

No, the data do not follow a linear pattern.

23. $y = 1.5x + 3.5; r = 1$ **25.** No candy bar weighs 0 grams.   **27.** $2x + y = 3$ or $y = -2x + 3$

28. $\{x \mid x \neq -5, x \neq 5\}$ **29.** $(g - f)(x) = x^2 - 8x + 12$ **30.** $y = (x + 3)^2 - 4$ **31.** $\{2 - \sqrt{7}, 2 + \sqrt{7}\}$

32. $\left[-\dfrac{52}{7}, \infty\right)$ **33.** Even   **34.** x-intercept: 2; y-intercept: $-\dfrac{3}{4}$ **35.** $-\dfrac{1}{6 + \sqrt{x + 1}}$ **36.** $\dfrac{7x + 6}{(2 + x)^{1/2}}$

4.3 Assess Your Understanding (page 308)

7. parabola   **8.** axis or axis of symmetry   **9.** $-\dfrac{b}{2a}$ **10.** T   **11.** T   **12.** T   **13.** a   **14.** d   **15.** C   **17.** F   **19.** G   **21.** H

23. (a) Vertex: $(3, -2)$; Axis of symmetry: $x = 3$
(b) Concave up
(c)

25. (a) Vertex: $(3, 5)$; Axis of symmetry: $x = 3$
(b) Concave down
(c)

27. (a) Vertex: $(6, 3)$; Axis of symmetry: $x = 6$
(b) Concave up
(c)

29. (a) Vertex: $\left(\dfrac{1}{2}, -\dfrac{7}{6}\right)$; Axis of symmetry: $x = \dfrac{1}{2}$
(b) Concave down
(c)

31.

33.

35. $f(x) = (x + 2)^2 - 2$

37. $f(x) = 2(x - 1)^2 - 1$

39. $f(x) = -(x + 1)^2 + 1$

41. $f(x) = \dfrac{1}{2}(x + 1)^2 - \dfrac{3}{2}$

43. (a) Vertex: $(-1, -1)$; Axis of symmetry: $x = -1$; Concave up
(b) y-intercept: 0; x-intercepts: $-2, 0$
(c)

(d) Domain: $(-\infty, \infty)$ Range: $[-1, \infty)$
(e) Decreasing: $(-\infty, -1]$ Increasing: $[-1, \infty)$
(f) $f(x) > 0$ on $(-\infty, -2) \cup (0, \infty)$ or for $x < -2, x > 0$; $f(x) < 0$ on $(-2, 0)$ or for $-2 < x < 0$

45. (a) Vertex: $(-3, 9)$; Axis of symmetry: $x = -3$; Concave down
  (b) y-intercept: 0; x-intercepts: $-6, 0$
  (c)

  (d) Domain: $(-\infty, \infty)$; Range: $(-\infty, 9]$
  (e) Increasing: $(-\infty, -3]$; Decreasing: $[-3, \infty)$
  (f) $f(x) > 0$ on $(-6, 0)$ or for $-6 < x < 0$;
    $f(x) < 0$ on $(-\infty, -6) \cup (0, \infty)$ or for
    $x < -6, x > 0$

47. (a) Vertex: $(-1, -9)$; Axis of symmetry: $x = -1$; Concave up
  (b) y-intercept: -8; x-intercepts: $-4, 2$
  (c)

  (d) Domain: $(-\infty, \infty)$; Range: $[-9, \infty)$
  (e) Decreasing: $(-\infty, -1]$; Increasing: $[-1, \infty)$
  (f) $f(x) > 0$ on $(-\infty, -4) \cup (2, \infty)$ or for $x < -4, x > 2$;
    $f(x) < 0$ on $(-4, 2)$ or for $-4 < x < 2$

49. (a) Vertex: $(-1, 0)$; Axis of symmetry: $x = -1$; Concave up
  (b) y-intercept: 1; x-intercept: -1
  (c)

  (d) Domain: $(-\infty, \infty)$; Range: $[0, \infty)$
  (e) Decreasing: $(-\infty, -1]$; Increasing: $[-1, \infty)$
  (f) $f(x) > 0$ on $(-\infty, -1) \cup (-1, \infty)$ or for $x < -1, x > -1$;
    $f(x)$ is never negative.

51. (a) Vertex: $\left(\dfrac{1}{4}, \dfrac{15}{8}\right)$; Axis of symmetry: $x = \dfrac{1}{4}$; Concave up
  (b) y-intercept: 2; x-intercept: None
  (c)

  (d) Domain: $(-\infty, \infty)$; Range: $\left[\dfrac{15}{8}, \infty\right)$
  (e) Decreasing: $\left(-\infty, \dfrac{1}{4}\right]$; Increasing: $\left[\dfrac{1}{4}, \infty\right)$
  (f) $f(x) > 0$ on $(-\infty, \infty)$ or for all real numbers;
    $f(x)$ is never negative.

53. (a) Vertex: $\left(\dfrac{1}{2}, -\dfrac{5}{2}\right)$; Axis of symmetry: $x = \dfrac{1}{2}$; Concave down
  (b) y-intercept: -3; x-intercept: None
  (c)

  (d) Domain: $(-\infty, \infty)$; Range: $\left(-\infty, -\dfrac{5}{2}\right]$
  (e) Increasing: $\left(-\infty, \dfrac{1}{2}\right]$; Decreasing: $\left[\dfrac{1}{2}, \infty\right)$
  (f) $f(x)$ is never positive; $f(x) < 0$ on $(-\infty, \infty)$
    or for all real numbers

55. (a) Vertex: $(-1, -1)$; Axis of symmetry: $x = -1$; Concave up
  (b) y-intercept: 2; x-intercepts: $\dfrac{-3 - \sqrt{3}}{3}, \dfrac{-3 + \sqrt{3}}{3}$
  (c)

  (d) Domain: $(-\infty, \infty)$; Range: $[-1, \infty)$
  (e) Decreasing: $(-\infty, -1]$; Increasing: $[-1, \infty)$
  (f) $f(x) > 0$ on $\left(-\infty, \dfrac{-3 - \sqrt{3}}{3}\right) \cup \left(\dfrac{-3 + \sqrt{3}}{3}, \infty\right)$ or
    for $x < \dfrac{-3 - \sqrt{3}}{3}, x > \dfrac{-3 + \sqrt{3}}{3}$;
    $f(x) < 0$ on $\left(\dfrac{-3 - \sqrt{3}}{3}, \dfrac{-3 + \sqrt{3}}{3}\right)$ or
    for $\dfrac{-3 - \sqrt{3}}{3} < x < \dfrac{-3 + \sqrt{3}}{3}$

57. (a) Vertex: $\left(-\dfrac{3}{4}, \dfrac{17}{4}\right)$; Axis of symmetry: $x = -\dfrac{3}{4}$; Concave down
  (b) y-intercept: 2; x-intercepts: $\dfrac{-3 - \sqrt{17}}{4}, \dfrac{-3 + \sqrt{17}}{4}$
  (c)

  (d) Domain: $(-\infty, \infty)$; Range: $\left(-\infty, \dfrac{17}{4}\right]$
  (e) Increasing: $\left(-\infty, -\dfrac{3}{4}\right]$; Decreasing: $\left[-\dfrac{3}{4}, \infty\right)$
  (f) $f(x) > 0$ on $\left(\dfrac{-3 - \sqrt{17}}{4}, \dfrac{-3 + \sqrt{17}}{4}\right)$ or
    for $\dfrac{-3 - \sqrt{17}}{4} < x < \dfrac{-3 + \sqrt{17}}{4}$;
    $f(x) < 0$ on $\left(-\infty, \dfrac{-3 - \sqrt{17}}{4}\right) \cup \left(\dfrac{-3 + \sqrt{17}}{4}, \infty\right)$ or
    for $x < \dfrac{-3 - \sqrt{17}}{4}, x > \dfrac{-3 + \sqrt{17}}{4}$

59. $f(x) = (x + 1)^2 - 2 = x^2 + 2x - 1$
61. $f(x) = -(x + 3)^2 + 5 = -x^2 - 6x - 4$
63. $f(x) = 2(x - 1)^2 - 3 = 2x^2 - 4x - 1$
65. Minimum value; -48 **67.** Minimum value; -21
69. Maximum value; 8   **71.** Maximum value; 23   **73.** $a = 6, b = 0, c = 2$
75. (a), (c), (d)   **77.** (a), (c), (d)   **79.** (a), (c), (d)

(b) $\{-1, 3\}$ (b) $\{-1, 3\}$ (b) $\{-1, 2\}$

81. (a) $a = 1$: $f(x) = (x + 3)(x - 1) = x^2 + 2x - 3$
$a = 2$: $f(x) = 2(x + 3)(x - 1) = 2x^2 + 4x - 6$
$a = -2$: $f(x) = -2(x + 3)(x - 1) = -2x^2 - 4x + 6$
$a = 5$: $f(x) = 5(x + 3)(x - 1) = 5x^2 + 10x - 15$

(b) The value of a does not affect the x-intercepts, but it changes the y-intercept by a factor of a.   (c) The value of a does not affect the axis of symmetry. It is $x = -1$ for all values of a.

(d) The value of a does not affect the x-coordinate of the vertex. However, the y-coordinate of the vertex is multiplied by a.

(e) The mean of the x-intercepts is the x-coordinate of the vertex.

83. (a) $(-2, -25)$ (b) $-7, 3$ (c) $-4, 0$; $(-4, -21)$, $(0, -21)$
(d)

85. (a) $\dfrac{625}{16} \approx 39.1$ ft   (b) $\dfrac{7025}{32} \approx 219.5$ ft   (c) About 170 ft
(d)
(f) When the height is 100 ft, the projectile is about 135.7 ft from the cliff.

87. 500; $1{,}000{,}000$ **89.** (a) 70,000 players   (b) $2500
91. (a) 187 or 188 watches; $7031.20   (b) $P(x) = -0.2x^2 + 43x - 1750$
(c) 107 or 108 watches; $561.20
93. (a) 171 ft   (b) 49 mph   **95.** $x = \dfrac{a}{2}$ **97.** 10   **99.** $(2, 2)$

101. f is increasing on $\left(-\infty, -\dfrac{1}{3}\right)$, $(5, \infty)$. f is decreasing on $\left(-\dfrac{1}{3}, 5\right)$

105.

109. 0, 1, or 2   **110.** Symmetric with respect to the x-axis, the y-axis, and the origin   **111.** $\{x \mid x \le 4\}$ or $(-\infty, 4]$
112. Center $(5, -2)$; radius $= 3$ **113.** $y = \sqrt{-x}$ **114.** $y = -\dfrac{5}{7}x + 7$
115. Domain: $\{1, 2, 3, 4, 5\}$; Range: $\{-7, -6, -5, -4, -3\}$; Function   **116.** 0   **117.** x **118.** $\dfrac{28x^2 + 40x}{(3x + 5)^{2/3}}$ **119.** $x + c + 5$

4.4 Assess Your Understanding *(page 317)*

3. (a) $R(p) = -6p^2 + 600p$
(b) $\{p \mid 0 \le p \le 100\}$
(c) $50
(d) $15,000
(e) 300
(f)
(g) Between $30 and $70

5. (a) $R(p) = -5p^2 + 100p$
(b) $\{p \mid 0 \le p \le 20\}$
(c) $10
(d) $500
(e) 50
(f)
(g) Between $8 and $12

7. (a) $A(w) = -w^2 + 200w$
(b) A is largest when $w = 100$ yd.
(c) 10,000 yd^2
9. 2,000,000 m^2
11. 18.75 m
13. (a) 3 in.   (b) Between 2 in. and 4 in.
15. $\dfrac{750}{\pi} \approx 238.73$ m by 375 m

17. (a)
(b) $I(x) = -49.421x^2 + 4749.034x - 60,370.056$
(c) About 48 years of age
(d) Approximately $53,717
(e)
The data appear to follow a quadratic relation with $a < 0$.

19. (a)
(b) $R(x) = 1.337x + 936.781$
(c) $2107
The data appear to be linearly related with positive slope.

21. (a)
The data appear to follow a quadratic relation with $a < 0$.
(b) $B(a) = -0.561a^2 + 32.550a - 371.675$ (c) 80.35 births per 1000
23. $\dfrac{38}{3}$ **25.** $\dfrac{248}{3}$ **28.** 13   **29.** $(x + 6)^2 + y^2 = 7$ **30.** $\left\{\dfrac{-4 - \sqrt{31}}{5}, \dfrac{-4 + \sqrt{31}}{5}\right\}$
31. x-intercept: 28; y-intercept 20   **32.** $\left\{-\dfrac{8}{3}, \dfrac{22}{3}\right\}$ **33.** $x^2 - 4x + 7$; remainder 6
34. $\{x \mid x \ne -4, x \ne 0, x \ne 4\}$ **35.** $y = 2\sqrt{9 - (x + 3)^2} - 4$ **36.** $\dfrac{-3}{(x + h - 1)(x - 1)}$
37. $(x + 1)^4(x - 7)^3(9x - 31)$

4.5 Assess Your Understanding *(page 323)*

3. (a) $\{x | x < -2 \text{ or } x > 2\}$; $(-\infty, -2) \cup (2, \infty)$ **(b)** $\{x | -2 \le x \le 2\}$; $[-2, 2]$

5. (a) $\{x | -2 \le x \le 1\}$; $[-2, 1]$ **(b)** $\{x | x < -2 \text{ or } x > 1\}$; $(-\infty, -2) \cup (1, \infty)$

7. $\{x | -2 < x < 5\}$; $(-2, 5)$ **9.** $\{x | x < 0 \text{ or } x > 4\}$; $(-\infty, 0) \cup (4, \infty)$ **11.** $\{x | -3 < x < 3\}$; $(-3, 3)$

13. $\{x | x < -4 \text{ or } x > 3\}$; $(-\infty, -4) \cup (3, \infty)$ **15.** $\left\{x \left| -\dfrac{1}{2} < x < 3\right.\right\}$; $\left(-\dfrac{1}{2}, 3\right)$ **17.** No real solution **19.** No real solution

21. $\left\{x \left| x < -\dfrac{2}{3} \text{ or } x > \dfrac{3}{2}\right.\right\}$; $\left(-\infty, -\dfrac{2}{3}\right) \cup \left(\dfrac{3}{2}, \infty\right)$ **23. (a)** $\{-1, 1\}$ **(b)** $\{-1\}$ **(c)** $\{-1, 4\}$ **(d)** $\{x | x < -1 \text{ or } x > 1\}$; $(-\infty, -1) \cup (1, \infty)$

(e) $\{x | x \le -1\}$; $(-\infty, -1]$ **(f)** $\{x | x < -1 \text{ or } x > 4\}$; $(-\infty, -1) \cup (4, \infty)$ **(g)** $\{x | x \le -\sqrt{2} \text{ or } x \ge \sqrt{2}\}$; $(-\infty, -\sqrt{2}] \cup [\sqrt{2}, \infty)$

25. (a) $\{-1, 1\}$ **(b)** $\left\{-\dfrac{1}{4}\right\}$ **(c)** $\{-4, 0\}$ **(d)** $\{x | -1 < x < 1\}$; $(-1, 1)$ **(e)** $\left\{x \left| x \le -\dfrac{1}{4}\right.\right\}$; $\left(-\infty, -\dfrac{1}{4}\right]$ **(f)** $\{x | -4 < x < 0\}$; $(-4, 0)$ **(g)** $\{0\}$

27. (a) $\{-2, 2\}$ **(b)** $\{-2, 2\}$ **(c)** $\{-2, 2\}$ **(d)** $\{x | x < -2 \text{ or } x > 2\}$; $(-\infty, -2) \cup (2, \infty)$ **(e)** $\{x | x \le -2 \text{ or } x \ge 2\}$; $(-\infty, -2] \cup [2, \infty)$

(f) $\{x | x < -2 \text{ or } x > 2\}$; $(-\infty, -2) \cup (2, \infty)$ **(g)** $\{x | x \le -\sqrt{5} \text{ or } x \ge \sqrt{5}\}$; $(-\infty, -\sqrt{5}] \cup [\sqrt{5}, \infty)$

29. (a) $\{-1, 2\}$ **(b)** $\{-2, 1\}$ **(c)** $\{0\}$ **(d)** $\{x | x < -1 \text{ or } x > 2\}$; $(-\infty, -1) \cup (2, \infty)$ **(e)** $\{x | -2 \le x \le 1\}$; $[-2, 1]$ **(f)** $\{x | x < 0\}$; $(-\infty, 0)$

(g) $\left\{x \left| x \le \dfrac{1 - \sqrt{13}}{2} \text{ or } x \ge \dfrac{1 + \sqrt{13}}{2}\right.\right\}$; $\left(-\infty, \dfrac{1 - \sqrt{13}}{2}\right] \cup \left[\dfrac{1 + \sqrt{13}}{2}, \infty\right)$ **31.** $\{x | x \le -4 \text{ or } x \ge 4\}$; $(-\infty, -4] \cup [4, \infty)$

33. (a) 5 s **(b)** The ball is more than 96 ft above the ground for time t between 2 and 3 s, $2 < t < 3$.

35. (a) $\$0, \$1000$ **(b)** The revenue is more than $\$800,000$ for prices between $\$276.39$ and $\$723.61$, $\$276.39 < p < \$723.61$.

37. (a) $\{c \mid 0.112 < c < 81.907\}$; $(0.112, 81.907)$ **(b)** It is possible to hit a target 75 km away if $c = 0.651$ or $c = 1.536$. **44.** $\{x | x \le 5\}$ **45.** Odd

46. (a) $(9, 0), (0, -6)$

(b)

47. $d = 58t$ **48.** $\{-3 - \sqrt{17}, -3 + \sqrt{17}\}$ **49.** $(0, 25), \left(-\dfrac{5}{2}, 0\right), \left(\dfrac{5}{2}, 0\right)$ **50.** $-x^2 + x + 3$ **51.** $3x^3 + 2x^2 - 29x + 28$

52. $6x + 3h - 5$ **53.** $\dfrac{x^4(2x + 35)}{(2x + 7)^5}$ or $\dfrac{2x^5 + 35x^4}{(2x + 7)^5}$

Review Exercises *(page 325)*

1. (a) $m = 2; b = -5$ **(b)** 2

(c)

(d) Increasing

2. (a) $m = \dfrac{4}{5}; b = -6$ **(b)** $\dfrac{4}{5}$

(c)

(d) Increasing

3. (a) $m = 0; b = 4$ **(b)** 0

(c)

(d) Constant

4. Linear; Slope: 5

5. Nonlinear

6.

7.

8.

9. (a)

(b) Domain: $(-\infty, \infty)$
Range: $[2, \infty)$
(c) Decreasing: $(-\infty, 2]$
Increasing: $[2, \infty)$

10. (a)

(b) Domain: $(-\infty, \infty)$
Range: $[-16, \infty)$
(c) Decreasing: $(-\infty, 0]$
Increasing: $[0, \infty)$

11. (a)

(b) Domain: $(-\infty, \infty)$
Range: $(-\infty, 1]$
(c) Increasing: $\left(-\infty, \dfrac{1}{2}\right]$
Decreasing: $\left[\dfrac{1}{2}, \infty\right)$

12. (a)

(b) Domain: $(-\infty, \infty)$
Range: $\left[\dfrac{1}{2}, \infty\right)$
(c) Decreasing: $\left(-\infty, -\dfrac{1}{3}\right]$
Increasing: $\left[-\dfrac{1}{3}, \infty\right)$

13. (a)

(b) Domain: $(-\infty, \infty)$
Range: $\left[-\dfrac{7}{3}, \infty\right)$
(c) Decreasing: $\left(-\infty, -\dfrac{2}{3}\right]$
Increasing: $\left[-\dfrac{2}{3}, \infty\right)$

14. Minimum value; 1 **15.** Maximum value; 12 **16.** Maximum value; 16 **17.** $\{x | -8 < x < 2\}$; $(-8, 2)$

18. $\left\{x \left| x \le -\dfrac{1}{3} \text{ or } x \ge 5\right.\right\}$; $\left(-\infty, -\dfrac{1}{3}\right] \cup [5, \infty)$ **19.** $y = -3x^2 + 12x - 16$ **20.** $y = x^2 + 2x + 3$

21. (a) $S(x) = 0.01x + 25,000$ **(b)** $\$35,000$ **(c)** $\$7,500,000$ **(d)** $x > \$12,500,000$

22. (a) $R(p) = -10p^2 + 1500p$ **(b)** $\{p | 0 \le p \le 150\}$ **(c)** $\$75$ **(d)** $\$56,250$ **(e)** 750 units **(f)** Between $\$70$ and $\$80$

23. 50 ft by 50 ft   **24.** 4,166,666.7 m² **25.** The side with the semicircles should be $\dfrac{50}{\pi}$ ft; the other side should be 25 ft.

26. (a) 63 clubs   **(b)** $151.90   **27.** $A(x) = -x^2 + 10x$; 25 sq. units   **28.** 3.6 ft

29. (a)

(b) yes
(c) $y = 1.3902\,x + 1.1140$
(d) 37.95 mm

30. (a) Quadratic, $a < 0$
(b) About $26.5 thousand
(c) $6408 thousand
(e)

Chapter Test (page 327)

1. (a) Slope: -4; y-intercept: 3
(b) -4 **(c)** Decreasing
(d)

2. $\left(-\dfrac{4}{3}, 0\right)$, $(2, 0)$, $(0, -8)$

3. $\left(\dfrac{2 - \sqrt{6}}{2}, 0\right)$, $\left(\dfrac{2 + \sqrt{6}}{2}, 0\right)$, $(0, 1)$

4. (a) $\{-1, 3\}$
(b)

(c) $\{x \mid -1 < x < 3\}$; $(-1, 3)$

5.

6. (a) Concave up
(b) $(2, -8)$
(c) $x = 2$
(d) x-intercepts: $\dfrac{6 - 2\sqrt{6}}{3}, \dfrac{6 + 2\sqrt{6}}{3}$; y-intercept: 4

(e)

7. Maximum value; 21
8. $\{x \mid x \le 4 \text{ or } x \ge 6\}$; $(-\infty, 4] \cup [6, \infty)$
9. (a) $C(m) = 0.15m + 129.50$ **(b)** $258.50   **(c)** 562 miles
10. (a) $R(p) = -10p^2 + 10{,}000p$ **(b)** $\{p \mid 0 \le p \le 1000\}$
(c) $500   **(d)** $2,500,000   **(e)** 5000
(f) Between $200 and $800

Cumulative Review (page 328)

1. $5\sqrt{2}$; $\left(\dfrac{3}{2}, \dfrac{1}{2}\right)$ **2.** $(-2, -1)$ and $(2, 3)$ are on the graph.

3. $\left\{x \,\middle|\, x \ge -\dfrac{3}{5}\right\}$; $\left[-\dfrac{3}{5}, \infty\right)$

4. $y = -2x + 2$

5. $y = -\dfrac{1}{2}x + \dfrac{13}{2}$

6. $(x - 2)^2 + (y + 4)^2 = 25$

7. Yes   **8. (a)** -3 **(b)** $x^2 - 4x - 2$ **(c)** $x^2 + 4x + 1$ **(d)** $-x^2 + 4x - 1$ **(e)** $x^2 - 3$ **(f)** $2x + h - 4$ **9.** $\left\{z \,\middle|\, z \ne \dfrac{7}{6}\right\}$

10. Yes   **11. (a)** No   **(b)** -1; $(-2, -1)$ is on the graph.   **(c)** -8; $(-8, 2)$ is on the graph.   **12.** Neither
13. Local maximum value is 5.30 and occurs at $x = -1.29$. Local minimum value is -3.30 and occurs at $x = 1.29$.
Increasing: $[-4, -1.29]$ and $[1.29, 4]$; Decreasing: $[-1.29, 1.29]$ **14. (a)** -4 **(b)** $\{x \mid x > -4\}$ or $(-4, \infty)$
15. (a) Domain: $\{x \mid -4 \le x \le 4\}$; Range: $\{y \mid -1 \le y \le 3\}$ **(b)** $(-1, 0)$, $(0, -1)$, $(1, 0)$ **(c)** y-axis   **(d)** 1   **(e)** -4 and 4   **(f)** $\{x \mid -1 < x < 1\}$
(g)   **(h)**   **(i)**   **(j)** Even   **(k)** $[0, 4]$

CHAPTER 5 Polynomial and Rational Functions

5.1 Assess Your Understanding (page 343)

6. smooth; continuous   **7.** b   **8.** $(-1, 1)$; $(0, 0)$; $(1, 1)$ **9.** r is a real zero of f; r is an x-intercept of the graph of f; $x - r$ is a factor of f.
10. turning points   **11.** $y = 3x^4$ **12.** ∞; $-\infty$ **13.** b   **14.** d   **15.** Polynomial function; degree 3; $f(x) = x^3 + 4x$; leading term: x^3; constant term: 0
17. Polynomial function; degree 2; $g(x) = \dfrac{3}{5}x^2 + \dfrac{2}{5}$; leading term: $\dfrac{3}{5}x^2$; constant term: $\dfrac{2}{5}$ **19.** Not a polynomial function; x is raised to the -1 power.

21. Not a polynomial function; x is raised to non-integer powers.

23. Polynomial function; degree 4; $F(x) = 5x^4 - \pi x^3 + \dfrac{1}{2}$; leading term: $5x^4$; constant term: $\dfrac{1}{2}$

25. Polynomial function; degree 4; $G(x) = 2x^4 - 4x^3 + 4x^2 - 4x + 2$; leading term: $2x^4$; constant term: 2

27.

29.

31.

33.

35.

37.

39.

41. $f(x) = (x + 1)(x - 1)(x - 3)$ for $a = 1$ **43.** $f(x) = x(x + 5)(x - 6)$ for $a = 1$

45. $f(x) = (x + 5)(x + 2)(x - 3)(x - 5)$ for $a = 1$ **47.** $f(x) = (x + 1)(x - 3)^2$ for $a = 1$

49. $f(x) = 3(x + 2)(x - 3)(x - 5)$ **51.** $f(x) = 16x(x + 2)(x - 1)(x - 3)$

53. $f(x) = 3(x + 3)(x - 1)(x - 4)$ **55.** $f(x) = 5(x + 1)^2(x - 1)^2$

57. $f(x) = -2(x + 5)^2(x - 2)(x - 4)$

59. (a) 7, multiplicity 1; -3, multiplicity 2   **(b)** Graph touches the x-axis at -3 and crosses it at 7.   **(c)** 2   **(d)** $y = 3x^3$

61. (a) 5, multiplicity 3   **(b)** Graph crosses x-axis at 5.   **(c)** 6   **(d)** $y = 7x^7$

63. (a) $-\dfrac{1}{2}$, multiplicity 2; -4, multiplicity 3   **(b)** Graph touches the x-axis at $-\dfrac{1}{2}$ and crosses at -4.   **(c)** 4   **(d)** $y = -2x^5$

65. (a) 5, multiplicity 3; -4, multiplicity 2   **(b)** Graph touches the x-axis at -4 and crosses it at 5.   **(c)** 4   **(d)** $y = x^5$

67. (a) No real zeros   **(b)** Graph neither crosses nor touches the x-axis.   **(c)** 5   **(d)** $y = 2x^6$

69. (a) 0, multiplicity 2; $-\sqrt{2}$, $\sqrt{2}$, multiplicity 1   **(b)** Graph touches the x-axis at 0 and crosses at $-\sqrt{2}$ and $\sqrt{2}$.   **(c)** 3   **(d)** $y = -2x^4$

71. Could be; zeros: $-1, 1, 2$; Least degree is 3.   **73.** Cannot be the graph of a polynomial; gap at $x = -1$

75. $f(x) = x(x - 1)(x - 2)$ **77.** $f(x) = -\dfrac{1}{2}(x + 1)(x - 1)^2(x - 2)$ **79.** $f(x) = 0.2(x + 4)(x + 1)^2(x - 3)$

81. $f(x) = -x(x + 3)^2(x - 3)^2$ **83. (a)** $-3, 2$ **(b)** $-6, -1$ **85.** -3, multiplicity 2; -1, multiplicity 3; 1, multiplicity 1

87. No, every polynomial function is defined at 0 so has a y-intercept; yes, the graph of a polynomial function can be completely above or below the

x-axis (e.g. $y = x^2 + 1$)   **89.** a, b, c, d   **95.** $y = -\dfrac{2}{5}x - \dfrac{11}{5}$ **96.** $\{x \mid x \neq -5\}$ **97.** $\dfrac{-2 - \sqrt{7}}{2}, \dfrac{-2 + \sqrt{7}}{2}$ **98.** $\left\{-\dfrac{4}{5}, 2\right\}$ **99.** Decreasing

100. $y = (x + 2)^2 + 5$ **101.** $(-\infty, 0) \cup (1, \infty)$ **102.** $-\dfrac{1}{3}$ **103.** $(8, -14)$ **104.** Quotient: $4x - 7$; remainder: $4x - 2$

5.2 Assess Your Understanding *(page 351)*

5. Step 1: $y = x^3$
  Step 2: x-intercepts: $0, 3$; y-intercept: 0
  Step 3: 0: multiplicity 2, touches; 3: multiplicity 1, crosses
  Step 4: At most 2 turning points
  Step 5: $f(-1) = -4$; $f(2) = -4$; $f(4) = 16$

7. Step 1: $y = -x^3$
  Step 2: x-intercepts: $-4, 1$; y-intercept: 16
  Step 3: -4: multiplicity 2, touches; 1: multiplicity 1, crosses
  Step 4: At most 2 turning points
  Step 5: $f(-5) = 6$; $f(-2) = 12$; $f(2) = -36$

9. Step 1: $y = -2x^4$
  Step 2: x-intercepts: $-2, 2$; y-intercept: 32
  Step 3: -2: multiplicity 1, crosses; 2: multiplicity 3, crosses
  Step 4: At most 3 turning points
  Step 5: $f(-3) = -250$; $f(-1) = 54$; $f(3) = -10$

11. Step 1: $y = x^3$
  Step 2: x-intercepts: $-4, -1, 2$; y-intercept: -8
  Step 3: $-4, -1, 2$: multiplicity 1, crosses
  Step 4: At most 2 turning points
  Step 5: $f(-5) = -28$; $f(-2) = 8$; $f(1) = -10$; $f(3) = 28$

13. Step 1: $y = x^3$
  Step 2: x-intercepts: $0, 1, 2$; y-intercept: 0
  Step 3: $0, 1, 2$: multiplicity 1, crosses
  Step 4: At most 2 turning points
  Step 5: $f\left(-\dfrac{1}{2}\right) = -\dfrac{15}{8}$; $f\left(\dfrac{5}{2}\right) = \dfrac{15}{8}$

15. Step 1: $y = x^4$
  Step 2: x-intercepts: $-1, 2$; y-intercept: 4
  Step 3: $-1, 2$: multiplicity 2, touches
  Step 4: At most 3 turning points
  Step 5: $f(-2) = 16$; $f(1) = 4$; $f(3) = 16$

17. Step 1: $y = -2x^4$
  Step 2: x-intercepts: $-4, 1, 4$; y-intercept: 32
  Step 3: $-4, 4$: multiplicity 1, crosses; 1: multiplicity 2, touches
  Step 4: At most 3 turning points
  Step 5: $f(-3) = 224; f(-1) = 120; f(2) = 24$

19. Step 1: $y = 5x^4$
  Step 2: x-intercepts: $-3, -2, 0, 2$; y-intercept: 0
  Step 3: $-3, -2, 0, 2$: multiplicity 1, crosses
  Step 4: At most 3 turning points
  Step 5: $f(-4) = 240$; $f(-2.5) \approx -14.1$;
    $f(-1) = 30$; $f(1) = -60$; $f(3) = 450$

21. Step 1: $y = x^5$
  Step 2: x-intercepts: $0, 2$; y-intercept: 0
  Step 3: 2: multiplicity 1, crosses; 0: multiplicity 2, touches
  Step 4: At most 4 turning points
  Step 5: $f(-1) = -12; f(1) = -4; f(3) = 108$

23. Step 1: $y = x^3$
  Step 2:
  Step 3: x-intercepts: $-1.26, -0.20, 1.26$; y-intercept: -0.31752
  Step 4:
  Step 5: $(-0.80, 0.57); (0.66, -0.99)$
  Step 6:
  Step 7: Range: $(-\infty, \infty)$
  Step 8: Increasing on $(-\infty, -0.80]$ and $[0.66, \infty)$
    Decreasing on $[-0.80, 0.66]$

25. Step 1: $y = x^3$
  Step 2:
  Step 3: x-intercepts: $-3.56, 0.50$; y-intercept: 0.89
  Step 4:
  Step 5: $(-2.21, 9.91); (0.50, 0)$
  Step 6:
  Step 7: Range: $(-\infty, \infty)$
  Step 8: Increasing on $(-\infty, -2.21]$ and $[0.50, \infty)$
    Decreasing on $[-2.21, 0.50]$

27. Step 1: $y = x^4$
  Step 2:
  Step 3: x-intercepts: $-1.5, -0.5, 0.5, 1.5$; y-intercept: 0.5625
  Step 4:
  Step 5: $(-1.12, -1); (1.12, -1); (0, 0.56)$
  Step 6:
  Step 7: Range: $[-1, \infty)$
  Step 8: Increasing on $[-1.12, 0]$ and $[1.12, \infty)$
    Decreasing on $(-\infty, -1.12]$ and $[0, 1.12]$

29. Step 1: $y = 2x^4$
  Step 2:
  Step 3: x-intercepts: $-1.07, 1.62$; y-intercept: -4
  Step 4:
  Step 5: $(-0.42, -4.64)$
  Step 6:
  Step 7: Range: $[-4.64, \infty)$
  Step 8: Increasing on $[-0.42, \infty)$
    Decreasing on $(-\infty, -0.42]$

31. $f(x) = -x(x + 2)(x - 2)$
   Step 1: $y = -x^3$
   Step 2: x-intercepts: $-2, 0, 2$; y-intercept: 0
   Step 3: $-2, 0, 2$: multiplicity 1, crosses
   Step 4: At most 2 turning points
   Step 5: $f(-3) = 15$; $f(-1) = -3$;
       $f(1) = 3$; $f(3) = -15$

33. $f(x) = x(x + 4)(x - 3)$
   Step 1: $y = x^3$
   Step 2: x-intercepts: $-4, 0, 3$; y-intercept: 0
   Step 3: $-4, 0, 3$: multiplicity 1, crosses
   Step 4: At most 2 turning points
   Step 5: $f(-5) = -40$; $f(-2) = 20$;
       $f(2) = -12$; $f(4) = 32$

35. $f(x) = 2x(x + 6)(x - 2)(x + 2)$
   Step 1: $y = 2x^4$
   Step 2: x-intercepts: $-6, -2, 0, 2$;
       y-intercept: 0
   Step 3: $-6, -2, 0, 2$: multiplicity 1, crosses
   Step 4: At most 3 turning points
   Step 5: $f(-7) = 630$; $f(-4) = -192$;
       $f(-1) = 30$; $f(1) = -42$;
       $f(3) = 270$

37. $f(x) = -x^2(x + 1)^2(x - 1)$
   Step 1: $y = -x^5$
   Step 2: x-intercepts: $-1, 0, 1$; y-intercept: 0
   Step 3: 1: multiplicity 1, crosses;
       $-1, 0$: multiplicity 2, touches
   Step 4: At most 4 turning points
   Step 5: $f(-1.5) = 1.40625$; $f(-0.54) = 0.10$;
       $f(0.74) = 0.43$; $f(1.2) = -1.39392$

39. $f(x) = 5x^3(3x + 4)(x + 4)$
   Step 1: $y = 15x^5$
   Step 2: x-intercepts: $-4, -\dfrac{4}{3}, 0$;
       y-intercept: 0
   Step 3: $-4, -\dfrac{4}{3}$: multiplicity 1, crosses;
       0: multiplicity 3, crosses
   Step 4: At most 4 turning points
   Step 5: $f(-4.2) = -637.157$; $f(-3) = 675$
       $f(-1) = -15$; $f(1) = 175$

41. $f(x) = \dfrac{1}{5}(x + 5)(x - 4)(x - 5)$
   Step 1: $y = \dfrac{1}{5}x^3$
   Step 2: x-intercepts: $-5, 4, 5$;
       y-intercept: 20
   Step 3: $-5, 4, 5$: multiplicity 1, crosses
   Step 4: At most 2 turning points
   Step 5: $f(-5.5) = -9.975$; $f(-3) = 22.4$;
       $f(2) = 8.4$; $f(4.5) = -0.475$;
       $f(6) = 4.4$

43. (a)

The relation appears to be cubic.
(b) $H(x) = 0.3948x^3 - 5.9563x^2 + 26.1965x - 7.4127$
(c) ≈ 24
(d)

(e) ≈ 54; no

45. (a)

The relation appears to be cubic.
(b) $0.7°/\text{h}$ **(c)** $-0.65°/\text{h}$
(d) $T(x) = -0.0079x^3 + 0.2930x^2 - 2.6481x + 47.6857$;
   $T(17) \approx 48.32°\text{F}$
(e)

(f) The predicted temperature at midnight is $47.7°\text{F}$.

47. (a)

(b)

(c)

(d) As more terms are added, the values of the polynomial function get closer to the values of f. The approximations near 0 are better than those near -1 or 1.

49. (a) Vertical scale may vary.   **(b)** $(-c, 0)$ and $(0, b)$ **(c)** $-c$ and 0

(d) (b, ∞) **(e)** $(-\infty, -b - 4)$

(f) Decreasing

50. $\left\{ x \,\middle|\, x < -\dfrac{2}{3} \text{ or } x > \dfrac{4}{3} \right\}$, or $\left(-\infty, -\dfrac{2}{3} \right) \cup \left(\dfrac{4}{3}, \infty \right)$

51. $y = -\dfrac{1}{3}\sqrt{x}$ **52.** $\left(\dfrac{7}{4}, \dfrac{25}{8} \right)$ **53.** 140   **54.** $\dfrac{17}{4}$

55. $[4, \infty)$ **56.** 3   **57.** Center: $(-2, 1)$; radius: 4

58. even   **59.** 6.25 years

5.3 Assess Your Understanding *(page 362)*

5. F   **6.** horizontal asymptote   **7.** vertical asymptote   **8.** proper   **9.** T   **10.** F   **11.** $y = 0$ **12.** T   **13.** d   **14.** a

15. All real numbers except 7; $\{x | x \neq 7\}$ **17.** All real numbers except 2 and -4; $\{x | x \neq 2, x \neq -4\}$

19. All real numbers except $-\dfrac{3}{2}$ and 4; $\left\{ x \,\middle|\, x \neq -\dfrac{3}{2}, x \neq 4 \right\}$ **21.** All real numbers except 4; $\{x | x \neq 4\}$

23. All real numbers   **25.** All real numbers except -2 and 2; $\{x | x \neq -2, x \neq 2\}$

27. (a) Domain: $\{x | x \neq 2\}$; range: $\{y | y \neq 1\}$ **(b)** $(0,0)$ **(c)** $y = 1$ **(d)** $x = 2$ **(e)** None

29. (a) Domain: $\{x | x \neq 0\}$; range: all real numbers   **(b)** $(-1, 0)$, $(1, 0)$ **(c)** None   **(d)** $x = 0$ **(e)** $y = 2x$

31. (a) Domain: $\{x | x \neq -2, x \neq 2\}$; range: $\{y | y \leq 0, y > 1\}$ **(b)** $(0, 0)$ **(c)** $y = 1$ **(d)** $x = -2, x = 2$ **(e)** None

33. (a)

(b) Domain: $\{x | x \neq 0\}$; range: $\{y | y \neq 2\}$
(c) Vertical asymptote: $x = 0$;
   horizontal asymptote: $y = 2$

35. (a)

(b) Domain: $\{x | x \neq 1\}$;
   range: $\{y | y > 0\}$
(c) Vertical asymptote: $x = 1$;
   horizontal asymptote: $y = 0$

37. (a)

(b) Domain: $\{x | x \neq -1\}$;
   range: $\{y | y \neq 0\}$
(c) Vertical asymptote: $x = -1$;
   horizontal asymptote: $y = 0$

39. (a)

(b) Domain: $\{x | x \neq -2\}$; range:
   $\{y | y < 0\}$
(c) Vertical asymptote: $x = -2$;
   horizontal asymptote: $y = 0$

41. (a)

(b) Domain: $\{x | x \neq 3\}$; range:
   $\{y | y > 1\}$
(c) Vertical asymptote: $x = 3$;
   horizontal asymptote: $y = 1$

43. (a)

(b) Domain: $\{x | x \neq 0\}$; range:
   $\{y | y < 1\}$
(c) Vertical asymptote: $x = 0$;
   horizontal asymptote: $y = 1$

45. Vertical asymptote: $x = -4$; horizontal asymptote: $y = 3$ **47.** Vertical asymptote: $x = 3$; oblique asymptote: $y = x + 5$

49. Vertical asymptotes: $x = 1, x = -1$; horizontal asymptote: $y = 0$ **51.** Vertical asymptote: $x = -\dfrac{1}{3}$; horizontal asymptote: $y = \dfrac{2}{3}$

53. Vertical asymptote: none; no horizontal or oblique asymptote   **55.** Vertical asymptote: $x = 0$; no horizontal or oblique asymptote

57. (a) 9.8208 m/sec^2 **(b)** 9.8195 m/sec^2 **(c)** 9.7936 m/sec^2 **(d)** h-axis   **(e)** $\varnothing$

59. (a)

(b) Horizontal: $R_{\text{tot}} = 10$; as the resistance
   of R_2 increases without bound, the total
   resistance approaches 10 ohms, the
   resistance R_1.
(c) $R_1 \approx 103.5$ ohms

61. (a) $R(x) = 2 + \dfrac{5}{x - 1} = 5\left(\dfrac{1}{x - 1} \right) + 2$

(b)

(c) Vertical asymptote: $x = 1$;
   horizontal asymptote: $y = 2$

67. $x = 5$ **68.** $\left\{ -\dfrac{4}{19} \right\}$

69. x-axis symmetry
70. $(-3, 11)$, $(2, -4)$
71. $(6, 0)$, $(0, -3)$ **72.** $f(-3) = 19$
73. $\left(-\dfrac{2}{5}, 3 \right)$ **74.** Odd   **75.** $\dfrac{9 - 2x}{x^2 - 9}$ **76.** $(\infty, -2)$

5.4 Assess Your Understanding *(page 376)*

2. False   **3.** c   **4.** T   **5. (a)** $\{x | x \neq 2\}$ **(b)** 0   **6.** a

7. 1. Domain: $\{x | x \neq 0, x \neq -4\}$ **2.** R is in lowest terms   **3.** no y-intercept; x-intercept: -1

   4. R is in lowest terms; vertical asymptotes: $x = 0, x = -4$ **5.** Horizontal asymptote: $y = 0$, intersected at $(-1, 0)$

6.

Interval	$(-\infty, -4)$	$(-4, -1)$	$(-1, 0)$	$(0, \infty)$
Number Chosen	-5	-2	$-\frac{1}{2}$	1
Value of R	$R(-5) = -\frac{4}{5}$	$R(-2) = \frac{1}{4}$	$R\left(-\frac{1}{2}\right) = -\frac{2}{7}$	$R(1) = \frac{2}{5}$
Location of Graph	Below x-axis	Above x-axis	Below x-axis	Above x-axis
Point on Graph	$\left(-5, -\frac{4}{5}\right)$	$\left(-2, \frac{1}{4}\right)$	$\left(-\frac{1}{2}, -\frac{2}{7}\right)$	$\left(1, \frac{2}{5}\right)$

7.

9. 1. $R(x) = \dfrac{3(x + 1)}{2(x + 2)}$; Domain: $\{x \mid x \neq -2\}$ **2.** R is in lowest terms   **3.** y-intercept: $\dfrac{3}{4}$; x-intercept: -1

4. R is in lowest terms; vertical asymptote: $x = -2$ **5.** Horizontal asymptote: $y = \dfrac{3}{2}$, not intersected

6.

Interval	$(-\infty, -2)$	$(-2, -1)$	$(-1, \infty)$
Number Chosen	-3	$-\frac{3}{2}$	0
Value of R	$R(-3) = 3$	$R\left(-\frac{3}{2}\right) = -\frac{3}{2}$	$R(0) = \frac{3}{4}$
Location of Graph	Above x-axis	Below x-axis	Above x-axis
Point on Graph	$(-3, 3)$	$\left(-\frac{3}{2}, -\frac{3}{2}\right)$	$\left(0, \frac{3}{4}\right)$

7.

11. 1. $R(x) = \dfrac{3}{(x + 2)(x - 2)}$; Domain: $\{x \mid x \neq -2, x \neq 2\}$ **2.** R is in lowest terms   **3.** y-intercept: $-\dfrac{3}{4}$; no x-intercept

4. R is in lowest terms; vertical asymptotes: $x = 2, x = -2$ **5.** Horizontal asymptote: $y = 0$, not intersected

6.

Interval	$(-\infty, -2)$	$(-2, 2)$	$(2, \infty)$
Number Chosen	-3	0	3
Value of R	$R(-3) = \frac{3}{5}$	$R(0) = -\frac{3}{4}$	$R(3) = \frac{3}{5}$
Location of Graph	Above x-axis	Below x-axis	Above x-axis
Point on Graph	$\left(-3, \frac{3}{5}\right)$	$\left(0, -\frac{3}{4}\right)$	$\left(3, \frac{3}{5}\right)$

7.

13. 1. $P(x) = \dfrac{(x^2 + x + 1)(x^2 - x + 1)}{(x + 1)(x - 1)}$; Domain: $\{x \mid x \neq -1, x \neq 1\}$ **2.** P is in lowest terms   **3.** y-intercept: -1; no x-intercept

4. P is in lowest terms; vertical asymptotes: $x = -1, x = 1$ **5.** No horizontal or oblique asymptote

6.

Interval	$(-\infty, -1)$	$(-1, 1)$	$(1, \infty)$
Number Chosen	-2	0	2
Value of P	$P(-2) = 7$	$P(0) = -1$	$P(2) = 7$
Location of Graph	Above x-axis	Below x-axis	Above x-axis
Point on Graph	$(-2, 7)$	$(0, -1)$	$(2, 7)$

7.

15. 1. $H(x) = \dfrac{(x - 1)(x^2 + x + 1)}{(x + 3)(x - 3)}$; Domain: $\{x \mid x \neq -3, x \neq 3\}$ **2.** H is in lowest terms   **3.** y-intercept: $\dfrac{1}{9}$; x-intercept: 1

4. H is in lowest terms; vertical asymptotes: $x = 3, x = -3$ **5.** Oblique asymptote: $y = x$, intersected at $\left(\dfrac{1}{9}, \dfrac{1}{9}\right)$

6.

Interval	$(-\infty, -3)$	$(-3, 1)$	$(1, 3)$	$(3, \infty)$
Number Chosen	-4	0	2	4
Value of H	$H(-4) \approx -9.3$	$H(0) = \frac{1}{9}$	$H(2) = -1.4$	$H(4) = 9$
Location of Graph	Below x-axis	Above x-axis	Below x-axis	Above x-axis
Point on Graph	$(-4, -9.3)$	$\left(0, \frac{1}{9}\right)$	$(2, -1.4)$	$(4, 9)$

7.

17. 1. $R(x) = \dfrac{x^2}{(x + 3)(x - 2)}$; Domain: $\{x \neq -3, x \neq 2\}$ **2.** R is in lowest terms   **3.** y-intercept: 0; x-intercept: 0

4. R is in lowest terms; vertical asymptotes: $x = 2, x = -3$ **5.** Horizontal asymptote: $y = 1$, intersected at $(6, 1)$

6.

Interval	$(-\infty, -3)$	$(-3, 0)$	$(0, 2)$	$(2, \infty)$
Number Chosen	-6	-1	1	3
Value of R	$R(-6) = 1.5$	$R(-1) = -\frac{1}{6}$	$R(1) = -0.25$	$R(3) = 1.5$
Location of Graph	Above x-axis	Below x-axis	Below x-axis	Above x-axis
Point on Graph	$(-6, 1.5)$	$\left(-1, -\frac{1}{6}\right)$	$(1, -0.25)$	$(3, 1.5)$

7.

19. 1. $G(x) = \dfrac{x}{(x+2)(x-2)}$; Domain: $\{x | x \neq -2, x \neq 2\}$ **2.** G is in lowest terms **3.** y-intercept: 0; x-intercept: 0

4. G is in lowest terms; vertical asymptotes: $x = -2, x = 2$ **5.** Horizontal asymptote: $y = 0$, intersected at $(0,0)$

6.

Interval	$(-\infty, -2)$	$(-2, 0)$	$(0, 2)$	$(2, \infty)$
Number Chosen	-3	-1	1	3
Value of G	$G(-3) = -\frac{3}{5}$	$G(-1) = \frac{1}{3}$	$G(1) = -\frac{1}{3}$	$G(3) = \frac{3}{5}$
Location of Graph	Below x-axis	Above x-axis	Below x-axis	Above x-axis
Point on Graph	$\left(-3, -\frac{3}{5}\right)$	$\left(-1, \frac{1}{3}\right)$	$\left(1, -\frac{1}{3}\right)$	$\left(3, \frac{3}{5}\right)$

7.

21. 1. $R(x) = \dfrac{3}{(x-1)(x+2)(x-2)}$; Domain: $\{x | x \neq 1, x \neq -2, x \neq 2\}$ **2.** R is in lowest terms **3.** y-intercept: $\dfrac{3}{4}$; no x-intercept

4. R is in lowest terms; vertical asymptotes: $x = -2, x = 1, x = 2$ **5.** Horizontal asymptote: $y = 0$, not intersected

6.

Interval	$(-\infty, -2)$	$(-2, 1)$	$(1, 2)$	$(2, \infty)$
Number Chosen	-3	0	1.5	3
Value of R	$R(-3) = -\frac{3}{20}$	$R(0) = \frac{3}{4}$	$R(1.5) = -\frac{24}{7}$	$R(3) = \frac{3}{10}$
Location of Graph	Below x-axis	Above x-axis	Below x-axis	Above x-axis
Point on Graph	$\left(-3, -\frac{3}{20}\right)$	$\left(0, \frac{3}{4}\right)$	$\left(1.5, -\frac{24}{7}\right)$	$\left(3, \frac{3}{10}\right)$

7.

23. 1. $H(x) = \dfrac{(x+1)(x-1)}{(x^2+4)(x+2)(x-2)}$; Domain: $\{x | x \neq -2, x \neq 2\}$ **2.** H is in lowest terms **3.** y-intercept: $\dfrac{1}{16}$; x-intercepts: $-1, 1$

4. H is in lowest terms; vertical asymptotes: $x = -2, x = 2$ **5.** Horizontal asymptote: $y = 0$, intersected at $(-1, 0)$ and $(1, 0)$

6.

Interval	$(-\infty, -2)$	$(-2, -1)$	$(-1, 1)$	$(1, 2)$	$(2, \infty)$
Number Chosen	-3	-1.5	0	1.5	3
Value of H	$H(-3) \approx 0.12$	$H(-1.5) \approx -0.11$	$H(0) = \frac{1}{16}$	$H(1.5) \approx -0.11$	$H(3) \approx 0.12$
Location of Graph	Above x-axis	Below x-axis	Above x-axis	Below x-axis	Above x-axis
Point on Graph	$(-3, 0.12)$	$(-1.5, -0.11)$	$\left(0, \frac{1}{16}\right)$	$(1.5, -0.11)$	$(3, 0.12)$

7.

25. 1. $F(x) = \dfrac{(x+1)(x-4)}{x+2}$; Domain: $\{x | x \neq -2\}$ **2.** F is in lowest terms **3.** y-intercept: -2; x-intercept: $-1, 4$

4. F is in lowest terms; vertical asymptote: $x = -2$ **5.** Oblique asymptote: $y = x - 5$, not intersected

6.

Interval	$(-\infty, -2)$	$(-2, -1)$	$(-1, 4)$	$(4, \infty)$
Number Chosen	-3	-1.5	0	5
Value of F	$F(-3) = -14$	$F(-1.5) = 5.5$	$F(0) = -2$	$F(5) \approx 0.86$
Location of Graph	Below x-axis	Above x-axis	Below x-axis	Above x-axis
Point on Graph	$(-3, -14)$	$(-1.5, 5.5)$	$(0, -2)$	$(5, 0.86)$

7.

27. 1. $R(x) = \dfrac{(x+4)(x-3)}{x-4}$; Domain: $\{x | x \neq 4\}$ **2.** R is in lowest terms **3.** y-intercept: 3; x-intercepts: $-4, 3$

4. R is in lowest terms; vertical asymptote: $x = 4$ **5.** Oblique asymptote: $y = x + 5$, not intersected

6.

Interval	$(-\infty, -4)$	$(-4, 3)$	$(3, 4)$	$(4, \infty)$
Number Chosen	-5	0	3.5	5
Value of R	$R(-5) = -\frac{8}{9}$	$R(0) = 3$	$R(3.5) = -7.5$	$R(5) = 18$
Location of Graph	Below x-axis	Above x-axis	Below x-axis	Above x-axis
Point on Graph	$\left(-5, -\frac{8}{9}\right)$	$(0, 3)$	$(3.5, -7.5)$	$(5, 18)$

7.

29. 1. $F(x) = \dfrac{(x+4)(x-3)}{x+2}$; Domain: $\{x | x \neq -2\}$ **2.** F is in lowest terms **3.** y-intercept: -6; x-intercepts: $-4, 3$

4. F is in lowest terms; vertical asymptote: $x = -2$ **5.** Oblique asymptote: $y = x - 1$, not intersected

6.

Interval	$(-\infty, -4)$	$(-4, -2)$	$(-2, 3)$	$(3, \infty)$
Number Chosen	-5	-3	0	4
Value of F	$F(-5) = -\frac{8}{3}$	$F(-3) = 6$	$F(0) = -6$	$F(4) = \frac{4}{3}$
Location of Graph	Below x-axis	Above x-axis	Below x-axis	Above x-axis
Point on Graph	$\left(-5, -\frac{8}{3}\right)$	$(-3, 6)$	$(0, -6)$	$\left(4, \frac{4}{3}\right)$

7.

31. 1. Domain: $\{x \mid x \neq -3\}$ **2.** R is in lowest terms   **3.** y-intercept: 0; x-intercepts: 0, 1   **4.** Vertical asymptote: $x = -3$
5. Horizontal asymptote: $y = 1$, not intersected
6.

Interval	$(-\infty, -3)$	$(-3, 0)$	$(0, 1)$	$(1, \infty)$
Number Chosen	-4	-1	$\frac{1}{2}$	2
Value of R	$R(-4) = 100$	$R(-1) = -0.5$	$R\left(\frac{1}{2}\right) \approx 0.003$	$R(2) = 0.016$
Location of Graph	Above x-axis	Below x-axis	Above x-axis	Above x-axis
Point on Graph	$(-4, 100)$	$(-1, -0.5)$	$\left(\frac{1}{2}, 0.003\right)$	$(2, 0.016)$

7.

33. 1. $R(x) = \dfrac{(x+4)(x-3)}{(x+2)(x-3)}$; Domain: $\{x \mid x \neq -2, x \neq 3\}$ **2.** In lowest terms, $R(x) = \dfrac{x+4}{x+2}$ **3.** y-intercept: 2; x-intercept: -4
4. Vertical asymptote: $x = -2$; hole at $\left(3, \dfrac{7}{5}\right)$ **5.** Horizontal asymptote: $y = 1$, not intersected
6.

Interval	$(-\infty, -4)$	$(-4, -2)$	$(-2, 3)$	$(3, \infty)$
Number Chosen	-5	-3	0	4
Value of R	$R(-5) = \frac{1}{3}$	$R(-3) = -1$	$R(0) = 2$	$R(4) = \frac{4}{3}$
Location of Graph	Above x-axis	Below x-axis	Above x-axis	Above x-axis
Point on Graph	$\left(-5, \frac{1}{3}\right)$	$(-3, -1)$	$(0, 2)$	$\left(4, \frac{4}{3}\right)$

7.

35. 1. $R(x) = \dfrac{(3x+1)(2x-3)}{(x-2)(2x-3)}$; Domain: $\left\{x \mid x \neq \dfrac{3}{2}, x \neq 2\right\}$ **2.** In lowest terms, $R(x) = \dfrac{3x+1}{x-2}$ **3.** y-intercept: $-\dfrac{1}{2}$; x-intercept: $-\dfrac{1}{3}$
4. Vertical asymptote: $x = 2$; hole at $\left(\dfrac{3}{2}, -11\right)$ **5.** Horizontal asymptote: $y = 3$, not intersected
6.

Interval	$\left(-\infty, -\frac{1}{3}\right)$	$\left(-\frac{1}{3}, \frac{3}{2}\right)$	$\left(\frac{3}{2}, 2\right)$	$(2, \infty)$
Number Chosen	-1	0	1.7	6
Value of R	$R(-1) = \frac{2}{3}$	$R(0) = -\frac{1}{2}$	$R(1.7) \approx -20.3$	$R(6) = 4.75$
Location of Graph	Above x-axis	Below x-axis	Below x-axis	Above x-axis
Point on Graph	$\left(-1, \frac{2}{3}\right)$	$\left(0, -\frac{1}{2}\right)$	$(1.7, -20.3)$	$(6, 4.75)$

7.

37. 1. $R(x) = \dfrac{(x+3)(x+2)}{x+3}$; Domain: $\{x \mid x \neq -3\}$ **2.** In lowest terms, $R(x) = x + 2$ **3.** y-intercept: 2; x-intercept: -2
4. Vertical asymptote: none; hole at $(-3, -1)$ **5.** Oblique asymptote: $y = x + 2$ intersected at all points except $x = -3$
6.

Interval	$(-\infty, -3)$	$(-3, -2)$	$(-2, \infty)$
Number Chosen	-4	$-\frac{5}{2}$	0
Value of R	$R(-4) = -2$	$R\left(-\frac{5}{2}\right) = -\frac{1}{2}$	$R(0) = 2$
Location of Graph	Below x-axis	Below x-axis	Above x-axis
Point on Graph	$(-4, -2)$	$\left(-\frac{5}{2}, -\frac{1}{2}\right)$	$(0, 2)$

7.

39. 1. $H(x) = \dfrac{-3(x-2)}{(x-2)(x+2)}$; Domain: $\{x \mid x \neq -2, x \neq 2\}$ **2.** In lowest terms, $H(x) = \dfrac{-3}{x+2}$ **3.** y-intercept: $-\dfrac{3}{2}$; no x-intercept
4. Vertical asymptote: $x = -2$; hole at $\left(2, -\dfrac{3}{4}\right)$ **5.** Horizontal asymptote: $y = 0$; not intersected
6.

Interval	$(-\infty, -2)$	$(-2, 2)$	$(2, \infty)$
Number Chosen	-3	0	3
Value of H	$H(-3) = 3$	$H(0) = -\frac{3}{2}$	$H(3) = -\frac{3}{5}$
Location of Graph	Above x-axis	Below x-axis	Below x-axis
Point on Graph	$(-3, 3)$	$\left(0, -\frac{3}{2}\right)$	$\left(3, -\frac{3}{5}\right)$

7.

41. 1. $F(x) = \dfrac{(x-1)(x-4)}{(x-1)^2}$; Domain: $\{x \mid x \neq 1\}$ **2.** In lowest terms, $F(x) = \dfrac{x-4}{x-1}$ **3.** y-intercept: 4; x-intercept: 4
4. Vertical asymptote: $x = 1$ **5.** Horizontal asymptote: $y = 1$; not intersected
6.

Interval	$(-\infty, 1)$	$(1, 4)$	$(4, \infty)$
Number Chosen	0	2	5
Value of F	$F(0) = 4$	$F(2) = -2$	$F(5) = \frac{1}{4}$
Location of Graph	Above x-axis	Below x-axis	Above x-axis
Point on Graph	$(0, 4)$	$(2, -2)$	$\left(5, \frac{1}{4}\right)$

7.

43. 1. $G(x) = \dfrac{x}{(x+2)^2}$; Domain: $\{x \mid x \neq -2\}$ **2.** G is in lowest terms **3.** y-intercept: 0; x-intercept: 0

4. Vertical asymptote: $x = -2$ **5.** Horizontal asymptote: $y = 0$; intersected at $(0, 0)$

6.

Interval	$(-\infty, -2)$	$(-2, 0)$	$(0, \infty)$
Number Chosen	-3	-1	1
Value of G	$G(-3) = -3$	$G(-1) = -1$	$G(1) = \frac{1}{9}$
Location of Graph	Below x-axis	Below x-axis	Above x-axis
Point on Graph	$(-3, -3)$	$(-1, -1)$	$\left(1, \frac{1}{9}\right)$

7.

45. 1. $f(x) = \dfrac{x^2 + 1}{x}$; Domain: $\{x \mid x \neq 0\}$ **2.** f is in lowest terms **3.** no y-intercept; no x-intercepts

4. f is in lowest terms; vertical asymptote: $x = 0$ **5.** Oblique asymptote: $y = x$, not intersected

6.

Interval	$(-\infty, 0)$	$(0, \infty)$
Number Chosen	-1	1
Value of f	$f(-1) = -2$	$f(1) = 2$
Location of Graph	Below x-axis	Above x-axis
Point on Graph	$(-1, -2)$	$(1, 2)$

7.

47. 1. $f(x) = \dfrac{x^3 + 1}{x} = \dfrac{(x+1)(x^2 - x + 1)}{x}$; Domain: $\{x \mid x \neq 0\}$ **2.** f is in lowest terms **3.** no y-intercept; x-intercept: -1

4. f is in lowest terms; vertical asymptote: $x = 0$ **5.** No horizontal or oblique asymptote

6.

Interval	$(-\infty, -1)$	$(-1, 0)$	$(0, \infty)$
Number Chosen	-2	$-\frac{1}{2}$	1
Value of f	$f(-2) = \frac{7}{2}$	$f\left(-\frac{1}{2}\right) = -\frac{7}{4}$	$f(1) = 2$
Location of Graph	Above x-axis	Below x-axis	Above x-axis
Point on Graph	$\left(-2, \frac{7}{2}\right)$	$\left(-\frac{1}{2}, -\frac{7}{4}\right)$	$(1, 2)$

7.

49. 1. $f(x) = \dfrac{x^4 + 1}{x^3}$; Domain: $\{x \mid x \neq 0\}$ **2.** f is in lowest terms **3.** no y-intercept; no x-intercepts

4. f is in lowest terms; vertical asymptote: $x = 0$ **5.** Oblique asymptote: $y = x$, not intersected

6.

Interval	$(-\infty, 0)$	$(0, \infty)$
Number Chosen	-1	1
Value of f	$f(-1) = -2$	$f(1) = 2$
Location of Graph	Below x-axis	Above x-axis
Point on Graph	$(-1, -2)$	$(1, 2)$

7.

51. One possibility: $R(x) = \dfrac{x^2}{x^2 - 4}$ **53.** One possibility: $R(x) = \dfrac{(x-1)(x-3)\left(x^2 + \dfrac{4}{3}\right)}{(x+1)^2(x-2)^2}$ **55.**

The likelihood of your ball not being chosen increases very quickly and approaches 1 as the number of attendees, x, increases.

57. (a) t-axis; $C(t) \to 0$

(b)

(c) 0.71 h after injection

59. (a) $C(x) = 16x + \dfrac{5000}{x} + 100$

(b) $x > 0$

(c)

(d) Approximately 17.7 ft by 56.6 ft (longer side parallel to river)

61. (a) $S(x) = 2x^2 + \dfrac{40,000}{x}$

(b)

(c) 2784.95 in.^{2}

(d) 21.54 in. $\times$ 21.54 in. $\times$ 21.54 in.

(e) To minimize the cost of materials needed for construction

63. (a) $C(r) = 12\pi r^2 + \dfrac{4000}{r}$

(b)

The cost is smallest when $r = 3.76$ cm.

65. (a) 0.8126

(b) 0.7759; a player serving, with probability 0.62 winning a point on a serve, has probability 0.7759 of winning the game.

(c) $x \approx 0.7$

(d)

67. (a) $R(x) = \begin{cases} \dfrac{x^2 + x - 12}{x^2 - x - 6} & \text{if } x \neq 3 \\ \dfrac{7}{5} & \text{if } x = 3 \end{cases}$

(b) $R(x) = \begin{cases} \dfrac{6x^2 - 7x - 3}{2x^2 - 7x + 6} & \text{if } x \neq \dfrac{3}{2} \\ -11 & \text{if } x = \dfrac{3}{2} \end{cases}$

69. No. Each function is a quotient of polynomials, but it is not written in lowest terms. Each function is undefined for $x = 1$; each graph has a hole at $x = 1$.

75. If there is a common factor between the numerator and the denominator, then the graph will have a hole.

76. $4x^3 - 5x^2 + 2x - 2$ **77.** $\left\{-\dfrac{1}{10}\right\}$ **78.** $\dfrac{17}{2}$ **79.** $(2, -5)$

80. $y = |x| - 4$ **81.** $g(3) = 6$ **82.** $x^2 - x - 4$

83. perpendicular **84.** $\{9\}$ **85.** $\{-\sqrt{2}, \sqrt{2}\}$

5.5 Assess Your Understanding *(page 384)*

3. c **4.** F **5. (a)** $\{x | 0 < x < 1 \text{ or } x > 2\}$; $(0, 1) \cup (2, \infty)$ **(b)** $\{x | x \le 0 \text{ or } 1 \le x \le 2\}$; $(-\infty, 0] \cup [1, 2]$

7. (a) $\{x | -1 < x < 0 \text{ or } x > 1\}$; $(-1, 0) \cup (1, \infty)$ **(b)** $\{x | x < -1 \text{ or } 0 \le x < 1\}$; $(-\infty, -1) \cup [0, 1)$

9. $\{x | x < 0 \text{ or } 0 < x < 3\}$; $(-\infty, 0) \cup (0, 3)$ **11.** $\{x | x \le 1\}$; $(-\infty, 1]$ **13.** $\{x | x \le -2 \text{ or } x \ge 2\}$; $(-\infty, -2] \cup [2, \infty)$

15. $\{x | -4 < x < -1 \text{ or } x > 0\}$; $(-4, -1) \cup (0, \infty)$ **17.** $\{x | -2 < x \le -1\}$; $(-2, -1]$ **19.** $\{x | x < -6\}$; $(-\infty, -6)$

21. $\{x | x > 4\}$; $(4, \infty)$ **23.** $\{x | -4 < x < 0 \text{ or } x > 0\}$; $(-4, 0) \cup (0, \infty)$ **25.** $\{x | x \le -2 \text{ or } 4 \le x \le 6\}$; $(-\infty, -2] \cup [4, 6]$

27. $\{x | -2 < x < 0 \text{ or } x > 6\}$; $(-2, 0) \cup (6, \infty)$ **29.** $\{x | x < -1 \text{ or } x > 1\}$; $(-\infty, -1) \cup (1, \infty)$

31. $\{x | x < -1 \text{ or } x > 1\}$; $(-\infty, -1) \cup (1, \infty)$ **33.** $\{x | x < 2\}$; $(-\infty, 2)$ **35.** $\{x | x < -1 \text{ or } x > 1\}$; $(-\infty, -1) \cup (1, \infty)$

37. $\{x | x \le -2 \text{ or } 0 < x \le 2\}$; $(-\infty, -2] \cup (0, 2]$ **39.** $\{x | x < -2 \text{ or } x > 2\}$; $(-\infty, -2) \cup (2, \infty)$ **41.** $\{x | x < 2\}$; $(-\infty, 2)$

43. $\{x | -2 < x \le 9\}$; $(-2, 9]$ **45.** $\{x | x < 3 \text{ or } x \ge 7\}$; $(-\infty, 3) \cup [7, \infty)$ **47.** $\{x | x < 2 \text{ or } 3 < x < 5\}$; $(-\infty, 2) \cup (3, 5)$

49. $\{x | x < -5 \text{ or } -4 \le x \le -3 \text{ or } x = 0 \text{ or } x > 1\}$; $(-\infty, -5) \cup [-4, -3] \cup \{0\} \cup (1, \infty)$

51. $\left\{x \left| -\dfrac{1}{2} < x < 1 \text{ or } x > 3\right.\right\}$; $\left(-\dfrac{1}{2}, 1\right) \cup (3, \infty)$ **53.** $\left\{x \left| x < -\dfrac{2}{3} \text{ or } 0 < x < \dfrac{3}{2}\right.\right\}$; $\left(-\infty, -\dfrac{2}{3}\right) \cup \left(0, \dfrac{3}{2}\right)$

55. (a)

(b) $(-\infty, -6] \cup [1, 2) \cup (2, \infty)$

57. (a)

(b) $[-4, -2) \cup [-1, 3) \cup (3, \infty)$

59. $\{x | x > 4\}$; $(4, \infty)$

61. $\{x | x \le -2 \text{ or } x \ge 2\}$; $(-\infty, -2] \cup [2, \infty)$

63. $\{x | x < -4 \text{ or } x \ge 2\}$; $(-\infty, -4) \cup [2, \infty)$

65.

$f(x) \le g(x)$ if $-1 \le x \le 1$

67.

$f(x) \le g(x)$ if $-2 \le x \le 2$

69. $(-\infty, -3) \cup (-2, 2) \cup (3, \infty)$

71. Produce at least 250 bicycles

73. (a) The stretch is less than 39 ft.

(b) The ledge should be at least 84 ft above the ground for a 150-lb jumper.

79. $\left[\dfrac{4}{3}, \infty\right)$ **80.** $3x^2y^4(x + 2y)(2x - 3y)$ **81.** $y = \dfrac{2}{3}\sqrt{x}$

82. $(f \cdot g)(x) = \sqrt{9x^2 - 1}$; Domain: $\left[\dfrac{1}{3}, \infty\right)$ **83.** x **84.** $C = \dfrac{1}{L\omega^2}$ **85.** x-axis **86.** $(0, 4), (1.33, 2.81)$ **87.** $\left\{-\dfrac{1}{3}, 4\right\}$

88. Quotient: $2x - \dfrac{3}{2}$; Remainder: $\dfrac{13}{2}$

Historical Problems *(page 398)*

1.

$$\left(x - \dfrac{b}{3}\right)^3 + b\left(x - \dfrac{b}{3}\right)^2 + c\left(x - \dfrac{b}{3}\right) + d = 0$$

$$x^3 - bx^2 + \dfrac{b^2 x}{3} - \dfrac{b^3}{27} + bx^2 - \dfrac{2b^2 x}{3} + \dfrac{b^3}{9} + cx - \dfrac{bc}{3} + d = 0$$

$$x^3 + \left(c - \dfrac{b^2}{3}\right)x + \left(\dfrac{2b^3}{27} - \dfrac{bc}{3} + d\right) = 0$$

Let $p = c - \dfrac{b^2}{3}$ and $q = \dfrac{2b^3}{27} - \dfrac{bc}{3} + d$. Then $x^3 + px + q = 0$.

2.

$$(H + K)^3 + p(H + K) + q = 0$$

$$H^3 + 3H^2 K + 3HK^2 + K^3 + pH + pK + q = 0$$

Let $3HK = -p$.

$$H^3 - pH - pK + K^3 + pH + pK + q = 0,$$

$$H^3 + K^3 = -q$$

3.

$$3HK = -p$$

$$K = -\frac{p}{3H}$$

$$H^3 + \left(-\frac{p}{3H}\right)^3 = -q$$

$$H^3 - \frac{p^3}{27H^3} = -q$$

$$27H^6 - p^3 = -27qH^3$$

$$27H^6 + 27qH^3 - p^3 = 0$$

$$H^3 = \frac{-27q \pm \sqrt{(27q)^2 - 4(27)(-p^3)}}{2 \cdot 27}$$

$$H^3 = \frac{-q}{2} \pm \sqrt{\frac{27^2 q^2}{2^2 (27^2)} + \frac{4(27)p^3}{2^2 (27^2)}}$$

$$H^3 = \frac{-q}{2} \pm \sqrt{\frac{q^2}{4} + \frac{p^3}{27}}$$

$$H = \sqrt[3]{\frac{-q}{2} + \sqrt{\frac{q^2}{4} + \frac{p^3}{27}}}$$

Choose the positive root for now.

4. $H^3 + K^3 = -q$

$$K^3 = -q - H^3$$

$$K^3 = -q - \left[\frac{-q}{2} + \sqrt{\frac{q^2}{4} + \frac{p^3}{27}}\right]$$

$$K^3 = \frac{-q}{2} - \sqrt{\frac{q^2}{4} + \frac{p^3}{27}}$$

$$K = \sqrt[3]{\frac{-q}{2} - \sqrt{\frac{q^2}{4} + \frac{p^3}{27}}}$$

5. $x = H + K$

$$x = \sqrt[3]{\frac{-q}{2} + \sqrt{\frac{q^2}{4} + \frac{p^3}{27}}} + \sqrt[3]{\frac{-q}{2} - \sqrt{\frac{q^2}{4} + \frac{p^3}{27}}}$$

(Note that if we had used the negative root in 3, the result would have been the same.)

6. $x = 3$ **7.** $x = 2$ **8.** $x = 2$

5.6 Assess Your Understanding (*page 398*)

5. a **6.** $f(c)$ **7.** b **8.** F **9.** 0 **10.** T **11.** $R = f(2) = 8$; no **13.** $R = f(2) = -82$; no **15.** $R = f(-4) = 0$; yes

17. $R = f(-4) = 1$; no **19.** $R = f\left(\frac{1}{2}\right) = 0$; yes **21.** 7; 3 or 1 positive; 2 or 0 negative **23.** 6; 2 or 0 positive; 2 or 0 negative

25. 3; 2 or 0 positive; 1 negative **27.** 4; 2 or 0 positive; 2 or 0 negative **29.** 5; 0 positive; 3 or 1 negative **31.** 6; 1 positive; 1 negative

33. $\pm 1, \pm\frac{1}{3}$ **35.** $\pm 1, \pm 5$ **37.** $\pm 1, \pm 3, \pm\frac{1}{9}, \pm\frac{1}{3}$ **39.** $\pm 1, \pm 3, \pm 9, \pm\frac{1}{2}, \pm\frac{1}{3}, \pm\frac{1}{6}, \pm\frac{3}{2}, \pm\frac{9}{2}$

41. $\pm 1, \pm 2, \pm 3, \pm 4, \pm 6, \pm 12, \pm\frac{1}{2}, \pm\frac{3}{2}$ **43.** $\pm 1, \pm 2, \pm 4, \pm 5, \pm 10, \pm 20, \pm\frac{1}{2}, \pm\frac{5}{2}, \pm\frac{1}{3}, \pm\frac{2}{3}, \pm\frac{4}{3}, \pm\frac{5}{3}, \pm\frac{10}{3}, \pm\frac{20}{3}, \pm\frac{1}{6}, \pm\frac{5}{6}$

45. $-3, -1, 2; f(x) = (x + 3)(x + 1)(x - 2)$ **47.** $\frac{1}{2}; f(x) = 2\left(x - \frac{1}{2}\right)(x^2 + 1)$ **49.** $2, \sqrt{5}, -\sqrt{5}; f(x) = 2(x - 2)(x - \sqrt{5})(x + \sqrt{5})$

51. $-1, \frac{1}{2}, \sqrt{3}, -\sqrt{3}; f(x) = 2(x + 1)\left(x - \frac{1}{2}\right)(x - \sqrt{3})(x + \sqrt{3})$ **53.** 1, multiplicity 2; $-2, -1; f(x) = (x + 2)(x + 1)(x - 1)^2$

55. $-1, -\frac{1}{4}; f(x) = 4(x + 1)\left(x + \frac{1}{4}\right)(x^2 + 2)$ **57.** $\{-1, 2\}$ **59.** $\left\{\frac{2}{3}, -1 + \sqrt{2}, -1 - \sqrt{2}\right\}$ **61.** $\left\{\frac{1}{3}, \sqrt{5}, -\sqrt{5}\right\}$ **63.** $\{-3, -2\}$

65. $\left\{-\frac{1}{3}\right\}$ **67.** $\left\{\frac{1}{2}, 2, 5\right\}$ **69.** LB $= -2$; UB $= 2$ **71.** LB $= -1$; UB $= 1$ **73.** LB $= -2$; UB $= 2$ **75.** LB $= -1$; UB $= 1$

77. LB $= -2$; UB $= 3$ **79.** $f(0) = -1; f(1) = 10$ **81.** $f(-5) = -58; f(-4) = 2$ **83.** $f(1.4) = -0.17536; f(1.5) = 1.40625$

85. 0.21 **87.** -4.04 **89.** 1.15 **91.** 2.53

93. **95.** **97.** **99.**

101. **103.** **105.** $-8, -4, -\frac{7}{3}$ **107.** $k = 5$ **109.** -7

111. If $f(x) = x^n - c^n$, then $f(c) = c^n - c^n = 0$, so $x - c$ is a factor of f.

113. 5 **115.** 7 in.

117. All the potential rational zeros are integers, so r either is an integer or is not a rational zero (and is therefore irrational). **119.** b

121. Define $f(x) = (x^3) - (1 - x^2) = x^3 + x^2 - 1$. Since $f(0) = -1$ and $f(1) = 1$, there is at least one point in the interval where the two functions have the same y-value, so there is at least one intersection point in the interval.

123. No; by the Rational Zeros Theorem, $\frac{1}{3}$ is not a potential rational zero. **125.** No; by the Rational Zeros Theorem, $\frac{2}{3}$ is not a potential rational zero.

126. $f(x) = -3(x - 5)^2 + 71$ **127.** $[3, 8)$ **128.** $y = \frac{2}{5}x - \frac{3}{5}$ **129.** No real solutions. The complex solutions are $3 \pm 3i\sqrt{2}$

130. $[-3, 2]$ and $[5, \infty)$ **131.** $(-\infty, -3]$ and $[2, 5]$ **132.** -5 and -1 **133.** $(-5, 0), (-1, 0), (0, 3)$ **134.** $(-3, -2), (2, 6),$ and $(5, 1)$

135. Absolute minimum: $f(-3) = -2$; no absolute maximum

5.7 Assess Your Understanding (*page 406*)

5. one **6.** $3 - 4i$ **7.** T **8.** F **9.** $4 + i$ **11.** $-i, 3 - i$ **13.** $-i, -5i$ **15.** $-i$ **17.** $4 - 9i, -7 + 2i$ **19.** $f(x) = x^4 - 14x^3 + 77x^2 - 200x + 208; a = 1$

21. $f(x) = x^5 - 4x^4 + 7x^3 - 8x^2 + 6x - 4; a = 1$ **23.** $f(x) = x^4 - 6x^3 + 10x^2 - 6x + 9; a = 1$ **25.** $-3i, 5$ **27.** $4i, -2, \frac{1}{4}$

29. $2 + 5i, -3, 6$ **31.** $2i, -\sqrt{7}, \sqrt{7}, -\dfrac{2}{3}$ **33.** $1, -\dfrac{1}{2} - \dfrac{\sqrt{3}}{2}i, -\dfrac{1}{2} + \dfrac{\sqrt{3}}{2}i; f(x) = (x - 1)\left(x + \dfrac{1}{2} + \dfrac{\sqrt{3}}{2}i\right)\left(x + \dfrac{1}{2} - \dfrac{\sqrt{3}}{2}i\right)$

35. $2, 3 - 2i, 3 + 2i; f(x) = (x - 2)(x - 3 + 2i)(x - 3 - 2i)$ **37.** $-i, i, -2i, 2i; f(x) = (x + i)(x - i)(x + 2i)(x - 2i)$

39. $-5i, 5i, -3, 1; f(x) = (x + 5i)(x - 5i)(x + 3)(x - 1)$ **41.** $-4, \dfrac{1}{3}, 2 - 3i, 2 + 3i; f(x) = 3(x + 4)\left(x - \dfrac{1}{3}\right)(x - 2 + 3i)(x - 2 - 3i)$

43. 130 **45.** **(a)** $f(x) = (x^2 - \sqrt{2}x + 1)(x^2 + \sqrt{2}x + 1)$ **(b)** $-\dfrac{\sqrt{2}}{2} - \dfrac{\sqrt{2}}{2}i, -\dfrac{\sqrt{2}}{2} + \dfrac{\sqrt{2}}{2}i, \dfrac{\sqrt{2}}{2} - \dfrac{\sqrt{2}}{2}i, \dfrac{\sqrt{2}}{2} + \dfrac{\sqrt{2}}{2}i$

47. Zeros that are complex numbers must occur in conjugate pairs; or a polynomial with real coefficients of odd degree must have at least one real zero.
49. If the remaining zero were a complex number, its conjugate would also be a zero, creating a polynomial of degree 5.

51. 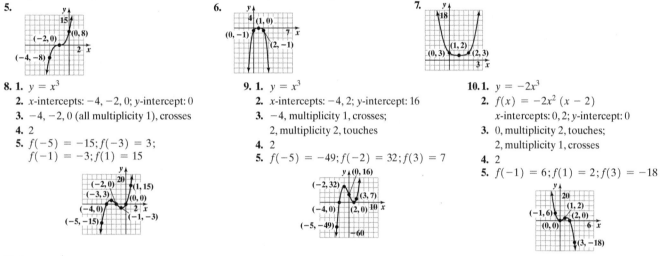 **52.** -22 **53.** $6x^3 - 13x^2 - 13x + 20$ **54.** $A = 9\pi$ ft^2 (≈ 28.274 ft^2); $C = 6\pi$ ft (≈ 18.850 ft)

55. $(g/f)(x) = \dfrac{x(3x - 2)}{x + 1}$; Domain: $x \neq -1$ and $x \neq 0$ **56.** $y = (x + 5)^2 - 3$ **57.** $[0, \infty)$

58. $(0, -2\sqrt{3}), (0, 2\sqrt{3}), (4, 0)$ **59.** $\dfrac{x - 9}{(x + 7)(\sqrt{x} + 3)}$ **60.** $3x^2 + 3xh + h^2$

Review Exercises *(page 410)*

1. Polynomial of degree 5 **2.** Rational **3.** Neither **4.** Polynomial of degree 0

5. **6.** **7.**

8. 1. $y = x^3$
2. x-intercepts: $-4, -2, 0$; y-intercept: 0
3. $-4, -2, 0$ (all multiplicity 1), crosses
4. 2
5. $f(-5) = -15; f(-3) = 3;$
$f(-1) = -3; f(1) = 15$

9. 1. $y = x^3$
2. x-intercepts: $-4, 2$; y-intercept: 16
3. -4, multiplicity 1, crosses;
2, multiplicity 2, touches
4. 2
5. $f(-5) = -49; f(-2) = 32; f(3) = 7$

10. 1. $y = -2x^3$
2. $f(x) = -2x^2(x - 2)$
x-intercepts: $0, 2$; y-intercept: 0
3. 0, multiplicity 2, touches;
2, multiplicity 1, crosses
4. 2
5. $f(-1) = 6; f(1) = 2; f(3) = -18$

11. 1. $y = x^4$
2. x-intercepts: $-3, -1, 1$; y-intercept: 3
3. $-3, -1$ (both multiplicity 1), crosses; 1, multiplicity 2, touches
4. 3
5. $f(-4) = 75; f(-2) = -9; f(2) = 15$

12. Domain: $\{x \mid x \neq -3, x \neq 3\}$; horizontal asymptote: $y = 0$;
vertical asymptotes: $x = -3, x = 3$

13. Domain: $\{x \mid x \neq 2\}$; oblique asymptote: $y = x + 2$; vertical asymptote: $x = 2$

14. Domain: $\{x \mid x \neq -2\}$; horizontal asymptote: $y = 1$; vertical asymptote: $x = -2$

15. 1. $R(x) = \dfrac{2(x - 3)}{x}$; domain: $\{x \mid x \neq 0\}$ **2.** R is in lowest terms **3.** no y-intercept; x-intercept: 3
4. R is in lowest terms; vertical asymptote: $x = 0$ **5.** Horizontal asymptote: $y = 2$; not intersected

6.

Interval	$(-\infty, 0)$	$(0, 3)$	$(3, \infty)$
Number Chosen	-2	1	4
Value of R	$R(-2) = 5$	$R(1) = -4$	$R(4) = \frac{1}{2}$
Location of Graph	Above x-axis	Below x-axis	Above x-axis
Point on Graph	$(-2, 5)$	$(1, -4)$	$\left(4, \frac{1}{2}\right)$

7.

16. 1. Domain: $\{x \mid x \neq 0, x \neq 2\}$ **2.** H is in lowest terms **3.** no y-intercept; x-intercept: -2
4. H is in lowest terms; vertical asymptotes: $x = 0, x = 2$ **5.** Horizontal asymptote: $y = 0$; intersected at $(-2, 0)$

6.

Interval	$(-\infty, -2)$	$(-2, 0)$	$(0, 2)$	$(2, \infty)$
Number Chosen	-3	-1	1	3
Value of H	$H(-3) = -\frac{1}{15}$	$H(-1) = \frac{1}{3}$	$H(1) = -3$	$H(3) = \frac{5}{3}$
Location of Graph	Below x-axis	Above x-axis	Below x-axis	Above x-axis
Point on Graph	$\left(-3, -\frac{1}{15}\right)$	$\left(-1, \frac{1}{3}\right)$	$(1, -3)$	$\left(3, \frac{5}{3}\right)$

7.

17. 1. $R(x) = \dfrac{(x+3)(x-2)}{(x-3)(x+2)}$; domain: $\{x \mid x \neq -2, x \neq 3\}$ **2.** R is in lowest terms **3.** y-intercept: 1; x-intercepts: $-3, 2$
4. R is in lowest terms; vertical asymptotes: $x = -2, x = 3$ **5.** Horizontal asymptote: $y = 1$; intersected at $(0, 1)$

6.

Interval	$(-\infty, -3)$	$(-3, -2)$	$(-2, 2)$	$(2, 3)$	$(3, \infty)$
Number Chosen	-4	$-\frac{5}{2}$	0	$\frac{5}{2}$	4
Value of R	$R(-4) = \frac{3}{7}$	$R\left(-\frac{5}{2}\right) = -\frac{9}{11}$	$R(0) = 1$	$R\left(\frac{5}{2}\right) = -\frac{11}{9}$	$R(4) = \frac{7}{3}$
Location of Graph	Above x-axis	Below x-axis	Above x-axis	Below x-axis	Above x-axis
Point on Graph	$\left(-4, \frac{3}{7}\right)$	$\left(-\frac{5}{2}, -\frac{9}{11}\right)$	$(0, 1)$	$\left(\frac{5}{2}, -\frac{11}{9}\right)$	$\left(4, \frac{7}{3}\right)$

7.

18. 1. $F(x) = \dfrac{x^3}{(x+2)(x-2)}$; domain: $\{x \mid x \neq -2, x \neq 2\}$ **2.** F is in lowest terms **3.** y-intercept: 0; x-intercept: 0
4. F is in lowest terms; vertical asymptotes: $x = -2, x = 2$ **5.** Oblique asymptote: $y = x$; intersected at $(0, 0)$

6.

Interval	$(-\infty, -2)$	$(-2, 0)$	$(0, 2)$	$(2, \infty)$
Number Chosen	-3	-1	1	3
Value of F	$F(-3) = -\frac{27}{5}$	$F(-1) = \frac{1}{3}$	$F(1) = -\frac{1}{3}$	$F(3) = \frac{27}{5}$
Location of Graph	Below x-axis	Above x-axis	Below x-axis	Above x-axis
Point on Graph	$\left(-3, -\frac{27}{5}\right)$	$\left(-1, \frac{1}{3}\right)$	$\left(1, -\frac{1}{3}\right)$	$\left(3, \frac{27}{5}\right)$

7.

19. 1. Domain: $\{x \mid x \neq 1\}$ **2.** R is in lowest terms **3.** y-intercept: 0; x-intercept: 0
4. R is in lowest terms; vertical asymptote: $x = 1$ **5.** No oblique or horizontal asymptote

6.

Interval	$(-\infty, 0)$	$(0, 1)$	$(1, \infty)$
Number Chosen	-2	$\frac{1}{2}$	2
Value of R	$R(-2) = \frac{32}{9}$	$R\left(\frac{1}{2}\right) = \frac{1}{2}$	$R(2) = 32$
Location of Graph	Above x-axis	Above x-axis	Above x-axis
Point on Graph	$\left(-2, \frac{32}{9}\right)$	$\left(\frac{1}{2}, \frac{1}{2}\right)$	$(2, 32)$

7.

20. 1. $G(x) = \dfrac{(x+2)(x-2)}{(x+1)(x-2)}$; domain: $\{x \mid x \neq -1, x \neq 2\}$ **2.** In lowest terms, $G(x) = \dfrac{x+2}{x+1}$ **3.** y-intercept: 2; x-intercept: -2
4. Vertical asymptote: $x = -1$; hole at $\left(2, \dfrac{4}{3}\right)$ **5.** Horizontal asymptote: $y = 1$, not intersected

6.

Interval	$(-\infty, -2)$	$(-2, -1)$	$(-1, 2)$	$(2, \infty)$
Number Chosen	-3	$-\frac{3}{2}$	0	3
Value of G	$G(-3) = \frac{1}{2}$	$G\left(-\frac{3}{2}\right) = -1$	$G(0) = 2$	$G(3) = \frac{5}{4}$
Location of Graph	Above x-axis	Below x-axis	Above x-axis	Above x-axis
Point on Graph	$\left(-3, \frac{1}{2}\right)$	$\left(-\frac{3}{2}, -1\right)$	$(0, 2)$	$\left(3, \frac{5}{4}\right)$

7.

21. $\{x \mid x < -2 \text{ or } -1 < x < 2\}$;
$(-\infty, -2) \cup (-1, 2)$

22. $\{x \mid -4 \leq x \leq -1 \text{ or } x \geq 1\}$
$[-4, -1] \cup [1, \infty)$

23. $\{x \mid x < 1 \text{ or } x > 2\}$; $(-\infty, 1) \cup (2, \infty)$

24. $\{x \mid 1 \leq x \leq 2 \text{ or } x > 3\}$; $[1, 2] \cup (3, \infty)$

25. $\{x \mid x < -4 \text{ or } 2 < x < 4 \text{ or } x > 6\}$; $(-\infty, -4) \cup (2, 4) \cup (6, \infty)$

26. $R = 10$; g is not a factor of f. **27.** $R = 0$; g is a factor of f. **28.** $f(4) = 47{,}105$ **29.** 4, 2, or 0 positive; 2 or 0 negative **30.** 1 positive; 2 or 0 negative
31. $\pm 1, \pm 3, \pm\dfrac{1}{2}, \pm\dfrac{3}{2}, \pm\dfrac{1}{3}, \pm\dfrac{1}{4}, \pm\dfrac{3}{4}, \pm\dfrac{1}{6}, \pm\dfrac{1}{12}$ **32.** $-2, 1, 4; f(x) = (x+2)(x-1)(x-4)$
33. $\dfrac{1}{2}$, multiplicity 2; -2; $f(x) = 4\left(x - \dfrac{1}{2}\right)^2(x+2)$ **34.** 2, multiplicity 2; $f(x) = (x-2)^2(x^2+5)$ **35.** $\{-3, 2\}$ **36.** $\left\{-3, -1, -\dfrac{1}{2}, 1\right\}$
37. LB: -2; UB: 3 **38.** LB: -3; UB: 5 **39.** $f(0) = -1$; $f(1) = 1$ **40.** $f(0) = -1$; $f(1) = 1$ **41.** 1.52 **42.** 0.93
43. $4 - i; f(x) = x^3 - 14x^2 + 65x - 102$ **44.** $-i, 1 - i; f(x) = x^4 - 2x^3 + 3x^2 - 2x + 2$ **45.** $-2, 1, 4; f(x) = (x+2)(x-1)(x-4)$
46. $-2, \dfrac{1}{2}$ (multiplicity 2); $f(x) = 4(x+2)\left(x - \dfrac{1}{2}\right)^2$ **47.** 2 (multiplicity 2), $-\sqrt{5}i, \sqrt{5}i; f(x) = (x + \sqrt{5}i)(x - \sqrt{5}i)(x-2)^2$
48. $-3, 2, -\dfrac{\sqrt{2}}{2}i, \dfrac{\sqrt{2}}{2}i; f(x) = 2(x+3)(x-2)\left(x + \dfrac{\sqrt{2}}{2}i\right)\left(x - \dfrac{\sqrt{2}}{2}i\right)$

49. Step 1: $y = x^3$
Step 2:

Step 3: x-intercepts: $-1.93, 1.14, 3.16$
y-intercept: 6.93
Step 4:

Step 5: $(-0.69, 8.70), (2.27, -4.21)$
Step 6:

Step 7: Range: $(-\infty, \infty)$
Step 8: Increasing on $(-\infty, -0.69]$ and
$[2.27, \infty)$;
Decreasing on $[-0.69, 2.27]$

50. (a) $A(r) = 2\pi r^2 + \dfrac{500}{r}$
(b) 223.22 cm²
(c) 257.08 cm²
(d)

A is smallest when $r \approx 3.41$ cm.

51. (a)

The relation appears to be cubic.
(b) $P(t) = 0.803t^3 - 7.460t^2 + 24.511t + 201.064; \approx \$503,000$
(c)

Chapter Test (page 411)

1.

2. (a) 3
(b) $\dfrac{p}{q}: \pm\dfrac{1}{2}, \pm 1, \pm\dfrac{3}{2}, \pm\dfrac{5}{2}, \pm 3, \pm 5, \pm\dfrac{15}{2}, \pm 15$
(c) $-5, -\dfrac{1}{2}, 3; g(x) = (x+5)(2x+1)(x-3)$
(d) y-intercept: -15; x-intercepts: $-5, -\dfrac{1}{2}, 3$
(e) Crosses at $-5, -\dfrac{1}{2}, 3$
(f) $y = 2x^3$
(g)

3. $4, -5i, 5i$ **4.** $\left\{1, \dfrac{5 - \sqrt{61}}{6}, \dfrac{5 + \sqrt{61}}{6}\right\}$ **5.** Domain: $\{x | x \neq -10, x \neq 4\}$; asymptotes: $x = -10, y = 2$

6. Domain: $\{x | x \neq -1\}$; asymptotes: $x = -1, y = x + 1$

7.

9. Answers may vary. One possibility is $r(x) = \dfrac{2(x-9)(x-1)}{(x-4)(x-9)}$.

10. $f(0) = 8; f(4) = -36$
Since $f(0) = 8 > 0$ and $f(4) = -36 < 0$, the Intermediate Value Theorem guarantees that there is at least one real zero between 0 and 4.

11. $\{x | x < 3 \text{ or } x > 8\}; (-\infty, 3) \cup (8, \infty)$

8. Answers may vary. One possibility is
$f(x) = x^4 - 4x^3 - 2x^2 + 20x$

12. $\{x | x \leq -3 \text{ or } -2 \leq x \leq 0\}; (-\infty, -3] \cup [-2, 0]$

Cumulative Review (page 412)

1. $\sqrt{26}$ **2.** $\{x | x \leq 0 \text{ or } x \geq 1\}$;
$(-\infty, 0] \text{ or } [1, \infty)$

3. $\{x | -1 < x < 4\}; (-1, 4)$ **4.** $f(x) = -3x + 1$ **5.** $y = 2x - 1$ **6.**

7. Not a function; 3 has two images.   **8.** $\{0, 2, 4\}$ **9.** $\left\{x \Big| x \geq \dfrac{3}{2}\right\}; \left[\dfrac{3}{2}, \infty\right)$

10. Center: $(-2, 1)$; radius: 3

11. x-intercepts: $-3, 0, 3$; y-intercept: 0; symmetric with respect to the origin

12. $y = -\dfrac{2}{3}x + \dfrac{17}{3}$ **13.** Not a function; it fails the vertical-line test.

14. (a) 22   **(b)** $x^2 - 5x - 2$ **(c)** $-x^2 - 5x + 2$ **(d)** $9x^2 + 15x - 2$ **(e)** $2x + h + 5$

15. (a) $\{x | x \neq 1\}$ **(b)** No; $(2, 7)$ is on the graph.   **(c)** 4; $(3, 4)$ is on the graph.   **(d)** $\dfrac{7}{4}; \left(\dfrac{7}{4}, 9\right)$ is on the graph.
(e) Rational

16.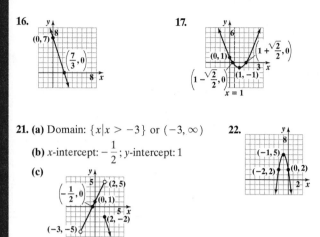

17.

18. $6; y = 6x - 1$
19. (a) x-intercepts: $-5, -1, 5$; y-intercept: -3
(b) No symmetry
(c) Neither
(d) Increasing: $(-\infty, -3]$ and $[2, \infty)$; decreasing: $[-3, 2]$
(e) A local maximum value of 5 occurs at $x = -3$.
(f) A local minimum value of -6 occurs at $x = 2$.
20. Odd

21. (a) Domain: $\{x | x > -3\}$ or $(-3, \infty)$

(b) x-intercept: $-\dfrac{1}{2}$; y-intercept: 1

(c)

22.

23. (a) $(f + g)(x) = x^2 - 9x - 6$; domain: all real numbers

(b) $\left(\dfrac{f}{g}\right)(x) = \dfrac{x^2 - 5x + 1}{-4x - 7}$; domain: $\left\{x \,\middle|\, x \neq -\dfrac{7}{4}\right\}$

24. (a) $R(x) = -\dfrac{1}{10}x^2 + 150x$

(b) \$14,000
(c) 750; \$56,250
(d) \$75

(d) Range: $\{y | y < 5\}$ or $(-\infty, 5)$

CHAPTER 6 Exponential and Logarithmic Functions

6.1 Assess Your Understanding *(page 419)*

4. composite function; $f(g(x))$ **5.** F   **6.** c   **7.** a   **8.** F   **9.** **(a)** -1 **(b)** -1 **(c)** 8   **(d)** 0   **(e)** 8   **(f)** -7 **11. (a)** 4   **(b)** 5   **(c)** -1 **(d)** -2

13. (a) 98   **(b)** 49   **(c)** 4   **(d)** 4   **15. (a)** 197   **(b)** $-\dfrac{835}{2}$ **(c)** 197   **(d)** $-\dfrac{3}{2}$ **17. (a)** $2\sqrt{5}$ **(b)** $5\sqrt{2}$ **(c)** 1   **(d)** 0   **19. (a)** $\dfrac{1}{25}$ **(b)** $\dfrac{1}{13}$ **(c)** 1

(d) $\dfrac{81}{730}$ **21. (a)** $\dfrac{3}{\sqrt[3]{4} + 1}$ **(b)** 1   **(c)** $\dfrac{6}{5}$ **(d)** 0   **23. (a)** $(f \circ g)(x) = 8x + 3$; all real numbers   **(b)** $(g \circ f)(x) = 8x + 12$; all real numbers

(c) $(f \circ f)(x) = 4x + 9$; all real numbers   **(d)** $(g \circ g)(x) = 16x$; all real numbers   **25. (a)** $(f \circ g)(x) = 3x^2 - 1$; all real numbers

(b) $(g \circ f)(x) = 9x^2 - 6x + 1$; all real numbers   **(c)** $(f \circ f)(x) = 9x - 4$; all real numbers   **(d)** $(g \circ g)(x) = x^4$; all real numbers

27. (a) $(f \circ g)(x) = x^4 + 8x^2 + 16$; all real numbers   **(b)** $(g \circ f)(x) = x^4 + 4$; all real numbers   **(c)** $(f \circ f)(x) = x^4$; all real numbers

(d) $(g \circ g)(x) = x^4 + 8x^2 + 20$; all real numbers   **29. (a)** $(f \circ g)(x) = \dfrac{3x}{2 - x}$; $\{x | x \neq 0, x \neq 2\}$ **(b)** $(g \circ f)(x) = \dfrac{2(x - 1)}{3}$; $\{x | x \neq 1\}$

(c) $(f \circ f)(x) = \dfrac{3(x - 1)}{4 - x}$; $\{x | x \neq 1, x \neq 4\}$ **(d)** $(g \circ g)(x) = x$; $\{x | x \neq 0\}$ **31. (a)** $(f \circ g)(x) = \dfrac{4}{4 + x}$; $\{x | x \neq -4, x \neq 0\}$

(b) $(g \circ f)(x) = \dfrac{-4(x - 1)}{x}$; $\{x | x \neq 0, x \neq 1\}$ **(c)** $(f \circ f)(x) = x$; $\{x | x \neq 1\}$ **(d)** $(g \circ g)(x) = x$; $\{x | x \neq 0\}$

33. (a) $(f \circ g)(x) = \sqrt{2x + 5}$; $\left\{x \,\middle|\, x \geq -\dfrac{5}{2}\right\}$ **(b)** $(g \circ f)(x) = 2\sqrt{x} + 5$; $\{x | x \geq 0\}$ **(c)** $(f \circ f)(x) = \sqrt[4]{x}$; $\{x | x \geq 0\}$

(d) $(g \circ g)(x) = 4x + 15$; all real numbers   **35. (a)** $(f \circ g)(x) = x$; $\{x | x \geq 7\}$ **(b)** $(g \circ f)(x) = |x|$; all real numbers

(c) $(f \circ f)(x) = x^4 + 14x^2 + 56$; all real numbers   **(d)** $(g \circ g)(x) = \sqrt{\sqrt{x - 7} - 7}$; $\{x | x \geq 56\}$ **37. (a)** $(f \circ g)(x) = -\dfrac{4x - 17}{2x - 1}$; $\left\{x \,\middle|\, x \neq 3; x \neq \dfrac{1}{2}\right\}$

(b) $(g \circ f)(x) = -\dfrac{3x - 3}{2x + 8}$; $\{x | x \neq -4; x \neq -1\}$ **(c)** $(f \circ f)(x) = -\dfrac{2x + 5}{x - 2}$; $\{x | x \neq -1; x \neq 2\}$ **(d)** $(g \circ g)(x) = -\dfrac{3x - 4}{2x - 11}$; $\left\{x \,\middle|\, x \neq \dfrac{11}{2}; x \neq 3\right\}$

39. $(f \circ g)(x) = f(g(x)) = f\left(\dfrac{1}{2}x\right) = 2 \cdot \dfrac{1}{2}x = x$; $(g \circ f)(x) = g(f(x)) = g(2x) = \dfrac{1}{2} \cdot 2x = x$

41. $(f \circ g)(x) = f(g(x)) = f(\sqrt[3]{x}) = (\sqrt[3]{x})^3 = x$; $(g \circ f)(x) = g(f(x)) = g(x^3) = \sqrt[3]{x^3} = x$

43. $(f \circ g)(x) = f(g(x)) = f\left(\dfrac{1}{9}(x + 6)\right) = 9 \cdot \dfrac{1}{9}(x + 6) - 6 = x + 6 - 6 = x$; $(g \circ f)(x) = g(f(x)) = g(9x - 6) = \dfrac{1}{9}(9x - 6 + 6) = \dfrac{1}{9} \cdot 9x = x$

45. $(f \circ g)(x) = f(g(x)) = f\left(\dfrac{1}{a}(x - b)\right) = a\left[\dfrac{1}{a}(x - b)\right] + b = x$; $(g \circ f)(x) = g(f(x)) = g(ax + b) = \dfrac{1}{a}(ax + b - b) = x$

47. $f(x) = x^4; g(x) = 2x + 3$ (Other answers are possible.)   **49.** $f(x) = \sqrt{x}; g(x) = x^2 + 1$ (Other answers are possible.)

51. $f(x) = |x|; g(x) = 2x + 1$ (Other answers are possible.)   **53.** $(f \circ g)(x) = 11$; $(g \circ f)(x) = 2$ **55.** $-3, 3$ **57. (a)** $(f \circ g)(x) = acx + ad + b$

(b) $(g \circ f)(x) = acx + bc + d$ **(c)** The domains of both $f \circ g$ and $g \circ f$ are all real numbers.   **(d)** $f \circ g = g \circ f$ when $ad + b = bc + d$

59. $S(t) = \dfrac{16}{9}\pi t^6$ **61.** $C(t) = 15{,}000 + 800{,}000t - 40{,}000t^2$ **63.** $C(p) = \dfrac{2\sqrt{100 - p}}{25} + 600, 0 \leq p \leq 100$ **65.** $V(r) = 2\pi r^3$

67. (a) $f(x) = 0.8101x$ **(b)** $g(x) = 132.317x$ **(c)** $g(f(x)) = 107.1900017x$ **(d)** 107,190.0017 yen   **69. (a)** $f(p) = p - 200$

(b) $g(p) = 0.8p$ **(c)** $(f \circ g)(p) = 0.8p - 200$; $(g \circ f)(p) = 0.8p - 160$; The 20% discount followed by the \$200 rebate is the better deal.

71. $-2\sqrt{2}, -\sqrt{3}, 0, \sqrt{3}, 2\sqrt{2}$ **73.** 15

75. f is an odd function, so $f(-x) = -f(x)$. g is an even function, so $g(-x) = g(x)$. Then $(f \circ g)(-x) = f(g(-x)) = f(g(x)) = (f \circ g)(x)$.

So $f \circ g$ is even. Also, $(g \circ f)(-x) = g(f(-x)) = g(-f(x)) = g(f(x)) = (g \circ f)(x)$, so $g \circ f$ is even.   **77.** -5

79. $(f + g)(x) = 4x + 3$; Domain: all real numbers; $(f - g)(x) = 2x + 13$; Domain: all real numbers; $(f \cdot g)(x) = 3x^2 - 7x - 40$;

Domain: all real numbers; $\left(\dfrac{f}{g}\right)(x) = \dfrac{3x + 8}{x - 5}$; Domain: $\{x | x \neq 5\}$

80. $\frac{1}{4}$, 4   **81.** Domain: $\{x \mid x \neq 3\}$; Vertical asymptote: $x = 3$; Oblique asymptote: $y = x + 9$ **82.** Vertex: $(3, 8)$; Axis of symmetry: $x = 3$; Concave down

83. $-1 \leq x \leq 7$ **84.** $b = \sqrt{3}$ **85.** $(-4, 11), (-1, 5)$ **86.** $5\sqrt{10}$ **87.** $-\dfrac{3}{(x+1)(c+1)}$ **88.** $\{-\sqrt{6}, 0, \sqrt{6}\}$

6.2 Assess Your Understanding *(page 431)*

5. $f(x_1) \neq f(x_2)$ **6.** one-to-one   **7.** 3   **8.** $y = x$ **9.** $[4, \infty)$ **10.** T   **11.** a   **12.** d   **13.** one-to-one   **15.** not one-to-one
17. not one-to-one   **19.** one-to-one   **21.** one-to-one   **23.** not one-to-one   **25.** one-to-one

27.

29.

31.

33. $f(g(x)) = f\left(\dfrac{1}{3}(x-4)\right) = 3 \cdot \dfrac{1}{3}(x-4) + 4 = x - 4 + 4 = x$

  $g(f(x)) = g(3x+4) = \dfrac{1}{3}[(3x+4)-4] = \dfrac{1}{3} \cdot 3x = x$

35. $f(g(x)) = f\left(\dfrac{x}{4}+2\right) = 4\left(\dfrac{x}{4}+2\right) - 8 = x + 8 - 8 = x$

  $g(f(x)) = g(4x-8) = \dfrac{4x-8}{4} + 2 = x - 2 + 2 = x$

37. $f(g(x)) = f(\sqrt[3]{x+8}) = (\sqrt[3]{x+8})^3 - 8 = x + 8 - 8 = x$

  $g(f(x)) = g(x^3 - 8) = \sqrt[3]{(x^3-8)+8} = \sqrt[3]{x^3} = x$

39. $f(g(x)) = f\left(\dfrac{1}{x}\right) = \dfrac{1}{\dfrac{1}{x}} = x; x \neq 0, g(f(x)) = g\left(\dfrac{1}{x}\right) = \dfrac{1}{\dfrac{1}{x}} = x, x \neq 0$

41. $f(g(x)) = f\left(\dfrac{4x-3}{2-x}\right) = \dfrac{2 \cdot \dfrac{4x-3}{2-x} + 3}{\dfrac{4x-3}{2-x} + 4} = \dfrac{2(4x-3) + 3(2-x)}{4x-3 + 4(2-x)} = \dfrac{5x}{5} = x, x \neq 2$

  $g(f(x)) = g\left(\dfrac{2x+3}{x+4}\right) = \dfrac{4 \cdot \dfrac{2x+3}{x+4} - 3}{2 - \dfrac{2x+3}{x+4}} = \dfrac{4(2x+3) - 3(x+4)}{2(x+4) - (2x+3)} = \dfrac{5x}{5} = x, x \neq -4$

43. (a) $f^{-1}(x) = \dfrac{1}{3}x$

  $f(f^{-1}(x)) = f\left(\dfrac{1}{3}x\right) = 3 \cdot \dfrac{1}{3}x = x$

  $f^{-1}(f(x)) = f^{-1}(3x) = \dfrac{1}{3} \cdot 3x = x$

(b) Domain of f = Range of f^{-1} = All real numbers
  Range of f = Domain of f^{-1} = All real numbers

(c)

45. (a) $f^{-1}(x) = \dfrac{x}{4} - \dfrac{1}{2}$

  $f(f^{-1}(x)) = f\left(\dfrac{x}{4} - \dfrac{1}{2}\right) = 4\left(\dfrac{x}{4} - \dfrac{1}{2}\right) + 2 = x - 2 + 2 = x$

  $f^{-1}(f(x)) = f^{-1}(4x+2) = \dfrac{4x+2}{4} - \dfrac{1}{2} = x + \dfrac{1}{2} - \dfrac{1}{2} = x$

(b) Domain of f = Range of f^{-1} = All real numbers
  Range of f = Domain of f^{-1} = All real numbers

(c)

47. (a) $f^{-1}(x) = \sqrt[3]{x+1}$

  $f(f^{-1}(x)) = f(\sqrt[3]{x+1}) = (\sqrt[3]{x+1})^3 - 1 = x$

  $f^{-1}(f(x)) = f^{-1}(x^3 - 1) = \sqrt[3]{(x^3-1)+1} = x$

(b) Domain of f = Range of f^{-1} = All real numbers
  Range of f = Domain of f^{-1} = All real numbers

(c)

49. (a) $f^{-1}(x) = \sqrt{x-4}, x \geq 4$

  $f(f^{-1}(x)) = f(\sqrt{x-4}) = (\sqrt{x-4})^2 + 4 = x$

  $f^{-1}(f(x)) = f^{-1}(x^2 + 4) = \sqrt{(x^2+4)-4} = \sqrt{x^2} = x, x \geq 0$

(b) Domain of f = Range of f^{-1} = $\{x \mid x \geq 0\}$
  Range of f = Domain of f^{-1} = $\{x \mid x \geq 4\}$

(c)

51. (a) $f^{-1}(x) = \dfrac{4}{x}$

  $f(f^{-1}(x)) = f\left(\dfrac{4}{x}\right) = \dfrac{4}{\dfrac{4}{x}} = x$

  $f^{-1}(f(x)) = f^{-1}\left(\dfrac{4}{x}\right) = \dfrac{4}{\dfrac{4}{x}} = x$

(b) Domain of f = Range of f^{-1} = $\{x \mid x \neq 0\}$
  Range of f = Domain of f^{-1} = $\{x \mid x \neq 0\}$

(c)

53. (a) $f^{-1}(x) = \dfrac{2x+1}{x}$

$$f(f^{-1}(x)) = f\left(\dfrac{2x+1}{x}\right) = \dfrac{1}{\dfrac{2x+1}{x} - 2} = \dfrac{x}{2x+1-2x} = x$$

$$f^{-1}(f(x)) = f^{-1}\left(\dfrac{1}{x-2}\right) = \dfrac{2\cdot\dfrac{1}{x-2}+1}{\dfrac{1}{x-2}} = \dfrac{2+x-2}{1} = x$$

(b) Domain of f = Range of f^{-1} = $\{x\,|\,x \neq 2\}$

Range of f = Domain of f^{-1} = $\{x\,|\,x \neq 0\}$

(c)

55. (a) $f^{-1}(x) = \dfrac{2-3x}{x}$

$$f(f^{-1}(x)) = f\left(\dfrac{2-3x}{x}\right) = \dfrac{2}{3+\dfrac{2-3x}{x}} = \dfrac{2x}{3x+2-3x} = \dfrac{2x}{2} = x$$

$$f^{-1}(f(x)) = f^{-1}\left(\dfrac{2}{3+x}\right) = \dfrac{2-3\cdot\dfrac{2}{3+x}}{\dfrac{2}{3+x}} = \dfrac{2(3+x)-3\cdot2}{2} = \dfrac{2x}{2} = x$$

(b) Domain of f = Range of f^{-1} = $\{x\,|\,x \neq -3\}$

Range of f = Domain of f^{-1} = $\{x\,|\,x \neq 0\}$

57. (a) $f^{-1}(x) = \dfrac{-2x}{x-3}$

$$f(f^{-1}(x)) = f\left(\dfrac{-2x}{x-3}\right) = \dfrac{3\cdot\dfrac{-2x}{x-3}}{\dfrac{-2x}{x-3}+2} = \dfrac{3(-2x)}{-2x+2(x-3)} = \dfrac{-6x}{-6} = x$$

$$f^{-1}(f(x)) = f^{-1}\left(\dfrac{3x}{x+2}\right) = \dfrac{-2\cdot\dfrac{3x}{x+2}}{\dfrac{3x}{x+2}-3} = \dfrac{-2\cdot3x}{3x-3(x+2)} = \dfrac{-6x}{-6} = x$$

(b) Domain of f = Range of f^{-1} = $\{x\,|\,x \neq -2\}$; Range of f = Domain of f^{-1} = $\{x\,|\,x \neq 3\}$

59. (a) $f^{-1}(x) = \dfrac{x}{3x-2}$

$$f(f^{-1}(x)) = f\left(\dfrac{x}{3x-2}\right) = \dfrac{2\cdot\dfrac{x}{3x-2}}{3\cdot\dfrac{x}{3x-2}-1} = \dfrac{2x}{3x-(3x-2)} = \dfrac{2x}{2} = x$$

$$f^{-1}(f(x)) = f^{-1}\left(\dfrac{2x}{3x-1}\right) = \dfrac{\dfrac{2x}{3x-1}}{3\cdot\dfrac{2x}{3x-1}-2} = \dfrac{2x}{6x-2(3x-1)} = \dfrac{2x}{2} = x$$

(b) Domain of f = Range of f^{-1} = $\left\{x\,\middle|\,x \neq \dfrac{1}{3}\right\}$; Range of f = Domain of f^{-1} = $\left\{x\,\middle|\,x \neq \dfrac{2}{3}\right\}$

61. (a) $f^{-1}(x) = \dfrac{3x+4}{2x-3}$

$$f(f^{-1}(x)) = f\left(\dfrac{3x+4}{2x-3}\right) = \dfrac{3\cdot\dfrac{3x+4}{2x-3}+4}{2\cdot\dfrac{3x+4}{2x-3}-3} = \dfrac{3(3x+4)+4(2x-3)}{2(3x+4)-3(2x-3)} = \dfrac{17x}{17} = x$$

$$f^{-1}(f(x)) = f^{-1}\left(\dfrac{3x+4}{2x-3}\right) = \dfrac{3\cdot\dfrac{3x+4}{2x-3}+4}{2\cdot\dfrac{3x+4}{2x-3}-3} = \dfrac{3(3x+4)+4(2x-3)}{2(3x+4)-3(2x-3)} = \dfrac{17x}{17} = x$$

(b) Domain of f = Range of f^{-1} = $\left\{x\,\middle|\,x \neq \dfrac{3}{2}\right\}$; Range of f = Domain of f^{-1} = $\left\{x\,\middle|\,x \neq \dfrac{3}{2}\right\}$

63. (a) $f^{-1}(x) = \dfrac{-2x+3}{x-2}$

$$f(f^{-1}(x)) = f\left(\dfrac{-2x+3}{x-2}\right) = \dfrac{2\cdot\dfrac{-2x+3}{x-2}+3}{\dfrac{-2x+3}{x-2}+2} = \dfrac{2(-2x+3)+3(x-2)}{-2x+3+2(x-2)} = \dfrac{-x}{-1} = x$$

$$f^{-1}(f(x)) = f^{-1}\left(\dfrac{2x+3}{x+2}\right) = \dfrac{-2\cdot\dfrac{2x+3}{x+2}+3}{\dfrac{2x+3}{x+2}-2} = \dfrac{-2(2x+3)+3(x+2)}{2x+3-2(x+2)} = \dfrac{-x}{-1} = x$$

(b) Domain of f = Range of f^{-1} = $\{x\,|\,x \neq -2\}$; Range of f = Domain of f^{-1} = $\{x\,|\,x \neq 2\}$

65. (a) $f^{-1}(x) = \dfrac{2}{\sqrt{1 - 2x}}$

$$f(f^{-1}(x)) = f\left(\dfrac{2}{\sqrt{1 - 2x}}\right) = \dfrac{\dfrac{4}{1 - 2x} - 4}{2 \cdot \dfrac{4}{1 - 2x}} = \dfrac{4 - 4(1 - 2x)}{2 \cdot 4} = \dfrac{8x}{8} = x$$

$$f^{-1}(f(x)) = f^{-1}\left(\dfrac{x^2 - 4}{2x^2}\right) = \dfrac{2}{\sqrt{1 - 2 \cdot \dfrac{x^2 - 4}{2x^2}}} = \dfrac{2}{\sqrt{\dfrac{4}{x^2}}} = \sqrt{x^2} = x, \text{ since } x > 0$$

(b) Domain of f = Range of f^{-1} = $\{x | x > 0\}$; Range of f = Domain of f^{-1} = $\left\{x \middle| x < \dfrac{1}{2}\right\}$

67. (a) $f^{-1}(x) = (x + 4)^{\frac{3}{2}}, x \geq -4$

$$f(f^{-1}(x)) = f\left((x + 4)^{\frac{3}{2}}\right) = \left((x + 4)^{\frac{3}{2}}\right)^{\frac{2}{3}} - 4 = x + 4 - 4 = x$$

$$f^{-1}(f(x)) = f^{-1}(x^{\frac{2}{3}} - 4) = \left((x^{\frac{2}{3}} - 4) + 4\right)^{\frac{3}{2}} = (x^{\frac{2}{3}})^{\frac{3}{2}} = x, \text{ since } x \geq 0$$

(b) Domain of f = Range of f^{-1} = $\{x | x \geq 0\}$; Domain of f^{-1} = Range of f = $\{x | x \geq -4\}$

69. (a) $f^{-1}(x) = \sqrt[5]{x^3 + 2}$

$$f(f^{-1}(x)) = f(\sqrt[5]{x^3 + 2}) = \sqrt[3]{(\sqrt[5]{x^3 + 2})^5 - 2} = \sqrt[3]{x^3 + 2 - 2} = \sqrt[3]{x^3} = x$$

$$f^{-1}(f(x)) = f^{-1}(\sqrt[3]{x^5 - 2}) = \sqrt[5]{(\sqrt[3]{x^5 - 2})^3 + 2} = \sqrt[5]{x^5 - 2 + 2} = \sqrt[5]{x^5} = x$$

(b) Domain of f = Range of f^{-1} = All Real Numbers; Domain of f^{-1} = Range of f = All Real Numbers

71. (a) $f^{-1}(x) = 3\sqrt{x - 2} + 1, x \geq 2$

$$f(f^{-1}(x)) = f(3\sqrt{x - 2} + 1) = \dfrac{1}{9}\left[(3\sqrt{x - 2} + 1) - 1\right]^2 + 2 = \dfrac{1}{9}[3\sqrt{x - 2}]^2 + 2 = \dfrac{1}{9} \cdot 9(x - 2) + 2 = x - 2 + 2 = x$$

$$f^{-1}(f(x)) = f^{-1}\left(\dfrac{1}{9}(x - 1)^2 + 2\right) = 3\sqrt{\dfrac{1}{9}(x - 1)^2 + 2 - 2} + 1 = 3\sqrt{\dfrac{1}{9}(x - 1)^2} + 1 = 3 \cdot \dfrac{1}{3}(x - 1) + 1 = x - 1 + 1 = x$$

(b) Domain of f = Range of f^{-1} = $\{x | x \geq 1\}$; Domain of f^{-1} = Range of f = $\{x | x \geq 2\}$

73. (a) 0 **(b)** 2 **(c)** 0 **(d)** 1 **75.** 7 **77.** Domain of f^{-1}: $[-2, \infty)$; range of f^{-1}: $[5, \infty)$ **79.** Domain of g^{-1}: $[0, \infty)$; range of g^{-1}: $(-\infty, 0]$

81. Increasing on the interval $[f(0), f(5)]$ **83.** $f^{-1}(x) = \dfrac{1}{m}(x - b), m \neq 0$ **85.** Quadrant I

87. Possible answer: $f(x) = |x|, x \geq 0$, is one-to-one; $f^{-1}(x) = x, x \geq 0$

89. (a) $r(d) = \dfrac{d + 90.39}{6.97}$

(b) $r(d(r)) = \dfrac{6.97r - 90.39 + 90.39}{6.97} = \dfrac{6.97r}{6.97} = r$

$d(r(d)) = 6.97 \cdot \dfrac{d + 90.39}{6.97} - 90.39 = d + 90.39 - 90.39 = d$

(c) 56 miles per hour

91. (a) 77.6 kg

(b) $h(W) = \dfrac{W - 50}{2.3} + 60 = \dfrac{W + 88}{2.3}$

(c) $h(W(h)) = \dfrac{50 + 2.3(h - 60) + 88}{2.3} = \dfrac{2.3h}{2.3} = h$

$W(h(W)) = 50 + 2.3\left(\dfrac{W + 88}{2.3} - 60\right) = 50 + W + 88 - 138 = W$

(d) 73 inches

93. (a) $\{g | 38,700 < g \leq 82,500\}$

(b) $\{T | 4453.50 < T \leq 14,089.50\}$

(c) $g(T) = \dfrac{T - 4453.50}{0.22} + 38,700$

Domain: $\{T | 4453.50 < T \leq 14,089.50\}$

Range: $\{g | 38,700 < g \leq 82,500\}$

95. (a) t represents time, so $t \geq 0$.

(b) $t(H) = \sqrt{\dfrac{H - 100}{-4.9}} = \sqrt{\dfrac{100 - H}{4.9}}$

(c) 2.02 seconds

97. $f^{-1}(x) = \dfrac{-dx + b}{cx - a}; f = f^{-1}$ if $a = -d$

99. (a) Domain: $(-\infty, \infty)$; Range: $(-\infty, 3) \cup [4, \infty)$

(b) $f^{-1}(x) = \begin{cases} \dfrac{x - 3}{2} & \text{if } x < 3 \\ \dfrac{x - 4}{3} & \text{if } x \geq 4 \end{cases}$

(c) Domain: $(-\infty, 3) \cup [4, \infty)$; Range: $(-\infty, \infty)$

103. No **107.** $6xh + 3h^2 - 7h$

108. Zeros: $\dfrac{-5 - \sqrt{13}}{6}, \dfrac{-5 + \sqrt{13}}{6}$;

x-intercepts: $\dfrac{-5 - \sqrt{13}}{6}, \dfrac{-5 + \sqrt{13}}{6}$; Vertex: $\left(-\dfrac{5}{6}, -\dfrac{13}{12}\right)$;

minimum; concave up

109.

110. Domain: $\left\{x \middle| x \neq -\dfrac{3}{2}, x \neq 2\right\}$; Vertical asymptote: $x = -\dfrac{3}{2}$, Horizontal asymptote: $y = 3$

111. $(x + 3)^2 + (y - 5)^2 = 49$ **112.** $y = -2x - 7$ **113.** Even **114.** $D = \dfrac{y - 2x}{2y - x}$

115. -16 **116.** $\dfrac{2}{\sqrt{2x + 2h + 3} + \sqrt{2x + 3}}$

6.3 Assess Your Understanding *(page 446)*

8. exponential function; growth factor; initial value   **9.** a **10.** T   **11.** T   **12.** $\left(-1, \frac{1}{a}\right)$; $(0, 1)$; $(1, a)$ **13.** 4   **14.** F   **15.** b   **16.** c

17. (a) 8.815   **(b)** 8.821   **(c)** 8.824   **(d)** 8.825   **19. (a)** 21.217   **(b)** 22.217   **(c)** 22.440   **(d)** 22.459   **21.** 1.265   **23.** 0.347   **25.** 3.320   **27.** 149.952
29. Neither   **31.** Exponential; $H(x) = 4^x$ **33.** Exponential; $f(x) = 3 \cdot 2^x$ **35.** Linear; $H(x) = 2x + 4$ **37.** B   **39.** D   **41.** A   **43.** E

45.
Domain: All real numbers
Range: $\{y \mid y > 1\}$ or $(1, \infty)$
Horizontal asymptote: $y = 1$
y-intercept: 2

47.
Domain: All real numbers
Range: $\{y \mid y > 0\}$ or $(0, \infty)$
Horizontal asymptote: $y = 0$
y-intercept: $\frac{1}{3}$

49.
Domain: All real numbers
Range: $\{y \mid y > 0\}$ or $(0, \infty)$
Horizontal asymptote: $y = 0$
y-intercept: 3

51.
Domain: All real numbers
Range: $\{y \mid y > -2\}$ or $(-2, \infty)$
Horizontal asymptote: $y = -2$
y-intercept: -1

53.
Domain: All real numbers
Range: $\{y \mid y > 2\}$ or $(2, \infty)$
Horizontal asymptote: $y = 2$
y-intercept: $\frac{9}{4}$

55.
Domain: All real numbers
Range: $\{y \mid y > 2\}$ or $(2, \infty)$
Horizontal asymptote: $y = 2$
y-intercept: 3

57.
Domain: All real numbers
Range: $\{y \mid y > 0\}$ or $(0, \infty)$
Horizontal asymptote: $y = 0$
y-intercept: 1

59.
Domain: All real numbers
Range: $\{y \mid y > 0\}$ or $(0, \infty)$
Horizontal asymptote: $y = 0$
y-intercept: e^2

61.
Domain: All real numbers
Range: $\{y \mid y < 5\}$ or $(-\infty, 5)$
Horizontal asymptote: $y = 5$
y-intercept: 4

63.
Domain: All real numbers
Range: $\{y \mid y < 2\}$ or $(-\infty, 2)$
Horizontal asymptote: $y = 2$
y-intercept: 1

65. $\{5\}$ **67.** $\{-4\}$ **69.** $\{2\}$ **71.** $\left\{\frac{7}{2}\right\}$ **73.** $\{-\sqrt{2}, 0, \sqrt{2}\}$

75. $\{3\}$ **77.** $\{-1, 7\}$ **79.** $\{-4, 2\}$ **81.** $\{-4\}$ **83.** $\{1, 2\}$ **85.** $\frac{1}{49}$

87. $\frac{1}{4}$ **89.** 5   **91.** $f(x) = 3^x$ **93.** $f(x) = -6^x$ **95.** $f(x) = 3^x + 2$

97. (a) 16; $(4, 16)$ **(b)** -4; $\left(-4, \frac{1}{16}\right)$ **99. (a)** $\frac{9}{4}$; $\left(-1, \frac{9}{4}\right)$ **(b)** 3; $(3, 66)$

101. (a) 60; $(-6, 60)$ **(b)** -4; $(-4, 12)$ **(c)** -2

103.
Domain: $(-\infty, \infty)$
Range: $[1, \infty)$
Intercept: $(0, 1)$

105.
Domain: $(-\infty, \infty)$
Range: $[-1, 0)$
Intercept: $(0, -1)$

107. (a) 74%   **(b)** 47%   **(c)** Each pane allows only 97% of light to pass through.
109. (a) \$16,231   **(b)** \$8626   **(c)** As each year passes, the sedan is worth 90% of its value the previous year.   **111. (a)** 30%   **(b)** 9%   **(c)** Each year only 30% of the previous survivors survive again.   **113.** 3.35 mg; 0.45 mg
115. (a) 0.632   **(b)** 0.982   **(c)** 1   **(d)**   **(e)** About 7 min

117. (a) 0.0516   **(b)** 0.0888   **119. (a)** 70.95%   **(b)** 72.62%   **(c)** 100%
121. (a) 5.41 amp, 7.59 amp, 10.38 amp   **(b)** 12 amp
(d) 3.34 amp, 5.31 amp, 9.44 amp
(e) 24 amp
(c), (f)

123. 36

125.

Final Denominator	Value of Expression	Compare Value to $e \approx$ 2.718281828
$1 + 1$	2.5	$2.5 < e$
$2 + 2$	2.8	$2.8 > e$
$3 + 3$	2.7	$2.7 < e$
$4 + 4$	2.721649485	$2.721649485 > e$
$5 + 5$	2.717770035	$2.717770035 < e$
$6 + 6$	2.718348855	$2.718348855 > e$

127. $f(A + B) = a^{A+B} = a^A \cdot a^B = f(A) \cdot f(B)$ **129.** $f(\alpha x) = a^{\alpha x} = (a^x)^\alpha = [f(x)]^\alpha$

131. (a) $f(-x) = \dfrac{1}{2}(e^{-x} + e^{-(-x)})$

$= \dfrac{1}{2}(e^{-x} + e^{x})$

$= \dfrac{1}{2}(e^{x} + e^{-x})$

$= f(x)$

(b)
$y = \dfrac{1}{2}(e^{x} + e^{-x})$

(c) $(\cosh x)^2 - (\sinh x)^2$

$= \left[\dfrac{1}{2}(e^{x} + e^{-x})\right]^2 - \left[\dfrac{1}{2}(e^{x} - e^{-x})\right]^2$

$= \dfrac{1}{4}[e^{2x} + 2 + e^{-2x} - e^{2x} + 2 - e^{-2x}]$

$= \dfrac{1}{4}\cdot 4 = 1$

133. $\{1, 2\}$ **135.** 59 minutes   **139.** $a^{-x} = (a^{-1})^x = \left(\dfrac{1}{a}\right)^x$ **140.** $(-\infty, -5]\cup[-2, 2]$ **141.** $(2, \infty)$ **142.** $f(x) = -2x^2 + 12x - 13$

143. (a)

(-3, 0) (1, 0) (-2, -3) (0, -3) (-1, -4) $x = -1$

(b) Domain: $(-\infty, \infty)$; Range: $[-4, \infty)$

(c) Decreasing: $(-\infty, -1]$; Increasing: $[-1, \infty)$

144. $\left\{\dfrac{3}{2}\right\}$ **145.** $x^2 + y^2 = 1$ **146.** $\{16, 144\}$

147. \$1050   **148.** $-3 \le x \le 3$ **149.** $4x + 2h - 7$

6.4 Assess Your Understanding *(page 460)*

3. $\{x | x > 0\}$ or $(0, \infty)$ **4.** $\left(\dfrac{1}{a}, -1\right)$, $(1, 0)$, $(a, 1)$ **5.** 1   **6.** F   **7.** T   **8.** a   **9.** c   **10.** b   **11.** $2 = \log_3 9$ **13.** $2 = \log_a 1.6$ **15.** $x = \log_2 7.2$

17. $x = \ln 8$ **19.** $2^3 = 8$ **21.** $a^6 = 3$ **23.** $3^x = 2$ **25.** $e^x = 4$ **27.** 0   **29.** 2   **31.** -3 **33.** $\dfrac{1}{2}$ **35.** 4   **37.** $\dfrac{1}{2}$ **39.** $\{x | x > 3\}$; $(3, \infty)$

41. All real numbers except 0; $\{x | x \neq 0\}$; $(-\infty, 0)\cup(0, \infty)$ **43.** $\{x | x > 10\}$; $(10, \infty)$ **45.** $\{x | x > -1\}$; $(-1, \infty)$

47. $\{x | x < -1 \text{ or } x > 0\}$; $(-\infty, -1)\cup(0, \infty)$ **49.** $\{x | x \ge 1\}$; $[1, \infty)$ **51.** 0.511   **53.** 30.099   **55.** 2.303   **57.** -53.991 **59.** $\sqrt{2}$

65. B   **67.** D   **69.** A   **71.** E

61.

$f(x) = 3^x$ $y = x$ (1, 3) $\left(-1, \dfrac{1}{3}\right)$ (0, 1)   (3, 1) $f^{-1}(x) = \log_3 x$ (1, 0) $\left(\dfrac{1}{3}, -1\right)$

63.

$f(x) = \left(\dfrac{1}{2}\right)^x$ $\left(\dfrac{1}{2}, 1\right)$ $y = x$ (-1, 2) $\left(1, \dfrac{1}{2}\right)$ $f^{-1}(x) = \log_{1/2} x$ (2, -1)

73. (a) Domain: $(-4, \infty)$

(b)

(-3, 0) $x = -4$

(c) Range: $(-\infty, \infty)$

Vertical asymptote: $x = -4$

(d) $f^{-1}(x) = e^x - 4$

(e) Domain of f^{-1}: $(-\infty, \infty)$

Range of f^{-1}: $(-4, \infty)$

(f)

(0, -3) $y = -4$

75. (a) Domain: $(0, \infty)$

(b)

(1, 2) $x = 0$

(c) Range: $(-\infty, \infty)$

Vertical asymptote: $x = 0$

(d) $f^{-1}(x) = e^{x-2}$

(e) Domain of f^{-1}: $(-\infty, \infty)$

Range of f^{-1}: $(0, \infty)$

(f)

$y = 0$ (2, 1)

77. (a) Domain: $(0, \infty)$

(b)

8   $\left(\dfrac{1}{2}, -3\right)$ $x = 0$

(c) Range: $(-\infty, \infty)$

Vertical asymptote: $x = 0$

(d) $f^{-1}(x) = \dfrac{1}{2}e^{x+3}$

(e) Domain of f^{-1}: $(-\infty, \infty)$

Range of f^{-1}: $(0, \infty)$

(f)

8   $\left(-3, \dfrac{1}{2}\right)$ $y = 0$

79. (a) Domain: $(4, \infty)$

(b)

$x = 4$ (5, 2)

(c) Range: $(-\infty, \infty)$

Vertical asymptote: $x = 4$

(d) $f^{-1}(x) = 10^{x-2} + 4$

(e) Domain of f^{-1}: $(-\infty, \infty)$

Range of f^{-1}: $(4, \infty)$

(f)

(2, 5) $y = 4$

81. (a) Domain: $(0, \infty)$

(b)

$\left(5, \dfrac{1}{2}\right)$ $\left(\dfrac{1}{2}, 0\right)$ $x = 0$

(c) Range: $(-\infty, \infty)$

Vertical asymptote: $x = 0$

(d) $f^{-1}(x) = \dfrac{1}{2}\cdot 10^{2x}$

(e) Domain of f^{-1}: $(-\infty, \infty)$

Range of f^{-1}: $(0, \infty)$

(f)

$\left(\dfrac{1}{2}, 5\right)$ $\left(0, \dfrac{1}{2}\right)$ $y = 0$

83. (a) Domain: $(-2, \infty)$

(b)

$x = -2$ (-1, 3)   (1, 4)

(c) Range: $(-\infty, \infty)$

Vertical asymptote: $x = -2$

(d) $f^{-1}(x) = 3^{x-3} - 2$

(e) Domain of f^{-1}: $(-\infty, \infty)$

Range of f^{-1}: $(-2, \infty)$

(f)

(4, 1) $y = -2$ (3, -1)

85. (a) Domain: $(-\infty, \infty)$
(b)

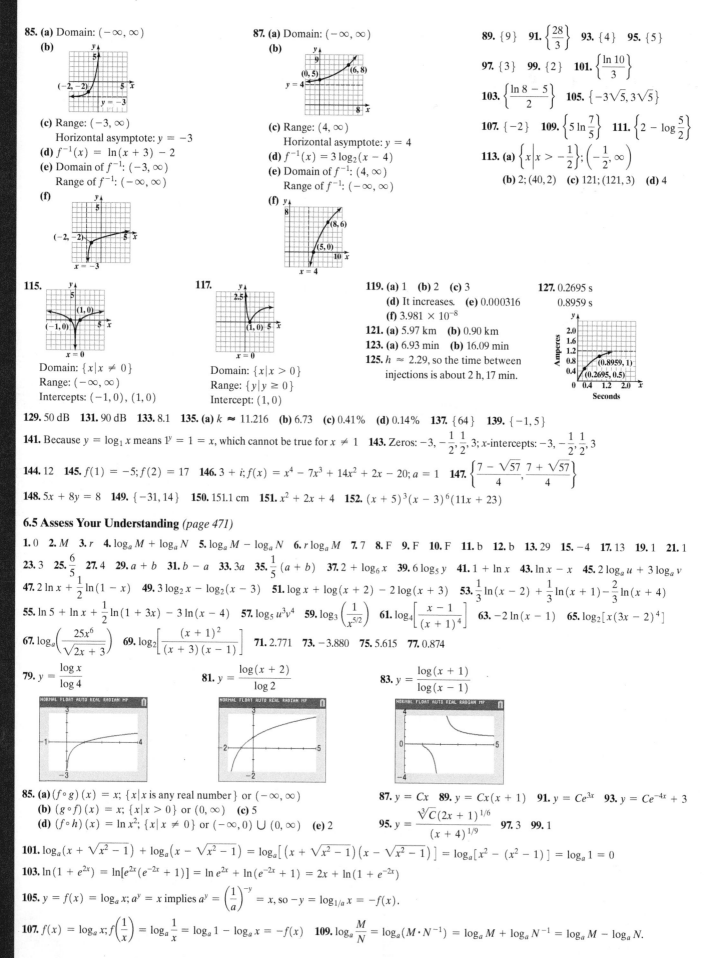

(c) Range: $(-3, \infty)$
Horizontal asymptote: $y = -3$
(d) $f^{-1}(x) = \ln(x + 3) - 2$
(e) Domain of f^{-1}: $(-3, \infty)$
Range of f^{-1}: $(-\infty, \infty)$
(f)

87. (a) Domain: $(-\infty, \infty)$
(b)

(c) Range: $(4, \infty)$
Horizontal asymptote: $y = 4$
(d) $f^{-1}(x) = 3\log_2(x - 4)$
(e) Domain of f^{-1}: $(4, \infty)$
Range of f^{-1}: $(-\infty, \infty)$
(f)

89. $\{9\}$ **91.** $\left\{\dfrac{28}{3}\right\}$ **93.** $\{4\}$ **95.** $\{5\}$

97. $\{3\}$ **99.** $\{2\}$ **101.** $\left\{\dfrac{\ln 10}{3}\right\}$

103. $\left\{\dfrac{\ln 8 - 5}{2}\right\}$ **105.** $\{-3\sqrt{5}, 3\sqrt{5}\}$

107. $\{-2\}$ **109.** $\left\{5\ln\dfrac{7}{5}\right\}$ **111.** $\left\{2 - \log\dfrac{5}{2}\right\}$

113. (a) $\left\{x\,\middle|\,x > -\dfrac{1}{2}\right\}; \left(-\dfrac{1}{2}, \infty\right)$
(b) $2; (40, 2)$ **(c)** $121; (121, 3)$ **(d)** 4

115.

Domain: $\{x \mid x \ne 0\}$
Range: $(-\infty, \infty)$
Intercepts: $(-1, 0), (1, 0)$

117.

Domain: $\{x \mid x > 0\}$
Range: $\{y \mid y \ge 0\}$
Intercept: $(1, 0)$

119. (a) 1 **(b)** 2 **(c)** 3
(d) It increases. **(e)** 0.000316
(f) 3.981×10^{-8}
121. (a) 5.97 km **(b)** 0.90 km
123. (a) 6.93 min **(b)** 16.09 min
125. $h \approx 2.29$, so the time between injections is about 2 h, 17 min.

127. 0.2695 s
0.8959 s

129. 50 dB **131.** 90 dB **133.** 8.1 **135. (a)** $k \approx 11.216$ **(b)** 6.73 **(c)** 0.41% **(d)** 0.14% **137.** $\{64\}$ **139.** $\{-1, 5\}$

141. Because $y = \log_1 x$ means $1^y = 1 = x$, which cannot be true for $x \ne 1$ **143.** Zeros: $-3, -\dfrac{1}{2}, \dfrac{1}{2}, 3$; x-intercepts: $-3, -\dfrac{1}{2}, \dfrac{1}{2}, 3$

144. 12 **145.** $f(1) = -5; f(2) = 17$ **146.** $3 + i; f(x) = x^4 - 7x^3 + 14x^2 + 2x - 20; a = 1$ **147.** $\left\{\dfrac{7 - \sqrt{57}}{4}, \dfrac{7 + \sqrt{57}}{4}\right\}$

148. $5x + 8y = 8$ **149.** $\{-31, 14\}$ **150.** 151.1 cm **151.** $x^2 + 2x + 4$ **152.** $(x + 5)^3(x - 3)^6(11x + 23)$

6.5 Assess Your Understanding *(page 471)*

1. 0 **2.** M **3.** r **4.** $\log_a M + \log_a N$ **5.** $\log_a M - \log_a N$ **6.** $r \log_a M$ **7.** 7 **8.** F **9.** F **10.** F **11.** b **12.** b **13.** 29 **15.** -4 **17.** 13 **19.** 1 **21.** 1

23. 3 **25.** $\dfrac{6}{5}$ **27.** 4 **29.** $a + b$ **31.** $b - a$ **33.** $3a$ **35.** $\dfrac{1}{5}(a + b)$ **37.** $2 + \log_6 x$ **39.** $6 \log_5 y$ **41.** $1 + \ln x$ **43.** $\ln x - x$ **45.** $2 \log_a u + 3 \log_a v$

47. $2 \ln x + \dfrac{1}{2}\ln(1 - x)$ **49.** $3 \log_2 x - \log_2(x - 3)$ **51.** $\log x + \log(x + 2) - 2\log(x + 3)$ **53.** $\dfrac{1}{3}\ln(x - 2) + \dfrac{1}{3}\ln(x + 1) - \dfrac{2}{3}\ln(x + 4)$

55. $\ln 5 + \ln x + \dfrac{1}{2}\ln(1 + 3x) - 3\ln(x - 4)$ **57.** $\log_5 u^3 v^4$ **59.** $\log_3\left(\dfrac{1}{x^{5/2}}\right)$ **61.** $\log_4\left[\dfrac{x - 1}{(x + 1)^4}\right]$ **63.** $-2\ln(x - 1)$ **65.** $\log_2[x(3x - 2)^4]$

67. $\log_a\left(\dfrac{25x^6}{\sqrt{2x + 3}}\right)$ **69.** $\log_2\left[\dfrac{(x + 1)^2}{(x + 3)(x - 1)}\right]$ **71.** 2.771 **73.** -3.880 **75.** 5.615 **77.** 0.874

79. $y = \dfrac{\log x}{\log 4}$

81. $y = \dfrac{\log(x + 2)}{\log 2}$

83. $y = \dfrac{\log(x + 1)}{\log(x - 1)}$

85. (a) $(f \circ g)(x) = x; \{x \mid x \text{ is any real number}\}$ or $(-\infty, \infty)$
(b) $(g \circ f)(x) = x; \{x \mid x > 0\}$ or $(0, \infty)$ **(c)** 5
(d) $(f \circ h)(x) = \ln x^2; \{x \mid x \ne 0\}$ or $(-\infty, 0) \cup (0, \infty)$ **(e)** 2

87. $y = Cx$ **89.** $y = Cx(x + 1)$ **91.** $y = Ce^{3x}$ **93.** $y = Ce^{-4x} + 3$

95. $y = \dfrac{\sqrt[3]{C}(2x + 1)^{1/6}}{(x + 4)^{1/9}}$ **97.** 3 **99.** 1

101. $\log_a\left(x + \sqrt{x^2 - 1}\right) + \log_a\left(x - \sqrt{x^2 - 1}\right) = \log_a\left[\left(x + \sqrt{x^2 - 1}\right)\left(x - \sqrt{x^2 - 1}\right)\right] = \log_a[x^2 - (x^2 - 1)] = \log_a 1 = 0$

103. $\ln(1 + e^{2x}) = \ln[e^{2x}(e^{-2x} + 1)] = \ln e^{2x} + \ln(e^{-2x} + 1) = 2x + \ln(1 + e^{-2x})$

105. $y = f(x) = \log_a x; a^y = x$ implies $a^y = \left(\dfrac{1}{a}\right)^{-y} = x$, so $-y = \log_{1/a} x = -f(x)$.

107. $f(x) = \log_a x; f\left(\dfrac{1}{x}\right) = \log_a \dfrac{1}{x} = \log_a 1 - \log_a x = -f(x)$ **109.** $\log_a \dfrac{M}{N} = \log_a(M \cdot N^{-1}) = \log_a M + \log_a N^{-1} = \log_a M - \log_a N$.

111. $\log_a b = \dfrac{\log_b b}{\log_b a} = \dfrac{1}{\log_b a}$ **113.** $\log_{a^n} b^m = \dfrac{\log_a b^m}{\log_a a^n} = \dfrac{m\log_a b}{n} = \dfrac{m}{n}\log_a b$

119. $\{-1.78, 1.29, 3.49\}$ **120.** A repeated real solution (double root)   **121.** $-2, \dfrac{1}{5}, \dfrac{-5 - \sqrt{21}}{2}, \dfrac{-5 + \sqrt{21}}{2}$

122.

123. $\left(-\infty, \dfrac{3}{5}\right]$ **124.** $\{x \mid -9 < x < 7\}$ or $(-9, 7)$ **125.** Vertex: $(4, 13)$; concave down

126. Center: $(5, -2)$; radius: 8   **127.** 7   **128.** Odd

Domain: $\{x \mid x \le 2\}$ or $(-\infty, 2]$
Range: $\{y \mid y \ge 0\}$ or $[0, \infty)$

6.6 Assess Your Understanding *(page 478)*

5. $\{16\}$ **7.** $\left\{\dfrac{16}{5}\right\}$ **9.** $\{11\}$ **11.** $\{-64, 64\}$ **13.** $\{-6, 7\}$ **15.** $\{64\}$ **17.** $\left\{\dfrac{1}{3}\right\}$ **19.** $\{7\}$ **21.** $\{5\}$ **23.** $\left\{\dfrac{16}{3}\right\}$ **25.** $\{-6\}$ **27.** $\{-2\}$

29. $\{-1 + \sqrt{1 + e^4}\}$ **31.** $\left\{\dfrac{-15 + 5\sqrt{13}}{2}\right\}$ **33.** $\{2\}$ **35.** $\left\{\dfrac{9}{2}\right\}$ **37.** $\{7\}$ **39.** $\{-2 + 4\sqrt{2}\}$ **41.** $\{-\sqrt{3}, \sqrt{3}\}$

43. $\left\{\dfrac{1}{9}, 243\right\}$ **45.** $\{8\}$ **47.** $\{\log_2 10\} = \left\{\dfrac{1}{\log 2}\right\}$ **49.** $\{-\log_8 1.2\} = \left\{-\dfrac{\ln 1.2}{\ln 8}\right\}$ **51.** $\left\{\dfrac{1}{3}\log_2 \dfrac{8}{5}\right\} = \left\{\dfrac{\ln \frac{8}{5}}{3 \ln 2}\right\}$ **53.** $\left\{\dfrac{\ln 3}{2 \ln 3 + \ln 4}\right\}$

55. $\left\{\dfrac{\ln 7}{\ln 0.6 + \ln 7}\right\}$ **57.** $\{0\}$ **59.** $\left\{\dfrac{\ln \pi}{1 + \ln \pi}\right\}$ **61.** $\left\{\dfrac{\ln 3}{\ln 2}\right\}$ **63.** $\{0\}$ **65.** $\{\log_4(-2 + \sqrt{7})\}$ **67.** $\{\log_5 4\}$ **69.** No real solution

71. $\{\log_4 5\}$ **73.** $\{2.79\}$ **75.** $\{-0.57\}$ **77.** $\{-0.70\}$ **79.** $\{0.57\}$ **81.** $\{0.39, 1.00\}$ **83.** $\{1.32\}$ **85.** $\{1.31\}$

87. (a) $\{5\}$; $(5, 3)$ **(b)** $\{5\}$; $(5, 4)$ **(c)** $\{1\}$; yes, at $(1, 2)$ **(d)** $\{5\}$ **(e)** $\left\{-\dfrac{1}{11}\right\}$

89. (a), (b)

(c) $\{x \mid x > 0.710\}$ or $(0.710, \infty)$

91. (a), (b), (c)

93. (a), (b), (c)

95. (a)

(b) 2   **(c)** $\{x \mid x < 2\}$ or $(-\infty, 2)$

97. (a) 2052   **(b)** 2070   **99. (a)** After 2.4 yr   **(b)** After 3.8 yr   **(c)** After 9.4 yr   **101.** $\{1\}$ **103.** $\left\{\dfrac{1}{4}, 4\right\}$ **105.** $\{3^{\log 3}\}$ **107.** $\left\{-3, \dfrac{1}{4}, 2\right\}$

108. one-to-one   **109.** $(f \circ g)(x) = \dfrac{x + 5}{-x + 11}$; $\{x \mid x \ne 3, x \ne 11\}$ **110.** $\{x \mid x \ge 1\}$ or $[1, \infty)$ **111.** $\{9\}$ **112.** 6   **113.** $\dfrac{x^2 + 7x - 10}{(x - 2)(x + 2)}$

114. $4\sqrt{10}$ **115.** $\dfrac{1}{6}$ **116.** $\dfrac{1}{\sqrt{x + 6} + \sqrt{x}}$

6.7 Assess Your Understanding *(page 487)*

3. principal   **4.** I; Prt; simple interest   **5.** effective rate of interest   **6.** a   **7.** \$108.29   **9.** \$969.56   **11.** \$1394.19   **13.** \$1246.08   **15.** \$88.72   **17.** \$1444.79

19. \$713.53   **21.** \$102.00   **23.** 5.095%   **25.** 4.081%   **27.** $6\dfrac{1}{4}$% compounded annually   **29.** 9% compounded monthly   **31.** 25.992%   **33.** 24.573%

35. (a) About 8.69 yr   **(b)** About 8.66 yr   **37.** 6.823%   **39.** 10.15 yr; 10.14 yr   **41.** 15.27 yr or 15 yr, 3 mo   **43.** \$104,335   **45.** \$12,910.62

47. About \$30.17 per share or \$3017   **49.** Not quite. Jim will have \$1057.60. The second bank gives a better deal, since Jim will have \$1060.62 after 1 yr.

51. Will has \$11,632.73; Henry has \$10,947.89.   **53. (a)** \$70,473   **(b)** \$49,167   **55.** 64.9 yr   **57.** \$940.90   **59.** 2.53%

61. 34.31 yr   **63. (a)** \$3686.45   **(b)** \$3678.79   **65.** \$6439.28

67. (a) 11.90 yr   **(b)** 22.11 yr   **(c)** $mP = P\left(1 + \dfrac{r}{n}\right)^{nt}$

$$m = \left(1 + \dfrac{r}{n}\right)^{nt}$$

$$\ln m = \ln\left(1 + \dfrac{r}{n}\right)^{nt} = nt \ln\left(1 + \dfrac{r}{n}\right)$$

$$t = \dfrac{\ln m}{n \ln\left(1 + \dfrac{r}{n}\right)}$$

69. (a) 1.23%   **(b)** In 2034 or after 22 yr   **71.** 22.7 yr   **76.** $R = 0$; yes

77. $f^{-1}(x) = \dfrac{2x}{x - 1}$ **78.** $-2, 5; f(x) = (x + 2)^2(x - 5)(x^2 + 1)$ **79.** $\{6\}$

80. $2x(x + 3)(x - 5)(x + 5)$ **81.** $(f \circ g)(x) = 45x^2 - 18x - 7$

82. Domain: $(-\infty, \infty)$; Range: $(-\infty, 9]$

83. Vertical: $x = 7$; horizontal: none; oblique: $y = 2x + 9$

84. $y = 4x - 18$ **85.** 3

6.8 Assess Your Understanding *(page 499)*

1. (a) 500 insects   **(b)** $0.02 = 2\%$ per day   **(c)** About 611 insects   **(d)** After about 23.5 days   **(e)** After about 34.7 days

3. (a) $-0.0244 = -2.44\%$ per year   **(b)** About 391.7 g   **(c)** After about 9.1 yr   **(d)** 28.4 yr   **5. (a)** $N(t) = N_0 e^{kt}$ **(b)** 5832   **(c)** 3.9 days

7. (a) $N(t) = N_0 e^{kt}$ **(b)** 25,198   **9.** 9.797 g   **11.** 9953 yr ago   **13. (a)** 5:18 PM   **(b)** About 14.3 min   **(c)** The temperature of the pizza approaches 70°F.

15. 18.63°C; 25.1°C   **17.** 1.7 ppm; 7.17 days, or 172 hr   **19.** 0.26 M; 6.58 hr, or 395 min   **21.** 26.6 days

23. (a) In 1984, 91.8% of households did not own a personal computer.

(b)

(c) 70.6%   **(d)** During 2011

25. (a)

(b) 0.78, or 78%
(c) 50 people
(d) As n increases, the probability decreases.

27. (a) $P(0) \approx 48$; In 1900, about 48 invasive species were present in the Great Lakes.   **(b)** 1.7%

(c)

(d) About 176   **(e)** During 1999

29. (a) $P(t) = 50(3)^{t/20}$ **(b)** 661   **(c)** In 48 days   **(d)** $P(t) = 50e^{0.055t}$

31. $f(x) = -\dfrac{3}{2}x + 7$ **32.** Neither   **33.** $2\ln x + \dfrac{1}{2}\ln y - \ln z$

34. $\{x \mid x \neq -4, x \neq 2\}$ **35.** $(g - f)(x) = \dfrac{x^2 - 2x - 13}{(x-3)(x-4)}$

36. x-intercepts: $\dfrac{5 - \sqrt{17}}{4}, \dfrac{5 + \sqrt{17}}{4}$; y-intercept: 1   **37.** $\left\{-\dfrac{\sqrt{2}}{2}, \dfrac{\sqrt{2}}{2}\right\}$

38. $y = -1.0714x + 3.9048$; $r = -0.985$

39.

Local minima: $f(-1.37) = -5.85$ and $f(1) = -1$
Local maximum: $f(0.37) = -0.65$
Increasing: $[-1.37, 0.37] \cup [1, 3]$
Decreasing: $[-3, -1.37] \cup [0.37, 1]$

40. $\dfrac{40x + 45}{3(2x+3)^{2/3}}$

6.9 Assess Your Understanding *(page 506)*

1. (a)

(b) $y = 0.0903(1.3384)^x$
(c) $N(t) = 0.0903e^{0.2915t}$

(d)

(e) 0.69
(f) After about 7.26 hr

3. (a)

(b) $y = 118.7226(0.7013)^x$
(c) $A(t) = 118.7226e^{-0.3548t}$

(d)

(e) 28.7%
(f) $k = -0.3548 = -35.48\%$ is the exponential growth rate. It represents the rate at which the percentage of patients surviving advanced-stage breast cancer is decreasing.

5. (a)

(b) $y = 344.5217 - 37.2566\ln x$

(c)

(d) 188 billion pounds
(e) Under by 2 billion pounds

7. (a)

(b) $y = \dfrac{762,176,844.4}{1 + 8.7428e^{-0.0162x}}$

(c)

(d) 762,176,844
(e) Approximately 315,203,288   **(f)** 2023

9. (a)

(b) Cubic
(c) $y = 0.0607x^3 - 0.5533x^2 + 4.1390x + 13.1560$

(d)

(e) About $186.4 billion

11. (a)

(b) Exponential
(c) $y = 115.5779(0.9012)^x$

(d)

(e) 5.1%

12. $f(x) = \dfrac{1}{3}(x+3)(x+1)^2(x-2)$ **13.** $\sqrt{3}$ **14.**

15. $(-3, -2)$ **16.** $f(x) = -\sqrt{x-4}$ **17.** $\left\{\dfrac{2 - \sqrt{19}}{3}, \dfrac{2 + \sqrt{19}}{3}\right\}$

18. $\{x \mid -5 < x \leq -1 \text{ or } x > 5\}$ or $(-5, -1] \cup (5, \infty)$

19. $R = f(-3) = 0$; yes   **20.** 6

21. The function f is a polynomial so it is a continuous function, and $f(-2) = -21$; $f(-1) = 3$; zero: -1.32

Review Exercises *(page 511)*

1. (a) -26 **(b)** -241 **(c)** 16   **(d)** -1 **2. (a)** $\sqrt{11}$ **(b)** 1   **(c)** $\sqrt{\sqrt{6} + 2}$ **(d)** 19   **3. (a)** e^4 **(b)** $3e^{-2} - 2$ **(c)** e^{e^4} **(d)** -17
4. $(f \circ g)(x) = 1 - 3x$, all real numbers; $(g \circ f)(x) = 7 - 3x$, all real numbers; $(f \circ f)(x) = x$, all real numbers; $(g \circ g)(x) = 9x + 4$, all real numbers
5. $(f \circ g)(x) = \sqrt{3 + 3x + 3x^2}$, all real numbers; $(g \circ f)(x) = 1 + \sqrt{3x} + 3x$, $\{x \mid x \geq 0\}$; $(f \circ f)(x) = \sqrt{3\sqrt{3x}}$, $\{x \mid x \geq 0\}$; $(g \circ g)(x) = 3 + 3x + 4x^2 + 2x^3 + x^4$, all real numbers

6. $(f \circ g)(x) = \dfrac{1+x}{1-x}, \{x \mid x \neq 0, x \neq 1\}; (g \circ f)(x) = \dfrac{x-1}{x+1}, \{x \mid x \neq -1, x \neq 1\}; (f \circ f)(x) = x, \{x \mid x \neq 1\}; (g \circ g)(x) = x, \{x \mid x \neq 0\}$

7. (a) one-to-one **(b)** $\{(2,1), (5,3), (8,5), (10,6)\}$ **8.**

$y = x$; $(0,2)$; $(3,3)$; $(-2,0)$; $(-3,-1)$

9. $f(g(x)) = f\left(\dfrac{1}{5}x + 2\right) = 5\left(\dfrac{1}{5}x + 2\right) - 10 = x + 10 - 10 = x$

$g(f(x)) = g(5x - 10) = \dfrac{1}{5}(5x - 10) + 2 = x - 2 + 2 = x$

10. $f(g(x)) = f\left(\dfrac{4}{1-x}\right) = \dfrac{\dfrac{4}{1-x} - 4}{4} = \dfrac{4 - 4(1-x)}{4} = \dfrac{4 - 4 + 4x}{4} = \dfrac{4x}{4} = x, x \neq 1$ **11.** $f^{-1}(x) = \dfrac{2x+3}{5x-2}$

$g(f(x)) = g\left(\dfrac{x-4}{x}\right) = \dfrac{4}{1 - \dfrac{x-4}{x}} = \dfrac{4x}{x - (x-4)} = \dfrac{4x}{4} = x, x \neq 0$

$f(f^{-1}(x)) = \dfrac{2\left(\dfrac{2x+3}{5x-2}\right) + 3}{5\left(\dfrac{2x+3}{5x-2}\right) - 2} = x$

12. $f^{-1}(x) = \dfrac{x+1}{x}$

$f(f^{-1}(x)) = \dfrac{1}{\dfrac{x+1}{x} - 1} = x$

$f^{-1}(f(x)) = \dfrac{\dfrac{1}{x-1} + 1}{\dfrac{1}{x-1}} = x$

$f^{-1}(f(x)) = \dfrac{2\left(\dfrac{2x+3}{5x-2}\right) + 3}{5\left(\dfrac{2x+3}{5x-2}\right) - 2} = x$

Domain of f = range of f^{-1} = all real numbers except $\dfrac{2}{5}$

Range of f = domain of f^{-1} = all real numbers except $\dfrac{2}{5}$

Domain of f = range of f^{-1} = all real numbers except 1
Range of f = domain of f^{-1} = all real numbers except 0

13. $f^{-1}(x) = x^2 + 2, x \geq 0$
$f(f^{-1}(x)) = \sqrt{x^2 + 2 - 2} = |x| = x, x \geq 0$
$f^{-1}(f(x)) = \left(\sqrt{x-2}\right)^2 + 2 = x$
Domain of f = range of f^{-1} = $[2, \infty)$
Range of f = domain of f^{-1} = $[0, \infty)$

14. $f^{-1}(x) = (x-1)^3$;
$f(f^{-1}(x)) = ((x-1)^3)^{1/3} + 1 = x$
$f^{-1}(f(x)) = (x^{1/3} + 1 - 1)^3 = x$
Domain of f = range of f^{-1} = $(-\infty, \infty)$
Range of f = domain of f^{-1} = $(-\infty, \infty)$

15. (a) 81 **(b)** 2 **(c)** $\dfrac{1}{9}$ **(d)** -3

16. $\log_5 z = 2$ **17.** $5^{13} = u$

18. $\left\{x \mid x > \dfrac{2}{3}\right\}; \left(\dfrac{2}{3}, \infty\right)$

19. $\{x \mid x < 1 \text{ or } x > 2\}; (-\infty, 1) \cup (2, \infty)$

20. -3 **21.** $\sqrt{2}$ **22.** 0.4

23. $\log_3 u + 2\log_3 v - \log_3 w$

24. $8\log_2 a + 2\log_2 b$

25. $2\log x + \dfrac{1}{2}\log(x^3 + 1)$ **26.** $2\ln(2x+3) - 2\ln(x-1) - 2\ln(x-2)$ **27.** $\dfrac{25}{4}\log_4 x$ **28.** $-2\ln(x+1)$ **29.** $\ln\left[\dfrac{16\sqrt{x^2 + 1}}{\sqrt{x}(x-4)}\right]$ **30.** 2.124

31.

NORMAL FLOAT AUTO REAL RADIAN MP

32. (a) Domain of f: $(-\infty, \infty)$
(b)

$(4,2)$; $(3,1)$; $y = 0$

(c) Range of f: $(0, \infty)$
Horizontal asymptote: $y = 0$
(d) $f^{-1}(x) = 3 + \log_2 x$
(e) Domain of f^{-1}: $(0, \infty)$
Range of f^{-1}: $(-\infty, \infty)$

(f)

$(2,4)$; $(1,3)$; $x = 0$

33. (a) Domain of f: $(-\infty, \infty)$
(b)

$(-1,4)$; $(0,2)$; $y = 1$

(c) Range of f: $(1, \infty)$
Horizontal asymptote: $y = 1$
(d) $f^{-1}(x) = -\log_3(x-1)$
(e) Domain of f^{-1}: $(1, \infty)$
Range of f^{-1}: $(-\infty, \infty)$
(f)

$(2,0)$; $(4,-1)$; $x = 1$

34. (a) Domain of f: $(-\infty, \infty)$
(b)

$(2,3)$; $y = 0$; $\left(1, \dfrac{3}{e}\right)$

(c) Range of f: $(0, \infty)$
Horizontal asymptote: $y = 0$
(d) $f^{-1}(x) = 2 + \ln\left(\dfrac{x}{3}\right)$
(e) Domain of f^{-1}: $(0, \infty)$
Range of f^{-1}: $(-\infty, \infty)$
(f)

$(3,2)$; $\left(\dfrac{3}{e}, 1\right)$; $x = 0$

35. (a) Domain of f: $(-3, \infty)$
(b)

$(-2,0)$; $\left(0, \dfrac{1}{2}\ln 3\right)$; $x = -3$

(c) Range of f: $(-\infty, \infty)$
Vertical asymptote: $x = -3$
(d) $f^{-1}(x) = e^{2x} - 3$
(e) Domain of f^{-1}: $(-\infty, \infty)$
Range of f^{-1}: $(-3, \infty)$
(f)

$\left(\dfrac{1}{2}\ln 3, 0\right)$; $(0,-2)$; $y = -3$

36. $\left\{-\dfrac{16}{9}\right\}$ **37.** $\left\{\dfrac{-1-\sqrt{3}}{2}, \dfrac{-1+\sqrt{3}}{2}\right\}$ **38.** $\left\{\dfrac{1}{4}\right\}$ **39.** $\left\{\dfrac{2\ln 3}{\ln 5 - \ln 3}\right\}$ **40.** $\{-2, 6\}$ **41.** $\{83\}$ **42.** $\left\{\dfrac{1}{2}, -3\right\}$

43. $\{1\}$ **44.** $\{-1\}$ **45.** $\{1 - \ln 5\}$ **46.** $\left\{\log_3\left(-2 + \sqrt{7}\,\right)\right\} = \left\{\dfrac{\ln\left(-2 + \sqrt{7}\,\right)}{\ln 3}\right\}$

47. (a), (e)

(b) $3; (6, 3)$
(c) $10; (10, 4)$
(d) $\left\{x \,\middle|\, x > \dfrac{5}{2}\right\}$ or $\left(\dfrac{5}{2}, \infty\right)$
(e) $f^{-1}(x) = 2^{x-1} + 2$

48. (a) 37.3 W **(b)** 6.9 dB **49. (a)** 11.77 **(b)** 9.56 in.
50. (a) 9.85 yr **(b)** 4.27 yr **51.** $20,398.87; 4.04\%; 17.5 yr
52. $41,668.97 **53.** 24,765 yr ago **54.** 55.22 min, or 55 min, 13 sec
55. 8,153,581,530 **56.** 7.204 g; 0.519 g

57. (a) 0.3 **(b)** 0.8
(c)

(d) In 2026

58. (a)

(b) $y = 3610.2684\,(1.0406)^x$
(c) $A(t) = 3610.2684e^{0.0398t}$
(d)

(e) 2027–28

59. (a)

(b) $y = 18.921 - 7.096\ln x$
(c)

(d) Approximately $-3°$F

60. (a)

(b) $C = \dfrac{46.9292}{1 + 21.2733e^{-0.7306t}}$
(c)

(d) About 47 people; 50 people
(e) 2.4 days; during the tenth hour of day 3
(f) 9.5 days

Chapter Test *(page 514)*

1. (a) $f \circ g = \dfrac{2x + 7}{2x + 3}$; domain: $\left\{x \,\middle|\, x \neq -\dfrac{3}{2}\right\}$ **(b)** $(g \circ f)(-2) = 5$ **(c)** $(f \circ g)(-2) = -3$

2. (a) The function is not one-to-one. **(b)** The function is one-to-one.

3. $f^{-1}(x) = \dfrac{2 + 5x}{3x}$; domain of $f = \left\{x \,\middle|\, x \neq \dfrac{5}{3}\right\}$, range of $f = \{y \,|\, y \neq 0\}$; domain of $f^{-1} = \{x \,|\, x \neq 0\}$; range of $f^{-1} = \left\{y \,\middle|\, y \neq \dfrac{5}{3}\right\}$

4. The point $(-5, 3)$ must be on the graph of f^{-1}. **5.** $x = 5$ **6.** $b = 4$ **7.** $x = 625$ **8.** -2 **9.** 4 **10.** 125 **11.** 7

12. (a) Domain of f: $\{x \,|\, -\infty < x < \infty\}$ or $(-\infty, \infty)$
(b)

(c) Range of f: $\{y \,|\, y > -2\}$ or $(-2, \infty)$
   Horizontal asymptote: $y = -2$
(d) $f^{-1}(x) = \log_4(x + 2) - 1$
(e) Domain of f^{-1}: $\{x \,|\, x > -2\}$ or $(-2, \infty)$
   Range of f^{-1}: $\{y \,|\, -\infty < y < \infty\}$ or $(-\infty, \infty)$
(f)

13. (a) Domain of f: $\{x \,|\, x > 2\}$ or $(2, \infty)$
(b)

(c) Range of f: $\{y \,|\, -\infty < y < \infty\}$ or $(-\infty, \infty)$
   Vertical asymptote: $x = 2$
(d) $f^{-1}(x) = 5^{1-x} + 2$
(e) Domain of f^{-1}: $\{x \,|\, -\infty < x < \infty\}$ or $(-\infty, \infty)$
   Range of f^{-1}: $\{y \,|\, y > 2\}$ or $(2, \infty)$
(f)

14. $\{1\}$ **15.** $\{91\}$ **16.** $\{-\ln 2\}$ **17.** $\left\{\dfrac{1 - \sqrt{13}}{2}, \dfrac{1 + \sqrt{13}}{2}\right\}$ **18.** $\left\{\dfrac{3\ln 7}{1 - \ln 7}\right\}$

19. $\{2\sqrt{6}\}$ **20.** $2 + 3\log_2 x - \log_2(x - 6) - \log_2(x + 3)$ **21.** About 250.39 days **22. (a)** $1033.82 **(b)** $963.42 **(c)** 11.9 yr
23. (a) About 83 dB **(b)** The pain threshold will be exceeded if 31,623 people shout at the same time.

Cumulative Review *(page 515)*

1. Yes; no **2. (a)** 10 **(b)** $2x^2 + 3x + 1$ **(c)** $2x^2 + 4xh + 2h^2 - 3x - 3h + 1$ **3.** $\left(\frac{1}{2}, \frac{\sqrt{3}}{2}\right)$ is on the graph. **4.** $\{-26\}$

5.   **6. (a)**   **(b)** $\{x \mid -\infty < x < \infty\}$ **7.** $f(x) = 2(x-4)^2 - 8 = 2x^2 - 16x + 24$ **8.**

9. $f(g(x)) = \dfrac{4}{(x-3)^2} + 2$; domain: $\{x \mid x \neq 3\}$; 3

10. (a) $y = 3x^4$
  (b) x-intercepts: $-2, 0, 2, 5$; y-intercept: 0
  (c) $-2, 0, 2, 5$: multiplicity 1, crosses
  (d) At most 3 turning points
  (e)

11. (a), (c)

Domain g = range $g^{-1} = (-\infty, \infty)$
Range g = domain $g^{-1} = (2, \infty)$
(b) $g^{-1}(x) = \log_3(x-2)$

12. $\left\{-\dfrac{3}{2}\right\}$ **13.** $\{2\}$

14. (a) $\{-1\}$ **(b)** $\{x \mid x > -1\}$ or $(-1, \infty)$
  (c) $\{25\}$

15. (a)

(b) Logarithmic; $y = 49.293 - 10.563 \ln x$
(c) Highest value of $|r|$

CHAPTER 7 Analytic Geometry

7.2 Assess Your Understanding *(page 525)*

7. parabola; axis of symmetry **8.** latus rectum **9.** c **10.** $(3, 2)$ **11.** d **12.** F **13.** B **15.** E **17.** H **19.** C

21. $y^2 = 16x$ **23.** $x^2 = -12y$ **25.** $y^2 = -8x$ **27.** $x^2 = 2y$

29. $x^2 = \dfrac{4}{3}y$ **31.** $(x-2)^2 = -8(y+3)$ **33.** $(y+2)^2 = 4(x+1)$ **35.** $(x+3)^2 = 4(y-3)$

37. $(y+2)^2 = -8(x+1)$ **39.** Vertex: $(0, 0)$; focus: $(0, 1)$; directrix: $y = -1$ **41.** Vertex: $(0, 0)$; focus: $(-4, 0)$; directrix: $x = 4$ **43.** Vertex: $(-1, 2)$; focus: $(1, 2)$; directrix: $x = -3$

45. Vertex: $(3, -1)$; focus: $\left(3, -\dfrac{5}{4}\right)$; directrix: $y = -\dfrac{3}{4}$ **47.** Vertex: $(2, -3)$; focus: $(4, -3)$; directrix: $x = 0$ **49.** Vertex: $(0, 2)$; focus: $(-1, 2)$; directrix: $x = 1$ **51.** Vertex: $(-4, -2)$; focus: $(-4, -1)$; directrix: $y = -3$

53. Vertex: $(-1,-1)$; focus: $\left(-\dfrac{3}{4},-1\right)$;

directrix: $x = -\dfrac{5}{4}$

55. Vertex: $(2,-8)$;

focus: $\left(2,-\dfrac{31}{4}\right)$;

directrix: $y = -\dfrac{33}{4}$

57. $(y-1)^2 = x$ **59.** $(y-1)^2 = -(x-2)$

61. $x^2 = 4(y-1)$ **63.** $y^2 = \dfrac{1}{2}(x+2)$

65. 20 ft **67.** 24.31 ft, 18.75 ft, 7.64 ft

69. 1.5625 ft from the base of the dish, along the axis of symmetry

71. 1 in. from the vertex, along the axis of symmetry

73. 0.78125 ft

75. 4.17 in. from the base, along the axis of symmetry

77. (a) $y = -\dfrac{2}{315}x^2 + 630$

(b) 567 ft: 119.7 ft; 478 ft: 267.3 ft; 308 ft: 479.4 ft **(c)** No

79. $Cy^2 + Dx = 0, C \neq 0, D \neq 0$

$Cy^2 = -Dx$

$y^2 = -\dfrac{D}{C}x$

This is the equation of a parabola with vertex at $(0,0)$ and axis of symmetry the x-axis.

The focus is $\left(-\dfrac{D}{4C},0\right)$; the directrix is the line $x = \dfrac{D}{4C}$. The parabola opens to the

right if $-\dfrac{D}{C} > 0$ and to the left if $-\dfrac{D}{C} < 0$.

81. $Cy^2 + Dx + Ey + F = 0, C \neq 0$

$Cy^2 + Ey = -Dx - F$

$y^2 + \dfrac{E}{C}y = -\dfrac{D}{C}x - \dfrac{F}{C}$

$\left(y + \dfrac{E}{2C}\right)^2 = -\dfrac{D}{C}x - \dfrac{F}{C} + \dfrac{E^2}{4C^2}$

$\left(y + \dfrac{E}{2C}\right)^2 = -\dfrac{D}{C}x + \dfrac{E^2 - 4CF}{4C^2}$

(a) If $D \neq 0$, then the equation may be written as

$$\left(y + \dfrac{E}{2C}\right)^2 = -\dfrac{D}{C}\left(x - \dfrac{E^2 - 4CF}{4CD}\right).$$

This is the equation of a parabola with vertex at $\left(\dfrac{E^2 - 4CF}{4CD}, -\dfrac{E}{2C}\right)$ and axis of symmetry parallel to the x-axis.

(b)–(d) If $D = 0$, the graph of the equation contains no points if $E^2 - 4CF < 0$, is a single horizontal line if $E^2 - 4CF = 0$, and is two horizontal lines if $E^2 - 4CF > 0$.

83. $(0,2), (0,-2), (-36,0)$; symmetric with respect to the x-axis. **84.** $\{5\}$ **85.** $\dfrac{11\sqrt{13}}{6}$ **86.** $(x+12)^2 + (y-7)^2 = 6$ **87.** \$6600

88. $\dfrac{\ln 2}{4}$ **89.** $\{-1,2,3,6\}$

7.3 Assess Your Understanding (page 535)

7. ellipse **8.** b **9.** $(0,-5); (0,5)$ **10.** $5; 3; x$ **11.** $(-2,-3); (6,-3)$ **12.** a **13.** C **15.** B

17. Vertices: $(-5,0), (0,5)$

Foci: $(-\sqrt{21},0), (\sqrt{21},0)$

19. Vertices: $(0,-5), (0,5)$

Foci: $(0,-4), (0,4)$

21. $\dfrac{x^2}{4} + \dfrac{y^2}{16} = 1$

Vertices: $(0,-4), (0,4)$

Foci: $(0,-2\sqrt{3}), (0,2\sqrt{3})$

23. $\dfrac{x^2}{8} + \dfrac{y^2}{2} = 1$

Vertices: $(-2\sqrt{2},0), (2\sqrt{2},0)$

Foci: $(-\sqrt{6},0), (\sqrt{6},0)$

25. $\dfrac{x^2}{16} + \dfrac{y^2}{16} = 1$

Vertices: $(-4,0), (4,0),$

$(0,-4), (0,4)$; Focus: $(0,0)$

27. $\dfrac{x^2}{25} + \dfrac{y^2}{16} = 1$

29. $\dfrac{x^2}{9} + \dfrac{y^2}{25} = 1$

31. $\dfrac{x^2}{9} + \dfrac{y^2}{5} = 1$

33. $\dfrac{x^2}{25} + \dfrac{y^2}{9} = 1$

35. $\dfrac{x^2}{4} + \dfrac{y^2}{13} = 1$

37. $x^2 + \dfrac{y^2}{16} = 1$

39. $\dfrac{(x+1)^2}{4} + (y-1)^2 = 1$

41. $(x-1)^2 + \dfrac{y^2}{4} = 1$

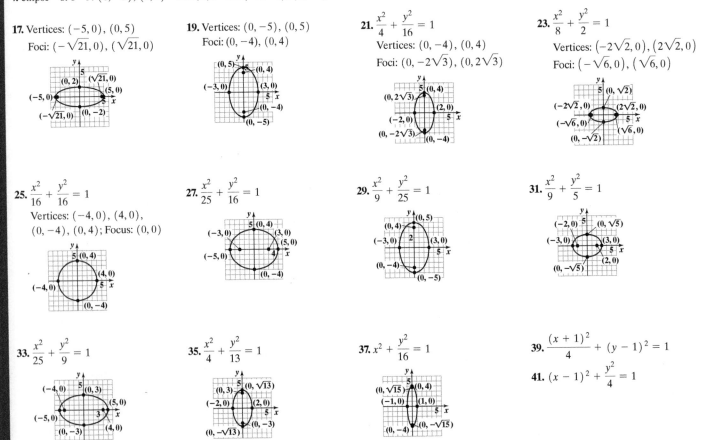

43. Center: $(3, -1)$; vertices: $(3, -4)$, $(3, 2)$; foci: $\left(3, -1 - \sqrt{5}\right), \left(3, -1 + \sqrt{5}\right)$

45. $\dfrac{(x + 5)^2}{16} + \dfrac{(y - 4)^2}{4} = 1$
Center: $(-5, 4)$; vertices: $(-9, 4)$, $(-1, 4)$; foci: $\left(-5 - 2\sqrt{3}, 4\right), \left(-5 + 2\sqrt{3}, 4\right)$

47. $\dfrac{(x + 2)^2}{4} + (y - 1)^2 = 1$
Center: $(-2, 1)$; vertices: $(-4, 1)$, $(0, 1)$; foci: $\left(-2 - \sqrt{3}, 1\right), \left(-2 + \sqrt{3}, 1\right)$

49. $\dfrac{(x - 2)^2}{3} + \dfrac{(y + 1)^2}{2} = 1$
Center: $(2, -1)$; vertices: $\left(2 - \sqrt{3}, -1\right)$; $\left(2 + \sqrt{3}, -1\right)$; foci: $(1, -1)$, $(3, -1)$

51. $\dfrac{(x - 1)^2}{4} + \dfrac{(y + 2)^2}{9} = 1$
Center: $(1, -2)$; vertices: $(1, -5)$, $(1, 1)$; foci: $\left(1, -2 - \sqrt{5}\right), \left(1, -2 + \sqrt{5}\right)$

53. $x^2 + \dfrac{(y + 2)^2}{4} = 1$
Center: $(0, -2)$; vertices: $(0, -4)$, $(0, 0)$; foci: $\left(0, -2 - \sqrt{3}\right), \left(0, -2 + \sqrt{3}\right)$

55. $\dfrac{(x - 2)^2}{25} + \dfrac{(y + 2)^2}{21} = 1$

57. $\dfrac{(x - 4)^2}{5} + \dfrac{(y - 6)^2}{9} = 1$

59. $\dfrac{(x - 2)^2}{16} + \dfrac{(y - 1)^2}{7} = 1$

61. $\dfrac{(x - 1)^2}{10} + (y - 2)^2 = 1$

63. $\dfrac{(x - 1)^2}{9} + \dfrac{(y - 2)^2}{9} = 1$

65.

67.

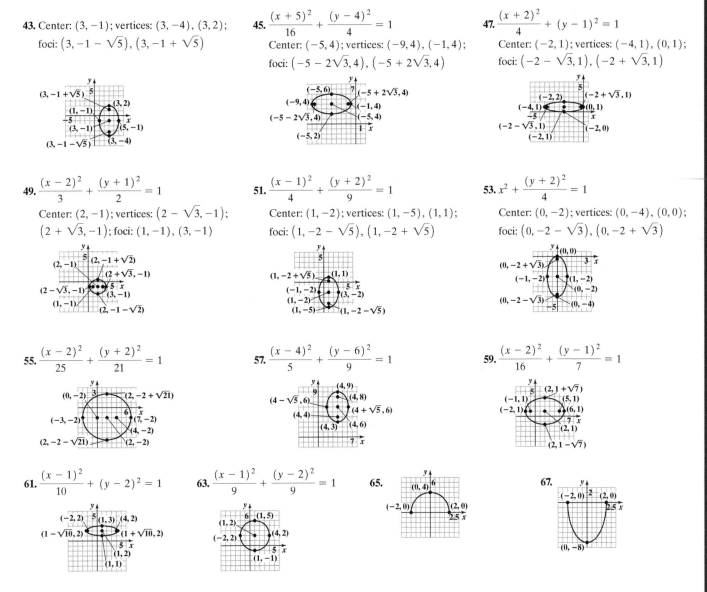

69. $\dfrac{x^2}{100} + \dfrac{y^2}{36} = 1$ **71.** 43.3 ft **73.** 24.65 ft, 21.65 ft, 13.82 ft **75.** 30 ft **77.** The elliptical hole will have a major axis of length $2\sqrt{41}$ in. and

a minor axis of length 8 in. **79.** 91.5 million mi; $\dfrac{x^2}{(93)^2} + \dfrac{y^2}{8646.75} = 1$ **81.** Perihelion: 460.6 million mi; mean distance: 483.8 million mi;

$\dfrac{x^2}{(483.8)^2} + \dfrac{y^2}{233,524.2} = 1$ **83.** 35 million mi **85.** $8\sqrt{14}$ cm $\approx$ 29.93 cm **87.** $5\sqrt{5} - 4$

89. (a) $Ax^2 + Cy^2 + F = 0$ If A and C are of the same sign and F is of opposite sign, then the equation takes the form
$Ax^2 + Cy^2 = -F$ $\dfrac{x^2}{\left(-\dfrac{F}{A}\right)} + \dfrac{y^2}{\left(-\dfrac{F}{C}\right)} = 1$, where $-\dfrac{F}{A}$ and $-\dfrac{F}{C}$ are positive. This is the equation of an ellipse with center at $(0, 0)$.

(b) If $A = C$, the equation may be written as $x^2 + y^2 = -\dfrac{F}{A}$. This is the equation of a circle with center at $(0, 0)$ and radius equal to $\sqrt{-\dfrac{F}{A}}$.

92. Zeros: $5 - 2\sqrt{3}, 5 + 2\sqrt{3}$; x-intercepts: $5 - 2\sqrt{3}, 5 + 2\sqrt{3}$ **93.** Domain: $\{x \mid x \neq 5\}$; Horizontal asymptote: $y = 2$; Vertical asymptote: $x = 5$

94. $c = 2, B = 30°$ **95.** $\dfrac{10}{7}$ **96.** $\{-0.5397\}$ **97.** $4x - 7$ **98.** $\{164\}$ **99.** $\left\{-3 - 2\sqrt{5}, -3 + 2\sqrt{5}\right\}$

7.4 Assess Your Understanding *(page 548)*

7. hyperbola   **8.** transverse axis   **9.** b   **10.** $(2,4);(2,-2)$ **11.** $(2,6);(2,-4)$ **12.** c   **13.** $2;3;x$ **14.** $y=-\dfrac{4}{9}x;\ y=\dfrac{4}{9}x$ **15.** B   **17.** A

19. $x^2-\dfrac{y^2}{8}=1$

21. $\dfrac{y^2}{16}-\dfrac{x^2}{20}=1$

23. $\dfrac{x^2}{9}-\dfrac{y^2}{16}=1$

25. $\dfrac{y^2}{36}-\dfrac{x^2}{9}=1$

27. $\dfrac{x^2}{8}-\dfrac{y^2}{8}=1$

29. $\dfrac{x^2}{25}-\dfrac{y^2}{9}=1$

Center: $(0,0)$
Transverse axis: x-axis
Vertices: $(-5,0),(5,0)$
Foci: $\left(-\sqrt{34},0\right),\left(\sqrt{34},0\right)$
Asymptotes: $y=\pm\dfrac{3}{5}x$

31. $\dfrac{x^2}{4}-\dfrac{y^2}{16}=1$

Center: $(0,0)$
Transverse axis: x-axis
Vertices: $(-2,0),(2,0)$
Foci: $\left(-2\sqrt{5},0\right),\left(2\sqrt{5},0\right)$
Asymptotes: $y=\pm2x$

33. $\dfrac{y^2}{9}-x^2=1$

Center: $(0,0)$
Transverse axis: y-axis
Vertices: $(0,-3),(0,3)$
Foci: $\left(0,-\sqrt{10}\right),\left(0,\sqrt{10}\right)$
Asymptotes: $y=\pm3x$

35. $\dfrac{y^2}{25}-\dfrac{x^2}{25}=1$

Center: $(0,0)$
Transverse axis: y-axis
Vertices: $(0,-5),(0,5)$
Foci: $\left(0,-5\sqrt{2}\right),\left(0,5\sqrt{2}\right)$
Asymptotes: $y=\pm x$

37. $x^2-y^2=1$

39. $\dfrac{y^2}{36}-\dfrac{x^2}{9}=1$

41. $\dfrac{(x-4)^2}{4}-\dfrac{(y+1)^2}{5}=1$

43. $\dfrac{(y+4)^2}{4}-\dfrac{(x+3)^2}{12}=1$

45. $(x-5)^2-\dfrac{(y-7)^2}{3}=1$

47. $\dfrac{(x-1)^2}{4}-\dfrac{(y+1)^2}{9}=1$

49. $\dfrac{(x-2)^2}{4}-\dfrac{(y+3)^2}{9}=1$

Center: $(2,-3)$
Transverse axis: parallel to x-axis
Vertices: $(0,-3),(4,-3)$
Foci: $\left(2-\sqrt{13},-3\right),\left(2+\sqrt{13},-3\right)$
Asymptotes: $y+3=\pm\dfrac{3}{2}(x-2)$

51. $\dfrac{(y-2)^2}{4} - (x+2)^2 = 1$

Center: $(-2, 2)$
Transverse axis: parallel to y-axis
Vertices: $(-2, 0), (-2, 4)$
Foci: $(-2, 2 - \sqrt{5}), (-2, 2 + \sqrt{5})$
Asymptotes: $y - 2 = \pm 2(x + 2)$

53. $\dfrac{(x+1)^2}{4} - \dfrac{(y+2)^2}{4} = 1$

Center: $(-1, -2)$
Transverse axis: parallel to x-axis
Vertices: $(-3, -2), (1, -2)$
Foci: $(-1 - 2\sqrt{2}, -2), (-1 + 2\sqrt{2}, -2)$
Asymptotes: $y + 2 = \pm(x + 1)$

55. $(x-1)^2 - (y+1)^2 = 1$

Center: $(1, -1)$
Transverse axis: parallel to x-axis
Vertices: $(0, -1), (2, -1)$
Foci: $(1 - \sqrt{2}, -1), (1 + \sqrt{2}, -1)$
Asymptotes: $y + 1 = \pm(x - 1)$

57. $\dfrac{(y-2)^2}{4} - (x+1)^2 = 1$

Center: $(-1, 2)$
Transverse axis: parallel to y-axis
Vertices: $(-1, 0), (-1, 4)$
Foci: $(-1, 2 - \sqrt{5}), (-1, 2 + \sqrt{5})$
Asymptotes: $y - 2 = \pm 2(x + 1)$

59. $\dfrac{(x-3)^2}{4} - \dfrac{(y+2)^2}{16} = 1$

Center: $(3, -2)$
Transverse axis: parallel to x-axis
Vertices: $(1, -2), (5, -2)$
Foci: $(3 - 2\sqrt{5}, -2), (3 + 2\sqrt{5}, -2)$
Asymptotes: $y + 2 = \pm 2(x - 3)$

61. $\dfrac{(y-1)^2}{4} - (x+2)^2 = 1$

Center: $(-2, 1)$
Transverse axis: parallel to y-axis
Vertices: $(-2, -1), (-2, 3)$
Foci: $(-2, 1 - \sqrt{5}), (-2, 1 + \sqrt{5})$
Asymptotes: $y - 1 = \pm 2(x + 2)$

63.

65.

67. Center: $(3, 0)$
Transverse axis: parallel to x-axis
Vertices: $(1, 0), (5, 0)$
Foci: $(3 - \sqrt{29}, 0), (3 + \sqrt{29}, 0)$
Asymptotes: $y = \pm \dfrac{5}{2}(x - 3)$

69. Vertex: $(0, 3)$; focus: $(0, 7)$;
directrix: $y = -1$

71. $\dfrac{(x-5)^2}{9} + \dfrac{y^2}{25} = 1$

Center: $(5, 0)$; vertices: $(5, 5), (5, -5)$;
foci: $(5, -4), (5, 4)$

73. $(x-3)^2 = 8(y+5)$
Vertex: $(3, -5)$; focus: $(3, -3)$;
directrix: $y = -7$

75. The fireworks display is 50,138 ft north of the person at point A.   **77.** The tower is 592.4 ft tall.   **79. (a)** $y = \pm x$ **(b)** $\dfrac{x^2}{100} - \dfrac{y^2}{100} = 1, x \geq 0$

81. $\dfrac{y^2}{81} - \dfrac{x^2}{88} = 1$ **83.** If the eccentricity is close to 1, the "opening" of the hyperbola is very small. As e increases, the opening gets bigger.

85. $\dfrac{x^2}{4} - y^2 = 1$; asymptotes $y = \pm \dfrac{1}{2}x$

$y^2 - \dfrac{x^2}{4} = 1$; asymptotes $y = \pm \dfrac{1}{2}x$

87. $Ax^2 + Cy^2 + F = 0$
$Ax^2 + Cy^2 = -F$

If A and C are of opposite sign and $F \neq 0$, this equation may be written as $\dfrac{x^2}{\left(-\dfrac{F}{A}\right)} + \dfrac{y^2}{\left(-\dfrac{F}{C}\right)} = 1$,

where $-\dfrac{F}{A}$ and $-\dfrac{F}{C}$ are opposite in sign. This is the equation of a hyperbola with center at $(0, 0)$. The transverse axis is the x-axis if $-\dfrac{F}{A} > 0$; the transverse axis is the y-axis if $-\dfrac{F}{A} < 0$.

89. $\{5\}$ **90.** 8.33 **91.** 8 **92.** $y = -5$ **93.** $f^{-1}(x) = \ln\left(\dfrac{x-4}{3}\right) + 1$ **94.** $\dfrac{9\pi}{4} - \dfrac{9}{2} \approx 2.57$ sq units   **95.** $\{-4\}$ **96.** $\left(\dfrac{1}{2}, -\dfrac{3}{2}\right)$

Review Exercises *(page 552)*

1. Parabola; vertex $(0,0)$, focus $(-4,0)$, directrix $x = 4$ **2.** Hyperbola; center $(0,0)$, vertices $(5,0)$ and $(-5,0)$, foci $\left(\sqrt{26},0\right)$ and $\left(-\sqrt{26},0\right)$, asymptotes $y = \dfrac{1}{5}x$ and $y = -\dfrac{1}{5}x$ **3.** Ellipse; center $(0,0)$, vertices $(0,5)$ and $(0,-5)$, foci $(0,3)$ and $(0,-3)$

4. $x^2 = -4(y-1)$: Parabola; vertex $(0,1)$, focus $(0,0)$, directrix $y = 2$ **5.** $\dfrac{x^2}{2} - \dfrac{y^2}{8} = 1$: Hyperbola; center $(0,0)$, vertices $\left(\sqrt{2},0\right)$ and $\left(-\sqrt{2},0\right)$, foci $\left(\sqrt{10},0\right)$ and $\left(-\sqrt{10},0\right)$, asymptotes $y = 2x$ and $y = -2x$ **6.** $(x-2)^2 = 2(y+2)$: Parabola; vertex $(2,-2)$, focus $\left(2,-\dfrac{3}{2}\right)$, directrix $y = -\dfrac{5}{2}$

7. $\dfrac{(y-2)^2}{4} - (x-1)^2 = 1$: Hyperbola; center $(1,2)$, vertices $(1,4)$ and $(1,0)$, foci $\left(1, 2 + \sqrt{5}\right)$ and $\left(1, 2 - \sqrt{5}\right)$, asymptotes $y - 2 = \pm 2(x-1)$

8. $\dfrac{(x-2)^2}{9} + \dfrac{(y-1)^2}{4} = 1$: Ellipse; center $(2,1)$, vertices $(5,1)$ and $(-1,1)$, foci $\left(2 + \sqrt{5}, 1\right)$ and $\left(2 - \sqrt{5}, 1\right)$

9. $(x-2)^2 = -4(y+1)$: Parabola; vertex $(2,-1)$, focus $(2,-2)$, directrix $y = 0$

10. $\dfrac{(x-1)^2}{4} + \dfrac{(y+1)^2}{9} = 1$: Ellipse; center $(1,-1)$, vertices $(1,2)$ and $(1,-4)$, foci $\left(1, -1 + \sqrt{5}\right)$ and $\left(1, -1 - \sqrt{5}\right)$

11. $y^2 = -8x$

12. $\dfrac{y^2}{4} - \dfrac{x^2}{12} = 1$

13. $\dfrac{x^2}{16} + \dfrac{y^2}{7} = 1$

14. $(x-2)^2 = -4(y+3)$

15. $(x+2)^2 - \dfrac{(y+3)^2}{3} = 1$

16. $\dfrac{(x+4)^2}{16} + \dfrac{(y-5)^2}{25} = 1$

17. $\dfrac{(x+1)^2}{9} - \dfrac{(y-2)^2}{7} = 1$

18. $\dfrac{(x-3)^2}{9} - \dfrac{(y-1)^2}{4} = 1$

19. $\dfrac{1}{4}$ ft or 3 in.   **20.** 19.72 ft, 18.86 ft, 14.91 ft   **21.** 450 ft

Chapter Test *(page 552)*

1. Hyperbola; center: $(-1,0)$; vertices: $(-3,0)$ and $(1,0)$; foci: $\left(-1 - \sqrt{13}, 0\right)$ and $\left(-1 + \sqrt{13}, 0\right)$; asymptotes: $y = -\dfrac{3}{2}(x+1)$ and $y = \dfrac{3}{2}(x+1)$

2. Parabola; vertex: $\left(1, -\dfrac{1}{2}\right)$; focus: $\left(1, \dfrac{3}{2}\right)$; directrix: $y = -\dfrac{5}{2}$

3. Ellipse; center: $(-1,1)$; foci: $\left(-1 - \sqrt{3}, 1\right)$ and $\left(-1 + \sqrt{3}, 1\right)$; vertices: $(-4,1)$ and $(2,1)$

4. $(x+1)^2 = 6(y-3)$

5. $\dfrac{x^2}{7} + \dfrac{y^2}{16} = 1$

6. $\dfrac{(y-2)^2}{4} - \dfrac{(x-2)^2}{8} = 1$

7. The microphone should be located $\dfrac{2}{3}$ ft from the base of the reflector, along its axis of symmetry.

Cumulative Review *(page 553)*

1. $-6x + 5 - 3h$ **2.** $\left\{-5, -\dfrac{1}{3}, 2\right\}$ **3.** $\{x \mid -3 \le x \le 2\}$ or $[-3, 2]$

4. (a) Domain: $(-\infty, \infty)$; range: $(2, \infty)$ **(b)** $y = \log_3(x - 2)$; domain: $(2, \infty)$; range: $(-\infty, \infty)$ **5. (a)** $\{18\}$ **(b)** $(2, 18]$

6. (a) $y = 2x - 2$ **(b)** $(x - 2)^2 + y^2 = 4$ **(c)** $\dfrac{x^2}{9} + \dfrac{y^2}{4} = 1$ **(d)** $y = 2(x - 1)^2$ **(e)** $y^2 - \dfrac{x^2}{3} = 1$ **(f)** $y = 4^x$

CHAPTER 8   Systems of Equations and Inequalities

8.1 Assess Your Understanding *(page 566)*

3. F   **4.** consistent; independent   **5.** $(3, -2)$ **6.** consistent; dependent   **7.** b   **8.** a   **9.** $\begin{cases} 2 \cdot 2 - (-1) = 5 \\ 5 \cdot 2 + 2(-1) = 8 \end{cases}$ **11.** $\begin{cases} 3 \cdot 2 - 4 \cdot \dfrac{1}{2} = 4 \\ \dfrac{1}{2} \cdot 2 - 3 \cdot \dfrac{1}{2} = -\dfrac{1}{2} \end{cases}$

13. $\begin{cases} 4 - 1 = 3 \\ \dfrac{1}{2} \cdot 4 + 1 = 3 \end{cases}$ **15.** $\begin{cases} 3 \cdot 1 + 3(-1) + 2 \cdot 2 = 4 \\ 1 - (-1) - 2 = 0 \\ 2(-1) - 3 \cdot 2 = -8 \end{cases}$ **17.** $\begin{cases} 3 \cdot 2 + 3(-2) + 2 \cdot 2 = 4 \\ 2 - 3(-2) + 2 = 10 \\ 5 \cdot 2 - 2(-2) - 3 \cdot 2 = 8 \end{cases}$

19. $x = 6, y = 2$; $(6, 2)$ **21.** $x = 3, y = -6$; $(3, -6)$ **23.** $x = 8, y = -4$; $(8, -4)$

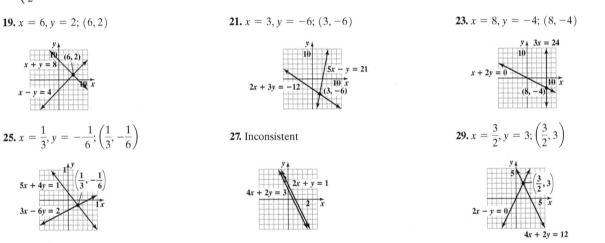

25. $x = \dfrac{1}{3}, y = -\dfrac{1}{6}$; $\left(\dfrac{1}{3}, -\dfrac{1}{6}\right)$ **27.** Inconsistent   **29.** $x = \dfrac{3}{2}, y = 3$; $\left(\dfrac{3}{2}, 3\right)$

31. $\{(x, y) \mid x = 4 - 2y, y \text{ is any real number}\}$, or $\left\{(x, y) \mid y = \dfrac{4 - x}{2}, x \text{ is any real number}\right\}$ **33.** $x = 1, y = 1$; $(1, 1)$ **35.** $x = \dfrac{3}{2}, y = 1$; $\left(\dfrac{3}{2}, 1\right)$

37. $x = 4, y = 3$; $(4, 3)$ **39.** $x = \dfrac{2}{3}, y = -\dfrac{5}{6}$; $\left(\dfrac{2}{3}, -\dfrac{5}{6}\right)$ **41.** $x = \dfrac{1}{5}, y = \dfrac{1}{3}$; $\left(\dfrac{1}{5}, \dfrac{1}{3}\right)$ **43.** $x = 6, y = 3, z = -1$; $(6, 3, -1)$

45. $x = 2, y = -1, z = 1$; $(2, -1, 1)$ **47.** Inconsistent   **49.** $\{(x, y, z) \mid x = 5z - 2, y = 4z - 3; z \text{ is any real number}\}$ **51.** Inconsistent

53. $x = 1, y = 3, z = -2$; $(1, 3, -2)$ **55.** $x = -3, y = \dfrac{1}{2}, z = 1$; $\left(-3, \dfrac{1}{2}, 1\right)$ **57.** Length 30 ft; width 15 ft   **59.** Commercial: 292; noncommercial: 177

61. 22.5 lb   **63.** Smartphone: \$775; tablet: \$890   **65.** Average wind speed 25 mph; average airspeed 175 mph   **67.** 80 \$25 sets and 120 \$45 sets

69. \$9.96   **71.** Mix 50 mg of first compound with 75 mg of second.   **73.** $a = \dfrac{4}{3}, b = -\dfrac{5}{3}, c = 1$ **75.** $Y = 9000, r = 0.06$ **77.** $I_1 = \dfrac{10}{71}, I_2 = \dfrac{65}{71}, I_3 = \dfrac{55}{71}$

79. 100 orchestra, 210 main, and 190 balcony seats   **81.** 1.5 chicken, 1 corn, 2 milk

83. If x = price of hamburgers, y = price of fries, and z = price of colas,

then $x = 2.75 - z, y = \dfrac{41}{60} + \dfrac{1}{3}z, \$0.60 \le z \le \$0.90$.

There is not sufficient information:

x	\$2.13	\$2.01	\$1.86
y	\$0.89	\$0.93	\$0.98
z	\$0.62	\$0.74	\$0.89

85. It will take Beth 30 h, Dan 24 h, and Edie 40 h.   **87.** $x = \dfrac{c}{a}, y = \dfrac{a}{b}, z = \dfrac{b}{c}$; $\left(\dfrac{c}{a}, \dfrac{a}{b}, \dfrac{b}{c}\right)$

91.

92. (a) $2(2x - 3)^3(x^3 + 5)(10x^3 - 9x^2 + 20)$ **(b)** $7(3x - 5)^{-1/2}(x + 3)^{-3/2}$

93. $\{x \mid -1 < x < 4\}$ or $(-1, 4)$ **94.** $f^{-1}(x) = 2 + \log_5(x - 3)$

95. $\{6, 12, 18, 24, 30\}$ **96.** $\dfrac{x^2}{100} + \dfrac{y^2}{36} = 1$ **97.** Maximum; 58

98. \$4709.29   **99.** 37   **100.** $\{x \mid -2 \le x \le 4\}$ or $[-2, 4]$

8.2 Assess Your Understanding *(page 580)*

1. matrix **2.** augmented **3.** third; fifth **4.** b **5.** T **6.** c **7.** $\begin{bmatrix} 1 & -5 & | & 5 \\ 4 & 3 & | & 6 \end{bmatrix}$ **9.** $\begin{bmatrix} 2 & 3 & | & 6 \\ 4 & -6 & | & -2 \end{bmatrix}$ **11.** $\begin{bmatrix} 0.01 & -0.03 & | & 0.06 \\ 0.13 & 0.10 & | & 0.20 \end{bmatrix}$

13. $\begin{bmatrix} 1 & -1 & 1 & | & 10 \\ 3 & 3 & 0 & | & 5 \\ 1 & 1 & 2 & | & 2 \end{bmatrix}$ **15.** $\begin{bmatrix} 1 & 1 & -1 & | & 2 \\ 3 & -2 & 0 & | & 2 \\ 5 & 3 & -1 & | & 1 \end{bmatrix}$ **17.** $\begin{bmatrix} 1 & -1 & -1 & | & 10 \\ 2 & 1 & 2 & | & -1 \\ -3 & 4 & 0 & | & 5 \\ 4 & -5 & 1 & | & 0 \end{bmatrix}$ **19.** $\begin{cases} x - 3y = -2 & (1) \\ 2x - 5y = 5 & (2) \end{cases}$; $\begin{bmatrix} 1 & -3 & | & -2 \\ 0 & 1 & | & 9 \end{bmatrix}$

21. $\begin{cases} x - 3y + 4z = 3 & (1) \\ 3x - 5y + 6z = 6 & (2) \\ -5x + 3y + 4z = 6 & (3) \end{cases}$; $\begin{bmatrix} 1 & -3 & 4 & | & 3 \\ 0 & 4 & -6 & | & -3 \\ 0 & -12 & 24 & | & 21 \end{bmatrix}$ **23.** $\begin{cases} x - 3y + 2z = -6 & (1) \\ 2x - 5y + 3z = -4 & (2) \\ -3x - 6y + 4z = 6 & (3) \end{cases}$; $\begin{bmatrix} 1 & -3 & 2 & | & -6 \\ 0 & 1 & -1 & | & 8 \\ 0 & -15 & 10 & | & -12 \end{bmatrix}$

25. $\begin{cases} 5x - 3y + z = -2 & (1) \\ 2x - 5y + 6z = -2 & (2) \\ -4x + y + 4z = 6 & (3) \end{cases}$; $\begin{bmatrix} 1 & 7 & -11 & | & 2 \\ 2 & -5 & 6 & | & -2 \\ 0 & -9 & 16 & | & 2 \end{bmatrix}$ **27.** $\begin{cases} x = 5 \\ y = -1 \end{cases}$
Consistent; $x = 5, y = -1$ or $(5, -1)$ **29.** $\begin{cases} x = 1 \\ y = 2 \\ 0 = 3 \end{cases}$
Inconsistent

31. $\begin{cases} x + 2z = -1 \\ y - 4z = -2 \\ 0 = 0 \end{cases}$
Consistent:
$\begin{cases} x = -1 - 2z \\ y = -2 + 4z \\ z \text{ is any real number or} \end{cases}$
$\{(x, y, z) | x = -1 - 2z,$
$y = -2 + 4z, z \text{ is any real}$
number$\}$

33. $\begin{cases} x_1 = 1 \\ x_2 + x_4 = 2 \\ x_3 + 2x_4 = 3 \end{cases}$
Consistent:
$\begin{cases} x_1 = 1, x_2 = 2 - x_4 \\ x_3 = 3 - 2x_4 \\ x_4 \text{ is any real number or} \end{cases}$
$\{(x_1, x_2, x_3, x_4) | x_1 = 1,$
$x_2 = 2 - x_4, x_3 = 3 - 2x_4,$
$x_4 \text{ is any real number}\}$

35. $\begin{cases} x_1 + 4x_4 = 2 \\ x_2 + x_3 + 3x_4 = 3 \\ 0 = 0 \end{cases}$
Consistent:
$\begin{cases} x_1 = 2 - 4x_4 \\ x_2 = 3 - x_3 - 3x_4 \\ x_3, x_4 \text{ are any real numbers or} \end{cases}$
$\{(x_1, x_2, x_3, x_4) | x_1 = 2 - 4x_4,$
$x_2 = 3 - x_3 - 3x_4, x_3, x_4 \text{ are}$
any real numbers$\}$

37. $\begin{cases} x_1 + x_4 = -2 \\ x_2 + 2x_4 = 2 \\ x_3 - x_4 = 0 \\ 0 = 0 \end{cases}$
Consistent:
$\begin{cases} x_1 = -2 - x_4 \\ x_2 = 2 - 2x_4 \\ x_3 = x_4 \\ x_4 \text{ is any real number or} \end{cases}$
$\{(x_1, x_2, x_3, x_4) | x_1 = -2 - x_4,$
$x_2 = 2 - 2x_4, x_3 = x_4, x_4 \text{ is any}$
real number$\}$

39. $x = 6, y = 2; (6, 2)$ **41.** $x = \dfrac{1}{3}, y = \dfrac{5}{6}; \left(\dfrac{1}{3}, \dfrac{5}{6}\right)$ **43.** $x = 4 - 2y, y$ is any real number; $\{(x, y) \mid x = 4 - 2y, y \text{ is any real number}\}$

45. $x = \dfrac{3}{2}, y = 1; \left(\dfrac{3}{2}, 1\right)$ **47.** $x = \dfrac{4}{3}, y = \dfrac{1}{5}; \left(\dfrac{4}{3}, \dfrac{1}{5}\right)$ **49.** $x = 8, y = 2, z = 0; (8, 2, 0)$ **51.** $x = 1, y = 2, z = -1; (1, 2, -1)$

53. Inconsistent **55.** $x = 5z - 2, y = 4z - 3$, where z is any real number; $\{(x, y, z) \mid x = 5z - 2, y = 4z - 3, z \text{ is any real number}\}$

57. Inconsistent **59.** $x = 1, y = 3, z = -2; (1, 3, -2)$ **61.** $x = -3, y = \dfrac{1}{2}, z = 1; \left(-3, \dfrac{1}{2}, 1\right)$ **63.** $x = \dfrac{1}{3}, y = \dfrac{2}{3}, z = 1; \left(\dfrac{1}{3}, \dfrac{2}{3}, 1\right)$

65. $x = 1, y = 2, z = 0, w = 1; (1, 2, 0, 1)$ **67.** $y = 0, z = 1 - x, x$ is any real number; $\{(x, y, z) \mid y = 0, z = 1 - x, x \text{ is any real number}\}$

69. $x = 2, y = z - 3, z$ is any real number; $\{(x, y, z) \mid x = 2, y = z - 3, z \text{ is any real number}\}$ **71.** $x = \dfrac{13}{9}, y = \dfrac{7}{18}, z = \dfrac{19}{18}; \left(\dfrac{13}{9}, \dfrac{7}{18}, \dfrac{19}{18}\right)$

73. $x = \dfrac{7}{5} - \dfrac{3}{5}z - \dfrac{2}{5}w, y = -\dfrac{8}{5} + \dfrac{7}{5}z + \dfrac{13}{5}w$, where z and w are any real numbers;

$\left\{(x, y, z, w) \left| x = \dfrac{7}{5} - \dfrac{3}{5}z - \dfrac{2}{5}w, y = -\dfrac{8}{5} + \dfrac{7}{5}z + \dfrac{13}{5}w, z \text{ and } w \text{ are any real numbers}\right.\right\}$

75. $y = -2x^2 + x + 3$ **77.** $f(x) = 3x^3 - 4x^2 + 5$ **79.** 1.5 salmon steak, 2 baked eggs, 1 acorn squash

81. \$4000 in Treasury bills, \$4000 in Treasury bonds, \$2000 in corporate bonds **83.** 8 Deltas, 5 Betas, 10 Sigmas **85.** $I_1 = 1, I_2 = 3, I_3 = 8, I_4 = 4$

87. (a)

Amount Invested At		
7%	9%	11%
0	10,000	10,000
1000	8000	11,000
2000	6000	12,000
3000	4000	13,000
4000	2000	14,000
5000	0	15,000

(b)

Amount Invested At		
7%	9%	11%
12,500	12,500	0
14,500	8500	2000
16,500	4500	4000
18,750	0	6250

(c) All the money invested at 7% provides \$2100, more than what is required.

89.

First Supplement	Second Supplement	Third Supplement
50 mg	75 mg	0 mg
36 mg	76 mg	8 mg
22 mg	77 mg	16 mg
8 mg	78 mg	24 mg

94. $\{x \mid -1 < x < 6\}$, or $(-1, 6)$

95.

96. $\{x \mid x \text{ is any real number}\}$, or $(-\infty, \infty)$ **97.** $\dfrac{8x^3}{27y^{12}}$ **98.** $\dfrac{(y-5)^2}{16} - \dfrac{(x-4)^2}{9} = 1$ **99.** \$3007.44   **100.** $-\dfrac{1}{5}$ **101.** $\dfrac{2x+h}{x^2(x+h)^2}$

8.3 Assess Your Understanding (*page 592*)

1. $ad - bc$ **2.** $\begin{vmatrix} 5 & 3 \\ -3 & -4 \end{vmatrix}$ **3.** F   **4.** F   **5.** F   **6.** a   **7.** 22   **9.** -2 **11.** 10   **13.** -26 **15.** $x = 6, y = 2; (6, 2)$ **17.** $x = 3, y = 2; (3, 2)$

19. $x = 8, y = -4; (8, -4)$ **21.** $x = -3, y = 5; (-3, 5)$ **23.** Not applicable   **25.** $x = \dfrac{1}{2}, y = \dfrac{3}{4}; \left(\dfrac{1}{2}, \dfrac{3}{4}\right)$ **27.** $x = \dfrac{1}{10}, y = \dfrac{2}{5}; \left(\dfrac{1}{10}, \dfrac{2}{5}\right)$

29. $x = \dfrac{3}{2}, y = 1; \left(\dfrac{3}{2}, 1\right)$ **31.** $x = \dfrac{4}{3}, y = \dfrac{1}{5}; \left(\dfrac{4}{3}, \dfrac{1}{5}\right)$ **33.** $x = 1, y = 3, z = -2; (1, 3, -2)$ **35.** $x = -2, y = \dfrac{1}{3}, z = 1; \left(-2, \dfrac{1}{3}, 1\right)$

37. Not applicable   **39.** $x = 0, y = 0, z = 0; (0, 0, 0)$ **41.** Not applicable   **43.** -4 **45.** 12   **47.** 8   **49.** 8   **51.** -5 **53.** $\dfrac{13}{11}$ **55.** 0 or -9

57. $(y_1 - y_2)x - (x_1 - x_2)y + (x_1 y_2 - x_2 y_1) = 0$

$$(y_1 - y_2)x + (x_2 - x_1)y = x_2 y_1 - x_1 y_2$$
$$(x_2 - x_1)y - (x_2 - x_1)y_1 = (y_2 - y_1)x + x_2 y_1 - x_1 y_2 - (x_2 - x_1)y_1$$
$$(x_2 - x_1)(y - y_1) = (y_2 - y_1)x - (y_2 - y_1)x_1$$
$$y - y_1 = \frac{y_2 - y_1}{x_2 - x_1}(x - x_1)$$

59. The triangle has an area of 5 square units.   **61.** 50.5 square units   **63.** $(x - 3)^2 + (y + 2)^2 = 25$

65. If $a = 0$, we have

$$by = s$$
$$cx + dy = t$$

Thus, $y = \dfrac{s}{b}$ and

$$x = \frac{t - dy}{c} = \frac{tb - ds}{bc}$$

Using Cramer's Rule, we get

$$x = \frac{sd - tb}{-bc} = \frac{tb - sd}{bc}$$
$$y = \frac{-sc}{-bc} = \frac{s}{b}$$

If $b = 0$, we have

$$ax = s$$
$$cx + dy = t$$

Since $D = ad \neq 0$, then $a \neq 0$ and $d \neq 0$.

Thus, $x = \dfrac{s}{a}$ and

$$y = \frac{t - cx}{d} = \frac{ta - cs}{ad}$$

Using Cramer's Rule, we get

$$x = \frac{sd}{ad} = \frac{s}{a}$$
$$y = \frac{ta - cs}{ad}$$

If $c = 0$, we have

$$ax + by = s$$
$$dy = t$$

Since $D = ad \neq 0$, then $a \neq 0$ and $d \neq 0$.

Thus, $y = \dfrac{t}{d}$ and

$$x = \frac{s - by}{a} = \frac{sd - bt}{ad}$$

Using Cramer's Rule, we get

$$x = \frac{sd - bt}{ad}$$
$$y = \frac{at}{ad} = \frac{t}{d}$$

If $d = 0$, we have

$$ax + by = s$$
$$cx = t$$

Since $D = -bc \neq 0$, then $b \neq 0$ and $c \neq 0$.

Thus, $x = \dfrac{t}{c}$ and

$$y = \frac{s - ax}{b} = \frac{sc - at}{bc}$$

Using Cramer's Rule, we get

$$x = \frac{-tb}{-bc} = \frac{t}{c}$$
$$y = \frac{at - sc}{-bc} = \frac{sc - at}{bc}$$

67. $\begin{vmatrix} a_{11} & a_{12} & a_{13} \\ ka_{21} & ka_{22} & ka_{23} \\ a_{31} & a_{32} & a_{33} \end{vmatrix} = -ka_{21}(a_{12}a_{33} - a_{32}a_{13}) + ka_{22}(a_{11}a_{33} - a_{31}a_{13}) - ka_{23}(a_{11}a_{32} - a_{31}a_{12})$

$$= k[-a_{21}(a_{12}a_{33} - a_{32}a_{13}) + a_{22}(a_{11}a_{33} - a_{31}a_{13}) - a_{23}(a_{11}a_{32} - a_{31}a_{12})] = k\begin{vmatrix} a_{11} & a_{12} & a_{13} \\ a_{21} & a_{22} & a_{23} \\ a_{31} & a_{32} & a_{33} \end{vmatrix}$$

69. $\begin{vmatrix} a_{11} + ka_{21} & a_{12} + ka_{22} & a_{13} + ka_{23} \\ a_{21} & a_{22} & a_{23} \\ a_{31} & a_{32} & a_{33} \end{vmatrix} = (a_{11} + ka_{21})(a_{22}a_{33} - a_{32}a_{23}) - (a_{12} + ka_{22})(a_{21}a_{33} - a_{31}a_{23}) + (a_{13} + ka_{23})(a_{21}a_{32} - a_{31}a_{22})$

$$= a_{11}a_{22}a_{33} - a_{11}a_{32}a_{23} + \overline{ka_{21}a_{22}a_{33}} - \overline{ka_{21}a_{32}a_{23}} - a_{12}a_{21}a_{33} + a_{12}a_{31}a_{23}$$
$$- \overline{ka_{22}a_{21}a_{33}} + \overline{ka_{22}a_{31}a_{23}} + a_{13}a_{21}a_{32} - a_{13}a_{31}a_{22} + \overline{ka_{23}a_{21}a_{32}} - \overline{ka_{23}a_{31}a_{22}}$$
$$= a_{11}a_{22}a_{33} - a_{11}a_{32}a_{23} - a_{12}a_{21}a_{33} + a_{12}a_{31}a_{23} + a_{13}a_{21}a_{32} - a_{13}a_{31}a_{22}$$

$$= a_{11}(a_{22}a_{33} - a_{32}a_{23}) - a_{12}(a_{21}a_{33} - a_{31}a_{23}) + a_{13}(a_{21}a_{32} - a_{31}a_{22}) = \begin{vmatrix} a_{11} & a_{12} & a_{13} \\ a_{21} & a_{22} & a_{23} \\ a_{31} & a_{32} & a_{33} \end{vmatrix}$$

70. 16.718 g   **71.** $\pm\dfrac{1}{2}, \pm\dfrac{5}{2}, \pm1, \pm2, \pm5, \pm10$

72.

73. $f^{-1}(x) = 5^{x-3} + 1$ **74.** $4\sqrt{5}$ **75.** $8x^3 - 60x^2 + 150x - 125$

76. $\dfrac{x - 93}{x\left(\sqrt{x + 7} + 10\right)}$ **77.** $y = \dfrac{5}{2}x - 22$

Historical Problems *(page 608)*

1. (a) $2 - 5i \longleftrightarrow \begin{bmatrix} 2 & -5 \\ 5 & 2 \end{bmatrix}, 1 + 3i \longleftrightarrow \begin{bmatrix} 1 & 3 \\ -3 & 1 \end{bmatrix}$ **(b)** $\begin{bmatrix} 2 & -5 \\ 5 & 2 \end{bmatrix}\begin{bmatrix} 1 & 3 \\ -3 & 1 \end{bmatrix} = \begin{bmatrix} 17 & 1 \\ -1 & 17 \end{bmatrix}$ **(c)** $17 + i$ **(d)** $17 + i$

2. $\begin{bmatrix} a & b \\ -b & a \end{bmatrix}\begin{bmatrix} a & -b \\ b & a \end{bmatrix} = \begin{bmatrix} a^2 + b^2 & 0 \\ 0 & b^2 + a^2 \end{bmatrix}$; the product is a real number.

3. (a) $x = k(ar + bs) + l(cr + ds) = r(ka + lc) + s(kb + ld)$ **(b)** $A = \begin{bmatrix} ka + lc & kb + ld \\ ma + nc & mb + nd \end{bmatrix}$
$y = m(ar + bs) + n(cr + ds) = r(ma + nc) + s(mb + nd)$

8.4 Assess Your Understanding *(page 608)*

1. square   **2.** T   **3.** F   **4.** inverse   **5.** T   **6.** $A^{-1}B$ **7.** a   **8.** d   **9.** $\begin{bmatrix} 4 & 4 & -5 \\ -1 & 5 & 4 \end{bmatrix}$ **11.** $\begin{bmatrix} 0 & 12 & -20 \\ 4 & 8 & 24 \end{bmatrix}$ **13.** $\begin{bmatrix} -8 & 7 & -15 \\ 7 & 0 & 22 \end{bmatrix}$ **15.** $\begin{bmatrix} 28 & -9 \\ 4 & 23 \end{bmatrix}$

17. $\begin{bmatrix} 1 & 14 & -14 \\ 2 & 22 & -18 \\ 3 & 0 & 28 \end{bmatrix}$ **19.** Not defined   **21.** $\begin{bmatrix} 15 & 21 & -16 \\ 22 & 34 & -22 \\ -11 & 7 & 22 \end{bmatrix}$ **23.** $\begin{bmatrix} 25 & -9 \\ 4 & 20 \end{bmatrix}$ **25.** $\begin{bmatrix} -13 & 7 & -12 \\ -18 & 10 & -14 \\ 17 & -7 & 34 \end{bmatrix}$ **27.** $\begin{bmatrix} -2 & 4 & 2 & 8 \\ 2 & 1 & 4 & 6 \end{bmatrix}$ **29.** $\begin{bmatrix} 5 & 14 \\ 9 & 16 \end{bmatrix}$

31. Not defined   **33.** $\begin{bmatrix} 9 & 2 \\ 34 & 13 \\ 47 & 20 \end{bmatrix}$ **35.** $\begin{bmatrix} 1 & -1 \\ -1 & 2 \end{bmatrix}$ **37.** $\begin{bmatrix} 1 & -\frac{5}{2} \\ -1 & 3 \end{bmatrix}$ **39.** $\begin{bmatrix} 1 & -\frac{1}{a} \\ -1 & \frac{2}{a} \end{bmatrix}$ **41.** $\begin{bmatrix} 3 & -3 & 1 \\ -2 & 2 & -1 \\ -4 & 5 & -2 \end{bmatrix}$ **43.** $\begin{bmatrix} -\frac{5}{7} & \frac{1}{7} & \frac{3}{7} \\ \frac{9}{7} & \frac{1}{7} & -\frac{4}{7} \\ \frac{3}{7} & -\frac{2}{7} & \frac{1}{7} \end{bmatrix}$

45. $x = -4, y = 7; (-4, 7)$ **47.** $x = -5, y = 10; (-5, 10)$ **49.** $x = 2, y = -1; (2, -1)$ **51.** $x = \dfrac{1}{2}, y = 2; \left(\dfrac{1}{2}, 2\right)$ **53.** $x = -2, y = 1; (-2, 1)$

55. $x = \dfrac{2}{a}, y = \dfrac{3}{a}; \left(\dfrac{2}{a}, \dfrac{3}{a}\right)$ **57.** $x = 5, y = -2, z = -3; (5, -2, -3)$ **59.** $x = \dfrac{1}{2}, y = -\dfrac{1}{2}, z = 1; \left(\dfrac{1}{2}, -\dfrac{1}{2}, 1\right)$

61. $x = -\dfrac{34}{7}, y = \dfrac{85}{7}, z = \dfrac{12}{7}; \left(-\dfrac{34}{7}, \dfrac{85}{7}, \dfrac{12}{7}\right)$ **63.** $x = \dfrac{1}{3}, y = 1, z = \dfrac{2}{3}; \left(\dfrac{1}{3}, 1, \dfrac{2}{3}\right)$ **65.** $\begin{bmatrix} 4 & 2 & | & 1 & 0 \\ 2 & 1 & | & 0 & 1 \end{bmatrix} \rightarrow \begin{bmatrix} 1 & \frac{1}{2} & | & \frac{1}{4} & 0 \\ 2 & 1 & | & 0 & 1 \end{bmatrix} \rightarrow \begin{bmatrix} 1 & \frac{1}{2} & | & \frac{1}{4} & 0 \\ 0 & 0 & | & -\frac{1}{2} & 1 \end{bmatrix}$

67. $\begin{bmatrix} 15 & 3 & | & 1 & 0 \\ 10 & 2 & | & 0 & 1 \end{bmatrix} \rightarrow \begin{bmatrix} 1 & \frac{1}{5} & | & \frac{1}{15} & 0 \\ 10 & 2 & | & 0 & 1 \end{bmatrix} \rightarrow \begin{bmatrix} 1 & \frac{1}{5} & | & \frac{1}{15} & 0 \\ 0 & 0 & | & -\frac{2}{3} & 1 \end{bmatrix}$

69. $\begin{bmatrix} -3 & 1 & -1 & | & 1 & 0 & 0 \\ 1 & -4 & -7 & | & 0 & 1 & 0 \\ 1 & 2 & 5 & | & 0 & 0 & 1 \end{bmatrix} \rightarrow \begin{bmatrix} 1 & 2 & 5 & | & 0 & 0 & 1 \\ 1 & -4 & -7 & | & 0 & 1 & 0 \\ -3 & 1 & -1 & | & 1 & 0 & 0 \end{bmatrix} \rightarrow \begin{bmatrix} 1 & 2 & 5 & | & 0 & 0 & 1 \\ 0 & -6 & -12 & | & 0 & 1 & -1 \\ 0 & 7 & 14 & | & 1 & 0 & 3 \end{bmatrix}$

$\rightarrow \begin{bmatrix} 1 & 2 & 5 & | & 0 & 0 & 1 \\ 0 & 1 & 2 & | & 0 & -\frac{1}{6} & \frac{1}{6} \\ 0 & 1 & 2 & | & \frac{1}{7} & 0 & \frac{3}{7} \end{bmatrix} \rightarrow \begin{bmatrix} 1 & 2 & 5 & | & 0 & 0 & 1 \\ 0 & 1 & 2 & | & 0 & -\frac{1}{6} & \frac{1}{6} \\ 0 & 0 & 0 & | & \frac{1}{7} & \frac{1}{6} & \frac{11}{42} \end{bmatrix}$ **71.** $\begin{bmatrix} -0.02 & 0.02 & 0.41 \\ 0.06 & 0.02 & -0.57 \\ 0.16 & 0.12 & -2.05 \end{bmatrix}$ **73.** $\begin{bmatrix} 0.02 & -0.04 & -0.01 & 0.01 \\ -0.02 & 0.05 & 0.03 & -0.03 \\ 0.02 & 0.01 & -0.04 & 0.00 \\ -0.02 & 0.06 & 0.07 & 0.06 \end{bmatrix}$

75. $x = 4.54, y = -6.42, z = -24.08; (4.54, -6.42, -24.08)$ **77.** $x = -1.10, y = 2.54, z = 8.30; (-1.10, 2.54, 8.30)$

79. $x = -5, y = 7; (-5, 7)$ **81.** $x = -4, y = 2, z = \dfrac{5}{2}; \left(-4, 2, \dfrac{5}{2}\right)$ **83.** Inconsistent; $\varnothing$

85. $x = -\dfrac{1}{5}z + \dfrac{1}{5}, y = \dfrac{1}{5}z - \dfrac{6}{5}$, where z is any real number; $\left\{ (x, y, z) \,\middle|\, x = -\dfrac{1}{5}z + \dfrac{1}{5}, y = \dfrac{1}{5}z - \dfrac{6}{5}, z \text{ is any real number} \right\}$

87. (a) $A = \begin{bmatrix} 6 & 9 \\ 3 & 12 \end{bmatrix}$; $B = \begin{bmatrix} 148.00 \\ 404.40 \end{bmatrix}$ **(b)** $AB = \begin{bmatrix} 4527.60 \\ 5296.80 \end{bmatrix}$; Nikki's total tuition is \$4527.60, and Joe's total tuition is \$5296.80.

89. (a) $\begin{bmatrix} 500 & 350 & 400 \\ 700 & 500 & 850 \end{bmatrix}$; $\begin{bmatrix} 500 & 700 \\ 350 & 500 \\ 400 & 850 \end{bmatrix}$ **(b)** $\begin{bmatrix} 15 \\ 8 \\ 3 \end{bmatrix}$ **(c)** $\begin{bmatrix} 11{,}500 \\ 17{,}050 \end{bmatrix}$ **(d)** $[0.10 \quad 0.05]$ **(e)** \$2002.50

91. (a) $K^{-1} = \begin{bmatrix} 1 & 0 & -1 \\ -1 & 1 & 1 \\ 0 & -1 & 1 \end{bmatrix}$ **(b)** $M = \begin{bmatrix} 13 & 1 & 20 \\ 8 & 9 & 19 \\ 6 & 21 & 14 \end{bmatrix}$ **(c)** Math is fun. **93.** $X = \begin{bmatrix} 4 & -3 & 5 \\ 6 & -2 & -7 \end{bmatrix}$

95. (a) $B_3 = A + A^2 + A^3 = \begin{bmatrix} 2 & 4 & 5 & 2 & 3 \\ 5 & 3 & 2 & 5 & 4 \\ 4 & 2 & 2 & 4 & 2 \\ 2 & 2 & 3 & 2 & 3 \\ 1 & 3 & 2 & 1 & 2 \end{bmatrix}$; Yes, all pages can reach every other page within 3 clicks. **(b)** Page 3

97. (a) $(3 - 2\sqrt{3}, 3\sqrt{3} + 2)$ **(b)** $R^{-1} = \begin{bmatrix} \dfrac{1}{2} & \dfrac{\sqrt{3}}{2} & 0 \\ -\dfrac{\sqrt{3}}{2} & \dfrac{1}{2} & 0 \\ 0 & 0 & 1 \end{bmatrix}$; This is the rotation matrix needed to get the translated coordinates back to the original coordinates.

99. $a = -\dfrac{1}{2}, b = -\dfrac{1}{2}; a = -\dfrac{1}{2}, b = \dfrac{1}{2}; a = 0, b = 0; a = -1, b = 0$ **104.** $x^6 - 4x^5 - 3x^4 + 18x^3$

105. $\{0, 3\}$ **106.** $\{-6\}$ **107.** $\dfrac{x^2 + 8x - 9}{x^2 - 9}$ or $\dfrac{(x + 9)(x - 1)}{(x - 3)(x + 3)}$ **108.** $\{x \mid x \le 5, x \ne -3\}$ **109.** $3x(x + 4)(x + 6)(x - 6)$

110. $2\pi + 4 \approx 10.28$ square units **111.** $y = -\dfrac{5}{2}x + \dfrac{21}{2}$

8.5 Assess Your Understanding *(page 619)*

5. Proper **7.** Improper; $1 + \dfrac{9}{x^2 - 4}$ **9.** Improper; $x - 1 + \dfrac{5x - 6}{x^2 + 2x - 15}$ **11.** Proper **13.** Improper; $5x + \dfrac{22x - 1}{x^2 - 4}$

15. Improper; $1 + \dfrac{-2(x - 6)}{(x + 4)(x - 3)}$ **17.** $\dfrac{-4}{x} + \dfrac{4}{x - 1}$ **19.** $\dfrac{1}{x} + \dfrac{-x}{x^2 + 1}$ **21.** $\dfrac{-1}{x - 1} + \dfrac{2}{x - 2}$ **23.** $\dfrac{\frac{1}{4}}{x + 1} + \dfrac{\frac{3}{4}}{x - 1} + \dfrac{\frac{1}{2}}{(x - 1)^2}$

25. $\dfrac{\frac{1}{12}}{x - 2} + \dfrac{-\frac{1}{12}(x + 4)}{x^2 + 2x + 4}$ **27.** $\dfrac{\frac{1}{4}}{x - 1} + \dfrac{\frac{1}{4}}{(x - 1)^2} + \dfrac{-\frac{1}{4}}{x + 1} + \dfrac{\frac{1}{4}}{(x + 1)^2}$ **29.** $\dfrac{-5}{x + 2} + \dfrac{5}{x + 1} + \dfrac{-4}{(x + 1)^2}$ **31.** $\dfrac{\frac{1}{4}}{x} + \dfrac{1}{x^2} + \dfrac{-\frac{1}{4}(x + 4)}{x^2 + 4}$

33. $\dfrac{\frac{2}{3}}{x + 1} + \dfrac{\frac{1}{3}(x + 1)}{x^2 + 2x + 4}$ **35.** $\dfrac{\frac{2}{7}}{3x - 2} + \dfrac{\frac{1}{7}}{2x + 1}$ **37.** $\dfrac{\frac{3}{4}}{x + 3} + \dfrac{\frac{1}{4}}{x - 1}$ **39.** $\dfrac{1}{x^2 + 4} + \dfrac{2x - 1}{(x^2 + 4)^2}$ **41.** $\dfrac{-1}{x} + \dfrac{2}{x - 3} + \dfrac{-1}{x + 1}$

43. $\dfrac{4}{x - 2} + \dfrac{-3}{x - 1} + \dfrac{-1}{(x - 1)^2}$ **45.** $\dfrac{x}{(x^2 + 16)^2} + \dfrac{-16x}{(x^2 + 16)^3}$ **47.** $\dfrac{-\frac{8}{7}}{2x + 1} + \dfrac{\frac{4}{7}}{x - 3}$ **49.** $\dfrac{-\frac{2}{9}}{x} + \dfrac{-\frac{1}{3}}{x^2} + \dfrac{\frac{1}{6}}{x - 3} + \dfrac{\frac{1}{18}}{x + 3}$

51. $x - 2 + \dfrac{10x - 11}{x^2 + 3x - 4}; \dfrac{\frac{51}{5}}{x + 4} + \dfrac{-\frac{1}{5}}{x - 1}; x - 2 + \dfrac{\frac{51}{5}}{x + 4} - \dfrac{\frac{1}{5}}{x - 1}$ **53.** $x - \dfrac{x}{x^2 + 1}$

55. $x^2 - 4x + 7 + \dfrac{-11x - 32}{x^2 + 4x + 4}; \dfrac{-11}{x + 2} + \dfrac{-10}{(x + 2)^2}; x^2 - 4x + 7 - \dfrac{11}{x + 2} - \dfrac{10}{(x + 2)^2}$

57. $x + 1 + \dfrac{2x^3 + x^2 - x + 1}{x^4 - 2x^2 + 1}; \dfrac{1}{x + 1} + \dfrac{\frac{1}{4}}{(x + 1)^2} + \dfrac{1}{x - 1} + \dfrac{\frac{3}{4}}{(x - 1)^2}; x + 1 + \dfrac{1}{x + 1} + \dfrac{\frac{1}{4}}{(x + 1)^2} + \dfrac{1}{x - 1} + \dfrac{\frac{3}{4}}{(x - 1)^2}$

59. $\dfrac{2}{e^x + 2} + \dfrac{1}{e^x - 1}$ **61.** 3.85 years **62.** -2 **63.** Odd **64.** $f^{-1}(x) = \log_8(x + 4) + 3$ **65.** $\dfrac{y^2}{25} - \dfrac{x^2}{144} = 1$

66. $\left\{ x \mid x \ge \dfrac{6}{5} \right\}$ or $\left[\dfrac{6}{5}, \infty \right)$ **67.** $D = \dfrac{2x - 4y}{4x - 2y + 1}$ **68.** $y = \dfrac{1}{2}x - \dfrac{11}{2}$

Historical Problem *(page 626)*

$x = 6$ units, $y = 8$ units

8.6 Assess Your Understanding *(page 626)*

5. **7.** **9.** **11.**

13. **15.** **17.** **19.**

21. No points of intersection   **23.**

25. $x = 1, y = 4; x = -1, y = -4; x = 2\sqrt{2}, y = \sqrt{2}; x = -2\sqrt{2}, y = -\sqrt{2}$ or $(1, 4), (-1, -4), (2\sqrt{2}, \sqrt{2}), (-2\sqrt{2}, -\sqrt{2})$

27. $x = 0, y = 2; x = -1, y = -1$ or $(0, 2), (-1, -1)$ **29.** $x = 0, y = -1; x = \dfrac{5}{2}, y = -\dfrac{7}{2}$ or $(0, -1), \left(\dfrac{5}{2}, -\dfrac{7}{2}\right)$

31. $x = \dfrac{1}{3}, y = \dfrac{9}{2}; x = 3, y = \dfrac{1}{2}$ or $\left(\dfrac{1}{3}, \dfrac{9}{2}\right), \left(3, \dfrac{1}{2}\right)$

33. $x = 3, y = 2; x = 3, y = -2; x = -3, y = 2; x = -3, y = -2$ or $(3, 2), (3, -2), (-3, 2), (-3, -2)$

35. $x = \dfrac{1}{2}, y = \dfrac{3}{2}; x = \dfrac{1}{2}, y = -\dfrac{3}{2}; x = -\dfrac{1}{2}, y = \dfrac{3}{2}; x = -\dfrac{1}{2}, y = -\dfrac{3}{2}$ or $\left(\dfrac{1}{2}, \dfrac{3}{2}\right), \left(\dfrac{1}{2}, -\dfrac{3}{2}\right), \left(-\dfrac{1}{2}, \dfrac{3}{2}\right), \left(-\dfrac{1}{2}, -\dfrac{3}{2}\right)$

37. $x = \sqrt{2}, y = 2\sqrt{2}; x = -\sqrt{2}, y = -2\sqrt{2}$ or $(\sqrt{2}, 2\sqrt{2}), (-\sqrt{2}, -2\sqrt{2})$ **39.** No real solution exists.

41. $x = \dfrac{8}{3}, y = \dfrac{2\sqrt{10}}{3}; x = -\dfrac{8}{3}, y = \dfrac{2\sqrt{10}}{3}; x = \dfrac{8}{3}, y = -\dfrac{2\sqrt{10}}{3}; x = -\dfrac{8}{3}, y = -\dfrac{2\sqrt{10}}{3}$ or $\left(\dfrac{8}{3}, \dfrac{2\sqrt{10}}{3}\right), \left(-\dfrac{8}{3}, \dfrac{2\sqrt{10}}{3}\right), \left(\dfrac{8}{3}, -\dfrac{2\sqrt{10}}{3}\right),$

$\left(-\dfrac{8}{3}, -\dfrac{2\sqrt{10}}{3}\right)$ **43.** $x = 1, y = \dfrac{1}{2}; x = -1, y = \dfrac{1}{2}; x = 1, y = -\dfrac{1}{2}; x = -1, y = -\dfrac{1}{2}$ or $\left(1, \dfrac{1}{2}\right), \left(-1, \dfrac{1}{2}\right), \left(1, -\dfrac{1}{2}\right), \left(-1, -\dfrac{1}{2}\right)$

45. No real solution exists.

47. $x = \sqrt{3}, y = \sqrt{3}; x = -\sqrt{3}, y = -\sqrt{3}; x = 2, y = 1; x = -2, y = -1$ or $(\sqrt{3}, \sqrt{3}), (-\sqrt{3}, -\sqrt{3}), (2, 1), (-2, -1)$

49. $x = 0, y = -2; x = 0, y = 1; x = 2, y = -1$ or $(0, -2), (0, 1), (2, -1)$ **51.** $x = 2, y = 8$ or $(2, 8)$ **53.** $x = 81, y = 3$ or $(81, 3)$

55.   **57.** $x = 0.48, y = 0.62$ **59.** $x = -1.65, y = -0.89$

61. $x = 0.58, y = 1.86; x = 1.81, y = 1.05; x = 0.58, y = -1.86; x = 1.81, y = -1.05$

63. $x = 2.35, y = 0.85$

65. **67.** **69.**

71. 3 and 1; -1 and -3 **73.** 2 and 2; -2 and -2 **75.** $\dfrac{1}{2}$ and $\dfrac{1}{3}$ **77.** 5   **79.** 5 in. by 3 in.   **81.** 2 cm and 4 cm   **83.** tortoise: 7 m/h, hare: $7\dfrac{1}{2}$ m/h

85. 12 cm by 18 cm   **87.** $x = 60$ ft; $y = 30$ ft   **89.** $y = 4x - 4$ **91.** $y = 2x + 1$ **93.** $y = -\dfrac{1}{3}x + \dfrac{7}{3}$ **95.** $y = 2x - 3$

97. $r_1 = \dfrac{-b + \sqrt{b^2 - 4ac}}{2a}; r_2 = \dfrac{-b - \sqrt{b^2 - 4ac}}{2a}$ **99.** $l = \dfrac{P + \sqrt{P^2 - 16A}}{4}; w = \dfrac{P - \sqrt{P^2 - 16A}}{4}$ **103.** $\left\{\dfrac{-3 - \sqrt{65}}{7}, \dfrac{-3 + \sqrt{65}}{7}\right\}$

104. $y = -\dfrac{2}{5}x - 3$ **105.** $(x + 3)^2 + (y - 4)^2 = 100$ **106.** $\{x | -3 < x < 7\}$ or $(-3, 7)$ **107.** $y = -\sqrt{25 - (x - 4)^2}$ **108.** $2x^2 - 20x + 49$

109. $-\dfrac{24}{(x - 8)(x + h - 8)}$ **110.** $-\dfrac{48x + 15}{(2x - 5)^{10}}$

8.7 Assess Your Understanding (*page 635*)

7. dashes; solid   **8.** half-planes   **9.** F   **10.** b

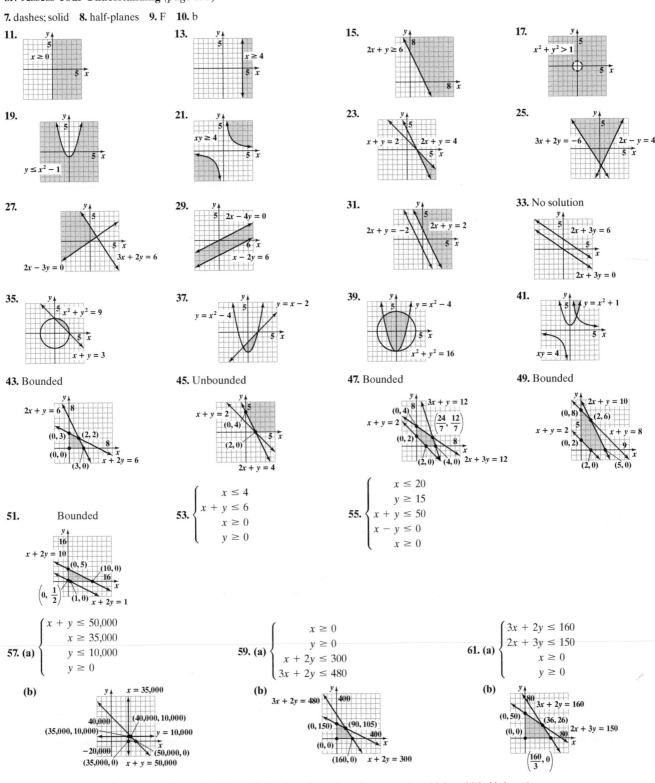

43. Bounded   **45.** Unbounded   **47.** Bounded   **49.** Bounded

51. Bounded

53. $\begin{cases} x \le 4 \\ x + y \le 6 \\ x \ge 0 \\ y \ge 0 \end{cases}$

55. $\begin{cases} x \le 20 \\ y \ge 15 \\ x + y \le 50 \\ x - y \le 0 \\ x \ge 0 \end{cases}$

57. (a) $\begin{cases} x + y \le 50{,}000 \\ x \ge 35{,}000 \\ y \le 10{,}000 \\ y \ge 0 \end{cases}$

59. (a) $\begin{cases} x \ge 0 \\ y \ge 0 \\ x + 2y \le 300 \\ 3x + 2y \le 480 \end{cases}$

61. (a) $\begin{cases} 3x + 2y \le 160 \\ 2x + 3y \le 150 \\ x \ge 0 \\ y \ge 0 \end{cases}$

(b)

63. $\{-1 - 2i, -1 + 2i\}$ **64.** $f(-1) = -5; f(2) = 28$; So, there is a value x between -1 and 2 for which $f(x) = 0$.

65. $\{x \mid -3 \le x \le 1\}$ or $[-3, 1]$ **66.** \$8823.30   **67.** 18.75 horsepower   **68.** $5^y = x$ **69.** $\{x \mid x \ge -2, x \ne 23\}$

70. Increasing: $\left(-\infty, -\dfrac{1}{3}\right)$, $(5, \infty)$; Decreasing: $\left(-\dfrac{1}{3}, 5\right)$

8.8 Assess Your Understanding (*page 642*)

1. objective function   **2.** T   **3.** Maximum value is 11; minimum value is 3.   **5.** Maximum value is 65; minimum value is 4.
7. Maximum value is 67; minimum value is 20.   **9.** The maximum value of z is 12, and it occurs at the point $(6, 0)$.
11. The minimum value of z is 4, and it occurs at the point $(2, 0)$.   **13.** The maximum value of z is 20, and it occurs at the point $(0, 4)$.
15. The minimum value of z is 8, and it occurs at the point $(0, 2)$.   **17.** The maximum value of z is 50, and it occurs at the point $(10, 0)$.

19. Produce 8 downhill and 24 cross-country; $1760; $1920 which is the profit when producing 16 downhill and 16 cross-country.
21. Rent 15 rectangular tables and 16 round tables for a minimum cost of $1252.   **23. (a)** $10,000 in a junk bond and $10,000 in Treasury bills
   (b) $12,000 in a junk bond and $8000 in Treasury bills   **25.** 120 lb of ground beef should be mixed with 60 lb of pork.
27. Manufacture 10 racing skates and 15 figure skates.   **29.** Order 2 metal samples and 4 plastic samples; $34
31. (a) Configure with 10 first class seats and 120 coach seats.   **(b)** Configure with 15 first class seats and 120 coach seats.   **34.** $\left\{-\dfrac{1}{32}, 1\right\}$

35. $-4, 4, -2i, 2i$ **36.** 89.1 years   **37.** $y = 3x + 7$ **38.** 5 and 11   **39.** $13,303.89   **40.** $\dfrac{x^2}{16} + \dfrac{y^2}{25} = 1$ **41.** $f^{-1}(x) = -\dfrac{x + 7}{5x - 2}$

42. $15x^2(x - 7)^{1/2}(3x - 14)$ **43.** Concave up: $(3, \infty)$; concave down: $(-\infty, 3)$

Review Exercises *(page 646)*

1. $x = 2, y = -1$ or $(2, -1)$ **2.** $x = 2, y = \dfrac{1}{2}$ or $\left(2, \dfrac{1}{2}\right)$ **3.** $x = 2, y = -1$ or $(2, -1)$ **4.** $x = \dfrac{11}{5}, y = -\dfrac{3}{5}$ or $\left(\dfrac{11}{5}, -\dfrac{3}{5}\right)$ **5.** Inconsistent

6. $x = 2, y = 3$ or $(2, 3)$ **7.** $y = -\dfrac{2}{5}x + 2$, where x is any real number, or $\left\{(x, y) \mid y = -\dfrac{2}{5}x + 2, x \text{ is any real number}\right\}$

8. $x = -1, y = 2, z = -3$ or $(-1, 2, -3)$

9. $x = \dfrac{7}{4}z + \dfrac{39}{4}, y = \dfrac{9}{8}z + \dfrac{69}{8}$, where z is any real number, or $\left\{(x, y, z) \mid x = \dfrac{7}{4}z + \dfrac{39}{4}, y = \dfrac{9}{8}z + \dfrac{69}{8}, z \text{ is any real number}\right\}$ **10.** Inconsistent

11. $\begin{cases} 3x + 2y = 8 \\ x + 4y = -1 \end{cases}$ **12.** $\begin{cases} x + 2y + 5z = -2 \\ 5x - 3z = 8 \\ 2x - y = 0 \end{cases}$ **13.** $\begin{bmatrix} 4 & -4 \\ 3 & 9 \\ 4 & 4 \end{bmatrix}$ **14.** $\begin{bmatrix} 6 & 0 \\ 12 & 24 \\ -6 & 12 \end{bmatrix}$ **15.** $\begin{bmatrix} 4 & -3 & 0 \\ 12 & -2 & -8 \\ -2 & 5 & -4 \end{bmatrix}$ **16.** $\begin{bmatrix} 9 & -31 \\ -6 & -3 \end{bmatrix}$

17. $\begin{bmatrix} \dfrac{1}{2} & -1 \\ -\dfrac{1}{6} & \dfrac{2}{3} \end{bmatrix}$ **18.** $\begin{bmatrix} -\dfrac{5}{7} & \dfrac{9}{7} & \dfrac{3}{7} \\ \dfrac{1}{7} & \dfrac{1}{7} & -\dfrac{2}{7} \\ \dfrac{3}{7} & -\dfrac{4}{7} & \dfrac{1}{7} \end{bmatrix}$ **19.** Singular   **20.** $x = \dfrac{2}{5}, y = \dfrac{1}{10}$ or $\left(\dfrac{2}{5}, \dfrac{1}{10}\right)$ **21.** $x = 9, y = \dfrac{13}{3}, z = \dfrac{13}{3}$ or $\left(9, \dfrac{13}{3}, \dfrac{13}{3}\right)$

22. Inconsistent   **23.** $x = -\dfrac{1}{2}, y = -\dfrac{2}{3}, z = -\dfrac{3}{4}$, or $\left(-\dfrac{1}{2}, -\dfrac{2}{3}, -\dfrac{3}{4}\right)$

24. $z = -1, x = y + 1$, where y is any real number, or $\left\{(x, y, z) \mid x = y + 1, z = -1, y \text{ is any real number}\right\}$

25. $x = 4, y = 2, z = 3, t = -2$ or $(4, 2, 3, -2)$ **26.** 5   **27.** 108   **28.** -100 **29.** $x = 2, y = -1$ or $(2, -1)$ **30.** $x = 2, y = 3$ or $(2, 3)$

31. $x = -1, y = 2, z = -3$ or $(-1, 2, -3)$ **32.** 16   **33.** -8 **34.** $\dfrac{-\dfrac{3}{2}}{x} + \dfrac{\dfrac{3}{2}}{x - 4}$ **35.** $\dfrac{-3}{x - 1} + \dfrac{3}{x} + \dfrac{4}{x^2}$ **36.** $\dfrac{-\dfrac{1}{10}}{x + 1} + \dfrac{\dfrac{1}{10}x + \dfrac{9}{10}}{x^2 + 9}$

37. $\dfrac{x}{x^2 + 4} + \dfrac{-4x}{(x^2 + 4)^2}$ **38.** $\dfrac{\dfrac{1}{2}}{x^2 + 1} + \dfrac{\dfrac{1}{4}}{x - 1} + \dfrac{-\dfrac{1}{4}}{x + 1}$ **39.** $x = -\dfrac{2}{5}, y = -\dfrac{11}{5}; x = -2, y = 1$ or $\left(-\dfrac{2}{5}, -\dfrac{11}{5}\right), (-2, 1)$

40. $x = 2\sqrt{2}, y = \sqrt{2}; x = -2\sqrt{2}, y = -\sqrt{2}$ or $(2\sqrt{2}, \sqrt{2}), (-2\sqrt{2}, -\sqrt{2})$

41. $x = 0, y = 0; x = -3, y = 3; x = 3, y = 3$ or $(0, 0), (-3, 3), (3, 3)$

42. $x = \sqrt{2}, y = -\sqrt{2}; x = -\sqrt{2}, y = \sqrt{2}; x = \dfrac{4}{3}\sqrt{2}, y = -\dfrac{2}{3}\sqrt{2}; x = -\dfrac{4}{3}\sqrt{2}, y = \dfrac{2}{3}\sqrt{2}$ or $(\sqrt{2}, -\sqrt{2}), (-\sqrt{2}, \sqrt{2}), \left(\dfrac{4}{3}\sqrt{2}, -\dfrac{2}{3}\sqrt{2}\right),$
$\left(-\dfrac{4}{3}\sqrt{2}, \dfrac{2}{3}\sqrt{2}\right)$ **43.** $x = 1, y = -1$ or $(1, -1)$

44.

45.

46. Unbounded

47. Bounded

48. Bounded

49.

50.

51. The maximum value is 32 when $x = 0$ and $y = 8$.   **52.** The minimum value is 3 when $x = 1$ and $y = 0$.   **53.** 10   **54.** A is any real number, $A \neq 10$.

55. $y = -\dfrac{1}{3}x^2 - \dfrac{2}{3}x + 1$ **56.** Mix 70 lb of $6.00 coffee and 30 lb of $9.00 coffee.   **57.** Buy 1 small, 5 medium, and 2 large.

58. (a) $\begin{cases} x \geq 0 \\ y \geq 0 \\ 4x + 3y \leq 960 \\ 2x + 3y \leq 576 \end{cases}$ **(b)**

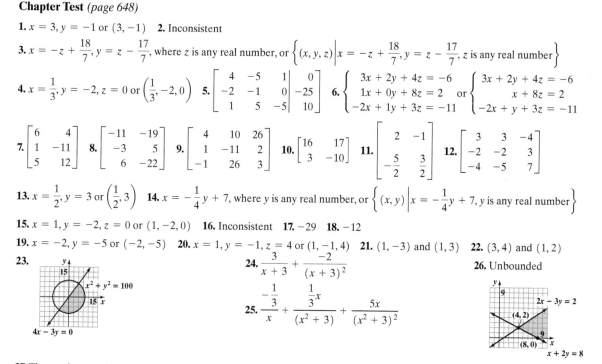

59. Speedboat: 36.67 km/h; Aguarico River: 3.33 km/h
60. Bruce: 4 h; Bryce: 2 h; Marty: 8 h
61. Produce 35 gasoline engines and 15 diesel engines; the factory is producing an excess of 15 gasoline engines and 0 diesel engines.

Chapter Test (page 648)

1. $x = 3, y = -1$ or $(3, -1)$ **2.** Inconsistent

3. $x = -z + \dfrac{18}{7}, y = z - \dfrac{17}{7}$, where z is any real number, or $\left\{ (x, y, z) \,\middle|\, x = -z + \dfrac{18}{7}, y = z - \dfrac{17}{7}, z \text{ is any real number} \right\}$

4. $x = \dfrac{1}{3}, y = -2, z = 0$ or $\left(\dfrac{1}{3}, -2, 0 \right)$ **5.** $\begin{bmatrix} 4 & -5 & 1 & | & 0 \\ -2 & -1 & 0 & | & -25 \\ 1 & 5 & -5 & | & 10 \end{bmatrix}$ **6.** $\begin{cases} 3x + 2y + 4z = -6 \\ 1x + 0y + 8z = 2 \\ -2x + 1y + 3z = -11 \end{cases}$ or $\begin{cases} 3x + 2y + 4z = -6 \\ x + 8z = 2 \\ -2x + y + 3z = -11 \end{cases}$

7. $\begin{bmatrix} 6 & 4 \\ 1 & -11 \\ 5 & 12 \end{bmatrix}$ **8.** $\begin{bmatrix} -11 & -19 \\ -3 & 5 \\ 6 & -22 \end{bmatrix}$ **9.** $\begin{bmatrix} 4 & 10 & 26 \\ 1 & -11 & 2 \\ -1 & 26 & 3 \end{bmatrix}$ **10.** $\begin{bmatrix} 16 & 17 \\ 3 & -10 \end{bmatrix}$ **11.** $\begin{bmatrix} 2 & -1 \\ -\dfrac{5}{2} & \dfrac{3}{2} \end{bmatrix}$ **12.** $\begin{bmatrix} 3 & 3 & -4 \\ -2 & -2 & 3 \\ -4 & -5 & 7 \end{bmatrix}$

13. $x = \dfrac{1}{2}, y = 3$ or $\left(\dfrac{1}{2}, 3 \right)$ **14.** $x = -\dfrac{1}{4}y + 7$, where y is any real number, or $\left\{ (x, y) \,\middle|\, x = -\dfrac{1}{4}y + 7, y \text{ is any real number} \right\}$

15. $x = 1, y = -2, z = 0$ or $(1, -2, 0)$ **16.** Inconsistent **17.** -29 **18.** -12

19. $x = -2, y = -5$ or $(-2, -5)$ **20.** $x = 1, y = -1, z = 4$ or $(1, -1, 4)$ **21.** $(1, -3)$ and $(1, 3)$ **22.** $(3, 4)$ and $(1, 2)$

23.

24. $\dfrac{3}{x+3} + \dfrac{-2}{(x+3)^2}$

25. $\dfrac{-\dfrac{1}{3}}{x} + \dfrac{\dfrac{1}{3}x}{(x^2+3)} + \dfrac{5x}{(x^2+3)^2}$

26. Unbounded

27. The maximum value of z is 64, and it occurs at the point $(0, 8)$. **28.** Jeans cost $24.50, camisoles cost $8.50, and T-shirts cost $6.00.

Cumulative Review (page 649)

1. $\left\{ 0, \dfrac{1}{2} \right\}$ **2.** $\{5\}$ **3.** $\left\{ -1, -\dfrac{1}{2}, 3 \right\}$ **4.** $\{-2\}$ **5.** $\left\{ \dfrac{5}{2} \right\}$ **6.** $\left\{ \dfrac{1}{\ln 3} \right\}$ **7.** Odd; symmetric with respect to the origin

8. Center: $(1, -2)$; radius $= 4$

9. Domain: all real numbers
Range: $\{y \,|\, y > 1\}$
Horizontal asymptote: $y = 1$

10. $f^{-1}(x) = \dfrac{5}{x} - 2$
Domain of f: $\{x \,|\, x \neq -2\}$
Range of f: $\{y \,|\, y \neq 0\}$
Domain of f^{-1}: $\{x \,|\, x \neq 0\}$
Range of f^{-1}: $\{y \,|\, y \neq -2\}$

12. (a) $;-2.28$ **(b)** Local maximum of 7 at $x = -1$;
local minimum of 3 at $x = 1$
(c) $(-\infty, -1], [1, \infty)$

CHAPTER 9 Sequences; Induction; the Binomial Theorem

9.1 Assess Your Understanding (page 658)

3. sequence **4.** True **5.** True **6.** b **7.** summation **8.** b **9.** 3,628,800 **11.** 504 **13.** 190,080 **15.** $s_1 = 1, s_2 = 2, s_3 = 3, s_4 = 4, s_5 = 5$

17. $a_1 = \dfrac{1}{3}, a_2 = \dfrac{1}{2}, a_3 = \dfrac{3}{5}, a_4 = \dfrac{2}{3}, a_5 = \dfrac{5}{7}$ **19.** $c_1 = 1, c_2 = -4, c_3 = 9, c_4 = -16, c_5 = 25$ **21.** $s_1 = \dfrac{3}{5}, s_2 = \dfrac{9}{7}, s_3 = \dfrac{27}{11}, s_4 = \dfrac{81}{19}, s_5 = \dfrac{243}{35}$

23. $t_1 = -\dfrac{1}{6}, t_2 = \dfrac{1}{12}, t_3 = -\dfrac{1}{20}, t_4 = \dfrac{1}{30}, t_5 = -\dfrac{1}{42}$ **25.** $b_1 = \dfrac{1}{e}, b_2 = \dfrac{2}{e^2}, b_3 = \dfrac{3}{e^3}, b_4 = \dfrac{4}{e^4}, b_5 = \dfrac{5}{e^5}$ **27.** $a_n = \dfrac{n}{n+1}$ **29.** $a_n = \dfrac{1}{2^{n-1}}$

31. $a_n = (-1)^{n+1}$ **33.** $a_n = (-1)^{n+1}n$ **35.** $a_1 = 2, a_2 = 5, a_3 = 8, a_4 = 11, a_5 = 14$ **37.** $a_1 = -2, a_2 = 0, a_3 = 3, a_4 = 7, a_5 = 12$

39. $a_1 = 4, a_2 = 12, a_3 = 36, a_4 = 108, a_5 = 324$ **41.** $a_1 = 3, a_2 = \dfrac{3}{2}, a_3 = \dfrac{1}{2}, a_4 = \dfrac{1}{8}, a_5 = \dfrac{1}{40}$ **43.** $a_1 = 1, a_2 = 2, a_3 = 2, a_4 = 4, a_5 = 8$

45. $a_1 = A, a_2 = A + d, a_3 = A + 2d, a_4 = A + 3d, a_5 = A + 4d$

47. $a_1 = \sqrt{2}, a_2 = \sqrt{2 + \sqrt{2}}, a_3 = \sqrt{2 + \sqrt{2 + \sqrt{2}}},$ **49.** $3 + 4 + \cdots + (n + 2)$ **51.** $\dfrac{1}{2} + 2 + \dfrac{9}{2} + \cdots + \dfrac{n^2}{2}$ **53.** $1 + \dfrac{1}{3} + \dfrac{1}{9} + \cdots + \dfrac{1}{3^n}$

$a_4 = \sqrt{2 + \sqrt{2 + \sqrt{2 + \sqrt{2}}}},$

$a_5 = \sqrt{2 + \sqrt{2 + \sqrt{2 + \sqrt{2 + \sqrt{2}}}}}$ **55.** $\dfrac{1}{3} + \dfrac{1}{9} + \cdots + \dfrac{1}{3^n}$ **57.** $\ln 2 - \ln 3 + \ln 4 - \cdots + (-1)^n \ln n$ **59.** $\displaystyle\sum_{k=1}^{20} k$

61. $\displaystyle\sum_{k=1}^{13} \dfrac{k}{k+1}$ **63.** $\displaystyle\sum_{k=0}^{6} (-1)^k \left(\dfrac{1}{3^k}\right)$ **65.** $\displaystyle\sum_{k=1}^{n} \dfrac{3^k}{k}$ **67.** $\displaystyle\sum_{k=1}^{n} (a + kd)$ or $\displaystyle\sum_{k=1}^{n+1} [a + (k-1)d]$ **69.** 200 **71.** 820 **73.** 1110 **75.** 1560 **77.** 3570

79. 44,000 **81.** \$2930 **83.** \$18,058.03 **85.** 21 pairs **87.** Fibonacci sequence **89. (a)** 3.630170833 **(b)** 3.669060828 **(c)** 3.669296668 **(d)** 12

91. (a) $a_1 = 0.4; a_2 = 0.7; a_3 = 1; a_4 = 1.6; a_5 = 2.8; a_6 = 5.2; a_7 = 10; a_8 = 19.6$

(b) Except for term 5, which has no match, Bode's formula provides excellent approximations for the mean distances of the planets from the sun.

(c) The mean distance of Ceres from the sun is approximated by $a_5 = 2.8$, and that of Uranus is $a_8 = 19.6$.

(d) $a_9 = 38.8; a_{10} = 77.2$ **(e)** Pluto's distance is approximated by a_9, but no term approximates Neptune's mean distance from the sun.

(f) According to Bode's Law, the mean orbital distance of Eris will be 154 AU from the sun.

93. (a) $a_n = 0.95^n \cdot I_0$ **(b)** 77 **95.** $a_0 = 2; a_5 = 2.236067977; 2.236067977$ **97.** $a_0 = 4; a_5 = 4.582575695; 4.582575695$

99. 1, 3, 6, 10, 15, 21, 28 **101.** $u_n = 1 + 2 + 3 + \cdots + n = \displaystyle\sum_{k=1}^{n} k = \dfrac{n(n+1)}{2}$, and from Problem 100, $u_{n+1} = \dfrac{(n+1)(n+2)}{2}$.

$u_{n+1} + u_n = \dfrac{(n+1)(n+2)}{2} + \dfrac{n(n+1)}{2} = \dfrac{(n+1)[(n+2)+n]}{2} = (n+1)^2$

105. \$2654.39 **106.** $27x^3 - 54x^2 + 36x - 8$ **107.** $(y-4)^2 = 16(x+3)$ **108.** $y = 0$ **109.** $-2, \dfrac{2}{5}$ **110.** $-2, 6$ **111.** $6x^2 + 6xh + 2h^2$

9.2 Assess Your Understanding (page 666)

1. arithmetic **2.** F **3.** 17 **4.** T **5.** d **6.** c **7.** $s_n - s_{n-1} = (n+4) - [(n-1)+4] = n+4 - (n+3) = n+4-n-3 = 1$, a constant;

$d = 1; s_1 = 5, s_2 = 6, s_3 = 7, s_4 = 8$

9. $a_n - a_{n-1} = (2n - 5) - [2(n-1) - 5] = 2n - 5 - (2n - 2 - 5) = 2n - 5 - (2n - 7) = 2n - 5 - 2n + 7 = 2$, a constant;

$d = 2; a_1 = -3, a_2 = -1, a_3 = 1, a_4 = 3$

11. $c_n - c_{n-1} = (6 - 2n) - [6 - 2(n-1)] = 6 - 2n - (6 - 2n + 2) = 6 - 2n - (8 - 2n) = 6 - 2n - 8 + 2n = -2$, a constant;

$d = -2; c_1 = 4, c_2 = 2, c_3 = 0, c_4 = -2$

13. $t_n - t_{n-1} = \left(\dfrac{1}{2} - \dfrac{1}{3}n\right) - \left[\dfrac{1}{2} - \dfrac{1}{3}(n-1)\right] = \dfrac{1}{2} - \dfrac{1}{3}n - \left(\dfrac{1}{2} - \dfrac{1}{3}n + \dfrac{1}{3}\right) = \dfrac{1}{2} - \dfrac{1}{3}n - \left(\dfrac{5}{6} - \dfrac{1}{3}n\right) = \dfrac{1}{2} - \dfrac{1}{3}n - \dfrac{5}{6} + \dfrac{1}{3}n = -\dfrac{1}{3}$, a constant;

$d = -\dfrac{1}{3}; t_1 = \dfrac{1}{6}, t_2 = -\dfrac{1}{6}, t_3 = -\dfrac{1}{2}, t_4 = -\dfrac{5}{6}$

15. $s_n - s_{n-1} = \ln 3^n - \ln 3^{n-1} = n \ln 3 - (n-1) \ln 3 = n \ln 3 - (n \ln 3 - \ln 3) = n \ln 3 - n \ln 3 + \ln 3 = \ln 3$, a constant;

$d = \ln 3; s_1 = \ln 3, s_2 = 2 \ln 3, s_3 = 3 \ln 3, s_4 = 4 \ln 3$

17. $a_n = 3n - 1; a_{51} = 152$ **19.** $a_n = 15 - 7n; a_{51} = -342$ **21.** $a_n = \dfrac{1}{2}(n - 1); a_{51} = 25$ **23.** $a_n = \sqrt{2}n; a_{51} = 51\sqrt{2}$ **25.** 200 **27.** -531 **29.** $\dfrac{83}{2}$

31. $a_1 = -13; d = 3; a_n = a_{n-1} + 3; a_n = -16 + 3n$ **33.** $a_1 = -53; d = 6; a_n = a_{n-1} + 6; a_n = -59 + 6n$

35. $a_1 = 28; d = -2; a_n = a_{n-1} - 2; a_n = 30 - 2n$ **37.** $a_1 = 25; d = -2; a_n = a_{n-1} - 2; a_n = 27 - 2n$ **39.** n^2 **41.** $\dfrac{n}{2}(9 + 5n)$ **43.** 1260

45. 195 **47.** -9312 **49.** 10,036 **51.** 12,240 **53.** -1925 **55.** 15,960 **57.** $-\dfrac{3}{2}$ **59.** 24 terms **61.** 1185 seats **63.** 2160 seats **65.** 8 yr

67. 210 beige and 190 blue **69.** $\{T_n\} = \{-5.5n + 67\}; T_5 = 39.5°F$ **71.** 14 **75.** 16.42% **76.** Yes

77. Ellipse: Center: $(0, 0)$; Vertices: $(0, -5), (0, 5)$; Foci: $(0, -\sqrt{21}), (0, \sqrt{21})$,

78. $\begin{bmatrix} \dfrac{1}{2} & 0 \\ \dfrac{3}{2} & -1 \end{bmatrix}$ **79.** $\dfrac{1}{x-1} - \dfrac{x-1}{x^2 + x + 1}$ **80.** $[-3, 11]$

81. $0, \sqrt{1 + \sqrt{2}}, -\sqrt{1 + \sqrt{2}}$ **82.** $\left\{-\dfrac{17}{8}\right\}$

Historical Problems (*page 676*)

1. $1\frac{2}{3}$ loaves, $10\frac{5}{6}$ loaves, 20 loaves, $29\frac{1}{6}$ loaves, $38\frac{1}{3}$ loaves   **2. (a)** 1 person   **(b)** 2401 kittens   **(c)** 2800

9.3 Assess Your Understanding (*page 677*)

3. Geometric   **4.** $\dfrac{a}{1-r}$ **5.** b   **6.** T   **7.** F   **8.** T   **9.** $r = 4; s_1 = 4, s_2 = 16, s_3 = 64, s_4 = 256$

11. $r = \dfrac{1}{2}; a_1 = -\dfrac{3}{2}, a_2 = -\dfrac{3}{4}, a_3 = -\dfrac{3}{8}, a_4 = -\dfrac{3}{16}$ **13.** $r = 2; c_1 = \dfrac{1}{4}, c_2 = \dfrac{1}{2}, c_3 = 1, c_4 = 2$ **15.** $r = 7^{1/4}; e_1 = 7^{1/4}, e_2 = 7^{1/2}, e_3 = 7^{3/4}, e_4 = 7$

17. $r = \dfrac{3}{2}; t_1 = \dfrac{1}{2}, t_2 = \dfrac{3}{4}, t_3 = \dfrac{9}{8}, t_4 = \dfrac{27}{16}$ **19.** $a_5 = 162; a_n = 2 \cdot 3^{n-1}$ **21.** $a_5 = 5; a_n = 5 \cdot (-1)^{n-1}$ **23.** $a_5 = 0; a_n = 0$

25. $a_5 = 9\sqrt{3}; a_n = (\sqrt{3})^n$ **27.** $a_7 = \dfrac{1}{64}$ **29.** $a_9 = 1$ **31.** $a_8 = 0.00000004$ **33.** $a_n = 6 \cdot 3^{n-1}$ or $a_n = 2 \cdot 3^n$ **35.** $a_n = -3 \cdot \left(-\dfrac{1}{3}\right)^{n-1} = \left(-\dfrac{1}{3}\right)^{n-2}$

37. $a_n = -(-3)^{n-1}$ **39.** $a_n = \dfrac{7}{15} \cdot 15^{n-1} = 7 \cdot 15^{n-2}$ **41.** $-\dfrac{1}{4}(1 - 2^n)$ **43.** $2\left[1 - \left(\dfrac{2}{3}\right)^n\right]$ **45.** $1 - 2^n$

47.
NORMAL FLOAT AUTO REAL RADIAN MP
(1/4)sum(seq(2ⁿ⁻¹,n,1,15,1)
                          8191.75

49.
NORMAL FLOAT AUTO REAL RADIAN MP
sum(seq((2/3)ⁿ,n,1,15,1)
                          1.995432683

51.
NORMAL FLOAT AUTO REAL RADIAN MP
-1sum(seq(2ⁿ⁻¹,n,1,15,1)
                          -32767

53. Converges; $\dfrac{3}{2}$ **55.** Converges; 16   **57.** Converges; $\dfrac{8}{5}$ **59.** Diverges   **61.** Converges; $\dfrac{20}{3}$ **63.** Diverges   **65.** Converges; $\dfrac{18}{5}$ **67.** Converges; 6

69. Arithmetic; $d = 1; 1375$ **71.** Neither   **73.** Arithmetic; $d = -\dfrac{2}{3}; -700$ **75.** Neither   **77.** Geometric; $r = \dfrac{2}{3}; 2\left[1 - \left(\dfrac{2}{3}\right)^{50}\right]$

79. Geometric; $r = -2; -\dfrac{1}{3}[1 - (-2)^{50}]$ **81.** Geometric; $r = 3^{1/2}; -\dfrac{\sqrt{3}}{2}(1 + \sqrt{3})(1 - 3^{25})$ **83.** -4 **85.** \$47,271.37

87. (a) 0.775 ft   **(b)** 8th   **(c)** 15.88 ft   **(d)** 20 ft   **89.** \$149,035.94   **91.** \$68,059.40   **93.** \$348.60   **95.** 1.845×10^{19} **97.** 10   **99.** \$72.67 per share
101. December 20; \$9999.92   **103.** 18   **105.** Option A results in a higher salary in 5 years (\$50,499 versus \$49,522); option B results in a higher 5-year total
   (\$225,484 versus \$233,602).   **107.** Option 2 results in the most: \$16,038,304; option 1 results in the least: \$14,700,000.
109. Yes. A constant sequence is both arithmetic and geometric. For example, 3, 3, 3, $\ldots$ is an arithmetic sequence with $a_1 = 3$ and $d = 0$ and is a geometric

   sequence with $a = 3$ and $r = 1$.   **113.** 2.121   **114.** $\dfrac{x^2}{4} - \dfrac{y^2}{12} = 1$ **115.** 54   **116.** $f(x) = -\dfrac{1}{8}(x + 2)(x - 1)(x - 4)^2$

117. $-16t - 13$ **118.** $g(x) = 7\sqrt{x + 5}$ **119.** $(x - 5)(x + 5)(x - 2)(x + 2)$ **120.** $[-1.95, 0.61]$

9.4 Assess Your Understanding (*page 683*)

1. (I) $n = 1: 2 \cdot 1 = 2$ and $1(1 + 1) = 2$
   (II) If $2 + 4 + 6 + \cdots + 2k = k(k + 1)$, then $2 + 4 + 6 + \cdots + 2k + 2(k + 1) = (2 + 4 + 6 + \cdots + 2k) + 2(k + 1)$
   $= k(k + 1) + 2(k + 1) = k^2 + 3k + 2 = (k + 1)(k + 2) = (k + 1)[(k + 1) + 1]$.

3. (I) $n = 1: 1 + 2 = 3$ and $\dfrac{1}{2} \cdot 1 \cdot (1 + 5) = \dfrac{1}{2} \cdot 6 = 3$

   (II) If $3 + 4 + 5 + \cdots + (k + 2) = \dfrac{1}{2}k(k + 5)$, then $3 + 4 + 5 + \cdots + (k + 2) + [(k + 1) + 2]$

   $= [3 + 4 + 5 + \cdots + (k + 2)] + (k + 3) = \dfrac{1}{2}k(k + 5) + k + 3 = \dfrac{1}{2}(k^2 + 7k + 6) = \dfrac{1}{2}(k + 1)(k + 6) = \dfrac{1}{2}(k + 1)[(k + 1) + 5]$.

5. (I) $n = 1: 3(1) - 1 = 2$ and $\dfrac{1}{2} \cdot 1 \cdot [3 \cdot 1 + 1] = \dfrac{1}{2} \cdot 4 = 2$

   (II) If $2 + 5 + 8 + \cdots + (3k - 1) = \dfrac{1}{2}k(3k + 1)$, then $2 + 5 + 8 + \cdots + (3k - 1) + [3(k + 1) - 1]$

   $= [2 + 5 + 8 + \cdots + (3k - 1)] + (3k + 2) = \dfrac{1}{2}k(3k + 1) + (3k + 2) = \dfrac{1}{2}(3k^2 + 7k + 4) = \dfrac{1}{2}(k + 1)(3k + 4)$

   $= \dfrac{1}{2}(k + 1)[3(k + 1) + 1]$.

7. (I) $n = 1: 2^{1-1} = 1$ and $2^1 - 1 = 1$
   (II) If $1 + 2 + 2^2 + \cdots + 2^{k-1} = 2^k - 1$, then $1 + 2 + 2^2 + \cdots + 2^{k-1} + 2^{(k+1)-1} = (1 + 2 + 2^2 + \cdots + 2^{k-1}) + 2^k$
   $= 2^k - 1 + 2^k = 2 \cdot 2^k - 1 = 2^{k+1} - 1$.

9. (I) $n = 1: 4^{1-1} = 1$ and $\dfrac{1}{3}(4^1 - 1) = \dfrac{1}{3} \cdot 3 = 1$

   (II) If $1 + 4 + 4^2 + \cdots + 4^{k-1} = \dfrac{1}{3}(4^k - 1)$, then $1 + 4 + 4^2 + \cdots + 4^{k-1} + 4^{(k+1)-1} = (1 + 4 + 4^2 + \cdots + 4^{k-1}) + 4^k$

   $= \dfrac{1}{3}(4^k - 1) + 4^k = \dfrac{1}{3}[4^k - 1 + 3 \cdot 4^k] = \dfrac{1}{3}[4 \cdot 4^k - 1] = \dfrac{1}{3}(4^{k+1} - 1)$.

11. (I) $n = 1: \dfrac{1}{1 \cdot 2} = \dfrac{1}{2}$ and $\dfrac{1}{1 + 1} = \dfrac{1}{2}$

(II) If $\dfrac{1}{1 \cdot 2} + \dfrac{1}{2 \cdot 3} + \dfrac{1}{3 \cdot 4} + \cdots + \dfrac{1}{k(k + 1)} = \dfrac{k}{k + 1}$, then $\dfrac{1}{1 \cdot 2} + \dfrac{1}{2 \cdot 3} + \dfrac{1}{3 \cdot 4} + \cdots + \dfrac{1}{k(k + 1)} + \dfrac{1}{(k + 1)[(k + 1) + 1]}$

$= \left[\dfrac{1}{1 \cdot 2} + \dfrac{1}{2 \cdot 3} + \dfrac{1}{3 \cdot 4} + \cdots + \dfrac{1}{k(k + 1)} \right] + \dfrac{1}{(k + 1)(k + 2)} = \dfrac{k}{k + 1} + \dfrac{1}{(k + 1)(k + 2)} = \dfrac{k(k + 2) + 1}{(k + 1)(k + 2)}$

$= \dfrac{k^2 + 2k + 1}{(k + 1)(k + 2)} = \dfrac{(k + 1)^2}{(k + 1)(k + 2)} = \dfrac{k + 1}{k + 2} = \dfrac{k + 1}{(k + 1) + 1}$.

13. (I) $n = 1: 1^2 = 1$ and $\dfrac{1}{6} \cdot 1 \cdot 2 \cdot 3 = 1$

(II) If $1^2 + 2^2 + 3^2 + \cdots + k^2 = \dfrac{1}{6}k(k + 1)(2k + 1)$, then $1^2 + 2^2 + 3^2 + \cdots + k^2 + (k + 1)^2$

$= (1^2 + 2^2 + 3^2 + \cdots + k^2) + (k + 1)^2 = \dfrac{1}{6}k(k + 1)(2k + 1) + (k + 1)^2 = \dfrac{1}{6}(2k^3 + 9k^2 + 13k + 6)$

$= \dfrac{1}{6}(k + 1)(k + 2)(2k + 3) = \dfrac{1}{6}(k + 1)[(k + 1) + 1][2(k + 1) + 1]$.

15. (I) $n = 1: 5 - 1 = 4$ and $\dfrac{1}{2} \cdot 1 \cdot (9 - 1) = \dfrac{1}{2} \cdot 8 = 4$

(II) If $4 + 3 + 2 + \cdots + (5 - k) = \dfrac{1}{2}k(9 - k)$, then $4 + 3 + 2 + \cdots + (5 - k) + [5 - (k + 1)]$

$= [4 + 3 + 2 + \cdots + (5 - k)] + 4 - k = \dfrac{1}{2}k(9 - k) + 4 - k = \dfrac{1}{2}(9k - k^2 + 8 - 2k) = \dfrac{1}{2}(-k^2 + 7k + 8)$

$= \dfrac{1}{2}(k + 1)(8 - k) = \dfrac{1}{2}(k + 1)[9 - (k + 1)]$.

17. (I) $n = 1: 1 \cdot (1 + 1) = 2$ and $\dfrac{1}{3} \cdot 1 \cdot 2 \cdot 3 = 2$

(II) If $1 \cdot 2 + 2 \cdot 3 + 3 \cdot 4 + \cdots + k(k + 1) = \dfrac{1}{3}k(k + 1)(k + 2)$, then $1 \cdot 2 + 2 \cdot 3 + 3 \cdot 4 + \cdots + k(k + 1)$

$+ (k + 1)[(k + 1) + 1] = [1 \cdot 2 + 2 \cdot 3 + 3 \cdot 4 + \cdots + k(k + 1)] + (k + 1)(k + 2)$

$= \dfrac{1}{3}k(k + 1)(k + 2) + \dfrac{1}{3} \cdot 3(k + 1)(k + 2) = \dfrac{1}{3}(k + 1)(k + 2)(k + 3) = \dfrac{1}{3}(k + 1)[(k + 1) + 1][(k + 1) + 2]$.

19. (I) $n = 1: 1^2 + 1 = 2$, which is divisible by 2.

(II) If $k^2 + k$ is divisible by 2, then $(k + 1)^2 + (k + 1) = k^2 + 2k + 1 + k + 1 = (k^2 + k) + 2k + 2$. Since $k^2 + k$ is divisible by 2 and $2k + 2$ is divisible by 2, $(k + 1)^2 + (k + 1)$ is divisible by 2.

21. (I) $n = 1: 1^2 - 1 + 2 = 2$, which is divisible by 2.

(II) If $k^2 - k + 2$ is divisible by 2, then $(k + 1)^2 - (k + 1) + 2 = k^2 + 2k + 1 - k - 1 + 2 = (k^2 - k + 2) + 2k$. Since $k^2 - k + 2$ is divisible by 2 and $2k$ is divisible by 2, $(k + 1)^2 - (k + 1) + 2$ is divisible by 2.

23. (I) $n = 1$: If $x > 1$, then $x^1 = x > 1$.

(II) Assume, for an arbitrary natural number k, that if $x > 1$ then $x^k > 1$. Multiply both sides of the inequality $x^k > 1$ by x. If $x > 1$, then $x^{k+1} > x > 1$.

25. (I) $n = 1: a - b$ is a factor of $a^1 - b^1 = a - b$.

(II) If $a - b$ is a factor of $a^k - b^k$, then $a^{k+1} - b^{k+1} = a(a^k - b^k) + b^k(a - b)$.
Since $a - b$ is a factor of $a^k - b^k$ and $a - b$ is a factor of $a - b$, then $a - b$ is a factor of $a^{k+1} - b^{k+1}$.

27. (a) $n = 1: (1 + a)^1 = 1 + a \geq 1 + 1 \cdot a$

(b) Assume that there is an integer k for which the inequality holds. So $(1 + a)^k \geq 1 + ka$. We need to show that $(1 + a)^{k+1} \geq 1 + (k + 1)a$.
$(1 + a)^{k+1} = (1 + a)^k (1 + a) \geq (1 + ka)(1 + a) = 1 + ka^2 + a + ka = 1 + (k + 1)a + ka^2 \geq 1 + (k + 1)a$.

29. If $2 + 4 + 6 + \cdots + 2k = k^2 + k + 2$, then $2 + 4 + 6 + \cdots + 2k + 2(k + 1) = (2 + 4 + 6 + \cdots + 2k) + 2k + 2$
$= k^2 + k + 2 + 2k + 2 = k^2 + 3k + 4 = (k^2 + 2k + 1) + (k + 1) + 2 = (k + 1)^2 + (k + 1) + 2$.
But $2 \cdot 1 = 2$ and $1^2 + 1 + 2 = 4$. The fact is that $2 + 4 + 6 + \cdots + 2n = n^2 + n$, not $n^2 + n + 2$ (Problem 1).

31. (I) $n = 1: [a + (1 - 1)d] = a$ and $1 \cdot a + d\dfrac{1 \cdot (1 - 1)}{2} = a$.

(II) If $a + (a + d) + (a + 2d) + \cdots + [a + (k - 1)d] = ka + d\dfrac{k(k - 1)}{2}$, then

$a + (a + d) + (a + 2d) + \cdots + [a + (k - 1)d] + [a + ((k + 1) - 1)d] = ka + d\dfrac{k(k - 1)}{2} + a + kd$

$= (k + 1)a + d\dfrac{k(k - 1) + 2k}{2} = (k + 1)a + d\dfrac{(k + 1)(k)}{2} = (k + 1)a + d\dfrac{(k + 1)[(k + 1) - 1]}{2}$.

33. (I) $n = 3$: The sum of the angles of a triangle is $(3 - 2) \cdot 180° = 180°$.

(II) Assume that for some $k \geq 3$, the sum of the angles of a convex polygon of k sides is $(k - 2) \cdot 180°$. A convex polygon of $k + 1$ sides consists of a convex polygon of k sides plus a triangle (see the illustration). The sum of the angles is
$(k - 2) \cdot 180° + 180° = (k - 1) \cdot 180° = [(k + 1) - 2] \cdot 180°$.

35. (a) 7; 15 **(b)** $c_n = 2^n - 1$ **(c) (I)** $n = 1$: one fold results in 1 crease and $c_1 = 2^1 - 1 = 1$.
(II) If $c_k = 2^k - 1$, then $c_{k+1} = 2c_k + 1 = 2(2^k - 1) + 1 = 2^{k+1} - 2 + 1 = 2^{k+1} - 1$
(d) Each fold doubles the thickness so the stack thickness will be $2^{25} \cdot 0.02 \, \text{mm} = 671,088.64 \, \text{mm}$ (or about 671 meters).

37. $\{251\}$ **38.** $x = \dfrac{1}{2}, y = -3; \left(\dfrac{1}{2}, -3\right)$ **39.** $\left\{\dfrac{21}{4}\right\}$ **40.** $\begin{bmatrix} 7 & -3 \\ -7 & 8 \end{bmatrix}$ **41.** $\dfrac{2}{x + 2} + \dfrac{1}{x - 1}$

42 $\left\{\dfrac{\ln 4 + 7}{3} \approx 2.795\right\}$ **43.** $x = 1$

9.5 Assess Your Understanding (page 690)

1. Pascal triangle **2.** $1; n$ **3.** F **4.** Binomial Theorem **5.** 10 **7.** 21 **9.** 50 **11.** 1 **13.** $\approx 1.8664 \times 10^{15}$ **15.** $\approx 1.4834 \times 10^{13}$
17. $x^5 + 5x^4 + 10x^3 + 10x^2 + 5x + 1$ **19.** $x^6 - 12x^5 + 60x^4 - 160x^3 + 240x^2 - 192x + 64$ **21.** $81x^4 + 108x^3 + 54x^2 + 12x + 1$
23. $x^{10} + 5x^8y^2 + 10x^6y^4 + 10x^4y^6 + 5x^2y^8 + y^{10}$ **25.** $x^3 + 6\sqrt{2}x^{5/2} + 30x^2 + 40\sqrt{2}x^{3/2} + 60x + 24\sqrt{2}x^{1/2} + 8$
27. $a^5x^5 + 5a^4bx^4y + 10a^3b^2x^3y^2 + 10a^2b^3x^2y^3 + 5ab^4xy^4 + b^5y^5$ **29.** 17,010 **31.** $-101,376$ **33.** 41,472 **35.** $2835x^3$ **37.** $314,928x^7$
39. 495 **41.** 3360 **43.** 1.00501

45. $\dbinom{n}{n-1} = \dfrac{n!}{(n-1)! \, [n - (n-1)]!} = \dfrac{n!}{(n-1)! \, 1!} = \dfrac{n \cdot (n-1)!}{(n-1)!} = n; \dbinom{n}{n} = \dfrac{n!}{n! \, (n-n)!} = \dfrac{n!}{n! \, 0!} = \dfrac{n!}{n!} = 1$

47. $2^n = (1 + 1)^n = \dbinom{n}{0}1^n + \dbinom{n}{1}(1)^{n-1}(1) + \cdots + \dbinom{n}{n}1^n = \dbinom{n}{0} + \dbinom{n}{1} + \cdots + \dbinom{n}{n}$ **49.** 1 **51.** 165

53. $\left\{\dfrac{\ln 5}{\ln 6 - \ln 5}\right\} \approx \{8.827\}$ **54.** $x = 1, y = 3, z = -2; (1, 3, -2)$

55. Bounded **56.** $g(f(x)) = \sqrt{x^2 - 4}$; Domain: $(-\infty, -2] \cup [2, \infty)$ **57.** $C = -46$

58. $\dfrac{1 - 8x^3}{3x^{2/3}(x^3 + 1)^2}$ **59.** $x = 3, x = -1$ **60.** $f(-2) = 5; (-2, 5)$

Review Exercises (page 693)

1. $a_1 = -\dfrac{4}{3}, a_2 = \dfrac{5}{4}, a_3 = -\dfrac{6}{5}, a_4 = \dfrac{7}{6}, a_5 = -\dfrac{8}{7}$ **2.** $c_1 = 2, c_2 = 1, c_3 = \dfrac{8}{9}, c_4 = 1, c_5 = \dfrac{32}{25}$ **3.** $a_1 = 3, a_2 = 2, a_3 = \dfrac{4}{3}, a_4 = \dfrac{8}{9}, a_5 = \dfrac{16}{27}$

4. $a_1 = 2, a_2 = 0, a_3 = 2, a_4 = 0, a_5 = 2$ **5.** $6 + 10 + 14 + 18 = 48$ **6.** $\displaystyle\sum_{k=1}^{13} (-1)^{k+1}\dfrac{1}{k}$ **7.** Arithmetic; $d = 1; S_n = \dfrac{n}{2}(n + 11)$ **8.** Neither

9. Geometric; $r = 8; S_n = \dfrac{8}{7}(8^n - 1)$ **10.** Arithmetic; $d = 4; S_n = 2n(n - 1)$ **11.** Geometric; $r = \dfrac{1}{2}; S_n = 6\left[1 - \left(\dfrac{1}{2}\right)^n\right]$ **12.** Neither

13. 9515 **14.** -1320 **15.** $\dfrac{1093}{2187} \approx 0.49977$ **16.** 682 **17.** 35 **18.** $\dfrac{1}{10^{10}}$ **19.** $9\sqrt{2}$ **20.** $\{a_n\} = \{5n - 4\}$ **21.** $\{a_n\} = \{n - 10\}$ **22.** Converges; $\dfrac{9}{2}$

23. Converges; $\dfrac{4}{3}$ **24.** Diverges **25.** Converges; 8

26. (I) $n = 1: 3 \cdot 1 = 3$ and $\dfrac{3 \cdot 1}{2}(1 + 1) = 3$

(II) If $3 + 6 + 9 + \cdots + 3k = \dfrac{3k}{2}(k + 1)$, then $3 + 6 + 9 + \cdots + 3k + 3(k + 1) = (3 + 6 + 9 + \cdots + 3k) + (3k + 3)$

$= \dfrac{3k}{2}(k + 1) + (3k + 3) = \dfrac{3k^2}{2} + \dfrac{3k}{2} + \dfrac{6k}{2} + \dfrac{6}{2} = \dfrac{3}{2}(k^2 + 3k + 2) = \dfrac{3}{2}(k + 1)(k + 2) = \dfrac{3(k + 1)}{2}[(k + 1) + 1]$.

27. (I) $n = 1: 2 \cdot 3^{1-1} = 2$ and $3^1 - 1 = 2$
(II) If $2 + 6 + 18 + \cdots + 2 \cdot 3^{k-1} = 3^k - 1$, then $2 + 6 + 18 + \cdots + 2 \cdot 3^{k-1} + 2 \cdot 3^{(k+1)-1} = (2 + 6 + 18 + \cdots + 2 \cdot 3^{k-1}) + 2 \cdot 3^k$
$= 3^k - 1 + 2 \cdot 3^k = 3 \cdot 3^k - 1 = 3^{k+1} - 1$.

28. (I) $n = 1: (3 \cdot 1 - 2)^2 = 1$ and $\dfrac{1}{2} \cdot 1 \cdot [6 \cdot 1^2 - 3 \cdot 1 - 1] = 1$

(II) If $1^2 + 4^2 + 7^2 + \cdots + (3k - 2)^2 = \dfrac{1}{2}k(6k^2 - 3k - 1)$, then

$1^2 + 4^2 + 7^2 + \cdots + (3k - 2)^2 + [3(k + 1) - 2]^2$

$= [1^2 + 4^2 + 7^2 + \cdots + (3k - 2)^2] + (3k + 1)^2 = \dfrac{1}{2}k(6k^2 - 3k - 1) + (3k + 1)^2 = \dfrac{1}{2}(6k^3 - 3k^2 - k) + (9k^2 + 6k + 1)$

$= \dfrac{1}{2}(6k^3 + 15k^2 + 11k + 2) = \dfrac{1}{2}(k + 1)(6k^2 + 9k + 2) = \dfrac{1}{2}(k + 1)[6(k + 1)^2 - 3(k + 1) - 1]$.

29. 10 **30.** $x^5 + 10x^4 + 40x^3 + 80x^2 + 80x + 32$ **31.** $81x^4 - 432x^3 + 864x^2 - 768x + 256$ **32.** 144 **33.** 84

34. (a) 8 bricks **(b)** 1100 bricks **35.** 360 **36. (a)** $20\left(\dfrac{3}{4}\right)^3 = \dfrac{135}{16}$ ft **(b)** $20\left(\dfrac{3}{4}\right)^n$ ft **(c)** 13 times **(d)** 140 ft **37.** $171,647.33 **38.** $58,492.93

Chapter Test (page 694)

1. $0, \dfrac{3}{10}, \dfrac{8}{11}, \dfrac{5}{4}, \dfrac{24}{13}$ **2.** 4, 14, 44, 134, 404 **3.** $2 - \dfrac{3}{4} + \dfrac{4}{9} = \dfrac{61}{36}$ **4.** $-\dfrac{1}{3} - \dfrac{14}{9} - \dfrac{73}{27} - \dfrac{308}{81} = -\dfrac{680}{81}$ **5.** $\displaystyle\sum_{k=1}^{10} (-1)^k\left(\dfrac{k + 1}{k + 4}\right)$ **6.** Neither

7. Geometric; $r = 4; S_n = \dfrac{2}{3}(1 - 4^n)$ **8.** Arithmetic; $d = -8; S_n = n(2 - 4n)$ **9.** Arithmetic; $d = -\dfrac{1}{2}; S_n = \dfrac{n}{4}(27 - n)$

10. Geometric; $r = \dfrac{2}{5}$; $S_n = \dfrac{125}{3}\left[1 - \left(\dfrac{2}{5}\right)^n\right]$ **11.** Neither **12.** Converges; $\dfrac{1024}{5}$ **13.** $243m^5 + 810m^4 + 1080m^3 + 720m^2 + 240m + 32$

14. First we show that the statement holds for $n = 1$. $\left(1 + \dfrac{1}{1}\right) = 1 + 1 = 2$. The equality is true for $n = 1$, so Condition I holds. Next we assume

that $\left(1 + \dfrac{1}{1}\right)\left(1 + \dfrac{1}{2}\right)\left(1 + \dfrac{1}{3}\right) \cdots \left(1 + \dfrac{1}{k}\right) = k + 1$ is true for some k, and we determine whether the formula then holds for $k + 1$.

$$\left(1 + \dfrac{1}{1}\right)\left(1 + \dfrac{1}{2}\right)\left(1 + \dfrac{1}{3}\right) \cdots \left(1 + \dfrac{1}{k}\right)\left(1 + \dfrac{1}{k+1}\right) = \left[\left(1 + \dfrac{1}{1}\right)\left(1 + \dfrac{1}{2}\right)\left(1 + \dfrac{1}{3}\right) \cdots \left(1 + \dfrac{1}{k}\right)\right]\left(1 + \dfrac{1}{k+1}\right)$$

$$= (k+1)\left(1 + \dfrac{1}{k+1}\right) = (k+1) \cdot 1 + (k+1) \cdot \dfrac{1}{k+1} = k + 1 + 1 = k + 2$$

Condition II also holds. So, the statement holds true for all natural numbers.
15. After 10 years, the car will be worth $6103.11. **16.** The weightlifter will have lifted a total of 8000 pounds after 5 sets.

Cumulative Review *(page 694)*

1. $\{-3, 3, -3i, 3i\}$ **2. (a)**

(b) $\left\{\left(\sqrt{\dfrac{-1 + \sqrt{3601}}{18}}, \dfrac{-1 + \sqrt{3601}}{6}\right), \left(-\sqrt{\dfrac{-1 + \sqrt{3601}}{18}}, \dfrac{-1 + \sqrt{3601}}{6}\right)\right\}$

(c) The circle and the parabola intersect at

$$\left(\sqrt{\dfrac{-1 + \sqrt{3601}}{18}}, \dfrac{-1 + \sqrt{3601}}{6}\right), \left(-\sqrt{\dfrac{-1 + \sqrt{3601}}{18}}, \dfrac{-1 + \sqrt{3601}}{6}\right)$$

3. $\left\{\ln\left(\dfrac{5}{2}\right)\right\}$ **4.** $y = 5x - 10$ **5.** $(x + 1)^2 + (y - 2)^2 = 25$ **6. (a)** 5 **(b)** 13 **(c)** $\dfrac{6x + 3}{2x - 1}$ **(d)** $\left\{x \middle| x \neq \dfrac{1}{2}\right\}$ **(e)** $\dfrac{7x - 2}{x - 2}$ **(f)** $\{x \mid x \neq 2\}$

(g) $g^{-1}(x) = \dfrac{1}{2}(x - 1)$; all reals **(h)** $f^{-1}(x) = \dfrac{2x}{x - 3}$; $\{x \mid x \neq 3\}$ **7.** $\dfrac{x^2}{7} + \dfrac{y^2}{16} = 1$ **8.** $(x + 1)^2 = 4(y - 2)$

CHAPTER 10 Counting and Probability

10.1 Assess Your Understanding *(page 701)*

5. subset; $\subseteq$ **6.** finite **7.** T **8.** b
9. $\varnothing$, $\{a\}$, $\{b\}$, $\{c\}$, $\{d\}$, $\{a, b\}$, $\{a, c\}$, $\{a, d\}$, $\{b, c\}$, $\{b, d\}$, $\{c, d\}$, $\{a, b, c\}$, $\{b, c, d\}$, $\{a, c, d\}$, $\{a, b, d\}$, $\{a, b, c, d\}$
11. 25 **13.** 40 **15.** 25 **17.** 37 **19.** 18 **21.** 5 **23.** 15 different arrangements **25.** 9000 numbers **27.** 175; 125
29. (a) 15 **(b)** 15 **(c)** 15 **(d)** 25 **(e)** 40 **31. (a)** 14.2 million **(b)** 78.4 million **33.** 480 portfolios

36.

$(x-2)^2 + (y+1)^2 = 9$

37. $2, 5, -2$ **38.** $\left\{\dfrac{1}{18}\right\}$ **39.** $\left\{-6\sqrt{2}, 0, 6\sqrt{2}\right\}$ **40.** $x = 3, y = -2$; $x = 8, y = 3$ or $(3, -2)$, $(8, 3)$

41. $6x^3 - 31x^2 + 43x - 28$ **42.** Converges; 10 **43. (a)** $x = \dfrac{15}{8}$ **(b)** at $x = 2$

44. $\dfrac{5}{x} - \dfrac{2}{x + 1} + \dfrac{7}{(x + 1)^2}$

10.2 Assess Your Understanding *(page 709)*

3. permutation **4.** combination **5.** $\dfrac{n!}{(n - r)!}$ **6.** $\dfrac{n!}{(n - r)!r!}$ **7.** 30 **9.** 24 **11.** 1 **13.** 1680 **15.** 28 **17.** 35 **19.** 1 **21.** 10,400,600
23. { abc, abd, abe, acb, acd, ace, adb, adc, ade, aeb, aec, aed, bac, bad, bae, bca, bcd, bce, bda, bdc, bde, bea, bec, bed, cab, cad, cae, cba, cbd, cbe, cda, cdb, cde, cea, ceb, ced, dab, dac, dae, dba, dbc, dbe, dca, dcb, dce, dea, deb, dec, eab, eac, ead, eba, ebc, ebd, eca, ecb, ecd, eda, edb, edc }; 60
25. { 123, 124, 132, 134, 142, 143, 213, 214, 231, 234, 241, 243, 312, 314, 321, 324, 341, 342, 412, 413, 421, 423, 431, 432 }; 24
27. { abc, abd, abe, acd, ace, ade, bcd, bce, bde, cde }; 10 **29.** { 123, 124, 134, 234 }; 4 **31.** 16 **33.** 8 **35.** 24 **37.** 60 **39.** 18,278 **41.** 35 **43.** 1024
45. 120 **47.** 132,860 **49.** 336 **51.** 90,720 **53. (a)** 63 **(b)** 35 **(c)** 1 **55.** 1.157×10^{76} **57.** 362,880 **59.** 660 **61.** 15
63. (a) 125,000; 117,600 **(b)** A better name for a *combination* lock would be a *permutation* lock because the order of the numbers matters.
68. $(3, 8]$ or $\{x \mid 3 < x \leq 8\}$ **69.** $(g \circ f)(x) = 4x^2 - 2x - 2$ **70.** $a_5 = 80$ **71.** $x^5 + 10x^4 y + 40x^3 y^2 + 80x^2 y^3 + 80xy^4 + 32 y^5$

72. $x = 3, y = -1$ or $(3, -1)$ **73.** $\begin{bmatrix} 6 & -6 \\ 14 & 5 \end{bmatrix}$ **74.** $\dfrac{5}{x^2 + 2} + \dfrac{3x + 4}{(x^2 + 2)^2}$ **75.** $\dfrac{16x - 30}{(x - 3)^{2/5}}$

Historical Problem *(page 718)*

1. (a) { AAAA, AAAB, AABA, AABB, ABAA, ABAB, ABBA, ABBB, BAAA, BAAB, BABA, BABB, BBAA, BBAB, BBBA, BBBB }

(b) $P(A \text{ wins}) = \dfrac{C(4, 2) + C(4, 3) + C(4, 4)}{2^4} = \dfrac{6 + 4 + 1}{16} = \dfrac{11}{16}$; $P(B \text{ wins}) = \dfrac{C(4, 3) + C(4, 4)}{2^4} = \dfrac{4 + 1}{16} = \dfrac{5}{16}$

10.3 Assess Your Understanding (*page 718*)

1. equally likely **2.** complement **3.** F **4.** T **5.** $0, 0.01, 0.35, 1$ **7.** Probability model **9.** Not a probability model

11. (a) $S = \{HH, HT, TH, TT\}$ **(b)** $P(HH) = \dfrac{1}{4}, P(HT) = \dfrac{1}{4}, P(TH) = \dfrac{1}{4}, P(TT) = \dfrac{1}{4}$

13. (a) $S = \{HH1, HH2, HH3, HH4, HH5, HH6, HT1, HT2, HT3, HT4, HT5, HT6, TH1, TH2, TH3, TH4, TH5, TH6, TT1, TT2, TT3, TT4, TT5, TT6\}$
(b) Each outcome has the probability of $\dfrac{1}{24}$.

15. (a) $S = \{HHH, HHT, HTH, HTT, THH, THT, TTH, TTT\}$ **(b)** Each outcome has the probability of $\dfrac{1}{8}$.

17. $S = \{$1 Yellow, 1 Red, 1 Green, 2 Yellow, 2 Red, 2 Green, 3 Yellow, 3 Red, 3 Green, 4 Yellow, 4 Red, 4 Green $\}$; each outcome has the probability
of $\dfrac{1}{12}$; thus, $P(2\text{ Red}) + P(4\text{ Red}) = \dfrac{1}{12} + \dfrac{1}{12} = \dfrac{1}{6}$.

19. $S = \{$1 Yellow Forward, 1 Yellow Backward, 1 Red Forward, 1 Red Backward, 1 Green Forward, 1 Green Backward, 2 Yellow Forward, 2 Yellow
Backward, 2 Red Forward, 2 Red Backward, 2 Green Forward, 2 Green Backward, 3 Yellow Forward, 3 Yellow Backward, 3 Red Forward, 3 Red
Backward, 3 Green Forward, 3 Green Backward, 4 Yellow Forward, 4 Yellow Backward, 4 Red Forward, 4 Red Backward, 4 Green Forward,
4 Green Backward$\}$; each outcome has the probability of $\dfrac{1}{24}$; thus, $P(1\text{ Red Backward}) + P(1\text{ Green Backward}) = \dfrac{1}{24} + \dfrac{1}{24} = \dfrac{1}{12}$.

21. $S = \{$11 Red, 11 Yellow, 11 Green, 12 Red, 12 Yellow, 12 Green, 13 Red, 13 Yellow, 13 Green, 14 Red, 14 Yellow, 14 Green, 21 Red, 21 Yellow,
21 Green, 22 Red, 22 Yellow, 22 Green, 23 Red, 23 Yellow, 23 Green, 24 Red, 24 Yellow, 24 Green, 31 Red, 31 Yellow, 31 Green, 32 Red, 32 Yellow,
32 Green, 33 Red, 33 Yellow, 33 Green, 34 Red, 34 Yellow, 34 Green, 41 Red, 41 Yellow, 41 Green, 42 Red, 42 Yellow, 42 Green, 43 Red,
43 Yellow, 43 Green, 44 Red, 44 Yellow, 44 Green$\}$; each outcome has the probability of $\dfrac{1}{48}$; thus, $E = \{$22 Red, 22 Green, 24 Red, 24 Green $\}$;
$P(E) = \dfrac{n(E)}{n(S)} = \dfrac{4}{48} = \dfrac{1}{12}$.

23. A, B, C, F **25.** B **27.** $P(H) = \dfrac{4}{5}$; $P(T) = \dfrac{1}{5}$ **29.** $P(1) = P(3) = P(5) = \dfrac{2}{9}$; $P(2) = P(4) = P(6) = \dfrac{1}{9}$ **31.** $\dfrac{3}{10}$ **33.** $\dfrac{1}{2}$ **35.** $\dfrac{1}{6}$ **37.** $\dfrac{1}{8}$

39. $\dfrac{1}{4}$ **41.** $\dfrac{1}{6}$ **43.** $\dfrac{1}{18}$ **45.** 0.55 **47.** 0.70 **49.** 0.30 **51.** 0.867 **53.** 0.62 **55.** 0.74 **57.** $\dfrac{17}{20}$ **59.** $\dfrac{11}{20}$ **61.** $\dfrac{1}{2}$ **63.** $\dfrac{19}{50}$ **65.** $\dfrac{9}{20}$

67. (a) 0.57 **(b)** 0.95 **(c)** 0.83 **(d)** 0.38 **(e)** 0.29 **(f)** 0.05 **(g)** 0.78 **(h)** 0.71

69. (a) $\dfrac{25}{33}$ **(b)** $\dfrac{25}{33}$ **71.** 0.167 **73.** $\dfrac{1}{25,989,600} \approx 0.0000000385$ **75.** 2; left; 3; down **76.** $\{22\}$ **77.** $(2, -3, -1)$ **78.** -40

79. $10\sqrt{3}$ **80.** 48 mph **81.** 593 **82.** $8\pi + 12$ square units **83.** $\dfrac{4}{x - 2} + \dfrac{3x - 7}{x^2 + 2x + 4}$

Review Exercises (*page 722*)

1. $\varnothing$, $\{$Dave$\}$, $\{$Joanne$\}$, $\{$Erica$\}$, $\{$Dave, Joanne$\}$, $\{$Dave, Erica$\}$, $\{$Joanne, Erica$\}$, $\{$Dave, Joanne, Erica$\}$ **2.** 17 **3.** 24 **4.** 29 **5.** 34 **6.** 7
7. 45 **8.** 25 **9.** 7 **10.** 336 **11.** 56 **12.** 60 **13.** 128 **14.** 3024 **15.** 1680 **16.** 91 **17.** 1,600,000 **18.** 216,000
19. 256 (allowing numbers with initial zeros, such as 011) **20.** 12,600 **21. (a)** 381,024 **(b)** 1260 **22. (a)** $8.634628387 \times 10^{45}$ **(b)** 0.6531 **(c)** 0.3469
23. (a) 0.038 **(b)** 0.962 **24.** $\dfrac{4}{9}$ **25.** 0.2; 0.26 **26. (a)** 0.68 **(b)** 0.58 **(c)** 0.32

Chapter Test (*page 723*)

1. 22 **2.** 3 **3.** 8 **4.** 45 **5.** 5040 **6.** 151,200 **7.** 462 **8.** There are 54,264 ways to choose 6 different colors from the 21 available colors.
9. There are 840 distinct arrangements of the letters in the word REDEEMED. **10.** There are 56 different exacta bets for an 8-horse race.
11. There are 155,480,000 possible license plates using the new format. **12. (a)** 0.95 **(b)** 0.30 **13. (a)** 0.25 **(b)** 0.55 **14.** 0.19 **15.** 0.000033069

16. $P(\text{exactly 2 fours}) = \dfrac{625}{3888} \approx 0.1608$

Cumulative Review (*page 724*)

1. $\left\{\dfrac{1}{3} - \dfrac{\sqrt{2}}{3}i, \dfrac{1}{3} + \dfrac{\sqrt{2}}{3}i\right\}$ **2.**   **3.**   **4.** $\{x\,|\,3.99 \le x \le 4.01\}$ or $[3.99, 4.01]$
5. $\left\{-\dfrac{1}{2} + \dfrac{\sqrt{7}}{2}i, -\dfrac{1}{2} - \dfrac{\sqrt{7}}{2}i, -\dfrac{1}{5}, 3\right\}$

6.   Domain: all real numbers
Range: $\{y\,|\,y > 5\}$
Horizontal asymptote: $y = 5$

7. 2 **8.** $\left\{\dfrac{8}{3}\right\}$ **9.** $x = 2, y = -5, z = 3$ **10.** 125; 700

APPENDIX Graphing Utilities

1 Exercises (page A2)

1. $(-1, 4)$; II **3.** $(3, 1)$; I **5.** $X\mathrm{min} = -6, X\mathrm{max} = 6, X\mathrm{scl} = 2, Y\mathrm{min} = -4, Y\mathrm{max} = 4, Y\mathrm{scl} = 2$
7. $X\mathrm{min} = -6, X\mathrm{max} = 6, X\mathrm{scl} = 2, Y\mathrm{min} = -1, Y\mathrm{max} = 3, Y\mathrm{scl} = 1$ **9.** $X\mathrm{min} = 3, X\mathrm{max} = 9, X\mathrm{scl} = 1, Y\mathrm{min} = 2, Y\mathrm{max} = 10, Y\mathrm{scl} = 2$
11. $X\mathrm{min} = -11, X\mathrm{max} = 5, X\mathrm{scl} = 1, Y\mathrm{min} = -3, Y\mathrm{max} = 6, Y\mathrm{scl} = 1$
13. $X\mathrm{min} = -30, X\mathrm{max} = 50, X\mathrm{scl} = 10, Y\mathrm{min} = -90, Y\mathrm{max} = 50, Y\mathrm{scl} = 10$
15. $X\mathrm{min} = -10, X\mathrm{max} = 110, X\mathrm{scl} = 10, Y\mathrm{min} = -10, Y\mathrm{max} = 160, Y\mathrm{scl} = 10$

2 Exercises (page A4)

3 Exercises (page A6)

1. -3.41 **3.** -1.71 **5.** -0.28 **7.** 3.00 **9.** 4.50 **11.** $1.00, 23.00$

5 Exercises (page A8)

1. No **3.** Yes **5.** Answers may vary. A possible answer is $Y\mathrm{min} = 0, Y\mathrm{max} = 10$, and $Y\mathrm{scl} = 1$.

Photo Credits

Cover   AlexRoz/Shutterstock

Chapter R   Page 1, Christo/Shutterstock; Page 31, Feraru Nicolae/Shutterstock; Page 38, Ben Buckwold/Red River Press

Chapter 1   Page 81, James Flint/123RF GB Ltd; Page 91, A.R. Monko/Shutterstock; Page 103, Peter Kotoff/Shutterstock; Page 103, Iain Masterton/Alamy Stock Photo; Page 133, Nancy R. Cohen/Getty Images; Page 146, Hemera Technologies/Photos.com/Thinkstock; Page 147, James Flint/123RF GB Ltd

Chapter 2   Page 149, Andrey_Popov/Shutterstock; Page 161, Alex Staroseltsev/Shutterstock; Page 167, Stockyimages/Fotolia; Page 168, U.S. Department of Energy; Page 182, Dmitry Kalinovsky/123RF GB Ltd; Page 190, Aneese/Shutterstock; Page 190, Pajor Pawel/Shutterstock; Page 201, Andrey_Popov/Shutterstock

Chapter 3   Page 202, Leigh Prather/Shutterstock; Page 218, NASA; Page 226, Exactostock-1557/Superstock; Page 278, Leigh Prather/Shutterstock

Chapter 4   Page 280, Andrey_Popov/Shutterstock; Page 318, Geno EJ Sajko Photography/Shutterstock; Page 329, Andrey_Popov/Shutterstock

Chapter 5   Page 330, FXQuadro/Shutterstock; Page 378, Perry Mastrovito/Getty Images; Page 413, FXQuadro/Shutterstock

Chapter 6   Page 414, zamollxis/123RF GB Ltd; Page 443, Pearson Education; Page 471, Pearson Education; Page 480, Grzegorz Czapski/Shutterstock; Page 485, Lacroix Serge/iStock/360/Getty Images; Page 500, Jupiterimages/Photos.com/360/Getty Images; Page 516, zamollxis/123RF GB Ltd

Chapter 7   Page 517, MarcelClemens/Shutterstock; Page 535, J. B. Spector/Museum of Science and Industry, Chicago/Getty Images; Page 550, cosma/Shutterstock; Page 553, MarcelClemens/Shutterstock

Chapter 8   Page 555, Rawpixel.com/Shutterstock; Page 608, Library of Congress (Photoduplication); Page 650, Rawpixel.com/Shutterstock

Chapter 9   Page 651, Arthimedes/Shutterstock; Page 661, frankie's/Shutterstock; Page 676, Pearson Education, Inc.; Page 690, Pearson Education, Inc.; Page 695, Arthimedes/Shutterstock

Chapter 10 Page 696, urbanbuzz/Alamy Stock Photo; Page 717, Georgios Kollidas/Getty Images; Page 724, urbanbuzz/Alamy Stock Photo

All Texas Instruments TI-84 Plus C Graphing Calulator graphs © copyright by Texas Instruments Education Technology.

All Desmos graphs courtesy of Desmos, Inc.

Subject Index

CONICS

Parabola

$$y^2 = 4ax \qquad y^2 = -4ax \qquad x^2 = 4ay \qquad x^2 = -4ay$$

Ellipse

$$\frac{x^2}{a^2} + \frac{y^2}{b^2} = 1, \quad a > b, \quad c^2 = a^2 - b^2 \qquad\qquad \frac{x^2}{b^2} + \frac{y^2}{a^2} = 1, \quad a > b, \quad c^2 = a^2 - b^2$$

Hyperbola

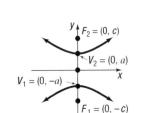

$$\frac{x^2}{a^2} - \frac{y^2}{b^2} = 1, \quad c^2 = a^2 + b^2 \qquad\qquad \frac{y^2}{a^2} - \frac{x^2}{b^2} = 1, \quad c^2 = a^2 + b^2$$

$$\text{Asymptotes:} \quad y = \frac{b}{a}x, \quad y = -\frac{b}{a}x \qquad\qquad \text{Asymptotes:} \quad y = \frac{a}{b}x, \quad y = -\frac{a}{b}x$$

PROPERTIES OF LOGARITHMS

$$\log_a (MN) = \log_a M + \log_a N$$

$$\log_a \left(\frac{M}{N}\right) = \log_a M - \log_a N$$

$$\log_a M^r = r \log_a M$$

$$\log_a M = \frac{\log M}{\log a} = \frac{\ln M}{\ln a}$$

$$a^r = e^{r \ln a}$$

PERMUTATIONS/COMBINATIONS

$$0! = 1 \qquad 1! = 1$$

$$n! = n(n-1) \cdot \ldots \cdot 3 \cdot 2 \cdot 1$$

$$P(n, r) = \frac{n!}{(n-r)!}$$

$$C(n, r) = \binom{n}{r} = \frac{n!}{(n-r)!\, r!}$$

BINOMIAL THEOREM

$$(a + b)^n = a^n + \binom{n}{1}ba^{n-1} + \binom{n}{2}b^2 a^{n-2}$$
$$+ \cdots + \binom{n}{n-1}b^{n-1}a + b^n$$

ARITHMETIC SEQUENCE

$$a_1 + (a_1 + d) + (a_1 + 2d) + \cdots + [a_1 + (n-1)d]$$
$$= \frac{n}{2}[2a_1 + (n-1)d] = \frac{n}{2}(a_1 + a_n)$$

GEOMETRIC SEQUENCE

$$a_1 + a_1 r + a_1 r^2 + \cdots + a_1 r^{n-1} = a_1 \cdot \frac{1 - r^n}{1 - r}$$

GEOMETRIC SERIES

$$\text{If } |r| < 1, \quad a_1 + a_1 r + a_1 r^2 + \cdots = \sum_{k=1}^{\infty} a_1 r^{k-1}$$
$$= \frac{a_1}{1 - r}$$

FORMULAS/EQUATIONS

Distance Formula

If $P_1 = (x_1, y_1)$ and $P_2 = (x_2, y_2)$, the distance from P_1 to P_2 is

$$d(P_1, P_2) = \sqrt{(x_2 - x_1)^2 + (y_2 - y_1)^2}$$

Standard Equation of a Circle

The standard equation of a circle of radius r with center (h, k) is

$$(x - h)^2 + (y - k)^2 = r^2$$

Slope Formula

The slope m of the line containing the points $P_1 = (x_1, y_1)$ and $P_2 = (x_2, y_2)$ is

$$m = \frac{y_2 - y_1}{x_2 - x_1} \qquad \text{if } x_1 \neq x_2$$

$$m \text{ is undefined} \qquad \text{if } x_1 = x_2$$

Point-Slope Equation of a Line

The equation of a line with slope m containing the point (x_1, y_1) is

$$y - y_1 = m(x - x_1)$$

Slope-Intercept Equation of a Line

The equation of a line with slope m and y-intercept b is

$$y = mx + b$$

Quadratic Formula

The solutions of the equation $ax^2 + bx + c = 0, a \neq 0$, are

$$x = \frac{-b \pm \sqrt{b^2 - 4ac}}{2a}$$

If $b^2 - 4ac > 0$, there are two unequal real solutions.
If $b^2 - 4ac = 0$, there is a repeated real solution.
If $b^2 - 4ac < 0$, there are two complex solutions that are not real.

GEOMETRY FORMULAS

Circle

$r = $ Radius, $A = $ Area, $C = $ Circumference
$$A = \pi r^2 \qquad C = 2\pi r$$

Triangle

$b = $ Base, $h = $ Altitude (Height), $A = $ area
$$A = \tfrac{1}{2} bh$$

Rectangle

$l = $ Length, $w = $ Width, $A = $ area, $P = $ perimeter
$$A = lw \qquad P = 2l + 2w$$

Rectangular Box (closed)

$l = $ Length, $w = $ Width, $h = $ Height, $V = $ Volume, $S = $ Surface area
$$V = lwh \qquad S = 2lw + 2lh + 2wh$$

Sphere

$r = $ Radius, $V = $ Volume, $S = $ Surface area
$$V = \tfrac{4}{3}\pi r^3 \qquad S = 4\pi r^2$$

Right Circular Cylinder (closed)

$r = $ Radius, $h = $ Height, $V = $ Volume, $S = $ Surface area
$$V = \pi r^2 h \qquad S = 2\pi r^2 + 2\pi rh$$